Nonparametric Statistical Methods

WILEY SERIES IN PROBABILITY AND STATISTICS
TEXTS AND REFERENCES SECTION

Established by WALTER A. SHEWHART and SAMUEL S. WILKS

Editors: *Vic Barnett, Noel A. C. Cressie, Nicholas I. Fisher, Iain M. Johnstone, J. B. Kadane, David G. Kendall, David W. Scott, Bernard W. Silverman, Adrian F. M. Smith, Jozef L. Teugels; Ralph A. Bradley, Emeritus, J. Stuart Hunter, Emeritus*

A complete list of the titles in this series appears at the end of this volume.

Nonparametric Statistical Methods

SECOND EDITION

MYLES HOLLANDER
Department of Statistics
The Florida State University
Tallahassee, Florida

DOUGLAS A. WOLFE
Department of Statistics
The Ohio State University
Columbus, Ohio

A Wiley-Interscience Publication
JOHN WILEY & SONS, INC.

New York   •   Chichester   •   Weinheim   •   Brisbane   •   Singapore   •   Toronto

This book is printed on acid-free paper. ∞

Copyright © 1999 by John Wiley & Sons, Inc. All rights reserved.

Published simultaneously in Canada.

No part of this publication may be reproduced, stored in a retrieval system or transmitted in any form or by any means, electronic, mechanical, photocopying, recording, scanning or otherwise, except as permitted under Sections 107 or 108 of the 1976 United States Copyright Act, without either the prior written permission of the Publisher, or authorization through payment of the appropriate per-copy fee to the Copyright Clearance Center, 222 Rosewood Drive, Danvers, MA 01923, (978) 750-8400, fax (978) 750-4744. Requests to the Publisher for permission should be addressed to the Permissions Department, John Wiley & Sons, Inc., 605 Third Avenue, New York, NY 10158-0012, (212) 850-6011, fax (212) 850-6008. E-Mail: PERMREQ@WILEY.COM.

For ordering and customer service, call 1-800-CALL-WILEY.

Library of Congress Cataloging-in-Publication Data:

Hollander, Myles.
    Nonparametric statistical methods / Myles Hollander, Douglas A. Wolfe. — 2nd ed.
       p.   cm. — (Wiley series in probability and statistics.
    Applied probability and statistics section)
    "A Wiley-Interscience publication."
    Includes bibliographical references and indexes.
    ISBN 0-471-19045-4 (cloth : alk. paper)
    1. Nonparametric statistics. I. Wolfe, Douglas A. II. Title.
III. Series: Wiley series in probability and statistics. Applied probability and statistics.
QA278.8.H65   1999
519.5—dc21   98-3314
                                                         CIP

Printed in the United States of America

10 9 8 7 6 5 4 3

To our wives,
Glee and Marilyn

Contents

Preface   **xiii**

1 Introduction   **1**

    1.1 Advantages of Nonparametric Methods, 1
    1.2 The Distribution-Free Property, 2
    1.3 Some Real-World Applications, 2
    1.4 Scope, 4
    1.5 Format and Organization, 8
    1.6 Computing Packages, 10
    1.7 Historical Background, 11
    1.8 Guide, 13

2 The Dichotomous Data Problem   **20**

    2.1 A Binomial Test, 20
    2.2 An Estimator for the Probability of Success, 29
    2.3 A Confidence Interval for the Probability of Success (Clopper-Pearson), 31

3 The One-Sample Location Problem   **35**

    Paired Replicates Analyses By Way of Signed Ranks
    3.1 A Distribution-Free Signed Rank Test (Wilcoxon), 36
    3.2 An Estimator Associated with Wilcoxon's Signed Rank Statistic (Hodges-Lehmann), 51
    3.3 A Distribution-Free Confidence Interval Based on Wilcoxon's Signed Rank Test (Tukey), 56
    Paired Replicate Analyses By Way of Signs
    3.4 A Distribution-Free Sign Test (Fisher), 60
    3.5 An Estimator Associated with the Sign Statistic (Hodges-Lehmann), 72
    3.6 A Distribution-Free Confidence Interval Based on the Sign Test (Thompson, Savur), 75

One-Sample Data

3.7 Procedures Based on the Signed Rank Statistic, 79
3.8 Procedures Based on the Sign Statistic, 83
3.9 An Asymptotically Distribution-Free Test of Symmetry (Randles-Fligner-Policello-Wolfe, Davis-Quade), 87

Bivariate Data

3.10 A Distribution-Free Test for Bivariate Symmetry (Hollander), 94
3.11 Efficiencies of Paired Replicates and One-Sample Location Procedures, 104

4 The Two-Sample Location Problem   106

4.1 A Distribution-Free Rank Sum Test (Wilcoxon, Mann and Whitney), 106
4.2 An Estimator Associated with Wilcoxon's Rank Sum Statistic (Hodges-Lehmann), 125
4.3 A Distribution-Free Confidence Interval Based on Wilcoxon's Rank Sum Test (Moses), 132
4.4 A Robust Rank Test for the Behrens-Fisher Problem (Fligner-Policello), 135
4.5 Efficiencies of Two-Sample Location Procedures, 139

5 The Two-Sample Dispersion Problem and Other Two-Sample Problems   141

5.1 A Distribution-Free Rank Test for Dispersion—Medians Equal (Ansari-Bradley), 142
5.2 An Asymptotically Distribution-Free Test for Dispersion Based on the Jackknife—Medians Not Necessarily Equal (Miller), 158
5.3 A Distribution-Free Rank Test for Either Location or Dispersion (Lepage), 169
5.4 A Distribution-Free Test for General Differences in Two Populations (Kolmogorov-Smirnov), 178
5.5 Efficiencies of Two-Sample Dispersion and Broad Alternatives Procedures, 187

6 The One-Way Layout   189

6.1 A Distribution-Free Test for General Alternatives (Kruskal-Wallis), 190
6.2 A Distribution-Free Test for Ordered Alternatives (Jonckheere, Terpstra), 202
6.3 Distribution-Free Tests for Umbrella Alternatives (Mack-Wolfe), 212
6.3.A A Distribution-Free Test for Umbrella Alternatives, Peak Known (Mack-Wolfe), 213
6.3.B A Distribution-Free Test for Umbrella Alternatives, Peak Unknown (Mack-Wolfe), 226
6.4 A Distribution-Free Test for Treatments versus a Control (Fligner-Wolfe), 234
6.5 Distribution-Free Two-Sided All-Treatments Multiple Comparisons Based on Pairwise Rankings—General Configuration (Dwass, Steel, Critchlow-Fligner), 240

6.6 Distribution-Free One-Sided All-Treatments Multiple Comparisons Based on Pairwise Rankings–Ordered Treatment Effects (Hayter-Stone), 249

6.7 Distribution-Free One-Sided Treatments versus Control Multiple Comparisons Based on Joint Rankings (Nemenyi, Damico-Wolfe), 254

6.8 Contrast Estimation Based on Hodges-Lehmann Two-Sample Estimators (Spjøtvoll), 260

6.9 Simultaneous Confidence Intervals for All Simple Contrasts (Critchlow-Fligner), 264

6.10 Efficiencies of One-Way Layout Procedures, 268

7 The Two-Way Layout 270

7.1 A Distribution-Free Test for General Alternatives in a Randomized Complete Block Design (Friedman, Kendall-Babington Smith), 272

7.2 A Distribution-Free Test for Ordered Alternatives in a Randomized Complete Block Design (Page), 284

7.3 Distribution-Free Two-Sided All-Treatments Multiple Comparisons Based on Friedman Rank Sums—General Configuration (Wilcoxon, Nemenyi, McDonald-Thompson), 295

7.4 Distribution-Free One-Sided Treatments versus Control Multiple Comparisons Based on Friedman Rank Sums (Nemenyi, Wilcoxon-Wilcox, Miller), 300

7.5 Contrast Estimation Based on One-Sample Median Estimators (Doksum), 305

Incomplete Block Data—Two-way Layout with Zero or One Observation per Treatment-Block Combination

7.6 A Distribution-Free Test for General Alternatives in a Randomized Balanced Incomplete Block Design (BIBD) (Durbin, Skillings-Mack), 309

7.7 Asymptotically Distribution-Free Two-Sided All-Treatments Multiple Comparisons for Balanced Incomplete Block Designs (Skillings-Mack), 317

7.8 A Distribution-Free Test for General Alternatives for Data from an Arbitrary Incomplete Block Design (Skillings-Mack), 319

Replications—Two-way Layout with at Least One Observation for Every Treatment-Block Combination

7.9 A Distribution-Free Test for General Alternatives in a Randomized Block Design with an Equal Number c (> 1) of Replications per Treatment-Block Combination (Mack-Skillings), 329

7.10 Asymptotically Distribution-Free Two-Sided All-Treatments Multiple Comparisons for a Two-Way Layout with an Equal Number of Replications in Each Treatment-Block Combination (Mack-Skillings), 340

Analyses Associated with Signed Ranks

7.11 A Test Based on Wilcoxon Signed Ranks for General Alternatives in a Randomized Complete Block Design (Doksum), 343

7.12 A Test Based on Wilcoxon Signed Ranks for Ordered Alternatives in a Randomized Complete Block Design (Hollander), 348

7.13 Approximate Two-Sided All-Treatments Multiple Comparisons Based on Signed Ranks (Nemenyi), 351

7.14 Approximate One-Sided Treatments versus Control Multiple Comparisons Based on Signed Ranks (Hollander), 353

7.15 Contrast Estimation Based on One-Sample Hodges-Lehmann Estimators (Lehmann), 357

7.16 Efficiencies of Two-Way Layout Procedures, 361

8 The Independence Problem   363

8.1 A Distribution-Free Test for Independence Based on Signs (Kendall), 363

8.2 An Estimator Associated with the Kendall Statistic (Kendall), 382

8.3 An Asymptotically Distribution-Free Confidence Interval Based on the Kendall Statistic (Samara-Randles, Fligner-Rust, Noether), 383

8.4 An Asymptotically Distribution-Free Confidence Interval Based on Efron's Bootstrap, 388

8.5 A Distribution-Free Test for Independence Based on Ranks (Spearman), 394

8.6 A Distribution-Free Test for Independence Against Broad Alternatives (Hoeffding), 408

8.7 Efficiencies of Independence Procedures, 413

9 Regression Problems   415

One Regression Line

9.1 A Distribution-Free Test for the Slope of the Regression Line (Theil), 416

9.2 A Slope Estimator Associated with the Theil Statistic (Theil), 421

9.3 A Distribution-Free Confidence Interval Associated with the Theil Test (Theil), 424

9.4 An Intercept Estimator Associated with the Theil Statistic and Use of the Estimated Linear Relationship for Prediction (Hettmansperger-McKean-Sheather), 426

$k(\geq 2)$ Regression Lines

9.5 An Asymptotically Distribution-Free Test for the Parallelism of Several Regression Lines (Sen, Adichie), 429

General Multiple Linear Regression

9.6 Asymptotically Distribution-Free Rank-Based Tests for General Multiple Linear Regression (Jaeckel, Hettmansperger-McKean), 438

Nonparametric Regression Analysis

9.7 An Introduction to Non-Rank-Based Approaches to Nonparametric Regression Analysis, 453

9.8 Efficiencies of Regression Procedures, 456

10 Comparing Two Success Probabilities   458

10.1 Approximate Tests and Confidence Intervals for the Difference between Two Success Probabilities (Pearson), 459

10.2 An Exact Test for the Difference between Two Success Probabilities (Fisher), 473

10.3 Inference for the Odds Ratio (Fisher, Cornfield), 477

10.4   Inference for k Strata of 2×2 Tables (Mantel and Haenszel), 484
   10.5   Efficiencies, 493

11   Life Distributions and Survival Analysis — 495

   11.1   A Test of Exponentiality versus IFR Alternatives (Epstein), 495
   11.2   A Test of Exponentiality versus NBU Alternatives (Hollander-Proschan), 504
   11.3   A Test of Exponentiality versus DMRL Alternatives (Hollander-Proschan), 513
   11.4   A Test of Exponentiality versus a Trend Change in Mean Residual Life (Guess-Hollander-Proschan), 520
   11.5   A Confidence Band for the Distribution Function (Kolmogorov), 526
   11.6   An Estimator of the Distribution Function When the Data Are Censored (Kaplan-Meier), 535
   11.7   A Two-Sample Test for Censored Data (Mantel), 550
   11.8   Efficiencies, 557

Appendix A   Tables and Charts — 563

Bibliography — 745

Answers to Selected Problems — 767

Author Index — 771

Subject Index — 779

Preface

In the 60+ years since its origins in the mid-1930s, nonparametric statistical methods have flourished and have emerged as the preferred methodology for statisticians and other scientists doing data analysis. There are a number of reasons for the success of the nonparametric approach (see Section 1.1). The primary reason is that the approach provides highly efficient techniques applicable to a wide variety of situations—techniques that do not require restrictive distributional assumptions about the underlying populations from which the data are drawn.

We hope that this second edition of *Nonparametric Statistical Methods* will satisfy the faithful audience and users of the first edition while also attracting a new generation of users. The major improvements to our book with respect to the first edition include expanded coverage of nonparametric methods for experimental designs in Chapters 6 and 7, new coverage of the bootstrap (Section 8.4), expanded coverage of nonparametric regression methods in Chapter 9, a new chapter on contingency tables and the odds ratio (Chapter 10), and a new chapter on life distributions and survival analysis (Chapter 11). In addition, all of the chapters of the first edition have been completely rewritten with more procedures, more real-world data sets, and more problems. Furthermore, whereas the first edition did not use software, in the second edition our example sections typically use nonparametric programs from `Minitab` and `StatXact`. For a detailed description of the contents of the second edition, see Section 1.4 and the guide in Section 1.8.

As in the first edition, the second edition is centered around actual experiments. The examples include problems from agricultural science, astronomy, biology, criminology, education, engineering, environmental science, geology, home economics, medicine, oceanography, physics, psychology, sociology, and space science.

We recommend the following timetable for a one-semester fifteen-week applied course meeting three hours per week: Chapter 2 (1 week), Chapter 3 (2 weeks), Chapters 4 and 5 (1 week each), Chapters 6 and 7 (2 weeks each), Chapters 8, 9, and 10 (1 week each) and Chapter 11 (2 weeks). This leaves two or three class periods for exams, not including finals week. If there is not sufficient time for complete coverage, the instructor can consider choosing between Chapters 10 and 11 to conclude the course.

At both Florida State University and Ohio State University, the first edition, and more recently various iterations of the notes upon which the second edition is based, have been used for an upper-level undergraduate/first-year graduate course in applied nonparametric statistics. The students typically come from departments of accounting, biology, biophysics, criminology, industrial engineering, oceanography, plant biology, sociology, statistics, and zoology. We aim to equip students with an arsenal of nonparametric techniques and to help them develop the insight they will need to choose appropriate procedures for various situations.

The format and organization of the book are described in Section 1.5. Appendix B (Computer Programs) and Appendix C (Glossary) can be found in the Solutions Manual, the Instructor's Manual, and on the Wiley website (http://www.wiley.com).

We have benefitted from the assistance of many friends and colleagues. Lora Bohn, Doug Critchlow, Mike Fligner, Joe McKean, Guohua Pan, and John Skillings provided classroom testing and helpful comments on earlier versions of this manuscript. Tom Fleming and Edsel Peña answered queries on survival analysis. Jay Aubuchon and Cyrus Mehta provided information about Minitab and StatXact, respectively. Bill Brown, Glee Hollander, and Duane Meeter led us to examples from the medical literature, the psychology literature, and statistical consulting projects, respectively. Alan Agresti furnished helpful references. S-plus programs to produce figures were written by Glen Laird (Figure 11.2), Chris Leddy (Figure 11.2), Edsel Peña (Figures 11.5, 11.6 and 11.7), Mourad Tighiouart (Figure 8.1), and Jie Yang (Figure 11.8). Mourad also helped with Appendix B and Bill Walton provided computer systems assistance. Chung-Lin (Joan) Hu and Shanggang Zhou worked examples, looked up references, solved problems, and found errors in earlier versions. Pirouz Foroutan, Lihong Qi, and William Turechek also found errors in early-stage drafts.

Chung-Lin Hu and Shen Zhang wrote the solutions for the Solutions Manual and the Instructor's Manual.

Melissa Elaine Smith and Peg Steigerwald provided expert, timely, and dedicated typesetting of the TeX/LaTeX manuscript.

Pam McGhee helped with permissions and the author index.

Steve Quigley, executive editor at Wiley, provided clear direction and good advice during the writing process.

Alison Bory, Steve's editorial assistant, helped with many aspects including permissions and coordinating the Solutions Manual and the Instructor's Manual. Lisa Van Horn, the production editor, provided an efficient and supportive production process, with the highest standards.

To all these friends and colleagues: It was a pleasure to have your assistance. We are very grateful.

MYLES HOLLANDER
DOUGLAS A. WOLFE

Tallahassee, Florida
Columbus, Ohio
October, 1998

Nonparametric Statistical Methods

CHAPTER 1

Introduction

1.1. ADVANTAGES OF NONPARAMETRIC METHODS

Roughly speaking, a nonparametric procedure is a statistical procedure that has certain desirable properties that hold under relatively mild assumptions regarding the underlying populations from which the data are obtained. The rapid and continuous development of nonparametric statistical procedures over the past six decades is due to the following advantages enjoyed by nonparametric techniques:

1. Nonparametric methods require few assumptions about the underlying populations from which the data are obtained. In particular, nonparametric procedures forgo the traditional assumption that the underlying populations are normal.
2. Nonparametric procedures enable the user to obtain exact P-values for tests, exact coverage probabilities for confidence intervals, exact experimentwise error rates for multiple comparison procedures, and exact coverage probabilities for confidence bands without relying on assumptions that the underlying populations are normal.
3. Nonparametric techniques are often (although not always) easier to apply than their normal theory counterparts.
4. Nonparametric procedures are often quite easy to understand.
5. Although at first glance most nonparametric procedures seem to sacrifice too much of the basic information in the samples, theoretical investigations have shown that this is not the case. Usually the nonparametric procedures are only slightly less efficient than their normal theory competitors when the underlying populations are normal (the home court of normal theory methods), and they can be mildly or wildly more efficient than these competitors when the underlying populations are not normal.
6. Nonparametric methods are relatively insensitive to outlying observations.
7. Nonparametric procedures are applicable in many situations where normal theory procedures cannot be utilized. Many nonparametric procedures require just the ranks of the observations, rather than the actual magnitude of the observations, whereas the parametric procedures require the magnitudes.
8. The Quenoulli–Tukey jackknife (Quenoulli (1949), Tukey (1958, 1962)) and Efron's computer-intensive (1979) bootstrap enable nonparametric approaches to be used in many complicated situations where the distribution theory needed to support parametric methods is intractable.
9. The development of computer software has facilitated fast computation of exact P-values for conditional nonparametric tests. Such exact P-values were, in principle, always available.

Before the advent of convenient software, however, users typically avoided the excessive computation required for exact conditional tests and instead relied on large-sample results, which yield only approximate P-values.

1.2. THE DISTRIBUTION-FREE PROPERTY

The term *nonparametric*, introduced in Section 1.1, is imprecise. The related term *distribution-free* has a precise meaning. The distribution-free property is a key aspect of many nonparametric procedures. In this section we informally introduce the concept of a distribution-free test statistic. The related notions of a distribution-free confidence interval, distribution-free multiple comparison procedure, distribution-free confidence band, asymptotically distribution-free test statistic, asymptotically distribution-free multiple comparison procedure, and asymptotically distribution-free confidence band are defined in the glossary (Appendix C) and introduced at appropriate points in the text.

Distribution-Free Test Statistic

We introduce the concept of a distribution-free test statistic by referring to the two-sample Wilcoxon rank sum statistic, which you will encounter in Section 4.1.

The data consist of a random sample of m observations from a population with continuous probability distribution F_1 and an independent random sample of n observations from a second population with continuous probability distribution F_2. The null hypothesis to be tested is

$$H_0 : F_1 = F_2 = F, \quad F \text{ unspecified.}$$

The null hypothesis asserts that the two random samples can be viewed as a single sample of size $N = m + n$ from a common population with unknown distribution F. The Wilcoxon (1945) statistic W is obtained by ranking the combined sample of N observations jointly from least to greatest. The test statistic is W, the sum of the ranks obtained by the Y's in the joint ranking.

When H_0 is true, the distribution of W does not depend on F. That is, when H_0 is true, for all a-values, the probability that $W \leq a$, denoted by $P_0(W \leq a)$, does not depend on F.

$$P_0(W \leq a) \text{ does not depend on } F. \tag{1.1}$$

The distribution-free property given by (1.1) enables one to table the distribution of W under H_0 without specifying the underlying F. It further enables one to exactly specify the type I error probability (the probability of rejecting H_0 when H_0 is true) without making distributional assumptions, such as the assumption that F is a normal distribution; the latter assumption is required by the parametric t-test.

The details concerning how to perform the Wilcoxon test are given in Section 4.1.

1.3. SOME REAL-WORLD APPLICATIONS

This book stresses the application of nonparametric techniques to real data. The following six examples are a sample of the type of problems you will learn to analyze using nonparametric methods.

Example 1.1: *Dose-Response Relationship.* In many situations a dose-response relationship may not be monotonic in the dosage. For example, with *in vitro* mutagenicity assays, experimental organisms may not survive the toxic side effects of high doses of the test agent, so there may be a reduction of the number of organisms at risk of mutation. This would lead to a downturn (that is, an umbrella pattern) in the dose-response curve. The data in Table 6.10 were considered by Simpson and Margolin (1986) in a discussion of the analysis of Ames test results. Plates containing Salmonella bacteria of strain TA98 were exposed to various doses of Acid Red 114. Table 6.10 gives the number of visible revertant colonies on the 18 plates in the study, 3 plates for each of the six doses (in μg/ml): 0, 100, 333, 1,000, 3,333, and 10,000. How can we test the hypothesis of equal population median numbers at each dose against the alternative that the peak of the dose response curve occurs at 1,000 μg/ml? How can we determine which particular pairs of doses, if any, significantly differ from one another in the number of revertant colonies? Which particular doses, out of 100, 333, 1000, 3,333, and 10,000, differ significantly from the 0 dose in terms of the number of revertant colonies? For doses that significantly differ, how can we estimate the magnitude of the difference? How can we simultaneously estimate all 15 "contrasts," $\tau_1 - \tau_2, \tau_1 - \tau_3, \tau_1 - \tau_4, \tau_1 - \tau_5, \tau_1 - \tau_6, \tau_2 - \tau_3, \tau_2 - \tau_4, \tau_2 - \tau_5, \tau_2 - \tau_6, \tau_3 - \tau_4, \tau_3 - \tau_5, \tau_3 - \tau_6, \tau_4 - \tau_5, \tau_4 - \tau_6, \tau_5 - \tau_6$, where, for example, $\tau_1 - \tau_2$ denotes the difference between the population medians at dose 0 and dose 100. The methods of Chapter 6 can be used to answer these questions.

Example 1.2: *Shelterbelts.* Shelterbelts are long rows of tree plantings across the direction of prevailing winds. They are used in developed countries to protect crops and livestock from the effects of the wind. A study was performed by Ujah and Adeoye (1984) to see if shelterbelts would limit severe losses from droughts regularly experienced in the arid and semiarid zones of Nigeria. The droughts are considered to be a leading factor in declining food production in Nigeria and in neighboring countries. Ujah and Adeoye studied the effect of shelterbelts on a number of factors relating to drought conditions, including wind velocity, air and soil temperatures, and soil moisture. Their experiment was conducted at two locations about $3\frac{1}{2}$ km apart, near Dambatta. Table 7.7 presents the wind velocity data, averaged over the two locations, at various distances leeward of the shelterbelt. The data are given as percent wind speed reduction relative to the wind velocity on the windward side of the shelterbelt. The data are given for 9 months (data were not available for July, November, and December) and five leeward distances, namely, 20, 40, 100, 150, and 250 m from the shelterbelt. Does the percent reduction in average wind speed tend to decrease as the leeward distance from a shelterbelt increases? Which particular leeward distances, if any, significantly differ from one another in percent reduction in average wind speed? How can the difference in percent reduction for two leeward distances be estimated? Chapter 7 presents nonparametric methods that will enable you to analyze the data and answer these questions.

Example 1.3: *Nasal Brushing.* In order to study the effects of pharmaceutical and chemical agents on mucociliary clearance, doctors often use the ciliary beat frequency (CBF) as an index of ciliary activity. One accepted way to measure CBF in a subject is through the collection and analysis of an endobronchial forceps biopsy specimen. This technique is, however, a rather invasive method for measuring CBF. In a study designed to assess the effectiveness of less invasive procedures for measuring CBF, Low, Luk, Dulfano, and Finch (1984) considered the alternative technique of nasal brushing. The data in Table 8.10 are a subset of the data collected by Low et al. during their investigation. The subjects in the study were all men undergoing bronchoscopies for diagnoses of a variety of pulmonary problems. The CBF values reported in Table 8.10 are averages of ten consecutive measurements on each subject.

How can we test the hypothesis of independence versus the alternative that the CBF measurements corresponding to nasal brushing and endobronchial forceps are positively associated? If there is evidence that the alternative is true, this would support the notion that nasal brushing is an acceptable substitute (for measuring CBF) for the more invasive endobronchial forceps biopsy technique. How can we obtain an estimate of a measure of the strength of association between the two techniques' CBF values? How can we compute confidence intervals for such a measure? These questions can be answered by the methods described in Chapter 8.

Example 1.4: *Coastal Sediments.* Coastal sediments are an important reservoir for organic nitrogen (ON). The degradation of ON is bacterially mediated. The mineralization of ON involves several distinct steps, and it is possible to measure the rates of these processes at each step. During the first stage of ON remineralization, ammonium is generated by heterotrophic bacteria during a process called *ammonification*. Ammonium can then be released to the environment or be microbially transformed to other nitrogenous species. The data in Table 9.4 are due to Mortazavi (1997) and are based on four sediment cores that were collected in Apalachicola Bay, Florida, in April 1995 and brought back to the main campus at Florida State University for analysis. The flux of ammonium to the overlying water was measured in each core during a 6-h incubation period. It is desired to know if there is a significant difference in ammonium flux between the cores. This is a regression problem, and it can be studied using the methods of Chapter 9.

Example 1.5: *Care Patterns for Black and White Patients with Breast Cancer.* Diehr et al. (1989) point out that it is well known that the survival rate of women with breast cancer tends to be lower in blacks than whites. Diehr and her colleagues sought to determine if these survival differences could be accounted for by differences in diagnostic methods and treatments. Diehr et al. reported on various breast cancer patterns; one pattern of interest was *liver scan*. Did patients with local or regional disease have a liver scan or CT scan of the liver? The data are given in Table 10.14. The data are for the 19 hospitals (out of 107 hospitals participating in the study) that had enough black patients for individual analysis. How can we determine, for a specific hospital, if there was a significant difference between the chance of a white patient receiving a scan and the chance of a black patient receiving a scan? How can the data from the 19 hospitals be utilized to get an overall assessment? The methods of Chapter 10 provide the means to answer these questions.

Example 1.6: *Times to First Review.* The data in Table 11.18, from Hollander, McKeague, and Yang (1997), relate to 432 manuscripts submitted to the Theory and Methods Section of the *Journal of the American Statistical Association* in the period January 1, 1994, to December 13, 1994. Of interest is the time (in days) to first review. When the data were studied on December 13, 1994, 157 papers had not yet received a first review. For example, for a paper received by *JASA* on November 1, 1994, and still awaiting a first review on December 13, 1994, we know on December 13 that its time to review is greater than 33 days, but we do not know at that point the actual time to review. The observation is said to be *censored*. How can we use the censored observations and uncensored observations (i.e., the ones for which we know the exact times to first review) to estimate the distribution of the time to first review? Section 11.6 shows how to estimate distributions when some of the data are censored.

1.4. SCOPE

This book emphasizes applications. The nonparametric methods presented are illustrated with examples based on real-world situations. A sample of these situations is given in Section 1.3. Tables and charts that facilitate implementation of the procedures are given in Appendix A. Computer packages that we use to obviate tedious computations are described in Section 1.6,

1.4. SCOPE

and some additional programs are given in Appendix B. An introductory nonmathematical statistics course should suffice as background, although at times we use terminology not generally presented in a first course. When confronted with unfamiliar concepts or terms, the reader should refer to the glossary in Appendix C for definitions. Appendices B and C can be found in the Solutions Manual, in the Instructors Manual, and on the Wiley website (http://www.wiley.com).

Hollander and Wolfe (1973), the precursor to this book, was the first book on nonparametric statistics that was not restricted to hypothesis tests. We continue that approach here and present not only test procedures, but also point estimators, confidence intervals, multiple comparison procedures, and confidence bands. The specific problems covered are outlined in the following paragraphs. A tabular guide to the procedures and problems is given in Section 1.8.

Chapter 2 (The Dichotomous Data Problem). In Chapter 2 the data consist of the outcomes of n independent repeated Bernoulli trials, each trial having the same success probability p. Section 2.1 describes the binomial test of the hypothesis $p = p_0$ (where p_0 is specified). Section 2.2 contains a point estimator for the unknown parameter p, and Section 2.3 gives confidence intervals for p.

Chapter 3 (The One-Sample Location Problem). In Sections 3.1–3.7 the data are in the form of paired replicates. The first observation in each pair may be conveniently viewed as a pretreatment observation and the second observation, as a posttreatment observation. The basic question is: has the application of the treatment caused a shift in location? The procedures in Sections 3.1–3.3 are associated with signed ranks. Section 3.1 presents a distribution-free test of the hypothesis that the treatment effect θ is equal to zero, Section 3.2 contains a point estimator of θ, and Section 3.3 gives a distribution-free confidence interval for θ.

The procedures in Sections 3.4–3.6 are analogous to those of Sections 3.1–3.3 but utilize signs rather than signed ranks. Sections 3.7 and 3.8 contain applications of the sign and signed rank procedures to one-sample data. Section 3.9 presents an asymptotically distribution-free test of the hypothesis that the one-sample data emanate from a symmetric population. Section 3.10 presents a distribution-free test of exchangeability for paired replicates.

Chapter 4 (The Two-Sample Location Problem). In Chapter 4, the data consist of two random samples, a sample from the control population and a sample from the treatment population. The basic hypothesis states that the two samples may be viewed as a single sample from one population. The alternative of interest is that the treatment population has a different location (median) than the control population. We let Δ (the treatment effect) denote the difference in locations. Section 4.1 presents a distribution-free rank test of $\Delta = 0$, Section 4.2 gives a point estimator of Δ, and Section 4.3 contains a distribution-free confidence interval for Δ. In Section 4.4 we present a test for location differences that allows the population dispersions to differ.

Chapter 5 (The Two-Sample Dispersion Problem and Other Two-Sample Problems). Here the basic data are as in Chapter 4, but the alternatives of interest are not restricted to be location-difference alternatives. Letting γ denote the unknown ratio of scale parameters for the two populations, Section 5.1 contains a distribution-free rank test of $\gamma = 1$. The test assumes that the two underlying populations have the same median. Section 5.2 presents an asymptotically distribution-free test of $\gamma = 1$ that does not require the assumption of equal medians. The section also gives point estimators and confidence intervals for γ^2. Section 5.3 contains a distribution-free rank test designed to detect location and dispersion differences. Section 5.4 contains a distribution-free rank test that will detect *all* alternatives to the null hypothesis of equivalent populations.

Chapter 6 (The One-Way Layout). In Chapter 6 the two-sample location problem of Chapter 4 is generalized to the case involving k treatment populations. The data consist of a random sample from each of the populations. The basic hypothesis H_0 is that of no treatment differences, so that when H_0 is true the k samples can be viewed as a single sample from one population. Section 6.1 presents a distribution-free test of H_0 that is designed to detect general location alternatives. A distribution-free test of H_0 designed to detect specified monotonically ordered alternatives is given in Section 6.2. Section 6.3 extends the idea of Section 6.2 and presents distribution-free tests of H_0 designed to detect a broader class of alternatives than the ordered alternatives of Section 6.2. The broader class consists of "umbrella" alternatives. Section 6.3.A considers the case where the position of the "peak" is known and Section 6.3.B considers the case where the position of the peak is unknown. In Section 6.4 a distribution-free test procedure is presented for the simultaneous comparison of $k - 1$ treatments with a control. Sections 6.5–6.7 contain multiple comparison procedures designed to detect which particular populations, if any, differ from one another. Section 6.8 presents estimators of contrasts in the treatment effects and Section 6.9 contains simultaneous confidence intervals for simple contrasts.

Chapter 7 (The Two-Way Layout). Chapter 7 may be viewed as a generalization of Chapter 3 to the case of k treatments. We consider experimental designs involving two factors, each at two or more levels. We are interested in location effects of one of these factors (the treatment factor) within the various levels of the second factor (the blocking factor). The basic null hypothesis asserts there are no differences in the location effects of the k treatments within each of the blocks. Sections 7.1–7.5 consider the case of one observation per treatment-block cell. This is commonly known as the randomized complete block design. Section 7.1 presents a distribution-free test designed to detect general alternatives. Section 7.2 considers a distribution-free test designed to detect ordered alternatives. Section 7.3 contains multiple comparison procedures for comparing all pairs of treatments, whereas Section 7.4 considers treatments versus control multiple comparison procedures. Section 7.5 presents estimators of contrasts in the treatment effects. Sections 7.6–7.8 deal with situations where certain treatment-block combinations have no observations. Section 7.6 presents a distribution-free test designed to detect general alternatives when the data arise from a balanced incomplete block design (BIBD). Section 7.7 contains an all-treatments multiple comparison procedure for BIBD data. Section 7.8 contains a distribution-free test designed to detect general alternatives when the data arise from an arbitrary incomplete block design. Sections 7.9 and 7.10 consider the setting where there is at least one observation per cell and some cells have multiple observations (replications). Section 7.9 contains a distribution-free test for this replications setting with an emphasis on the special case where there is an equal number (> 1) of replications in each cell. Section 7.10 presents an all-treatments multiple comparison procedure for the situation of an equal number of replications.

The procedures of Sections 7.1–7.10 are based on within-blocks rankings and are extensions of the paired-replicates sign procedures of Sections 3.4–3.6. Sections 7.11–7.15 present signed-rank extensions of the procedures in Sections 3.1–3.3. The signed-rank extensions are not distribution-free when $k > 2$, but they are asymptotically distribution-free. For simplicity, we present conservative approximations to these asymptotically distribution procedures. The approximations are, however, nearly asymptotically distribution-free and differ only slightly from their asymptotically distribution-free counterparts. Sections 7.11–7.15 consider randomized block designs with k treatments and n blocks. The test of Section 7.11 is directed to general alternatives, and the test of Section 7.12 is designed for ordered alternatives. Section 7.13 contains an all treatments multiple comparisons procedure and Section 7.14 contains

1.4. SCOPE

a treatments versus control multiple comparison procedure. Section 7.15 presents contrast estimators.

Chapter 8 (The Independence Problem). In Chapter 8 the basic data are a random sample of (X, Y) observations from a bivariate population. The independence hypothesis H_0 asserts that the X and Y variables are independent. The chapter is devoted to tests of H_0 and point and interval estimators of the strength of the association between X and Y. Sections 8.1 and 8.5 present distribution-free tests of H_0 that are based on sample correlation coefficients. Section 8.6 presents a distribution-free test of H_0 that detects a much broader class of alternatives to independence than the classes of alternatives detected by the tests of Sections 8.1 and 8.5. Section 8.2 contains a point estimator of a parameter τ that measures association between X and Y. Sections 8.3 and 8.4 present asymptotically distribution-free confidence intervals for τ.

Chapter 9 (Regression Problems). Chapter 9 is devoted to nonparametric regression techniques. Sections 9.1–9.4 consider a single linear relationship between a dependent (response) variable and an independent (predictor) variable. Section 9.1 presents a distribution-free test of the hypothesis that the slope of the regression line is a specified value. Section 9.2 provides a point estimator of the slope parameter. Section 9.3 describes a confidence interval for the slope parameter. Section 9.4 considers an estimator of the intercept of the regression line and shows how the estimated linear relationship can be used to predict dependent variable responses to additional values of the predictor variable. Section 9.5 describes an asymptotically distribution-free test of the hypothesis that k regression lines have the same slope, that is, the k regression lines are parallel. Section 9.6 is an introduction to rank-based regression analysis for more complicated regression relationships than that of a straight line. Section 9.7 provides short introductions to some recent developments in non-rank-based nonparametric regression methods.

Chapter 10 (Comparing Two Success Probabilities). Chapter 2 concerns inferential procedures for a single unknown success probability, p, based on the observed rate, $\hat{p}$, of success. In Chapter 10 we describe inferential procedures for comparing two unknown success probabilities, p_1 and p_2, on the basis of the corresponding rates of success in independent samples. Approximate tests and confidence intervals for $p_1 - p_2$ are presented in Section 10.1. Section 10.2 contains an exact test of the hypothesis $p_1 = p_2$. Section 10.3 introduces the odds ratio and presents tests, estimators, and confidence intervals for the odds ratio. Section 10.4 describes tests, estimators, and confidence intervals for analyzing k strata of 2×2 contingency tables.

Chapter 11 (Life Distributions and Survival Analysis). This chapter considers survival methods where the data are times to an endpoint event. Sections 11.1–11.4 consider various nonparametric classes of life distributions used to model aging. The classes include the increasing failure rate (IFR) class, the increasing failure rate average (IFRA) class, the new better than used (NBU) class, the new better than used in expectation (NBUE) class, and the decreasing mean residual life (DMRL) class. Each of the aforementioned classes can be used to model deleterious aging, and with each class is associated a dual class that is used to model beneficial aging. Sections 11.1–11.4 use a random sample from an unknown life distribution to test the hypothesis that the underlying distribution is an exponential distribution. Section 11.1 considers tests against IFR and IFRA alternatives. Section 11.2 considers tests against NBU and NBUE alternatives. Section 11.3 considers tests against DMRL alternatives and also presents a confidence band for the mean residual life function. Section 11.4 contains tests to detect alternatives corresponding to a turning point (from increasing to decreasing or from decreasing to increasing) in the mean residual life. Two cases are considered, namely, when the turning point is known and when it is not known. Section 11.5 presents a confidence band for an arbitrary distribution, not restricted to be a life distribution. Section 11.5 also considers goodness-of-fit tests that the underlying distribution is a specified distribution or a member of a specified parametric family. Section 11.6 presents an estimator of a survival function for the case when the data are randomly right-censored. Section 11.6 also contains confidence bands for the survival function and confidence bands for the quantile function. In Section 11.7 the data consist of two independent randomly right-censored right samples. We present a test of the hypothesis that the two underlying populations are equivalent.

1.5. FORMAT AND ORGANIZATION

The basic data, assumptions, and procedures are described precisely in each chapter according to the following format. *Data* and *Assumptions* are specified prior to the group of particular procedures discussed. Then, for each technique, we include (when applicable) the following subsections: *Procedure*, *Large-Sample Approximation*, *Ties*, *Example*, *Comments*, *Properties*, and *Problems*. We now describe the purpose of each subsection.

Procedure

This subsection contains a description of how to apply the procedure under discussion. If the procedure uses values from a special table, such as an exact null distribution table for a statistic used in a hypothesis test, the reader is referred to the appropriate table in Appendix A for the necessary values.

Large-Sample Approximation

This subsection contains an approximation to the method described in *Procedure*. The approximation is intended for use when the sample size (or sample sizes, as the case may be) is large. The importance of such an approximation depends on the particular procedure. If exact tables or software needed to implement the procedure are limited or unavailable, the approximation is useful because we can make a decision based on the approximation. Even if the procedure is such that tables or software are available, the large sample approximation may be preferred for convenience. Furthermore, by applying the approximation in situations where we also have exact results, we obtain information on the accuracy of the approximation.

Ties

A common assumption in the development of nonparametric procedures is that the underlying population(s) is (are) continuous. This assumption implies that the probability of obtaining tied observations is zero. Nevertheless, tied observations do occur in practice. These ties may arise when the underlying population is not continuous. They may even arise if the continuity assumption is valid. We simply may be unable, owing to inaccuracies in measurement, to distinguish between two very close observations (temperatures, lengths, etc.) that emanate from a continuous population. The *Ties* subsection contains a prescription to adjust the necessary steps in the *Procedure* in order that we may treat tied observations. The adjusted procedure should then be viewed as an approximation.

Example

This subsection is basic to our text. We present a problem in which the procedure under discussion is applicable. The reader has a set of data he or she may use to apply each step of the *Procedure*, to become familiar with our notation, and to gain familiarity in performing the method. In many examples, computations are done using `Minitab` or `StatXact` (see Section 1.6). In addition to practice, the example provides the first step toward developing an appreciation for the simplicity (difficulty) of the procedure and toward developing an intuitive feeling of how the procedure summarizes the data. The enthusiastic reader can seek out the journal article upon which the example is based to obtain a more detailed specification of the experiment (in some cases our descriptions of the experiments are simplified so that the examples can be easily explained) and to question whether the *Assumptions* underlying the nonparametric method are indeed satisfied.

Comments

The comments supplement the text. In the comments we may discuss the underlying assumptions, give an intuitive motivation for the method being considered, relate the method to other procedures in different parts of the book, provide helpful computational hints, or single out certain references.

Properties

This subsection is primarily intended as a set of directions for the reader who wishes to probe the theoretical aspects of the subject and, in particular, the theory of the procedure under discussion. No theory is presented, but the citations guide the reader to sources furnishing the basic properties and their derivations.

Problems

Typically, the first problem of each *Problems* subsection provides practice in applying the procedure just introduced. Some problems require a comparison of an exact procedure with its large-sample approximation. Other problems are more thought-provoking. We sometimes ask the reader to find or create an example that illustrates a desirable or undesirable property of the procedure under discussion.

There are occasional deviations from the format. For example, in many of the sections devoted to estimators and confidence intervals, there is no need for a *Ties* subsection, because the procedures described are well defined even when ties observations occur. In some chapters the *Assumptions* are given prior to the particular (group of) sections that contain procedures based on those *Assumptions*.

A few more remarks about the organization of this book are necessary. How do the nonparametric procedures we present compare with their classical competitors, which are based on parametric assumptions such as the assumption of normality for the underlying populations? The answer depends on the particular problem and procedures under consideration. When possible, we indicate a partial answer in an efficiency section at the end of each chapter.

We have already mentioned the appendices. Appendix A contains tables that are used to implement the nonparametric procedures we present. The specific table for a particular method is cited, by number, in the *Procedure* subsection describing the method. Appendix B contains some additional computer programs in the `S-plus` language (see Appendix B and the references therein). The programs can be used to implement some procedures described in Chapters 8 and 11. Appendix C contains a glossary. We explain some symbols that appear frequently in the text and define terms and concepts that may, as indicated previously, be unfamiliar to the reader who has had only an introductory nonmathematical statistics course.

We have relegated such definitions to an appendix in order to keep the body of the text directed to applications. Appendices B and C can be found in the Solutions Manual, the Instructors Manual, and on the Wiley website (http://www.wiley.com).

1.6. COMPUTING PACKAGES

In many of our *Example* subsections, we not only illustrate the direct computation of the procedure, but are also provide the output obtained by using various commands in standard statistical computing packages. In our text we have found it convenient to use `StatXact` and `Minitab`. `StatXact` is primarily devoted to nonparametric methods, whereas `Minitab` is a more general statistics computing package with some nonparametric programs.

StatXact

`StatXact` is a specialized software package for the exact analysis of small-sample categorical and nonparametric data. The term *small-sample* as used here applies not only to data sets with just a few observations but also to large but unbalanced data sets and to contingency tables with zeros and small cell counts in some of the cells but large cell counts in other cells. In these settings, `StatXact` produces exact P-values and exact confidence intervals instead of relying on possibly unreliable large-sample theory for its inferences. The inference is based on generating permutation distributions of the appropriate test statistics in a conditional reference set.

At the time of this writing `StatXact` offers exact P-values and confidence intervals for one-, two-, and k-sample problems, 2×2, $2 \times c$, and $r \times c$ contingency tables, and measures of association. The data may be either unstratified or stratified. Both independent and blocked samples are accommodated. `StatXact` has inference procedures that cater explicitly to binomial data, nominal categorical data, ordered categorical data, continuous complete data, and continuous right-censored data.

`StatXact` runs on all Microsoft Windows platforms as a stand-alone product. In addition, a special version, `PROC-StatXact` for `SAS` users, is available as external `SAS` procedures for both the Microsoft Windows and UNIX operating systems. It is available from Cytel Software Corporation, Cambridge, MA 02139 (http://www.cytel.com).

The underpinnings and improvements for `StatXact` are based on a series of papers by Mehta, Patel and their colleagues. These include Mehta and Patel (1983, 1986), Mehta, Patel, and Tsiatis (1984), Mehta, Patel, and Gray (1985), Mehta, Patel, and Senchaudhuri (1988), Agresti, Mehta, and Patel (1990), Strawderman and Mehta (1992), Mehta and Hilton (1993), and Senchaudhuri, Mehta, and Patel (1995).

Minitab

`Minitab` is a general-purpose statistical package that provides a wide range of data analysis capabilities. At the time of this writing, its nonparametric procedures include, among others, the one-sample sign test and associated confidence interval, the Wilcoxon signed rank test and associated confidence interval, the two-sample Mann-Whitney-Wilcoxon test and associated confidence interval, the Kruskal-Wallis test, and Friedman's test for randomized blocks. In general, `Minitab` uses a normal or chi-square approximation to calculate P-values.

`Minitab`'s general capabilities include, among others, basic statistics, regression, analysis of variance, design of experiments, multivariate analysis, time series, cross-tabulations and quality control methods.

Most analyses include graphical as well as printed output. In addition, `Minitab` has a very flexible capability that allows you to create your own graphs, from simple plots and histograms to highly sophisticated presentation-quality graphs. All graphs can be customized and edited. A powerful macro facility allows you to "program" your own commands. `Minitab` provides both a menu–dialog box interface and a command-line interface.

`Minitab` is available for all Microsoft platforms as stand-alone and network products, as well as for 30 other platforms including Macintosh, DOS, UNIX, VAX/MVS, and Solaris. `Minitab` also offers a student version that has capabilities for teaching at an introductory level. `Minitab` is distributed by Minitab Inc., 3081 Enterprise Drive, State College, PA 16801 (http://www.minitab.com).

Other nonparametric computing packages are `DISFREE`, which is tied in with the book by Krauth (1988), and `TESTIMATE`. Other standard statistical packages that include some nonparametric programs are `BMDP` (http://www.StatSolUSA.com), `SAS` (http://www.sas.com/rnd), `SPSS` (http://www.spss.com), `Stata` (http://www.stata.com), and `STATISTICA` (http://www.statsoft.com).

1.7. HISTORICAL BACKGROUND

Binomial probability calculations were used early in the eighteenth century by the British physician Arbuthnott (1710) (see Comment 2.13). Nevertheless, Savage (1953) (also see Savage (1962)) designated 1936 as the true beginning of the subject of nonparametric statistics, marked by the publication of the Hotelling and Pabst (1936) paper on rank correlation. Scheffé (1943), in a survey paper, pointed to (among others) the articles by Pearson (1900, 1911) and the presence of the sign test in Fisher's first edition of *Statistical Methods for Research Workers* (1925). Other important papers, in the late thirties, include those by Friedman (1937), Kendall (1938), and Smirnov (1939). Wilcoxon (1945), in a paper that is brief, yet elegant in its simplicity and usefulness, introduced his now-famous two-sample rank sum test and paired-sample signed rank test. The rank sum test was given by Wilcoxon only for equal sample sizes, but Mann and Whitney (1947) treated the general case. Wilcoxon's procedures played a major role in stimulating the development of rank-based procedures in the 1950s and 1960s, including rank procedures for multivariate situations (see Puri and Sen (1971)). Further momentum was provided by Pitman (1948), Hodges and Lehmann (1956), and Chernoff and Savage (1958), who showed that nonparametric rank tests have desirable efficiency properties relative to parametric competitors. An important advance that enabled nonparametric methods to be used in a variety of situations was the jackknife, introduced by Quenouille (1949) as a bias-reduction technique and extended by Tukey (1958, 1962) to provide approximate significance tests and confidence intervals.

There was major nonparametric research in the 1960s, and the most important contribution was that of Hodges and Lehmann (1963). They showed how to derive estimators from rank tests and established that these estimators have desirable properties. Their work paved the way for the nonparametric approach to be used to derive estimators in experimental design settings and for nonparametric testing and estimation in regression. Two seminal papers in the 1970s are those of Cox (1972) and Ferguson (1973). Cox's paper sparked research on nonparametric models and methods for survival analysis. The development of such methods is a very active area. See, for example, the advanced books by Fleming and Harrington (1991) and Andersen et al. (1993). Ferguson (1973) presented an approach (based on his Dirichlet process prior) to nonparametric Bayesian methods that combines the advantages of the nonparametric approach and the use of prior information incorporated in Bayesian procedures. Susarla and van Ryzin (1976) used Ferguson's approach to derive nonparametric

Bayesian estimators of survival curves. Dykstra and Laud (1981) used a different prior, the gamma process, to develop a Bayesian nonparametric approach to reliability. Hjort (1990b) proposed nonparametric Bayesian estimators based on using beta processes to model the cumulative hazard function. In the late 1980s and the 1990s there has been a surge of activity in Bayesian methods due to Markov chain Monte Carlo (MCMC) methods. See, for example, Gelfand and Smith (1990), Gamerman (1991), West (1992), Smith and Roberts (1993), and Arjas and Gasbarra (1994). Gilks, Richardson, and Spiegelhalter (1996) give a practical review. Key algorithms for developing and implementing modern Bayesian methods include the Metropolis-Hastings-Green algorithm (see Metropolis et al. (1953), Hastings (1970), and Green (1995)) and the Tanner-Wong (1987) data augmentation algorithm.

The most important advance in nonparametric statistics in the last two decades is Efron's (1979) bootstrap. Efron's computer-intensive method makes use of the (ever increasing) computational power of computers to provide standard errors and confidence intervals in many settings, including complicated situations where it is difficult, if not impossible, to use a parametric approach. Bootstrapping has been a subject of vigorous research in the 1980s and 1990s and is likely to continue to receive considerable attention. See the books by Efron and Tibshirani (1993) and Davison and Hinkley (1997).

The rapid development of nonparametric statistics over a period of 60 years is mirrored in the numerous books devoted to the subject. The first textbook on ranking methods was Kendall's 1948 book on rank correlation methods (see the fifth edition by Kendall and Gibbons (1990)). Elementary books in the 1950s included a 1956 book by Siegel (see the second edition by Siegel and Castellan (1988)) and Tate and Clelland (1957). The first theoretical book on nonparametric statistics was written by Fraser (1957). In the 1960s elementary books were written by Walsh (1962, 1965, 1968), and Kraft and van Eeden (1968). Noether (1967a) and Hájek (1969) wrote intermediate-level texts. Hájek and Šidák (1967) is an advanced theoretical text. Elementary books in the 1970s include a 1971 book by Conover (see the second edition, by Conover (1980)), Hollander and Wolfe (1973), a 1976 book by Gibbons (see the third edition, Gibbons (1997)), Marasuilo and McSweeny (1977), a 1978 book by Daniel (see the second edition, by Daniel (1990)), and Leach (1979). Intermediate-level books include a text by Gibbons in 1971 (see the third edition by Gibbons and Chakraborti (1992)), Lehmann (1975), and Randles and Wolfe (1979). Elementary books in the 1980s, not previously cited in this section are Krauth (1988), Neave and Worthington (1988), and a 1989 book by Sprent (see the second edition by Sprent (1993)). Intermediate-level books in the 1980s were written by Maritz (1981), Pratt and Gibbons (1981), and Hettmansperger (1984). New entries in the 1990s include a specialized theoretical book by Nitikin (1995) on asymptotic efficiency of nonparametric tests and a theoretical book by Hettmansperger and McKean (1998) on robust nonparametric statistical methods. Updated editions in the 1990s already mentioned are Daniel (1990), Kendall and Gibbons (1990), Gibbons and Chakraborti (1992), Sprent (1993), and Gibbons (1997).

Books on bootstrapping include Efron's (1982) monograph on the bootstrap, the jackknife and other resampling methods, Hall's (1992) theoretical book, Efron and Tibshirani's (1993) introductory book, Davison and Hinkley's (1997) intermediate-level book, and Manly's (1997) second edition on bootstrapping methods in biology. Efron and Tibshirani (1993) and Davison and Hinkley (1997) contain many practical illustrations.

Specialized works on subsets of the nonparametric field include the intermediate-level volume by Critchlow (1980) on metric methods for analyzing partially ranked data, the advanced book by Sen (1981) on the theory of sequential nonparametric methods, and the volume of papers edited by Fligner and Verducci (1993) on probability models and statistical analyses for ranking data. Books dealing with various aspects of nonparametric regression methods include Eubank (1988), Müller (1988), Carroll and Ruppert (1988), Härdle (1990),

Wahba (1990), Green and Silverman (1994), Wand and Jones (1995), Fan and Gijbels (1996), Hart (1997), and Hettmansperger and McKean (1998).

The books by Fleiss (1981), Agresti (1984), Agresti (1990) and Everitt (1992) are devoted to categorical data and contain some nonparametrical methods. Survival analysis and reliability books with nonparametric techniques include those of Kalbfleisch and Prentice (1980), Miller (1981b), Nelson (1982), Cox and Oakes (1984), Crowder, Kimber, Smith and Sweeting (1991), Fleming and Harrington (1991), Zacks (1992), Andersen, Borgan, Gill and Keiding (1993), and Klein and Moeschberger (1997).

There are many reasons why the development of nonparametric techniques—and the exposition of those techniques in books—continues at a vigorous pace. We delineated advantages of the nonparametric approach in Section 1.1. In addition to those practical advantages, the theory supporting nonparametric methods is elegant, and researchers find it challenging to advance the theory. The primary reasons for the success of nonparametric methods are the wide applicability and desirable efficiency properties of the procedures and the realization that it is sound statistical practice to use methods that do not depend on restrictive parametric assumptions because such assumptions often fail to be valid.

1.8. GUIDE

Table 1.1 provides a guide to the various procedures.

TABLE 1.1. A Guide for Finding the Desired Nonparametric Procedures

Topic or Problem	Data	Procedure	In This Section We Present
Binomial	Outcomes of independent Bernoulli trials	Binomial test (2.1)	A hypothesis test concerning p, the unknown success probability
		Binomial estimator (2.2)	An estimator of p
		Clopper-Pearson confidence interval (2.3)	A confidence interval for p
One-sample location	Paired replicates $(X_1, Y_1), \ldots, (X_n, Y_n)$	Wilcoxon signed rank test (3.1)	A hypothesis test concerning the treatment effect θ, the unknown median of $Y - X$
		Fisher sign test (3.4)	
		Hodges-Lehmann estimator based on median of Walsh averages (3.2)	An estimator of θ
		Estimator based on sample median (3.5)	
		Tukey confidence interval	A confidence interval for θ (3.3)
		Thompson, Savur confidence interval (3.6)	
	One sample $Z_1, \ldots, Z_n$ (the Z's can, but need not, be of the form $Z = Y - X$)	Signed rank procedures (3.7)	A test, estimator, and confidence interval for θ, the unknown median of Z
		Sign procedures (3.8)	
		Randles-Fligner-Policello-Wolfe, Davis-Quade symmetry test (3.9)	A test of the hypothesis that the Z population is symmetric about 0
	Bivariate sample $(X_1, Y_1), \ldots, (X_n, Y_n)$	Hollander bivariate symmetry test (3.10)	A test of the hypothesis of exchangeability
Two-sample location	Two samples, $X_1, \ldots, X_m$ and $Y_1, \ldots, Y_n$	Wilcoxon, Mann-Whitney rank sum test (4.1)	A test of the hypothesis that the X and Y samples are from the same population against the location alternative that the Y's are shifted an amount Δ (difference in medians between the Y and X populations)
		Hodges-Lehmann two-sample estimator (4.2)	An estimator of Δ
		Moses confidence interval (4.3)	A confidence interval for Δ
		Fligner-Policello test (4.4)	A rank test for the Behrens-Fisher problem
Two-sample dispersion	Two samples, $X_1, \ldots, X_m$ and $Y_1, \ldots, Y_n$	Ansari-Bradley test (5.1)	A test of the hypothesis that the X and Y samples are from the same population against the dispersion alternative that the unknown squared ratio of scale parameters $\gamma^2 = (\sigma_1^2/\sigma_2^2)$ differs from 1
		Miller jackknife test (5.2)	

	Lepage test (5.3)	A test of the hypothesis that the X and Y samples are from the same population against location and dispersion alternatives
	Kolmogorov-Smirnov test (5.4)	A test of the hypothesis that the X and Y samples are from the same population against broad alternatives that the populations differ
One-way layout (k samples)	Kruskal-Wallis test (6.1)	A test of the hypothesis that the k samples are from the same population ($\tau_1 = \cdots = \tau_k$) against location alternatives for which the τ's (the treatment effects) are not all equal
N X's are observed, where $X_{ij} = \mu + \tau_j + e_{ij}, i = 1,\ldots,n_j,$ $j = 1,\ldots,k$ $N = \sum_{j=1}^{k} n_j$	Jonckheere, Terpstra test (6.2)	A test of the hypothesis that the k samples are from the same population against the ordered alternatives $\tau_1 \leq \tau_2 \leq \cdots \leq \tau_k$
	Mack-Wolfe test, peak known (6.3A)	A test of the hypothesis $\tau_1 = \cdots = \tau_k$ against $\tau_1 \leq \cdots \leq \tau_p \geq \tau_{p+1} \geq \cdots \geq \tau_k$ for a specific p
	Mack-Wolfe test, peak known (6.3B)	A test of the hypothesis $\tau_1 = \cdots = \tau_k$ against $\tau_1 \leq \cdots \leq \tau_p \geq \tau_{p+1} \geq \cdots \geq \tau_k$ for some $p \in \{1, 2,\ldots, k\}$
	Fligner-Wolfe test (6.4)	A test of the hypothesis $\tau_i = \tau_1, i = 2,\ldots,k$
	Dwass, Steel, Critchlow-Fligner all-treatment multiple comparisons (6.5)	Multiple comparison procedures to select those pairs of treatment effects (τ_u, τ_v) for which $\tau_u \neq \tau_v$
	Hayter-Stone all-treatment multiple comparisons (6.6)	Multiple comparison procedures to select those pairs of treatment effects (τ_u, τ_v) with $\tau_v > \tau_u$ in a setting of ordered alternatives
	Nemenyi, Damico-Wolfe treatments versus control multiple comparisons (6.7)	Multiple comparison procedures to select those treatment effects τ_j, $j = 2,\ldots,k$, for which $\tau_j > \tau_1$ (where τ_1 corresponds to the control)

(*continued*)

TABLE 1.1. A Guide for Finding the Desired Nonparametric Procedures (*continued*)

Topic or Problem	Data	Procedure	In This Section We Present
		Spjøtvoll contrast estimator (6.8)	An estimator of contrasts in the τ's
		Critchlow-Fligner simultaneous confidence intervals (6.9)	Simultaneous confidence intervals for the collection C of all $\binom{k}{2}$ simple contrasts given by $C = \{\Delta_{uv} : \Delta_{uv} = \tau_v - \tau_u, 1 \leq u < v \leq k\}$
Two-way layout	$N = \sum_{i=1}^{n} \sum_{j=1}^{k} c_{ij}$ X's are observed $X_{ij} = \theta + \beta_i + \tau_j + e_{ijt}, i = 1, \ldots, n;$ $j = 1, \ldots, k; t = 1, \ldots, c_{ij}$	Friedman, Kendall-Babington Smith test (7.1) Doksum test (7.11)	A test, for the case where each $c_{ij} = 1$, of the hypothesis $\tau_1 = \cdots = \tau_k$ against location alternatives for which the τ's are not all equal
		Page test (7.2) Hollander test (7.12)	A test, for the case where each $c_{ij} = 1$, of the hypothesis $\tau_1 = \cdots = \tau_k$ against the ordered alternatives $\tau_1 \leq \tau_2 \leq \cdots \leq \tau_k$
		Wilcoxon, Nemenyi, McDonald-Thompson all-treatment multiple comparisons based on Friedman rank sums (7.3) Nemenyi all-treatment multiple comparisons based on Wilcoxon signed ranks (7.13)	Multiple comparison procedures, for the case where each $c_{ij} = 1$, to select those pairs of treatment effects (τ_u, τ_v) for which $\tau_u \neq \tau_v$
		Nemenyi, Wilcoxon-Wilcox, Miller treatment versus control multiple comparisons based on Friedman rank sums (7.4) Hollander treatment versus control multiple comparisons based on Wilcoxon signed ranks (7.14)	Multiple comparison procedures, for the case where each $c_{ij} = 1$, to select those treatment effects τ_j for which $j = 2, \ldots, k$ for which $\tau_j > \tau_1$ (where τ_1 corresponds to the control)
		Doksum estimator (7.5) Lehmann estimator (7.15)	An estimator, for the case where each $c_{ij} = 1$, of contrasts in the τ's
		Durbin, Skillings-Mack test (7.6)	A test of $\tau_1 = \cdots = \tau_k$, against general alternatives, in balanced incomplete block designs
		Skillings-Mack all-treatment multiple comparisons (7.7)	Multiple comparison procedures, to select those pairs of treatment effects (τ_u, τ_v) for which $\tau_u \neq \tau_v$, in balanced incomplete block designs

		Skillings–Mack test (7.8)	A test of $\tau_1 = \cdots = \tau_k$, against general alternatives, in an arbitrary incomplete block design
		Mack–Skillings test (7.9)	A test of $\tau_1 = \cdots = \tau_k$, against general alternatives, in a randomized block design with an equal number (>1) of replications per treatment-block combination
		Mack–Skillings all-treatment multiple comparisons (7.10)	Multiple comparison procedures, to select those pairs of treatment effects (τ_u, τ_v) for which $\tau_u \neq \tau_v$, in a two-way layout with an equal number of replications in each treatment-block combination
Independence	Bivariate observations $(X_1, Y_1), \ldots, (X_n, Y_n)$	Kendall test (8.1) Spearman test (8.5) Hoeffding test (8.6)	A test of the hypothesis that the X and Y variables are independent
		Estimator associated with Kendall's statistic (8.2)	An estimator of the parameter $\tau = 2P\{(X_1 - X_2)(Y_1 - Y_2) > 0\}$
		Samara–Randles, Fligner-Rust, Noether confidence interval (8.3)	A confidence interval for τ
		Efron bootstrap confidence interval (8.4)	
Regression One regression line	Observations Y_i at fixed points x_i, $i = 1, \ldots, n$, where $Y_i = \alpha + \beta x_i + e_i$	Theil test (9.1)	A hypothesis test concerning β, the slope of the regression line
		Theil estimator (9.2)	An estimator of β
		Theil confidence interval (9.3)	A confidence interval for β
		Intercept estimator and prediction (9.4)	An estimator of α, the intercept of the regression line, and use of the full estimated regression line for prediction
Several regression lines	Observations Y_{ij} at fixed points x_{ij}, $i = 1, \ldots, k, j = 1, \ldots, n_i$, where $Y_{ij} = \alpha_i + \beta_i x_{ij} + e_{ij}$	Sen, Adichie test (9.5)	A test of the hypothesis $\beta_1 = \cdots = \beta_k$, that is, the k regression lines are parallel

(*continued*)

TABLE 1.1. A Guide for Finding the Desired Nonparametric Procedures (*continued*)

Topic or Problem	Data	Procedure	In This Section We Present
General multiple linear regression	Observations Y_i at vectors of fixed points $x_i' = (x_{i1}, \ldots, x_{pi})$, $i = 1, \ldots, n$ where $Y_i = \xi + \beta_1 x_{1i} + \cdots + \beta_p x_{pi} + e_i$ and $\beta' = (\beta_1, \ldots, \beta_p)$ is a row vector of regression coefficients	Jaeckel, Hettmansperger-McKean test (9.6)	A hypothesis test of $\beta_q' = (\beta_1, \ldots, \beta_q) = (0, \ldots, 0)$, with $\beta_{p-q}' = (\beta_{q+1}, \ldots, \beta_p)$ and ξ unspecified, that is, that the fixed independent variables $x_1, \ldots, x_q$ do not play significant roles in determining the value of the dependent variable Y
Nonparametric regression analysis	Observations Y_i at fixed points x_i, $i = 1, \ldots, n$, where $Y_i = \mu(x_i) + e_i$ and $\mu(\cdot)$ is the arbitrary regression function relating the dependent variable Y to the independent variable x	(9.7)	General discussion of a variety of modern nonparametric approaches to the estimation of the regression function $\mu(\cdot)$
Comparing two rates	Outcomes of two independent sets of Bernoulli trials	Pearson test (10.1) Fisher test (10.2)	A test of the hypothesis that $p_1 = p_2$
		Inference for the odds ratio (Fisher, Cornfield) (10.3)	Tests, estimators, and confidence intervals for the odds ratio parameter
	k strata; within each stratum a 2×2 contingency table	Mantel-Haenszel test (10.4)	A test of the hypothesis that within each stratum, the success probabilities are equal
Life distributions and survival analysis	One sample $X_1, \ldots, X_n$ from a life distribution	Epstein IFR test (11.1) Hollander-Proschan NBU test (11.2)	A test of the hypothesis that the sample is from an exponential distribution
		Hollander-Proschan DMRL test (11.3) Guess-Hollander-Proschan turning point test (turning point known), Hawkins-Kochar-Loader turning point test (turning point unknown) (11.4)	
		Hall-Wellner confidence bands (11.3)	Confidence bands for the mean residual life function

One sample $X_1, \ldots, X_n$ from an arbitrary distribution	Kolmogorov confidence band (11.5)	A confidence band for the distribution function
One censored sample from a life distribution	Kaplan-Meier estimator (11.6)	An estimator of the survival function
	Hall-Wellner confidence bands, Hollander-Peña confidence bands (11.6)	Confidence bands for the survival function
	Li-Hollander-McKeague-Yang confidence bands (11.6)	Confidence bands for the quantile function
Two independent censored samples, each from a life distribution	Mantel test (11.7)	A test of the hypothesis that the two life distributions are equal

CHAPTER 2

The Dichotomous Data Problem

INTRODUCTION

In this chapter the data consist of the dichotomous outcomes of independent repeated Bernoulli trials having constant probability of success p. On the basis of these outcomes, we wish to make inferences about the parameter p.

Section 2.1 presents a binomial test for the hypothesis that $p = p_0$, where p_0 is a specified number; Section 2.2 gives a point estimator for p; and Section 2.3 gives a class of confidence intervals for p.

Data. We observe the outcomes of n independent repeated Bernoulli trials.

Assumptions

A1. The outcome of each trial can be classified as a *success* or a *failure*.
A2. The probability of a success, denoted by p, remains constant from trial to trial.
A3. The n trials are independent.

2.1. A BINOMIAL TEST

Procedure

To test

$$H_0 : p = p_0, \tag{2.1}$$

where p_0 is some specified number, $0 < p_0 < 1$, set

$$B = \text{number of successes.} \tag{2.2}$$

a. One-Sided Upper-Tail Test. To test

$$H_0 : p = p_0$$

versus

$$H_1 : p > p_0$$

at the α level of significance,

$$\text{Reject } H_0 \text{ if } B \geq b_\alpha; \text{ otherwise do not reject,} \qquad (2.3)$$

where the constant b_α is chosen to make the type I error probability equal to α. Values of b_α are given in Table A.2; b_α is the upper α percentile point of the binomial distribution with sample size n and success probability p_0.

b. One-Sided Lower-Tail Test. To test

$$H_0 : p = p_0$$

versus

$$H_2 : p < p_0$$

at the α level of significance,

$$\text{Reject } H_0 \text{ if } B \leq c_\alpha; \text{ otherwise do not reject.} \qquad (2.4)$$

Values of c_α can also be determined from Table A.2. This is illustrated in Comment 3. Here, c_α is the lower α percentile point of the binomial distribution with sample size n and success probability p_0. For the special case of testing $p = \frac{1}{2}$,

$$c_\alpha = n - b_\alpha. \qquad (2.5)$$

Equation (2.5) is explained in Comment 4.

c. Two-Sided Test. To test

$$H_0 : p = p_0$$

versus

$$H_3 : p \neq p_0$$

at the α level of significance,

$$\text{Reject } H_0 \text{ if } B \geq b_{\alpha_1} \text{ or } B \leq c_{\alpha_2}; \text{ otherwise do not reject,} \qquad (2.6)$$

where b_{α_1} is the upper α_1 percentile point, c_{α_2} is the lower α_2 percentile point, and $\alpha_1 + \alpha_2 = \alpha$.

Large-Sample Approximation

The large-sample approximation is based on the asymptotic normality of B, suitably standardized. To standardize, we need to know the mean and variance of B when the null hypothesis is true. When H_0 is true, the mean and variance of B are, respectively,

$$E_{p_0}(B) = np_0, \qquad (2.7)$$

$$\text{var}_{p_0}(B) = np_0(1 - p_0). \qquad (2.8)$$

Comment 8 gives the derivations for (2.7) and (2.8).

The standardized version of B is

$$B^* = \frac{B - E_{p_0}(B)}{\{\text{var}_{p_0}(B)\}^{1/2}} = \frac{B - np_0}{\{np_0(1-p_0)\}^{1/2}}. \quad (2.9)$$

When H_0 is true, B^* has, as n tends to infinity, an asymptotic $N(0, 1)$ distribution. Let z_α denote the upper α percentile point of the $N(0, 1)$ distribution. Values of z_α are given in Table A.1.

The normal approximation to procedure (2.3) is

$$\text{Reject } H_0 \text{ if } B^* \geq z_\alpha; \text{ otherwise do not reject.} \quad (2.10)$$

The normal approximation to procedure (2.4) is

$$\text{Reject } H_0 \text{ if } B^* \leq -z_\alpha; \text{ otherwise do not reject.} \quad (2.11)$$

The normal approximation to procedure (2.6), with $\alpha_1 = \alpha_2 = \alpha/2$, is

$$\text{Reject } H_0 \text{ if } |B^*| \geq z_{\alpha/2}; \text{ otherwise do not reject.} \quad (2.12)$$

Example 2.1: *Canopy Gap Closure.* Dickinson, Putz, and Canham (1993) investigated canopy gap closure in thickets of the clonal shrub *Cornus racemosa*. Shrubs often form dense clumps where tree abundance has been kept artificially low (for example, on power-line right-of-ways). These shrub clumps then retard reinvasion of the sites by trees. Individual clumps may persist for many years. Clumps outlast the lives of the individual stems of which they are formed; stems die and leave temporary holes in the canopies of the clumps. Closure of the hole (gap) left by dead stems occurs in part by the lateral growth of branches of stems that surround the hole. Opening of the gap often occurs when individual branches of hole-edge stems die. Between sample dates, more branches in six out of seven gaps in clumps, at a site with nutrient-poor and dry soil, died than lived. Let us say we have a success if more branches die than live in the gaps in clumps. Let p denote the corresponding probability of success. We suppose that the success probability for sites that are nutrient rich with moist soil has been established by previous studies to be 15%. Do the nutrient-poor and dry soil sites have the same success probability as the nutrient rich and moist soil sites or is it larger? This reduces to the hypothesis-testing problem

$$H_0 : p = .15$$

versus

$$H_1 : p > .15.$$

Our sample size is $n = 7$ and we observe $B = 6$ successes. Suppose we wish to perform a test with type I error probability $\alpha = .0121$. From Table A.2 entered at $n = 7$, $p = .15$, we find $b_{.0121} = 4$. Thus the $\alpha = .0121$ test is

$$\text{Reject } H_0 \text{ if } B \geq 4; \text{ otherwise do not reject.}$$

Since $B = 6$, we reject H_0 at $\alpha = .0121$. To find the P-value, we use Table A.2 to determine $P_{.15}\{B \geq 6\} = .0001$. (The notation $P_{.15}\{B \geq 6\}$ is shorthand for the probability that $B \geq 6$, computed under the assumption that the true success probability is .15.) Thus the P-value

2.1. A BINOMIAL TEST   23

is .0001. This is the smallest significance level at which we can reject H_0 (in favor of the alternative $p > .15$) with our observed value of B. We conclude that there is strong evidence against H_0 favoring the alternative. For more on the P-value, see Comment 9.

Example 2.2: *Sensory Difference Tests.* Sensory difference tests are often used in quality control and quality evaluation. The triangle test (cf. Bradley (1963)) is a sensory difference test that provides a useful application of the binomial model. In its simplest form, the triangle procedure is as follows. To each of n panelists three test samples are presented in a randomized order. Two of the samples are known to be identical; the third is different. The panelist is then supposed to select the odd sample, perhaps on the basis of a specified sensory attribute. If the panelists are homogeneous trained judges, the experiment can be viewed as n independent repeated Bernoulli trials, where a success corresponds to a correct identification of the odd sample. (If the panelists are not homogeneous trained judges, we may question the validity of Assumption A2.) Under the hypothesis that there is no basis for discrimination, the probability p of success is $\frac{1}{3}$, whereas a basis for discrimination would correspond to values of p that exceed $\frac{1}{3}$.

Byer and Abrams (1953) considered triangular bitterness tests in which each taster received three glasses, two containing the same quinine solution and the third a different quinine solution. In their first bitterness test, the solutions contained .0075% and .0050%, respectively, of quinine sulfate. The six presentation orders, LHH, HLH, HHL, HLL, LHL, LLH (L denotes the lower concentration, H the higher concentration), were randomly distributed among the tasters. Out of 50 trials, there were 25 correct selections and 25 incorrect selections.

We consider the binomial test of $H_0 : p = \frac{1}{3}$ versus the one-sided alternative $p > \frac{1}{3}$, and use the large-sample approximation to (2.3). We set $\alpha = .05$ for purposes of illustration. From Table A.1 we find $z_{.05} = 1.645$ and thus approximation (2.10), at the $\alpha = .05$ level, reduces to:

Reject H_0 if $B^* \geq 1.645$; otherwise do not reject.

From the data we have $n = 50$ and B (the number of correct identifications) $= 25$. Thus from (9), with $p_0 = \frac{1}{3}$, we obtain

$$B^* = \frac{25 - 50(\frac{1}{3})}{\{50(\frac{1}{3})(\frac{2}{3})\}^{1/2}} = 2.5.$$

Since $B^* = 2.5 > 1.645$, we reject $H_0 : p = \frac{1}{3}$ in favor of $p > \frac{1}{3}$ at the approximate $\alpha = .05$ level. Thus there is evidence of a basis for discrimination in the taste bitterness test. (From Table A.1 we also note that $z_{.0062} = 2.5$. Thus the smallest significance level at which we reject H_0 in favor of $p > \frac{1}{3}$ is .0062.)

Comments

1. *Binomial Test Procedures.* Assumptions A1–A3 are the general assumptions underlying a binomial experiment. Research problems possessing these assumptional underpinnings are common, and thus the binomial test procedures find frequent use. A particularly important special case in which procedures (2.3), (2.4), and (2.6) are applicable occurs when we wish to test hypotheses about the unknown median, θ, of a population. The application of binomial theory to this problem leads to a test statistic, B, that counts the number of sample observations larger than a specified null hypothesis value of θ, say θ_0. For this particular special case, the statistic B is referred to as the sign statistic, and the associated test procedures are referred to

as sign test procedures. See Sections 3.4 and 3.8 for a more detailed discussion of the sign test procedures corresponding to (2.3), (2.4), and (2.6).

2. *Distribution-Free Test.* The critical constant b_α of (2.3) is chosen so that the probability of rejecting H_0, when H_0 is true, is α. We can control this type I error because Assumptions A1–A3 and a specification of p (the null hypothesis specifies p to be equal to p_0) determine, without further assumptions regarding the underlying populations from which the dichotomous data emanate, the probability distribution of B. Thus under Assumptions A1–A3, the test in (2.3) is said to be a distribution-free test of H_0. The same statement can be made for tests (2.4) and (2.6).

3. *Use of Binomial Tables.* Suppose $n = 8$ and we wish to test $H_0 : p = .4$ versus $p > .4$ via procedure (2.3). From Table A.2 we find

b	0	1	2	3	4	5	6	7	8
$P_{.4}\{B \geq b\}$	1	.9832	.8936	.6846	.4059	.1737	.0498	.0085	.0007

(Recall that the $P_{.4}$ notation indicates that the probabilities are computed under the assumption that $p = .4$.) Hence we can find constants b_α that satisfy the equation $P_{.4}\{B \geq b_\alpha\} = \alpha$ only for certain values of α. For $\alpha = .0085$, $b_{.0085} = 7$. For $\alpha = .0498$, $b_{.0498} = 6$. As α increases, the critical constant b_α decreases. Thus when we increase α it is easier to reject H_0; hence we increase the power or, equivalently, decrease the probability of a type II error for our test (against a particular alternative). Similarly, if we lower α, we raise the probability of a type II error. This is illustrated in Comment 9.

Again consider the case $n = 8$ and suppose we want to test $p = .4$ versus the alternative $p < .4$. We can use the lower-tail test described by (2.4). For example, suppose we want $\alpha = .1064$. From Table A.2 we find $P_{.4}\{B \geq 2\} = .8936$. Hence $P_{.4}\{B \leq 1\} = 1 - .8936 = .1064$. Thus in (2.4), $c_{.1064} = 1$ and this yields the $\alpha = .1064$ test; namely, reject H_0 if $B \leq 1$ and accept H_0 if $B > 1$.

We close this comment with an example of the two-sided test described by (2.6). For convenience, we stay with the case $n = 8$ and test $H_0 : p = .4$. Since 6 is the upper .0498 percentile point of the null distribution of B and since 1 is the lower .1064 percentile point, the test that rejects H_0 when $B \geq 6$ or when $B \leq 1$ and accepts H_0 when $1 < B < 6$ is an $\alpha = .0498 + .1064 = .1562$ two-tailed test.

4. *Binomial Distribution.* The statistic B has been defined as the number of successes in n independent Bernoulli trials, each trial having a success probability equal to p. The distribution of the random variable B is known as the binomial distribution with parameters n and p.

For the special case when $p = \frac{1}{2}$, it can be shown that the distribution of B is symmetric about its mean $n/2$. This implies that

$$P_{.5}\{B \geq x\} = P_{.5}\{B \leq (n - x)\} \quad \text{for} \quad x = 0, \ldots, n. \tag{2.13}$$

Equation (2.13) implies that the lower α percentile point of the binomial distribution, with $p = .5$, is equal to n minus the upper α percentile point. That result was expressed as equation (2.5) after we introduced the lower-tail test given by (2.4).

5. *Motivation for the Test Based on B.* The statistic B/n is an estimator (see Section 2.2) of the true unknown parameter p. Thus if $p > p_0$, B/n will tend to be larger than p_0. This suggests rejecting $H_0 : p = p_0$ in favor of $p > p_0$ for large values of B and serves as partial motivation for (2.3).

6. *An Example of the Exact Distribution of B.* The exact distribution of B can be obtained from the equation

2.1. A BINOMIAL TEST

$$B = \sum_{i=1}^{n} \psi_i, \tag{2.14}$$

where

$$\psi_i = \begin{cases} 1, & \text{if the } i\text{th Bernoulli trial is a success,} \\ 0, & \text{if the } i\text{th Bernoulli trial is a failure.} \end{cases}$$

We consider the 2^n possible outcomes of the configurations $(\psi_1, \ldots, \psi_n)$ and use the fact that under H_0 any outcome with b 1s and $(n - b)$ 0s has probability $p_0^b(1 - p_0)^{n-b}$. For example, in the case $n = 2$, $p_0 = \frac{1}{4}$, the $2^2 = 4$ possible outcomes for (ψ_1, ψ_2) and associated values of B are as follows:

(ψ_1, ψ_2)	$P_{.25}\{(\psi_1, \psi_2)\}$	$B = \psi_1 + \psi_2$
$(0,0)$	$\left(\frac{1}{4}\right)^0 \left(\frac{3}{4}\right)^{2-0} = \frac{9}{16}$	0
$(0,1)$	$\left(\frac{1}{4}\right)^1 \left(\frac{3}{4}\right)^{2-1} = \frac{3}{16}$	1
$(1,0)$	$\left(\frac{1}{4}\right)^1 \left(\frac{3}{4}\right)^{2-1} = \frac{3}{16}$	1
$(1,1)$	$\left(\frac{1}{4}\right)^2 \left(\frac{3}{4}\right)^{2-2} = \frac{1}{16}$	2

Thus, for example, $P_{.25}\{B \geq 1\} = P_{.25}\{B = 1\} + P_{.25}\{B = 2\} = \frac{6}{16} + \frac{1}{16} = \frac{7}{16}$. (See the $p = .25, n = 2, b = 1$ entry of Table A.2.)

7. *The Exact Distribution of B.* By methods similar to the particular case illustrated in Comment 6, it can be shown that for each of the $n + 1$ possible values of B (namely, $b = 0, \ldots, n$), we have, when $p = p_0$,

$$P\{B = b\} = \binom{n}{b} p_0^b (1 - p_0)^{n-b}. \tag{2.15}$$

In (2.15) the symbol $\binom{n}{b}$ (read "binomial n, b") is given by

$$\binom{n}{b} = \frac{n!}{b!(n-b)!}, \tag{2.16}$$

where the symbol $m!$ (read "m factorial") is, for positive integers, defined as $m! = m(m - 1)(m - 2) \cdots (3)(2)(1)$, and 0! is defined to be equal to 1. The number $\binom{n}{b}$ is known as the number of combinations of n things taken b at a time. It is equal to the number of subsets of size b that may be formed from the members of a set of size n.

8. *The Asymptotic Distribution of B.* Using representation (2.14), we find, when $p = p_0$, the mean of B is

$$E_{p_0}(B) = E_{p_0}\left(\sum_{i=1}^{n} \psi_i\right) = \sum_{i=1}^{n} E_{p_0}(\psi_i) = np_0$$

where we have used the calculation

$$E_{p_0}(\psi_i) = 1 \cdot P(\psi_i = 1) + 0 \cdot P(\psi_i = 0) = 1 \cdot p_0 + 0 \cdot (1 - p_0) = p_0.$$

Then, using the fact that $\psi_1, \psi_2, \ldots, \psi_n$ are independent,

$$\text{var}_{p_0}(B) = \text{var}_{p_0}\left(\sum_{i=1}^n \psi_i\right) = \sum_{i=1}^n \text{var}_{p_0}(\psi_i). \tag{2.17}$$

The variance of any one of the indicator random variables ψ_i is determined as follows. Since $\psi_i^2 = \psi_i$,

$$E_{p_0}(\psi_i^2) = E_{p_0}(\psi_i) = p_0,$$

and thus

$$\text{var}_{p_0}(\psi_i) = E_{p_0}(\psi_i^2) - \{E_{p_0}(\psi_i)\}^2 = p_0 - p_0^2 = p_0(1 - p_0).$$

Hence, from (2.17),

$$\text{var}_{p_0}(B) = \sum_{i=1}^n p_0(1 - p_0) = np_0(1 - p_0).$$

Since B is a sum of independent and identically distributed random variables, the central limit theorem (cf. Casella and Berger (1990, 216)) establishes that, as $n \to \infty$, $(B - np_0)/\sqrt{np_0(1 - p_0)}$ has a limiting $N(0, 1)$ distribution.

9. *The P-Value.* Rather than specify an α level, and report whether the test rejects at that specific α level, it is more informative to state the lowest significance level at which we can reject with the observed data. This is called the *P*-value. Consider the $\alpha = .0085$ test (test T_1, say) and the $\alpha = .0498$ test (T_2) of $H_0 : p = .4$ versus $p > .4$ for the case $n = 8$. Suppose in an actual experiment that our observed value of B is 7. Then with test T_2 we reject H_0 since the critical region for test T_2 consists of the values $\{B = 6, B = 7, B = 8\}$ and our observed value 7 is in the critical region. Thus, it is correct for us to state that the value $B = 7$ is significant at the $\alpha = .0498$ level. But the value $B = 7$ is also significant at the $\alpha = .0085$ level. If we simply state that we reject H_0 at the .0498 level, we do not convey the additional information that, with the value $B = 7$, we also can reject H_0 at the .0085 level. To remedy this, the following approach is suggested.

Suppose, as in the previous example, large values of some statistic S (say) lead to rejection of the null hypothesis. Let s denote the observed value of S. Compute $P_0\{S \geq s\}$, the probability, under the null hypothesis, that S will be greater than or equal to s. This is the lowest level at which we can reject H_0. The observation $S = s$ will be significant at all levels greater than or equal to $P_0\{S \geq s\}$ and not significant at levels less than $P_0\{S \geq s\}$.

To further illustrate this point, consider the test of $p = \frac{1}{3}$ versus $p > \frac{1}{3}$ of Example 2.2. We apply procedure (2.10), based on the large-sample approximation to the null distribution of B. The (approximate) $\alpha = .05$ test rejects if $B^* \geq 1.645$ and accepts otherwise. Since the observed value of B^* is $B^* = 2.5$, we can reject $p = \frac{1}{3}$ in favor of $p > \frac{1}{3}$ at the .05 level. From Table A.1, however, we find $z_{.0062} = 2.5$. Thus the smallest significance level at which we can reject is approximately .0062, and this statement is more informative than the statement that the .05 test leads to rejection.

10. *Calculating Power.* Take $n = 8$, and consider the following two tests of $H_0 : p = .4$ versus $p > .4$, based on (2.3). Test T_1, corresponding to $\alpha = .0085$, rejects H_0 if $B \geq 7$ and accepts otherwise. Test T_2, corresponding to $\alpha = .0498$, rejects H_0 if $B \geq 6$ and accepts otherwise. Suppose, in fact, that the alternative $p = .5$ is true. Let R_1 denote the power of test

2.1. A BINOMIAL TEST

T_1 (for this alternative) and let R_2 denote the power of test T_2. Thus, R_1 is the probability of rejecting H_0 with test T_1 and R_2 is the probability of rejecting H_0 with test T_2. These powers are to be calculated when the alternative $p = .5$ is true. Entering Table A.2 at $p = .5$, we find

$$R_1 = P_{.5}\{B \geq 7\} = .0352, \qquad R_2 = P_{.5}\{B \geq 6\} = .1445.$$

For the alternative $p = .5$, let β_1 denote the probability of a type II error using test T_1 and let β_2 denote the probability of a type II error using test T_2. We find

$$\beta_1 = 1 - R_1 = .9648, \qquad \beta_2 = 1 - R_2 = .8555.$$

Test T_1 has a lower probability of a type I error than test T_2, but the probability of a type II error for test T_1 exceeds that of test T_2. Incidentally, the reader should not be shocked at the very high values of β_1 and β_2. The alternative $p = .5$ is quite close to the null hypothesis value $p = .4$, and a sample of size 8 is simply not large enough to make a better (in terms of power) distinction between the hypothesis and alternative.

11. *More Power Calculations.* We return to Example 2 concerning sensory difference tests. Suppose we have $n = 50$ and we decide to employ the approximate $\alpha = .05$ level test of $H_0 : p = \frac{1}{3}$ versus $H_1 : p > \frac{1}{3}$. Recall that test rejects H_0 if

$$\frac{B - n\left(\frac{1}{3}\right)}{\left\{n\left(\frac{1}{3}\right)\left(\frac{2}{3}\right)\right\}^{1/2}} > 1.645$$

and accepts H_0 otherwise. What is the power of this test if in fact $p = .6$? We approximate the power using the asymptotic normality of B, suitably standardized. If $p = .6$, then

$$\frac{B - n(.6)}{\{n(.6)(.4)\}^{1/2}},$$

has an approximate $N(0, 1)$ distribution. Using this we find

$$\text{Power} = P_{.6}\left\{\frac{B - n\left(\frac{1}{3}\right)}{\left\{n\left(\frac{1}{3}\right)\left(\frac{2}{3}\right)\right\}^{1/2}} > 1.645\right\}$$

$$= P_{.6}\{B > [(1.645)\{n\left(\tfrac{1}{3}\right)\left(\tfrac{2}{3}\right)\}^{1/2}] + n\left(\tfrac{1}{3}\right)\}$$

$$= P_{.6}\left\{\frac{B - n(.6)}{\{n(.6)(.4)\}^{1/2}} > \left[\frac{(1.645)\{n\left(\tfrac{1}{3}\right)\left(\tfrac{2}{3}\right)\}^{1/2} + n\left(\tfrac{1}{3}\right) - n(.6)}{\{n(.6)(.4)\}^{1/2}}\right]\right\}$$

$$\doteq P\{Z > -2.27\}$$

where $Z = \{B - n(.6)\}/\{n(.6)(.4)\}^{1/2}$ is approximately a $N(0, 1)$ random variable and -2.27 is the value, when $n = 50$, of the term in large square brackets. From Table A.1 we thus find

$$\text{Power} \doteq P\{Z > -2.27\} = .9884.$$

12. *Counting Failures instead of Successes.* Define B^- to be the number of failures in the n Bernoulli trials. Note that B^- could be defined by (2.13) with ψ_i replaced by $(1 - \psi_i)$, for $i = 1, \ldots, n$. Test procedures (2.3), (2.4), and (2.6) could equivalently be based on B^-, since $B^- = n - B$.

13. *Some History.* The binomial distribution has been utilized for statistical inferences about dichotomous data for more than 250 years. Binomial probability calculations were used by the British physician Arbuthnott (1710) in the early eighteenth century as an argument for the sexual balance maintained by Divine Providence and against the practice of polygamy. Bernoulli trials are so named in honor of Jacques Bernoulli. His book *Ars Conjectandi* (1713) contains a profound study of such trials and is viewed as a milestone in the history of probability theory. (LeCam and Neyman (1965) reported that the original Latin edition was followed by several in modern languages; the last republication, in German, appeared in 1899 in No. 107 and No. 108 of the series *Ostwald's Klassiker der exakten Wissenschaften*, Wilhelm Engelman, Leipzig.) Today the binomial procedures remain one of the easiest and most useful sets of procedures in the statistical catalog.

Properties

1. *Consistency.* Test procedures (2.3), (2.4), and (2.6) will be consistent against alternatives for which $p >, <$, and $\neq p_0$, respectively.

PROBLEMS

1. Stanton (1969) investigated the problem of paroling criminal offenders. He studied the behavior of all male criminals paroled from New York's correctional institutions to original parole supervision during 1958 and 1959 (exclusive of those released to other warrants or to deportation). The parolees were observed for 3 years following their releases or until they exhibited some delinquent parole behavior. In a study involving a very large number of subjects, Stanton considered criminals convicted of crimes other than first- or second-degree murder. He found that approximately 60% of these parolees did not have any delinquent parole behavior during the 3 years following their releases.

    During the same period, Stanton found that 56 of the 65 paroled murderers (first- or second-degree murderers who were also original parolees) in the study had no delinquent parole behavior. Let a success correspond to a male murderer on original parole who does not exhibit any delinquent parole behavior in the 3-year observation period. Note that we could question Assumption A2 in this context—parolees convicted of first-degree murder may have a different success probability than parolees convicted of second-degree murder. Even the parolees in the first-degree (or second-degree) group may have different individual success probabilities. For pedagogical purposes, we proceed as if Assumption A2 is valid and denote the common success probability by p.

    It is of interest to investigate whether murderers are better risks as original parolees than are criminals convicted of lesser crimes. This suggests testing $H_0: p = .6$ against the alternative $p > .6$. Perform this test using the large-sample approximation to procedure (2.3).

2. Describe a situation in which Assumptions A1 and A2 hold but Assumption A3 is violated.
3. Describe a situation in which Assumptions A1 and A3 hold but Assumption A2 is violated.
4. Suppose that ten Bernoulli trials satisfying Assumptions A1–A3 result in eight successes. Investigate the accuracy of the large-sample approximation by comparing the smallest significance level at which we would reject $H_0: p = \frac{1}{2}$ in favor of $p > \frac{1}{2}$ when using procedure (2.3) with the corresponding smallest significance level for the large-sample approximation to procedure (2.3) given by (2.10).
5. Return to the $\alpha = .0121$ test of Example 2.1. Recall that test of $H_0: p = .15$ versus $H_1: p > .15$ rejects H_0 if in $n = 7$ trials there are 4 or more successes and accepts H_0 if there are 3 or fewer successes. What is the power of that test when (a) $p = .4$; when (b) $p = .6$; and when (c) $p = .8$?
6. A standard surgical procedure has a success rate of .7. A surgeon claims a new technique improves the success rate. In 20 applications of the new technique, there are 18 successes. Is there evidence to support the surgeon's claim?
7. A multiple-choice quiz contains ten questions. For each question there are one correct answer and four incorrect answers. A student gets three correct answers on the quiz. Test the hypothesis that the student is guessing.

8. Return to Example 2.2 and, in the case of $n = 50$, approximate the power of the $\alpha = .05$ test when $p = .5$

9. Forsman and Lindell (1993) studied swallowing performance of adders (snakes). Captive snakes were fed with dead field voles (rodents) of differing body masses and the number of successful swallowing attempts was recorded. Out of 67 runs resulting in swallowing attempts, 58 were successful and 9 failed. (A failure was easy to detect because the fur of a partly swallowed and regurgitated vole is slick and sticks to the anterior part of the body.) Test the hypothesis that $p = .6$ against the alternative $p > .6$.

2.2. AN ESTIMATOR FOR THE PROBABILITY OF SUCCESS

Procedure

The estimator of the probability of success p, associated with the statistic B is

$$\hat{p} = \frac{B}{n}. \qquad (2.18)$$

Example 2.3: *Example 2.2 Continued.* Consider the triangle test data of Example 2.2. Then $\hat{p} = B/n = (25/50) = .5$. Thus our point estimate of p, the probability of correctly identifying the odd sample, is $\hat{p} = .5$.

Comments

14. *Observed Relative Frequency of Success.* The statistic $\hat{p}$ is simply the observed relative frequency of success in n Bernoulli trials satisfying Assumptions A1–A3. Thus $\hat{p}$ qualifies as a natural estimator of p, the unknown probability of success in a single Bernoulli trial. That is, we estimate the true unknown probability of success by the observed frequency of success.

15. *Standard Deviation of $\hat{p}$.* We have shown in Comment 8 that the variance of B is $np(1-p)$, where p is the success probability. It follows that the variance of $\hat{p}$ is

$$\text{var}(\hat{p}) = \frac{p(1-p)}{n}. \qquad (2.19)$$

The standard deviation of $\hat{p}$ is

$$sd(\hat{p}) = \sqrt{\frac{p(1-p)}{n}}. \qquad (2.20)$$

Note that $sd(\hat{p})$ cannot be computed unless we know the value of p, but it can be estimated by substituting $\hat{p}$ for p in (2.20). This quantity, which we denote as $\widehat{sd}(\hat{p})$, is a consistent estimator of $sd(\hat{p})$. The quantity $\widehat{sd}(\hat{p})$ is also known as the standard error of $\hat{p}$. We have

$$\widehat{sd}(\hat{p}) = \sqrt{\frac{\hat{p}(1-\hat{p})}{n}}. \qquad (2.21)$$

Rather than simply stating the value of $\hat{p}$ when reporting an observed relative frequency of success, it is important to also report the value of $\widehat{sd}(\hat{p})$, which (as does $\text{var}(\hat{p})$) measures the variability of the estimate.

Thus, for the adder data of Problem 9, we could report

$$\hat{p} = \frac{58}{67} = .87; \quad \widehat{sd}(\hat{p}) = \sqrt{\frac{\left(\frac{58}{67}\right)\left(\frac{9}{67}\right)}{67}} = .04.$$

Alternatively, we could use a confidence interval for p. See Section 2.3.

16. *Sample Size Determination.* Suppose we want to choose the sample size n so that $\hat{p}$ is within a distance D of p, with probability $1 - \alpha$. That is, we want

$$P_p(-D < \hat{p} - p < D) = 1 - \alpha.$$

This is equivalent to

$$P_p\left(\frac{-D}{\sqrt{\frac{p(1-p)}{n}}} < \frac{\hat{p} - p}{\sqrt{\frac{p(1-p)}{n}}} < \frac{D}{\sqrt{\frac{p(1-p)}{n}}}\right) = 1 - \alpha.$$

Since $(\hat{p} - p)/\sqrt{p(1-p)/n}$ has an asymptotic $N(0, 1)$ distribution we know

$$P\left(-z_{\alpha/2} < \frac{\hat{p} - p}{\sqrt{\frac{p(1-p)}{n}}} < z_{\alpha/2}\right) \doteq 1 - \alpha.$$

From the two previous equations we see that

$$\frac{D}{\sqrt{\frac{p(1-p)}{n}}} \doteq z_{\alpha/2}.$$

Solving for n yields

$$n \doteq \frac{(z_{\alpha/2})^2 p(1-p)}{D^2}. \tag{2.22}$$

Expression (2.22) requires a guess or estimate for p, since p is not known. Since $p(1 - p)$ is maximized at $p = \frac{1}{2}$ and decreases to zero as p approaches 0 or 1, we obtain the most conservative sample size by substituting $\frac{1}{2}$ for p in (2.22). This yields

$$n = \frac{(z_{\alpha/2})^2}{4D^2}. \tag{2.23}$$

17. *Competing Estimators.* An estimator for p derived by way of statistical decision theory by Hodges and Lehmann (1950) (who acknowledge priority to H. Rubin) is $\tilde{p} = [n^{1/2}\hat{p} + (\frac{1}{2})]/[1 + n^{1/2}]$, where $\hat{p}$ is defined by (2.18). When the parameter p is not equal to $\frac{1}{2}$, the estimator $\hat{p}$ has asymptotic relative efficiency greater than 1 with respect to $\tilde{p}$. However, when $p = \frac{1}{2}$, the asymptotic relative efficiency of $\hat{p}$ with respect to $\tilde{p}$ is 1, and further considerations are necessary to determine which of the two estimators is superior. See Hodges and Lehmann

(1970). For a Bayes estimator of p that uses prior information, see Casella and Berger (1990, 298). Also see Skibinsky and Cote (1963) for an estimator that uses prior information. Other methods for estimating p are described in Chew (1971).

Properties

1. Standard Deviation of $\widehat{p}$. See Comment 14.
2. Asymptotic Normality. See Casella and Berger (1990, 216).

PROBLEMS

10. Calculate $\widehat{p}$ for the parolee data of Problem 1 and obtain an estimate of the standard deviation of $\widehat{p}$.
11. Obtain an estimate for the standard deviation of the estimate $\widehat{p}$ calculated in Example 2.1.
12. Suppose $n = 7$. What are the possible values for $\widehat{p}$ and what are the possible values for $\widetilde{p}$ defined in Comment 17?
13. Suppose you are designing a study to estimate a success probability p. Determine the sample size n so that $\widehat{p}$ is within a distance .05 of p with probability .99.

2.3. A CONFIDENCE INTERVAL FOR THE PROBABILITY OF SUCCESS (CLOPPER-PEARSON)

Procedure

For a two-sided confidence interval for p, with confidence coefficient at least $1 - \alpha$, obtain the values $p_L(\alpha)$ and $p_U(\alpha)$ from Table A.3. The $1 - \alpha$ confidence interval for p is then $(p_L(\alpha), p_U(\alpha))$, and we have

$$P_p\{p_L(\alpha) < p < p_U(\alpha)\} \geq 1 - \alpha. \tag{2.24}$$

Large-Sample Approximation

For large n the values $p_L(\alpha)$ and $p_U(\alpha)$ may be approximated by

$$p_L(\alpha) \approx \widehat{p} - z_{\alpha/2}\left[\frac{\widehat{p}(1 - \widehat{p})}{n}\right]^{1/2} \tag{2.25}$$

and

$$p_U(\alpha) \approx \widehat{p} + z_{\alpha/2}\left[\frac{\widehat{p}(1 - \widehat{p})}{n}\right]^{1/2}, \tag{2.26}$$

where $\widehat{p} = B/n$.

Example 2.4: *Example 2.1 Continued.* Consider the canopy gap–closure problem of Example 2.1. There $n = 7$ and $B = 6$. From Table A.3, with $\alpha = .02$, we find $p_L(.02) = .3566$ and $p_U(.02) = .9986$. Thus the confidence interval for p, with confidence $\geq .98$, is (.3566, .9986). If, instead, we choose $\alpha = .05$, $p_L(.05) = .4213$, $p_U(.05) = .9964$ with confidence $\geq .95$. Note the 95% interval is shorter than the 98% interval but provides less confidence; i.e., the coverage probability is lower. This is the usual trade-off in statistics. If we desire more confidence, then we must settle for longer intervals.

StatXact will give the point estimate and an exact confidence interval. After using the TA command and entering the data, you escape to the command prompt, and type AD; then after the menu appears type ON and finally BI. For the data with seven trials and six successes, the StatXact output is:

```
ESTIMATION OF BINOMIAL PARAMETER (PI)
      NUMBER OF TRIALS = 7
      NUMBER OF SUCCESSES = 6
   Point Estimate of PI = 0.8571
   95.00% Confidence Interval for PI = (0.4213,0.9964)
```

Example 2.5: *Example 2.2 Continued.* We again consider the triangle test data of Example 2.2. From Table A.1 we find $z_{.025} = 1.96$. Then from (2.25) and (2.26) we have

$$p_L(.05) \approx \left(\frac{1}{2}\right) - 1.96\left[\frac{\left(\frac{1}{2}\right)\left(\frac{1}{2}\right)}{50}\right]^{1/2} = .36$$

and

$$p_U(.05) \approx \left(\frac{1}{2}\right) + 1.96\left[\frac{\left(\frac{1}{2}\right)\left(\frac{1}{2}\right)}{50}\right]^{1/2} = .64,$$

Thus (.36, .64) is an approximate 95% confidence interval for p.

Comments

18. *Large Sample Approximation Intervals Centered at $\hat{p}$.* We note that $\hat{p}$ is the midpoint of the approximate confidence intervals for p defined by (2.25) and (2.26). However, this is not true in general for the confidence intervals for p given by Table A.3.

19. *Graphical procedure.* A graphical procedure for finding the confidence intervals given by Table A.3 was presented by Clopper and Pearson (1934).

20. *Interval Endpoints.* In this comment we denote $p_L(\alpha)$ by $p_L^\alpha(n, B)$ and $p_U(\alpha)$ by $p_U^\alpha(n, B)$ to indicate their dependence on n and B. It can be shown that the confidence limits given in Table A.3 satisfy the equations

$$p_L^\alpha(n, B) = \frac{B}{B + (n - B + 1)f_{\alpha/2, 2(n-B+1), 2B}} \tag{2.27}$$

and

$$p_U^\alpha(n, B) = 1 - p_L^\alpha(n, n - B), \tag{2.28}$$

where B is the number of successes in the n Bernoulli trials and f_{γ, n_1, n_2} is the upper γth percentile for the F distribution with n_1 degrees of freedom in the numerator and n_2 degrees of freedom in the denominator.

21. *Equivariance.* Binomial confidence interval procedures that satisfy equation (2.28) are said to be equivariant (Casella 1986). The motivation for the term *equivariance* is that the binomial distribution is invariant under the transformations $B \to n - B$ and $p \to 1 - p$. See Casella (1986) for further details. The Clopper-Pearson intervals are equivariant but they are

not the only ones which enjoy the equivariance property. Casella (1986) gives a method for refining equivariant binomial confidence procedures to obtain new intervals with uniformly shorter lengths for the same confidence coefficient.

22. *Partial Historical Development.* Procedures for obtaining the confidence intervals given by Table A.3 were presented both numerically and graphically by Clopper and Pearson (1934). A modification of their scheme was proposed by Crow (1956). Other competitors of the Clopper-Pearson procedure include one proposed by Sterne (1954) and a modification of Sterne's procedure due to Crow (1956), the latter resulting in generally shorter-length confidence intervals than those obtained by the Clopper-Pearson technique. Anderson and Burstein (1967, 1968) also considered approximate one-sided confidence intervals for p. A basic paper in the general theory of confidence intervals is that of Neyman (1937). Louis (1981) discusses the special case of confidence intervals for p when the observed number of successes is zero. Casella (1986) presents a method for refining equivariant binomial confidence interval procedures to obtain narrower intervals. Casella's refinement construction is equivalent to a continuous version of a construction by Blyth and Still (1983). Duffy and Santner (1987) provide an algorithm associated with multistage hypothesis tests for computing confidence intervals for a binomial parameter. Applied to single-stage data, Santner (1998) notes that the Duffy-Santner algorithm produces confidence intervals for p that attain at least their nominal confidence levels for any number of trials n and true value of p.

Properties

1. For Bernoulli trials satisfying Assumptions A1–A3, (2.22) holds. Thus $(p_L(\alpha), p_U(\alpha))$ is a confidence interval for p with confidence coefficient at least $1 - \alpha$ for a very large class of experimental situations.

PROBLEMS

14. For the parolee data of Problem 1, obtain a confidence interval for p with approximate confidence coefficient of .96.

15. Shlafer and Karow (1971) considered some of the problems involved with cardiac preservation. In particular, they were interested in the morphological and physiological injury occurring in hearts that had been frozen to various temperatures without the benefit of a cryoprotectant. Hearts from adult rats were perfused with a balanced salt solution *in vitro* for 20 min, and during this time contractions were noted. After disconnection from the perfusion apparatus, each heart, surrounded by a plastic shield, was inserted into a metal canister and chilled by an acetone bath (maintained at $-20°C$ by addition of dry ice) until the lowest desired temperature was attained. The individual hearts were then thawed (in 1 min or less) by removing the metal canister and running $35°C$ tap water over the plastic shields, being careful to prevent the water from flowing directly over the hearts. After thawing, the hearts were again perfused with the balanced salt solution. Hearts spontaneously resuming coordinated atrioventricular contractions within 20 min of thawing were considered to be "survivors" of the freeze-thawing process.

    The authors conducted experiments where the lowest attained temperatures were $-10, -12, -17,$ and $-20°C$, respectively. We focus here on the data for the $-12°C$ investigation, in which the authors found that of six hearts frozen to $-12°C$, three were survivors. If we let success denote survival, then p represents the probability that a rat heart frozen to $-12°C$ will spontaneously resume coordinated atrioventricular contractions within 20 min after thawing and perfusion with a balanced salt solution. Find, from Table A.3, an approximate 95% confidence interval for p.

16. Many materials that are satisfactory for use in air will ignite and burn in pure oxygen when subjected to mechanical impact. This problem is of vital concern to the aerospace industry, which uses enormous amounts of both liquid and gaseous oxygen. In particular, there is a need for guidelines

to aid the designer in selecting materials to be employed in pressurized oxygen systems. In order to provide an appropriate method for determining gaseous oxygen-material compatibility, the Kennedy Space Center developed a gaseous oxygen impact test procedure. Jamison (1971) reported on the use of this testing scheme to analyze the gaseous oxygen-material compatibility for 33 Apollo spacecraft test materials. One such material tested, silicone elastomer 342, failed the gaseous oxygen impact test (i.e., the material ignited) on 4 out of 20 trials. Let p denote the probability of ignition for silicone elastomer 342 when subjected to the conditions employed in the gaseous oxygen impact test. Obtain a confidence interval for p with approximate confidence coefficient .96.

17. Suppose the value of $\hat{p}$ obtained in Example 2.3 had been the result of 30 instead of 50 trials. How would this affect the length of the approximate 95% confidence interval for p obtained in Example 2.5?

18. Suppose that ten Bernoulli trials satisfying Assumptions A1–A3 result in six successes. Investigate the accuracy of the large-sample approximation by comparing the endpoints of the approximate 95% confidence interval for p obtained from Table A.3 with the corresponding endpoints of the approximate 95% confidence interval for p obtained using (2.25) and (2.26).

19. Consider the adder data of Problem 9. Determine an approximate 96% confidence interval for p.

20. In Example 2.4, we determined Clopper-Pearson confidence intervals for p with $n = 7$ and $B = 6$. With confidence at least 95%, the interval was found to be (.4216, .9964). Use (2.25) and (2.26) to obtain an approximate 95% interval based on the large-sample approximation and compare it with the Clopper-Pearson interval.

21. Fix the value of n and the value of α. Use Table A.3 to investigate if the inequalities

$$p_L^\alpha(n, B+1) > p_L^\alpha(n, B)$$

and

$$p_U^\alpha(n, B+1) > p_U^\alpha(n, B)$$

hold for $B = 0, 1, \ldots, n-1$. Explain why the inequalities being satisfied should be viewed as a desirable property.

22. Fix $n = 8$ and $\alpha = .05$. Compute the expected length of the Clopper-Pearson interval if, in fact, $p = .3$. *Big hint:* Letting ℓ denote the random length, we have

$$E_p(\ell) = \sum_{b=0}^{n} (p_U^\alpha(n, B) - p_L^\alpha(n, B)) \binom{n}{b} p^b (1-p)^{n-b}.$$

23. Ehlers (1995) performed a 1-year follow-up study of panic disorder. As partial motivation for the study, the author offered the following quote from Wolfe and Maser (1994, 241): "Little is known about the long-term course of the disorder. The limited findings to date suggest that in most cases it is a chronic disorder that waxes and wanes in severity. However, some people may have a limited period of dysfunction that never recurs, while others tend to have a more severe and complicated course." In this study diagnoses were made by trained interviewers (either the author, A. Ehlers, or trained graduate students) according to the criteria of the revised third edition of the *Diagnostic and Statistical Manual of Mental Disorders* (DSM–III–R; American Psychiatric Association, 1987). One year after initial assessments, participants were mailed a questionnaire for the purpose of assessing their current symptoms. In this problem, we give one small portion of the data that were obtained. Out of 46 people who were initially diagnosed as "infrequent panickers," 23 experienced panic attacks during follow-up. Give a 96% confidence interval for p, the probability that an infrequent panicker will experience panic attacks during a 1-year follow-up period.

24. Consider the study in Problem 23 and describe the inherent sources for error when one uses a mailed questionnaire.

25. Consider the study in Problem 23 and discuss the possibility of unintentional bias entering the study, since some of the diagnoses were made by the author of the study.

CHAPTER 3

The One-Sample Location Problem

INTRODUCTION

The procedures of this chapter are designed for statistical analyses in which primary interest is centered on the location (median) of a population. We encounter two types of data for which such analyses are important. The first of these, referred to as paired replicates data, represents pairs of "pretreatment" and "posttreatment" observations; here we are concerned with a shift in location due to the application of the "treatment." The second type of data, referred to as one-sample data, consists of observations from a single population about whose location we wish to make inferences.

In Sections 3.1 through 3.3, procedures are considered for analyzing paired replicates data using signed ranks. In particular, Section 3.1 presents a distribution-free signed rank test, Section 3.2, a point estimator associated with the signed rank statistic, and Section 3.3, a related distribution-free confidence interval. In Section 3.7 these procedures are applied to some one-sample data. An asymptotically distribution-free test for symmetry of the underlying population (one of the assumptions in Sections 3.1–3.3 and 3.7) is considered in Section 3.9. A distribution-free test for exchangeability of the paired replicate data is discussed in Section 3.10.

Procedures for analyzing paired replicates data using signs are discussed in Sections 3.4–3.6. A distribution-free sign test is considered in Section 3.4, a point estimator associated with the sign statistic, in Section 3.5, and a related distribution-free confidence interval in Section 3.6. These sign procedures are applied to some one-sample data in Section 3.8.

The asymptotic relative efficiencies for translation alternatives of the procedures based on the signed rank statistic and those based on the sign statistic with respect to their normal theory counterparts based on the sample mean are discussed in Section 3.11.

PAIRED REPLICATES ANALYSES BY WAY OF SIGNED RANKS

Data. We obtain $2n$ observations, two observations on each of n subjects (blocks, patients, etc.).

Subject i	X_i	Y_i
1	X_1	Y_1
2	X_2	Y_2
.	.	.
.	.	.
.	.	.
n	X_n	Y_n

Assumptions

A1. We let $Z_i = Y_i - X_i$, for $i = 1, \ldots, n$. The differences $Z_1, \ldots, Z_n$ are mutually independent.

A2. Each $Z_i, i = 1, \ldots, n$, comes from a continuous population (not necessarily the same one) that is symmetric about a common median θ. If F_i represents the distribution function for $Z_i, i = 1, \ldots, n$, this assumption requires that

$$F_i(\theta + t) + F_i(\theta - t) = 1, \text{ for every } t \text{ and } i = 1, \ldots, n.$$

The parameter θ is referred to as the *treatment effect*.

3.1. A DISTRIBUTION-FREE SIGNED RANK TEST (WILCOXON)

Hypothesis

The null hypothesis of interest here is that of zero shift in location due to the treatment, namely,

$$H_0 : \theta = 0. \qquad (3.1)$$

This null hypothesis asserts that **each of the distributions (not necessarily the same)** for the differences (posttreatment minus pretreatment observations) is symmetrically distributed about 0, corresponding to no shift in location due to the treatment.

Procedure

To compute the Wilcoxon signed rank statistic T^+, form the absolute values $|Z_1|, \ldots, |Z_n|$ of the differences and order them from least to greatest. Let R_i denote the rank of $|Z_i|, i = 1, \ldots, n$, in this ordering. Define indicator variables $\psi_i, i = 1, \ldots, n$, where

$$\psi_i = \begin{cases} 1, & \text{if } Z_i > 0, \\ 0, & \text{if } Z_i < 0, \end{cases} \qquad (3.2)$$

and obtain the n products $R_1\psi_1, \ldots, R_n\psi_n$. The product $R_i\psi_i$ is known as the **positive signed rank of Z_i**. It takes on the value zero if Z_i is negative and is equal to the rank of $|Z_i|$ when Z_i is positive. The Wilcoxon signed rank statistic T^+ is then the sum of the positive signed ranks, namely,

$$T^+ = \sum_{i=1}^{n} R_i \psi_i. \qquad (3.3)$$

a. One-Sided Upper-Tail Test. To test

$$H_0 : \theta = 0$$

versus

$$H_1 : \theta > 0$$

at the α level of significance,

$$\text{Reject } H_0 \text{ if } T^+ \geq t_\alpha; \text{ otherwise do not reject,} \qquad (3.4)$$

where the constant t_α is chosen to make the type I error probability equal to α. Values of t_α are given in Table A.4.

b. One-Sided Lower-Tail Test. To test

$$H_0 : \theta = 0$$

versus

$$H_2 : \theta < 0$$

at the α level of significance,

$$\text{Reject } H_0 \text{ if } T^+ \leq \frac{n(n+1)}{2} - t_\alpha; \text{ otherwise do not reject.} \qquad (3.5)$$

c. Two-Sided Test. To test

$$H_0 : \theta = 0$$

versus

$$H_3 : \theta \neq 0$$

at the α level of significance,

$$\text{Reject } H_0 \text{ if } T^+ \geq t_{\alpha/2} \text{ or } T^+ \leq \frac{n(n+1)}{2} - t_{\alpha/2}; \text{ otherwise do not reject.} \qquad (3.6)$$

This two-sided procedure is the two-sided symmetric test with $\alpha/2$ probability in each tail of the null distribution of T^+.

Large-Sample Approximation

The large-sample approximation is based on the asymptotic normality of T^+, suitably standardized. We first need to know the expected value and variance of T^+ when the null hypothesis is true. When H_0 is true, the expected value and variance of T^+ are

$$E_0(T^+) = \frac{n(n+1)}{4} \qquad (3.7)$$

and

$$\text{var}_0(T^+) = \frac{n(n+1)(2n+1)}{24}, \tag{3.8}$$

respectively. These expressions for $E_0(T^+)$ and $\text{var}_0(T^+)$ are verified by direct calculations in Comment 6 for the special case of $n = 3$. General derivations of both expressions are presented in Comment 7.

The standardized version of T^+ is

$$T^* = \frac{T^+ - E_0(T^+)}{\{\text{var}_0(T^+)\}^{1/2}} = \frac{T^+ - \{\frac{n(n+1)}{4}\}}{\{n(n+1)(2n+1)/24\}^{1/2}}. \tag{3.9}$$

When H_0 is true, T^* has, as n tends to infinity, an asymptotic $N(0, 1)$ distribution. (See Comment 7 for indications of the proof.) The normal theory approximation for procedure (3.4) is

$$\text{Reject } H_0 \text{ if } T^* \geq z_\alpha; \text{ otherwise do not reject;} \tag{3.10}$$

the normal theory approximation for procedure (3.5) is

$$\text{Reject } H_0 \text{ if } T^* \leq -z_\alpha; \text{ otherwise do not reject;} \tag{3.11}$$

and the normal theory approximation for procedure (3.6) is

$$\text{Reject } H_0 \text{ if } |T^*| \geq z_{\alpha/2}; \text{ otherwise do not reject.} \tag{3.12}$$

Ties

If there are zero values among the Z's, discard the zero values and redefine n to be the number of nonzero Z's. If there are ties among the (nonzero) $|Z|$'s, assign each of the observations in a tied group the average of the integer ranks that are associated with the tied group. After computing T^+ with these average ranks for nonzero Z's, use procedure (3.4), (3.5), or (3.6) and refer the value of T^+ to Table A.4. Note, however, that this test associated with tied $|Z|$'s is only approximately, and not exactly, of significance level α. (To get an exact level α test even in this tied setting, see Comment 11.)

When applying the large-sample approximation, an additional factor must be taken into account. Although ties in the nonzero $|Z|$'s do not affect the null expected value of T^+, its null variance is reduced to

$$\text{var}_0(T^+) = (24)^{-1} \left[n(n+1)(2n+1) - \frac{1}{2} \sum_{j=1}^{g} t_j(t_j - 1)(t_j + 1) \right], \tag{3.13}$$

where in equation (3.13) g denotes the number of tied groups of nonzero $|Z|$'s and t_j is the size of tied group j. We note that an untied observation is considered to be a tied "group" of size 1. In particular, if there are no ties among the $|Z|$'s, then $g = n$ and $t_j = 1$ for $j = 1, \ldots, n$. In this case each term in (3.13) of the form $t_j(t_j - 1)(t_j + 1)$ reduces to zero, and the variance expression in (3.13) reduces to the usual null variance of T^+ when there are no ties, as given in equation (3.8). Note that the term $(48)^{-1} \sum_{j=1}^{g} t_j(t_j - 1)(t_j + 1)$ represents the reduction in the null variance of T^+ due to the presence of tied nonzero Z's.

TABLE 3.1. Hamilton Depression Scale Factor IV Values

Patient i	X_i	Y_i
1	1.83	0.878
2	0.50	0.647
3	1.62	0.598
4	2.48	2.05
5	1.68	1.06
6	1.88	1.29
7	1.55	1.06
8	3.06	3.14
9	1.30	1.29

Source: D. S. Salsburg (1970).

As a consequence of the effect that ties have on the null variance of T^+, the following modification is needed to apply the large-sample approximation when there are tied nonzero Z's. Compute T^+ using average ranks and set

$$T^* = \frac{T^+ - \{\frac{n(n+1)}{4}\}}{\{\mathrm{var}_0(T^+)\}^{1/2}}, \tag{3.14}$$

where $\mathrm{var}_0(T^+)$ is now given by display (3.13). With this modified value of T^*, approximations (3.10), (3.11), or (3.12) can be applied.

Example 3.1: *Hamilton Depression Scale Factor IV.* The data in Table 3.1 are a portion of the data obtained by Salsburg (1970). These data, based on nine patients who received tranquilizer T, were taken from a double-blind clinical trial involving two tranquilizers. The measure used was the Hamilton (1960) depression scale factor IV (the "suicidal" factor). The X(pre) value was obtained at the first patient visit after initiation of therapy, whereas the Y(post) value was obtained at the second visit after initiation of therapy. The patients had been diagnosed as having mixed anxiety and depression.

In this example, an improvement due to tranquilizer T corresponds to a reduction in factor IV values. Hence, we apply test (3.5), which is designed to detect the alternative $\theta < 0$. For purpose of illustration we take the significance level to be $\alpha = .049$. From Table A.4 we find $t_{.049} = 37$. Since

$$\frac{n(n+1)}{2} - t_{.049} = \frac{9(10)}{2} - 37 = 8,$$

procedure (3.5) reduces to

$$\text{Reject } H_0 \text{ if } T^+ \leq 8.$$

Now, we illustrate the computations leading to the sample value of T^+ (3.3) in tabular form.

We then have $T^+ = \sum_{i=1}^{9} R_i \psi_i = 5$. Since this value of T^+ is less than the critical value 8, we reject H_0 in favor of $\theta < 0$ at the $\alpha = .049$ level.

Since the one-sided P-value for these data is the lowest significance level at which we can reject H_0 in favor of $\theta < 0$ with the observed value of the test statistic $T^+ = 5$, we use

i	Z_i	$\|Z_i\|$	R_i	ψ_i	$R_i\psi_i$
1	−.952	.952	8	0	0
2	.147	.147	3	1	3
3	−1.022	1.022	9	0	0
4	−.430	.430	4	0	0
5	−.620	.620	7	0	0
6	−.590	.590	6	0	0
7	−.490	.490	5	0	0
8	.080	.080	2	1	2
9	−.010	.010	1	0	0

equation (3.17) in Comment 8 with $n = 9$ to obtain

$$P_0(T^+ \leq 5) = P_0(T^+ \geq 40) = .020,$$

where the P-value .020 is obtained by entering Table A.4 with $n = 9$ and $x = 40$.

For the large-sample approximation we find (since there are no ties) from equation (3.9) that

$$T^* = \frac{5 - (9(10)/4)}{\{9(10)(19)/24\}^{1/2}} = -2.07.$$

Thus, the smallest significance level at which we can reject H_0 in favor of $\theta < 0$ using the normal approximation is .0192. Both the exact test and the large sample approximation indicate that there is strong evidence that tranquilizer T does lead to patient improvement, as measured by a reduction in Hamilton scale factor IV values.

Both the exact P-value and the approximate P-value based on the large sample approximation can also be obtained by using StatXact's WI/EX command. The StatXact output for the factor IV data of Table 3.1 is as follows.

Summary of exact distribution of WILCOXON SIGNED RANK statistic:

Min	Max	Mean	Std-dev	Observed	Standardized
00.00	45.00	22.50	8.441	5.000	−2.073

Asymptotic Inference:
One-sided P-value: Pr{Test Statistic. LE. Observed} = .0191
Two-sided P-value: 2 ∗ One-sided = .0382
Exact Inference:
One-sided P-value: Pr {Test Statistic. LE. Observed} = .0195
Point Probability:Pr {Test Statistic. EQ. Observed} = .0059
Two-sided P-value: 2 ∗ One-sided = .0391

Under the heading "Exact Inference" we see that StatXact provides .0195 as the one-sided P-value, agreeing with what we found from Table A.4. In addition, we note that the approximate one-sided P-value provided by StatXact under "Asymptotic Inference" is .0191, also in agreement with what we previously obtained using the large-sample approximation. (We note that Minitab's WTEST command can also be used to obtain an approximate P-value based on the large-sample approximation with a continuity correction.)

Example 3.2: *Government versus Private-Sector Salaries.* In an annual survey to determine whether federal pay scales were commensurate with private-sector salaries, government

3.1. A DISTRIBUTION-FREE SIGNED RANK TEST (WILCOXON)

TABLE 3.2. Annual Salaries

Pair i	Private	Government
1	12,500	11,750
2	22,300	20,900
3	14,500	14,800
4	32,300	29,900
5	20,800	21,500
6	19,200	18,400
7	15,800	14,500
8	17,500	17,900
9	23,300	21,400
10	42,100	43,200
11	16,800	15,200
12	14,500	14,200

Source: J.T. McClave and G. Benson (1978).

and private workers were matched as closely as possible (with respect to type of job, educational background, years experience, etc.) and the salaries of the matched pairs were obtained. The data in Table 3.2 are the annual salaries (in dollars) for 12 such matched pairs, as reported in McClave and Benson (1978).

Letting X correspond to the government worker's salary and Y to the matched private-sector salary, the tabular presentation of the associated positive signed ranks (using average ranks to break ties) is as follows:

i	Z_i	$|Z_i|$	R_i	ψ_i	$R_i \psi_i$
1	750	750	5	1	5
2	1,400	1,400	9	1	9
3	−300	300	1.5	0	0
4	2,400	2,400	12	1	12
5	−700	700	4	0	0
6	800	800	6	1	6
7	1,300	1,300	8	1	8
8	−400	400	3	0	0
9	1,900	1,900	11	1	11
10	−1,100	1,100	7	0	0
11	1,600	1,600	10	1	10
12	300	300	1.5	1	1.5

To test H_0 versus the alternative that government workers are generally paid less than their counterparts in the private sector, we use the signed rank test of $H_0 : \theta = 0$ versus $H_1 : \theta > 0$. From the signed rank computational array, we see that

$$T^+ = 5 + 9 + 12 + 6 + 8 + 11 + 10 + 1.5 = 62.5.$$

Entering Table A.4 with $n = 12$, we find that the smallest significance level at which these data lead to rejection of $H_0 : \theta = 0$ in favor of $H_1 : \theta > 0$ (i.e., the one-sided P-value) is $\alpha = .039$. (Note that the P-value corresponds to the upper-tail probability associated with the critical value 62 and not that associated with 63, since 62 is the largest such critical value for which we would reject H_0.) Thus, we would reject H_0 in favor of $H_1 : \theta > 0$ for any significance level greater than or equal to this P-value .039. Hence, there is moderate evidence to indicate that federal government workers (at least in the type of jobs considered in this survey) are, indeed, paid less than their private sector counterparts. (We point out that the P-value for these data is only approximate, due to the tied $300 differences. For a discussion of how to obtain the exact conditional P-value in this case, see Comment 11.)

For the normal approximation with the data in Table 3.2, we need to use the ties-corrected version of T^* given in equation (3.14). For the salary data, we have $g = 11$ and (arbitrarily labeling the tied groups in order of increasing ranks) $t_1 = 2, t_2 = t_3 = \cdots = t_{10} = t_{11} = 1$. Using the ties-corrected formula (3.13) for $\text{var}_0(T^+)$, we obtain

$$T^* = \frac{62.5 - \frac{12(13)}{4}}{\left\{\frac{12(12+1)(2(12)+1) - \frac{1}{2}(2)(1)(3)}{24}\right\}^{1/2}} = \frac{62.5 - 39}{\left\{\frac{3897}{24}\right\}^{1/2}} = 1.84.$$

To find the P-value associated with this normal approximation, we refer $T^* = 1.84$ to Table A.1. The approximate P-value is then the area under the standard normal curve to the right of 1.84. From Table A.1 we find this area to be .0329, which is in good agreement with the value of .039 obtained without using the normal approximation.

Comments

1. *Motivation for the Test.* When θ is greater than 0, there will tend to be a large proportion of positive Z differences and they will tend to have the larger absolute values. Hence, when θ is greater than 0, we would expect a higher proportion of positive signed ranks with relatively large sizes, leading to a big value of T^+. This suggests rejecting H_0 in favor of $\theta > 0$ for large values of T^+ and motivates procedures (3.4) and (3.10). Similar rationales lead to procedures (3.5), (3.6), (3.11), and (3.12).

2. *Assumptions.* There is no requirement that the individual X_i and Y_i be independent, only that the pairs $(X_1, Y_1), \ldots, (X_n, Y_n)$, and therefore the resulting differences $Z_1, \ldots, Z_n$, be mutually independent. Indeed, in most applications the individual X_i and Y_i are dependent. For paired replicates data, the symmetry part of Assumption A2 is often inherently satisfied. In particular, if each X_i and $Y_i, i = 1, \ldots, n$, arise from populations differing only in location (i.e., the only treatment "effect" is a change in location), then the $(Z_i - \theta)$'s come from populations that are symmetric about zero. (This is, in fact, true under more general conditions.)

3. *Testing θ Equal to Some Specified Nonzero Value.* Procedures (3.4), (3.5), (3.6) and the corresponding normal approximations in (3.10), (3.11), (3.12) are for testing θ equal to zero. To test $\theta = \theta_0$, where θ_0 is some specified nonzero number, subtract θ_0 from each of the differences $Z_1, \ldots, Z_n$ to form a modified sample $Z'_1 = Z_1 - \theta_0, \ldots, Z'_n = Z_n - \theta_0$. Then compute T^+ as the sum of the positive signed ranks for these Z'_i's. Procedures (3.4), (3.5), (3.6) and their corresponding large-sample approximations in (3.10), (3.11), (3.12) are then applied as previously described.

4. *Equivalent Form.* It may appear that some of the information in the ranking of the sample Z-differences is being lost by using only the positive signed ranks to compute T^+. Such is not the case. If we define T^- to be the sum of ranks (of the absolute values) corresponding to the negative Z observations, then $T^- = \sum_{i=1}^{n}(1 - \psi_i)R_i$. It follows that $T^+ + T^- = \sum_{i=1}^{n} R_i = n(n+1)/2$. Thus, the test procedures defined in (3.4), (3.5), (3.6) and (3.10), (3.11), (3.12) could equivalently be based on $T^- = [n(n+1)/2] - T^+$.

5. *Derivation of the Distribution of T^+ under H_0 (No Ties Case).* Let B be the number of positive Z's and let $r_1 < \cdots < r_B$ denote the ordered ranks of the absolute values of these

3.1. A DISTRIBUTION-FREE SIGNED RANK TEST (WILCOXON)

positive Z's. Then the null (H_0) distribution can be obtained directly from the representation $T^+ = \sum_{i=1}^{B} r_i$. Under the assumption that the underlying Z_i distributions are all continuous, the probabilities are zero that there are ties among the absolute values of the Z's or that any of the Z's are exactly zero. In addition, under H_0 these underlying Z_i distributions are all symmetric about $\theta = 0$. It follows that under H_0 each of the 2^n possible outcomes for the ordered configuration $(r_1, \ldots, r_B)$ occurs with equal probability $(\frac{1}{2})^n$. For example, in the case of $n = 3$, the $2^3 = 8$ possible outcomes for $(r_1, \ldots, r_B)$ and associated values of T^+ are given in the following table.

B	$(r_1, r_2, \ldots, r_B)$	Probability under H_0	$T^+ = \sum_{i=1}^{B} r_i$
0		$\frac{1}{8}$	0
1	$r_1 = 1$	$\frac{1}{8}$	1
1	$r_1 = 2$	$\frac{1}{8}$	2
1	$r_1 = 3$	$\frac{1}{8}$	3
2	$r_1 = 1, r_2 = 2$	$\frac{1}{8}$	3
2	$r_1 = 1, r_2 = 3$	$\frac{1}{8}$	4
2	$r_1 = 2, r_2 = 3$	$\frac{1}{8}$	5
3	$r_1 = 1, r_2 = 2, r_3 = 3$	$\frac{1}{8}$	6

Thus, for example, the probability is $\frac{2}{8}$ under H_0 that T^+ is equal to 3, since $T^+ = 3$ when either of the exclusive outcomes $B = 1, r_1 = 3$ or $B = 2, (r_1 = 1, r_2 = 2)$ occurs and each of these outcomes has null probability $\frac{1}{8}$. Simplifying, we obtain the null distribution.

Possible value of T^+	Probability under H_0
0	$\frac{1}{8}$
1	$\frac{1}{8}$
2	$\frac{1}{8}$
3	$\frac{2}{8}$
4	$\frac{1}{8}$
5	$\frac{1}{8}$
6	$\frac{1}{8}$

The probability, under H_0, that T^+ is greater than or equal to 5, for example, is therefore

$$P_0(T^+ \geq 5) = P_0(T^+ = 5) + P_0(T^+ = 6)$$
$$= .125 + .125 = .25.$$

This agrees with the upper-tail probability entry for $n = 3$ and the value $T^+ = 5$ in Table A.4.

Note that we have derived the null distribution of T^+ without specifying the forms of the underlying Z populations under H_0 beyond the point of requiring that they be continuous and symmetric about zero. This is why the test procedures based on T^+ are called distribution-free procedures. From the null distribution of T^+ we can determine the critical value t_α and control the probability α of falsely rejecting H_0 when H_0 is true, and this error probability does not depend on the specific forms of the underlying continuous and symmetric (about 0) Z distributions.

6. *Calculation of the Mean and Variance of T^+ under the Null Hypothesis H_0.* In displays (3.7) and (3.8) we presented formulas for the mean and variance of T^+ when the null hypothesis is true. In this comment we illustrate a direct calculation of $E_0(T^+)$ and $\text{var}_0(T^+)$ in the particular case of $n = 3$, using the null distribution of T^+ obtained in Comment 5. (Later, in Comment 7, we present general derivations of $E_0(T^+)$ and $\text{var}_0(T^+)$.) The null mean, $E_0(T^+)$, is obtained by multiplying each possible value of T^+ with its probability under H_0. Thus,

$$E_0(T^+) = 0(.125) + 1(.125) + 2(.125) + 3(.25) + 4(.125) + 5(.125) + 6(.125) = 3.$$

This is in agreement with what we obtain using equation (3.7), namely,

$$E_0(T^+) = \frac{n(n+1)}{4} = \frac{3(3+1)}{4} = 3.$$

A check on the expression for $\text{var}_0(T^+)$ is also easily performed, using the well-known fact that

$$\text{var}_0(T^+) = E_0[(T^+)^2] - \{E_0(T^+)\}^2.$$

The value of $E_0[(T^+)^2]$, the second moment of the null distribution of T^+, is again obtained by multiplying possible values (in this case, of $(T^+)^2$) by the corresponding probabilities under H_0. We find

$$E_0[(T^+)^2] = [(0 + 1 + 4)(.125) + 9(.25) + (16 + 25 + 36)(.125)] = 12.5.$$

Thus,

$$\text{var}_0(T^+) = 12.5 - (3)^2 = 3.5,$$

which agrees with what we obtain using equation (3.8) directly, namely,

$$\text{var}_0(T^+) = \frac{3(3+1)(2(3)+1)}{24} = 3.5.$$

7. *Large-Sample Approximation.* In view of the representation $T^+ = \sum_{i=1}^{B} r_i$, it follows from the discussion in Comment 5 that $T^+ \stackrel{d}{=} \sum_{i=1}^{n} V_i$, where the symbol $\stackrel{d}{=}$ means "has the same distribution as" and $V_1, \ldots, V_n$ are mutually independent dichotomous random variables with probability distributions

$$P(V_i = i) = P(V_i = 0) = \tfrac{1}{2},$$

for $i = 1, \ldots, n$. From this distributionally equivalent form, we can immediately use well-known expressions for the mean and variance of a sum of mutually independent random variables to obtain

$$E_0(T^+) = E\left[\sum_{i=1}^{n} V_i\right] = \sum_{i=1}^{n} E[V_i] \qquad (3.15)$$

3.1. A DISTRIBUTION-FREE SIGNED RANK TEST (WILCOXON)

and

$$\text{var}_0(T^+) = \text{var}\left(\sum_{i=1}^n V_i\right) = \sum_{i=1}^n \text{var}(V_i). \tag{3.16}$$

Since V_i is a dichotomous variable, we have, for $i = 1, \ldots, n$, that

$$E_0(V_i) = i\left(\frac{1}{2}\right) + 0\left(\frac{1}{2}\right) = \frac{i}{2}$$

and

$$\text{var}_0(V_i) = E_0(V_i^2) - [E_0(V_i)]^2 = \left[i^2\left(\frac{1}{2}\right) + 0^2\left(\frac{1}{2}\right)\right] - \left[\frac{i}{2}\right]^2 = \frac{i^2}{2} - \frac{i^2}{4} = \frac{i^2}{4}.$$

Using these results, together with closed-form expressions for the sum of the first n positive integers and the sum of the squares of the first n positive integers, in equations (3.15) and (3.16), we obtain

$$E_0(T^+) = \frac{1}{2}\sum_{i=1}^n i = \frac{1}{2}\left[\frac{n(n+1)}{2}\right] = \frac{n(n+1)}{4}$$

and

$$\text{var}_0(T^+) = \frac{1}{4}\sum_{i=1}^n i^2 = \frac{1}{4}\left[\frac{n(n+1)(2n+1)}{6}\right] = \frac{n(n+1)(2n+1)}{24},$$

which agree with the general expressions stated in equations (3.7) and (3.8).

Also using the distributional equality between T^+ and $\sum_{i=1}^n V_i$, the asymptotic normality of the standardized form

$$T^* = \frac{T^+ - E_0(T^+)}{\{\text{var}_0(T^+)\}^{1/2}} = \frac{T^+ - \dfrac{n(n+1)}{4}}{\left\{\dfrac{n(n+1)(2n+1)}{24}\right\}^{1/2}}$$

follows from standard theory for sums of mutually independent, but not identically distributed, random variables, such as the Liapounov central limit theorem (cf. Randles and Wolfe (1979, 423)). Asymptotic normality results are also obtainable under general alternatives to H_0. See, for example, Hoeffding's (1948a) U-statistic theorem as stated and applied to the Wilcoxon signed rank statistic on pages 82–85 of Randles and Wolfe (1979).

8. *Symmetry of the Distribution of T^+ under the Null Hypothesis.* When H_0 is true, the distribution of T^+ is symmetric about its mean $n(n+1)/4$. (See Comment 5 for verification of this when $n = 3$.) This implies that

$$P_0(T^+ \leq x) = P_0\left(T^+ \geq \frac{n(n+1)}{2} - x\right), \tag{3.17}$$

for $x = 0, 1, \ldots, n(n+1)/2$.

Equation (3.17) is directly used to convert upper-tail probabilities, as presented in Table A.4, to lower-tail probabilities. We have already used this relationship in equation (3.17) in Example 1.

9. *Zero Z Values.* We have recommended dealing with zero values among the Z's by discarding them and redefining n to be the number of nonzero Z's. This approach is satisfactory as long as the zero values do not represent a sizable percentage of the Z differences. If, however, there is a relatively large number of zero Z's, it would be advisable to consider an appropriate statistical procedure designed for analyzing such discrete data. See, for example, Chapter 10 or a book on categorical data analysis such as Agresti (1990).

We should also point out that there are methods other than elimination that have been proposed for dealing with zero Z values. One could use individual randomization (e.g., flipping a fair coin) to decide whether each of the zero Z values is to be counted as positive or negative in the construction of T^+. (Although this approach maintains many of the nice properties of T^+ that hold when there are no zeros, it introduces extraneous randomness that could quite easily have a direct effect on the outcome of any subsequent inferences based on such a modified T^+.) A second alternative approach in the case of the one-sided test procedures in (3.4), (3.5), (3.10), and (3.11) is to be conservative about rejecting the null hypothesis H_0; that is, we could count all the zero Z values as if they were in favor of not rejecting H_0. Thus, for example, in applying either procedure (3.4) or (3.10) to test H_0 against the alternative $\theta > 0$, we would treat all of the zero Z's as if they were negative (in favor of not rejecting H_0) in the calculation of T^+. (In the case of procedures (3.5) and (3.11), zero Z's would be considered positive in the calculation of T^+.) Any rejection of H_0 with this conservative approach to dealing with zero Z values could then be viewed as providing strong evidence in favor of the appropriate alternative. For a more detailed discussion of methods for handling zero observations, see Pratt (1959).

10. *Tied Nonzero Absolute Z Values.* Methods for dealing with tied nonzero absolute Z values other than using average ranks have been discussed in the literature. These include analogues to the randomization and conservative approaches mentioned in Comment 9 with regard to zero Z values. For further discussion of these alternative methods for dealing with tied nonzero absolute Z's, see Pratt (1959).

11. *Exact Conditional Distribution of T^+ with Ties Among the Nonzero Absolute Z Values.* To have a test with exact significance level even in the presence of tied absolute Z's (assuming there are no zero Z values or they have been discarded and n reduced accordingly), one considers all 2^n possible outcomes for the ordered configuration $(r_1, \ldots, r_B)$, where B represents the number of positive Z's as in Comment 5 but where $r_1 < \cdots < r_B$ now denote the ordered ranks of the absolute values of the positive Z's using average ranks to break the ties. As in Comment 5, it still follows that under H_0 each of the 2^n possible outcomes for the ordered configurations $(r_1, \ldots, r_B)$ based on using average ranks to break ties occurs with probability $(\frac{1}{2})^n$. For each such configuration, the value of T^+ is computed and the results tabulated. We illustrate this construction for $n = 4$ and the data $Z_1 = -12$, $Z_2 = -10$, $Z_3 = 10$, $Z_4 = 12$. Using average ranks to break ties, the associated absolute value ranks are $R_1 = 3.5, R_2 = 1.5, R_3 = 1.5,$ and $R_4 = 3.5$. Thus, $B = 2$ and the ordered ties-broken-ranks for the positive Z's are $r_1 = 1.5$ and $r_2 = 3.5$, leading to an attained value of $T^+ = 5$. To assess the significance of T^+, we obtain its conditional distribution by considering the $2^4 = 16$ equally likely (under H_0) possible values of $(r_1, \ldots, r_B)$ for the given tied rank vector $(1.5, 1.5, 3.5, 3.5)$. These 16 values of $(r_1, \ldots, r_B)$ and associated values of T^+ are:

3.1. A DISTRIBUTION-FREE SIGNED RANK TEST (WILCOXON)

B	$(r_1, r_2, \ldots, r_B)$	Probability under H_0	Value of T^+
0		$\frac{1}{16}$	0
1	$r_1 = 1.5$	$\frac{1}{16}$	1.5
1	$r_1 = 1.5$	$\frac{1}{16}$	1.5
1	$r_1 = 3.5$	$\frac{1}{16}$	3.5
1	$r_1 = 3.5$	$\frac{1}{16}$	3.5
2	$r_1 = 1.5, r_2 = 1.5$	$\frac{1}{16}$	3
2	$r_1 = 1.5, r_2 = 3.5$	$\frac{1}{16}$	5
2	$r_1 = 1.5, r_2 = 3.5$	$\frac{1}{16}$	5
2	$r_1 = 1.5, r_2 = 3.5$	$\frac{1}{16}$	5
2	$r_1 = 1.5, r_2 = 3.5$	$\frac{1}{16}$	5
2	$r_1 = 3.5, r_2 = 3.5$	$\frac{1}{16}$	7
3	$r_1 = 1.5, r_2 = 1.5, r_3 = 3.5$	$\frac{1}{16}$	6.5
3	$r_1 = 1.5, r_2 = 1.5, r_3 = 3.5$	$\frac{1}{16}$	6.5
3	$r_1 = 1.5, r_2 = 3.5, r_3 = 3.5$	$\frac{1}{16}$	8.5
3	$r_1 = 1.5, r_2 = 3.5, r_3 = 3.5$	$\frac{1}{16}$	8.5
4	$r_1 = 1.5, r_2 = 1.5, r_3 = 3.5, r_4 = 3.5$	$\frac{1}{16}$	10

This yields the null tail probabilities

$$P_0(T^+ \geq 10) = \tfrac{1}{16},$$
$$P_0(T^+ \geq 8.5) = \tfrac{3}{16},$$
$$P_0(T^+ \geq 7) = \tfrac{4}{16},$$
$$P_0(T^+ \geq 6.5) = \tfrac{6}{16},$$
$$P_0(T^+ \geq 5) = \tfrac{10}{16},$$
$$P_0(T^+ \geq 3.5) = \tfrac{12}{16},$$
$$P_0(T^+ \geq 3) = \tfrac{13}{16},$$
$$P_0(T^+ \geq 1.5) = \tfrac{15}{16},$$
$$P_0(T^+ \geq 0) = 1.$$

This distribution is called the conditional distribution or the permutation distribution of T^+, given the set of tied ranks $\{1.5, 1.5, 3.5, 3.5\}$. For the particular observed value $T^+ = 5$, we have $P_0(T^+ \geq 5) = \tfrac{10}{16}$, so that such a value does not indicate a deviation from H_0 in the direction of $\theta > 0$.

For a given data set of Z values, StatXact will generate the null permutation distribution of T^+. After entering the signed ranks of the data (using average ranks to break ties in absolute Z values) with the case command CA, the commands

```
SE/FI/TDIST.PRN (enter)
PE/EX (enter)
```

will generate the permutation distribution and save it in a file named TDIST.PRN. The output will include

One-sided P-value: Pr {Test Statistic. LE. Observed} $\qquad = .6250$

and

Point probability: Pr {Test Statistic. EQ. Observed} $\qquad = .2500.$

This yields $P_0(T^+ \geq 5) = (1 - .6250) + .2500 = .6250$, which agrees with what we found previously in this comment by direct enumeration.

12. *Some Power Results for the Wilcoxon Signed Rank Test.* We consider the upper-tail α-level test of $H_0 : \theta = 0$ versus $H_1 : \theta > 0$ given by procedure (3.4). Under the additive shift model (see Assumption A2) and common underlying distribution $F_1 \equiv F_2 \equiv \cdots \equiv F_n \equiv F$ for the Z differences, the power, or probability of correctly rejecting H_0, for median θ_0 values "near" the null hypothesis value of 0 can be approximated by

$$\text{Power} \doteq \Phi(A_F), \qquad (3.18)$$

where $\Phi(A_F)$ is the area under a standard normal density to the left of the point

$$A_F = \left\{ \frac{n(n-1)f^\star(0) + nf(0)}{[n(n+1)(2n+1)/24]^{1/2}} \right\} \theta - z_\alpha, \qquad (3.19)$$

where $f(0)$ is the common density function, evaluated at 0, for the Z differences and $f^\star(0)$ is the density function, also evaluated at 0, of the sum of two independent random variables drawn from the Z-population having distribution F (cf. Lehmann (1975, 167 and 403)).

When F is normal with standard deviation σ, we have $f(0) = (\sigma\sqrt{2\pi})^{-1}$ and $f^\star(0) = (2\sigma\sqrt{\pi})^{-1}$. Under this setting, A_F in (3.19) reduces to

$$A_{\text{normal}} = \left\{ \frac{(n(n-1)/2) + n/\sqrt{2}}{[n(n+1)(2n+1)/24]^{1/2}} \right\} \frac{\theta}{\sigma\sqrt{\pi}} - z_\alpha. \qquad (3.20)$$

Thus, when F is normal the approximate power for the additive shift model depends on θ and σ only through their ratio θ/σ. This implies, for example, that the approximate power for the pair ($\theta = .5, \sigma = 4$) is the same as the approximate power for the pair ($\theta = 1, \sigma = 8$).

For purposes of illustration, suppose that the additive shift model holds with the common underlying population F taken to be normal with variance $\sigma^2 = 4$ and treatment effect $\theta = 1.5$. For the case where $n = 10$ and $\alpha = .053$, the test rejects H_0 if and only if $T^+ \geq 44$. Substituting the appropriate values in equation (3.20), we obtain

$$A_{\text{normal}} = \left\{ \frac{10(9)/2 + 10/\sqrt{2}}{[10(11)(21)/24]^{1/2}} \right\} \frac{1.5}{2\sqrt{\pi}} - 1.62$$

$$= \left\{ \frac{45 + 7.07}{9.81} \right\} .42 - 1.62 = .61.$$

Thus, from Table A.1, the approximate power of this test at $\theta = 1.5$ (and $\sigma^2 = 4$) is

$$\text{Power} \doteq \Phi(.61) = 1 - .27 = .73.$$

This compares with the exact power of .70 as given in Table 1 of Klotz (1963). Additional exact power values for the one-sided Wilcoxon signed rank test and sample sizes 5(1)10 can be found in Klotz (1963) for normal shift alternatives and in Arnold (1965) for shifted t-distributions with $\nu = \frac{1}{2}, 1, 2$, and 4 degrees of freedom.

13. *Sample Size Determination.* The Wilcoxon signed rank test detects a more general class of alternatives than the location-shift alternatives associated with model Assumption A2. When $Z_1, \ldots, Z_n$ are a random sample from a single continuous, symmetric population F, then the one-sided upper-tail test defined by procedure (3.4) is consistent (that is, has power tending to 1 as n tends to infinity) against those F populations for which $\eta > \frac{1}{2}$, with

$$\eta = P(Z_1 + Z_2 > 0), \qquad (3.21)$$

where Z_1, Z_2 are independent and identically distributed as F. The parameter η (3.21) is the probability that a Z_1 randomly selected from the continuous and symmetric F will be greater than the negative of a second independent Z_2 also randomly selected from the same distribution F.

Noether (1987) shows how to determine an approximate sample size n so that the α-level one-sided test given by procedure (3.4) will have approximate power $1 - \beta$ against an alternative value of η (3.21) greater than $\frac{1}{2}$. This approximate value of n is

$$n \doteq \frac{(z_\alpha + z_\beta)^2}{3(\eta - \frac{1}{2})^2}. \qquad (3.22)$$

As an illustration of the use of equation (3.22), suppose we are testing H_0 and we desire to have an upper-tail level $\alpha = .025$ test with power $1 - \beta$ of at least .95 against an alternative for which $\eta = P(Z_1 + Z_2 > 0) = .8$. (Recall that under H_0, $\eta = .5$.) Since $z_\alpha = z_{.025} = 1.96$ and $z_\beta = z_{.05} = 1.65$, we find that the approximate required sample size for the alternative $\eta = .8$ is

$$n \doteq \frac{(1.96 + 1.65)^2}{3(.8 - .5)^2} = 48.3.$$

To be conservative, we would take $n = 49$.

14. *Consistency of the T^+ Test.* Under the assumption that $Z_1, \ldots, Z_n$ is a random sample from a single continuous population F, the consistency of the tests based on T^+ depends on the parameter

$$\eta^\star = P(Z_1 + Z_2 > 0) - \frac{1}{2},$$

where Z_1, Z_2 are independent and identically distributed as F. The test procedures defined by (3.4), (3.5) and (3.6) are consistent against the classes of alternatives corresponding to $\eta^\star >, <,$ and $\neq 0$, respectively.

Properties

1. *Consistency.* For our consistency statement we strengthen Assumption A2 to require that each Z has the same continuous population that is symmetric about θ. Then the tests defined by (3.4), (3.5), and (3.6) are consistent against the alternatives $\theta >, <,$ and $\neq 0$, respectively. See also Comment 14.
2. *Asymptotic Normality.* See Randles and Wolfe (1979, 83–85).
3. *Efficiency.* See Section 3.11.

PROBLEMS

1. The data in Table 3.3 are a subset of the data obtained by Kaneto, Kosaka, and Nakao (1967). The experiment investigated the effect of vagal nerve stimulation on insulin secretion. The subjects were mongrel dogs with varying body weights. Table 3.3 gives the amount of immunoreactive insulin in pancreatic venous plasma just before stimulation of the left vagus nerve (X) and the amount measured 5 min after stimulation (Y) for seven dogs. Test the hypothesis of no effect against the alternative that stimulation of the vagus nerve increases the blood level of immunoreactive insulin.

TABLE 3.3. Blood Levels of Immunoreactive Insulin (μU/ml)

Dog i	X_i	Y_i
1	350	480
2	200	130
3	240	250
4	290	310
5	90	280
6	370	1450
7	240	280

Source: A. Kaneto, K. Kosaka, and K. Nakao (1967).

2. Change the value of X_3, in Table 3.1, from 1.62 to 16.2. What effect does this outlying observation have on the calculations performed in Example 3.1? What does this suggest about the relative insensitivity of the signed rank tests to outliers? Construct an example in which changing one observation has a marked effect on the final decision regarding rejection or acceptance of H_0.

3. Let $T^- = \sum_{i=1}^{n} R_i(1 - \psi_i)$, where $\psi_i = 1$ if $Z_i > 0$, and 0 otherwise. Verify directly, or illustrate using the data of Table 3.1, the equation $T^+ + T^- = n(n + 1)/2$.

4. August, Hung, and Houck (1974) studied collagen metabolism in children deficient in growth hormone before and after growth hormone therapy. The data in Table 3.4 are the values of heat-insoluble hydroxyproline in the skin of children before and 3 months after growth hormone therapy. Can we conclude on the basis of these data that growth hormone therapy increases heat-insoluble hydroxyproline in the skin?

TABLE 3.4. Heat-Insoluble Hydroxyproline Micromoles per Gram of Dry Weight

Child i	Before	After
1	349	425
2	400	533
3	520	362
4	490	628
5	574	463
6	427	427
7	435	449

Source: G. P. August, W. Hung, and J. C. Houck (1974).

5. Assume that the additive shift model (see Assumption A2) holds with common underlying distribution $F_1 \equiv F_2 \equiv \cdots \equiv F_n \equiv F$. If we have 15 observations and F is normal with variance 16, what is the approximate power of the level $\alpha = .076$ test of $H_0 : \theta = 0$ versus the alternative $\theta > 0$ when the treatment effect is $\theta = 1.25$?

6. For arbitary number of observations n, what are the smallest and largest possible values for T^+? Justify your answers.

7. Suppose $n = 14$. Compare the critical region for the exact level $\alpha = .039$ test of $H_0 : \theta = 0$ versus $H_1 : \theta > 0$ based on T^+ with the critical region for the corresponding nominal level $\alpha = .039$ test based on the large-sample approximation. What is the exact significance level of this .039 nominal level test based on the large-sample approximation?

8. Consider a level $\alpha = .05$ test of $H_0 : \theta = 0$ versus the alternative $\theta > 0$ based on T^+ and let η be as given in equation (3.21). If our data $Z_1, \ldots, Z_n$ are a random sample from a single continuous, symmetric distribution $F(\cdot)$, how many observations n will we need to collect in order to have approximate power at least .84 against an alternative for which $\eta = .7$?

9. Suppose $n = 5$ and we observe the data $Z_1 = -1.3, Z_2 = 2.4, Z_3 = 1.3, Z_4 = 1.3$, and $Z_5 = 2.4$. What is the conditional probability distribution of T^+ under $H_0 : \theta = 0$ when average ranks are used to break ties among the absolute values of the Z's? How extreme is the observed value of T^+ in this conditional null distribution? Compare this fact with that obtained by taking the observed value of T^+ to the (incorrect) unconditional null distribution of T^+ given in Table A.4.

10. Apply the large-sample approximation test of $H_0 : \theta = 5$ versus $H_1 : \theta > 5$ based on T^+ to the beak-clapping data in Table 3.5.

11. Consider procedure (3.6) with n observations for testing $H_0 : \theta = 0$ versus $H_1 : \theta \neq 0$. If your critical region consists of the four values $T^+ = 0, 1, [n(n + 1)/2] - 1, n(n + 1)/2$, what is the significance level for your test?

12. Apply the appropriate form of the test based on T^+ to the data on Stanford Profile Scales of hypnotic susceptibility in Table 3.6.

13. For the case of $n = 5$ untied Z observations, use the representation for T^+ discussed in Comment 5 to obtain the form of the exact null (H_0) distribution of T^+.

14. Let Z_1 and Z_2 be independent, identically distributed continuous random variables with a common probability distribution that is symmetric about 0. What is the value of η^* in Comment 14 for this setting?

15. Consider the test of $H_0 : \theta = 0$ versus $H_1 : \theta > 0$ based on T^+ for the following $n = 10$ Z observations: $Z_1 = 2.5, Z_2 = 3.7, Z_3 = 0, Z_4 = -0.6, Z_5 = 4.7, Z_6 = 0, Z_7 = 1.4, Z_8 = 0, Z_9 = 1.9, Z_{10} = 5.2$. Compute the P-values for the competing T^+-procedures based on either (i) discarding the zero Z values and reducing n accordingly, as recommended in the ties portion of this section, or (ii) treating the zero Z values in a conservative manner, as presented in Comment 9. Discuss the results.

16. Use the computer software StatXact to verify the small-sample and large-sample approximation P-values for the test of $H_0 : \theta = 0$ versus $H_1 : \theta > 0$ based on T^+ for the salary data in Example 3.2.

17. Use the computer software StatXact to generate the conditional permutation distribution of T^+, given the set of tied ranks for the salary data in Example 3.2. From this conditional permutation distribution of T^+, obtain the exact conditional P-value, $P_0(T^+ \geq 62.5)$, for the corresponding test of $H_0 : \theta = 0$ versus $H_1 : \theta > 0$. Compare this exact conditional P-value with the approximate small-sample p-value of .039 obtained in Example 3.2 from Table A.4.

3.2. AN ESTIMATOR ASSOCIATED WITH WILCOXON'S SIGNED RANK STATISTIC (HODGES-LEHMANN)

Procedure

To estimate the treatment effect θ, form the $M = n(n + 1)/2$ averages $(Z_i + Z_j)/2$, for $i \leq j = 1, \ldots, n$. The estimator of θ associated with the Wilcoxon signed rank statistic T^+

(see Comment 15) is

$$\hat{\theta} = \text{median}\left\{\frac{Z_i + Z_j}{2}, \quad i \leq j = 1, \ldots, n\right\}. \tag{3.23}$$

Let $W^{(1)} \leq \cdots \leq W^{(M)}$ denote the ordered values of $(Z_i + Z_j)/2$. Then if M is odd, say $M = 2k + 1$, we have $k = (M - 1)/2$ and

$$\hat{\theta} = W^{(k+1)}, \tag{3.24}$$

the value that occupies position $k + 1$ in the list of the ordered $(Z_i + Z_j)/2$ averages. If M is even, say $M = 2k$, then $k = M/2$ and

$$\hat{\theta} = \frac{W^{(k)} + W^{(k+1)}}{2}. \tag{3.25}$$

That is, when M is even, $\hat{\theta}$ is the average of the two $(Z_i + Z_j)/2$ values that occupy positions k and $k + 1$ in the ordered list of the M $(Z_i + Z_j)/2$ averages.

Example 3.3: *Continuation of Example 3.1.* To estimate θ for the Hamilton depression scale factor IV data in Table 3.1, we first calculate the $M = 9(10)/2 = 45$ $(Z_i + Z_j)/2$ averages. To facilitate this computation and eventual ordering of these averages, we first obtain the individual ordered Z values, denoted by $Z^{(1)} \leq \cdots \leq Z^{(9)}$, and place them as the headings for a 9×9 upper triangular array. The (i, j)th entry in this array then corresponds to the value of the sum $(Z^{(i)} + Z^{(j)})$, for $i \leq j = 1, 2, \ldots, 9$. This representation for the Hamilton depression scale factor IV data is as follows:

	−1.022	−.952	−.620	−.590	−.490	−.430	−.010	.080	.147
−1.022	−2.044	−1.974	−1.642	−1.612	−1.512	−1.452	−1.032	−.942	−.875
−.952		−1.904	−1.572	−1.542	−1.442	−1.382	−.962	−.872	−.805
−.620			−1.240	−1.210	−1.110	−1.050	−.630	−.540	−.473
−.590				−1.180	−1.080	−1.020	−.600	−.510	−.443
−.490					−.980	−.920	−.500	−.410	−.343
−.430						−.860	−.440	−.350	−.283
−.010							−.020	.070	.137
.080								.160	.227
.147									.294

Thus, for example, the entry in the fourth row and sixth column of the array ($i = 4, j = 6$) is $Z^{(4)} + Z^{(6)} = -.590 - .430 = -1.020$. The remaining 44 entries are calculated similarly. The ordered $Z^{(i)} + Z^{(j)}$ sums are then obtained by observation in this array, moving carefully from the upper left (the $(1, 1)$ entry) across and down the array to the lower right (the $(9,9)$ entry). These ordered $Z^{(i)} + Z^{(j)}$ sums for these data are: $-2.044, -1.974, -1.904, -1.642,$ $-1.612, -1.572, -1.542, -1.512, -1.452, -1.442, -1.382, -1.240, -1.210, -1.180,$ $-1.110, -1.080, -1.050, -1.032, -1.020, -.980, -.962, -.942, -.920, -.875, -.872,$ $-.860, -.805, -.630, -.600, -.540, -.510, -.500, -.473, -.443, -.440, -.410, -.350,$ $-.343, -.283, -.020, .070, .137, .160, .227,$ and $.294$. The ordered values of the $(Z_i + Z_j)/2$ averages, namely, $W^{(1)} \leq \cdots \leq W^{(45)}$, then correspond to these ordered $Z^{(i)} + Z^{(j)}$ sums divided by 2. Since $M = 45$ is odd, we use equation (3.24) with $k = (45 - 1)/2 = 22$ to obtain the estimate $\hat{\theta} = W^{(23)} = -.920/2 = -.460$ for the treatment effect θ. Thus, we

3.2. AN ESTIMATOR ASSOCIATED WITH WILCOXON'S SIGNED RANK STATISTIC

estimate that a typical patient of the type included in this study will have a drop in Hamilton depression scale factor IV value of roughly .460 due to treatment with tranquilizer T.

We can also use Minitab's WINTERVAL command to compute the Hodges-Lehmann estimator for the Hamilton depression scale factor IV data, as follows:

```
MTB > SET C1
DATA > -.952  .147  -1.022  -.430  -.620  -.590  -.490  .080  -.010
DATA > END
MTB > WINTERVAL .96 C1
```

The output is, in Minitab's notation:

	N	ESTIMATED MEDIAN	ACHIEVED CONFIDENCE	CONFIDENCE INTERVAL
C1	9	-.460	96	(-.786, -.010)

The point estimate is $\hat{\theta} = -.460$, as we found previously. We explain the confidence interval portion of Minitab's output in Section 3.3.

We can also get the $M = n(n + 1)/2$ averages $(Z_i + Z_j)/2$, for $i \leq j = 1, \ldots, n$, using Minitab's WALSH command. These averages, in addition to their importance in finding $\hat{\theta}$, are also necessary for determining the confidence interval for θ described in Section 3.3. Again we illustrate the Minitab WALSH command with the Hamilton depression scale factor IV data. The form of the command is

WALSH averages for C, put into C [put indices into C C].

After setting the ordered Z differences in Column C1, use

```
MTB > WALSH C1  C2  C3  C4.
```

For given $i, j \in \{1, 2, \ldots, n\}$, the individual average $(Z^{(i)} + Z^{(j)})/2$ appears in Column C2 corresponding to index i in Column C3 and index j in Column C4. To get the ordered values of the $(Z^{(i)} + Z^{(j)})/2$ averages from Minitab, use the SORT command. Thus, to put the $(Z^{(i)} + Z^{(j)})/2$ averages from Column C2 in order in Column C5, use

```
MTB > SORT C2  C5.
```

The Minitab output for Columns C2–C5 for the Hamilton depression scale factor IV data then consists of the following four columns:

C2 $(Z^{(i)} + Z^{(j)})/2$ average	C3 Index i	C4 Index j	C5 ordered $(Z^{(i)} + Z^{(j)})/2$ averages
-1.022	1	1	-1.022
-0.987	1	2	-0.987
-0.821	1	3	-0.952
-0.806	1	4	-0.821
-0.756	1	5	-0.806
-0.726	1	6	-0.786
-0.516	1	7	-0.771
-0.471	1	8	-0.756
-0.4375	1	9	-0.726
-0.952	2	2	-0.721
-0.86	2	3	-0.691
-0.71	2	4	-0.620
-0.721	2	5	-0.605
-0.691	2	6	-0.590
-0.481	2	7	-0.555
-0.436	2	8	-0.540

(continued)

C2 $(Z^{(i)} + Z^{(j)})/2$ average	C3 Index i	C4 Index j	C5 ordered $(Z^{(i)} + Z^{(j)})/2$ averages
−0.4025	2	9	−0.525
−0.620	3	3	−0.516
−0.605	3	4	−0.510
−0.555	3	5	−0.490
−0.525	3	6	−0.481
−0.315	3	7	−0.471
−0.270	3	8	−0.460
−0.2365	3	9	−0.4375
−0.590	4	4	−0.436
−0.540	4	5	−0.430
−0.510	4	6	−0.4025
−0.300	4	7	−0.315
−0.255	4	8	−0.300
−0.2215	4	9	−0.270
−0.490	5	5	−0.255
−0.460	5	6	−0.250
−0.250	5	7	−0.2365
−0.205	5	8	−0.2215
−0.1715	5	9	−0.220
−0.430	6	6	−0.205
−0.220	6	7	−0.175
−0.175	6	8	−0.1715
−0.1415	6	9	−0.1415
−0.010	7	7	−0.010
0.035	7	8	0.035
0.0685	7	9	0.0685
0.080	8	8	0.080
0.1135	8	9	0.1135
0.147	9	9	0.147

Comments

15. *Motivation for the Hodges-Lehmann Estimator.* The Hodges-Lehmann estimator $\hat{\theta}$, defined by equation (3.23), is associated with the Wilcoxon signed rank test. When $\theta = 0$, the distribution of the statistic T^+ is symmetric about its mean, $n(n + 1)/4$ (see Comment 8). A natural estimator of θ is the amount $\hat{\theta}$ (say) that should be subtracted from each Z_i so that the value of T^+, when applied to the shifted sample $Z_1 - \hat{\theta}, \ldots, Z_n - \hat{\theta}$, is as close to $n(n + 1)/4$ as possible. Roughly speaking, we estimate θ by the amount $(\hat{\theta})$ that the Z sample should be shifted in order that $Z_1 - \hat{\theta}, \ldots, Z_n - \hat{\theta}$ appears (when "viewed" by the signed rank statistic T^+) as a sample from a population with median 0. (Under Assumptions A1 and A2, each of the $Z_1 - \theta, \ldots, Z_n - \theta$ variables is from a population with median 0.)

The Hodges-Lehmann method can be applied to a large class of statistics containing T^+. However, the forms of the resulting estimators for other members of this class are not always as convenient for calculation as is $\hat{\theta}$. See Hodges and Lehmann (1983) for an expository article on their method.

16. *Sensitivity to Gross Errors.* The estimator $\hat{\theta}$ is relatively insensitive to outliers. This is not the case with the classical estimator $\bar{Z} = \sum_{i=1}^{n} Z_i/n$. Thus the use of $\hat{\theta}$ provides protection against gross errors.

17. *Walsh Averages.* Each of the $n(n + 1)/2$ averages $(Z_i + Z_j)/2, i \leq j = 1, \ldots, n$, is called a Walsh average (see Walsh (1949)). If we define W^+ to be the number of positive

Walsh averages, then (when there are no ties among the $|Z|$'s and none of the Z's is zero) the statistic W^+ is identical to T^+ (3.3) (See Problem 22). This result is due to Tukey (1949).

18. *Zero and Tied Absolute Z's.* Note that in calculating the estimator $\hat{\theta}$ we use *all* of the Z differences in computing the $(Z_i + Z_j)/2$ averages. Although we recommend (see Ties in Section 3.1) discarding the zero Z values (and reducing n accordingly) prior to applying the signed rank test to the data, it is not necessary to do so when calculating $\hat{\theta}$. In fact, the zero Z values contain important information about the magnitude of the treatment effect. This is also the case when we consider (Section 3.3) confidence intervals and bounds for θ.

19. *Pseudomedian.* A pseudomedian (cf. Høyland (1965)) of a distribution F is defined to be a median of the distribution of $(Z_1 + Z_2)/2$, where Z_1 and Z_2 are independent, each with the same distribution F. We assume here that our F is such that both the median and the pseudomedian of F are unique. The estimator $\hat{\theta}$ (3.23) is a consistent estimator of the pseudomedian, which in general may differ from the median θ. However, when F is symmetric as assumed in this section, the median and the pseudomedian coincide.

Properties

1. *Standard Deviation of $\hat{\theta}$.* For the asymptotic standard deviation of $\hat{\theta}$ (3.23), see Hodges and Lehmann (1963), Lehmann (1963c), and Comment 24.
2. *Asymptotic Normality.* See Hodges and Lehmann (1963) and Ramachandramurty (1966a).
3. *Efficiency.* See Hodges and Lehmann (1963), Bickel (1965), Høyland (1968), Gastwirth and Rubin (1969), and Section 3.11.

PROBLEMS

18. Consider the data of Table 3.2. Using the X and Y associations from Example 3.2, estimate θ for the salary data of that example.
19. Estimate θ for the blood level data of Table 3.3.
20. Change the value of X_3, as given in Table 3.1, from 1.62 to 16.2. How does this affect the value of $\bar{Z} = \sum_{i=1}^{9} Z_i/9$? How does it affect the estimate of θ given by $\hat{\theta}$? Interpret these calculations in light of Comment 16.
21. Estimate θ for the heat-insoluble hydroxyproline data of Table 3.4.
22. Verify directly, or illustrate using the data of Table 3.1, that (when there are no ties among the absolute values of the Z's and none of the Z's is zero) T^+ is equal to the number of positive Walsh averages W^+ (see Comment 17).
23. (a) What happens to $\hat{\theta}$ when we add a number b to each of the sample values $Z_1, \ldots, Z_n$?
    (b) What happens to $\hat{\theta}$ when we multiply each sample value $Z_1, \ldots, Z_n$ by a number d?
    (c) Let k be a positive integer such that $n > 2k$. What happens to $\hat{\theta}$ when we discard the k largest and the k smallest Z values from the sample?
24. (a) Do we need to calculate all of the $n(n+1)/2$ Walsh averages in order to compute the value of $\hat{\theta}$? Explain.
    (b) If you were to write a computer program to calculate $\hat{\theta}$, how would you proceed?
25. Use the computer software Minitab to compute the value of $\hat{\theta}$ for the salary data of Table 3.2. (Use the X and Y labels as assigned in Example 3.2.)
26. Use the computer software Minitab to obtain the $M = 78$ ordered Walsh averages for the salary data of Table 3.2. (Use the X and Y labels as assigned in Example 3.2.)

3.3. A DISTRIBUTION-FREE CONFIDENCE INTERVAL BASED ON WILCOXON'S SIGNED RANK TEST (TUKEY)

Procedure

For a symmetric two-sided confidence interval for θ, with confidence coefficient $1 - \alpha$, first obtain the upper $(\alpha/2)$th percentile point $t_{\alpha/2}$ of the null distribution of T^+ from Table A.4. Set

$$C_\alpha = \frac{n(n+1)}{2} + 1 - t_{\alpha/2}. \tag{3.26}$$

The $100(1 - \alpha)\%$ confidence interval (θ_L, θ_U) for θ that is associated with the two-sided Wilcoxon signed rank test (see Comment 20) of $H_0 : \theta = 0$ is then given by

$$\theta_L = W^{(C_\alpha)}, \quad \theta_U = W^{(M+1-C_\alpha)} = W^{(t_{\alpha/2})}, \tag{3.27}$$

where $M = n(n+1)/2$ and $W^{(1)} \leq \cdots \leq W^{(M)}$ are the ordered values of the $(Z_i + Z_j)/2$ averages, $1 \leq i \leq j \leq n$, used in computing the point estimator $\hat{\theta}$ (3.23). That is, θ_L is the $(Z_i + Z_j)/2$ average (i.e., Walsh average; see Comment 17) that occupies position C_α in the list of M ordered $(Z_i + Z_j)/2$ averages. The upper endpoint θ_U is the $(Z_i + Z_j)/2$ average that occupies position $M + 1 - C_\alpha = t_{\alpha/2}$ in this ordered list. With θ_L and θ_U given by display (3.27), we have

$$P_\theta(\theta_L < \theta < \theta_U) = 1 - \alpha \text{ for all } \theta. \tag{3.28}$$

(For upper or lower confidence bounds for θ associated with appropriate one-sided Wilcoxon signed rank tests of $H_0 : \theta = 0$, see Comment 21.)

Large-Sample Approximation

For large n the integer C_α may be approximated by

$$C_\alpha \approx \frac{n(n+1)}{4} - z_{\alpha/2} \left\{ \frac{n(n+1)(2n+1)}{24} \right\}^{1/2}. \tag{3.29}$$

In general the value of the right-hand side of equation (3.29) is not an integer. To be conservative, take C_α to be the largest integer that is less than or equal to the right-hand side of equation (3.29).

Example 3.4: *Continuation of Examples 3.1 and 3.3.* Consider the Hamilton depression scale factor IV data of Table 3.1. We illustrate how to obtain the 96% confidence interval for θ. With $1 - \alpha = .96$ (so that $\alpha = .04$), we see from Table A.4 with $n = 9$ that $t_{\alpha/2} = t_{.02} = 40$. From equation (3.26) it follows that

$$C_{.04} = \left[\frac{9(9+1)}{2} + 1 - 40 \right] = 6.$$

Using these values of $C_{.04} = 6$ and $t_{.02} = 40$ in display (3.27), we see that

$$\theta_L = W^{(6)} = -.786 \quad \text{and} \quad \theta_U = W^{(40)} = -.010.$$

3.3. CONFIDENCE INTERVAL FOR TREATMENT EFFECT (TUKEY)

The value $\theta_L = -.786$ is found by dividing the sixth smallest ordered $Z_i + Z_j$ value in the list presented in Example 3.3, namely, -1.572, by 2. Similarly, the value $\theta_U = -.010$ corresponds to dividing the sixth largest (or 40th ordered) $Z_i + Z_j$ value in that list, namely, $-.020$, by 2.

If we choose to apply the large-sample approximation, we find from approximation (3.29) that

$$C_{.04} \approx \left[\frac{9(9+1)}{4}\right] - 2.05 \left\{\frac{9(9+1)(2(9)+1)}{24}\right\}^{1/2} = 5.2.$$

Thus, with a conservative approach and the large-sample approximation, we set $C_{.04} = 5$ and find that

$$(\theta_L, \theta_U) = (W^{(5)}, W^{(41)}) = (-.806, .035)$$

is the approximate 96% confidence interval for θ.

Recall that in Example 3.3 the output from Minitab's WINTERVAL command included the presentation of a confidence interval for θ. There, with approximate confidence coefficient $1 - \alpha = .96$, Minitab gives $(-.786, -.010)$ as the approximate (using the large-sample approximation with continuity correction) 96% confidence interval for θ, agreeing with the exact 96% confidence interval we previously obtained using display (3.27) and Table A.4.

Comments

20. *Relationship of Confidence Interval to Two-Sided Test.* The $100(1 - \alpha)$% confidence interval for θ given by display (3.27) can be obtained from the two-sided signed rank test as follows. The confidence interval (θ_L, θ_U) consists of those θ_0 values for which the two-sided α-level test of $\theta = \theta_0$ (see Comment 3) does not reject the hypothesis $\theta = \theta_0$. The confidence interval given by display (3.27) was defined by way of a graphical procedure by Lincoln Moses (who attributed it to John Tukey) in Chapter 18 of Walker and Lev (1953). See Lehmann (1986, 90) for a general result relating confidence intervals and acceptance regions of tests, and see Lehmann (1963c) for the specific result involving the signed rank test.

21. *Confidence Bounds.* In many settings we are interested only in making one-sided confidence statements about the parameter θ; that is, we wish to assert with specified confidence that θ is no larger (or, in other settings, no smaller) than some upper (lower) confidence bound based on the sample data. To obtain such one-sided confidence bounds for θ, we proceed as follows. For specified confidence coefficient $1 - \alpha$, find the upper αth (not $(\alpha/2)$nd, as for the confidence interval) percentile point t_α of the null distribution of T^+ from Table A.4. Set

$$C_\alpha^* = \frac{n(n+1)}{2} + 1 - t_\alpha. \tag{3.30}$$

The $100(1 - \alpha)$% *lower* confidence bound θ_L^* for θ that is associated with the one-sided Wilcoxon signed rank test of $H_0 : \theta = 0$ against the alternative $H_1 : \theta > 0$ is then given by

$$(\theta_L^*, \infty) = (W^{(C_\alpha^*)}, \infty), \tag{3.31}$$

where, as before, $M = n(n+1)/2$ and $W^{(1)} \leq \cdots \leq W^{(M)}$ are the ordered values of the $(Z_i + Z_j)/2$ averages, $1 \leq i \leq j \leq n$. With θ_L^* given by display (3.31), we have

$$P_\theta(\theta_L^* < \theta < \infty) = 1 - \alpha \text{ for all } \theta. \tag{3.32}$$

The corresponding $100(1 - \alpha)\%$ *upper* confidence bound θ_U^* for θ that is associated with the one-sided Wilcoxon signed rank test of $H_0 : \theta = 0$ against the alternative $H_1 : \theta < 0$ is given by

$$(-\infty, \theta_U^*) = \left(-\infty, W^{(M+1-C_\alpha^*)}\right) = (-\infty, W^{(t_\alpha)}), \qquad (3.33)$$

where C_α^* is given in (3.30). It follows that

$$P_\theta(-\infty < \theta < \theta_U^*) = 1 - \alpha \quad \text{for all } \theta. \qquad (3.34)$$

For large n the integer C_α^* may be approximated by

$$C_\alpha^* \approx \frac{n(n+1)}{4} - z_\alpha \left\{\frac{n(n+1)(2n+1)}{24}\right\}^{1/2}. \qquad (3.35)$$

As with C_α (3.29) and the confidence interval for θ, the value of the right-hand side of equation (3.35) is not an integer. To be conservative, take C_α^* to be the largest integer that is less than or equal to the right-hand side of equation (3.35).

The $100(1 - \alpha)\%$ lower and upper confidence bounds θ_L^* (3.31) and θ_U^* (3.33) are related to the acceptance regions of the one-sided Wilcoxon signed rank tests of $H_0 : \theta = \theta_0$ against the alternatives $\theta > \theta_0$ and $\theta < \theta_0$, respectively, in the same way that the confidence interval (θ_L, θ_U) is related to the acceptance region of the two-sided Wilcoxon signed rank test of $H_0 : \theta = \theta_0$ (see Comment 20).

22. *Zero and Tied Absolute Z's.* Note that in calculating the confidence interval (θ_L, θ_U) from display (3.27) or the confidence bounds θ_L^* (3.31) or θ_U^* (3.33) for θ, we use *all* the Z differences in computing the $(Z_i + Z_j)/2$ averages. This is in common with our recommendation (see Comment 18) for computing the point estimator $\hat{\theta}$ (3.23), but different from the recommended policy (see Ties in Section 3.1) of discarding the zero Z values (and reducing n accordingly) prior to applying the signed rank test to the data. However, if there are zero Z's in the data, the equivalence (discussed in Comments 20 and 21) between the acceptance regions of the one-sided and two-sided signed rank tests and the appropriate confidence bound and confidence interval, respectively, are no longer valid. In addition, in cases with tied absolute Z's, the nominal confidence coefficient $1 - \alpha$ used in displays (3.27), (3.31), and (3.33) is not exact anymore, since in such cases the exact conditional null distribution of T^+ is no longer given by Table A.4. (See Comment 11.)

23. *Midpoint of Confidence Interval as an Estimator.* The midpoint of the interval (3.27), namely, $[W^{(C_\alpha)} + W^{(M+1-C_\alpha)}]/2$, suggests itself as a reasonable estimator of θ. (Note that this actually yields a class of estimators depending on the value of α.) In general this midpoint is not the same as $\hat{\theta}$ (3.23). Lehmann (1963c) has also derived an asymptotically distribution-free confidence interval centered at $\hat{\theta}$. This asymptotically distribution-free confidence interval is based on the assumption that each of the n Z_i's comes from the *same* continuous population that is symmetric about θ. This assumption is more restrictive than Assumption A2.

24. *Estimating the Asymptotic Standard Deviation of $\hat{\theta}$.* Replace Assumption A2 by the stronger Assumption A2': each Z comes from the *same* continuous population that is symmetric about θ. Then, it follows from Lehmann (1963c) that the statistic $(\theta_U - \theta_L)/2z_{\alpha/2}$, where (θ_L, θ_U) is the $100(1 - \alpha)\%$ confidence interval for θ defined by display (3.27), is a consistent estimator for the asymptotic standard deviation of the point estimator $\hat{\theta}$ (3.23).

3.3. CONFIDENCE INTERVAL FOR TREATMENT EFFECT (TUKEY)

Properties

1. *Distribution-Freeness.* For populations satisfying Assumptions A1 and A2, equation (3.28) holds. Hence, we can control the coverage probability to be $1 - \alpha$ without having more specific knowledge about the forms of the underlying Z distributions. Thus (θ_L, θ_U) is a distribution-free confidence interval for θ over a very large class of populations.
2. *Efficiency.* See Lehmann (1963c) and Section 3.11.

PROBLEMS

27. For the blood-level data of Table 3.3, obtain a confidence interval for θ with exact confidence coefficient .954.
28. For the heat-insoluble hydroxyproline data of Table 3.4, obtain a confidence interval for θ with exact confidence coefficient .922.
29. For the blood-level data of Table 3.3 and $\alpha = .078$, calculate the point estimator of θ defined in Comment 23. Compare with the value of $\hat{\theta}$ obtained in Problem 19.
30. Use the results of Example 3.4 to obtain an estimate of the asymptotic standard deviation of $\hat{\theta}$ for the Hamilton depression scale factor IV data of Table 3.1. (See Comment 24.)
31. For the Hamilton depression scale factor IV data of Table 3.1, find an upper confidence bound for θ with exact confidence coefficient .973. (See Comment 21.)
32. For the salary data of Table 3.2, use equation (3.31) in Comment 21 and find a lower confidence bound for θ with approximate confidence coefficient .936. Why is the confidence coefficient only approximate and not exact?
33. Consider the $1 - \alpha$ confidence interval for θ defined by display (3.27). Let $Z_{(1)} \leq \cdots \leq Z_{(n)}$ be the ordered Z's. Show that when $\alpha = 2/2^n$,

$$\theta_L = Z_{(1)} \quad \text{and} \quad \theta_U = Z_{(n)}.$$

34. Consider the $1 - \alpha$ upper confidence bound for θ given in equation (3.33) in Comment 21. If $Z_{(1)} \leq \cdots \leq Z_{(n)}$ denote the ordered Z's and $\alpha = 2/2^n$, show that

$$\theta_U^* = \frac{Z_{(n-1)} + Z_{(n)}}{2}.$$

35. Consider the $1 - \alpha$ confidence interval for θ defined by display (3.27). Let $Z_{(1)} \leq \cdots \leq Z_{(n)}$ be the ordered Z's. If $\alpha = 4/2^n$, express the length $(\theta_U - \theta_L)$ of the confidence interval in terms of $Z_{(1)}, \ldots, Z_{(n)}$.
36. How does varying α affect the length of the confidence interval defined by display (3.27)? How does it affect the point estimator defined in Comment 23?
37. Consider the blood level data of Table 3.3. Obtain an approximate 95% confidence interval for θ using the large-sample approximation of this section. Compare this approximate confidence interval with the exact 95.4% confidence interval obtained in Problem 27.
38. Consider the salary data of Table 3.2. Use the large-sample approximation of this section to obtain an approximate 90% confidence interval for θ.
39. Consider the heat-insoluble hydroxyproline data of Table 3.4. Use the large-sample approximation to obtain an approximate 99% lower confidence bound for θ. (See Comment 21.)
40. Consider the case $n = 10$ and compare the length of the exact 95.2% confidence interval for θ given by display (3.27) with the length of the approximate 95.2% confidence interval for θ obtained using the large-sample approximation of this section.
41. Consider the case $n = 15$ and compare the exact 96.8% upper confidence bound for θ given by equation (3.33) with the approximate 96.8% upper confidence bound for θ obtained from the large-sample approximation in Comment 21.
42. Use the computer software `Minitab` to obtain a 93.6% confidence interval for θ for the salary data of Table 3.2. (Use the X and Y labels as assigned in Example 3.2.)

PAIRED REPLICATE ANALYSES BY WAY OF SIGNS

Data. We obtain $2n$ observations, two observations on each of n subjects (blocks, patients, etc.)

Subject i	X_i	Y_i
1	X_1	Y_1
2	X_2	Y_2
$\vdots$	$\vdots$	$\vdots$
n	X_n	Y_n

Assumptions

B1. We let $Z_i = Y_i - X_i$, for $i = 1, \ldots, n$. The differences $Z_1, \ldots, Z_n$ are mutually independent.

B2. Each Z_i, $i = 1, \ldots, n$, comes from a continuous population (not necessarily the same) that has a common median θ. If F_i represents the distribution function for Z_i, $i = 1, \ldots, n$, this assumption requires that

$$F_i(\theta) = P(Z_i \leq \theta) = P(Z_i > \theta) = 1 - F_i(\theta), \quad \text{for } i = 1, \ldots, n. \tag{3.36}$$

The parameter θ is referred to as the unknown treatment effect.

3.4. A DISTRIBUTION-FREE SIGN TEST (FISHER)

Hypothesis

The null hypothesis of interest here is that of zero shift in location due to the treatment, namely,

$$H_0 : \theta = 0. \tag{3.37}$$

This null hypothesis asserts that each of the distributions (not necessarily the same) for the differences (posttreatment minus pretreatment observations) has median 0, corresponding to no shift in location due to the treatment.

Procedure

To compute the sign statistic B, define indicator variables ψ_i, $i = 1, \ldots, n$, where

$$\psi_i = \begin{cases} 1, & \text{if } Z_i > 0 \\ 0, & \text{if } Z_i < 0, \end{cases} \tag{3.38}$$

and set

$$B = \sum_{i=1}^{n} \psi_i. \tag{3.39}$$

The sign statistic B is the number of positive Z's.

a. One-Sided Upper-Tail Test. To test

$$H_0 : \theta = 0$$

versus

$$H_1 : \theta > 0,$$

at the α level of significance,

$$\text{Reject } H_0 \text{ if } B \geq b_{\alpha,1/2}; \text{ otherwise do not reject,} \quad (3.40)$$

where the constant $b_{\alpha,1/2}$ is chosen to make the type I error probability equal to α and is the upper αth percentile point for the binomial distribution with sample size n and $p = \frac{1}{2}$. Values of $b_{\alpha,1/2}$ are given in Table A.2.

b. One-Sided Lower-Tail Test. To test

$$H_0 : \theta = 0$$

versus

$$H_2 : \theta < 0,$$

at the α level of significance,

$$\text{Reject } H_0 \text{ if } B \leq n - b_{\alpha,1/2}; \text{ otherwise do not reject.} \quad (3.41)$$

c. Two-Sided Test. To test

$$H_0 : \theta = 0$$

versus

$$H_3 : \theta \neq 0,$$

at the α level of significance,

$$\text{Reject } H_0 \text{ if } B \geq b_{\alpha/2,1/2} \text{ or } B \leq n - b_{\alpha/2,1/2}; \text{ otherwise do not reject.} \quad (3.42)$$

This two-sided procedure is the two-sided symmetric test with $\alpha/2$ probability in each tail of the null distribution of B.

Large-Sample Approximation

The large-sample approximation is based on the asymptotic normality of B, suitably standardized. Since the distribution of B under the null hypothesis $H_0 : \theta = 0$ is binomial with parameters n and $p = \frac{1}{2}$, we know that

$$E_0(B) = \frac{n}{2} \quad (3.43)$$

and

$$\text{var}_0(B) = \frac{n}{4}. \quad (3.44)$$

The standardized version of B is then

$$B^* = \frac{B - E_0(B)}{\{\text{var}_0(B)\}^{1/2}} = \frac{B - (n/2)}{\{n/4\}^{1/2}}. \quad (3.45)$$

When H_0 is true, B^* has, as n tends to infinity, an asymptotic $N(0, 1)$ distribution. (See Comment 32 for indications of the proof.) The normal approximation for procedure (3.40) is

$$\text{Reject } H_0 \text{ if } B^* \geq z_\alpha; \text{ otherwise do not reject,} \qquad (3.46)$$

the normal theory approximation for procedure (3.41) is

$$\text{Reject } H_0 \text{ if } B^* \leq -z_\alpha; \text{ otherwise do not reject,} \qquad (3.47)$$

and the normal theory approximation for procedure (3.42) is

$$\text{Reject } H_0 \text{ if } |B^*| \geq z_{\alpha/2}; \text{ otherwise do not reject.} \qquad (3.48)$$

Ties

If there are zero values among the Z's, discard the zero values and redefine n to be the number of nonzero Z's.

Example 3.5: *Beak-Clapping Counts*. The data in Table 3.5 are a subset of the data obtained by Oppenheim (1968) in an experiment investigating light responsivity in chick embryos. The subjects were white leghorn chick embryos, and the behavioral response measured in the investigation was beak-clapping (i.e., the rapid opening and closing of the beak that occurs during the latter one-third of incubation in chick embryos). (Gottlieb (1965) had previously shown that changes in the rate of beak-clapping constituted a sensitive indicator of auditory responsiveness in chick embryos.) The embryos were placed in a dark chamber 30 min before the intitiation of testing. Then ten 1-min readings were taken in the dark, and at the end of this 10-min period, a single reading was obtained for a 1-min period of illumination. Table 3.5 gives the average number of claps per minute during the dark period (X) and the corresponding rate during the period of illumination (Y) for 25 chick embryos.

Since responsivity of a chick embryo to a light stimulus is expected to correspond to positive Z differences, we apply procedure (3.40) which is designed to detect the alternative $\theta > 0$. For purpose of illustration, we take the significance level to be $\alpha = .0216$. From Table A.2 with $n = 25$ and $p = .5$, we find $b_{.0216, 1/2} = 18$. Procedure (3.40) is then given by

$$\text{Reject } H_0 \text{ if } B \geq 18.$$

Now, the sample value of B can be obtained directly from the indicator variables $\psi_1, \ldots, \psi_{25}$ listed in Table 3.5.

We find that $B = \sum_{i=1}^{25} \psi_i =$ (number of positive Z's) $= 21$. Since this value of B is greater than the critical value 18, we reject H_0 in favor of $\theta > 0$ at the $\alpha = .0216$ level.

The one-sided P-value for these data is the smallest significance level at which we can reject H_0 in favor of $\theta > 0$. With the observed value of the test statistic $B = 21$ and $n = 25$, we see that

$$P_0(B \geq 21) = .0005,$$

where this P-value of .0005 is obtained by entering Table A.2 with $n = 25, p = .5$, and $b = 21$. (We note that the actual magnitudes of the Z differences are not needed to calculate B. We require only the information as to whether or not Y_i is larger than X_i, for $i = 1, \ldots, n$, and this information is contained entirely in the indicator variables $\psi_1, \ldots, \psi_n$. However, the actual

3.4. A DISTRIBUTION-FREE SIGN TEST (FISHER)

TABLE 3.5. Beak-Clapping Counts per Minute

Embryo i	X_i (Dark Period)	Y_i (Illumination)	$Z_i = Y_i - X_i$	ψ_i
1	5.8	5	−0.8	0
2	13.5	21	7.5	1
3	26.1	73	46.9	1
4	7.4	25	17.6	1
5	7.6	3	−4.6	0
6	23.0	77	54.0	1
7	10.7	59	48.3	1
8	9.1	13	3.9	1
9	19.3	36	16.7	1
10	26.3	46	19.7	1
11	17.5	9	−8.5	0
12	17.9	25	7.1	1
13	18.3	59	40.7	1
14	14.2	38	23.8	1
15	55.2	70	14.8	1
16	15.4	36	20.6	1
17	30.0	55	25.0	1
18	21.3	46	24.7	1
19	26.8	25	−1.8	0
20	8.1	30	21.9	1
21	24.3	29	4.7	1
22	21.3	46	24.7	1
23	18.2	71	52.8	1
24	22.5	31	8.5	1
25	31.1	33	1.9	1

Source: R. W. Oppenheim (1968).

magnitude of the Z_i's will be necessary in Sections 3.5 and 3.6 to obtain point and interval estimates, respectively, of θ associated with the sign test.)

For the large-sample approximation, we find from equation (3.45) that

$$B^* = \frac{21 - \left(\frac{25}{2}\right)}{\left(\frac{25}{4}\right)^{1/2}} = 3.40.$$

Thus, the smallest significance level at which we can reject H_0 in favor of $\theta > 0$ using the normal approximation (i.e., the approximate P-value) is .0003. Clearly, both the exact test and the large-sample approximation indicate that there is strong evidence that chick embryos are indeed responsive to a light stimulus, as measured by an increase in the frequency of beak-claps.

The exact P-value for the sign test and the beak-clapping data can also be obtained by using Minitab's STEST command, as follows

```
MTB > SET C1
DATA > -.8  7.5  46.9  17.6  -4.6  54.0  48.3  3.9  16.7  19.7  -8.5
DATA > 7.1  40.7  23.8  14.8  20.6  25.0  24.7  -1.8  21.9  4.7  24.7
DATA > 52.8  8.5  1.9
DATA > END
MTB > STEST 0  C1;
SUBC > ALTERNATIVE = 1.
```

The output is, in Minitab's notation:

```
SIGN TEST OF MEDIAN = 0 VERSUS G.T. 0

      N    BELOW   EQUAL   ABOVE   P VALUE   MEDIAN
C1   25      4       0       21     0.0005    17.6
```

Under the heading "P VALUE" we see that Minitab provides .0005 as the one-sided P-value, agreeing with what we previously found from Table A.2.

Comments

25. *Motivation for the Test.* When θ is greater than 0, there will tend to be a large number of positive Z differences, leading to a big value of B. This suggests rejecting H_0 in favor of $\theta > 0$ for large values of B and motivates the procedures (3.40) and (3.46). Similar rationales lead to procedures (3.41), (3.42), (3.47), and (3.48).

26. *Assumptions.* Assumption B2 is implied by Assumption A2, but the converse is not true. Thus, Assumption B2 is less stringent than Assumption A2—an advantage of the sign test over the signed rank test. We can, when testing $\theta = 0$, weaken Assumption B2 further to the Assumption B2', namely: $P(Z_i < 0) = P(Z_i > 0) = \frac{1}{2}, i = 1, \ldots, n$, when θ is the hypothesized value 0. When testing $\theta = \theta_0$ (see Comment 28), for $\theta_0 \neq 0$, Assumption B2 can be replaced by the weaker Assumption B_2'', namely: $P(Z_i < \theta_0) = P(Z_i > \theta_0) = \frac{1}{2}, i = 1, \ldots, n$, when θ is the hypothesized value θ_0.

We also note that there is no requirement that the individual X_i and Y_i be independent, only that the pairs $(X_1, Y_1), \ldots, (X_n, Y_n)$, and therefore the resulting differences $Z_1, \ldots, Z_n$, be mutually independent. Indeed, in most applications the individual X_i and Y_i are dependent.

27. *Binomial Test.* The test procedures based on the sign statistic B are actually special cases of the general binomial test procedures considered in Chapter 2. The sign test procedures are simply binomial procedures with "success" corresponding to a positive Z difference, "failure" corresponding to a negative Z difference, and $p = P(\text{"success"}) = P(Z_i > 0)$ assuming the value $p_0 = \frac{1}{2}$ when the null hypothesis $H_0 : \theta = 0$ is true.

28. *Testing θ Equal to Some Specified Nonzero Value.* Procedures (3.40), (3.41), and (3.42) and the corresponding normal approximations in (3.46), (3.47), and (3.48) are for testing θ equal to zero. To test $H_0 : \theta = \theta_0$, where θ_0 is some specified nonzero number, subtract θ_0 from each of the differences $Z_1, \ldots, Z_n$ to form a modified sample $Z_1' = Z_1 - \theta_0, \ldots, Z_n' = Z_n - \theta_0$. Then compute B as the number of these Z_i''s that are positive. Procedures (3.40), (3.41), and (3.42) and their corresponding large-sample approximations in (3.46), (3.47), and (3.48) are then applied as previously described.

29. *Equivalent Form.* The statistic B (3.39) is the number of positive Z differences. If we define B^- to be the number of negative Z differences, then $B^- = \sum_{i=1}^n (1 - \psi_i) = n - \sum_{i=1}^n \psi_i = n - B$. Thus, the test procedures (3.40), (3.41), (3.42) and (3.46), (3.47), (3.48) could equivalently be based on $B^- = (n - B)$. (We point out that B^- also (as does B) has a binomial distribution with sample size n and $p = \frac{1}{2}$ when $H_0 : \theta = 0$ is true.)

30. *Derivation of the Distribution of B under H_0 (When There Are No Zero Z Values).* The null (H_0) distribution of B can be obtained directly from the representation $B = \sum_{i=1}^n \psi_i$. Under the assumption that the underlying Z_i distributions are all continuous, the probabilities are zero that any of the Z_i's are zero. Hence, under H_0 each of the 2^n possible outcomes for the configuration $(\psi_1, \ldots, \psi_n)$ occurs with probability $(\frac{1}{2})^n$. For example, in the case of $n = 3$, the $2^3 = 8$ possible outcomes for (ψ_1, ψ_2, ψ_3) and associated values of B are given in the following table.

3.4. A DISTRIBUTION-FREE SIGN TEST (FISHER)

(ψ_1, ψ_2, ψ_3)	Probability under H_0	$B = \sum_{i=1}^{3} \psi_i$
(0, 0, 0)	$\frac{1}{8}$	0
(0, 0, 1)	$\frac{1}{8}$	1
(0, 1, 0)	$\frac{1}{8}$	1
(1, 0, 0)	$\frac{1}{8}$	1
(0, 1, 1)	$\frac{1}{8}$	2
(1, 0, 1)	$\frac{1}{8}$	2
(1, 1, 0)	$\frac{1}{8}$	2
(1, 1, 1)	$\frac{1}{8}$	3

Thus, for example, the probability is $\frac{3}{8}$ under H_0 that B is equal to 2, since $B = 2$ when any of the three exclusive outcomes $(\psi_1, \psi_2, \psi_3) = (0, 1, 1), (1, 0, 1)$, or $(1, 0, 1)$ occurs, and each of these outcomes has null probability $\frac{1}{8}$. Simplifying, we obtain the null distribution.

Possible Value of B	Probability under H_0
0	$\frac{1}{8}$
1	$\frac{3}{8}$
2	$\frac{3}{8}$
3	$\frac{1}{8}$

The probability, under H_0, that B is greater than or equal to 2, for example, is therefore

$$P_0(B \geq 2) = P_0(B = 2) + P_0(B = 3)$$
$$= .375 + .125 = .50.$$

This agrees with the upper-tail probability entry for $p = .5, n = 3$ and the value $B = 2$ in Table A.2. (We note that this null distribution of B could alternatively be obtained from the binomial probability distribution in equation (2.15) in Comment 2.7 by taking $p_0 = \frac{1}{2}$.)

Note that we have derived the null distribution of B without specifying the forms of the underlying Z populations under H_0 beyond the requirement that they be continuous and have common median 0. That is why the test procedures based on B are called distribution-free procedures. From the null distribution of B we can determine the critical value $b_{\alpha/2, 1/2}$ and control the probability α of falsely rejecting H_0 when H_0 is true, and this error probability does not depend on the specific forms of the underlying continuous Z distributions with common median 0.

31. *Calculation of the Mean and Variance of B under the Null Hypothesis H_0.* In displays (3.43) and (3.44) we presented formulas for the mean and variance of B when the null hypothesis is true. In this comment we illustrate a direct calculation of $E_0(B)$ and $\text{var}_0(B)$ in the particular case of $n = 3$, using the null distribution of B obtained in Comment 30. (Later, in Comment 32, we present general derivations of $E_0(B)$ and $\text{var}_0(B)$.) The null mean, $E_0(B)$, is obtained by multiplying each possible value of B with its probability under H_0. Thus

$$E_0(B) = 0(.125) + 1(.375) + 2(.375) + 3(.125) = 1.5.$$

This is in agreement with what we obtain using equation (3.43), namely,

$$E_0(B) = \frac{n}{2} = \frac{3}{2} = 1.5.$$

A check on the expression for $\text{var}_0(B)$ is also easily performed, using the well-known fact that

$$\text{var}_0(B) = E_0(B^2) - \{E_0(B)\}^2.$$

The value of $E_0(B^2)$, the second moment of the null distribution of B, is again obtained by multiplying possible values (in this case, of B^2) by the corresponding probabilities under H_0. We find

$$E_0(B^2) = 0(.125) + 1(.375) + 4(.375) + 9(.125) = 3.0.$$

Thus,

$$\text{var}_0(B) = 3.0 - (1.5)^2 = 0.75,$$

which agrees with what we obtain using equation (3.44) directly, namely,

$$\text{var}_0(B) = \frac{n}{4} = \frac{3}{4} = 0.75.$$

32. Large-Sample Approximation. Under Assumption $B1$, the variables $Z_1, \ldots, Z_n$ are mutually independent. Since ψ_i is a function of Z_i only, for $i = 1, \ldots, n$, it follows that $\psi_1, \ldots, \psi_n$ are also mutually independent variables. In view of the representation $B = \sum_{i=1}^{n} \psi_i$ in (3.39), we can immediately use well-known expressions for the mean and variance of a sum of mutually independent random variables to obtain

$$E_0(B) = E_0\left[\sum_{i=1}^{n} \psi_i\right] = \sum_{i=1}^{n} E_0(\psi_i) \tag{3.49}$$

and

$$\text{var}_0(B) = \text{var}_0\left(\sum_{i=1}^{n} \psi_i\right) = \sum_{i=1}^{n} \text{var}_0(\psi_i). \tag{3.50}$$

Now, under H_0, the ψ_i's are also identically distributed, each following a Bernoulli probability distribution with $p = \frac{1}{2}$. Thus, for $i = 1, \ldots, n$, we see that

$$E_0(\psi_i) = 0\left(\frac{1}{2}\right) + 1\left(\frac{1}{2}\right) = \frac{1}{2}$$

and

$$\text{var}_0(\psi_i) = E_0(\psi_i^2) - [E_0(\psi_i)]^2$$

$$= 0^2\left(\frac{1}{2}\right) + 1^2\left(\frac{1}{2}\right) - \left(\frac{1}{2}\right)^2 = \frac{1}{2} - \frac{1}{4} = \frac{1}{4}.$$

Using these results in equations (3.49) and (3.50), we obtain

$$E_0(B) = \sum_{i=1}^{n}\left(\frac{1}{2}\right) = \frac{n}{2}$$

3.4. A DISTRIBUTION-FREE SIGN TEST (FISHER)

and

$$\text{var}_0(B) = \sum_{i=1}^{n} \left(\frac{1}{4}\right) = \frac{n}{4},$$

which agree with the general expressions stated in equations (3.43) and (3.44).

The asymptotic normality of the standardized form

$$B^* = \frac{B - E_0(B)}{\{\text{var}_0(B)\}^{1/2}} = \frac{B - \frac{n}{2}}{\left\{\frac{n}{4}\right\}^{1/2}}$$

follows from standard central limit theory for sums of mutually independent, identically distributed random variables (cf. Randles and Wolfe (1979, 421)). Asymptotic normality results are also obtainable under general alternatives to H_0 (see Comment 35).

33. *Symmetry of the Distribution of B under the Null Hypothesis.* When H_0 is true, the distribution of B is symmetric about its mean $n/2$. (See Comment 30 for verification of this when $n = 3$.) This implies that

$$P_0(B \leq x) = P_0(B \geq n - x), \tag{3.51}$$

for $x = 0, 1, \ldots, n$.

Equation (3.51) is used directly to convert upper-tail probabilities, as presented in Table A.2, to lower-tail probabilities.

34. *Zero Z Values.* We have recommended discarding zero Z values and redefining n to be the number of nonzero Z's. This approach is satisfactory as long as the zero values do not represent a sizable percentage of the total number of Z differences. If, however, there is a relatively large number of zero Z's, it would be advisable to consider an appropriate statistical procedure designed specifically for analyzing such discrete data. See, for example, Chapter 10 or a book on categorical data analysis such as Agresti (1990).

We should also point out that there are methods other than elimination that have been proposed for dealing with zero Z values. One could use individual randomization (e.g., flipping a fair coin) to decide whether each of the zero Z values is to be counted as positive or negative in the computation of B. (Although this approach maintains many of the nice theoretical properties of B that hold when there are no zeros, it introduces extraneous randomness that could quite easily have a direct effect on the outcome of any subsequent inferences based on such a modified B.) A second alternative approach in the case of the one-sided test procedures in (3.40), (3.41), (3.46), and (3.47) is to be conservative about rejecting the null hypothesis H_0; that is, we could count all the zero Z values as if they were in favor of not rejecting H_0. Thus, for example, in applying either procedure (3.41) or (3.47) to test H_0 against the alternative $\theta < 0$, we would treat all the zero Z's as if they were "positive" (in favor of not rejecting H_0) in the calculation of B. (In the case of procedures (3.40) and (3.46), zero Z values would be considered "negative" in the calculation of B.) Any rejection of H_0 with this conservative approach to dealing with zero Z values could be viewed as providing strong evidence in favor of the appropriate alternative. For a more detailed discussion of methods for handling zero observations, see Pratt (1959).

35. *Some Power Results for the Sign Test.* We consider the upper-tail α-level test of $H_0 : \theta = 0$ versus $H_1 : \theta > 0$ given by procedure (3.40). When we have a common underlying distribution with median θ and distribution function $F_1 \equiv F_2 \equiv \cdots \equiv F_n \equiv F$ for the Z differences, the sign statistic B has a binomial distribution with parameters n and

$p_\theta = P_\theta(Z_1 > 0) = 1 - F(0)$. It follows (see equation (2.15) in Comment 7 of Chapter 2) that the exact power of the sign test procedure (3.40) against the alternative $\theta > 0$ is given by the expression

$$\text{Power}_\theta = \sum_{t=b_{\alpha,1/2}}^{n} \binom{n}{t} p_\theta^t (1-p_\theta)^{n-t} = \sum_{t=b_{\alpha,1/2}}^{n} \binom{n}{t} [1-F(0)]^t [F(0)]^{n-t}. \qquad (3.52)$$

(Since $p_\theta = 1 - F(0) > 1 - F(\theta) = 1 - \frac{1}{2} = \frac{1}{2}$ for all $\theta > 0$, it follows that $\text{Power}_\theta > \text{Power}_0 = \alpha$ for all alternatives $\theta > 0$.) Evaluation of Power_θ for a moderate sample size n and particular value of $\theta > 0$ (and associated $p_\theta = 1 - F(0) > \frac{1}{2}$) can thus be accomplished through the use of readily available binomial distribution tables or by direct computation.

For large sample sizes, we can make use of the standard central limit theorem for sums of mutually independent and identically distributed random variables to conclude that

$$\frac{B - np_\theta}{[np_\theta(1-p_\theta)]^{1/2}} = \frac{B - n(1 - F(0))}{[n(1 - F(0))(F(0))]^{1/2}} \qquad (3.53)$$

has an asymptotic ($n \longrightarrow \infty$) standard normal distribution. Thus, for large n, we can approximate the exact power in (3.52) by

$$\text{Power}_\theta \approx 1 - \Phi\left(\frac{b_{\alpha,1/2} - np_\theta}{[np_\theta(1-p_\theta)]^{1/2}}\right)$$

$$= 1 - \Phi\left(\frac{b_{\alpha,1/2} - n(1 - F(0))}{[n(1 - F(0))(F(0))]^{1/2}}\right), \qquad (3.54)$$

where $\Phi(t)$ is the area under a standard normal density to the left of t.

We note that both the exact power (3.52) and the approximate power (3.54) against an alternative $\theta > 0$ depend on the common distribution only through the value of its distribution function $F(z)$ at $z = 0$. Thus, if two distributions have a common median $\theta > 0$ and distribution functions F_1 and F_2 such that $F_1(0) = F_2(0)$, then the exact power (3.52) of the sign test against the alternative $\theta > 0$ will be the same for both distributions F_1 and F_2. (The same is, of course, true for the approximate power in (3.54).)

For purposes of illustration, consider the case where $n = 10$ and $\alpha = .0547$. Then from Table A.2 we see that $b_{.0547, 1/2} = 8$ and the test (3.40) rejects H_0 if and only if $B \geq 8$. If F is the distribution function for a probability distribution with median $\theta = 2$ (i.e., $F(2) = \frac{1}{2}$) and $F(0) = \frac{1}{4}$, then from equation (3.52) and Table A.2, the exact power of the sign test in this setting is

$$\text{Power}_{\theta=2} = \sum_{t=8}^{10} \binom{10}{t} \left(\frac{3}{4}\right)^t \left(\frac{1}{4}\right)^{10-t} = 1 - .4744 = .5256.$$

The approximate power for the same setting is seen from (3.54) to be

$$\text{Power}_{\theta=2} \approx 1 - \Phi\left(\frac{8 - 10\left(\frac{3}{4}\right)}{\left[10\left(\frac{3}{4}\right)\left(\frac{1}{4}\right)\right]^{1/2}}\right)$$

$$= 1 - \Phi\left(\frac{.5}{\left[\frac{30}{16}\right]^{1/2}}\right) = 1 - \Phi(.365),$$

3.4. A DISTRIBUTION-FREE SIGN TEST (FISHER)

which, using Table A.1, yields

$$\text{Power}_{\theta=2} \approx .3575.$$

Thus, for n as small as 10 and $p_\theta = \frac{3}{4}$, the agreement between the exact power from (3.52) and the approximate power from (3.54) is not too good. (We note that for an underlying normal distribution F with mean $\theta = 2$ and variance σ^2, the condition $F(0) = \frac{1}{4}$ corresponds to $\Phi((0 - 2)/\sigma) = \frac{1}{4}$, which in turn corresponds to $(-2/\sigma) = z_{.75} = -.675$, or $\sigma = (2/.675) = 2.96$. More generally, the test in (3.40) for $n = 10$ and $\alpha = .0547$ has exact power of .5256 against any normal distribution with mean θ and variance σ^2 for which $-(\theta/\sigma) = z_{.75} = -.675$, or $\theta = .675\sigma$.)

36. *Sample Size Determination.* When $Z_1, \ldots, Z_n$ are a random sample from a single continuous F, then the one-sided upper-tail test defined by procedure (3.40) is consistent (that is, has power tending to 1 as n tends to infinity) against those F populations for which $p > \frac{1}{2}$, with

$$p = P(Z > 0), \tag{3.55}$$

where Z is distributed as F.

Noether (1987), among others, shows how to determine an approximate sample size n so that the α-level one-sided test given by procedure (3.40) will have approximate power $1 - \beta$ against an alternative value of p (3.55) greater than $\frac{1}{2}$. This approximate value of n is

$$n \doteq \frac{(z_\alpha + z_\beta)^2}{4(p - \frac{1}{2})^2}. \tag{3.56}$$

As an illustration of the use of equation (3.56), suppose we are testing H_0 and we wish to have an upper-tail level $\alpha = .04$ test with power $1 - \beta$ at least .975 against an alternative for which $p = P(Z > 0) = .7$. (Recall that $p = .5$ under H_0.) Since $z_\alpha = z_{.04} = 1.75$ and $z_\beta = z_{.025} = 1.96$, we find that the required sample size for the alternative $p = .7$ is

$$n \doteq \frac{(1.75 + 1.96)^2}{4(.7 - .5)^2} = 86.03.$$

To be conservative, we take $n = 87$.

37. *Consistency of the B Test.* Under the assumption that $Z_1, \ldots, Z_n$ is a random sample from a single continuous population F, the consistency of the tests based on B depends on the parameter

$$p^\star = P(Z_1 > 0) - \tfrac{1}{2}. \tag{3.57}$$

The test procedures defined by (3.40), (3.41), and (3.42) are consistent against the classes of alternatives corresponding to $p^\star >, <,$ and $\neq 0$, respectively.

Properties

1. *Consistency.* For our consistency statement we strengthen Assumption B2 to require that each Z has the same continuous population with median θ. Then the test procedures defined by (3.40), (3.41), and (3.42) are consistent against the alternatives $\theta >, <,$ and $\neq 0$, respectively. See also Comment 37.

2. *Asymptotic Normality.* Under a strengthened Assumption B2 that requires that each Z has the same continuous population with median θ, the asymptotic normality of the standardized form of the B statistic follows from the standard central limit theorem for sums of mutually independent and identically distributed random variables. Also see Comment 35.
3. *Efficiency.* See Section 3.11.

PROBLEMS

43. The data in Table 3.6 are a portion of the data obtained by Cooper et al. (1967). The purpose of their investigation was to determine whether hypnotic susceptibility as measured on objective scales can be changed with practice and training. The objective measures used were the Stanford Profile Scales of Hypnotic Susceptibility, forms I and II (Hilgard, Lauer, and Morgan (1963)). The subjects were administered these Profile Scales, both forms I and II, by a hypnotist other than the experimenter. Each subject was then seen by one of the authors for an extensive period of "hypnotic training." After these sessions were concluded, each subject was retested by a different hypnotist (again not the experimenter), using equivalent forms of the Profile Scales, forms I' and II'. Table 3.6 gives the average score obtained on forms I and II prior to hypnotic training (X) and the corresponding average score obtained on forms I' and II' after the training (Y) for the six subjects. Note that a high (low) score on the Profile Scales indicates a high (low) degree of hypnotic susceptibility.

    Test the hypothesis of no change in hypnotic susceptibility versus the alternative that hypnotic susceptibility (as measured by the Profile Scales) can be increased with practice and training.

TABLE 3.6. Average Scores on the Stanford Profile Scales of Hypnotic Susceptibility

Subject i	X_i	Y_i
1	10.5	18.5
2	19.5	24.5
3	7.5	11.0
4	4.0	2.5
5	4.5	5.5
6	2.0	3.5

Source: L. M. Cooper, E. Schubot, S. A. Banford, and C. T. Tart (1967).

44. Change the value of Y_3 in Table 3.5 from 73 to 173. What effect does this outlying observation have on the calculations performed in Example 3.5? What does this suggest about the relative insensitivity of the sign tests to outliers? Construct an example in which changing one observation has an effect on the final decision regarding rejection or acceptance of H_0.
45. Suppose $n = 25$. Compare the critical region for the exact level $\alpha = .0539$ test of $H_0 : \theta = 0$ versus $H_1 : \theta < 0$ based on B with the critical region for the corresponding nominal level $\alpha = .0539$ test based on the large-sample approximation. What is the exact significance level of this .0539 nominal level test based on the large-sample approximation?
46. In an investigation to determine the effect of aspirin on bleeding time and platelet adhesion, Bick, Adams, and Schmalhorst (1976) studied the reactions of normal subjects to aspirin. A subset of their data is presented in Table 3.7, where the X observation for each subject is the bleeding time (in seconds) before ingestion of 600 mg aspirin and the Y observation is the bleeding time (again in seconds) 2h after administration of the aspirin.

    Find the P-value for an appropriate test of the hypothesis that a 600-mg dose of aspirin has no effect on bleeding time versus the alternative that it typically leads to an increase in bleeding time.

TABLE 3.7. Bleeding Times (in seconds)

Subject i	X_i	Y_i
1	270	525
2	150	570
3	270	190
4	420	395
5	202	370
6	255	210
7	165	490
8	220	250
9	305	360
10	210	285
11	240	630
12	300	385
13	300	195
14	70	295

Source: R. L. Bick, T. Adams, and W. R. Schmalhorst (1976).

47. Assume that we have a common underlying distribution $F_1 \equiv F_2 \equiv \cdots \equiv F_n \equiv F$ (in Assumption B2). If we have 20 observations, what is the exact power of the level $\alpha = .0577$ test of $H_0 : \theta = 0$ versus the alternative $\theta > 0$ when $F(0) = .3$?

48. Assume that we have a common underlying distribution $F_1 \equiv F_2 \equiv \cdots \equiv F_n \equiv F$ (in Assumption B2). If we have 18 observations and F is normal with variance 4, what is the exact power of the level $\alpha = .0154$ test of $H_0 : \theta = 0$ versus the alternative $\theta < 0$ when the treatment effect is $\theta = -2$?

49. Consider a level $\alpha = .025$ test of $H_0 : \theta = 0$ versus the alternative $\theta > 0$ based on B. If our data $Z_1, \ldots, Z_n$ are a random sample from a single, continuous distribution $F(\cdot)$, how many observations n will we need to collect in order to have approximate power at least .75 against an alternative for which $F(0) = .20$?

50. Apply the appropriate form of the test based on B to the Hamilton depression scale factor IV data in Table 3.1.

51. Assume that we have a common underlying distribution $F_1 \equiv F_2 \equiv \cdots \equiv F_n \equiv F$ (in Assumption B2). If we have 20 observations, what is the approximate power of the level $\alpha = .0577$ test of $H_0 : \theta = 0$ versus the alternative $\theta > 0$ when $F(0) = .3$? Compare this approximate power with the exact power from Problem 47.

52. Apply the large-sample approximation test of $H_0 : \theta = 1,000$ versus $H_1 : \theta > 1,000$ based on B to the salary data in Table 3.2.

53. For the case of $n = 5$ nonzero Z values, use the approach discussed in Comment 30 to obtain the form of the exact null (H_0) distribution of B. Verify numerically that this null distribution is, indeed, the binomial distribution with parameters $n = 5$ and $p_0 = .5$.

54. Consider the test of $H_0 : \theta = 0$ versus $H_1 : \theta > 0$ based on B for the following $n = 15$ Z observations: $Z_1 = 2.5, Z_2 = 0, Z_3 = 3.7, Z_4 = -0.6, Z_5 = 1.7, Z_6 = 0, Z_7 = 5.9, Z_8 = 4.6, Z_9 = 0, Z_{10} = -1.4, Z_{11} = 5.4, Z_{12} = 4.6, Z_{13} = 3.1, Z_{14} = -2.0$, and $Z_{15} = 6.3$. Compute the P-values for the competing B-procedures based on either (i) discarding the zero Z values and reducing n accordingly, as recommended in the ties portion of this section, or (ii) treating the zero Z values in a conservative manner, as presented in Comment 34. Discuss the results.

55. Consider the same setting as in Problem 54. Suppose that you had decided to use randomization to deal with the three zero Z values in the data (see Comment 34). Consider the various possible outcomes for this randomization process and compute the associated P-value for each of these outcomes. Discuss the implications of these findings in conjunction with the results of Problem 54.

56. Use the computer software `Minitab` to obtain the exact P-value for the appropriate test based on B for the bleeding times data in Table 3.7.

57. Use the computer software `Minitab` to obtain the exact P-value for the test of $H_0 : \theta = 1,000$ versus $H_1 : \theta > 1,000$ based on B for the salary data in Table 3.2.

3.5. AN ESTIMATOR ASSOCIATED WITH THE SIGN STATISTIC (HODGES-LEHMANN)

Procedure

To estimate the treatment effect θ, order the sample observations and let $Z^{(1)} \leq \cdots \leq Z^{(n)}$ denote these ordered items. The estimator of θ associated with the sign statistic (see Comment 38) is

$$\tilde{\theta} = \text{median}\{Z_i, \ 1 \leq i \leq n\}. \tag{3.58}$$

Thus, if n is odd, say $n = 2k + 1$, we have $k = (n-1)/2$ and

$$\tilde{\theta} = Z^{(k+1)}, \tag{3.59}$$

the value that occupies position $k + 1$ in the list of the ordered Z_i values. If n is even, say $n = 2k$, then $k = n/2$ and

$$\tilde{\theta} = \frac{Z^{(k)} + Z^{(k+1)}}{2}. \tag{3.60}$$

That is, when n is even $\tilde{\theta}$ is the average of the two Z_i values that occupy positions k and $k + 1$ in the ordered list of the n data values.

Example 3.6: *Continuation of Example 3.5.* To estimate θ for the beak-clapping data in Table 3.5, we first form the $n = 25$ ordered Z values, namely, $Z^{(1)} \leq \cdots \leq Z^{(25)}$: -8.5, -4.6, -1.8, -0.8, 1.9, 3.9, 4.7, 7.1, 7.5, 8.5, 14.8, 16.7, 17.6, 19.7, 20.6, 21.9, 23.8, 24.7, 24.7, 25.0, 40.7, 46.9, 48.3, 52.8, 54.0. Since $n = 25$ is odd, we use equation (3.59) with $k = (25-1)/2 = 12$ to obtain the estimate $\tilde{\theta} = Z^{(13)} = 17.6$ for the treatment effect θ. Thus, we estimate that a typical chick embryo of the type included in this study will produce 17.6 more beak-claps per minute during periods of illumination than during periods of darkness.

We can also use Minitab's SINTERVAL command to compute the estimator $\tilde{\theta}$ for the beak-clapping data, as follows:

```
MTB > SET C1
DATA > -.8  7.5  46.9  17.6  -4.6  54.0  48.3  3.9  16.7  19.7  -8.5
DATA > 7.1  40.7  23.8  14.8  20.6  25.0  24.7  -1.8  21.9  4.7  24.7
DATA > 52.8  8.5  1.9
DATA > END
MTB > SINTERVAL .8922  C1
```

The output is, in Minitab's notation:

SIGN CONFIDENCE INTERVAL FOR MEDIAN

	N	MEDIAN	ACHIEVED CONFIDENCE	CONFIDENCE INTERVAL	POSITION
C1	25	17.6	.8922	(7.5 , 23.8)	9

The point estimate is $\tilde{\theta} = 17.6$ as we found previously. We explain the confidence interval portion of Minitab's output in Section 3.6.

Comments

38. *Motivation for the Hodges-Lehmann Estimator.* The estimator $\tilde{\theta}$ defined by equation (3.58) is associated with the sign test in the same way as the estimator $\hat{\theta}$ (3.23) is associated

3.5. AN ESTIMATOR ASSOCIATED WITH THE SIGN STATISTIC (HODGES-LEHMANN)

with the signed rank test (see Comment 15). When $\theta = 0$, the distribution of the statistic B (3.39) is symmetric about its mean $n/2$ (see Comment 33). A natural estimator of θ is the amount $\tilde{\theta}$ (say) that should be subtracted from each Z_i so that the value of B, when applied to the shifted sample $Z_1 - \tilde{\theta}, \ldots, Z_n - \tilde{\theta}$, is as close to $n/2$ as possible. Intuitively, we estimate θ by the amount ($\tilde{\theta}$) that the Z sample should be shifted in order that $Z_1 - \tilde{\theta}, \ldots, Z_n - \tilde{\theta}$ appears (when "viewed" by the sign statistic B) as a sample from a population with median 0. (Under Assumptions B1 and B2, each of the $Z_1 - \theta, \ldots, Z_n - \theta$ variables is from a population with median 0.)

The Hodges-Lehmann method can be applied to a large class of statistics containing both B and T^+ (3.3). However, the forms of the resulting estimator for other members of this class are not always as convenient for calculation as are $\tilde{\theta}$ (3.58) or $\hat{\theta}$ (3.23). See Hodges and Lehmann (1983) for an expository article on their method of estimation.

39. *Simplicity.* One of the virtues of $\tilde{\theta}$ (3.58) is its simplicity. Whereas many estimators associated with distribution-free test statistics are tedious to compute (e.g., $\hat{\theta}$ (3.23) requires computing the median of $n(n + 1)/2$ values), $\tilde{\theta}$ requires only that we find the median of the n Z observations. However, although the signs of the Z differences provide sufficient information to conduct a sign test, the magnitudes of these differences are needed in order to obtain the value of the estimator $\tilde{\theta}$.

40. *Sensitivity to Gross Errors.* The estimator $\tilde{\theta}$ (3.58) is even less sensitive to outliers than is the estimator $\hat{\theta}$ (3.23) associated with the signed rank statistic T^+ (3.3). (See Comment 16 and Problems 20 and 60.) As a result, $\tilde{\theta}$ protects well against gross errors. However, all the information contained in the collected sample is not utilized in computing $\tilde{\theta}$. Consequently, $\tilde{\theta}$ is rather inefficient for many populations.

41. *Zero Z Values.* Note that in calculating the estimator $\tilde{\theta}$ we use *all* the Z differences. Although we recommend (see Ties in Section 3.4) discarding the zero Z values (and reducing n accordingly) prior to applying the sign test to the data, it is not necessary to do so when calculating $\tilde{\theta}$. In fact, the zero Z values contain important information about the magnitude of the treatment effect. This is also the case when we consider (Section 3.6) confidence intervals and bounds for θ.

42. *Historical Perspective.* The use of the estimator $\tilde{\theta}$ predates most of the recent unified developments in the field of nonparametric statistics. A. T. Craig (1932) first found the sampling distribution of $\tilde{\theta}$, and its asymptotic properties were developed shortly thereafter by Smirnov (1935).

43. *Quasimedians.* Let $Z^{(1)} \leq \cdots \leq Z^{(n)}$ be the ordered sample observations, as in step 1 of the Procedure. Hodges and Lehmann (1967) defined the sample quasimedians by

$$\tilde{\theta}_i = \begin{cases} \dfrac{Z^{(k+1-i)} + Z^{(k+1+i)}}{2}, & \text{if } n = 2k + 1, \\ \dfrac{Z^{(k-i)} + Z^{(k+1+i)}}{2}, & \text{if } n = 2k, \end{cases}$$

for $i = 0, 1, \ldots, k$ if $n = 2k + 1$, or $i = 0, 1, \ldots, k - 1$ if $n = 2k$; that is, each quasimedian $\tilde{\theta}_i$ is an average of two symmetrically situated, ordered Z observations. (Note that this definition of a quasimedian generalizes the concept of a sample median, since the sample median $\tilde{\theta}$ (3.58) is equal to $\tilde{\theta}_0$.) These quasimedians are natural estimators for the parameter θ (see Comment 52) and were considered by Hodges and Lehmann (1967), who investigated some of the asymptotic properties of this class of statistics.

44. *Linear Combinations of Order Statistics.* Let $Z^{(1)} \leq \cdots \leq Z^{(n)}$ be the ordered sample observations, as in step 1 of the Procedure. Under the additional assumption that we have

a common underlying distribution $F_1 \equiv F_2 \equiv \cdots \equiv F_n \equiv F$ (in Assumption B2), the n variables $Z^{(1)}, \ldots, Z^{(n)}$ are called the order statistics for the random sample $Z_1, \ldots, Z_n$. The estimator $\tilde{\theta}$ (3.58) is a special case of a general class of estimators of θ based on linear combinations of these sample order statistics, corresponding to estimators of the form

$$\tilde{\theta}_\mathbf{b} = \sum_{i=1}^{n} b_i Z^{(i)}, \qquad (3.61)$$

where $\mathbf{b} = (b_1, \ldots, b_n)$ is a vector of n nonnegative constants such that $\sum_{i=1}^{n} b_i = 1$. For more detailed discussion about estimators of the form $\tilde{\theta}_\mathbf{b}$ (3.61), see David (1981) or Arnold, Balakrishnan, and Nagaraja (1992), for example.

45. *Variance Approximation.* Hodges and Lehmann (1967) obtained an approximation for the variance of the estimator $\tilde{\theta}$ (3.58) under the additional assumption that we have a common underlying distribution $F_1 \equiv F_2 \equiv \cdots \equiv F_n \equiv F$ (in Assumption B2). (See equation (1.4) of their paper.) They point out that, up to the accuracy of their approximation, it is not wise to compute the sample median $\tilde{\theta}$ using an odd number of observations, say $n = 2k + 1$. The next smaller even number, $n = 2k$, yields a sample median that is just as accurate. This conclusion does not depend on the shape of the underlying population except that it be symmetric, although the degree of accuracy of the approximation is affected by the shape.

46. *Estimating the Asymptotic Standard Deviation of $\tilde{\theta}$.* Assume that we have a common underlying distribution $F_1 \equiv F_2 \equiv \cdots \equiv F_n \equiv F$ (in Assumption B2) and set

$$D = \sum_{i=1}^{n} a_i,$$

where

$$a_i = \begin{cases} 1, & \text{if } [\tilde{\theta} - (n)^{-1/5}] \leq Z_i \leq [\tilde{\theta} + (n)^{-1/5}], \\ 0, & \text{otherwise,} \end{cases}$$

for $i = 1, \ldots, n$. Let $A = \text{maximum } \{1, D\}$. Under the additional assumption on the common distribution F that the probability of obtaining a Z observation in any (sufficiently) small interval I centered at the median θ is greater than or equal to some fixed constant (not depending on I) times the length of I, the statistic $C = n^{3/10} A^{-1}$ is a consistent estimator of the asymptotic standard deviation of the point estimator $\tilde{\theta}$ (3.58). The statistic C is related to general classes of estimators of probability density functions considered by Rosenblatt (1956), Parzen (1962), and Gupta (1967). The consistency of C follows directly from the results in Korwar (1971).

47. *Relative Merits of $\hat{\theta}$ and $\tilde{\theta}$.* The point estimator $\tilde{\theta}$ (3.58) associated with the sign test statistic B is to be preferred to the point estimator $\hat{\theta}$ (3.23) associated with the signed rank test statistic T^+ when ease of computation is a consideration (see Comment 39). Generally (but not always) $\hat{\theta}$ is more efficient than $\tilde{\theta}$ (see Comment 40 and Section 3.11).

Properties

1. *Standard Deviation of $\tilde{\theta}$.* For the asymptotic standard deviation of $\tilde{\theta}$ (3.58), see Fisz (1963, 383) and Comment 46.
2. *Asymptotic Normality.* See Fisz (1963, 383).
3. *Efficiency.* See Hodges and Lehman (1963) and Section 3.11.

PROBLEMS

58. Using the designated X and Y associations, estimate θ for the average Profile Scales data of Table 3.6.
59. Estimate θ for the bleeding time data of Table 3.7.
60. Change the value of Y_3 in Table 3.5 from 73 to 173. What effect does this have on the value of $\bar{Z} = \sum_{i=1}^{25} Z_i/25$? What is the new value of $\tilde{\theta}$ (3.58)? Interpret these calculations (see Comment 40).
61. Calculate $\tilde{\theta}$ for the heat-insoluble hydroxyproline data of Table 3.4. Compare with the value of $\hat{\theta}$ obtained in Problem 21.
62. (a) What happens to $\tilde{\theta}$ when we add a number b to each of the sample values $Z_1, \ldots, Z_n$?
    (b) What happens to $\tilde{\theta}$ when we multiply each sample value by the number d?
    (c) What happens to $\tilde{\theta}$ when we discard the k largest and k smallest values from the sample (assume $n > 2k$)? Compare your answers with the corresponding answers to Problem 23.
63. Calculate $\tilde{\theta}$ for the blood level data of Table 3.3. Compare with the value of $\hat{\theta}$ obtained in Problem 19.
64. Calculate $\tilde{\theta}$ for the salary data in Table 3.2. Compare with the value of $\hat{\theta}$ obtained in Problem 18.
65. Calculate $\tilde{\theta}$ for the Hamilton depression scale factor IV data in Table 3.1. Compare with the value of $\hat{\theta}$ obtained in Example 3.3.
66. Find the vector $\mathbf{b} = (b_1, \ldots, b_n)$ to show that $\tilde{\theta}$ can be written as a linear combination of the sample order statistics $Z^{(1)} \leq \cdots \leq Z^{(n)}$, as discussed in Comment 44.
67. Show that the class of quasimedian estimators of θ (see Comment 43) is a subset of the class of estimators of θ based on linear combinations of the sample order statistics $Z^{(1)} \leq \cdots \leq Z^{(n)}$, as discussed in Comment 44.
68. Use the computer software Minitab to compute the value of $\tilde{\theta}$ for the salary data of Table 3.2. (Use X and Y labels as assigned in Example 3.2.)
69. Use the computer software Minitab to compute the value of $\tilde{\theta}$ for the bleeding time data of Table 3.7.

3.6. A DISTRIBUTION-FREE CONFIDENCE INTERVAL BASED ON THE SIGN TEST (THOMPSON, SAVUR)

Procedure

For a symmetric two-sided confidence interval for θ, with confidence coefficient $1 - \alpha$, first obtain the upper $(\alpha/2)$nd percentile point $b_{\alpha/2,1/2}$ of the null distribution of B from Table A.2. Set

$$C_\alpha = n + 1 - b_{\alpha/2,1/2}. \tag{3.62}$$

The $100(1 - \alpha)\%$ confidence interval (θ_L, θ_U) for θ that is associated with the two-sided sign test (see Comment 48) of $H_0 : \theta = 0$ is then given by

$$\theta_L = Z^{(C_\alpha)}, \quad \theta_U = Z^{(n+1-C_\alpha)} = Z^{\left(b_{\alpha/2,1/2}\right)}, \tag{3.63}$$

where $Z^{(1)} \leq \cdots \leq Z^{(n)}$ are the ordered sample observations. That is, θ_L is the sample observation that occupies position C_α in the list of ordered sample data. The upper endpoint θ_U is the sample observation that occupies position $n + 1 - C_\alpha = b_{\alpha/2,1/2}$ in this ordered list. With θ_L and θ_U given by display (3.63), we have

$$P_\theta(\theta_L < \theta < \theta_U) = 1 - \alpha \quad \text{for all } \theta. \tag{3.64}$$

(For upper or lower confidence bounds for θ associated with appropriate one-sided sign tests of $H_0 : \theta = 0$, see Comment 49.)

Large-Sample Approximation

For large n the integer C_α may be approximated by

$$C_\alpha \approx \frac{n}{2} - z_{\alpha/2} \left(\frac{n}{4}\right)^{1/2}. \tag{3.65}$$

In general the value of the right-hand side of equation (3.65) is not an integer. To be conservative, take C_α to be the largest integer that is less than or equal to the right-hand side of equation (3.65).

Example 3.7: *Continuation of Examples 3.5 and 3.6.* Consider the beak-clapping data in Table 3.5. We illustrate how to obtain the 89.22% confidence interval for θ. With $1 - \alpha = .8922$ (so that $\alpha = .1078$), we see from Table A.2 with $n = 25$ and $p = \frac{1}{2}$ that $b_{\alpha/2,1/2} = b_{.0539,1/2} = 17$. From equation (3.62) it follows that

$$C_{.1078} = 25 + 1 - 17 = 9.$$

Using these values of $C_{.1078} = 9$ and $b_{.0539,1/2}$ in display (3.63), we see that

$$\theta_L = Z^{(9)} = 7.5 \quad \text{and} \quad \theta_U = Z^{(17)} = 23.8,$$

so that our 89.22% confidence interval for θ is

$$(\theta_L, \theta_U) = (7.5, 23.8).$$

If we choose instead to apply the large-sample approximation, we find from equation (3.65) that

$$C_{.1078} \approx \frac{25}{2} - 1.608 \left(\frac{25}{4}\right)^{1/2} = 8.48.$$

Thus, with a conservative approach and the large-sample approximation, we set $C_{.1078} = 8$ and find that

$$(\theta_L, \theta_U) = (Z^{(8)}, Z^{(18)}) = (7.1, 24.7)$$

is the approximate (conservative) 89.22% confidence interval for θ.

In Example 3.6 we discussed the output from Minitab's SINTERVAL applied to the beak-clapping data. This included the presentation of a confidence interval for θ. There, with confidence coefficient $1 - \alpha = .8922$, Minitab gives (7.5, 23.8) as the exact 89.22% confidence interval for θ, in agreement with the result we previously obtained using display (3.63) and Table A.2.

Comments

48. *Relationship of Confidence Interval to Two-Sided Test.* The $100(1 - \alpha)\%$ confidence interval for θ given by display (3.63) can be obtained from the two-sided sign test as follows.

3.6. DISTRIBUTION-FREE CONFIDENCE INTERVAL BASED ON THE SIGN TEST

The confidence interval (θ_L, θ_U) consists of those θ_0 values for which the two-sided α-level test of $\theta = \theta_0$ (see Comment 28) does not reject the hypothesis $\theta = \theta_0$.

49. Confidence Bounds. Often we are interested only in making one-sided confidence statements about the parameter θ; that is, we wish to assert with specified confidence that θ is no larger (or, in other settings, no smaller) than some upper (lower) confidence bound based on the sample data. To obtain such one-sided confidence bounds for θ, we proceed as follows. For specified confidence coefficient $1 - \alpha$, find the upper αth (not $(\alpha/2)$nd, as for the confidence interval) percentile point $b_{\alpha,1/2}$ of the null distribution of B from Table A.2. Set

$$C_\alpha^\star = n + 1 - b_{\alpha,1/2}. \tag{3.66}$$

The $100(1 - \alpha)\%$ *lower* confidence bound $\theta_L^\star$ for θ that is associated with the one-sided sign test of $H_0 : \theta = \theta_0$ against the alternative $H_1 : \theta > \theta_0$ is then given by

$$(\theta_L^\star, \infty) = (Z^{(C_\alpha^\star)}, \infty), \tag{3.67}$$

where, as before, $Z^{(1)} \leq \cdots \leq Z^{(n)}$ are the ordered sample observations. With $\theta_L^\star$ given by display (3.67), we have

$$P_\theta(\theta_L^\star < \theta < \infty) = 1 - \alpha \text{ for all } \theta. \tag{3.68}$$

The corresponding $100(1 - \alpha)\%$ *upper* confidence bound $\theta_U^\star$ for θ that is associated with the one-sided sign test of $H_0 : \theta = \theta_0$ against the alternative $H_1 : \theta < \theta_0$ is given by

$$(-\infty, \theta_U^\star) = \left(-\infty, Z^{(n+1-C_\alpha^\star)}\right) = \left(-\infty, Z^{(b_{\alpha,1/2})}\right), \tag{3.69}$$

where $C_\alpha^\star$ is given in (3.66). It follows that

$$P_\theta(-\infty < \theta < \theta_U^\star) = 1 - \alpha \text{ for all } \theta. \tag{3.70}$$

For large n the integer $C_\alpha^\star$ may be approximated by

$$C_\alpha^\star \approx \frac{n}{2} - z_\alpha \left(\frac{n}{4}\right)^{1/2}. \tag{3.71}$$

As with C_α (3.65) and the confidence interval for θ, the value of the right-hand side of equation (3.71) is not an integer. To be conservative, take $C_\alpha^\star$ to be the largest integer that is less than or equal to the right-hand side of equation (3.71).

The $100(1 - \alpha)\%$ lower and upper confidence bounds $\theta_L^\star$ (3.67) and $\theta_U^\star$ (3.69) are related to the acceptance regions of the one-sided sign tests of $H_0 : \theta = \theta_0$ against the alternatives $\theta > \theta_0$ and $\theta < \theta_0$, respectively, in the same way that the confidence interval (θ_L, θ_U) is related to the acceptance region of the two-sided sign test of $H_0 : \theta = \theta_0$ (see Comment 48).

50. Zero Z Values. Note that in calculating the confidence interval (θ_L, θ_U) from display (3.63) or the confidence bounds $\theta_L^\star$ (3.67) or $\theta_U^\star$ (3.69) for θ, we use *all* the Z differences. This is in common with our recommendation (see Comment 41) for computing the point estimator $\tilde{\theta}$ (3.58), but differs from the recommended policy (see Ties in Section 3.4) of discarding the zero Z values (and reducing n accordingly) prior to applying the sign test to the data. However, if there are zero Z's in the data, the equivalence (discussed in Comments 48 and 49) between the acceptance regions of the one-sided and two-sided sign tests and the appropriate confidence bound and confidence interval, respectively, are no longer valid.

51. *Necessity of Magnitudes.* The confidence interval and bounds (see Comment 49) for θ based on the sign tests are simple to compute, since the end points depend only on the ordered sample Z observations. However, for such a computation, knowledge of the signs of the Z differences is no longer sufficient as it was for the computation of B (3.39) for the various sign tests. We need the observation magnitudes to obtain θ_L, θ_U, $\theta_L^\star$, or $\theta_U^\star$.

52. *Midpoint of the Confidence Interval as an Estimator.* The midpoint of the interval (3.63), namely, $[Z^{(C_\alpha)} + Z^{(n+1-C_\alpha)}]/2$, is also a natural estimator of θ. (Note that this actually yields a class of estimators, depending on the value of α.) In general this midpoint is not the same as $\tilde{\theta}$ (3.58). (See Hodges and Lehmann (1967) and Comment 43 for additional discussion of this midpoint class of estimators.)

53. *Comparison of Sign and Signed Rank Confidence Intervals for θ.* The confidence interval (3.63) for θ associated with the sign test and based on the n ordered Z differences is easier to compute than the confidence interval (3.27) for θ associated with the signed rank test and based on the $n(n+1)/2$ ordered Walsh averages (see Comment 17). However, the signed rank confidence interval (3.27) is generally (but not always) more efficient than the sign confidence interval (3.63). (See Section 3.11.)

54. *Extension to Discrete Distributions.* Consider the closed version $[\theta_L, \theta_U] = [Z^{(C_\alpha)}, Z^{(n+1-C_\alpha)}]$ of the $100(1-\alpha)\%$ confidence interval for θ given in display (3.63) under the alternative (to Assumptions B1 and B2) assumption that $Z_1, \ldots, Z_n$ are a random sample from an underlying distribution $F(\cdot)$ with *unique* median θ. Suppose that this common distribution $F(\cdot)$ is such that in any bounded interval of the real line there are at most a finite number (could be zero) of values having positive probability. (If $F(\cdot)$ is continuous, this is trivially satisfied since in that case no real number has positive probability. However, the large majority of discrete probability distributions also satisfy this mild assumption.) Under these weakened conditions on the common $F(\cdot)$, the closed interval $[\theta_L, \theta_U]$ remains a conservative $100(1-\alpha)\%$ confidence interval for θ in the sense that

$$P_\theta(\theta_L \leq \theta \leq \theta_U) \geq 1 - \alpha \quad \text{for all } \theta$$

is guaranteed for every such $F(\cdot)$. The closed versions of the upper and lower confidence bounds (see Comment 49), namely, $(-\infty, \theta_U^\star] = (-\infty, Z^{(b_{\alpha,1/2})}]$ and $[\theta_L^\star, \infty) = [Z^{(C_\alpha^\star)}, \infty)$, respectively, also remain conservative $100(1-\alpha)\%$ bounds over this expanded class of common distributions $F(\cdot)$. (For more details on the extension of these confidence intervals and bounds to common discrete distributions, see Scheffé and Tukey (1945) and Noether (1967a).)

Properties

1. *Distribution-Freeness.* For populations satisfying Assumptions B1 and B2, equation (3.64) holds. Hence, we can control the coverage probability to be $1 - \alpha$ without having more specific knowledge about the forms of the underlying Z distributions. Thus, (θ_L, θ_U) is a distribution-free confidence interval for θ over a very large class of populations. (See also Comment 54.)

2. *Efficiency.* See Section 3.11.

PROBLEMS

70. For the Profile Scales data of Table 3.6, obtain a confidence interval for θ with exact confidence coefficient .9688.

3.7. PROCEDURES BASED ON THE SIGNED RANK STATISTIC

71. For the bleeding time data in Table 3.7, obtain a confidence interval for θ with exact confidence coefficient .9426.
72. For the beak-clapping data of Table 3.5, obtain an estimate for the asymptotic standard deviation of $\tilde{\theta}$ (see Comment 46).
73. For the beak-clapping data of Table 3.5 and $\alpha = .1078$, calculate the point estimator of θ defined in Comment 52. Compare with the value of $\tilde{\theta}$ obtained in Example 3.6.
74. For the Hamilton depression scale factor IV data of Table 3.1, find a confidence interval for θ with exact confidence coefficient .9610.
75. For the bleeding time data in Table 3.7, obtain an approximate 94.26% confidence interval for θ using the large-sample approximation of this section. Compare this approximate confidence interval with the exact 94.26% confidence interval obtained in Problem 71.
76. How does varying α affect the length of the confidence interval defined by display (3.63)? How does it affect the point estimator of θ defined in Comment 52?
77. For the beak-clapping data of Table 3.5, find a lower confidence bound for θ with exact confidence coefficient .9461. (See Comment 49.)
78. Consider the Stanford Profile Scores data of Table 3.6. Obtain an upper confidence bound for θ with exact confidence coefficient .8906. (See Comment 49.)
79. For the salary data in Table 3.2, find a lower confidence bound for θ with exact confidence coefficient .9270. How does this compare with the approximate 93.6% lower confidence bound for θ obtained in Problem 32?
80. Consider the beak-clapping data of Table 3.5. Use the large-sample approximation to obtain an approximate 95% lower confidence bound for θ (see Comment 49). Compare this approximate bound with the exact 94.61% lower confidence bound obtained in Problem 77.
81. Consider the bleeding time data of Table 3.7. Use the large-sample approximation to find an approximate 92% upper confidence bound for θ (see Comment 49).
82. Consider the case $n = 15$ and compare the length of the exact 96.48% confidence interval for θ given by display (3.63) with the length of the approximate 96.48% confidence interval for θ obtained using the large-sample approximation of this section.
83. Consider the case $n = 25$ and compare the exact 94.61% lower confidence bound for θ given by equation (3.67) with the approximate 94.61% lower confidence bound for θ obtained from the large-sample approximation in Comment 49.
84. Use the computer software `Minitab` to obtain a 98.7% confidence interval for θ for the bleeding time data of Table 3.7. (Use the X and Y labels given in Table 3.7.)
85. Use the computer software `Minitab` to obtain a 95.68% confidence interval for θ for the beak-clapping data in Table 3.5.
86. Discuss how the computer software `Minitab` could be used to obtain lower or upper confidence bounds for the parameter θ. Illustrate this application of `Minitab` by using it to obtain a 92.70% lower confidence bound for the salary data of Table 3.2.

ONE-SAMPLE DATA*

3.7. PROCEDURES BASED ON THE SIGNED RANK STATISTIC

Data. We obtain n observations $Z_1, \ldots, Z_n$.

Assumptions

C1. The Z's are mutually independent.
C2. Each Z comes from a population (not necessarily the same) that is continuous and symmetric about θ.

*Sections 3.7 to 3.10 are optional. The contents of these sections are not used in the sequel.

TABLE 3.8. Estimated Values of θ from Mariner and Pioneer Spacecraft

Spacecraft	θ
Mariner 2 (Venus)	81.3001
Mariner 4 (Mars)	81.3015
Mariner 5 (Venus)	81.3006
Mariner 6 (Mars)	81.3011
Mariner 7 (Mars)	81.2997
Pioneer 6	81.3005
Pioneer 7	81.3021

Source: J. D. Anderson, L. Efron, and S. K. Wong (1970).

Procedures

To test $H_0 : \theta = \theta_0$, where θ_0 is some specified number, we create the modified observations $Z_i' = Z_i - \theta_0$, for $i = 1, \ldots, n$. Then we apply any of the test procedures of Section 3.1 to these modified Z' observations.

To obtain a point estimator of θ or a confidence interval for θ, we apply the procedures of Sections 2 and 3 directly to the Z observations without modification.

Example 3.8: *Mariner and Pioneer Spacecraft Data.* The data in Table 3.8 were reported by Anderson, Efron, and Wong (1970). The seven observations represent average measurements of θ, the ratio of the mass of the earth to that of the moon, obtained from seven different spacecraft.

On the basis of previous (2 to 3 years earlier) Ranger spacecraft findings, scientists had considered the value of θ to be approximately 81.3035. Thus, with the data of Table 3.8, we are interested in testing $H_0 : \theta = 81.3035$ versus the alternative $\theta \neq 81.3035$, and we perform test procedure (3.6). With $\alpha = .078$, we see from Table A.4 that $t_{.039} = 25$.

Now, we form the modified Z' observations as follows.

i	Z_i	$Z_i' = Z_i - 81.3035$
1	81.3001	−.0034
2	81.3015	−.0020
3	81.3006	−.0029
4	81.3011	−.0024
5	81.2997	−.0038
6	81.3005	−.0030
7	81.3021	−.0014

Using the computational setup of Section 3.1 on the Z' observations, we calculate $T^+ = 0$. Thus we reject $H_0 : \theta = 81.3035$ at the $\alpha = .078$ level, since $T^+ = 0 < [28 - t_{.039}] = 3$. From Table A.4 we note that $[28 - t_{.008}] = (28 - 28) = 0$. Hence the smallest significance level at which we could reject H_0 in favor of $\theta \neq 81.3035$ by using a symmetric test based on T^+ is .016.

For the large-sample approximation, we see from (3.9) that

$$T^* = \frac{0 - [7(8)/4]}{[7(8)(15)/24]^{1/2}} = -2.366.$$

3.7. PROCEDURES BASED ON THE SIGNED RANK STATISTIC

Thus the smallest significance level at which we could reject H_0 by using a symmetric test based on the normal approximation is .018. This means that both the exact test and the large-sample approximation indicate the existence of strong evidence to reject the findings of the earlier Ranger spacecraft that $\theta = 81.3035$.

The ordered values of $(Z_i + Z_j)/2$ are $W^{(1)} \leq \cdots \leq W^{(28)}$: 81.2997, 81.2999, 81.3001, 81.3001, 81.30015, 81.3003, 81.30035, 81.3004, 81.3005, 81.30055, 81.3006, 81.3006, 81.3006, 81.3008, 81.3008, 81.30085, 81.3009, 81.3010, 81.30105, 81.3011, 81.3011, 81.3013, 81.3013, 81.30135, 81.3015, 81.3016, 81.3018, 81.3021. Since $M = 7(8)/2 = 28$, we see that $M = 2k$ with $k = 14$. Thus from (3.25) we have

$$\hat{\theta} = \frac{W^{(14)} + W^{(15)}}{2} = \frac{81.3008 + 81.3008}{2} = 81.3008.$$

From Table A.4 with $n = 7$ and $\alpha = .046$ we find that $t_{\alpha/2} = t_{.023} = 26$. Thus $C_{.046} = \{7(8)/2\} + 1 - t_{.023} = 28 + 1 - 26 = 3$.

From (3.27) it follows that

$$\theta_L = W^{(3)} = 81.3001 \quad \text{and} \quad \theta_U = W^{(26)} = 81.3016$$

so that our 95.4% confidence interval for θ is

$$(\theta_L, \theta_U) = (81.3001, 81.3016).$$

Applying the large-sample approximation, we find from (3.29) that

$$C_{.046} \approx [7(8)/4] - 1.996[7(8)(15)/24]^{1/2} \approx 2,$$

and since $W^{(2)} = 81.2999$ and $W^{(27)} = 81.3018$, the approximate 95.4% confidence interval for θ is (81.2999, 81.3018).

It is important to comment that in applying the procedures based on the signed rank statistic T^+ (3.3), we made the assumption that the population of average θ measurements for each of the satellites was symmetric about θ. (For a test of this basic assumption, see Section 3.9.) We also note that this set of data provides an example in which the populations of the Z observations are probably not the same (see Assumption C2).

Comments

55. *Assumptions.* Note that Assumption A1 for the paired replicates procedures based on the signed rank statistic is not necessary for the one-sample data because these data need not consist of differences for paired observations.

Properties

1. The properties of the one-sample procedures based on the signed rank statistic are essentially the same as those of the corresponding paired replicates procedures. An exception occurs in the efficiencies of the procedures and is due to the difference in the type of data for the two problems. See Section 3.11 for a discussion of the difference in efficiencies of the procedures of Sections 3.1, 3.2, and 3.3 when they are applied to single-sample problems.

PROBLEMS

87. The data in Table 3.9 are a subset of the data reported by Ijzermans (1970) from an investigation into the susceptibility to corrosion of 18Cr_10Ni_2Mo stainless steel (i.e., stainless steel containing 18% chromium, 10% nickel, and 2% molybdenum by weight).

    Twelve specimens of steel were selected for use in the corrosion study. Although Ijzermans' experiment was directed toward corrosion, we are concerned here with the quality of the steel from which the stainless steel samples were chosen. Table 3.9 gives the percentage of chromium in the 12 samples used by Ijzermans.

    Test the hypothesis that the median percentage of chromium content (θ) of the steel is 18% against the alternative that it is not 18%. Obtain a point estimate of θ and find a confidence interval for θ with confidence coefficient .936.

88. For the percentage of chromium data in Table 3.9, obtain a point estimate of θ from the midpoint of the confidence interval calculated in Problem 87 (see Comment 23). Compare with the point estimate obtained in Problem 87.

89. Compute $\hat{\theta}$ for the settling velocity data of Table 3.12 and compare with the value of $\tilde{\theta}$ obtained in Example 3.9.

90. Lamp (1976) studied the age distribution of a common mayfly species, *Stenacron interpunctatum*, among various habitats in Big Darby Creek, Ohio. One of the measurements considered was head width (in micrometer divisions, 1 division = .0345 mm); a subset of Lamp's data from the mayflies in habitat A is presented in Table 3.10.

    Test the hypothesis that the median head width for mayflies from habitat A (θ) is 22 μm divisions against the alternative that it is greater than 22. Obtain a point estimate of θ and find a lower confidence bound (see Comment 21) for θ with confidence coefficient .976.

TABLE 3.9. Percentage of Chromium in the Stainless Steel Samples

Steel Sample	% Cr
1	17.4
2	17.9
3	17.6
4	18.1
5	17.6
6	18.9
7	16.9
8	17.5
9	17.8
10	17.4
11	24.6
12	26.0

Source: A. B. Ijzermans (1970).

TABLE 3.10. Mayfly Head Width, Habitat A (Micrometer Divisions)

Mayfly i	Z_i
1	36
2	31
3	30
4	27
5	20
6	33
7	27
8	18
9	19
10	28

Source: W. O. Lamp (1976).

91. The data in Table 3.11 are a subset of the data obtained by Poland et al. (1970) in an experiment concerned with the effect of occupational exposure to DDT on human drug and steroid metabolism. The DDT-exposed subjects were employees of the Montrose Chemical Corporation who had been working in the DDT plant at Torrance, California for more than 5 years. All these individuals had received moderate to intense occupational exposure to DDT, and all were in good health. One of the measures used in the study was the 24-h urinary excretion of 6β-hydroxycortisol.

    Test the hypothesis that the median 6β-hydroxycortisol excretion rate for subjects with occupational exposure to DDT similar to the workers in this study (θ) is 175 μg/24 h against the alternative that it is greater than 175. Obtain a point estimate of θ and find a confidence interval for θ with confidence coefficient .916.

TABLE 3.11. 6β-hydroxycortisol Excretion ($\mu g/24$ hours)

Worker i	Z_i
1	254
2	171
3	345
4	134
5	190
6	447
7	106
8	173
9	449
10	198

Source: A. Poland, D. Smith, R. Kuntzman, M. Jacobson, and A. H. Conney (1970).

92. Consider the oxidant content of dew water data in Table 3.13. Use the computer software Minitab to test the hypothesis that the median oxidant content of dew water (θ) was .25 against the alternative that it was less than .25. Also use Minitab to obtain a point estimate of θ and find an upper confidence bound (see Comment 21) for θ with confidence coefficient .961. Compare with the answers to Problem 94.

93. Consider the settling velocity data of Table 3.12. Use the computer software Minitab to test the hypothesis that the median settling velocity for the Middle Ground sand ridge (θ) was 14 cm/s against the alternative that it was not equal to 14. Also use Minitab to obtain a point estimate of θ and find a confidence interval for θ with confidence coefficient .890. Compare with the results obtained in Example 3.9.

3.8. PROCEDURES BASED ON THE SIGN STATISTIC

Data. We obtain n observations $Z_1, \ldots, Z_n$.

Assumptions

D1. The Z's are mutually independent.

D2. Each Z comes from the same continuous population with median θ, so that $P(Z_i < \theta) = P(Z_i > \theta) = \frac{1}{2}$, $i = 1, \ldots, n$.

Procedures

To test $H_0 : \theta = \theta_0$, where θ_0 is some specified number, we form the modified observations $Z_i' = Z_i - \theta_0$, for $i = 1, \ldots, n$. Then we can apply any of the test procedures of Section 3.4 to these modified Z' observations. (In the test of H_0, we can weaken Assumption D2 to D2', namely, that each Z comes from a population, not necessarily the same population, such that $P(Z_i < \theta_0) = P(Z_i > \theta_0) = \frac{1}{2}, i = 1, \ldots, n$, when θ is equal to the hypothesized value θ_0.)

To obtain a point estimator of θ or a confidence interval for θ, we apply the procedures of Sections 3.5 and 3.6 directly to the Z observations without modification.

Example 3.9: *Sediment Settling Velocities.* The data in Table 3.12 are a subset of the data obtained by Smith (1969) in an experiment investigating the geomorphology of the Middle Ground sand ridge, which lies in Vineyard Sound, Massachusetts.

TABLE 3.12. Settling Velocities at 22°C

Sample i	Z_i (cm/s)
1	12.9
2	13.7
3	14.5
4	13.3
5	12.8
6	13.8
7	13.4

Source: J. D. Smith (1969).

Seven samples were obtained from a particular portion of the ridge by using a van Veen grab. One of the objective measurements reported by Smith was the settling velocity of the sediment at 22°C. For sediment from a sand-wave crest section of a sand ridge, the settling velocity has a typical value of 14 cm/s. Table 3.12 gives the settling velocities for the seven sediment samples collected from a particular portion of the Middle Ground sand ridge.

We would like to detect whether the seven sediment samples came from a sand-wave crest section of the Middle Ground sand ridge. Let θ denote the median settling velocity for the population of sediment samples from this portion of Middle Ground. Then we are interested in testing $H_0 : \theta = 14$ cm/s versus the alternative $\theta \neq 14$ cm/s, and we perform test procedure (3.42). With $\alpha = .0156$, we see from Table A.2 that $b_{.0078,1/2} = 7$.

Now, we create the modified Z' observations by using the following setup.

i	Z_i	$Z'_i = Z_i - 14$
1	12.9	-1.1
2	13.7	-0.3
3	14.5	0.5
4	13.3	-0.7
5	12.8	-1.2
6	13.8	-0.2
7	13.4	-0.6

Using the computational setup of Section 3.4 on the Z' observations, we calculate $B = 1$. Thus we accept $H_0 : \theta = 14$ cm/s at the $\alpha = .0156$ level, since $[7 - b_{.0078,1/2}] = 0 < B < 7 = b_{.0078,1/2}$. From Table A.2 we note that $[n - b_{.0625,1/2}] = (7 - 6) = 1$. Hence the smallest significance level at which we could reject H_0 in favor of $\theta \neq 14$ cm/s by using a symmetric test based on B is .1250.

For the large-sample approximation, we see from (3.45) that

$$B^* = \frac{1 - \left(\frac{7}{2}\right)}{\left(\frac{7}{4}\right)^{1/2}} \approx -1.89.$$

Thus the smallest significance level at which we could reject H_0 by using a symmetric test based on the normal approximation is .0588.

The ordered Z observations are $Z^{(1)} \leq \cdots \leq Z^{(7)}$: 12.8, 12.9, 13.3, 13.4, 13.7, 13.8, 14.5. Since $n = (2k + 1)$ with $k = 3$, we see from (3.59) that

$$\tilde{\theta} = Z^{(4)} = 13.4.$$

3.8. PROCEDURES BASED ON THE SIGN STATISTIC

From Table A.2 with $n = 7$ and $\alpha = .1250$ we find that $b_{\alpha/2, 1/2} = b_{.0625, 1/2} = 6$. Thus $C_{.1250} = 7 + 1 - 6 = 2$. From (3.63) it follows that

$$\theta_L = Z^{(2)} = 12.9 \quad \text{and} \quad \theta_U = Z^{(6)} = 13.8,$$

so that our 87.50% confidence interval for θ is

$$(\theta_L, \theta_U) = (12.9, 13.8).$$

Applying the large-sample approximation, we find from (3.65) that

$$C_{.1250} \approx \left(\tfrac{7}{2}\right) - 1.534 \left(\tfrac{7}{4}\right)^{1/2} \approx 1,$$

and since $Z^{(1)} = 12.8$ and $Z^{(n+1-1)} = Z^{(7)} = 14.5$, the approximate 87.50% confidence interval for θ is (12.8, 14.5).

Comments

56. *Assumptions.* Note that Assumption B1 for the paired replicates procedures based on the sign statistic is not necessary for the one-sample data because these data do not consist of differences for paired observations.

57. *Procedures for Population Quantiles Other Than the Median.* For one-sample data the theory underlying the sign statistic can also be used to construct distribution-free test procedures for population quantiles other than the median. Such test procedures are similar to procedures (3.40) to (3.42), but they have different p values in the null hypothesis binomial distribution. For example, let $Z_1, \ldots, Z_n$ be a random sample from a population Π. Define μ_ξ to be the unknown ξ quantile of the population. (For convenience, let us assume that μ_ξ is unique.) Consider the problem of testing $H_0 : \mu_\xi = \mu_0$ (specified) versus the one-sided alternative $\mu_\xi > \mu_0$. Define B to be the number of Z's that are greater than μ_0. Under H_0, B has the binomial distribution with parameters n and $p = 1 - \xi$. Large values of B indicate that $\mu_\xi > \mu_0$, so an appropriate one-sided α-level test is to reject H_0 in favor of $\mu_\xi > \mu_0$ if $B \geq b_{\alpha, 1-\xi}$ and accept H_0 if $B < b_{\alpha, 1-\xi}$, where the $b_{\alpha, 1-\xi}$ constant is to be obtained from Table A.2. One-sided tests against $\mu_\xi < \mu_0$ and two-sided tests for alternatives $\mu_\xi \neq \mu_0$ are constructed in a similar manner. The natural point estimator of the parameter $P(Z > \mu_0)$ is the statistic B/n. Approximate confidence intervals for μ_ξ can also be obtained (cf. Conover (1971)).

Properties

1. The properties of the one-sample procedures based on the sign statistic are essentially the same as those of the corresponding paired replicates procedures. An exception occurs in the efficiencies of the procedures and is due to the difference in the type of data for the two problems. See Section 3.11 for a discussion of the difference in efficiencies for the procedures of Sections 3.4 to 3.6 when they are applied to single-sample problems.

PROBLEMS

94. The data in Table 3.13 are a subset of the data obtained by Cole and Katz (1966). They were investigating the relation between ozone concentrations and weather fleck damage to tobacco

TABLE 3.13. Oxidant Content of Dew Water, Port Burwell, 1960

Sample i	Z_i (ppm ozone)
1	.32
2	.21
3	.28
4	.15
5	.08
6	.22
7	.17
8	.35
9	.20
10	.31
11	.17
12	.11

Source: A. F. W. Cole and M. Katz (1966).

crops in southern Ontario, Canada. One of the objective measurements reported was oxidant content of dew water in parts per million (ppm) ozone. Twelve samples of dew were collected during the period of August 25–30, 1960, at Port Burwell, Ontario; the resulting oxidant contents are given in Table 3.13.

Test the hypothesis that the median oxidant content (θ) of dew water was .25 against the alternative that it was less than .25. Obtain a point estimate of θ and find a confidence interval for θ with confidence coefficient .9614.

95. For the oxidant content data of Table 3.13, obtain a point estimate of θ from the midpoint of the confidence interval calculated in Problem 94 (see Comment 52). Compare with $\tilde{\theta}$ obtained in Problem 94.

96. Compute $\tilde{\theta}$ for the mass ratio data of Table 3.8 and compare with the value of $\hat{\theta}$ obtained in Example 3.8.

97. Maxson (1977) studied the activity patterns of female ruffed grouse with broods. Using surveillance techniques he recorded the movements of seven female ruffed grouse with broods over a fixed period of time. The percentage of time that these grouse were in active movement are recorded in Table 3.14.

Test the hypothesis that the median percentage time active for female ruffed grouse with broods (θ) is 50% against the alternative that it is greater than 50%. Obtain a point estimate of θ and find a lower confidence bound (see Comment 49) for θ with confidence coefficient .99.

TABLE 3.14. Ruffed Grouse, Percentage Time in Active Movement

Grouse i	Z_i (% Time Active)
1	52.7
2	51.5
3	58.4
4	56.9
5	58.5
6	54.4
7	47.1

Source: S. J. Maxson (1977).

TABLE 3.15. Net Oxygen Consumption (cc)

Patient i	Z_i
1	339
2	349
3	387
4	159
5	579
6	586
7	519
8	275

Source: A. M. Flores and L. R. Zohman (1970).

98. The data in Table 3.15 are a subset of the data obtained by Flores and Zohman (1970) in an experiment investigating the effect that the method of bed-making has on the oxygen consumption for patients assigned to complete or modified bed rest. The subjects were inpatients of the Rehabilitation Medicine Service, Montefiore Hospital and Medical Center, Bronx, New York. The measure used was net oxygen consumption for the patients during bed-making. The data in Table 3.15 are the net oxygen consumptions (in cc) for the eight patients in the study during a cardiac top-to-bottom bed-making procedure, consisting of moving the patient to a sitting position and changing the sheets from the top to the bottom of the bed.

    Test the hypothesis that the median oxygen consumption rate during cardiac bed-making for patients assigned to complete or modified bed rest (θ) is 350 cc against the alternative that it is not 350. Obtain a point estimate of θ and find a confidence interval for θ with confidence coefficient .95.

99. Consider the 6β-hydroxycortisol excretion data in Table 3.11. Use the computer software Minitab to test the hypothesis that the median 6β-hydroxycortisol excretion rate for subjects with occupational exposure to DDT similar to the workers in the Poland et al. (1970) study (θ) is $175/\mu g/24$ h against the alternative that it is greater than 175. Obtain a point estimate of θ and find a confidence interval for θ with confidence coefficient .925. Compare with the answers to Problem 91.

100. Consider the mayfly head width data in Table 3.10. Let $\mu_{.75}$ be the 75th percentile for the distribution of mayfly head widths in habitat A studied by Lamp (1976). Test the hypothesis that $\mu_{.75} = 25$ against the alternative that $\mu_{.75}$ is greater than 25. (See Comment 57.)

3.9. AN ASYMPTOTICALLY DISTRIBUTION-FREE TEST OF SYMMETRY (RANDLES-FLIGNER-POLICELLO-WOLFE, DAVIS-QUADE)

Data. We obtain n observations $Z_1, \ldots, Z_n$.

Assumptions

E1. The Z's are mutually independent.

E2. Each Z comes from the same continuous population having distribution function F and unknown median θ. This assumption requires that $F(\theta) = \frac{1}{2}$.

Hypothesis

The null hypothesis of interest here is that the common underlying distribution for the Z observations is symmetric about θ. This hypothesis of symmetry can be written as

$$H_0 : [F(\theta + b) + F(\theta - b) = 1, \quad \text{for every } b], \tag{3.72}$$

and it is equivalent to the statement that $P(0 < Z - \theta < b) = P(-b < Z - \theta < 0)$ for all $b > 0$.

Procedure

For each triple of observations (Z_i, Z_j, Z_k), $1 \leq i < j < k \leq n$, obtain the value of

$$f^*(Z_i, Z_j, Z_k) = [\text{sign}(Z_i + Z_j - 2Z_k) + \text{sign}(Z_i + Z_k - 2Z_j) + \text{sign}(Z_j + Z_k - 2Z_i)], \tag{3.73}$$

where sign $(t) = -1, 0, 1$ as $t <, =, > 0$. (Note that there are $n(n-1)(n-2)/6$ distinct triples in the sample.) We say that (Z_i, Z_j, Z_k) forms a **right triple** (looks skewed to the right) if $f^*(Z_i, Z_j, Z_k) = 1$. (Note that being a right triple is equivalent to the middle *ordered* observation in (Z_i, Z_j, Z_k) being closer to the smallest of the three observations than it is to the largest of them.) Conversely, (Z_i, Z_j, Z_k) is said to be a **left triple** (looks skewed to the left) if $f^*(Z_i, Z_j, Z_k) = -1$ (i.e., the middle *ordered* observation in (Z_i, Z_j, Z_k) is closer to the largest than to the smallest of the three observations). Finally, when $f^*(Z_i, Z_j, Z_k) = 0$, the triple (Z_i, Z_j, Z_k) is neither right nor left.

For the data $Z_1, \ldots, Z_n$, set

$$T = \sum\sum\sum_{1 \leq i < j < k \leq n} f^*(Z_i, Z_j, Z_k)$$

$$= \{[\text{number of right triples}] - [\text{number of left triples}]\}. \tag{3.74}$$

For each fixed $t = 1, \ldots, n$, let

$B_t = \{[\text{number of right triples involving } Z_t] - [\text{number of left triples involving } Z_t]\}$

$$= \left[\sum_{j=t+1}^{n-1} \sum_{k=j+1}^{n} f^*(Z_t, Z_j, Z_k) + \sum_{j=1}^{t-1} \sum_{k=t+1}^{n} f^*(Z_j, Z_t, Z_k) + \sum_{j=1}^{t-2} \sum_{k=j+1}^{t-1} f^*(Z_j, Z_k, Z_t) \right]. \tag{3.75}$$

For each fixed integer pair (s, t) such that $1 \leq s < t \leq n$, define

$B_{s,t} = \{[\text{number of right triples involving } Z_s \text{ and } Z_t]$

$\quad - [\text{number of left triples involving } Z_s \text{ and } Z_t]\}$

$$= \left[\sum_{j=1}^{s-1} f^*(Z_j, Z_s, Z_t) + \sum_{j=s+1}^{t-1} f^*(Z_s, Z_j, Z_t) + \sum_{j=t+1}^{n} f^*(Z_s, Z_t, Z_j) \right]. \tag{3.76}$$

Using the expressions for B_t (3.75), $B_{s,t}$ (3.76), and the triple statistic T (3.74), set

$$V = T/\hat{\sigma}, \tag{3.77}$$

where

$$\hat{\sigma}^2 = \left[\frac{(n-3)(n-4)}{(n-1)(n-2)} \sum_{t=1}^{n} B_t^2 + \frac{(n-3)}{(n-4)} \sum_{s=1}^{n-1} \sum_{t=s+1}^{n} B_{s,t}^2 \right.$$
$$\left. + \frac{n(n-1)(n-2)}{6} - \left\{ 1 - \frac{(n-3)(n-4)(n-5)}{n(n-1)(n-2)} \right\} T^2 \right]. \tag{3.78}$$

3.9. AN ASYMPTOTICALLY DISTRIBUTION-FREE TEST OF SYMMETRY

When H_0 is true and the underlying distribution is symmetric, V has, as n tends to infinity, an asymptotic $N(0, 1)$ distribution. (In order for this normal approximation to be reasonably effective, the sample size n should be at least ten. For further discussion along these lines, see Comment 60.)

To test H_0 (3.72), corresponding to symmetry of the underlying distribution, versus the general alternative of asymmetry, corresponding to

$$H_1 : [P(Z \leq \theta + b) + P(Z \leq \theta - b) \neq 1 \text{ for at least one } b], \quad (3.79)$$

at the approximate (n large) α level of significance,

$$\text{Reject } H_0 \text{ if } |V| \geq z_{\alpha/2}; \quad \text{otherwise do not reject.} \quad (3.80)$$

Ties

The test procedure in (3.80) is well-defined when zeros occur in the $(Z_i + Z_j - 2Z_k)$ variables and further adjustments are not necessary.

Example 3.10: *Percentage Chromium in Stainless Steel.* In order to clearly illustrate the details of the rather involved calculations necessary to obtain the value of the test statistic V (3.77), we consider the application of the test for symmetry to the first five (i.e., $n = 5$) percentage chromium data values in Table 3.9, namely, $Z_1 = 17.4, Z_2 = 17.9, Z_3 = 17.6, Z_4 = 18.1$, and $Z_5 = 17.6$. (We emphasize that this application is for illustrative purposes only. The test for symmetry is totally ineffective at detecting asymmetry for sample sizes as small as $n = 5$. See Comment 60 for related discussion.) We must calculate $n(n-1)(n-2)/6 = 5(4)(3)/6 = 10$ values of the triple indicator $f^*(Z_i, Z_j, Z_k)$ given by equation (3.73). We have that

$$f^*(Z_1, Z_2, Z_3) = [\text{sign } (17.4 + 17.9 - 2(17.6)) + \text{sign } (17.4 + 17.6 - 2(17.9))$$
$$+ \text{sign } (17.9 + 17.6 - 2(17.4))]$$
$$= [\text{sign } (.1) + \text{sign } (-.8) + \text{sign } (.7)] = 1 - 1 + 1 = 1. \quad (3.81)$$

Similarly, we obtain

$$f^*(Z_1, Z_2, Z_5) = f^*(Z_1, Z_3, Z_4) = f^*(Z_1, Z_4, Z_5)$$
$$= f^*(Z_2, Z_3, Z_5) = f^*(Z_3, Z_4, Z_5) = 1 \quad (3.82)$$

and

$$f^*(Z_1, Z_2, Z_4) = f^*(Z_1, Z_3, Z_5) = f^*(Z_2, Z_3, Z_4) = f^*(Z_2, Z_4, Z_5) = -1. \quad (3.83)$$

Hence, from equation (3.74) we have that

$$T = \sum\sum\sum_{1 \leq i < j < k \leq 5} f^*(Z_i, Z_j, Z_k) = 6 - 4 = 2. \quad (3.84)$$

For the calculation of $\hat{\sigma}^2$, we first need to obtain the values of $B_1, \ldots, B_5$ and $B_{s,t}$, for $1 \leq s < t \leq 5$. From equations (3.75), (3.81), (3.82), and (3.83), we have that

$$B_1 = \sum_{j=2}^{4} \sum_{k=j+1}^{5} f^*(Z_1, Z_j, Z_k) = [1 - 1 + 1 + 1 - 1 + 1] = 2,$$

$$B_2 = \left[\sum_{j=3}^{4} \sum_{k=j+1}^{5} f^*(Z_2, Z_j, Z_k) + \sum_{k=3}^{5} f^*(Z_1, Z_2, Z_k) \right] = [(-1 + 1 - 1) + (1 - 1 + 1)] = 0,$$

$$B_3 = \left[f^*(Z_3, Z_4, Z_5) + \sum_{j=1}^{2} \sum_{k=4}^{5} f^*(Z_j, Z_3, Z_k) + f^*(Z_1, Z_2, Z_3) \right]$$
$$= [1 + (1 - 1 - 1 + 1) + 1] = 2,$$

$$B_4 = \left[\sum_{j=1}^{3} f^*(Z_j, Z_4, Z_5) + \sum_{j=1}^{2} \sum_{k=j+1}^{3} f^*(Z_j, Z_k, Z_4) \right] = [(1 - 1 + 1) + (-1 + 1 - 1)] = 0,$$

and

$$B_5 = \sum_{j=1}^{3} \sum_{k=j+1}^{4} f^*(Z_j, Z_k, Z_5) = [1 - 1 + 1 + 1 - 1 + 1] = 2.$$

It follows that

$$\sum_{t=1}^{5} B_t^2 = [2^2 + 0^2 + 2^2 + 0^2 + 2^2] = 12. \qquad (3.85)$$

Furthermore, using equations (3.76), (3.81), (3.82), and (3.83), we obtain

$$B_{1,2} = \sum_{j=3}^{5} f^*(Z_1, Z_2, Z_j) = [1 - 1 + 1] = 1,$$

$$B_{1,3} = f^*(Z_1, Z_2, Z_3) + \sum_{j=4}^{5} f^*(Z_1, Z_3, Z_j) = [1 + (1 - 1)] = 1,$$

$$B_{1,4} = \sum_{j=2}^{3} f^*(Z_1, Z_j, Z_4) + f^*(Z_1, Z_4, Z_5) = [(-1 + 1) + 1] = 1,$$

$$B_{1,5} = \sum_{j=2}^{4} f^*(Z_1, Z_j, Z_5) = [1 - 1 + 1] = 1,$$

$$B_{2,3} = f^*(Z_1, Z_2, Z_3) + \sum_{j=4}^{5} f^*(Z_2, Z_3, Z_j) = [1 + (-1 + 1)] = 1,$$

$$B_{2,4} = f^*(Z_1, Z_2, Z_4) + f^*(Z_2, Z_3, Z_4) + f^*(Z_2, Z_4, Z_5) = [-1 - 1 - 1] = -3,$$

$$B_{2,5} = f^*(Z_1, Z_2, Z_5) + \sum_{j=3}^{4} f^*(Z_2, Z_j, Z_5) = [1 + (1 - 1)] = 1,$$

3.9. AN ASYMPTOTICALLY DISTRIBUTION-FREE TEST OF SYMMETRY

$$B_{3,4} = \sum_{j=1}^{2} f^*(Z_j, Z_3, Z_4) + f^*(Z_3, Z_4, Z_5) = [(1-1) + 1] = 1,$$

$$B_{3,5} = \sum_{j=1}^{2} f^*(Z_j, Z_3, Z_5) + f^*(Z_3, Z_4, Z_5) = [(-1+1) + 1] = 1,$$

and

$$B_{4,5} = \sum_{j=1}^{3} f^*(Z_j, Z_4, Z_5) = [1 - 1 + 1] = 1.$$

These $B_{s,t}$ values yield

$$\sum_{s=1}^{4} \sum_{t=s+1}^{5} B_{s,t}^2 = [1^2 + 1^2 + 1^2 + 1^2 + 1^2 + (-3)^2 + 1^2 + 1^2 + 1^2 + 1^2] = 18. \quad (3.86)$$

Using the computational results from equations (3.84), (3.85), and (3.86) in the formula for $\hat{\sigma}^2$ (3.78), we obtain

$$\hat{\sigma}^2 = \left[\frac{2(1)}{4(3)}(12) + \frac{2}{1}(18) + \frac{5(4)(3)}{6} - \left\{ 1 - \frac{2(1)(0)}{5(4)(3)} \right\} (2)^2 \right] = [2 + 36 + 10 - 4] = 44.$$

Finally, from equation (3.77), we have

$$V = \frac{T}{\hat{\sigma}} = \frac{2}{(44)^{1/2}} = .30.$$

With significance level $\alpha = .05$, we see from Table A.1 that $z_{.025} = 1.96$. Since $|V| = .30$ is less than 1.96, we cannot reject the null hypothesis of symmetry for the underlying distribution. In fact, the smallest significance level at which we could reject this distributional symmetry (i.e., the two-sided P-value for these data) is

$$\alpha = 2P(\text{standard normal variable exceeds } .30)$$
$$= 2(.3821) = .7642,$$

clearly indicating that there is virtually no evidence in this subset of the percentage chromium data to indicate asymmetry in the underlying probability distribution. (Remember, however, that this subset was a sample of only five observations. These are simply not sufficient data to detect asymmetry even if it were present. See Comment 60.)

Comments

58. *Motivation.* A right triple is indicative of skewness to the right and a left triple is indicative of skewness to the left. The absolute value of the statistic T (3.74) is the difference between the numbers of right and left triples among the $n(n-1)(n-2)/6$ triples in the sample. When the null hypothesis H_0 (3.72) of symmetry is true, we would expect half of the sample triples to be right triples and the other half to be left triples. Thus, when H_0 is true we would expect T to be near zero. A substantial deviation in either direction from zero for T is therefore

indicative of asymmetry in the population and serves as partial motivation for the procedure defined in (3.80).

59. *Asymptotic Distribution-Freeness.* Asymptotically (i.e., for infinitely large samples) the true level of the test defined by (3.80) will agree with the nominal level. Subject to Assumptions E1 and E2, this asymptotic result does not depend on the underlying population of the Z's. More precisely, subject to Assumptions E1 and E2, V has an asymptotic $N(0, 1)$ distribution when H_0 is true. Since this asymptotic distribution does not depend on the underlying population of the Z's, we say that the test based on V is asymptotically distribution-free. Of course, in practice, we do not have the luxury of infinite samples. Thus in any particular case, with n large, we hope the level of a test based on V is close to the nominal level α but it may not be exactly equal to α. The closeness of the approximation depends on n and α and, for fixed α, the closeness generally improves as n increases. In the case of the V test, the reader is warned that the question of how large n should be, in order for the approximation to be good, is unanswered. Exact null distribution tables for V cannot be provided since, for a specified value of n, the exact null distribution of V depends on the underlying Z population; thus exact critical points would vary with the form of the Z population. The procedure in (3.80) based on V, therefore, is not (strictly) distribution-free.

60. *Sample Size Requirement.* As noted in Comment 59, the test of symmetry described in equation (3.80) is not an exact distribution-free procedure. The nominal significance level α is guaranteed only asymptotically, as the number of observations, n, becomes infinite. In addition, symmetry is a rather complex property of a probability distribution. It is, therefore, virtually impossible to deny its presence without at least a moderate sample size. It is simply difficult to "see" asymmetry in a small number of sample observations. Both Randles et al. (1980) and Davis and Quade (1978) found this to be the case. They concluded that the symmetry test (3.80) is not effective at detecting asymmetry in the underlying population unless the sample size (n) is at least 20.

61. *One-Sided Tests for Right-Skewness or Left-Skewness.* The test procedure in (3.80) is a two-sided test of symmetry against a very general class of asymmetric alternatives. However, one-sided tests of symmetry versus specific classes of right-skewed (or left-skewed) asymmetric alternatives can also be based on the statistic V (3.77). In particular, a one-sided (approximate) level α test of H_0 (3.72) (symmetry) versus the specific class of right-skewed alternatives satisfying

$$F(\theta + b) \leq [1 - F(\theta - b)] \quad \text{for every } b > 0,$$

with strict inequality for at least one positive b, \hfill (3.87)

is given by

$$\text{Reject } H_0 \text{ if } V \geq z_\alpha; \text{ otherwise do not reject.} \hfill (3.88)$$

Similarly, a one-sided (approximate) level α test of H_0 (3.72) versus the specific class of left-skewed alternatives satisfying

$$F(\theta + b) \geq [1 - F(\theta - b)], \quad \text{for every } b > 0,$$

with strict inequality for at least one positive b, \hfill (3.89)

is given by

$$\text{Reject } H_0 \text{ if } V \leq -z_\alpha; \text{ otherwise do not reject.} \hfill (3.90)$$

These one-sided hypothesis tests in (3.88) and (3.90) are asymptotically distribution-free in the same sense as the two-sided test given by (3.80). See Comment 59 for further discussion of this property.

62. *Signed Rank Procedures.* One of the critical assumptions permitting the application of signed rank procedures to one-sample data is that of underlying distributional symmetry (see Assumption C2 in Section 3.7). Under this symmetry *assumption*, procedures based on the signed rank statistic T^+ (3.3) for one-sample data are used to make inferences about the median of a population. Procedure (3.80), on the other hand, is used to test for the symmetry of a population and is not directly concerned with the numerical value of the median of the population. Therefore, in an appropriate one-sample location problem we might wish to apply procedure (3.80) [to check the symmetry assumption] prior to using the signed rank procedures of Section 3.7 for making inferences about the actual value of the unknown median of the population. Procedure (3.80) is appropriate, but the *known* median test mentioned in Comment 63 is inappropriate as a pretest in this situation. (For the paired-replicates data in Sections 3.1–3.3, we remind the reader that the symmetry assumption is most often inherently satisfied through the nature of the pairing. See Comment 2.)

63. *Case of Known Median.* For the situation when the median of the underlying population is *known* to be a specified value θ_0 (say), Gupta (1967) proposed a procedure for testing the hypothesis of symmetry about θ_0. However, situations in which the median of the underlying population is known but the symmetry of the distribution is not known are encountered considerably less frequently than situations in which both the median and the symmetry are not known (see Comment 62). (Gupta (1967) also proposed a test for symmetry when the underlying median is not known. His procedure in this case is a competitor to the test given by (3.80). He investigated the loss of efficiency that results from using his test for symmetry with unknown median when his known median procedure is applicable.)

64. *Alternative Determination of Right and Left Triples.* The original definitions of right and left triples in this section involve the sign function $f^*(Z_i, Z_j, Z_k)$ in equation (3.73). A more intuitive interpretation is associated with the comparison of two common sample measures of location. For a triple (Z_i, Z_j, Z_k), let $\bar{Z} = (Z_i + Z_j + Z_k)/3$ and $\tilde{Z} = \text{median}\{Z_i, Z_j, Z_k\}$ be the average and median, respectively, for the observations in the triple. Then the triple (Z_i, Z_j, Z_k) is a right triple if $\bar{Z} > \tilde{Z}$ and it is a left triple if $\bar{Z} < \tilde{Z}$. (It is neither right nor left if $\bar{Z} = \tilde{Z}$.) This formulation provides a very natural interpretation of what it means to be a right or left triple, since we know that the population mean is greater than or less than the population median according to whether the population is skewed to the right or left, respectively. If the population is symmetric, its mean and median are equal and it would be a toss up as to which of $\bar{Z}$ or $\tilde{Z}$ would be greater. This should lead to about an equal number of right and left triples in the sample.

65. *Consistent Estimator of the Asymptotic Variance of $\sqrt{n}T$.* In order to insure that V (3.77) is asymptotically distribution-free, $n\hat{\sigma}^2$ (3.78) is taken to be a consistent estimator of the asymptotic null variance of $n^{1/2}T$. The consistency of this estimator $n\hat{\sigma}^2$ follows from a standard body of theory about a class of statistics introduced by Hoeffding (1948a) and referred to as U-statistics. (For more details about U-statistics and their application in the triples test, see Randles and Wolfe (1979).) The asymptotic normality (and, thereby, the asymptotic distribution-freeness) for V (3.77) follows from standard U-statistics theory and Slutsky's theorem (see, for example, Theorem A.3.13 in Randles and Wolfe (1979)).

66. *Consistency of the V Test.* Under Assumptions E1 and E2, the consistency of the tests based on V depend on the parameter

$$p^* = P(Z_1 + Z_2 - 2Z_3 > 0) - \tfrac{1}{2}. \qquad (3.91)$$

The two-sided test defined by (3.80) is consistent against the class of asymmetric alternatives corresponding to $p^* \neq 0$. We point out that while asymmetry of a probability distribution implies that $p^* \neq 0$ for that distribution, the converse is not necessarily true. That is, there are asymmetric probability distributions for which $p^* = 0$ and against which, therefore, the two-sided test (3.80) based on V will not be consistent. Randles et al. (1980) note, however, that the class of distributions with this property is quite small. (The one-sided tests discussed in Comment 61 and defined by (3.88) and (3.90) are consistent against the classes of asymmetric alternatives corresponding to $p^* > 0$ and < 0, respectively.)

Properties

1. *Consistency.* See Comment 66 and Randles et al. (1980).
2. *Asymptotic Normality.* See Randles and Wolfe (1979, 99–101).

PROBLEMS

101. Consider the percentage chromium data in Table 3.9. Test the hypothesis of symmetry versus general asymmetry. (Note that some of the necessary calculations for this test have been completed in Example 3.10.)
102. Show that a triple (Z_1, Z_2, Z_3) is a right triple if and only if $\bar{Z} = (Z_1 + Z_2 + Z_3)/3$ is greater than $\tilde{Z} = $ median (Z_1, Z_2, Z_3). (See also Comment 64.)
103. Consider the oxidant content data of Table 3.13. Test the hypothesis of symmetry versus general asymmetry.
104. What effect does the addition of a number b to each of the Z observations have on the value of the V (3.77) statistic? Comment on this as a desirable property for a test of population symmetry.
105. What effect does the multiplication of each of the Z observations by a number b have on the absolute value of the V (3.77) statistic? Comment on this as a desirable property for a test of population symmetry versus general asymmetry.
106. Consider the settling velocity data in Table 3.12. Test the hypothesis of symmetry against the alternative that the population of settling velocities is skewed to the right (see Comment 61).
107. Consider the Z differences for the beak-clapping data in Table 3.5. Test the hypothesis of symmetry against the alternative that the population of beak-clapping differences is skewed to the left (see Comment 61).
108. For n observations $Z_1, \ldots, Z_n$, what is the maximum possible value for T (3.74)? What is the minimum possible value for T? For $n = 4$, construct examples where these extreme values for T are achieved.
109. Consider the four observations $Z_1 = 2$, $Z_2 = 2.4$, $Z_3 = 3$, and $Z_4 = 3.5$. Compute the value of T (3.74) for these data. Indicate how to change only one of the sample observations in such a way that T achieves its maximum value (see Problem 108) on the altered data. Similarly, indicate how to change only one of the sample observations in such a way that T achieves its minimum value (see Problem 108) on the altered data.

BIVARIATE DATA

3.10. A DISTRIBUTION-FREE TEST FOR BIVARIATE SYMMETRY (HOLLANDER)

Data. We obtain $2n$ observations, two observations on each of n subjects.

3.10. A DISTRIBUTION-FREE TEST FOR BIVARIATE SYMMETRY (HOLLANDER)

Subject i	X_i	Y_i
1	X_1	Y_1
2	X_2	Y_2
$\vdots$	$\vdots$	$\vdots$
n	X_n	Y_n

Assumptions

F1. The n bivariate observations $(X_1, Y_1), \ldots, (X_n, Y_n)$ are mutually independent.

F2. Each $(X_i, Y_i), i = 1, \ldots, n$, comes from the same bivariate population with joint distribution function $F(x, y)$.

Hypothesis

The null hypothesis of interest here is that the X and Y variables are exchangeable or, equivalently, that there is no treatment effect (see Comment 67). This hypothesis of exchangeability can be written as

$$H_0 : [F(x, y) = F(y, x), \text{ for all } (x, y)]. \tag{3.92}$$

(Another way to state this exchangeability property is that the pairs (X, Y) and (Y, X) have the same joint bivariate distribution.)

Procedure

For each observation pair $(X_i, Y_i), i = 1, \ldots, n$, let

$$a_i = \min(X_i, Y_i), \quad b_i = \max(X_i, Y_i), \tag{3.93}$$

where, without loss of generality, we take $a_1 \leq a_2 \leq \cdots \leq a_n$. (We may simply relabel the n (X, Y) pairs so that the a's defined by (3.93) are increasing.) Define the n (observed) r values $r_1, r_2, \ldots, r_n$ by

$$r_i = \begin{cases} 1, & \text{if } X_i = a_i < b_i = Y_i \\ 0, & \text{if } X_i = b_i \geq a_i = Y_i. \end{cases} \tag{3.94}$$

That is, r_i is defined to be 1 if $X_i < Y_i$ and 0 if $X_i \geq Y_i$. (Note that the designation that $r_i = 0$ for those cases where $X_i = Y_i$ is purely arbitrary, since such a tied situation makes no contribution to the overall test statistic to be defined by (3.98).)

Define the n^2 values d_{ij}, for $i, j = 1, \ldots, n$, by

$$d_{ij} = \begin{cases} 1, & \text{if } a_j < b_i \leq b_j \text{ and } a_i \leq a_j, \\ 0, & \text{otherwise.} \end{cases} \tag{3.95}$$

For $j = 1, \ldots, n$, set

$$T_j = \sum_{i=1}^{n} s_i d_{ij}, \tag{3.96}$$

where d_{ij} is given by (3.95) and

$$s_i = 2r_i - 1. \tag{3.97}$$

Let A_{obs}, to be read as "A observed," be defined as

$$A_{\text{obs}} = \sum_{j=1}^{n} \frac{T_j^2}{n^2}. \tag{3.98}$$

Now, in addition to our observed r configuration $(r_1, \ldots, r_n)$ defined by (3.94), there are $2^n - 1$ other possible r configurations, corresponding to the cases in which each r_i can be either 0 or 1 and excluding the observed configuration (see Comment 68). For each of these $2^n - 1$ additional r configurations, calculate a corresponding value of A using (3.96) to (3.98). It is important to note that the d's defined by (3.95) remain the same for each of these additional calculations of A_{obs}.

Let

$$A^{(1)} \leq A^{(2)} \leq \cdots \leq A^{(2^n)} \tag{3.99}$$

denote the 2^n ordered values of the A's. (Note that A_{obs} will be one of these ordered A's.) Set

$$m = 2^n - [[2^n \alpha]], \tag{3.100}$$

where $[[2^n \alpha]]$ is the greatest integer less than or equal to $2^n \alpha$. Define M_1 to be the number of ordered values $A^{(1)} \leq \cdots \leq A^{(2^n)}$ that are greater than $A^{(m)}$, and take M_2 to be the number of the $A^{(1)} \leq \cdots \leq A^{(2^n)}$ values that are equal to $A^{(m)}$, where $A^{(m)}$ is determined by (3.99) and (3.100).

To test H_0 (3.92), corresponding to exchangeability of the X and Y variables, versus the general (two-sided) alternative that they are not exchangeable, corresponding to

$$H_1 : [F(x, y) \neq F(y, x) \text{ for at least one } (x, y)], \tag{3.101}$$

at the exact α level of significance,

$$\text{Reject } H_0 \text{ if } A_{\text{obs}} > A^{(m)}; \text{ do not reject } H_0 \text{ if } A_{\text{obs}} < A^{(m)}; \tag{3.102}$$

and

If $A_{\text{obs}} = A^{(m)}$, make a randomized decision to reject H_0

with probability q and to not reject H_0 with probability $1 - q$,

where

$$q = \frac{2^n \alpha - M_1}{M_2}. \tag{3.103}$$

Large-Sample Approximation

Koziol (1979) derived the asymptotic (n tending to infinity) conditional (see Comment 68) null (H_0) distribution of A_{obs} under the additional assumption that the bivariate distribution $F(x, y)$ is continuous. Under this mild regularity condition of continuity, the large sample approximation to the exact α-level conditional test procedure given in (3.102) is

$$\text{Reject } H_0 \text{ if } \frac{A_{\text{obs}}}{n} \geq a_\alpha; \text{ otherwise do not reject,} \quad (3.104)$$

where the constant a_α is the upper αth percentile point for the asymptotic conditional null distribution of A_{obs}/n. Values of a_α are given in Table A.5. The experimenter may also set $a_\alpha = A_{\text{obs}}/n$ for his or her data and then use Table A.5 to approximate the smallest significance level at which the observed data would lead to rejection of H_0 (3.92) with the approximate conditional procedure (3.104).

Ties

No adjustment for ties is necessary. The calculation of A_{obs} (3.98) is well-defined when ties occur. As a result, the associated test procedures in (3.102) and (3.104) handle ties automatically.

Example 3.11: *Inulin Clearance in Kidney Transplants.* The data in Table 3.16 are a subset of the data obtained by Shelp et al. (1970) in a study of renal transplants. Part of their study dealt with living related donor kidneys and pertained to a comparison of clearance capacity of the donor and recipient after the transplant was done. Table 3.16 gives inulin clearance values for seven recipients and their corresponding donors. (We note that Assumption F2 may not be satisfied since the subjects are not homogeneous. They differ in various factors that may be pertinent to clearance, such as the basic disease, age when the transplant was performed, age of the donor, and sex of the donor. In order to illustrate the bivariate symmetry test, we neglect this heterogeneity of subjects.)

From (3.93) and Table 3.16, we find

$$\begin{aligned} a_1 &= 61.4, \quad a_2 = 63.3, \quad a_3 = 63.7, \quad a_4 = 67.1, \quad a_5 = 77.3, \\ & \qquad a_6 = 84.0, \quad a_7 = 88.1 \\ b_1 &= 70.8, \quad b_2 = 89.2, \quad b_3 = 65.8, \quad b_4 = 80.0, \quad b_5 = 87.3, \\ & \qquad b_6 = 85.1, \quad b_7 = 105.0. \end{aligned} \quad (3.105)$$

TABLE 3.16. Inulin Clearance of Living Donors and Recipients of Their Kidneys

Patient†	Inulin Clearance (ml/min)	
	Recipient, X_i	Donor, Y_i
1'	61.4	70.8
2'	63.3	89.2
3'	63.7	65.8
4'	80.0	67.1
5'	77.3	87.3
6'	84.0	85.1
7'	105.0	88.1

Source: W. D. Shelp, F. H. Bach, W. A. Kisken, M. Newton, R. E. Rieselbach, and A. B. Weinstein (1970).
†The primes on the patient numbers indicate that our numbering is different from that in the study. We have renumbered so that the a's defined by (3.93) are in the order $a_1 < a_2 < a_3 < a_4 < a_5 < a_6 < a_7$.

Our observed r configuration is, from (3.94),

$$r_1 = 1, \quad r_2 = 1, \quad r_3 = 1, \quad r_4 = 0, \quad r_5 = 1, \quad r_6 = 1, \quad r_7 = 0. \tag{3.106}$$

We next calculate the $7^2 = 49$ d values, using (3.95). We have

$$\begin{aligned}
&d_{11} = 1, &&d_{12} = 1, &&d_{13} = 0, &&d_{14} = 1, &&d_{15} = 0, &&d_{16} = 0, &&d_{17} = 0,\\
&d_{21} = 0, &&d_{22} = 1, &&d_{23} = 0, &&d_{24} = 0, &&d_{25} = 0, &&d_{26} = 0, &&d_{27} = 1,\\
&d_{31} = 0, &&d_{32} = 0, &&d_{33} = 1, &&d_{34} = 0, &&d_{35} = 0, &&d_{36} = 0, &&d_{37} = 0,\\
&d_{41} = 0, &&d_{42} = 0, &&d_{43} = 0, &&d_{44} = 1, &&d_{45} = 1, &&d_{46} = 0, &&d_{47} = 0,\\
&d_{51} = 0, &&d_{52} = 0, &&d_{53} = 0, &&d_{54} = 0, &&d_{55} = 1, &&d_{56} = 0, &&d_{57} = 0,\\
&d_{61} = 0, &&d_{62} = 0, &&d_{63} = 0, &&d_{64} = 0, &&d_{65} = 0, &&d_{66} = 1, &&d_{67} = 0,\\
&d_{71} = 0, &&d_{72} = 0, &&d_{73} = 0, &&d_{74} = 0, &&d_{75} = 0, &&d_{76} = 0, &&d_{77} = 1.
\end{aligned} \tag{3.107}$$

From (3.97), and (3.106), we have

$$s_1 = 1, \quad s_2 = 1, \quad s_3 = 1, \quad s_4 = -1, \quad s_5 = 1, \quad s_6 = 1, \quad s_7 = -1 \tag{3.108}$$

From (3.96), (3.107), and (3.108) we obtain

$$\begin{aligned}
T_1 &= d_{11}s_1 + d_{21}s_2 + d_{31}s_3 + d_{41}s_4 + d_{51}s_5 + d_{61}s_6 + d_{71}s_7\\
&= 1(1) + 0(1) + 0(1) + 0(-1) + 0(1) + 0(1) + 0(-1) = 1,\\
T_2 &= d_{12}s_1 + d_{22}s_2 + d_{32}s_3 + d_{42}s_4 + d_{52}s_5 + d_{62}s_6 + d_{72}s_7\\
&= 1(1) + 1(1) + 0(1) + 0(-1) + 0(1) + 0(1) + 0(-1) = 2,\\
T_3 &= d_{13}s_1 + d_{23}s_2 + d_{33}s_3 + d_{43}s_4 + d_{53}s_5 + d_{63}s_6 + d_{73}s_7\\
&= 0(1) + 0(1) + 1(1) + 0(-1) + 0(1) + 0(1) + 0(-1) = 1,\\
T_4 &= d_{14}s_1 + d_{24}s_2 + d_{34}s_3 + d_{44}s_4 + d_{54}s_5 + d_{64}s_6 + d_{74}s_7\\
&= 1(1) + 0(1) + 0(1) + 1(-1) + 0(1) + 0(1) + 0(-1) = 0,\\
T_5 &= d_{15}s_1 + d_{25}s_2 + d_{35}s_3 + d_{45}s_4 + d_{55}s_5 + d_{65}s_6 + d_{75}s_7\\
&= 0(1) + 0(1) + 0(1) + 1(-1) + 1(1) + 0(1) + 0(-1) = 0,\\
T_6 &= d_{16}s_1 + d_{26}s_2 + d_{36}s_3 + d_{46}s_4 + d_{56}s_5 + d_{66}s_6 + d_{76}s_7\\
&= 0(1) + 0(1) + 0(1) + 0(-1) + 0(1) + 1(1) + 0(-1) = 1,\\
T_7 &= d_{17}s_1 + d_{27}s_2 + d_{37}s_3 + d_{47}s_4 + d_{57}s_5 + d_{67}s_6 + d_{77}s_7\\
&= 0(1) + 1(1) + 0(1) + 0(-1) + 0(1) + 0(1) + 1(-1) = 0.
\end{aligned} \tag{3.109}$$

Equation (3.98) then yields

$$\begin{aligned}
A_{\text{obs}} &= \frac{T_1^2 + T_2^2 + T_3^2 + T_4^2 + T_5^2 + T_6^2 + T_7^2}{49}\\
&= \frac{1 + 4 + 1 + 0 + 0 + 1 + 0}{49} = \frac{7}{49}.
\end{aligned} \tag{3.110}$$

3.10. A DISTRIBUTION-FREE TEST FOR BIVARIATE SYMMETRY (HOLLANDER)

Now, to apply the exact procedure given by (3.102), we need to obtain the additional $2^7 - 1 = 127$ A values, corresponding to the other 127 possible r configurations. The 128 possible r configurations, including r observed, which is given by (3.106), are displayed in Table 3.17.

TABLE 3.17. The 128 Possible r Configurations and Corresponding Values of 49A

r_1	r_2	r_3	r_4	r_5	r_6	r_7	(49A)
1	1	1	1	1	1	1	(19)
1	1	1	1	1	1	0	(15)
1	1	1	1	1	0	1	(19)
1	1	1	1	1	0	0	(15)
1	1	1	1	0	1	1	(15)
1	1	1	1	0	1	0	(11)
1	1	1	1	0	0	1	(15)
1	1	1	1	0	0	0	(11)
1	1	1	0	1	1	1	(11)
1	1	1	0	1	0	1	(11)
1	1	1	0	1	1	0*	(7)
1	1	1	0	1	0	0	(7)
1	1	1	0	0	1	1	(15)
1	1	1	0	0	0	1	(15)
1	1	1	0	0	1	0	(11)
1	1	1	0	0	0	0	(11)
1	1	0	1	1	1	1	(19)
1	1	0	1	1	1	0	(15)
1	1	0	1	1	0	1	(19)
1	1	0	1	1	0	0	(15)
1	1	0	1	0	1	1	(15)
1	1	0	1	0	1	0	(11)
1	1	0	1	0	0	1	(15)
1	1	0	1	0	0	0	(11)
1	1	0	0	1	1	1	(11)
1	1	0	0	1	0	1	(11)
1	1	0	0	1	1	0	(7)
1	1	0	0	1	0	0	(7)
1	1	0	0	0	1	1	(15)
1	1	0	0	0	0	1	(15)
1	1	0	0	0	1	0	(11)
1	1	0	0	0	0	0	(11)
1	0	1	1	1	1	1	(11)
1	0	1	1	1	1	0	(15)
1	0	1	1	1	0	1	(11)
1	0	1	1	1	0	0	(15)
1	0	1	1	0	1	1	(7)
1	0	1	1	0	1	0	(11)
1	0	1	1	0	0	1	(7)
1	0	1	1	0	0	0	(11)
1	0	1	0	1	1	1	(3)
1	0	1	0	1	0	1	(3)
1	0	1	0	1	1	0	(7)
1	0	1	0	1	0	0	(7)

*Note that (1, 1, 1, 0, 1, 1, 0) was our observed configuration (see (3.106)).

(continued)

TABLE 3.17. The 128 Possible r Configurations and Corresponding Values of 49A (*continued*)

r_1	r_2	r_3	r_4	r_5	r_6	r_7	(49A)
1	0	1	0	0	1	1	(7)
1	0	1	0	0	0	1	(7)
1	0	1	0	0	1	0	(11)
1	0	1	0	0	0	0	(11)
1	0	0	1	1	1	1	(11)
1	0	0	1	1	1	0	(15)
1	0	0	1	1	0	1	(11)
1	0	0	1	1	0	0	(15)
1	0	0	1	0	1	1	(7)
1	0	0	1	0	1	0	(11)
1	0	0	1	0	0	1	(7)
1	0	0	1	0	0	0	(11)
1	0	0	0	1	1	1	(3)
1	0	0	0	1	0	1	(3)
1	0	0	0	1	1	0	(7)
1	0	0	0	1	0	0	(7)
1	0	0	0	0	1	1	(7)
1	0	0	0	0	0	1	(7)
1	0	0	0	0	1	0	(11)
1	0	0	0	0	0	0	(11)
0	1	1	1	1	1	1	(11)
0	1	1	1	1	1	0	(7)
0	1	1	1	1	0	1	(11)
0	1	1	1	1	0	0	(7)
0	1	1	1	0	1	1	(7)
0	1	1	1	0	1	0	(3)
0	1	1	1	0	0	1	(7)
0	1	1	1	0	0	0	(3)
0	1	1	0	1	1	1	(11)
0	1	1	0	1	0	1	(11)
0	1	1	0	1	1	0	(7)
0	1	1	0	1	0	0	(7)
0	1	1	0	0	1	1	(15)
0	1	1	0	0	0	1	(15)
0	1	1	0	0	1	0	(11)
0	1	1	0	0	0	0	(11)
0	1	0	1	1	1	1	(11)
0	1	0	1	1	1	0	(7)
0	1	0	1	1	0	1	(11)
0	1	0	1	1	0	0	(7)
0	1	0	1	0	1	1	(7)
0	1	0	1	0	1	0	(3)
0	1	0	1	0	0	1	(7)
0	1	0	1	0	0	0	(3)
0	1	0	0	1	1	1	(11)
0	1	0	0	1	0	1	(11)
0	1	0	0	1	1	0	(7)
0	1	0	0	1	0	0	(7)
0	1	0	0	0	1	1	(15)
0	1	0	0	0	0	1	(15)

TABLE 3.17. (continued)

r_1	r_2	r_3	r_4	r_5	r_6	r_7	(49A)
0	1	0	0	0	1	0	(11)
0	1	0	0	0	0	0	(11)
0	0	1	1	1	1	1	(11)
0	0	1	1	1	1	0	(15)
0	0	1	1	1	0	1	(11)
0	0	1	1	1	0	0	(15)
0	0	1	1	0	1	1	(7)
0	0	1	1	0	1	0	(11)
0	0	1	1	0	0	1	(7)
0	0	1	1	0	0	0	(11)
0	0	1	0	1	1	1	(11)
0	0	1	0	1	0	1	(11)
0	0	1	0	1	1	0	(15)
0	0	1	0	1	0	0	(15)
0	0	1	0	0	1	1	(15)
0	0	1	0	0	0	1	(15)
0	0	1	0	0	1	0	(19)
0	0	1	0	0	0	0	(19)
0	0	0	1	1	1	1	(11)
0	0	0	1	1	1	0	(15)
0	0	0	1	1	0	1	(11)
0	0	0	1	1	0	0	(15)
0	0	0	1	0	1	1	(7)
0	0	0	1	0	1	0	(11)
0	0	0	1	0	0	1	(7)
0	0	0	1	0	0	0	(11)
0	0	0	0	1	1	1	(11)
0	0	0	0	1	0	1	(11)
0	0	0	0	1	1	0	(15)
0	0	0	0	1	0	0	(15)
0	0	0	0	0	1	1	(15)
0	0	0	0	0	0	1	(15)
0	0	0	0	0	1	0	(19)
0	0	0	0	0	0	0	(19)

The parenthetical values to the right of each r configuration in Table 3.17 are the corresponding values of $49A$. These values are calculated in the same way that we calculated A_{obs} in (3.107) to (3.110). The s's corresponding to (3.108) must be recalculated for each r configuration for use in the T_j equations, but the d's remain the same for each calculation. The ordered A's defined by (3.99) are $A^{(1)} \leq \cdots \leq A^{(128)}$.

We now list the ordered values of $49A$: $49A^{(1)} = \cdots = 49A^{(8)} = 3$, $49A^{(9)} = \cdots = 49A^{(40)} = 7$, $49A^{(41)} = \cdots = 49A^{(88)} = 11$, $49A^{(89)} = \cdots = 49A^{(120)} = 15$, $49A^{(121)} = \cdots = 49A^{(128)} = 19$.

Let us illustrate the $\alpha = \frac{8}{128} = .0625$ test. The value of m (3.100) is

$$m = 2^7 - \left[\left[2^7 \left(\tfrac{8}{128}\right)\right]\right] = 128 - 8 = 120,$$

and thus $A^{(m)} = A^{(120)} = \left(\frac{15}{49}\right)$. We then have

$$M_1 = \text{number of } A \text{ values greater than } \tfrac{15}{49} = 8,$$
$$M_2 = \text{number of } A \text{ values equal to } \tfrac{15}{49} = 32,$$

and

$$q_1 = \left(128\left(\tfrac{8}{128}\right) - 8\right) = 0.$$

Hence, procedure (3.102) reduces to, at the $\alpha = .0625$ level,

$$\text{Reject } H_0 \text{ if } A_{\text{obs}} > \tfrac{15}{49}. \tag{3.111}$$

Since $A_{\text{obs}} = \tfrac{7}{49}$, we do not reject the hypothesis of bivariate symmetry at the .0625 level. Furthermore, since there are 120 configurations (including the one corresponding to A_{obs}) that yield a value greater than or equal to A_{obs}, the lowest level at which we can reject using a nonrandomized test based on A is $\tfrac{120}{128} = .9375$.

For the large-sample approximation, we set $a_\alpha = (A_{\text{obs}}/n) = \{(\tfrac{7}{49})/7\} = .02$. Taking this value of a_α to Table A.5, we find that the approximate P-value based on the large-sample approximation is greater than .95. Thus the large-sample approximation is in close agreement with the exact P-value and the same conclusion (i.e., the data do not support rejecting bivariate symmetry) is reached if one uses the large-sample approximation.

Comments

67. *Motivation.* The hypothesis H_0 (3.92) is a natural one when an experimenter is testing for a treatment effect and finds it convenient (or necessary) to have the same subjects receive the treatment and also act as controls. Since (X_i, Y_i) then represent two observations on the same subject, it is unrealistic to assume that X_i and Y_i are independent. The hypothesis of no treatment effect is precisely H_0. Terms used by various workers to describe H_0 include exchangeability, interchangeability, and bivariate symmetry. (See Hollander (1971).)

68. *Conditional Nature of the Test.* The hypothesis H_0 implies that the r's defined by (3.94) are independent and identically distributed, each r_i assuming values 1 and 0 with probabilities $\tfrac{1}{2}$ and $\tfrac{1}{2}$, respectively. This leads to a conditional distribution P_c that assigns probability $(\tfrac{1}{2})^n$ to each of the A-values associated with each of the possible 2^n r-configurations. (In the foregoing statement we implicitly distinguish between all A-values, although, as we see in Example 3.11, two different r's may yield the same value of A.) The test defined by (3.102) investigates how large A_{obs} is with respect to this conditional distribution. For further information on conditional tests of this nature (which are known as permutation tests) see Hoeffding (1952), Box and Andersen (1955), Lehmann (1959), and Scheffé (1959).

69. *Alternative Computation of the d's.* In computing the d's defined by (3.95), life can be made easier by observing:

(i) $d_{ij} = 0$ for all j, if $a_i = b_i$.
(ii) $d_{ii} = 1$ if $a_i \neq b_i$, and $d_{ii} = 0$ if $a_i = b_i$.
(iii) When $i > j$, if $a_i \neq a_j$, then $d_{ij} = 0$.

70. *Parametric Representation of the Null Hypothesis* (H_0). Consider (3.92) and define

$$A^*(x, y) = P(X \leq x \text{ and } Y \leq y) - P(X \leq y \text{ and } Y \leq x). \tag{3.112}$$

The hypothesis H_0 (3.92) is true if and only if $A^*(x, y) = 0$ for all (x, y). The statistic (A/n) estimates the parameter

$$\Delta(F) = E_F\{A^*(X', Y')\}^2, \qquad (3.113)$$

where (X', Y') is a random member from the underlying bivariate population with distribution F. We may view $A^*(x, y)$ as a measure of the deviation from H_0 at the point (x, y), and $\Delta(F)$ (3.113) as the average value of the square of this deviation.

71. *Consistency: Comparison of A Test and Signed Rank Test.* The A test was designed by Hollander (1971) to detect a broad class of alternatives to the hypothesis of no treatment effect. Thus, although the A test will detect alternatives of the form associated with nonzero ($\theta \neq 0$) treatment effects as discussed for paired replicates data in Section 3.1, it will also be sensitive to differences in dispersion in the (marginal) X and Y populations, as well as to more general deviations from H_0. Of course, a price must be paid for this more general type of protection. Namely, we cannot expect the A test to have power as good as that of, say, the Wilcoxon signed rank test (3.6) when the location model of Section 3.1 is true, since the signed rank test is directed to location changes. On the other hand, there are many alternatives to H_0 for which the signed rank test will have power remaining at α (for any sample size), whereas the A test will have power tending to 1 (as n tends to infinity). In fact, under mild conditions on the nature of the underlying bivariate population F, the A test is consistent when H_0 is false.

Properties

1. *Consistency.* The test defined by (3.102) is consistent against populations for which the parameter $\Delta(F)$ defined by (3.113) is positive. For conditions on F insuring that $\Delta(F)$ will be positive, see Hollander (1971).
2. *Asymptotic Distribution.* See Koziol (1979).

PROBLEMS

110. Cain, Mayer, and Jones (1970) have studied albumin and fibrinogen metabolism by using the carbonate-^{14}C method to measure the synthetic rate of liver-produced plasma proteins before and after a 13-day course of prednisolone. The eight subjects were patients with hepatocellular disease as established by needle biopsy. Part of the study related to changes in the intravascular albumin pool. Table 3.18 is based on a subset of the Cain-Mayer-Jones data.

Find the exact conditional P-value for these data and the test procedure given by (3.102).

TABLE 3.18. Intravascular Albumin Pool Before and After Prednisolone

	Intravascular Albumin Pool (g)	
Patient	Before, X_i	After, Y_i
1	74.4	83.8
2	100.0	97.5
3	82.5	77.4
4	84.3	87.2
5	91.4	116.2
6	92.8	88.2
7	104.2	115.1
8	58.3	50.5

Source: G. D. Cain, G. Mayer, and E. A. Jones (1970).

111. Consider the intravascular albumin data in Table 3.18. Find the approximate P-value for these data and the test procedure in (3.104) based on the large-sample approximation. Compare with the exact conditional P-value for these data as found in Problem 110.
112. Verify directly, or illustrate with a numerical example, remarks (i)–(iii) of Comment 69.
113. Consider the immunoreactive insulin blood level data of Table 3.3. Find the exact conditional P-value for those data and the test procedure given by (3.102).
114. Consider the immunoreactive insulin blood level data of Table 3.3. Find the approximate P-value for those data and the test procedure in (3.104) based on the large-sample approximation. Compare with the exact conditional P-value for these data as found in Problem 113.

3.11. EFFICIENCIES OF PAIRED REPLICATES AND ONE-SAMPLE LOCATION PROCEDURES

Recall the normal theory one-sample t-test based on the statistic

$$V = \frac{\sqrt{n}\,\bar{Z}}{S_z}, \tag{3.114}$$

where $\bar{Z} = \sum_{i=1}^{n} Z_i/n$ and $S_z^2 = \sum_{i=1}^{n}(Z_i - \bar{Z})^2/(n-1)$. The Pitman asymptotic relative efficiency of the one-sample test procedure (one- or two-sided) based on the signed rank statistic T^+ (3.3) with respect to the corresponding normal theory test based on V is

$$e(T^+, V) = 12\,\sigma_F^2 \left\{ \int_{-\infty}^{\infty} f^2(u)\,du \right\}^2, \tag{3.115}$$

where σ_F^2 is the variance of the common (continuous and symmetric) distribution $F(\cdot)$ of $Z_1, \ldots, Z_n$ and $f(\cdot)$ is the probability density function corresponding to $F(\cdot)$. The parameter $\int_{-\infty}^{\infty} f^2(u)\,du$ is the area under the curve associated with $f^2(\cdot)$, the square of the common probability density function.

The expression in equation (3.115) was first obtained by Pitman (1948) in the context of hypothesis testing. Hodges and Lehmann (1963) showed that the same expression, $e(T^+, V)$, also pertains for the asymptotic relative efficiency of the point estimator $\hat{\theta}$ [see equation (3.23)] with respect to $\bar{\theta} = \bar{Z}$. Finally, Lehmann (1963c) established that equation (3.115) also provides the asymptotic relative efficiency of the confidence interval (or bound) for θ derived from T^+ (see Section 3.3) relative to the corresponding confidence interval (or bound) for θ associated with the one-sample t-test based on V (3.114).

Hodges and Lehmann (1956) demonstrated that within the class of continuous and symmetric $F(\cdot)$, $e(T^+, V)$ is always at least .864. Thus, in this class of distributions the most efficiency that can be lost when employing a procedure (test, point estimator, or confidence interval/bound) based on T^+ instead of the corresponding normal theory procedure associated with V (3.114) is about 14%. Even when $F(\cdot)$ is normal (the proper setting for procedures based on V), $e(T^+, V) = .955$ and there is only a minor loss (4.5%) in efficiency from using a T^+-based procedure rather than the optimal procedure based on V. On the other hand, $e(T^+, V)$ exceeds 1 for many populations and it can be infinite (for example, when $F(\cdot)$ is Cauchy). Some values of $e(T^+, V)$ for selected $F(\cdot)$ are:

F :	Normal	Uniform	Logistic	Double Exponential	Cauchy	
$e(T^+, V)$:	.955	1.000	1.097	1.500	∞	(3.116)

The Pitman asymptotic relative efficiency of the one-sample test procedure (one- or two-sided) basd on the sign statistic B (3.39) with respect to the corresponding normal theory test based on V (3.114) is

$$e(B, V) = 4\, \sigma_F^2 f^2(0), \tag{3.117}$$

where σ_F^2 is the variance and $f(\cdot)$ is the probability density function for the common (continuous and symmetric) distribution $F(\cdot)$ of the Z observations.

Pitman (1948) established the general efficiency expression in (3.117) for the hypothesis tests based on B and V, although Cochran (1937) had previously obtained the particular efficiency value of .637 for the case of an underlying (F) normal distribution. Hodges and Lehmann (1963) showed that the expression $e(B, V)$ also holds for the asymptoptic relative efficiency of the point estimator $\tilde{\theta}$ (see equation (3.58)) with respect to $\bar{\theta} = \bar{Z}$, and the results in Lehmann (1963c) lead to the same conclusion for the confidence interval (or bound) for θ based on B (see Section 3.6) relative to the corresponding confidence interval (or bound) for θ associated with the one-sample t-test based on V (3.114).

Hodges and Lehmann (1956) found that within a certain class of populations, $e(B, V)$ is always at least $\frac{1}{3}$ and it can be infinite. Some values of $e(B, V)$ for selected $F(\cdot)$ are:

	Normal	Uniform	Logistic	Double Exponential	Cauchy
F:	Normal	Uniform	Logistic	Double Exponential	Cauchy
$e(B,V)$:	.637	.333	.822	2.000	∞

(3.118)

We note that for the paired replicates problem each Z is actually a difference of two observations. For the efficiency calculation the common F in the parameters $e(T^+, V)$ and $e(B, V)$ is a distribution for a difference of two independent and identically distributed random variables. Since neither all continuous distributions nor all continuous and unimodal distributions can be distributions for such a difference, the lower bounds for $e(T^+, V)$ and $e(B, V)$ for paired replicates data are obtained over smaller classes of distributions than for the one-sample data. In particular, in the paired case, Hollander (1967a) proved that the lower bound of .864 for $e(T^+, V)$ is no longer attainable. Similarly, Puri and Sen (1968) demonstrated that the lower bound of $\frac{1}{3}$ for $e(B, V)$ is not attainable in the paired case.

For the paired replicate data, the values of $e(T^+, V)$ and $e(B, V)$ remain the same as given in expressions (3.116) and (3.118), respectively, for an underlying (F) normal, logistic, double exponential, or Cauchy distribution. However, the uniform distribution cannot be a distribution for a difference of two independent and identically distributed random variables (see Puri and Sen (1968)).

We do not know of any results for the asymptotic efficiencies of the Randles et al. test for distributional symmetry (Section 3.9) or Hollander's bivariate symmetry test (Section 3.10).

CHAPTER 4

The Two-Sample Location Problem

In this chapter the data consist of two random samples, a sample from the control population and an independent sample from the treatment population. On the basis of these samples we wish to investigate the presence of a treatment effect that results in a shift of location. The basic hypothesis is that of no treatment effect; that is, the samples can be thought of as a single sample from one population.

Section 4.1 presents a distribution-free rank sum test for the hypothesis of no treatment effect; Section 4.2, a point estimator associated with the rank sum statistic; and Section 4.3, a related distribution-free confidence interval that emanates from the rank sum test. The basic model for Sections 4.1, 4.2, and 4.3 assumes the populations differ only by a location shift. In Section 4.4 we present a test for location differences that allows the population dispersions to differ. Section 4.5 considers the asymptotic relative efficiencies for translation alternatives of the procedures based on the rank sum statistic with respect to their normal theory counterparts based on sample means.

Data. We obtain $N = m + n$ observations $X_1, \ldots, X_m$ and $Y_1, \ldots, Y_n$.

Assumptions

A1. The observations $X_1, \ldots, X_m$ are a random sample from population 1. That is, the X's are independent and identically distributed. The observations $Y_1, \ldots, Y_n$ are a random sample from population 2. That is, the Y's are independent and identically distributed.

A2. The X's and Y's are mutually independent. Thus, in addition to assumptions of independence within each sample, we also assume independence between the two samples.

A3. Populations 1 and 2 are continuous populations.

4.1. A DISTRIBUTION-FREE RANK SUM TEST (WILCOXON, MANN AND WHITNEY)

Hypothesis

Let F be the distribution function corresponding to population 1 and let G be the distribution function corresponding to population 2.
 The null hypothesis is

$$H_0 : F(t) = G(t), \quad \text{for every } t. \tag{4.1}$$

The null hypothesis asserts that the X variable and the Y variable have the same probability distribution, but the common distribution is not specified.

The alternative hypothesis in a two-sample location problem typically specifies that Y tends to be larger (or smaller) than X. One model that is useful to describe such alternatives is the translation model—also called the location-shift model. The location-shift model is

$$G(t) = F(t - \Delta), \quad \text{for every } t. \tag{4.2}$$

Model (4.2) says population 2 is the same as population 1 except it is shifted by the amount Δ. Another way of writing this is

$$Y \stackrel{d}{=} X + \Delta$$

where the symbol $\stackrel{d}{=}$ means "has the same distribution as." The parameter Δ is called the location shift. It is also known as the treatment effect. If X is a randomly selected value from population 1, the control population, and Y is a randomly selected value from population 2, the treatment population, then Δ is the expected effect due to the treatment. If Δ is positive, it is the expected increase due to the treatment and if Δ is negative, it is the expected decrease due to the treatment. If the mean $E(X)$ of population 1 exists, then letting $E(Y)$ denote the mean of population 2,

$$\Delta = E(Y) - E(X),$$

the difference in population means. In terms of the location-shift model, the null hypothesis H_0 reduces to

$$H_0 : \Delta = 0,$$

the hypothesis that asserts the population means are equal or, equivalently, that the treatment has no effect.

We note that although we find it convenient to use the "treatment" and "control" terminology, many situations will arise in which we want to compare two random samples, neither one of which can be described as a sample from a control population. The procedures of this chapter are applicable even when there are no natural control or treatment designations.

Procedure

To compute the Wilcoxon two-sample rank sum statistic W, order the combined sample of $N = m + n$ X-values and Y-values from least to greatest. Let S_1 denote the rank of $Y_1, \ldots, S_n$ denote the rank of Y_n in this joint ordering. W is the sum of the ranks assigned to the Y-values. That is,

$$W = \sum_{j=1}^{n} S_j. \tag{4.3}$$

a. One-Sided Upper-Tail Test. To test

$$H_0 : \Delta = 0$$

versus

$$H_1 : \Delta > 0$$

at the α level of significance,

$$\text{Reject } H_0 \text{ if } W \geq w_\alpha; \quad \text{otherwise do not reject,} \tag{4.4}$$

where the constant w_α is chosen to make the type I error probability equal to α. Values of w_α are given in Table A.6. Table A.6 is entered with the sum of the ranks corresponding to the smaller sample size, so when naming the samples call the Y-sample the one with the smaller sample size, i.e., $n \leq m$. If $n = m$, either sample can be designated the Y-sample for use of Table A.6.

b. One-Sided Lower-Tail Test. To test

$$H_0 : \Delta = 0$$

versus

$$H_2 : \Delta < 0$$

at the α level of significance,

$$\text{Reject } H_0 \text{ if } W \leq n(m + n + 1) - w_\alpha; \quad \text{otherwise do not reject.} \tag{4.5}$$

c. Two-Sided Test. To test

$$H_0 : \Delta = 0$$

versus

$$H_3 : \Delta \neq 0$$

at the α level of significance,

$$\text{Reject } H_0 \text{ if } W \geq w_{\alpha/2} \text{ or if } W \leq n(m + n + 1) - w_{\alpha/2}; \quad \text{otherwise do not reject.} \tag{4.6}$$

The two-sided procedure given by (4.6) is the two-sided symmetric test with $\alpha/2$ probability in each tail of the distribution.

Large-Sample Approximation

The large-sample approximation is based on the asymptotic normality of W, suitably standardized. We first need to know the mean and variance of W when the null hypothesis is true. When H_0 is true, the mean and variance of W are, respectively,

$$E_0(W) = \frac{n(m + n + 1)}{2} \tag{4.7}$$

$$\text{var}_0(W) = \frac{mn(m + n + 1)}{12}. \tag{4.8}$$

Comment 4 gives direct calculations of $E_0(W)$ and $\text{var}_0(W)$ in the special case where $m = 3$, $n = 2$. Comment 6 gives general derivations.

The standardized version of W is

$$W^* = \frac{W - E_0(W)}{\{\text{var}_0(W)\}^{1/2}} = \frac{W - \{n(m+n+1)/2\}}{\{mn(m+n+1)/12\}^{1/2}}. \quad (4.9)$$

When H_0 is true, W^* has, as $\min(m, n)$ tends to infinity, an asymptotic $N(0, 1)$ distribution.

The normal theory approximation to procedure (4.4) is

$$\text{Reject } H_0 \text{ if } W^* \geq z_\alpha; \quad \text{otherwise do not reject.} \quad (4.10)$$

The normal approximation to procedure (4.5) is

$$\text{Reject } H_0 \text{ if } W^* \leq -z_\alpha; \quad \text{otherwise do not reject.} \quad (4.11)$$

The normal approximation to procedure (4.6) is

$$\text{Reject } H_0 \text{ if } |W^*| \geq z_{\alpha/2}; \quad \text{otherwise do not reject.} \quad (4.12)$$

Ties

If there are ties, give tied observations the average of the ranks for which those observations are competing. After computing W using average ranks, use procedures (4.4), (4.5) or (4.6) and refer the value of W to Table A.6. Now, however, the test is approximate rather than exact. (To get an exact test, even in the tied case, see Comment 5.)

When applying the large-sample approximation, the following modification should be made. When there are ties, the null mean of W is unaffected, but the null variance is reduced to

$$\text{var}_0(W) = \frac{mn}{12}\left[m + n + 1 - \frac{\sum_{j=1}^{g}(t_j - 1)t_j(t_j + 1)}{(m+n)(m+n-1)}\right], \quad (4.13)$$

or, equivalently,

$$\text{var}_0(W) = \frac{mn(N+1)}{12} - \left\{\frac{mn}{12N(N-1)} \cdot \sum_{j=1}^{g}(t_j - 1)t_j(t_j + 1)\right\}. \quad (4.14)$$

In displays (4.13) and (4.14) g denotes the number of tied groups and t_j is the size of tied group j. Furthermore, an untied observation is considered to be a tied "group" of size 1. In particular, if there are no tied observations, $g = N$, $t_j = 1$ for $j = 1, \ldots, N$, and thus each term of the form $(t_j - 1)(t_j)(t_j + 1)$ reduces to 0 and $\text{var}_0(W)$ reduces to $mn(m+n+1)/12$, the null variance of W when there are no ties. Note also that the term in curly braces on the right-hand side of display (4.14) measures the reductions in the null variance due to the presence of ties.

To apply the large-sample approximation when ties are present, compute W using average ranks, and compute

$$W^* = \frac{W - [n(m+n+1)/2]}{\{\text{var}_0(W)\}^{1/2}}$$

where $\text{var}_0(W)$ is given by display (4.13). With this modified value of W^*, approximations (4.10), (4.11) and (4.12) can be applied.

TABLE 4.1. Tritiated Water Diffusion Across Human Chorioamnion

$Pd(10^{-4}$ cm/s)	
At Term	12–26 Weeks Gestational Age
0.80	1.15
0.83	0.88
1.89	0.90
1.04	0.74
1.45	1.21
1.38	
1.91	
1.64	
0.73	
1.46	

Source: S.J. Lloyd, K.D. Garlid, R.C. Reba and A.E. Seeds (1969).

Example 4.1: *Water Transfer in Placental Membrane.* The data in Table 4.1 are a portion of the data obtained by Lloyd et al. (1969). Among other things, these authors investigated whether there is a difference in the transfer of tritiated water (water containing tritium, a radioactive isotope of hydrogen) across the tissue layers in the term human chorioamnion (a placental membrane) and in the human chorioamnion between 3 and 6 months gestation age. The objective measure used was the permeability constant Pd of the human chorioamnion to water. The tissues used for the study were obtained within 5 min of delivery from the placentas of healthy, uncomplicated pregnancies in the following two gestational age categories: (a) between 12 and 26 weeks following termination of pregnancy via abdominal hysterotomy (surgical incision of the uterus) for psychiatric indications; and (b) term, uncomplicated vaginal deliveries. Tissues from ten term pregnancies and five terminated pregnancies were used in the experiment. Table 4.1 gives the average permeability constant (in units of 10^{-4} cm/s) for six measurements on each of the 15 tissues in the study.

In this example, the alternative of interest is greater permeability of the human chorioamnion for the term pregnancy. Thus, if we let X correspond to the Pd values of tissues from term pregnancies and Y to the Pd values of tissues from terminated pregnancies, we perform test (4.5), which is designed to detect the alternative $\Delta < 0$.

For purpose of illustration we choose α to be .082. From Table A.6 we find $w_{.082} = 52$.

Now we list the combined sample in increasing order to facilitate the joint ranking. The ranks are given in parentheses

X	Y	X	X	Y	Y	X	Y
.73	.74	.80	.83	.88	.90	1.04	1.15
(1)	(2)	(3)	(4)	(5)	(6)	(7)	(8)

Y	X	X	X	X	X	X
1.21	1.38	1.45	1.46	1.64	1.89	1.91
(9)	(10)	(11)	(12)	(13)	(14)	(15)

We see that the Y-ranks are 2, 5, 6, 8, 9 and thus

$$W = 2 + 5 + 6 + 8 + 9 = 30.$$

Since
$$n(m + n + 1) - w_\alpha = 5(10 + 5 + 1) - 52 = 28,$$

procedure (4.5) reduces to

$$\text{Reject } H_0 \text{ if } W \leq 28; \quad \text{otherwise do not reject.}$$

Since $W = 30$, we accept H_0 at the $\alpha = .082$ level.

Recall that the one-sided P-value is the lowest significance level at which we can reject H_0 in favor of $\Delta < 0$ with the observed test statistic value $W = 30$. To find this P-value we make use of equation (4.20), Comment 8. With $n = 5$, $m = 10$, and $x = 30$, equation (4.20) yields

$$P(W \leq 30) = P(W \geq 50) = .127,$$

where the P-value .127 is obtained by entering Table A.6 at $n = 5$, $m = 10$, $x = 50$.

For the large-sample approximation we find, from equation (4.7),

$$W^* = \frac{30 - \{5(16)/2\}}{\{5(10)(16)/12\}^{1/2}} = -1.225.$$

Thus the smallest significance level at which we can reject H_0 with the lower-tail one-sided test using the normal approximation is .11. This is the one-sided P-value. The two-sided P-value is .22.

Both the exact test and the large-sample approximation indicate that there is not sufficiently strong evidence to support the hypothesis that human chorioamnion is more permeable to water transfer at term than at 12 to 26 weeks gestational age.

The exact P-value and approximate P-value based on the large-sample approximation can also be found using StatXact's WI/EX command. The StatXact output for the water transfer data of Table 4.1 is

```
Summary of Exact distribution of WILCOXON RANK SUM statistic:

             Min    Max   Mean   Std-dev   Observed   Standardized
            15.00  65.00  40.00   8.165     30.00        -1.225

Mann-Whitney Statistic =15.00

      Asymptotic Inference:
      One-sided P-value:  Pr{Test Statistic .GE. Observed}              =.1103
      Two-sided P-value:  2*One-sided                                    =.2207

      Exact Inference:
      One-sided P-value:  Pr{Test Statistic .GE. Observed}              =.1272
      Point probability:  Pr{Test Statistic .EQ. Observed}              =.0240
      Two-sided P-value:  Pr{|Test Statistic-Mean|.GE.|Observed-Mean|}  =.2544
      Two-sided P-value:  2*One-sided                                    =.2544
```

Under the heading "Exact Inference" you see that StatXact gives the one-sided P-value as .1272, agreeing with what we found in Table A.6. You will note that StatXact also gives the two-sided P-value based on the large-sample approximations as .2207, agreeing with what we found earlier when we referred W^* to Table A.1. StatXact also gives the value of the Mann-Whitney statistic, which is linearly related to W. See Comment 7.

TABLE 4.2. Alcohol Intake for 1 Year (cl of Pure Alcohol)

Control		SST	
1,042	(13)	874	(9)
1,617	(23)	389	(2)
1,180	(18)	612	(4)
973	(12)	798	(7)
1,552	(22)	1,152	(17)
1,251	(19)	893	(10)
1,151	(16)	541	(3)
1,511	(21)	741	(6)
728	(5)	1,064	(14)
1,079	(15)	862	(8)
951	(11)	213	(1)
1,319	(20)		

Source: L. Eriksen, S. Björnstad, and K. G. Götestam (1986).

Example 4.2: *Alcohol Intakes.* Eriksen, Björnstad, and Götestam (1986) studied a social skills training program for alcoholics. Twenty-four "alcohol-dependent" male inpatients at an alcohol treatment center were randomly assigned to two groups. The control group patients were given a traditional treatment program. The treatment group patients were given the traditional treatment program plus a class in social skills training (SST). After being discharged from the program, each patient reported—in 2-week intervals—the quantity of alcohol consumed, the number of days prior to his first drink, the number of sober days, the days worked, the times admitted to an institution, and the nights slept at home. Reports were verified by other sources (wives or family members). (Such data can be unreliable!) One patient in the SST group, discovered to be an opiate addict, disappeared after discharge and submitted no reports. The remaining 23 patients reported faithfully for a year. The results for alcohol intake are given in Table 4.2. The ranks in the joint ranking of the 23 observations are given in parentheses in Table 4.2.

To test H_0 versus the alternative that the SST group tends to have lower alcohol intakes, we need to test $H_0 : \Delta = 0$ versus $H_2 : \Delta < 0$. Suppose, for example, we choose $\alpha = .05$. Then $z_{.05} = 1.645$ and the normal approximation given by display (4.11) is

Reject H_0 if $W^* \leq -1.645$; otherwise do not reject.

From Table 4.2, we find the sum of the SST ranks is

$$W = 9 + 2 + 4 + 7 + 17 + 10 + 3 + 6 + 14 + 8 + 1 = 81.$$

Then from equation (4.9) we obtain

$$W^* = \frac{81 - \frac{11(12 + 11 + 1)}{2}}{\left\{\frac{12(11)(12 + 11 + 1)}{12}\right\}^{1/2}} = \frac{81 - 132}{\{264\}^{1/2}} = -3.14.$$

Since $W^* = -3.14$ is less than -1.645, we reject H_0 and conclude that there is evidence that the SST class in combination with the traditional treatment program tends to lower alcohol intake in alcoholics. In fact, the evidence is quite strong. To find the P-value we refer $W^* = -3.14$ to Table A.1. The P-value is the area under the standard normal curve to the left of -3.14, which is equal to the area under the standard normal curve to the right of 3.14. From Table A.1 this area is found to be .0008.

Comments

1. *Motivation for the Test.* When Δ is greater than 0, the Y-values will tend to be larger than the X-values, and thus the Y-ranks will tend to be larger than the X-ranks. Hence the value of W will tend to be large. This suggests rejecting H_0 in favor of $\Delta > 0$ for large values of W and motivates procedure (4.4). An analogous motivation leads to procedure (4.5).

The test based on W was introduced by Wilcoxon in 1945. An equivalent test based on the number of X before Y occurrences in the jointly ordered sample (see Comment 7) was proposed by Mann and Whitney (1947). Kruskal (1957) gives a detailed history of the Wilcoxon statistic dating back to 1914.

2. *Testing Δ is Equal to Some Specified Nonzero Value.* Procedures (4.4), (4.5), and (4.6) and the corresponding large-sample approximations given by procedures (4.10), (4.11), and (4.12) are for testing if Δ is equal to zero. To test $\Delta = \Delta_0$, where Δ_0 is some specified nonzero number, subtract Δ_0 from each Y-value to form a pseudosample, namely, $Y_1' = Y_1 - \Delta_0$, $Y_2' = Y_2 - \Delta_0, \ldots, Y_n' = Y_n - \Delta_0$. Then compute W as the sum of the Y'-ranks in the joint ranking of the m X-values and the n Y'-values. Then procedures (4.4), (4.5), and (4.6), and their corresponding large-sample approximations given by displays (4.10), (4.11), and (4.12), can be applied as described earlier.

3. *Derivation of the Distribution of W under H_0 (No-ties Case).* Assume that the underlying distribution under H_0 is continuous so that ties have probability zero of occurring. Then under H_0, all $\binom{N}{n}$ possible assignments for the Y-ranks are equally likely, each having probability $1/\binom{N}{n}$. For example, in the case $m = 3$, $n = 2$, the $\binom{5}{2} = 10$ possible outcomes for the ranks attained by the two Y-observations and the corresponding values of W are given in the following table.

Y-ranks	Probability	W
1, 2	$\frac{1}{10}$	3
1, 3	$\frac{1}{10}$	4
1, 4	$\frac{1}{10}$	5
1, 5	$\frac{1}{10}$	6
2, 3	$\frac{1}{10}$	5
2, 4	$\frac{1}{10}$	6
2, 5	$\frac{1}{10}$	7
3, 4	$\frac{1}{10}$	7
3, 5	$\frac{1}{10}$	8
4, 5	$\frac{1}{10}$	9

Thus, for example, under H_0, the probability is $\frac{2}{10}$ that W is equal to 5, because $W = 5$ when either Y-rank configuration $\{1, 4\}$ or Y-rank configuration $\{2, 3\}$ occurs, each has a $\frac{1}{10}$ chance of occurring (and, of course, they cannot both occur simultaneously). Simplifying, we obtain for the null distribution,

Possible Value of W	Probability of Value
3	.1
4	.1
5	.2
6	.2
7	.2
8	.1
9	.1

Thus, for example, under H_0, the probability that W is greater than or equal to 7 is

$$P_0(W \geq 7) = P_0(W = 7) + P_0(W = 8) + P_0(W = 9)$$
$$= .2 + .1 + .1 = .4.$$

Note that this agrees with the tail probability entry corresponding to the value $W = 7$ in Table A.6 with $m = 3$, $n = 2$.

Observe that we have derived the null distribution of W without specifying the common underlying continuous distribution of the two populations. This is why the procedures based on W are called distribution-free procedures. From the null distribution of W we can determine the critical values w_α and control the probability α of falsely rejecting H_0 when H_0 is true, and this error probability does not depend on the common underlying distribution.

4. *Calculation of the Mean and Variance of W under the Null Hypothesis.* In displays (4.7) and (4.8), we presented formulas for the mean and variance of W when the null hypothesis is true. In this comment, we illustrate a direct calculation of $E_0(W)$ and $\text{var}_0(W)$ in a particular case. We use the null distribution of W obtained in Comment 3. (Later, in Comment 6, we present general derivations of $E_0(W)$ and $\text{var}_0(W)$.) Comment 3 treated the case where $m = 3$, $n = 2$. The null mean of W, $E_0(W)$, is obtained by multiplying each possible value of W with its probability under H_0. Thus

$$E_0(W) = 3(.1) + 4(.1) + 5(.2) + 6(.2) + 7(.2) + 8(.1) + 9(.1) = 6.$$

This is in agreement with what we obtain using equation (4.7), namely,

$$E_0(W) = \frac{n(m + n + 1)}{2} = \frac{2(3 + 2 + 1)}{2} = 6.$$

A check on the expression for $\text{var}_0(W)$ is also easily performed. Recall

$$\text{var}_0(W) = E_0(W^2) - \{E_0(W)\}^2,$$

where $E_0(W^2)$, the second moment of the distribution of W, is again obtained by multiplying possible values (in this case, of W^2) by the corresponding probabilities under H_0. We find

$$E_0(W^2) = 9(.1) + 16(.1) + 25(.2) + 36(.2) + 49(.2) + 64(.1) + 81(.1) = 39.$$

Thus

$$\text{var}_0(W) = 39 - (6)^2 = 39 - 36 = 3.$$

This agrees with what we obtain using equation (4.8) directly, namely,

$$\text{var}_0(W) = \frac{3(2)(3+2+1)}{12} = 3.$$

5. *Exact Conditional Distribution of W with Ties.* To get an exact test in the presence of ties, we consider all $\binom{N}{n}$ possible assignments of the N observations with n observations serving as Y's and m observations serving as X's. For each such assignment, we compute a value of W. Then we see how extreme our observed value of W is in this "built-up" conditional distribution. To keep computations simple, we illustrate for the $n = 2, m = 3$ data

Y	X	X	Y	X
.7	1.2	1.7	1.7	2.8
(1)	(2)	(3.5)	(3.5)	(5)

Note the two tied 1.7 values get the average ranks 3.5. We then find W, the sum of the Y-ranks, is

$$W = 1 + 3.5 = 4.5.$$

To assess the significance of W, we obtain a conditional distribution by considering the $\binom{5}{2} = 10$ possible assignments of the observations

$$.7, \quad 1.2, \quad 1.7, \quad 1.7, \quad 2.8$$

to serve as three X-values and two Y-values, or, equivalently, the 10 possible assignments of the ranks

$$1, \quad 2, \quad 3.5, \quad 3.5, \quad 5$$

to serve as three X-ranks and two Y-ranks. These ten assignments and the corresponding values of W are:

Y-ranks	Probability	W
1, 2	$\frac{1}{10}$	3
1, 3.5	$\frac{1}{10}$	4.5
1, 3.5	$\frac{1}{10}$	4.5
1, 5	$\frac{1}{10}$	6
2, 3.5	$\frac{1}{10}$	5.5
2, 3.5	$\frac{1}{10}$	5.5
2, 5	$\frac{1}{10}$	7
3.5, 3.5	$\frac{1}{10}$	7
3.5, 5	$\frac{1}{10}$	8.5
3.5, 5	$\frac{1}{10}$	8.5

Then, for the tail probabilities, we obtain

$$P_0(W \geq 8.5) = \tfrac{2}{10},$$

$$P_0(W \geq 7) = \tfrac{4}{10},$$

$$P_0(W \geq 6) = \tfrac{5}{10},$$
$$P_0(W \geq 5.5) = \tfrac{7}{10},$$
$$P_0(W \geq 4.5) = \tfrac{9}{10},$$
$$P_0(W \geq 3) = 1.$$

This distribution is called the conditional distribution or the permutation distribution of W. For the particular observed value $W = 4.5$, we see $P_0(W \leq 4.5) = 1 - P_0(W \geq 5.5) = \tfrac{3}{10}$, and such a value would not indicate a deviation from H_0.

StatXact will generate the permutation distribution of W. After entering the data with the case command CA, the commands

SE FI WDIST.PRN⟨enter⟩
WI/EX⟨enter⟩

will generate the permutation distribution and save it in a file named WDIST.PRN. The output will include

One-sided p-value: Pr{Test Statistic .LE. Observed}=.3000

That agrees with what we found in this comment by direct enumeration.

6. *Large-Sample Approximation.* The statistic W/n is the average of the Y-ranks. Since all $\binom{N}{n}$ possible outcomes of the Y-ranks are equally likely under H_0, the null distribution of W/n is the same as the distribution of the sample mean of a random sample of size n drawn without replacement from the finite population $\{1, 2, \ldots, N\}$ of the first N integers. Next we use results (i) and (ii), which are basic results from finite population theory concerning the mean and variance of the distribution of the sample mean of a sample of size n drawn without replacement from a finite population of N elements:

(i) The mean is equal to the mean μ_{pop} of the finite population.
(ii) The variance is equal to

$$\frac{\sigma_{\text{pop}}^2}{n} \times \frac{N-n}{N-1},$$

where σ_{pop}^2 denotes the variance of the finite population and the factor $(N-n)/(N-1)$ is the finite-population correction factor.

For the finite population $\{1, 2, \ldots, N\}$, direct calculations establish

(iii) $\mu_{\text{pop}} = \dfrac{1 + 2 + \cdots + N}{N} = \dfrac{N+1}{2},$

(iv) $\sigma_{\text{pop}}^2 = \dfrac{1}{N}\{1^2 + 2^2 + \cdots + N^2\} - \left(\dfrac{N+1}{2}\right)^2 = \dfrac{(N-1)(N+1)}{12}.$

From (i), (ii), (iii), and (iv) we then obtain

$$E_0\left(\frac{W}{n}\right) = \frac{N+1}{2},$$

$$\text{var}_0\left(\frac{W}{n}\right) = \frac{(N-1)(N+1)}{12n} \times \frac{N-n}{N-1} = \frac{m(N+1)}{12n}$$

and it follows that

$$\text{var}_0(W) = \frac{mn(N+1)}{12}.$$

Asymptotic normality of

$$W^* = \frac{W - \frac{n(N+1)}{2}}{\sqrt{\frac{mn(N+1)}{12}}} = \frac{W - E_0(W)}{\sigma_0(W)}$$

follows from standard theory for the mean of a sample from a finite population (cf. Wilks, 1962, 268).

Asymptotic normality results are also obtainable under general alternatives. See, for example, Lehmann's (1951) extension of Hoeffding's (1948a) U-statistic theorem as stated and applied to the Wilcoxon statistic on pages 92–94 of Randles and Wolfe (1979).

7. *The Mann-Whitney U Statistic.* For testing the hypothesis $H_0 : \Delta = 0$, Mann and Whitney (1947) proposed the statistic

$$U = \sum_{i=1}^{m} \sum_{j=1}^{n} \phi(X_i, Y_j), \tag{4.15}$$

where

$$\phi(X_i, Y_j) = \begin{cases} 1, & \text{if } X_i < Y_j, \\ 0, & \text{otherwise}. \end{cases}$$

The statistic U can be computed as follows. For each pair of values X_i and Y_j, observe which is smaller. If the X_i value is smaller, score one for that pair; if the Y_j value is smaller, score 0 for that pair. Add up the 0s and 1s and call the sum U. Mann and Whitney showed that, in the case of no ties,

$$W = U + \frac{n(n+1)}{2}. \tag{4.16}$$

This implies that tests based on U are equivalent to tests based on W.

To establish equation (4.16), write

$$W = \sum_{j=1}^{n} R(Y_j), \tag{4.17}$$

where $R(Y_j)$ denotes the rank of Y_j in the joint ranking of the $m + n$ X's and Y's. Since the rank of Y_j is equal to the number of X's less than Y_j plus the number of Y's less than Y_j plus 1, write

$$R(Y_j) = \sum_{i=1}^{m} \varphi(X_i, Y_j) + \sum_{j'=1}^{n} \varphi(Y_{j'}, Y_j) + 1. \tag{4.18}$$

Substituting (4.18) into (4.17) yields

$$W = \sum_{j=1}^{n} \sum_{i=1}^{m} \varphi(X_i, Y_j) + \sum_{j=1}^{n} \sum_{j'=1}^{n} \varphi(Y_{j'}, Y_j) + n. \tag{4.19}$$

In equation (4.19), the first term on the right is U. The second term on the right is equal to the number of Y's less than the smallest Y plus the number of Y's less than the second smallest

Y plus ... plus the number of Y's less than the largest Y, that is $0 + 1 + \cdots + n - 1$. Thus

$$W = U + \{1 + 2 + \cdots + n - 1\} + n = U + \frac{n(n + 1)}{2},$$

recalling that the sum of the first n integers is equal to $n(n + 1)/2$.

We illustrate the computation of U for the diffusion data of Table 4.1. The first row below counts the number of X-values less than 1.15, the second row counts the number of X-values less than .88, the third row the number of X-values less than .90, the fourth row the number of X-values less than .74, and the fifth row the number of X-values less than 1.21. The counts are given in braces; each count is simply the number of 1s in the particular row. U is then obtained by summing the counts, which is equivalent to summing all the 1s.

$$\begin{aligned}
U = &\ 1 + 1 + 0 + 1 + 0 + 0 + 0 + 0 + 1 + 0 && [4] \\
&+ 1 + 1 + 0 + 0 + 0 + 0 + 0 + 0 + 1 + 0 && [3] \\
&+ 1 + 1 + 0 + 0 + 0 + 0 + 0 + 0 + 1 + 0 && [3] \\
&+ 0 + 0 + 0 + 0 + 0 + 0 + 0 + 0 + 1 + 0 && [1] \\
&+ 1 + 1 + 0 + 1 + 0 + 0 + 0 + 0 + 1 + 0 && [4] \\
= &\ 15.
\end{aligned}$$

Recall that in Example 4.1 we found $W = 30$. We could alternatively obtain W by computing U as before, and then use equation (4.16) to find

$$W = 15 + \frac{5(6)}{2} = 30.$$

There is a generalization of equation (4.16) that holds when there are ties. If W is computed using average ranks and U is computed via

$$U = \sum_{i=1}^{m} \sum_{j=1}^{n} \varphi^*(X_i, Y_j)$$

where

$$\varphi^*(X_i, Y_j) = \begin{cases} 1, & \text{if } X_i < Y_j \\ \frac{1}{2}, & \text{if } X_i = Y_j \\ 0, & \text{if } X_i > Y_j, \end{cases}$$

then we still have $W = U + n(n + 1)/2$. In other words, when there are ties, instead of scoring 1 if X is less than Y and 0 otherwise, compute U by scoring 1 if X is less than Y, $\frac{1}{2}$ if X equals Y, and 0 if X is greater than Y.

Bohn and Wolfe (1992, 1994) developed statistical procedures based on an analogue of the Mann-Whitney statistic U (4.15) for data obtained under the structure of ranked-set sampling. This form of data collection is a preferable alternative to simple random sampling when the actual sample measurements are costly and/or difficult to obtain, but ranking a small set of items is relatively easy and inexpensive.

Bohn (1996) provides a nice review of the general concept of ranked-set sampling, as well as an overview of the releated nonparametric literature in this area of research.

8. *Symmetry of the Distribution of W under the Null Hypothesis.* When H_0 is true, the distribution of W is symmetric about its mean. This implies that when H_0 is true,

$$P(W \leq x) = P(W \geq n(m + n + 1) - x) \quad (4.20)$$

for $x = n(n + 1)/2, \ldots, n(2m + n + 1)/2$.

Equation (4.20) is useful for converting upper-tail probabilities given in Table A.6 to lower-tail probabilities. We have already used equation (4.20) in Example 4.1.

9. *Some Power Results for the Wilcoxon Test.* We consider the upper-tail α-level test of $H_0 : \Delta = 0$ versus $H_1 : \Delta > 0$ given by procedure (4.4). Suppose that the Y-population is the X-population shifted by an amount Δ, so that model (4.2) holds. Recall that

Power = probability of rejecting H_0, given that H_0 is false.

Then for Δ values "near" the null hypothesis value of 0, the power can be approximated as

$$\text{Power} \doteq \Phi(A_F), \quad (4.21)$$

where $\Phi(A_F)$ is the area under a standard normal density to the left of the point

$$A_F = \left[\left(\frac{12mn}{N+1} \right)^{1/2} \cdot f^*(0) \cdot \Delta \right] - z_\alpha \quad (4.22)$$

where $f^*(0)$ is the density function, evaluated at 0, of the difference between two independent values drawn from the X-population having distribution F (cf. Lehmann (1975, 72, 403)).

When F is normal with standard deviation σ, $f^*(0) = 1/\{2\sigma(\pi)^{1/2}\}$ and A_F reduces to

$$A_{\text{normal}} = \left(\sqrt{\frac{3mn}{(N+1)\pi}} \cdot \frac{\Delta}{\sigma} \right) - z_\alpha. \quad (4.23)$$

Equation (4.23) shows that when F is normal, the approximate power depends on Δ and σ only through their ratio Δ/σ. (This is also true of the exact power.) Thus, for example, the power for the pair $(\Delta = 1, \sigma = 2)$ is the same as the power for the pair $(\Delta = .5, \sigma = 1)$.

Exact power values for the one-sided Wilcoxon test for model (4.2) when F is normal are given in Table B-1 of Milton (1970). Exact power values for the two-sided Wilcoxon test when F is normal are given in Table B-2 of Milton (1970). Milton's tables give power values for all sample sizes $2 \leq n \leq m \leq 7$ that yield nontrivial results. If the sample of size m (or n) is from a normal population with mean μ_1 (or μ_2), $\mu_2 > \mu_1$, and variance σ^2, the location shift alternative is defined in terms of $d = \{(\mu_2 - \mu_1)/\sigma\} = \Delta/\sigma$. Values are given for $d = .2(.2)1.0, 1.5, 2.0, 3.0$. Entries in the tables are ordered according to increasing values of $m + n$, from $2 \leq m + n \leq 14$. In Tables B-1 and B-2, the nominal levels of α are $\alpha = .25, .10, .05, .025, .01, .005$. The α's appearing in the tables are the attainable levels of significance nearest to but less than the nominal α's.

We suppose, for purposes of illustration, that model (4.2) holds with the underlying population F taken to be normal with variance $\sigma^2 = 16$ and the treatment effect $\Delta = 4$. Suppose further that we wish to determine, in a case where $m = 7$ and $n = 7$, the power of the $\alpha = .082$ test that rejects H_0 if $W \geq 64$ and accepts H_0 if $W < 64$. Substituting into equation (4.23) yields

$$A_{\text{normal}} = \left\{\left(\sqrt{\frac{3(7)(7)}{(15)\pi}} \cdot \left(\frac{4}{4}\right) - 1.39\right)\right\} = .376$$

and thus the power is approximately

$$\text{Power} \doteq \Phi(.376) = 1 - .35 = .65.$$

The exact power in this case is found from Table B-1 of Milton (1970) to be .635.

10. *Sample-Size Determination.* The Wilcoxon rank sum test detects a more general class of alternatives than the location-shift alternatives described by model (4.2). The one-sided upper-tail test defined by procedure (4.4) is consistent (that is, has power tending to 1 as m, n tend to infinity) against those (F, G) populations for which $\delta > \frac{1}{2}$, where

$$\delta = P(X < Y). \tag{4.24}$$

The parameter δ defined by equation (4.24) is the probability that an X randomly selected from the distribution F will be less than an independent Y randomly selected from the distribution G. We say more about δ in Comment 18.

Noether (1987) shows how to determine an approximate total sample size N so that the α-level one-sided test given by procedure (4.4) will have approximate power $1 - \beta$ against an alternative value δ, where δ is greater than $\frac{1}{2}$. With $m = cN$, the approximate value of N is

$$N \doteq \frac{(z_\alpha + z_\beta)^2}{12c(1-c)(\delta - \frac{1}{2})^2} \tag{4.25}$$

We illustrate the use of equation (4.25). Suppose we are testing H_0 and we desire to use an upper-tail $\alpha = .05$ test with power $= 1 - \beta$ at least .90 against an alternative where $\delta = P(X < Y) = .7$. (Recall that under H_0, $\delta = .5$.) For simplicity, we take $m = n$ so that $c = .5$. From equation (4.25) with $z_\alpha = z_{.05} = 1.65$, $z_\beta = z_{.10} = 1.28$, and $\delta = .7$, we find

$$N \doteq \frac{(1.65 + 1.28)^2}{12(.5)(.5)(.7 - .5)^2} = 71.54, \qquad m = n = \frac{N}{2} = 35.8.$$

To be conservative take $m = n = 36$ rather than 35.

11. *Robustness of Level.* The significance level of the rank sum test is not preserved if the two populations differ in dispersion or shape. This is also the case for the normal theory two-sample t-test. For the effect of shape differences between the populations on the level of the rank sum test and other two-sample location procedures, see Pratt (1964). For a test of location differences which does not assume equal dispersions, see Fligner and Policello (1981) and Section 4.

The level of the rank sum test is not preserved if dependencies exist among the X's or among the Y's, or if the X's are not independent of the Y's. Recall we have assumed the N X's and Y's are mutually independent. For the effect on the level when this assumption is relaxed so that dependencies are allowed, see Serfling (1968), Hollander, Pledger, and Lin (1974), and Pettitt and Siskind (1981).

There are other situations and designs in which the exact conditional randomization distribution of the Wilcoxon statistic is different than the usual Wilcoxon null distribution and different approaches need to be used to obtain a P-value for comparing two treatments. See, for example, Efron's (1971) biased coin design and other restricted randomization designs considered by Hollander and Peña (1988) and Mehta, Patel and Wei (1988).

12. *Van der Waerden's Test.* Van der Waerden's rank statistic is

$$c = \sum_{j=1}^{n} \Phi^{-1}\left(\frac{S_j}{N+1}\right) \tag{4.26}$$

where, as before, $S_1, \ldots, S_n$ are the Y-ranks and $\Phi^{-1}(t)$ is the tth percentile of the $N(0, 1)$ distribution. That is, $\Phi^{-1}(t)$ is the point such that the area under a $N(0, 1)$ curve to the left of $\Phi^{-1}(t)$ is equal to t. The test of H_0 based on c has competitive efficiency properties versus the test based on W (see Section 4.5) and therefore is a popular competitor of W. To test $H_0 : \Delta = 0$ versus $H_1 : \Delta > 0$, reject H_0 for significantly large values of c. To test $H_0 : \Delta = 0$ versus $H_1 : \Delta < 0$, reject H_0 for significantly small values of c. To test $H_0 : \Delta = 0$ versus $H_3 : \Delta \neq 0$, reject H_0 for significantly large values of $|c|$. Under H_0, the distribution of c is symmetric about 0. Tables of critical values are given by van der Waerden and Nievergelt (1956). Exact P-values can be obtained from StatXact's NO/EX command.

The large-sample approximation is easy to perform. Under H_0, c has mean 0 and variance

$$\mathrm{var}_0(c) = \frac{mn\left[\sum_{i=1}^{N}\{\Phi^{-1}(i/(N+1))\}^2\right]}{N(N-1)}. \tag{4.27}$$

The normal approximation to the distribution of

$$c^* = \frac{c}{\sqrt{\mathrm{var}_0(c)}} \tag{4.28}$$

treats c^* as an approximate $N(0, 1)$ random variable for large m, n.

We illustrate the large-sample test based on c^* using the chorioamnion permeability data of Table 4.1 for which $m = 10$, $n = 5$, and $N = 15$. From Table A.1 or, for example, using Minitab's INVCDF command in cooperation with the NORMAL subcommand, we find the values of $\Phi^{-1}(i/16)$. From the symmetry of the normal distribution, we note $\Phi^{-1}(i/16) = -\Phi^{-1}((16-i)/16)$ for $i = 1, \ldots, 7$, and $\Phi^{-1}(\frac{8}{16}) = \Phi^{-1}(\frac{1}{2}) = 0$. The values of $\Phi^{-1}(i/16)$ are:

i:	1	2	3	4	5	6	7	8
$\Phi^{-1}(i/16)$:	-1.534	-1.150	$-.887$	$-.674$	$-.488$	$-.318$	$-.157$	0

i:	9	10	11	12	13	14	15
$\Phi^{-1}(i/16)$:	.157	.318	.488	.674	.887	1.150	1.534

Recall that the Y-ranks for the data of Table 4.1 are 2, 5, 6, 8, and 9. From equation (4.26) we obtain

$$c = \Phi^{-1}(\tfrac{2}{16}) + \Phi^{-1}(\tfrac{5}{16}) + \Phi^{-1}(\tfrac{6}{16}) + \Phi^{-1}(\tfrac{8}{16}) + \Phi^{-1}(\tfrac{9}{16})$$
$$= -1.150 - .488 - .318 + 0 + .157 = -1.80.$$

From equation (4.27) we obtain

$$\mathrm{var}_0(c) = \frac{10(5)}{15(14)}\{(-1.534)^2 + (-1.150)^2 + (-.887)^2 + (-.674)^2$$
$$+ (-.488)^2 + (-.318)^2 + (-.157)^2 + (.157)^2 + (.318)^2 + (.488)^2$$
$$+ (.674)^2 + (.887)^2 + (1.150)^2 + (1.534)^2\} = 2.51.$$

Then from equation (4.28) we find

$$c^* = \frac{-1.80}{\sqrt{2.51}} = -1.14$$

with a one-sided P-value of .13. StatXact gives the exact P-value, $P = P_0(c \leq -1.8) = .1345$.

Note that the results based on c are very close to those based on W that we found in Example 4.1. The large-sample approximation based on W gave a one-sided P-value of .11 and the exact P-value from W is $P = P_0(W \leq 30) = .127$.

A test that is asymptotically equivalent to the test based on c is the Fisher-Yates-Terry-Hoeffding (cf. Terry (1952), Hoeffding (1951)) test based on

$$c_1 = \sum_{j=1}^{n} E(V^{(S_j)})$$

where $V^{(1)} < V^{(2)} < \cdots < V^{(N)}$ are the order statistics of a sample of size N from a $N(0, 1)$ distribution and $S_1, \ldots, S_n$ are the Y-ranks. Values of $E(V^{(i)})$, $i = 1, \ldots, N$ for $N \leq 100$ and some larger sizes are given in Harter (1961). Exact tables can be found in Terry (1952) and Klotz (1964). Under H_0, the distribution of c_1 is symmetric about 0. The large-sample normal approximation treats

$$c_1^* = \frac{c_1}{\sqrt{\text{var}_0(c_1)}}$$

as a $N(0, 1)$ random variable under H_0, where

$$\text{var}_0(c_1) = \frac{mn \sum_{i=1}^{N} \{E(V^{(i)})\}^2}{N(N-1)}.$$

Since $E(V^{(i)}) \doteq \Phi^{-1}(i/(N+1))$, it can be shown that tests based on c and c_1 are asymptotically equivalent. Both tests are often referred to as the normal scores test. Exact power values for the one-sided and two-sided tests based on c_1 for model (4.2) when F is normal are given in Tables B-3 and B-4 of Milton (1970).

13. *The Location-Shift Function*. Model (4.2) implies that the treatment effect is the same constant value Δ for each possible value of X. In some instances, it will be more appropriate to use a model that allows the treatment effect to be a function $\Delta(X)$ that is allowed to vary with X. For example, the treatment effect may be the expected increase (decrease) in systolic blood pressure due to taking a tranquilizer. In such a case, $\Delta(X)$ would depend on the patient's pretranquilizer blood pressure level X. This suggests the model

$$Y \stackrel{d}{=} X + \Delta(X) \qquad (4.29)$$

where Y is systolic blood pressure after taking the tranquilizer. Model (4.29) was introduced by Lehmann (1975, p. 68). The function $\Delta(X)$ is called the location-shift function. Properties of $\Delta(X)$ were developed by Doksum (1974) and Switzer (1976). Doksum and Sievers (1976) derive simultaneous confidence bands for $\Delta(X)$. Hollander and Korwar (1982) and Wells and Tiwari (1989) extended the results of Doksum (1974) and Switzer (1976) to a nonparametric Bayesian framework. Lu, Wells, and Tiwari (1994) studied the location-shift function when the two samples are censored.

14. *Consistency of the W Test.* Under assumptions A1, A2, and A3, the consistency of the tests based on W depends on the parameter

$$\delta^* = P(X < Y) - \tfrac{1}{2}.$$

The test procedures defined by (4.4), (4.5), and (4.6) are consistent against the alternatives for which $\delta^* >, <,$ and $\neq 0$, respectively.

Properties

1. *Consistency.* For the location-shift model defined by equation (4.2), the tests defined by (4.4), (4.5), and (4.6) are consistent against the alternatives $\Delta >, <,$ and $\neq 0$, respectively. Also see Comment 14.
2. *Asymptotic Normality.* See Lehmann (1975, 365–366).
3. *Efficiency.* See Section 4.5.

PROBLEMS

1. The data in Table 4.3 are a subset of the data obtained by Thomas and Simmons (1969), who investigated the relation of sputum histamine levels to inhaled irritants or allergens. The histamine content was reported in micrograms per gram dry weight of sputum. The subjects for this portion of the study consisted of 22 smokers; 9 of them were allergics and the remaining 13 were asymptomatic (nonallergic) individuals. Care was taken to avoid people who carried out part of their daily work in an atmosphere of noxious gases or other respiratory toxicants. Table 4.3 gives the ordered sputum histamine levels for the 22 individuals in the study.

    Test the hypothesis of equal levels versus the alternative that allergic smokers have higher sputum histamine levels than nonallergic smokers. Use the large-sample approximation.

2. Let W' be the sum of the ranks of the X observations. Verify directly, or illustrate using the chorioamnion permeability data of Table 4.1, the equation $W + W' = (m + n)(m + n + 1)/2$.

3. Suppose a sixth Y observation is added to the five Y's of Table 4.1, and assume that the value $W = 30$ based on the original ten X's and five Y's has already been calculated. How would you calculate the new value of W? Compare the method of reranking (to obtain new Y ranks) with a method based on using the Mann-Whitney statistic U in conjunction with the equation relating U

TABLE 4.3. Sputum Histamine Levels ($\mu g/g$ Dry Weight Sputum)

Allergics	Nonallergics
1651.0	48.1
1112.0	48.0
102.4	45.5
100.0	41.7
67.6	35.4
65.9	34.3
64.7	32.4
39.6	29.1
31.0	27.3
	18.9
	6.6
	5.2
	4.7

Source: H. V. Thomas and E. Simmons (1969).

and W (see Comment 7). Generalize the problem to different m and n values and make the same comparison.

4. Let U' denote the number of (X_i, Y_j) pairs for which $X_i > Y_j$. Assume there are no $X = Y$ ties, and either establish directly, or illustrate with the chorioamnion permeability data of Table 4.1, the relation $U' + U = mn$.

5. Molitor (1989) conducted a study to see if children who watched TV or film violence were significantly more tolerant of "real-life" violent behavior than children who instead watched a nonviolent TV show or film. Half of the 42 children in the study were shown violent TV (an edited version of *The Karate Kid*), whereas the other half watched exciting but nonviolent sports (highlights from the 1984 Summer Olympic Games). Each child was asked to "watch over" two younger children, supposedly in the next room, via a television monitor. Each child was instructed to go and get the research assistant (who stated she had to leave for an emergency) if the younger children "got into trouble." What each child witnessed, while alone, was actually a videotaped sequence depicting two small children first play with blocks and then progressively get more violent. That is, they called each other names, then pushed each other, chased each other, fought, and then supposedly broke a video camera while fighting.

TABLE 4.4. Seconds Spent in Room after Witnessing Violence

Olympics Watchers	*Karate Kid* Watchers
12	37
44	39
34	30
14	7
9	13
19	139
156	45
23	25
13	16
11	146
47	94
26	16
14	23
33	1
15	290
62	169
5	62
8	145
0	36
154	20
146	13

Source: F. T. Molitor (1989).

Toleration of violence was measured by the time (in seconds) each child stayed in the room after he or she witnessed the two younger children's first act of violence. As soon as the subject child left the room, the timing clock was stopped. Each child was subsequently assured that an adult had entered the room where the two children were and that they were not hurt and the video camera was not damaged.

Do the data of Table 4.4 indicate that the children who viewed the violent TV tend to take longer to seek help (were more tolerant) than the children who viewed the nonviolent sports-action TV? Use Wilcoxon's W.

6. Assume that model (4.2) holds and that F is normal with variance 13. We have eight X-observations and eight Y-observations. If we use the $\alpha = .065$ test of $H_0 : \Delta = 0$ versus the alternative $\Delta > 0$, what is the approximate power of this test when the treatment effect is $\Delta = 2$?

7. For testing $H_0 : \Delta = 0$ versus the alternative $\Delta > 0$, you choose to use a type I error probability $\alpha = .10$. Using equal sample sizes, what should the common value of m, n be to have power at least .88 against an alternative where $\delta = .8$?

8. We observe $X_1 = 2.1, X_2 = 1.9, X_3 = 2.6, X_4 = 3.3, Y_1 = 1.9, Y_2 = 2.6, Y_3 = 3.7$. What is the conditional distribution of W obtained by considering all $\binom{7}{3}$ possible choices of three data points to serve as the Y-values? How extreme is the observed value of W in this conditional distribution?

9. Apply van der Waerden's test based on c to the data of Table 4.4. Compare your result with that obtained in Problem 5 using Wilcoxon's W.

10. Apply the test based on W to the plasma glucose data of Table 4.6.

11. Apply the test based on c to the plasma glucose data of Table 4.6. Compare with the results obtained in Problem 10.

12. Show directly, or illustrate via an example, that the maximum value of W is $n(2m + n + 1)/2$. What is the minimum value of W?

13. Suppose you reject H_0 if $W = n(2m + n + 1)/2$ or if $W = n(n + 1)/2$, and you accept H_0 otherwise. What is α for this test?

14. Suppose $m = n = 7$. Compare the exact $\alpha = .049$ test of $H_0 : \Delta = 0$ versus $H_1 : \Delta > 0$ based on W with its corresponding test based on large-sample approximation. What is the exact α value of the test based on the large-sample approximation whose nominal α value is .049?

4.2. AN ESTIMATOR ASSOCIATED WITH WILCOXON'S RANK SUM STATISTIC (HODGES-LEHMANN)

Procedure

To estimate Δ of model (4.2), form the mn differences $Y_j - X_i$, for $i = 1, \ldots, m$ and $j = 1, \ldots, n$. The estimator of Δ associated with the Wilcoxon rank sum statistic (see Comment 16) is

$$\widehat{\Delta} = \text{median}\{(Y_j - X_i), i = 1, \ldots, m; j = 1, \ldots, n\}. \tag{4.30}$$

Let $U^{(1)} \leq \cdots \leq U^{(mn)}$ denote the ordered values of $Y_j - X_i$. Then if mn is odd, say $mn = 2k + 1$, we have $k = (mn - 1)/2$ and

$$\widehat{\Delta} = U^{(k+1)}, \tag{4.31}$$

the value that occupies position $k + 1$ in the list of the ordered $Y - X$ differences. If mn is even, say $mn = 2k$, then $k = mn/2$ and

$$\widehat{\Delta} = \frac{U^{(k)} + U^{(k+1)}}{2}. \tag{4.32}$$

That is, $\widehat{\Delta}$ is the average of the two $Y - X$ differences that occupy positions k and $k + 1$ in the ordered list of the mn differences.

Example 4.3: *Continuation of Example 4.1* To estimate Δ for the chorioamnion permeability data of Table 4.1, we first calculate the 50 $Y_j - X_i$ differences. These values are given in column 1 of Table 4.5. For example, the first two values in column 1 of Table 4.5 are

$$Y_1 - X_1 = 1.15 - .80 = .35, \qquad Y_1 - X_2 = 1.15 - .83 = .32.$$

The remaining 48 values in column 1 are similarly calculated. Column 2 (the "j" column) and column 3 (the "i" column) of Table 4.5 indicate which $Y_j - X_i$ value appears in the corresponding position in column 1. Column 4 of Table 4.5 gives the ordered values of $Y_j - X_i$. These are, in our notation, $U^{(1)}, U^{(2)}, \ldots, U^{(50)}$.

Since $mn = 50$ is even, we use equation (4.32) with $k = \frac{50}{2} = 25$ to obtain

$$\widehat{\Delta} = \frac{U^{(25)} + U^{(26)}}{2} = \frac{-.31 - .30}{2} = -.305.$$

We can also use `Minitab`'s `MANN-WHITNEY` command to compute the Hodges-Lehmann estimator for the chorioamnion permeability data.

```
MTB>SET C1
DATA>1.15 .88 .90 .74 1.21
DATA>END
MTB>SET C2 .80 .83 1.89 1.04 1.45 1.38 1.91 1.64 .73 1.46
DATA>END
MTB>MANN-WHITNEY C1 C2
```

The output is, in `Minitab`'s notation:

```
Mann-Whitney Confidence Interval and Test
C1  N=5  MEDIAN=0.9000
C2  N=10  MEDIAN=1.4150
POINT ESTIMATE FOR ETA1-ETA2 IS -.3050
95.7 PCT C.I. FOR ETA1-ETA2 IS (-0.7602, 0.1499)
TEST OF ETA1=ETA2 VS. ETA1 N.E. ETA2 IS SIGNIFICANT AT 0.2446
CANNOT REJECT AT ALPHA=0.05
```

The point estimate is $-.3050$, as we have already found. The significance level of the test is calculated using a normal approximation with a continuity correction. We do not employ a continuity correction, and thus our two-sided P-value, .22, found in Section 4.1 using W^* is slightly different than `Minitab`'s two-sided .2446. We explain the confidence interval portion of `Minitab`'s output in Section 4.3.

We can also get the mn $Y - X$ differences of Table 4.5 using `Minitab`'s `WDIFF` command. These differences, in addition to their importance in finding $\widehat{\Delta}$ of this section, are also useful for determining the confidence interval for Δ described in Section 4.3. Again using the chorioamnion permeability data, proceed as follows. The form of the command is

```
WDIFF C C, put into C [put indices into C and C]
```

After setting the Y's in `C1` and the X's in `C2`, use

```
MTB>WDIFF C1 C2 C3 C4 C5
```

To get the ordered values of $Y - X$ from `Minitab` use the `SORT` command. Thus to put the $Y - X$ differences from column `C3` in order in Column `C6`, use

```
MTB>SORT C3 C6
```

The `Minitab` output for columns `C3–C6` consists, respectively, of the values we give in columns 1–4 of Table 4.5.

Comments

15. *Motivation for the Hodges-Lehmann Estimator.* The Hodges-Lehmann (1963) estimator $\widehat{\Delta}$ defined by equation (4.30) is associated with the Wilcoxon rank sum test. When $\Delta = 0$,

4.2. HODGES-LEHMANN ESTIMATOR

TABLE 4.5. Unordered and Ordered $Y - X$ Differences for the Chorioamnion Permeability Data

$Y_j - X_i$	j	i	Ordered $Y_j - X_i$ values
0.35	1	1	−1.17
0.32	1	2	−1.15
−0.74	1	3	−1.03
0.11	1	4	−1.01
−0.30	1	5	−1.01
−0.23	1	6	−0.99
−0.76	1	7	−0.90
−0.49	1	8	−0.76
0.42	1	9	−0.76
−0.31	1	10	−0.74
0.08	2	1	−0.74
0.05	2	2	−0.72
−1.01	2	3	−0.71
−0.16	2	4	−0.70
−0.57	2	5	−0.68
−0.50	2	6	−0.64
−1.03	2	7	−0.58
−0.76	2	8	−0.57
0.15	2	9	−0.56
−0.58	2	10	−0.55
0.10	3	1	−0.50
0.07	3	2	−0.49
−0.99	3	3	−0.48
−0.14	3	4	−0.43
−0.55	3	5	−0.31
−0.48	3	6	−0.30
−1.01	3	7	−0.30
−0.74	3	8	−0.25
0.17	3	9	−0.24
−0.56	3	10	−0.23
−0.06	4	1	−0.17
−0.09	4	2	−0.16
−1.15	4	3	−0.14
−0.30	4	4	−0.09
−0.71	4	5	−0.06
−0.64	4	6	0.01
−1.17	4	7	0.05
−0.90	4	8	0.07
0.01	4	9	0.08
−0.72	4	10	0.10
0.41	5	1	0.11
0.38	5	2	0.15
−0.68	5	3	0.17
0.17	5	4	0.17
−0.24	5	5	0.32
−0.17	5	6	0.35
−0.70	5	7	0.38
−0.43	5	8	0.41
0.48	5	9	0.42
−0.25	5	10	0.48

the distribution of the statistic W is symmetric about its mean $n(m + n + 1)/2$ (see Comment 8). A reasonable estimator of Δ is the amount $\widehat{\Delta}$ (say) that should be subtracted from each Y_j so that the value of W, when applied to the aligned samples $X_1,\ldots,X_m, Y_1 - \widehat{\Delta},\ldots,Y_n - \widehat{\Delta}$, is $n(m + n + 1)/2$. Roughly speaking, we estimate Δ by the amount $(\widehat{\Delta})$ that the Y sample should be shifted in order that $X_1,\ldots,X_m$ and $Y_1 - \widehat{\Delta},\ldots,Y_n - \widehat{\Delta}$ appear (when "viewed" by the rank sum statistic W) as two samples from the same population. (Under Assumptions A1–A3, the variables $X_1,\ldots,X_m$ and $Y_1 - \Delta,\ldots,Y_n - \Delta$ can be taken as a single sample of size $N = m + n$ from the underlying population.)

The Hodges-Lehmann method can be applied to large classes of statistics which, for example, include van der Waerden's V. The forms of the resulting estimators are not always as convenient for calculation as in the case of $\widehat{\Delta}$. See Hodges and Lehmann (1983) for an expository article on their method. See McKean and Ryan (1977) for an algorithm for computing $\widehat{\Delta}$.

16. *Sensitivity to Gross Errors.* The estimator $\widehat{\Delta}$ is less sensitive to gross errors than its normal theory analog $\overline{Y} - \overline{X}$, the difference of the sample averages.

17. *Competing Estimators.* Observe that the estimator $\widehat{\Delta}$ cannot be written as a difference of a statistic based on the Y observations only and a second statistic based on the X observations only. The classical estimator $\overline{\Delta} = \overline{Y} - \overline{X}$ can be written as such a difference. When the underlying population is symmetric, Lehmann (1963a) proposed to estimate Δ by

$$\widehat{\widehat{\Delta}} = \widehat{\theta}_2 - \widehat{\theta}_1,$$

where $\widehat{\theta}_1(\widehat{\theta}_2)$ is the estimator (3.10) associated with the signed rank statistic T^+ for estimating the location of the population corresponding to the $X(Y)$ observations. That is,

$$\widehat{\widehat{\Delta}} = \text{median}\left\{\frac{Y_i + Y_j}{2}, 1 \leq i \leq j \leq n\right\} - \text{median}\left\{\frac{X_i + X_j}{2}, 1 \leq i \leq j \leq m\right\}. \quad (4.33)$$

The standard deviation of $\widehat{\widehat{\Delta}}$ can be estimated by

$$\widehat{\sigma}_{\widehat{\widehat{\Delta}}} = \left\{\left(\frac{\theta_{2U} - \theta_{2L}}{2z_{\alpha_2/2}}\right)^2 + \left(\frac{\theta_{1U} - \theta_{1L}}{2z_{\alpha_1/2}}\right)^2\right\}^{1/2} \quad (4.34)$$

where θ_{2U}, θ_{2L} are the upper and lower endpoints of the $100(1 - \alpha_2)\%$ confidence interval obtained from the method of Section 3.3 by replacing the Z's of Section 3.3 by the n Y's of sample 2. Similarly, θ_{1U}, θ_{1L} are the endpoints of the $100(1 - \alpha_1)\%$ confidence interval obtained by the method of Section 3.3 by replacing the Z's of Section 3.3 by the m X's of sample 1.

An approximate confidence interval for Δ, with confidence coefficient $1 - \alpha$, is

$$\Delta_\ell = \widehat{\widehat{\Delta}} - z_{\alpha/2}\widehat{\sigma}_{\widehat{\widehat{\Delta}}}, \qquad \Delta_u = \widehat{\widehat{\Delta}} + z_{\alpha/2}\widehat{\sigma}_{\widehat{\widehat{\Delta}}}. \quad (4.35)$$

Lehmann (1963a), Høyland (1965), and Ramachandramurty (1966a) investigated the properties of $\overline{\Delta}$, $\widehat{\Delta}$ and $\widehat{\widehat{\Delta}}$ for various deviations from the assumptions, including asymmetry and nonlocation differences between the populations.

Other competing estimators of Δ include those in classes initiated by Serfling (1984), Akritas (1986), and Serfling (1992).

18. *The Probability That X Is Less Than Y.* A quantity of interest in the two-sample location problem is the parameter $\delta = P(X_1 < Y_1)$, where X_1 is a random member from the X population, Y_1 is a random member from the Y population, and X_1, Y_1 are independent; that is, δ is the probability that a single Y observation will be larger than a single X observation. Pitman (1948) and Birnbaum (1956) discussed a point estimator for δ given by $\widehat{\delta} = U/mn$, where U is the Mann-Whitney form of the rank sum statistic (see Comment 7). Upper bounds for the variance of U, which are useful when using $\widehat{\delta}$ as a point estimator for δ, were obtained in terms of δ by van Dantzig (1951). See Birnbaum and Klose (1957) for lower bounds. Lehmann (1951) showed that $\widehat{\delta}$ is the uniform minimum variance unbiased estimator of δ over the class of continuous populations. (Also see Blyth (1950).) For the use of the sign statistic in obtaining a point estimator for δ, see Saxena (1969).

Many statisticians, including Wolfe and Hogg (1971), have emphasized the importance of natural parameters such as δ. Consider a medical application in which X represents the response to treatment A and Y is the response to treatment B. Let μ_1, μ_2 be the respective means of the X and Y populations and let σ denote the (assumed) common standard deviation. Then $P(X < Y) = .76$ will usually make more sense to a doctor than the statement $\{(\mu_2 - \mu_1)/\sigma\} = 1$. (If X and Y are normal, and independent, with means μ_1, μ_2 and common standard deviation σ, then $\{(\mu_2 - \mu_1)/\sigma\} = 1$ implies $P(X < Y) = .76$.) Furthermore, we are often more interested in the probability that X is less than Y than, say, the difference between the Y and X means. This is true in a good deal of biological research, where, for example, a large liver is a large liver, but how large it is makes little difference except possibly in comparison with other livers (rather than in comparison with scale measurements on a weighing machine). In situations such as these, the estimator $\widehat{\delta}$ may be more useful than the estimator $\widehat{\Delta}$.

Birnbaum (1956) and Birnbaum and McCarty (1958) considered a distribution-free upper confidence bound for $\delta = P(X_1 < Y_1)$ based on the Mann-Whitney U when the underlying populations are continuous. Owen, Craswell, and Hanson (1964) extended this to discrete populations, and Govindarajulu (1968) sharpened the Birnbaum-McCarty upper bound and provided corresponding two-sided distribution-free confidence intervals for δ. Sen (1967) and Govindarajulu (1968) considered asymptotically distribution-free confidence bounds for δ based on consistent estimators of the variance of the Mann-Whitney U. Saxena (1969) discussed distribution-free confidence bounds for δ based on the sign statistic.

The parameter δ also arises naturally in reliability. Let X be the stress on a component and let Y be the strength of the component. Then $\delta = P(X < Y)$ is the probability that the component functions properly. Johnson (1988) surveys many of the methods referenced in this comment in the context of reliability. His focus is on getting estimators and confidence bounds on system reliability in reliability systems such as "k out of n" systems.

Sen's (1967) asymptotic nonparametric interval for δ is relatively easy to obtain. Sen's interval is based on the asymptotic normality of $\sqrt{n_0}(\widehat{\delta} - \delta)/s$, where $n_0 = mn/(m+n)$. Here s is a consistent estimator of the standard deviation of $\sqrt{n_0}\widehat{\delta}$. Many estimators are available. A particularly convenient one defined by Sen is

$$s^2 = \frac{nS_{10}^2 + mS_{01}^2}{m+n},$$

where

$$S_{10}^2 = \frac{\sum_{i=1}^m (R_i - i)^2 - m\left(\overline{R} - (m+1)/2\right)^2}{(m-1)n^2}$$

and

$$S_{01}^2 = \frac{\sum_{j=1}^{n}(S_j - j)^2 - n\left(\overline{S} - (n+1)/2\right)^2}{(n-1)m^2}.$$

Here R_i is the rank of $X_{(i)}$ in the joint ranking of the X's and Y's, S_j is the rank of $Y_{(j)}$ in the joint ranking of the X's and Y's, $\overline{R} = \sum_{i=1}^{m} R_i/m$, and $\overline{S} = \sum_{j=1}^{n} S_j/n$. Recall $X_{(1)} \leq \cdots \leq X_{(m)}$ are the ordered X-values and $Y_{(1)} \leq \cdots \leq Y_{(n)}$ are the ordered Y-values. The lower and upper endpoints, δ_L and δ_U, respectively, of the asymptotic $1 - \alpha$ confidence interval are

$$\delta_L^S = \widehat{\delta} - z_{\alpha/2}\sqrt{\frac{nS_{10}^2 + mS_{01}^2}{mn}},$$

$$\delta_U^S = \widehat{\delta} + z_{\alpha/2}\sqrt{\frac{nS_{10}^2 + mS_{01}^2}{mn}}. \quad (4.36)$$

A competing interval has been proposed by Halperin, Gilbert, and Lachin (1987). Their $1 - \alpha$ confidence interval is

$$\delta_L^H = \frac{A - B}{C}, \quad \delta_U^H = \frac{A + B}{C}, \quad (4.37)$$

where

$$A = \widehat{\delta} + \frac{\gamma z_{\alpha/2}^2}{2mn},$$

$$B = \left(\frac{(\widehat{\delta}(1 - \widehat{\delta})\gamma z_{\alpha/2}^2 + \gamma^2 z_{\alpha/2}^4/4mn)}{mn}\right)^{1/2},$$

$$C = 1 + \frac{\gamma z_{\alpha/2}^2}{mn},$$

$$\gamma = \widehat{\theta}(m + n - 2) + 1,$$

$$\widehat{\theta} = \frac{\frac{\widehat{K} + 2(n-1)\widehat{\delta}}{m + n - 2} - \widehat{\delta}^2}{\widehat{\delta}(1 - \widehat{\delta})},$$

$$\widehat{K} = \left\{\frac{\sum_{j=1}^{n} r_{1j}(r_{1j} - 1)}{mn}\right\} + \left\{\frac{\sum_{i=1}^{m} s_{1i}(s_{1i} - 1)}{mn}\right\} - (n - 1)$$

where r_{1j} is the number of X-observations that are less than $Y_{(j)}$ and s_{1i} is the number of Y-observations that are less than $X_{(i)}$.

Halperin, Gilbert, and Lachin point out that δ_U^H is less than 1 and δ_L^U is greater than 0. Also $\widehat{\theta} \leq 1$ if $\widehat{\delta}$ is neither 0 nor 1, but for some samples $\widehat{\theta}$ may be less than 0. If that happens, take $\widehat{\theta} = 0$ in the definition of γ. If $\widehat{\delta} = 0$ or 1, take $\widehat{\theta} = 1$.

Halperin, Gilbert, and Lachin did simulations that indicated their method generally yields coverage probabilities closer to the nominal $1 - \alpha$ than does the Sen method.

4.2. HODGES-LEHMANN ESTIMATOR

For the chorioamnion permeability data of Table 4.1, the approximate 95% Sen confidence interval for δ and the approximate 95% Halperin-Gilbert-Lachin confidence interval for δ are as follows. Recall that for these data we have found (see Comment 7) $U = 15$ and thus

$$\hat{\delta} = \frac{15}{10(5)} = .3.$$

For the Sen interval,

$$S_{10}^2 = .171, \qquad S_{01}^2 = .015.$$

From display (4.36) we obtain, with $\alpha = .05$,

$$\delta_L^S = .02, \qquad \delta_U^S = .58.$$

For the Halperin-Gilbert-Lachin interval, with $\alpha = .05$, we find

$$\hat{K} = -.76, \qquad \hat{\theta} = .172, \qquad \gamma = 3.24$$
$$A = .424, \qquad B = .260, \qquad C = 1.25.$$

From display (4.37) we obtain

$$\delta_L^H = .13, \qquad \delta_U^H = .55.$$

Properties

1. *Standard Deviation of* $\hat{\Delta}$. For the asymptotic standard deviation of $\hat{\Delta}$, see Hodges and Lehmann (1963), Lehmann (1963c), and Comment 21.
2. *Asymptotic Normality.* See Hodges and Lehmann (1963) and Ramachandramurty (1966a).
3. *Efficiency.* See Hodges and Lehmann (1963), Høyland (1965), Ramachandramurty (1966a), and Section 4.5.

PROBLEMS

15. Consider the data of Table 4.3. Associate the Y's (X's) with the allergies (nonallergies) and estimate Δ of model (4.2) using $\hat{\Delta}$.
16. Again consider the data of Table 4.3. Estimate Δ using $\hat{\hat{\Delta}}$ and compare your estimate with $\hat{\Delta}$ obtained in Problem 15.
17. Consider the data of Table 4.3. Use display (4.35) to obtain an approximate 95% confidence interval for Δ.
18. Consider the data of Table 4.3. Estimate $\delta = P(X < Y)$ and determine an approximate 90% confidence interval for δ.
19. Consider the data of Table 4.4. Estimate Δ of model (4.2) using $\hat{\Delta}$.
20. Consider the data of Table 4.4. Estimate Δ using $\hat{\hat{\Delta}}$ and compare your estimate with $\hat{\Delta}$ obtained in Problem 19.
21. Consider the data of Table 4.4. Use Comment 17 to obtain an approximate 93% confidence interval for Δ.

22. Consider the data of Table 4.4. Estimate $\delta = P(X < Y)$ and determine (a) an approximate 93% confidence interval for δ using Sen's interval, and (b) an approximate 93% confidence interval for δ using the Halperin-Gilbert-Lachin interval.

23. Change the value 102.4, appearing in Table 4.3, to 1024. How does this affect the estimate of Δ given by $\widehat{\Delta}$? How does this affect the estimate of Δ given by $\overline{\Delta} = \overline{Y} - \overline{X}$?

24. (a) What happens to $\widehat{\Delta}$ when we add a number b to each of the m X values and a number c to each of the n Y values? In particular, what happens when $b = c$?

    (b) What happens to $\widehat{\Delta}$ when we multiply each of the X and Y values by the same number d?

25. Answer parts (a) and (b) of Problem 24 with $\widehat{\Delta}$ replaced by $\widehat{\widehat{\Delta}}$.

26. (a) Do you need to calculate the values of all mn $Y - X$ differences in order to compute the value of $\widehat{\Delta}$? Explain.

    (b) In writing a computer program to calculate $\widehat{\Delta}$, how would you proceed?

4.3. A DISTRIBUTION-FREE CONFIDENCE INTERVAL BASED ON WILCOXON'S RANK SUM TEST (MOSES)

Procedure

For a symmetric two-sided confidence interval for Δ, with confidence coefficient $1 - \alpha$, determine the upper $\alpha/2$ percentile point $w_{\alpha/2}$ of the null distribution of W. These percentile points can be obtained from Table A.6.

Set

$$C_\alpha = \frac{n(2m + n + 1)}{2} + 1 - w_{\alpha/2}. \qquad (4.38)$$

The $1 - \alpha$ confidence interval (Δ_L, Δ_U) is given by

$$\Delta_L = U^{(C_\alpha)}, \qquad \Delta_U = U^{(mn+1-C_\alpha)}. \qquad (4.39)$$

That is, Δ_L is the $Y - X$ difference that occupies position C_α in the list of the mn ordered $Y - X$ differences. The upper endpoint Δ_U is the $Y - X$ difference that occupies position $mn + 1 - C_\alpha$ in the ordered list. With Δ_L and Δ_U given by display (4.39), we have, for all Δ,

$$P_\Delta(\Delta_L < \Delta < \Delta_U) = 1 - \alpha. \qquad (4.40)$$

Large-Sample Approximation

For large m and n, the integer C_α may be approximated by

$$C_\alpha \approx \frac{mn}{2} - z_{\alpha/2}\left\{\frac{mn(m + n + 1)}{12}\right\}^{1/2}. \qquad (4.41)$$

In general the value of the right-hand side of equation (4.41) is not an integer. To be conservative, take C_α to be the largest integer that is less than or equal to the right-hand side of equation (4.41).

Example 4.4: *Continuation of Example 4.1* Consider the chorioamnion permeability data of Table 4.1. We will illustrate how to obtain the 96% confidence interval for Δ. With $1 - \alpha =$

4.3. CONFIDENCE INTERVAL FOR TREATMENT EFFECT (MOSES)

.96, so that $\alpha = .04$, from Table A.6 with $m = 10$ and $n = 5$ we find $w_{\alpha/2} = w_{.02} = 57$. From equation (4.38) we then obtain

$$C_{.04} = \left\{\frac{5(20+5+1)}{2}\right\} + 1 - 57 = 9.$$

From display (4.39) we see that

$$\Delta_L = U^{(9)} = -.76, \qquad \Delta_U = U^{(42)} = .15.$$

The value $-.76$ is found from the list of the ordered $Y - X$ values. It is in column 4, row 9 of Table 4.5. Similarly, the value .15 is in column 4 of that same table on the ninth row up from the bottom.

Applying the large-sample approximation, we find from approximation (4.41)

$$C_{.04} \approx \frac{10(5)}{2} - 2.05 \left\{\frac{10(5)(10+5+1)}{12}\right\}^{1/2} = 8.3.$$

Thus, with the large sample approximation, we set $C_{.04}$ equal to 8 and

$$\Delta_L = U^{(8)} = -.76, \qquad \Delta_U = U^{(43)} = .17.$$

Recall that in Example 4.2 we found the 96% confidence interval using Minitab's MANN-WHITNEY command. Minitab gives $(-.76, .15)$ as the 95.7% confidence interval, agreeing with our findings based on Table 4.5 and equations (4.38) and (4.39).

Comments

19. *Relationship of Confidence Interval to Test.* The $1 - \alpha$ confidence interval given by display (4.39) can be obtained from the two-sided rank sum test as follows. The confidence interval (Δ_L, Δ_U) consists of those Δ_0 values for which the two-sided α-level test of $\Delta = \Delta_0$ (see Comment 2) accepts the hypothesis $\Delta = \Delta_0$. The confidence interval given by display (4.39) was defined by way of a graphical procedure by Lincoln Moses in Chapter 18 of Walker and Lev (1953). See Lehmann (1986, 90) for a general result relating confidence intervals and acceptance regions of tests, and see Lehmann (1963c) for the specific result involving the rank sum test.

20. *Midpoint of Confidence Interval as an Estimator.* The midpoint of the interval (4.39), namely, $\{U^{(C_\alpha)} + U^{(mn+1-C_\alpha)}\}/2$, suggests itself as a reasonable estimator of Δ. (Note that this actually yields a class of estimators depending on the value of α.) In general this midpoint is not the same as $\widehat{\Delta}$. Lehmann (1963b) has also dealt with an asymptotically distribution-free confidence interval for Δ centered at $\widehat{\Delta}$, and Lehmann (1963c) has shown that the asymptotically distribution-free confidence interval has the same asymptotic behavior as the distribution-free confidence interval given by display (4.39).

21. *Estimating the Asymptotic Standard Deviation of $\widehat{\Delta}$.* The quantity $(\Delta_U - \Delta_L)/(2z_{\alpha/2})$, where (Δ_L, Δ_U) is the $1 - \alpha$ confidence interval defined by display (4.39), provides us with a consistent estimator for the asymptotic standard deviation of the point estimator $\widehat{\Delta}$. See Lehmann (1963c).

22. *Confidence Bounds.* To obtain a lower confidence bound for Δ, with confidence coefficient $1 - \alpha$, set

$$C_\alpha^* = \frac{n(2m+n+1)}{2} + 1 - w_\alpha \qquad (4.42)$$

where w_α, the upper α percentile point of the null distribution of W, is obtained from Table A.6. The $100(1-\alpha)\%$ lower confidence bound Δ_L^* for Δ that is associated with the one-sided Wilcoxon rank sum test of $H_0 : \Delta = 0$ against the alternative $H_1 : \Delta > 0$ is given by

$$(\Delta_L^*, \infty) = (U^{(C_\alpha^*)}, \infty) \tag{4.43}$$

where $U^{(1)} \leq \cdots \leq U^{(mn)}$ are the ordered values of $Y_j - X_i$. With Δ_L^* defined by (4.43), we have, for all Δ,

$$P_\Delta(\Delta_L^* < \Delta < \infty) = 1 - \alpha. \tag{4.44}$$

The $100(1-\alpha)\%$ upper confidence bound Δ_U^* for Δ that is associated with the one-sided Wilcoxon rank sum test of $H_0 : \Delta = 0$ against the alternative $H_1 : \Delta < 0$ is given by

$$(-\infty, \Delta_U^*) = (-\infty, U^{(mn+1-C_\alpha^*)}), \tag{4.45}$$

where C_α^* is given by (4.42). With Δ_U^* defined by (4.45), we have, for all Δ,

$$P_\Delta(-\infty < \Delta < \Delta_U^*) = 1 - \alpha. \tag{4.46}$$

For large m, n the integer C_α^* can be approximated by

$$C_\alpha^* \cong \frac{mn}{2} - z_\alpha \left\{ \frac{mn(m+n+1)}{12} \right\}^{1/2}. \tag{4.47}$$

Properties

1. Under Assumptions A1 to A3 and model (4.2), equation (4.40) holds. Hence, we can control the coverage probability to be $1 - \alpha$ without having more specific knowledge about the form of the underlying distribution. Thus (Δ_L, Δ_U) is a distribution-free confidence interval for Δ over a very large class of populations.
2. *Efficiency.* See Lehmann (1963c) and Section 4.5.

PROBLEMS

27. Refer to Problem 15 and obtain a confidence interval for Δ with approximate confidence coefficient .95.
28. For the chorioamnion permeability data of Table 4.1, compute an estimate of Δ utilizing the estimator defined in Comment 20. Compare with the value of $\widehat{\Delta}$ obtained in Example 4.3.
29. Use the results of Example 4.4 to obtain an estimate for the asymptotic standard deviation of $\widehat{\Delta}$ (see Comment 21).
30. Consider the $1 - \alpha$ confidence interval defined by display (4.39). Show that when $\alpha = 2/\binom{N}{n}$,

$$\Delta_L = Y_{(1)} - X_{(m)}, \qquad \Delta_U = Y_{(n)} - X_{(1)},$$

where $X_{(1)} \leq \ldots \leq X_{(m)}$ are the ordered X's and $Y_{(1)} \leq \cdots \leq Y_{(n)}$ are the ordered Y's.
31. Consider the $1 - \alpha$ confidence interval defined by display (4.39). Show that when $\alpha = 4/\binom{N}{n}$,

$$\Delta_L = \text{minimum}\{Y_{(2)} - X_{(m)}, Y_{(1)} - X_{(m-1)}\},$$
$$\Delta_U = \text{maximum}\{Y_{(n)} - X_{(2)}, Y_{(n-1)} - X_{(1)}\}.$$

32. Consider the data of Table 4.3. Obtain an approximate 95% confidence interval for Δ using the large-sample approximation of this section. Compare your result with the approximate 95% confidence interval obtained in Problem 17.

33. Consider the data of Table 4.2 and obtain an approximate 90% confidence interval for Δ using the large-sample approximation of this section.

34. Consider the data of Table 4.4 and obtain an approximate 99% confidence interval for Δ using the large-sample approximation of this section.

35. Consider the case $m = n = 8$ and compare the exact 91.8% confidence interval given by display (4.39) with that obtained by the large-sample approximation.

36. Consider the case $m = n = 10$ and compare the exact 91% confidence interval given by display (4.39) with that obtained by the large-sample approximation.

4.4. A ROBUST RANK TEST FOR THE BEHRENS-FISHER PROBLEM (FLIGNER-POLICELLO)

Hypothesis

In this section we introduce new assumptions. Let $X_1, \ldots, X_m$ and $Y_1, \ldots, Y_n$ be independent random samples from continuous distributions that are symmetric about the population medians θ_x and θ_y, respectively. Note that we do not require the X and Y populations to have the same distributional form, nor do we assume that the variances of the two populations are equal. We are interested in testing $H_0' : \theta_x = \theta_y$ versus $\theta_x < \theta_y$ [or $\theta_x > \theta_y$ or $\theta_x \neq \theta_y$]. This problem of testing $H_0' : \theta_x = \theta_y$ without assuming equal variances is often referred to as the Behrens-Fisher problem.

Procedure

Let

$$P_i = [\text{number of sample } Y \text{ observations less than } X_i], \tag{4.48}$$

for $i = 1, \ldots, m$. Similarly, set

$$Q_j = [\text{number of sample } X \text{ observations less than } Y_j], \tag{4.49}$$

for $j = 1, \ldots, n$. We call P_i and Q_j the *placements* of X_i and Y_j, respectively. Compute

$$\overline{P} = \frac{1}{m} \sum_{i=1}^{m} P_i = \text{average } X \text{ sample placement} \tag{4.50}$$

and

$$\overline{Q} = \frac{1}{n} \sum_{j=1}^{n} Q_j = \text{average } Y \text{ sample placement.} \tag{4.51}$$

Let

$$V_1 = \sum_{i=1}^{m}(P_i - \overline{P})^2 \quad \text{and} \quad V_2 = \sum_{j=1}^{n}(Q_j - \overline{Q})^2 \tag{4.52}$$

and set

$$\widehat{U} = \frac{\sum_{j=1}^{n} Q_j - \sum_{i=1}^{m} P_i}{2(V_1 + V_2 + \overline{P}\,\overline{Q})^{1/2}}. \tag{4.53}$$

a. One-Sided Upper-Tail Test. For a one-sided test of $H'_0 : \theta_x = \theta_y$ versus the one-sided alternative $H'_1 : \theta_y > \theta_x$ at the approximate α level of significance,

$$\text{Reject } H'_0 \text{ if } \widehat{U} \geq u_\alpha; \quad \text{otherwise do not reject,} \tag{4.54}$$

where u_α is a constant satisfying $P_0(\widehat{U} \geq u_\alpha) \approx \alpha$. Selected values of u_α can be obtained from Table A.7. Note that to use this table we must label the sample with the fewer number of observations to be the Y-sample. If $m = n$, the sample can be applied with either sample designated as the Y-sample.

b. One-Sided Lower-Tail Test. For a one-sided test of $H'_0 : \theta_x = \theta_y$ versus the alternative $H'_2 : \theta_y < \theta_x$ at the approximate α level of significance, we

$$\text{Reject } H'_0 \text{ if } \widehat{U} \leq -u_\alpha; \quad \text{otherwise do not reject.} \tag{4.55}$$

c. Two-Sided Test. For a two-sided test of $H'_0 : \theta_x = \theta_y$ versus the alternative $H'_3 : \theta_y \neq \theta_x$ at the approximate α level of significance, we

$$\text{Reject } H'_0 \text{ if } |\widehat{U}| \geq u_{\alpha/2}; \quad \text{otherwise do not reject.} \tag{4.56}$$

Large-Sample Approximation

When $H'_0 : \theta_x = \theta_y$ is true, the statistic $\widehat{U}$ has an asymptotic ($\min(m, n)$ tending to infinity) $N(0, 1)$ distribution. Thus the normal theory approximations to procedures (4.54), (4.55), and (4.56) are obtained by replacing u_α and $u_{\alpha/2}$ by z_α and $z_{\alpha/2}$, respectively.

Ties

If there are ties among the N sample observations, replace the placement formulas (4.48) and (4.49) by

$$P_i = \{[\text{number of sample } Y \text{ observations less than } X_i] + \tfrac{1}{2}[\text{number of sample } Y \text{ observations equal to } X_i]\} \tag{4.57}$$

and

$$Q_j = \{[\text{number of sample } X \text{ observations less than } Y_j] \tag{4.58}$$
$$+ \tfrac{1}{2}[\text{number of sample } X \text{ observations equal to } Y_j]\},$$

respectively.

Example 4.5: *Plasma Glucose in Geese.* March et al. (1976) were interested in, among other things, examining the differences between healthy (normal) and lead-poisoned Canadian geese. In particular, one of the measures examined was plasma glucose (in mg/100 ml

4.4. A ROBUST RANK TEST FOR THE BEHRENS-FISHER PROBLEM (FLIGNER-POLICELLO)

TABLE 4.6. **Plasma Glucose Values**

Healthy Geese	Lead-Poisoned Geese
297	293
340	291
325	289
227	430
277	510
337	353
250	318
290	

Source: G. L. March, T. M. John, B. A. McKeown, L. Sileo and J. C. George (1976).

plasma). The data they obtained for eight healthy and seven lead-poisoned geese are given in Table 4.6.

Labeling the lead-poisoned geese as the Y-sample (since there are fewer lead-poisoned observations), the authors were interested in testing $H_0' : \theta_x = \theta_y$ versus $H_1' : \theta_y > \theta_x$; that is, do lead-poisoned Canadian geese tend to have larger plasma glucose values than healthy geese? Computing the placements for the X and Y observations, we obtain:

$$P_1 = 3, \quad P_2 = 4, \quad P_3 = 4, \quad P_4 = 0, \quad P_5 = 0, \quad P_6 = 4, \quad P_7 = 0, \quad P_8 = 1$$

and

$$Q_1 = 4, \quad Q_2 = 4, \quad Q_3 = 3, \quad Q_4 = 8, \quad Q_5 = 8, \quad Q_6 = 8, \quad Q_7 = 5.$$

Thus,

$$\overline{P} = \frac{3 + 4 + 4 + 0 + 0 + 4 + 0 + 1}{8} = \frac{16}{8} = 2$$

and

$$\overline{Q} = \frac{4 + 4 + 3 + 8 + 8 + 8 + 5}{7} = \frac{40}{7}.$$

Using the values in equation (4.52), we have

$$V_1 = [(3-2)^2 + (4-2)^2 + (4-2)^2 + (0-2)^2 + (0-2)^2$$
$$+ (4-2)^2 + (0-2)^2 + (1-2)^2]$$
$$= 1 + 4 + 4 + 4 + 4 + 4 + 4 + 1 = 26$$

and

$$V_2 = \left[\left(4 - \frac{40}{7}\right)^2 + \left(4 - \frac{40}{7}\right)^2 + \left(3 - \frac{40}{7}\right)^2 + \left(8 - \frac{40}{7}\right)^2\right.$$
$$\left. + \left(8 - \frac{40}{7}\right)^2 + \left(8 - \frac{40}{7}\right)^2 + \left(5 - \frac{40}{7}\right)^2\right]$$

$$= \frac{144}{49} + \frac{144}{49} + \frac{361}{49} + \frac{256}{49} + \frac{256}{49} + \frac{256}{49} + \frac{25}{49}$$

$$= \frac{1442}{49} = \frac{206}{7}.$$

Combining these quantities, we obtain

$$\widehat{U} = \frac{(40 - 16)}{2 \left[26 + \frac{206}{7} + 2 \left(\frac{40}{7} \right) \right]^{1/2}}$$

$$= \frac{12}{[468/7]^{1/2}} = \frac{12}{8.177} = 1.468.$$

From Table A.7 we find $u_{.05} = 1.807$ and $u_{.10} = 1.310$. Thus, for these data, the P-value obtained for testing $H_0' : \theta_x = \theta_y$ versus $H_1' : \theta_y > \theta_x$ is between .05 and .10. Thus we would reject H_0 for $\alpha = .10$ but not for $\alpha = .05$.

Comments

23. *Relationship of $\widehat{U}$ to U*. The statistic $\widehat{U}$ defined by equation (4.53) is of the form

$$\widehat{U} = \frac{n^{1/2} \{ (U/mn) - \frac{1}{2} \}}{\widehat{\sigma}} \tag{4.59}$$

where U is the Mann-Whitney statistic defined by equation (4.15) and

$$\widehat{\sigma}^2 = \frac{\sum_{j=1}^{n} (Q_j - \overline{Q})^2 + \sum_{i=1}^{m} (P_i - \overline{P})^2 + \overline{P} \, \overline{Q}}{m^2 n}.$$

Fligner and Policello (1981) point out that when written in the form (4.53), namely,

$$\widehat{U} = \frac{\sum_{j=1}^{n} Q_j - \sum_{i=1}^{m} P_i}{2 \{ V_1 + V_2 + \overline{P} \, \overline{Q} \}^{1/2}},$$

$\widehat{U}$ resembles Welch's t statistic (Welch 1937, 1947) for the normal theory Behrens-Fisher problem.

24. *Symmetry of the Distribution of $\widehat{U}$*. When H_0: [Identical X and Y distributions] is true, the distribution of $\widehat{U}$ is symmetric about its mean 0, which implies that

$$P_0(\widehat{U} \geq x) = P_0(\widehat{U} \leq -x)$$

for every x. From this it follows that the lower αth percentile for the null H_0 distribution of $\widehat{U}$ is $-u_\alpha$; hence, its use in the test of H_0' versus H_2' defined by (4.55).

25. *Maintaining Levels*. The test procedures in (4.54), (4.55) and (4.56) have *exact* significance levels equal to α for testing H_0: [Identical X and Y distributions]. However, they also maintain *approximate* level α for the more general null hypothesis $H_0' : \theta_x = \theta_y$, without requiring equal variances or identical distributional forms for the two underlying populations.

26. *Consistency of the Test Based on $\widehat{U}$*. Fligner and Policello (1981) consider the consistency of their test based on $\widehat{U}$. To test $H_0 : \theta_x = \theta_y$ versus $\theta_x < \theta_y$, it is necessary to impose conditions on F and G to ensure that whenever $\theta_x = \theta_y$, we have $P(X < Y) = \frac{1}{2}$ and

whenever $\theta_x < \theta_y$, we have $P(X < Y) > \frac{1}{2}$. Fligner and Policello point out that a sufficient condition is that F and G be symmetric.

Properties

1. *Consistency.* Assuming F, G are symmetric, the tests defined by (4.54), (4.55), and (4.56) are consistent against the alternatives for which $\theta_x < \theta_y$, $\theta_x > \theta_y$, and $\theta_x \neq \theta_y$, respectively.
2. *Asymptotic Normality.* See Fligner and Policello (1981).
3. *Efficiency.* See Fligner and Policello (1981) and Section 4.5.

PROBLEMS

37. Apply the test based on $\widehat{U}$ to the data of Table 4.1. Compare your results with those of Example 4.1.
38. Apply the test based on $\widehat{U}$ to the data of Table 4.2. Compare your results with those of Example 4.2.
39. Apply the test based on $\widehat{U}$ to the data of Table 4.3. Compare your results with those of Problem 1.
40. Apply the test based on $\widehat{U}$ to the data of Table 4.4. Compare your results with those of Problem 5.
41. Establish equation (4.59) directly, or illustrate it using an example.
42. Show that $\widehat{U}$ is a rank statistic. That is, show that you can compute $\widehat{U}$ from knowledge of $S_1, \ldots, S_n$ where S_j = rank of Y_j in the joint ranking of the N X's and Y's.

4.5. EFFICIENCIES OF TWO-SAMPLE LOCATION PROCEDURES

Recall the normal theory t-test based on

$$t = \frac{\overline{Y} - \overline{X}}{s_p \sqrt{\frac{m+n}{mn}}}$$

where

$$s_p^2 = \frac{\sum_{i=1}^{m}(X_i - \overline{X})^2 + \sum_{j=1}^{n}(Y_j - \overline{Y})^2}{m + n - 2}$$

is the pooled variance. The Pitman asymptotic relative efficiency of the test based on W versus the test based on t is

$$E(W, t) = 12\sigma_F^2 \left\{ \int f^2 \right\}^2. \tag{4.60}$$

In equation (4.60), σ_F^2 is the variance of the population with distribution F and f is the probablity density corresponding to F. The parameter $\int f^2$ is the area under the curve of f^2.

Equation (4.60) was derived by Pitman (1948) in the testing context and shown by Hodges and Lehmann (1963) to hold also for the asymptotic relative efficiency of the point estimator $\widehat{\Delta}$ (see equation (4.30)) with respect to $\overline{\Delta} = \overline{Y} - \overline{X}$. Lehmann (1963c) showed equation (4.60) also gives the asymptotic relative efficiency of the confidence interval derived from W to that of the confidence interval derived from $\overline{Y} - \overline{X}$.

Hodges and Lehmann (1956) showed that for all populations, $E(W, t)$ is at least .864. Thus the most efficiency one can lose when employing the Wilcoxon test instead of the t-test is about 14%. When F is the normal (the home turf of the t-test), $E(W, t) = .955$. For many populations, $E(W, t)$ exceeds 1, and it can be infinite, as it is in the case when F is Cauchy. Some values of $E(W, t)$ are:

F	Normal	Uniform	Logistic	Double Exponential	Cauchy	Exponential
$E(W, t)$	.955	1.000	1.097	1.500	∞	3.00

Some asymptotic relative efficiency values of van der Waerden's c test relative to the test based on W are

F	Normal	Uniform	Logistic	Double Exponential	Cauchy	Exponential
$E(c, W)$	1.047	∞	.955	.847	.708	∞

These values are also the values of the asymptotic relative efficiency $E(c_1, W)$, where c_1 is the Fisher-Yates-Terry-Hoeffding statistic.

Chernoff and Savage (1958) showed that for all populations, the asymptotic relative efficiency of the Fisher-Yates-Terry-Hoeffding test with respect to the t-test is always greater than or equal to 1. It equals 1 when F is normal.

For model (4.2), the asymptotic relative efficiency of the Fligner-Policello test based on $\widehat{U}$ with respect to W is 1 for all F.

CHAPTER 5

The Two-Sample Dispersion Problem and Other Two-Sample Problems

INTRODUCTION

In this chapter the data once again consist of two independent random samples, one sample from each of two underlying populations. This is the same as the data setting considered in Chapter 4, where we discussed procedures designed for statistical analyses in which primary interest was on possible differences in the locations (medians) of the populations. In this chapter we deal with statistical procedures designed to make inferences about possible difference other than location between two populations.

In Section 5.1 we present a distribution-free rank test for the hypothesis of equal scale parameters when the two underlying populations have a common median. Section 5.2 is devoted to an asymptotically distribution-free test for equality of scale parameters when the assumption of common medians is not justified. In Section 5.3 we consider a distribution-free rank test for the dual hypothesis of equal location and equal scale parameters for the underlying populations. Section 5.4 contains a distribution-free test of the general hypothesis that two populations are identical in all respects. Some aspects of the asymptotic relative efficiencies of the procedures in this chapter with respect to their normal theory counterparts are discussed in Section 5.5.

Data. We obtain $N = m + n$ observations $X_1, \ldots, X_m$ and $Y_1, \ldots, Y_n$.

Assumptions

A1. The observations $X_1, \ldots, X_m$ are a random sample from a continuous population 1. That is, the X's are mutually independent and identically distributed. The observations $Y_1, \ldots, Y_n$ are a random sample from a continuous population 2, so that the Y's are also mutually independent and identically distributed.

A2. The X's and Y's are mutually independent. Thus, in addition to assumptions of independence within each sample, we also assume independence between the two samples.

5.1. A DISTRIBUTION-FREE RANK TEST FOR DISPERSION—MEDIANS EQUAL (ANSARI–BRADLEY)

Hypothesis

Let F and G be the distribution functions corresponding to populations 1 and 2, respectively. The null hypothesis of interest here is that the X and Y variables have the same probability distribution but that their common distribution is not specified. Formally stated, this null hypothesis is

$$H_0 : [F(t) = G(t), \text{ for every } t]. \tag{5.1}$$

The typical alternative hypothesis in a two-sample dispersion problem specifies that the Y population has greater (or less) variability associated with it than does the X population. One model that is often used to describe such alternatives is the location-scale parameter model. In our two-sample setting, this location-scale parameter model corresponds to taking

$$F(t) = H\left(\frac{t - \theta_1}{\eta_1}\right) \quad \text{and} \quad G(t) = H\left(\frac{t - \theta_2}{\eta_2}\right), \quad -\infty < t < \infty, \tag{5.2}$$

where $H(u)$ is the distribution function for a continuous distribution with median 0, so that $F(\theta_1) = G(\theta_2) = \frac{1}{2}$. Thus, θ_1 and θ_2 are the population medians for the X and Y distributions, respectively. Moreover, η_1 and η_2 are the scale parameters associated with the X and Y distributions, respectively. Model (5.2) states that the Y population has the same general form as the X population, but they could have different medians and scale parameters. Another way to express this is to write

$$\frac{X - \theta_1}{\eta_1} \stackrel{d}{=} \frac{Y - \theta_2}{\eta_2}, \tag{5.3}$$

where the symbol $\stackrel{d}{=}$ means "has the same distribution as."

This two-sample location-scale problem will be further discussed in this most general context in Sections 5.2 and 5.3. In this section, however, we impose the further restriction that $\theta_1 = \theta_2$; that is, we also assume

A3. The median (θ_1) of the X population is equal to the median (θ_2) of the Y population.

Under this additional assumption, A3, the equal-in-distribution statement in (5.3) simplifies to

$$\frac{X - \theta}{\eta_1} \stackrel{d}{=} \frac{Y - \theta}{\eta_2}, \tag{5.4}$$

where θ is the common median and the only possible difference between the X and Y populations is in their respective scale parameters, as illustrated in Figure 5.1. (If the medians θ_1 and θ_2 of the X and Y populations are not necessarily equal but are known, the shifted variables $X_1 - \theta_1, \ldots, X_m - \theta_1$ and $Y_1 - \theta_2, \ldots, Y_n - \theta_2$ will satisfy Assumptions A1, A2, and A3. In such a situation, the procedures of this section can be applied to the shifted $(X - \theta_1)$ and $(Y - \theta_2)$ sample observations. For more about this known medians setting, see Comment 1.)

Under Assumptions A1–A3, the parameter of interest in this section is the ratio of the scale parameters, $\gamma = (\eta_1/\eta_2)$. (See Comment 3.) If the variance of population 1, Var(X), exists

5.1. A DISTRIBUTION-FREE RANK TEST FOR DISPERSION—MEDIANS EQUAL

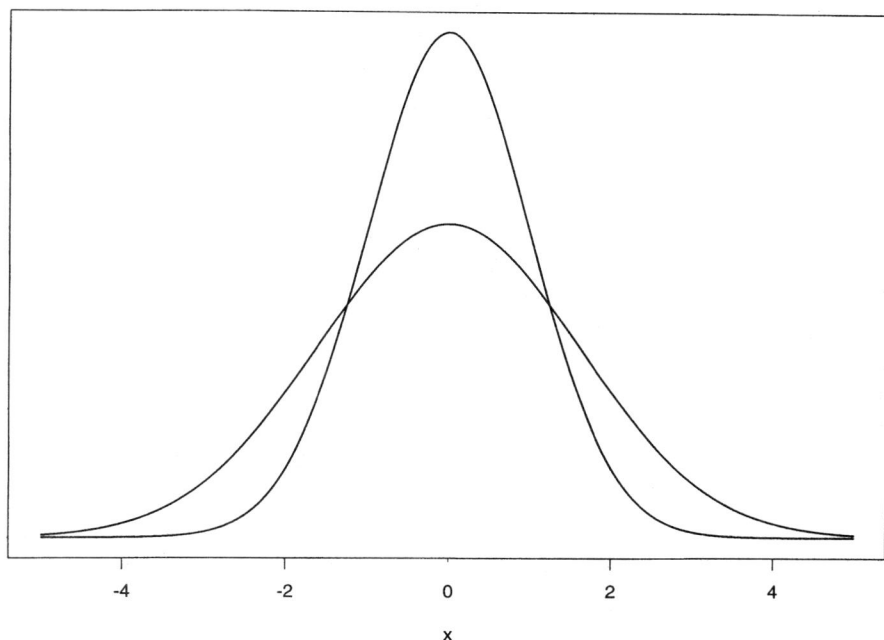

FIGURE 5.1. Probability distributions with the same general form and equal medians but different scale parameters.

and equation (5.4) is satisfied, then the variance of population 2, Var(Y), also exists and

$$\gamma^2 = \left[\frac{\text{Var}(X)}{\text{Var}(Y)}\right], \tag{5.5}$$

the ratio of population variances (also see Comment 7). In terms of this location-scale parameter model with equal location parameters, as given in (5.4), the null hypothesis H_0 (5.1) reduces to $H_0 : \gamma^2 = 1$, corresponding to the assertion that the population scale parameters are equal.

Procedure

To compute the Ansari-Bradley two-sample scale statistic C, order the combined sample of $N = (m + n)$ X-values and Y-values from least to greatest. Assign the score 1 to both the smallest and largest observations in this combined sample, assign the score 2 to the second smallest and second largest, and continue in this manner. If N is an even integer, the array of assigned scores is $1, 2, 3, \ldots, N/2, N/2, \ldots, 3, 2, 1$. If N is an odd integer, the array of assigned scores is $1, 2, 3, \ldots, (N-1)/2, (N+1)/2, (N-1)/2, \ldots, 3, 2, 1$. Let R_j denote the score asigned in this manner to Y_j, for $j = 1, \ldots, n$, and set

$$C = \sum_{j=1}^{n} R_j. \tag{5.6}$$

Thus the statistic C is the sum of the scores assigned via this scheme to the Y observations.

a. One-Sided Upper-Tail Test. To test

$$H_0 : \gamma^2 = 1$$

versus

$$H_1 : \gamma^2 > 1,$$

at the α level of significance,

$$\text{Reject } H_0 \text{ if } C \geq c_\alpha; \quad \text{otherwise do not reject,} \tag{5.7}$$

where the constant c_α is chosen to make the type I error probability equal to α. Values of c_α are given in Table A.8. We enter Table A.8 with the sum of the scores assigned to the sample with the fewer number of observations. Thus, in view of the way C is defined in (5.6), we need to label the sample with the fewer number of observations to be the Y sample (i.e., take $n \leq m$).

b. One-Sided Lower-Tail Test. To test

$$H_0 : \gamma^2 = 1$$

versus

$$H_2 : \gamma^2 < 1,$$

at the α level of significance,

$$\text{Reject } H_0 \text{ if } C \leq [c_{1-\alpha} - 1]; \quad \text{otherwise do not reject,} \tag{5.8}$$

where, as with the upper-tail test in (5.4), the appropriate value of $c_{1-\alpha}$ is obtained from Table A.8. (See Comment 10.)

c. Two-Sided Test. To test

$$H_0 : \gamma^2 = 1$$

versus

$$H_3 : \gamma^2 \neq 1,$$

at the α level of significance,

$$\text{Reject } H_0 \text{ if } C \geq c_{\alpha_1} \text{ or } C \leq [c_{1-\alpha_2} - 1]; \quad \text{otherwise do not reject,} \tag{5.9}$$

where $\alpha_1 + \alpha_2 = \alpha$ and the appropriate values of c_{α_1} and $c_{1-\alpha_2}$ are obtained from Table A.8. We note that the null distribution of C is symmetric when $N = (m + n)$ is an even number (see Comment 5). In such a case, it is most natural to place an equal amount of probability in each tail of the null distribution of C, corresponding to setting $\alpha_1 = \alpha_2 = \alpha/2$. Thus, when N is even, the two-sided symmetric version of procedure (5.9) uses the critical values $c_{\alpha/2}$ and $[c_{1-(\alpha/2)} - 1]$ from Table A.8.

Large-Sample Approximation

The large-sample approximation is based on the asymptotic normality of C, suitably standardized. For this purpose we first need to know the expected value and variance of C when the null hypothesis is true. Since the set of scores being assigned to the jointly ranked sample X and Y observations (see Procedure) depends on whether N is an even or odd integer, it is not surprising that the form of the mean and variance for C also depends on whether N is even or odd. When H_0 is true and $N = m + n$ is an even number, the expected value and variance of C are

$$E_0(C) = \frac{n(N + 2)}{4} \tag{5.10}$$

and

$$\text{Var}_0(C) = \frac{mn(N + 2)(N - 2)}{48(N - 1)}, \tag{5.11}$$

respectively. When N is an odd integer, the null expected value and variance of C are

$$E_0(C) = \frac{n(N + 1)^2}{4N} \tag{5.12}$$

and

$$\text{var}_0(C) = \frac{mn(N + 1)(3 + N^2)}{48N^2}, \tag{5.13}$$

respectively. These expressions for $E_0(C)$ and $\text{var}_0(C)$ are verified by direct calculations in Comment 8 for the special cases of $m = n = 2$ (where $N = 4$ is even) and $m = 3, n = 2$ (where $N = 5$ is odd). General derivations of the null expected value and variance expressions in equations (5.10)–(5.13) are presented in Comment 9.

For general N (even or odd), the standardized version of C is given by

$$C^* = \frac{C - E_0(C)}{\{\text{var}_0(C)\}^{1/2}}, \tag{5.14}$$

where $E_0(C)$ and $\text{var}_0(C)$ correspond to expressions (5.10) and (5.11), respectively, if N is even or to expressions (5.12) and (5.13), respectively, if N is odd. In either case, when H_0 is true, C^* has, as $\min(m, n)$ tends to infinity, an asymptotic $N(0, 1)$ distribution. (See Comment 9 for indications of the proof.) The normal theory approximation for procedure (5.7) is

$$\text{Reject } H_0 \text{ if } C^* \geq z_\alpha; \quad \text{otherwise do not reject}, \tag{5.15}$$

the normal theory approximation for procedure (5.8) is

$$\text{Reject } H_0 \text{ if } C^* \leq -z_\alpha; \quad \text{otherwise do not reject}, \tag{5.16}$$

and the normal theory approximation for procedure (5.9) is

$$\text{Reject } H_0 \text{ if } |C^*| \geq z_{\alpha/2}; \quad \text{otherwise do not reject}. \tag{5.17}$$

Ties

If there are ties among the X and/or Y observations, assign each of the observations in a tied group the average of the integer scores that are associated with the tied group. After computing

C with these average scores for tied observations, use procedure (5.7), (5.8), or (5.9) and refer the value of C to Table A.8. Note, however, that this test associated with tied X and/or Y observations is only approximately, and not exactly, of significance level α. (To get an exact level α test even in this tied setting, see Comment 11.)

When applying the large-sample approximation, an additional factor must be taken into account. Although ties among the X and/or Y observations do not affect the null expected value of C, its null variance is reduced to

$$\text{var}_0(C) = \frac{mn \left[16 \sum_{j=1}^{g} t_j r_j^2 - (N)(N+2)^2\right]}{16N(N-1)} \qquad (5.18)$$

in the presence of ties when N is even and to

$$\text{var}_0(C) = \frac{mn \left[16N \sum_{j=1}^{g} t_j r_j^2 - (N+1)^4\right]}{16N^2(N-1)} \qquad (5.19)$$

when N is odd, where in equations (5.18) and (5.19) g denotes the number of tied groups among the N sample observations, t_j is the size of tied group j, and r_j is the average score associated with the observations in tied group j. We note that an untied observation is considered to be a tied "group" of size 1. In particular, if there are no ties among the X's and/or Y's, then $g = N$ and $t_j = 1$ for $j = 1, \ldots, N$. In this case of no tied sample observations, we have

$$\sum_{j=1}^{g} t_j r_j^2 = 2 \sum_{j=1}^{N/2} j^2 = \frac{2 \left(\frac{N}{2}\right) \left(\frac{N}{2}+1\right) \left(2\left(\frac{N}{2}\right)+1\right)}{6} = \frac{N(N+1)(N+2)}{12},$$

when N is an even integer, and

$$\sum_{j=1}^{g} t_j r_j^2 = 2 \sum_{j=1}^{(N-1)/2} j^2 + \left(\frac{N+1}{2}\right)^2$$

$$= \frac{2}{6} \left(\frac{N-1}{2}\right) \left(\frac{N-1}{2}+1\right) \left(2\left(\frac{N-1}{2}\right)+1\right) + \left(\frac{N+1}{2}\right)^2$$

$$= \left(\frac{N+1}{12}\right)(N^2 + 2N + 3),$$

when N is an odd integer. Using these expressions for $\sum_{j=1}^{g} t_j r_j^2$, the associated ties-adjusted expressions for $\text{var}_0(C)$ given in equations (5.18) and (5.19) reduce to the corresponding untied null variances in equations (5.11) and (5.13), respectively, in the case of no tied observations.

As a consequence of the effect that ties have on the null variance of C, the following modification is needed to apply the large-sample approximation when there are ties among the X and/or Y observations. Compute C using average scores and set

$$C^* = \frac{C - E_0(C)}{\{\text{var}_0(C)\}^{1/2}}, \qquad (5.20)$$

where $E_0(C)$ and $\text{var}_0(C)$ are now given by displays (5.10) and (5.18), respectively, if N is even or by displays (5.12) and (5.19), respectively, if N is odd. With this modified form of C^*, approximations (5.15), (5.16), or (5.17) can be applied.

5.1. A DISTRIBUTION-FREE RANK TEST FOR DISPERSION—MEDIANS EQUAL

TABLE 5.1. Serum Iron (μg/100 ml) Determination Using Hyland Control Sera

Ramsay Method	Jung-Parekh Method
111	107
107	108
100	106
99	98
102	105
106	103
109	110
108	105
104	104
99	100
101	96
96	108
97	103
102	104
107	114
113	114
116	113
113	108
110	106
98	99

Source: D. H. Jung and A. C. Parekh (1970).

Example 5.1: *Serum Iron Determination.* The data in Table 5.1 are a portion of the data obtained by Jung and Parekh (1970) in a study concerned with techniques for direct determination of serum iron. In particular, they attempted to eliminate some of the problems associated with other commonly used methods, which often result in turbidity of the analyzed serum, as well as requiring large samples and slow, tedious analyses. To accomplish this, the authors proposed an improved method for serum iron determination based on a different detergent. One of the purposes of their investigation was to study the accuracy of their method for serum iron determination in comparison to a method due to Ramsay (1957). Twenty duplicate analyses were made, each by the proposed method and by the method of Ramsay, using Hyland control sera containing 105 μg of serum iron per 100 ml. Table 5.1 gives the serum iron detected (in μg/100 ml) for the 40 analyses in the study.

From the point of view of procedural technique, the Jung-Parekh method competes favorably with the Ramsay method for serum iron determination. An additional concern, however, is whether there is a loss of accuracy when the Jung-Parekh procedure is used instead of the Ramsay procedure. As a result, the alternative of interest in this example is greater dispersion or variation for the Jung-Parekh method of serum iron determination than for the method of Ramsay. Hence, letting Y correspond to the Ramsay determinations and X to the Jung-Parekh determinations, we are interested in a one-sided test designed to detect the alternative $H_1 : \gamma^2 > 1$. Since there are ties among the X and Y sample observations and $N = m + n = 20 + 20 = 40$ is an even integer, we will apply the large-sample approximation (with ties), as detailed in equations (5.15) and (5.20), to procedure (5.7).

For purposes of illustration, we consider the approximate level $\alpha = .05$. Hence, from Table A.1, we have $z_{.05} = 1.645$, and the large-sample approximation to procedure (5.7) is given by

$$\text{Reject } H_0 \text{ if } \frac{C - E_0(C)}{\{\text{var}_0(C)\}^{1/2}} \geq 1.645,$$

where $E_0(C)$ and $\text{var}_0(C)$ are given by expressions (5.10) and (5.18), respectively.

To calculate C (5.6) we need the Ansari-Bradley ranks of the 20 Y (Ramsay) observations. In the following display we list in order (from least to greatest) the combined sample of 40 X (Jung-Parekh) and Y (Ramsay) values and assign the ranks according to the Ansari-Bradley scheme.

Ansari-Bradley Ranking Scheme for the Data of Table 4.1

Y	X	Y	Y	X	Y	Y	X	X	Y
96	96	97	98	98	99	99	99	100	100
1.5	1.5	3	4.5	4.5	7	7	7	9.5	9.5
Y	Y	Y	X	X	X	Y	X	X	X
101	102	102	103	103	104	104	104	105	105
11	12.5	12.5	14.5	14.5	17	17	17	19.5	19.5
X	X	Y	Y	X	Y	Y	X	X	X
106	106	106	107	107	107	108	108	108	108
19	19	19	16	16	16	12.5	12.5	12.5	12.5
Y	X	Y	Y	X	Y	Y	X	X	Y
109	110	110	111	113	113	113	114	114	116
10	8.5	8.5	7	5	5	5	2.5	2.5	1

Thus $C = \sum_{i=1}^{20} R_i = 185.5$. In order to calculate C^* we need to evaluate expressions (5.10) and (5.18). We illustrate the calculation of $\sum_{j=1}^{g} t_j r_j^2$ in the following table, where for our data there are $g = 19$ tied groups.

Tied Group	t_j	r_j^2	$t_j r_j^2$
1	2	2.25	4.5
2	1	9	9
3	2	20.25	40.5
4	3	49	147
5	2	90.25	180.5
6	1	121	121
7	2	156.25	312.5
8	2	210.25	420.5
9	3	289	867
10	2	380.25	760.5
11	3	361	1083
12	3	256	768
13	4	156.25	625
14	1	100	100
15	2	72.25	144.5
16	1	49	49
17	3	25	75
18	2	6.25	12.5
19	1	1	1

Thus we have $\sum_{j=1}^{19} t_j r_j^2 = 5721$, and from (5.10) and (5.18) we obtain

$$C^* = \frac{185.5 - [20(42)/4]}{\{(20)(20)[16(5721) - 40(42)^2]/[16(40)(39)]\}^{1/2}} = -1.34,$$

which tells us not to reject H_0 at the approximate $\alpha = .05$ level, since $C^* = -1.34 < 1.645 = z_{.05}$. Hence, there is not sufficient evidence to indicate loss of accuracy when the Jung-Parekh method is used instead of the Ramsay method.

Since the one-sided P-value for these data is the lowest significance level at which we can reject H_0 in favor of $\gamma^2 > 1$ with the observed value of the test statistic C^*, we see from Table A.1 that the P-value for these data is approximately $(1 - .0901) = .9099$. Thus there is absolutely no evidence in the sample data to indicate any loss of accuracy with the Jung-Parekh method. In fact, the C^* value of -1.34 actually provides evidence pointing in the other direction, namely, $\gamma^2 < 1$, corresponding to improved accuracy with the Jung-Parekh method.

Comments

1. *Known Population Medians.* When the population median, θ_2, for the Y observations is known to be equal to $\theta_1 + \xi$, where θ_1 is the population median for the X observations and ξ is a known constant, we can create modified observations $X_i' = X_i + \xi, i = 1, \ldots, m$, and apply the Ansari-Bradley procedures of this section to the modified X' observations and the unchanged Y observations.

2. *Testing γ^2 Equal to Some Specified Value Other Than One.* To test the hypothesis $\gamma^2 = \gamma_0^2$, where γ_0^2 is some specified positive number different from 1, when the common median for the underlying X and Y populations has known value θ_0, we obtain the modified observations $X_i' = (X_i - \theta_0)/\gamma_0$, for $i = 1, \ldots, m$, and $Y_j' = (Y_j - \theta_0)$, for $j = 1, \ldots, n$, and compute C (5.6) using the X''s and Y''s (instead of the X's and Y's). Procedures (5.7), (5.8), or (5.9) or the corresponding large-sample approximations (5.15), (5.16), or (5.17) may then be applied as described.

3. *Motivation for the Test.* Under Assumptions A1 to A3, the X and Y populations have the same median. Suppose, for example, that γ^2 is greater than 1. Then the X values would tend to be more spread out than the Y values. Thus the Y's would tend to get larger scores than the X's from the scheme described in the Procedure and C (5.6) would tend to be larger. (Visualize an extreme sample where the sample values, when ordered, fall in the pattern $XXYYYXX$.) This serves as partial motivation for the one-sided upper-tail test procedure given in (5.7).

4. *Derivation of the Distribution of C under H_0 (No-Ties Case).* Under H_0 (5.1), each of the $\binom{N}{n}$ possible "meshings" of the X's and Y's has probability $1/\binom{N}{n}$. This fact can be used to obtain the null distribution of C (5.6). We illustrate the steps involved in constructing this null distribution for the two cases $m = 3, n = 2$ (where $N = 5$ is odd) and $m = 2, n = 2$ (where $N = 4$ is even). First, for $m = 3$ and $n = 2$, we use the set of scores $\{1, 1, 2, 2, 3\}$. Let $R^{(1)} < R^{(2)}$ denote the ordered Y scores so that $C = R_1 + R_2 = R^{(1)} + R^{(2)}$. The $\binom{5}{2} = 10$ possible meshings and associated values of $(R^{(1)}, R^{(2)})$ and C are given in the following table.

Meshing	Probability	$(R^{(1)}, R^{(2)})$	$C = R^{(1)} + R^{(2)}$
YYXXX	$\frac{1}{10}$	(1, 2)	3
YXYXX	$\frac{1}{10}$	(1, 3)	4
YXXYX	$\frac{1}{10}$	(1, 2)	3

(continued)

Meshing	Probability	$(R^{(1)}, R^{(2)})$	$C = R^{(1)} + R^{(2)}$
YXXXY	$\frac{1}{10}$	(1, 1)	2
XYYXX	$\frac{1}{10}$	(2, 3)	5
XYXYX	$\frac{1}{10}$	(2, 2)	4
XYXXY	$\frac{1}{10}$	(1, 2)	3
XXYYX	$\frac{1}{10}$	(2, 3)	5
XXYXY	$\frac{1}{10}$	(1, 3)	4
XXXYY	$\frac{1}{10}$	(1, 2)	3

Thus, for example, the probability is $\frac{3}{10}$ under H_0 that C is equal to 4, since $C = 4$ when either of the exclusive outcomes $(R^{(1)}, R^{(2)}) = (1, 3)$ or $(R^{(1)}, R^{(2)}) = (2, 2)$ occurs. These two outcomes for $(R^{(1)}, R^{(2)})$ are associated with three mutually exclusive meshings, each with null probability $\frac{1}{10}$. Hence, it follows that $P_0(C = 4) = 3(\frac{1}{10})$. Proceeding in the same manner for all possible values for C and simplifying, we obtain the null distribution.

Possible Value of C	Probability under H_0
2	$\frac{1}{10}$
3	$\frac{4}{10}$
4	$\frac{3}{10}$
5	$\frac{2}{10}$

The probability, under H_0, that C is greater than or equal to 4, for example, is therefore

$$P_0(C \geq 4) = P_0(C = 4) + P_0(C = 5) = .3 + .2 = .5.$$

This agrees with the upper-tail probability entry for $m = 3, n = 2$, and the value $C = 4$ in Table A.8.

For the case of $m = n = 2$ (where $N = 4$ is even), we use the set of scores $\{1, 1, 2, 2\}$. The $\binom{4}{2} = 6$ possible meshings, as well as the associated ordered Y-scores $(R^{(1)}, R^{(2)})$ and values of C are given in the following table.

Meshing	Probability	$(R^{(1)}, R^{(2)})$	$C = R^{(1)} + R^{(2)}$
XXYY	$\frac{1}{6}$	(1, 2)	3
XYXY	$\frac{1}{6}$	(1, 2)	3
YXXY	$\frac{1}{6}$	(1, 1)	2
XYYX	$\frac{1}{6}$	(2, 2)	4
YXYX	$\frac{1}{6}$	(1, 2)	3
YYXX	$\frac{1}{6}$	(1, 2)	3

Proceeding as for the previous case of $m = 3, n = 2$, we obtain the null distribution for C.

Possible Value of C	Probability under H_0
2	$\frac{1}{6}$
3	$\frac{4}{6}$
4	$\frac{1}{6}$

Note that we have derived the null distribution of C without specifying the form of the common (under H_0) underlying X and Y populations beyond the point of requiring that they be continuous. This is why the test procedures based on C are called distribution-free procedures. From the null distribution of C we can determine the critical value c_α and control the probability α of falsely rejecting H_0 when H_0 is true, and this error probability does not depend on the specific form of the common underlying continuous distribution for the X and Y observations.

5. *Symmetry of the Distribution of C under the Null Hypothesis When $N = m + n$ Is Even.* When H_0 is true and $N = m + n$ is an even integer, the distribution of C is symmetric about its mean $n(N+2)/4$. (See Comment 4 for verification of this when $m = n = 2$.) This implies that when N is even

$$P_0(C \leq x) = P_0\left(C \geq \frac{n(N+2)}{2} - x\right), \tag{5.21}$$

for every possible value of x.

Equation (5.21) is directly used to convert upper-tail probabilities, as presented in Table A.8, to lower-tail probabilities when N is even. Thus, the lower-tail critical value $[c_{1-\alpha} - 1]$ used in test procedures (5.8) or (5.9) can be expressed in terms of the upper-tail critical value c_α by

$$[c_{1-\alpha} - 1] = \left[\frac{n(N+2)}{2} - c_\alpha\right], \tag{5.22}$$

when N is even.

6. *Equivalent Form.* The statistic C (5.6) is the sum of the scores assigned to the Y observations by the Ansari-Bradley scoring scheme described in the Procedure. Test procedures (5.7), (5.8), and (5.9) could equivalently be based on the statistic $C' =$ [sum of the scores assigned by this scheme to the X observations], since $C' = [N(N+2)/4] - C$ when $N = m + n$ is even and $C' = [(N-1)^2/4] - C$ when N is odd (see Problem 6).

7. *Assumptions.* We can use the Ansari-Bradley test procedures in (5.7), (5.8), or (5.9) without even requiring that the variances for the X and Y populations exist. Indeed, our Assumptions A1 to A3 for this Section do not specify anything about the existence of even the first moments of the X and Y populations. However, when the first two moments (and, therefore, the variance) for the underlying distributional model $H(u)$ in (5.2) exist, we see from the equal-in-distribution statement in (5.3) that

$$\text{var}\left(\frac{X}{\eta_1}\right) = \text{var}\left(\frac{Y}{\eta_2}\right),$$

which, in turn, implies that

$$[\text{var}(X)]/\eta_1^2 = [\text{var}(Y)]/\eta_2^2.$$

Thus, when the variances exist, we see that $\gamma^2 = [\eta_1^2/\eta_2^2] = [\text{var}(X)/\text{var}(Y)]$.

Assumptions A1 to A3 do imply that the only possible difference between the X and Y populations is a difference in scale parameters. In particular, these assumptions imply that the two populations do not differ in location, as they have a common median θ. (See Comment 1 for a slight relaxation of this condition.) Whereas the requirement of equal medians is not necessary for the classical $\mathcal{F}$-test based on the ratio of the X and Y sample variances, this requirement is essential for the Ansari-Bradley test. For example, suppose that $m = 5, n = 4$, and the X and Y probability distributions are such that $P(X < Y) = 0$. Then, for *all* possible X and Y samples, the joint ordering of the five X observations and four Y observations would *always* result in a value of $C = 10$, regardless of the scale parameters for the two populations. That is, in such a setting, *no* information about γ^2 can be obtained from the joint ranking and the Ansari-Bradley scoring scheme.

Moses (1963) has emphasized this bizarre behavior of tests for dispersion based on joint rankings of the sample X and Y observations and has shown that such tests are inadequate unless strong assumptions (such as equal or known medians) are made concerning the locations of the X and Y populations. For an asymptotically distribution-free test that does not require equal or known medians, see Section 5.2.

8. *Calculation of the Mean and Variance of C under the Null Hypothesis, H_0.* In equations (5.10) and (5.11) we presented formulas for the mean and variance of C when the null hypothesis is true and $N = (m + n)$ is an even number. The corresponding expressions for the null mean and variance of C when N is an odd number are given in equations (5.12) and (5.13). In this comment, we illustrate a direct calculation of $E_0(C)$ and $\text{var}_0(C)$ in the particular cases of $m = 3, n = 2$ (where $N = 5$ is odd) and $m = n = 2$ (where $N = 4$ is even), using the null distributions of C obtained in Comment 4. (Later, in Comment 9, we present general derivations of $E_0(C)$ and $\text{var}_0(C)$.) The null mean, $E_0(C)$, is obtained by multiplying each possible value of C by its probability under H_0. Thus, for $m = 3, n = 2$, we have

$$E_0(C) = 2(.1) + 3(.4) + 4(.3) + 5(.2) = 3.6.$$

This is in agreement with what we obtain using equation (5.12), namely,

$$E_0(C) = \frac{n(N+1)^2}{4N} = \frac{2(5+1)^2}{4(5)} = 3.6.$$

Similarly, for $m = n = 2$, we have by direct computation from the null distribution of C in Comment 4 that

$$E_0(C) = 2\left(\tfrac{1}{6}\right) + 3\left(\tfrac{4}{6}\right) + 4\left(\tfrac{1}{6}\right) = 3,$$

in agreement with the value obtained from equation (5.10), namely,

$$E_0(C) = \frac{n(N+2)}{4} = \frac{2(4+2)}{4} = 3.$$

Checks on the expressions for $\text{var}_0(C)$ are also easily performed, using the well-known fact that

$$\text{var}_0(C) = E_0(C^2) - \{E_0(C)\}^2.$$

The required values of $E_0(C^2)$, the second moment of the null distribution of C, are again obtained by multiplying the possible values of C^2 by the corresponding probabilities under

H_0. For the case of $m = 3, n = 2$, we find that

$$E_0(C^2) = 2^2(.1) + 3^2(.4) + 4^2(.3) + 5^2(.2) = 13.8,$$

yielding

$$\text{var}_0(C) = 13.8 - (3.6)^2 = 13.8 - 12.96 = .84,$$

which is in agreement with the value obtained from equation (5.13), namely,

$$\text{var}_0(C) = \frac{mn(N+1)(3+N^2)}{48N^2}$$

$$= \frac{3(2)(5+1)(3+5^2)}{48(5)^2} = .84.$$

Similary, for $m = n = 2$, we have by direct computation from the null distribution in Comment 4 that

$$E_0(C^2) = 2^2 \left(\tfrac{1}{6}\right) + 3^2 \left(\tfrac{4}{6}\right) + 4^2 \left(\tfrac{1}{6}\right) = \tfrac{56}{6},$$

yielding

$$\text{var}_0(C) = \tfrac{56}{6} - (3)^2 = \tfrac{1}{3},$$

which is in agreement with the value obtained from equation (5.11), namely,

$$\text{Var}_0(C) = \frac{mn(N+2)(N-2)}{48(N-1)}$$

$$= \frac{2(2)(4+2)(4-2)}{48(4-1)} = \frac{1}{3}.$$

9. *Large-Sample Approximation.* The statistic C/n is the average of the scores assigned to the Y observations. Since all $\binom{N}{n}$ possible distributions of the appropriate scores (depending on whether N is even or odd) to the X and Y observations are equally likely under H_0, the null distribution of C/n is the same as the distribution of the sample mean for a random sample of size n drawn without replacement from the finite population of scores S_N, where $S_N = \{1, 2, 3, \ldots, N/2, N/2, \ldots, 3, 2, 1\}$ if N is an even number and $S_N = \{1, 2, 3, \ldots, (N-1)/2, (N+1)/2, (N-1)/2, \ldots, 3, 2, 1\}$ if N is odd.

From basic results for a random sample of size n drawn without replacement from a finite population of N elements, we know that:

(i) The expected value of the sample average is equal to the average, μ_{pop}, of the finite population, and
(ii) The variance of the sample average is equal to

$$\frac{\sigma^2_{pop}}{n} \left(\frac{N-n}{N-1}\right),$$

where σ^2_{pop} is the variance of the finite population and the factor $(N-n)/(N-1)$ is known as the finite population correction factor.

For the case of N even and the finite population $S_N = \{1, 2, 3, \ldots, N/2, N/2, \ldots, 3, 2, 1\}$, we see that

(iii)
$$\mu_{\text{pop}} = \frac{2}{N}\sum_{i=1}^{N/2} i = \frac{(N/2)[(N/2)+1]}{2(N/2)} = \frac{N+2}{4}$$

and

(iv)
$$\sigma^2_{\text{pop}} = \left[\frac{2}{N}\left(\sum_{i=1}^{N/2} i^2\right) - \left(\frac{N+2}{4}\right)^2\right]$$
$$= \left[\frac{(N/2)[(N/2)+1][2(N/2)+1]}{6(N/2)} - \left(\frac{N+2}{4}\right)^2\right]$$
$$= \left[\frac{(N+2)(N+1)}{12} - \frac{(N+2)(N+2)}{16}\right]$$
$$= \frac{(N+2)(N-2)}{48}.$$

From (i), (ii), (iii), and (iv), it follows that

$$E_0\left(\frac{C}{n}\right) = \frac{N+2}{4}$$

and

$$\text{var}_0\left(\frac{C}{n}\right) = \left[\frac{(N+2)(N-2)}{48n}\right]\left[\frac{N-n}{N-1}\right] = \frac{m(N+2)(N-2)}{48n(N-1)}.$$

Thus,

$$E_0(C) = nE_0\left(\frac{C}{n}\right) = \frac{n(N+2)}{4}$$

and

$$\text{var}_0(C) = n^2 \text{var}_0\left(\frac{C}{n}\right) = \frac{mn(N+2)(N-2)}{48(N-1)},$$

in agreement with the formulas in equations (5.10) and (5.11). The corresponding expressions for $E_0(C)$ and $\text{var}_0(C)$ when N is an odd integer, as given in equations (5.12) and (5.13), respectively, can be similarly obtained using the expressions in (i) and (ii) and the finite population

$$S_N = \left\{1, 2, 3, \ldots, \frac{N-1}{2}, \frac{N+1}{2}, \frac{N-1}{2}, \ldots, 3, 2, 1\right\}.$$

For any N (even or odd), the asymptotic normality under H_0 of the standardized

$$C^* = \frac{C - E_0(C)}{\sqrt{\text{var}_0(C)}}$$

follows from standard theory for the mean of a sample from a finite population (cf. Wilks 1962, 268). Asymptotic normality results for C^* are also available under general alternatives

5.1. A DISTRIBUTION-FREE RANK TEST FOR DISPERSION—MEDIANS EQUAL

to H_0. See, for example, Ansari and Bradley (1960), Randles and Wolfe (1979), or Hájek and Šidák (1967).

10. *Lower-Tail Critical Values.* In the expression for the one-sided lower-tail test in (5.9), the critical value is given to be $c_{1-\alpha} - 1$, where $c_{1-\alpha}$ is the upper $(1 - \alpha)$th percentile of the null distribution of C as presented in Table A.8. This means that

$$P_0(C \leq c_{1-\alpha} - 1) = 1 - P_0(C > c_{1-\alpha} - 1) = 1 - P_0(C \geq c_{1-\alpha}),$$

where the last equality follows from the fact that C is a discrete random variable assuming only positive integer values. Since $c_{1-\alpha}$ is the upper $(1 - \alpha)$th percentile for the null distribution of C, it follows that

$$P_0(C \leq c_{1-\alpha} - 1) = 1 - (1 - \alpha) = \alpha.$$

Hence, $c_{1-\alpha} - 1$ is, indeed, the *lower* αth percentile for the null distribution of C, as required for the level α one-sided lower-tail test procedure in expression (5.8).

When N is an even integer, we have already noted in Comment 5 that the null distribution of C is symmetric about its mean, $n(N + 2)/4$. It follows that $[c_{1-\alpha} - 1] = [\{n(N + 2)/2\} - c_\alpha]$ when N is even.

11. *Exact Conditional Distribution of C with Ties.* To have a test with exact significance level even in the presence of ties among the X's and/or Y's, we need to consider all $\binom{N}{n}$ possible assignments of the N observations with n observations serving as Y's and m observations serving as X's. As in Comment 4, it still follows that, under H_0 (5.1), each of the $\binom{N}{n}$ possible "meshings" of the X's and Y's has probability $1/\binom{N}{n}$. The only difference in the case of ties is that we now use average scores in the computation of C for each of these $\binom{N}{n}$ "meshings" leading to the tabulation of the null distribution. We illustrate this construction for N odd (a similar approach will work for N even) and the following $m = 3, n = 2$ data: $X_1 = 3.2, X_2 = 5.7, X_3 = 6.3, Y_1 = 1.9, Y_2 = 6.3$. The associated average scores assignments (taking into account the tie between X_3 and Y_2) are 2, 3, 1.5, 1, and 1.5, respectively, and the corresponding value of C, the sum of the scores for the Y observations, is $C = 1.5 + 1 = 2.5$. To assess the significance of this value of C, we obtain its conditional distribution by considering the $\binom{5}{2} = 10$ possible assignments of the observations 1.9, 3.2, 5.7, 6.3, and 6.3 to serve as three X observations and two Y observations, or, equivalently the 10 possible assignments of the average scores 1, 1.5, 1.5, 2, and 3 to serve as three X scores and two Y scores. These ten assignments and the corresponding values of C are as follows.

Y Scores	Probability under H_0	Value of C
1, 1.5	$\frac{1}{10}$	2.5
1, 1.5	$\frac{1}{10}$	2.5
1, 2	$\frac{1}{10}$	3
1, 3	$\frac{1}{10}$	4
1.5, 1.5	$\frac{1}{10}$	3
1.5, 2	$\frac{1}{10}$	3.5
1.5, 3	$\frac{1}{10}$	4.5
1.5, 2	$\frac{1}{10}$	3.5
1.5, 3	$\frac{1}{10}$	4.5
2, 3	$\frac{1}{10}$	5

This yields the null tail probabilities

$$P_0(C \geq 5) = \tfrac{1}{10},$$
$$P_0(C \geq 4.5) = \tfrac{3}{10},$$
$$P_0(C \geq 4) = \tfrac{4}{10},$$
$$P_0(C \geq 3.5) = \tfrac{6}{10},$$
$$P_0(C \geq 3) = \tfrac{8}{10},$$
$$P_0(C \geq 2.5) = 1.$$

This distribution is called the conditional null distribution or the permutation null distribution of C, given the set of tied scores $\{1, 1.5, 1.5, 2, 3\}$. For the particular observed value $C = 2.5$, we have $P_0(C \geq 2.5) = 1$, so that such a value does not indicate a deviation from H_0 in the direction of $\gamma^2 > 1$ (although it would provide marginal support for the alternative $\gamma^2 < 1$).

For a given set of X and Y observations, StatXact will generate the null permutation distribution of C. After entering the tie-adjusted average scores for the data with the case command CA, the commands

SE/FI/ ABDIST.PRN (enter)
PE/EX (enter)

will generate the null permutation distribution of C and save it in a file named ABDIST.PRN. For our sample example, the output will include

One-sided P-value: Pr {Test Statistic. LE. Observed} = .2000

and

Point probability: Pr {Test Statistic. EQ. Observed} = .2000.

This yields $P_0(C \geq 2.5) = (1 - .2000) + .2000 = 1$, which agrees with what we found previously in this comment by direct enumeration.

12. *Confidence Intervals, Confidence Bounds, and Point Estimators for γ^2.* The Ansari-Bradley statistic, C (5.6), is a member of a large class of rank statistics (referred to as *linear rank statistics* in the literature; see, for example, Section 9.3 of Randles and Wolfe (1979)) that can be used to test for equality of scale parameters under the strict assumption of equal or known medians for the X and Y populations. Bauer (1972) has shown how to invert some of these linear rank tests of $\gamma^2 = 1$, including the Ansari-Bradley procedure, to obtain point estimators and confidence intervals or bounds for γ^2 in such a setting.

13. *Unequal and Unknown Medians.* If the medians of the X and Y populations are not known and it is questionable whether or not they are equal, Ansari and Bradley (1960) suggested the following modification to their test procedures. Define the adjusted observations $X'_i = X_i - \tilde{X}, i = 1, \ldots, m$, and $Y'_j = Y_j - \tilde{Y}, j = 1, \ldots, n$, where $\tilde{X}$ and $\tilde{Y}$ are the sample medians for the X and Y observations, respectively. Let C' be C (5.6) calculated for these adjusted X' and Y' observations. Depending on the alternative to $H_0 : \gamma^2 = 1$ that is of interest, the appropriate procedure (5.7), (5.8), or (5.9), or the corresponding large-sample approximation, can then be applied directly to the modified statistic C' instead of C. Such tests based on C' are no longer strictly distribution-free. However, Gross (1966) has given sufficient conditions under which such procedures are asymptotically distribution-free. Under such conditions, the various tests based on C' maintain an approximate (both m and n large) significance level α over a large class of continuous underlying distributions.

14. *Consistency of the C-Tests.* Under Assumptions A1, A2, and A3 the consistency of the tests based on C depends on the parameter

$$\Delta^* = \left[P(X > Y > \theta) + P(X < Y < \theta) - \tfrac{1}{4}\right].$$

The test procedures defined by (5.7), (5.8), and (5.9) are consistent against the alternatives corresponding to $\Delta^* >, <$, and $\neq 0$, respectively.

15. *More General Alternatives.* In many two-sample situations we are interested in simultaneously detecting either location or scale differences between the X and Y populations. One solution to this broader problem is to use a test procedure designed to detect quite general alternatives. One such test procedure based on the two-sample Kolmogorov (1933)–Smirnov (1939) statistic is discussed in Section 5.4. A second approach is to conduct simultaneously a test such as the Wilcoxon rank sum procedure based on W(4.3) for detecting differences in location and a second test such as the Ansari-Bradley procedure based on C (5.6) for detecting differences in scale. One such simultaneous testing approach, due to Lepage (1971, 1973), for dealing with general alternatives is the topic of Section 5.3. Randles and Hogg (1971) have shown that in such a situation, W and C are uncorrelated and, in fact, asymptotically independent when $H_0(1)$ is true. This implies, among other things, that if we conduct the Wilcoxon rank sum test at a significance level α_1, and the Ansari-Bradley test at a significance level α_2, then the probability of incorrectly rejecting with at least one of the two tests, given that H_0 (5.1) is true, is approximately $\alpha_1 + \alpha_2 - \alpha_1\alpha_2$.

Properties

1. *Consistency.* For our statement we consider the more stringent location-scale parameter model described in (5.4). Then the tests defined by (5.7), (5.8), and (5.9) are consistent against the alternatives $\gamma^2 >, <$, and $\neq 1$, respectively. See also Comment 14.
2. *Asymptotic Normality.* See Randles and Wolfe (1979, 315–320).
3. *Efficiency.* See Section 5.5.

PROBLEMS

1. Consider the chorioamnion permeability data in Table 4.1. In Section 4.1 we saw that a test procedure based on the Wilcoxon rank sum statistic did not reject the null hypothesis that the human chorioamnion is as permeable to water transfer at 12 to 26 weeks gestational age as it is at term. With this in mind and using the same data, test the hypothesis of equal dispersions versus the alternative that the variation in tritiated water diffusion across human chorioamnion is different at term than at 12 to 26 weeks gestational age.
2. Find or construct an example in which there exists a level α and a constant d such that when C (5.6) is computed for the original data $X_1,\ldots,X_m, Y_1,\ldots,Y_n$, the level α procedure in (5.7) does not lead to rejection of $H_0 : \gamma^2 = 1$, but when C is computed for the values $X_1,\ldots,X_m, Y_1 + d,\ldots,Y_n + d$, the level α procedure in (5.7) does lead to rejection of H_0. Note that such an example exposes an undesirable aspect of the test procedures based on C. Let Y be a random member from a population Π. Then the population Π^* formed by adding the constant d to each member of Π must, by any reasonable definition of dispersion, have the same dispersion as the Π population. Thus the difference in dispersions between the X and Y populations must be the same as the differences in dispersions between the X and $Y + d$ populations. Yet the tests based on C, applied to the data $X_1,\ldots,X_m, Y_1 + d,\ldots,Y_n + d$, can yield a decision that differs from the one that results from applying the same C-test to $X_1,\ldots,X_m, Y_1,\ldots,Y_n$. (For a related discussion, see Comment 7.)
3. Consider the television-viewing behavior data in Table 4.4. For these data, find the approximate P-value for an appropriate test of the hypothesis of equal dispersions versus the alternative that

there is more variability in the time spent in the room after witnessing the violent behavior for those children who had previously watched the *Karate Kid* than for those children who had previously watched parts of the 1984 Summer Olympic Games. Comment on the importance of the results of Problem 4.5 in relationship to this dispersion test.

4. Verify the expressions for $E_0(C)$ and $\text{var}_0(C)$ in equations (5.12) and (5.13), respectively, when $N = (m + n)$ is an odd integer. (See Comment 9 for guidance.)

5. Consider the following two-sample data for $m = 3, n = 3$: $X_1 = -3.7, X_2 = 4.6, X_3 = 1.5, Y_1 = 1.5, Y_2 = 4.6, Y_3 = 1.5$. Here, $N = 3 + 3 = 6$ is an even integer. Using the approach discussed in Comment 11, find the exact conditional null distribution of the Ansari-Bradley statistic, C (5.6). Compare and contrast this conditional null distribution with the null distribution of C given in Table A.8 for $m = n = 3$ and no tied observations.

6. Let C' be the sum of the scores assigned to the X observations by the Ansari-Bradley scoring scheme described in the Procedure. Verify directly, or illustrate using the serum iron determination data in Table 5.1, that $C' = [N(N + 2)/4] - C$, when $N = (m + n)$ is even.

7. For an arbitrary total number of observations $N = (m + n)$, find an expression for the smallest and largest possible values of C. Consider the two cases of N even and N odd.

8. Let X and Y be independent, identically distributed continuous random variables with a common probability distribution with median θ. What is the value of Δ^* in Comment 14 for this setting?

9. Consider the alcoholic intake data in Table 4.2. In Example 4.2 the Wilcoxon rank sum test procedure led to rejection of $H_0 : \Delta = 0$ in favor of $H_1 : \Delta < 0$. What does this result imply about the appropriateness of the Ansari-Bradley procedure in (5.9) or its approximate large-sample counterpart in (5.17) as a test of $H_0 : \gamma^2 = 1$ versus $H_1 : \gamma^2 \neq 1$ for these data? In view of this fact, find the approximate P-value for an appropriate modification of the large-sample procedure in (5.17) to test for possible differences in dispersions between the control and SST data. (See Comment 13.)

✓ 10. Consider the television-viewing behavior data in Table 4.4. *Without* the assumption of equal medians, find the approximate P-value for an appropriate test (see Comment 13) of the hypothesis of equal dispersions versus the alternative that there is more variability in the time spent in the room after witnessing the violent behavior for those children who had previously watched the *Karate Kid* than for those children who had previously watched parts of the 1984 Summer Olympic Games. Compare the P-value obtained here with the one found in Problem 3. Interpret the similarity or lack thereof between the two P-values.

✓ 11. Suppose $m = n = 10$. Compare the critical region for the exact level $\alpha = .056$ test of $H_0 : \gamma^2 = 1$ versus $H_1 : \gamma^2 > 1$ based on C with the critical region for the corresponding nominal level $\alpha = .056$ test based on the large-sample approximation. What is the exact significance level of this .056 nominal level test based on the large-sample approximation?

12. Use the computer software StatXact to generate the conditional permutation distribution of C (see Comment 11), given the set of tied values for the serum iron data in Example 5.1. From this conditonal permutation distribution of C, obtain the exact conditional P-value, $P_0(C \geq 185.5)$, for the corresponding test of $H_0 : \gamma^2 = 1$ versus $H_1 : \gamma^2 > 1$. Compare this exact conditional P-value with the approximate P-value for the large-sample test procedure applied to these data in Example 5.1.

5.2. AN ASYMPTOTICALLY DISTRIBUTION-FREE TEST FOR DISPERSION BASED ON THE JACKKNIFE—MEDIANS NOT NECESSARILY EQUAL (MILLER)

Hypothesis

Let $X_1, \ldots, X_m$ and $Y_1, \ldots, Y_n$ be independent random samples satisfying Assumptions A1 and A2 from continuous populations with distribution functions F and G, respectively, satisfying the location-scale parameter model relationship in equations (5.2) and (5.3). In addition,

5.2. AN ASYMPTOTICALLY DISTRIBUTION-FREE TEST FOR DISPERSION

we assume that the continuous distribution associated with the distribution function $H(\cdot)$ in equation (5.2) has finite fourth moment; that is, we assume

A4. If V is a continuous random variable with distribution function H, then $E(V^4) < \infty$.

Under Assumptions A1, A2, and A4, but without the equal median Assumption A3, we are once again interested in the ratio of scale parameters $\gamma = (\eta_1/\eta_2)$. In view of Assumption A4 (see Comment 7), we note that $\gamma^2 = [\text{var}(X)/\text{var}(Y)]$, the ratio of population variances. We are interested in testing the null hypothesis H_0 (5.1), which reduces to $H_0 : \gamma^2 = 1$, corresponding to the assertion that the population variances are equal, under the location-scale parameter model (5.2).

Procedure

Consider the X sample data with the first observation deleted and set

$$\bar{X}_1 = \sum_{s=2}^{m} \frac{X_s}{m-1} \quad \text{and} \quad D_1^2 = \sum_{s=2}^{m} \frac{(X_s - \bar{X}_1)^2}{m-2}. \tag{5.23}$$

Thus, $\bar{X}_1$ and D_1^2 are the sample average and sample variance for the data $X_2,\ldots,X_m$, corresponding to the X sample less X_1. Similarly, let

$$\bar{X}_i = \sum_{s \neq i}^{m} \frac{X_s}{m-1} \quad \text{and} \quad D_i^2 = \sum_{s \neq i}^{m} \frac{(X_s - \bar{X}_i)^2}{m-2} \tag{5.24}$$

be the sample average and sample variance, respectively, for the data $X_1,\ldots,X_{i-1},X_{i+1},\ldots,X_m$, corresponding to the X sample less X_i, for $i = 1,\ldots,m$. In the same fashion, let

$$\bar{Y}_j = \sum_{t \neq j}^{n} \frac{Y_t}{n-1} \quad \text{and} \quad E_j^2 = \sum_{t \neq j}^{n} \frac{(Y_t - \bar{Y}_j)^2}{n-2} \tag{5.25}$$

be the sample average and sample variance, respectively, for the data $Y_1,\ldots,Y_{j-1},Y_{j+1},\ldots,Y_n$, corresponding to the Y sample less Y_j, for $j = 1,\ldots,n$. Define $S_1,\ldots,S_m$ and $T_1,\ldots,T_n$ by

$$S_i = \ln D_i^2, \quad i = 1,\ldots,m, \tag{5.26}$$

and

$$T_j = \ln E_j^2, \quad j = 1,\ldots,n. \tag{5.27}$$

In addition, let

$$S_0 = \ln \left[\sum_{s=1}^{m} \frac{(X_s - \bar{X}_0)^2}{m-1} \right] \tag{5.28}$$

and

$$T_0 = \ln \left[\sum_{t=1}^{n} \frac{(Y_t - \bar{Y}_0)^2}{n-1} \right], \tag{5.29}$$

where $\bar{X}_0 = \sum_{s=1}^{m} X_s/m$ and $\bar{Y}_0 = \sum_{t=1}^{n} Y_t/n$, be the corresponding statisics for the complete samples $X_1, \ldots, X_m$ and $Y_1, \ldots, Y_n$, respectively. Compute

$$A_i = mS_0 - (m-1)S_i, \quad \text{for } i = 1, \ldots, m, \tag{5.30}$$

and

$$B_j = nT_0 - (n-1)T_j, \quad \text{for } j = 1, \ldots, n. \tag{5.31}$$

(This is what is referred to as the "jackknifing process," as applied to the sample variance.) Set

$$\bar{A} = \sum_{i=1}^{m} \frac{A_i}{m} \quad \text{and} \quad \bar{B} = \sum_{j=1}^{n} \frac{B_j}{n}, \tag{5.32}$$

and compute

$$V_1 = \sum_{i=1}^{m} \frac{(A_i - \bar{A})^2}{m(m-1)} \tag{5.33}$$

and

$$V_2 = \sum_{j=1}^{n} \frac{(B_j - \bar{B})^2}{n(n-1)}. \tag{5.34}$$

Finally, set

$$Q = \frac{\bar{A} - \bar{B}}{\sqrt{V_1 + V_2}}. \tag{5.35}$$

a. One-Sided Upper-Tail Test. To test

$$H_0 : \gamma^2 = 1$$

versus

$$H_1 : \gamma^2 > 1,$$

at the approximate α level of significance,

$$\text{Reject } H_0 \text{ if } Q \geq z_\alpha; \quad \text{otherwise do not reject,} \tag{5.36}$$

where, as previously, z_α is the upper αth percentile for the standard normal distribution.

b. One-Sided Lower-Tail Test. To test

$$H_0 : \gamma^2 = 1$$

versus

$$H_1 : \gamma^2 < 1,$$

at the approximate α level of significance,

$$\text{Reject } H_0 \text{ if } Q \leq -z_\alpha; \quad \text{otherwise do not reject.} \qquad (5.37)$$

c. Two-Sided Test. To test

$$H_0 : \gamma^2 = 1$$

versus

$$H_1 : \gamma^2 \neq 1,$$

at the approximate α level of significance,

$$\text{Reject } H_0 \text{ if } |Q| \geq z_{\alpha/2}; \quad \text{otherwise do not reject.} \qquad (5.38)$$

This two-sided procedure is the two-sided symmetric test with $\alpha/2$ probability in each tail of the approximating standard normal distribution.

When m and n are small and equal, the approximate level α test procedures given by (5.36), (5.37), and (5.38) can be improved slightly by replacing z_α and $z_{\alpha/2}$ by $t_{m+n-2,\alpha}$ and $t_{m+n-2,\alpha/2}$, respectively, where $t_{m+n-2,\alpha}$ is the upper α percentile point of the t distribution with $m+n-2$ degrees of freedom. Values of $t_{m+n-2,\alpha}$ can be obtained from Chart A.1.

Ties

The jackknife procedures are well defined when ties within or between the X's and Y's occur and further adjustments are not necessary.

Example 5.2: *Southern Armyworm and Pokeweed.* Burnett and Jones (1973) investigated the idea of coevolution between the southern armyworm and pokewood. They suspected that armyworms might have developed a greater resistance to the toxins from pokeweed populations that lie within their geographic range than to the toxins of pokeweeds found in other areas of the country. Pokeweed plants from Florida populations (within the range of southern armyworms) and Kentucky populations (well north of the range of southern armyworms) were raised under similar conditions in greenhouses for the study. Larval southern armyworms were then used in feeding experiments to determine whether they would eat less of the Kentucky pokeweed possessing toxins to which they are not resistant. Five samples of Kentucky pokeweed and five samples of Florida pokeweed were used, with each such sample being exposed to ten separate southern armyworm larvae. (There were 100 *different* larvae used in the experiment.) Following an individual larva's 24-h feeding period (in darkness at $25 \pm 1°C$) on a moist filter paper in a disposable petri dish, the fecal material of the larva was dried overnight in an oven and weighed the following day. This was then used as a measure of the quantity of plant material ingested by the armyworm larva during its feeding. The data in Table 5.2 are the average (over the ten armyworm larvae replications) dry feces weights (in mg) for the five Kentucky pokeweed and five Florida pokeweed plant samples.

It is clear from the data in Table 5.2 that the southern armyworm larvae had a tendency to eat more (on the average) of the Florida pokeweed than the Kentucky pokeweed. As a result, if we are interested in assessing whether there is any difference in the variability or dispersion of the southern armyworm's consumption of the two pokeweed varieties, it would not be appropriate to directly apply one of the Ansari–Bradley procedures discussed in Section 5.1, since they require equality of the respective population medians. However, the jackknifed

TABLE 5.2. Average Dry Feces Weight (mg)

Kentucky Pokeweed	Florida Pokeweed
6.2	9.5
5.9	9.8
8.9	9.5
6.5	9.6
8.6	10.3

Source: W. C. Burnett, Jr., and S. B. Jones, Jr. (1973).

variances procedure of this section makes no such assumption and can be applied directly to the sample data.

Letting X correspond to the Kentucky pokeweed observations and Y to the Florida pokeweed data, we consider testing the null hypothesis of no difference in dispersion against the alternative that the variability is greater for Kentucky pokeweed; that is, we want to use procedure (5.36) to test $H_0 : \gamma^2 = 1$ against the alternative $H_1 : \gamma^2 > 1$.

The five Kentucky pokeweed subgroups of four observations each, corresponding to the five different ways to delete a single measurement, are given by:

$$G_1 = \{6.2, 5.9, 8.9, 6.5\}, \quad G_2 = \{6.2, 5.9, 8.9, 8.6\}, \quad G_3 = \{6.2, 5.9, 6.5, 8.6\},$$

$$G_4 = \{6.2, 8.9, 6.5, 8.6\} \quad \text{and} \quad G_5 = \{5.9, 8.9, 6.5, 8.6\}.$$

Following (5.23), the sample average and sample variance associated with subgroup G_1 is

$$\bar{X}_1 = \frac{6.2 + 5.9 + 8.9 + 6.5}{4} = 6.875 \tag{5.39}$$

and

$$D_1^2 = \frac{(6.2 - 6.875)^2 + (5.9 - 6.875)^2 + (8.9 - 6.875)^2 + (6.5 - 6.875)^2}{3} = 1.8825. \tag{5.40}$$

In a similar manner, it follows from (5.24) that the sample averages and sample variances for the other four Kentucky pokeweed subgroups are

Subgroup G_i	$\bar{X}_i$	D_i^2
G_2	7.4	2.46
G_3	6.8	1.50
G_4	7.55	1.95
G_5	7.475	2.2425

(5.41)

Proceeding in the same fashion with the Florida pokeweed data, we obtain the five deleted-observation subgroups

$$H_1 = \{9.5, 9.8, 9.5, 9.6\}, \quad H_2 = \{9.5, 9.8, 9.5, 10.3\}, \quad H_3 = \{9.5, 9.8, 9.6, 10.3\},$$

$$H_4 = \{9.5, 9.5, 9.6, 10.3\}, \quad \text{and} \quad H_5 = \{9.8, 9.5, 9.6, 10.3\}.$$

Using equation (5.25), we obtain the associated subgroup sample means and sample variances to be

Subgroup H_j	$\bar{Y}_j$	E_j^2
H_1	9.6	.02
H_2	9.775	.1425
H_3	9.8	.1267
H_4	9.725	.1492
H_5	9.8	.1267

(5.42)

Taking natural logarithms of the D_i^2's (in (5.40) and (5.41)) and the E_j^2's (in (5.42)), it follows from (5.26) and (5.27) that

$$S_1 = .6326, \quad S_2 = .9002, \quad S_3 = .4055, \quad S_4 = .6678, \quad S_5 = .8076 \quad (5.43)$$

and

$$T_1 = -3.9120, \quad T_2 = -1.9484, \quad T_3 = -2.0662, \quad T_4 = -1.9027, \quad T_5 = -2.0662.$$

(5.44)

Finally, using all five of the X sample observations we see from (5.28) that

$$\bar{X}_0 = 7.22 \quad \text{and} \quad S_0 = \ln\left[\sum_{s=1}^{5} \frac{(X_s - 7.22)^2}{4}\right] = \ln 2.007 = .6966.$$

Similarly, using all five of the Y sample observations, it follows from (5.29) that

$$\bar{Y}_0 = 9.74 \quad \text{and} \quad T_0 = \ln\left[\sum_{t=1}^{5} \frac{(Y_t - 9.74)^2}{4}\right] = \ln .113 = -2.1804.$$

Combining these complete sample values with the subgroup calculations in (5.43) and (5.44) via the jackknifing process in (5.30) and (5.31), we obtain

$$A_1 = 5(.6966) - 4(.6326) = .9526, \quad A_2 = 5(.6966) - 4(.9002) = -.1178,$$
$$A_3 = 5(.6966) - 4(.4055) = 1.861, \quad (5.45)$$
$$A_4 = 5(.6966) - 4(.6678) = .8118, \quad A_5 = 5(.6966) - 4(.8076) = .2526$$

and

$$B_1 = 5(-2.1804) - 4(-3.9120) = 4.746,$$
$$B_2 = 5(-2.1804) - 4(-1.9484) = -3.1084,$$
$$B_3 = 5(-2.1804) - 4(-2.0662) = -2.6372, \quad (5.46)$$
$$B_4 = 5(-2.1804) - 4(-1.9027) = -3.2912,$$
$$B_5 = 5(-2.1804) - 4(-2.0662) = -2.6372.$$

From (5.32), (5.33), and (5.45), we see that

$$\bar{A} = .7520 \quad \text{and} \quad V_1 = \sum_{i=1}^{5} \frac{(A_i - \bar{A})^2}{5(4)} = .1140.$$

Similarly, from (5.32), (5.34), and (5.46), we have

$$\bar{B} = -1.3856 \quad \text{and} \quad V_2 = \sum_{j=1}^{5} \frac{(B_j - \bar{B})^2}{5(4)} = 2.3664.$$

It then follows from (5.35) that

$$Q = \frac{.7520 - (-1.3856)}{(.1140 + 2.3664)^{1/2}} = 1.36.$$

Hence, from (5.36) and Table A.1, we see that the lowest significance level at which we can reject H_0 in favor of $\gamma^2 > 1$ with the observed value of the test statistic Q (i.e., the one-sided P-value) is approximately .0869. Thus, there is only mild evidence in the sample data to indicate greater variability for the Kentucky pokeweed population.

Comments

16. *Assumptions.* Note that Assumptions A1, A2, and A4 do not impose the severe condition that the two underlying populations have equal medians. Although these assumptions do require that the two underlying populations have finite fourth moments, this is not a serious restriction for most common data collection settings. This means that the Miller procedures are applicable in more general settings than the Ansari-Bradley procedures based on C (5.6). (See Comment 7.)

17. *Testing γ^2 Equal to Some Specified Value Other Than One.* To test the hypothesis $\gamma^2 = \gamma_0^2$, where γ_0^2 is some specific positive number different from 1, we obtain the modified observations $Y_j' = Y_j/\gamma_0$, $j = 1, \ldots, n$, and compute Q (5.35) using the X's and the Y''s (instead of the X's and the Y's). The appropriate procedure (5.36), (5.37), or (5.38) may then be applied as described.

18. *Asymptotic Distribution-Freeness.* Asymptotically (i.e., for infinitely large samples) the true level of the tests defined by (5.36), (5.37), and (5.38) will agree with the nominal level α. Subject to Assumptions A1, A2, and A4, this asymptotic result does not depend on the underlying populations of the X's and Y's. More precisely, subject to Assumptions A1, A2, and A4, Q (5.35) has an asymptotic $N(0, 1)$ distribution when H_0 is true. Since this asymptotic distribution does not depend on the underlying populations of the X's and Y's, we say that the tests based on Q are asymptotically distribution-free. Of course, in practice, we do not have the luxury of infinite samples. Thus, in any particular setting with m and n large, although the level of any of the tests based on Q is not necessarily exactly equal to the nominal level α, we hope it is close to it. The closeness of this approximation depends on m, n, α, and the underlying populations, but, for fixed α, the closeness generally improves as m and n increase, regardless of the underlying populations. In the case of the Q tests, the reader is cautioned that the question of how large m and n should be, in order for the normal approximation to be good, is relatively unanswered. Exact null distribution tables for Q cannot be provided since, for specified values of m and n, the exact null distribution of Q depends on the underlying X and Y populations; thus exact critical points would vary with the forms of the X and Y

populations. The procedures (5.36), (5.37), and (5.38) based on Q, therefore, are not (strictly) distribution-free.

19. *Alternative Method of Calculation.* For $i = 1, \ldots, m$, S_i is the natural log of the sample variance for the $(m - 1)$ X observations $X_1, \ldots, X_{i-1}, X_{i+1}, \ldots, X_m$. Similarly, for $j = 1, \ldots, n$, T_j is the natural log of the sample variance for the $(n - 1)$ Y observations $Y_1, \ldots, Y_{j-1}, Y_{j+1}, \ldots, Y_n$. The following equivalent formulas for D_i^2 (5.24) and E_j^2 (5.25), namely,

$$D_i^2 = \frac{\sum_{s \neq i}^{m} X_s^2 - \frac{(\sum_{s \neq i}^{m} X_s)^2}{m - 1}}{m - 2}$$

and

$$E_j^2 = \frac{\sum_{t \neq j}^{n} Y_t^2 - \frac{(\sum_{t \neq j}^{n} Y_t)^2}{n - 1}}{n - 2},$$

are computationally more convenient than the definitions given in (5.24) and (5.25), respectively.

20. *General Jackknife Technique.* The jackknife technique applied in this section to the problem of testing two-sample dispersion hypotheses is a tool that can be used successfully in certain statistical problems to accomplish two goals: (a) reducing the bias of point estimators and (b) generating broadly applicable and reasonably powerful test procedures for problems where classical test procedures are sensitive to nonnormality of the underlying populations. Although the jackknife technique is not always effective in achieving these goals (see Miller (1964)), it performs well in the two-sample dispersion problem, providing us with asymptotically distribution-free test procedures for a problem where deviations from normality of the underlying populations can be disastrous for the classical $\mathcal{F}$-test for equal variances (cf. Box (1953), Shorack (1969), and Comment 26) and where distribution-free rank tests for the problem are limited in their applicability (see Comment 7 and Moses (1963)). Miller (1968) discussed in detail the advantages gained by applying the jackknife technique to this two-sample dispersion problem. See also Comment 21.

21. *Motivation.* The jackknife is an extension of an idea due to Quenouille (1949) and is designed to reduce the bias of an estimator. Suppose we have a sample of N independent observations, each from the same distribution which depends on an unknown parameter θ. Assume that we have a general method for estimating θ and let $\hat{\theta}$ denote this estimator based on all N observations. Divide the data into n groups of size k. Let $\hat{\theta}_{-i}, i = 1, \ldots, n$, denote the estimator of θ obtained by deleting the ith group and estimating θ from the remaining $(n - 1)k$ observations. Define $\tilde{\theta}_i = n\hat{\theta} - (n - 1)\hat{\theta}_{-i}$. The jackknife estimator of θ is $\tilde{\theta} = \sum_{i=1}^{n} \tilde{\theta}_i / n$. In certain situations the jackknife can be shown to be less biased than the estimator $\hat{\theta}$. Tukey (1958, 1962) extended the jackknife to construct approximate significance tests and confidence intervals for θ.

The traditional estimator of the variance of $\tilde{\theta}$, in the case of $k = 1$, is

CORRECTION

$$\hat{V}^2 = \frac{1}{N(N - 1)} \sum_{i=1}^{N} (\hat{\theta}_{-i} - \hat{\theta})^2.$$

$$\tilde{\theta} = \sum_{i=1}^{N} \frac{\tilde{\theta}_i}{N}$$

$$\hat{V}^2 = \frac{1}{N(N-1)} \sum_{i=1}^{N} (\tilde{\theta}_i - \tilde{\theta})^2$$

Asymptotic $100(1 - \alpha)\%$ confidence intervals for θ are then

$$(\tilde{\theta} - z_{\alpha/2}\widehat{V}, \tilde{\theta} + z_{\alpha/2}\widehat{V}).$$

Under certain conditions (i.e., when $\widehat{\theta}$ is not "sufficiently smooth"), $\widehat{V}^2$ may be inconsistent. One such situation is when $\widehat{\theta}$ is a sample quantile (see, for example, Miller, 1974). To overcome this difficulty, Shao (1988), Shao and Wu (1989), and Wu (1990) have studied a "delete-d" jackknife variance estimator. See these references and Maesono (1996) for further details.

In the dispersion problem, Miller jackknifed the natural logs of the sample variances rather than the sample variances themselves because the natural log transformation tends to stabilize the variance and create a distribution that is "closer" to the normal distribution. The statistic $\bar{B}$ (5.32) is an estimator of $\ln\{\text{var}(Y)\}$, the statistic $\bar{A}$ (5.32) is an estimator of $\ln\{\text{var}(X)\}$, and $\bar{A} - \bar{B}$ estimates $\ln\{\text{var}(X)/\text{var}(Y)\} = \ln \gamma^2$. The quantity $(V_1 + V_2)^{1/2}$ in the denominator of Q (5.35) is an estimator of the standard deviation of $\bar{A} - \bar{B}$. If, for example, the X's are more disperse than the Y's, $\bar{A} - \bar{B}$ would tend to be large, and this is partial motivation for procedure (5.36).

22. *Generalization.* In its most general formulation (see Comment 21) the jackknife process can be applied to any randomly selected partition of the data set into subsets of size k each, where k can be any positive integer that is a factor of the number of observations in the data set. In fact, Miller (1968) discussed the test for dispersion based on jackknifing the natural logs of the sample variances in the context of this most general formulation. However, for any integer $k > 1$ the associated Miller jackknifed variances procedures have the rather severe deficiency that it is possible for two different people to arrive at different conclusions when analyzing the same data set with the same test and at the same significance level. This possibility arises because of the variety of ways that the data set could be randomly partitioned into subsets of size k each. To avoid this undesirable feature, we have chosen to discuss the jackknifed variances procedure only for $k = 1$. In this case there is no flexibility in partitioning a data set and the associated Miller test procedures are unambiguous in their conclusions.

23. *t Distribution Approximation.* The standard normal percentiles used in (5.36) to (5.38) should be replaced by the corresponding percentile points for a t distribution with $m + n - 2$ degrees of freedom only when m and n are small and equal. For other situations, the matter of which t distribution (i.e., what degrees of freedom) should be used to find the approximating percentile is somewhat ambiguous.

24. *Point Estimators and Confidence Intervals and Bounds for γ^2.* Point estimators of and approximate confidence intervals and bounds for γ^2 can be readily obtained from the jackknife procedures. In particular, the estimator for γ^2 associated with the jackknifed variances procedures is

$$\tilde{\gamma}^2 = e^{\bar{A} - \bar{B}}. \tag{5.47}$$

Moreover, an asymptotically distribution-free confidence interval for γ^2, with approximate confidence coefficient $1 - \alpha$, based on the jackknifed variances procedures is given by

$$(\gamma_L^2, \gamma_U^2), \tag{5.48}$$

where

$$\gamma_L^2 = e^{[(\bar{A} - \bar{B}) - z_{\alpha/2}(V_1 + V_2)^{1/2}]} \tag{5.49}$$

and

$$\gamma_U^2 = e^{[(\bar{A}-\bar{B})+z_{\alpha/2}(V_1+V_2)^{1/2}]}. \qquad (5.50)$$

With γ_L^2 and γ_U^2 given by (5.49) and (5.50), we have

$$P_{\gamma^2}\{\gamma_L^2 < \gamma^2 < \gamma_U^2\} \approx 1 - \alpha. \qquad (5.51)$$

The corresponding asymptotically distribution-free approximate $100(1 - \alpha)\%$ *lower* and *upper* confidence bounds for γ^2 based on the jackknifed variances procedures are

$$\gamma_L^{*2} = e^{[(\bar{A}-\bar{B})-z_\alpha(V_1+V_2)^{1/2}]} \qquad (5.52)$$

and

$$\gamma_U^{*2} = e^{[(\bar{A}-\bar{B})+z_\alpha(V_1+V_2)^{1/2}]}, \qquad (5.53)$$

respectively, satisfying

$$P_{\gamma^2}\{\gamma_L^{*2} < \gamma^2\} \approx 1 - \alpha \quad \text{and} \quad P_{\gamma^2}\{\gamma^2 < \gamma_U^{*2}\} \approx 1 - \alpha. \qquad (5.54)$$

For the armyworm/pokeweed data in Example 5.2, the point estimate of γ^2 is $\tilde{\gamma}^2 = e^{[.7520-(-1.3856)]} = e^{2.1376} = 8.479$ and, with $\alpha = .0548$, the approximate 94.52% lower confidence bound for γ^2 is

$$\gamma_L^{*2} = e^{[(.7520-(-1.3856))-1.6(.1140+2.3664)^{1/2}]}$$
$$= e^{-.3823} = .6823.$$

25. *Asymptotic Coverage Probability.* Asymptotically (i.e., for infinitely large samples), the true coverage probabilities of the confidence interval defined by (5.48) and the confidence bounds defined by (5.52) and (5.53) will agree with the nominal confidence coefficient $1 - \alpha$. Subject to Assumptions A1, A2, and A4, this asymptotic result does not depend on the form of the distribution function $H(\cdot)$ in equation (5.2). Hence, we say that the interval given by (5.48) and the bounds given by (5.52) and (5.53) are an asymptotically distribution-free confidence interval and asymptotically distribution-free confidence bounds, respectively, for γ^2.

The interval (5.48) has also been defined so that it is "asymptotically symmetric." Here, the word symmetric refers to the equal-tail probabilities of $\alpha/2$. The $1 - \alpha$ confidence interval for γ^2 defined by (5.48) can be called asymptotically symmetric, since it is constructed so that $P_{\gamma^2}(\gamma_U^2 \leq \gamma^2) \approx P_{\gamma^2}(\gamma_L^2 \geq \gamma^2) \approx \alpha/2$. The approximation is a result of approximating the true distribution of the statistic $\bar{A} - \bar{B}$ by its asymptotic normal distribution.

26. *Lack of Robustness of the Classical $\mathcal{F}$-Test for Equal Variances.* The classical normal theory $\mathcal{F}$-test for equality of variances is not robust with respect to the assumption of normality in the sense that when the underlying populations are not normal, the true level of an $\mathcal{F}$-test that is supposed to be of size α may be quite far from α. Box (1953) gave examples in which the level of the $\mathcal{F}$-test is specified to be .05, although the actual level is as large as .166 or as small as .0056. Furthemore, there exist nonnormal populations in which, even with large samples, the level of the $\mathcal{F}$-test will not be what it is supposed to be. This nonrobustness, which was pointed out as early as 1931 by Pearson (1931), has been emphasized by Box (1953) and more recently by Miller (1968) and Shorack (1969).

Properties

1. *Consistency.* The tests given by (5.36), (5.37), and (5.38) are consistent against the alternatives $\gamma^2 >, <,$ and $\neq 1$, respectively.
2. *Asymptotic Normality.* See Miller (1968).
3. *Efficiency.* See Miller (1968) and Section 5.5.

PROBLEMS

13. The data in Table 5.3 are a portion of those collected by Bugyi et al. (1969) in a study concerned with ascertaining sodium ion content in erythrocytes (red blood cells). Such determinations are helpful in the diagnoses of certain diseases, where merely knowing the sodium ion content in plasma does not provide sufficient information. However, erythrocyte sodium ion determination is extremely variable and subject to error. This prompted the authors to propose using the flame photometric method to determine sodium ion content in erythrocytes, with the hope of providing better accuracy than can be obtained with the inefficient procedures currently employed.

    One of the ways to assess the accuracy of the proposed method is to compare the variation in erythrocyte sodium ion measurements with the variation in plasma sodium ion determinations, where it is known that the measurement variation for the flame photometric method is acceptably low. Sodium ion determinations were obtained by the flame photometric method on each of 10 plasma and 10 erythrocyte samples. Table 5.3 gives the sodium ion content in mequiv/l for the 20 samples.

    TABLE 5.3. Sodium Ion Content (mequiv/l)

Plasma	Erythrocytes
147.0	10.3
147.0	12.2
146.0	16.5
145.0	19.3
146.5	8.3
161.0	15.2
141.0	27.0
146.5	26.3
145.0	17.5
153.5	21.7

    Source: H. I. Bugyi, E. Magnier, W. Joseph, and G. Frank (1969).

    Use a Miller jackknife procedure to test the hypothesis of equal dispersions for the plasma and erythrocyte sodium ion measurements against the alternative of interest in the study. Find the approximate P-value for the test.

14. Consider the chorioamnion permeability data given in Table 4.1. Find the approximate P-value for the Miller jackknife test of the hypothesis of equal dispersions versus the alternative that the variation in tritiated water diffusion across human chorioamnion is different at term than at 12 to 26 weeks gestational age. Compare your findings with those obtained in Problem 1 using an Ansari-Bradley procedure to analyze the data.

15. Consider the television-viewing behavior data in Table 4.4. Find the approximate P-value for the Miller jackknife test of equal dispersions versus the alternative that there is more variability in the time spent in the room after witnessing the violent behavior for those children who had previously watched the *Karate Kid* than for those children who had previously watched parts of the 1984 Summer Olympic Games. Comment on your analysis in conjunction with the findings of Problems 3 and 4.5.

16. Consider the alcoholic intake data in Table 4.2. Find the approximate P-value for the Miller jackknife test of whether there is any difference in dispersions for the control and SST data. Comment on your analysis relative to the findings in Problem 9 and Example 4.2.

17. For the sodium ion determination data of Table 5.3, compute the value of the estimator $\tilde{\gamma}^2$ defined in expression (5.47).

18. For the chorioamnion permeability data given in Table 4.1, obtain the value of the estimator $\tilde{\gamma}^2$ defined in expression (5.47).

19. Obtain the value of the estimator $\tilde{\gamma}^2$ defined in expression (5.47) for the television-viewing behavior data in Table 4.4.

20. Compute the value of the estimator $\tilde{\gamma}^2$ defined in expression (5.47) for the alcoholic intake data in Table 4.2.

21. With respect to the chorioamnion permeability data given in Table 4.1, find an approximate 96.6% confidence interval for γ^2 utilizing the procedure discussed in Comment 24.

22. Consider the alcoholic intake data in Table 4.2. Using the procedure discussed in Comment 24, find an approximate 93.72% confidence interval for γ^2.

23. Consider the television-viewing behavior data in Table 4.4. Labeling the Olympic watchers data as the X sample, use the procedure discussed in Comment 24 to find an approximate 98.96% upper confidence bound for γ^2.

24. Consider the sodium ion determination data of Table 5.3. Labeling the erythrocyte sodium ion measurements as the X sample, use the procedure discussed in Comment 24 to find an approximate 91.92% lower confidence bound for γ^2.

5.3. A DISTRIBUTION-FREE RANK TEST FOR EITHER LOCATION OR DISPERSION (LEPAGE)

Hypothesis

Let $X_1, \ldots, X_m$ and $Y_1, \ldots, Y_n$ be independent random samples satisfying Assumptions A1 and A2 from continuous populations with distribution functions F and G, respectively, satisfying the location-scale parameter model relationships in equations (5.2) and (5.3).

Under these assumptions, we are interested in assessing whether there are differences in *either* the location parameters (i.e., medians) θ_1 and θ_2 *or* the scale parameters η_1 and η_2 for the X and Y populations. Thus, we are interested in testing the null hypothesis H_0 (5.1) versus the general alternative $H_1 : [\theta_1 \neq \theta_2 \text{ and/or } \eta_1 \neq \eta_2]$. Note that under the location-scale parameter model, as stated in equations (5.2) and (5.3), the null hypotheses H_0 (5.1) reduces to $H_0 : [\theta_1 = \theta_2 \text{ and } \eta_1 = \eta_2]$, corresponding to the assertion that both the population location parameters and the population scale parameters are equal.

Procedure

To compute the Lepage two-sample location-scale statistic D, order the combined sample of $N = (m+n)$ X-values and Y-values from least to greatest. Let S_j denote the combined samples rank of Y_j, for $j = 1, \ldots, n$, and let $W = \sum_{j=1}^{n} S_j$ be the Wilcoxon rank sum statistic defined in equation (4.3). In addition, for $j = 1, \ldots, n$, let R_j be the score assigned to Y_j by the Ansari-Bradley scoring scheme discussed in the Procedure of Section 5.2 and let $C = \sum_{j=1}^{n} R_j$ be the Ansari-Bradley scale statistic defined in equation (5.6). The Lepage rank statistic is then defined by

$$D = \frac{[W - E_0(W)]^2}{\text{var}_0(W)} + \frac{[C - E_0(C)]^2}{\text{var}_0(C)}, \quad (5.55)$$

where $E_0(W)$ and $\text{var}_0(W)$ are the expected value and variance of W under H_0(5.1), as given in equations (4.7) and (4.8), respectively, and $E_0(C)$ and $\text{var}_0(C)$ are the corresponding expected value and variance of C under H_0 (5.1), as stated in equations (5.10) and (5.11), respectively, when $N = (m + n)$ is an even number, or in equations (5.12) and (5.13), respectively, when N is odd. Thus, if we let W^* (4.9) and C^* (5.20) represent the standardized forms for the Wilcoxon rank sum statistic and Ansari-Bradley scale statistic, respectively, then the Lepage statistic D can be written as

$$D = (W^*)^2 + (C^*)^2. \tag{5.56}$$

To test H_0 (5.1), corresponding to equality of both the location and scale parameters for the X and Y populations, versus the general alternatives that either the location parameters are different or the scale parameters are different or both, corresponding to

$$H_1 : [\theta_1 \neq \theta_2 \text{ and/or } \eta_1 \neq \eta_2], \tag{5.57}$$

at the α level of significance,

$$\text{Reject } H_0 \text{ if } D \geq d_\alpha; \text{ otherwise do not reject}, \tag{5.58}$$

where the constant d_α is chosen to make the type I error probability equal to α. Values of d_α are given in Table A.9. We enter Table A.9 with the X and Y labels assigned to the samples in such a way that there are at least as many Y observations as there are X observations; that is, the sample X and Y labels are assigned so that $m \leq n$. (We note that this labeling of the X and Y samples so that $m \leq n$ for the Lepage procedure and the use of Table A.9 is opposite to the manner in which we labeled the X and Y samples for the use of Tables A.6 and A.8 for the Wilcoxon rank sum and Ansari-Bradley test procedures, respectively. For the use of Tables A.6 and A.8 with those procedures, we labeled the X and Y samples so that $n \leq m$. However, the formulas for the Wilcoxon rank sum and Ansari-Bradley statistics and their null means and null variances, as used in the computation of the Lepage test statistic D (5.56), are valid whether $m \leq n$ or $n \leq m$. The sample labeling, such that $m \leq n$, is required solely to use the null distribution critical values in Table A.9 for the Lepage test procedure, since they correspond to the form of the test statistic D (5.56) associated with summing the Wilcoxon and Ansari-Bradley ranks for the sample (Y) with the *greater* number (n) of observations.)

Large-Sample Approximation

The large-sample approximation is based on the fact that when H_0 (5.1) is true, the statistic D has, as $\min(m, n)$ tends to infinity, a chi-square distribution with two degrees of freedom. (See Comment 31 for indications of the proof.) The large-sample approximation for the exact level α procedure in (5.58) is

$$\text{Reject } H_0 \text{ if } D \geq \chi^2_{2,\alpha}; \text{ otherwise do not reject}, \tag{5.59}$$

where $\chi^2_{2,\alpha}$ is the upper α percentile point of the chi-square distribution with two degrees of freedom. Values of $\chi^2_{2,\alpha}$ can be found from Chart A.2.

Ties

If there are ties among the X and/or Y observations, we modify the standardized Wilcoxon rank sum statisic W^* and the standardized Ansari-Bradley scale statistic C^* in the manners prescribed for the large-sample approximations in the Ties portions of Sections 4.1 and 5.1, respectively. When applying either the small-sample procedure in (5.58) or the large-sample approximation in (5.59), the Lepage statistic D should be computed using these ties-modified versions of W^* and C^*. The corresponding modified version of procedure (5.58) in the case of ties among the X and/or Y observations is only approximately, and not exactly, of significance level α. (To get an exact level α test even in this tied setting, see Comment 32.)

Example 5.3: *Effect of Maternal Steroid Therapy on Platelet Counts of Newborn Infants.* Autoimmune thrombocytopenic purpura (ATP) is a disease in which the patient produces antibodies to her own platelets. Because of transplacental passage of antiplatelet antibodies during pregnancy, children of women with ATP are often born with low platelet counts. For this reason, there is medical concern that a vaginal delivery for a mother with (ATP) could result in intracranial hemorrhage for the infant. However, the proper obstetrical management of pregnant women with ATP is controversial. Most doctors have advocated cesarean section as the preferable method of delivery for mothers with ATP. Others suggest that cesarean section, with its obvious complications for both mother and infant, be avoided unless there is some additional obstetrical reason for it. Karpatkin, Porges, and Karpatkin (1981) studied the effect of administering the corticosteroid prednisone to pregnant women with ATP with the intent of raising the infants' platelet counts to safe levels during their deliveries. The rationale for this treatment is the fact that steroids, in general, increase the platelet counts in patients with ATP by blocking splenic destruction of antibody-coated platelets. In theory then, the corticosteroid prednisone should cross the placenta, enter the infant's circulation, and prevent splenic removal of those infant's platelets which are coated by the mother's antibodies.

The data in Table 5.4 are a subset of the data obtained by Karpatkin et al. in their study of the effect that administration of prednisone to pregnant women with ATP had on their infants' platelet counts. All the infants included in this example were delivered vaginally. Table 5.4 gives the platelet counts (per cubic millimeter) of 10 infants whose mothers received the steroid prednisone prior to delivery and 6 infants whose mothers were not treated with prednisone prior to delivery. All 16 mothers in the study were diagnosed with ATP.

The primary interest in the study is in whether or not the predelivery administration of prednisone typically leads to an increased newborn platelet count. Thus the principal statistical issue in the study is that of a possible difference in locations for the prednisone and nonprednisone populations. However, there is some concern that the administration of predelivery prednisone could also lead to a rather large increase in variability in the newborn platelet

TABLE 5.4. Platelet Counts of Newborn Infants (per mm^3)

Mothers Given Prednisone	Mothers Not Given Prednisone
120,000	12,000
124,000	20,000
215,000	112,000
90,000	32,000
67,000	60,000
95,000	40,000
190,000	
180,000	
135,000	
399,000	

Source: M. Karpatkin, R. F. Porges, and S. Karpatkin (1981).

counts. (Such a finding would certainly affect our interpretation of any possible increase in typical platelet count resulting from the prednisone.) As a result, we will apply the Lepage test procedure to test H_0 (5.1) versus the general alternative H_1 (5.57). For purposes of illustration, we consider the exact procedure (5.58) with level of significance $\alpha = .02$. In order to use Table A.9, we take the infant platelet count data for mothers given prednisone to be the Y sample ($n = 10$) and the corresponding control (nonprednisone) data to be the X sample ($m = 6$). Thus, we see from Table A.9 that the level $\alpha = .02$ application of procedure (5.58) is

$$\text{Reject } H_0 \text{ if } D \geq 6.9025. \tag{5.60}$$

To calculate D, we need first to calculate the standardized versions of the Wilcoxon rank sum and Ansari-Bradley statistics. Proceeding as in equation (4.3), we note that the combined samples ranks for the ten Y observations are 10, 11, 16, 7, 6, 8, 14, 13, 12, and 15, yielding a value of

$$W = 10 + 11 + 16 + 7 + 6 + 8 + 14 + 13 + 12 + 15 = 112$$

for the Wilcoxon rank sum statistic. Since there are no ties among the sixteen X and Y observations, it follows from equations (4.7), (4.8), and (4.9) that the standardized form of W for the data in Table 5.4 is

$$W^* = \frac{112 - \{10(6+10+1)/2\}}{\{6(10)(6+10+1)/12\}^{1/2}} = 2.929. \tag{5.61}$$

For the calculation of the standardized Ansari-Bradley statistic C^*, we observe that the Ansari-Bradley scores (as defined in the Procedure of Section 5.1) for the ten Y observations are 7, 6, 2, 7, 6, 8, 3, 4, 5, and 1. From equation (5.6), this produces a value of

$$C = 7 + 6 + 2 + 7 + 6 + 8 + 3 + 4 + 5 + 1 = 49.$$

Since $N = 6 + 10 = 16$ is an even number and there are no ties among the sixteen X and Y observations, it follows from (5.10), (5.11), and (5.14) that the standardized form of C for the data in Table 5.4 is

$$C^* = \frac{49 - \{10(16+2)/4\}}{\left\{\frac{10(6)(16+2)(16-2)}{48(16-1)}\right\}^{1/2}} = .873. \tag{5.62}$$

Using these values of W^* (5.61) and C^* (5.62) in equation (5.56) yields

$$D = (2.929)^2 + (.873)^2 = 9.341, \tag{5.63}$$

which, in view of expression (5.60), tells us to reject H_0 at the $\alpha = .02$ level, since $D = 9.341 > d_{.02} = 6.9025$. Hence, there is rather strong evidence that there are differences in locations or scales (or both) between the prednisone and control infant platelet count populations. In fact, we see from Table A.9 that the P-value for these data with observed value $C^* = 9.341$ is less than .01 (since $d_{.01} = 8.0275$), providing even a stronger statement in favor of the alternative H_1 (5.57).

For the large-sample approximation we see from equation (5.59) that the approximate P-value for these data is

$$P\text{-value} \approx P(Q \geq 9.341),$$

where Q has a chi-square distribution with two degrees of freedom. From Chart A.2 we find that this approximate P-value is roughly .009, in good agreement with the exact upper bound of .01 previously obtained from Table A.9.

We conclude this example by noting that the large value of D is due primarily to a large value of W^*. This would suggest intuitively that the rejection of H_0 is due primarily to a difference in locations between the infant platelet counts for the prednisone and control populations. However, we emphasize that such a conclusion is not statistically justified through the application of the general Lepage procedure (5.58). The only valid conclusion based on the Lepage procedure is that of the general alternative H_1 (5.57). (If you do, however, apply the Wilcoxon rank sum procedure of Section 4.1 to the data in Table 5.4, you would be able to conclude that there is, indeed, a difference in locations between the infant platelet counts for the prednisone and control populations. In view of this fact, would it be legitimate to then apply the Ansari-Bradley procedure of Section 5.1 directly to the data in Table 5.4 to test for possible scale differences in the two populations?)

Comments

27. *Motivation for the Test*. From Section 4.1 we know that a large value of $(W^*)^2$ is indicative of a possible difference in locations for the X and Y populations. We also know from Section 5.1 that a large value of $(C^*)^2$ is indicative of a possible difference in dispersions for the X and Y populations. Since D (5.56) will be large if and only if either $(W^*)^2$ is large or $(C^*)^2$ is large or both, then such a large value of D is indicative of either $\theta_1 \neq \theta_2$ or $\eta_1 \neq \eta_2$ or both. This serves as partial motivation for the test procedure given by (5.58).

28. *Derivation of the Distribution of D under H_0 (No-Ties Case)*. Under H_0 (5.1) each of the $\binom{N}{n}$ possible "meshings" of the X's and Y's has probability $1/\binom{N}{n}$. This fact can be used to obtain the null distribution of D (5.56). We illustrate the steps involved in constructing this null distribution for the simple case $m = 2, n = 2$. Since $N = 4$, we must consider $\binom{4}{2} = 6$ possible meshings of the X and Y observations. For this setting, it follows from equations (4.7) and (4.8) that $E_0(W) = 2(2 + 1)/2 = 5$ and $\text{var}_0(W) = 2(2)(2 + 1)/12 = \frac{5}{3}$. Similarly, from equations (5.10) and (5.11) we have $E_0(C) = 2(4 + 2)/4 = 3$ and $\text{var}_0(C) = [2(2)(4 + 2)(4 - 2)]/48(4 - 1) = \frac{1}{3}$. Thus, for $m = n = 2$ we have $(W^*)^2 = 3(W - 5)^2/5$ and $(C^*)^2 = 3(C - 3)^2$. Using these facts and the same approach taken in Comments 4.3 and 5.4 for the calculations of W and C, respectively, the values of D for these six meshings are given in the following table.

Meshing	Probability	$D = (W^*)^2 + (C^*)^2$
XXYY	$\frac{1}{6}$	2.4
XYXY	$\frac{1}{6}$	.6
YXXY	$\frac{1}{6}$	3.0
XYYX	$\frac{1}{6}$	3.0
YXYX	$\frac{1}{6}$	.6
YYXX	$\frac{1}{6}$	2.4

Thus, for example, the probability is $\frac{1}{3}$ under H_0 that D is equal to .6, since $D = .6$ when either of the exclusive meshings $XYXY$ or $YXYX$ occurs and each of these meshings has null probability $\frac{1}{6}$. Proceeding in the same manner for all possible values for D and simplifying, we obtain the null distribution.

Possible Value of D	Probability under H_0
.6	$\frac{1}{3}$
2.4	$\frac{1}{3}$
3.0	$\frac{1}{3}$

Thus, for example, the probability under H_0 that D is greater than or equal to 3 is, therefore, $P_0(D \geq 3) = \frac{1}{3}$, which implies that $d_{1/3} = 3$ for the setting $m = n = 2$.

Note that we have derived the null distribution of D without specifying the form of the common (under H_0) underlying X and Y populations beyond the point of requiring that they be continuous. This is why the test procedure based on D is called a distribution-free procedure. From the null distribution of D we can determine the critical value d_α and control the probability α of falsely rejecting H_0 when H_0 is true, and this error probability does not depend on the specific form of the common underlying distribution for the X and Y populations.

29. *Equivalent Form.* In computing the Wilcoxon rank sum statistic W (4.3) we use the combined samples ranks of the Y observations. In computing the Ansari-Bradley statistic C (5.6) we used the scores assigned to the Y observations by the Ansari-Bradley outside-in scoring scheme. However, both W and C, and therefore D (5.55), can be computed solely from knowledge of the combined samples ranks of the Y observations. This follows directly from the fact that the Ansari-Bradley statistic can also be represented (in the case of no tied X and/or Y observations) as

$$C = \frac{n(N+1)}{2} - \sum_{j=1}^{n} \left| S_j - \frac{N+1}{2} \right|, \qquad (5.64)$$

where, as in the calculation of W, S_j is the combined samples rank of Y_j, for $j = 1, \ldots, n$. In fact, both W and C are members of a very large class of statistics based solely on the combined samples ranks in a special way. This collection is referred to as the class of two-sample linear rank statistics, and they have been extensively studied in the literature (see, for example, Randles and Wolfe (1979)).

Thus, although the Ansari-Bradley scoring scheme is useful in helping to motivate the statistic C as one appropriate for assessing possible scale differences in the X and Y populations, we could, in view of (5.64), just as easily have initially defined C in terms of the combined samples ranks as we did W (4.3). This means, of course, that D (5.55) is also a function of the X and Y observations only through their combined samples ranks.

30. *Assumptions.* We can use the Lepage test procedure in (5.58) without even requiring that the variances for the X and Y populations exist. Indeed, neither Assumptions A1 and A2 nor the location-scale parameter model in (5.2) and (5.3) specify anything about the existence of even the first moments of the X and Y populations. However, when the first two moments (and, therefore, the variance) for the underlying distributional model $H(u)$ in (5.2) exist, we see from the equal-in-distribution statement in (5.3) that

$$\text{var}\left(\frac{X}{\eta_1}\right) = \text{var}\left(\frac{Y}{\eta_2}\right),$$

which, in turn, implies that

$$\frac{\text{var}(X)}{\eta_1^2} = \frac{\text{var}(Y)}{\eta_2^2}.$$

Thus, when the variances exist, we see that $\gamma^2 = [\eta_1^2/\eta_2^2] = [\text{var}(X)/\text{var}(Y)]$.

It also follows from (5.3) that

$$E\left[\frac{X - \theta_1}{\eta_1}\right] = E\left[\frac{Y - \theta_2}{\eta_2}\right],$$

provided only that the first moment exists for $H(u)$ in (5.2). Thus, if this first moment exists, we have the relationship

$$E[Y] - \theta_2 = \frac{\eta_2}{\eta_1}\{E[X] - \theta_1\}.$$

As a result, if $\gamma^2 = \eta_1^2/\eta_2^2 = 1$, then $E[Y] - E[X] = \theta_2 - \theta_1$, corresponding to the standard interpretation of a location-only difference between two populations. This is the setting previously considered in Chapter 4 with the identification $\Delta = \theta_2 - \theta_1$. (We emphasize, however, that the existence of the first moment is not a necessary assumption for any of the statistical procedures developed in Chapter 4.)

31. *Large-Sample Approximation.* We have previously seen that both W^* (see Comment 4.6) and C^* (see Comment 9) have asymptotic standard normal distributions under H_0 (5.1) as $\min(m, n)$ becomes infinite. Moreover, it can be shown (see, for example, Lepage (1971)) that W^* and C^* are asymptotically independent under H_0 (5.1) as $\min(m, n)$ becomes infinite. The conclusion that the statistic D (5.56) has, as $\min(m, n)$ tends to infinity, an asymptotic distribution under H_0 (5.1) that is chi-square with two degrees of freedom then follows from the properties (i) the square of a standard normal variable has a chi-square distribution with one degree of freedom and (ii) the sum of independent chi-square variables with degrees of freedom f_1 and f_2 has a chi-square distribution with $f_1 + f_2$ degrees of freedom.

32. *Exact Conditional Distribution of D with Ties.* To have a test with exact significance level even in the presence of ties among the X's and/or Y's, we need to consider all $\binom{N}{n}$ possible assignments of the N observations, with n observations serving as Y's and m observations serving as X's. As in Comment 28, it still follows that under H_0 (5.1), each of the $\binom{N}{n}$ possible "meshings" of the X's and Y's has probability $1/\binom{N}{n}$. The only difference in the case of ties (see Ties in this section) is that we now use average scores and the appropriately modified $\text{var}_0(C)$ in the computation of C^* and average ranks and the appropriately modified $\text{var}_0(W)$ in the computation of W^* to calculate the value of D for each of these $\binom{N}{n}$ meshings leading to the tabulation of the exact conditional null distribution of D.

An example illustrating how to obtain such a conditional null distribution of D for a specific case of tied observations is not included here, because the details are much the same as those provided in Comments 4.5 and 11 for the conditional null distributions of W and C, respectively, in the presence of tied observations. As for those two settings, the computer software `StatXact` can be helpful in obtaining the exact conditional null distribution of D when there are ties.

33. *More General Alternatives.* In his original discussion of the test procedure based on D, Lepage (1971) considered a slightly more general setting than that dictated by the location-scale parameter model in (5.2). In addition to Assumptions A1 and A2, he required that the X

distribution function F and the Y distribution function G be related by the equation

$$G(t) = F(at + b), \quad \text{for every } t, \tag{5.65}$$

for some constants $a > 0$ and $-\infty < b < \infty$. He then considered tests of $H_0^* : [a = 1, b = 0]$ versus $H_1^* : [a \neq 1 \text{ or } b \neq 0 \text{ or both}]$. His null hypothesis H_0^* is, of course, identical to H_0 (5.1) considered in this section. However, his alternative H_1^* is more general than the location-scale parameter alternative H_1 (5.57) discussed here. The alternative H_1 (5.57) represents a slightly reduced subset of H_1^* corresponding to the identifications $a = \eta_1/\eta_2$ and $b = (\eta_2 \theta_1 - \eta_1 \theta_2)/\eta_2$.

34. *Consistency of the D Tests.* Let $\delta^* = [P(X < Y) - \frac{1}{2}]$ and $\Delta_\theta^* = [P(X > Y > \theta) + P(X < Y < \theta) - \frac{1}{4}]$. Under the minimal Assumptions A1 and A2 only, the test procedure (5.58) based on D is consistent if either $\delta^* \neq 0$ or $\theta_1 = \theta_2 = \theta$ and $\Delta_\theta^* \neq 0$. (See Comments 4.15 and 14.)

Under Assumptions A1, A2, and the additional general distributional relationship given by (5.65), the test procedure (5.58) based on D is consistent against any alternative for which either $a \neq 1$ or $b \neq 0$.

Properties

1. *Consistency.* For our statement we consider the more stringent location-scale parameter model described in (5.2). Then the test defined by (5.58) is consistent against alternatives for which either $\eta_1 \neq \eta_2$ or $\theta_1 \neq \theta_2$. See also Comment 34.
2. *Asymptotic Chi-Squareness.* See Lepage (1971) and Comment 31.
3. *Efficiency.* See Section 5.5.

PROBLEMS

25. It has long been generally accepted by medical doctors that exercise tends to stimulate the release of growth hormones in adolescents. However, little previous research had been directed toward assessment of possible effects that various medications might have on this phenomenon. This fact led Falkner et al. (1981) to investigate whether the use of the drug clonidine to treat hypertension in adolescents has any effect on this exercise-induced release of growth hormones. Two groups of adolescents were involved in the study. The first was a control group consisting of 10 teenagers who had been diagnosed as hypertensive but were not being treated with clonidine. (Note that the "control" group considered by Falkner et al. included an additional 7 nonhypertensive teenagers. In order not to possibly confound the effects of hypertension itself and the treatment clonidine on the release of the growth hormone during exercise, these 7 subjects are not included in the control group presented in the problem. In addition, 2 subjects studied both as controls and again later after clonidine treatment are included here only in the control sample.) The second treatment group consisted of 13 hypertensive teenagers who were being treated with clonidine.

    The experiment proceeded as follows. First the basal level of growth hormone in the blood was measured for each of the subjects prior to exercising. Then each subject exercised on a treadmill until attaining a heart rate of 180 to 200 beats per minute, at which time the blood level of growth hormone was once again obtained. The data in Table 5.5 represent these pre- and postexercise growth hormone blood levels (in ng/ml) for the 23 subjects in the study.

    Use an appropriate nonparametric test procedure to assess whether there are significant location or dispersion differences between the control hypertension population and the clonidine treated population in their increases in growth hormone levels following exercise. Find the approximate P-value for the test.

TABLE 5.5. Growth Hormone Level (ng/ml)

	Preexercise	Postexercise
Control		
1	1.3	19.0
2	1.3	40.0
3	5.8	3.8
4	2.0	6.5
5	2.7	16.0
6	1.7	13.0
7	1.8	18.0
8	1.7	2.6
9	1.8	18.0
10	4.7	5.8
Clonidine-treated		
1	1.2	5.1
2	1.2	7.2
3	5.8	14.0
4	0.3	4.0
5	3.3	25.0
6	2.2	15.0
7	4.1	10.0
8	1.2	7.6
9	6.4	10.0
10	1.8	10.0
11	1.8	8.0
12	5.2	40.0
13	1.3	21.0

Source: B. Falkner, G. Onesti, T. Moshang, Jr., and D. T. Lowenthal (1981).

26. In Example 5.3 we used a Lepage test procedure to assess whether or not the administration of the corticosteroid prednisone to pregnant women with ATP (autoimmune thrombocytopenic purpura) resulted in any location or dispersion changes in the platelet counts of their newborn infants. It would also be of interest to know whether there were any baseline (predelivery) differences in the platelet counts of those mothers in the study who were given the prednisone and those who served as the no-prednisone control group. The platelet count (per mm^3) data for the mothers are given in Table 5.6.

Find the approximate P-value for an appropriate nonparametric test procedure to assess whether there are any significant location or dispersion differences in the predelivery maternal platelet counts for the control and prednisone-treated groups.

TABLE 5.6. Maternal Platelet Counts (per mm^3)

Mothers Given Prednisone	Mothers Not Given Prednisone
12,000	15,000
25,000	44,000
30,000	52,000
38,000	64,000
50,000	65,000
80,000	80,000
85,000	
126,000	
130,000	
180,000	

Source: M. Karpatkin, R. F. Porges, and S. Karpatkin (1981).

27. When there are no tied X and/or Y observations, show that the representation for C given in equation (5.64) in Comment 29 is indeed equivalent to the original definition of C in equation (5.6).
28. Generate the exact null distribution of D for the setting $m = 2, n = 3$. (See Comment 28.)
29. Consider the general relationship between the distribution functions for the X and Y populations prescribed in equation (5.65) of Comment 33. Verify that the location-scale parameter model relationship given in equation (5.2) corresponds to the special case of (5.65) with $a = \eta_1/\eta_2$ and $b = (\eta_2\theta_1 - \eta_1\theta_2)/\eta_2$.
30. Consider the television-viewing behavior data in Table 4.4. For these data find the approximate P-value for an appropriate test of whether there are either location or dispersion differences in the time spent in the room after witnessing the violent behavior for those children who had previously watched the *Karate Kid* versus those children who had previously watched parts of the 1984 Summer Olympic Games. Comment on your finding in view of the results of Problems 3 and 4.5.
31. Consider the alcoholic intake data in Table 4.2. For these data find the approximate P-value for an appropriate test of whether there are either location or dispersion differences between the control and SST data. Discuss the result in conjunction with the previous findings in Example 4.2 and Problem 9.
32. Consider the following two-sample data for $m = 3, n = 3$: $X_1 = -3.7, X_2 = 4.6, X_3 = 1.5, Y_1 = 1.5, Y_2 = 4.6, Y_3 = 1.5$. Using the approach discussed in Comment 32, find the exact conditional null distribution of the Lepage statistic D (5.55). Compare and contrast the upper $\alpha = .10$ percentile for this exact conditional null distribution with the corresponding upper $\alpha = .10$ percentile for the null distribution of D given in Table A.9 for $m = n = 3$ and no tied observations.

5.4. A DISTRIBUTION-FREE TEST FOR GENERAL DIFFERENCES IN TWO POPULATIONS (KOLMOGOROV-SMIRNOV)

Hypothesis

Let $X_1, \ldots, X_m$ and $Y_1, \ldots, Y_n$ be independent random samples satisfying Assumptions A1 and A2 from continuous populations with distribution functions F and G, respectively. Under these assumptions we are interested in assessing whether there are *any* differences whatsoever between the X and Y probability distributions. Thus, we are interested in testing the null hypothesis H_0 (5.1) against the most general alternative possible, namely,

$$H_1 : [F(t) \neq G(t) \text{ for at least one } t]. \tag{5.66}$$

Procedure

To compute the two-sided two-sample Kolmogorov-Smirnov general alternative statistic J, we first need to obtain the empirical distribution functions for the X and Y samples. For every real number t, let

$$F_m(t) = \frac{\text{number of sample } X\text{'s} \leq t}{m} \tag{5.67}$$

and

$$G_n(t) = \frac{\text{number of sample } Y\text{'s} \leq t}{n}. \tag{5.68}$$

(The functions $F_m(t)$ and $G_n(t)$ are called the *empirical distribution functions* for the X and Y samples, respectively.) Let

$$d = \text{greatest common divisor of } m \text{ and } n \tag{5.69}$$

and set

$$J = \frac{mn}{d} \max_{(-\infty < t < \infty)} \{|F_m(t) - G_n(t)|\}. \tag{5.70}$$

The statistic J is the two-sided two-sample Kolmogorov-Smirnov statistic. To actually calculate J for given X and Y samples, we use the fact that $F_m(t)$ and $G_n(t)$ are step functions changing functional values only at the observed X and Y sample observations, respectively. Thus, if we let $Z_{(1)} \leq \cdots \leq Z_{(N)}$ denote the $N = (m + n)$ ordered values for the combined sample of $X_1, \ldots, X_m$ and $Y_1, \ldots, Y_n$, then we can rewrite J (5.70) in the computational form

$$J = \frac{mn}{d} \max_{i=1,\ldots,N} \{|F_m(Z_{(i)}) - G_n(Z_{(i)})|\}. \tag{5.71}$$

To test H_0 (5.1), corresponding to identical X and Y probability distributions, versus the general alternative H_1 (5.66), corresponding to *any* possible difference between the X and Y probability distributions, at the α level of significance,

$$\text{Reject } H_0 \text{ if } J \geq j_\alpha; \quad \text{otherwise do not reject,} \tag{5.72}$$

where the constant j_α is chosen to make the type I error probability equal to α. Values of j_α are given in Table A.10. Enter Table A.10 with the X and Y labels assigned to the samples in such a way that there are at least as many Y observations as there are X observations; that is, the sample X and Y labels are assigned so that $m \leq n$.

Large-Sample Approximation

The large-sample approximation is based on the asymptotic distribution of J, suitably normalized, as $\min(m, n)$ tends to infinity. Set

$$J^* = \left(\frac{mn}{N}\right)^{1/2} \max_{i=1,\ldots,N} \{|F_m(Z_{(i)}) - G_n(Z_{(i)})|\} = \frac{d}{(mnN)^{1/2}} J. \tag{5.73}$$

As $\min(m, n)$ tends to infinity,

$$P_0(J^* < s) \longrightarrow \sum_{k=-\infty}^{\infty} (-1)^k e^{-2k^2 s^2}, 0 \quad \text{for} \quad s >, \leq 0. \tag{5.74}$$

Table A.11 lists values of the function $Q(s)$ defined by

$$Q(s) = 1 - \sum_{k=-\infty}^{\infty} (-1)^k e^{-2k^2 s^2}, \quad s > 0. \tag{5.75}$$

The large-sample approximation to procedure (5.72) based on (5.74) and (5.75) is

$$\text{Reject } H_0 \text{ if } J^* \geq q_\alpha^*; \quad \text{otherwise do not reject,} \tag{5.76}$$

where q_α^* is defined by

$$Q(q_\alpha^*) = \alpha, \tag{5.77}$$

and can be obtained from Table A.11.

5. THE TWO-SAMPLE DISPERSION PROBLEM AND OTHER TWO-SAMPLE PROBLEMS

TABLE 5.7. Mean Drop Differences

Feedback Group	No-Feedback Group
−.15	2.55
8.60	12.07
5.00	.46
3.71	.35
4.29	2.69
7.74	−.94
2.48	1.73
3.25	.73
−1.15	−.35
8.38	−.37

Source: F. C. Delse and B. W. Feather (1968).

Ties

The empirical distribution functions $F_m(t)$ and $G_n(t)$, given by (5.67) and (5.68), respectively, are well-defined in the case of ties and no adjustments are necessary in the calculation of J (5.70). (See Comment 39.) The test is then conducted using the same critical point j_α (5.72) as specified for the untied case. This approach is conservative; it yields a test with a significance level that does not exceed the nominal level α (see Hájek and Šidák (1967, 123), Noether (1963), and Walsh (1963)). For different methods of treating ties when using the Kolmogorov-Smirnov statistic J, see Hájek (1969, 134, 145).

Example 5.4: *Effect of Feedback on Salivation Rate.* The effect of enabling a subject to hear himself salivate while trying to increase or decrease his salivary rate has been studied by Delse and Feather (1968). Two groups of subjects were told to attempt to increase their salivary rates upon observing a light to the left, and decrease their salivary rates upon observing a light to the right. The apparatus for collecting and recording the amounts of saliva was described by Delse and Feather (1968) and also Feather and Wells (1966). Members of the feedback group received a 0.2-s, 1000-cps tone for each drop collected, whereas members of the no-feedback group did not receive any indication of their salivary rates. Table 5.7 gives differences of the form mean number of drops over 13 increase signals − mean number of drops over 13 decrease signals for the feedback group and the no-feedback group, each group consisting of 10 subjects.

Since the sample sizes are both equal to 10, we arbitrarily choose to label the feedback group data as the X sample and the no-feedback group data as the Y sample. Thus we have $m = n = 10, N = (10 + 10) = 20$, and $d = 10$. We simultaneously illustrate the calculation of the values of the empirical distribution functions $F_{10}(t)$ and $G_{10}(t)$ at the ordered combined sample values $Z_{(1)} \leq \cdots \leq Z_{(20)}$ from Table 5.7, as well as the absolute differences $|F_{10}(Z_{(i)}) - G_{10}(Z_{(i)})|$, in the following display.

i	$Z_{(i)}$	$F_{10}(Z_{(i)})$	$G_{10}(Z_{(i)})$	$\|F_{10}(Z_{(i)}) - G_{10}(Z_{(i)})\|$
1	−1.15	$\frac{1}{10}$	$\frac{0}{10}$	$\frac{1}{10}$
2	−.94	$\frac{1}{10}$	$\frac{1}{10}$	0
3	−.37	$\frac{1}{10}$	$\frac{2}{10}$	$\frac{1}{10}$
4	−.35	$\frac{1}{10}$	$\frac{3}{10}$	$\frac{2}{10}$

(continued)

5.4. A DISTRIBUTION-FREE TEST FOR GENERAL DIFFERENCES

i	$Z_{(i)}$	$F_{10}(Z_{(i)})$	$G_{10}(Z_{(i)})$	$\|F_{10}(Z_{(i)}) - G_{10}(Z_{(i)})\|$
5	−.15	$\frac{2}{10}$	$\frac{3}{10}$	$\frac{1}{10}$
6	.35	$\frac{2}{10}$	$\frac{4}{10}$	$\frac{2}{10}$
7	.46	$\frac{2}{10}$	$\frac{5}{10}$	$\frac{3}{10}$
8	.73	$\frac{2}{10}$	$\frac{6}{10}$	$\frac{4}{10}$
9	1.73	$\frac{2}{10}$	$\frac{7}{10}$	$\frac{5}{10}$
10	2.48	$\frac{3}{10}$	$\frac{7}{10}$	$\frac{4}{10}$
11	2.55	$\frac{3}{10}$	$\frac{8}{10}$	$\frac{5}{10}$
12	2.69	$\frac{3}{10}$	$\frac{9}{10}$	$\frac{6}{10}$
13	3.25	$\frac{4}{10}$	$\frac{9}{10}$	$\frac{5}{10}$
14	3.71	$\frac{5}{10}$	$\frac{9}{10}$	$\frac{4}{10}$
15	4.29	$\frac{6}{10}$	$\frac{9}{10}$	$\frac{3}{10}$
16	5.00	$\frac{7}{10}$	$\frac{9}{10}$	$\frac{2}{10}$
17	7.74	$\frac{8}{10}$	$\frac{9}{10}$	$\frac{1}{10}$
18	8.38	$\frac{9}{10}$	$\frac{9}{10}$	0
19	8.60	$\frac{10}{10}$	$\frac{9}{10}$	$\frac{1}{10}$
20	12.07	$\frac{10}{10}$	$\frac{10}{10}$	0

For example, consider the evaluation of $F_{10}(Z_{(4)})$. We must count the number of X's less than or equal to $Z_{(4)} = -.35$, and divide this count by 10. From Table 5.7 we find that only one of the X values (−1.15) is less than −.35, none is equal to −.35, and thus $F_{10}(Z_{(4)}) = \frac{1}{10}$. Similarly, $G_{10}(Z_{(4)})$ is equal to {the number of Y's that are less than or equal to $-.35$}/10. From Table 5.7 we find two Y-values (−.94 and −.37) that are less than −.35 and one Y-value that is equal to −.35; thus $G_{10}(Z_{(4)}) = \frac{3}{10}$. From this computational Table for the $|F_{10}(Z_{(i)}) - G_{10}(Z_{(i)})|$ values we find

$$\max_{i=1,\ldots,20}\{|F_{10}(Z_{(i)}) - G_{10}(Z_{(i)})|\} = \tfrac{6}{10},$$

corresponding to $Z_{(12)}$. It follows from equation (5.71) that $J = [(10)(10)/10](6/10) = 6$.

From Table A.10 we find $P_0(J \geq 6) = .0524$. That is, in the notation of (5.72) with $m = n = 10$, we have $j_{.0524} = 6$. Thus the lowest level at which we can reject H_0 (5.1) with our observed value of $J = 6$ (i.e., the P-value for the data) using procedure (5.72) is .0524, indicating some marginal evidence in the samples that feedback might have an effect on salivation rate.

To perform the large-sample approximation, we compute J^* (5.73). We find that $J^* = \{10/[10(10)(20)]^{1/2}\}(6) = 1.34$. From Table A.11 we find $Q(1.34) = .0551$. Thus the smallest significance level at which we reject H_0, using the large-sample approximation to the Kolmogorov-Smirnov test, is approximately .0551.

Comments

35. *Motivation for the Test.* The empirical distribution functions $F_m(t)$ (5.67) and $G_n(t)$ (5.68) are estimators of the underlying distribution functions $F(t) = P\{X \leq t\}$ and $G(t) = P\{Y \leq t\}$, respectively. Thus dJ/mn may be viewed as an estimator of $\max_{-\infty < t < \infty} |F(t) - G(t)| = \max_{-\infty < t < \infty} |P\{X \leq t\} - P\{Y \leq t\}|$, and this parameter is zero when H_0 (5.1) is true. Hence,

large J values indicate a deviation from H_0 in the direction of the general alternative specified by (5.66).

36. *Equivalent Form.* In the case of no ties among the N combined $Z_{(i)}$ values, there is an alternative counting formulation for the test statistic J (5.70). Define the variables $\delta_i, i = 1, \ldots, N$, by

$$\delta_i = \begin{cases} 1, & \text{if } Z_{(i)} \text{ is an } X \text{ observation,} \\ 0, & \text{if } Z_{(i)} \text{ is a } Y \text{ observation.} \end{cases} \qquad (5.78)$$

Set

$$s_j = \left[\frac{jm}{N} - \delta_1 - \cdots - \delta_j\right], \quad j = 1, \ldots, N. \qquad (5.79)$$

Then the Kolmogorov-Smirnov statistic J (5.70) can also be expressed as

$$J = (N/d) \max\{|s_1|, \ldots, |s_N|\}. \qquad (5.80)$$

(We note that, unlike expression (5.70), the formulation in (5.80) is not well defined in the case of ties among the $Z_{(i)}$'s.)

37. *Equal Sample Sizes.* In settings where $m = n$, the computational expression for J (5.71) can be simplified to

$$J = \max_{i=1,\ldots,N} |Q(Z_{(i)}) - S(Z_{(i)})|, \qquad (5.81)$$

where, for every real number t,

$$Q(t) = mF_m(t) = [\text{number of sample } X\text{'s } \leq t] \qquad (5.82)$$

and

$$S(t) = nG_n(t) = [\text{number of sample } Y\text{'s } \leq t]. \qquad (5.83)$$

Thus, for the salivation data in Example 5.4 we have

$$J = |Q(Z_{(12)}) - S(Z_{(12)})| = |3 - 9| = 6,$$

in agreement with the value obtained via equation (5.71).

38. *Derivation of the Distribution of J under H_0 (No-Ties Case).* The null (H_0) distribution of J in the case of no ties can be obtained by using the fact that under H_0 (5.1) all possible $\binom{N}{n}$ meshings of the X's and Y's are equally likely, each having probability $1/\binom{N}{n}$. In the ensuing illustration we derive the null distribution of J (5.70) for the sample sizes $m = 1$, $n = 3$. Here $N = 4, d = 1$, and thus $J = 3 \max_{i=1,\ldots,4} |F_1(Z_{(i)}) - G_3(Z_{(i)})|$. We now list the $\binom{4}{1} = 4$ possible meshings, and for each of these meshings we give the associated values of $(F_1(Z_{(1)}), \ldots, F_1(Z_{(4)}))$, the associated values of $(G_3(Z_{(1)}), \ldots, G_3(Z_{(4)}))$, and finally the values of J. Thus $P_0\{J = 2\} = (\frac{2}{4}) = .5$ and $P_0\{J = 3\} = .5$, the latter agreeing with the associated entry in Table A.10.

5.4. A DISTRIBUTION-FREE TEST FOR GENERAL DIFFERENCES

Meshings	$(F_1(Z_{(1)}), \ldots, F_1(Z_{(4)}))$	$(G_3(Z_{(1)}), \ldots, G_3(Z_{(4)}))$	J
XYYY	(1,1,1,1)	$(0, \frac{1}{3}, \frac{2}{3}, 1)$	3
YXYY	(0,1,1,1)	$(\frac{1}{3}, \frac{1}{3}, \frac{2}{3}, 1)$	2
YYXY	(0,0,1,1)	$(\frac{1}{3}, \frac{2}{3}, \frac{2}{3}, 1)$	2
YYYX	(0,0,0,1)	$(\frac{1}{3}, \frac{2}{3}, 1, 1)$	3

39. *Ties.* To illustrate how the computational formula for J given in expression (5.71) is well defined in the case of ties, we consider the following artificial set of tied data: $X_1 = 3, X_2 = 3, X_3 = 5, X_4 = 7, X_5 = 9$ and $Y_1 = 3, Y_2 = 4, Y_3 = 4, Y_4 = 6, Y_5 = 7, Y_6 = 8, Y_7 = 10, Y_8 = 10, Y_9 = 11, Y_{10} = 12$. Here we have $m = 5, n = 10, N = 15$, and $d = 5$. Following the tabular approach of Example 5.4 for computation of J we obtain:

i	$Z_{(i)}$	$F_5(Z_{(i)})$	$G_{10}(Z_{(i)})$	$|F_5(Z_{(i)}) - G_{10}(Z_{(i)})|$
1	3	$\frac{2}{5}$	$\frac{1}{10}$	$\frac{3}{10}$
2	3	$\frac{2}{5}$	$\frac{1}{10}$	$\frac{3}{10}$
3	3	$\frac{2}{5}$	$\frac{1}{10}$	$\frac{3}{10}$
4	4	$\frac{2}{5}$	$\frac{3}{10}$	$\frac{1}{10}$
5	4	$\frac{2}{5}$	$\frac{3}{10}$	$\frac{1}{10}$
6	5	$\frac{3}{5}$	$\frac{3}{10}$	$\frac{3}{10}$
7	6	$\frac{3}{5}$	$\frac{4}{10}$	$\frac{3}{10}$
8	7	$\frac{4}{5}$	$\frac{5}{10}$	$\frac{3}{10}$
9	7	$\frac{4}{5}$	$\frac{5}{10}$	$\frac{3}{10}$
10	8	$\frac{4}{5}$	$\frac{6}{10}$	$\frac{2}{10}$
11	9	$\frac{5}{5}$	$\frac{6}{10}$	$\frac{4}{10}$
12	10	$\frac{5}{5}$	$\frac{8}{10}$	$\frac{2}{10}$
13	10	$\frac{5}{5}$	$\frac{8}{10}$	$\frac{2}{10}$
14	11	$\frac{5}{5}$	$\frac{9}{10}$	$\frac{1}{10}$
15	12	$\frac{5}{5}$	$\frac{10}{10}$	0

Thus, the empirical distribution function $F_5(t)$ for the X sample jumps from 0 to $\frac{2}{5}$ at $Z_{(1)} = Z_{(2)} = Z_{(3)} = 3$, since two of the five X values are 3's. Similarly, the empirical distribution function $G_{10}(t)$ for the Y sample jumps from $\frac{1}{10}$ to $\frac{3}{10}$ and $\frac{6}{10}$ to $\frac{8}{10}$ at $Z_{(4)} = Z_{(5)} = 4$ and $Z_{(12)} = Z_{(13)} = 10$, respectively, since there are two 4s and two 10s among the Y observations. For these tied data we find

$$\max_{i=1,\ldots,15} |F_5(Z_{(i)}) - G_{10}(Z_{(i)})| = |F_5(Z_{(11)}) - G_{10}(Z_{(11)})| = \tfrac{4}{10}.$$

From equation (5.71) it then follows (as in the case of no ties) that $J = [\frac{5(10)}{5}](\frac{4}{10}) = 4$.

40. *Exact Conditional Distribution of J with Ties.* To have a test with exact significance level even in the presence of ties among the X's and/or Y's, we need to consider all $\binom{N}{n}$ possible assignments of the N observations with n observations serving as Y's and m observations serving as X's. As in Comment 38, it still follows that, under H_0 (5.1), each of the $\binom{N}{n}$ possible meshings of the X's and Y's has probability $1/\binom{N}{n}$. The only difference in the case of ties

is that now in the computation of J for each of these $\binom{N}{n}$ meshings, the jumps in the X and Y empirical distribution functions can occur at common observations and the sizes of these jumps can be greater than $1/m$ or $1/n$, respectively. We illustrate this construction for the following $m = 2, n = 3$ data: $X_1 = 3.2, X_2 = 6.3, Y_1 = 1.9, Y_2 = 1.9, Y_3 = 6.3$. The associated ordered $Z_{(i)}$ values are: $Z_{(1)} = Z_{(2)} = 1.9 < Z_{(3)} = 3.2 < Z_{(4)} = Z_{(5)} = 6.3$ and the corresponding value of J (5.71) is $\frac{2(3)}{1}|F_2(Z_{(1)}) - G_3(Z_{(1)})| = 6|F_2(Z_{(2)}) - G_3(Z_{(2)})| = 6|F_2(1.9) - G_3(1.9)| = 6|0 - \frac{2}{3}| = 4$. To assess the significance of this value of J, we obtain its conditional distribution by considering the $\binom{5}{3} = 10$ possible assignments of the observations 1.9, 1.9, 3.2, 6.3, and 6.3 to serve as two X observations and three Y observations. These ten assignments and the corresponding values of J are:

X Observations	Y Observations	Probability under H_0	Value of J
1.9, 1.9	3.2, 6.3, 6.3	$\frac{1}{10}$	6
1.9, 3.2	1.9, 6.3, 6.3	$\frac{1}{10}$	4
1.9, 3.2	1.9, 6.3, 6.3	$\frac{1}{10}$	4
1.9, 6.3	1.9, 3.2, 6.3	$\frac{1}{10}$	1
1.9, 6.3	1.9, 3.2, 6.3	$\frac{1}{10}$	1
1.9, 6.3	1.9, 3.2, 6.3	$\frac{1}{10}$	1
1.9, 6.3	1.9, 3.2, 6.3	$\frac{1}{10}$	1
3.2, 6.3	1.9, 1.9, 6.3	$\frac{1}{10}$	4
3.2, 6.3	1.9, 1.9, 6.3	$\frac{1}{10}$	4
6.3, 6.3	1.9, 1.9, 3.2	$\frac{1}{10}$	6

This yields the null tail probabilities

$$P_0(J \geq 6) = \tfrac{2}{10}, \quad P_0(J \geq 4) = \tfrac{6}{10}, \quad P_0(J \geq 1) = 1.$$

This distribution is called the conditional null distribution or the permutation null distribution of J, given the set of tied observations $\{1.9, 1.9, 3.2, 6.3, 6.3\}$. For the particular observed value $J = 4$, we have that $P_0(J \geq 4) = \tfrac{6}{10}$. (Note that the particular observed X and Y sample values are not important to the calculation of this conditional null distribution of J. It is critical only that the two smallest observations are tied in value, the middle ordered value is untied, and the two largest observations are tied. Thus, for example, the two sets of sample observations $\{X_1 = 3.2, X_2 = 6.3, Y_1 = 1.9, Y_2 = 1.9, Y_3 = 6.3\}$ and $\{X_1 = -12.1, X_2 = 13.7, Y_1 = -12.1, Y_2 = 0, Y_3 = 13.7\}$ yield the same exact conditional null distribution of J.) The computer software StatXact can be helpful in obtaining the exact conditional null distribution of J when there are ties among the X and/or Y observations.

41. *Large-Sample Approximation.* Smirnov (1939) derived the asymptotic (min (m, n) tending to infinity) distribution of the standardized Kolmogorov-Smirnov statistic J^* (5.73) using the work of Kolmogorov (1933) on the asymptotic (m tending to infinity) distribution of the one-sample statistic

$$J_0 = \sqrt{m} \max_{-\infty < a < \infty} |F_m(a) - F_0(a)|, \tag{5.84}$$

where $F_m(\cdot)$ is the empirical distribution function for a random sample of size m from the (assumed) continuous distribution with distribution function $F(a) = P(X \leq a)$ and $F_0(a)$ is a completely specified distribution function. The statistic J_0 can be used to test the goodness-of-fit hypothesis that the random sample $X_1, \ldots, X_m$ has been drawn from a population with distribution function F_0, namely,

$$H_0': [P(X \leq a) = F_0(a) \text{ for all } -\infty < a < \infty], \tag{5.85}$$

versus the broad alternative that the population from which the sample was drawn does not have distribution function F_0.

42. *Test Based on the One-Sample Limit of the Wilcoxon Rank Sum Statistic.* It is of interest to note that the two-sample Wilcoxon test discussed in Section 4.1 can be reduced to a test of H_0' (5.85) by allowing one of the sample sizes, say n, to become infinite. Moses (1964) showed how this leads to a test based on $W_0 = \sum_{j=1}^{m} F_0(X_j)$. (The normal approximation to W_0 treats $[W_0 - (m/2)]/(m/12)^{1/2}$ as an approximate $N(0, 1)$ random variable under H_0'.) Moses pointed out that a test based on W_0 is particularly convenient when F_0 is known but is specified by tabular data, such as demographic data on age of death distributions, rather than being given by a mathematical expression.

43. *Consistency of the J Tests.* Define the class $\mathcal{C}$ of pairs of distribution functions F and G by

$$\mathcal{C} = \{(F, G) : F(x) \neq G(x) \text{ for at least one } x\}. \tag{5.86}$$

Under the minimal Assumptions A1 and A2 only, the test procedure (5.72) is consistent for any $(F, G) \in \mathcal{C}$; that is, the test is consistent against *any* differences between the F and G distributions (i.e., *whenever* H_0 (5.1) is false). In gaining this extra protection against all differences, we do, however, sacrifice power against specific subclasses of alternatives (such as location shifts or differences in dispersions).

Properties

1. *Consistency.* See Comment 43.
2. *Asymptotic Distribution.* See Smirnov (1939) and Comment 41.
3. *Efficiency.* See Capon (1965), Ramachandramurty (1966b), Yu (1971), and Section 5.5.

PROBLEMS

33. The data in Table 5.8 are a subset of the data obtained by Friedman et al. (1971) in an experiment comparing the average concentrations of human plasma growth hormone both resting and after arginine hydrochloride infusion in relatively coronary-prone subjects (persons with type A behavior patterns) with the corresponding concentrations of relatively coronary-resistant individuals (subjects with type B behavior patterns). Type A behavior is characterized by an excessive sense of time urgency, drive, and competitiveness; type B denotes a converse type of behavior. Earlier studies (cf. Friedman and Rosenman (1959)) indicated that type A individuals may be more prone to coronary heart disease than type B individuals.
    Find the P-value for an appropriate test of whether there is any difference between the probability distribution of peak level human plasma growth hormone (after arginine hydrochloride infusion) for type A subjects and that for type B subjects.
34. Consider the alcoholic intake data in Table 4.2. For these data find the approximate P-value for an appropriate test of whether there are *any* differences between the control and SST probability

TABLE 5.8. Peak Levels of Human Plasma Growth Hormone After Arginine Hydrochloride Infusion (Initial Test, ng/ml)

Type A Subjects	Type B Subjects
3.6	16.2
2.6	17.4
4.7	8.5
8.0	15.6
3.1	5.4
8.8	9.8
4.6	14.9
5.8	16.6
4.0	15.9
4.6	5.3
	10.5

Source: M. Friedman, S. O. Byers, R. H. Rosenman, and R. Neuman (1971).

distributions. Discuss this result in conjunction with the previous findings in Example 4.2, Problem 9, and Problem 31.

35. Verify directly, or illustrate with a numerical example, that representations (5.71) and (5.80) for J are indeed equivalent.

36. When $m = n$, show that both representations (5.71) and (5.80) for J are equivalent to the expression

$$J = \max\{|t_1|, |t_2|, \ldots, |t_N|\}, \tag{5.87}$$

where

$$t_j = (1 - 2\delta_1) + (1 - 2\delta_2) + \cdots + (1 - 2\delta_j), \tag{5.88}$$

and the δ's are given by (5.78).

37. Calculate the value of J for the salivation data in Table 5.7 using the equivalent (when $m = n$) expression in (5.87).

38. Apply the two-sided Wilcoxon rank sum test procedure from Section 4.1 to the salivation data in Table 5.7 by finding the appropriate P-value. Compare the conclusion indicated by this Wilcoxon rank sum procedure with that indicated by the Kolmogorov-Smirnov procedure in Example 5.4. Comment on your findings.

39. Generate the exact null distribution of J (5.70) for the setting $m = 3, n = 3$. (See Comment 38.)

40. Consider the growth hormone level data found in Table 5.5. Use the Kolmogorov-Smirnov test procedure to assess whether there are significant differences of *any* kind between the control hypertension population and the clonidine treated population in their increases in growth hormone levels following exercise. Find the appropriate P-value for the test and compare it with the P-value obtained in Problem 25.

41. Consider the following two-sample data for $m = 3, n = 3$: $X_1 = -3.7, X_2 = 4.6, X_3 = 1.5, Y_1 = 1.5, Y_2 = 4.6, Y_3 = 1.5$. Using the approach discussed in Comment 40, find the exact conditional null distribution of the Kolmogorov-Smirnov statistic J (5.70). Compare and contrast this exact conditional null distribution with the corresponding null distribution of J for $m = n = 3$ and no tied observations, as obtained in Problem 39.

42. Consider the serum iron data in Table 5.1. Use the Kolmogorov-Smirnov test procedure to assess whether there are significant differences of *any* kind between the distribution of serum iron values obtained by the Ramsay method and the distribution of serum iron values obtained by the Jung-Parekh method. Find the appropriate P-value for the test and compare it with the results discussed in Example 5.1.

5.5. EFFICIENCIES OF TWO-SAMPLE DISPERSION AND BROAD ALTERNATIVES PROCEDURES

Recall the classical normal theory $\mathcal{F}$-test for equality of variances based on the statistic

$$D = \frac{S_x^2}{S_y^2}, \tag{5.89}$$

where $S_x^2 = \sum_{i=1}^{m}(X_i - \bar{X})^2/(m-1)$, $S_y^2 = \sum_{j=1}^{n}(Y_j - \bar{Y})^2/(n-1)$, $\bar{X} = \sum_{i=1}^{m} X_i/m$, and $\bar{Y} = \sum_{j=1}^{n} Y_j/n$. The significance level of this $\mathcal{F}$-test is extremely sensitive to nonnormality. (See Comment 26.) This is also true of the coverage probability of the confidence intervals for σ_2^2/σ_1^2 that are based on the ratio of sample variances and are derived from the $\mathcal{F}$-test. The Box-Andersen (1955) test "adjusts" the $\mathcal{F}$-test to remedy this difficulty. Since this Box-Andersen approach has desirable properties, we report asymptotic efficiencies of the test procedures of Sections 5.1 and 5.2, as well as the point estimators and confidence intervals/bounds associated with the jackknife approach (see Comment 24), with respect to the corresponding Box-Andersen procedures. (The specific Box-Andersen procedures that we refer to are (a) the APF test of Shorack (1969), which is a slight variation of the test used by Box and Andersen for the case where the parameters θ_1 and θ_2 of model (5.2) are known; (b) an associated estimator given by Shorack (1965); and (c) the associated confidence interval and bounds discussed in Shorack (1969).)

The Pitman asymptotic relative efficiency for scale alternatives of the Ansari-Bradley test based on C (5.6) relative to the Box-Andersen adjusted $\mathcal{F}$-test based on D (5.89) is

$$e(C,D) = 12(\beta_G - 1)\left[\int_{-\infty}^{0} xg^2(x)\,dx - \int_{0}^{\infty} xg^2(x)\,dx\right]^2, \tag{5.90}$$

where

$$\beta_G = \frac{\int_{-\infty}^{\infty}(x-\mu)^4 g(x)\,dx}{\{\int_{-\infty}^{\infty}(x-\mu)^2 g(x)\,dx\}^2}$$

is the kurtosis and $\mu = \int_{-\infty}^{\infty} xg(x)\,dx$ is the mean of the population with distribution function $G(\cdot)$ and probability density function $g(\cdot)$.

The expression in (5.90) was obtained by Ansari and Bradley (1960). Some values of $e(C,D)$ for selected $G(\cdot)$ are:

G	Normal	Uniform	Double Exponential
$e(C,D)$	.61	.60	.94

Miller (1968) pointed out that the asymptotic relative efficiency of the jackknife procedures (tests, point estimators, and confidence intervals/bounds) with respect to the Box-Andersen procedures has the value 1 for *any* underlying distribution $F(\cdot)$; that is, $e(Q,D) \equiv 1$, where Q is given by (5.35) and D represents the Box-Andersen adjusted $\mathcal{F}$-test procedures.

We do not know of any results for the asymptotic efficiencies of the Lepage test for location or scale differences (Section 5.3).

The determination of asymptotic relative efficiencies for the Kolmogorov-Smirnov test based on J (5.70) is difficult, owing to the complicated form of the asymptotic distribution of the Kolmogorov-Smirnov statistic. Capon (1965) obtained lower bounds for the asymptotic relative efficiency of the Kolmogorov-Smirnov test. In particular, for normal translation alternatives, Capon derived the lower bound of .637 for the asymptotic relative efficiency of the Kolmogorov-Smirnov test with respect to the normal theory two-sample t test (see Section 4.5). See also Ramachandramurty (1966b) and Yu (1971). For related efficiency results using different notions of asymptotic efficiency, see Klotz (1967), Hájek and Šidák (1967, 272) and Anděl (1967).

CHAPTER 6

The One-Way Layout

INTRODUCTION

The procedures of this chapter are designed for statistical analyses in which primary interest is centered on the relative locations (medians) of three or more populations. This development represents a direct generalization of the two-sample location problem (discussed in Chapter 4) to situations in which the data consist of $k (\geq 3)$ random samples, one sample from each of k populations. The basic null hypothesis of interest is that of no differences in locations (medians), under which the k samples can be treated as a single (combined) sample from one population. The alternatives considered here correspond to a variety of restricted nonnull relationships between the locations (medians). We encounter two types of data for which such analyses are important. The first of these corresponds to a general setting of k populations (referred to as *treatments* for convenience) with no additional conditions. The second deals with the setting where one of the *treatments* represents a *control* (or placebo) population and we are interested in detecting which, if any, of the other $(k-1)$ treatments are different from this control.

Section 6.1 presents a distribution-free test directed at general alternatives for the setting of k treatments. A distribution-free test designed for detecting ordered alternatives among k treatments is considered in Section 6.2 and generalized in Section 6.3 to the broader class of umbrella alternatives. In Section 6.4 a distribution-free test procedure is presented for the simultaneous comparison of $(k-1)$ treatments with a control. In Sections 6.5–6.7 we introduce multiple comparison procedures designed to detect which particular populations, if any, differ from one another. Sections 6.5 and 6.6 are devoted to procedures for making the total of $\binom{k}{2}$ pairwise comparisons between all k treatments in the general and ordered alternatives settings, respectively. Section 6.7 presents multiple comparisons procedures based on simple random samples for deciding which, if any, of $(k-1)$ treatments are different from a control. Section 6.8 considers estimators of contrasts in the treatment effects and Section 6.9 deals with simultaneous confidence intervals for simple contrasts. The asymptotic relative efficiencies for translation alternatives of the procedures discussed in this chapter with respect to their normal theory counterparts based on sample averages are discussed in Section 6.10.

Data. The data consist of $N = \sum_{j=1}^{k} n_j$ observations, with n_j observations from the jth treatment, $j = 1, \ldots, k$.

	Treatments		
1	2	...	k
X_{11}	X_{12}	...	X_{1k}
X_{21}	X_{22}	...	X_{2k}
$\vdots$	$\vdots$		$\vdots$
$X_{n_1 1}$	$X_{n_2 2}$	...	$X_{n_k k}$

Assumptions

A1. The N random variables $\{X_{1j}, X_{2j}, \ldots, X_{n_j j}\}$, $j = 1, \ldots, k$, are mutually independent.

A2. For each fixed $j \in \{1, \ldots, k\}$, the n_j random variables $\{X_{1j}, X_{2j}, \ldots, X_{n_j j}\}$ are a random sample from a continuous distribution with distribution function F_j.

A3. The distribution functions $F_1, \ldots, F_k$ are connected through the relationship

$$F_j(t) = F(t - \tau_j), \quad -\infty < t < \infty, \tag{6.1}$$

for $j = 1, \ldots, k$, where F is a distribution function for a continuous distribution with unknown median θ and τ_j is the unknown treatment effect for the jth population.

We note that Assumptions A1–A3 correspond directly to the usual one-way layout model commonly associated with normal theory assumptions; that is, Assumptions A1–A3 are equivalent to the representation

$$X_{ij} = \theta + \tau_j + e_{ij}, \quad i = 1, \ldots, n_j, \quad j = 1, \ldots, k,$$

where θ is the overall median, τ_j is the *treatment j effect,* and the N e's form a random sample from a continuous distribution with median 0. (Under the additional assumption of normality, the medians θ and 0 are, of course, also the respective means.)

Hypothesis

The null hypothesis of interest in Sections 6.1–6.4 of this chapter is that of no differences among the treatment effects $\tau_1, \ldots, \tau_k$, namely,

$$H_0 : [\tau_1 = \cdots = \tau_k]. \tag{6.2}$$

This null hypothesis asserts that each of the underlying distributions $F_1, \ldots, F_k$ is the same, corresponding to $F_1 \equiv F_2 \equiv \cdots \equiv F_k \equiv F$ in equation (6.1).

6.1. A DISTRIBUTION-FREE TEST FOR GENERAL ALTERNATIVES (KRUSKAL-WALLIS)

In this section we present a procedure for testing H_0 (6.2) against the general alternative that at least two of the treatment effects are not equal, namely,

$$H_1 : [\tau_1, \ldots, \tau_k \text{ not all equal}]. \tag{6.3}$$

Procedure

To compute the Kruskal-Wallis statistic, H, we first combine all N observations from the k samples and order them from least to greatest. Let r_{ij} denote the rank of X_{ij} in this joint ranking and set

$$R_j = \sum_{i=1}^{n_j} r_{ij} \quad \text{and} \quad R_{.j} = \frac{R_j}{n_j}, \quad j = 1, \ldots, k. \tag{6.4}$$

Thus, for example, R_1 is the sum of the joint ranks received by the treatment 1 observations and $R_{.1}$ is the average rank for these same observations. The Kruskal-Wallis statistic H is then given by

$$H = \frac{12}{N(N+1)} \sum_{j=1}^{k} n_j \left(R_{.j} - \frac{N+1}{2} \right)^2$$

$$= \left(\frac{12}{N(N+1)} \sum_{j=1}^{k} \frac{R_j^2}{n_j} \right) - 3(N+1), \tag{6.5}$$

where $(N+1)/2 = \left(\sum_{j=1}^{k} \sum_{i=1}^{n_j} r_{ij}/N \right)$ is the average rank assigned in the joint ranking.

To test

$$H_0 : [\tau_1 = \cdots = \tau_k]$$

versus the general alternative

$$H_1 : [\tau_1, \ldots, \tau_k \text{ not all equal}],$$

at the α level of significance,

$$\text{Reject } H_0 \text{ if } H \geq h_\alpha; \quad \text{otherwise do not reject}, \tag{6.6}$$

where the constant h_α is chosen to make the type I error probability equal to α. Values of h_α are given in Table A.12.

Large-Sample Approximation

When H_0 is true, the statistic H has, as $\min(n_1, \ldots, n_k)$ tends to infinity, an asymptotic chi-square (χ^2) distribution with $k - 1$ degrees of freedom. (See Comment 9 for indications of the proof.) The chi-square approximation for procedure (6.6) is

$$\text{Reject } H_0 \text{ if } H \geq \chi^2_{k-1,\alpha}; \quad \text{otherwise do not reject}, \tag{6.7}$$

where $\chi^2_{k-1,\alpha}$ is the upper α percentile point of a chi-square distribution with $k - 1$ degrees of freedom. Values of $\chi^2_{k-1,\alpha}$ can be obtained from Chart A.2.

Ties

If there are ties among the N X's, assign each of the observations in a tied group the average of the integer ranks that are associated with the tied group and compute H with these average ranks. As a consequence of the effect that ties have on the null distribution of H, the following modification is needed to apply either procedure (6.6) or the large-sample approximation in (6.7) when there are tied X's. In either of these procedures, we replace H by

$$H' = \frac{H}{1 - \left(\sum_{j=1}^{g}(t_j^3 - t_j)/[N^3 - N]\right)}, \quad (6.8)$$

where, in equation (6.8), H is computed using average ranks, g denotes the number of tied X groups, and t_j is the size of tied group j. We note that an untied observation is considered to be a tied group of size 1. In particular, if there are no ties among the X's, then $g = N$ and $t_j = 1$ for $j = 1, \ldots, N$. In this case, each term in (6.8) of the form $t_j^3 - t_j$ reduces to zero, the denominator of the right-hand side of expression (6.8) reduces to 1, and H' (6.8) reduces to H, as given in (6.5).

We note that even procedure (6.6) using the critical values from Table A.12 is only approximately, and not exactly, of significance level α in the presence of tied X observations. To get an exact level α test in this tied setting, see Comment 8.

Example 6.1: *Half-Time of Mucociliary Clearance.* Thomson and Short (1969) have assessed mucociliary efficiency from the rate of removal of dust in normal subjects, subjects with obstructive airway disease, and subjects with asbestosis. Table 6.1 is based on a subset of the Thomson-Short data. The joint ranks (r_{ij}'s) of the observations are given in Table 6.1 in parentheses after the data values and the treatment rank sums (R_1, R_2, and R_3) are provided at the bottom of the columns.

We are interested in using procedure (6.6) to test if there are any differences in median mucociliary clearance half-times for the three subject populations. For purpose of illustration we take the significance level to be $\alpha = .0502$. From Table A.12 with sample sizes 4, 5, and 5 (see Comment 7) we see that $h_{.0502} = 5.643$ and procedure (6.6) reduces to

Reject H_0 if $H \geq 5.643$.

Now, we illustrate the computations leading to the sample value of H (6.5). For these data, we have $n_1 = n_3 = 5, n_2 = 4$, and $N = 14$. Combining these facts with the treatment rank sums in Table 6.1, we find from (6.5) that

$$H = \frac{12}{14(14+1)} \left(\frac{(36)^2}{5} + \frac{(36)^2}{4} + \frac{(33)^2}{5}\right) - 3(14+1) = .771.$$

TABLE 6.1. Half-Time of Mucociliary Clearance (h)

	Subjects with	
Normal Subjects	Obstructive Airways Disease	Asbestosis
2.9 (8)	3.8 (13)	2.8 (7)
3.0 (9)	2.7 (6)	3.4 (11)
2.5 (4)	4.0 (14)	3.7 (12)
2.6 (5)	2.4 (3)	2.2 (2)
3.2 (10)		2.0 (1)
$R_1 = 36$	$R_2 = 36$	$R_3 = 33$

Source: M. L. Thomson and M. D. Short (1969).

Since this value of H is less than the critical value 5.643, we do not reject H_0 at the $\alpha = .0502$ level. In fact, we find from Table A.12, with $x = .771$ and sample sizes 4, 5, and 5, that $P_0(H \geq .771) > .1009$. Thus the lowest significance level at which we can reject H_0 in favor of H_1 with the observed value of the test statistic $H = .771$ is greater than .1009.

For the large-sample approximation, we compare the value of H (since there are no ties) to the chi-square distribution with $k - 1 = 2$ degrees of freedom. From Chart A.2, we see that the observed value of $H = .771$ is approximately the .64 upper percentile for the chi-square distribution with 2 degrees of freedom. Thus the approximate P-value for these data and test procedure (6.7) is .64. Both the exact test and large-sample approximation indicate that there is virtually no sample evidence in support of significant differences in mucociliary clearance half-times for the three subject populations.

A Monte Carlo estimate and an approximate 99% confidence interval for the exact P-value as well as the approximate P-value based on the large-sample approximation can also be obtained by using StatXact's KW/MO command. The StatXact output for the mucociliary clearance data of Table 6.1 is

$$KW(X) = \text{Kruskal-Wallis test statistic} = .7714.$$

Asymptotic P-value (based on Chi-square distribution with 2 df)

$$\Pr\{KW(X).GE..7714\} = .6800$$

Monte Carlo estimate of P-value:

$$\Pr\{KW(X).GE..7714\} = .7147$$

$$99.00\% \text{ Confidence Interval} = (.7031, .7263).$$

Under the heading "Monte Carlo estimate of P-value" we see that StatXact provides .7147 as the Monte Carlo estimate of the exact P-value. In addition, StatXact provides (.7031, .7263) as a 99% confidence interval for the exact P-value. Finally, we note that the approximate P-value provided by StatXact under "Asymptotic P-value (based on Chi-square distribution with 2 df)" is .6800, also in agreement with what we previously obtained from Chart A.2 using the large-sample approximation.

We can also use Minitab's KRUSKAL-WALLIS command to compute the value of H and provide an approximate P-value based on the large-sample approximation. The Minitab setup for the mucociliary data in Table 6.1 is as follows:

```
MTB > READ C1    C2
DATA > 2.9    1
DATA > 3.0    1
DATA > 2.5    1
DATA > 2.6    1
DATA > 3.2    1
DATA > 3.8    2
DATA > 2.7    2
DATA > 4.0    2
DATA > 2.4    2
DATA > 2.8    3
DATA > 3.4    3
DATA > 3.7    3
DATA > 2.2    3
DATA > 2.0    3
DATA > END
MTB > KRUSKAL-WALLIS C1    C2
```

The output is, in Minitab's notation:

LEVEL	NOBS	MEDIAN	AVE.RANK	Z VALUE
1	5	2.9	7.2	−0.20
2	4	3.25	9	0.85
3	5	2.8	6.6	−0.60
OVERALL	14		7.5	

$H = .77$ d.f. $= 2$ $p = 0.680$
$H = .77$ d.f. $= 2$ $p = 0.680$ (adj. for ties).

The approximate P-value provided by Minitab is $p = 0.680$ without the ties correction given in equation (6.8) and $p = 0.680$ with the ties correction in (6.8). Since there are no ties in the mucociliary data of Table 6.1, these two approximate P-values in the Minitab output are identical, and both are in good agreement with what we previously obtained from Chart A.2.

Comments

1. *More General Setting.* We could replace Assumptions A1–A3 and H_0 (6.2) with the more general null hypothesis that all $N! / \left(\prod_{j=1}^{k} n_j! \right)$ assignments of n_1 ranks to the treatment 1 observations, n_2 ranks to the treatment 2 observations , ..., n_k ranks to the treatment k observations, are equally likely.

2. *Motivation for the Test.* Under Assumptions A1–A3 and H_0 (6.2), the rank vector $\mathbf{R}^* = (r_{11}, \ldots, r_{n_1 1}, r_{22}, \ldots, r_{n_2 2}, \ldots, r_{1k}, \ldots, r_{n_k k})$ has a uniform distribution over the set of all $N!$ permutations of the vector of integers $(1, 2, \ldots, N)$. It follows that

$$E_0(r_{ij}) = \frac{1}{N!}(N-1)! \sum_{i=1}^{N} i = \frac{N+1}{2},$$

the average rank being assigned in the joint ranking. Thus,

$$E_0(R_{\cdot j}) = E_0 \left(\frac{1}{n_j} \sum_{i=1}^{n_j} r_{ij} \right) = \frac{1}{n_j} \sum_{i=1}^{n_j} E_0(r_{ij}) = \frac{n_j(N+1)}{2n_j} = \frac{N+1}{2}, \text{ for } j = 1, 2, \ldots, k,$$

and we would expect the $R_{\cdot j}$'s to be close to $(N+1)/2$ when H_0 is true. Since the test statistic H (6.5) is a constant times a weighted sum of squared differences between the observed treatment average ranks, $R_{\cdot j}$, and their null expected values, $E_0(R_{\cdot j}) = (N+1)/2$, small values of H represent agreement with H_0 (6.2). When the τ's are not all equal, we would expect a portion of the associated treatment average ranks to differ from their common null expectation, $(N+1)/2$, with some tending to be larger and some smaller. The net result (after squaring the observed differences to obtain the $(R_{\cdot j} - (N+1)/2)^2$ terms) would be a large value of H. This suggests rejecting H_0 in favor of H_1 (6.3) for large values of H and motivates procedures (6.6) and (6.7). (See also Comment 3.)

3. *Connection to Normal Theory Test.* The Kruskal-Wallis test can also be motivated by considering the usual analysis of variance $\mathcal{F}$ statistic calculated using the ranks, rather than the original observations. The $\mathcal{F}$ statistic can be written as $\mathcal{F} = c(\text{SSB})/(\text{SST} - \text{SSB})$, where c is a constant depending only on the sample sizes, SST is the total sum of squares, and SSB is the between sum of squares. The statistic SSB reduces to $\sum_{j=1}^{k} n_j (R_{\cdot j} - (N+1)/2)^2$ when applied to the ranks rather than the original observations and SST becomes a fixed constant

when calculated on the ranks. Using these facts it can be shown that when $\mathcal{F}$ is calculated for the ranks, $\mathcal{F}$ is an increasing function of H.

4. *Assumptions.* It is important to point out that Assumption A3 stipulates that the k treatment distributions $F_1, \ldots, F_k$ can differ at most in their locations (medians). In particular, Assumption A3 requires that the k underlying distributions belong to the same general family (F) and that they do not differ in scale parameters (variability). (For a discussion of methodology designed for a more general setting where differences in scale parameters are permitted, see Comment 11.)

5. *Special Case of Two Treatments.* For the case of $k = 2$ treatments, the procedures in (6.6) and (6.7) are equivalent to the exact and large-sample approximation forms, respectively, of the two-sided Wilcoxon rank sum test, as discussed in Section 4.1.

6. *Derivation of the Distribution of H under H_0 (No-Ties Case).* The null distribution of H (6.5) can be obtained by using the fact that under H_0 (6.2), all $N!/\left(\prod_{j=1}^{k} n_j!\right)$ assignments of n_1 ranks to the treatment 1 observations, n_2 ranks to the treatment 2 observations, ..., n_k ranks to the treatment k observations are equally likely. We illustrate how the null distribution can be derived in the particular case $k = 3$, $n_1 = n_2 = n_3 = 2$. In this case we have $H = [\{12/[6(7)]\}\{(R_1^2 + R_2^2 + R_3^2)/2\} - 21] = [(A/7) - 21]$, where $A = R_1^2 + R_2^2 + R_3^2$. We next enumerate 15 of the total possible $\{6!/[(2!)(2!)(2!)]\} = 90$ rank assignments and their corresponding values of A and H.

	I	II	III				I	II	III	
(a)	1	3	5	$A = 179$		(b)	1	3	4	$A = 173$
	2	4	6	$H = 4.57$			2	5	6	$H = 3.71$
(c)	1	3	4	$A = 171$		(d)	1	2	5	$A = 173$
	2	6	5	$H = 3.43$			3	4	6	$H = 3.71$
(e)	1	2	4	$A = 165$		(f)	1	2	4	$A = 161$
	3	5	6	$H = 2.57$			3	6	5	$H = 2$
(g)	1	2	3	$A = 155$		(h)	1	2	5	$A = 171$
	4	5	6	$H = 1.14$			4	3	6	$H = 3.43$
(i)	1	2	3	$A = 153$		(j)	1	2	4	$A = 161$
	4	6	5	$H = .86$			5	3	6	$H = 2$
(k)	1	2	3	$A = 153$		(l)	1	2	3	$A = 149$
	5	4	6	$H = .86$			5	6	4	$H = .29$
(m)	1	2	4	$A = 155$		(n)	1	2	3	$A = 149$
	6	3	5	$H = 1.14$			6	4	5	$H = .29$
(o)	1	2	4	$A = 147$						
	6	5	3	$H = 0$						

For each of the foregoing rank configurations, there are five other configurations (corresponding to the six permutations of the names of the samples I, II, and III) which yield the same value of H. This covers the complete total of 90 possible rank assignments. Thus

$$P_0\{H = 4.57\} = 1/15, \quad P_0\{H = 3.71\} = 2/15, \quad P_0\{H = 3.43\} = 2/15,$$
$$P_0\{H = 2.57\} = 1/15, \quad P_0\{H = 2\} = 2/15, \quad P_0\{H = 1.14\} = 2/15,$$
$$P_0\{H = .86\} = 2/15, \quad P_0\{H = .29\} = 2/15, \quad P_0\{H = 0\} = 1/15.$$

The probability, under H_0, that H is greater than or equal to 3.71, for example, is therefore

$$P_0\{H \geq 3.71\} = P_0\{H = 3.71\} + P_0\{H = 4.57\}$$
$$= \tfrac{1}{15} + \tfrac{2}{15} = .20.$$

This agrees with the upper-tail probability entry for $k = 3, n_1 = n_2 = n_3 = 2$ in Table A.12.

Note that we have derived the null distribution of H without specifying the common form (F) of the underlying distribution function for the X's under H_0 beyond the point of requiring that it be continuous. This is why the test procedure (6.6) based on H is called a distribution-free procedure. From the null distribution of H we can determine the critical value h_α and control the probability α of falsely rejecting H_0 when H_0 is true, and this error probability does not depend on the specific form of the common underlying continuous X distribution.

7. *Sample Sizes Not in Increasing Order.* Table A.12 gives critical values for $n_1 \leq \cdots \leq n_k$ situations. However, critical points for $n_1 \leq \cdots \leq n_k$ configurations not in this order are obtained by simply putting the sample sizes in increasing order and then entering Table A.12. (This occurs because relabeling the samples does not change the value of H.) Thus, as in Example 6.1, to find critical values for the case $n_1 = 5, n_2 = 4, n_3 = 5$, we enter Table A.12 at $n_1 = 4, n_2 = 5, n_3 = 5$.

8. *Exact Conditional Distribution of H with Ties among the X-Values.* To have a test with exact significance level even in the presence of tied X's, we need to consider all $N!/\left(\prod_{j=1}^{k} n_j!\right)$ assignments of n_1 ranks to the treatment 1 observations, n_2 ranks to the treament 2 observations, ..., n_k ranks to the treatment k observations, where now these joint ranks are obtained by using average ranks to break the ties. As in Comment 6, it still follows that under H_0 each of these $N!/\left(\prod_{j=1}^{k} n_j!\right)$ assignments is equally likely. For each such assignment, the value of H is computed and the results are tabulated. We illustrate this construction for $k = 3$ and $n_1 = n_2 = 2, n_3 = 1$ and the data $X_{11} = 1.3, X_{21} = 1.7, X_{12} = 1.3, X_{22} = 2.0$, and $X_{13} = 2.0$. Using average ranks to break ties, the observed rank vector is $(r_{11}, r_{21}, r_{12}, r_{22}, r_{13})$ $= (1.5, 3, 1.5, 4.5, 4.5)$. Thus, $R_1 = 4.5, R_2 = 6, R_3 = 4.5$, and the attained value of H is

$$H = \left[\frac{12}{5(6)}\left\{\frac{(4.5)^2}{2} + \frac{(6)^2}{2} + \frac{(4.5)^2}{1}\right\} - 3(6)\right] = 1.35.$$

To assess the significance of H, we obtain its conditional null distribution by considering the $[5!/(2! \, 2! \, 1!)] = 30$ equally likely (under H_0) possible assignments of the observed rank vector $(1.5, 3, 1.5, 4.5, 4.5)$ to the three treatments. These 30 assignments and associated values of H are:

I	II	III		I	II	III	
1.5	4.5	1.5		1.5	4.5	1.5	
3	4.5		$H = 3.15$	3	4.5		$H = 3.15$

I	II	III		I	II	III	
1.5	3	1.5		1.5	3	1.5	
4.5	4.5		$H = 1.35$	4.5	4.5		$H = 1.35$
1.5	3	1.5		1.5	3	1.5	
4.5	4.5		$H = 1.35$	4.5	4.5		$H = 1.35$
3	1.5	1.5		3	1.5	1.5	
4.5	4.5		$H = 1.35$	4.5	4.5		$H = 1.35$
3	1.5	1.5		3	1.5	1.5	
4.5	4.5		$H = 1.35$	4.5	4.5		$H = 1.35$
4.5	1.5	1.5		4.5	1.5	1.5	
4.5	3		$H = 3.15$	4.5	3		$H = 3.15$
1.5	4.5	3		1.5	1.5	3	
1.5	4.5		$H = 3.60$	4.5	4.5		$H = 0$
1.5	1.5	3		1.5	1.5	3	
4.5	4.5		$H = 0$	4.5	4.5		$H = 0$
1.5	1.5	3		4.5	1.5	3	
4.5	4.5		$H = 0$	4.5	1.5		$H = 3.60$
1.5	3	4.5		1.5	3	4.5	
1.5	4.5		$H = 3.15$	1.5	4.5		$H = 3.15$
1.5	1.5	4.5		1.5	1.5	4.5	
3	4.5		$H = 1.35$	3	4.5		$H = 1.35$
1.5	1.5	4.5		1.5	1.5	4.5	
3	4.5		$H = 1.35$	3	4.5		$H = 1.35$
1.5	1.5	4.5		1.5	1.5	4.5	
4.5	3		$H = 1.35$	4.5	3		$H = 1.35$
1.5	1.5	4.5		1.5	1.5	4.5	
4.5	3		$H = 1.35$	4.5	3		$H = 1.35$
3	1.5	4.5		3	1.5	4.5	
4.5	1.5		$H = 3.15$	4.5	1.5		$H = 3.15$

Since each of these values for H has null probability $\frac{1}{30}$, it follows that

$$P_0(H = 3.60) = \frac{2}{30} \qquad P_0(H = 1.35) = \frac{16}{30}$$
$$P_0(H = 3.15) = \frac{8}{30} \qquad P_0(H = 0) = \frac{4}{30}.$$

This distribution is called the conditional distribution or the permutation distribution of H, given the set of tied ranks $\{1.5, 1.5, 3, 4.5, 4.5\}$. For the particular observed value $H = 1.35$, we have $P_0(H \geq 1.35) = \frac{28}{30}$, so that such a value does not indicate a deviation from H_0.

9. *Large-Sample Approximation.* Define the random variables $T_j = R_{.j} - E_0(R_{.j}) = R_{.j} - (N+1)/2$, for $j = 1, 2, \ldots, k$. Since each $R_{.j} = \sum_{i=1}^{n_j} r_{ij}/n_j$ is an average, it is not surprising (see Kruskal and Wallis (1952) for example, for justification) that a properly standardized version of the vector $\mathbf{T}^* = (T_1, \ldots, T_{k-1})$ has an asymptotic $(\min(n_1, \ldots, n_k)$ tending to infinity) $(k-1)$-variate normal distribution with mean vector $\mathbf{0} = (0, \ldots, 0)$ and appropriate covariance matrix $\mathbf{\Sigma}$ when the null hypothesis H_0 is true. (Note that $\mathbf{T}^*$ does not include $T_k = R_{.k} - (N+1)/2$, since T_k can be expressed as a linear combination of $T_1, \ldots, T_{k-1}$. This is the reason that the asymptotic normal distribution is $(k-1)$-variate and not k-variate.) Since the test statistic H (6.5) is a quadratic form in the variables $(T_1, \ldots, T_{k-1})$,

it is therefore quite natural that H has an asymptotic $(\min(n_1, \ldots, n_k)$ tending to infinity) chi-square distribution with $k - 1$ degrees of freedom.

10. *Family Monotonicity.* Gabriel (1969) introduced a desirable property of a testing family called monotonicity and pointed out that the H statistic does not enjoy the property. We refer the interested user to Gabriel's paper, but we briefly mention here that the problem arises because it is possible that the H statistic computed for a subset can exceed the H statistic computed for a set containing the subset. Gabriel gave the following example. The sample 1 ranks are 8, 9, 10, 11, the sample 2 ranks are 1, 2, 6, 7, and the sample 3 ranks are 3, 4, 5, 12. Then H based on samples 1 and 2 ($k = 2$) is 5.33, whereas H based on samples 1, 2, and 3 ($k = 3$) is 4.77. The same anomaly can arise with Friedman's statistic (Section 7.1).

11. *k-Sample Behrens-Fisher Problem.* Two of the implicit requirements associated with Assumptions A1–A3 are that the underlying distributions belong to the same common family (F) and that they differ within this family at most in their medians. The less restrictive setting, where these assumptions are relaxed to permit the possibility of differences in scale parameters as well as medians (but still requiring the same common family F), is generally referred to as the k-sample Behrens-Fisher problem. (Note that this is a direct k-sample extension of the corresponding two-sample Behrens-Fisher problem considered in Section 4.4.) The Kruskal-Wallis procedure (6.6) is no longer distribution-free under these relaxed assumptions permitting unequal scale parameters. Rust and Fligner (1984) proposed a modification of the Kruskal-Wallis statistic H (6.5) to deal with this broader Behrens-Fisher setting. Their procedure is designed as a test for the less restrictive null and alternative hypotheses

$$H_0^* : [\delta_{ij} = \tfrac{1}{2} \text{ for all } i \neq j = 1, \ldots, k] \tag{6.9}$$

and

$$H_1^* : [\delta_{ij} \neq \tfrac{1}{2} \text{ for at least one } i \neq j = 1, \ldots, k], \tag{6.10}$$

respectively, where

$$\delta_{ij} = P(X_{1i} > X_{1j}), \quad \text{for } i \neq j = 1, \ldots, k.$$

The Rust-Fligner modification of the Kruskal-Wallis statistic provides a test procedure that is still exactly distribution-free under the more restrictive null hypothesis H_0 (6.2). However, their modified procedure is also asymptotically $(\min(n_1, \ldots, n_k)$ tending to infinity) distribution-free under the considerably broader null hypothesis H_0^* (6.9) so long as the underlying populations (not necessarily of the same form) are all symmetric. In the special case of $k = 2$ populations, the Rust-Fligner procedure reduces approximately to the Fligner-Policello modifications to the Mann-Whitney-Wilcoxon two-sample test procedure discussed in Section 4.4.

12. *Pairwise Rankings.* The Kruskal-Wallis statistic H (6.5) is based on the treatment rank sums $R_1, \ldots, R_k$ associated with the *joint* ranking of all N sample observations. As an alternative approach, one could just as well choose to compare the k treatments through a combination of all $k(k - 1)/2$ *pairwise* rankings. Fligner (1985) proposed such a pairwise ranking analogue of the Kruskal-Wallis statistic and demonstrated that the associated pairwise ranking test procedure has some nice efficiency properties. Such pairwise rankings (as opposed to joint rankings) have also proved useful in certain multiple comparisons settings (see Sections 6.5 and 6.10 for more in this regard).

13. *Consistency of the H Test.* Replace Assumptions A1–A3 by the less restrictive Assumptions A1': the X's are mutually independent and A2': $X_{1j}, \ldots, X_{n_j j}$ come from the same continuous population $\Pi_j, j = 1, \ldots, k$, but where $\Pi_1, \ldots, \Pi_k$ are not assumed to be identical.

Then Kruskal and Wallis (1952) pointed out that (roughly speaking) the test defined by (6.6) is consistent if (and only if) "... there be at least one of the populations for which the limiting probability is not one-half that a random observation from this population is greater than an independent random member of the N sample observations."

Properties

1. *Consistency.* Under Assumptions A1–A3 and equal sample sizes ($n_1 = \cdots = n_k$), the test defined by (6.6) is consistent against the alternative for which $\tau_i \neq \tau_j$ for at least one $i \neq j = 1, \ldots, k$. For arbitrary sample sizes, see Kruskal (1952) and Comment 13.
2. *Asymptotic Chi-Squareness.* See Kruskal and Wallis (1952) and Hettmansperger (1984, 184–185).
3. *Efficiency.* See Andrews (1954), Hodges and Lehmann (1956), and Section 6.10.

PROBLEMS

1. Pretherapy training of clients has been shown to have beneficial effects on the process and outcome of counseling and psychotherapy. Sauber (1971) investigated four different approaches to pretherapy training:
   1. Control (no treatment).
   2. Therapeutic reading (TR) (indirect learning).
   3. Vicarious therapy pretraining (VTP) (videotaped, vicarious learning).
   4. Group, role induction interview (RII) (direct learning).

   Treatment conditions 2 to 4 were expected to enhance the outcome of counseling and psychotherapy as compared with a control group, those subjects receiving no prior set of structuring procedures. One of the major variables of the study was that of "psychotherapeutic attraction." The basic data in Table 6.2 consist of the raw scores for this measure according to each of the four experimental conditions. Apply procedure (6.6), with the correction for ties given by (6.8).

   TABLE 6.2. Raw Scores Indicating the Degree of Psychotherapeutic Attraction for Each Experimental Condition

Control	Reading (TR)	Videotape (VTP)	Group (RII)
0	0	0	1
1	6	5	5
3	7	8	12
3	9	9	13
5	11	11	19
10	13	13	22
13	20	16	25
17	20	17	27
26	24	20	29

   Source: S. R. Sauber (1971).

2. Show that the two expressions for H in (6.5) are indeed equivalent.
3. Show directly, or illustrate by means of an example, that the maximum value of H is $H_{max} = \{N^3 - \sum_{j=1}^{k} n_j^3\}/\{N(N+1)\}$. For what rank configurations is this maximum achieved?
4. To determine the number of gamefish to stock in a given system and to set appropriate catch limits, it is important for fishery managers to be able to assess potential growth and survival of gamefish

in that system. Such growth and survival rates are closely related to the availability of appropriately sized prey. Young-of-year (YOY) gizzard shad (*Dorosoma cepedianum*) are the primary food source for gamefish in many Ohio environments. However, because of their fast growth rate, YOY gizzard shad can quickly become too large for predators to swallow. Thus to be able to predict predator growth rates in such settings it is useful to know both the density and the size structure of the resident YOY shad populations. With this in mind, Johnson (1984) sampled the YOY gizzard shad population at four different sites in Kokosing Lake (Ohio) in summer 1984. The data in Table 6.3 are lengths (mm) for a subset of the YOY gizzard shad sampled by Johnson.

TABLE 6.3. Length of YOY Gizzard Shad from Kokosing Lake, Ohio, Sampled in Summer, 1984 (mm)

Site I	Site II	Site III	Site IV
46	42	38	31
28	60	33	30
46	32	26	27
37	42	25	29
32	45	28	30
41	58	28	25
42	27	26	25
45	51	27	24
38	42	27	27
44	52	27	30

Source: B. Johnson (1984).

Apply procedure (6.7), with the correction for ties given in (6.8), to assess whether there are any differences between the median lengths for the YOY gizzard shad populations in the four Kokosing Lake sites.

5. Suppose $k = 3$ and $n_1 = 2, n_2 = n_3 = 6$. Compare the critical region for the exact level $\alpha = .050$ test of H_0 (6.2) based on H with the critical region for the corresponding nominal level $\alpha = .050$ test based on the large-sample approximation. What is the exact significance level of this .050 nominal level test based on the large-sample approximation?

6. Suppose $k = 4, n_1 = n_2 = n_3 = 1$, and $n_4 = 2$. Obtain the form of the exact null (H_0) distribution of H for the case of no tied X observations.

7. Suppose $k = 3, n_1 = n_2 = n_3 = 2$, and we observe the data $X_{11} = 2.7, X_{21} = 3.4, X_{12} = 2.7$, $X_{22} = 4.5, X_{13} = 4.9$, and $X_{23} = 2.7$. What is the conditional probability distribution of H under H_0 (6.2) when average ranks are used to break ties among the X's? How extreme is the observed value of H in this conditional null distribution? Compare this fact with that obtained by taking the observed value of H to the (incorrect) unconditional null distribution of H given in Table A.12.

8. Leukemia is a disease characterized by proliferation of the white blood cells, or leukocytes. One form of chemotherapy used in the treatment of leukemia involves the administration of corticosteroids. Some researchers suggested that forms of leukemia characterized by leukocytes with a large number of glucocorticoid receptor (GR) sites per cell are more effectively controlled by corticosteroids. Other researchers questioned this relationship. In an effort to aid in the resolution of this controversy, Kontula, Andersson, et al. (1980) developed a method for determining more accurately the number of GR sites per cell. In this research and later work by Kontula, Paavonen et al. (1982), this new methodology was used to count the number of GR sites for samples of leukocyte cells from normal subjects, as well as patients with hairy-cell leukemia, chronic lymphatic leukemia, chronic myelocytic leukemia, or acute leukemia. The data in Table 6.4 are a subset of the data considered by the authors in these two publications.

Use these data to assess whether there are any differences between the median numbers of GR sites per leukocyte cell for the population of normal subjects and the populations of patients with hairy-cell leukemia, chronic lymphatic leukemia, chronic myelocytic leukemia, or acute leukemia.

TABLE 6.4. Number of Glucocorticoid Receptor (GR) Sites per Leukocyte Cell

Normal Subjects	Hairy-cell Anemia	Chronic Lymphatic Leukemia	Chronic Myelocytic Leukemia	Acute Leukemia
3,500	5,710	2,930	6,320	3,230
3,500	6,110	3,330	6,860	3,880
3,500	8,060	3,580	11,400	7,640
4,000	8,080	3,880	14,000	7,890
4,000	11,400	4,280		8,280
4,000		5,120		16,200
4,300				18,250
4,500				29,900
4,500				
4,900				
5,200				
6,000				
6,750				
8,000				

Source: K. Kontula, L. C. Andersson, T. Paavonen, G. Myllyla, L. Teerenhovi, and P. Vuopio (1980) and K. Kontula, T. Paavonen, P. Vuopio, and L. C. Andersson (1982).

9. Use the computer software StatXact to conduct the desired analysis (see Problem 4) of the gizzard shad data in Table 6.3.

10. Use the computer software Minitab to conduct the desired analysis (see Problem 8) of the leukocyte glucocorticoid receptor site data in Table 6.4.

11. Generate the conditional permutation distribution of H using only the last two sample lengths from each of the four sites for the gizzard shad data in Table 6.3. From this conditional permutation distribution of H, obtain the exact conditional P-value for a test of H_0 (6.2) versus H_1 (6.3) with this subset of data from Table 6.3. Compare this exact conditional P-value with the approximate P-value associated with taking the observed value of H for these tied data directly to Table A.12.

12. Habitat plays an important role in fish behavior, particularly feeding, spawning, and protection/security. One of the modern methods of fisheries management is habitat modification in large, constructed reservoirs. Previous studies have shown that the type of structure introduced is an important factor in such habitat modifications. Of particular relevance in many settings is the size openings or *interstices* in the introduced structure. The data in Table 6.5 represent a subset of that obtained

TABLE 6.5. Mean Interstitial Lengths (mm)

Scotch Pine	Blue Spruce	White Pine
52.2	46.7	75.2
56.4	60.5	63.7
57.1	58.9	73.2
46.9	82.9	66.2
49.1	65.8	67.4
52.5	93.3	69.4
63.0	66.9	70.4
52.0	70.9	72.3
61.1	73.7	63.6
55.3	65.8	61.9
46.2	90.2	74.4
57.2	68.9	70.1

Source: K. A. Kayle (1984).

by Kayle (1984) from Alum Creek Lake in Westerville, Ohio in a study to determine the relative effectiveness of three species of pine trees for habitat modification. The measurements in Table 6.5 are averages (mm) of interstitial lengths (distances between midpoints) of ten pairs of secondary branches for each of 12 scotch pine, 12 blue spruce, and 12 white pine trees. Use an appropriate procedure to test whether there are any differences in median interstitial lengths between secondary branches for the three studied species of pine.

6.2. A DISTRIBUTION-FREE TEST FOR ORDERED ALTERNATIVES (JONCKHEERE, TERPSTRA)

In many practical settings the treatments are such that the appropriate alternatives to no differences in treatment effects (H_0) are those of increasing (or decreasing) treatment effects according to some natural labeling for the treatments. Examples of such settings include "treatments" corresponding to degrees of knowledge of performance, quality or quantity of materials, severity of disease, amount of practice, drug dosage levels, intensity of a stimulus, and temperature. We note that the Kruskal-Wallis procedure (6.6) does not utilize any such partial prior information regarding a postulated alternative ordering. The statistic H (6.5) takes on the same value for all $k!$ possible labelings of the treatments. In this section we consider a procedure for testing H_0 (6.2) against the a priori ordered alternatives

$$H_2 : [\tau_1 \leq \tau_2 \leq \cdots \leq \tau_k, \text{ with at least one strict inequality}]. \tag{6.11}$$

The Jonckheere-Terpstra test of this section is preferred to the Kruskal-Wallis test in Section 6.1 when the treatments can be labeled a priori in such a way that the experimenter expects any deviation from H_0 (6.2) to be in the particular direction associated with H_2 (6.11). We emphasize, however, that the labeling of the treatments so that the ordered alternatives (6.11) are appropriate *cannot* depend on the observed sample observations. This labeling must correspond completely to a factor(s) implicit in the nature of the *experimental design* and *not* the *observed data*.

Procedure

First we must label the treatments so that they are in the expected order associated with the alternative H_2 (6.11). (This labeling must be done prior to data collection.) To compute the Jonckheere-Terpstra statistic, J, we calculate the $k(k-1)/2$ Mann-Whitney (see Comment 4.7) counts U_{uv} given by

$$U_{uv} = \sum_{i=1}^{n_u} \sum_{j=1}^{n_v} \phi(X_{iu}, X_{jv}), \quad 1 \leq u < v \leq k, \tag{6.12}$$

where $\phi(a, b) = 1$ if $a < b$, 0 otherwise. (Thus, U_{uv} is the number of sample u before sample v precedences.) The Jonckheere-Terpstra statistic, J, is then the sum of these $k(k-1)/2$ Mann-Whitney counts,

$$J = \sum_{u=1}^{v-1} \sum_{v=2}^{k} U_{uv}. \tag{6.13}$$

To test

$$H_0 : [\tau_1 = \cdots = \tau_k]$$

versus the ordered alternative

$$H_2 : [\tau_1 \leq \tau_2 \leq \cdots \leq \tau_k, \text{ with at least one strict inequality}],$$

at the α level of significance,

$$\text{Reject } H_0 \text{ if } J \geq j_\alpha; \quad \text{otherwise do not reject}, \tag{6.14}$$

where the constant j_α is chosen to make the type I error probability equal to α. Values of j_α are given in Table A.13.

Large-Sample Approximation

The large-sample approximation is based on the asymptotic $(\min(n_1, n_2, \ldots, n_k)$ tending to infinity) normality of J, suitably standardized. We first need to know the expected value and variance of J when the null hypothesis is true. Under H_0, the expected value and variance of J are

$$E_0(J) = \frac{N^2 - \sum_{j=1}^k n_j^2}{4} \tag{6.15}$$

and

$$\text{var}_0(J) = \frac{N^2(2N + 3) - \sum_{j=1}^k n_j^2(2n_j + 3)}{72}, \tag{6.16}$$

respectively. These expressions for $E_0(J)$ and $\text{var}_0(J)$ are verified by direct calculations in Comment 20 for the special case of $k = 3, n_1 = n_2 = 1, n_3 = 2$. General derivations of both expressions are outlined in Comment 21.

The standardized version of J is

$$J^* = \frac{J - E_0(J)}{\sqrt{\text{var}_0(J)}} = \frac{J - \left[\frac{N^2 - \sum_{j=1}^k n_j^2}{4}\right]}{\left\{\left[N^2(2N + 3) - \sum_{j=1}^k n_j^2(2n_j + 3)\right]/72\right\}^{1/2}}. \tag{6.17}$$

When H_0 is true, J^* has, as $\min(n_1, \ldots, n_k)$ tends to infinity, an asymptotic $N(0, 1)$ distribution. (See Comment 21 for indications of the proof.) The normal theory approximation for procedure (6.14) is

$$\text{Reject } H_0 \text{ if } J^* \geq z_\alpha; \quad \text{otherwise do not reject}. \tag{6.18}$$

Ties

If there are ties among the N X's, replace $\phi(a, b)$ in the calculation of the Mann-Whitney counts U_{uv} by $\phi^*(a, b) = 1, \frac{1}{2}, 0$ if $a <, =,$ or $> b$, respectively, so that for each between-sample comparison where there is a tie, the contribution to the appropriate Mann-Whitney count will be $\frac{1}{2}$. After computing J with these modified Mann-Whitney counts, use procedure (6.14) and refer the value of J to Table A.13. Note, however, that this test associated with tied X's is only approximately, and not exactly, of significance level α.

When applying the large-sample approximation, an additional factor must be taken into account. Although ties in the X's do not affect the null expected value of J, its null variance is reduced to

$$\text{var}_0(J) = \left\{ \frac{1}{72} \left[N(N-1)(2N+5) - \sum_{i=1}^{k} n_i(n_i-1)(2n_i+5) - \sum_{j=1}^{g} t_j(t_j-1)(2t_j+5) \right] \right.$$

$$+ \frac{1}{36N(N-1)(N-2)} \left[\sum_{i=1}^{k} n_i(n_i-1)(n_i-2) \right] \left[\sum_{j=1}^{g} t_j(t_j-1)(t_j-2) \right]$$

$$\left. + \frac{1}{8N(N-1)} \left[\sum_{i=1}^{k} n_i(n_i-1) \right] \left[\sum_{j=1}^{g} t_j(t_j-1) \right] \right\}, \qquad (6.19)$$

where, in equation (6.19), g denotes the number of tied X groups and t_j is the size of tied group j. We note that an untied observation is considered to be a tied group of size 1. In particular, if there are no ties among the X's, then $g = N$ and $t_j = 1$, for $j = 1,\ldots,N$. In this case each term in (6.19) that involves the factor $(t_j - 1)$ reduces to zero and (as you are asked to show in Problem 21) the variance expression in (6.19) reduces to the usual null variance of J when there are no ties, as given previously in equation (6.16).

As a consequence of the effect that ties have on the null variance of J, the following modification is needed to apply the large-sample approximation when there are tied X's. Compute J using the modified Mann-Whitney counts and set

$$J^* = \frac{J - \left[\frac{N^2 - \sum_{j=1}^{k} n_j^2}{4} \right]}{\{\text{var}_0(J)\}^{1/2}}, \qquad (6.20)$$

where $\text{var}_0(J)$ is now given by display (6.19). With this modified value of J^*, the approximation (6.18) can be applied.

Example 6.2: *Motivational Effect of Knowledge of Performance.* Hundal (1969) described a study designed to assess the purely motivational effects of knowledge of performance in a repetitive industrial task. The task was to grind a metallic piece to a specified size and shape. Eighteen male workers were divided randomly into three groups. The subjects in the control group, A, received no information about their output, subjects in group B were given a rough estimate of their output, and subjects in group C were given accurate information about their output and could check it further by referring to a figure that was placed before them. The basic data in Table 6.6 consist of the numbers of pieces processed by each subject in the experimental period.

We apply the Jonckheere-Terpstra test with the notion that a deviation from H_0 is likely to be in the direction of increased output with increased degree of knowledge of performance. Thus we are interested in using procedure (6.14) with the treatment labels $1 \equiv$ control (no information), $2 \equiv$ group B (rough information), and $3 \equiv$ group C (accurate information). For purpose of illustration we take the significance level to be $\alpha = .0490$. From Table A.13 with sample sizes 6, 6, and 6, we see that $j_{.0490} = 75$, and procedure (6.14) reduces to

Reject H_0 if $J \geq 75$.

We now illustrate the computations leading to the sample value of J (6.13). Since there are ties in the sample data, we use $\phi^*(a,b) = 1, \frac{1}{2}, 0$ if $a <, =,$ or $> b$, respectively, to compute

TABLE 6.6. Number of Pieces Processed

Control (No Information)	Group B (Rough Information)	Group C (Accurate Information)
40 (5.5)*	38 (2.5)	48 (18)
35 (1)	40 (5.5)	40 (5.5)
38 (2.5)	47 (17)	45 (15)
43 (10.5)	44 (13)	43 (10.5)
44 (13)	40 (5.5)	46 (16)
41 (8)	42 (9)	44 (13)

Source: P. S. Hundal (1969).
* Although we do not need to perform the joint ranking to compute the Jonckheere-Terpstra statistic, we give these ranks here for use in Sections 6.4 and 6.7.

the $3(2)/2 = 3$ Mann-Whitney counts. We obtain

$$U_{12} = 1.5 + 2.5 + 6 + 5.5 + 2.5 + 4 = 22,$$

$$U_{13} = 6 + 2.5 + 6 + 4.5 + 6 + 5.5 = 30.5,$$

and

$$U_{23} = 6 + 2 + 5 + 4 + 5 + 4.5 = 26.5.$$

From (6.13) it follows that

$$J = 22 + 30.5 + 26.5 = 79.$$

Since this value is greater than the critical value 75, we reject H_0 at the .0490 level. In fact, we find from Table A.13, with $k = 3$ and $n_1 = n_2 = n_3 = 6$, that $j_{.0231} = 79$. Thus the lowest significance level at which we can reject H_0 in favor of H_2 with the observed value of $J = 79$ is the P-value .0231.

For the large-sample approximation we need to compute the standardized form of J^* using equations (6.19) and (6.20), since there are ties in the data. The null expected value for J is $E_0(J) = [(18)^2 - (6^2 + 6^2 + 6^2)]/4 = 54$. For the ties-corrected null variance of J, we note that $g = 11$ and $t_1 = 1, t_2 = 2, t_3 = 4, t_4 = 1, t_5 = 1, t_6 = 2, t_7 = 3, t_8 = 1, t_9 = 1, t_{10} = 1, t_{11} = 1$ for the Hundal data. Hence, using the ties correction in equation (6.19), we have

$$\mathrm{var}_0(J) = \left\{ \frac{1}{72}[18(17)(41) - 3(6)(5)(17) - 2(2)(1)(9) - 3(2)(11) - 4(3)(13)] \right.$$

$$+ \frac{1}{36(18)(17)(16)}[3(6)(5)(4)][3(2)(1) + 4(3)(2)]$$

$$\left. + \frac{1}{8(18)(17)}[3(6)(5)][2(2)(1) + 1(3)(2) + 1(4)(3)] \right\} = 150.29,$$

from which it follows that the ties-corrected value of J^* (6.20) is

$$J^* = \frac{79 - 54}{\{150.29\}^{1/2}} = 2.04.$$

Thus, using the approximate procedure (6.18) with the ties-corrected value of $J^* = 2.04$, we see from Table A.1 that the approximate P-value for these data is .0207. Both the exact test and large-sample approximation indicate strong evidence in support of increased output with increase in degree of knowledge of performance for the task considered by Hundal.

The approximate P-value based on the large-sample approximation as well as a Monte Carlo estimate and approximate 99% confidence interval for the exact P-value can also be obtained by using StatXact's JT/MO command. The StatXact output for the Hundal motivational data in Table 6.6 is:

Mean	Std − dev	Observed($JT(x)$)	Standardized($JT^*(x)$)
54	12.26	79	2.039

Asymptotic P-values:

One-sided: $Pr\{JT^*(X).\text{GE}.2.039\} = .0207$

Monte Carlo estimate of P-values:

One-sided: $Pr\{JT^*(X).\text{GE}.2.039\} = .0198$

99% Confidence Interval = $(.0162, .0234)$.

Under the heading "Monte Carlo estimate of P-values" we see that StatXact provides .0198 as the Monte Carlo estimate of the exact P-value, in good agreement with what we found from Table A.13. In addition, StatXact provides $(.0162, 0234)$ as a 99% confidence interval for the exact P-value. Finally, we note that the approximate P-value provided by StatXact under "Asymptotic P-values" is .0207, also in agreement with what we previously obtained from Table A.1 using the large-sample approximation. (Note that StatXact also provides some output for a two-sided version of the Jonckheere-Terpstra test procedure. However, this output is not relevant to the one-sided procedures given in (6.14) and (6.18).)

Comments

14. *More General Setting.* As with the Kruskal-Wallis procedure in Section 6.1, we could replace Assumptions A1–A3 and H_0 (6.2) with the more general null hypothesis that all $N!/\left(\prod_{j=1}^{k} n_j!\right)$ assignments of n_1 joint ranks to the treatment 1 observations, n_2 joint ranks to the treatment 2 observations, ..., n_k joint ranks to the treatment k observations are equally likely.

15. *Motivation for the Test.* Consider J (6.13) and note that the term $\sum_{u=1}^{v-1} \sum_{v=2}^{k} U_{uv}$ takes the postulated ordering into account. Consider, for simplicity, the case $k = 3$. Then $\sum_{u=1}^{v-1} \sum_{v=2}^{3} U_{uv} = U_{12} + U_{13} + U_{23}$, and if $\tau_1 < \tau_2 < \tau_3$, U_{12} would tend to be larger than $n_1 n_2/2$ (its null expectation); U_{13} would tend to be larger than $n_1 n_3/2$; U_{23} would tend to be larger than $n_2 n_3/2$ and, consequently, $J = U_{12} + U_{13} + U_{23}$ would tend to be larger than its null expectation $(n_1 n_2 + n_1 n_3 + n_2 n_3)/2 = \{[N^2 - (n_1^2 + n_2^2 + n_3^2)]/4\}$. This serves as partial motivation for the J test.

16. *Assumptions.* It is once again (as with the Kruskal-Wallis procedure in Section 6.1) important to point out that Assumption A3 stipulates that the k treatment distributions $F_1, \ldots, F_k$ can differ at most in their locations (medians). (See also Comment 4.)

17. *Special Case of Two Treatments.* When there are only two treatments, the procedures in (6.14) and (6.18) are equivalent to the exact and large-sample approximation forms, respectively, of the one-sided upper-tail Wilcoxon rank sum test, as discussed in Section 4.1.

18. *Derivation of the Distribution of J under H_0 (No-Ties Case).* A little thought will convince the reader that J can be computed from the joint ranking of all $N = \sum_{j=1}^{k} n_j$ observations. That is, although we do not need to perform this joint ranking in order to compute J, given the ranking we can, without knowledge of the actual X_{ij} values, retrieve the value of J. Thus one way to obtain the null distribution of J is to follow the method of Comment 6; namely, use the fact that under H_0 (6.2) all $N!/\left(\prod_{j=1}^{k} n_j!\right)$ rank assignments are equally likely, and compute the associated value of J for each possible ranking. Consider how this would work in the small-sample-size case of $k = 3$, $n_1 = 1$, $n_2 = 1$, and $n_3 = 2$. The $4!/[1!\ 1!\ 2!] = 12$ possible assignments of the joint ranks 1, 2, 3, and 4 to the three treatments and their associated values of J (6.13) are as follows:

(a) I II III   (b) I II III   (c) I II III
     1   2   3   2   1   3   1   3   2
                 4   4   4
     $J = 5$ $J = 4$ $J = 4$

(d) I II III   (e) I II III   (f) I II III
     3   1   2   1   4   2   4   1   2
                 4   3   3
     $J = 3$ $J = 3$ $J = 2$

(g) I II III   (h) I II III   (i) I II III
     2   3   1   3   2   1   2   4   1
                 4   4   3
     $J = 3$ $J = 2$ $J = 2$

(j) I II III   (k) I II III   (l) I II III
     4   2   1   3   4   1   4   3   1
                 3   2   2
     $J = 1$ $J = 1$ $J = 0$

Thus the null distribution for J for $n_1 = 1, n_2 = 1, n_3 = 2$, and $k = 3$ is given by

$$P_0\{J = 0\} = \tfrac{1}{12},\ P_0\{J = 1\} = \tfrac{2}{12},\ P_0\{J = 2\} = \tfrac{3}{12},$$
$$P_0\{J = 3\} = \tfrac{3}{12},\ P_0\{J = 4\} = \tfrac{2}{12},\ P_0\{J = 5\} = \tfrac{1}{12}.$$

The probability, under H_0, that J is greater than or equal to 4, for example, is therefore

$$P_0\{J \geq 4\} = P_0\{J = 4\} + P_0\{J = 5\}$$
$$= \tfrac{1}{12} + \tfrac{2}{12} = .25.$$

Note that we have derived the null distribution of J without specifying the common form (F) of the underlying distribution function for the X's under H_0 beyond the requirement that it be continuous. This is why the test procedure (6.14) based on J is called a distribution-free procedure. From the null distribution of J we can determine the critical value j_α and control the probability α of falsely rejecting H_0 when H_0 is true, and this error probability does not depend on the specific form of the common underlying continuous X distribution.

19. *Sample Sizes Not in Increasing Order.* Table A.13 gives critical values for $n_1 \leq n_2 \leq n_3$ situations. However, critical points for (n_1, n_2, n_3) configurations not in this order can be obtained by simply putting the three sample sizes in increasing order and then entering Table A.13. (This is a consequence of certain symmetry properties of the distribution of J.) Thus, for example, to find critical values for the case $n_1 = 4, n_2 = 6, n_3 = 2$, enter Table A.13 at $n_1 = 2, n_2 = 4, n_3 = 6$.

20. *Calculation of the Mean and Variance of J under the Null Hypothesis H_0.* In displays (6.15) and (6.16) we presented formulas for the mean and variance of J when the null hypothesis is true. In this comment we illustrate a direct calculation of $E_0(J)$ and $\text{var}_0(J)$ in the particular case of $k = 3$ and $n_1 = n_2 = 1, n_3 = 2$ and no tied observations, using the null distribution of J obtained in Comment 18. (Later, in Comment 21, we present arguments for the general derivations of $E_0(J)$ and $\text{var}_0(J)$.) The null mean, $E_0(J)$, is obtained by multiplying each possible value of J with its probability under H_0. Thus

$$E_0(J) = 0\left(\tfrac{1}{12}\right) + 1\left(\tfrac{2}{12}\right) + 2\left(\tfrac{3}{12}\right) + 3\left(\tfrac{3}{12}\right) + 4\left(\tfrac{2}{12}\right) + 5\left(\tfrac{1}{12}\right) = 2.5.$$

This is in agreement with what we obtain using equation (6.15), namely,

$$E_0(J) = \frac{4^2 - \{1^2 + 2^2 + 1^2\}}{4} = 2.5.$$

A check on the expression for $\text{var}_0(J)$ is also easy, using the well-known fact that

$$\text{var}_0(J) = E_0(J^2) - \{E_0(J)\}^2.$$

The value of $E_0(J^2)$, the second moment of the null distribution of J, is again obtained by multiplying possible values (in this case, of J^2) by the corresponding probabilities under H_0. We find

$$E_0(J^2) = 0^2\left(\tfrac{1}{12}\right) + 1^2\left(\tfrac{2}{12}\right) + 2^2\left(\tfrac{3}{12}\right) + 3^2\left(\tfrac{3}{12}\right) + 4^2\left(\tfrac{2}{12}\right) + 5^2\left(\tfrac{1}{12}\right) = \tfrac{49}{6}.$$

Thus,

$$\text{var}_0(J) = \tfrac{49}{6} - (2.5)^2 = \tfrac{23}{12} = 1.92,$$

which agrees with what we obtain using equation (6.16) directly, namely,

$$\text{var}_0(J) = \frac{\{4^2(2(4) + 3) - [1^2(2(1) + 3) + 1^2(2(1) + 3) + 2^2(2(2) + 3)]\}}{72}$$
$$= 1.92.$$

21. *Large-Sample Approximation.* From the definition of J (6.13) and U_{uv} (6.12) we see that

$$E(J) = E\left[\sum_{u=1}^{v-1}\sum_{v=2}^{k} U_{uv}\right] = E\left[\sum_{u=1}^{v-1}\sum_{v=2}^{k}\sum_{i=1}^{n_u}\sum_{j=1}^{n_v} \phi(X_{iu}, X_{jv})\right]$$

$$= \sum_{u=1}^{v-1}\sum_{v=2}^{k}\sum_{i=1}^{n_u}\sum_{j=1}^{n_v} E[\phi(X_{iu}, X_{jv})]$$

6.2. A DISTRIBUTION-FREE TEST FOR ORDERED ALTERNATIVES

$$= \sum_{u=1}^{v-1}\sum_{v=2}^{k}\sum_{i=1}^{n_u}\sum_{j=1}^{n_v} P(X_{iu} < X_{jv})$$

$$= \sum_{u=1}^{v-1}\sum_{v=2}^{k} n_u n_v P(X_{1u} < X_{1v}). \qquad (6.21)$$

Under the null hypothesis H_0 (6.2), $P_0(X_{1u} < X_{1v}) = \frac{1}{2}$ for every $1 \leq u < v \leq k$. It follows that

$$E_0(J) = \sum_{u=1}^{v-1}\sum_{v=2}^{k} \frac{(n_u n_v)}{2} = \frac{1}{4}\sum_{\substack{u=1 \\ u \neq v}}^{k}\sum_{v=1}^{k} n_u n_v$$

$$= \frac{1}{4}\left[\sum_{u=1}^{k}\sum_{v=1}^{k} n_u n_v - \sum_{t=1}^{k} n_t^2\right]$$

$$= \frac{1}{4}\left[N^2 - \sum_{t=1}^{k} n_t^2\right],$$

which agrees with the general expression stated in equation (6.15).

It also follows from equations (6.12) and (6.13) that

$$\mathrm{var}(J) = \mathrm{var}\left(\sum_{u=1}^{v-1}\sum_{v=2}^{k} U_{uv}\right)$$

$$= \sum_{u=1}^{v-1}\sum_{v=2}^{k} \mathrm{var}(U_{uv}) + \sum_{u=1}^{v-1}\sum_{v=2}^{k}\sum_{\substack{s=1 \\ (u,v) \neq (s,t)}}^{t-1}\sum_{t=2}^{k} \mathrm{cov}(U_{uv}, U_{st}). \qquad (6.22)$$

Under H_0 (6.2), it can be shown (we will not here) that

$$\mathrm{var}_0(U_{uv}) = \frac{n_u n_v (n_u + n_v + 1)}{12}, \quad \text{for } 1 \leq u < v \leq k, \qquad (6.23)$$

$$\mathrm{cov}_0(U_{uv}, U_{st}) = 0, \quad \text{for all distinct } u, v, s, t \text{ in } \{1, \ldots, k\}, \qquad (6.24)$$

$$\mathrm{cov}_0(U_{uv}, U_{ut}) = \frac{n_u n_v n_t}{12}, \quad \text{for } 1 \leq u < v, t \leq k, \ v \neq t, \qquad (6.25)$$

$$\mathrm{cov}_0(U_{uv}, U_{su}) = \frac{-n_s n_u n_v}{12}, \quad \text{for } 1 \leq s < u < v \leq k, \qquad (6.26)$$

$$\mathrm{cov}_0(U_{uv}, U_{vt}) = \frac{-n_u n_v n_t}{12}, \quad \text{for } 1 \leq u < v < t \leq k, \qquad (6.27)$$

$$\mathrm{cov}_0(U_{uv}, U_{sv}) = \frac{n_u n_v n_s}{12}, \quad \text{for } 1 \leq u, s < v \leq k, \ u \neq s. \qquad (6.28)$$

Combining the results in equations (6.23)–(6.28) with the expression for var(J) in equation (6.22), it follows after significant algebraic manipulation that

$$\operatorname{var}_0(J) = \frac{N^2(2N+3) - \sum_{j=1}^{k} n_j^2(2n_j+3)}{72},$$

which agrees with the general expression stated in equation (6.16).

The null asymptotic normality of the standardized form

$$J^* = \frac{J - E_0(J)}{\{\operatorname{var}_0(J)\}^{1/2}} = \frac{J - \left[\frac{N^2 - \sum_{j=1}^{k} n_j^2(2n_j+3)}{4}\right]}{\left\{\left[N^2(2N+1) - \sum_{t=1}^{k} n_t^2(2n_t+3)\right]/72\right\}^{1/2}}$$

follows from the fact that J can be expressed as a sum of certain mutually independent combined-samples Mann-Whitney statistics and standard theory for such sums of mutually independent, but not necessarily identically distributed, random variables (see, for example, Terpstra (1952) or Section 12.1 of Randles and Wolfe (1979)). Asymptotic normality results for J under general alternatives to H_0 are obtainable from standard results in k-sample U-statistics theory (see, for example, Lehmann (1975, 401–402)).

22. *Power of the Jonckheere-Terpstra Test.* The Jonckheere-Terpstra procedures (6.14) and (6.18) are quite superior to the Kruskal-Wallis procedures in (6.6) and (6.7) when the conjectured ordering of the treatment effects ($\tau_1 \leq \tau_2 \leq \cdots \leq \tau_k$) is, indeed, appropriate. In addition, small violations in the conjectured ordering for τ_i and τ_j do not seriously affect the power of the Jonckheere-Terpstra tests if i and j correspond to treatment labels near the middle of the conjectured orderings. However, if i and j are both near 1 or k, the affect of such violations can be rather substantial, especially if the magnitude of the difference $|\tau_j - \tau_i|$ is fairly large. Mack and Wolfe (1981) present the results of a small-sample power study that illustrate this phenomenon about the power of the Jonckheere-Terpstra procedures. In Section 6.3 we will discuss test procedures designed to deal with this possibility of early or late violations of the conjectured orderings $\tau_1 \leq \tau_2 \leq \cdots \leq \tau_k$. The Jonckheere-Terpstra procedures will turn out to be special cases of this class of tests designed for the more general form of alternatives $\tau_1 \leq \tau_2 \leq \cdots \leq \tau_{p-1} \leq \tau_p \geq \tau_{p+1} \geq \cdots \geq \tau_k$, known in the literature as "umbrella orderings" for the pictoral shape of the graphed treatment effects.

23. *k-Sample Behrens-Fisher Problem.* Two of the implicit requirements associated with Assumptions A1–A3 are that the underlying distributions belong to the same common family (F) and that they differ within this family at most in their medians. The less restrictive setting where these assumptions are relaxed to permit the possibility of differences in scale parameters as well as medians within the common family F is referred to as the k-sample Behrens-Fisher problem. The Jonckheere-Terpstra procedure (6.14) is no longer distribution-free under this more relaxed Behrens-Fisher setting. Chen and Wolfe (1990a) suggested a modification of the Jonckheere-Terpstra statistic J (6.13) to deal with this less restrictive setting. Their approach is similar to that used by Rust and Fligner (1984) to modify the Kruskal-Wallis statistic H for the same setting (see Comment 11).

24. *Consistency of the J Test.* Replace Assumptions A1–A3 by the less restrictive Assumptions A1': the X's are mutually independent and A2' : $X_{1j}, \ldots, X_{n_j j}$ comes from the same continuous population $\Pi_j, j = 1, \ldots, k$. The populations $\Pi_1, \ldots, \Pi_k$ need not be identical, but we do assume that

$$\delta_{ij} = P(X_{1j} > X_{1i}) \geq \tfrac{1}{2}, \quad \text{for } 1 \leq i < j \leq k.$$

Then, roughly speaking, the test defined by (6.14) is consistent if and only if there is at least one pair (i, j), with $i < j$, such that $\delta_{ij} > \frac{1}{2}$.

Properties

1. *Consistency.* The condition n_j/N tends to $\lambda_j, 0 < \lambda_j < 1, j = 1, \ldots, k$, is sufficient to ensure that the test defined by (6.14) is consistent against the H_2 (6.11) alternatives. For a more general consistency statement, see Terpstra (1952) and Comment 24.
2. *Asymptotic Normality.* See Randles and Wolfe (1979, 396–397) and Lehmann (1975, 401–402).
3. *Efficiency.* See Puri (1965) and Section 6.10.

PROBLEMS

13. Apply the Jonckheere-Terpstra test to the psychotherapeutic attraction data of Table 6.2 using the postulated ordering $\tau_1 \leq \tau_2 \leq \tau_3 \leq \tau_4$. Compare and contrast this result with that obtained for the Kruskal-Wallis test in Problem 1.
14. The statistic J can be computed either from (a) the joint ranking of the $N = \sum_{j=1}^{k} n_j$ observations or from (b) $k(k-1)/2$ "two-sample" rankings. Explain.
15. What are the minimum and maximum values for J? Justify your answer.
16. Suppose $k = 3$ and $n_1 = 4, n_2 = 7, n_3 = 8$. Compare the critical region for the exact level $\alpha = .0444$ test of H_0 (6.2) based on J with the critical region for the corresponding nominal level $\alpha = .0444$ test based on the large-sample approximation. What is the nominal probability of a type I error assigned by the large-sample approximation to the exact level $\alpha = .0444$ critical region?
17. Suppose $k = 4, n_1 = n_2 = n_3 = 1$, and $n_4 = 2$. Obtain the form of the exact null (H_0) distribution of J for the case of no tied observtions.
18. Use equations (6.23)–(6.28) to show that the expression for $\text{var}_0(J)$ in equation (6.16) follows, under H_0, from the general expression for $\text{var}(J)$ in equation (6.22).
19. In a project designed to study stand density (i.e., number of trees in a fixed area) and its relationship to other important features of a timber area such as tree growth, wood quality, and total wood production, Dale (1984) collected data on a quantity (related to yearly growth increment in a tree) known as basal area increment (BAI) for 16 stands of mixed species of oak trees in southeastern Ohio. The 16 stands were grouped according to the value of a second factor called growing site index. This index ranges in value from the low 50s to 100 for oak species, and as the value of the site index increases, the growing environment becomes more favorable for a stand of trees. The data in Table 6.7 are a subset of the data obtained by Dale and represent average BAI values for the 16 stands in his study. The BAI data are grouped into five distinct categories according to the associated growing site index values.

TABLE 6.7. Average Basal Area Increment (BAI) Values for Oak Stands in Southeastern Ohio

Growing Site Index Interval				
66–68	69–71	72–74	75–77	78–80
1.91	2.44	2.45	2.52	2.78
1.53		2.04	2.36	2.88
2.08		1.60	2.73	2.10
1.71		2.37		1.66

Source: M. Dale (1984).

Use an appropriate test procedure to evaluate the conjecture that the average basal area increment for a given stand of oak trees is an increasing function of the value of the stand's growing site index.

20. Apply the Kruskal-Wallis test to the knowledge of performance data in Table 6.6. Compare and contrast this result with that obtained by the Jonckheere-Terpstra test in Example 6.2.
21. Show that the expression given in equation (6.19) for the null variance of J in the case of tied X observations reduces to the usual null variance of J when there are no ties, as given in equation (6.16).
22. Use the computer software StatXact to conduct the desired analysis (see Problem 6.19) of the basal area increment data in Table 6.7.

6.3. DISTRIBUTION-FREE TESTS FOR UMBRELLA ALTERNATIVES (MACK-WOLFE)

In Section 6.2 we introduced the idea of designing test procedures to be especially effective against a restricted class of alternatives. There we considered the special class of monotonically ordered alternatives. In this section we extend that idea to a broader class of alternatives, which includes the ordered alternatives of Section 6.2 as a special case.

Let $p \in \{1, 2, \ldots, k\}$ be a fixed treatment label. In this section we consider procedures for testing H_0 (6.2) against the class of umbrella alternatives corresponding to

$$H_3 : [\tau_1 \leq \tau_2 \leq \cdots \leq \tau_{p-1} \leq \tau_p \geq \tau_{p+1} \geq \cdots \geq \tau_k,$$

with at least one strict inequality]. (6.29)

(The label *umbrella* was given to these alternatives by Mack and Wolfe (1981) because of the pictoral configuration of the τ's.) The umbrella in (6.29) is said to have a peak at population p. (Note that the ordered alternatives of Section 6.2 are simply a special case of umbrella alternatives with peak at $p = k$.) These umbrella alternatives are one-way layout analogs to a quadratic regression setting and are appropriate, for example, in evaluating marginal gain in performance efficiency as a function of time, crop yield as a function of fertilizer applied, reaction to increasing drug dosage levels where a downturn in effect may occur after the optimal dosage is exceeded, effect of age on responses to certain stimuli, etc. (These umbrella alternatives can be effectively used in place of ordered alternatives when one is concerned about possible violations of the monotonic ordering at either the beginning or the end of the sequence of treatment effects. See Comment 22 for further discussion along these lines.)

In Section 6.3A we present a procedure specifically designed to test H_0 (6.2) against the umbrella alternatives H_3 (6.29), where the peak, p, of the conjectured umbrella is known *prior* to data collection. This procedure is preferred to the general alternatives Kruskal-Wallis test in Section 6.1 when the treatments can be labeled a priori in such a way that the experimenter expects any deviation from H_0 (6.2) to be in the particular direction of H_3 (6.29) with known p. In Section 6.3B we extend the idea of umbrella alternatives to the more practical setting where it is not necessary to specify the peak, p, of the umbrella configuration prior to data collection. Here we present a procedure designed to test H_0 (6.2) against the class of umbrella alternatives with peak (p) unspecified, namely,

$$H_4 : [\tau_1 \leq \cdots \leq \tau_{p-1} \leq \tau_p \geq \tau_{p+1} \geq \cdots \geq \tau_k,$$

with at least one strict inequality, for some $p \in \{1, 2, \ldots, k\}$]. (6.30)

The Mack-Wolfe procedure in Section 6.3B is preferred to the "peak-known" procedure presented in Section 6.3A for the more common settings when umbrella alternatives are appropriate but where there is some uncertainty about the treatment at which the maximum effect is expected to occur if H_0 (6.2) is not true.

As with the ordered alternatives in Section 6.2, we emphasize that the labeling of the treatments so that either of the umbrella alternatives H_3 (6.29) or H_4 (6.30) is appropriate *cannot* depend on the observed sample values. This labeling must correspond to a factor (s) associated with the *experimental design* and *not* on the *sample data*. In Section 6.3B, however, the peak of the conjectured umbrella need not be specified prior to data collection.

6.3.A. A DISTRIBUTION-FREE TEST FOR UMBRELLA ALTERNATIVES, PEAK KNOWN (MACK-WOLFE)

In this subsection we present a procedure for testing H_0 (6.2) against the peak-known (at p) umbrella alternatives given by H_3 (6.29).

Procedure

First we must label the treatments so that they are in the prescribed ordered relationships to the known peak, p, corresponding to the umbrella configuration in H_3 (6.29). To calculate the known-peak umbrella statistic, A_p, we first compute the $p(p-1)/2$ Mann-Whitney counts U_{uv} (6.12) for every pair of treatments with labels less than or equal to the hypothesized peak (i.e., for $1 \leq u < v \leq p$). In addition, we compute the $(k-p+1)(k-p)/2$ reverse Mann-Whitney counts U_{vu} (6.12) for every pair of treatments with labels greater than or equal to the hypothesized peak (i.e., for $p \leq u < v \leq k$). (Thus, U_{vu} is the number of sample v before sample u precedences. Note that if there are no ties between the uth sample and vth sample observations, $p \leq u < v \leq k$, then $U_{vu} = n_u n_v - U_{uv}$.) The Mack-Wolfe peak-known statistic, A_p, is then the sum of the Mann-Whitney counts to the left of the peak and the reverse Mann-Whitney counts to the right of the peak (as appropriate for the umbrella alternatives H_3 (6.29)), namely,

$$A_p = \sum_{u=1}^{v-1} \sum_{v=2}^{p} U_{uv} + \sum_{u=p}^{v-1} \sum_{v=p+1}^{k} U_{vu}. \tag{6.31}$$

To test

$$H_0 : [\tau_1 = \cdots = \tau_k]$$

versus the peak-known (at $p \in \{1,\ldots,k\}$) umbrella alternative

$$H_3 : [\tau_1 \leq \tau_2 \leq \cdots \leq \tau_{p-1} \leq \tau_p \geq \tau_{p+1} \geq \cdots \geq \tau_k,$$

with at least one strict inequality],

at the α level of significance,

$$\text{Reject } H_0 \text{ if } A_p \geq a_{p,\alpha}; \quad \text{otherwise do not reject,} \tag{6.32}$$

where the constant $a_{p,\alpha}$ is chosen to make the type I error probability equal to α. Values of $a_{p,\alpha}$ are given in Table A.14 for selected p and k combinations.

Large-Sample Approximation

The large-sample approximation is based on the asymptotic $(\min(n_1,\ldots,n_k)$ tending to infinity) normality of A_p, suitably standardized. For this purpose we need to know the expected value and variance of A_p when the null hypothesis is true. Under H_0, the expected value and variance of A_p are

$$E_0(A_p) = \frac{N_1^2 + N_2^2 - \sum_{i=1}^{k} n_i^2 - n_p^2}{4} \tag{6.33}$$

and

$$\text{var}_0(A_p) = \frac{1}{72}\left\{2(N_1^3 + N_2^3) + 3(N_1^2 + N_2^2) - \sum_{i=1}^{k} n_i^2(2n_i + 3)\right.$$

$$\left. - n_p^2(2n_p + 3) + 12n_p N_1 N_2 - 12n_p^2 N\right\}, \tag{6.34}$$

respectively, with $N_1 = \sum_{i=1}^{p} n_i$ and $N_2 = \sum_{i=p}^{k} n_i$. (Note that $N = N_1 + N_2 - n_p$, since the observations in the peak treatment p are counted in both N_1 and N_2.) These expressions for $E_0(A_p)$ and $\text{var}_0(A_p)$ are verified by direct calculations in Comment 29 for the special case of $k = 4$, $p = 3$, $n_1 = n_2 = n_4 = 1, n_3 = 2$. General derivations of both expressions are outlined in Comment 30.

The standardized version of A_p is

$$A_p^* = \frac{A_p - E_0(A_p)}{\sqrt{\text{var}_0(A_p)}}$$

$$= \left\{A_p - \left[\frac{N_1^2 + N_2^2 - \sum_{i=1}^{k} n_i^2 - n_p^2}{4}\right]\right\}$$

$$\div \left\{\left[2(N_1^3 + N_2^3) + 3(N_1^2 + N_2^2) - \sum_{i=1}^{k} n_i^2(2n_i + 3)\right.\right.$$

$$\left.\left. - n_p^2(2n_p + 3) + 12n_p N_1 N_2 - 12n_p^2 N\right]/72\right\}^{1/2}. \tag{6.35}$$

When H_0 is true, A_p^* has, as $\min(n_1,\ldots,n_k)$ tends to infinity, an asymptotic $N(0, 1)$ distribution. (See Comment 30 for indications of the proof.) The normal theory approximation to procedure (6.32) is

$$\text{Reject } H_0 \text{ if } A_p^* \geq z_\alpha; \text{ otherwise do not reject.} \tag{6.36}$$

Ties

If there are ties among either the N_1 X's in treatments $1,\ldots,p$ or the N_2 X's in treatments $p,\ldots,k$, replace $\phi(a,b)$ in the calculations of the appropriate Mann-Whitney counts U_{uv} or reverse Mann-Whitney counts U_{vu} by $\phi^*(a,b) = 1, \frac{1}{2}, 0$ if $a <, =,$ or $> b$, respectively, so that for each between-sample comparison where there is a tie, the contribution to the appropriate Mann-Whitney or reverse Mann-Whitney count will be $\frac{1}{2}$. After computing A_p with these modified counts, use procedure (6.32) and refer the value of A_p to Table A.14. Note, however, that this test associated with tied X's is only approximately, and not exactly, of significance level α.

When applying the large-sample approximation, an additional factor should be taken into account. Although ties in the X's do not affect the null expected value of A_p, its true null variance is smaller in the case of ties than the numerical value given by the expression in (6.34). However, the appropriate expression for the exact variance of A_p in the case of ties is not available. Therefore, in the case of tied X's and large sample sizes, we recommend computing A_p using the modified Mann-Whitney counts and then A_p^* via equation (6.35). With this modified value of A_p^* the approximation (6.36) can be applied. However, the associated approximate P-value will be larger than what we would obtain if the appropriate expression for the ties-corrected null variance of A_p were available to use in the computation of A_p^*.

Example 6.3: *Fasting Metabolic Rate of White-Tailed Deer.* Seasonal energy requirements of deer is an important consideration when evaluating wildlife plans for certain habitats. Both nutritional quality of the range and the physiological demands of the deer must be studied in order to prevent starvation during critical seasons and to select optimum harvest strategies. Some aspects of the energy demand were considered by Silver et al. (1969) as they studied the fasting metabolic rate (FMR) of white-tailed deer. In particular, one of the questions of interest was whether or not FMR is an increasing function of environmental temperature, for which they collected the data in Table 6.8.

For these data we expect any deviation from H_0 (6.2) to be in the direction of increasing FMR values from the January–February period up through the warmest 2-month period, July–August, with declining FMR values from July–August through the November–December period. Thus we are interested in testing H_0 against the peak-known umbrella alternatives (6.29) with treatment labels $1 \equiv$ January–February, $2 \equiv$ March–April, $3 \equiv$ May–June, $4 \equiv$ July–August, $5 \equiv$ September–October, $6 \equiv$ November–December and known umbrella peak at $p = 4$, corresponding to the warmest (July–August) 2-month period. Since exact critical values are not available in Table A.14 for the setting $k = 6, p = 4, n_1 = 7, n_2 = 3, n_3 = 5, n_4 = 4, n_5 = 4$, and $n_6 = 3$, we will need to use the large-sample approximation and procedure (6.36). For purpose of illustration, we consider the approximate significance level $\alpha = .0274$. From Table A.1 we have $z_{.0274} = 1.92$ and procedure (6.36) is given by

$$\text{reject } H_0 \text{ if } A_4^* \geq 1.92.$$

TABLE 6.8. Fasting Metabolic Rate (FMR) for White-Tailed Deer (kcal/kg/day)

Two-Month Period					
Jan.–Feb.	Mar.–Apr.	May–June	July–Aug.	Sept.–Oct.	Nov.–Dec.
36.0	39.9	44.6	53.8	44.3	31.7
33.6	29.1	54.4	53.9	34.1	22.1
26.9	43.4	48.2	62.5	35.7	30.7
35.8		55.7	46.6	35.6	
30.1		50.0			
31.2					
35.3					

Source: H. Silver, N. F. Colovos, J. B. Holter, and H. H. Hayes (1969).

We now illustrate the computations leading to the sample value of A_4^* (6.35). For that purpose we first need to compute the $4(3)/2 = 6$ Mann-Whitney counts U_{uv}, for $1 \leq u < v \leq 4$, and the $3(2)/2 = 3$ reverse Mann-Whitney counts U_{vu}, for $4 \leq u < v \leq 6$. We obtain

$$U_{12} = 7 + 1 + 7 = 15, \quad U_{13} = 7 + 7 + 7 + 7 + 7 = 35, \quad U_{14} = 7 + 7 + 7 + 7 = 28,$$
$$U_{23} = 3 + 3 + 3 + 3 + 3 = 15, \, U_{24} = 3 + 3 + 3 + 3 = 12, \, U_{34} = 3 + 3 + 5 + 1 = 12,$$
$$U_{54} = 4 + 4 + 4 + 4 = 16, \quad U_{64} = 3 + 3 + 3 + 3 = 12, \quad U_{65} = 3 + 3 + 3 + 3 = 12.$$

From (6.31) it follows that

$$A_4 = U_{12} + U_{13} + U_{14} + U_{23} + U_{24} + U_{34} + U_{65} + U_{64} + U_{54}$$
$$= 15 + 35 + 28 + 15 + 12 + 12 + 12 + 12 + 16 = 157,$$

and $n_1 = 7, n_2 = 3, n_3 = 5, n_4 = 4, n_5 = 4, n_6 = 3$. Hence, we have $N_1 = (7 + 3 + 5 + 4) = 19$, $N_2 = 3 + 4 + 4 = 11$ and $N = (7 + 3 + 5 + 4 + 4 + 3) = 26$. Using these figures in expressions (6.33) and (6.34) for $E_0(A_4)$ and $\text{var}_0(A_4)$, respectively, we see that

$$E_0(A_4) = \frac{(19)^2 + (11)^2 - [(7)^2 + (3)^2 + (5)^2 + (4)^2 + (4)^2 + (3)^2 + (4)^2]}{4}$$
$$= 85.5$$

and

$$\text{var}_0(A_4) = \tfrac{1}{72}\{2[(19)^3 + (11)^3] + 3[(19)^2 + (11)^2]$$
$$- [(7)^2(2(7) + 3) + (3)^2(2(3) + 3)$$
$$+ (5)^2(2(5) + 3) + (4)^2(2(4) + 3)$$
$$+ (4)^2(2(4) + 3) + (3)^2(2(3) + 3)]$$
$$- (4)^2(2(4) + 3) + 12(4)(19)(11) - 12(4)^2(26)\}$$
$$= \frac{21,018}{72} = 291.92.$$

Thus, from (6.35), we obtain

$$A_4^* = \frac{A_4 - E_0(A_4)}{\sqrt{\text{var}_0(A_4)}} = \frac{157 - 85.5}{\sqrt{291.92}} = 4.18.$$

Since this value is greater than the approximate critical value 1.92, we reject H_0 at the approximate .0274 level. In fact, from Table A.1 we see that even $z_{.0002} = 3.49$ is smaller than the observed value $A_4^* = 4.18$. Thus the smallest approximate level at which we can reject H_0 in favor of H_3 with the observed value of $A_4^* = 4.18$ (i.e., the approximate P-value) is less than .0002 and there is very strong evidence in support of the claim that FMR for white-tailed deer is an increasing function of environmental temperature. (We note that the Jonckheere-Terpstra procedure from Section 6.2 would not be appropriate for these FMR data even with relabeled treatments, since it would be difficult to properly order the temperatures of the March–April and September–October periods, for example.)

Comments

25. *Motivation for the Test*. Notice that the statistic A_p can be viewed as the simple sum of two Jonckheere-Terpstra statistics, one (J_{up}) on treatments 1 through p with the postulated ordering $\tau_1 \leq \cdots \leq \tau_p$ and the second (J_{down}) on treatments k through p with the postulated reverse ordering $\tau_k \leq \tau_{k-1} \leq \cdots \leq \tau_p$. Thus the statistic $A_p = J_{\text{up}} + J_{\text{down}}$ will be large if either J_{up} or J_{down} (or both) is large. In view of Comment 15, this serves as partial motivation for the A_p test.

26. *Computer Software*. The umbrella statistic A_p can be represented as a sum of two Jonckheere-Terpstra statistics (see Comment 25). Thus the `StatXact JT/MO` command can be used twice to compute the numerical value of A_p and its null expected value. However, neither the null standard deviation for A_p (and, therefore, the standardized form A_p^*) nor the approximate P-value for the test based on A_p (or A_p^*) can be obtained from these two `JT/MO` outputs.

27. *Special Case of Three Treatments*. When there are only $k = 3$ treatments, the umbrella statistic A_p can be viewed in a special way. If $p = 3$, then $A_3 = U_{12} + U_{13} + U_{23}$ is just the usual Jonckheere-Terpstra statistic for the ordered alternatives $\tau_1 \leq \tau_2 \leq \tau_3$. If $p = 1$, we have $A_1 = U_{31} + U_{32} + U_{21}$, which is the Jonckheere-Terpstra statistic for the reverse ordered alternatives $\tau_3 \leq \tau_2 \leq \tau_1$. In either of these cases, all the properties of the Jonckheere-Terpstra test procedure (including null distribution and critical values) discussed in Section 6.2 apply directly to tests based on A_1 or A_3, as appropriate. For the third umbrella setting with $p = 2$, we see that $A_2 = U_{12} + U_{32}$, which is the same as a *single* Mann-Whitney statistic comparing the peak sample (treatment 2) with the combined set of data from treatments 1 and 3. (Thus, A_2 is the number of sample 1 *or* sample 3 before sample 2 precedences.) As a result, if $p = 2$ and $k = 3$ the procedures in (6.32) and (6.36) for sample sizes n_1, n_2, and n_3 are equivalent to the exact and large-sample approximation forms, respectively, of the one-sided upper-tail two-sample Wilcoxon rank sum test (as discussed in Section 4.1) for sample sizes $m = n_1 + n_3$ and $n = n_2$.

28. *Derivation of the Distribution of A_p under H_0 (No Ties)*. As with the Jonckheere-Terpstra statistic J (see Comment 18), it is clear that the umbrella peak-known statistic A_p can be computed from the joint ranking of all $N = \sum_{i=1}^{k} n_i$ observations. Thus one way to obtain the null distribution of A_p is to follow the method of Comments 6 and 18, namely, to compute the value of A_p for each of the $N! / \left(\prod_{j=1}^{k} n_j! \right)$ equally likely (under H_0) rank assignments. We illustrate how this works in the small-sample-size case of $k = 4, p = 3, n_1 = n_2 = n_4 = 1, n_3 = 2$. The $5!/[1!\ 1!\ 1!\ 2!] = 60$ possible arrangements of the joint ranks 1, 2, 3, 4, and 5 to the four treatments and their associated values of A_3 (6.31) are as follows:

	I	II	III	IV		I	II	III	IV
1.	1	2	4 5	3	2.	2	1	4 5	3
			$A_3 = 7$					$A_3 = 6$	
3.	1	3	4 5	2	4.	3	1	4 5	2
			$A_3 = 7$					$A_3 = 6$	
5.	2	3	4 5	1	6.	3	2	4 5	1
			$A_3 = 7$					$A_3 = 6$	

7. | I | II | III | IV
    | 1 | 2 | 3 | 4
    | | | 5 |
    | | | $A_3 = 6$ |

8. | I | II | III | IV
    | 2 | 1 | 3 | 4
    | | | 5 |
    | | | $A_3 = 5$ |

9. | I | II | III | IV
    | 1 | 4 | 3 | 2
    | | | 5 |
    | | | $A_3 = 6$ |

10. | I | II | III | IV
    | 4 | 1 | 3 | 2
    | | | 5 |
    | | | $A_3 = 5$ |

11. | I | II | III | IV
    | 2 | 4 | 3 | 1
    | | | 5 |
    | | | $A_3 = 6$ |

12. | I | II | III | IV
    | 4 | 2 | 3 | 1
    | | | 5 |
    | | | $A_3 = 5$ |

13. | I | II | III | IV
    | 1 | 3 | 2 | 4
    | | | 5 |
    | | | $A_3 = 5$ |

14. | I | II | III | IV
    | 3 | 1 | 2 | 4
    | | | 5 |
    | | | $A_3 = 4$ |

15. | I | II | III | IV
    | 1 | 4 | 2 | 3
    | | | 5 |
    | | | $A_3 = 5$ |

16. | I | II | III | IV
    | 4 | 1 | 2 | 3
    | | | 5 |
    | | | $A_3 = 4$ |

17. | I | II | III | IV
    | 3 | 4 | 2 | 1
    | | | 5 |
    | | | $A_3 = 5$ |

18. | I | II | III | IV
    | 4 | 3 | 2 | 1
    | | | 5 |
    | | | $A_3 = 4$ |

19. | I | II | III | IV
    | 2 | 3 | 1 | 4
    | | | 5 |
    | | | $A_3 = 4$ |

20. | I | II | III | IV
    | 3 | 2 | 1 | 4
    | | | 5 |
    | | | $A_3 = 3$ |

21. | I | II | III | IV
    | 2 | 4 | 1 | 3
    | | | 5 |
    | | | $A_3 = 4$ |

22. | I | II | III | IV
    | 4 | 2 | 1 | 3
    | | | 5 |
    | | | $A_3 = 3$ |

23. | I | II | III | IV
    | 3 | 4 | 1 | 2
    | | | 5 |
    | | | $A_3 = 4$ |

24. | I | II | III | IV
    | 4 | 3 | 1 | 2
    | | | 5 |
    | | | $A_3 = 3$ |

25. | I | II | III | IV
    | 1 | 2 | 3 | 5
    | | | 4 |
    | | | $A_3 = 5$ |

26. | I | II | III | IV
    | 2 | 1 | 3 | 5
    | | | 4 |
    | | | $A_3 = 4$ |

27. | I | II | III | IV
    | 1 | 5 | 3 | 2
    | | | 4 |
    | | | $A_3 = 5$ |

28. | I | II | III | IV
    | 5 | 1 | 3 | 2
    | | | 4 |
    | | | $A_3 = 4$ |

	I	II	III	IV		I	II	III	IV
29.	2	5	3 4	1 $A_3 = 5$	30.	5	2	3 4	1 $A_3 = 4$
31.	1	3	2 4	5 $A_3 = 4$	32.	3	1	2 4	5 $A_3 = 3$
33.	1	5	2 4	3 $A_3 = 4$	34.	5	1	2 4	3 $A_3 = 3$
35.	3	5	2 4	1 $A_3 = 4$	36.	5	3	2 4	1 $A_3 = 3$
37.	2	3	1 4	5 $A_3 = 3$	38.	3	2	1 4	5 $A_3 = 2$
39.	2	5	1 4	3 $A_3 = 3$	40.	5	2	1 4	3 $A_3 = 2$
41.	3	5	1 4	2 $A_3 = 3$	42.	5	3	1 4	2 $A_3 = 2$
43.	1	4	2 3	5 $A_3 = 3$	44.	4	1	2 3	5 $A_3 = 2$
45.	1	5	2 3	4 $A_3 = 3$	46.	5	1	2 3	4 $A_3 = 2$
47.	4	5	2 3	1 $A_3 = 3$	48.	5	4	2 3	1 $A_3 = 2$
49.	2	4	1 3	5 $A_3 = 2$	50.	4	2	1 3	5 $A_3 = 1$

51.
I	II	III	IV
2	5	1	4
		3	

$A_3 = 2$

52.
I	II	III	IV
5	2	1	4
		3	

$A_3 = 1$

53.
I	II	III	IV
4	5	1	2
		3	

$A_3 = 2$

54.
I	II	III	IV
5	4	1	2
		3	

$A_3 = 1$

55.
I	II	III	IV
3	4	1	5
		2	

$A_3 = 1$

56.
I	II	III	IV
4	3	1	5
		2	

$A_3 = 0$

57.
I	II	III	IV
3	5	1	4
		2	

$A_3 = 1$

58.
I	II	III	IV
5	3	1	4
		2	

$A_3 = 0$

59.
I	II	III	IV
4	5	1	3
		2	

$A_3 = 1$

60.
I	II	III	IV
5	4	1	3
		2	

$A_3 = 0$

Thus the null distribution for A_3 when $k = 3, n_1 = n_2 = n_4 = 1$, and $n_3 = 2$ is given by

$$P_0\{A_3 = 0\} = \tfrac{3}{60}, \quad P_0\{A_3 = 1\} = \tfrac{6}{60}, \quad P_0\{A_3 = 2\} = \tfrac{9}{60}$$
$$P_0\{A_3 = 3\} = \tfrac{12}{60}, \quad P_0\{A_3 = 4\} = \tfrac{12}{60}, \quad P_0\{A_3 = 5\} = \tfrac{9}{60}$$
$$P_0\{A_3 = 6\} = \tfrac{6}{60}, \quad P_0\{A_3 = 7\} = \tfrac{3}{60}.$$

The probability, under H_0, that A_3 is greater than or equal to 5, for example, is

$$P_0\{A_3 \geq 5\} = P_0\{A_3 = 5\} + P_0\{A_3 = 6\} + P_0\{A_3 = 7\} = \frac{9 + 6 + 3}{60} = .3.$$

Note that we have derived the null distribution of A_3 without specifying the common form (F) of the underlying distribution function for the X's under H_0 beyond the requirement that it be continuous. This is why the test procedure (6.32) based on A_p is called a distribution-free procedure. From the null distribution of A_p we can determine the critical value $a_{p,\alpha}$ and control the probability α of falsely rejecting H_0 when H_0 is true, and this error probability does not depend on the specific form of the common underlying continuous X distribution.

29. *Calculation of the Mean and Variance of A_p under the Null Hypothesis H_0.* In displays (6.33) and (6.34) we presented formulas for the mean and variance of A_p when the null hypothesis is true. In this comment we provide a direct calculation of $E_0(A_p)$ and $\text{var}_0(A_p)$ in the specific case of $k = 4, p = 3, n_1 = n_2 = n_4 = 1, n_3 = 2$ and no tied observations using the null distribution of A_3 obtained in Comment 28. (Later, in Comment 30, we discuss general derivations of $E_0(A_p)$ and $\text{var}_0(A_p)$.) From the null distribution provided in Comment

28, we see that

$$E_0(A_3) = \left[0\left(\tfrac{3}{60}\right) + 1\left(\tfrac{6}{60}\right) + 2\left(\tfrac{9}{60}\right) + 3\left(\tfrac{12}{60}\right) + 4\left(\tfrac{12}{60}\right)\right.$$
$$\left. + 5\left(\tfrac{9}{60}\right) + 6\left(\tfrac{6}{60}\right) + 7\left(\tfrac{3}{60}\right)\right]$$
$$= 3.5.$$

This is in agreement with what we obtain using equation (6.33), namely,

$$E_0(A_3) = \frac{\{(1+1+2)^2 + (2+1)^2 - [1^2 + 1^2 + 2^2 + 1^2 + 2^2]\}}{4}$$
$$= \frac{16 + 9 - 11}{4} = 3.5.$$

Again using the null distribution in Comment 28, we have

$$E_0(A_3^2) = \left[0^2\left(\tfrac{3}{60}\right) + 1^2\left(\tfrac{6}{60}\right) + 2^2\left(\tfrac{9}{60}\right) + 3^2\left(\tfrac{12}{60}\right) + 4^2\left(\tfrac{12}{60}\right)\right.$$
$$\left. + 5^2\left(\tfrac{9}{60}\right) + 6^2\left(\tfrac{6}{60}\right) + 7^2\left(\tfrac{3}{60}\right)\right]$$
$$= 15.5.$$

Using the well-known expression for $\text{var}_0(A_3)$, it follows that

$$\text{var}_0(A_3) = E_0(A_3^2) - \{E_0(A_3)\}^2 = 15.5 - (3.5)^2 = 3.25,$$

which agrees with what we obtain using equation (6.34) directly, namely,

$$\text{var}_0(A_3) = \{2(4^3 + 3^3) + 3(4^2 + 3^2) - [(3)(1)^2(2(1) + 3) + 2(2)^2(2(2) + 3)]$$
$$+ 12(2)(4)(3) - 12(2)^2(5)\}/72$$
$$= \frac{182 + 75 - 71 + 288 - 240}{72} = 3.25.$$

30. *Large-Sample Approximation.* As noted in Comment 25, the umbrella statistic A_p can be expressed as $A_p = J_{\text{up}} + J_{\text{down}}$, where J_{up} is the Jonckheere-Terpstra statistic on treatments 1 through p with the postulated ordering $\tau_1 \leq \cdots \leq \tau_p$ and J_{down} is the Jonckheere-Terpstra statistic on treatments k through p with the postulated ordering $\tau_k \leq \tau_{k-1} \leq \cdots \leq \tau_p$. Thus, using the previous development for the Jonckheere-Terpstra statistic in Comment 21, we see that

$$E_0(A_p) = E_0(J_{\text{up}}) + E_0(J_{\text{down}})$$
$$= \frac{1}{4}\left[N_1^2 - \sum_{t=1}^{p} n_t^2\right] + \frac{1}{4}\left[N_2^2 - \sum_{t=p}^{k} n_t^2\right]$$
$$= \frac{1}{4}\left[N_1^2 + N_2^2 - \sum_{t=1}^{k} n_t^2 - n_p^2\right],$$

which agrees with the general expression stated in equation (6.33).

It also follows from the representation $A_p = J_{\text{up}} + J_{\text{down}}$ that

$$\text{var}_0(A_p) = \text{var}_0(J_{\text{up}} + J_{\text{down}})$$
$$= \text{var}_0(J_{\text{up}}) + \text{var}_0(J_{\text{down}}) + 2\text{cov}_0(J_{\text{up}}, J_{\text{down}}). \qquad (6.37)$$

Now,

$$\text{cov}_0(J_{\text{up}}, J_{\text{down}}) = \text{cov}_0\left(\sum_{u=1}^{v-1}\sum_{v=2}^{p} U_{uv}, \sum_{s=p}^{t-1}\sum_{t=p+1}^{k} U_{ts}\right)$$

$$= \text{cov}_0\left(\sum_{u=1}^{v-1}\sum_{v=2}^{p-1} U_{uv} + \sum_{u=1}^{p-1} U_{up},\right.$$

$$\left.\sum_{s=p+1}^{t-1}\sum_{t=p+2}^{k} U_{ts} + \sum_{t=p+1}^{k} U_{tp}\right)$$

$$= \left[\text{cov}_0\left(\sum_{u=1}^{v-1}\sum_{v=2}^{p-1} U_{uv}, \sum_{s=p+1}^{t-1}\sum_{t=p+2}^{k} U_{ts}\right)\right.$$

$$+ \text{cov}_0\left(\sum_{u=1}^{v-1}\sum_{v=2}^{p-1} U_{uv}, \sum_{t=p+1}^{k} U_{tp}\right)$$

$$+ \text{cov}_0\left(\sum_{u=1}^{p-1} U_{up}, \sum_{s=p+1}^{t-1}\sum_{t=p+2}^{k} U_{ts}\right)$$

$$\left.+ \text{cov}_0\left(\sum_{u=1}^{p-1} U_{up}, \sum_{t=p+1}^{k} U_{tp}\right)\right]. \qquad (6.38)$$

The term $\sum_{u=1}^{v-1}\sum_{v=2}^{p-1} U_{uv}$ involves only X observations from the first $(p-1)$ samples, whereas the terms $\sum_{s=p+1}^{t-1}\sum_{t=p+2}^{k} U_{ts}$ and $\sum_{t=p+1}^{k} U_{tp}$ involve only X observations from samples $p+1, p+2, \ldots, k$ and $p, p+1, \ldots, k$, respectively. Since the X observations are mutually independent, it follows that

$$\text{cov}_0\left(\sum_{u=1}^{v-1}\sum_{v=2}^{p-1} U_{uv}, \sum_{s=p+1}^{t-1}\sum_{t=p+2}^{k} U_{ts}\right) = \text{cov}_0\left(\sum_{u=1}^{v-1}\sum_{v=2}^{p-1} U_{uv}, \sum_{t=p+1}^{k} U_{tp}\right) = 0. \qquad (6.39)$$

Similarly, the term $\sum_{u=1}^{p-1} U_{up}$ involves only X observations from the first p samples, and the term $\sum_{s=p+1}^{t-1}\sum_{t=p+2}^{k} U_{ts}$ involves only X observations from samples $p+1, p+2, \ldots, k$, leading to

$$\text{cov}_0\left(\sum_{u=1}^{p-1} U_{up}, \sum_{s=p+1}^{t-1}\sum_{t=p+2}^{k} U_{ts}\right) = 0. \qquad (6.40)$$

(Note that equations (6.39) and (6.40) are a consequence of the fact that the sample observations from the peak treatment p are the only data used in both J_{up} and J_{down}.) Combining equations (6.38), (6.39), and (6.40) with a well-known result about covariances of sums, we obtain

$$\operatorname{cov}_0(J_{\text{up}}, J_{\text{down}}) = \operatorname{cov}_0\left(\sum_{u=1}^{p-1} U_{up}, \sum_{t=p+1}^{k} U_{tp}\right)$$

$$= \sum_{u=1}^{p-1} \sum_{t=p+1}^{k} \operatorname{cov}_0\left(U_{up}, U_{tp}\right). \quad (6.41)$$

From equation (6.28), it follows that

$$\operatorname{cov}_0(J_{\text{up}}, J_{\text{down}}) = \frac{n_p}{12} \sum_{u=1}^{p-1} \sum_{t=p+1}^{k} n_u n_t = \frac{n_p}{12}\left(\sum_{u=1}^{p-1} n_u\right)\left(\sum_{t=p+1}^{k} n_t\right)$$

$$= \frac{n_p(N_1 - n_p)(N_2 - n_p)}{12}. \quad (6.42)$$

Combining equations (6.37) and (6.42), we see that

$$\operatorname{var}_0(A_p) = \operatorname{var}_0(J_{\text{up}}) + \operatorname{var}_0(J_{\text{down}}) + \frac{n_p(N_1 - n_p)(N_2 - n_p)}{6}. \quad (6.43)$$

Using the expression in (6.16) for both $\operatorname{var}_0(J_{\text{up}})$ and $\operatorname{var}_0(J_{\text{down}})$, it follows from (6.43) after some algebraic manipulation (Problem 29) that

$$\operatorname{var}_0(A_p) = \frac{1}{72}\left\{2(N_1^3 + N_2^3) + 3(N_1^2 + N_2^2) - \sum_{i=1}^{k} n_i^2(2n_i + 3)\right.$$

$$\left. - n_p^2(2n_p + 3) + 12n_p N_1 N_2 - 12n_p^2 N\right\},$$

which agrees with the general expression stated in equation (6.34).

The null asymptotic normality of the standardized form

$$A_p^* = \frac{A_p - E_0(A_p)}{\{\operatorname{var}_0(A_p)\}^{1/2}}$$

follows from the fact that A_p can be expressed as a sum of certain mutually independent combined-samples Mann-Whitney statistics and standard theory for such sums of mutually independent, but not necessarily identically distributed, random variables (see, for example, Mack and Wolfe (1981)). Asymptotic normality results for A_p under general alternatives to H_0 follow directly from work by Archambault, Mack, and Wolfe (1977) on a large class of k-sample statistics.

31. *k-Sample Behrens-Fisher Problem.* Two of the implicit requirements associated with Assumptions A1–A3 are that the underlying distributions belong to the same common family (F) and that they differ within this family at most in their medians. The less restrictive setting where these assumptions are relaxed to permit the possibility of differences in scale parameters

as well as medians within the common family F is referred to as the k-sample Behrens-Fisher problem. The Mack-Wolfe procedure (6.32) is no longer distribution-free under this more relaxed Behrens-Fisher setting. Chen and Wolfe (1990a) suggested a modification of the Mack-Wolfe statistic A_p (6.31) to deal with this less restrictive setting. Their approach is similar to that used by Rust and Fligner (1984) to modify the Kruskal-Wallis statistic H for the same setting (see Comment 11).

32. *Consistency of the A_p Test.* Replace Assumptions A1–A3 by the less restrictive Assumptions A1': the X's are mutually independent and A2' : $X_{1j}, \ldots, X_{n_j j}$ come from the same continuous population $\Pi_j, j = 1, \ldots, k$. The populations $\Pi_1, \ldots, \Pi_k$ need not be identical, but they are restricted to conform with the umbrella alternatives. Letting $\delta_{ij} = P(X_{1j} > X_{1i})$, for $1 \leq i < j \leq k$, we do assume that

$$\delta_{ij} \geq \tfrac{1}{2}, \quad \text{for } 1 \leq i < j \leq p$$

$$\delta_{ij} \leq \tfrac{1}{2}, \quad \text{for } p \leq i < j \leq k, \tag{6.44}$$

with no restrictions on δ_{ij} for $i < p$ and $j > p$. Under these conditions on $\Pi_1, \ldots, \Pi_k$, the test defined by (6.32) is, roughly speaking, consistent if and only if at least one of the inequalities in (6.44) is strict.

Properties

1. *Consistency.* The condition n_j/N tends to $\lambda_j, 0 < \lambda_j < 1, j = 1, \ldots, k$, is sufficient to insure that the test defined by (6.32) is consistent against the umbrella alternatives H_3 (6.29). For a more general consistency statement, see Mack and Wolfe (1981) and Comment 32.
2. *Asymptotic Normality.* See Mack and Wolfe (1981).
3. *Efficiency.* See Mack and Wolfe (1981) and Section 6.10.

PROBLEMS

23. Survival of stocked tiger muskellunge (*Esox masquinongy*), like other stocked sportfish, is variable to poor in Ohio reservoirs. Previous research with this species suggests three possible reasons for poor survival: (i) predation by largemouth bass, (ii) inability to forage, and (iii) stress-related mortality associated with the stocking process. Among other things, Mather (1984) studied the effect on mortality of three components of the stocking process: netting, confinement, and temperature increase. One portion of her study dealt with the glucose response to the stress of an increase in temperature. A sample of 40 tiger muskellunge were transferred from a 15°C holding tank into a test tank (also held at 15°C) and allowed 24 h to recover. (This is the period of time previous experimenters have found to be necessary for the fish's plasma glucose level to return to normal after a dipnet stressor.) Then, a random sample of eight fish were removed from the tank, anesthetized, blood collected, and plasma glucose determined. These data serve as a baseline or control sample. Next the stressor (a 12°C temperature increase) was applied to the test tank and blood samples were collected (in the way previously described) for random samples of eight additional fish at each of the time periods 1 h, 4 h, 24 h, and 96 h after the temperature increase. These plasma glucose measurements (mg%) are given in Table 6.9 for the 40 fish in the study.

    In anticipation that a 24 h period is also necessary for a tiger muskellunge's plasma glucose level to recover from the 12°C temperature increase stressor, test the hypothesis of interest using an approximate significance level of .0281. What is the approximate P-value for these data?

24. (a) The statistic A_p can be computed from the joint ranking of all N observations. Explain.
    (b) The statistic A_p can also be computed from pairwise *two-sample* rankings. Explain.
    (c) How many different two-sample rankings are required in (b) to compute A_p?

6.3.A. A DISTRIBUTION-FREE TEST FOR UMBRELLA ALTERNATIVES, PEAK KNOWN

TABLE 6.9. Plasma Glucose (mg%)

		Hours after 12°C Temperature Increase		
0	1	4	24	96
61.08	95.45	205.96	67.74	61.76
86.21	169.19	82.55	79.84	69.12
90.15	216.16	116.60	78.23	77.45
72.91	141.92	107.23	90.23	73.45
83.74	116.16	103.83	64.92	71.08
76.35	172.22	96.60	65.73	52.45
91.63	126.26	112.77	49.60	71.57
56.65	177.78	140.85	77.42	54.90

Source: M. Mather (1984).

25. In Example 6.2, we used the Jonckheere-Terpstra procedure to analyze the knowledge of performance data. It is quite reasonable to postulate that "too much information" (e.g., supervisor looking over your shoulder commenting at each step of the process) might actually lead to a downturn in the number of satisfactory pieces produced. Suppose that the following data were collected under such a too-much-information scenario.

Group D (Too Much Information)
38
41
37
46
39
42

Use both the Jonckheere-Terpstra procedure and the Mack-Wolfe procedure with $p = 3$ to analyze the performance data with these added group D data. Discuss your findings.

26. What are the minimum and maximum values for A_p? Justify your answers.
27. Notice that the statistic A_p (6.31) does not include any Mann-Whitney comparisons between samples from pairs of treatments on opposite sides of the peak treatment p. Discuss the pros and cons of this fact in relation to the Mack-Wolfe test procedure based on A_p.
28. Consider the umbrella statistic A_p for k treatments.
    (a) Which value(s) of p requires computation of the maximum number of Mann-Whitney statistics? How many Mann-Whitney statistics are required?
    (b) Which value(s) of p requires computation of the fewest number of Mann-Whitney statistics? How many Mann-Whitney statistics are required?
29. Suppose $k = 4, n_1 = n_3 = n_4 = 1$, and $n_2 = 2$. Obtain the form of the exact null (H_0) distribution of A_2 for the case of no tied observations. Compare the null distribution of A_2 for $k = 4, n_1 = n_3 = n_4 = 1, n_2 = 2$ with the null distribution of A_3 for $k = 4, n_1 = n_2 = n_4 = 1, n_3 = 2$, as obtained in Comment 28. Discuss the differences.
30. Suppose $k = 4, n_1 = n_4 = 1$, and $n_2 = n_3 = 2$. Obtain the form of the exact null (H_0) distribution of A_2 for the case of no tied observations.
31. Show that the expression for the null variance (no ties) of A_p given in equation (6.43) is indeed the same as that stated in equation (6.34).
32. In many settings a dose-response relationship need not be monotonic in the dosage. In in vitro mutagenicity assays, for example, experimental organisms may not survive the toxic side effects of high doses of the test agent, thereby actually reducing the number of organisms at risk of mutation

and leading to a downturn (i.e., umbrella pattern) in the dose-response curve. The data in Table 6.10 are a subset of the data considered by Simpson and Margolin (1986) in a discussion of the analysis of Ames test results. Plates containing Salmonella bacteria of strain TA98 were exposed to various doses of Acid Red 114. The tabled observations are the numbers of visible revertant colonies on the 18 plates in the study.

TABLE 6.10. Number of Revertant Colonies of Salmonella Bacteria of Strain TA98 under Exposure to Various Doses of Acid Red 114, with Hamster Liver Activation

\multicolumn{6}{c}{Dose (μg/ml)}
0	100	333	1000	3333	10,000
22	60	98	60	22	23
23	59	78	82	44	21
35	54	50	59	33	25

Source: D. G. Simpson and B. H. Margolin (1986).

Test the null hypothesis H_0 (6.2) against the alternative that the peak of the dose response curve for Salmonella bacteria of strain TA98 exposure to Acid Red 114 occurs at dosage level 1000μg/ml.

33. For the Salmonella bacteria strain TA98 data in Table 6.10, test the null hypothesis H_0 (6.2) against the alternative that the peak of the dose response curve for Salmonella bacteria of strain TA98 exposure to Acid Red 114 occurs at dosage level 333 μg/ml. Compare the result with that from Problem 32.

34. For the Salmonella bacteria strain TA98 data in Table 6.10, test the null hypothesis H_0 (6.2) against the alternative that the number of revertant colonies of the bacteria is a monotone increasing function of the dose level of Acid Red 114 over the range of exposure in Table 6.10. Compare this result with those obtained in Problems 32 and 33.

35. For the Salmonella bacteria strain TA98 data in Table 6.10, use the Kruska-Wallis procedure to test H_0 (6.2) against the general alternatives H_1 (6.3). Compare this result with those obtained in Problems 32, 33, and 34.

6.3.B. A DISTRIBUTION-FREE TEST FOR UMBRELLA ALTERNATIVES, PEAK UNKNOWN (MACK-WOLFE)

In this section we present a procedure for testing H_0 (6.2) against the general peak unknown umbrella alternatives H_4 (6.30).

Procedure

We label the treatments so that they are in the proper umbrella relationship to the unknown peak treatment p. To calculate the Mack-Wolfe statistic for this unknown peak setting, we first use the sample data to estimate which of the treatments is most likely to correspond to the peak of the umbrella; that is, we first estimate p from the sample data. To accomplish this, we calculate k combined-samples Mann-Whitney statistics

$$U_{.q} = \sum_{i \neq q} U_{iq}, \quad \text{for } q = 1, \ldots, k, \tag{6.45}$$

where U_{iq} = (number of ith sample observations that precede qth sample observations) is the usual Mann-Whitney statistic for the ith and qth samples. Thus, $U_{.q}$ is itself simply a single Mann-Whitney statistic computed between the qth sample and the remaining $(k-1)$ samples combined (i.e., it equals the number of times an observation from the qth sample exceeds an observation from the other $(k-1)$ combined samples). Next, we standardize each of the $U_{.q}$'s by subtracting off its expected value under the null hypothesis H_0 (6.2) and dividing by its null standard deviation (see Comment 39) to obtain

$$U_{.q}^* = \frac{U_{.q} - E_0(U_{.q})}{\{\mathrm{var}_0(U_{.q})\}^{1/2}} = \frac{U_{.q} - [n_q(N-n_q)/2]}{\left\{\frac{n_q(N-n_q)(N+1)}{12}\right\}^{1/2}}, \quad q = 1,\ldots,k. \qquad (6.46)$$

Let r equal the number of treatments that are tied for having the maximum $U_{.q}^*$ value and let B be the subset of $\{1, 2, \ldots, k\}$ that corresponds to the r treatments tied for the maximum $U_{.q}^*$ value. (Since $U_{.1}^*, \ldots, U_{.k}^*$ are discrete random variables, there are sample size configurations for which the probability is positive that r will be greater than 1. See also Comment 34 and Problem 38.) The Mack-Wolfe peak-unknown statistic is then given by

$$A_{\hat{p}}^* = \frac{1}{r}\sum_{j \in B} \left[\frac{A_j - E_0(A_j)}{\{\mathrm{var}_0(A_j)\}^{1/2}}\right], \qquad (6.47)$$

where A_j (6.31) is the peak-known statistic with the peak at the jth treatment group and $E_0(A_j)$ and $\mathrm{var}_0(A_j)$ are the null expected value and null variance of A_j given by equations (6.33) and (6.34), respectively. (Thus $A_{\hat{p}}^*$ is equal to the average of the r standardized peak-known statistics corresponding to peaks at each of the r samples tied for the maximum $U_{.q}^*$. In most cases $r = 1$ and $A_{\hat{p}}^*$ is equal to the single standardized peak-known statistic with the peak at the indicated treatment group.)

To test

$$H_0 : [\tau_1 = \cdots = \tau_k]$$

versus the peak-unknown umbrella alternatives

$$H_4 : [\tau_1 \leq \cdots \leq \tau_{p-1} \leq \tau_p \geq \tau_{p+1} \geq \cdots \geq \tau_k,$$

with at least one strict inequality, for some $p \in \{1, 2, \ldots, k\}]$,

at the α level of significance,

$$\text{Reject } H_0 \text{ if } A_{\hat{p}}^* \geq a_{\hat{p},\alpha}^*; \quad \text{otherwise do not reject}, \qquad (6.48)$$

where the constant $a_{\hat{p},\alpha}^*$ is chosen to make the type I error probability equal to α. Values of $a_{\hat{p},\alpha}^*$ are given in Table A.15 for $k = 3(1)10$ and equal sample sizes $n_1 = \cdots = n_k = 2(1)10$. For an unequal sample-size configuration $(n_1, \ldots, n_k)$, Mack and Wolfe (1981) suggest approximating $a_{\hat{p},\alpha}^*$ by the corresponding critical value for equal sample sizes $n_1 = \cdots = n_k = m$, where m is the integer closest to the average sample size $(n_1 + \cdots + n_k)/k$. If this value of m is greater than 10, use the critical value for $m = 10$. Finally, if $k > 10$, they suggest using the appropriate $k = 10$ equal sample size critical value.

Ties

If there are ties among the N X's, replace $\phi(a, b)$ in the calculation of the associated Mann-Whitney counts U_{uv} or reverse Mann-Whitney counts U_{vu} by $\phi^*(a,b) = 1, \frac{1}{2}, 0$ if $a <, =,$ or $> b$, respectively, so that for each between-sample comparison where there is a tie, the contribution to the appropriate Mann-Whitney or reverse Mann-Whitney count will be $\frac{1}{2}$. After computing the $U_{.q}^*$'s (6.46) and $A_{\hat{p}}^*$ (6.47) with these modified counts, use procedure (6.48)

TABLE 6.11. Wechsler Adult Intelligence Scale (WAIS) Values

		Age Group		
16–19	20–34	35–54	55–69	≥ 70
8.62	9.85	9.98	9.12	4.80
9.94	10.43	10.69	9.89	9.18
10.06	11.31	11.40	10.57	9.27

Source: R. D. Norman (1966).

and refer the value of $A_{\hat{p}}^*$ to Table A.15. Note, however, that this test associated with tied X's is only approximately, and not exactly, of significance level α.

Example 6.4: *Learning Comprehension and Age.* It is generally believed that the ability to comprehend ideas and learn is an increasing function of age up to a certain point, and then it declines with increasing age. The data in Table 6.11 are values in the range typically obtained on the Wechsler Adult Intelligence Scale (WAIS) by males of various ages. (Actually, the averages of the five samples agree with the corresponding age-group means in Norman (1966).)

With $k = 5$ and $n_1 = \cdots = n_5 = 3$, we wish to test

$$H_0 \text{ (6.2) versus } H_4 : [\tau_1 \leq \cdots \leq \tau_p \geq \tau_{p+1} \geq \cdots \geq \tau_5,$$

with at least one strict inequality, for some $p \in \{1, 2, \ldots, 5\}]$,

where the five age groups are numbered as treatments in order of increasing age. For purpose of illustration, we consider the approximate significance level $\alpha = .05$. From Table A.15 we have $a_{\hat{p},.05}^* = 2.239$, and procedure (6.48) is given by

$$\text{Reject } H_0 \text{ if } A_{\hat{p}}^* \geq 2.239.$$

We now illustrate the computations leading to the sample value of $A_{\hat{p}}^*$ (6.47). First, we compute all of the $5(4)/2 = 10$ possible Mann-Whitney statistics, obtaining

$$U_{12} = 1 + 3 + 3 = 7, U_{13} = 2 + 3 + 3 = 8, \ U_{14} = 1 + 1 + 3 = 5, U_{15} = 0 + 1 + 1 = 2$$
$$U_{23} = 1 + 2 + 3 = 6, U_{24} = 0 + 1 + 2 = 3, \ U_{25} = 0 + 0 + 0 = 0, U_{34} = 0 + 0 + 1 = 1$$
$$U_{35} = 0 + 0 + 0 = 0, \ U_{45} = 0 + 1 + 1 = 2.$$

In order to estimate the age group at which WAIS values peak we next need to compute the combined-samples Mann-Whitney statistics $U_{.q}$ (6.45), for $q = 1, \ldots, 5$. Using the fact that $U_{vu} = n_u n_v - U_{uv}$ (since there are no ties in the data), for $u, v = 1, \ldots, 5$, we find that

$$U_{.1} = U_{21} + U_{31} + U_{41} + U_{51}$$
$$= \{[3(3) - U_{12}] + [3(3) - U_{13}] + [3(3) - U_{14}] + [3(3) - U_{15}]\}$$
$$= (9 - 7) + (9 - 8) + (9 - 5) + (9 - 2) = 14,$$

6.3.B. A DISTRIBUTION-FREE TEST FOR UMBRELLA ALTERNATIVES, PEAK UNKNOWN

$$U_{.2} = U_{12} + U_{32} + U_{42} + U_{52}$$
$$= U_{12} + [3(3) - U_{23}] + [3(3) - U_{24}] + [3(3) - U_{25}]$$
$$= 7 + (9 - 6) + (9 - 3) + (9 - 0) = 25,$$
$$U_{.3} = U_{13} + U_{23} + U_{43} + U_{53}$$
$$= U_{13} + U_{23} + [3(3) - U_{34}] + [3(3) - U_{35}]$$
$$= 8 + 6 + (9 - 1) + (9 - 0) = 31,$$
$$U_{.4} = U_{14} + U_{24} + U_{34} + U_{54}$$
$$= U_{14} + U_{24} + U_{34} + [3(3) - U_{45}]$$
$$= 5 + 3 + 1 + (9 - 2) = 16,$$

and

$$U_{.5} = U_{15} + U_{25} + U_{35} + U_{45} = 2 + 0 + 0 + 2 = 4.$$

For this study we have equal sample sizes $n_1 = \cdots = n_5 = 3$. This implies that each of the combined-samples Mann-Whitney statistics has the same null mean and null variance; that is, for $q = 1, \ldots, 5$, we have

$$E_0(U_{.q}) = \frac{3(15-3)}{2} = 18, \quad \text{var}_0(U_{.q}) = \frac{3(15-3)(15+1)}{12} = 48.$$

As a result, for this equal-sample-sizes setting we do not need to compute the standardized forms $U_{.q}^*$ (6.46), as the treatment group with the largest $U_{.q}$ value will also be the one with the largest $U_{.q}^*$ (see also Comment 34). Therefore, the third age group (35–54) is estimated to be the unique peak group (i.e., $\hat{p} = 3$ and $r = 1$), since

$$U_{.3} = \max\{U_{.1}, U_{.2}, U_{.3}, U_{.4}, U_{.5}\} = 31.$$

The Mack-Wolfe peak-unknown statistic $A_{\hat{p}}^*$ (6.47) with $r = 1$ and $\hat{p} = 3$ becomes

$$A_{\hat{p}}^* = \frac{A_3 - E_0(A_3)}{\{\text{var}_0(A_3)\}^{1/2}}.$$

Using the computational formula (6.35) for the peak-known setting in Section 6.3A, we obtain

$$A_3 = 45, \quad E_0(A_3) = 27, \quad \text{var}_0(A_3) = 58.5,$$

which yields

$$A_{\hat{p}}^* = \frac{45 - 27}{\sqrt{58.5}} = 2.353.$$

Since this value is greater than the critical value 2.239, we reject H_0 at the approximate .05 level and conclude that there is sufficient evidence in support of the claim that the ability to comprehend ideas and learn is an increasing function of age up through the age group 35–54, from which point it declines with further age.

Comments

33. *Motivation for the Test.* The combined-samples Mann-Whitney statistic $U_{\cdot q}$ represents the number of times an observation from the qth sample exceeds an observation from the other $(k-1)$ combined samples. If the sample sizes are all equal and $\tau_1 < \tau_2 < \cdots < \tau_{p-1} < \tau_p > \tau_{p+1} > \cdots > \tau_k$, then we would expect $U_{\cdot p}$ to be the largest of the k combined-samples Mann-Whitney statistics. Such an outcome would lead to the selection of the pth treatment as the peak group and to $A_{\hat{p}}^* = [A_p - E_0(A_p)]/\{\text{var}_0(A_p)\}^{1/2}$. In view of Comment 25, this provides partial motivation for the $A_{\hat{p}}^*$ test when we have equal sample sizes. (See also Comment 34.)

34. *Equal versus Unequal Sample Sizes.* The number of individual comparisons required to produce the value of $U_{\cdot q}$ (6.45) is $n_q(N - n_q)$. If the sample sizes are not all equal, then we will have differing numbers of comparisons leading to the various $U_{\cdot q}$ values. This leads to the undesirable situation where even under the null hypothesis (H_0) those treatments with more sample observations are more likely to be selected as the estimated peak if we use the $U_{\cdot q}$ statistics directly. One way to address this problem is to first standardize the $U_{\cdot q}$'s by subtracting off their null expected values and then dividing by their null standard derivations. The use of these standardized $U_{\cdot q}^*$ statistics to select the peak results in each treament having as nearly as possible an equal chance of being selected as the peak under H_0.

If the sample sizes are all equal, say $n_1 = \cdots = n_k = n^*$, then we have

$$E_0(U_{\cdot q}) = \frac{n^*(N - n^*)}{2} \quad \text{and} \quad \text{var}_0(U_{\cdot q}) = \frac{n^*(N - n^*)(N + 1)}{12}$$

for every $q = 1, \ldots, k$. Thus in order to obtain the standardized $U_{\cdot q}^*$ in such a setting we would be subtracting the same quantity from each $U_{\cdot q}$ and dividing each of the resulting differences by the same value. The rank order of the resulting $U_{\cdot q}^*$'s would be identical with the rank order of the original $U_{\cdot q}$'s; that is, if we have equal sample sizes and t is such that

$$U_{\cdot t} = \text{maximum } \{U_{\cdot 1}, \ldots, U_{\cdot k}\}$$

then it is also true that

$$U_{\cdot t}^* = \text{maximum } \{U_{\cdot 1}^*, \ldots, U_{\cdot k}^*\}.$$

As a result, the standardization to obtain the $U_{\cdot q}^*$'s is not necessary in the case of all equal sample sizes as the $U_{\cdot q}$'s themselves can be directly used to select the peak $\hat{p}$.

35. *More General Setting.* As with the other procedures in this chapter we could replace Assumptions A1–A3 and H_0 (6.2) for the Mack-Wolfe umbrella procedures (both peak-known and peak-unknown) with the more general null hypothesis that all $N!/(\prod_{j=1}^{k} n_j!)$ assignments of n_1 joint ranks to the treatment 1 observations, n_2 joint ranks to the treatment 2 observations, $\ldots, n_k$ joint ranks to the treatment k observations are equally likely.

36. *Assumptions.* As with the other procedures in this chapter, it is important to point out that for the Mack-Wolfe umbrella procedures (both the peak-known and peak-unknown) the k treatment distributions $F_1, \ldots, F_k$ can differ at most in their locations (medians). (See also Comment 4.)

37. *Computer Software.* As far as we are aware there is no commercial computer software designed specifically to calculate $A_{\hat{p}}^*$. However, either the StatXact WI/EX command or Minitab's Mann-Whitney command can be used to compute the numerical values of the k combined-samples Mann-Whitney statistics $U_{\cdot q}$ used in selecting the umbrella peak. Once

38. *Estimation of the Umbrella Peak.* In situations where there is a unique, single treatment label, say t, for which

$$U^*_{.t} = \text{maximum } \{U^*_{.1}, \ldots, U^*_{.k}\},$$

then $r = 1$ in equation (6.47) and

$$A^*_{\hat{p}} = \frac{A_t - E_0(A_t)}{\{\text{var}_0(A_t)\}^{1/2}}.$$

In this setting, t also provides us with a point estimator for the unknown peak p (i.e., $\hat{p} = t$).

Pan (1996) developed a distribution-free confidence procedure designed to identify those treatments that yield the optimal effects in a one-way layout with umbrella configuration. It utilizes the theory of U-statistics and isotonic regression to provide a random confidence subset of the treatments that contains all the unknown peaks (optimal treatments) within an umbrella ordering with prespecified confidence level.

39. *Null Mean and Variance of Combined-Samples Mann-Whitney Statistics.* The combined-samples statistic $U_{.q}$ (6.45) can be viewed as a single Mann-Whitney statistic between the qth sample with n_q observations and the remaining $k - 1$ samples combined with $N - n_q$ observations. Thus, from the standard formulas for the null mean and null variance of a Mann-Whitney statistic (see the derivation in Comment 21, for example, particularly equation (6.23)), we see that

$$E_0(U_{.q}) = \frac{n_q(N - n_q)}{2} \quad \text{and} \quad \text{var}_0(U_{.q}) = \frac{n_q(N - n_q)(N + 1)}{12},$$

which agree with the expressions used in equation (6.46).

40. *Derivation of the Distribution of $A^*_{\hat{p}}$ under H_0 (No Ties).* As with the peak-known statistic A_p, the peak-unknown statistic $A^*_{\hat{p}}$ can also be computed from the joint ranking of all $N = \sum_{i=1}^{k} n_i$ observations. Thus one way to obtain the null distribution of $A^*_{\hat{p}}$ is to follow the method of Comments 6, 18, and 28, namely, to compute the value of $A^*_{\hat{p}}$ for each of the $N!/\left(\Pi_{j=1}^{k} n_j!\right)$ equally likely (under H_0) rank assignments. We illustrate this development in the specific case of $k = 3, n_1 = 1, n_2 = 2$ and $n_3 = 1$. The $4!/[1! \ 2! \ 1!] = 12$ possible assignments of the joint ranks 1, 2, 3, and 4 to the three treatments and the associated values of $A^*_{\hat{p}}$ are as follows:

1.	I	II	III	2.	I	II	III	
	1	2	4		4	2	1	
		3				3		
			$A^*_{\hat{p}} = 1.806$				$A^*_{\hat{p}} = 1.806$	
3.	I	II	III	4.	I	II	III	
	1	2	3		3	2	1	
		4				4		
			$A^*_{\hat{p}} = 0.775$				$A^*_{\hat{p}} = 0.775$	

5.	I	II	III	6.	I	II	III	
	3	1	4		4	1	3	
		2				2		
			$A_{\hat{p}}^* = 0.361$				$A_{\hat{p}}^* = 0.361$	

7.	I	II	III	8.	I	II	III	
	2	1	4		4	1	2	
		3				3		
			$A_{\hat{p}}^* = 1.084$				$A_{\hat{p}}^* = 1.084$	

9.	I	II	III	10.	I	II	III	
	2	1	3		3	1	2	
		4				4		
			$A_{\hat{p}}^* = 0.361$				$A_{\hat{p}}^* = 0.361$	

11.	I	II	III	12.	I	II	III	
	1	3	2		2	3	1	
		4				4		
			$A_{\hat{p}}^* = 1.549$				$A_{\hat{p}}^* = 1.549$	

Thus the null distribution for $A_{\hat{p}}^*$ when $k = 3$, $n_1 = n_3 = 2$, and $n_2 = 1$ is given by

$$P_0\{A_{\hat{p}}^* = 0.361\} = \tfrac{4}{12}, \quad P_0\{A_{\hat{p}}^* = 0.775\} = \tfrac{2}{12}, \quad P_0\{A_{\hat{p}}^* = 1.084\} = \tfrac{2}{12}$$

$$P_0\{A_{\hat{p}}^* = 1.549\} = \tfrac{2}{12}, \quad P_0\{A_{\hat{p}}^* = 1.806\} = \tfrac{2}{12}.$$

The probability, under H_0, that $A_{\hat{p}}^*$ is greater than or equal to 1.549, for example, is therefore

$$P_0\{A_{\hat{p}}^* \geq 1.549\} = P_0\{A_{\hat{p}}^* = 1.549\} + P_0\{A_{\hat{p}}^* = 1.806\} = \tfrac{2+2}{12} = \tfrac{1}{3}.$$

Note that we have derived the null distribution of $A_{\hat{p}}^*$ without specifying the common form (F) of the underlying distribution function for the X's under H_0 beyond the requirement that it be continuous. This is why the test procedure (6.48) based on $A_{\hat{p}}^*$ is called a distribution-free procedure. From the null distribution of $A_{\hat{p}}^*$ we can determine the critical value $a_{\hat{p},\alpha}^*$ and control the probability α of falsely rejecting H_0 when H_0 is true, and this error probability does not depend on the specific form of the common underlying continuous X distribution.

41. *Powers of the Mack-Wolfe Umbrella Tests.* The Mack-Wolfe unknown-peak umbrella procedure (6.48) based on $A_{\hat{p}}^*$ is generally much superior to the Kruskal-Wallis procedures in (6.6) and (6.7) when the treatment effects do, indeed, follow an umbrella pattern. When the peak is known a priori to be at treatment p, then the peak-known test (6.32) based on A_p has even better power properties. However, if there is serious uncertainty concerning the location of the true peak, the $A_{\hat{p}}^*$ procedure is preferable since the power of the A_p test can be somewhat diminished when p is not the correct peak. Mack and Wolfe (1981) present the results of a small-sample power study comparing the relative performances of the Kruskal-Wallis, Jonckheere-Terpstra, and the two Mack-Wolfe procedures for settings where umbrella alternatives pertain.

42. *Inverted Umbrella Alternatives.* The Mack-Wolfe procedures in this section can easily be adapted to provide tests for "inverted umbrella" alternatives of the form $\tau_1 \geq \cdots \geq \tau_{p-1} \geq \tau_p \leq \tau_{p+1} \leq \cdots \leq \tau_k$, with at least one strict inequality, for both p-known and p-unknown situations. To test for such inverted umbrella alternatives, simply redefine the peak-known

statistics to be

$$A_p = \sum_{u=1}^{v-1} \sum_{v=2}^{p} U_{vu} + \sum_{u=p}^{v-1} \sum_{v=p+1}^{k} U_{uv}, \quad \text{for } p = 1, \ldots, k,$$

and for the peak-unknown case, redefine the peak selectors to be "valley" selectors of the form

$$U_{q.} = \sum_{i \neq q} U_{qi}, \quad q = 1, 2, \ldots, k.$$

Everything else remains unchanged, including the necessary null distribution tables.

43. *k-Sample Behrens-Fisher Problem.* Two of the implicit requirements associated with Assumptions A1–A3 are that the underlying distributions belong to the same common family (F) and that they differ within this family at most in their medians. The less restrictive setting where these assumptions are relaxed to permit the possibility of differences in scale parameters as well as medians within the common F is referred to as the k-sample Behrens-Fisher problem. The Mack-Wolfe peak-unknown procedure (6.48) is no longer distribution-free under this more relaxed Behrens-Fisher setting. Chen and Wolfe (1990a) propose a modification of the Mack-Wolfe statistic A_p^* (6.47) to deal with this less restrictive setting. Their approach is similar to that used by Rust and Fligner (1984) to modify the Kruskal-Wallis statistic H for the same setting (see Comment 11).

44. *Ordered versus Umbrella Alternatives.* In this section and Section 6.2 we have considered procedures for testing the null hypothesis H_0 (6.2) of no differences in treatment effects against either ordered or, more generally, umbrella alternatives. In some settings, however, what is actually of interest is the ability to distinguish *directly between* a strictly upward trend (ordered alternatives) and an early upward trend with an eventual downturn (umbrella alternatives). This is frequently the case with dose-response data. Simpson and Margolin (1986) propose a recursive procedure based on the Jonckheere-Terpstra statistic for dealing with such problems.

45. *An alternative Approach Based on Maximums.* The Mack-Wolfe approach to the setting of umbrella alternatives with unknown peak is to first use the data to estimate the unknown peak and then to base the test of H_0 (6.2) on the peak-known statistic with peak at this estimated value. An alternative approach would be to bypass the first step of estimating the unknown peak and simply assess directly which of the treatments provides the most evidence of an umbrella alternative. To this effect, Chen and Wolfe (1990b) studied competitor test procedures to procedure (6.48) based on the extreme statistic $A_{\max} = \max\{A_1^*, \ldots, A_k^*\}$, with A_p^* given by (6.35). Hettmansperger and Norton (1987) considered similar competitors to (6.48) based on the maximum of certain linear rank statistics. The results of a substantial small sample power study of these competitors (as well as the Simpson-Margolin (1986) procedure mentioned in Comment 44) is provided in Chen and Wolfe (1990b).

PROBLEMS

36. Consider the tiger muskellunge data in Table 6.9. Test the hypothesis of no differences in the plasma glucose values over time against a general umbrella alternative using an approximate significance level of .01. Compare your result with that obtained in Problem 21.
37. Consider the fasting metabolic rate (FMR) data on white-tailed deer in Table 6.8. Test the hypothesis of no difference in FMR over the 2-month periods against a general umbrella alternative. Use an approximate significance level of .01. Compare your result with that obtained in Example 6.3.

38. (a) The statistic $A_{\hat{p}}^*$ can be computed from the joint ranking of all N observations. Explain.
    (b) The statistic $A_{\hat{p}}^*$ can also be computed from pairwise two-sample rankings. Explain.
39. Suppose $k = 3, n_1 = n_2 = 1$, and $n_3 = 2$. Obtain the form of the exact null (H_0) distribution of $A_{\hat{p}}^*$ for the case of no tied observations. Compare this null distribution with that of $A_{\hat{p}}^*$ for $k = 3, n_1 = n_3 = 1$, and $n_2 = 2$, as obtained in Comment 40.
40. Construct a set of data with no tied observations for which $r > 1$ in the definition of $A_{\hat{p}}^*$ (6.47). Discuss the implications this has for estimation of the umbrella peak.
41. Consider the Acid Red 114 revertant colonies data in Table 6.10. Test the hypothesis of no differences in the number of revertant colonies over the dosage levels against a general umbrella alternative. Use an approximate significance level of .05. Compare this result with those obtained in Problems 32, 33, 34, and 35.

6.4. A DISTRIBUTION-FREE TEST FOR TREATMENTS VERSUS A CONTROL (FLIGNER-WOLFE)

In this section we discuss a test procedure specifically designed for the setting where one of the treatments corresponds to a control or baseline set of conditions and we are interested in assessing which, if any, of the treatments are better than the control. Without loss of generality, we label the treatments so that the control corresponds to treatment 1. In this setting, the null hypothesis of interest is still H_0 (6.2), but now it corresponds to the statement that none of the treatments $2, \ldots, k$ is different from the control (treatment 1). This is usually expressed as

$$H_0 : [\tau_i = \tau_1, \ i = 2, \ldots, k]. \tag{6.49}$$

(Note that the expression in (6.49) is, indeed, equivalent to the original expression for H_0 (6.2).)

Procedure

To compute the Fligner-Wolfe statistic FW, we first combine all N observations from the k samples and order them from least to greatest. Letting r_{ij} denote the rank of X_{ij} in this joint ranking, the Fligner-Wolfe statistic FW is then the sum of these joint ranks for the noncontrol treatments, namely,

$$FW = \sum_{j=2}^{k} \sum_{i=1}^{n_j} r_{ij}. \tag{6.51}$$

a. One-Sided Upper-Tail Test. To test

$$H_0 : [\tau_i = \tau_1, \ \text{for } i = 2, \ldots, k]$$

versus

$$H_5 : [\tau_i \geq \tau_1, \ \text{for } i = 2, \ldots, k, \ \text{with at least one strict inequality}], \tag{6.52}$$

at the α level of significance,

$$\text{Reject } H_0 \text{ if } FW \geq f_\alpha; \quad \text{otherwise do not reject,} \tag{6.53}$$

where the constant f_α is chosen to make the type I error probability equal to α. In order to determine the critical value f_α we note that the statistic FW can be viewed as a two-sample Wilcoxon rank sum statistic (see Section 4.1) computed for the n_1 control treatment observations (playing the role of the X's in the two-sample setting) and the $N^* = \sum_{j=2}^{k} n_j$ combined observations from treatments $2, \ldots, k$ (playing the role of the Y's in the two-sample setting). As a result, the null distribution of FW is the same as that of the Wilcoxon rank sum statisic W with sample sizes $m = n_1$ and $n = N^*$. Thus, if $N^* \leq n_1$ the critical value f_α is just the upper αth percentile w_α for the null distribution of the Wilcoxon rank sum statistic with sample sizes $m = n_1$ and $n = N^*$. Values of $f_\alpha = w_\alpha$ in this case ($N^* \leq n_1$) can be obtained directly from Table A.6. In the case of $N^* > n_1$, the appropriate relationship between f_α and Wilcoxon rank sum critical values (see Problem 43) is $f_\alpha = w'_\alpha + [(N - 2n_1)(N + 1)/2]$, where w'_α is the upper αth percentile for the null distribution of the Wilcoxon rank sum statistic with sample sizes $m = N^*$ and $n = n_1$. Again, w'_α and, therefore, $f_\alpha = w'_\alpha + [(N - 2n_1)(N + 1)/2]$ can be found directly from Table A.6.

b. One-Sided Lower-Tail Test. To test

$$H_0 : [\tau_i = \tau_1, \text{ for } i = 2, \ldots, k]$$

versus

$$H_6 : [\tau_i \leq \tau_1, \text{ for } i = 2, \ldots, k, \text{ with at least one strict inequality}], \quad (6.54)$$

at the α level of significance,

$$\text{Reject } H_0 \text{ if } FW \leq N^*(N + 1) - f_\alpha; \quad \text{otherwise do not reject.} \quad (6.55)$$

Large-Sample Approximation

As previously noted, when H_0 is true, the statistic FW has the same probability distribution as the null distribution of the two-sample Wilcoxon rank sum statisic W with sample sizes $m = n_1$ and $n = N^*$. Hence, it follows directly from the Large-Sample Approximation discussion of Section 4.1 that the standardized version of FW, namely,

$$FW^* = \frac{FW - E_0(FW)}{\{\text{var}_0(FW)\}^{1/2}} = \frac{FW - \{N^*(N + 1)/2\}}{\{n_1 N^*(N + 1)/12\}^{1/2}} \quad (6.56)$$

has, as $\min(n_1, N^*)$ tends to infinity, an asymptotic $N(0, 1)$ distribution when H_0 is true. The normal theory approximation for procedure (6.53) is

$$\text{Reject } H_0 \text{ if } FW^* \geq z_\alpha; \quad \text{otherwise do not reject,} \quad (6.57)$$

and the normal theory approximation to procedure (6.55) is

$$\text{Reject } H_0 \text{ if } FW^* \leq -z_\alpha; \quad \text{otherwise do not reject.} \quad (6.58)$$

Ties

If there are ties among the X's, assign each of the observations in a tied group the average of the integer ranks that are associated with the tied group. After computing FW with these average ranks, use procedure (6.53) or (6.55) with the appropriate critical values from Table A.6.

Note, however, that this test associated with tied X's is only approximately, and not exactly, of significance level α. (To get an exact level α test even in this tied setting, see Comment 49.)

When applying the large-sample approximation, an additional factor must be taken into account. Although ties in the X's do not affect the null expected value of FW, its null variance is reduced to

$$\text{var}_0(FW) = \frac{n_1 N^*}{12}\left[N + 1 - \frac{\sum_{j=1}^{g} t_j(t_j - 1)(t_j + 1)}{N(N-1)}\right], \qquad (6.59)$$

where in display (6.59) g denotes the number of tied groups and t_j is the size of tied group j. We note that an untied observation is considered to be a tied group of size 1. In particular, if there are no ties among the X's, then $g = N$ and $t_j = 1$ for $j = 1, \ldots, N$. In this case each term in (6.59) of the form $t_j(t_j - 1)(t_j + 1)$ reduces to zero and the variance expression in (6.59) reduces to the usual null variance of FW when there are no ties, as given previously in equation (6.56). Note that the term $[n_1 N^*/12N(N-1)]\sum_{j=1}^{g} t_j(t_j - 1)(t_j + 1)$ represents the reduction in the null variance of FW due to the presence of the tied X's.

As a consequence of the effect that ties have on the null variance of FW, the following modification is needed to apply the large-sample approximation when there are tied X's. Compute FW using average ranks and set

$$FW^* = \frac{FW - \left\{\frac{N^*(N+1)}{2}\right\}}{\{\text{var}_0(FW)\}^{1/2}}, \qquad (6.60)$$

where $\text{var}_0(FW)$ is now given by display (6.59). With this modified value of FW^*, approximation (6.57) or (6.58) can be applied.

Example 6.5: *Motivational Effect of Knowledge of Performance–Example 6.2 Continued.* For Hundal's (1969) study to assess the motivational effects of knowledge of performance, the no information category clearly serves as a control population, and it is very natural to ask if additional performance information of either type (rough or accurate) leads to improved performance as measured by an increase in the number of pieces processed. Thus, we will apply the Fligner-Wolfe procedure (6.52) to the data in Table 6.6 to assess whether there is a deviation from H_0 in the direction of $\tau_2 > \tau_1$ and/or $\tau_3 > \tau_1$, where the treatment numbers are the same as those taken in Example 6.2. For purpose of illustration we take the significance level to be $\alpha = .0415$. From Table A.6 with $m = N^* = n_2 + n_3 = 6 + 6 = 12$ and $n = n_1 = 6$, we find that $w'_{.0415} = 76$. It then follows (since $N^* > n_1$) that $f_{.0415} = 76 + [(18 - 2(6))(18 + 1)/2] = 133$ and procedure (6.52) reduces to

Reject H_0 if $FW \geq 133$.

Using the joint ranks provided in parentheses beside the data in Table 6.6, we see that

$$FW = [2.5 + 5.5 + 17 + 13 + 5.5 + 9 + 18 + 5.5 + 15 + 10.5 + 16 + 13] = 130.5.$$

Since this value of FW is smaller than the critical value 133, we do not reject H_0 at the .0415 level. (This example illustrates the added power of the Jonckheere-Terpstra procedure relative to that of the Fligner-Wolfe procedure when we are able to utilize the additional piece of information that $\tau_3 \geq \tau_2$. From Example 6.2, the P-value for the Jonckheere-Terpstra procedure applied to these Hundal data is .0231, indicating rejection of H_0 at $\alpha = .0415$.)

6.4. A DISTRIBUTION-FREE TEST FOR TREATMENTS VERSUS A CONTROL

For the large-sample approximation we need to compute the standardized form of FW^* using equation (6.60) since there are ties in the data. The null expected value for FW is $E_0(FW) = 12(18+1)/2 = 114$. For the ties-corrected null variance of FW, we note that $g = 11$ and $t_1 = 1, t_2 = 2, t_3 = 4, t_4 = 1, t_5 = 1, t_6 = 2, t_7 = 3, t_8 = 1, t_9 = 1, t_{10} = 1$, and $t_{11} = 1$ for the Hundal data. Hence, using the ties correction in equation (6.59), we have that

$$\text{var}_0(FW) = \frac{6(12)}{12}\left\{18 + 1 - \left[\frac{2(2)(1)(3) + 3(2)(4) + 4(3)(5)}{18(17)}\right]\right\}$$

$$= 6\left(19 - \frac{16}{51}\right) = 112.12,$$

from which it follows that the ties-corrected vaue of FW^* (6.60) is

$$FW^* = \frac{130.5 - 114}{\{112.12\}^{1/2}} = 1.56.$$

Thus, using the approximate procedure (6.57) with the ties-corrected value of $FW^* = 1.56$, we see from Table A.1 that the approximate P-value for these data is .0594. Thus we have marginal evidence from the Fligner-Wolfe treatments versus control procedure that additional performance knowledge (either rough or accurate) leads to an increase in the number of pieces produced.

Comments

46. *More General Setting*. As with the other procedures of this chapter, we could replace Assumptions A1–A3 and H_0 (6.2) with the more general null hypothesis that all $N!/\left(\prod_{j=1}^{k} n_j!\right)$ assignments of n_1 joint ranks to the control observations, n_2 joint ranks to the treatment 2 observations, ..., n_k joint ranks to the treatment k observations are equally likely.

47. *Motivation for the Test*. The statistic FW (6.51) is the sum of the joint ranks assigned to the noncontrol treatments. When some of the τ_i's are strictly greater than the control effect τ_1 we would expect the joint ranks for the observations from those treatments to be larger than the joint ranks for the control observations. The net result would be a larger value of FW. This suggests rejecting H_0 in favor of H_5 (6.52) for large values of FW and motivates procedures (6.53) and (6.57). A similar motivation leads to procedures (6.55) and (6.58). (See also Comment 51.)

48. *Assumptions*. As with the other test procedures of this chapter, Assumption A3 requires that the control and the $(k - 1)$ treatment distributions $F_1, \ldots, F_k$ can differ at most in their locations (medians). (See also Comments 4 and 54.)

49. *Exact Conditional Distribution of FW with Ties among the Data*. To get an exact level α test in the presence of ties, we rely on the fact that the null distribution of FW conditional on the observed configuration of joint tied ranks is the same as the corresponding conditional tied ranks null distribution of the Wilcoxon rank sum statistic W with sample sizes $m = n_1$ and $n = N^*$. Therefore, the approach discussed and illustrated in Comment 4.5 can be used to get the exact conditional null distribution of FW and associated exact level α test in the case of ties among the data.

50. *Two-Sided Test*. We note that we have not discussed a test based on the FW statistic that is designed for a two-sided alternative. The "natural" two-sided alternative for this treatment versus control setting corresponds to [either $\tau_i \geq \tau_1$ for all $i = 2, \ldots, k$ or $\tau_i \leq \tau_1$ for all

$i = 2, \ldots, k$, with at least one strict inequality]. We feel that it is rather unlikely that we would find ourselves in such a setting where either all the treatments are better than the control or all the treatments are worse than the control, but we have no idea of which of the two cases pertains. As a result, a two-sided test based on FW is not presented in this section.

51. *Limitations.* The test procedures in (6.53) and (6.55) deal with very restricted alternatives where *all* the treatments are either at least as good as the control (i.e., $\tau_i \geq \tau_1$ for all $i = 2, \ldots, k$) or *all* the treatments are no better than the control (i.e., $\tau_i \leq \tau_1$ for all $i = 2, \ldots, k$), respectively. They are not appropriate tests when the possibility exists that some of the treatments might be better ($\tau_i > \tau_1$) and some might be worse ($\tau_i < \tau_1$) than the control. For such mixed alternatives, one would need to use the general alternatives Kruskal-Wallis procedure presented in Section 6.1.

52. *Comparisons Between Treatments.* The Fligner-Wolfe procedure (6.53) is a test designed to decide if *any* of treatments $2, \ldots, k$ are better (i.e., $\tau_i > \tau_1$) than the control. It involves no direct comparisons between the various treatments observations themselves. In order to reach conclusions about whether there are any differences among the treatment effects $\tau_2, \ldots, \tau_k$, one would need to apply the Kruskal-Wallis procedure of Section 6.1 (or, if appropriate, the Jonckheere-Terpstra ordered alternatives or Mack-Wolfe umbrella alternatives procedures of Sections 6.2 and 6.3, respectively) to the sample data from treatments $2, \ldots, k$. Under the null hypothesis H_0 (6.2), the Fligner-Wolfe statistic FW is independent of the Kruskal-Wallis statistic H (and also of the Jonckheere-Terpstra statistic J and the Mack-Wolfe statistics A_p and A_p^*). This implies, for example, that if we conduct the Fligner-Wolfe test (6.53) at a significance level α_1 and the Kruskal-Wallis test (6.6) (or Jonckheere-Terpstra test (6.14), Mack-Wolfe peak-known test (6.32), or Mack-Wolfe peak-unknown test (6.48)) on treatments $2, \ldots, k$ at significance level α_2, then the probability of incorrectly rejecting H_0 when it is true with at least one of the two tests is exactly $\alpha_1 + \alpha_2 - \alpha_1 \alpha_2$. A similar comment applies to procedure (6.55).

53. *Multiple Comparisons.* If test procedure (6.53) leads to rejection of H_0 (6.2), we are led to the conclusion that at least one treatment has a greater effect than the control. However, procedure (6.53) does not address the question of exactly how many treatment effects are greater than that of the control, nor does it provide us information as to which specific treatments are better than the control. For answers to such questions we turn to treatments versus control multiple comparisons procedures, as discussed in Sections 6.7 and 6.8. Similar comments apply to the lower-tail test procedure in (6.55).

54. *Treatments versus Control Behrens-Fisher Problem.* Two of the implicit requirements imposed by Assumptions A1–A3 are that the underlying distributions belong to the same common family (F) and that they differ within this family at most in their medians. The less restrictive setting where these assumptions are relaxed to permit the possibility of differences in scale parameters as well as medians within the common family F is referred to as the k-sample treatments versus control Behrens-Fisher problem. The Fligner-Wolfe procedures (6.53) and (6.55) are no longer distribution-free under this more general Behrens-Fisher setting. If we replace Assumption A3 by the less restrictive Assumption A3*: [The treatments' distribution functions $F_2, \ldots, F_k$ are connected through the relationship

$$F_i(t) = F^*(t - \tau_j), \quad -\infty < t < \infty,$$

for $i = 2, \ldots, k$, where F^* is a distribution function for a continuous distribution that is symmetric about its median θ and, in addition, the control distribution (F_1) is continuous and symmetric about its median $\theta + \tau_1$.], then the Fligner-Policello two-sample robust rank procedure discussed in Section 4.4 can be adapted to provide distribution-free tests of H_0 (6.2) against either H_5 (6.52) or H_6 (6.54) under these more general treatments versus control Behrens-Fisher Assumptions A1, A2, and A3*.

55. *Treatments versus Control under Umbrella Configurations.* In many settings where we are interested in comparing a number of treatments with a control, we will have additional a priori information regarding the relative magnitude of the treatment effects. One such piece of information might be that the treatment effects are known to follow an umbrella pattern (see Section 6.3) $\tau_1 \leq \cdots \leq \tau_{p-1} \leq \tau_p \geq \tau_{p+1} \geq \cdots \geq \tau_k$ with either known or unknown peak p. (Remember that the ordered pattern of Section 6.2 corresponds to $p = k$ or 1.) In a drug study, for instance, increasing dosage levels may be compared with a zero-dose control. If the treatment effects are not identical to that of the control, then it is often reasonable to assume that the higher the dose of the drug applied, the better (say, higher) will be the resulting effect on a patient, corresponding to monotonically ordered treatment effects. However, it may also be the case that a subject might potentially succumb to toxic effects at high doses, thereby actually decreasing the associated treatment effects. Such a setting would correspond to an ordering in the treatment effects that is monotonically increasing up to a point, followed by a monotonic decrease; that is, an umbrella pattern on the treatment effects. Chen and Wolfe (1993) consider a test procedure designed specifically to compare a number of treatments with a single control under this basic umbrella pattern for the treatment effects. Their test requires an equal number of observations in each of the treatments (i.e., $n_2 = \cdots = n_k = n^*$), but permits a differing number (n_1) of observations from the control setting. The necessary null distribution critical values are provided for a variety of k, n_1, and n^* combinations and the results of a substantial Monte Carlo simulation power study are presented. (An example of the type of data for which this Chen-Wolfe procedure would be appropriate is provided by the muskellunge plasma glucose data in Table 6.9.)

56. *Consistency of the FW Test.* Replace Assumptions A1–A3 by the less restrictive Assumptions A1': The X's are mutually independent and A2' : $X_{1j}, \ldots, X_{n_j j}$ come from the same continuous population Π_j, $j = 1, \ldots, k$. The populations $\Pi_1, \ldots, \Pi_k$ need not be identical, but we do assume that

$$\delta_{ij} = P(X_{1j} > X_{11}) \geq \tfrac{1}{2}, \quad \text{for } j = 2, \ldots, k.$$

Then, roughly speaking, the test defined by (6.53) is consistent if and only if there is at least one $j \in \{2, 3, \ldots, k\}$ for which $\delta_{1j} > \tfrac{1}{2}$.

Properties

1. *Consistency.* The condition n_j/N tends to λ_j, $0 < \lambda_j < 1$, $j = 1, \ldots, k$, is sufficient to ensure that the tests defined by (6.53) and (6.55) are consistent against the H_5 (6.52) and H_6 (6.54) alternatives, respectively. For a more general consistency statement, see Comment 56.
2. *Asymptotic Normality.* See Fligner and Wolfe (1982).
3. *Efficiency.* See Fligner and Wolfe (1982) and Section 6.10.

PROBLEMS

42. Apply the appropriate Fligner-Wolfe test to the psychotherapeutic attraction data of Table 6.2. Compare and contrast this result with that obtained for the Kruskal-Wallis test in Problem 1.
43. Apply the appropriate Fligner-Wolfe procedure to the glucocorticoid receptor data for the leukemia patients in Table 6.4, using the normal subjects as the control. Compare and contrast with the result obtained from the Kruskal-Wallis test in Problem 8.

44. Apply the appropriate Fligner-Wolfe test to the muskellunge plasma glucose data in Table 6.9. Compare and contrast with the result obtained from the Mack-Wolfe test in Problem 23. (See also Comment 55.)

45. Suppose $N^* > n_1$, where n_1 is the number of control observations and N^* is the combined number of observations in treatments $2, \ldots, k$. Let f_α be the upper αth percentile for the null distribution of FW (6.51) for this setting. Show that $f_\alpha = w'_\alpha + [(N - 2n_1)(N + 1)/2]$, where w'_α is the upper αth percentile for the null distribution of the Wilcoxon rank sum statistic with sample sizes $m = N^*$ and $n = n_1$.

RATIONALE FOR MULTIPLE COMPARISON PROCEDURES

In Sections 6.1–6.4 of this chapter we have discussed procedures designed to test the null hypothesis H_0 (6.2) against a variety of alternative hypotheses. Upon rejection of H_0 with one of these test procedures for a given set of data, our conclusions range from the general statement that there are some unspecified differences among the treatment effects (associated with the Kruskal-Wallis procedure discussed in Section 6.1) to the more informative relationships between the treatment effects associated with test procedures designed for ordered or umbrella alternatives or the treatments versus control setting. However, in none of these test procedures are our conclusions pair-specific; that is, the tests in Sections 6.1–6.4 are not designed to enable us to reach conclusions about specific pairs of treatment effects. The relative sizes of the specific treatment effects τ_1 and τ_2, for example, cannot be inferred from the conclusions reached by any of the test procedures in Sections 6.1–6.4. To elicit such pairwise specific information we turn to the class of multiple comparison procedures. In Section 6.5 we present such two-sided all-treatments multiple comparison procedures for the omnibus setting corresponding to the general alternatives H_1 (6.3). In Section 6.6 we deal with one-sided all-treatments multiple comparison procedures associated with the restricted ordered alternatives H_2 (6.11). Finally, in Section 6.7 we discuss an approach for making treatments versus control multiple comparison decisions.

6.5. DISTRIBUTION-FREE TWO-SIDED ALL-TREATMENTS MULTIPLE COMPARISONS BASED ON PAIRWISE RANKINGS—GENERAL CONFIGURATION (DWASS, STEEL, CRITCHLOW-FLIGNER)

In this section we present a multiple comparison procedure based on pairwise two-sample rankings that is designed to make decisions about individual differences between pairs of treatment effects (τ_i, τ_j), for $i < j$, in a setting where general alternatives H_1 (6.3) are of interest. Thus, the multiple comparison procedure of this section would generally be applied to one-way layout data *after* rejection of H_0 (6.2) with the Kruskal-Wallis procedure from Section 6.1. In this setting it is important to reach conclusions about all $\binom{k}{2} = k(k-1)/2$ pairs of treatment effects, and these conclusions are naturally two-sided in nature.

Procedure

For each pair of treatments (i, j), let

$$W_{ij} = \sum_{b=1}^{n_j} R_{ib}, \quad \text{for } 1 \leq i < j \leq k, \tag{6.61}$$

where $R_{i1},\ldots,R_{in_j}$ are the ranks of $X_{1j},\ldots,X_{n_jj}$, respectively, among the combined ith and jth samples; that is, W_{ij} is the Wilcoxon rank sum of the jth sample ranks in the joint two-sample ranking of the ith and jth sample observations. Compute

$$W_{ij}^* = \sqrt{2}\left[\frac{W_{ij} - E_0(W_{ij})}{\{\text{var}_0(W_{ij})\}^{1/2}}\right] = \frac{W_{ij} - \frac{n_j(n_i+n_j+1)}{2}}{\{n_in_j(n_i+n_j+1)/24\}^{1/2}}, \quad \text{for } 1 \leq i < j \leq k. \tag{6.62}$$

(Thus, W_{ij}^* is the standardized (under H_0) version of W_{ij} multiplied by $\sqrt{2}$.)

At an experimentwise error rate of α, the Steel-Dwass-Critchlow-Fligner two-sided all-treatments multiple comparison procedure reaches its $k(k-1)/2$ pairwise decisions, corresponding to each (τ_u, τ_v) pair, $1 \leq u < v \leq k$, by the criterion

$$\text{Decide } \tau_u \neq \tau_v \text{ if } |W_{uv}^*| \geq w_\alpha^*; \quad \text{otherwise decide } \tau_u = \tau_v, \tag{6.63}$$

where the constant w_α^* is chosen to make the experimentwise error rate equal to α; that is, w_α^* satisfies the restriction

$$P_0(|W_{uv}^*| < w_\alpha^*, u = 1,\ldots,k-1; v = u+1,\ldots,k) = 1 - \alpha, \tag{6.64}$$

where the probability $P_0(\cdot)$ is computed under H_0 (6.2). Equation (6.64) stipulates that the $k(k-1)/2$ inequalities $|W_{uv}^*| < w_\alpha^*$, corresponding to all pairs (u, v) of treatments with $u < v$, hold simultaneously with probability $1 - \alpha$ when H_0 (6.2) is true. Selected values of w_α^* can be found in Table A.16.

Large-Sample Approximation

When H_0 is true, the $[k(k-1)/2]$-component vector $(W_{12}^*, W_{13}^*, \ldots, W_{k-1,k}^*)$ has, as $\min(n_1,\ldots,n_k)$ tends to infinity, an asymptotic multivariate normal distribution with mean vector $\mathbf{0}$. It then follows (see Comment 62 for indications of the proof) that w_α^* can be approximated for large sample sizes by q_α, where q_α is the upper αth percentile point for the distribution of the range of k independent $N(0,1)$ variables. Thus the large-sample approximation for procedure (6.63) is

$$\text{Decide } \tau_u \neq \tau_v \text{ if } |W_{uv}^*| \geq q_\alpha; \quad \text{otherwise decide } \tau_u = \tau_v. \tag{6.65}$$

Values of q_α for $\alpha = .0001, .0005, .001, .005, .01, .025, .05, .10, .20$ and $k = 2(1)20(2)40(10)100$ are given in Table A.17.

Ties

If there are ties among the X observations, use average ranks in computing the individual Wilcoxon rank sum statistics W_{ij} (6.61). In addition, replace the term $\text{var}_0(W_{ij})/2 = n_in_j(n_i+n_j+1)/24$ in the denominator of W_{ij}^* (6.62) by

$$\frac{\text{var}_0(W_{ij})}{2} = \frac{n_in_j}{24}\left[n_i + n_j + 1 - \frac{\sum_{b=1}^{g_{ij}}(t_b-1)t_b(t_b+1)}{(n_i+n_j)(n_i+n_j-1)}\right], \tag{6.66}$$

where, for $1 \leq i < j \leq k$, g_{ij} denotes the number of tied groups in the joint ranking of the ith and jth sample observations and t_b is the size of tied group b in this joint ranking. Furthermore, an untied observation is considered to be a tied group of size 1. In particular, if there are no tied observations in the ith and jth combined samples, then $g_{ij} = n_i + n_j$ and $t_b = 1$ for $b = 1, \ldots, n_i + n_j$, in which case each term of the form $(t_b - 1)t_b(t_b + 1)$ reduces to 0 and $\text{var}_0(W_{ij})/2$ reduces to $n_i n_j (n_i + n_j + 1)/24$, the appropriate expression when there are no ties in the ith and jth combined samples.

Example 6.6: *Length of YOY Gizzard Shad.* Consider the length of YOY gizzard shad data discussed in Problem 4. Applying the Kruskal-Wallis procedure to the length data from the four sites in Kokosing Lake yields highly significant differences between the median YOY lengths at the four sites. To examine which particular sites differ in median YOY lengths, we apply the approximate procedure (6.65) with the appropriate corrections for ties given in (6.66). For this study we have $k = 4, n_1 = n_2 = n_3 = n_4 = 10$, and we must compute $k(k-1)/2 = 4(3)/2 = 6$ standardized W_{ij}^* statistics. For sake of illustration, we take our experimentwise error rate to be $\alpha = .01$. From Table A.17 with $k = 4$ we find $q_{.01} = 4.403$ and procedure (6.65) reduces to

$$\text{Decide } \tau_u \neq \tau_v \text{ if } |W_{uv}^*| \geq 4.403.$$

Next, we compute the six W_{ij}^* statistics. For the sample observations from sites I and II (populations 1 and 2, respectively), the combined-samples ranking yields the sum of ranks for the site II data to be

$$W_{12} = 9.5 + 20 + 3.5 + 9.5 + 13.5 + 19 + 1 + 17 + 9.5 + 18 = 120.5.$$

For this pair of samples there are tied observations, and we have $g_{12} = 14$ and $t_1 = t_2 = 1, t_3 = 2, t_4 = t_5 = t_6 = 1, t_7 = 4, t_8 = 1, t_9 = t_{10} = 2$, and $t_{11} = t_{12} = t_{13} = t_{14} = 1$. From (6.66) we find

$$\frac{\text{var}_0(W_{12})}{2} = \frac{10(10)}{24} \left[10 + 10 + 1 - \frac{3(1)(2)(3) + (3)(4)(5)}{(10 + 10)(10 + 10 - 1)} \right]$$

$$= \frac{25}{6} \left[\frac{7,980 - 78}{380} \right] = 86.64.$$

Using this result in equation (6.62), we obtain

$$W_{12}^* = \frac{[120.5 - 10(21)/2]}{\sqrt{86.64}} = 1.67.$$

For the other five population pairs, similar calculations yield the following:

Site I and Site III

$$W_{13} = 13.5 + 11 + 2.5 + 1 + 8 + 8 + 2.5 + 5 + 5 + 5 = 61.5,$$

$$g_{13} = 13, \quad t_1 = 1, \quad t_2 = 2, \quad t_3 = t_4 = 3, \quad t_5 = t_6 = t_7 = 1, \quad t_8 = 2,$$

$$t_9 = t_{10} = t_{11} = t_{12} = 1, \quad t_{13} = 2,$$

$$\text{var}_0(W_{13}) = \frac{10(10)}{24} \left[10 + 10 + 1 - \frac{3(1)(2)(3) + 2(2)(3)(4)}{(10 + 10)(10 + 10 - 1)} \right] = 86.78,$$

$$W_{13}^* = \frac{[61.5 - 10(21)/2]}{\sqrt{86.78}} = -4.67.$$

6.5. DISTRIBUTION-FREE TWO-SIDED ALL-TREATMENTS MULTIPLE COMPARISONS

Site I and Site IV

$$W_{14} = 11 + 9 + 4.5 + 7 + 9 + 2.5 + 2.5 + 1 + 4.5 + 9 = 60,$$
$$g_{14} = 15, \quad t_1 = 1, \quad t_2 = t_3 = 2, \quad t_4 = t_5 = 1, \quad t_6 = 3,$$
$$t_7 = t_8 = t_9 = t_{10} = t_{11} = t_{12} = t_{13} = t_{14} = 1, \quad t_{15} = 2,$$
$$\text{var}_0(W_{14}) = \frac{10(10)}{24}\left[10 + 10 + 1 - \frac{3(1)(2)(3) + 2(3)(4)}{(10 + 10)(10 + 10 - 1)}\right] = 87.04,$$
$$W_{14}^* = \frac{[60 - 10(21)/2]}{\sqrt{87.04}} = -4.82.$$

Site II and Site III

$$W_{23} = 12 + 11 + 2.5 + 1 + 8.5 + 8.5 + 2.5 + 5.5 + 5.5 + 5.5 = 62.5,$$
$$g_{23} = 13, \quad t_1 = 1, \quad t_2 = 2, \quad t_3 = 4, \quad t_4 = 2, \quad t_5 = t_6 = t_7 = 1, \quad t_8 = 3,$$
$$t_9 = t_{10} = t_{11} = t_{12} = t_{13} = 1,$$
$$\text{var}_0(W_{23}) = \frac{10(10)}{24}\left[10 + 10 + 1 - \frac{2(1)(2)(3) + 2(3)(4) + 3(4)(5)}{(10 + 10)(10 + 10 - 1)}\right] = 86.45,$$
$$W_{23}^* = \frac{[62.5 - 10(21)/2]}{\sqrt{86.45}} = -4.57.$$

Site II and Site IV

$$W_{24} = 11 + 9 + 5 + 7 + 9 + 2.5 + 2.5 + 1 + 5 + 9 = 61,$$
$$g_{24} = 13, \quad t_1 = 1, \quad t_2 = 2, \quad t_3 = 3, \quad t_4 = 1, \quad t_5 = 3, \quad t_6 = t_7 = 1, \quad t_8 = 3,$$
$$t_9 = t_{10} = t_{11} = t_{12} = t_{13} = 1,$$
$$\text{var}_0(W_{24}) = \frac{10(10)}{24}\left[10 + 10 + 1 - \frac{1(2)(3) + 3(2)(3)(4)}{(10 + 10)(10 + 10 - 1)}\right] = 86.64,$$
$$W_{24}^* = \frac{[61 - 10(21)/2]}{\sqrt{86.64}} = -4.73.$$

Site III and Site IV

$$W_{34} = 18 + 16 + 9 + 14 + 16 + 3 + 3 + 1 + 9 + 16 = 105,$$
$$g_{34} = 10, \quad t_1 = 1, \quad t_2 = 3, \quad t_3 = 2, \quad t_4 = 5, \quad t_5 = 2, \quad t_6 = 1,$$
$$t_7 = 3, \quad t_8 = t_9 = t_{10} = 1,$$
$$\text{var}_0(W_{34}) = \frac{10(10)}{24}\left[10 + 10 + 1 - \frac{2(1)(2)(3) + 2(2)(3)(4) + 4(5)(6)}{(10 + 10)(10 + 10 - 1)}\right] = 85.53,$$
$$W_{34}^* = \frac{[105 - 10(21)/2]}{\sqrt{85.53}} = 0.$$

Taking absolute values and referring them to the critical value $q_{.01} = 4.403$, we see that:

$$|W^*_{12}| = 1.67 < 4.403 \implies \text{decide } \tau_1 = \tau_2,$$
$$|W^*_{13}| = 4.67 > 4.403 \implies \text{decide } \tau_1 \neq \tau_3,$$
$$|W^*_{14}| = 4.82 > 4.403 \implies \text{decide } \tau_1 \neq \tau_4,$$
$$|W^*_{23}| = 4.57 > 4.403 \implies \text{decide } \tau_2 \neq \tau_3,$$
$$|W^*_{24}| = 4.73 > 4.403 \implies \text{decide } \tau_2 \neq \tau_4,$$
$$|W^*_{34}| = 0 < 4.403 \implies \text{decide } \tau_3 = \tau_4.$$

Thus, at an experimentwise error rate of .01, the six multiple comparison decisions can be summarized by the statement $(\tau_1 = \tau_2) \neq (\tau_3 = \tau_4)$. This multiple comparison procedure provides more detailed information about the lengths of the YOY gizzard shad populations in Kokosing Lake. We now know that sites I and II may be viewed as providing similar living environments for gizzard shad. The same conclusion holds for sites III and IV. However, we also know that the common living environment at sites I and II is significantly different from the common living environment at sites III and IV.

Comments

57. *Rationale for Multiple Comparison Procedures.* We think of the methods of this section as multiple comparison procedures. The aim of applying such procedures goes beyond the point of deciding whether the treatments are equivalent to the (often more important) problem of selecting which, if any, treatments differ from one another. Thus the user makes $k(k-1)/2$ decisions, one for each pair of treatments. Equation (6.64) states that the probability of making all correct decisions when H_0 is true is controlled to be $1 - \alpha$. That is, when using procedure (6.63), the probability of at least one incorrect decision, when H_0 is true, is controlled to be α. This error rate is derived under the assumption that H_0 is true, but it does not depend on the particular underlying distribution F. This is why we call (6.63) a distribution-free multiple comparison procedure.

Multiple comparison procedures can be interpreted as hypothesis tests. If we consider the test that rejects H_0 if the inequality of (6.63) holds for at least one (u, v) pair and accepts H_0 if, for every (u, v) pair, the inequality of (6.63) is not satisfied, this is a distribution-free test of size α for H_0 (6.2).

58. *Experimentwise Error Rate.* The use of an experimentwise error rate represents a very conservative approach to multiple comparisons. We are insisting that the probability of making only correct decisions be $1 - \alpha$ when the hypothesis H_0 (6.2) of treatment equivalence is true. Thus we have a high degree of protection when H_0 is true, but we often apply such techniques when we have evidence (perhaps based on a priori information or perhaps obtained by applying the Kruskal-Wallis test, as in Example 6.6) that H_0 is not true.

This protection under H_0 also makes it harder for the procedure to judge treatments as differing significantly when in fact H_0 is false, and this difficulty becomes more severe as k increases. We justify our use of an experimentwise error rate in much the same way as Kurtz et al. (1965). The rate provides a precise measure of a level of uncertainty, and statements at higher or lower levels are readily obtained.

Anscombe (1965), although not advocating the use of such rates, mentioned an interesting hypothetical situation (which he attributed to Richard Olshen) in defense of such a conservative approach. Anscombe was commenting on simultaneous confidence intervals proposed by Kurtz et al., but his statements would also apply to multiple comparison procedures of the type discussed here. We quote from his comments. "A panacea manufacturer advertises on television that trials have shown his product to be more effective than any other leading brand. Such trials (if they are not a downright fabrication) certainly seem to present a situation of the third type.* Their objective is not to help the manufacturer reach a decision, but hopefully to permit him to make a multiple-comparison statement that will impress the public and boost sales. He could appropriately use the simultaneous confidence intervals of this paper; indeed, the Food and Drug Administration could appropriately require him to do so. The more equally ineffective other leading brands there were, the harder would it be for him to obtain the evidence he needed, and the more trials he would have to conduct and suppress before achieving a favorable one. Thus would Statistics and Economics go hand in hand to protect the public."

59. *Critical Values* w_α^*. The w_α^* values of Table A.16 can be obtained by using the fact that under H_0 (6.2), all $N!/\left(\prod_{j=1}^k n_j!\right)$ *joint* (of all N sample observations) rank assignments of n_1 ranks to the treatment 1 observations, n_2 ranks to the treatment 2 observations, ..., n_k ranks to the treatment k observations are equally likely. (Although the standardized pairwise Wilcoxon statistics W_{ij}^* (6.62) are formally defined in terms of pairwise two-sample ranks, it is clear that all $k(k-1)/2$ W_{ij}^* values can also be computed from the joint ranks of all N observations.) Thus to obtain the probability, under H_0, that $|W_{uv}^*| < c$, for all $u < v$, we simply count the number of configurations for which the event $A = \{|W_{uv}^*| < c, \text{ for all } u < v\}$ occurs, and divide this count by $N!/\left[\prod_{j=1}^k n_j!\right]$. For an illustration, we return to Comment 6 and use the 15 joint rank configurations displayed there for the case $k = 3$ and $n_1 = n_2 = n_3 = 2$. (Again, we can reduce the number of configurations that need to be considered from 90 to 15 by the same reasoning as in Comment 6.) For each of these 15 configurations we now display the values of $|W_{12}^*|, |W_{13}^*|,$ and $|W_{23}^*|$.

(a) $|W_{12}^*| = 2.1909$
$|W_{13}^*| = 2.1909$
$|W_{23}^*| = 2.1909$

(b) $|W_{12}^*| = 2.1909$
$|W_{13}^*| = 2.1909$
$|W_{23}^*| = 1.0954$

(c) $|W_{12}^*| = 2.1909$
$|W_{13}^*| = 2.1909$
$|W_{23}^*| = 0$

(d) $|W_{12}^*| = 1.0954$
$|W_{13}^*| = 2.1909$
$|W_{23}^*| = 2.1909$

(e) $|W_{12}^*| = 1.0954$
$|W_{13}^*| = 2.1909$
$|W_{23}^*| = 1.0954$

(f) $|W_{12}^*| = 1.0954$
$|W_{13}^*| = 2.1909$
$|W_{23}^*| = 0$

(g) $|W_{12}^*| = 1.0954$
$|W_{13}^*| = 1.0954$
$|W_{23}^*| = 1.0954$

(h) $|W_{12}^*| = 0$
$|W_{13}^*| = 2.1909$
$|W_{23}^*| = 2.1909$

(i) $|W_{12}^*| = 1.0954$
$|W_{13}^*| = 1.0954$
$|W_{23}^*| = 0$

(j) $|W_{12}^*| = 0$
$|W_{13}^*| = 1.0954$
$|W_{23}^*| = 2.1909$

(k) $|W_{12}^*| = 0$
$|W_{13}^*| = 1.0954$
$|W_{23}^*| = 1.0954$

(l) $|W_{12}^*| = 1.0954$
$|W_{13}^*| = 0$
$|W_{23}^*| = 0$

(m) $|W_{12}^*| = 0$
$|W_{13}^*| = 0$
$|W_{23}^*| = 2.1909$

(n) $|W_{12}^*| = 0$
$|W_{13}^*| = 0$
$|W_{23}^*| = 1.0954$

(o) $|W_{12}^*| = 0$
$|W_{13}^*| = 0$
$|W_{23}^*| = 0$

*The term *third type* is used by Anscombe to refer to experiments intended to give fundamental knowledge or insight into some phenomenon but not to aid in a particular job of decision making.

Thus, for example,

$$P_0\{|W^*_{uv}| < 2.1909, (u,v) = (1,2), (1,3), (2,3)\}$$
$$= P_0\{|W^*_{12}| < 2.1909; |W^*_{13}| < 2.1909; |W^*_{23}| < 2.1909\}$$
$$= \tfrac{6}{15} = 1 - .6,$$

since for 6 of the 15 configurations—[(g), (i), (k), (l), (n), and (o)]—the event $\{|W^*_{12}| < 2.1909; |W^*_{13}| < 2.1909; |W^*_{23}| < 2.1909\}$ occurs. Similarly, $P_0\{|W^*_{uv}| < 1.0954, (u,v) = (1,2), (1,3), (2,3)\} = \tfrac{1}{15} = 1 - .9333$, as the event $\{|W^*_{12}| < 1.0954; |W^*_{13}| < 1.0954; |W^*_{23}| < 1.0954\}$ occurs only for the single configuration (o). Hence, for $k = 3$ and $n_1 = n_2 = n_3 = 2$, we have $w^*_{.6000} = 2.1909$ and $w^*_{.9333} = 1.0954$, and the values .6000 and .9333 are the only available experimentwise error rates for the Dwass-Steel-Critchlow-Fligner procedure (6.63) in this setting.

60. *Historical Development.* The multiple comparison procedures (6.63) and (6.65) based on Wilcoxon rank sum statistics were first proposed independently by Steel (1960, 1961) and Dwass (1960) for the setting of equal sample sizes $n_1 = \cdots = n_k$. Critchlow and Fligner (1991) presented a natural generalization of these Steel-Dwass procedures when the n_i are not all equal and provided the exact critical values w^*_α given in Table A.16 for $k = 3$ and $2 \leq n_1 \leq n_2 \leq n_3 \leq 7$.

61. *Maximum Type I Error Rate.* The multiple comparison procedure (6.63) is designed so that the experimentwise error rate (see Comment 58) is controlled to be equal to α; that is, the probability of falsely declaring any pair of treatment effects to be different, when in fact *all* of the treatment effects are the same, is equal to α. However, it also satisfies the more stringent *maximum type I error rate* requirement that the probability of falsely declaring any pair of treatment effects to be different, regardless of the values of the other $k - 2$ treatment effects, is no larger than the stated α. This requires controlling the probability of making false declarations about treatment effect differences even in situations when *not all* of the treatment effects are the same. For example, if $\tau_1 < \tau_2 = \tau_3$ the probability of incorrectly deciding that $\tau_2 \neq \tau_3$ is still controlled to be α by multiple comparison procedure (6.63). Similar comments apply to the approximate procedure in (6.65).

62. *Large-Sample Approximation.* Let $\mathbf{W}^* = (W^*_{12}, W^*_{13}, \ldots, W^*_{k-1,k})$, where W^*_{ij} is given by equation (6.62) for $1 \leq i < j \leq k$. Then it can be shown that $\mathbf{W}^*$ has, as $\min(n_1, \ldots, n_k)$ tends to infinity, an asymptotic multivariate normal distribution with mean vector $\mathbf{0}$ and appropriate covariance matrix $\mathbf{\Sigma}$ (see Miller (1981a) for further details). It follows directly from this result (again, see Miller (1981a)) that the procedure in (6.65) has an asymptotic experimentwise error rate equal to α when $n_1 = n_2 = \cdots = n_k$. Critchlow and Fligner (1991) used a result by Hayter (1984) to establish the fact that the asymptotic experimentwise error rate for procedure (6.65) is also bounded above by α when we have unequal sample sizes.

When H_0 is true and $n_1 = n_2 = \cdots = n_k$, the asymptotic correlation matrix $\mathbf{\Sigma}_1$ (say) of the $\binom{k}{2}$ W_{ij}'s is the same as the correlation matrix $\mathbf{\Sigma}_2$ (say) of the $\binom{k}{2}$ differences $Z_i - Z_j, 1 \leq i < j \leq k$, where $Z_1, \ldots, Z_k$ are independent $N(0,1)$ random variables (cf. Miller (1966), 155–156). It follows that the asymptotic distribution of

$$\sqrt{2} \max_{1 \leq i < j \leq k} \left\{ \frac{|W_{ij} - E_0(W_{ij})|}{[\mathrm{var}_0(W_{ij})]^{1/2}} \right\} = \max_{1 \leq i < j \leq k} |W^*_{ij}|$$

can be approximated by the distribution of

$$\max_{1 \le i < j \le k} |Z_i - Z_j| = \text{range } (Z_1, \ldots, Z_k).$$

The $\sqrt{2}$ occurs because the variance of $Z_i - Z_j$ equals 2. This justifies the use of approximation (6.65) in the equal-sample-size case. When the sample sizes are unequal, the asymptotic correlation matrix of the $\binom{k}{2} W_{ij}$'s will not in general agree with Σ_2, but (6.65) can be justified via a Tukey-Kramer approximation (see, for example, Tukey (1953), Kramer (1956, 1957) and pages 91–93 of Hochberg and Tamhane (1987)).

63. *Joint Ranking Approach.* The multiple comparison procedure discussed in this section is based on $k(k-1)/2$ separate two-sample rankings. However it is also reasonable to consider all treatments multiple comparisons based on a single joint ranking of all N observations. Let $R_{.j}$ (6.4), $j = 1, \ldots, k$, be the average rank for the jth treatment sample in the joint ranking of all N observations. The joint ranking analogue to procedure (6.63) is then given by

$$\text{Decide } \tau_u \ne \tau_v \text{ if } N^*|R_{.u} - R_{.v}| \ge y_\alpha; \quad \text{otherwise decide } \tau_u = \tau_v, \tag{6.67}$$

where N^* is the least common multiple of $n_1, \ldots, n_k$ and the constant y_α is chosen to make the experimentwise error rate equal to α; that is, y_α satisfies the restriction

$$P_0(N^*|R_{.u} - R_{.v}| < y_\alpha, u = 1, \ldots, k-1; \; v = u+1, \ldots, k) = 1 - \alpha, \tag{6.68}$$

where the probability $P_0(\cdot)$ is computed under H_0 (6.2). As with the multiple comparison procedures based on pairwise rankings, equation (6.68) stipulates that the $k(k-1)/2$ inequalities $N^*|R_{.u} - R_{.v}| < y_\alpha$, corresponding to all pairs (u, v) of treatments with $u < v$, hold simultaneously with probability $1 - \alpha$ when H_0 (6.2) is true.

Nemenyi (1963) first proposed procedure (6.68) for the special case of equal sample sizes, in which case $N^*|R_{.u} - R_{.v}| = |R_u - R_v|$, where R_j (6.4) is the sum of the joint ranks for the treatment j observations. The general form of (6.68) for arbitrary sample sizes was considered by Damico and Wolfe (1987).

The y_α critical values can be obtained in exactly the same way as the w_α^* values of Table A.16. Proceeding as in Comment 59, we simply count the number of joint rank configurations for which the event $B = \{N^*|R_{.u} - R_{.v}| < c, \text{ for all } u < v\}$ occurs and divide this count by $N! / \left[\prod_{j=1}^{k} n_j!\right]$ to obtain the probability, under H_0, that $N^*|R_{.u} - R_{.v}| < c$ for all $u < v$. For an illustration, we again return to Comment 6 and use the 15 joint rank configurations displayed there for the case $k = 3$ and $n_1 = n_2 = n_3 = 2$. (For this setting, $N^*|R_{.u} - R_{.v}| = |R_u - R_v|$, and we can once again reduce the number of configurations that need to be considered from 90 to 15 by the same reasoning as in Comment 6.) For each of these 15 configurations we now display the values of $|R_1 - R_2|, |R_1 - R_3|,$ and $|R_2 - R_3|$.

(a) $|R_1 - R_2| = 4$
$|R_1 - R_3| = 8$
$|R_2 - R_3| = 4$

(b) $|R_1 - R_2| = 5$
$|R_1 - R_3| = 7$
$|R_2 - R_3| = 2$

(c) $|R_1 - R_2| = 6$
$|R_1 - R_3| = 6$
$|R_2 - R_3| = 0$

(d) $|R_1 - R_2| = 2$
$|R_1 - R_3| = 7$
$|R_2 - R_3| = 5$

(e) $|R_1 - R_2| = 3$
$|R_1 - R_3| = 6$
$|R_2 - R_3| = 3$

(f) $|R_1 - R_2| = 4$
$|R_1 - R_3| = 5$
$|R_2 - R_3| = 1$

(g) $|R_1 - R_2| = 2$
$|R_1 - R_3| = 4$
$|R_2 - R_3| = 2$

(h) $|R_1 - R_2| = 0$
$|R_1 - R_3| = 6$
$|R_2 - R_3| = 6$

(i) $|R_1 - R_2| = 3$
$|R_1 - R_3| = 3$
$|R_2 - R_3| = 0$

(j) $|R_1 - R_2| = 1$
$|R_1 - R_3| = 4$
$|R_2 - R_3| = 5$

(k) $|R_1 - R_2| = 0$
$|R_1 - R_3| = 3$
$|R_2 - R_3| = 3$

(l) $|R_1 - R_2| = 2$
$|R_1 - R_3| = 1$
$|R_2 - R_3| = 1$

(m) $|R_1 - R_2| = 2$
$|R_1 - R_3| = 2$
$|R_2 - R_3| = 4$

(n) $|R_1 - R_2| = 1$
$|R_1 - R_3| = 1$
$|R_2 - R_3| = 2$

(o) $|R_1 - R_2| = 0$
$|R_1 - R_3| = 0$
$|R_2 - R_3| = 0$

Thus, for example,

$$P_0\{|R_u - R_v| < 8, (u, v) = (1, 2), (1, 3), (2, 3)\}$$
$$= P_0\{|R_1 - R_2| < 8; |R_1 - R_3| < 8; |R_2 - R_3| < 8\}$$
$$= \tfrac{14}{15} = 1 - .067,$$

since for 14 of the configurations—all but configuration (a)—the event $\{|R_1 - R_2| < 8; |R_1 - R_3| < 8; |R_2 - R_3| < 8\}$ occurs. Similarly, $P_0\{|R_u - R_v| < 7; (u, v) = (1, 2), (1, 3), (2, 3)\} = \tfrac{12}{15} = .80$, because the event $\{|R_1 - R_2| < 7; |R_1 - R_3| < 7; |R_2 - R_3| < 7\}$ occurs for 12 of the configurations—all but (a), (b), and (d). Hence, for $k = 3$ and $n_1 = n_2 = n_3 = 2$, we have $y_{.067} = 8$ and $y_{.200} = 7$. Values of y_α are available in Damico and Wolfe (1987) for available experimentwise error rates (α) closest to but not exceeding .001, .005, .01 (.005) .05 (.01) .15 and most useful combinations of either $k = 3, 1 \leq n_1 \leq n_2 \leq n_3 \leq 6$ or $k = 4, 1 \leq n_1 \leq n_2 \leq n_3 \leq n_4 \leq 6$. For the special cases of equal sample sizes, these tabled values agree with those previously given by Nemenyi (1963) and McDonald and Thompson (1967). An approximation to y_α for large common sample size is discussed in Miller (1966). A related approximate procedure based on joint ranks and appropriate for large unequal sample sizes is suggested by Dunn (1964).

The joint ranking multiple comparison procedure given by (6.67) is a good deal simpler computationally than the corresponding pairwise ranking multiple comparison procedure in (6.63). Both procedures maintain the designated experimentwise error rate α. However, the joint ranking procedure does not provide the additional maximum type I error rate protection level α guarantee associated with the pairwise ranking procedure (see Comment 61). A second drawback for the joint ranking procedure is the fact that the absolute differences $|R_{.u} - R_{.v}|$ depend on the values of the observations from the other $k - 2$ treatments, in addition to the observations from treatments u and v. Thus, in the case of $k = 3$ the decision concerning treatments 1 and 2, for example, depends on the treatment 3 observations. This difficulty is discussed in Miller (1966) and Gabriel (1969).

Properties

1. *Asymptotic Multivariate Normality.* See Hayter (1984) and Critchlow and Fligner (1991).

2. *Efficiency.* See Sherman (1965) and Section 6.10.

PROBLEMS

46. Apply procedure (6.63) to the mean interstitial length data of Table 6.5.
47. Procedure (6.63) is defined specifically in terms of the $k(k-1)/2$ pairwise two-sample rankings. However, it can be applied to settings where only the joint ranks of all N observations are available. Explain.
48. Apply procedure (6.63) to the half-time of mucociliary clearance data of Table 6.1.
49. Apply the approximate procedure (6.65) to the glucocorticoid receptor data of Table 6.4.
50. For the case $k = 3, \alpha = .05$, and $n_1 = n_2 = n_3 = 6$, compare procedures (6.63) and (6.65).
51. Apply the approximate procedure (6.65) to the psychotherapeutic attraction data of Table 6.2.
52. Find the totality of all available experimentwise error rates α and the associated critical values w_α^* for procedure (6.63) when $k = 4, n_1 = 1$, and $n_2 = n_3 = n_4 = 2$.
53. Consider the joint ranking procedure (6.67) discussed in Comment 63. Find the totality of all available experimentwise error rates α and the associated critical values y_α for this procedure when $k = 4, n_1 = 1$ and $n_2 = n_3 = n_4 = 2$.
54. Consider the YOY gizzard shad data discussed in Example 6.6. Find the smallest (available) approximate experimentwise error rate at which the most significant difference in treatment effects (i.e., that between site I and site IV) would be detected.
55. Consider the mean interstitial length data in Table 6.5. Find the smallest (available) approximate experimentwise error rate at which we would declare that the typical mean interstitial length for white pines is different from that for Scotch pines.

6.6. DISTRIBUTION-FREE ONE-SIDED ALL-TREATMENTS MULTIPLE COMPARISONS BASED ON PAIRWISE RANKINGS–ORDERED TREATMENT EFFECTS (HAYTER-STONE)

In this section we discuss a multiple comparison procedure based on pairwise two-sample rankings that is designed to make decisions about individual differences between pairs of treatment effects (τ_i, τ_j), for $i < j$, in a setting where ordered alternatives H_2 (6.11) are of interest. Thus, the multiple comparison procedure of this section would be appropriate for one-way layout data *after* rejection of H_0 (6.2) with the Jonckheere-Terpstra procedure from Section 6.2. As with the procedure for general alternatives discussed in Section 6.5, we will once again reach conclusions about all $\binom{k}{2} = k(k-1)/2$ pairs of treatment effects. However, here these conclusions are naturally one-sided, in accordance with the ordered alternatives setting.

Procedure

For each pair of treatments $(i, j), 1 \leq i < j \leq k$, let W_{ij} be defined by expression (6.61); that is, W_{ij} is the Wilcoxon rank sum of the jth sample ranks in the two-sample ranking of the ith and jth sample observations. Compute the standardized form W_{ij}^* given in equation (6.62) for each treatment pair combination (i, j) with $i < j$.

At an experimentwise error rate of α, the Hayter-Stone one-sided all-treatments multiple comparison procedure reaches its $k(k-1)/2$ pairwise decisions, corresponding to each (τ_u, τ_v) pair, $1 \leq u < v \leq k$, by the criterion

$$\text{Decide } \tau_v > \tau_u \text{ if } W_{uv}^* \geq c_\alpha^*; \quad \text{otherwise decide } \tau_u = \tau_v, \tag{6.69}$$

where the constant c_α^* is chosen to make the experimentwise error rate equal to α; that is, c_α^* satisfies the restriction

$$P_0(W_{uv}^* < c_\alpha^*, u = 1, \ldots, k-1; v = u+1, \ldots, k) = 1 - \alpha, \tag{6.70}$$

where the probability $P_0(\cdot)$ is computed under H_0 (6.2). Equation (6.70) requires that the $k(k-1)/2$ inequalities $W_{uv}^* < c_\alpha^*$, corresponding to all pairs (u, v) of treatments with $u < v$, hold simultaneously with probability $1 - \alpha$ when H_0 (6.2) is true. Selected values of c_α^* can be found in Table A.18. (See also Comment 66.)

Large-Sample Approximation

When H_0 is true, the $k(k-1)/2$ component vector $(W_{12}^*, W_{13}^*, \ldots, W_{k-1,k}^*)$ has, as $\min(n_1, \ldots, n_k)$ tends to infinity, an asymptotic multivariate normal distribution with mean vector **0**. It then follows (see Hayter and Stone (1991) for example, for an indication of the proof) that c_α^* can be approximated for large sample sizes by d_α, where d_α is the upper αth percentile point for the distribution of

$$D = \max_{1 \leq i < j \leq k} \left[\frac{Z_j - Z_i}{\left\{ \frac{n_i + n_j}{2n_i n_j} \right\}^{1/2}} \right],$$

where $Z_1, \ldots, Z_k$ are mutually independent and Z_i has a $N(0, 1/n_i)$ distribution, for $i = 1, \ldots, k$. Thus, the large-sample approximation for procedure (6.69) is

$$\text{Decide } \tau_v > \tau_u \text{ if } W_{uv}^* \geq d_\alpha; \quad \text{otherwise decide } \tau_u = \tau_v. \tag{6.71}$$

Values of d_α for large equal sample sizes $n_1 = \cdots = n_k = n, k = 3(1)9$, and experimentwise error rate $\alpha = .01, .05$, and $.10$ are given in Table A.19. Approximate critical values d_α for large, but unequal, sample sizes $n_1, \ldots, n_k$ are not presently available in the literature. (See also Comment 68.)

Ties

If there are ties among the X observations, use average ranks in computing the individual Wilcoxon rank sum statistics W_{ij} (6.61). In addition, replace the term $\text{var}_0(W_{ij})/2 = n_i n_j (n_i + n_j + 1)/24$ in the denominator of W_{ij}^* (6.62) by the expression in equation (6.66).

Example 6.7: *Motivational Effect of Knowledge of Performance–Example 6.2 Continued.* For Hundal's (1969) study to assess the motivational effects of knowledge of performance, we found in Example 6.2 (using the Jonckheere-Terpstra test procedure) that there was sufficient evidence in the sample data to conclude that $\tau_1 \leq \tau_2 \leq \tau_3$ with at least one strict inequality. To examine which of the types of information (none, rough, or accurate) lead to differences in median numbers of pieces processed, we apply procedure (6.69) with the appropriate corrections for ties, as given in (6.66). For this study we have $k = 3, n_1 = n_2 = n_3 = 6$, and we must compute $k(k-1)/2 = 3(2)/2 = 3$ standardized W_{ij}^* statistics. For sake of illustration we take our experimentwise error rate to be $\alpha = .0553$. From Table A.18 with $k = 3$ and $n_1 = n_2 = n_3 = 6$, we find $c_{.0553}^* = 2.9439$ and procedure (6.69) reduces to

$$\text{Decide } \tau_u > \tau_v \text{ if } W_{uv}^* \geq 2.9439.$$

Next we compute the three W_{ij}^* statistics. For the control (no information) and group B (partial information) sample observations, the combined-samples ranking yields the sum of ranks for the group B data to be

$$W_{12} = 2.5 + 5 + 12 + 10.5 + 5 + 8 = 43.$$

For this pair of samples there are tied observations and we have $g_{12} = 8$ and $t_1 = 1, t_2 = 2, t_3 = 3, t_4 = t_5 = t_6 = 1, t_7 = 2$, and $t_8 = 1$. From (6.66) we obtain

$$\frac{\text{var}_0(W_{12})}{2} = \frac{6(6)}{24}\left[6 + 6 + 1 - \frac{2(1)(2)(3) + 2(3)(4)}{(6+6)(6+6-1)}\right]$$

$$= \frac{3}{2}\left[\frac{1,716 - 36}{132}\right] = 19.09.$$

Using this result in equation (6.62), we find

$$W_{12}^* = \frac{[43 - 6(13)/2]}{\sqrt{19.09}} = .92.$$

For the other two population pairs, similar calculations lead to the following.

Control (No Information) and Group C (Accurate Information)

$$W_{13} = 12 + 3.5 + 10 + 6.5 + 11 + 8.5 = 51.5,$$

$$g_{13} = 9, \quad t_1 = t_2 = 1, \quad t_3 = 2, \quad t_4 = 1, \quad t_5 = t_6 = 2, \quad t_7 = t_8 = t_9 = 1,$$

$$\text{var}_0(W_{13}) = \frac{6(6)}{24}\left[6 + 6 + 1 - \frac{3(1)(2)(3)}{(6+6)(6+6-1)}\right] = 19.30,$$

$$W_{13}^* = \frac{[51.5 - 6(13)/2]}{\sqrt{19.30}} = 2.85.$$

Group B (Partial Information) and Group C (Accurate Information)

$$W_{23} = 12 + 3 + 9 + 6 + 10 + 7.5 = 47.5,$$

$$g_{23} = 9, \quad t_1 = 1, \quad t_2 = 3, \quad t_3 = t_4 = 1, \quad t_5 = 2, \quad t_6 = t_7 = t_8 = t_9 = 1,$$

$$\text{var}_0(W_{23}) = \frac{6(6)}{24}\left[6 + 6 + 1 - \frac{1(2)(3) + 2(3)(4)}{(6+6)(6+6-1)}\right] = 19.16,$$

$$W_{23}^* = \frac{[47.5 - 6(13)/2]}{\sqrt{19.16}} = 1.94.$$

Referring these W_{ij}^* values to the critical point $c_{.0553}^* = 2.9439$, we see that

$$W_{12}^* = .92 < 2.9439 \quad \Rightarrow \quad \text{decide } \tau_1 = \tau_2,$$

$$W_{13}^* = 2.85 < 2.9439 \quad \Rightarrow \quad \text{decide } \tau_1 = \tau_3,$$

$$W_{23}^* = 1.94 < 2.9439 \quad \Rightarrow \quad \text{decide } \tau_2 = \tau_3.$$

Thus, at an experimentwise error rate of .0553, we have reached the conclusion that $\tau_1 = \tau_2 = \tau_3$ (i.e., there are no differences in median numbers of pieces processed between the different levels of information), in contradiction with the conclusion from the Jonckheere-Terpstra test that $\tau_1 \leq \tau_2 \leq \tau_3$ with at least one strict inequality. Even though the P-value for the Jonckheere-Terpstra test procedure for these data is .0231, we are not able to detect any individual differences between treatment effects with the multiple comparison procedure (6.69), even with an experimentwise error rate as high as .0553. Such occurrences are, unfortunately, rather common in practice because of the conservative nature of the multiple comparison procedures (see Comment 58). For this reason, we often conduct our multiple comparison procedure at an experimentwise error rate that is higher than a typical significance level (such as .01 or .05) for a hypothesis test. If we have previously conducted a hypothesis test (such as the Jonckheere-Terpstra test in the example) and rejected H_0, we would at least like to know the *most* significant difference between pairs of treatment effects. For this reason, it is always informative in such cases to find the smallest experimentwise error rate at which the first pairwise difference in treatment effects would become significant. For the Hundal data, that corresponds to treatments 1 (no information) and 3 (accurate information) with an observed value $W_{13}^* = 2.85$. Referring to Table A.18 we see that $c_{.0850}^* = 2.7175 < W_{13}^* = 2.85 < 2.9439 = c_{.0553}^*$. Thus the smallest experimentwise error rate (among the limited number available) at which we would decide $\tau_3 > \tau_1$ (and thus conclude that accurate information is more effective than no information) is .0850.

Comments

64. *Rationale for Multiple Comparison Procedures.* The general rationale for the multiple comparison procedures of this section is the same as that given in Comment 57 for the two-sided all-treatments multiple comparison procedures of Section 6.5. The only additional factor here is that the procedures of this section yield decisions that are one-sided by nature in line with their association with the ordered restriction ($\tau_1 \leq \cdots \leq \tau_k$) on the treatment effects.

65. *Experimentwise Error Rate.* The use of an experimentwise error rate represents a very conservative approach to multiple comparisons. We are insisting that the probability of making only correct decisions be $1 - \alpha$ when the hypothesis H_0 (6.2) of treatment equivalence is true. Thus we have a high degree of protection when H_0 is true, but we often apply the techniques of this section when we have evidence (perhaps based on a priori information or perhaps obtained by applying the Jonckheere-Terpstra test, as in Example 6.7) that H_0 is not true. (For additional general remarks about experimentwise error rates, see Comment 58.)

66. *Critical Values c_α^*.* The c_α^* values of Table A.18 can be obtained by using the fact that under H_0 (6.2), all $N!/\left(\prod_{j=1}^{k} n_j!\right)$ joint (of all N sample observations) rank assignments of n_1 ranks to the treatment 1 observations, n_2 ranks to the treatment 2 observations,..., n_k ranks to the treatment k observations are equally likely. (Although the standardized pairwise Wilcoxon statistics W_{ij}^* (6.62) are formally defined in terms of pairwise two-sample ranks, it is clear that all $k(k-1)/2$ W_{ij}^* statistics can also be computed from the joint ranks of all N observations.) Thus to obtain the probability, under H_0, that $W_{uv}^* < t$, for all $u < v$, we simply count the number of configurations for which the event $A = \{W_{uv}^* < t, \text{ for all } u < v\}$ occurs and divide this count by $N!/\left[\prod_{j=1}^{k} n_j!\right]$. For an illustration, we return to Comment 18 and use the 12 joint rank configurations displayed there for the case $k = 3, n_1 = 1, n_2 = 1$, and $n_3 = 2$. For each of these 12 configurations we now display the values of W_{12}^*, W_{13}^*, and W_{23}^*.

(a) $W_{12}^* = 1.4142$
$W_{13}^* = 1.7321$
$W_{23}^* = 1.7321$

(b) $W_{12}^* = -1.4142$
$W_{13}^* = 1.7321$
$W_{23}^* = 1.7321$

(c) $W_{12}^* = 1.4142$
$W_{13}^* = 1.7321$
$W_{23}^* = 0$

(d) $W_{12}^* = -1.4142$
$W_{13}^* = 0$
$W_{23}^* = 1.7321$

(e) $W_{12}^* = 1.4142$
$W_{13}^* = 1.7321$
$W_{23}^* = -1.7321$

(f) $W_{12}^* = -1.4142$
$W_{13}^* = -1.7321$
$W_{23}^* = 1.7321$

(g) $W_{12}^* = 1.4142$
$W_{13}^* = 0$
$W_{23}^* = 0$

(h) $W_{12}^* = -1.4142$
$W_{13}^* = 0$
$W_{23}^* = 0$

(i) $W_{12}^* = 1.4142$
$W_{13}^* = 0$
$W_{23}^* = -1.7321$

(j) $W_{12}^* = -1.4142$
$W_{13}^* = -1.7321$
$W_{23}^* = 0$

(k) $W_{12}^* = 1.4142$
$W_{13}^* = -1.7321$
$W_{23}^* = -1.7321$

(l) $W_{12}^* = -1.4142$
$W_{13}^* = -1.7321$
$W_{23}^* = -1.7321$

Thus, for example,

$$P_0\{W_{uv}^* < 1.7321, (u,v) = (1,2), (1,3), (2,3)\}$$
$$= P_0\{W_{12}^* < 1.7321; W_{13}^* < 1.7321; W_{23}^* < 1.7321\}$$
$$= \tfrac{6}{12} = 1 - .5,$$

since for 6 of the 12 configurations—(g), (h), (i), (j), (k), and (l)—the event $\{W_{12}^* < 1.7321; W_{13}^* < 1.7321; W_{23}^* < 1.7321\}$ occurs. Similarly, $P_0\{W_{uv}^* < 1.4142, (u,v) = (1,2), (1,3), (2,3)\} = \tfrac{3}{15} = 1 - .8$, because the event $\{W_{12}^* < 1.4142; W_{13}^* < 1.4142; W_{23}^* < 1.4142\}$ occurs only for the three configurations, (h), (j), and (l). Finally, $P_0\{W_{uv}^* < 0, (u,v) = (1,2), (1,3), (2,3)\} = \tfrac{1}{15} = 1 - .9333$, corresponding to the single configuration (l). Hence, for $k = 3, n_1 = 1, n_2 = 1$, and $n_3 = 2$, we have $c_{.5000}^* = 1.7321, c_{.8000}^* = 1.4142$, and $c_{.9333}^* = 0$, and the values .5000, .8000, and .9333 are the only available experimentwise error rates for the Hayter-Stone procedure (6.69) in this setting.

67. *Maximum Type I Error Rate.* The multiple comparison procedure (6.69) is designed so that the experimentwise error rate (see Comment 65) is controlled to be equal to α; that is, the probability of falsely declaring any pair of treatment effects to be different, when in fact *all* the treatment effects are the same, is equal to α. However, it also satisfies the more stringent *maximum type I error rate* requirement that the probability of falsely declaring any pair of treatment effects to be different, regardless of the values of the other $k - 2$ treatment effects, is no larger than the stated α. This requires controlling the probability of making false declarations about treatment effect differences even in situations when *not all* the treatment effects are the same. For example, if $\tau_1 < \tau_2 = \tau_3$ the probability of incorrectly deciding that $\tau_2 \neq \tau_3$ is still controlled to be α by multiple comparison procedure (6.69). Similar comments apply to the approximate procedure in (6.71).

68. *Large and Unequal Sample Sizes.* In order to obtain the large-sample approximate critical values d_α for use in procedure (6.71) when we have an unbalanced setting (i.e., where the sample sizes are not all equal), we must evaluate a $(k - 1)$-dimensional integral expression. In view of this difficulty (even with the availability of high-speed computers) and the fact that there is a large number of possible unequal-sample-size combinations for each fixed k and N, the evaluation of these approximate critical values is practically feasible only for a small percentage of the necessary cases. To make matters even more complicated, the use of an appropriate equal-sample-size asymptotic critical value when we actually have

unequal sample sizes does not result in a conservative procedure, as it does for the Dwass-Steel-Critchlow-Fligner two-sided all-treatments multiple comparison procedure in Section 6.5 (see, for example, Comment 62). In the case of the Hayter-Stone one-sided procedure (6.71), use of a particular equal-sample-size asymptotic critical value d_α may result in either a conservative or liberal (i.e. experimentwise error rate $\leq \alpha$ or $> \alpha$, respectively) procedure, depending on the particular unequal sample size configurations involved. Thus, for k and unequal $(n_1, \ldots, n_k)$ configurations beyond those for which the exact critical values are given in Table A.18, Hayter and Stone (1991) recommend that computer simulation techniques be used to obtain appropriate asymptotic critical values.

Properties

1. *Asymptotic Multivariate Normality.* See Hayter (1984) and Hayter and Stone (1991).

PROBLEMS

56. Apply procedure (6.71) to the psychotherapeutic attraction data of Table 6.2 using the postulated order $\tau_1 \leq \tau_2 \leq \tau_3 \leq \tau_4$.
57. Procedure (6.69) is defined specifically in terms of the $k(k-1)/2$ pairwise two-sample rankings. However, it can be applied to settings where only the joint ranks of all N observations are available. Explain.
58. Apply procedure (6.69) to the average basal area increment data in Table 6.7. Use only the growing site index intervals 72–74, 75–77, and 78–80 with the postulated ordering $\tau_{72-74} \leq \tau_{75-77} \leq \tau_{78-80}$.
59. For the case $k = 3$, $\alpha = .05$, and $n_1 = n_2 = n_3 = 6$, compare procedures (6.69) and (6.71).
60. Find the totality of all available experimentwise error rates α and the associated critical values c_α^* for procedure (6.69) when $k = 4$, $n_1 = 1$, and $n_2 = n_3 = n_4 = 2$.
61. Consider the psychotherapeutic attraction data of Table 6.2 with the postulated ordering $\tau_1 \leq \tau_2 \leq \tau_3 \leq \tau_4$. Find the smallest (available) approximate experimentwise error rate at which the most significant difference in treatment effects would be detected.
62. Consider the average basal area increment data in Table 6.7. Using only the growing site index intervals 72–74, 75–77, and 78–80 with the postulated ordering $\tau_{72-74} \leq \tau_{75-77} \leq \tau_{78-80}$, find the smallest available experimentwise error rate at which we would declare $\tau_{78-80} > \tau_{72-74}$.

6.7. DISTRIBUTION-FREE ONE-SIDED TREATMENTS VERSUS CONTROL MULTIPLE COMPARISONS BASED ON JOINT RANKINGS (NEMENYI, DAMICO-WOLFE)

In this section our attention turns to a multiple comparison procedure designed to make decisions about individual differences between the median effect for a single, baseline control population and the median effects for each of the the remaining $k-1$ treatments. This treatment versus control multiple comparison procedure is based on the joint ranking of all N sample observations and can be applied to one-way layout data containing a single control sample *after* rejection of H_0 (6.2) with any of the test procedures in Sections 6.1 to 6.4. Its application leads to conclusions about the differences between each of the $k-1$ treatment effects and the control effect, and these conclusions are naturally one-sided in nature.

Procedure

For simplicity of notation, we let treatment 1 assume the role of the single baseline control. In addition, let N^* be the least common multiple of the sample sizes $n_1, \ldots, n_k$. Jointly rank all N

of the sample observations and let $R_{.1},\ldots,R_{.k}$ be the averages of these joint ranks associated with treatments $1,\ldots,k$, respectively. (Thus, $R_{.1},\ldots,R_{.k}$ are as originally defined in (6.4) in conjunction with the Kruskal-Wallis statistic.) For each of the $k - 1$ noncontrol treatments, calculate the difference $R_{.u} - R_{.1}$, $u = 2,\ldots,k$.

At an experimentwise error rate of α, the Nemenyi-Damico-Wolfe one-sided treatments versus control multiple comparison procedure (see Comment 69) reaches its $k - 1$ pairwise decisions, corresponding to each (τ_1, τ_u) pair, $u = 2,\ldots,k$, by the criterion

$$\text{Decide } \tau_u > \tau_1 \text{ if } N^*(R_{.u} - R_{.1}) \geq y_\alpha^*; \quad \text{otherwise decide } \tau_u = \tau_1, \quad (6.72)$$

where the constant y_α^* is chosen to make the experimentwise error rate equal to α; that is, y_α^* satisfies the restriction

$$P_0\{N^*(R_{.u} - R_{.1}) < y_\alpha^*, u = 2,\ldots,k\} = 1 - \alpha, \quad (6.73)$$

where the probability $P_0(\cdot)$ is computed under H_0 (6.2). Equation (6.73) stipulates that the $k - 1$ inequalities $N^*(R_{.u} - R_{.1}) < y_\alpha^*$, corresponding to all pairs $(1, u)$ of noncontrol treatments $(u = 2,\ldots,k)$ with the control treatment 1, hold simultaneously with probability $1 - \alpha$ when H_0 (6.2) is true. Selected values of y_α^* can be found in Table A.20 for $k = 3$ and most useful combinations of $n_1 = 1(1)6$ and $1 \leq n_2, n_3 \leq 6$ and for a variety of $k = 4$ settings with $n_2 = n_3 = n_4$, but not necessarily equal to n_1.

Large-Sample Approximations

When H_0 is true, the $(k - 1)$ component vector $(R_{.2} - R_{.1}, R_{.3} - R_{.1},\ldots,R_{.k} - R_{.1})$ has, as $\min(n_1,\ldots,n_k)$ tends to infinity, an asymptotic $(k - 1)$-variate normal distribution with mean vector $\mathbf{0}$. (For an indication of the proof, see Miller (1966).) For the special case of $n_1 = b$ and $n_2 = \cdots = n_k = n$, with both n and b large, the critical value y_α^* can be approximated by $[N(N + 1)/12]^{1/2}[(1/b) + (1/n)]^{1/2}N^*m_\alpha^*$, where m_α^* is the upper αth percentile point for the distribution of the maximum of $(k - 1)$ $N(0, 1)$ variables with common correlation $\rho = n/(b + n)$. Thus the large-sample approximation for procedure (6.72) when we have equal treatment sample sizes $n_2 = \cdots = n_k = n$ (possibly different from $b = n_1$) is

$$\text{Decide } \tau_u > \tau_1 \text{ if } (R_{.u} - R_{.1}) \geq m_\alpha^* \left[\frac{N(N + 1)}{12}\right]^{1/2} \left(\frac{1}{b} + \frac{1}{n}\right)^{1/2};$$

$$\text{otherwise decide } \tau_u = \tau_1, u = 2,\ldots,k. \quad (6.74)$$

Selected values of m_α^* for $k = 3(1)13$ and $\rho = .1, .125, .2, .25, .3, \frac{1}{3}, .375, .4, .5, .6, .625, \frac{2}{3}, .7, .75, .8, .875,$ and $.9$ are given in Table A.21.

For the general setting of arbitrary (not necessarily equal) treatments sample sizes, Dunn (1964) used Bonferroni's Inequality to provide the large-sample approximation to procedure (6.72) given by

$$\text{Decide } \tau_u > \tau_1 \text{ if } (R_{.u} - R_{.1}) \geq z_{\alpha^*} \left[\frac{N(N + 1)}{12}\right]^{1/2} \left(\frac{1}{n_1} + \frac{1}{n_u}\right)^{1/2};$$

$$\text{otherwise decide } \tau_u = \tau_1, u = 2,\ldots,k, \quad (6.75)$$

where $\alpha^* = \alpha/(k - 1)$. (We note that this general approximate procedure can often be quite conservative in practice, as a direct result of the conservative nature of the Bonferroni Inequality.)

Ties

If there are ties among the X observations, use average ranks in computing the individual treatment sums of ranks $R_1, \ldots, R_k$.

Example 6.8: *Motivational Effect of Knowledge of Performance—Example 6.2 Continued.*
Once again we consider Hundal's (1969) study to assess the motivational effects of knowledge of performance. We previously found in Example 6.2 (using the Jonckheere–Terpstra test procedure) that there was sufficient evidence in the sample data to conclude that $\tau_1 \leq \tau_2 \leq \tau_3$ with at least one strict inequality. To further investigate which (if either) of the two types of additional information (rough or accurate) lead to differences in median numbers of pieces processed relative to the no information control (treatment 1), we apply procedure (6.72). Here we have $k = 3$ and $n_1 = n_2 = n_3 = N^* = 6$. For sake of illustration we take our experimentwise error rate to be $\alpha = .0554$. From Table A.20 with $k = 3$ and $n_1 = n_2 = n_3 = 6$, we find $y^*_{.0554} = 35$ and procedure (6.72) reduces to

$$\text{Decide } \tau_u > \tau_1 \text{ if } 6(R_{.u} - R_{.1}) = (R_u - R_1) \geq 35.$$

Using the joint ranks (with average ranks to break ties among the observations) provided in parentheses beside the data in Table 6.6, we see that

$$R_1 = 5.5 + 1 + 2.5 + 10.5 + 13 + 8 = 40.5,$$
$$R_2 = 2.5 + 5.5 + 17 + 13 + 5.5 + 9 = 52.5,$$

and
$$R_3 = 18 + 5.5 + 15 + 10.5 + 16 + 13 = 78.$$

Thus $(R_2 - R_1) = (52.5 - 40.5) = 12$ and $(R_3 - R_1) = (78 - 40.5) = 37.5$. Referring these rank sum differences to the critical point $y^*_{.0554} = 35$, we see that

$$(R_2 - R_1) = 12 < 35 \quad \Rightarrow \quad \text{decide } \tau_2 = \tau_1,$$
$$(R_3 - R_1) = 37.5 \geq 35 \quad \Rightarrow \quad \text{decide } \tau_3 > \tau_1.$$

Thus at an experimentwise error rate of .0554, we have reached the conclusion that accurate information leads to significantly more pieces processed than the no information control. (We note that the smallest experimentwise error rate at which we would reach this conclusion is .0426, since $y^*_{.0426} = 37$ and $y^*_{.0371} = 38$.)

For sake of illustration for the associated large-sample approximation (with equal sample sizes) procedure in (6.74), we note that $\rho = n/(b + n) = 6/(6 + 6) = \frac{1}{2}$ (which is always the case with equal sample sizes in the control and the noncontrol treatments). Using an approximate experimentwise error rate of $\alpha = .05183$ with $(k - 1) = 2$, we see from Table A.21 that $m^*_{.05183} = 1.90$. Thus we have that

$$\left[\frac{N(N+1)}{12}\right]^{1/2} \left(\frac{1}{b} + \frac{1}{n}\right)^{1/2} m^*_{.05183} = \left[\frac{18(19)}{12}\right]^{1/2} \left(\frac{1}{6} + \frac{1}{6}\right)^{1/2} (1.90) = 5.856$$

and procedure (6.74) becomes

$$\text{Decide } \tau_u > \tau_1 \text{ if } (R_{.u} - R_{.1}) \geq 5.856$$

or, equivalently,

$$\text{Decide } \tau_u > \tau_1 \text{ if } (R_u - R_1) = 6(R_{.u} - R_{.1}) \geq 6(5.856) = 35.14.$$

Thus, for $k = 3$ and $n_1 = n_2 = n_3 = 6$, the exact procedure (6.72) and the large-sample approxmation (6.74) are virtually identical and lead to the same conclusions that $\tau_2 = \tau_1$ and $\tau_3 > \tau_1$.

We note that the treatment-versus-control procedure (6.72) yields the conclusion that $\tau_3 > \tau_1$ at a considerably smaller experimentwise error rate (as low as .0426) than is the case with the Hayter-Stone one-sided all-treatments multiple comparison procedure (as detailed in Example 6.7), where the smallest experimentwise error rate leading to this conclusion is .0850. This situation is due primarily to the fact that the Hayter-Stone procedure is required to make an additional decision about the relative magnitude of τ_2 and τ_3, which, for these data, do not appear to be significantly different.

Comments

69. *Rationale for Treatments versus Control Multiple Comparison Procedures.* The general rationale for the multiple comparison procedures of this section is the same as that given in Comment 57 for the two-sided all-treatments multiple comparison procedures of Section 6.5. The only additional factor here is that the treatment-versus-control procedures of this section do not compare all treatments, but only each noncontrol treatment with the control on a directional basis. This situation arises, for example, in drug screening in the examination of many new treatments in hopes of improving on a standard, and there is no initial reason to perform between treatment comparisons. Of course, similar comparisons would be carried out later between treatments that were selected as being better than the control.

70. *Experimentwise Error Rate.* The use of an experimentwise error rate represents a very conservative approach to multiple comparisons. We are insisting that the probability of making only correct decisions be $1 - \alpha$ when the hypothesis H_0 (6.2) of treatment equivalence is true. Thus we have a high degree of protection when H_0 is true, but we often apply the techniques of this section when we have evidence (perhaps based on a priori information or perhaps obtained by applying a previous test procedure, as in Example 6.8) that H_0 is not true. (For additional general remarks about experimentwise error rates, see Comment 58.)

71. *Opposite Direction Decisions.* Procedures (6.72), (6.74) and (6.75) are designed for the one-sided case where the decisions are $\tau_u > \tau_1$ versus $\tau_u = \tau_1, u = 2, \ldots, k$. To handle the analogous one-sided situation where the decisions involve $\tau_u < \tau_1$ versus $\tau_u = \tau_1, u = 2, \ldots, k$. we use (6.72), (6.74), and (6.75) with $(R_{.u} - R_{.1})$ replaced by $(R_{.1} - R_{.u})$ for $u = 2, \ldots, k$.

72. *Critical Values y_α^*.* The y_α^* values in Table A.20 can be obtained by using the fact that under H_0 (6.2), all $N! / \left[\prod_{j=1}^{k} n_j!\right]$ rank assignments are equally likely. However, in this one-sided treatments versus control setting we must work a little harder than in the two-sided all-treatments case (see Comments 59 and 63 in Section 6.5) since the values $R_{.u} - R_{.1}, u = 2, \ldots, k$, are, in general, changed when we relabel the control treatment. (In either of the previous two-sided all-treatments cases, the relevant statistic is unaffected by treatment relabelings.) As a result, we will have to take the complete enumeration approach employed in Comment 66 for the one-sided all-treatments setting, where the relevant statistic is also not invariant with respect to treatment relabelings.

For an illustration, we return to Comment 18 and use the 12 rank configurations displayed there for the case $k = 3, n_1 = 1, n_2 = 1$, and $n_3 = 2$. (Here, $N^* = 2$.) For each of these 12 configurations we now display the values of $2(R_{.2} - R_{.1})$ and $2(R_{.3} - R_{.1})$.

(a) $2(R_{.2} - R_{.1}) = 2$
     $2(R_{.3} - R_{.1}) = 5$

(b) $2(R_{.2} - R_{.1}) = -2$
     $2(R_{.3} - R_{.1}) = 3$

(c) $2(R_{.2} - R_{.1}) = 4$
     $2(R_{.3} - R_{.1}) = 4$

(d) $2(R_{.2} - R_{.1}) = -4$
     $2(R_{.3} - R_{.1}) = 0$

(e) $2(R_{.2} - R_{.1}) = 6$
     $2(R_{.3} - R_{.1}) = 3$

(f) $2(R_{.2} - R_{.1}) = -6$
     $2(R_{.3} - R_{.1}) = -3$

(g) $2(R_{.2} - R_{.1}) = 2$
     $2(R_{.3} - R_{.1}) = 1$

(h) $2(R_{.2} - R_{.1}) = -2$
     $2(R_{.3} - R_{.1}) = -1$

(i) $2(R_{.2} - R_{.1}) = 4$
     $2(R_{.3} - R_{.1}) = 0$

(j) $2(R_{.2} - R_{.1}) = -4$
     $2(R_{.3} - R_{.1}) = -4$

(k) $2(R_{.2} - R_{.1}) = 2$
     $2(R_{.3} - R_{.1}) = -3$

(l) $2(R_{.2} - R_{.1}) = -2$
     $2(R_{.3} - R_{.1}) = -5$

Thus, for example,

$$P_0\{2(R_{.u} - R_{.1}) < 6, u = 2, 3\}$$
$$= P_0\{2(R_{.2} - R_{.1}) < 6 \text{ and } 2(R_{.3} - R_{.1}) < 6\}$$
$$= \tfrac{11}{12} = 1 - .0833,$$

since the event $\{2(R_{.2} - R_{.1}) < 6 \text{ and } 2(R_{.3} - R_{.1}) < 6\}$ occurs for all but configuration (e). Similarly, $P_0\{2(R_{.u} - R_{.1}) < 2, u = 2, 3\} = \tfrac{5}{12} = 1 - .5833$, because the event $\{2(R_{.2} - R_{.1}) < 2 \text{ and } 2(R_{.3} - R_{.1}) < 2\}$ occurs only for the five configurations (d), (f), (h), (j), and (l). Hence, for $k = 3, n_1 = 1, n_2 = 1$, and $n_3 = 2$, we have $y^*_{.0833} = 6$ (agreeing with the value in Table A.20) and $y^*_{.5833} = 2$. The other possible experimentwise error rates (there are ten, including 1, all together) and the associated critical values for this setting are obtained through the same type of enumeration.

73. *Interpretation as Hypothesis Tests.* Procedures (6.72), (6.74), and (6.75) can also be interpreted as hypothesis tests of H_0 (6.2). (For example, the procedure that rejects H_0 if at least one of the $k - 1$ inequalities of (6.72) holds is a distribution-free test of level α for H_0.) However, theyare generally more effective at detecting differences between individual treatment effects when applied to data for which the null hypothesis H_0 has previously been rejected by one of the test procedures in Sections 6.1 through 6.4.

74. *Dependence on Observations from Other Noninvolved Treatments.* The differences $(R_{.u} - R_{.1})$ depend on the values of the observations from the other $k - 2$ treatments, in addition to the observations from the control and treatment u. Thus the multiple comparison procedures in (6.72), (6.74), and (6.75) all have the disadvantage that the decision concerning treatment u and the control can be affected by changes only in the observations from one or more of the other $k - 2$ non-involved treatments. This difficulty has been emphasized by Miller (1966) and Gabriel (1969).

75. *Two-Sided Treatments versus Control Multiple Comparison Procedures.* All the multiple comparison procedures of this section are one-sided by nature, resulting in decisions between $\tau_u = \tau_1$ and $\tau_u > \tau_1$ for every $u = 2, \ldots, k$ (or between $\tau_u = \tau_1$ and $\tau_u < \tau_1$ for every $u = 2, \ldots, k$, as noted in Comment 71). We view such one-sided comparisons to be the most natural approach for treatments versus control settings. In such situations we are generally interested in seeing which, if any, of the proposed new treatments are better than a standard control or placebo. In most practical applications, *better* is synonymous with one-sided comparisons (all in one direction or all in the other)—thus our emphasis on such procedures in this section. However, a two-sided treatments-versus-control analogue to procedure (6.72) has

6.7. ONE-SIDED TREATMENTS VERSUS CONTROL MULTIPLE COMPARISONS

been developed in the literature and corresponds to the criterion

$$\text{Decide } \tau_u \neq \tau_1 \text{ if } N^*|R_{.u} - R_{.1}| \geq y_\alpha^{**}; \quad \text{otherwise decide } \tau_u = \tau_1, \quad (6.76)$$

where the constant y_α^{**} is chosen to make the experimentwise error rate equal to α; that is,

$$P_0\{N^*|R_{.u} - R_{.1}| < y_\alpha^{**}, u = 2, \ldots, k\} = 1 - \alpha,$$

where the probability $P_0(\cdot)$ is computed under H_0 (6.2). However, the required tables of y_α^{**} are available only in a very limited fashion. Leach (1972) has provided such critical values y_α^{**} for the very special case of $k = 3$ and equal sample sizes $n_1 = n_2 = n_3 = 2(1)6$. To our knowledge, this is the extent of the tabulated critical values y_α^{**} for the two-sided exact procedure (6.76). Associated large-sample approximations to (6.76) for equal and unequal noncontrol treatment sample sizes have been considered by Miller (1966) and Dunn (1964), respectively. For further discussion of these two-sided treatments versus control multiple comparison procedures, see Miller (1966).

76. *Pairwise Ranking Approach.* The treatments versus control multiple comparison procedures discussed in this section are based on the joint ranking of all N of the sample observations. They suffer from the same drawbacks as do other one-way layout multiple comparison procedures based on joint rankings. For example, they do not provide the maximum type I error rate protection level α guarantee and decisions between treatment u and the control depend on the values of the observations from the other $k - 2$ treatments (for more details, see, for example, Fligner (1984)).

Steel (1959) developed a competitor of these Nemenyi-Damico-Wolfe procedures that takes the pairwise ranking approach discussed in Sections 6.5 and 6.6. His procedure is based on $k - 1$ separate two-sample rankings between the control sample and each of the $k - 1$ noncontrol samples and has form

$$\text{Decide } \tau_u > \tau_1 \text{ if } W_{1u}^* \geq b_\alpha^*; \quad \text{otherwise decide } \tau_u = \tau_1, u = 2, \ldots, k, \quad (6.77)$$

where $W_{12}^*, \ldots, W_{1k}^*$ are defined by equation (6.62) and b_α^* is chosen to make the experimentwise error rate equal to α. This pairwise ranking treatments-versus-control procedure has many of the nice properties of the analogous pairwise rankings all-treatments multiple comparison procedures discussed in Sections 6.5 and 6.6, including proper control of the maximum type I error rate (see Comments 61 and 67). However, the lack of existing tables of the critical values b_α^* (or of computer software to generate them) makes the practical use of procedure (6.77) difficult.

Properties

1. *Asymptotic Multivariate Normality.* See Miller (1966).
2. *Efficiency.* See Sherman (1965) and Section 6.10.

PROBLEMS

63. Apply the approximate procedure (6.74) to the psychotherapeutic attraction data in Table 6.2.
64. For the case $k = 3, \alpha = .01, n_1 = n_2 = n_3 = 6$, compare procedures (6.72) and (6.74).
65. For the psychotherapeutic attraction data in Table 6.2, find the smallest approximate experimentwise error rate at which we would decide $\tau_4 > \tau_1$ using procedure (6.74).

66. Consider the mucociliary clearance data in Table 6.1. Use procedure (6.72) to decide whether or not either obstructive airways disease or asbestosis (or both) lead to a deterioration (slowdown) in median mucociliary clearance half-times.
67. Apply the approximate procedure (6.75) to the glucocorticoid receptor site data in Table 6.4.
68. For the glucocorticoid receptor site data in Table 6.4, find the smallest approximate experimentwise error rate at which we would decide $\tau_5 > \tau_1$ using procedure (6.75).
69. Apply the approximate procedure (6.74) to the plasma glucose data in Table 6.9.
70. For the plasma glucose data in Table 6.9, find the smallest approximate experimentwise error rate at which we would decide $\tau_3 > \tau_1$ using procedure (6.74).
71. For the plasma glucose data in Table 6.9, find the smallest approximate experimentwise error rate at which we would decide $\tau_5 < \tau_1$ with an appropriate treatments versus control multiple comparison procedure (see Comment 71).
72. Apply the approximate procedure (6.74) to the revertant colonies data in Table 6.10.
73. For the revertant colonies data in Table 6.10, find the smallest approximate experimentwise error rate at which we would decide $\tau_4 > \tau_1$ using procedure (6.74).
74. Consider the revertant colonies data in Table 6.10. Find the smallest approximate experimentwise error rate at which the most significant difference in treatment (dosage) effects would be detected.
75. Find the totality of all available experimentwise error rates α and the associated critical values y_α^* for procedure (6.72) when $k = 4, n_1 = 1$, and $n_2 = n_3 = n_4 = 2$.

6.8. CONTRAST ESTIMATION BASED ON HODGES-LEHMANN TWO-SAMPLE ESTIMATORS (SPJØTVOLL)

In this section we discuss a method for the point estimation of certain linear combinations of treatment effects known in the literature as *contrasts*. We define such a contrast in the treatment effects $\tau_1, \ldots, \tau_k$ to be any linear combination of the form

$$\theta = \sum_{i=1}^{k} a_i \tau_i, \tag{6.78}$$

where $a_1, \ldots, a_k$ are any specified set of constants such that $\sum_{i=1}^{k} a_i = 0$. Equivalently, we can write θ in terms of the individual differences in treatment effects (known in the literature as *simple contrasts*)

$$\Delta_{hj} = \tau_h - \tau_j, \quad h = 1, \ldots, k; j = 1, \ldots, k, \tag{6.79}$$

by noting that

$$\theta = \sum_{h=1}^{k} \sum_{j=1}^{k} d_{hj} \Delta_{hj}, \tag{6.80}$$

where

$$d_{hj} = \frac{a_h}{k}, h = 1, \ldots, k; \quad j = 1, \ldots, k. \tag{6.81}$$

For a given setting, decisions about which contrasts to estimate can be related either to a priori interest in particular linear combinations of the τ's or the results of one of the multiple comparison procedures discussed in Sections 6.5–6.7.

6.8. CONTRAST ESTIMATION BASED ON HODGES-LEHMANN TWO-SAMPLE ESTIMATORS

Procedure

For each pair of treatments $(h, j), h \neq j = 1, \ldots, k$, define the pairwise estimators

$$Z_{hj} = \text{median } \{X_{\alpha h} - X_{\beta j}, \alpha = 1, \ldots, n_h; \beta = 1, \ldots, n_j\}. \tag{6.82}$$

Since $Z_{hj} = -Z_{jh}$, we need to calculate only the $k(k-1)/2$ estimators Z_{hj} corresponding to $h < j$. We refer to Z_{hj} as the raw or unadjusted estimator of the simple contrast $\Delta_{hj} = \tau_h - \tau_j$. (Note that Z_{hj} is exactly the Hodges-Lehmann two-sample estimator defined in Section 4.2, as applied to the hth sample (playing the role of the Y's) and the jth sample (playing the role of the X's). For example, Z_{13} is simply the median of the $n_1 n_3$ differences $X_{\alpha 1} - X_{\beta 3}$ obtained from the treatments 1 and 3 observations.) Next, we obtain the set $\overline{\Delta}_1, \ldots, \overline{\Delta}_k$ of individual weighted averages of these unadjusted estimators Z_{hj} corresponding to

$$\overline{\Delta}_h = \sum_{j=1}^{k} \left(\frac{n_j}{N}\right) Z_{hj}, h = 1, \ldots, k, \tag{6.83}$$

where we note that $Z_{hh} = 0$ for $h = 1, \ldots, k$.

The weighted-adjusted estimator of the contrast θ (6.78) is given by

$$\widehat{\theta} = \sum_{i=1}^{k} a_i \overline{\Delta}_i, \tag{6.84}$$

or, equivalently,

$$\widehat{\theta} = \sum_{h=1}^{k} \sum_{j=1}^{k} d_{hj}(\overline{\Delta}_h - \overline{\Delta}_j) = \sum_{h=1}^{k} \sum_{j=1}^{k} d_{hj} W_{hj}, \tag{6.85}$$

where

$$W_{hj} = \overline{\Delta}_h - \overline{\Delta}_j = \widehat{\Delta}_{hj} \tag{6.86}$$

is the weighted-adjusted estimator of the simple contrast $\Delta_{hj} = \tau_h - \tau_j$. (We note that in the special case $n_1 = n_2 = \cdots = n_k$, $\overline{\Delta}_h$ (6.83) reduces to

$$Z_{h.} = \frac{\sum_{j=1}^{k} Z_{hj}}{k}, \quad h = 1, \ldots, k, \tag{6.87}$$

and $W_{hj} = \overline{\Delta}_h - \overline{\Delta}_j$ (6.86) can be simplified to

$$W_{hj} = Z_{h.} - Z_{j.}, \quad h \neq j = 1, \ldots, k. \tag{6.88}$$

Example 6.9: *Motivational Effect of Knowledge of Performance—Examples 6.2 and 6.8 Continued.* Consider the Hundal knowledge of performance data originally presented in Example 6.2. In the application of the Nemenyi-Damico-Wolfe one-sided treatments versus control multiple comparison procedure (Example 6.8) to these data, we concluded that the group receiving accurate information about their output produced significantly more (experimentwise error rate .0554) pieces than the group that received no information. Thus, it is of interest to use the knowledge of performance data in Table 6.6 to estimate the simple contrast

$\theta = \tau_{\text{accurate information}} - \tau_{\text{no information}} = \tau_3 - \tau_1$, thereby providing an idea of the increased output that might be expected for this task by providing accurate information to the workers.

From Table 6.6 and equation (6.82), the three pairwise estimators are

$$Z_{12} = \text{median } \{2, 0, -7, -4, 0, -2, -3, -5, -12, -9, -5, -7, 0, -2, -9,$$
$$-6, -2, -4, 5, 3, -4, -1, 3, 1, 6, 4, -3, 0, 4, 2, 3, 1, -6, -3, 1, -1\}$$
$$= -1.5,$$
$$Z_{13} = \text{median } \{-8, 0, -5, -3, -6, -4, -13, -5, -10, -8, -11, -9, -10, -2, -7,$$
$$-5, -8, -6, -5, 3, -2, 0, -3, -1, -4, 4, -1, 1, -2, 0, -7, 1, -4, -2, -5, -3\}$$
$$= -4,$$

and

$$Z_{23} = \text{median } \{-10, -2, -7, -5, -8, -6, -8, 0, -5, -3, -6, -4, -1, 7, 2,$$
$$4, 1, 3, -4, 4, -1, 1, -2, 0, -8, 0, -5, -3, -6, -4, -6, 2, -3, -1, -4, -2\}$$
$$= -3.$$

From expression (6.83), or equivalently (since $n_1 = n_2 = n_3 = 6$) from (6.87), we have

$$\overline{\Delta}_1 = \frac{Z_{11} + Z_{12} + Z_{13}}{3} = \frac{0 - 1.5 - 4}{3} = -\frac{11}{6},$$
$$\overline{\Delta}_2 = \frac{Z_{21} + Z_{22} + Z_{23}}{3} = \frac{1.5 + 0 - 3}{3} = -.5,$$

and

$$\overline{\Delta}_3 = \frac{Z_{31} + Z_{32} + Z_{33}}{3} = \frac{4 + 3 + 0}{3} = \frac{7}{3}.$$

(Note that in calculating $\overline{\Delta}_2$ and $\overline{\Delta}_3$ we have used the fact that $Z_{21} = -Z_{12}, Z_{31} = -Z_{13}$, and $Z_{32} = -Z_{23}$.) The weighted-adjusted estimator of $\theta = \tau_3 - \tau_1$ is now obtained from (6.84) with $a_1 = -1, a_2 = 0$, and $a_3 = 1$. We find

$$\widehat{\theta} = W_{31} = \overline{\Delta}_3 - \overline{\Delta}_1 = \tfrac{7}{3} - \left(-\tfrac{11}{6}\right) = \tfrac{25}{6} = 4.17 \text{ pieces.}$$

(We note that the values of the raw estimator Z_{31} and the classical estimator $\overline{X}_3 - \overline{X}_1$ are 4.00 and 4.17, respectively, for these data.)

Comments

77. *Ambiguities with the Unadjusted Estimators.* The unadjusted estimators Z_{hj} (6.82) lead to ambiguities in contrast estimation because they do not satisfy the linear relations that are satisfied by the contrasts they estimate. For example, $\Delta_{13} = \tau_1 - \tau_3 = (\tau_1 - \tau_2) + (\tau_2 - \tau_3) = \Delta_{12} + \Delta_{23}$, but in general $Z_{13} \neq Z_{12} + Z_{23}$. Thus, the two "reasonable" estimators Z_{13} and $Z_{12} + Z_{23}$ of $\Delta_{13} = \tau_1 - \tau_3$ can give different estimates. This was pointed out by Lehmann (1963a), who called the unadjusted estimators incompatible.

78. *Compatible, but Inconsistent Estimators.* Lehmann (1963a) removed the incompatibility difficulty discussed in Comment 77 by using the estimators $W_{hj} = Z_{h.} - Z_{j.}$ (6.88). These

estimators are obtained by minimizing the sum of squares $\sum\sum_{h\neq j}[Z_{hj} - (\tau_h - \tau_j)]^2$. Although these estimators are compatible, Lehmann also pointed out that two additional difficulties now arise. First, the estimator $Z_{h.} - Z_{j.}$ of $\Delta_{hj} = \tau_h - \tau_j$ depends, in addition to the observations from samples h and j, on the observations from the other $k - 2$ samples. Furthermore, in the case of $k = 3$, for example, the estimator $Z_{1.} - Z_{2.}$ (of $\tau_1 - \tau_2$) is not consistent when n_1 and n_2 tend to infinity unless n_3 also tends to infinity.

79. *Consistency.* Spjøtvoll (1968) removed the nonconsistency difficulty by obtaining the weighted-adjusted estimators $W_{hj} = \overline{\Delta}_h - \overline{\Delta}_j$ (6.86). These estimators minimize the sum of squares $N^{-2} \sum\sum_{h\neq j} n_h n_j [Z_{hj} - (\tau_h - \tau_j)]^2$. Spjøtvoll's estimators do, however, retain the disadvantage that the estimator of $\tau_h - \tau_j$ depends on unrelated observations from the other samples.

80. *Competitor Contrast Estimator.* Spjøtvoll (1968) also proposed weighted-adjusted estimators that minimize

$$\sum\sum_{h\neq j} \left(\frac{N}{n_h} + \frac{N}{n_j}\right)^{-1} [Z_{hj} - (\tau_h - \tau_j)]^2, \qquad (6.89)$$

using the asymptotic variances of the Z_{hj}'s as weights in the sum of squares. These estimators are more difficult to compute than the estimators W_{hj} (6.86). Furthermore, Spjøtvoll showed that the weighted-adjusted estimators W_{hj} (6.86) and those obtained by minimizing (6.89) have the same asymptotic properties when n_j tends to infinity in such a way that (n_j/N) tends to λ_j with $0 < \lambda_j < 1$, for $j = 1, \ldots, k$.

81. *Equivalence with Equal Sample Sizes.* Spjøtvoll pointed out that the estimator of Δ_{hj} obtained by minimizing (6.89) and the estimator W_{hj} (6.86) both reduce to Lehmann's estimator $Z_{h.} - Z_{j.}$ (6.88) when $n_1 = n_2 = \cdots = n_k$.

Properties

1. *Standard Deviation of $\widehat{\theta}$ (6.84).* For the asymptotic standard deviation of $\widehat{\theta}$ (6.84), see Spjøtvoll (1968).
2. *Asymptotic Normality.* See Spjøtvoll (1968) and Lehmann (1963a).
3. *Efficiency.* See Spjøtvoll (1968), Lehmann (1963a), and Section 6.10.

PROBLEMS

76. Estimate the simple contrast $\theta = \tau_4 - \tau_1$ for the psychotherapeutic attraction data in Table 6.2.
77. Estimate the simple contrasts $\theta_1 = \tau_2 - \tau_1$, $\theta_2 = \tau_4 - \tau_1$, and $\theta_3 = \tau_5 - \tau_1$ for the glucocorticoid receptor sites data in Table 6.4.
78. Estimate all possible simple contrasts for the mean interstitial lengths data in Table 6.5.
79. Estimate the simple contrasts $\tau_2 - \tau_1$, $\tau_4 - \tau_1$, and $\tau_5 - \tau_1$ for the BAI data in Table 6.7.
80. Take $k = 4$ and construct a data example where $Z_{13} + Z_{24} \neq Z_{14} + Z_{23}$. (Note that $\Delta_{13} + \Delta_{24} = \Delta_{14} + \Delta_{23}$. See also Comment 77.)
81. As suggested by the application of the Dwass-Steel-Critchlow-Fligner multiple comparison procedure (in Example 6.6), estimate the contrast $\theta = \left[\frac{1}{2}(\tau_1 + \tau_2) - \frac{1}{2}(\tau_3 + \tau_4)\right]$ for the gizzard shad data in Table 6.3.
82. Estimate the contrast $\theta = \left[\frac{1}{3}(\tau_2 + \tau_3 + \tau_4) - \frac{1}{3}(\tau_1 + \tau_5 + \tau_6)\right]$ for the revertant colonies data in Table 6.10.

83. Estimate the simple contrasts $\tau_2 - \tau_1$ and $\tau_3 - \tau_1$ for the plasma glucose data in Table 6.9.
84. Estimate all contrasts found to be of interest in Problem 63 for the psychotherapeutic attraction data in Table 6.2.

6.9. SIMULTANEOUS CONFIDENCE INTERVALS FOR ALL SIMPLE CONTRASTS (CRITCHLOW-FLIGNER)

A contrast θ (6.78) is said to be a simple contrast if it involves only two treatment effects (i.e., all but two of the a_i coefficients are zero). In this section we present a method for obtaining simultaneous confidence intervals for the entire collection, C, of all $\binom{k}{2}$ simple contrasts given by

$$C = \{\Delta_{uv} : \Delta_{uv} = \tau_v - \tau_u, \ 1 \leq u < v \leq k\}. \tag{6.90}$$

Procedure

For each pair of treatments $(u, v), u \neq v = 1, \ldots, k$, define the sample differences

$$D_{ij}^{uv} = X_{jv} - X_{iu}, \quad i = 1, \ldots, n_u; \quad j = 1, \ldots, n_v. \tag{6.91}$$

Let $D_{(1)}^{uv} \leq D_{(2)}^{uv} \leq \cdots \leq D_{(n_u n_v)}^{uv}$ denote the ordered values of the $n_u n_v$ D_{ij}^{uv} differences, for $u \neq v = 1, \ldots, k$. Let w_α^* be the upper αth percentile for the distribution of maximum $\{|W_{uv}^*|, u \neq v = 1, \ldots, k\}$ under H_0 (6.2), where W_{uv}^* is the standardized two-sample rank sum statistic (multiplied by $\sqrt{2}$) for the uth and vth samples, as defined previously in equation (6.62) for the two-sided all-treatments multiple comparisons setting. Selected values of w_α^* can be found in Table A.16.

For $1 \leq u < v \leq k$, set

$$a_{uv} = \frac{n_u n_v}{2} - w_\alpha^* \left[\frac{n_u n_v (n_u + n_v + 1)}{24}\right]^{1/2} + 1 \tag{6.92}$$

and

$$b_{uv} = a_{uv} - 1. \tag{6.93}$$

The simultaneous $100(1 - \alpha)\%$ confidence intervals for the collection C (6.90) of all simple contrasts are then

$$\{[D_{(\langle a_{uv}\rangle)}^{uv}, D_{(n_u n_v - \langle b_{uv}\rangle)}^{uv}), 1 \leq u < v \leq k\}, \tag{6.94}$$

where $\langle t \rangle$ denotes the greatest integer less than or equal to t. This set of intervals satisfies the condition

$$P_{\tau_1,\ldots,\tau_k}(D_{(\langle a_{uv}\rangle)}^{uv} \leq \tau_v - \tau_u < D_{(n_u n_v - \langle b_{uv}\rangle)}^{uv}, \text{ for } 1 \leq u < v \leq k)$$
$$= 1 - \alpha, \quad \text{for all } -\infty < \tau_i < \infty, i = 1, \ldots, k. \tag{6.95}$$

(For simultaneous lower confidence bounds for the collection C that are appropriate under the ordered alternatives setting of Section 6.2, see Comment 83.)

6.9. SIMULTANEOUS CONFIDENCE INTERVALS FOR ALL SIMPLE CONTRASTS

Large-Sample Approximation

When H_0 is true, the $[k(k-1)/2]$-component vector $(W^*_{12}, W^*_{13}, \ldots, W^*_{k-1,k})$ has, as min $(n_1, \ldots, n_k)$ tends to infinity, an asymptotic multivariate normal distribution with mean vector $\mathbf{0}$. It then follows (see Comment 62) that w^*_α can be approximated for large sample sizes by q_α, where q_α is the upper αth percentile point for the distribution of the range of k independent $N(0, 1)$ variables. Thus the large-sample approximate simultaneous $100(1-\alpha)\%$ confidence intervals for C (6.90) are given by (6.94) with w^*_α replaced by q_α in the expressions for a_{uv} (6.92) and b_{uv} (6.93). Values of q_α for $\alpha = .0001, .0005, .001, .005, .01, .025, .05, .10, .20$, and $k = 2(1)20(2)40(10)(100)$ are given in Table A.17.

Example 6.10: *Motivational Effect of Knowledge of Performance—Examples 6.2, 6.8, and 6.9 Continued.* Consider the Hundal knowledge of performance data originally presented in Example 6.2. In this example, we wish to find simultaneous $100(1-\alpha)\%$ confidence intervals for the $3(2)/2 = 3$ simple contrasts

$$C = \{\tau_2 - \tau_1, \tau_3 - \tau_1, \tau_3 - \tau_2\}.$$

For sake of illustration, we take $\alpha = .1041$. From Table A.16, we see that $w^*_{.1041} = 2.9439$. For the Hundal study, we have $n_1 = n_2 = n_3 = 6$. It follows from expressions (6.92) and (6.93) that

$$a_{12} = a_{13} = a_{23} = \frac{6(6)}{2} - 2.944 \left[\frac{6(6)(6+6+1)}{24}\right]^{1/2} + 1 \approx 6.00$$

and

$$b_{12} = b_{13} = b_{23} = 6.00 - 1 = 5.00.$$

Thus the simultaneous 89.51% confidence intervals for the simple contrasts $\Delta_{12} = \tau_2 - \tau_1, \Delta_{13} = \tau_3 - \tau_1$, and $\Delta_{23} = \tau_3 - \tau_2$, correspond to $[D^{12}_{((6))}, D^{12}_{(36-\langle 5 \rangle)}) = [D^{12}_{(6)}, D^{12}_{(31)})$, $[D^{13}_{(6)}, D^{13}_{(31)})$, and $[D^{23}_{(6)}, D^{23}_{(31)})$, respectively. Using the individual differences already computed in Example 6.9 to obtain a point estimate of the contrast $\tau_3 - \tau_1 = \tau_{\text{accurate information}} - \tau_{\text{no information}}$, we see that the three sets of $n_u n_v = 36$ ordered $D^{uv}_{(t)}$'s are given by

$$D^{12}_{(t)} : \{-6, -5, -4, -4, -3, -3, -3, -2, -2, -1, -1, -1, 0, 0, 0, 0, 1, 1,$$
$$2, 2, 2, 3, 3, 3, 4, 4, 4, 5, 5, 6, 6, 7, 7, 9, 9, 12\},$$

$$D^{13}_{(t)} : \{-4, -3, -1, -1, 0, 0, 0, 1, 1, 2, 2, 2, 2, 3, 3, 3, 4, 4,$$
$$4, 5, 5, 5, 5, 5, 6, 6, 7, 7, 8, 8, 8, 9, 10, 10, 11, 13\},$$

and

$$D^{23}_{(t)} : \{-7, -4, -4, -3, -2, -2, -1, -1, 0, 0, 0, 1, 1, 1, 2, 2, 2, 3,$$
$$3, 3, 4, 4, 4, 4, 5, 5, 5, 6, 6, 6, 6, 7, 8, 8, 8, 10\}.$$

Hence the simultaneous 89.51% confidence intervals for the simple contrasts $\Delta_{12} = \tau_2 - \tau_1, \Delta_{13} = \tau_3 - \tau_1$ and $\Delta_{23} = \tau_3 - \tau_2$ for the Hundal data are

$$[D^{12}_{(6)}, D^{12}_{(31)}) = [-3, 6),$$

$$[D^{13}_{(6)}, D^{13}_{(31)}) = [0, 8),$$

and

$$[D^{23}_{(6)}, D^{23}_{(31)}) = [-2, 6),$$

respectively.

For sake of illustration for the associated large-sample approximation we take an approximate α value of .10. From Table A.17 with $k = 3$, we find $q_{.10} = 2.902$. The associated approximate values for the a_{uv}'s and b_{uv}'s are

$$a_{12} = a_{13} = a_{23} \approx \frac{6(6)}{2} - 2.902 \left[\frac{6(6)(6 + 6 + 1)}{24} \right]^{1/2} + 1 = 6.19$$

and

$$b_{12} = b_{13} = b_{23} \approx 6.19 - 1 = 5.19.$$

Since $\langle 6.19 \rangle = 6$ and $\langle 5.19 \rangle = 5$, we see that the approximate 90% simultaneous confidence intervals for the simple contrasts $\tau_2 - \tau_1$, $\tau_3 - \tau_1$, and $\tau_3 - \tau_2$ are identical with the exact 89.51% simultaneous confidence intervals for these Hundal data. This provides some indication that the common sample size of 6 observations is already large enough to enable the large-sample approximation to be effective for these simultaneous confidence intervals.

Comments

82. *Relationship of Simultaneous Confidence Intervals to Two-Sided All-Treatments Multiple Comparisons.* The simultaneous $100(1 - \alpha)\%$ confidence intervals (6.94) for the collection C (6.90) of all simple contrasts are directly related to the Dwass-Steel-Critchlow-Fligner two-sided all-treatments multiple comparison procedure (6.63) discussed in Section 6.5. In fact, for *every* (u, v) pair, $1 \leq u < v \leq k$, the two-sided multiple comparison procedure (6.63) yields the decision $\tau_u \neq \tau_v$ at an experimentwise error rate α if and only if 0 does not belong to the corresponding simultaneous $100(1 - \alpha)\%$ confidence interval $[D^{uv}_{(\langle a_{uv} \rangle)}, D^{uv}_{(n_u n_v - \langle b_{uv} \rangle)})$ for $\tau_v - \tau_u$. Thus each of the $2^{\binom{k}{2}}$ sets of possible multiple comparison decisions associated with procedure (6.63) at an experimentwise error rate α correponds to a collection of simultaneous $100(1 - \alpha)\%$ confidence intervals (6.94) for C (6.90) for which the particular (u, v) intervals not containing the value 0 match exactly with those treatment pairs for which procedure (6.63) leads to the decision $\tau_u \neq \tau_v$.

83. *Simultaneous $100(1 - \alpha)\%$ Lower Confidence Bounds.* In situations where an order restriction $\tau_1 \leq \tau_2 \leq \cdots \leq \tau_k$ on the treatment effects is appropriate (see Sections 6.2 and 6.6 for further details), it is more natural to seek out simultaneous $100(1 - \alpha)\%$ lower confidence bounds (rather than two-sided intervals) for the collection C (6.90) of simple contrasts. In such a setting, let c^*_α be the critical value for the Hayter-Stone one-sided all-treatments multiple comparison procedure (6.69) and set

$$h_{uv} = \frac{n_u n_v}{2} - c^*_\alpha \left[\frac{n_u n_v (n_u + n_v + 1)}{24} \right]^{1/2} + 1. \quad (6.96)$$

The simultaneous $100(1 - \alpha)\%$ lower confidence bounds for the collection C (6.90) suggested by Hayter and Stone (1991) are then given by

$$\{[D^{uv}_{(\langle h_{uv} \rangle)}, \infty), 1 \leq u < v \leq k\}, \tag{6.97}$$

where, once again, $\langle t \rangle$ denotes the greatest integer less than or equal to t and the ordered $D^{uv}_{(t)}$'s are as defined in the Procedure of this section. When either the number of treatments exceeds 3 or $k = 3$ and one or more of the sample sizes is larger than 7, Hayter and Stone (1991) suggest approximating c^*_α in expression (6.96) by d_α, the upper αth percentile point for the distribution of

$$D = \underset{1 \leq i < j \leq k}{\text{maximum}} \left[\frac{Z_j - Z_i}{\left\{ \frac{n_i + n_j}{2 n_i n_j} \right\}^{1/2}} \right],$$

where $Z_1, \ldots, Z_k$ are mutually independent and Z_i has a $N(0, 1/n_i)$ distribution, for $i = 1, \ldots, k$. Values of d_α for large equal sample sizes $n_1 = \cdots = n_k = n, k = 3(1)9$, and confidence level $1 - \alpha = .90, .95,$ and $.99$ are given in Table A.19. Approximate values d_α for large, but unequal, sample sizes $n_1, \ldots, n_k$ are not presently available in the literature.

The relationship between these simultaneous $100(1 - \alpha)\%$ lower confidence bounds (6.97) for C (6.90) and the Hayter-Stone one-sided all-treatments multiple comparison procedure (6.69) at experimentwise error rate α is identical to that described in Comment 82 for the simultaneous $100(1 - \alpha)\%$ confidence intervals (6.94) for C (6.90) and the two-sided all-treatments multiple comparison procedure (6.63) at experimentwise error rate α.

84. *Pairwise versus Joint Rankings.* In the latter portion of Comment 63 we discussed some of the pros and cons of pairwise rankings versus joint rankings in the one-way layout setting. The simultaneous $100(1 - \alpha)\%$ confidence intervals (6.94) for the collection of all simple contrasts (and the analogous simultaneous lower confidence bounds (6.97) discussed in Comment 83) are clearly associated with pairwise rankings. This provides an additional advantage to the use of pairwise rankings, since the joint ranking approach discussed in Comment 63 does not lead directly to such simultaneous confidence intervals or bounds for C (6.90).

Properties

1. *Distribution-Freeness.* For populations satisfying Assumptions A1, A2, and A3, equation (6.95) holds. Hence, we can control the simultaneous coverage probability to be $1 - \alpha$ without having more specific knowledge about the form of the underlying F. As a result the intervals in (6.94) are distribution-free simultaneous confidence intervals for the collection C (6.90) of all simple contrasts over a very large class of populations.
2. *Asymptotic Multivariate Normality.* See Hayter (1984) and Critchlow and Fligner (1991).

PROBLEMS

85. Consider the length of YOY gizzard shad data discussed in Problem 4. Find a set of approximate simultaneous 95% confidence intervals for the set of all simple contrasts.
86. Consider the Hundal knowledge of performance data originally discussed in Example 6.2. Find a set of simultaneous 88.11% lower confidence bounds for the three simple contrasts $\tau_2 - \tau_1, \tau_3 - \tau_1$, and $\tau_3 - \tau_2$. (See Comment 83.) Compare with the set of 89.59% simultaneous confidence intervals obtained in Example 6.10 for these same simple contrasts.

87. Consider the Acid Red 114 revertant colonies data in Table 6.10. Find a set of approximate simultaneous 90% confidence intervals for the set of all simple contrasts.

88. Consider the tiger muskellunge plasma glucose data in Table 6.9. Find a set of approximate simultaneous 95% confidence intervals for the set of all simple contrasts.

89. Consider the white-tailed deer fasting metabolic rate data in Table 6.8. Find a set of approximate simultaneous 80% confidence intervals for the set of all simple contrasts.

90. Consider the half-time of mucociliary clearance data in Table 6.1. Find a set of approximate simultaneous 91.81% confidence intervals for the set of all simple contrasts. Without further calculations, what decisions would be reached for these data by the multiple comparison procedure (6.63) at experimentwise error rate $\alpha = .0819$? (See Comment 82.)

91. Consider the average basal area increment data in Table 6.7. Find a set of approximate simultaneous 90% confidence intervals for the set of all simple contrasts. Do you have any concerns about the application of this procedure to these data?

92. Consider the Wechsler Adult Intelligence Scale data in Table 6.11. For the age groups 16–19, 20–34, 35–54, and 55–69 only, find a set of approximate simultaneous 90% lower confidence bounds for the set of all simple contrasts for these four age groups. Without further calculations, what decisions would be reached for these data by the multiple comparison procedure (6.71) at approximate experimentwise error rate $\alpha = .10$? (See Comments 82 and 83.)

6.10. EFFICIENCIES OF ONE-WAY LAYOUT PROCEDURES

The Pitman asymptotic relative efficiencies for translation alternatives of most of the nonparametric procedures discussed in this chapter with respect to the corresponding normal theory procedures are given by the expression

$$e_F = 12\sigma_F^2 \left\{ \int_{-\infty}^{\infty} f^2(u)\, du \right\}^2, \tag{6.98}$$

where σ_F^2 is the variance of the common underlying (continuous) distribution F (6.1) and $f(\cdot)$ is the probability density function corresponding to F. The parameter $\int_{-\infty}^{\infty} f^2(u)\, du$ is the area under the curve associated with $f^2(\cdot)$, the square of the common probability density function. We note that this same expression (6.98) also yields the corresponding Pitman efficiencies in the one-sample and two-sample location settings (see Sections 3.11 and 4.5).

In particular, the Pitman asymptotic relative efficiency of the Kruskal-Wallis test based on H (6.5) with respect to the normal theory one-way layout $\mathcal{F}$-test was found to be e_F (6.98) by Andrews (1954). The asymptotic relative efficiency of the Jonckheere-Terpstra test for ordered alternatives based on the statistic J (6.13) with respect to a suitable normal theory competitor was found by Puri (1965) to be e_F (6.98) as well. Mack and Wolfe (1981) found the same expression to hold for the asymptotic relative efficiency of their peak-known umbrella test procedure based on A_p (6.31) relative to an analogous normal theory procedure based on sample averages. Fligner and Wolfe (1982) found the same to be the case for the treatments versus control test based on FW (6.51).

Sherman (1965) obtained e_F (6.98) as the asymptotic relative efficiency of the two-sided all-treatments and the one-sided treatments versus control multiple comparison procedures discussed in Sections 6.5 and 6.7 with respect to the corresponding classical normal theory procedures based on sample means. Spjøtvoll (1968) showed that, when n_j/N tends to ρ_j, with $0 < \rho_j < 1$, the estimators W_{hj} (6.86) have the same asymptotic properties as the estimators $[Z_{h.} - Z_{j.}]$ (see Comment 78). It then follows from Lehmann's (1963a) results that e_F (6.98) is the asymptotic relative efficiency of the estimator $\widehat{\theta}$ (6.84) with respect to the least squares

6.10. EFFICIENCIES OF ONE-WAY LAYOUT PROCEDURES

estimator $\overline{\theta} = \sum_{h=1}^{k} \sum_{j=1}^{k} d_{hj}(\overline{X}_{.h} - \overline{X}_{.j})$, where

$$\overline{X}_{.t} = \sum_{i=1}^{n_t} \frac{X_{it}}{n_t}, \quad \text{for } t = 1, \ldots, k.$$

As noted in both Sections 3.11 and 4.5, the asymptotic relative efficiency e_F (6.98) is always greater than or equal to .864 and can be infinite. See expression (3.116) for the value of e_F (6.98) for a variety of underlying F populations.

We do not know of any results for the asymptotic relative efficiencies of the Mack-Wolfe peak-unknown umbrella test (Section 6.3B), the Hayter-Stone one-sided all-treatments multiple comparison procedure (Section 6.6), or the Critchlow-Fligner procedure for simultaneous confidence intervals for all simple contrasts (Section 6.9).

CHAPTER 7

The Two-Way Layout

INTRODUCTION

The procedures of this chapter are designed for statistical analyses of data collected under the auspices of an experimental design involving two factors, each at two or more levels. Our primary interest is in the relative location effects (medians) of the different levels of one of these factors, hereafter called the *treatment* factor, *within* the various levels of the second factor, hereafter called the *blocking* factor. This blocking factor is associated quite commonly with the experimental design where subjects are first divided into more homogeneous subgroups (called *blocks*) and then randomly assigned to the various treatment levels within these blocks. Such a design is called a randomized block design, and we will use this treatment/block terminology to describe the two-way layout setting throughout this chapter. In addition, we will refer, without loss of generality, to the k levels of a treatment as the k *treatments*. (In the case of a randomized complete block design, where the data consist of one observation on each of k treatments in each of n blocks, this represents a direct generalization of the paired replicates setting discussed in Chapter 3.)

The basic null hypothesis of interest is that of no differences in the location effects (medians) of the k treatments within each of the blocks. The alternatives considered here correspond to either general or ordered differences between the treatment effects (medians). As with the one-way layout setting in Chapter 6, we also differentiate between those cases where all of the k treatments represent "new" categories for study and those where one of the treatments corresponds to a control or baseline category. Finally, we must deal separately with a variety of different possibilities (and correspondingly different statistical procedures) for the number of observations available from each treatment-block combination (cell), ranging from 0 (missing data), 1, or more than 1 (replications).

Sections 7.1–7.5 are devoted to the case of one observation per treatment-block cell (commonly known as a randomized complete block design). Section 7.1 presents a distribution-free test directed at general alternatives. A distribution-free test designed specifically to detect ordered differences among the k treatments is discussed in Section 7.2. Multiple comparison procedures designed to detect which, if any, treatment effects differ from one another are presented in Section 7.3 (all-treatments comparisons) and 7.4 (treatments versus control comparisons). In Section 7.5 we present estimators of contrasts in the treatment effects.

Sections 7.6–7.8 deal with settings where certain treatment-block cells yield single observations, but there are also treatment-block combinations for which we have no observations; that is, we have either zero or one observation from each treatment-block cell. Sections 7.6 and 7.7 present a distribution-free hypothesis test for general alternatives and an all-treatments multiple comparison procedure, respectively, for the structured setting where the data arise from a balanced incomplete block design (BIBD). Section 7.8 discusses a distribution-free hypothesis test for general alternatives in a two-way layout with an arbitrary configuration of either zero or one observation per cell.

In Sections 7.9 and 7.10 we discuss procedures for the setting where there is at least one observation from each cell and there are some cells with multiple observations (replications). Section 7.9 presents a distribution-free hypothesis test for general alternatives for this replications setting, with an emphasis on the special case where we have an equal number (> 1) of replications in each cell. An all-treatments multiple comparison procedure for this setting of an equal number of replications is detailed in Section 7.10.

All of the procedures in Sections 7.1–7.10 are associated with within-blocks rankings (known as Friedman ranks) and represent direct extensions to the two-way layout of the paired replicates sign procedures discussed in Sections 3.4–3.6. The corresponding extensions to the two-way layout of the paired replicates Wilcoxon signed ranks procedures discussed in Sections 3.1–3.3 yield asymptotically (number of blocks tending to infinity) distribution-free test and multiple comparison procedures, and we present simplified conservative versions that are nearly asymptotically distribution-free. In Sections 7.11–7.15 we discuss these extensions associated with Wilcoxon signed ranks for data from a randomized complete block design with k treatments and n blocks. Section 7.11 contains a conservative signed ranks test directed at general alternatives, and Section 7.12 presents the corresponding conservative signed ranks test procedure designed for ordered alternatives. The associated approximate signed ranks multiple comparison procedures are given in Sections 7.13 (all-treatments comparisons) and 7.14 (treatments versus control comparisons). Section 7.15 contains the contrast estimators linked to the Wilcoxon signed ranks.

The asymptotic relative efficiencies for translation alternatives of the procedures discussed in this chapter with respect to their normal theory counterparts are discussed in Section 7.16.

Data. The data consist of $N = \sum_{i=1}^{n} \sum_{j=1}^{k} c_{ij}$ observations, with c_{ij} observations from the combination of the ith block with the jth treatment (i.e., the (i, j)th cell), for $i = 1, \ldots, n$ and $j = 1, \ldots, k$.

		Treatments		
Blocks	1	2	$\ldots$	k
1	X_{111}	X_{121}	$\ldots$	X_{1k1}
	$\vdots$	$\vdots$	$\ldots$	$\vdots$
	$X_{11c_{11}}$	$X_{12c_{12}}$	$\ldots$	$X_{1kc_{1k}}$
2	X_{211}	X_{221}	$\ldots$	X_{2k1}
	$\vdots$	$\vdots$	$\ldots$	$\vdots$
	$X_{21c_{21}}$	$X_{22c_{22}}$	$\ldots$	$X_{2kc_{2k}}$
$\vdots$	$\vdots$	$\vdots$	$\vdots$	$\vdots$
n	X_{n11}	X_{n21}	$\ldots$	X_{nk1}
	$\vdots$	$\vdots$	$\ldots$	$\vdots$
	$X_{n1c_{n1}}$	$X_{n2c_{n2}}$	$\ldots$	$X_{nkc_{nk}}$

Assumptions

A1. The N random variables $\{(X_{ij1}, \ldots, X_{ijc_{ij}}), i = 1, \ldots, n$ and $j = 1, \ldots, k\}$ are mutually independent.

A2. For each fixed (i, j), with $i \in \{1, \ldots, n\}$ and $j \in \{1, \ldots, k\}$, the c_{ij} random variables $(X_{ij1}, \ldots, X_{ijc_{ij}})$ are a random sample from a continuous distribution with distribution function F_{ij}.

A3. The distribution functions $F_{11}, \ldots, F_{1k}, \ldots, F_{n1}, \ldots, F_{nk}$ are connected through the relationship

$$F_{ij}(u) = F(u - \beta_i - \tau_j), \quad -\infty < u < \infty, \tag{7.1}$$

for $i = 1, \ldots, n$ and $j = 1, \ldots, k$, where F is a distribution function for a continuous distribution with unknown median θ, β_i is the unknown additive effect contributed by block i, and τ_j is the unknown additive treatment effect contributed by the jth treatment.

We note that Assumptions A1–A3 correspond directly to the usual two-way layout *additive* (see Comment 6) model associated with normal theory assumptions; that is, Assumptions A1–A3 are equivalent to the representation

$$X_{ijt} = \theta + \beta_i + \tau_j + e_{ijt}, \quad i = 1, \ldots, n; \quad j = 1, \ldots, k; \quad t = 1, \ldots, c_{ij},$$

where θ is the overall median, τ_j is the treatment j effect, β_i is the block i effect, and the N e's form a random sample from a continuous distribution with median 0. (Under the additional assumption of normality, the medians θ and 0 are, of course, also the respective means.)

Hypothesis

The null hypothesis of interest in Sections 7.1–7.2, 7.6, 7.8–7.9, and 7.11–7.12 is that of no differences among the additive treatment effects $\tau_1, \ldots, \tau_k$, namely,

$$H_0 : [\tau_1 = \cdots = \tau_k]. \tag{7.2}$$

The null hypothesis asserts that the underlying distributions $F_{i1}, \ldots, F_{ik}$ within block i are the same, for each fixed $i = 1, \ldots, n$; that is, $F_{i1} \equiv F_{i2} \equiv \cdots \equiv F_{ik} \equiv F_i$, for $i = 1, \ldots, n$, in equation (7.1).

In Sections 7.1–7.5 we consider the special case of one observation per treatment-block combination (commonly known as a randomized complete block design), corresponding to $c_{ij} = 1$ for every $i = 1, \ldots, n$ and $j = 1, \ldots, k$. For ease of notation in these five sections, we drop the third subscript on the X variables, since it is always equal to 1 in this setting.

7.1. A DISTRIBUTION-FREE TEST FOR GENERAL ALTERNATIVES IN A RANDOMIZED COMPLETE BLOCK DESIGN (FRIEDMAN, KENDALL-BABINGTON SMITH)

In this section we present a procedure for testing H_0 (7.2) against the general alternative that at least two of the treatment effects are not equal, namely,

$$H_1 : [\tau_1, \ldots, \tau_k \text{ not all equal}], \tag{7.3}$$

when $c_{ij} \equiv 1$, for $i = 1, \ldots, n$ and $j = 1, \ldots, k$.

Procedure

To compute the Friedman statistic S, we first order the k observations from least to greatest separately within each of the n blocks. Let r_{ij} denote the rank of X_{ij} in the joint ranking of the

7.1. A DISTRIBUTION-FREE TEST FOR GENERAL ALTERNATIVES

observations $X_{i1}, \ldots, X_{ik}$ in the ith block and set

$$R_j = \sum_{i=1}^{n} r_{ij} \quad \text{and} \quad R_{\cdot j} = \frac{R_j}{n}. \tag{7.4}$$

Thus, for example, R_2 is the sum (over the n blocks) of the within-blocks ranks received by the treatment 2 observations and $R_{\cdot 2}$ is the average within-blocks rank for these same observations. The Friedman statistic S is then given by

$$S = \frac{12n}{k(k+1)} \sum_{j=1}^{k} \left(R_{\cdot j} - \frac{k+1}{2} \right)^2$$

$$= \left[\frac{12}{nk(k+1)} \sum_{j=1}^{k} R_j^2 \right] - 3n(k+1), \tag{7.5}$$

where $(k+1)/2 = \sum_{i=1}^{n} \sum_{j=1}^{k} r_{ij}/nk$ is the average rank assigned via this within-blocks ranking scheme.

To test

$$H_0 : [\tau_1 = \cdots = \tau_k]$$

versus the general alternative

$$H_1 : [\tau_1, \ldots, \tau_k \text{ not all equal}],$$

at the α level of significance,

$$\text{Reject } H_0 \text{ if } S \geq s_\alpha; \quad \text{otherwise do not reject,} \tag{7.6}$$

where the constant s_α is chosen to make the type I error probability equal to α. Values of s_α are given in Table A.22.

Large-Sample Approximation

When H_0 is true, the statistic S has, as n tends to infinity, an asymptotic chi-square (χ^2) distribution with $k-1$ degrees of freedom. (See Comment 10 for indications of the proof.) The chi-square approximation for procedure (7.6) is

$$\text{Reject } H_0 \text{ if } S \geq \chi^2_{k-1,\alpha}; \quad \text{otherwise do not reject,} \tag{7.7}$$

where $\chi^2_{k-1,\alpha}$ is the upper α percentile point of a chi-square distribution with $k-1$ degrees of freedom. Values of $\chi^2_{k-1,\alpha}$ can be obtained from Chart A.2.

Ties

If there are ties among the k observations in a given block, assign each of the observations in a tied group the average of the within-blocks integer ranks that are associated with the tied group and compute S with these within-blocks average ranks. As a consequence of the effect that ties have on the null distribution of S, the following modification is required to apply either procedure (7.6) or the large-sample approximation in (7.7) when there are tied data values within any of the blocks. For either of these procedures, we replace S by

$$S' = \frac{12 \sum_{j=1}^{k} \left(R_j - \frac{n(k+1)}{2}\right)^2}{nk(k+1) - [1/(k-1)] \sum_{i=1}^{n} \left\{ \left(\sum_{j=1}^{g_i} t_{i,j}^3\right) - k \right\}} \quad (7.8)$$

$$= \frac{12 \sum_{j=1}^{k} R_j^2 - 3n^2 k(k+1)^2}{nk(k+1) - [1/(k-1)] \sum_{i=1}^{n} \left\{ \left(\sum_{j=1}^{g_i} t_{i,j}^3\right) - k \right\}},$$

where, in equation (7.8), g_i denotes the number of tied groups in the ith block and $t_{i,j}$ is the size of the jth tied group in the ith block. We note that an untied observation within a block is considered to be a tied group of size 1. In particular, if there are no ties among the X's in the ith block, then $g_i = k$, $t_{i,j} = 1$ for each $j = 1, \ldots, k$, and the correction term for the ith block becomes $\{(\sum_{j=1}^{g_i} t_{i,j}^3) - k\} = k - k = 0$. If each block is void of ties, then we have $\sum_{i=1}^{n} \{(\sum_{j=1}^{g_i} t_{i,j}^3) - k\} = 0$ and S' (7.8) reduces to S, as given in (7.5).

We note that even procedure (7.6) using the critical values from Table A.22 is only approximately, and not exactly, of significance level α in the presence of tied X observations within any of the blocks. To get an exact level α test in this tied setting, see Comment 9.

Example 7.1: *Rounding First Base.* The data in Table 7.1 were obtained by Woodward (1970) in a study to determine which, if any, of three methods of rounding first base is best, in

TABLE 7.1. Rounding First Base Times

	Methods		
Players	Round Out	Narrow Angle	Wide Angle
1	5.40 (1)	5.50 (2)	5.55 (3)
2	5.85 (3)	5.70 (1)	5.75 (2)
3	5.20 (1)	5.60 (3)	5.50 (2)
4	5.55 (3)	5.50 (2)	5.40 (1)
5	5.90 (3)	5.85 (2)	5.70 (1)
6	5.45 (1)	5.55 (2)	5.60 (3)
7	5.40 (2.5)	5.40 (2.5)	5.35 (1)
8	5.45 (2)	5.50 (3)	5.35 (1)
9	5.25 (3)	5.15 (2)	5.00 (1)
10	5.85 (3)	5.80 (2)	5.70 (1)
11	5.25 (3)	5.20 (2)	5.10 (1)
12	5.65 (3)	5.55 (2)	5.45 (1)
13	5.60 (3)	5.35 (1)	5.45 (2)
14	5.05 (3)	5.00 (2)	4.95 (1)
15	5.50 (2.5)	5.50 (2.5)	5.40 (1)
16	5.45 (1)	5.55 (3)	5.50 (2)
17	5.55 (2.5)	5.55 (2.5)	5.35 (1)
18	5.45 (1)	5.50 (2)	5.55 (3)
19	5.50 (3)	5.45 (2)	5.25 (1)
20	5.65 (3)	5.60 (2)	5.40 (1)
21	5.70 (3)	5.65 (2)	5.55 (1)
22	6.30 (2.5)	6.30 (2.5)	6.25 (1)
	$R_1 = 53$	$R_2 = 47$	$R_3 = 32$

Source: W. F. Woodward (1970).

7.1. A DISTRIBUTION-FREE TEST FOR GENERAL ALTERNATIVES

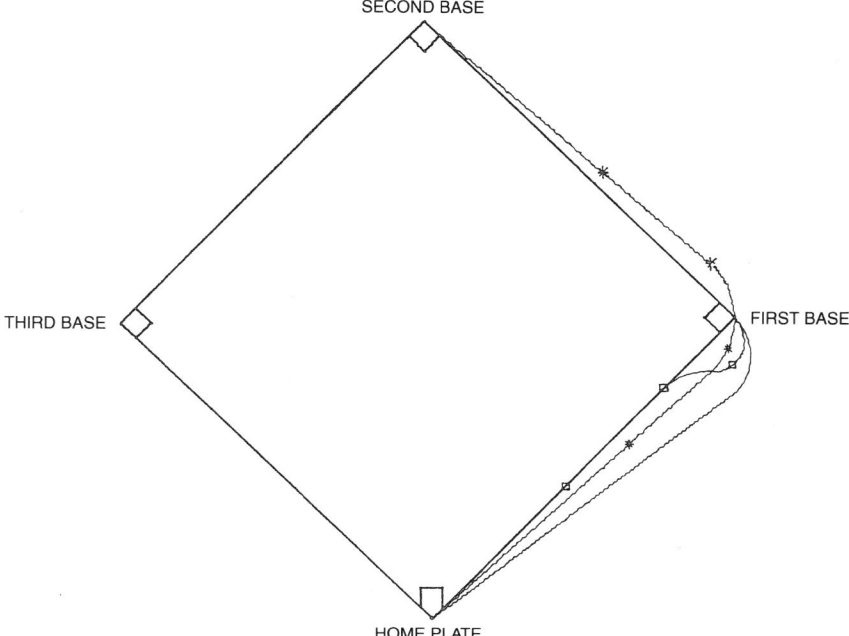

FIGURE 7.1. Three methods of rounding first base: ◇ path = round out method, ⋆ path = narrow angle method, solid path = wide angle method.

the sense that it minimizes, on the average, the time to reach second base. The three methods, "round out," "narrow angle," and "wide angle" are illustrated in Figure 7.1.

Twenty-two baseball players participated in the study, and each of them ran from home plate to second base six times. Using a randomized order, these six trials per player were evenly divided (two each) among the three methods (round out, narrow angle, and wide angle). The entries in Table 7.1 are average times of the two runs per method from a point on the first base line 35 ft from home plate to a point 15 ft short of second base. The within-blocks (players) ranks (r_{ij}'s) of the observations are also given in Table 7.1 in parentheses after the data values (using average ranks to break the ties) and the treatment (running method) rank sums (R_1, R_2, and R_3) are provided at the bottom of the columns.

Since ties exist in blocks 7, 15, 17, and 22, we use S' (7.8). The term in braces in the denominator of (7.8) is zero for each block i in which there are no tied observations. Thus we need evaluate that term only for $i = 7, 15, 17,$ and 22, corresponding to the blocks in which ties exist. In block 7 there is one tied group of size 2 (5.40, 5.40) and one tied group of size 1 (5.35). Thus $t_{7,1} = 2, t_{7,2} = 1, g_7 = 2$, and $\{(\sum_{j=1}^{g_7} t_{7,j}^3) - k\} = \{(2^3 + 1^3) - 3\} = 6$. In the same way, $\{(\sum_{j=1}^{g_i} t_{i,j}^3) - k\} = 6$ for $i = 15, 17,$ and 22. Hence from (7.8) we obtain

$$S' = \frac{12[(53-44)^2 + (47-44)^2 + (32-44)^2]}{22(3)(4) - (\frac{1}{2})(6+6+6+6)} = 11.1.$$

From the χ_2^2 distribution (Chart A.2) we see that the lowest level at which we reject, using the large-sample procedure (7.7) adjusted for ties, is approximately .004. Hence there is strong evidence here to reject the hypothesis that the methods are equivalent with respect to speed.

We can also use Minitab's FRIEDMAN command to compute the value of S' and provide an approximate P-value based on the large-sample approximation. The Minitab setup for the base-running data in Table 7.1 is as follows:

```
MTB > READ C1-C3
DATA > 1 1 5.40          DATA > 2 13 5.35
DATA > 1 2 5.85          DATA > 2 14 5.00
DATA > 1 3 5.20          DATA > 2 15 5.50
DATA > 1 4 5.55          DATA > 2 16 5.55
DATA > 1 5 5.90          DATA > 2 17 5.55
DATA > 1 6 5.45          DATA > 2 18 5.50
DATA > 1 7 5.40          DATA > 2 19 5.45
DATA > 1 8 5.45          DATA > 2 20 5.60
DATA > 1 9 5.25          DATA > 2 21 5.65
DATA > 1 10 5.85         DATA > 2 22 6.30
DATA > 1 11 5.25         DATA > 3 1 5.55
DATA > 1 12 5.65         DATA > 3 2 5.75
DATA > 1 13 5.60         DATA > 3 3 5.50
DATA > 1 14 5.05         DATA > 3 4 5.40
DATA > 1 15 5.50         DATA > 3 5 5.70
DATA > 1 16 5.45         DATA > 3 6 5.60
DATA > 1 17 5.55         DATA > 3 7 5.35
DATA > 1 18 5.45         DATA > 3 8 5.35
DATA > 1 19 5.50         DATA > 3 9 5.00
DATA > 1 20 5.65         DATA > 3 10 5.70
DATA > 1 21 5.70         DATA > 3 11 5.10
DATA > 1 22 6.30         DATA > 3 12 5.45
DATA > 2 1 5.50          DATA > 3 13 5.45
DATA > 2 2 5.70          DATA > 3 14 4.95
DATA > 2 3 5.60          DATA > 3 15 5.40
DATA > 2 4 5.50          DATA > 3 16 5.50
DATA > 2 5 5.85          DATA > 3 17 5.35
DATA > 2 6 5.55          DATA > 3 18 5.55
DATA > 2 7 5.40          DATA > 3 19 5.25
DATA > 2 8 5.50          DATA > 3 20 5.40
DATA > 2 9 5.15          DATA > 3 21 5.55
DATA > 2 10 5.80         DATA > 3 22 6.25
DATA > 2 11 5.20         DATA > END
DATA > 2 12 5.55         MTB > FRIEDMAN C3 C1 C2
```

The output is, in Minitab's notation:

$$S = 10.64 \quad \text{d.f.} = 2 \quad p = 0.005$$
$$S = 11.14 \quad \text{d.f.} = 2 \quad p = 0.004 \text{ (adjusted for ties)}$$

C1	N	Est. Median	Sum of RANKS
1	22	5.5667	53
2	22	5.5250	47
3	22	5.4333	32

Grand median = 5.5083

The approximate P-value provided by Minitab is $p = .005$ without the ties correction given in equation (7.8) and $p = .004$ with the ties correction in (7.8). Since the number of ties in these data is minimal, both of these approximate P-values are in good agreement with what we previously obtained from Chart A.2.

Comments

1. *Basic Model.* Model (7.1) is the most basic form of the two-way layout. There is just one observation per cell, and we assume that there is no interaction between the block and treatment factors.

7.1. A DISTRIBUTION-FREE TEST FOR GENERAL ALTERNATIVES 277

2. *More General Setting*. We could replace Assumptions A1–A3 and H_0 (7.2) with the more general null hypothesis that all possible $(k!)^n$ rank configurations for the r_{ij}'s are equally likely. Procedure (7.6) remains distribution-free for this more general hypothesis.

3. *Design Rationale*. The n blocks in this basic two-way layout design represent an effort to reduce experimental errors and prevent misleading comparisons of "apples and oranges." (We prefer to compare apples with apples.) Thus in Example 7.1 the 22 blocks correspond to 22 different baseball players. The treatments are to be assigned at random within each block (i.e., in each block, the order in which each player is assigned to run the three different rounding-first-base methods should be decided by a random mechanism, where each of the six possible orders has equal probability of being chosen, and the assignments in the different blocks are to be independent). Note that in the Procedure we rank only within each block. Thus in block 1, for example, the three treatment times of player 1 are compared. This is an attempt to eliminate a nuisance effect due to player 1's intrinsic speed. It would be foolish to compare round out times of player 1 with wide angle times of player 2 if player 1 is a (slow) 200-lb catcher and player 2 is a (speedy) 160-lb shortstop. In such a comparison, a difference in treatment effects would be confounded with the basic speed differences of the players, the latter being of little or no interest in this particular experiment.

4. *Motivation for the Test*. Under Assumptions A1–A3 and H_0 (7.2), each of the block rank vectors $\mathbf{R}_i^* = (r_{i1}, \ldots, r_{ik}), i = 1, \ldots, n$, has a uniform distribution over the set of all $k!$ permutations of the vector of integers $(1, 2, \ldots, k)$. It follows that

$$E_0(r_{ij}) = \frac{1}{k!}(k-1)! \sum_{t=1}^{k} t = \frac{k+1}{2},$$

the average rank being assigned separately in each of the blocks. Thus, we have

$$E_0(R_{.j}) = E_0\left(\frac{1}{n}R_j\right) = \frac{1}{n}E_0\left(\sum_{i=1}^{n} r_{ij}\right) = \frac{1}{n}\sum_{i=1}^{n} E_0(r_{ij})$$

$$= \frac{n(k+1)}{2n} = \frac{k+1}{2}, \quad \text{for } j = 1, \ldots, k,$$

and we would expect the $R_{.j}$'s to be close to $(k+1)/2$ when H_0 is true. Since the test statistic S (7.5) is a constant times a sum of squared differences between the observed treatment average ranks, $R_{.j}$, and their common null expected value, $E_0(R_{.j}) = (k+1)/2$, small values of S represent agreement with H_0 (7.2). When the τ's are not all equal, we would expect a portion of the associated treatment average ranks to differ from their common null expectation, $(k+1)/2$, with some tending to be smaller and some larger. The net result (after squaring the observed differences to obtain the $[R_{.j} - (k+1)/2]^2$ terms) would be a large value of S. This quite naturally suggests rejecting H_0 in favor of H_1 (7.3) for large values of S and motivates procedures (7.6) and (7.7). (See also Comment 5.)

5. *Connection to Normal Theory Test*. The Friedman S statistic also arises naturally if we apply the usual two-way layout $\mathcal{F}$ statistic to the ranks instead of the actual observations. Then S may be written as $S = [12/k(k+1)]$ SST, where SST is the treatment sum of squares applied to the ranks.

6. *Assumptions*. We emphasize that Assumption A3 stipulates that the nk cell distributions $F_{ij}, i = 1, \ldots, n$ and $j = 1, \ldots, k$, can differ at most in their locations (medians), and that these location differences (if any) must be a result of additive block and/or treatment effects (i.e., there is no interaction between the treatment and block factors). In particular, Assumption

A3 requires that the nk underlying distributions belong to the same general family (F) and that they do not differ in scale parameters (variability). We do note, however, that the test procedure (7.6) remains distribution-free under the less restrictive setting where Assumption A3 is replaced by the weaker condition

A3'. The distribution functions $F_{11}, \ldots, F_{1k}, \ldots, F_{n1}, \ldots, F_{nk}$ are connected through the relationship

$$F_{ij}(u) = F_i(u - \tau_j), \quad -\infty < u < \infty,$$

for $i = 1, \ldots, n$ and $j = 1, \ldots, k$, where $F_1, \ldots, F_n$ are arbitrary distribution functions for continuous distributions with unknown medians $\theta_1, \ldots, \theta_n$, respectively, and, as before, τ_j is the unknown additive treatment effect contributed by the jth treatment.

Assumption A3 then corresponds to Assumption A3' with the additional condition that $F_1 \equiv \cdots \equiv F_n$. (See also Comment 2.)

7. *Special Case of Two Treatments*. For the case of $k = 2$ treatments, the procedures in (7.6) and (7.7) are equivalent to the exact and large-sample approximation forms, respectively, of the two-sided sign test, as discussed in Section 3.4.

8. *Derivation of the Distribution of S under H_0 (No-Ties Case)*. The null distribution of S (7.5) can be obtained by using the fact that under H_0 (7.2), all possible $(k!)^n$ rank configurations for the r_{ij}'s are equally likely. We now take $k = 4, n = 2$ to illustrate how the null distribution can be derived. In this case, S (7.5) reduces to $S = (.3R^* - 30)$, where $R^* = R_1^2 + R_2^2 + R_3^2 + R_4^2$. We note that S does not vary with changes of the names of the blocks or with relabeling of the k samples. Thus, for example,

(a)

	I	II	III	IV
Block 1	1	2	3	4
Block 2	3	1	2	4

(b)

	I	II	III	IV
Block 1	3	1	2	4
Block 2	1	2	3	4

yield the same value of S, since (b) is obtained from (a) by reversing the roles of blocks 1 and 2. Similarly,

(c)

	I	II	III	IV
Block 1	1	2	3	4
Block 2	3	1	2	4

(d)

	I	II	III	IV
Block 1	2	1	3	4
Block 2	1	3	2	4

yield the same value of S, since (d) is obtained from (c) by reversing the roles of samples I and II. Instead of $(4!)^2$ rank configurations, therefore, we list only $4! = 24$ configurations (the 24 different configurations in block 2 corresponding to a fixed configuration 1, 2, 3, 4 in block 1) and their associated values of R^* and S.

(a)

I	II	III	IV
1	2	3	4
1	2	3	4

$R^* = 120, S = 6$

(b)

I	II	III	IV
1	2	3	4
1	2	4	3

$R^* = 118, S = 5.4$

(c)

I	II	III	IV
1	2	3	4
1	3	4	2

$R^* = 114, S = 4.2$

(d)

I	II	III	IV
1	2	3	4
1	3	2	4

$R^* = 118, S = 5.4$

(e)

I	II	III	IV
1	2	3	4
1	4	2	3

$R^* = 114, S = 4.2$

(f)

I	II	III	IV
1	2	3	4
1	4	3	2

$R^* = 112, S = 3.6$

7.1. A DISTRIBUTION-FREE TEST FOR GENERAL ALTERNATIVES

(g)

I	II	III	IV
1	2	3	4
2	1	3	4

$R^* = 118, S = 5.4$

(h)

I	II	III	IV
1	2	3	4
2	1	4	3

$R^* = 116, S = 4.8$

(i)

I	II	III	IV
1	2	3	4
2	3	4	1

$R^* = 108, S = 2.4$

(j)

I	II	III	IV
1	2	3	4
2	3	1	4

$R^* = 114, S = 4.2$

(k)

I	II	III	IV
1	2	3	4
2	4	1	3

$R^* = 110, S = 3$

(l)

I	II	III	IV
1	2	3	4
2	4	3	1

$R^* = 106, S = 1.8$

(m)

I	II	III	IV
1	2	3	4
3	1	2	4

$R^* = 114, S = 4.2$

(n)

I	II	III	IV
1	2	3	4
3	1	4	2

$R^* = 110, S = 3$

(o)

I	II	III	IV
1	2	3	4
3	2	4	1

$R^* = 106, S = 1.8$

(p)

I	II	III	IV
1	2	3	4
3	2	1	4

$R^* = 112, S = 3.6$

(q)

I	II	III	IV
1	2	3	4
3	4	1	2

$R^* = 104, S = 1.2$

(r)

I	II	III	IV
1	2	3	4
3	4	2	1

$R^* = 102, S = .6$

(s)

I	II	III	IV
1	2	3	4
4	1	2	3

$R^* = 108, S = 2.4$

(t)

I	II	III	IV
1	2	3	4
4	1	3	2

$R^* = 106, S = 1.8$

(u)

I	II	III	IV
1	2	3	4
4	2	1	3

$R^* = 106, S = 1.8$

(v)

I	II	III	IV
1	2	3	4
4	2	3	1

$R^* = 102, S = .6$

(w)

I	II	III	IV
1	2	3	4
4	3	1	2

$R^* = 102, S = .6$

(x)

I	II	III	IV
1	2	3	4
4	3	2	1

$R^* = 100, S = 0$

Thus we find

$$P_0\{S = 6\} = \tfrac{1}{24}, \quad P_0\{S = 5.4\} = \tfrac{3}{24}, \quad P_0\{S = 4.8\} = \tfrac{1}{24},$$
$$P_0\{S = 4.2\} = \tfrac{4}{24}, \quad P_0\{S = 3.6\} = \tfrac{2}{24}, \quad P_0\{S = 3\} = \tfrac{2}{24},$$
$$P_0\{S = 2.4\} = \tfrac{2}{24}, \quad P_0\{S = 1.84\} = \tfrac{4}{24}, \quad P_0\{S = 1.2\} = \tfrac{1}{24},$$
$$P_0\{S = .6\} = \tfrac{3}{24}, \quad P_0\{S = 0\} = \tfrac{1}{24}.$$

The probability, under H_0, that S is greater than or equal to 5.4, for example, is therefore

$$P_0\{S \geq 5.4\} = P_0\{S = 5.4\} + P_0\{S = 6\}$$
$$= \tfrac{3}{24} + \tfrac{1}{24} = \tfrac{1}{6}.$$

This agrees with the upper-tail probability entry for $k = 4$ and $n = 2$ in Table A.22.

Note that we have derived the null distribution of S without specifying the common form (F) of the underlying distribution function for the X's under H_0 beyond the point of requiring that it be continuous. This is why the test procedure (7.6) based on S is called a distribution-free procedure. From the null distribution of S we can determine the critical value s_α and control the probability α of falsely rejecting H_0 when H_0 is true, and this error probability does not depend on the specific form of the common underlying continuous X distribution.

9. *Exact Conditional Distribution of S with Ties among the X Values.* To have a test with exact significance level even in the presence of tied X's, we need to consider all $(k!)^n$ block rank configurations, where now these within-blocks ranks are obtained by using average ranks to break the ties. As in Comment 8, it still follows that under H_0 each of the $(k!)^n$ block rank configurations (now with these tied ranks) is equally likely. For each such configuration, the value of S is computed and the results are tabulated. We illustrate this construction only for the very limited case of $k = 3, n = 2$, and the tied data $X_{11} = 2.4, X_{12} = 3.0, X_{13} = 3.0$, $X_{21} = 4.0, X_{22} = 6.0$, and $X_{23} = 3.0$. Using average ranks to break within-blocks ties, the observed rank vector is $(r_{11}, r_{12}, r_{13}, r_{21}, r_{22}, r_{23}) = (1, 2.5, 2.5, 2, 3, 1)$. Thus, $R_1 = 3, R_2 = 5.5, R_3 = 3.5$, and the attained value of S is

$$S = \left[\frac{12}{2(3)(4)}\{(3)^2 + (5.5)^2 + (3.5)^2\} - 3(2)(4)\right] = 1.75.$$

To assess the significance of S, we obtain its conditional null distribution by considering the 36 equally likely (under H_0) possible rank configurations (i.e., permutation combinations) of the observed rank vector $(1, 2.5, 2.5, 2, 3, 1)$. These 36 configurations and associated values of S are as follows:

I	II	III		I	II	III	
1	2.5	2.5		1	2.5	2.5	
2	3	1	$S = 1.75$	2	3	1	$S = 1.75$
2.5	1	2.5		2.5	1	2.5	
2	3	1	$S = 0.25$	2	3	1	$S = 0.25$
2.5	2.5	1		2.5	2.5	1	
2	3	1	$S = 3.25$	2	3	1	$S = 3.25$
1	2.5	2.5		1	2.5	2.5	
2	1	3	$S = 1.75$	2	1	3	$S = 1.75$
2.5	1	2.5		2.5	1	2.5	
2	1	3	$S = 3.25$	2	1	3	$S = 3.25$
2.5	2.5	1		2.5	2.5	1	
2	1	3	$S = 0.25$	2	1	3	$S = 0.25$
1	2.5	2.5		1	2.5	2.5	
1	2	3	$S = 3.25$	1	2	3	$S = 3.25$
2.5	1	2.5		2.5	1	2.5	
1	2	3	$S = 1.75$	1	2	3	$S = 1.75$
2.5	2.5	1		2.5	2.5	1	
1	2	3	$S = 0.25$	1	2	3	$S = 0.25$
1	2.5	2.5		1	2.5	2.5	
3	2	1	$S = 0.25$	3	2	1	$S = 0.25$
2.5	1	2.5		2.5	1	2.5	
3	2	1	$S = 1.75$	3	2	1	$S = 1.75$
2.5	2.5	1		2.5	2.5	1	
3	2	1	$S = 3.25$	3	2	1	$S = 3.25$
1	2.5	2.5		1	2.5	2.5	
1	3	2	$S = 3.25$	1	3	2	$S = 3.25$
2.5	1	2.5		2.5	1	2.5	
1	3	2	$S = 0.25$	1	3	2	$S = 0.25$
2.5	2.5	1		2.5	2.5	1	
1	3	2	$S = 1.75$	1	3	2	$S = 1.75$

I	II	III			I	II	III	
1	2.5	2.5			1	2.5	2.5	
3	1	2	$S = 0.25$		3	1	2	$S = 0.25$
2.5	1	2.5			2.5	1	2.5	
3	1	2	$S = 3.25$		3	1	2	$S = 3.25$
2.5	2.5	1			2.5	2.5	1	
3	1	2	$S = 1.75$		3	1	2	$S = 1.75$

Since each of these values of S has null probability $\frac{1}{36}$, it follows that

$$P_0\{S = 0.25\} = P_0\{S = 1.75\} = P_0\{S = 3.25\} = \tfrac{1}{3}.$$

This distribution is called the conditional distribution or the permutation distribution of S, given the tied ranks $\{(1, 2.5, 2.5), (1, 2, 3)\}$. For the particular observed value $S = 1.75$, we have $P_0\{S \geq 1.75\} = \tfrac{2}{3}$.

10. *Large-Sample Approximation.* Define the random variables $T_j = R_{\cdot j} - E_0(R_{\cdot j}) = R_{\cdot j} - (k + 1)/2$, for $j = 1, \ldots, k$. Since each $R_{\cdot j} = \sum_{i=1}^{n} r_{ij}/n$ is an average, it is not surprising (see, for example, pages 388–9 of Lehmann (1975) for justification) that a properly standardized version of the vector $\mathbf{T}^* = (T_1, \ldots, T_{k-1})$ has an asymptotic (n tending to infinity) $(k-1)$-variate normal distribution with mean vector $\mathbf{0} = (0, \ldots, 0)$ and appropriate covariance matrix Σ when the null hypothesis H_0 is true. (Note that $\mathbf{T}^*$ does not include $T_k = R_{\cdot k} - (k + 1)/2$, since T_k can be expressed as a linear combination of $T_1, \ldots, T_{k-1}$. This is the reason that the asymptotic normal distribution is $(k-1)$-variate and not k-variate.) Since the test statistic S (7.5) is a quadratic form in the variables $(T_1, \ldots, T_{k-1})$, it is therefore quite natural that S has an asymptotic (n tending to infinity) chi-square distribution with $k - 1$ degrees of freedom.

11. *Competitor Based on Wilcoxon Signed Ranks.* The statistic S (7.5) utilizes the treatment observations only through comparisons within blocks. As noted in Comment 7, this provides a natural extension of the sign test procedure for paired data and it is this restriction to within blocks comparisons that leads directly to the distribution-free nature of procedure (7.6). An alternative approach would be to extend the (generally) more powerful signed rank test procedure, as discussed in Section 3.1. This approach utilizes between-block comparisons of the observations and is discussed further in Section 7.11. The associated test procedure utlizing between-blocks signed rank comparisons is (generally) more powerful than the Friedman test based on S (7.5). However, this two-way layout signed rank procedure is no longer exactly distribution-free for small numbers (n) of blocks and tests based on this approach require the use of a large-sample approximation.

12. *Consistency of the S Test.* Replace Assumptions A1–A3 by the less restrictive Assumptions A1': $X_{ij} = \beta_i + e_{ij}$, where the e's are mutually independent, and A2': $e_{1j}, \ldots, e_{nj}$ come from the same continuous population $\prod_j, j = 1, \ldots, k$, but where $\prod_1, \ldots, \prod_k$ are not assumed to be identical. Then the test defined by (7.6) is consistent against alternatives for which $\sum_{v=1}^{k}(1 - p_{uv}) \neq \sum_{v=1}^{k} p_{uv}$ for at least one $u \in \{1, \ldots, k\}$, where $p_{uv} = P(e_{iu} < e_{iv})$ with e_{iu} a random member from $\prod_u$ and e_{iv} a random member from $\prod_v$ that is independent of e_{iu}.

Properties

1. *Consistency.* See Noether (1967a, 54) and Comment 12.
2. *Asymptotic Chi-Squaredness.* See Lehmann (1975, 388–9).
3. *Efficiency.* See van Elteren and Noether (1959) and Section 7.16.

PROBLEMS

1. Goldsmith and Nadel (1969) have studied respiratory function following exposure to various levels of ozone for periods of 1 h. The subjects were four presumably healthy males employed by the California State Department of Public Health. The objective measurement used was airway resistance as evaluated by the body plethysmographic technique (see DuBois et al. (1956) and Comroe, Botelho, and DuBois (1959)). Goldsmith and Nadel reported average values for four consecutive measurements taken immediately prior to and again about 5 min after termination of each level of ozone exposure. Table 7.2 is based on a subset of the Goldsmith-Nadel data, where the tabled values are average airway resistance after ozone exposure minus average airway resistance prior to ozone exposure. Use procedure (7.6) to test H_0.

TABLE 7.2. Effect of Experimental Ozone Exposures on Airway Resistance (cm $H_2 0 \ell/s$)

Subject	After .1 ppm	After .6 ppm	After 1.0 ppm
1	−.08	.01	.06
2	.21	.17	.19
3	.50	−.11	.34
4	.14	.07	.14

Source: J. R. Goldsmith and J. A. Nadel (1969).

2. Show that the two expressions for S in (7.5) are, indeed, equivalent.
3. Could Friedman's test be applied to data from a one-way layout in which there are the same number, n, of observations from each of the k treatments? Explain. Should Friedman's test be applied to such data? Explain.
4. Show directly, or illustrate by means of an example, that the maximum value of S is $S_{max} = n(k-1)$. For what configuration is this maximum achieved?
5. Creatine phosphokinase (CPK) is a skeletal muscle isoenzyme that is often found to be elevated in the serum of acutely psychotic subjects during the initial stages of a psychotic episode. A number of variables known to affect serum CPK activity have been evaluated as possible causes of the serum CPK activity elevations observed during acute psychotic episodes. One such variable of interest is that of physical exercise, which is well known to increase serum CPK levels in normal subjects. In this regard, Goode and Meltzer (1976) studied the relationship between isometric exercises (designed to strengthen and tone muscle without lengthening and contracting the muscles themselves) and increased CPK levels in psychotic patients. In particular, they were interested in whether the elevation of CPK in the serum of psychiatric patients may be in part due to increased covert isometric motor activity. The subjects in their study were patients hospitalized on a research unit at the Illinois Psychiatric Institute. Fourteen such patients were isometrically exercised following remission of psychotic symptoms, usually 2 to 4 weeks after admission. The 60-min isometric exercise procedure involved stationary wall bars and required maximal use of all major muscle groups. The subjects described the exercises as extremely fatiguing and at or near the limits of their endurances.

    Table 7.3 contains the plasma CPK activity (mU/ℓ) levels for each of these 14 patients prior to the period of isometric exercises, as well as at 18 and 42 h after completion of such exercises. Also recorded for each patient is the peak plasma CPK activity exhibited during the period of psychosis immediately following admission to the Institute.

    Use these data to assess whether there are any differences in CPK activity between the four patient conditions considered in Table 7.3.

6. Suppose $k = 3$ and $n = 13$. Compare the critical region for the exact level $\alpha = .025$ test of H_0 (7.2) based on S with the critical region for the corresponding nominal level $\alpha = .025$ test based on the large-sample approximation. What is the exact significance level of this .025 nominal level test based on the large-sample approximation?

TABLE 7.3. Effect of Isometric Exercise on Serum Creatine Phosphokinase (CPK) Activity (mU/ℓ) in Psychotic Patients

Subject	Preexercise	19 h Postexercise	42 h Postexercise	Peak Psychotic Period
1	27	101	82	63
2	30	112	50	78
3	24	26	68	69
4	54	89	135	1,137
5	21	30	49	57
6	36	41	48	800
7	36	29	46	105
8	16	20	8	111
9	21	26	25	61
10	26	25	31	74
11	65	60	69	190
12	25	27	28	107
13	19	18	21	306
14	48	41	28	109

Source: D. J. Goode and H. Y. Meltzer (1976).

7. Suppose $k = 3$ and $n = 3$. Obtain the form of the exact null (H_0) distribution of S for the case of no tied observations.

8. Suppose $k = 4$ and $n = 8$. Compare the critical region for the exact level $\alpha = .005$ test of H_0 (7.2) based on S with the critical region for the corresponding nominal level $\alpha = .005$ test based on the large-sample approximation. What is the exact significance level of this .005 nominal level test based on the large-sample approximation?

9. Suppose $k = 3$ and $n = 2$, and we observe the data $X_{11} = 3.6, X_{12} = 3.6, X_{13} = 5.2, X_{21} = 4.3, X_{22} = 5.2$, and $X_{23} = 4.3$. What is the conditional probability distribution of S under H_0 (7.2) when average ranks are used to break ties among the X's? How extreme is the observed value of S in this conditional null distribution? Compare this fact with that obtained by taking the observed value of S to the (incorrect) unconditional null distribution of S given in Table A.22.

10. Consider the creatine phosphokinase (CPK) activity data in Table 7.3. Ignoring the patients' peak psychotic period data, assess the conjecture that isometric exercise has an effect on the CPK activity of psychotic patients.

11. Use the computer software Minitab to conduct the desired analysis (see Problem 5) of the complete creatine phosphokinase (CPK) activity data in Table 7.3.

12. Use the creatine phosphokinase (CPK) data in Table 7.3 and an appropriate nonparametric test procedure to assess whether there is any difference between peak CPK activity during the psychotic period and peak CPK activity over the combined pre/post exercise periods.

13. Nicholls and Ling (1982) conducted a study to assess the effectiveness of a system employing hand cues in the teaching of language to severely hearing-impaired children. In particular, they considered syllables presented to hearing-impaired children under the following seven conditions: (A) audition, (L) lip reading, (AL) audition and lip reading, (C) cued speech, (AC) audition and cued speech, (LC) lip reading and cued speech, and (ALC) audition, lip reading, and cued speech. The 18 subjects in the study were all severely hearing-impaired children who had been taught through the use of cued speech for at least 4 years. Syllables were presented to the subjects under each of the seven conditions (presented in random orders) and the subjects were asked in each case to identify the consonants in each syllable by writing down what they perceived them to be. The subjects' results were scored by marking properly identified consonants in the appropriate order as correct. After tallying the responses, an overall percentage correct was assigned to each participant under each experimental condition. These correct percentage data for the 18 children in the study are given in Table 7.4.

TABLE 7.4. Percentage Consonants Correctly Identified under Each of the Conditions: (A) Audition, (L) Lip Reading, (AL) Audition and Lip Reading, (C) Cued Speech, (AC) Audition and Cued Speech, (LC) Lip Reading and Cued Speech, and (ALC) Audition, Lip Reading, and Cued Speech

Subject	A	L	AL	C	AC	LC	ALC
1	1.1	36.9	52.4	42.9	31.0	83.3	63.0
2	1.1	33.3	34.5	34.5	41.7	77.3	81.0
3	13.0	28.6	40.5	33.3	44.0	81.0	76.1
4	0	23.8	22.6	33.3	33.3	69.0	65.5
5	11.9	40.5	57.1	35.7	46.4	98.8	96.4
6	0	27.4	46.4	42.9	47.4	78.6	77.4
7	5.0	20.2	22.6	35.7	37.0	69.0	73.8
8	4.0	29.8	42.9	13.0	33.3	95.2	91.7
9	0	27.4	38.0	42.9	45.2	89.3	85.7
10	1.1	26.2	31.0	31.0	32.1	70.2	71.4
11	2.4	29.8	38.0	34.5	46.4	86.9	92.9
12	0	21.4	21.4	41.7	33.3	67.9	59.5
13	0	32.1	33.3	44.0	34.5	86.9	82.1
14	0	28.6	23.8	32.1	39.3	85.7	72.6
15	1.1	28.6	29.8	41.7	35.7	81.0	78.6
16	1.1	36.9	33.3	25.0	31.0	95.2	95.2
17	0	27.4	26.1	40.5	44.0	91.7	89.3
18	0	41.7	35.7	42.9	45.2	95.2	95.2

Source: G. H. Nicholls and D. Ling (1982).

Use these data to assess whether there are any differences in the effectiveness of these seven conditions for teaching severely hearing-impaired children.

14. Consider the study with severely hearing-impaired children in Problem 13. Using the percentage correctly identified data from Table 7.4, assess whether there are any differences in the effectiveness of the three stand-alone conditions A, L, and C for teaching severely hearing-impaired children.

15. Use the computer software Minitab to conduct the desired analysis in Problem 13 of the percentage correctly identified data in Table 7.4.

16. Consider the study with severely hearing-impaired children in Problem 13. Using the percentage correctly identified data for only the first eight children (and the proper correction for ties), assess whether there are any differences in the teaching effectiveness from adding one or more of the factors L (lip reading) and C (cued speech) to the baseline A (audition) approach.

7.2. A DISTRIBUTION-FREE TEST FOR ORDERED ALTERNATIVES IN A RANDOMIZED COMPLETE BLOCK DESIGN (PAGE)

In many practical two-way layout settings where an additive model is appropriate it is also the case that the treatments are such that the appropriate alternatives to no differences in treatment effects (H_0) are those of increasing (or decreasing) treatment effects according to some natural labeling for the treatments. Examples of such settings include treatments corresponding to quality or quantity of materials, severity of disease, drug dosage levels, and intensity of a stimulus. We note that the Friedman procedure (7.6) does not utilize any such partial prior information regarding a postulated alternative ordering. The statistic S (7.5) takes on the same value for all possible $k!$ labelings of the treatments. In this section we consider a procedure for testing H_0 (7.2) against the a priori ordered alternatives.

$$H_2 : [\tau_1 \leq \tau_2 \leq \cdots \leq \tau_k, \text{with at least one strict inequality}]. \tag{7.9}$$

7.2. A DISTRIBUTION-FREE TEST FOR ORDERED ALTERNATIVES

The Page test of this section is preferred to the Friedman test in Section 7.1 when the treatments can be labeled a priori in such a way that the experimenter expects any deviation from H_0 (7.2) to be in the particular direction associated with H_2 (7.9). We emphasize, however, that the labeling of the treatments so that the ordered alternatives (7.9) are appropriate *cannot* depend on the observed sample values. This labeling must correspond completely to a factor (s) implicit in the nature of the *experimental design* and *not* the *observed data*.

Procedure

First we must label the treatments so that they are in the expected order associated with the alternative H_2 (7.9). (This labeling must be done prior to data collection.) To compute the Page statistic L we once again rank within blocks and compute the Friedman treatment sums of ranks $R_1, \ldots, R_k$ as defined in (7.4). The Page statistic L is then the weighted combination of these rank sums given by

$$L = \sum_{j=1}^{k} jR_j = R_1 + 2R_2 + \cdots + kR_k. \tag{7.10}$$

To test

$$H_0 : [\tau_1 = \cdots = \tau_k]$$

versus the ordered alternative

$$H_2 : [\tau_1 \leq \tau_2 \leq \cdots \leq \tau_k, \text{with at least one strict inequality}],$$

at the α level of significance,

$$\text{Reject } H_0 \text{ if } L \geq l_\alpha; \quad \text{otherwise do not reject,} \tag{7.11}$$

where the constant l_α is chosen to make the type I error probability equal to α. Approximate values of l_α are given in Table A.23.

Large-Sample Approximation

The large-sample approximation is based on the asymptotic (n tending to infinity) normality of L, suitably standardized. We first need to know the expected value and variance of L when the null hypothesis is true. Under H_0, the expected value and variance of L are

$$E_0(L) = \frac{nk(k+1)^2}{4} \tag{7.12}$$

and

$$\text{var}_0(L) = \frac{nk^2(k+1)(k^2-1)}{144}, \tag{7.13}$$

respectively. These expressions for $E_0(L)$ and $\text{var}_0(L)$ are verified by direct calculations in Comment 18 for the special case of $k = 3$ and $n = 2$. General derivations of both expressions are outlined in Comment 20.

The standardized version of L is

$$L^* = \frac{L - E_0(L)}{\sqrt{\text{var}_0(L)}} = \frac{L - \left[\frac{nk(k+1)^2}{4}\right]}{\left\{\frac{nk^2(k+1)(k^2-1)}{144}\right\}^{1/2}}. \tag{7.14}$$

When H_0 is true, L^* has, as n tends to infinity, an asymptotic $N(0, 1)$ distribution. (See Comment 20 for indications of the proof.) The normal theory approximation for procedure (7.11) is

$$\text{Reject } H_0 \text{ if } L^* \geq z_\alpha; \quad \text{otherwise do not reject.} \tag{7.15}$$

Ties

If there are ties among the k X's within any of the n blocks, assign each of the observations in a tied group the average of the integer ranks that are associated with the tied group and compute L with these average ranks.

We note that even procedure (7.11) using these average ranks to break ties and the critical values of Table A.23 is only approximately, and not exactly, of significance level α in the presence of tied X observations within any of the blocks. To get an exact level α test in this tied setting, see Comment 19. (See also Comment 21 regarding the use of the large-sample approximation in the case of within-blocks ties.)

Example 7.2: *Breaking Strength of Cotton Fibers.* An experiment reported in Cochran and Cox (1957, 108) considered the effect, in terms of breaking strength of cotton fibers, of the level of potash (K_2O) in the soil. Five levels of potash were applied ($k = 5$) in a randomized block pattern with three blocks ($n = 3$). The criterion used for the analysis was the Pressley strength index, obtained by measuring the breaking strength of a bundle of fibers of a given cross-sectional area. A single sample of cotton was taken from each plot, and four determinations were made on each sample. The main entries of Table 7.5 are the means of the four determinations and the parenthetical values are the within-block ranks. (No dimensions are associated with the data of Table 7.5, since the machine that measures the strength index is calibrated in arbitrary units.)

We are interested here in using procedure (7.11) to test the hypothesis of equivalent strengths versus the ordered alternative that specifies a trend of decreasing breaking strength with increasing levels of potash. For purpose of illustration we take the significance level to be $\alpha = .01$. From Table A.23, with $k = 5$ and $n = 3$, we see that $l_{.01} = 155$ and procedure (7.11) becomes

$$\text{Reject } H_0 \text{ if } L \geq 155.$$

TABLE 7.5. Strength Index of Cotton

Replications	Potash (lb/acre)				
	144	108	72	54	36
1	7.46 (2)	7.17 (1)	7.76 (4)	8.14 (5)	7.63 (3)
2	7.68 (2)	7.57 (1)	7.73 (3)	8.15 (5)	8.00 (4)
3	7.21 (1)	7.80 (3)	7.74 (2)	7.87 (4)	7.93 (5)
	$R_1 = 5$	$R_2 = 5$	$R_3 = 9$	$R_4 = 14$	$R_5 = 12$

Source: W. G. Cochran and G. M. Cox (1957).

Now, we illustrate the computations leading to the sample value of L (7.10). Using the treatment sums of within-block ranks given in Table 7.3, we see from equation (7.10) that

$$L = R_1 + 2R_2 + 3R_3 + 4R_4 + 5R_5$$
$$= 5 + 2(5) + 3(9) + 4(14) + 5(12) = 158.$$

Since the value of L is greater than the critical value 155, we can reject H_0 at the $\alpha = .01$ level, providing strong evidence (for the levels of potash considered) in favor of the trend of decreasing breaking strength with increasing level of potash.

For the large-sample approximation we need to compute the standardized form of L^* using equation (7.14). Since $k = 5$, $n = 3$, and the sample value of L is 158, we see from (7.14) that

$$L^* = \frac{158 - \left[\frac{3(5)(5+1)^2}{4}\right]}{\left\{\frac{3(5^2)(5+1)(5^2-1)}{144}\right\}^{1/2}} = \frac{158 - 135}{\sqrt{75}} = 2.66.$$

Thus, using the approximate procedure (7.15) with the value of $L^* = 2.66$, we see from Table A.1 that the approximate P-value for these data is .0039. This is in good agreement with our previous outcome using the exact test, even though n is only three.

Comments

13. *More General Setting.* As with the Friedman procedure in Section 7.1, we could replace Assumptions A1–A3 and H_0 (7.2) with the more general null hypothesis that all possible $(k!)^n$ rank configurations for the r_{ij}'s are equally likely. Procedure (7.11) remains distribution-free for this more general hypothesis.

14. *Motivation for the Test.* If the ordering $\tau_1 < \tau_2 < \cdots < \tau_k$ is true, then R_v will tend to be larger than R_u for $u < v$. Note that L (7.10) weights R_v by the integer v and R_u by the integer u. Thus L tends to be large when H_2 (7.9) is true, serving as partial motivation for the L test in (7.11).

15. *Assumptions.* As with the Friedman procedure in Section 7.1, we emphasize that Assumption A3 stipulates that the nk cell distributions F_{ij}, $i = 1, \ldots, n$ and $j = 1, \ldots, k$, can differ at most in their locations (medians) and that these location differences (if any) must be a result of additive block and/or treatment effects (i.e., there is no interaction between the treatment and block factors). In particular, Assumption A3 requires that the nk underlying distributions belong to the same general family (F) and that they do not differ in scale parameters (variability). We do note, however, that the test procedure (7.11) remains distribution-free under the less restrictive setting where Assumption A3 is replaced by the weaker condition A3' stated in Comment 6. (See also Comment 13.)

16. *Special Case of Two Treatments.* For the case of $k = 2$ treatments, the procedures in (7.11) and (7.15) are equivalent to the exact and large-sample approximation forms, respectively, of the one-sided sign test, as discussed in Section 3.4.

17. *Derivation of the Distribution of L under H_0 (No-Ties Case).* The null distribution of L (7.10) can be obtained by using the fact that under H_0 (7.2), all $(k!)^n$ possible rank configurations are equally likely. As is the case for S (7.5) (see Comment 8), L does not vary with changes of the names of the blocks; however, unlike S, since it is directed toward a particular ordered alternative, its values (in general) change with changes of names of the treatments. Thus building up the null distribution of L is more tedious than in the case of S. We illustrate this construction for the very special case of $k = 3$ and $n = 2$. In this case we

need to consider $(3!)^2 = 36$ block-treatment rank configurations. We list these configurations and their associated values of $L = R_1 + 2R_2 + 3R_3$.

(a)

	I	II	III
	1	2	3
	1	2	3

$L = 28$

(b)

	I	II	III
	1	2	3
	1	3	2

$L = 27$

(c)

	I	II	III
	1	2	3
	2	1	3

$L = 27$

(d)

	I	II	III
	1	2	3
	2	3	1

$L = 25$

(e)

	I	II	III
	1	2	3
	3	1	2

$L = 25$

(f)

	I	II	III
	1	2	3
	3	2	1

$L = 24$

(g)

	I	II	III
	1	3	2
	1	2	3

$L = 27$

(h)

	I	II	III
	1	3	2
	1	3	2

$L = 26$

(i)

	I	II	III
	1	3	2
	2	1	3

$L = 26$

(j)

	I	II	III
	1	3	2
	2	3	1

$L = 24$

(k)

	I	II	III
	1	3	2
	3	1	2

$L = 24$

(l)

	I	II	III
	1	3	2
	3	2	1

$L = 23$

(m)

	I	II	III
	2	1	3
	1	2	3

$L = 27$

(n)

	I	II	III
	2	1	3
	1	3	2

$L = 26$

(o)

	I	II	II
	2	1	3
	2	1	3

$L = 26$

(p)

	I	II	III
	2	1	3
	2	3	1

$L = 24$

(q)

	I	II	III
	2	1	3
	3	1	2

$L = 24$

(r)

	I	II	III
	2	1	3
	3	2	1

$L = 23$

(s)

	I	II	III
	2	3	1
	1	2	3

$L = 25$

(t)

	I	II	III
	2	3	1
	1	3	2

$L = 24$

(u)

	I	II	III
	2	3	1
	2	1	3

$L = 24$

(v)

	I	II	III
	2	3	1
	2	3	1

$L = 22$

(w)

	I	II	III
	2	3	1
	3	1	2

$L = 22$

(x)

	I	II	III
	2	3	1
	3	2	1

$L = 21$

(y)

	I	II	III
	3	1	2
	1	2	3

$L = 25$

(z)

	I	II	III
	3	1	2
	1	3	2

$L = 24$

(aa)

	I	II	III
	3	1	2
	2	1	3

$L = 24$

(bb)

	I	II	III
	3	1	2
	2	3	1

$L = 22$

(cc)

	I	II	III
	3	1	2
	3	1	2

$L = 22$

(dd)

	I	II	III
	3	1	2
	3	2	1

$L = 21$

(ee)

	I	II	III
	3	2	1
	1	2	3

$L = 24$

(ff)

	I	II	III
	3	2	1
	1	3	2

$L = 23$

(gg)

	I	II	III
	3	2	1
	2	1	3

$L = 23$

(hh)

	I	II	III
	3	2	1
	2	3	1

$L = 21$

(ii)

	I	II	III
	3	2	1
	3	1	2

$L = 21$

(jj)

	I	II	III
	3	2	1
	3	2	1

$L = 20$

Thus, we find

$$P_0\{L = 28\} = \tfrac{1}{36}, \quad P_0\{L = 27\} = \tfrac{4}{36}, \quad P_0\{L = 26\} = \tfrac{4}{36},$$
$$P_0\{L = 25\} = \tfrac{4}{36}, \quad P_0\{L = 24\} = \tfrac{10}{36}, \quad P_0\{L = 23\} = \tfrac{4}{36},$$
$$P_0\{L = 22\} = \tfrac{4}{36}, \quad P_0\{L = 21\} = \tfrac{4}{36}, \quad P_0\{L = 20\} = \tfrac{1}{36}.$$

The probability, under H_0, that L is greater than or equal to 27, for example, is therefore

$$P_0\{L \geq 27\} = P_0\{L = 27\} + P_0\{L = 28\} = \tfrac{4}{36} + \tfrac{1}{36} = \tfrac{5}{36} = .139 > .05.$$

Similarly, $P_0\{L \geq 28\} = P_0\{L = 28\} = \tfrac{1}{36} = .028$, which is less than .05 but greater than .01. This agrees with the upper-tail probability entry for $k = 3$ and $n = 2$ in Table A.23.

Since the null distribution for L has been derived without specifying the common form (F) of the underlying distribution function for the X's under H_0 beyond the point of requiring that it be continuous, the test procedure (7.11) based on L is a distribution-free procedure. From the null distribution of L we can determine the critical value l_α and control the probability α of falsely rejecting H_0 when H_0 is true, and this error probability does not depend on the specific form of the common underlying X distribution.

18. *Calculation of the Mean and Variance of L under the Null Hypothesis H_0.* In displays (7.12) and (7.13) we presented formulas for the mean and variance of L when the null hypothesis is true. In this comment we illustrate a direct calculation of $E_0(L)$ and $\text{var}_0(L)$ in the particular case of $k = 3, n = 2$ and no tied observations, using the null distribution of L obtained in Comment 17. (Later, in Comment 20, we present arguments for the general derivations of $E_0(L)$ and $\text{var}_0(L)$.) The expected value of the null distribution of L is obtained directly from multiplication of each possible value of L by its probability under H_0. Thus, using the null probability values from Comment 17, we obtain

$$E_0(L) = \tfrac{1}{36}(20 + 28) + \tfrac{4}{36}(21 + 22 + 23 + 25 + 26 + 27) + \tfrac{10}{36}(24) = 24.$$

This is in agreement with what we obtain using equation (7.12), namely,

$$E_0(L) = \frac{2(3)(3 + 1)^2}{4} = 24.$$

A direct check on the expression for $\text{var}_0(L)$ is also easy. Again using the null probabilities from Comment 17, we have

$$\begin{aligned}
\text{var}_0(L) &= E_0[\{L - E_0(L)\}^2] = E_0[\{L - 24\}^2] \\
&= \Big\{ \tfrac{1}{36}[(20 - 24)^2 + (28 - 24)^2] + \tfrac{4}{36}[(21 - 24)^2 \\
&\quad + (22 - 24)^2 + (23 - 24)^2 + (25 - 24)^2 \\
&\quad + (26 - 24)^2 + (27 - 24)^2] + \tfrac{10}{36}[(24 - 24)^2] \Big\} \\
&= \tfrac{1}{36}(16 + 16) + \tfrac{4}{36}(9 + 4 + 1 + 1 + 4 + 9) + \tfrac{10}{36}(0) = 4,
\end{aligned}$$

which agrees with what we obtain using equation (7.13) directly, namely,

$$\text{var}_0(L) = \frac{2(3)^2(3 + 1)(3^2 - 1)}{144} = 4.$$

19. *Exact Conditional Distribution of L under H_0 with Ties within the Blocks.* To have a test with exact significance level even in the presence of tied X's within some of the blocks, we need to consider all $(k!)^n$ possible rank configurations, where now the within-blocks ranks are

obtained by using average ranks to break the ties. As in Comment 17, it still follows that under H_0 each of these $(k!)^n$ configurations is equally likely. For each such configuration, the value of L is computed and the results are tabulated. As an example, consider the case of $k = 3, n = 2$, and the data $X_{11} = 2.4, X_{12} = 3.6, X_{13} = 2.4, X_{21} = 4.0, X_{22} = 5.9$, and $X_{23} = 1.7$. Using average ranks to break the tie in the first block, the observed block rank vectors are $(r_{11}, r_{12}, r_{13}) = (1.5, 3, 1.5)$ and $(r_{21}, r_{22}, r_{23}) = (2, 3, 1)$. Thus, $R_1 = 3.5, R_2 = 6, R_3 = 2.5$, and the attained value of L is $3.5 + 2(6) + 3(2.5) = 23$. To assess the significance of this value of L, we would need to obtain the entire conditional null distribution of L by computing its value for each of the $(3!)^2 = 36$ equally likely (under H_0) possible configurations of the observed block rank vectors $(1.5, 3, 1.5)$ and $(2, 3, 1)$. This would be accomplished in exactly the same manner as is illustrated for the no-ties case in Comment 17.

20. *Large-Sample Approximation*. We can rewrite the expression for L (7.10) to obtain

$$L = \sum_{j=1}^{k} jR_j = \sum_{j=1}^{k} j\left(\sum_{i=1}^{n} r_{ij}\right) = \sum_{i=1}^{n} \left(\sum_{j=1}^{k} jr_{ij}\right) = \sum_{i=1}^{n} Q_i,$$

with $Q_i = \sum_{j=1}^{k} jr_{ij}, i = 1, \ldots, n$. Moreover, from Assumptions A1 and A3, $Q_1, \ldots, Q_n$ are mutually independent and identically distributed random variables, regardless of whether or not the null hypothesis H_0 is true. The asymptotic normality, as n tends to infinity, of the standardized form

$$L^* = \frac{L - E(L)}{[\mathrm{var}(L)]^{1/2}} = \frac{L - nE[Q_1]}{[n\,\mathrm{var}(Q_1)]^{1/2}} \tag{7.16}$$

then follows at once from standard central limit theory for sums of mutually independent, identically distributed random variables (cf. Randles and Wolfe (1979, 421)).

The computation of $E(Q_1)$ and $\mathrm{var}(Q_1)$ is simplified by noting that

$$E(Q_1) = E\left(\sum_{j=1}^{k} jr_{1j}\right) = \sum_{j=1}^{k} jE(r_{1j}) \tag{7.17}$$

and

$$\mathrm{var}(Q_1) = \mathrm{var}\left(\sum_{j=1}^{k} jr_{1j}\right)$$

$$= \sum_{j=1}^{k} \mathrm{var}(jr_{1j}) + 2\sum_{u=1}^{v-1}\sum_{v=2}^{k} \mathrm{cov}(ur_{1u}, vr_{1v})$$

$$= \sum_{j=1}^{k} j^2 \mathrm{var}(r_{1j}) + 2\sum_{u=1}^{v-1}\sum_{v=2}^{k} uv\,\mathrm{cov}(r_{1u}, r_{1v}). \tag{7.18}$$

In particular, when H_0 is true, $(r_{11}, \ldots, r_{1k})$ is an exchangeable random vector. Thus, under H_0, we have

$$\mathrm{var}_0(r_{1j}) = \mathrm{var}_0(r_{11}), \quad \text{for } j = 2, \ldots, k$$

and

$$\mathrm{cov}_0(r_{1u}, r_{1v}) = \mathrm{cov}_0(r_{11}, r_{12}), \quad \text{for } 1 \le u < v \le k.$$

7.2. A DISTRIBUTION-FREE TEST FOR ORDERED ALTERNATIVES

Using these facts in (7.15) and (7.16), we obtain

$$E_0(Q_1) = E_0(r_{11}) \sum_{j=1}^{k} j = \frac{k(k+1)}{2} E_0(r_{11}) \tag{7.19}$$

and

$$\text{var}_0(Q_1) = \text{var}_0(r_{11}) \sum_{j=1}^{k} j^2 + 2\,\text{cov}_0(r_{11}, r_{12}) \sum_{u=1}^{v-1} \sum_{v=2}^{k} uv$$

$$= \frac{k(k+1)(2k+1)}{6} \text{var}_0(r_{11})$$

$$+ \text{cov}_0(r_{11}, r_{12}) \left[\sum_{u=1}^{k} \sum_{v=1}^{k} uv - \sum_{t=1}^{k} t^2 \right]$$

$$= \frac{k(k+1)(2k+1)}{6} [\text{var}_0(r_{11}) - \text{cov}_0(r_{11}, r_{12})]$$

$$+ \left[\frac{k(k+1)}{2} \right]^2 \text{cov}_0(r_{11}, r_{12}). \tag{7.20}$$

It can be shown (see Problem 24) that

$$E_0(r_{11}) = \frac{k+1}{2},\ \text{var}_0(r_{11}) = \frac{k^2-1}{12} \tag{7.21}$$

and

$$\text{cov}_0(r_{11}, r_{12}) = -\frac{(k+1)}{12}. \tag{7.22}$$

Using these results in expressions (7.19) and (7.20), we obtain

$$E_0(Q_1) = \frac{k(k+1)}{2} \frac{(k+1)}{2} = \frac{k(k+1)^2}{4} \tag{7.23}$$

and

$$\text{var}_0(Q_1) = \frac{k(k+1)(2k+1)}{6} \left[\frac{k^2-1}{12} + \frac{k+1}{12} \right]$$

$$+ \left[\frac{k(k+1)}{2} \right]^2 \left(-\frac{k+1}{12} \right),$$

which, after some straightforward algebra, yields

$$\text{var}_0(Q_1) = \frac{k^2(k+1)(k^2-1)}{144}. \tag{7.24}$$

Combining equations (7.17), (7.18), (7.23), and (7.24), we obtain

$$E_0(L) = nE_0(Q_1) = \frac{nk(k+1)^2}{4}$$

and

$$\text{var}_0(L) = \frac{nk^2(k+1)(k^2-1)}{144},$$

as stated in expressions (7.12) and (7.13), respectively. In conjunction wth equation (7.16), this provides the justification for the approximate α level procedure in (7.15).

21. *Conservative Nature of the Large-Sample Approximation when There Are Ties within Blocks.* In applications where tied X values are observed in one or more of the blocks and average ranks are used to deal with these ties, the null variance of L based on its exact conditional null distribution (see Comment 19) is always smaller than the value obtained from expression (7.13). (This fact is illustrated in Problems 22 and 23.) As a result, the approximate level α procedure in (7.15) is conservative in the presence of within-blocks ties in the following sense: If we reject H_0 using procedure (7.15) with $\text{var}_0(L)$ obtained from expression (7.13), then we would also reject H_0 if we were to more properly use the exact conditional null variance of L in computing the value of L^* (7.14).

22. *Relation to Rank Order Correlation.* The L test is directly related to Spearman's rank order correlation coefficient r_s (8.63). Let r_i denote Spearman's correlation coefficient computed between the observed order and the postulated order in block i, and set $\bar{r} = (\sum_{i=1}^{n} r_i/n)$. Then it can be shown that

$$\bar{r} = \left\{ \frac{12L}{nk(k^2-1)} - \frac{3(k+1)}{(k-1)} \right\}.$$

23. *Consistency of the L Test.* Replace Assumptions A1–A3 by the less restrictive Assumptions $\text{A1}' : X_{ij} = \beta_i + e_{ij}$, where the e's are mutually independent, and $\text{A}_2' : e_{1j}, \ldots, e_{nj}$ come from the same continuous population $\prod_j$, $j = 1, \ldots, k$, but where $\prod_1, \ldots, \prod_k$ are not assumed to be identical. Then the test defined by (7.11) is consistent against alternatives for which $\{\sum_{u<v}(v-u)p_{uv} > k(k-1)(k+1)/12\}$, where $p_{uv} = P(e_{iu} < e_{iv})$ with e_{iu} a random member from $\prod_u$ and e_{iv} a random member from $\prod_v$ that is independent of e_{iu} (see Hollander (1967a)). For those situations covered by Assumptions A1–A3, this consistency statement implies the consistency statement given in Property 1.

Properties

1. *Consistency.* The test defined by (7.11) is consistent against the H_2 (7.9) alternatives. See Hollander (1967a) and Comment 23.
2. *Asymptotic Normality.* See Comment 20 and Randles and Wolfe (1979, 421).
3. *Efficiency.* See Hollander (1967a) and Section 7.16.

PROBLEMS

17. Brady (1969) described an experiment concerning the influence of the rhythmicity of a metronome on the speech of stutterers. The subjects were 12 severe stutterers. Each subject spoke extemporaneously for 3 min under the three conditions $N, A,$ and R.

    N: Subject spoke unaided by a metronome.

    R: Subject spoke with a regular (rhythmic) metronome set at 120 ticks per minute and was instructed to pace one syllable of speech to each tick.

    A: Subject spoke with an arrhythmic metronome in which the intervals between ticks ranged randomly between 0.3 and 0.7 s but with an average of 120 ticks per minute. Again the subject was instructed to pace one syllable of speech to each tick.

TABLE 7.6. Influence of Rhythmicity of Metronome on Speech Fluency

Subject	Dysfluencies under Each Condition		
	R	A	N
1	3	5	15
2	3	3	11
3	1	3	18
4	5	4	21
5	2	2	6
6	0	2	17
7	0	2	10
8	0	3	8
9	0	2	13
10	1	0	4
11	2	4	11
12	2	1	17

Source: J. P. Brady (1969).

Table 7.6 gives the number of dysfluencies under each condition. On the basis of the conditions, and prior to looking at the data, we might expect a deviation from H_0 to be in the direction $\tau_R < \tau_A < \tau_N$. Perform Page's test using this postulated ordering.

18. Verify the relationship (see Comment 22) between L (7.10) and r_s (8.63).

19. Show directly, or illustrate by means of an example, that the maximum value of L is $L_{max} = nk(k+1)(2k+1)/6$. For what rank configuration is the maximum achieved?

20. Show directly, or illustrate by means of an example, that the minimum value of L is $L_{min} = nk(k+1)(k+2)/6$. For what rank configuration is this minimum achieved?

21. Shelterbelts (long rows of tree plantings across the direction of the prevailing winds) have been used extensively for some time in developed countries to protect crops and livestock from the effects of the wind. Ujah and Adeoye (1984) conducted a study to see if such shelterbelts could be used effectively to ameliorate the severe losses from droughts experienced almost annually in the arid and semiarid zones of Nigeria and considered to be a leading factor in the declining food production in Nigeria and many of its neighbors.

Ujah and Adeoye investigated the effect of shelterbelts on a variety of factors related to drought conditions, including wind velocity, air and soil temperatures, and soil moisture. The experiment was conducted at two locations about 3.5 km apart, near Dambatta. Table 7.7 presents the wind

TABLE 7.7. Percent Reduction in Average Wind Speed at Dambatta, 1980/81

Month	Leeward Distance from Shelterbelt (m)				
	20	40	100	150	200
January	22.1	20.7	15.4	12.3	6.9
February	19.2	18.7	14.9	9.3	6.5
March	21.5	21.9	14.3	9.9	7.1
April	21.5	21.2	11.1	9.4	6.2
May	21.3	20.9	11.2	9.4	7.7
June	20.9	19.6	16.9	11.6	7.0
August	19.3	18.7	14.4	12.5	7.0
September	20.1	19.6	15.6	12.6	7.5
October	23.7	20.4	14.6	12.4	8.5

Source: C. E. Ujah and K. B. Adeoye (1984).

velocity data (averaged over these two locations) at various distances leeward of the shelterbelt expressed as percent wind speed reduction relative to the wind velocity on the windward side of the shelterbelt. The data are monthly (except for July, November, and December, for which the data were not available) and at leeward distances of 20, 40, 100, 150, and 200 m from the shelterbelt.

Use these data to test the hypothesis of a negative relationship between percent reduction in average wind speed and the leeward distance from a shelterbelt.

22. Consider the case of $k = 3, n = 2$, and the tied data set $X_{11} = 2.4, X_{12} = 3.6, X_{13} = 2.4, X_{21} = 4.0, X_{22} = 5.9$, and $X_{23} = 1.7$. What is the conditional probability distribution of L under H_0 (7.2) when average ranks are used to break within-blocks ties among the X's? (See Comment 19.) How extreme is the observed value of $L = 23$ in this conditional null distribution?

23. Consider the tied data set in Problem 22 for the setting of $k = 3$ and $n = 2$. Use the conditional null probability distribution of L obtained in Problem 22 to compute the conditional null variance of L and compare this value with that of the unconditional null variance given by (7.13). Interpret these two numbers in view of the discussion in Comment 21.

24. Let $\mathbf{r}_1 = (r_{11}, \ldots, r_{1k})$ be a random vector of ranks that is uniformly distributed over the set of all $k!$ permutations of $(1, \ldots, k)$. Show that $E(r_{11}) = (k+1)/2$, $\text{var}(r_{11}) = (k^2 - 1)/12$, and $\text{cov}(r_{11}, r_{12}) = -(k+1)/12$.

25. Carry out the algebra to verify the final expression for $\text{var}_0(Q)$ in equation (7.24).

26. Suppose $k = 3$ and $n = 3$. Obtain the form of the exact null (H_0) distribution of L for the case of no tied observations.

27. In their study of shelterbelts (see Problem 21), Ujah and Adeoye (1984) also obtained measurements of the monthly maximum soil temperature (°C) at a 5-cm depth at leeward distances of 20, 40, 100, and 200 m from the shelterbelt. These data are presented in Table 7.8.

TABLE 7.8. Maximum Soil Temperature (°C) at 5-cm Depth at Dambatta, 1980/81

Month	Leeward Distance from Shelterbelt (m)			
	20	40	100	200
January	37.7	37.5	37.6	37.4
February	39.7	39.4	39.6	39.6
March	42.0	42.0	41.9	41.9
April	43.4	43.1	42.8	43.0
May	42.5	42.3	42.3	42.1
June	39.7	39.7	39.6	39.7
July	38.7	38.5	38.6	38.5
August	39.1	38.8	38.9	38.4
September	39.7	39.5	39.2	39.4
October	39.9	40.0	40.0	40.2
November	39.6	39.7	39.8	39.7

Source: J. E. Ujah and K. B. Adeoye (1984).

Use these data to test the hypothesis that there is a negative relationship between maximum soil temperature at a 5-cm depth and the leeward distance from a shelterbelt.

28. For the case of $k = 2$, show that procedure (7.11) is equivalent to the exact one-sided sign test, as discussed in Section 3.4.

29. Consider the data on percentage consonants correctly identified in Table 7.4 from the study on hearing-impaired children by Nicholls and Ling (1982). From previous studies, there is reason to believe that cued speech (C) is more effective as a stand-alone method for teaching language to hearing-impaired children than is lip reading (L), which, in turn, is thought to be more effective than audition (A) by itself. Find an approximate P-value using procedure (7.15) to test this conjecture.

RATIONALE FOR MULTIPLE COMPARISON PROCEDURES

In Sections 7.1 and 7.2 we have discussed procedures designed to test the null hypothesis H_0 (7.2) against either general or ordered alternatives. Upon rejection of H_0 with one of these test procedures for a given set of data, our conclusion is either that there are some unspecified differences among the treatment effects (associated with the Friedman procedure discussed in Section 7.1) or that the treatment effects follow an ordered pattern (associated with the Page procedure of Section 7.2). However, in neither of these test procedures is our conclusion pair-specific; that is, the tests in Sections 7.1 and 7.2 are not designed to enable us to reach conclusions about specific pairs of treatment effects. The relative sizes of the specific treatment effects τ_1 and τ_2, for example, cannot be inferred from the conclusions reached by either of the test procedures of Sections 7.1 or 7.2. To elicit such pairwise specific information we turn to the class of multiple comparison procedures. In Section 7.3 we present a two-sided all-treatments multiple comparison procedure for the omnibus setting corresponding to the general alternatives H_1 (7.3). In Section 7.4 we deal with treatments versus control multiple comparison decisions for settings where one of the treatments plays a special role as the study control.

7.3. DISTRIBUTION-FREE TWO-SIDED ALL-TREATMENTS MULTIPLE COMPARISONS BASED ON FRIEDMAN RANK SUMS—GENERAL CONFIGURATION (WILCOXON, NEMENYI, MCDONALD-THOMPSON)

In this section we present a multiple comparison procedure based on Friedman's within-blocks ranks that is designed to make decisions about individual differences between pairs of treatment effects (τ_i, τ_j), for $i < j$, in a setting where general alternatives H_1 (7.3) are of interest. Thus, the multiple comparison procedure of this section would generally be applied to two-way layout data (with one observation per cell) *after* rejection of H_0 (7.2) with the Friedman procedure from Section 7.1. In this setting it is important to reach conclusions about all $\binom{k}{2} = k(k-1)/2$ pairs of treatment effects and these conclusions are naturally two-sided in nature.

Procedure

Let $R_1, \ldots, R_k$ be the treatment sums of within-blocks ranks given by (7.4). Calculate the $k(k-1)/2$ absolute differences $|R_u - R_v|, 1 \leq u < v \leq k$. At an experimentwise error rate of α, the Wilcoxon-Nemenyi-McDonald-Thompson two-sided all-treatments multiple comparison procedure reaches its $k(k-1)/2$ pairwise decisions, corresponding to each (τ_u, τ_v) pair, $1 \leq u < v \leq k$, by the criterion

$$\text{Decide } \tau_u \neq \tau_v \text{ if } |R_u - R_v| \geq r_\alpha; \quad \text{otherwise decide } \tau_u = \tau_v, \qquad (7.25)$$

where the constant r_α is chosen to make the experimentwise error rate equal to α; that is, r_α satisfies the restriction

$$P_0(|R_u - R_v| < r_\alpha, u = 1, \ldots, k-1; v = u+1, \ldots, k) = 1 - \alpha, \qquad (7.26)$$

where the probability $P_0(.)$ is computed under H_0 (7.2). Equation (7.26) stipulates that the $k(k-1)/2$ inequalities $|R_u - R_v| < r_\alpha$, corresponding to all pairs (u, v) of treatments with $u < v$, hold simultaneously with probability $1 - \alpha$ when H_0 (7.2) is true. Selected approximate values r_α can be found in Table A.24 for all combinations of $k = 3(1)15$ and $n = 3(1)15$.

Large-Sample Approximation

When H_0 is true, the $[k(k-1)/2]$-component vector $(R_1, \ldots, R_k)$ has, as n tends to infinity, an asymptotic $(k-1)$-variate normal distribution with appropriate mean vector and covariance matrix (see Comment 29 for indications of the proof). It then follows that the critical value r_α can, when the number of blocks n is large, be approximated by $[nk(k+1)/12]^{1/2} q_\alpha$, where q_α is the upper αth percentile point for the distribution of the range of k independent $N(0,1)$ variables. Thus the large-sample approximation for procedure (7.25) is

$$\text{Decide } \tau_u \neq \tau_v \text{ if } |R_u - R_v| \geq q_\alpha \left[\frac{nk(k+1)}{12}\right]^{1/2} ; \quad \text{otherwise decide } \tau_u = \tau_v. \quad (7.27)$$

Values of q_α for $\alpha = .0001, .0005, .001, .005, .01, .025, .05, .10, .20$ and $k = 2(1)20(2)40(10)100$ are given in Table A.17.

Ties

If there are ties among the X observations within any of the blocks, use average ranks to break the ties and compute the individual treatment sums of ranks $R_1, \ldots, R_k$. In such cases, the experimentwise error rate associated with procedure (7.25) is only approximately equal to α.

Example 7.3: *Rounding First Base.* Consider the rounding first base data discussed in Example 7.1. There we had found (using the large-sample approximation for the Friedman procedure) that there is strong evidence to conclude that the three methods of running to first base are not equivalent with respect to time to reach second base. To determine which of the three running methods differ in median times to second base, we apply the approximate procedure (7.27), using average ranks to break the within-runners ties in computing R_1, R_2, and R_3. Here we have $k = 3$ and $n = 22$. For sake of illustration we take our approximate experimentwise error rate to be $\alpha = .01$. From Table A.17 with $k = 3$ we find $q_{.01} = 4.12$, and procedure (7.27) reduces to

$$\text{Decide } \tau_u \neq \tau_v \text{ if } |R_u - R_v| \geq (4.12)\left[\frac{22(3)(4)}{12}\right]^{1/2} = 19.3.$$

Using the treatments sums of within-runners ranks given in Table 7.1, we find that

$$|R_2 - R_1| = 6, \quad |R_3 - R_1| = 21, \quad \text{and} \quad |R_3 - R_2| = 15.$$

Referring these absolute value rank sum differences to the approximate critical value 19.3, we see that

$$|R_2 - R_1| = 6 < 19.3 \quad \Rightarrow \quad \text{decide } \tau_2 = \tau_1,$$
$$|R_3 - R_1| = 21 \geq 19.3 \quad \Rightarrow \quad \text{decide } \tau_3 \neq \tau_1,$$

and

$$|R_3 - R_2| = 15 < 19.3 \quad \Rightarrow \quad \text{decide } \tau_3 = \tau_2.$$

Thus at an approximate experimentwise error rate of .01, we have reached the conclusion that only the round out (treatment 1) and wide angle (treatment 3) running methods yield significantly different median times to second base. (We note that the smallest approximate experimentwise error rate available in Table A.17 at which we would reach this conclusion is .005, since $\sqrt{22}\,q_{.005} = \sqrt{22}\,(4.424) = 20.750 < 21 < \sqrt{22}\,q_{.001} = \sqrt{22}\,(5.063) = 23.748$.)

For sake of illustration for the exact procedure in (7.25), we consider the subset of the sample data associated with the first 15 baseball players in Table 7.1. For that subset we have $k = 3, n = 15$, and the three treatment sums of ranks are $R_1^* = 37, R_2^* = 31$, and $R_3^* = 22$. From Table A.24 with $k = 3, n = 15$, and experimentwise error rate $\alpha = .047$, we find $r_{.047} = 13$ and procedure (7.25) becomes

$$\text{Decide } \tau_u \neq \tau_v \text{ if } |R_u - R_v| \geq 13.$$

Since $|R_2^* - R_1^*| = 6, |R_3^* - R_1^*| = 15$, and $|R_3^* - R_2^*| = 9$, we see that our decisions for this subset of data using procedure (7.25) would be $\tau_2 = \tau_1, \tau_3 \neq \tau_1$, and $\tau_3 = \tau_2$, in agreement with what we found using the entire set of 22 baseball players and the approximate procedure (7.27). (Note, however, that for this smaller set of data we could no longer conclude that $\tau_3 \neq \tau_1$ at an experimentwise error rate as low as .01.)

Comments

24. *Rationale for Multiple Comparison Procedures.* We think of the methods of this section as multiple comparison procedures. The aim of applying such procedures goes beyond the point of deciding whether the treatments are equivalent to the (often more important) problem of selecting which, if any, treatments differ from one another. Thus the user makes $k(k-1)/2$ decisions, one for each pair of treatments. Equation (7.26) states that the probability of making all correct decisions when H_0 is true is controlled to be $1 - \alpha$. That is, when using procedure (7.25), the probability of at least one incorrect decision, when H_0 is true, is controlled to be α. This error rate is derived under the assumption that H_0 is true, but it does not depend on the particular underlying distributional form F. This is why we call (7.25) a distribution-free multiple comparison procedure.

The multiple comparison procedures of this section can also be interpreted as hypothesis tests. If we consider the procedure that rejects H_0 if and only if the inequality of (7.25) [or of (7.27)] holds for at least one (u, v) pair, $1 \leq u < v \leq k$, this is a distribution-free test of size α for H_0 (7.2).

25. *Experimentwise Error Rate.* The use of an experimentwise error rate represents a very conservative approach to multiple comparisons. We are insisting that the probability of making only correct decisions be $1 - \alpha$ when the null hypothesis H_0 (7.2) of treatment equivalence is true. Thus, although we have a high degree of protection when H_0 is true, we often apply such techniques when we have evidence (perhaps based on a priori information or perhaps obtained by applying the Friedman test, as in Example 7.3) that H_0 is not true. This protection under H_0 also makes it harder for the procedure to judge treatments as differing significantly when in fact H_0 is false, and this difficulty becomes more severe as k increases. See Comment 6.58 for additional discussion of experimentwise error rates.

26. *Critical Values r_α.* The r_α values in Table A.24 can be obtained by using the fact that under H_0 (7.2), all $(k!)^n$ rank configurations are equally likely. Thus to obtain the probability under H_0 that $|R_u - R_v| < c$ simultaneously for $u = 1, \ldots, k-1$ and $v = u+1, \ldots, k$, we can count the number of configurations for which the event $B = \{|R_u - R_v| < c, u = 1, \ldots, k-1; v = u+1, \ldots, k\}$ occurs and divide this number by $(k!)^n$. For an illustration, consider the 24 configurations of Comment 8, corresponding to the case $k = 4, n = 2$. (As in Comment 8, the same reasoning enables us to consider only 24 rather than $(4!)^2 = 576$ configurations.) For each configuration, we now display the values of $|R_1 - R_2|, |R_1 - R_3|, |R_1 - R_4|, |R_2 - R_3|, |R_2 - R_4|$, and $|R_3 - R_4|$.

(a) $|R_1 - R_2| = 2$
$|R_1 - R_3| = 4$
$|R_1 - R_4| = 6$
$|R_2 - R_3| = 2$
$|R_2 - R_4| = 4$
$|R_3 - R_4| = 2$

(b) $|R_1 - R_2| = 2$
$|R_1 - R_3| = 5$
$|R_1 - R_4| = 5$
$|R_2 - R_3| = 3$
$|R_2 - R_4| = 3$
$|R_3 - R_4| = 0$

(c) $|R_1 - R_2| = 3$
$|R_1 - R_3| = 5$
$|R_1 - R_4| = 4$
$|R_2 - R_3| = 2$
$|R_2 - R_4| = 1$
$|R_3 - R_4| = 1$

(d) $|R_1 - R_2| = 3$
$|R_1 - R_3| = 3$
$|R_1 - R_4| = 6$
$|R_2 - R_3| = 0$
$|R_2 - R_4| = 3$
$|R_3 - R_4| = 3$

(e) $|R_1 - R_2| = 4$
$|R_1 - R_3| = 3$
$|R_1 - R_4| = 5$
$|R_2 - R_3| = 1$
$|R_2 - R_4| = 1$
$|R_3 - R_4| = 2$

(f) $|R_1 - R_2| = 4$
$|R_1 - R_3| = 4$
$|R_1 - R_4| = 4$
$|R_2 - R_3| = 0$
$|R_2 - R_4| = 0$
$|R_3 - R_4| = 0$

(g) $|R_1 - R_2| = 0$
$|R_1 - R_3| = 3$
$|R_1 - R_4| = 5$
$|R_2 - R_3| = 3$
$|R_2 - R_4| = 5$
$|R_3 - R_4| = 2$

(h) $|R_1 - R_2| = 0$
$|R_1 - R_3| = 4$
$|R_1 - R_4| = 4$
$|R_2 - R_3| = 4$
$|R_2 - R_4| = 4$
$|R_3 - R_4| = 0$

(i) $|R_1 - R_2| = 2$
$|R_1 - R_3| = 4$
$|R_1 - R_4| = 2$
$|R_2 - R_3| = 2$
$|R_2 - R_4| = 0$
$|R_3 - R_4| = 2$

(j) $|R_1 - R_2| = 2$
$|R_1 - R_3| = 1$
$|R_1 - R_4| = 5$
$|R_2 - R_3| = 1$
$|R_2 - R_4| = 3$
$|R_3 - R_4| = 4$

(k) $|R_1 - R_2| = 3$
$|R_1 - R_3| = 1$
$|R_1 - R_4| = 4$
$|R_2 - R_3| = 2$
$|R_2 - R_4| = 1$
$|R_3 - R_4| = 3$

(l) $|R_1 - R_2| = 3$
$|R_1 - R_3| = 3$
$|R_1 - R_4| = 2$
$|R_2 - R_3| = 0$
$|R_2 - R_4| = 1$
$|R_3 - R_4| = 1$

(m) $|R_1 - R_2| = 1$
$|R_1 - R_3| = 1$
$|R_1 - R_4| = 4$
$|R_2 - R_3| = 2$
$|R_2 - R_4| = 5$
$|R_3 - R_4| = 3$

(n) $|R_1 - R_2| = 1$
$|R_1 - R_3| = 3$
$|R_1 - R_4| = 2$
$|R_2 - R_3| = 4$
$|R_2 - R_4| = 3$
$|R_3 - R_4| = 1$

(o) $|R_1 - R_2| = 0$
$|R_1 - R_3| = 3$
$|R_1 - R_4| = 1$
$|R_2 - R_3| = 3$
$|R_2 - R_4| = 1$
$|R_3 - R_4| = 2$

(p) $|R_1 - R_2| = 0$
$|R_1 - R_3| = 0$
$|R_1 - R_4| = 4$
$|R_2 - R_3| = 0$
$|R_2 - R_4| = 4$
$|R_3 - R_4| = 4$

(q) $|R_1 - R_2| = 2$
$|R_1 - R_3| = 0$
$|R_1 - R_4| = 2$
$|R_2 - R_3| = 2$
$|R_2 - R_4| = 0$
$|R_3 - R_4| = 2$

(r) $|R_1 - R_2| = 2$
$|R_1 - R_3| = 1$
$|R_1 - R_4| = 1$
$|R_2 - R_3| = 1$
$|R_2 - R_4| = 1$
$|R_3 - R_4| = 0$

(s) $|R_1 - R_2| = 2$
$|R_1 - R_3| = 0$
$|R_1 - R_4| = 2$
$|R_2 - R_3| = 2$
$|R_2 - R_4| = 4$
$|R_3 - R_4| = 2$

(t) $|R_1 - R_2| = 2$
$|R_1 - R_3| = 1$
$|R_1 - R_4| = 1$
$|R_2 - R_3| = 3$
$|R_2 - R_4| = 3$
$|R_3 - R_4| = 0$

(u) $|R_1 - R_2| = 1$
$|R_1 - R_3| = 1$
$|R_1 - R_4| = 2$
$|R_2 - R_3| = 0$
$|R_2 - R_4| = 3$
$|R_3 - R_4| = 3$

(v) $|R_1 - R_2| = 1$
$|R_1 - R_3| = 1$
$|R_1 - R_4| = 0$
$|R_2 - R_3| = 2$
$|R_2 - R_4| = 1$
$|R_3 - R_4| = 1$

(w) $|R_1 - R_2| = 0$
$|R_1 - R_3| = 1$
$|R_1 - R_4| = 1$
$|R_2 - R_3| = 1$
$|R_2 - R_4| = 1$
$|R_3 - R_4| = 2$

(x) $|R_1 - R_2| = 0$
$|R_1 - R_3| = 0$
$|R_1 - R_4| = 0$
$|R_2 - R_3| = 0$
$|R_2 - R_4| = 0$
$|R_3 - R_4| = 0$

Thus, for example,

$$P_0\{|R_u - R_v| < 6, u = 1, 2, 3; v = 2, 3, 4\}$$
$$= P_0\{|R_1 - R_2| < 6; |R_1 - R_3| < 6; |R_1 - R_4| < 6;$$
$$|R_2 - R_3| < 6; |R_2 - R_4| < 6; |R_3 - R_4| < 6\}$$
$$= \tfrac{22}{24} = 1 - .083,$$

since for 22 of the configurations—all but configurations (a) and (d)—the event $\{|R_1 - R_2| < 6; |R_1 - R_3| < 6; |R_1 - R_4| < 6; |R_2 - R_3| < 6; |R_2 - R_4| < 6; |R_3 - R_4| < 6\}$ occurs. This .083 probability agrees with the appropriate entry in Table A.24.

27. *Relationship to Range of Rank Sums.* Define the range of $R_1, \ldots, R_k$ as

$$\text{Range } [R_1, \ldots, R_k] = \max[R_1, \ldots, R_k] - \min[R_1, \ldots, R_k].$$

Then $|R_u - R_v|$ is less than c, for all $u < v$, if and only if range $[R_1, \ldots, R_k]$ is less than c. Thus in Comment 26, instead of computing the values of $|R_1 - R_2|, |R_1 - R_3|, |R_1 - R_4|$, $|R_2 - R_3|, |R_2 - R_4|, |R_3 - R_4|$ for each rank configuration, we need to have calculated only range $[R_1, \ldots, R_k]$ for each configuration. That is, we can obtain the critical constants r_α by computing only the range for each possible rank configuration. We do, however, need to compute the individual absolute differences $|R_u - R_v|$ in order to apply procedure (7.25) to a set of data.

28. *Historical Development.* The basic idea behind the multiple comparison procedures (7.25) and (7.27) based on Friedman rank sums is attributed by McDonald and Thompson (1967) to Wilcoxon who "...in 1956, in an unpublished notebook, carried out the first correct probability computation for 3 objects (treatments) and 3 judges (blocks)...." Nemenyi (1963) obtained a small number of exact critical values r_α for procedure (7.25) in his Ph.D. dissertation. McDonald and Thompson (1967) provided the additional critical values that appear in Table A.24.

29. *Large-Sample Approximation.* Let $\boldsymbol{R} = (R_1, \ldots, R_k)$ be the vector of Friedman rank sums. Then it can be shown that a properly standardized $\boldsymbol{R}$ has, as n tends to infinity, an asymptotic multivariate normal distribution with appropriate mean vector $\boldsymbol{\mu}$ and covariance matrix $\boldsymbol{\Sigma}$ (see Miller (1966) for details). It follows directly from this result (again, see Miller (1966)) that the procedure in (7.27) has an asymptotic experimentwise error rate equal to α.

30. *Dependence on Observations from Other Noninvolved Treatments.* The absolute difference $|R_u - R_v|$ depends on the values of the observations from the other $k - 2$ treatments, in addition to the observations from treatments u and v. Thus the multiple comparison procedures (7.25) and (7.27) both have the disadvantage that the decision concerning treatment u and treatment v can be affected by changes only in the observations from one or more of the other $k - 2$ treatments that are not directly involved. This difficulty has been emphasized by Miller (1966) and Gabriel (1969).

31. *Approximately Distribution-Free Multiple Comparison Procedures Based on Signed Ranks.* The multiple comparison procedures in (7.25) and (7.27) both utilize within-blocks Friedman ranking schemes, and, as a result, they are related to the sign procedures for paired replicates data (see Comments 7 and 16). Competitor all-treatments multiple comparison procedures based on signed ranks that utilize between-block information are discussed in Section 7.13. These signed rank procedures are, however, only asymptotically distribution-free.

Properties

1. *Asymptotic Multivariate Normality.* See Miller (1966).
2. *Efficiency.* See Section 7.16.

PROBLEMS

30. Livesey (1967) has compared the performance of rats, rabbits, and cats on the Hebb-Williams (1946) elevated pathway test (EPT). Table 7.9, based on a subset of the Livesey data, gives mean error scores by species for 12 problems. Using procedure (7.25), find the species (if any) that differ significantly.

TABLE 7.9. Error Scores by Species

Problem Number	Rats	Rabbits	Cats
1	1.5	1.7	0.3
2	1.1	1.5	1.0
3	1.8	8.1	3.6
4	1.9	1.3	0.0
5	4.3	4.0	0.6
6	2.0	4.6	5.5
7	8.4	4.0	1.0
8	6.6	5.1	3.1
9	2.4	2.5	0.1
10	6.5	6.9	1.6
11	2.6	2.5	4.3
12	6.5	6.8	1.0

Source: P. J. Livesey (1967).

31. For the case of $k = 3, n = 9$, and $\alpha = .05$, compare procedures (7.25) and (7.27).
32. Consider the rounding first base data discussed in Examples 7.1 and 7.3. Using the data for all 22 players, find the smallest (available) approximate experimentwise error rate at which we would declare that the median time to second base for the narrow angle method of running is different from that for the wide angle method of running.
33. Apply procedure (7.25) to the ozone exposure data of Table 7.2.
34. Apply the approximate procedure (7.27) to the percentage of correctly identified consonants data of Table 7.4.
35. Find the totality of all available experimentwise error rates α and the associated critical values r_α for procedure (7.25) when $k = 3$ and $n = 3$.
36. Illustrate the difficulty discussed in Comment 30 by means of a numerical example.
37. Consider the serum creatine phosphokinase data of Table 7.3 in Problem 5. Find the smallest (available) experimentwise error rate at which the most significant difference in treatment effects between the time measurements would be detected.
38. Consider the serum creatine phosphokinase data of Table 7.3 in Problem 5. Find the smallest (available) experimentwise error rate at which we would declare that the typical serum creatine phosphokinase activity at 19 h postexercise is different from that at 42 h postexercise.

7.4. DISTRIBUTION-FREE ONE-SIDED TREATMENTS VERSUS CONTROL MULTIPLE COMPARISONS BASED ON FRIEDMAN RANK SUMS (NEMENYI, WILCOXON-WILCOX, MILLER)

In this section we turn our attention to a multiple comparison procedure designed to make decisions about individual differences between the median effect for a single, baseline control population and the median effects of each of the remaining $k - 1$ treatments. This treatment versus control multiple comparison procedure can be applied *after* rejection of H_0 (7.2) with either the Friedman or Page test discussed in Sections 7.1 and 7.2, respectively. Its application leads to conclusions about the differences between each of the $k - 1$ treatment effects and the control effect and these conclusions are naturally one-sided in nature.

Procedure

For simplicity of notation, we let treatment 1 assume the role of the single baseline control. Let $R_1, \ldots, R_k$ be the treatment sums of within-blocks ranks given by (7.4). Calculate the

$k-1$ differences $(R_u - R_1), u = 2, \ldots, k$. At an experimentwise error rate of α, the Nemenyi-Wilcoxon-Wilcox-Miller one-sided treatments versus control multiple comparison procedure reaches its $k-1$ pairwise decisions, corresponding to each (τ_1, τ_u) pair, for $u = 2, \ldots, k$, by the criterion

$$\text{Decide } \tau_u > \tau_1 \text{ if } (R_u - R_1) \geq r_\alpha^*; \text{ otherwise decide } \tau_u = \tau_1, \qquad (7.28)$$

where the constant r_α^* is chosen to make the experimentwise error rate equal to α; that is, r_α^* satisfies the restriction

$$P_0((R_u - R_1) < r_\alpha^*, u = 2, \ldots, k) = 1 - \alpha, \qquad (7.29)$$

where the probability $P_0(.)$ is computed under H_0 (7.2). Equation (7.29) stipulates that the $k-1$ inequalities $(R_u - R_1) < r_\alpha^*$, corresponding to each treatment paired with the control, hold simultaneously with probability $1 - \alpha$ when H_0 (7.2) is true. Selected values of r_α^* can be found in Table A.25 for $k = 3, n = 2(1)18$; $k = 4(1)5, n = 2(1)8$; and $k = 6, n = 2(1)6$. (For discussion of how to adjust procedure (7.28) for settings where it is of interest to decide whether the treatment effects are *smaller* than the control effect, see Comment 34.)

Large-Sample Approximation

When H_0 is true, the $(k-1)$-component vector $(R_2 - R_1, \ldots, R_k - R_1)$ has, as n tends to infinity, an asymptotic $(k-1)$-variate normal distribution with mean vector **0** (see Comment 37 for an indication of the proof). It then follows that the critical value r_α^* can, when the number of blocks is large, be approximated by $[nk(k+1)/6]^{1/2} m_{\alpha,1/2}^*$, where $m_{\alpha,1/2}^*$ is the upper αth percentile point for the distribution of the maximum of $(k-1)$ $N(0, 1)$ random variables with common correlation $\rho = \frac{1}{2}$. Thus the large-sample approximation for procedure (7.28) is

$$\text{Decide } \tau_u > \tau_1 \text{ if } (R_u - R_1) \geq [nk(k+1)/6]^{1/2} m_{\alpha,1/2}^*;$$
$$\text{otherwise decide } \tau_u = \tau_1. \qquad (7.30)$$

Selected values of $m_{\alpha,1/2}^*$ can be found in Table A.21 with $\rho = \frac{1}{2}$ for $k = 3(1)13$.

Ties

If there are ties among the X observations within any of the blocks, use average ranks to break the ties and compute the individual treatment sums of ranks $R_1, \ldots, R_k$. In such cases, the experimentwise error rate associated with procedure (7.28) is only approximately equal to α.

Example 7.4: *Stuttering Adaptation.* Daly and Cooper (1967) considered the rate of stuttering adaptation under three conditions. Eighteen subjects (college-age stutterers) read each of three different passages five consecutive times. In one condition electroshock was administered during each moment of stuttering and in another condition electroshock was administered immediately following each stuttered word. The remaining condition was a control with no electroshock administered. The percentage of stuttering behavior during each reading was recorded, and Table 7.10 presents for each subject a rate of adaptation score under each condition. The score was found by using the residual measurement method suggested by Tate, Cullinan, and Ahlstrand (1961).

To determine if either of the two treatments yield improved (larger) median adaptation scores, we apply procedure (7.28), using average ranks to break the within-subjects ties in

TABLE 7.10. Adaptation Scores for College-Age Stutterers

Subject	Treatment		
	1 (No Shock)	2 (Shock Following)	3 (Shock During)
1	57 (3)	38 (1)	51 (2)
2	59 (3)	48 (1)	56 (2)
3	44 (1.5)	50 (3)	44 (1.5)
4	51 (2)	53 (3)	44 (1)
5	43 (1)	53 (3)	50 (2)
6	49 (1)	56 (3)	54 (2)
7	48 (2)	37 (1)	50 (3)
8	56 (2)	58 (3)	40 (1)
9	44 (1.5)	44 (1.5)	50 (3)
10	50 (2)	50 (2)	50 (2)
11	44 (1)	58 (3)	56 (2)
12	50 (3)	48 (2)	46 (1)
13	70 (2)	60 (1)	74 (3)
14	42 (1)	58 (3)	57 (2)
15	58 (1)	60 (2)	74 (3)
16	54 (3)	38 (1)	48 (2)
17	38 (1)	48 (2.5)	48 (2.5)
18	48 (2)	56 (3)	44 (1)
	$R_1 = 33$	$R_2 = 39$	$R_3 = 36$

Source: D. A. Daly and E. B. Cooper (1967).

computing R_1, R_2, and R_3. Here we have $k = 3$ and $n = 18$. For sake of illustration, we take our experimentwise error rate to be $\alpha = .0492$. From Table A.25 with $k = 3$ and $n = 18$, we find $r^*_{.0492} = 12$, and procedure (7.28) reduces to

$$\text{Decide } \tau_u > \tau_1 \text{ if } (R_u - R_1) \geq 12.$$

Using the treatments sums of within-subjects ranks given in Table 7.10, we find that

$$(R_2 - R_1) = 6 \quad \text{and} \quad (R_3 - R_1) = 3.$$

Referring these rank sum differences to the critical value 12, we see that

$$(R_2 - R_1) = 6 < 12 \quad \Rightarrow \quad \text{decide } \tau_2 = \tau_1,$$
$$(R_3 - R_1) = 3 < 12 \quad \Rightarrow \quad \text{decide } \tau_3 = \tau_1.$$

Thus, at an experimentwise error rate of .0492, we find no statistical evidence that either of the two electroshock treatments lead to an increase in median adaptation scores over the control setting. (In fact, we note that the smallest experimentwise error rate at which we would be able to declare a statistically significant increase in median adaptation scores for either of the two treatments is .2859, since the largest observed difference in rank sums is $(R_2 - R_1) = 6$ and, from Table A.25, $r^*_{.2859} = 6$.)

For sake of illustration in the use of Table A.21 with $\rho = \frac{1}{2}$ for the large-sample approximation (7.30), we simply note that $m^*_{.02002,1/2} = 2.30$ for $k = 3$ and $m^*_{.05410,1/2} = 2.20$ for $k = 6$.

Comments

32. *Rationale for Treatments versus Control Multiple Comparison Procedures.* The general rationale for the multiple comparison procedures of this section is the same as that given in Comment 24 for the two-sided all-treatments multiple comparison procedures of Section 7.3. The only additional factor here is that the treatments-versus-control procedures of this section do not compare all treatments, but only each noncontrol treatment with the control on a directional basis. This situation arises, for example, in drug screening in the examination of many new treatments in hopes of improving on a standard, and there is no initial reason to perform between treatment comparisons. Of course, similar comparisons between treatments that were selected as being better than the control would most likely be carried out later in a follow-up study.

The multiple comparison procedures of this section, which involve making $k - 1$ decisions, can also be interpreted as hypothesis tests. If we consider the procedure that rejects H_0 if and only if the inequality in (7.28) [or in (7.30)] holds for at least one $(1, u)$ pair, $u = 2, \ldots, k$, then this is a distribution-free test of level α for H_0 (7.2).

33. *Experimentwise Error Rate.* The use of an experimentwise error rate represents a very conservative approach to multiple comparisons. We are insisting that the probability of making only correct decisions be $1 - \alpha$ when the hypothesis H_0 (7.2) of treatment equivalence is true. Thus we have a high degree of protection when H_0 is true, but we often apply the techniques of this section when we have evidence (perhaps based on a priori information or perhaps obtained by applying a previous test procedure) that H_0 is not true. (For additional general remarks about experimentwise error rates, see Comment 6.58.)

34. *Opposite Direction Decisions.* Procedures (7.28) and (7.30) are designed for the one-sided case where the decisions are $\tau_u > \tau_1$ versus $\tau_u = \tau_1, u = 2, \ldots, k$. To handle the analogous one-sided situation, where the decisions involve $\tau_u < \tau_1$ versus $\tau_u = \tau_1, u = 2, \ldots, k$, use (7.28) and (7.30) with $(R_u - R_1)$ replaced by $(R_1 - R_u)$ for $u = 2, \ldots, k$.

35. *Critical Values r_α^*.* The r_α^* values of Table A.25 can be obtained by using the fact that under H_0 (7.2), all $(k!)^n$ rank configurations are equally likely. However, the computational effort is greater in this treatments-versus-control problem than in the all-treatments problem, since the values $R_u - R_1, u = 2, \ldots, k$, are in general changed when we relabel the control treatment. (In the all-treatments case, the relevant statistic range $[R_1, \ldots, R_k]$ is unaffected by treatment relabelings. See Comment 27.)

Let us now check a value in Table A.25 to illustrate the nature of the necessary computations. For simplicity, we take the case $n = 3, k = 3$. Here the largest possible value of $R_3 - R_1$ is 6, corresponding to the configuration

	I	II	III
(a)	1	2	3
	1	2	3
	1	2	3,

where $R_1 = 3, R_3 = 9$. Similarly, the largest possible value of $R_2 - R_1$ is 6, corresponding to the configuration

	I	II	III
(b)	1	3	2
	1	3	2
	1	3	2.

Since none of the other configurations can yield an $R_u - R_1$ difference as large as 6, we have $P_0\{(R_2 - R_1) \geq 6 \text{ or } (R_3 - R_1) \geq 6\} = 2/[(3!)^3] = 2/216 = .0093$. Thus in the notation of (7.28) we have $r^*_{.0093} = 6$, in agreement with the value given in Table A.25 for this setting.

36. *Historical Development.* The basic idea behind the treatments-versus-control multiple comparison procedures (7.28) and (7.30) based on Friedman rank sums was initially discussed by Nemenyi (1963), Wilcoxon and Wilcox (1964) and Miller (1966). Windham (1971) provided the exact critical values r^*_α for procedure (7.28) for the case of $k = 3, n = 2(1)18$ and for $k = 4, n = 2(1)5$. Additional values of r^*_α were obtained by Odeh (1977c) for the settings $k = 2(1)5, n = 2(1)8$ and $k = 6, n = 2(1)6$.

37. *Large-Sample Approximation.* Let $\boldsymbol{R}_d = (R_2 - R_1, \ldots, R_k - R_1)$ be the vector of differences between the treatment rank sums and the control rank sum. Then it can be shown that a properly standardized $\boldsymbol{R}_d$ has, as n tends to infinity, an asymptotic multivariate normal distribution with mean vector $\boldsymbol{0}$ and appropriate covariance matrix Σ (see Miller (1966) for details). It follows directly from this result (again, see Miller (1966)) that the procedure in (7.30) has an asymptotic experimentwise error rate equal to α.

38. *Dependence on Observations from Other Noninvolved Treatments.* The treatments-versus-control multiple comparison procedures of this section suffer from the same disadvantage mentioned in Comment 30 for the corresponding all-treatments multiple comparisons. The decision between treatment $u (> 1)$ and the control can be affected by changes only in the observations from one or more of the other $k - 2$ treatments that are not directly involved.

39. *Two-Sided Treatments-versus-Control Multiple Comparison Procedures.* The multiple comparison procedures of this section are both one-sided by nature, resulting in decisions between $\tau_u = \tau_1$ and $\tau_u > \tau_1$ for every $u = 2, \ldots, k$ (or between $\tau_u = \tau_1$ and $\tau_u < \tau_1$ for every $u = 2, \ldots, k$, as noted in Comment 34). We view such one-sided comparisons to be the most natural approach for treatments-versus-control settings. In such situations we are generally interested in seeing which, if any, of the proposed new treatments are better than a standard control or placebo. In most practical applications, *better* is synonymous with one-sided comparisons (all in one direction or all in the other), and thus our emphasis on such procedures in this section. However, a two-sided treatments-versus-control analogue to procedure (7.28) has been developed in the literature and corresponds to the criterion

$$\text{Decide } \tau_u \neq \tau_1 \text{ if } |R_u - R_1| \geq r^{**}_\alpha; \quad \text{otherwise decide } \tau_u = \tau_1, \qquad (7.31)$$

where the constant r^{**}_α is chosen to make the experimentwise error rate equal to α; that is,

$$P_0\{|R_u - R_1| < r^{**}_\alpha, u = 2, \ldots, k\} = 1 - \alpha,$$

where the probability $P_0(.)$ is computed under H_0 (7.2). Windham (1971) provided values of r^{**}_α for procedure (7.31) for the settings $k = 3, n = 2(1)18$ and $k = 4, n = 2(1)5$ (see also Hollander and Wolfe (1973)). A large-sample approximation to (7.31) is discussed in Miller (1966).

40. *Approximately Distribution-Free Multiple Comparison Procedures Based on Signed Ranks.* The multiple comparison procedures (7.28) and (7.30) both utilize within-blocks Friedman ranking schemes, and, as a result, they are related to the sign procedures for paired replicates data (see Comments 7 and 16). Competitor treatments-versus-control multiple comparison procedures based on signed ranks that utilize between-block information are discussed in Section 7.14. These signed rank procedures are, however, only asymptotically distribution-free.

Properties

1. *Asymptotic Multivariate Normality.* See Miller (1966).
2. *Efficiency.* See Section 7.16.

PROBLEMS

39. Consider the serum creatine phosphokinase activity data from Problem 5. Treating preexercise as a control and ignoring the peak psychotic period data, apply procedure (7.28) to decide if there is statistical evidence of increased serum creatine phosphokinase activity either 19 or 42 h after exercise.
40. Apply an appropriate one-sided multiple comparison procedure (see (7.28) and Comment 34) to the rhythmicity data of Table 7.6, letting the condition N (subject spoke unaided by a metronome) serve as the control.
41. For the case $k = 3, n = 18$, and $\alpha \approx .01$, compare procedures (7.28) and (7.30).
42. Consider the rounding first base data discussed in Examples 7.1 and 7.3. Using the data for all 22 players and treating the round out method of running to second base as the control, find the smallest approximate experimentwise error rate at which we would declare that the median time to second base for the wide angle method of running is smaller than that for the round out method.
43. Illustrate the difficulty discussed in Comment 38 by means of a numerical example.
44. Verify the available experimentwise error rates α and the associated r_α^* critical values in Table A.25 for procedure (7.28) when $k = 3$ and $n = 2$.
45. Consider the subset of the data on percentage correctly identified consonants in Table 7.4 corresponding to conditions A, AL, and AC. Treating condition A as a control, find the smallest experimentwise error rate for procedure (7.28) at which we would detect the condition (L or C) yielding the most improvement in performance when added to A in the syllable presentation.
46. Treating condition A as a control, apply procedure (7.30) to the data on percentage correctly identified consonants in Table 7.4.
47. Consider the maximum soil temperature data in Table 7.8. Apply the appropriate treatments versus control procedure to decide if maximum soil temperature is significantly warmer at 20, 40, or 100 m from the shelterbelt than at a distance of 200 m.

7.5. CONTRAST ESTIMATION BASED ON ONE-SAMPLE MEDIAN ESTIMATORS (DOKSUM)

In this section we discuss a method for point estimation of certain linear combinations of treatment effects known in the literature as *contrasts*. We define such a contrast in the treatment effects $\tau_1, \ldots, \tau_k$ to be any linear combination of the form

$$\theta = \sum_{i=1}^{k} a_i \tau_i, \tag{7.32}$$

where $a_1, \ldots, a_k$ are any specified constants such that $\sum_{i=1}^{k} a_i = 0$. Equivalently, we can write θ in terms of the individual differences in treatment effects (known in the literature as *simple contrasts*)

$$\Delta_{hj} = \tau_h - \tau_j, \quad h = 1, \ldots, k; \quad j = 1, \ldots, k, \tag{7.33}$$

by noting that

$$\theta = \sum_{h=1}^{k}\sum_{j=1}^{k} d_{hj}\Delta_{hj},\qquad(7.34)$$

where

$$d_{hj} = \frac{a_h}{k},\quad h = 1,\ldots,k;\quad j = 1,\ldots,k.\qquad(7.35)$$

For a given setting, decisions about which contrasts to estimate can be related either to a priori interest in particular linear combinations of the τ's or the results of one of the multiple comparison procedures discussed in Sections 7.3 and 7.4.

Procedure

For each pair of treatments $(u,v), u \neq v = 1,\ldots,k$, compute the differences

$$D_{uv}^{i} = X_{iu} - X_{iv}, i = 1,\ldots,n,\qquad(7.36)$$

between the treatment u and treatment v observations for each of the n blocks. For $1 \leq u \neq v \leq k$, let

$$Z_{uv} = \text{median } \{D_{uv}^{i}, i = 1,\ldots,n\}.\qquad(7.37)$$

Since $Z_{vu} = -Z_{uv}$, we need only calculate the $k(k-1)/2$ values Z_{uv} corresponding to $u < v$. We refer to Z_{uv} as the "unadjusted" estimator of the simple contrast $\Delta_{uv} = \tau_u - \tau_v$. (Note that Z_{uv} is just the median estimator of Section 3.5, applied here to the $X_{iu} - X_{iv}$ differences. For example, Z_{23} is the median of the $X_{i2} - X_{i3}$ differences, $i = 1,\ldots,n$, and is the "unadjusted" estimator of the simple contrast $\tau_2 - \tau_3$.) Next, we compute

$$Z_{u.} = \sum_{j=1}^{k} \frac{Z_{uj}}{k},\quad u = 1,\ldots,k,\qquad(7.38)$$

where we note that $Z_{uu} = 0$ for $u = 1,\ldots,k$. Setting

$$\tilde{\Delta}_{uv} = Z_{u.} - Z_{v.},\qquad(7.39)$$

the adjusted estimator of θ is given by

$$\tilde{\theta} = \sum_{j=1}^{k} a_j Z_{j.},\qquad(7.40)$$

or, equivalently,

$$\tilde{\theta} = \sum_{h=1}^{k}\sum_{j=1}^{k} d_{hj}\tilde{\Delta}_{hj}.\qquad(7.41)$$

7.5. CONTRAST ESTIMATION BASED ON ONE-SAMPLE MEDIAN ESTIMATORS

TABLE 7.11. Values of D_{uv} Differences for Data of Table 7.1

Player i	D^i_{12}	D^i_{13}	D^i_{23}
1	−.10	−.15	−.05
2	.15	.10	−.05
3	−.40	−.30	.10
4	.05	.15	.10
5	.05	.20	.15
6	−.10	−.15	−.05
7	.00	.05	.05
8	−.05	.10	.15
9	.10	.25	.15
10	.05	.15	.10
11	.05	.15	.10
12	.10	.20	.10
13	.25	.15	−.10
14	.05	.10	.05
15	.00	.10	.10
16	−.10	−.05	.05
17	.00	.20	.20
18	−.05	−.10	−.05
19	.05	.25	.20
20	.05	.25	.20
21	.05	.15	.10
22	.00	.05	.05

Example 7.5: *Rounding First Base*. Consider the rounding first base data originally presented in Table 7.1 of Example 7.1. We illustrate the Doksum contrast estimator $\tilde{\theta}$ (7.40) on the simple constrast $\theta = \tau_{\text{roundout}} - \tau_{\text{wide angle}} = \tau_1 - \tau_3$. In Example 7.3 we found the round out and wide angle methods differed significantly at the .01 experimentwise error rate. An estimate of $\tau_1 - \tau_3$ provides us with an idea of the time saved by running wide angle as opposed to round out.

From Table 7.11 and (7.37), we obtain $Z_{12} = .05, Z_{13} = .125$, and $Z_{23} = .10$. From (7.38) we have

$$Z_{1.} = \frac{Z_{11} + Z_{12} + Z_{13}}{3}$$
$$= \frac{0 + .05 + .125}{3} = .058,$$

$$Z_{2.} = \frac{Z_{21} + Z_{22} + Z_{23}}{3} = \frac{-.05 + 0 + .10}{3} = .017,$$

$$Z_{3.} = \frac{Z_{31} + Z_{32} + Z_{33}}{3} = \frac{-.125 - .10 + 0}{3} = -.075.$$

Note that in calculating $Z_{2.}$ and $Z_{3.}$ we have used the fact that $Z_{uv} = -Z_{vu}$.

The adjusted estimator of $\theta = \tau_1 - \tau_3$ is now obtained by using (7.32) with

$$a_1 = 1, \quad a_2 = 0, \quad a_3 = -1,$$

so that from (7.40) we have

$$\tilde{\theta} = Z_{1.} - Z_{3.} = .058 - (-.075) = .133.$$

Parenthetically, it should be noted that the quivalent form (7.34) is obtained with the identifications

$$d_{11} = d_{12} = d_{13} = \tfrac{1}{3},$$
$$d_{21} = d_{22} = d_{23} = 0,$$
$$d_{31} = d_{32} = d_{33} = -\tfrac{1}{3}.$$

Comments

41. *Unadjusted Estimator.* The unadjusted estimator Z_{uv} (7.37) of Δ_{uv} (7.33) is simply the estimator associated with the sign test and previously discussed in Section 3.5. However the Doksum adjusted estimator $\tilde{\theta}$ (7.40) is quite often different from this simple unadjusted estimator Z_{uv}. This is the case in Example 7.5, for instance, where $Z_{13} = .125$, but $\tilde{\theta} = \tau_1 - \tau_3 = .133$.

42. *Ambiguities with the Unadjusted Estimators.* The unadjusted estimators Z_{uv} (7.37) lead to ambiguities in contrast estimation because they do not satisfy the linear relations that are satisfied by the contrasts they estimate. For example, $\Delta_{13} = \tau_1 - \tau_3 = (\tau_1 - \tau_2) + (\tau_2 - \tau_3) = \Delta_{12} + \Delta_{23}$, but, in general, $Z_{13} \neq Z_{12} + Z_{23}$. Thus, the two "reasonable" estimators Z_{13} and $Z_{12} + Z_{23}$ of $\Delta_{13} = \tau_1 - \tau_3$ can give different estimates. We refer to this property as the incompatibility of the unadjusted estimators Z_{uv}.

43. *Efficiency.* The adjusted estimators $\tilde{\Delta}_{uv}$ (7.39) of Δ_{uv} (7.33) are always at least as efficient as the unadjusted ones and they are compatible. They do, however, have the disadvantage that the estimator of $\Delta_{uv} = \tau_u - \tau_v$ depends on the observations from the other $k - 2$ treatments.

44. *Contrast Estimator Associated with Signed Ranks.* As noted in Comment 41, the contrast estimator $\tilde{\theta}$ (7.40) is related to paired replicates estimators associated with the sign statistic, as discussed in Sections 3.4 and 3.5. A competitor contrast estimator related to the Hodges-Lehmann paired replicates estimator associated with the signed rank statistic and utilizing between-block information is discussed in Section 7.15.

Properties

1. *Standard Deviation of $\tilde{\theta}$ (7.40).* For the asymptotic standard deviation of $\tilde{\theta}$ (7.40), see Doksum (1967).
2. *Asymptotic Normality.* See Doksum (1967).
3. *Efficiency.* See Doksum (1967) and Section 7.16.

PROBLEMS

48. Estimate $2\tau_N - \tau_A - \tau_R$ for the metronome data of Table 7.6.
49. Illustrate, using a numerical example, the incompatibility of the unadjusted estimators Z_{uv} (see Comment 42).
50. Estimate the simple contrasts $\theta_1 = \tau_2 - \tau_1, \theta_2 = \tau_3 - \tau_1$, and $\theta_3 = \tau_3 - \tau_2$ for the creatine phosphokinase activity data in Table 7.3.

51. Estimate the contrast $3\tau_{ALC} - \tau_{AL} - \tau_{AC} - \tau_{LC}$ for the percentage consonants correctly identified data in Table 7.4.
52. Using the data of Table 7.4, estimate the simple contrast that represents the benefit from adding lip reading to audition in teaching severely hearing-impaired children.
53. Estimate the contrast $\tau_{AC} + \tau_{LC} - 2\tau_C$ for the percentage consonants correctly identified data in Table 7.4.
54. Estimate all contrasts found to be of interest in Problem 47 for the maximum soil temperature data in Table 7.8.
55. Estimate all possible simple contrasts for the ozone exposure data in Table 7.2.
56. Consider the percent average wind speed reduction data in Table 7.7. Use an appropriate all-treatments multiple comparison procedure (see Section 7.3) to decide which distances from the shelterbelt have significantly different reductions in average wind speed. Estimate all contrasts suggested to be important from this multiple comparison analysis.
57. Estimate the contrast $\tau_{\text{rats}} - \tau_{\text{cats}}$ for the Livesey EPT error score data of Table 7.9.

INCOMPLETE BLOCK DATA—TWO-WAY LAYOUT WITH ZERO OR ONE OBSERVATION PER TREATMENT-BLOCK COMBINATION

In two-way layout settings the most common form of data collection corresponds to the case of a single observation for every treatment-block combination. However, it is not uncommon to deal with two-way layout problems where certain treatment-block cells yield single observations, but where there are also treatment-block combinations for which we have no observations. This could be a result of a deliberate design to deal with data collection problems where it is not feasible (economically, time-constraints, etc.) to collect data from every treatment-block combination or it could be simply a result of missing data from what was intended to be a complete block design.

In the next three sections we discuss procedures developed for such incomplete block data sets. In Sections 7.6 and 7.7 we present a distribution-free hypothesis test for general alternatives and an all-treatments multiple comparison procedure, respectively, for the most commonly used design specifically structured to yield less than complete block data, namely, the balanced incomplete block design (BIBD). In Section 7.8 we detail a distribution-free hypothesis test for general alternatives that is applicable for two-way layout data representing an arbitrary configuration of either zero or one observation per cell.

Throughout these next three sections, we continue to operate under the general conditions of Assumptions A1–A3. However, in these sections we impose the additional constraint that each c_{ij} is either 0 or 1 and $N = \sum_{i=1}^{n} \sum_{j=1}^{k} c_{ij} \neq kn$; that is, we have incomplete block data. We again drop the third subscript on the X variables in Sections 7.6–7.8. This will not be problematic, however, as there are no cells with more than one observation.

7.6. A DISTRIBUTION-FREE TEST FOR GENERAL ALTERNATIVES IN A RANDOMIZED BALANCED INCOMPLETE BLOCK DESIGN (BIBD) (DURBIN, SKILLINGS-MACK)

In this section we present a procedure for testing H_0 (7.2) against the general alternatives H_1 (7.3) for incomplete block data that arise from a very structured randomized balanced incomplete block design (BIBD). Such a BIBD corresponds to a setting where we observe $s(< k)$ treatments in each of the n blocks, every pair of treatments occurs together in the same number, λ, of blocks, and each of the k treatments is observed a total of p times. These parameters of a BIBD must satisfy the restriction that $p(s - 1) = \lambda(k - 1)$, which, of course, forces additional constraints on the c_{ij}'s. (See Problems 59 and 62.)

Procedure

To compute the Durbin-Skillings-Mack statistic for such a balanced incomplete block design setting, we first order the available s observations from least to greatest separately within each of the n blocks. Let r_{ij} be this within-block rank of X_{ij} if there is an observation for the ith block-jth treatment combination; otherwise, let $r_{ij} = 0$. Set

$$R_j = \sum_{i=1}^{n} r_{ij}, \quad \text{for } j = 1, \ldots, k. \tag{7.42}$$

Thus, for example, R_3 is the sum (over the n blocks) of the within-blocks ranks received by the p available treatment 3 observations. (Note that each R_j will be the sum of exactly p nonzero within-blocks ranks.) The Durbin-Skillings-Mack statistic is then given by

$$\begin{aligned} D &= \left[\frac{12}{\lambda k(s + 1)}\right] \sum_{j=1}^{k} \left\{R_j - \frac{p(s + 1)}{2}\right\}^2 \\ &= \left\{\left[\frac{12}{\lambda k(s + 1)}\right] \sum_{j=1}^{k} R_j^2\right\} - \frac{3(s + 1)p^2}{\lambda}, \end{aligned} \tag{7.43}$$

where $(s + 1)/2 = \sum_{j=1}^{k} r_{ij}/k$ is the average within-blocks rank assigned for each of the n blocks. Since each treatment is observed p times, it follows that $p(s + 1)/2$ is the expected sum of within-blocks ranks for each of the k treatments when H_0 (7.2) is true.

To test

$$H_0 : [\tau_1 = \cdots = \tau_k]$$

versus the general alternative

$$H_1 : [\tau_1, \ldots, \tau_k \text{ not all equal}],$$

at the α level of significance,

$$\text{Reject } H_0 \text{ if } D \geq d_{\alpha,s}; \quad \text{otherwise do not reject}, \tag{7.44}$$

where the constant $d_{\alpha,s}$ is chosen to make the type I error probability equal to α. Selected values of $d_{\alpha,s}$ for a variety of BIBDs are given in Table A.26.

Large-Sample Approximation

When H_0 is true, the statistic D has, as n tends to infinity, an asymptotic chi-square (χ^2) distribution with $k - 1$ degrees of freedom. (See Comment 50 for indications of the proof.) The chi-square approximation for procedure (7.44) is

$$\text{Reject } H_0 \text{ if } D \geq \chi^2_{k-1,\alpha}; \quad \text{otherwise do not reject}, \tag{7.45}$$

where $\chi^2_{k-1,\alpha}$ is the upper α percentile of a chi-square distribution with $k-1$ degrees of freedom. Values of $\chi^2_{k-1,\alpha}$ can be obtained from Chart A.2.

Skillings and Mack (1981) noted that this approximate procedure (7.45) can be quite conservative when α is small (say, $\leq .01$) and either the number of blocks n or the number of common occurrences λ is small. In particular, they suggest that the approximation is conservative whenever λ is not at least 3. In such cases, they strongly recommend the use of the exact values of $d_{\alpha,s}$ from Table A.26 when available (or, if possible, the generation of such "exact" values $d_{\alpha,s}$ via simulation when $\alpha \leq .01$ and the BIBD is not one that is covered by Table A.26).

Ties

If there are ties among the X observations within any of the blocks, use average ranks to break the ties and compute the individual treatment sums of ranks $R_1, \ldots, R_k$. In such cases, the significance level associated with procedure (7.44) is only approximately equal to α. (See Comment 51 for discussion of how to construct a conditionally distribution-free test of H_0 even when there are ties within some of the blocks.)

Example 7.6: *Chemical Toxicity.* Moore and Bliss (1942) compared the toxicity of each of seven chemicals applied to *Aphis rumicis*, a black aphid found on nasturtiums. The logarithm of the dose required to kill 95% of the insects exposed to a chemical was the measurement reported. Since the experimenters could test only three chemicals in any given day, they used a balanced incomplete block design requiring seven days for completion of the experiment. The toxicities for the studied chemicals are shown in Table 7.12.

This experiment constitutes a BIBD with $k = 7$ treatments, of which $s = 3$ are observed in each of the $n = 7$ blocks, every pair of treatments occur together in $\lambda = 1$ of the blocks, and each of the treatments is observed a total of $p = 3$ times. We are interested in assessing whether there are any differences in the toxicities of the seven chemicals relative to *Aphis rumicis*. Since this BIBD is not one of those for which the exact critical values are available in Table A.26, we will use the approximate procedure (7.45) to test if there are any differences in the toxicities of the seven chemicals. For sake of illustration, we take the approximate significance level to be $\alpha = .05$. From Chart A.2 with $k - 1 = 6$ degrees of freedom, we find the value $\chi^2_{6,.05} \approx 12.5$ and procedure (7.45) reduces to

$$\text{Reject } H_0 \text{ if } D \geq 12.5.$$

TABLE 7.12. Logarithm of Toxic Dosages

Day	Chemical						
	A	B	C	D	E	F	G
1	.465	.343		.396			
2	.602		.873		.634		
3			.875	.325			.330
4	.423					.987	.426
5		.652	1.142			.989	
6		.536			.409		.309
7				.609	.417	.931	

Source: W. Moore and C. I. Bliss (1942).

Now, we illustrate the computations leading to the sample value of D (7.43). Ranking from 1 to 3 within each of the seven blocks (days) and summing across the blocks for each of the chemicals, we obtain the following treatment sums of ranks:

$R_1 = 3 + 1 + 1 = 5$, $R_2 = 1 + 1 + 3 = 5$, $R_3 = 3 + 3 + 3 = 9$, $R_4 = 2 + 1 + 2 = 5$,
$R_5 = 2 + 2 + 1 = 5$, $R_6 = 3 + 2 + 3 = 8$, $R_7 = 2 + 2 + 1 = 5$.

Hence, from (7.43) we find that

$$D = \left\{\left[\frac{12}{1(7)(4)}\right](5^2 + 5^2 + 9^2 + 5^2 + 5^2 + 8^2 + 5^2)\right\} - \frac{3(4)(3^2)}{1}$$
$$= \left\{\frac{3(270)}{7}\right\} - 108 = 7.71.$$

Comparison of this observed value of 7.71 with the critical value $\chi^2_{6,.05} \approx 12.5$ leads us to conclude that there is not strong sample evidence to indicate any significant difference between the seven studied chemicals with respect to their toxicities for *Aphis rumicis*. In fact, from Chart A.2, we see that the observed value of $D = 7.71$ is approximately the .26 upper percentile for the chi-square distribution with 6 degrees of freedom. Thus, the approximate P-value for these data and procedure (7.45) is .26, providing further evidence of the similarity of the studied chemicals with respect to their toxic effects on *Aphis rumicis*.

Comments

45. *More General Setting.* We could replace Assumptions A1–A3 and H_0 (7.2) with the more general null hypothesis that all possible $(s!)^n$ rank configurations for the nonzero r_{ij}'s are equally likely. Procedure (7.44) remains distribution-free for this more general hypothesis.

46. *Design Rationale.* In a two-way layout setting with no replications within block-treatment combinations, it is best to use a randomized complete block design (as discussed in Sections 7.1–7.5) whenever possible. However, there are times when experimental constraints such as fixed costs or limited time or facilities make it impossible to obtain an observation for every treatment-block combination. When this is the case, the use of a balanced incomplete block design is often a good alternative. Such a BIBD provides for sufficient data to be collected to permit comparison of each treatment with every other one. Moreover, the fact that the BIBD imposes a rigid structure on the missing observations within and across the blocks enables the associated data analysis to be both relatively simple and efficient.

47. *Motivation for the Test.* Under Assumptions A1–A3 and H_0 (7.2), each of the block rank vectors $\boldsymbol{R}_i^*$ for those s observations present in the ith block, $i = 1, \ldots, n$, has a uniform distribution over the set of all $s!$ permutations of the vector of integers $(1, 2, \ldots, s)$. If r_{ij} is nonzero, it follows that $E_0(r_{ij}) = (1/s!)[(s-1)!]\sum_{t=1}^{k} t = (s+1)/2$, the average rank being assigned separately to the partial data in each of the blocks. Thus, $E_0(R_j) = \sum_{i=1}^{n} E_0(r_{ij}) = p(s+1)/2$, for $j = 1, \ldots, k$, since there are observations in exactly p of the blocks for each of the k treatments. Therefore, we would expect each of the R_j's to be close to $p(s+1)/2$ when H_0 is true. Since the test statistic D (7.43) is a constant times a sum of squared differences between the observed treatment sums of ranks, R_j, and their common null expected value, $E_0(R_j) = p(s+1)/2$, small values of D represent agreement with H_0 (7.2). When the τ's are not all equal, we would expect a portion of the associated treatment sums of ranks to differ from their common null expectation, $p(s+1)/2$, with some tending to be smaller and some

7.6. TEST FOR GENERAL ALTERNATIVES IN A BALANCED INCOMPLETE BLOCK DESIGN

larger. The net result (after squaring the observed differences to obtain the $[R_j - p(s+1)/2]^2$ terms) would be a large value of D. This quite naturally suggests rejecting H_0 in favor of H_1 (7.3) for large values of D and motivates procedures (7.44) and (7.45).

48. *Assumptions.* We emphasize that Assumption A3 stipulates that the ns cell distributions F_{ij} for those treatment-block combinations where observations are collected can differ at most in their locations (medians) and that these location differences (if any) must be a result of additive block and/or treatment effects (i.e., there is no interaction between the treatment and block factors). In particular, Assumption A3 requires that the ns underlying distributions belong to the same general family (F) and that they do not differ in scale parameters (variability). We do note, however, that the test procedure (7.44) remains distribution-free under the less restrictive setting where Assumption A3 is replaced by the weaker condition

A3'. The distribution functions $F_{11}, \ldots, F_{1k}, \ldots, F_{n1}, \ldots, F_{nk}$ are connected through the relationship

$$F_{ij}(u) = F_i(u - \tau_j), \quad -\infty < u < \infty,$$

for $i = 1, \ldots, n$ and $j = 1, \ldots, k$, where $F_1, \ldots, F_n$ are arbitrary distribution functions for continuous distributions with unknown medians $\theta_1, \ldots, \theta_n$, respectively, and, as before, τ_j is the unknown additive treatment effect contributed by the jth treatment.

Assumption A3 then corresponds to Assumption A3' with the additional condition that $F_1 \equiv \cdots \equiv F_n$. (See also Comment 45.)

49. *Derivation of the Distribution of D under H_0 (No-Ties Case).* The null distribution of D (7.43) can be obtained by using the fact that under H_0 (7.2), all possible $(s!)^n$ rank configurations for the nonzero r_{ij}'s are equally likely. We now take the simplest (but not very useful in practice) balanced incomplete block design corresponding to $k = 3, s = 2, n = 3, p = 2$, and $\lambda = 1$ to illustrate how the null distribution can be derived. In this case, D (7.43) reduces to $D = (\frac{4}{3})R^* - 36$, where $R^* = R_1^2 + R_2^2 + R_3^2$.

The value of D for each of the $(2!)^3 = 8$ possible rank configurations for this setting are presented below.

I	II	III		I	II	III		I	II	III
1	2			2	1			1	2	
1		2		1		2		2		1
	1	2			1	2			1	2

$R^* = 29, D = 2.67 \qquad R^* = 29, D = 2.67 \qquad R^* = 27, D = 0$

I	II	III		I	II	III		I	II	III
2	1			1	2			1	2	
2		1		1		2		2		1
	1	2			2	1			2	1

$R^* = 29, D = 2.67 \qquad R^* = 29, D = 2.67 \qquad R^* = 29, D = 2.67$

I	II	III		I	II	III
2	1			2	1	
1		2		2		1
	2	1			2	1

$R^* = 27, D = 0 \qquad R^* = 29, D = 2.67$

Thus, we find

$$P_0\{D = 0\} = .25 \quad \text{and} \quad P_0\{D = 2.67\} = .75,$$

which agrees with the upper-tail probabilities listed in Table A.26 for $k = 3, s = 2, n = 3$, $p = 2$, and $\lambda = 1$.

Note that we have derived the null distribution of D without specifying the common form (F) of the underlying distribution function for the X's under H_0 beyond the point of requiring that it be continuous. This is why the test procedure (7.44) based on D is called a distribution-free procedure. From the null distribution of D we can determine the critical value $d_{\alpha,s}$ and control the probability α of falsely rejecting H_0 when it is true, and this error probability does not depend on the specific form of the common underlying continuous X distribution.

50. *Large-Sample Approximation.* Define the random variables $T_j = R_j - E_0(R_j) = R_j - p(s+1)/2$, for $j = 1,\ldots,k$. Since each R_j is a sum, it is not surprising (see, for example, Skillings and Mack (1981) for formal justification) that a properly standardized version of the vector $\boldsymbol{T}^* = (T_1,\ldots,T_{k-1})$ has an asymptotic (n tending to infinity) ($k-1$)-variate normal distribution with mean vector $\boldsymbol{0} = (0,\ldots,0)$ and appropriate covariance matrix Σ when the null hypothesis H_0 is true. (Note that $\boldsymbol{T}^*$ does not include $T_k = R_k - p(s+1)/2$, since T_k can be expressed as a linear combination of $T_1,\ldots,T_{k-1}$. This is the reason that the asymptotic normal distribution is $(k-1)$-variate and not k-variate.) Since the test statistic D (7.43) is a quadratic form in the variables $(T_1,\ldots,T_{k-1})$, it is therefore quite natural that D has an asymptotic (n tending to infinity) chi-square distribution with $k-1$ degrees of freedom.

51. *Exact Conditional Distribution of D with Ties among the Observed X Values within Blocks.* To have a test with exact significance level even in the presence of tied X's within some of the blocks, we need to consider all $(s!)^n$ block rank configurations, where now these within-blocks ranks are obtained using average ranks to break ties. As in Comment 49, it still follows that under H_0 each of the $(s!)^n$ block rank configurations (now with these tied ranks) is equally likely. For each such configuration, the value of D is computed and the results are tabulated. We illustrate this for the same setting as was used in the untied example in Comment 49 (namely, $k = 3, s = 2, n = 3, p = 2$, and $\lambda = 1$), except here we assume that the two observations within block one are tied in value. Thus the block ranks we are dealing with here are (1.5, 1.5), (1, 2), and (1, 2) for blocks 1, 2, and 3, respectively.

As in Comment 49, D (7.43) reduces to $D = (\frac{4}{3})R^* - 36$, where $R^* = R_1^2 + R_2^2 + R_3^2$. Since the rank configuration for the first block is always (1.5, 1.5), we need only compute the value of D for $(2!)^2 = 4$ possible rank configurations. The values of D for these 4 configurations are as follows:

I	II	III		I	II	III
1.5	1.5			1.5	1.5	
1		2		1		2
	1	2			2	1
$R^* = 28.5, D = 2$				$R^* = 27.5, D = 0.67$		

I	II	III		I	II	III
1.5	1.5			1.5	1.5	
2		1		2		1
	1	2			2	1
$R^* = 27.5, D = 0.67$				$R^* = 28.5, D = 2$		

Thus, we find

$$P_0\{D = 0.67\} = .50 \quad \text{and} \quad P_0\{D = 2\} = .50.$$

This distribution is called the conditional distribution or the permutation distribution of D, given the set of tied within-blocks ranks (1.5, 1.5), (1, 2), and (1, 2).

52. *Historical Development.* The test procedures (7.44) and (7.45) based on D were first proposed by Durbin (1951). Later, Skillings and Mack (1981) studied a more general procedure for arbitrary incomplete block data (see Section 7.8) and, for the first time, made available the exact critical values $d_{\alpha,s}$ for procedure (7.44) for a reasonable set of balanced incomplete block designs.

Properties

1. *Consistency.* See van Elteren and Noether (1959).
2. *Asymptotic Chi-Squaredness.* See Durbin (1951), Benard and van Elteren (1953) or Skillings and Mack (1981).
3. *Efficiency.* See van Elteren and Noether (1959) and Section 7.16.

PROBLEMS

58. Mendenhall (1968) discusses an experiment that was conducted to compare the effects of seven different chemical substances on the skin of male rats. The necessity to use relatively homogeneous patches of a rat's skin for the study restricted the experimenter to three experimental units (patches of skin) per animal. However, to avoid the confounding effect of rat-to-rat variability in the comparison of the seven chemicals, the experimenter was obligated to block on rats and any given rat could be treated with only three of the seven chemicals. This resulted in the use of a balanced incomplete block design with parameters $k = 7, n = 7, s = 3, p = 3$, and $\lambda = 1$. The experimental measurements for the study are presented in Table 7.13.

TABLE 7.13. Reactions of Male Rats to Chemical Substances

	Chemical Substance						
Rat	A	B	C	D	E	F	G
1	10.2	6.9		14.2			
2			9.9	12.9		14.1	
3		12.1	11.7		8.6		
4				14.3	9.1		7.7
5		8.8				16.3	8.6
6	13.1				9.2	15.2	
7	11.3		9.7				6.2

Source: W. Mendenhall (1968).

Apply procedure (7.45) with approximate significance level $\alpha \approx .05$ to these data to test the hypothesis of interest.

59. Verify that the relationship $p(s - 1) = \lambda(k - 1)$ must hold for a balanced incomplete block design.
60. Verify that the two representations for D (7.43) are, in fact, equivalent.
61. What are the maximum and minimum values for the test statistic D (7.43)? What rank configurations lead to these maximum and minimum values?

62. Consider the relationship $p(s-1) = \lambda(k-1)$ that must hold for a balanced incomplete block design. What constraints does this condition place on the c_{ij}'s for the data?

63. Kuehl (1994) described an experiment by J. Berry and A. Deutschman at the University of Arizona designed to study the effect of pressure on percent conversion of methyl glucoside to monovinyl isomers. The conversion is achieved by addition of acetylene to methyl glucoside in the presence of a base under high pressure. Five pressures were of interest in the study, but only three could be examined at any one time under identical experimental conditions because of limited laboratory space. This necessitated the use of a balanced incomplete block design. The data obtained in the experiment and design are given in Table 7.14.

TABLE 7.14. Percent Conversion of Methyl Glucoside to Monovinyl Isomers

	Pressure (psi)				
Experimental Run	250	325	400	475	550
1	16	18		32	
2	19			46	45
3		26	39		61
4			21	35	55
5		19		47	48
6	20		33	31	
7	13	13	34		
8	21		30		52
9	24	10			50
10		24	31	37	

Source: R. O. Kuehl (1994).

State the parameters for the BIBD employed in this chemical conversion study. Apply procedure (7.44) to these data to assess whether pressure (at the levels included in the study) has any effect on the percent conversion of methyl glucoside to monovinyl isomers.

64. Consider the BIBD corresponding to $k=5, n=10, s=3, p=6$, and $\lambda=3$. Compare the critical region for the exact level $\alpha = .0499$ test of H_0 (7.2) based on D with the critical region for the corresponding nominal level $\alpha = .0499$ test based on the large-sample approximation.

65. Consider the BIBD corresponding to $k=4, n=6, s=2, p=3$, and $\lambda=1$. Obtain the form of the exact null H_0 distribution of D (7.43) for the case of no tied observations.

66. Consider the BIBD corresponding to $k=5, n=20, s=2, p=8$, and $\lambda=2$. Compare the critical region for the exact level $\alpha = .0685$ test of H_0 (7.2) based on D with the critical region for the corresponding nominal level $\alpha = .0685$ test based on the large-sample approximation.

67. Consider the BIBD corresponding to $k=4, n=6, s=2, p=3$, and $\lambda=1$. Suppose that the two observations in each of blocks 3 and 5 are tied. Obtain the conditional exact probability distribution of D under H_0 (7.2) when average ranks are used to break these two sets of within-blocks ties. Compare this conditional null distribution of D with the null distribution for D obtained in Problem 65 when there are no ties.

68. Consider the percentage consonants correctly identified data in Table 7.4 for conditions AL, AC, LC, and ALC only. Using the null distributions for D available in Table A.26 as a guide, what BIBD could have been utilized in this study for these four treatments to reduce the number of conditions under which each severely hearing-impaired child had to be observed? Using a random mechanism for deciding how to apply the BIBD in question to the existing data set, analyze the corresponding data subset to assess whether there are any differences in the effectiveness of the conditions AL, AC, LC, and ALC for teaching severely hearing-impaired children.

69. Consider the percentage consonants correctly identified data in Table 7.4 for conditions A, L, and C only. Using the null distributions for D available in Table A.26 as a guide, what BIBD could have

been utilized in this study for these three treatments to reduce the number of conditions under which each severely hearing-impaired child had to be observed? Using a random mechanism for deciding how to apply the BIBD in question to the existing data set, analyze the corresponding data subset to assess whether there are any differences in the effectiveness of the conditions A, L, and C for teaching severely hearing-impaired children.

7.7. ASYMPTOTICALLY DISTRIBUTION-FREE TWO-SIDED ALL-TREATMENTS MULTIPLE COMPARISONS FOR BALANCED INCOMPLETE BLOCK DESIGNS (SKILLINGS-MACK)

In this section we present an asymptotically distribution-free multiple comparison procedure using Friedman within-blocks ranks that is designed to make decisions about individual differences between pairs of treatment effects (τ_i, τ_j), for $i < j$, for data obtained from a balanced incomplete block design. The multiple comparison procedure of this section would generally be applied to BIBD data *after* rejection of H_0 (7.2) with the Durbin-Skillings-Mack procedure from Section 7.6. In this setting, we will reach conclusions about all $k(k-1)/2$ pairs of treatment effects and these conclusions are naturally two-sided in nature.

Procedure

Let $R_1, \ldots, R_k$ be the treatment sums of within-blocks ranks given by (7.42). Calculate the $k(k-1)/2$ absolute differences $|R_u - R_v|$, $1 \leq u < v \leq k$.

When H_0 (7.2) is true, the $k(k-1)/2$-component vector $(R_1, \ldots, R_k)$ has, when properly standardized and as n tends to infinity, an asymptotic $(k-1)$-variate normal distribution with appropriate mean vector and covariance matrix (see Skillings and Mack (1981) for details of the proof). At an approximate experimentwise error rate of α, the Skillings-Mack two-sided all-treatments multiple comparison procedure reaches its $k(k-1)/2$ pairwise decisions, corresponding to each (τ_u, τ_v) pair, $1 \leq u < v \leq k$, by the criterion

$$\text{Decide } \tau_u \neq \tau_v \text{ if } |R_u - R_v| \geq [(s+1)(ps - p + \lambda)/12]^{1/2} q_\alpha;$$
$$\text{otherwise decide } \tau_u = \tau_v, \quad (7.46)$$

where q_α is the upper αth percentile for the distribution of the range of k independent $N(0, 1)$ variables. Values of q_α for $\alpha = .0001, .0005, .001, .005, .01, .025, .05, .10, .20$ and $k = 2(1)20(2)40(10)100$ are given in Table A.17. (See also Comment 55.)

Ties

If there are ties among the X observations within any of the blocks, use average ranks to break the ties and compute the individual treatment sums of ranks $R_1, \ldots, R_k$.

Example 7.7: *Chemical Toxicity*. For sake of illustration, we apply procedure (7.46) to the chemical toxicity data relative to the black aphid, *Aphis rumicis*, as previously discussed in Example 7.6, even though the Durbin-Skillings-Mack procedure did not find any significant differences (approximate P-value of .26) between the treatment effects. Taking our approximate experimentwise error rate to be $\alpha \approx .05$, we see from Table A.17 that $q_{.05} = 4.170$ for $k = 7$ and procedure (7.46) reduces to

$$\text{Decide } \tau_u \neq \tau_v \text{ if } |R_u - R_v| \geq [4(9 - 3 + 1)/12]^{1/2}(4.170) = 6.370.$$

Using the treatments sums of within-blocks ranks obtained in Example 7.6, we find that

$|R_2 - R_1| = 0,$ $|R_3 - R_1| = 4,$ $|R_4 - R_1| = 0,$ $|R_5 - R_1| = 0,$ $|R_6 - R_1| = 3,$
$|R_7 - R_1| = 0,$ $|R_3 - R_2| = 4,$ $|R_4 - R_2| = 0,$ $|R_5 - R_2| = 0,$ $|R_6 - R_2| = 3,$
$|R_7 - R_2| = 0,$ $|R_4 - R_3| = 4,$ $|R_5 - R_3| = 4,$ $|R_6 - R_3| = 1,$ $|R_7 - R_3| = 4,$
$|R_5 - R_4| = 0,$ $|R_6 - R_4| = 3,$ $|R_7 - R_4| = 0,$ $|R_6 - R_5| = 3,$ $|R_7 - R_5| = 0,$
$|R_7 - R_6| = 3.$

Since all these absolute differences are less than the critical value, 6.370, we see that the $7(6)/2 = 21$ decisions at this approximate experimentwise error rate of .05 are that $\tau_u = \tau_v$, for $1 \le u < v \le 7$. This is, of course, not at all surprising, since, in Example 7.6, the Durbin-Skillings-Mack test procedure found no support for rejecting H_0 (7.2) for these data.

Comments

53. *Rationale for Multiple Comparison Procedure*. The rationale behind the multiple comparison procedure of this section for data from a balanced incomplete block design is similar to that behind the two-sided multiple comparison procedures for data from a complete randomized block design. For further discussion, see Comment 24.

54. *Experimentwise Error Rate*. The use of an experimentwise error rate represents a very conservative approach to multiple comparisons. We are insisting that the probability of making only correct decisions be $1 - \alpha$ when the null hypothesis H_0 (7.2) of treatment equivalence is true. Thus we have a high degree of protection when H_0 is true, but we often apply such techniques when we have evidence (perhaps based on a priori information or perhaps obtained by applying the Durbin-Skillings-Mack test, as in Example 7.7) that H_0 is not true. The protection under H_0 also makes it harder for the procedure to judge treatments as differing significantly when, in fact, H_0 is false, and this difficulty becomes more severe as k increases. See Comment 6.58 for additional discussion of experimentwise error rates.

55. *Conservative Procedure*. Skillings and Mack (1981) also proposed a conservative multiple comparison procedure that guarantees an upper bound on the experimentwise error rate. Let $R_1, \ldots, R_k$ be the treatment sums of within-blocks ranks given by (7.42). At an experimentwise error rate *no greater* than α, the Skillings-Mack conservative two-sided all-treatments multiple comparison procedure reaches its $k(k-1)/2$ decisions through the criterion

$$\text{Decide } \tau_u \ne \tau_v \text{ if } |R_u - R_v| \ge [k\lambda d_{\alpha,s}(s+1)/6]^{1/2};$$

$$\text{otherwise decide } \tau_u = \tau_v, \qquad (7.47)$$

where $d_{\alpha,s}$ is the upper α percentile for the null distribution of the Durbin-Skillings-Mack statistic D (7.43). Values of $d_{\alpha,s}$ are available in Table A.26 for a variety of BIBD's. Skillings and Mack (1981) note that although procedure (7.47) does not require a large number of blocks, it is, nevertheless, rather conservative since it is based on the projection procedure of Scheffé; that is, the true experimentwise error rate might be considerably smaller than the bound α provided by (7.47). As a result, they recommend using the approximation (7.46) whenever the number of blocks is reasonably large.

56. *Contrast Estimators for BIBD's*. Greenberg (1966) proposed a method of contrast estimation for general (including balanced) incomplete block designs where the number of observations in a block is smaller than the number of treatments to be compared.

57. *Dependence on Observations from Other Noninvolved Treatments*. The all-treatments multiple comparison procedure of this section suffers from the same disadvantage as do the

other two-way layout multiple comparison procedures of this chapter. The decision between treatment u and treatment v can be affected by changes only in the observations from one or more of the other $k - 2$ treatments which are not directly involved.

Properties

1. *Asymptotic Multivariate Normality.* See Skillings and Mack (1981).
2. *Efficiency.* See Section 7.16.

PROBLEMS

70. Apply procedure (7.46) to the chemical substance effect data of Table 7.13 in Problem 58.
71. Illustrate the difficulty discussed in Comment 57 by means of a numerical example.
72. Apply procedure (7.46) to the percent conversion data of Table 7.14 in Problem 63.
73. Consider the chemical toxicity data of Table 7.12 in Example 7.6. Find the smallest approximate experimentwise error rate at which the most significant difference(s) in black aphid (*Aphis rumicis*) toxicity between the studied substances would be detected by procedure (7.46).
74. Consider the chemical substance effect data of Table 7.13 in Problem 58. Find the smallest approximate experimentwise error rate at which procedure (7.46) would declare that chemical substances F and G have differing effects on the skin of male rats.
75. Consider the percent conversion data of Table 7.14 in Problem 63. Find the smallest approximate experimentwise error rate at which the most significant difference in the effects of the various pressures on the percent conversion of methyl glucoside to monovinyl isomers would be detected by procedure (7.46).

7.8. A DISTRIBUTION-FREE TEST FOR GENERAL ALTERNATIVES FOR DATA FROM AN ARBITRARY INCOMPLETE BLOCK DESIGN (SKILLINGS-MACK)

Not every set of data resulting from less than a randomized complete block design satisfies the necessary constraints (see Section 7.6) to be analyzed by the Durbin-Skillings-Mack procedure for balanced incomplete block designs. In this section we present a general procedure for analyzing data from a two-way layout where there are either zero or one observation for each treatment-block combination but where there is not necessarily any nice pattern to the particular combinations for which we do not have observations. Such an incomplete data configuration could, of course, be intentionally designed this way, but it could also be the consequence of missing observations from an experiment that was intended to yield data for a randomized complete block design.

For this general two-way layout setting, let s_i denote the number of treatments for which an observation is present in block i, for $i = 1, \ldots, n$. (If $s_i = 1$ for block i, we remove that block from the analysis. Therefore, throughout this section, n will denote the number of blocks for which $s_i \geq 2$.) We discuss a distribution-free procedure for testing H_0 (7.2) against the general alternatives H_1 (7.3) when we are faced with such arbitrarily incomplete block data.

Procedure

To compute the Skillings-Mack statistic for arbitrarily incomplete block data, we first rank the s_i observed data values in block i from least to greatest, for each block $i = 1, \ldots, n$. Thus, in the ith block, we will be assigning ranks $1, 2, \ldots, s_i$. For $i = 1, \ldots, n$ and $j = 1, \ldots, k$, let

r_{ij} = rank of X_{ij} among the observations in block i, if $c_{ij} = 1$,
$= (s_i + 1)/2$, if $c_{ij} = 0$,

where $(s_i + 1)/2$ is the average of the ranks assigned to the observations present in the ith block. Set

$$A_j = \sum_{i=1}^{n} \left(\frac{12}{s_i + 1}\right)^{1/2} \left(r_{ij} - \frac{s_i + 1}{2}\right), \quad j = 1, \ldots, k. \tag{7.48}$$

Thus, A_j is the weighted sum of centered (around the block average) ranks for the observations from the jth treatment, with the block weighting factor $[12/(s_i + 1)]^{1/2}$ being inversely proportional to the square root of the number of observations present in the block (see Comment 63). Set

$$\mathbf{A} = (A_1, \ldots, A_{k-1}). \tag{7.49}$$

(Without loss of generality, we have chosen to omit A_k from the vector $\mathbf{A}$. The A_j's are linearly dependent, since a weighted linear combination of all k of them is a constant. We could omit any one of the A_j's in the definition of $\mathbf{A}$ (7.49) and the procedure we now describe would lead to the same value of the test statistic. For further discussion, see Skillings and Mack (1981).)

The covariance matrix for $\mathbf{A}$ under H_0 (7.2) is given by

$$\Sigma_0 = \begin{bmatrix} \sum_{t=2}^{k} \lambda_{1t} & -\lambda_{12} & -\lambda_{13} & \cdots & -\lambda_{1,k-1} \\ -\lambda_{12} & \sum_{t \neq 2}^{k} \lambda_{2t} & -\lambda_{23} & \cdots & -\lambda_{2,k-1} \\ \vdots & \vdots & \vdots & & \vdots \\ -\lambda_{1,k-1} & -\lambda_{2,k-1} & -\lambda_{3,k-1} & \cdots & \sum_{t \neq k-1}^{k} \lambda_{k-1,t} \end{bmatrix} \tag{7.50}$$

where, for $t \neq q = 1, \ldots, k$,

$\lambda_{qt} = \lambda_{tq} =$ [number of blocks in which both treatments q and t are observed]. (7.51)

Let Σ_0^- be any (see Comment 62) generalized inverse for Σ_0. The Skillings-Mack statistic is then given by

$$SM = \mathbf{A} \Sigma_0^- \mathbf{A}'. \tag{7.52}$$

(We note that if $\lambda_{qt} > 0$ for all $q \neq t$, then the rank of the covariance matrix Σ_0 (7.50) is $k - 1$, and we can simply use the ordinary inverse Σ_0^{-1} in the definition of SM (7.52).)

To test

$$H_0 : [\tau_1 = \cdots = \tau_k]$$

versus the general alternative

$$H_1 : [\tau_1, \tau_2, \ldots, \tau_k \text{ not all equal}],$$

at the α level of significance,

$$\text{Reject } H_0 \text{ if } SM \geq sm_\alpha; \quad \text{otherwise do not reject,} \tag{7.53}$$

where the constant sm_α is chosen to make the type I error probability equal to α.

To generate critical values sm_α, we would need to look at all possible rank configurations of the within-blocks r_{ij} values for the sample observations present in the study (i.e., for those cells where $c_{ij} = 1$). (For more details, see Comment 64.) However, in view of the substantial number of incomplete block data configurations that are possible, even for a fixed combination of k treatments and n blocks, tables of sm_α critical values are simply not available in the literature for arbitrary incomplete block configurations. The one exception is for the very special case where there is only a single missing observation from what would otherwise be a randomized complete block design. Selected values of sm_α for this latter setting were given by Skillings and Mack (1981) and are reproduced in Table A.27 for $k = 3(1)6$ and $n = 3(1)11$. (See also expression (7.55) in Comment 60 for a simplified computational form for SM when there is only a single missing observation.)

Large-Sample Approximation

When H_C (7.2) is true and $\lambda_{qt} > 0$ for every $q \neq t = 1, \ldots, k$ (that is, every pair of treatments occur together in at least one block), the statistic SM has, as n tends to infinity, an asymptotic chi-square (χ^2) distribution with $k - 1$ degrees of freedom. (See Comment 65 for indications of the proof.) The chi-square approximation for procedure (7.53) is

$$\text{Reject } H_0 \text{ if } SM \geq \chi^2_{k-1,\alpha}; \quad \text{otherwise do not reject,} \tag{7.54}$$

where $\chi^2_{k-1,\alpha}$ is the upper α percentile of a chi-square distribution with $k - 1$ degrees of freedom. Values of $\chi^2_{k-1,\alpha}$ can be obtained from Chart A.2.

As with the BIBD procedure discussed in Section 7.6, Skillings and Mack (1981) have pointed out that this approximate procedure (7.54) can be quite conservative when α is small (say, $\leq .01$) and the number of blocks, n, is not large. In such cases, it would, of course, be preferable to use the exact procedure (7.53). However, the lack of published exact critical values sm_α for most incomplete block designs makes use of the approximate procedure (7.54) the only available choice in many settings, unless one is willing to generate the exact null distribution of SM for the particular incomplete block configuration of interest (see Comment 64). (We should also point out that the approximate procedure (7.54) is simply not applicable if there are at least two treatments which do not have observations together in any of the blocks; that is, if $\lambda_{qt} = 0$ for at least one pair $q \neq t = 1, \ldots, k$.)

Ties

If there are ties among the X observations within any of the blocks, use average ranks to break the ties and compute the individual treatment weighted sums of centered ranks $A_1, \ldots, A_k$. In such cases, the significance level associated with procedure (7.53) is only approximately equal to α. (See Comment 66 for discussion of how to construct a conditionally distribution-free test of H_0 even when there are tied observations within some of the blocks.)

Example 7.8: *Effect of Rhythmicity of a Metronome on Speech Fluency.* Consider Table 7.15, which contains a subset (with subject labels renumbered) of the data in Table 7.6 obtained by Brady (1969) in his study of the influence of rhythmicity of a metronome on the speech of

TABLE 7.15. Subset of Data on the Influence of Rhythmicity of Metronome on Speech Fluency

Subject	Dysfluencies under Each Condition		
	R	A	N
1	3(1)	5(2)	15(3)
2	1(1)	3(2)	18(3)
3	5(2)	4(1)	21(3)
4	2(1)	—	6(2)
5	0(1)	2(2)	17(3)
6	0(1)	2(2)	10(3)
7	0(1)	3(2)	8(3)
8	0(1)	2(2)	13(3)

Source: J. P. Brady (1969).

stutterers, where the missing observation for subject 4 might be due to a malfunction in the arrhythmic metronome during her evaluation.

The within-blocks (subjects) ranks (r_{ij}'s) for the observations present are also given in Table 7.15 in parentheses after the data values. With respect to the missing value for subject 4 under condition A, the average rank $(2 + 1)/2 = 1.5$ for subject 4 is assigned as the value for r_{42}. From (7.48), we compute the weighted sums of centered ranks to be

$$A_1 = \left\{ \left[\frac{12}{(3+1)}\right]^{1/2} [(1-2) + (1-2) + (2-2) + (1-2) + (1-2) \right.$$
$$\left. + (1-2) + (1-2)] + \left[\frac{12}{(2+1)}\right]^{1/2} (1 - 1.5) \right\}$$
$$= 1.732(-6) + 2(-.5) = -11.392,$$

$$A_2 = \left\{ \left[\frac{12}{(3+1)}\right]^{1/2} [(2-2) + (2-2) + (1-2) + (2-2) + (2-2) \right.$$
$$\left. + (2-2) + (2-2)] + \left[\frac{12}{(2+1)}\right]^{1/2} (1.5 - 1.5) \right\}$$
$$= 1.732(-1) + 2(0) = -1.732,$$

and

$$A_3 = \left\{ \left[\frac{12}{(3+1)}\right]^{1/2} [(3-2) + (3-2) + (3-2) + (3-2) + (3-2) \right.$$
$$\left. + (3-2) + (3-2)] + \left[\frac{12}{(2+1)}\right]^{1/2} (2 - 1.5) \right\}$$
$$= 1.732(7) + 2(.5) = 13.124.$$

Thus, we obtain $\mathbf{A} = (-11.392, -1.732)$.

7.8. TEST FOR GENERAL ALTERNATIVES FOR ARBITRARY INCOMPLETE BLOCK DATA

With the single missing observation for subject 4 under condition A, the combination counts λ_{qt} (7.51) are $\lambda_{12} = 7, \lambda_{13} = 8$, and $\lambda_{23} = 7$. Hence, from representation (7.50), the null covariance matrix Σ_0 has form

$$\Sigma_0 = \begin{bmatrix} 15 & -7 \\ -7 & 14 \end{bmatrix}.$$

Since each of $\lambda_{12}, \lambda_{13}$, and λ_{23} is positive, the rank of Σ_0 is 2, and its ordinary inverse is

$$\Sigma_0^{-1} = (\tfrac{1}{161}) \begin{bmatrix} 14 & 7 \\ 7 & 15 \end{bmatrix} = \begin{bmatrix} .0870 & .0435 \\ .0435 & .0932 \end{bmatrix}.$$

From (7.52) we obtain

$$SM = A\Sigma_0^{-1}A'$$

$$= (-11.392, -1.732) \begin{bmatrix} .0870 & .0435 \\ .0435 & .0932 \end{bmatrix} \begin{pmatrix} -11.392 \\ -1.732 \end{pmatrix}$$

$$= 13.287.$$

Since these data represent a setting where we are missing only a single observation from what would otherwise be a randomized complete block design, we can use the exact procedure (7.53) with critical values from Table A.27. With $k = 3$ and $n = 8$, we see from Table A.27 that $sm_{.0097} = 8.528$. Since we have observed $SM = 13.287 > sm_{.0097} = 8.528$, the P-value for this test procedure is smaller than .0097. (In fact, from Chart A.2 with $k - 1 = 2$ degrees of freedom, we see that the approximate P-value for these data and this procedure is $\approx .0014$.) Thus, there is strong evidence to support the hypothesis that the rhythmicity of a metronome does, indeed, influence the speech of stutterers. (See also Comment 60 and Problem 78.)

Comments

58. *More General Setting.* We could replace Assumptions A1–A3 and H_0 (7.2) with the more general null hypothesis that all possible $\prod_{i=1}^{n} s_i!$ rank configurations for the within-blocks ranks, r_{i_j}, of the *observed* data are equally likely. Procedure (7.53) remains distribution-free for this more general hypothesis.

59. *Motivation for the Test.* Under Assumptions A1–A3 and H_0 (7.2), the block rank vector R_i^* for those s_i observations present in the ith block has a uniform distribution over the set of all $s_i!$ permutations of the vector of integers $(1, 2, \ldots, s_i)$, and this is true for all blocks, $i = 1, \ldots, n$. For those r_{ij} corresponding to observations present in the collected data set, it is then the case that $E_0(r_{ij}) = (s_i + 1)/2$, the average rank being assigned to the partial data present in the ith block, for every block $i = 1, \ldots, n$. Thus it follows from (7.48) and the definition of r_{ij} for an empty cell that $E_0(A_j) = 0$ for every $j = 1, \ldots, k$. Therefore, we would expect each of the A_j's to be close to zero when H_0 is true. Since the test statistic SM (7.52) is a quadratic form in $A_1, \ldots, A_k$, small values of SM represent agreement with H_0 (7.2). When the τ's are not all equal, we would expect a portion of the A_j's to differ from their common null expectation of zero, with some tending to be positive and some tending to be negative. The net effect would be a large value of the quadratic form SM. This quite naturally suggests rejecting H_0 in favor of H_1 (7.3) for large values of SM and motivates procedures (7.53) and (7.54).

60. *Special Cases.* When the configuration of observed data in each of the blocks satisfies the constraints of a BIBD (see Section 7.6), the procedures in (7.53) and (7.54) are equivalent to the exact and large-sample approximation forms, respectively, of the Durbin-Skillings-Mack test procedure given in (7.44) and (7.45), respectively, of Section 7.6. Moreover, when we have an observation present in every treatment-block combination (i.e., we have a randomized complete block design), the Skillings-Mack procedures in (7.53) and (7.54) are equivalent to the exact and large-sample approximation forms, respectively, of the Friedman test procedure presented in (7.6) and (7.7), respectively, of Section 7.1. Thus the Skillings-Mack procedures in (7.53) and (7.54) represent natural extensions of the most commonly used nonparametric procedures for randomized complete block and balanced incomplete block designs to the setting of arbitrary incompleteness of the blocks.

We note that there is also an alternative closed-form expression for the test statistic SM (7.52) when only a single treatment has missing data. Without loss of generality, suppose that treatment k is missing observations in blocks $t+1, t+2, \ldots, n$ (i.e., treatment k has only t observations). Then, it can be shown (see Problem 78) that SM can be written as

$$SM = [t + (k-1)n]^{-1} \left\{ \sum_{j=1}^{k-1} A_j^2 + \left[\frac{nA_k^2}{t}\right] \right\}. \tag{7.55}$$

61. *Assumptions.* We emphasize that Assumption A3 stipulates that the ns cell distributions F_{ij} for those treatment-block combinations where observations are collected can differ at most in their locations (medians) and that these location differences (if any) must be a result of additive block and/or treatment effects (i.e., there is no interaction between the treatment and block factors). In particular, Assumption A3 requires that the ns underlying distributions belong to the same general family (F) and that they do not differ in scale parameters (variability). We do note, however, that the test procedure (7.53) remains distribution-free under the less restrictive setting where Assumption A3 is replaced by the weaker condition Assumption A3' stated in Comment 43. Assumption A3 then corresponds to Assumption A3' with the additional condition that $F_1 \equiv \cdots \equiv F_n$. (See also Comment 58.)

62. *Use of Generalized Inverse.* We noted in the body of the text that *any* generalized inverse Σ_0^- can be used in the computation of SM (7.52). Skillings and Mack (1981) have shown that the value of SM is invariant with respect to the choice of generalized inverse, so that there is no ambiguity in the computation of SM and the associated test procedures (7.53) and (7.54) even if Σ_0 (7.50) is not of full rank. Of course, as we also noted previously, if $\lambda_{qt} > 0$ for all $q \neq t = 1, \ldots, k$, then the rank of the covariance matrix Σ_0 is, in fact, $k-1$ and we can simply use the ordinary inverse Σ_0^{-1} in the definition of SM.

63. *Weighting Factor.* In the computation of the weighted sums of centered ranks $A_1, \ldots, A_k$, Skillings and Mack (1981) chose to weight the within-blocks centered ranks $[r_{ij} - (s_i + 1)/2]$ by the factor $[12/(s_i + 1)]^{1/2}$. They noted that this weighting factor has several advantages over other alternatives. First, it leads to a simple null covariance structure Σ_0 (7.50), which is useful for computational purposes. Second, since the range of the $[r_{ij} - (s_i + 1)/2]$ values is less for blocks having fewer observations than for complete blocks, it is quite reasonable to adjust for this fact by using larger weights in those blocks with fewer observations. This has the effect of equalizing the contribution of each block when computing the A_j's. Prentice (1979) showed that a similar weighting scheme (using weights of $[s_i + 1]^{-1}$) leads to increased power over use of the unweighted forms of the treatment sums. Skillings and Mack (1981) also note that use of the alternative simple weights s_i^{-1} would minimize the null variance of the weighted sums of centered ranks $A_1, \ldots, A_k$. However, the use of these simple weights would also alter the simplicity of the null covariance matrix Σ_0, and

computation of the associated test statistic would be much more difficult that is the case for *SM* (7.52). Other weighting schemes have been considered by Benard and van Elteren (1953) and Brunden and Mohberg (1976).

64. *Derivation of the Distribution of SM under H_0 (No-Ties Case)*. The null distribution of *SM* (7.52) can be obtained by using the fact that under H_0 (7.2), all possible $s_1!s_2!\ldots s_k!$ rank configurations for the within-blocks ranks of the *observed* data are equally likely. We would simply compute the value of *SM* for each of these $s_1!s_2!\ldots s_k!$ block rank configurations and then tabulate the collective distribution of the values. Since the specifics of generating such a null distribution for *SM* are virtually identical with those for the Durbin-Skillings-Mack statistic *D* (7.43) for balanced incomplete block designs, the reader is referred to Comment 49 for illustration of the details of the process.

65. *Large-Sample Approximation*. Let A be the vector defined in (7.49). Since each A_j is a weighted sum of centered ranks, it is not surprising (see Skillings and Mack (1981) for more details) that a properly standardized version of A has an asymptotic (n tending to infinity) $(k-1)$-variate normal distribution with mean vector $\mathbf{0} = (0,\ldots,0)$ and covariance matrix $\sum_0$ (7.50) when the null hypothesis H_0 is true and $\lambda_{qt} > 0$, for every $q \neq t = 1,\ldots,k$. (Note once again that A does not include A_k, since A_k can be expressed as a weighted linear combination of $A_1,\ldots,A_{k-1}$. This is the reason that the asymptotic normal distribution is $(k-1)$-variate and not k-variate.) Since the test statistic *SM* (7.52) is a quadratic form in the variables $A_1,\ldots,A_{k-1}$, it is therefore quite natural that *SM* has an asymptotic (n tending to infinity) chi-square distribution with $k-1$ degrees of freedom when the null hypothesis H_0 is true and $\lambda_{qt} > 0$, for every $q \neq t = 1,\ldots,k$.

66. *Exact Conditional Null Distribution of SM with Ties among the Observed X Values within Blocks*. To have a test with exact significance level even in the presence of tied *X*'s within some of the blocks, we need to consider all $s_1!s_2!\ldots s_k!$ block rank configurations for the *observed* data, where now these within-blocks ranks are obtained using average ranks to break ties. As in Comment 64, it still follows that under H_0 each of the $s_1!s_2!\ldots s_k!$ observed block rank configurations (now with these tied ranks) is equally likely. For each such configuration, the value of *SM* (7.52) is computed and the results tabulated. Since the specifics of generating such a conditional null distribution for *SM* in the case of tied within-blocks observations are virtually identical with those for the case of tied observations with the Durbin-Skillings-Mack statistic *D* (7.43) for balanced incomplete block designs, the reader is referred to Comment 51 for illustration of the details of the process.

67. *Settings Where the Chi-Square Approximation Is Not Applicable*. We note that the sole condition (other than the number of blocks becoming large) for the chi-square approximation to be applicable is that each of the λ_{qt}'s, $q \neq t = 1,\ldots,k$, must be positive. In settings where at least one of the λ_{qt}'s is zero, the approximate procedure (7.54) is not applicable. For such cases, one could still use procedure (7.53) by generating the (exact or simulated) null distribution of *SM* (7.52) and obtaining the appropriate critical values. On the other hand, if $\lambda_{qt} = 0$ for a particular pair of treatments q and t (so that q and t never appear together in a block), then procedure (7.53) would not necessarily be effective in testing H_0 (7.2) even when τ_q and τ_t are quite different.

68. *Historical Development*. The test procedures (7.53) and (7.54) were proposed and studied by Skillings and Mack (1981). They provided some exact null distribution critical values for the special case of BIBD's (see Section 7.6) and for a second special case where we are only a single observation short of having complete block data.

Properties

1. *Asymptotic Chi-Squaredness*. See Skillings and Mack (1981).

PROBLEMS

76. In the data on percent reduction in average wind speed due to shelterbelts discussed in Problem 21, the month of November was omitted from the data in Table 7.7 because the percent reduction observation at 20 meters was missing. In Table 7.16 we again present these data with the month of November included.

TABLE 7.16. Percent Reduction in Average Wind Speed at Dambatta, 1980/81

Month	Leeward Distance from Shelterbelt (m)				
	20	40	100	150	200
January	22.1	20.7	15.4	12.3	6.9
February	19.2	18.7	14.9	9.3	6.5
March	21.5	21.9	14.3	9.9	7.1
April	21.5	21.2	11.1	9.4	6.2
May	21.3	20.9	11.2	9.4	7.7
June	20.9	19.6	16.9	11.6	7.0
August	19.3	18.7	14.4	12.5	7.0
September	20.1	19.6	15.6	12.6	7.5
October	23.7	20.4	14.6	12.4	8.5
November	19.5	18.4	13.8	8.4	—

Source: J. E. Ujah and K. B. Adeoye (1984).

Use the Skillings-Mack exact procedure (7.53) to test the hypothesis that there is a difference in percent reduction in average wind speed over the five leeward distances from a shelterbelt. (Note that the conclusion of the Skillings-Mack procedure applied to these data is not directional, as was the case with the decision in Problem 21 using the Page ordered alternatives procedure and the randomized complete block design data without the month of November. A corresponding ordered alternatives analogue to the Skillings-Mack procedure for arbitrary missing data is not available in the literature.)

77. Skillings and Mack (1981) consider the experiment evaluating four methods of assembling a product, where the blocking factor corresponds to the individual assembly workers. The data for this experiment are presented in Table 7.17, where the observations are assembly times in minutes. The missing observations are due to machinery breakdowns or employee absenteeism. Use these data to

TABLE 7.17. Assembly Times (m)

Workers	Assembly Methods			
	A	B	C	D
1	3.2	4.1	3.8	4.2
2	3.1	3.9	3.4	4.0
3	4.3	3.5	4.6	4.8
4	3.5	3.6	3.9	4.0
5	3.6	4.2	3.7	3.9
6	4.5	4.7	3.7	—
7	—	4.2	3.4	—
8	4.3	4.6	4.4	4.9
9	3.5	—	3.7	3.9

Source: J. H. Skillings and G. A. Mack (1981).

assess whether there are any differences among the assembly methods with regard to median time for assembly of the product.

78. Verify that expression (7.55) is an alternative way to compute the statistic SM (7.52) when only a single treatment has missing data.

79. Consider Table 7.18, which gives a subset of the rounding first base data in Table 7.1 obtained by Woodward (1970) in his study of the best method of rounding first base to minimize the time to second base. (The missing observations might be due to injury during one of the other runs.) Use these data to assess whether there are any differences among median times to second base for these three ways of rounding first base.

TABLE 7.18. Rounding First Base Times

	Methods		
Players	Round Out	Narrow Angle	Wide Angle
1	5.40	5.50	5.55
2	5.85	5.70	5.75
3	5.20	5.60	5.50
4	5.55	5.50	—
5	5.90	5.85	5.70
6	5.45	5.55	5.60
7	5.40	5.40	5.35
8	—	5.50	5.35
9	5.25	5.15	5.00
10	5.85	—	5.70
11	5.25	5.20	5.10
12	5.65	5.55	—
13	5.60	5.35	5.45
14	5.05	—	4.95
15	5.50	5.50	5.40
16	—	5.55	5.50
17	5.55	5.55	—
18	5.45	5.50	5.55
19	5.50	5.45	5.25
20	5.65	5.60	5.40
21	5.70	5.65	5.55
22	6.30	6.30	6.25

Source: W. F. Woodward (1970).

80. We noted that the value of the test statistic SM (7.52) does not depend on the form of the particular generalized inverse $\overline{\sum}_0$ used in the calculation. Illustrate this fact by computing SM using two different generalized inverses for a setting where the rank of $\sum_0$ is not $k - 1$.

81. Consider Table 7.19, which gives a subset of the serum creatine phosphokinase (CPK) activity data in Table 7.3 obtained by Goode and Meltzer (1976) in their study of the effect of isometric exercise on serum CPK levels. Use these data to assess whether there are any differences in median serum CPK activity (in mU/ℓ) for the three measurement periods.

82. Verify for the data in Table 7.15 of Example 7.8 that the value of $SM = 13.287$ would also be obtained if we take $\boldsymbol{A}$ to be (A_1, A_3) or (A_2, A_3) and make the corresponding changes in the definition of the null covariance matrix $\sum_0$ (7.50).

83. Verify that the Skillings-Mack statistic SM (7.52) simplifies to the closed form expression for the Friedman statistic S (7.5) when we have data from a randomized complete block design.

84. Verify that the Skillings-Mack statistic SM (7.52) simplifies to the closed form expression for the Durbin-Skillings-Mack statistic D (7.43) when we have data from a balanced incomplete block design.

TABLE 7.19. Effect of Isometric Exercise on Serum Creatine Phosphokinase (CPK) Activity (mU/ℓ) in Psychotic Patients

Subject	Preexercise	19 h Postexercise	42 h Postexercise
1	27	101	82
2	30	112	50
3	24	26	68
4	54	89	—
5	21	30	49
6	36	41	48
7	36	29	46
8	16	20	8
9	21	26	25

Source: D. J. Goode and H. Y. Meltzer (1976).

85. Consider the setting corresponding to $k = 4$ and $n = 10$ where we have only a single missing observation in one of the blocks. Compare the critical region for the exact level $\alpha = .0501$ test of H_0 (2) based on SM with the critical region for the corresponding nominal level $\alpha = .0501$ test based on the large-sample approximation.

86. Consider the setting corresponding to $k = 6$ and $n = 5$ where we have only a single missing observation in one of the blocks. Compare the critical region for the exact level $\alpha = .0499$ test of H_0 (7.2) based on SM with the critical region for the corresponding nominal level $\alpha = .0499$ test based on the large-sample approximation.

87. Consider the incomplete block data setting corresponding to $k = 3, n = 3, s_1 = s_2 = 3,$ and $s_3 = 2$. Obtain the form of the exact null H_0 distribution of SM (7.52) for the case of no tied observations.

88. Consider the incomplete block data setting corresponding to $k = 3, n = 3, s_1 = s_2 = 3,$ and $s_3 = 2$. Suppose the three observations in block 2 are tied at a single value. Obtain the conditional exact probability distribution of SM (7.52) under H_0 (7.2) when average ranks are used to break this set of within-block ties. Compare this conditional null distribution of SM with the null distribution of SM obtained in Problem 87 when there are no ties.

REPLICATIONS—TWO-WAY LAYOUT WITH AT LEAST ONE OBSERVATION FOR EVERY TREATMENT-BLOCK COMBINATION

It is often the case in two-way layout settings that we have more than one observation for some of the treatment-block combinations. These multiple observations in a given cell are referred to as replications for that treatment-block combination. Of course, permitting such replications opens the possibility of a much wider variety of data configurations for our two-way layout. There could be some cells with no observations, some with one observation, and some with multiple observations. In the next two sections, we emphasize nonparametric procedures for general alternatives in the setting where we have a common, equal number $c > 1$ of replications for every treatment-block combination. Direct extensions of these general alternatives procedures to less restrictive settings where the number of replications need not be equal but there are no empty cells are discussed in Comment 77. Nonparametric procedures that are valid for the most general two-way layout settings where there may be a mix of cells with more than one observation (i.e., replications), cells with a single observation, and empty cells with no observations are discussed in Comment 78 for the cases of general and ordered alternatives.

In Section 7.9 we present a distribution-free hypothesis test for general alternatives when we have an equal number (> 1) of replications for every treatment-block combination. In Section 7.10 we discuss an all-treatments multiple comparison procedure for the same setting.

Throughout these two sections, we continue to operate under the general conditions of Assumptions A1–A3. However, in Sections 7.9 and 7.10, we impose the additional constraint that each c_{ij} is equal to $c(> 1)$ and, thus, that

$$N = \sum_{i=1}^{n} \sum_{j=1}^{k} c_{ij} = nkc.$$

7.9. A DISTRIBUTION-FREE TEST FOR GENERAL ALTERNATIVES IN A RANDOMIZED BLOCK DESIGN WITH AN EQUAL NUMBER $c(> 1)$ OF REPLICATIONS PER TREATMENT-BLOCK COMBINATION (MACK-SKILLINGS)

In this section we present a procedure for testing H_0 (7.2) against the general alternatives H_1 (7.3) for block data where we have an equal number $c > 1$ replications for each of the treatment-block combinations. Here the total number of observations is $N = nkc$.

Procedure

To compute the Mack-Skillings statistic for this equal replications setting, we first rank the observations from least to greatest separately within each of the n blocks. Let r_{ijq} be this within-block rank of X_{ijq} (the qth replication from the jth treatment in the ith block) among the kc total observations present in the ith block, for $i = 1, \ldots, n$. Set

$$S_j = \sum_{i=1}^{n} \left[\sum_{q=1}^{c} r_{ijq}/c \right], \quad \text{for } j = 1, \ldots, k. \tag{7.56}$$

Thus, S_j is the sum (across blocks) of the cellwise averages of the within-blocks ranks assigned to the c observations from treatment j, for $j = 1, \ldots, k$. The Mack-Skillings statistic for equal replications is then given by

$$MS = \left[\frac{12}{k(N+n)} \right] \sum_{j=1}^{k} \left[S_j - \frac{N+n}{2} \right]^2,$$

$$= \left[\frac{12}{k(N+n)} \right] \left\{ \sum_{j=1}^{k} S_j^2 \right\} - 3(N+n), \tag{7.57}$$

where $(N+n)/2n = (kc+1)/2 = \sum_{j=1}^{k} \sum_{q=1}^{c} r_{ijq}/kc$ is the average within-blocks rank assigned for each of the n blocks. It follows that $n(N+n)/2n = (N+n)/2$ is the expected sum (across blocks) of the cellwise averages for each of the k treatments when H_0 (7.2) is true; that is, $(N+n)/2$ is the expected value of S_j, for each $j = 1, \ldots, k$, when the null hypothesis H_0 is true.

To test

$$H_0 : [\tau_1 = \cdots = \tau_k]$$

versus the general alternative

$$H_1 : [\tau_1, \ldots, \tau_k \text{ not all equal}],$$

at the α level of significance,

$$\text{Reject } H_0 \text{ if } MS \geq ms_\alpha; \quad \text{otherwise do not reject,} \tag{7.58}$$

where the constant ms_α is chosen to make the type I error probability equal to α. Selected values of ms_α are given in Table A.28 for $k = 2(1)5, n = 2(1)5$, and $c = 2(1)5$.

Large-Sample Approximation

When H_0 (7.2) is true, the statistic MS has, as the common number of observations on each treatment, nc, tends to infinity, an asymptotic chi-square (χ^2) distribution with $k - 1$ degrees of freedom. (See Comment 74 for indications of the proof.) The chi-square approximation for procedure (7.58) is

$$\text{Reject } H_0 \text{ if } MS \geq \chi^2_{k-1,\alpha}; \quad \text{otherwise do not reject,} \tag{7.59}$$

where $\chi^2_{k-1,\alpha}$ is the upper α percentile of a chi-square distribution with $k - 1$ degrees of freedom. Values of $\chi^2_{k-1,\alpha}$ can be obtained from Chart A.2.

Mack and Skillings (1980) have pointed out that this chi-square approximation is adequate when the significance level α is at least .05 and the number of replications c is at least 4, even though it is slightly conservative when the level nears .05. However, for significance levels as low as .01, they note that the conservative nature of the approximate procedure (7.59) can be somewhat severe unless the common number of replications c is rather large. Whenever possible they recommend the use of the exact procedure (7.58) for such small significance levels.

Ties

If there are ties among the X observations within any of the blocks, use average ranks to break the ties and compute the individual sums of cell-wise averages of the within-blocks ranks $S_1, \ldots, S_k$. In such cases, the significance level associated with procedure (7.58) is only approximately equal to α. (See Comment 75 for discussion of how to construct an exact conditionally distribution-free test of H_0 even when there are tied observations within some of the blocks.)

Example 7.9: *Determination of Niacin in Bran Flakes.* In a study to investigate the precision and homogeneity of a procedure for assessing the amount of niacin in bran flakes, Campbell and Pelletier (1962) prepared homogenized samples of bran flakes enriched with 0, 4, or 8 mg niacin per 100 g of cereal. Portions of the homogenized samples were sent to different laboratories, which were asked to carry out the specified procedure for each of three separate samples. The resulting data (in milligrams per 100 g bran flakes) for a subset (4 out of 12) of the laboratories included in the study are presented in Table 7.20.

7.9. TEST FOR GENERAL ALTERNATIVES FOR A BLOCK DESIGN WITH REPLICATIONS

TABLE 7.20. Amount of Niacin in Enriched Bran Flakes

Laboratory	Amount of Niacin Enrichment (Milligrams per 100 g Bran Flakes)		
	0	4	8
1	7.58 (3)	11.63 (7)	15.00 (2)
	7.87 (8)	11.87 (11)	15.92 (9)
	7.71 (6)	11.40 (3)	15.58 (4)
2	8.00 (9.5)	12.20 (12)	16.60 (12)
	8.27 (12)	11.70 (8.5)	16.40 (11)
	8.00 (9.5)	11.80 (10)	15.90 (7)
3	7.60 (4)	11.04 (2)	15.87 (6)
	7.30 (1)	11.50 (5.5)	15.91 (8)
	7.82 (7)	11.49 (4)	16.28 (10)
4	8.03 (11)	11.50 (5.5)	15.10 (3)
	7.35 (2)	10.10 (1)	14.80 (1)
	7.66 (5)	11.70 (8.5)	15.70 (5)

Source: J. A. Campbell and O. Pelletier (1962).

Of primary interest here is the precision of the laboratory procedure for determining niacin content in bran flakes. The actual amount of niacin enrichment in the prepared bran flakes serves only as a "nuisance" blocking factor in our evaluation of the consistency of the results across the four laboratories for which data are included in Table 7.20. Hence, we have data from a two-way layout with $k = 4$ treatments (laboratories), $n = 3$ blocks (amounts of niacin enrichment), and $c = 3$ replications (individual bran flake samples) per laboratory/enrichment combination. For purpose of illustration, we consider the significance level $\alpha = .0501$. From Table A.28 with $k = 4, n = 3$, and $c = 3$, we see that $ms_{.0501} = 7.479$ and procedure (7.58) becomes

$$\text{Reject } H_0 \text{ if } MS \geq 7.479.$$

Now, we illustrate the computations leading to the sample value of MS. The numbers in parentheses after the data values in Table 7.20 are the within-enrichment-levels (i.e., blocks) ranks (using average ranks to break ties) of the niacin content measurements obtained from the four laboratories. Using these block ranks, we obtain the following sums of cellwise averages for the four laboratories:

$$S_1 = \frac{3 + 8 + 6 + 7 + 11 + 3 + 2 + 9 + 4}{3} = 17.67,$$

$$S_2 = \frac{9.5 + 12 + 9.5 + 12 + 8.5 + 10 + 12 + 11 + 7}{3} = 30.5,$$

$$S_3 = \frac{4 + 1 + 7 + 2 + 5.5 + 4 + 6 + 8 + 10}{3} = 15.83,$$

and

$$S_4 = \frac{11 + 2 + 5 + 5.5 + 1 + 8.5 + 3 + 1 + 5}{3} = 14.$$

Hence, with $k = 4$, $n = 3$, and $N = 36$, we find from (7.57) that

$$MS = \left[\frac{12}{4(36 + 3)}\right] \{(17.67)^2 + (30.5)^2 + (15.83)^2 + (14)^2\} - 3(36 + 3)$$

$$= \left[\frac{1}{13}\right] \{312.23 + 930.25 + 250.59 + 196\} - 117 = 12.93.$$

Since the observed value of MS is greater than the critical value 7.479, we can reject H_0 at the $\alpha = .0501$ level, providing rather strong evidence that the studied process for assessing niacin content in bran flakes does not produce consistent results across a variety of laboratories and is therefore not reliable as an evaluative procedure. (In fact, from Table A.28, we note that we would even have rejected the null hypothesis at significance level $\alpha = .0099$, since the observed value $MS = 12.93$ is also larger than $ms_{.0099} = 10.573$. Using the chi-square approximation with $k - 1 = 3$ degrees of freedom, we also see from Chart A.2 that the P-value for these data is approximately .005.)

We should note in passing that there also appears to be an even more basic problem with this studied procedure for assessing niacin content in bran flakes and that is the accuracy (in addition to the lack of reproducibility) of the numerical values of the measurements. For example, for those samples enriched with 4 mg niacin per 100 g bran flakes, the values obtained by applying this procedure to the sample bran flakes ranged from 10.10 mg to 12.20 mg per 100 g bran flakes, well over the preestablished niacin content. (Similar comments apply to the 0 and 8 mg enrichment samples.) This clearly indicates a rather severe basic calibration problem with the assessment procedure, in addition to the lack of portability across laboratories detected by our application of the Mack-Skillings procedure to these data.

Comments

69. *More General Setting*. We could replace Assumptions A1–A3 and H_0 (7.2) with the more general null hypothesis that all possible $[(ck)!]^n$ configurations for the permutations of the within-blocks ranks (r_{ijq}'s) are equally likely. Procedure (7.58) remains distribution-free for this more general null hypothesis.

70. *Motivation for the Test*. Under Assumptions A1–A3 and H_0 (7.2), each of the block rank vectors $R_i^* = (r_{i11}, \ldots, r_{i1c}, r_{i21}, \ldots, r_{i2c}, \ldots, r_{ik1}, \ldots, r_{ikc})$, $i = 1, \ldots, n$, has a uniform distribution over the set of all $(ck)!$ permutations of the vector of integers $(1, 2, \ldots, ck)$ and this is true, independently, for each of the n blocks. It is then the case that $E_0(r_{ijq}) = (ck + 1)/2$ for every $i = 1, \ldots, n$; $j = 1, \ldots, k$; and $q = 1, \ldots, c$. It follows from (7.56) that $E_0(S_j) = nc(ck + 1)/2c = (nck + n)/2 = (N + n)/2$. Since the test statistic MS (7.57) is a constant times a sum of squared differences between the observed treatment sums of cellwise average ranks, S_j, and their common null expected value, $E_0(S_j) = (N + n)/2$, small values of MS represent agreement with H_0 (7.2). When the τ's are not all equal, we would expect a portion of the associated treatment sums of cellwise average ranks, S_j, to differ from their common null expectation, $(N + n)/2$, with some tending to smaller and some larger. The net result (after squaring the observed differences to obtain the $[S_j - (N + n)/2]^2$ terms) would be a large value of MS. This quite naturally suggests rejecting H_0 in favor of H_1 (7.3) for larger values of MS and motivates procedures (7.58) and (7.59).

71. *Special Case of $c = 1$*. When we have a single observation for every treatment-block combination (i.e., $c = 1$), we are dealing with data from a complete randomized block design. In this setting, the Mack-Skillings statistic MS (7.57) is equivalent to the Friedman statistic S (7.5). Thus the Mack-Skillings procedures (7.58) and (7.59) represent natural extensions of

the Friedman procedures (7.6) and (7.7), respectively, to the case of an equal number $c > 1$ of replications per cell.

72. *Assumptions.* We emphasize that Assumption A3 stipulates that the nk cell distributions F_{ij} can differ at most in their locations (medians) and that these location differences (if any) must be a result of additive block and/or treatment effects (i.e., there is no interaction between the treatment and block factors). In particular, Assumption A3 requires that the ns underlying distributions belong to the same general family (F) and that they do not differ in scale parameters (variability). We do note, however, that the test procedure (7.58) remains distribution-free under the less restrictive setting where Assumption A3 is replaced by the weaker condition Assumption A3' stated in Comment 43. Assumption A3 then corresponds to Assumption A3' with the additional condition that $F_1 \equiv \cdots \equiv F_n$. (Also see Comment 69.)

73. *Derivation of the Distribution of MS under H_0 (No-Ties Case).* The null distribution of *MS* (7.57) can be obtained by using the fact that under H_0 (7.2), all possible $[(ck)!]^n$ configurations for the permutations of the within-blocks ranks (r_{ijq}'s) are equally likely. Thus, to obtain the exact null distribution of *MS*, we compute its value for each of these $[(ck)!]^n$ block rank configurations and then tabulate the collected outcomes. We must point out, of course, that the number $[(ck)!]^n$ of configurations for which we need to compute the value of *MS* can get large rather quickly, as either k or c is moderately increased. Since the specifics of generating such a null distribution for *MS* are virtually identical with those for the Durbin-Skillings-Mack statistic D (7.43) for balanced incomplete block designs, the reader is referred to Comment 49 for illustration of the details of the process.

74. *Large-Sample Approximation.* Define the centered treatment sums of cellwise average ranks $S_j^* = S_j - E_0(S_j) = S_j - (N + n)/2$, for $j = 1, \ldots, k$, and set $\mathbf{S}^* = (S_1^*, \ldots, S_{k-1}^*)$. Since each S_j is an average, it is not surprising (see Mack and Skillings (1980) for more details) that a properly standardized version of $\mathbf{S}^*$ has an asymptotic (nc tending to infinity) $(k - 1)$-variate normal distribution with mean vector $\mathbf{0} = (0, \ldots, 0)$ and appropriate covariance matrix Σ^* when the null hypothesis H_0 is true. (Note that $\mathbf{S}^*$ does not include S_k^*, since S_k^* can be expressed as a weighted linear combination of $S_1^*, \ldots, S_{k-1}^*$. This is the reason that the asymptotic normal distribution is $(k - 1)$-variate and not k-variate.) Since the test statistic *MS* (7.57) is a quadratic form in the variables $(S_1^*, \ldots, S_{k-1}^*)$, it is therefore quite natural that *MS* has an asymptotic (nc tending to infinity) chi-square distribution with $k - 1$ degrees of freedom when the null hypothesis H_0 is true.

75. *Exact Conditional Null Distribution of MS with Ties among the X Values within Blocks.* To have a test with exact significance level even in the presence of tied X's within some of the blocks, we need to consider all $[(ck)!]^n$ block rank configurations for the observed data, where now these within-blocks ranks are obtained using average ranks to break ties. As in Comment 73, it still follows that under H_0 each of the $[(ck)!]^n$ observed block rank configurations (now with these tied ranks) is equally likely. For each such configuration, the value of *MS* (7.57) is computed and the results tabulated. Since the specifics of generating such a conditional null distribution for *MS* in the case of tied within-blocks observations are virtually identical with those for the case of tied observations with the Durbin-Skillings-Mack statistic D (7.43) for balanced incomplete block designs, the reader is referred to Comment 51 for illustration of the details of the process.

76. *Simple Competitor Procedure when the Number of Replications Is the Same for Every Cell.* As an alternative to the Mack-Skillings procedure (7.58), we could first compute the median of the c replications separately in each of the nk cells and then apply either the Friedman procedure (7.6) or the Page procedure (7.11), whichever is appropriate for the alternatives of interest, to these nk cell medians (which now represent data from a complete randomized block design). In general, this approach could result in substantial loss of information, especially

when the number of replications per cell, c, is large. However, it is simple and does provide the only available nonparametric procedure for dealing specifically with ordered alternatives when we have an equal number (> 1) of replications per cell. (We note, in passing, that any appropriate measure of central tendency, such as the cell means or the medians of the Walsh averages (see Comment 3.17) for the individual cell data, could be used instead of the cell medians to summarize the data prior to application of the Friedman or Page procedure.)

77. *Extension to Arbitrary Replication (≥ 1) Configurations.* We have described the Mack-Skillings procedure in detail for the setting where we have the same number of replications $c\ (\geq 1)$ for each of the treatment-block combinations. However, in their original work, Mack and Skillings (1980) proposed a more general test procedure that is appropriate for any two-way layout setting for which we have at least one replication for every treatment-block combination (i.e., there are no empty cells). We now present their procedure for this more general setting where the only stipulation is that $c_{ij} > 0$ for every $i = 1, \ldots, n$ and $j = 1, \ldots, k$.

For $i = 1, \ldots, n$, let $q_i = \sum_{j=1}^{k} c_{ij}$ be the total number of observations present in the ith block. Once again we rank the observations from least to greatest within each of the blocks and let r_{iju} denote the rank of X_{iju} within the q_i observations present in the ith block, for $u = 1, \ldots, c_{ij}; i = 1, \ldots, n;$ and $j = 1, \ldots, k$. For each treatment, compute the sum of cellwise weighted average ranks

$$V_j = \sum_{i=1}^{n} \sum_{u=1}^{c_{ij}} \frac{r_{iju}}{q_i}, \quad j = 1, \ldots, k. \tag{7.60}$$

Define the vector

$$\begin{aligned} \mathbf{V} &= (V_1 - E_0[V_1], \ldots, V_{k-1} - E_0[V_{k-1}]) \\ &= \left(V_1 - \sum_{i=1}^{n} \left[\frac{c_{i1}(q_i + 1)}{2q_i} \right], \ldots, V_{k-1} - \sum_{i=1}^{n} \left[\frac{c_{i,k-1}(q_i + 1)}{2q_i} \right] \right). \end{aligned} \tag{7.61}$$

Thus the components of $\mathbf{V}$ are the sums of cellwise weighted average ranks centered about their expected values under H_0. (Without loss of generality, we have chosen to omit the centered V_k from the vector $\mathbf{V}$. The V_j's are linearly dependent, since a weighted linear combination of all k of them is a constant. We could omit any one of the V_j's in the definition of $\mathbf{V}$ and the procedure we now describe would lead to the same value of the test statistic. For further discussion, see Mack and Skillings (1980).)

The covariance matrix for $\mathbf{V}$ under H_0 (7.2) has the form $\Sigma_{\mathbf{V},0} = ((\sigma_{s,t}))$, where

$$\begin{aligned} \sigma_{s,t} &= \sum_{i=1}^{n} \left[\frac{c_{is}(q_i - c_{is})(q_i + 1)}{12 q_i^2} \right], \quad \text{for } s = t = 1, \ldots, k-1 \\ &= -\sum_{i=1}^{n} \left[\frac{c_{is} c_{it}(q_i + 1)}{12 q_i^2} \right], \quad \text{for } s \neq t = 1, \ldots, k-1. \end{aligned} \tag{7.62}$$

The rank of the matrix $\Sigma_{\mathbf{V},0}$ is $k - 1$. Letting $\Sigma_{\mathbf{V},0}^{-1}$ denote the inverse of $\Sigma_{\mathbf{V},0}$, the Mack-Skillings test statistic for this general setting of unequal, but positive, numbers of replications in the various treatment-block combinations, is given by

$$MS_g = \mathbf{V} \Sigma_{\mathbf{V},0}^{-1} \mathbf{V}'. \tag{7.63}$$

7.9. TEST FOR GENERAL ALTERNATIVES FOR A BLOCK DESIGN WITH REPLICATIONS

To test

$$H_0 : [\tau_1 = \cdots = \tau_k]$$

versus the general alternative

$$H_1 : [\tau_1, \tau_2, \ldots, \tau_k \text{ not all equal}],$$

at the α level of significance, the Mack-Skillings general procedure is then to

$$\text{Reject } H_0 \text{ if } MS_g \geq ms_{g,\alpha}; \quad \text{otherwise do not reject,} \tag{7.64}$$

where the constant $ms_{g,\alpha}$ is chosen to make the type I error probability equal to α.

The critical values $ms_{g,\alpha}$ are available in the literature only for the setting where we have an equal number, c, of replications in each cell, in which case the general Mack-Skillings test procedure (7.64) is equivalent to the equal replications version given in equation (7.58). However, when H_0 (7.2) is true, the general form statistic MS_g has, as N tends to infinity in such a way that c_{ij}/N tends to $\rho_{ij} > 0$ for every $i = 1, \ldots, n$ and $j = 1, \ldots, k$, an asymptotic chi-square (χ^2) distribution with $k - 1$ degrees of freedom. Thus, when N is large, the chi-square approximation for the general Mack-Skillings procedure (7.64) is

$$\text{Reject } H_0 \text{ if } MS_g \geq \chi^2_{k-1,\alpha}; \quad \text{otherwise do not reject,} \tag{7.65}$$

where $\chi^2_{k-1,\alpha}$ is the upper α percentile of a chi-square distribution with $k - 1$ degrees of freedom. Values of $\chi^2_{k-1,\alpha}$ can be obtained from Chart A.2.

78. *Competitor Procedures Applicable for Most General Two-Way Layout Settings Where There Are Both Replications and Empty Cells.* Thus far in this chapter we have discussed procedures that are appropriate either for settings where we have either 0 or 1 observation for every treatment-block combination or for settings where we have at least one observation in every cell. None of these procedures are appropriate for the most general settings that represent a combination of these two structures, namely, those data sets where we have replications ($c_{ij} > 1$) for some treatment-block combinations and no observations ($c_{ij} = 0$) for others. We briefly discuss now two test procedures for such general two-way layout settings, one designed for general alternatives to H_0 (7.2) and the second specifically oriented toward detecting ordered alternatives.

Let k_i be the number of treatments in the ith block for which $c_{ij} > 0$, for $i = 1, \ldots, n$. (Once again, we discard any block i for which $k_i = 1$, as such a block contains no information relative to possible differences in the treatment effects. Notationally, then, n represents the number of blocks remaining after discarding blocks with observations on only a single treatment.)

General Alternatives. We first compute the one-way layout Kruskal-Wallis statistic H (6.5) separately in each of the n blocks. Letting H_i denote this Kruskal-Wallis statistic for the ith block, $i = 1, \ldots, n$, the statistic considered by Mack (1981) for this most general two-way layout setting is given by

$$H_{\text{tot}} = \sum_{i=1}^{n} H_i. \tag{7.66}$$

The level α test of H_0 (7.2) versus the general alternatives H_1 (7.3) studied by Mack (1981) is

$$\text{Reject } H_0 \text{ if } H_{\text{tot}} \geq h^*_\alpha; \quad \text{otherwise do not reject,} \tag{7.67}$$

where the constant h_α^* is chosen to make the type I error probability equal to α. Values of h_α^* are available in Mack (1981) for $k = 3, n = 2, 3$ and all combinations of replications $0 \leq c_{ij} \leq 3$, as well as for $k = 3, n = 4, 5$ and all combinations of replications $0 \leq c_{ij} \leq 2$. Additional values of h_α^* can be found in DeKroon and Van der Laan (1981) for $\alpha = .01, .05$ and various combinations of k, n, and equal number of replications c in the ranges $2 \leq k \leq 4, 1 \leq n \leq 10$, and $2 \leq c \leq 4$. (We note, in passing, that procedure (7.67) can be particularly sensitive to a large degree of interaction between the treatment and block factors. In the presence of such extensive interaction, it is possible that a rejection of H_0 with procedure (7.67) could be a direct consequence of this interaction, rather than because of any significant differences in the treatment effects $\tau_1, \ldots, \tau_k$.)

When H_0 (7.2) is true, the statistic H_{tot} has, as min (nonzero $c_{ij}, i = 1, \ldots, n; j = 1, \ldots, k$) tends to infinity, an asymptotic chi-square (χ^2) distribution with $d = (k_1 + k_2 + \cdots + k_n - n)$ degrees of freedom (see Mack (1981) for details). Thus, when the minimum nonzero c_{ij} is large, the chi-square approximation for procedure (7.67) is

$$\text{Reject } H_0 \text{ if } H_{\text{tot}} \geq \chi^2_{d,\alpha}; \quad \text{otherwise do not reject,} \tag{7.68}$$

where $\chi^2_{d,\alpha}$ is the upper α percentile point of a chi-square distribution with d degrees of freedom. Values of $\chi^2_{d,\alpha}$ can be obtained from Chart A.2.

Ordered Alternatives. If we are interested in ordered alternatives, H_2 (7.9), we first compute the one-way layout Jonckheere-Terpstra statistic J (6.13) separately in each of the n blocks. Letting J_i denote this Jonckheere-Terpstra statistic for the ith block, $i = 1, \ldots, n$, the statistic proposed by Skillings and Wolfe (1977, 1978) for this most general two-way layout ordered alternatives setting is given by

$$J_{\text{tot}} = \sum_{i=1}^{n} J_i. \tag{7.69}$$

The level α test of H_0 (7.2) versus the ordered alternatives H_2 (7.9) suggested by Skillings and Wolfe (1977, 1978) is

$$\text{Reject } H_0 \text{ if } J_{\text{tot}} \geq j_\alpha^*; \quad \text{otherwise do not reject,} \tag{7.70}$$

where the constant j_α^* is chosen to make the type I error probability equal to α. Values of j_α^* are available in Skillings (1980) for $k = 2(1)6, n = 2(1)5$ and selected configurations of the c_{ij}'s such that $c_{ij} = C_i$, for $i = 1, \ldots, n$ and $j = 1, \ldots, k$ (i.e., within a given block, each treatment has the same number of observations C_i, but $C_1, C_2, \ldots, C_n$ need not all be equal). (We note that procedure (7.70) does not have the same sensitivity to the presence of extensive interaction as does the general alternatives procedure (7.67). Rejection of H_0 with procedure (7.70) will always be indicative of the presence of an ordered structure on the treatment effects $\tau_1, \ldots, \tau_k$.)

When H_0 (7.2) is true, the standardized form

$$J_{\text{tot}}^* = \frac{J_{\text{tot}} - E_0(J_{\text{tot}})}{[\text{var}_0(J_{\text{tot}})]^{1/2}} \tag{7.71}$$

has, as min (nonzero $c_{ij}, i = 1, \ldots, n; j = 1, \ldots, k$) tends to infinity, an asymptotic $N(0, 1)$ distribution (see Skillings and Wolfe (1977, 1978) for details), where

$$E_0(J_{\text{tot}}) = \frac{\sum_{i=1}^{n} \left[q_i^2 - \sum_{j=1}^{k} c_{ij}^2 \right]}{4} \tag{7.72}$$

and

$$\text{var}_0(J_{\text{tot}}) = \frac{\sum_{i=1}^{n}\left[q_i^2(2q_i+3) - \sum_{j=1}^{k} c_{ij}^2(2c_{ij}+3)\right]}{72}, \qquad (7.73)$$

are the expected value and variance, respectively, of J_{tot} (7.69) under the null hypothesis H_0 and $q_i = c_{i1} + \cdots + c_{ik}$ is the total number of observations present in the ith block, $i = 1, \ldots, n$. Thus, when the minimum nonzero c_{ij} is large, the normal theory approximation for procedure (7.70) is

$$\text{Reject } H_0 \text{ if } J_{\text{tot}}^* \geq z_\alpha; \quad \text{otherwise do not reject.} \qquad (7.74)$$

79. *Historical Development.* Mack and Skillings (1980) proposed and studied a general test procedure for an arbitrary two-way layout setting where we have at least one observation for every treatment-block combination (see Comment 77). For the special case of an equal number of replications, c, in every cell, their general test procedure simplifies to the expression in (7.58) based on the test statistic MS. They also provided some exact null distribution critical values sm_α in this equal replications setting for a variety of combinations of k, n, and c.

Properties

1. *Asymptotic Chi-Squaredness.* See Mack and Skillings (1980).
2. *Efficiency.* See Mack and Skillings (1980) and Section 7.16.

PROBLEMS

89. Rice (1988) considered an experiment to determine whether two forms of iron, Fe^{2+} and Fe^{3+}, are retained differently, with the goal of comparing their potentials for use as dietary supplements. One hundred eight mice were randomly divided into 6 groups of 18 mice each. Three of these groups were given Fe^{2+} in the different concentrations, 10.2, 1.2, and .3 millimolar, and three groups were given Fe^{3+} in the same concentrations. The iron was radioactively labeled so that a counter could be used to accurately measure the initial amount given, and it was administered orally to the mice. At a later time, a second count was obtained on each mouse, and the percentage of iron retained was recorded. The data in Table 7.21 are the percentages retained by each of the 108 mice.

    Use the Mack-Skillings large-sample procedure (7.59) to test the hypothesis that there is a difference across the concentrations studied between the two forms of iron Fe^{2+} and Fe^{3+} in percentage iron retained.

90. Let V_j be as defined in expression (7.60), for $j = 1, \ldots, k$. Show that

$$E_0[V_j] = \sum_{i=1}^{n} \left[\frac{c_{ij}(q_i+1)}{2q_i}\right],$$

as noted in expression (7.61), where q_i is the number of observations present in the ith block, for $i = 1, \ldots, n$.

91. Show that for the special case of one replication per cell (i.e., $c = 1$), the Mack-Skillings procedures (7.58) and (7.59) are equivalent to the Friedman procedures (7.6) and (7.7), respectively. (See Comment 71.)

92. Anderson and McLean (1974) considered the data from an experiment measuring the strength of a weld in steel bars. The two factors of interest in the experiment were the total time of the automatic weld cycle and the distance the weld die travels during the automatic weld cycle. Two weld-strength observations were collected at each combination of five different weld cycle times

TABLE 7.21. Percentage of Iron Retained

Concentration	Form of Iron					
	Fe^{2+}			Fe^{3+}		
.3 millimolar	2.71	5.43	6.38	2.25	3.93	5.08
	6.38	8.32	9.04	5.82	5.84	6.89
	9.56	10.01	10.08	8.50	8.56	9.44
	10.62	13.80	15.99	10.52	13.46	13.57
	17.90	18.25	19.32	14.76	16.41	16.96
	19.87	21.60	22.25	17.56	22.82	29.13
1.2 millimolar	4.04	4.16	4.42	2.20	2.93	3.08
	4.93	5.49	5.77	3.49	4.11	4.95
	5.86	6.28	6.97	5.16	5.54	5.68
	7.06	7.78	9.23	6.25	7.25	7.90
	9.34	9.91	13.46	8.85	11.96	15.54
	18.40	23.89	26.39	15.89	18.30	18.59
10.2 millimolar	2.20	2.69	3.54	0.71	1.66	2.01
	3.75	3.83	4.08	2.16	2.42	2.42
	4.27	4.53	5.32	2.56	2.60	3.31
	6.18	6.22	6.33	3.64	3.74	3.74
	6.97	6.97	7.52	4.39	4.50	5.07
	8.36	11.65	12.45	5.26	8.15	8.24

Source: J. A. Rice (1988).

and three different weld die travel distances (gage bar settings). These weld-strength data are given in Table 7.22.

Use the Mack-Skillings procedure to test the hypothesis that weld cycle time has an effect on the strength of a weld, at least over the weld die travel distances considered in the study.

93. For the weld-strength data in Table 7.22, compute the median of the two observations in each of the gage bar setting/weld cycle time combinations. Apply the Friedman procedure (7.6) to the resulting medians to test the hypothesis that weld cycle time has an effect on the strength of a weld, at least over the weld die travel distances in the study. Compare with the result obtained in Problem 92. (See also Comment 76.)

TABLE 7.22. Strength of Weld

Gage Bar Setting	Weld Cycle Times				
	1	2	3	4	5
1	10 12	13 17	21 30	18 16	17 21
2	15 19	14 12	30 38	15 11	14 12
3	10 8	12 9	10 5	14 15	19 11

Source: V. L. Anderson and R. A. McLean (1974).

94. Consider the Mack-Skillings statistic MS_g (7.63) for the most general two-way layout setting with at least one replication for every treatment-block combination, as discussed in Comment 77. Show that the test procedure (7.64) based on MS_g is equivalent to the equal replications test procedure (7.58) based on MS (7.57) when, in fact, we have an equal number, c, of replications for every treatment-block combination.

95. One method for the determination of coal acidity is based on the use of ethanolic NaOH. In an effort to assess the effect of the ethanolic NaOH concentration on the obtained acidity values, Sternhell (1958) studied three different NaOH concentrations (.404N, .626N, and .786N) in conjunction with

7.9. TEST FOR GENERAL ALTERNATIVES FOR A BLOCK DESIGN WITH REPLICATIONS

three different types of coal (Morwell, Yallourn, and Maddingley). The data in Table 7.23 are the resulting acidity values determined under each of these three concentration levels for two different samples from each type of coal.

TABLE 7.23. Coal Acidity Value

Type of Coal	NaOH Concentration		
	.404N	.626N	.786N
Morwell	8.27 8.17	8.03 8.21	8.60 8.20
Yallourn	8.66 8.61	8.42 8.58	8.61 8.76
Maddingley	8.14 7.96	8.02 7.89	8.13 8.07

Source: S. Sternhell (1958).

Use the Mack-Skillings procedure to test the hypothesis that the NaOH concentration has an effect on the measured coal acidity values, at least over the three types of coal included in this study.

96. Consider the percentage retained iron data in Table 7.21. Test the hypothesis that the iron concentration affects the percentage iron retention, regardless of which form of iron is involved.

97. What is the maximum value for the Mack-Skillings statistic MS (7.57) when there are c replications per cell? For what rank configuration is this maximum achieved?

98. Consider the setting corresponding to $k = 4, n = 5$, and $c = 3$ replications per cell. Compare the critical region for the exact level $\alpha = .0100$ test of H_0 (7.2) based on MS with the critical region for the corresponding nominal level $\alpha = .0100$ test based on the large-sample approximation.

99. Consider the setting corresponding to $k = 2, n = 2$, and $c = 2$ replications per cell. Obtain the form of the exact null H_0 distribution of MS (7.57) for the case of no tied observations.

100. Consider the setting corresponding to $k = 2, n = 2$, and $c = 2$ replications per cell. Suppose that one of the observations in the first cell (block 1 and treatment 1) is tied in value with one of the observations in the second cell (block 1 and treatment 2). Obtain the conditional exact probability distribution of MS (7.57) under H_0 (7.2) when average ranks are used to break this pair of within-blocks ties. Compare this conditional null distribution of MS with the null distribution of MS obtained in Problem 99 when there are no ties.

101. Consider the setting corresponding to $k = 2, n = 2$, and $c = 2$ replications per cell. Suppose that one of the observations in the first cell (block 1 and treatment 1) is tied in value with the other observation in the same cell. Obtain the conditional exact probability distribution of MS (7.57) under H_0 (7.2) when average ranks are used to break this pair of within-cell ties. Compare this conditional null distribution of MS with the null distributions of MS obtained in Problems 99 and 100 when there are no ties and ties between cells, respectively.

102. Consider the setting corresponding to $k = 5, n = 4$, and $c = 4$ replications per cell. Compare the critical region for the exact level $\alpha = .0500$ test of H_0 (7.2) based on MS with the critical region for the corresponding nominal level $\alpha = .0500$ test based on the large-sample approximation.

103. In a study to determine the effect of light on the release of luteinizing hormone (LH), Rice (1988) compared data for male and female rats kept in constant light with similar animals exposed to a regime of 14 h of light and 10 h of darkness. Five different dosages of a luteinizing release factor (LRF) were considered in the study and the measurement obtained from the animals was the level of LH (in nanograms per milliliter of serum) in blood samples collected after exposure to one of the light regimes in combination with one of the LRF dosages. We consider data for the male rats only.

Sixty male rats were randomly allocated to the various experimental settings in such a way that 6 rats were exposed to each of the ten combinations of light regime and LRH dosage. The LH level data for these 60 rats are given in Table 7.24.

Use the Mack-Skillings large-sample procedure (7.59) to test the hypothesis that degree of exposure to light has an effect on serum levels of LH across the LRH dosages included in the study.

TABLE 7.24. Serum Level of LH (Nanograms per Milliliter of Serum)

	Light Regime					
LRF Dosage	Constant Light				14 h Light/ 10 h Dark	
0 ng (control)	72	64	78	212	27	68
	20	56	70	72	130	153
10 ng	74	82	40	32	98	148
	87	78	88	186	203	188
50 ng	130	187	133	294	306	234
	185	107	98	219	281	288
250 ng	159	167	193	515	340	348
	196	174	250	205	505	432
1250 ng	137	426	178	296	545	630
	208	196	251	418	396	227

Source: J. A. Rice (1988).

7.10. ASYMPTOTICALLY DISTRIBUTION-FREE TWO-SIDED ALL-TREATMENTS MULTIPLE COMPARISONS FOR A TWO-WAY LAYOUT WITH AN EQUAL NUMBER OF REPLICATIONS IN EACH TREATMENT-BLOCK COMBINATION (MACK-SKILLINGS)

In this section we present an asymptotically distribution-free multiple comparison procedure using within-blocks ranks that is designed to make two-sided decisions about individual differences between pairs of treatment effects (τ_i, τ_j), for $i < j$, for data obtained from a two-way layout design with an equal number of replications for every treatment-block combination. The multiple comparison procedure of this section would generally be applied to data from such a two-way layout with an equal number of replications *after* rejection of H_0 (7.2) with the Mack-Skillings procedure from Section 7.9. In this setting, we will reach conclusions about all $k(k-1)/2$ pairs of treatment effects and these conclusions are naturally two-sided in nature.

Procedure

Let $S_1, \ldots, S_k$ be the treatment sums of cellwise averages of within-blocks ranks given by (7.56). Calculate the $k(k-1)/2$ absolute differences $|S_u - S_v|$, $1 \leq u < v \leq k$.

When H_0 (7.2) is true, the $k(k-1)/2$-component vector $(S_1, \ldots, S_k)$ has, when properly standardized and as N tends to infinity, an asymptotic $(k-1)$-variate normal distribution with appropriate mean vector and covariance matrix (see Mack and Skillings (1980) for details of the proof). At an approximate experimentwise error rate of α, the Mack-Skillings two-sided all-treatments multiple comparison procedure reaches its $k(k-1)/2$ pairwise decisions, corresponding to each (τ_u, τ_v) pair, $1 \leq u < v \leq k$, by the criterion

$$\text{Decide } \tau_u \neq \tau_v \text{ if } |S_u - S_v| \geq [k(N+n)/12]^{1/2} q_\alpha; \quad \text{otherwise decide } \tau_u = \tau_v, \quad (7.75)$$

where q_α is the upper αth percentile for the distribution of the range of k independent $N(0,1)$ variables. Values of q_α for $\alpha = .0001, .0005, .001, .005, .01, .025, .05, .10, .20$ and $k = 2(1)20(2)40(10)100$ are given in Table A.17. (See also Comment 82.)

Ties

If there are ties among the X observations within any of the blocks, use average ranks to break the ties and compute the individual sums of cellwise averages of within-blocks ranks $S_1, \ldots, S_k$.

Example 7.10: *Determination of Niacin in Bran Flakes.* For sake of illustration, we apply procedure (7.75) to the niacin determination data discussed in Example 7.9. There we had found rather strong evidence that the studied process for assessing niacin content in bran flakes does not produce consistent results across a variety of laboratories. To determine which of the laboratories differ in median detected niacin content in the bran flakes, we consider procedure (7.75) with an approximate experimentwise error rate $\alpha \approx .025$. From Table A.17, we see that $q_{.025} = 3.984$ for $k = 4$ and procedure (7.75) reduces to

$$\text{Decide } \tau_u \neq \tau_v \text{ if } |S_u - S_v| \geq [4(36 + 3)/12]^{1/2}(3.984) = 14.365.$$

Using the treatments sums of cellwise averages of within-blocks ranks obtained in Example 7.9, we find that

$$|S_2 - S_1| = |30.5 - 17.67| = 12.83, \quad |S_3 - S_1| = |15.83 - 17.67| = 1.84,$$
$$|S_4 - S_1| = |14 - 17.67| = 3.67, \quad |S_3 - S_2| = |15.83 - 30.5| = 14.67,$$
$$|S_4 - S_2| = |14 - 30.5| = 16.5, \quad |S_4 - S_3| = |14 - 15.83| = 1.83.$$

Referring these differences to the approximate critical value 14.365, we see that

$$|S_2 - S_1| = 12.83 < 14.365 \quad \Rightarrow \quad \text{decide } \tau_2 = \tau_1,$$
$$|S_3 - S_1| = 1.84 < 14.365 \quad \Rightarrow \quad \text{decide } \tau_3 = \tau_1,$$
$$|S_4 - S_1| = 3.67 < 14.365 \quad \Rightarrow \quad \text{decide } \tau_4 = \tau_1,$$
$$|S_3 - S_2| = 14.67 > 14.365 \quad \Rightarrow \quad \text{decide } \tau_3 \neq \tau_2,$$
$$|S_4 - S_2| = 16.5 > 14.365 \quad \Rightarrow \quad \text{decide } \tau_4 \neq \tau_2,$$
$$\text{and} \quad |S_4 - S_3| = 1.83 < 14.365 \quad \Rightarrow \quad \text{decide } \tau_4 = \tau_3.$$

Thus, at an approximate experimentwise error rate of .025, we see that Laboratory 2 yielded significantly different median detected niacin content than either Laboratory 3 or Laboratory 4. These multiple comparison decisions help to focus the rationale for the original rejection of H_0 (7.2) by the Mack-Skillings procedure in Example 7.9, as it now seems reasonable to question the reliability of Laboratory 2 in conducting this niacin content process.

Comments

80. *Rationale for Multiple Comparison Procedure.* The rationale behind the multiple comparison procedure of this section for data from a two-way layout design with an equal number of replications is similar to that for the two-sided multiple comparison procedures for data from a complete randomized block design. For further discussion, see Comment 24.

81. *Experimentwise Error Rate.* The use of an experimentwise error rate represents a very conservative approach to multiple comparisons. We are insisting that the probability of making only correct decisions be $1 - \alpha$ when the null hypothesis H_0 (7.2) of treatment equivalence is true. Thus we have a high degree of protection when H_0 is true, but we often apply such techniques when we have evidence (perhaps based on a priori information or perhaps obtained by applying the Mack-Skillings test, as in Example 7.9) that H_0 is not true. The protection under H_0 also makes it harder for the procedure to judge treatments as differing significantly

when, in fact, H_0 is false, and this difficulty becomes more severe as k increases. See Comment 6.58 for additional discussion of experimentwise error rates.

82. *Conservative Procedure.* Mack and Skillings (1980) also proposed a conservative multiple comparison procedure that guarantees an upper bound on the experimentwise error rate. Let $S_1, \ldots, S_k$ be the treatment sums of cellwise averages of within-blocks ranks given by (7.56). At an experimentwise error rate *no greater* than α, the Mack-Skillings conservative two-sided all-treatments multiple comparison procedure reaches its $k(k-1)/2$ decisions through the criterion

$$\text{Decide } \tau_u \neq \tau_v \text{ if } |S_u - S_v| \geq [k(N+n)ms_\alpha/6]^{1/2};$$

$$\text{otherwise decide } \tau_u = \tau_v, \qquad (7.76)$$

where ms_α is the upper α percentile for the null distribution of the Mack-Skillings statistic MS (7.57). Values of ms_α are available in Table A.28 for $k = 2(1)5$, $n = 2(1)5$, and $c = 2(1)5$. Mack and Skillings (1980) note that although procedure (7.76) does not require a large number of blocks, it is, nevertheless, rather conservative since it is based on the projection procedure of Scheffé; that is, the true experimentwise error rate might be considerably smaller than the bound α provided by (7.76). As a result, they recommend using the approximation (7.75) whenever the number of blocks is reasonably large.

83. *Dependence on Observations from Other Noninvolved Treatments.* The all-treatments multiple comparison procedure of this section suffers from the same disadvantage as do the other two-way layout multiple comparison procedures of this chapter. The decision between treatment u and treatment v can be affected by changes only in the observations from one or more of the other $k - 2$ treatments that are not directly involved.

Properties

1. *Asymptotic Multivariate Normality.* See Mack and Skillings (1980).
2. *Efficiency.* See Section 7.16.

PROBLEMS

104. Apply procedure (7.75) to the weld-strength data of Table 7.22 in Problem 92.
105. Illustrate the difficulty discussed in Comment 83 by means of a numerical example.
106. Apply procedure (7.75) to the coal acidity data of Table 7.23 in Problem 95.
107. Consider the niacin content data of Table 7.20 in Example 7.9. Find the smallest approximate experimentwise error rate at which the most significant difference(s) in median bran flake niacin content between the four laboratories would be detected by procedure (7.75).
108. Consider the weld-strength data of Table 7.22 in Problem 92. Find the smallest approximate experimentwise error rate at which procedure (7.75) would declare that weld cycle times 1 and 3 have differing effects on the strength of a weld.
109. Consider the coal acidity data of Table 7.23 in Problem 95. Find the smallest approximate experimentwise error rate at which the most significant difference(s) in effects of the NaOH concentration on the measured coal acidity value would be detected by procedure (7.75).
110. Consider the coal acidity data of Table 7.23 in Problem 95. Find the smallest approximate experimentwise error rate at which procedure (7.75) would declare that there is a difference in median coal acidity level between the Morwell and Yallourn types of coal.

ANALYSES ASSOCIATED WITH SIGNED RANKS

The statistical procedures discussed in Sections 7.1–7.5 (for randomized block designs with a single observation on each treatment-block combination) utilize the treatment observations only through comparisons within blocks. It is this restriction to within-blocks comparisons that leads directly to many of these procedures being strictly distribution-free, even for small sample sizes. An alternative approach is to consider accessing between-blocks information via utilization of pairwise signed ranks in the construction of appropriate statistical procedures. Hypothesis test and multiple comparison procedures based on these pairwise signed ranks will no longer be exactly distribution-free for small numbers (n) of blocks and they require the use of large-sample approximations. However, improved efficiency can result in many cases from this use of between-blocks signed ranks.

In the next five sections we assume (as done in Sections 7.1–7.5) that we have data from a randomized complete block design satisfying Assumptions A1–A3 for the case of one observation per treatment-block combination, corresponding to $c_{ij} = 1$ for every $i = 1,\ldots,n$ and $j = 1,\ldots,k$. For ease of notation in these five sections, we once again drop the third subscript on the X variables, since it is always equal to 1 in this setting.

Section 7.11 contains a conservative signed ranks test procedure directed at general alternatives for randomized block designs with a single observation on each treatment-block combination, while Section 7.12 presents the corresponding conservative signed ranks test procedure designed for ordered alternatives. The associated approximate signed ranks multiple comparison procedures are given in Sections 7.13 (all-treatments comparisons) and 7.14 (treatments versus control comparisons). Section 7.15 contains the contrast estimators linked to the Wilcoxon signed ranks for this setting.

7.11. A TEST BASED ON WILCOXON SIGNED RANKS FOR GENERAL ALTERNATIVES IN A RANDOMIZED COMPLETE BLOCK DESIGN (DOKSUM)

In this section we present a conservative procedure based on pairwise signed ranks for testing H_0 (7.2) against the general alternative H_1 (7.3) that at least two of the treatment effects are not equal.

Procedure

For each of the $k(k-1)/2$ pairs of treatments (u, v), with $1 \leq u < v \leq k$, we form the n absolute differences

$$Y_{uv}^i = |X_{iu} - X_{iv}|, \quad i = 1,\ldots,n. \tag{7.77}$$

(Note that $Y_{uv}^i = |D_{uv}^i|$, where the D_{uv}^i are the same differences given in (7.36) and used in the contrast estimator discussed in Section 7.5.) For each pair of treatments (u, v), we let R_{uv}^i be the rank of Y_{uv}^i in the ranking from least to greatest of the n values $Y_{uv}^1,\ldots,Y_{uv}^n$. To compute the Doksum (1967) statistic D, set

$$T_{uv} = \sum_{i=1}^{n} R_{uv}^i \Psi_{uv}^i \quad \text{and} \quad B_{uv} = \sum_{i=1}^{n} \Psi_{uv}^i, \tag{7.78}$$

where

$$\Psi_{uv}^i = \begin{cases} 1, & \text{if } X_{iu} < X_{iv}, \\ 0, & \text{otherwise}. \end{cases} \quad (7.79)$$

Let

$$H_{uv} = \frac{2(T_{uv} - B_{uv})}{n(n-1)}, \quad 1 \leq u < v \leq k. \quad (7.80)$$

(We note that the statistics H_{uv} need be calculated directly only for $u < v$, since for $u > v$ we can use the relationship $H_{vu} = 1 - H_{uv}$.) Next we obtain the averages

$$H_{u.} = \sum_{j=1}^{k} \frac{H_{uj}}{k}, \quad u = 1, \ldots, k, \quad (7.81)$$

where we note that $H_{uu} = 0$, for $u = 1, \ldots, k$.

The common null variance of each of the $k(k-1)/2$ differences $H_{u.} - H_{v.}$, $1 \leq u < v \leq k$, is given by the expression

$$\text{var}_0(H_{u.} - H_{v.}) = \frac{2n - 1 + (k-2)[24(n-2)\lambda_F + 13 - 6n]}{3kn(n-1)}, \quad (7.82)$$

with

$$\lambda_F = P_0(X_1 < X_2 + X_3 - X_4 \text{ and } X_1 < X_5 + X_6 - X_7), \quad (7.83)$$

where $X_1, X_2, \ldots, X_7$ are independent and identically distributed according to the common continuous underlying distribution F in Assumption A3. Since the value of λ_F (7.83) depends on the particular form of the continuous F, we can not use the expression in (7.82) to construct a distribution-free procedure for testing H_0 (7.2). However, Lehmann (1964) showed that $\lambda_F \leq \frac{7}{24}$ for all continuous F. (See Comment 87.) Replacing λ_F in equation (7.82) by this upper bound of $\frac{7}{24}$ yields the expression

$$V_U = \frac{2n - 1 + (k-2)[7(n-2) + 13 - 6n]}{3kn(n-1)}. \quad (7.84)$$

The Doksum test statistic for the conservative test of H_0 (7.2) is then

$$D = \sum_{j=1}^{k} \frac{[H_{j.} - \{(k-1)/2k\}]^2}{(k-1)V_U/2k}. \quad (7.85)$$

For a conservative test (see Comment 85) of

$$H_0 : [\tau_1 = \cdots = \tau_k]$$

versus the general alternative

$$H_1 : [\tau_1, \tau_2, \ldots, \tau_k \text{ not all equal}],$$

7.11. A TEST BASED ON WILCOXON SIGNED RANKS FOR GENERAL ALTERNATIVES

at the approximate α level of significance,

$$\text{Reject } H_0 \text{ if } D \geq \chi^2_{k-1,\alpha}; \quad \text{otherwise do not reject,} \tag{7.86}$$

where $\chi^2_{k-1,\alpha}$ is the upper α percentile point of a chi-square distribution with $k-1$ degrees of freedom. Values of $\chi^2_{k-1,\alpha}$ can be obtained from Chart A.2.

Ties

For any Y^i_{uv} (7.77), $1 \leq u < v \leq k$, that is zero, compute T_{uv} and B_{uv} in (7.78) by replacing the associated Ψ^i_{uv} (7.79) with

$$\Psi^{*i}_{uv} = \begin{cases} 1, & \text{if } X_{iu} < X_{iv}, \\ \frac{1}{2}, & \text{if } X_{iu} = X_{iv}, \\ 0, & \text{if } X_{iu} > X_{iv}. \end{cases} \tag{7.87}$$

For ties among $Y^1_{uv}, \ldots, Y^n_{uv}$, use average ranks to compute T_{uv} (7.78).

Example 7.11: *Rounding First Base.* Consider once again the rounding first base data presented in Table 7.1 and discussed in Example 7.1. The reader should already be familiar with the calculations of the paired-data signed rank statistics T_{uv} and sign statistics B_{uv} (see Comment 86) from the materials in Sections 3.1 and 3.4, respectively. We include a detailed calculation of T_{12} and B_{12} in Table 7.25 to illustrate the method for handling zero differences and ties. (See Ties and Comment 88.)

TABLE 7.25. Calculation of T_{12} and B_{12} for Data in Table 7.1

j	$X_{j1} - X_{j2}$	Y^j_{12}	R^j_{12}	Ψ^{*j}_{uv}	$R^j_{uv} \Psi^{*j}_{uv}$
1	−.10	.10	17	1	17
2	.15	.15	20	0	0
3	−.40	.40	22	1	22
4	.05	.05	9.5	0	0
5	.05	.05	9.5	0	0
6	−.10	.10	17	1	17
7	.00	.00	2.5	$\frac{1}{2}$	1.25
8	−.05	.05	9.5	1	9.5
9	.10	.10	17	0	0
10	.05	.05	9.5	0	0
11	.05	.05	9.5	0	0
12	.10	.10	17	0	0
13	.25	.25	21	0	0
14	.05	.05	9.5	0	0
15	.00	.00	2.5	$\frac{1}{2}$	1.25
16	−.10	.10	17	1	17
17	.00	.00	2.5	$\frac{1}{2}$	1.25
18	−.05	.05	9.5	1	9.5
19	.05	.05	9.5	0	0
20	.05	.05	9.5	0	0
21	.05	.05	9.5	0	0
22	.00	.00	2.5	$\frac{1}{2}$	1.25
				$B_{12} = 8$	$T_{12} = 97$

The statistics B_{12} and T_{12} are obtained by summing the entries in the next-to-last and last columns, respectively, of Table 7.25. We obtain B_{13}, B_{23}, T_{13}, and T_{23} in a similar manner, and the results are

$$B_{13} = 5, \quad B_{23} = 5, \quad T_{13} = 54, \quad \text{and} \quad T_{23} = 30.5.$$

It then follows from (7.80) that

$$H_{12} = .385, \quad H_{13} = .212, \quad \text{and} \quad H_{23} = .110.$$

From (7.81) and the fact that $H_{uv} = 1 - H_{vu}$, we have

$$H_{1.} = \frac{H_{11} + H_{12} + H_{13}}{3}$$
$$= \frac{0 + .385 + .212}{3} = .199,$$

$$H_{2.} = \frac{H_{21} + H_{22} + H_{23}}{3}$$
$$= \frac{.615 + 0 + .110}{3} = .242,$$

$$H_{3.} = \frac{H_{31} + H_{32} + H_{33}}{3}$$
$$= \frac{.788 + .890 + 0}{3} = .559.$$

We next find V_U (7.84) to be

$$V_U = \frac{2(22) - 1 + [7(20) + 13 - 6(22)]}{3(3)(22)(21)} = .015.$$

Substituting these values for $H_{1.}, H_{2.}, H_{3.}$, and V_U into equation (7.43) yields

$$D = \frac{[.199 - (\frac{1}{3})]^2 + [.242 - (\frac{1}{3})]^2 + [.559 - (\frac{1}{3})]^2}{2(.015)/6} = 15.5.$$

Referring this value of D to the chi-square distribution with $k - 1 = 2$ degrees of freedom (Chart A.2), we find that the lowest significance level at which we would reject H_0 is less than .001. (Cf. Example 7.1.)

Comments

84. *Motivation for the Test.* Under H_0 (7.2), the $H_{j.}$'s (7.81) tend to be near $(k - 1)/2k$, their common null expectation, and thus the numerator of D (7.85) tends to be small. When the τ's are not all equal, we expect the $H_{j.}$'s to be more disparate, and thus (at least some of) the $[H_{j.} - \{(k - 1)/2k\}]^2$ terms tend to be large, yielding a large value of D. This provides partial motivation for procedure (7.86).

85. *Conservative Nature of the Test.* The test defined by (7.86) is neither distribution-free nor asymptotically $(n \to \infty)$ distribution-free. Rather, it is conservative in the sense that (asymptotically) the actual probability of rejecting H_0 (7.2) when it is true tends to be slightly

smaller than the nominal level α. This is a consequence of using an upper bound for the parameter λ_F (7.83). See also Comments 87 and 89.

86. *Pairwise Signed Rank and Sign Statistics.* For a given pair of treatments (u, v), $1 \leq u < v \leq k$, the statistics T_{uv} and B_{uv} (7.78) are simply the Wilcoxon signed rank and sign statistics, respectively, as discussed in Sections 3.1 and 3.4, respectively, applied to the paired data in treatments u and v. With this relationship in mind, we note that the difference $T_{uv} - B_{uv}$ in the numerator of H_{uv} (7.80) may equivalently be calculated as the number of Walsh averages $(X_{su} - X_{sv} + X_{tu} - X_{tv})/2$, with $1 \leq s < t \leq n$, that are negative. (See Comment 3.17.)

87. *Bounds for the Parameter λ_F.* The null correlation between two overlapping statistics H_{uv} and H_{uw} defined by (7.80), with $u \neq v, u \neq w$, and $v \neq w$, depends on the parameter λ_F (7.83). This, combined with the fact that λ_F varies with F (Lehmann (1964)), prevents the development of a distribution-free test procedure based on the numerator of D (7.85). Lehmann (1964) showed that $\lambda_F \leq \frac{7}{24}$ ($\approx .2917$) for all continuous F. Replacement of λ_F in expression (7.82) for the null variance of $H_{u.} - H_{v.}$ by the upper bound $\frac{7}{24}$ enables the development of the conservative procedure based on D (7.86). Spurrier (1991) established the lower bound $\lambda_F \geq \frac{89}{315}$ ($\approx .2825$) for all continuous F. Since the value of λ_F is so narrowly confined between .2825 and .2917 for all continuous F, replacing λ_F by its upper bound of $\frac{7}{24}$ in expression (7.82) sacrifices little to permit the construction of the conservative test procedure (7.86).

88. *Ties.* The reader may have noted that the method we advocate in Ties for dealing with zero differences, when computing the $T_{uv}(B_{uv})$ signed rank (sign) statistics for use in procedure (7.86), differs from the corresponding directions given for the signed rank (sign) statistic in Section 3.1 (Section 3.4). In Chapter 3 we recommended reducing the sample size by the number of zero differences. This change is initiated in the calculation of D (7.85) in order to keep all of the T_{uv}'s and B_{uv}'s based on the same sample size (n).

89. *Asymptotically Distribution-Free Competitor.* As an alternative to the conservative test procedure (7.86) based on the replacement of the unknown parameter λ_F (7.83) by its upper bound $\frac{7}{24}$, we could instead choose to estimate the value of λ_F from the sample data. Use of a consistent estimator of λ_F in this manner leads to an asymptotically ($n \to \infty$) distribution-free procedure for testing H_0 (7.2), rather than the conservative procedure in (7.86). Lehmann (1964) proposed the estimator $\widehat{\lambda}_F$ of λ_F, where $\widehat{\lambda}_F$ is the proportion of sample sextuples $(\alpha, \beta, \gamma; u, v, w)$ for which the simultaneous inequalities

$$(X_{\alpha u} < X_{\beta u} + X_{\alpha v} - X_{\beta v} \quad \text{and} \quad X_{\alpha u} < X_{\gamma u} + X_{\alpha w} - X_{\gamma w})$$

are satisfied. In practice, when estimating λ_F, it would normally suffice to check only a subset of the total number of such sample sextuples. Due to the closeness of the upper bound $\frac{7}{24}$ to all values of λ_F, procedure (7.86) is, for all practical purposes, virtually equivalent to Doksum's (1967) asymptotically distribution-free procedure based on estimating λ_F.

Properties

1. *Consistency.* See Doksum (1967) and Hollander and Wolfe (1973, 166).
2. *Asymptotic Chi-Square Distribution.* See Doksum (1967).
3. *Efficiency.* See Doksum (1967) and Section 7.16.

PROBLEMS

111. Apply procedure (7.86) to the adaptation score data of Table 7.10, Example 7.4.
112. The Doksum test procedure (7.86) uses between-block information, whereas Friedman's test procedure (7.6) uses only within-block information. Explain.

113. Apply procedure (7.86) to the serum creatine phosphokinase (CPK) activity data in Table 7.3, Problem 5.
114. Apply procedure (7.86) to the percentage correctly identified consonants data in Table 7.4, Problem 13.
115. Both the Doksum (7.86) and Friedman (7.6) procedures are appropriate for testing against general alternatives when we have data from a randomized complete block design with one observation per treatment-block combination. Discuss the relative advantages and disadvantages of the two competing procedures.

7.12. A TEST BASED ON WILCOXON SIGNED RANKS FOR ORDERED ALTERNATIVES IN A RANDOMIZED COMPLETE BLOCK DESIGN (HOLLANDER)

In this section we present a conservative procedure based on pairwise signed ranks for testing H_0 (7.2) against the a priori ordered alternatives H_2 (7.9), corresponding to $\tau_1 \leq \tau_2 \leq \cdots \leq \tau_k$, with at least one strict inequality.

Procedure

For each of the $k(k-1)/2$ pairs of treatments (u, v), with $1 \leq u < v \leq k$, we compute the signed rank statistic T_{uv}, as defined in (7.78). To compute the Hollander statistic Q, set

$$Y = \sum_{u=1}^{k-1} \sum_{v=u+1}^{k} T_{uv}. \tag{7.88}$$

The null expected value of Y is given by

$$E_0(Y) = \frac{nk(k-1)(n+1)}{8}, \tag{7.89}$$

but the null variance of Y is unknown (see Comment 93) and depends on the particular form of the underlying continuous distribution F in Assumption 3. Thus, a test of H_0 (7.2) based on Y will not be distribution-free. However, a conservative procedure can be developed by using an upper bound for this unknown null variance of Y. From Table A.29 we obtain the value of the upper bound ρ_U^n for the null correlation between two overlapping signed rank statistics based on n observations. An upper bound for the null variance of Y (7.88) is then given by

$$\text{var}_U(Y) = \frac{nk(n+1)(2n+1)(k-1)\{3 + 2(k-2)\rho_U^n\}}{144}. \tag{7.90}$$

The Hollander test statistic for the conservative test of H_0 (7.2) is then

$$Q = \frac{Y - E_0(Y)}{\{\text{var}_U(Y)\}^{1/2}}, \tag{7.91}$$

with the expressions for $E_0(Y)$ and $\text{var}_U(Y)$ given in equations (7.89) and (7.90), respectively. For a conservative test (see Comment 92) of

$$H_0 : [\tau_1 = \cdots = \tau_k]$$

7.12. A TEST BASED ON WILCOXON SIGNED RANKS FOR ORDERED ALTERNATIVES

versus the ordered alternatives

$$H_2 : [\tau_1 \leq \tau_2 \leq \cdots \leq \tau_k, \text{ with at least one strict inequality}],$$

at the approximate α level of significance,

$$\text{Reject } H_0 \text{ if } Q \geq z_\alpha; \quad \text{otherwise do not reject.} \tag{7.92}$$

Ties

See Ties of Section 7.11 and Comment 88.

Example 7.12: *Effect of Weight on Forearm Tremor Frequency.* The data in Table 7.26 are based on a subset of the data obtained by Fox and Randall (1970) in their study of forearm tremor. Each entry in the table is the mean of five experimental values of tremor frequency. We identify treatment 1 with 7.5 lb, treatment 2 with 5 lb, treatment 3 with 2.5 lb, treatment 4 with 1.25 lb, and treatment 5 with 0 lb, and use procedure (7.92) to test H_0 (7.2) versus the ordered alternatives H_2 (7.9), which specify that adding mass decreases the tremor frequency.

Calculations similar to those presented in Example 7.11 yield

$$\begin{array}{lllll} T_{12} = 18.5, & T_{13} = 21, & T_{14} = 21, & T_{15} = 21, & T_{23} = 20, \\ T_{24} = 21, & T_{25} = 21, & T_{34} = 21, & T_{35} = 21, & T_{45} = 21. \end{array} \tag{7.93}$$

From (7.88) we obtain

$$Y = T_{12} + T_{13} + T_{14} + T_{15} + T_{23} + T_{24} + T_{25} + T_{34} + T_{35} + T_{45} = 206.5.$$

From Table A.29 we find $\rho_U^6 = .452$, and evaluating (7.89) and (7.90) gives

$$E_0(Y) = \frac{5(4)(6)(7)}{8} = 105,$$

$$\text{Var}_U(Y) = \frac{6(7)(13)(5)(4)\{3 + 6(.452)\}}{144} = 433.2.$$

TABLE 7.26. Forearm Tremor Frequency (Hz) as a Function of Weight Applied at the Wrist

Treatment	1	2	3	4	5
			Weight (lb)		
Subject	7.5	5	2.5	1.25	0
1	2.58	2.63	2.62	2.85	3.01
2	2.70	2.83	3.15	3.43	3.47
3	2.78	2.71	3.02	3.14	3.35
4	2.36	2.49	2.58	2.86	3.10
5	2.67	2.96	3.08	3.32	3.41
6	2.43	2.50	2.85	3.06	3.07

Source: J R. Fox and J. E. Randall (1970).

From (7.91) we then have

$$Q = \frac{206.5 - 105}{[433.2]^{1/2}} = 4.88.$$

From Table A.1, we see that the lowest approximate level at which we would reject H_0 with these data is below .0002. Thus there is very strong evidence (over the range of weights considered in the study) that the tremor frequency does decrease as the applied weight increases.

Comments

90. *Motivation for the Test.* Note that the statistic Y (7.88) is designed to guard against the postulated ordered alternatives H_2 (7.9). Consider the case $k = 3$. Then $Y = T_{12} + T_{13} + T_{23}$, and if $\tau_1 < \tau_2 < \tau_3$, each of T_{12}, T_{13}, and T_{23} would tend to be larger than $n(n + 1)/4$, their common null expectation. Thus Y would tend to be large, as desired. Contrast this with a situation in which we suspect (and design the test for) the alternative $\tau_1 < \tau_2 < \tau_3$ but in actuality we have $\tau_3 < \tau_2 < \tau_1$. In this case, each of T_{12}, T_{13}, and T_{23} would tend to be small. This provides partial motivation for procedure (7.92).

91. *Non-Distribution-Free Property of Y (7.88).* Consider the Y (7.88) statistic for testing against ordered alternatives in the two-way layout (7.1) in relation to Jonckheere's J (6.13) statistic for testing against ordered alternatives in the one-way layout (6.1). The statistic J is the sum $\sum_{u<v}^{k} U_{uv}$ of two-sample Mann-Whitney statistics U_{uv} (or, equivalently, Wilcoxon rank sum statistics), where each U_{uv} is distribution-free under H_0 (6.2). The statistic Y is a sum $\sum_{u<v}^{k} T_{uv}$ of paired-sample Wilcoxon signed rank statistics T_{uv}, where each T_{uv} is distribution-free under H_0 (7.2). Although J itself is also distribution-free under H_0 (6.2), Y is not distribution-free under H_0 (7.2) when $k > 2$. (For $k = 2, Y$ reduces to T_{12}, which is distribution-free.) See Hollander (1967a) for details of the non-distribution-free character of Y.

92. *Conservative Nature of the Test.* The test defined in (7.91) is neither distribution-free nor asymptotically ($n \to \infty$) distribution-free. Rather, it is conservative in the sense that (asymptotically) the actual probability of rejecting H_0 (7.2) when it is true tends to be smaller than the nominal level α. This is a direct consequence of using an upper bound $\text{var}_U(Y)$ to replace the unknown null variance of Y. Also see Comment 94.

93. *Asymptotic Null Variance of Y.* Hollander (1967a) showed that the asymptotic ($n \to \infty$) null variance of Y (7.88) has the form

$$\text{var}_0(Y) = \frac{nk(n + 1)(2n + 1)(k - 1)\{3 + 2(k - 2)\rho^*\}}{144},$$

where ρ^* is the limiting ($n \to \infty$) null correlation between two overlapping signed rank statistics T_{12} and T_{13}. This limiting correlation can also be expressed as

$$\rho^* = 12\lambda_F - 3, \qquad (7.94)$$

where λ_F is defined by (7.83). In forming the Q test statistic (7.91) for the conservative test procedure (7.92), we replace ρ^* by its upper bound ρ_U^n.

94. *Asymptotically Distribution-Free Competitor.* As an alternative to the conservative test procedure (7.92) based on the use of the upper bound ρ_U^n, we could instead replace ρ^* (7.94) by a consistent estimator $\hat{\rho}$ based on the sample data. Use of a consistent estimator of ρ^* in this manner leads to an asymptotically ($n \to \infty$) distribution-free procedure for testing H_0 (7.2),

rather than the conservative procedure in (7.92). Hollander suggested such an approach to this problem based on the consistent estimator $\widehat{\rho} = 12\widehat{\lambda}_F - 3$, where $\widehat{\lambda}_F$ is defined in Comment 89. Due to the closeness of the upper bound $\frac{7}{24}$ to all values of λ_F, procedure (7.92) is, for all practical purposes, virtually equivalent to Hollander's (1967a) asympotically distribution-free procedure based on estimating λ_F.

Properties

1. *Consistency.* The test defined by (7.92) is consistent against the ordered alternatives (7.9). See Hollander (1967a) and Hollander and Wolfe (1973, 170).
2. *Asymptotic Normality.* See Hollander (1967a).
3. *Efficiency.* See Hollander (1967a) and Section 7.16.

PROBLEMS

116. Apply the Q (7.92) test to the metronome data of Table 7.6. Use the postulated ordering $\tau_R < \tau_A < \tau_N$.
117. The Hollander test procedure (7.92) uses between-block information, but Page's test procedure (7.11) uses only within-block information. Explain.
118. Apply procedure (7.92) to the shelterbelt data in Table 7.7, Problem 21.
119. Apply procedure (7.92) to the cotton strength index data in Table 7.5, Example 7.2. Compare with the result from the use of Page's test in Example 7.2.
120. Both the Hollander (7.92) and Page (7.11) procedures are appropriate for testing against ordered alternatives when we have data from a randomized complete block design with one observation per treatment-block combination. Discuss the relative advantages and disadvantages of the two competing procedures.

7.13. APPROXIMATE TWO-SIDED ALL-TREATMENTS MULTIPLE COMPARISONS BASED ON SIGNED RANKS (NEMENYI)

In this section we present a multiple comparison procedure based on Wilcoxon signed rank statistics that is designed to make decisions about individual differences between pairs of treatment effects (τ_u, τ_v), for $u < v$, in a setting where general alternatives H_1 (7.3) are of interest. Thus, the multiple comparison procedure of this section would generally be applied to two-way layout data (with one observation per cell) *after* rejection of H_0 (7.2) with the Doksum-Lehmann procedure from Section 7.11. In this setting it is important to reach conclusions about all $k(k-1)/2$ pairs of treatment effects and these conclusions are naturally two-sided in nature.

Procedure

For $1 \leq u < v \leq k$, let T_{uv} be the signed rank statistic (7.78) between treatments u and v. Calculate the $k(k-1)/2$ statistics

$$T'_{uv} = \max\{T_{uv}, [n(n+1)/2] - T_{uv}\}, 1 \leq u < v \leq k. \qquad (7.95)$$

At an approximate (see Comment 95) experimentwise error rate of α, the two-sided signed rank multiple comparison procedure reaches its $k(k-1)/2$ pairwise decisions, corresponding to each (τ_u, τ_v) pair, for $1 \leq u < v \leq k$, by the criterion

$$\text{Decide } \tau_u \neq \tau_v \text{ if } T'_{uv} \geq t'_\alpha; \quad \text{otherwise decide } \tau_u = \tau_v, \tag{7.96}$$

where the constant t'_α is chosen to make the experimentwise error rate approximately equal to α; that is, t'_α satisfies the restriction

$$P_0\{T'_{uv} < t'_\alpha, u = 1, \ldots, k-1 \quad \text{and} \quad v = u+1, \ldots, k\} \approx 1 - \alpha, \tag{7.97}$$

where the probability $P_0(.)$ is computed under H_0 (7.2). Equation (7.97) stipulates that the $k(k-1)/2$ inequalities $T'_{uv} < t'_\alpha$, corresponding to all pairs (u, v) of treatments with $u < v$, hold simultaneously with approximate probability $1 - \alpha$ when H_0 (7.2) is true. Selected approximate values of t'_α can be found from the relationship

$$t'_\alpha \approx \left[\frac{n(n+1)}{4}\right] + \left[\frac{n(n+1)(2n+1)}{48}\right]^{1/2} q_\alpha, \tag{7.98}$$

where q_α is the upper αth percentile point for the distribution of the range of k independent $N(0, 1)$ variables. Values of q_α for α = .0001, .0005, .001, .005, .01, .025, .05, .10, .20 and $k = 2(1)20(2)40(10)100$ are given in Table A.17.

Ties

See Ties of Section 7.11 and Comment 88.

Example 7.13: *Rounding First Base.* We illustrate procedure (7.96) using the approximation (7.98) with the rounding first base data of Table 7.1. In Example 7.11 we found

$$T_{12} = 97, \quad T_{13} = 54, \quad \text{and} \quad T_{23} = 30.5.$$

From (7.95) we obtain

$$T'_{12} = \max\{97, 253 - 97\} = 156, \quad T'_{13} = \max\{54, 253 - 54\} = 199,$$
$$T'_{23} = \max\{30.5, 253 - 30.5\} = 222.5.$$

With an experimentwise error rate of α = .01 and $k = 3$, we find $q_{.01} = 4.12$ from Table A.17. Thus, with approximation (7.98), the inequality in (7.96) reduces to

$$T'_{uv} \geq t'_{.01} \approx \left[\frac{22(23)}{4}\right] + \left[\frac{22(23)(45)}{48}\right]^{1/2} (4.12) = 216.2,$$

and procedure (7.96) becomes

$$\text{Decide } \tau_u \neq \tau_v \text{ if } T'_{uv} \geq 216.2, \quad 1 \leq u < v \leq 3.$$

Since $T'_{12} < 216.2, T'_{13} < 216.2$, and $T'_{23} \geq 216.2$, only the narrow angle (treatment 2) and wide angle (treatment 3) running methods differ significantly at the approximate .01 experimentwise error rate using the signed rank procedure (7.96).

At this point the reader may have noticed that, at the approximate .01 experimentwise error rate, the signed rank analysis in this example yields a conclusion different from the corresponding analysis based on Friedman rank sums performed in Example 7.3. Since the analyses are based on different rankings and different statistics, the reader should not be shocked. It is instructive to note that if, for example, the multiple comparisons were made at an approximate .10 experimentwise error rate, the two procedures would agree in the sense that differences between treatments 2 and 3 and between treatments 1 and 3 would be declared significant under both analyses.

Comments

95. *Non-Distribution-Free Property.* Procedure (7.96), using approximation (7.98), is neither distribution-free nor asymptotically distribution-free. Nemenyi (1963) proposed this procedure under the assumptions that (a) the statistic $\max\{T'_{uv}, 1 \leq u < v \leq k\}$ is distribution-free and (b) the limiting ($n \to \infty$) null correlation between T_{12} and T_{13} (say) is close to $\frac{1}{2}$. Assumption (a) is incorrect, but the reasonableness of assumption (b) is supported by the values of λ_F, for various distributions F, obtained by Lehmann (1964), Hollander (1966), and Obenchain (1969). (See also Comments 87 and 93.)

96. *Independence from Observations for Other Noninvolved Treatments.* The value of T'_{uv}, the statistic used in the decision relating to τ_u and τ_v, does not depend on the observation values from the other $k - 2$ treatments. Thus the signed ranks procedure (7.96) eliminates a difficulty encountered with the corresponding multiple comparison procedures (7.25) and (7.27) of Section 7.3 based on Friedman rank sums. (See Comment 30.)

Properties

1. *Efficiency.* See Section 7.16.

PROBLEMS

121. Apply procedure (7.96) to the serum creatine phosphokinase (CPK) activity data in Table 7.3, Problem 5.
122. Apply procedure (7.96) to the Hebb-Williams elevated pathway test data in Table 7.9, Problem 30.
123. Both procedures (7.27) and (7.96) are appropriate multiple comparison procedures when we have data from a randomized complete block design with one observation per treatment-block combination, and we are interested in two-sided comparisons between all treatments. Discuss the relative advantages and disadvantages of the two competing procedures.
124. Apply procedure (7.96) to the percentage correctly identified consonants data in Table 7.4, Problem 13.

7.14. APPROXIMATE ONE-SIDED TREATMENTS VERSUS CONTROL MULTIPLE COMPARISONS BASED ON SIGNED RANKS (HOLLANDER)

In this section we turn our attention to a multiple comparison procedure based on Wilcoxon signed rank statistics that is designed to make decisions about individual differences between the median effect for a single, baseline control population and the median effects of each of the remaining $k - 1$ treatments. This treatment versus control multiple comparison procedure can be applied to two-way layout data (with one observation per cell) *after* rejection of H_0 (7.2) with either the Doksum-Lehmann or Hollander procedure discussed in Sections 7.11 and 7.12, respectively. Its application leads to conclusions about the differences between each of the $k - 1$ treatment effects and the control effect and these conclusions are naturally one-sided in nature.

Procedure

For simplicity of notation, we let treatment 1 assume the role of the single, baseline control. For each of the $k - 1$ treatments $u = 2, \ldots, k$, we compute the signed rank statistic T_{1u} (7.78) between the control treatment 1 and treatment u. At an approximate (see Comment 98) experimentwise error rate of α, the one-sided treatments versus control signed rank multiple comparison procedure reaches its $k - 1$ pairwise decisions, corresponding to each (τ_1, τ_u) pair, for $u = 2, \ldots, k$, by the criterion

$$\text{Decide } \tau_u > \tau_1 \text{ if } T_{1u} \geq t_\alpha^*; \quad \text{otherwise decide } \tau_u = \tau_1, \tag{7.99}$$

where the constant t_α^* is chosen to make the experimentwise error rate approximately equal to α; that is, t_α^* satisfies the restriction

$$P_0\{T_{1u} < t_\alpha^*, u = 2, \ldots, k\} \approx 1 - \alpha, \tag{7.100}$$

where the probability $P_0(.)$ is computed under H_0 (7.2). Equation (7.100) stipulates that the $k - 1$ inequalities $T_{1u} < t_\alpha^*$, corresponding to each treatment paired with the control, hold simultaneously with approximate probability $1 - \alpha$ when H_0 (7.2) is true. Selected approximate values of t_α^* can be found from the relationship

$$t_\alpha^* \approx \left[\frac{n(n+1)}{4}\right] + \left[\frac{n(n+1)(2n+1)}{24}\right]^{1/2} m_\alpha^*, \tag{7.101}$$

where m_α^* is the upper αth percentile point for the distribution of the maximum of $(k-1)N(0, 1)$ variables with common correlation ρ^* equal to the upper bound ρ_U^n for the null correlation between two overlapping signed rank statistics based on n observations. Selected values of m_α^* for $k = 3(1)13$ and $\rho^* = .1, .125, .2, .25, .3, \frac{1}{3}, .375, .4, .5, .6, .625, \frac{2}{3}, .7, .75, .8, .875,$ and $.9$ are given in Table A.21. The upper bound ρ_U^n for signed rank statistics based on n observations is found in Table A.29. Linear interpolation in Table A.21 between common correlation ρ^* values adjacent to ρ_U^n is often required. (For a discussion of how to adjust procedure (7.99) for settings where it is of interest to decide whether the treatment effects are *smaller* than the control effect, see Comment 97.)

Ties

See Ties of Section 7.11 and Comment 88.

Example 7.14: *Effect of Weight on Forearm Tremor Frequency.* We use the tremor data of Table 7.26 to illustrate procedure (7.99) using the approximation (7.101). We relabel the treatments so that the no-weight (0 lb) treatment assumes the role of the control. To make this clear in the ensuing computations, we reproduce Table 7.26 as Table 7.26′ with the new treatment designations.

We illustrate the one-sided decisions of $\tau_u = \tau_1$ versus $\tau_u < \tau_1, u = 2, \ldots, 5$. We see from Comment 97 that our procedure is based on (7.99) with $T_{u1} = [n(n+1)/2] - T_{1u}$ replacing T_{1u} in the left-hand side of the inequality in (7.99). From the relabeling in Table 7.26′ and the basic computations in Example 7.12, we obtain

$$T_{12} = 0, \quad T_{13} = 0, \quad T_{14} = 0, \quad \text{and} \quad T_{15} = 0,$$

TABLE 7.26'. Forearm Tremor Frequency (Hz) as a Function of Weight Applied at the Wrist

Treatment	1	2	3	4	5
			Weight (lb)		
Subject	0	1.25	2.5	5	7.5
1	3.01	2.85	2.62	2.63	2.58
2	3.47	3.43	3.15	2.83	2.70
3	3.35	3.14	3.02	2.71	2.78
4	3.10	2.86	2.58	2.49	2.36
5	3.41	3.32	3.08	2.96	2.67
6	3.07	3.06	2.85	2.50	2.43

Source: J. R. Fox and J. E. Randall (1970).

which, in turn, implies

$$T_{21} = 21 - T_{12} = 21, \quad T_{31} = 21 - T_{13} = 21,$$
$$T_{41} = 21 - T_{14} = 21, \quad T_{51} = 21 - T_{15} = 21.$$

From Table A.29 we find $\rho_U^6 = .452$. Since Table A.21 does not contain m_α^* values for $k = 5$ and $\rho^* = .452$, we must interpolate between m_α^* values for $\rho^* = .400$ and $\rho^* = .500$. With an approximate experimentwise error rate of $\alpha = .10$, we find (using linear interpolation *within* the ρ^* values) that $m_{.10}^* \approx 1.87$ for $\rho^* = .400$ and $m_{.10}^* \approx 1.84$ for $\rho^* = .500$. Again using linear interpolation (but this time *across* the ρ^* values), we find $m_{.10}^* \approx 1.85$ for $\rho^* = .452$. Thus, with approximation (7.101), the inequality in (7.99) for our one-sided decisions reduces to

$$T_{u1} \geq t_{.10}^* \approx \left[\frac{6(7)}{4}\right] + \left[\frac{6(7)(13)}{24}\right]^{1/2} (1.85) = 19.3,$$

and procedure (7.99) for these one-sided decisions becomes

Decide $\tau_u < \tau_1$ if $T_{u1} \geq 19.3$, $u = 2, \ldots, 5$.

Since $T_{21} = T_{31} = T_{41} = T_{51} = 21 > 19.3$, the signed rank procedure (7.99) with an approximate .10 experimentwise error rate concludes that all four weight levels (treatments) yield significantly smaller forearm tremor frequencies than does the zero weight control.

Comments

97. *Opposite Direction Decisions.* Procedure (7.99) is designed for the one-sided situation in which the relevant decisions are $\tau_u = \tau_1$ versus $\tau_u > \tau_1, u = 2, \ldots, k$. To treat the analogous one-sided case of $\tau_u = \tau_1$ versus $\tau_u < \tau_1, u = 2, \ldots, k$, we simply replace T_{1u} by $T_{u1} = [n(n+1)/2] - T_{1u}$ in the left-hand side of the inequality in (7.99).

98. *Non-Distribution-Free Property.* Procedure (7.99), using approximation (7.101), is neither distribution-free nor asymptotically distribution-free. However, it is generally conservative in that the attained approximate experimentwise error rate associated with procedure (7.99) tends to be slightly lower than the nominally stipulated rate. Due to the closeness of the upper bound $\frac{7}{24}$ to all values of λ_F, procedure (7.99) is, for all practical purposes, virtually equivalent to Hollander's (1966) asymptotically distribution-free procedure based on estimating λ_F.

99. *Simplification of Approximation.* One of the disadvantages of the approximation to t_α^* provided in equation (7.101) is that it requires obtaining the value of m_α^* for common correlation ρ_U^n. With currently existing tables (Table A.21), this must usually be accomplished by way of linear interpolation between the two values of m_α^* corresponding to ρ_1 and ρ_2, with $\rho_1 < \rho_U^n < \rho_2$. To simplify matters, one could always use the further approximation associated with replacing ρ_U^n by its asymptotic ($n \to \infty$) limit of $\frac{1}{2}$. The approximation in (7.101) would then use the proper value of m_α^* for common correlation $\rho^* = \frac{1}{2}$.

100. *Asymptotically Distribution-Free Competitor.* As an alternative to the conservative one-sided multiple comparison procedure (7.99) based on the use of the upper bound ρ_U^n in approximation (7.101), we could instead use a consistent estimator $\hat{\rho}$ of the null correlation between two overlapping signed rank statistics based on n observations. The value of m_α^* used in approximation (7.101) would then correspond to this estimate ($\hat{\rho}$) of the null correlation rather than the upper bound ρ_U^n. Use of a consistent estimator $\hat{\rho}$ in this manner leads to an asymptotically ($n \to \infty$) distribution-free one-sided multiple comparison procedure, rather than the conservative procedure in (7.99). Hollander (1966) suggested such an approach based on the consistent estimator $\hat{\rho} = 12\hat{\lambda}_F - 3$, where $\hat{\lambda}_F$ is defined in Comment 89.

101. *Two-Sided Treatments versus Control Multiple Comparison Procedure.* The multiple comparison procedure (7.99) of this section is one-sided by nature, resulting in decisions between $\tau_u = \tau_1$ and $\tau_u > \tau_1$ for every $u = 2, \ldots, k$ (or between $\tau_u = \tau_1$ and $\tau_u < \tau_1$ for every $u = 2, \ldots, k$, as noted in Comment 97). We view such one-sided comparisons to be the most natural approach for treatments versus control settings. In such situations we are generally interested in seeing which, if any, of the proposed new treatments are better than a standard control or placebo. In most practical applications, *better* is synonymous with one-sided comparisons (all in one direction or all in the other) and thus our emphasis on such procedures in this section. However, a two-sided treatments versus control analogue to procedure (7.99) has been developed in the literature and corresponds to the criterion

$$\text{Decide } \tau_u \neq \tau_1 \text{ if } T'_{1u} \geq t_\alpha^{**}; \quad \text{otherwise decide } \tau_u = \tau_1, \quad (7.102)$$

where the T'_{1u}'s are defined by (7.95) and the constant t_α^{**} is chosen to make the experimentwise error rate approximately equal to α; that is,

$$P_0\{T'_{1u} < t_\alpha^{**}, u = 2, \ldots, k\} \approx 1 - \alpha,$$

where the probability $P_0(.)$ is computed under H_0 (7.2). One approximation for t_α^{**} sets

$$t_\alpha^{**} \approx \left[\frac{n(n+1)}{4}\right] + \left[\frac{n(n+1)(2n+1)}{24}\right]^{1/2} v_\alpha^*, \quad (7.103)$$

where v_α^* is the upper αth percentile of the maximum absolute value of $(k-1)$ $N(0, 1)$ random variables with common correlation $\frac{1}{2}$. Selected values of v_α^* can be obtained from Dunnett (1964).

102. *Independence from Observations for Other Noninvolved Treatments.* The value of T_{1u}, the statistic used in the decision relating to τ_u and τ_1, does not depend on the observation values from the other $k - 2$ treatments. Thus the signed ranks procedure (7.99) eliminates a difficulty encountered with the corresponding one-sided multiple comparison procedures (7.28) and (7.30) of Section 7.4 based on Friedman rank sums. (See Comment 38.)

Properties

1. *Efficiency.* See Section 7.16.

PROBLEMS

125. Apply an appropriate one-sided signed rank multiple comparison procedure (see (7.99) and Comment 97) to the rhythmicity data of Table 7.6 in Problem 17, letting the condition N serve as the control.

126. Consider the serum creatine phosphokinase activity data from Problem 5. Treating preexercise as a control and ignoring the peak psychotic period data, apply procedure (7.99) to decide if there is statistical evidence of increased serum creatine phosphokinase activity either 19 or 42 h after exercise.

127. Both procedures (7.30) and (7.99) are appropriate multiple comparison procedures when we have data from a randomized complete block design with one observation per treatment-block combination and we are interested in one-sided comparisons between $(k - 1)$ treatments and a single, baseline control. Discuss the relative advantages and disadvantages of the two competing procedures.

128. Treating condition A as a control, apply procedure (7.99) to the percentage correctly identified consonants data in Table 7.4, Problem 13.

7.15. CONTRAST ESTIMATION BASED ON ONE-SAMPLE HODGES-LEHMANN ESTIMATORS (LEHMANN)

In this section, we describe how to use Hodges-Lehmann estimators based on appropriate Walsh averages to construct estimators of a contrast θ (7.32, 7.34) in the treatment effects $\tau_1, \ldots, \tau_k$. For a given setting, decisions about which contrasts to estimate can be related either to a priori interest in particular linear combinations of the τ's or to the result of one of the multiple comparison procedures discussed in Sections 7.3, 7.4, 7.13, and 7.14.

Procedure

Let θ be an arbitrary contrast (7.32, 7.34) in the treatment effects $\tau_1, \ldots, \tau_k$. For each pair of treatments $(u, v), u \neq v = 1, \ldots, k$, compute the differences D_{uv}^i (7.36), $i = 1, \ldots, n$, between the treatment u and treatment v observations for each of the n blocks. For each (u, v) pair, obtain the values of the $n(n + 1)/2$ Walsh averages for these sample differences, namely,

$$\frac{D_{uv}^i + D_{uv}^j}{2}, \quad 1 \leq i \leq j \leq n. \tag{7.104}$$

Let W_{uv} be the median of the Walsh averages associated with the $u - v$ treatment differences; that is,

$$W_{uv} = \text{median}\left\{\frac{D_{uv}^i + D_{uv}^j}{2}, 1 \leq i \leq j \leq n\right\}, \quad u \neq v = 1, \ldots, k. \tag{7.105}$$

(Since $W_{vu} = -W_{uv}$, we need calculate only the $k(k - 1)/2$ values W_{uv} corresponding to $u < v$.) Note that each W_{uv} is a Hodges-Lehmann estimator of the form considered in Section 3.2, applied here to the $X_{iu} - X_{iv}$ differences. For example, W_{23} is the median of the $n(n+1)/2$ Walsh averages of the form $[D_{23}^i + D_{23}^j]/2, 1 \leq i \leq j \leq n$, and can be viewed as an "unadjusted" estimator (see Comments 7.103 and 7.104) of the simple contrast $\tau_2 - \tau_3$.

Next, we compute

$$W_{u.} = \sum_{j=1}^{k} \frac{W_{uj}}{k}, \quad u = 1, \ldots, k, \tag{7.106}$$

where we note that $W_{uu} = 0$ for $u = 1, \ldots, k$. Setting

$$\widehat{\Delta}_{uv} = W_{u.} - W_{v.}, \tag{7.107}$$

the adjusted estimator of θ is given by

$$\widehat{\theta} = \sum_{j=1}^{k} a_j W_{j.}, \tag{7.108}$$

or, equivalently,

$$\widehat{\theta} = \sum_{h=1}^{k} \sum_{j=1}^{k} d_{hj} \widehat{\Delta}_{hj}. \tag{7.109}$$

(See equation (7.35) for the relationship between the d's and the a's.)

Example 7.15: *Rounding First Base.* In Example 7.5, we obtained the Doksum estimator of the contrast $\theta = \tau_{\text{roundout}} - \tau_{\text{wide angle}} = \tau_1 - \tau_3$ relating to the rounding first base data of Table 7.1. We now use (7.108) to obtain the Lehmann estimator of the same contrast. To evaluate W_{12}, defined by (7.105), note that $W_{12} = \text{median} \{[D^i_{12} + D^j_{12}]/2, 1 \leq i \leq j \leq 22\}$, where D^i_{12} and D^j_{12} are defined by (7.36). The $D^1_{12}, \ldots, D^{22}_{12}$ values are exhibited in Table 7.11. Letting $F^{(1)}_{12} \leq \cdots \leq F^{(253)}_{12}$ denote the 253 ordered $[D^i_{12} + D^j_{12}]/2$ values, we find:

$F^{(1)}_{12} = -.4, \quad F^{(2)}_{12} = F^{(3)}_{12} = F^{(4)}_{12} = -.25, \quad F^{(5)}_{12} = F^{(6)}_{12} = -.225,$
$F^{(7)}_{12} = \cdots = F^{(10)}_{12} = -.2, \quad F^{(11)}_{12} = \cdots = F^{(18)}_{12} = -.175,$
$F^{(19)}_{12} = F^{(20)}_{12} = -.15, \quad F^{(21)}_{12} = -.125, F^{(22)}_{12} = \cdots = F^{(27)}_{12} = -.1,$
$F^{(28)}_{12} = \cdots = F^{(34)}_{12} = -.075, \quad F^{(35)}_{12} = \cdots = F^{(49)}_{12} = -.05,$
$F^{(50)}_{12} = \cdots = F^{(81)}_{12} = -.025, \quad F^{(82)}_{12} = \cdots = F^{(113)}_{12} = 0,$
$F^{(114)}_{12} = \cdots = F^{(152)}_{12} = .025, \quad F^{(153)}_{12} = \cdots = F^{(198)}_{12} = .05,$
$F^{(199)}_{12} = \cdots = F^{(221)}_{12} = .075, \quad F^{(222)}_{12} = \cdots = F^{(234)}_{12} = .1,$
$F^{(235)}_{12} = \cdots = F^{(240)}_{12} = .125, \quad F^{(241)}_{12} = \cdots = F^{(249)}_{12} = .15,$
$F^{(250)}_{12} = F^{(251)}_{12} = .175, \quad F^{(252)}_{12} = .2, F^{(253)}_{12} = .25.$

Thus

$$W_{12} = F^{(127)}_{12} = .025.$$

To evaluate W_{13}, we use the equation $W_{13} = \text{median} \{[D^i_{13} + D^j_{13}]/2, 1 \leq i \leq j \leq 22\}$, where D^i_{13} and D^j_{13} are defined by (7.36). The $D^1_{13}, \ldots, D^{22}_{13}$ values are exhibited in Table 7.11. Letting $F^{(1)}_{13} \leq \cdots \leq F^{(253)}_{13}$ denote the 253 ordered $[D^i_{13} + D^j_{13}]/2$ values, we have:

7.15. CONTRAST ESTIMATION BASED ON ONE-SAMPLE HODGES-LEHMANN ESTIMATORS

$$F_{13}^{(1)} = -.3, \quad F_{13}^{(2)} = F_{13}^{(3)} = -.225, \quad F_{13}^{(4)} = -.2, \quad F_{13}^{(5)} = -.175,$$
$$F_{13}^{(6)} = F_{13}^{(7)} = F_{13}^{(8)} = -.15, \quad F_{13}^{(9)} = \cdots = F_{13}^{(12)} = -.125,$$
$$F_{13}^{(13)} = \cdots = F_{13}^{(19)} = -.1, \quad F_{13}^{(20)} = \cdots = F_{13}^{(25)} = -.075,$$
$$F_{13}^{(26)} = \cdots = F_{13}^{(33)} = -.05, \quad F_{13}^{(34)} = \cdots = F_{13}^{(46)} = -.025,$$
$$F_{13}^{(47)} = \cdots = F_{13}^{(62)} = 0, \quad F_{13}^{(63)} = \cdots = F_{13}^{(77)} = .025,$$
$$F_{13}^{(78)} = \cdots = F_{13}^{(94)} = .05, \quad F_{13}^{(95)} = \cdots = F_{13}^{(108)} = .075,$$
$$F_{13}^{(109)} = \cdots = F_{13}^{(131)} = .1, \quad F_{13}^{(132)} = \cdots = F_{13}^{(157)} = .125,$$
$$F_{13}^{(158)} = \cdots = F_{13}^{(190)} = .15, \quad F_{13}^{(191)} = \cdots = F_{13}^{(217)} = .175,$$
$$F_{13}^{(218)} = \cdots = F_{13}^{(238)} = .2, \quad F_{13}^{(239)} = \cdots = F_{13}^{(247)} = .225,$$
$$F_{13}^{(248)} = \cdots = F_{13}^{(253)} = .25.$$

Thus

$$W_{13} = F_{13}^{(127)} = .1.$$

In the same way, we calculate W_{23} by using the $D_{23}^1, \ldots, D_{23}^{22}$ values in Table 7.11 and the fact that $W_{23} = \text{median } \{[D_{23}^i + D_{23}^j]/2, 1 \leq i \leq j \leq 22\}$. Letting $F_{23}^{(1)} \leq \cdots \leq F_{23}^{(253)}$ denote the 253 ordered $[D_{23}^i + D_{23}^j]/2$ values, we see that:

$$F_{23}^{(1)} = -.1, F_{23}^{(2)} = \cdots = F_{23}^{(5)} = -.075, \quad F_{23}^{(6)} = \cdots = F_{23}^{(15)} = -.05,$$
$$F_{23}^{(16)} = \cdots = F_{23}^{(19)} = -.025, \quad F_{23}^{(20)} = \cdots = F_{23}^{(42)} = 0,$$
$$F_{23}^{(43)} = \cdots = F_{23}^{(73)} = .025, \quad F_{23}^{(74)} = \cdots = F_{23}^{(98)} = .05,$$
$$F_{23}^{(99)} = \cdots = F_{23}^{(138)} = .075, \quad F_{23}^{(139)} = \cdots = F_{23}^{(178)} = .1,$$
$$F_{23}^{(179)} = \cdots = F_{23}^{(211)} = .125, \quad F_{23}^{(212)} = \cdots = F_{23}^{(238)} = .15,$$
$$F_{23}^{(239)} = \cdots = F_{23}^{(247)} = .175, \quad F_{23}^{(248)} = \cdots = F_{23}^{(253)} = .2.$$

Thus

$$W_{23} = F_{23}^{(127)} = .075.$$

From (7.106), we find

$$W_{1.} = \frac{W_{11} + W_{12} + W_{13}}{3}$$
$$= \frac{0 + .025 + .1}{3} = .0417,$$

$$W_{2.} = \frac{W_{21} + W_{22} + W_{23}}{3}$$
$$= \frac{-.025 + 0 + .075}{3} = .0167,$$

and

$$W_{3.} = \frac{W_{31} + W_{32} + W_{33}}{3} = \frac{-.1 - .075 + 0}{3} = -.0583.$$

Note that in calculating $W_{2.}$ and $W_{3.}$ we use the relationship $W_{uv} = -W_{vu}$.

The Lehmann estimator $\widehat{\theta}$ is now obtained from (7.108) by noting that $a_1 = 1, a_2 = 0$, and $a_3 = -1$, so that

$$\widehat{\theta} = W_{1.} - W_{3.} = .0417 - (-.0583) = .10.$$

For these data, the adjusted estimator $W_{1.} - W_{3.}$ agrees with the unadjusted estimator W_{13}. However, we do note that the value of the Lehmann estimator $\widehat{\theta} = .10$ differs from that of the Doksum estimator $\widetilde{\theta} = .133$ (see Example 7.5) for these rounding first base data.

Comments

103. *Unadjusted Estimator.* The unadjusted estimator W_{uv} (7.105) of $\Delta_{uv} = \tau_u - \tau_v$ is simply the estimator associated with the signed rank test and discussed in Section 3.2.

104. *Ambiguities with the Unadjusted Estimators.* The unadjusted estimators W_{uv} (7.105) are incompatible, leading to possible ambiguities in contrast estimation because they do not satisfy the linear relations that are satisfied by the contrasts they estimate. We have encountered this difficulty before (see Comments 6.77 and 7.42). The adjusted estimators $\widehat{\Delta}_{uv}$ (7.107) are compatible, but have the disadvantage that the estimator of $\Delta_{uv} = \tau_u - \tau_v$ depends on the observations from the other $k - 2$ treatments.

105. *Computational Difficulty.* Example 7.15 is a glaring illustration of the labor involved in computing W_{uv} when n is moderately large. It is necessary to obtain the median of $n(n+1)/2$ Walsh averages, whereas the estimator Z_{uv} (7.37) is based on the median of only n differences. Thus, Doksum's contrast estimator is preferred to Lehmann's contrast estimator in terms of ease of computation. On the other hand, asymptotic efficiencies generally (but not always) favor Lehmann's contrast estimator. (See Section 7.16.)

Properties

1. *Standard Deviation of $\widehat{\theta}$ (7.108).* For the asymptotic standard deviation of $\widehat{\theta}$ (7.108), see Lehmann (1964).
2. *Asymptotic Normality.* See Lehmann (1964).
3. *Efficiency.* See Lehmann (1964) and Section 7.16.

PROBLEMS

129. Calculate the Lehmann estimator of the contrast $2\tau_N - \tau_A - \tau_R$ for the metronome data of Table 7.6. Compare with the Doksum estimator from Problem 48.
130. Give an example illustrating the incompatibility of the unadjusted estimators W_{uv} (7.105). (See Comment 104.)
131. Calculate the Lehmann estimators for the simple contrasts $\theta_1 = \tau_2 - \tau_1, \theta_2 = \tau_3 - \tau_1$, and $\theta_3 = \tau_3 - \tau_2$ for the creatine phosphokinase activity data in Table 7.3.
132. Estimate the contrast $3\tau_{ALC} - \tau_{AL} - \tau_{AC} - \tau_{LC}$ for the percentage consonants correctly identified data in Table 7.4.
133. Using the data of Table 7.4, obtain Lehmann's estimator of the simple contrast that represents the benefit from adding lip reading to audition in teaching severely hearing-impaired children. Compare with the Doksum estimator from Problem 52.
134. Compute Lehmann's estimator for all contrasts found to be of interest in Problem 47 for the maximum soil temperature data in Table 7.8.
135. Calculate Lehmann's estimator of the contrast $\tau_{\text{rats}} - \tau_{\text{cats}}$ for the Livesey EPT error score data of Table 7.9.

7.16. EFFICIENCIES OF TWO-WAY LAYOUT PROCEDURES

We first consider the procedures of Sections 7.1–7.5, which are associated with Friedman rank sums for the case of one observation per treatment-block combination (i.e., a randomized complete block design). The Pitman asymptotic relative efficiencies (for translation alternatives) of these procedures with respect to the corresponding normal theory counterparts are given by the expression

$$e_F = \left[\frac{k}{(k+1)}\right] \left[12\sigma_F^2 \left\{\int_{-\infty}^{\infty} f^2(u)\, du\right\}^2\right], \qquad (7.110)$$

where σ_F^2 is the variance of the common underlying (continuous) distribution F (7.1) and $f(.)$ is the probability density function corresponding to F. The parameter $\int_{-\infty}^{\infty} f^2(u)\, du$ is the area under the curve associated with $f^2(.)$, the square of the common probability density function. We note that e_F (7.110) is simply $k/(k+1)$ times the corresponding Pitman efficiencies in the one-sample, two-sample, and k-sample location settings (see Sections 3.11, 4.5, and 6.10).

In particular, the Pitman asymptotic relative efficiency of the Friedman test based on S (7.5) with respect to the normal theory two-way layout F test was found to be e_F (7.110) by van Elteren and Noether (1959). The asymptotic relative efficiency of the Page test for ordered alternatives, based on the statistic L (7.10), with respect to a suitable normal theory competitor was found by Hollander (1967a) to be e_F (7.110) as well. Furthermore, methods analogous to those of Sherman (1965) lead to expression (7.110) as the asymptotic relative efficiency of both the all-treatments two-sided and the treatments-versus-control one-sided multiple comparison procedures in Sections 7.3 and 7.4, respectively, with respect to the classical normal theory procedures based on sample means. Finally, Doksum (1967) obtained (7.110) as the asymptotic relative efficiency of the estimator $\tilde{\theta}$ (7.40) with respect to the least-squares estimator $\overline{\theta} = \sum_{j=1}^{k} a_j X_{.j}$, where $X_{.j} = \sum_{i=1}^{n} X_{ij}/n$.

The efficiency e_F (7.110) is always greater than or equal to .576 and it can be infinite. Some values of e_F for various F and k combinations are given in Table 7.27.

We next turn to the procedures in Sections 7.6–7.8 that are designed for two-way layout data with zero or one observation per treatment-block combination. The Pitman asymptotic relative efficiency of the Durbin-Skillings-Mack test based on D (7.43) with respect to the standard normal theory procedure for a balanced incomplete block design was found to be e_F (7.110) by van Elteren and Noether (1959). Once again, methods analogous to those of Sherman (1965) lead to expression (7.110) as the asymptotic relative efficiency of the all-treatments two-sided multiple comparison procedures in Section 7.7. We do not know of any results for the asymptotic efficiencies of the general alternatives Skillings-Mack test in Section 7.8 for data from an arbitrary incomplete block design.

For the case of two-way layout data with at least one observation for every treatment-block combination, Mack and Skillings (1980) found that under certain conditions the asymptotic

TABLE 7.27. Values of e_F for Various Distributions and Numbers (k) of Treatments

k Distribution	2	3	4	5 e_F	10	20	50	∞
Normal	.637	.716	.764	.796	.868	.909	.936	.955
Uniform	.667	.750	.800	.833	.909	.952	.980	1.000
Double exponential	1.000	1.125	1.200	1.250	1.364	1.429	1.471	1.500

TABLE 7.28. Values of e_F^* for Various Distributions and Numbers (k) of Treatments

k Distribution	2	3	4	5 e_F^*	10	20	50	∞
Normal	.955	.966	.972	.975	.983	.987	.989	.990
Uniform	.889	.894	.897	.899	.902	.904	.905	.906
Exponential	1.500	1.528	1.543	1.552	1.570	1.579	1.585	1.588

relative efficiency of their test for general alternatives based on the statistic MS (7.57) with respect to a suitable normal theory competitor is, once again, given by e_F (7.110). Combining their results with methods analogous to those of Sherman (1965) yields expression (7.110) as the asymptotic relative efficiency of the all-treatments two-sided multiple comparison procedures in Section 7.10, as well.

Finally, we turn to the procedures in Sections 7.11–7.15 which are associated with Wilcoxon signed ranks. The asymptotic efficiencies of Doksum's conservative test of Section 7.11, based on replacing λ_F by its upper bound $\frac{7}{24}$, are very close to those of a related test proposed by Doksum (1967) in which λ_F is estimated. The expression for the asymptotic relative efficiency e_F^* of the related test, relative to the normal theory $\mathcal{F}$-test, is given by the right-hand side of equation (2.12) in Doksum (1967). The parameter e_F^* is always greater than .864 and can be infinite. In Table 7.28 we provide values of e_F^* for normal, uniform and exponential distributions and various numbers (k) of treatments. Similarly, the efficiencies of Hollander's conservative test of Section 7.12, based on replacing λ_F by its upper bound $\frac{7}{24}$, are very close to those of a related test proposed by Hollander (1967a), in which λ_F is estimated. The expression for the asymptotic relative efficiency e_F^{**} of this related test, with respect to a normal theory t-test for ordered alternatives, is given by the right-hand sign of equation (4.6) of Hollander (1967a). The parameter e_F^{**} is always greater than .864 and can be infinite. In Table 7.29 we provide values of e_F^{**} for normal, uniform and exponential distributions and various numbers (k) of treatments. The efficiencies in Table 7.29 are also close approximations to the efficiencies of the conservative multiple comparison procedures of Sections 7.13 and 7.14 with respect to normal theory competitors based on sample means. Lehmann (1964) obtained the asymptotic relative efficiency (for translation alternatives) of the contrast estimator (7.108) of Secton 7.15 with respect to the least-squares estimator based on the sample means. The asymptotic relative efficiency is given by e_F^*. See Table 7.28.

TABLE 7.29. Values of e_F^{**} for Various Distributions and Numbers (k) of Treatments

k Distribution	2	3	4	5 e_F^{**}	10	20	50	∞
Normal	.955	.963	.969	.972	.980	.985	.988	.990
Uniform	.889	.893	.895	.897	.901	.903	.905	.906
Exponential	1.500	1.521	1.534	1.543	1.563	1.575	1.583	1.588

CHAPTER 8

The Independence Problem

INTRODUCTION

The data in this chapter consist of a random sample from a bivariate population. Our basic interest here is in the statistical relationship between the two variables involved in the bivariate structure. In particular, we will discuss procedures for deciding whether or not these two variables are independent and, if not independent, for assessing both the type and degree of dependency that exists between them.

In Section 8.1 we present a distribution-free test for independence that is based on signs of appropriate products of differences. Section 8.2 presents an estimator of the measure of association τ defined by (8.2). Section 8.3 contains an asymptotically distribution-free confidence interval for τ. Section 8.4 uses Efron's bootstrap method to obtain a different asymptotically distribution-free confidence interval for τ. Section 8.5 presents a distribution-free test for independence based on ranks. Section 8.6 contains a distribution-free test of independence which is consistent against a broader class of alternatives than those classes of alternatives that can be detected by the tests of Sections 8.1 and 8.5. Section 8.7 considers the asymptotic relative efficiencies of the procedures in this chapter with respect to their normal theory counterparts.

Data. We obtain n bivariate observations $(X_1, Y_1), \ldots, (X_n, Y_n)$, one observation on each of n subjects.

Assumptions

A The n bivariate observations $(X_1, Y_1), \ldots, (X_n, Y_n)$ are a random sample from a continuous bivariate population. That is, the (X, Y) pairs are mutually independent and identically distributed according to some continuous bivariate population.

8.1. A DISTRIBUTION-FREE TEST FOR INDEPENDENCE BASED ON SIGNS (KENDALL)

Hypothesis

Let $F_{X,Y}$ be the joint distribution function for the common bivariate population of the (X, Y) pairs. Moreover, let $F_X(x)$ and $F_Y(y)$ be the distribution functions for the marginal X and Y

populations, respectively. The null hypothesis of interest here is that the X and Y random variables are independent. Formally stated, this null hypothesis is

$$H_0 : [F_{X,Y}(x,y) \equiv F_X(x)F_Y(y), \text{ for all } (x,y) \text{ pairs}]. \tag{8.1}$$

The alternative hypothesis to (8.1) will be a function of the type of dependence between the X and Y variables that is of principal interest. In this section we concentrate on a type of dependence measured by the Kendall population correlation coefficient

$$\tau = 2P\{(Y_2 - Y_1)(X_2 - X_1) > 0\} - 1. \tag{8.2}$$

We note that the event $\{(Y_2 - Y_1)(X_2 - X_1) > 0\}$ occurs if and only if either the event $\{X_2 > X_1 \text{ and } Y_2 > Y_1\}$ or the event $\{X_2 < X_1 \text{ and } Y_2 < Y_1\}$ occurs. Since these latter two events are mutually exclusive, we have that

$$P\{(Y_2 - Y_1)(X_2 - X_1) > 0\} = P(X_2 > X_1, Y_2 > Y_1) + P(X_2 < X_1, Y_2 < Y_1). \tag{8.3}$$

If X and Y are independent, it follows that

$$P(X_2 > X_1, Y_2 > Y_1) = P(X_2 > X_1)P(Y_2 > Y_1) = \left(\tfrac{1}{2}\right)\left(\tfrac{1}{2}\right) = \tfrac{1}{4}, \tag{8.4}$$

since X_1, X_2 are independent and identically distributed variables, as are Y_1, Y_2 (though not necessarily, of course, with the same distribution as the X's). Similarly, if X and Y are independent, we also have

$$P(X_2 < X_1, Y_2 < Y_1) = \tfrac{1}{4}.$$

Combining this result with (8.3) and (8.4), we see that the Kendall population correlation coefficient $\tau = 2(\tfrac{1}{4} + \tfrac{1}{4}) - 1 = 0$ if X and Y are independent. (It is important to point out that this is not an if and only if statement as $\tau = 0$ does not necessarily imply that X and Y are independent. See Comment 2 for more on this relationship.)

Procedure

To compute the Kendall sample correlation statistic K, we first calculate the values of the $n(n-1)/2$ paired sign statistics $Q((X_i, Y_i), (X_j, Y_j))$, for $1 \leq i < j \leq n$, where

$$Q((a,b),(c,d)) = \begin{cases} 1, & \text{if } (d-b)(c-a) > 0, \\ -1, & \text{if } (d-b)(c-a) < 0. \end{cases} \tag{8.5}$$

That is, for each pair of subscripts (i, j) with $i < j$, score 1 if $(Y_j - Y_i)(X_j - X_i)$ is positive and score -1 if $(Y_j - Y_i)(X_j - X_i)$ is negative. The Kendall statistic K is then

$$K = \sum_{i=1}^{n-1} \sum_{j=i+1}^{n} Q((X_i, Y_i), (X_j, Y_j)), \tag{8.6}$$

corresponding to adding up the 1s and -1s from the paired sign statistics.

a. One-Sided Upper-Tail Test. To test the null hypothesis of independence, namely,

$$H_0 : [F_{X,Y}(x, y) \equiv F_X(x)F_Y(y), \text{ for all } (x, y) \text{ pairs}]$$

(which implies $\tau = 0$) versus the alternative that X and Y are positively correlated (see Comment 2) corresponding to

$$H_1 : \tau > 0, \tag{8.7}$$

at the α-level of significance,

$$\text{Reject } H_0 \text{ if } K \geq k_\alpha; \quad \text{otherwise do not reject}, \tag{8.8}$$

where the constant k_α is chosen to make the type I error probability equal to α. Values of k_α are given in Table A.30.

b. One-Sided Lower-Tail Test. To test

$$H_0 : [F_{X,Y}(x, y) \equiv F_X(x)F_Y(y), \text{ for all } (x, y) \text{ pairs}]$$

versus the alternative that X and Y are negatively correlated (see Comment 2) corresponding to

$$H_2 : \tau < 0,$$

at the α-level of significance,

$$\text{Reject } H_0 \text{ if } K \leq -k_\alpha; \quad \text{otherwise do not reject.} \tag{8.9}$$

c. Two-Sided Test. To test

$$H_0 : [F_{X,Y}(x, y) \equiv F_X(x)F_Y(y), \text{ for all } (x, y) \text{ pairs}]$$

versus the general alternative that X and Y are dependent variables corresponding to

$$H_3 : \tau \neq 0,$$

at the α-level of significance,

$$\text{Reject } H_0 \text{ if } |K| \geq k_{\alpha/2}; \quad \text{otherwise do not reject.} \tag{8.10}$$

This two-sided procedure is the two-sided symmetric test with $\alpha/2$ probability in each tail of the null distribution of K.

Large-Sample Approximation

The large-sample approximation is based on the asymptotic normality of K, suitably standardized. For this standardization, we need to know the expected value and variance of K when the null hypothesis of independence is true. Under H_0, the expected value and variance of K are

$$E_0(K) = 0 \tag{8.11}$$

and

$$\text{var}_0(K) = \frac{n(n-1)(2n+5)}{18}, \tag{8.12}$$

respectively. These expressions for $E_0(K)$ and $\text{var}_0(K)$ are verified by direct calculations in Comment 7 for the special case of $n = 4$. General derivations of both expressions are presented in Comment 10.

The standardized version of K is

$$K^* = \frac{K - E_0(K)}{\{\text{var}_0(K)\}^{1/2}} = \frac{K}{\{n(n-1)(2n+5)/18\}^{1/2}}. \tag{8.13}$$

When H_0 is true, K^* has, an n tends to infinity, an asymptotic $N(0, 1)$ distribution. (See Comment 10 for indications of the proof.) The normal theory approximation for procedure (8.8) is

$$\text{Reject } H_0 \text{ if } K^* \geq z_\alpha; \quad \text{otherwise do not reject,} \tag{8.14}$$

the normal theory approximation for procedure (8.9) is

$$\text{Reject } H_0 \text{ if } K^* \leq -z_\alpha; \quad \text{otherwise do not reject,} \tag{8.15}$$

and the normal theory approximation for procedure (8.10) is

$$\text{Reject } H_0 \text{ if } |K^*| \geq z_{\alpha/2}; \quad \text{otherwise do not reject.} \tag{8.16}$$

Ties

If there are ties among the n X observations and/or separately among the n Y observations, replace the function $Q((a,b),(c,d))$ in the definition of K (8.6) by

$$Q^*((a,b),(c,d)) = \begin{cases} 1, & \text{if } (d-b)(c-a) > 0, \\ 0, & \text{if } (d-b)(c-a) = 0, \\ -1, & \text{if } (d-b) \end{cases} \tag{8.17}$$

(Thus, in the case of tied X values and/or tied Y values, zeros are assigned to the associated paired sign statistics.) After computing K with these modified paired sign statistics, use procedure (8.8), (8.9), or (8.10) and refer the value of K to Table A.30. Note, however, that this test associated with tied X's and/or Y's is only approximately, and not exactly, of significance level α.

When applying the large-sample approximation, however, the loss in variability due to the tied X's and/or tied Y's must also be taken into account. While these ties do not affect the null expected value of K, its null variance is reduced to

$$\text{var}_0(K) = \frac{\left\{n(n-1)(2n+5) - \sum_{i=1}^{g} t_i(t_i-1)(2t_i+5) - \sum_{j=1}^{h} u_j(u_j-1)(2u_j+5)\right\}}{18}$$

$$+ \frac{\left\{\sum_{i=1}^{g} t_i(t_i-1)(t_i-2)\right\} \left\{\sum_{j=1}^{h} u_j(u_j-1)(u_j-2)\right\}}{9n(n-1)(n-2)}$$

$$+ \frac{\left\{\sum_{i=1}^{g} t_i(t_i-1)\right\}\left\{\sum_{j=1}^{h} u_j(u_j-1)\right\}}{2n(n-1)} \tag{8.18}$$

in the presence of such ties, where in equation (8.18) g denotes the number of tied X groups, t_i is the size of tied X group i, h is the number of tied Y groups, and u_j is the size of the tied Y group j. We note that an untied $X(Y)$ observation is considered to be a tied $X(Y)$ "group" of size 1. In particular, if neither the collection of n X nor the collection of n Y observations contains tied observations, we have $g = h = n, t_i = u_j = 1, i = 1, \ldots, n$ and $j = 1, \ldots, n$. In this case of no tied X's and no tied Y's, each term involving either $(t_i - 1)$ or $(u_j - 1)$ or both reduces to zero and the variance expression in (8.18) reduces to the usual null variance of K, as given previously in equation (8.12).

As a consequence of the effect that ties have on the null variance of K, the following modification is needed to apply the large-sample approximation when there are tied X observations and/or tied Y observations. Compute K with the modified paired sign statistic using equation (8.17) and set

$$K^* = \frac{K}{\{\text{var}_0(K)\}^{1/2}}, \qquad (8.19)$$

where $\text{var}_0(K)$ is now given by display (8.18). With this modified form of K^*, approximations (8.14), (8.15), or (8.16) can be applied.

Example 8.1: *Tuna Lightness and Quality.* The data in Table 8.1 are a subset of the data obtained by Rasekh, Kramer, and Finch (1970) in a study designed to ascertain the relative importance of the various factors contributing to tuna quality and to find objective methods for determining quality parameters and consumer preference. Table 8.1 gives values of the Hunter L measure of lightness, along with panel scores for nine lots of canned tuna. The original consumer panel scores of excellent, very good, good, fair, poor, and unacceptable were converted to the numerical values of 6, 5, 4, 3, 2, and 1, respectively. The panel scores in Table 8.1 are averages of 80 such values. (The Y random variable is thus discrete, and hence the continuity portion of Assumption A is not satisfied. Nevertheless, since each Y is an average of 80 values, we need not be nervous about this departure from Assumption A.)

It is suspected that the Hunter L value is positively associated with the panel score. Thus, we will apply procedure (8.8) to test H_0 (8.1) versus $\tau > 0$. For purposes of illustration, we consider the significance level $\alpha = .090$. From Table A.30 we find $k_{.090} = 14$ and procedure (8.8) becomes

Reject H_0 if $K \geq 14$.

TABLE 8.1. Hunter L Values and Consumer Panel Scores for Nine Lots of Canned Tuna

Lot	Hunter L Value (X)	Panel Score (Y)
1	44.4	2.6
2	45.9	3.1
3	41.9	2.5
4	53.3	5.0
5	44.7	3.6
6	44.1	4.0
7	50.7	5.2
8	45.2	2.8
9	60.1	3.8

Source: J. Rasekh, A. Kramer, and R. Finch (1970).

TABLE 8.2. $Q((X_i, Y_i), (X_j, Y_j))$ Values for Canned Tuna Data

j\i	1	2	3	4	5	6	7	8
2	1							
3	1	1						
4	1	1	1					
5	1	−1	1	1				
6	−1	−1	1	1	−1			
7	1	1	1	−1	1	1		
8	1	1	1	1	−1	−1	1	
9	1	1	1	−1	1	−1	−1	1

We illustrate the computations of the paired sign statistics in (8.5) leading to the sample value of K (8.6) in Table 8.2.
Summing the $+1$ and -1 values in Table 8.2 we see that

$$K = \sum_{i=1}^{8} \sum_{j=i+1}^{9} Q((X_i, Y_i), (X_j, Y_j)) = 26 - 10 = 16.$$

Since this value of K is greater than the critical value 14, we reject H_0 in favor of $\tau > 0$ at the $\alpha = .090$ level.

Since the one-sided P-value for these data is the lowest significance level at which we can reject H_0 in favor of $\tau > 0$ with the observed value of the test statistic $K = 16$, we see from Table A.30 (entering the table with $n = 9$ and $x = 16$) that the P-value for these data is $P_0(K \geq 16) = .060$. Thus there is some evidence (though not overwhelming) that the Hunter L lightness values and the panel scores are positively correlated.

For the large-sample approximation we find (since there are no ties in the data) from equation (8.13) that

$$K^* = \frac{16}{\{9(8)(23)/18\}^{1/2}} = 1.67.$$

Thus, the smallest significance level at which we can reject H_0 in favor of $\tau > 0$ using the normal approximation is .0475, since $z_{.0475} = 1.67$ in Table A.1. This is in good agreement with the exact P-value of .060 found previously.

Comments

1. *Motivation for the Test*. The null hypothesis of this section is that the X and Y random variables are independent, which implies (see the discussion in the Procedure) that the Kendall population correlation coefficient τ is equal to 0. However, the alternatives are stated directly in terms of $\tau (>, <, \text{ or } \neq 0)$. When τ is greater than 0 (and thus $P((Y_2 - Y_1)(X_2 - X_1) > 0) > \frac{1}{2}$), there will tend to be a large number of positive paired sign statistics and fewer negative paired sign statistics. Hence, when τ is greater than 0, we would expect the sample to lead to a big, positive value for K. This suggests rejecting H_0 in favor of $\tau > 0$ for large values of K and motivates procedures (8.8) and (8.14). Similar rationales lead to procedures (8.9), (8.10), (8.15), and (8.16).

8.1. A DISTRIBUTION-FREE TEST FOR INDEPENDENCE BASED ON SIGNS (KENDALL)

2. *Interpretation of τ.* The Kendall correlation coefficient τ can also be written as $\tau = [P((Y_2 - Y_1)(X_2 - X_1) > 0) - P((Y_2 - Y_1)(X_2 - X_1) < 0)]$. We have already noted that if X and Y are independent, then $\tau = 0$. On the other hand, if $\tau > 0$, then it is more likely that $\{X_2 > X_1 \text{ and } Y_2 > Y_1\}$ or $\{X_2 < X_1 \text{ and } Y_2 < Y_1\}$ occurs than either of the complementary events $\{X_2 > X_1 \text{ and } Y_2 < Y_1\}$ or $\{X_2 < X_1 \text{ and } Y_2 > Y_1\}$. Thus, if $\tau > 0$, it is more likely that the change from X_1 to X_2 has the same (rather than opposite) sign as that from Y_1 to Y_2. It is reasonable to interpret this type of relationship between X and Y as indicative of a positive association (as measured by τ). Similarly, $\tau < 0$ may reasonably be interpreted as indicative of a negative association (as measured by τ) between X and Y.

3. *Concordant/Discordant Pairs.* Call the $(X_i, Y_i), (X_j, Y_j)$ pairs *concordant* if $(X_i - X_j)(Y_i - Y_j) > 0$ and *discordant* if $(X_i - X_j)(Y_i - Y_j) < 0$. Thus (X_i, Y_i) and (X_j, Y_j) are concordant if either (a) $X_i > X_j$ and $Y_i > Y_j$ or if (b) $X_i < X_j$ and $Y_i < Y_j$. Similarly, (X_i, Y_i) and (X_j, Y_j) are discordant if (c) $X_i < X_j$ and $Y_i > Y_j$ or if (d) $X_i > X_j$ and $Y_i < Y_j$. Now K (8.6) can be expressed as $K = K' - K''$, where

$$K' = \text{number of concordant pairs,}$$

$$K'' = \text{number of discordant pairs,}$$

and the count is taken over the $n(n-1)/2$ sets of pairs $(X_i, Y_i), (X_j, Y_j)$ with $i < j$. Note that $(X_i, Y_i), (X_j, Y_j)$ are concordant if the ordering of X_i, X_j agrees with that of Y_i, Y_j. We have discordance when these orderings do not agree. Thus $K/\{n(n-1)/2\}$ can be viewed as an average measure of agreement between the X's and the Y's, where agreement refers to order.

4. *Equivalent Expression When There Are No Ties.* Let K' = (number of concordant pairs) and K'' = (number of discordant pairs), as defined in Comment 3. If there are no ties among the X's and no ties among the Y's then $K' + K'' = n(n-1)/2$. Thus, with no ties we have $K = K' - K'' = K' - [n(n-1)/2 - K'] = 2K' - \{n(n-1)/2\}$. To illustrate, consider the tuna data in Example 8.1. Summing the 1s in Table 8.2 (corresponding to concordant pairs), we obtain $K' = 7 + 5 + 6 + 3 + 2 + 1 + 1 + 1 = 26$. Adding the 0s in Table 8.2 (corresponding to discordant pairs), we have $K'' = 1 + 2 + 0 + 2 + 2 + 2 + 1 + 0 = 10$. It follows that $K = K' - K'' = 26 - 10 = 2K' - \{n(n-1)/2\} = [2(26) - 9(8)/2] = 16$, in agreement with the value obtained directly in Example 8.1.

5. *Convenience Through Ordering.* It is convenient to compute the number of concordant pairs, K', by first rearranging the (X_i, Y_i) pairs so that the (new) X's are in increasing order. Then, after rearrangement, K' is equal to the number of pairs for which the corresponding Y's are in increasing order. For example, suppose our observations are

i	1	2	3	4	5
X_i	4.1	−2.4	−2.2	−5.6	5.5
Y_i	2.3	3.7	1.1	2.2	3.8

We arrange these so that the X's are in increasing order and obtain the following:

X	−5.6	−2.4	−2.2	4.1	5.5
Y	2.2	3.7	1.1	2.3	3.8

370   8. THE INDEPENDENCE PROBLEM

Then, proceeding from left to right, we find the Y pairs that are in increasing order to be: (2.2, 3.7), (2.2, 2.3), (2.2, 3.8), (3.7, 3.8), (1.1, 2.3), (1.1, 3.8), (2.3, 3.8). Thus, $K' = 7$ and $K = 2K' - \{5(4)/2\} = 4$.

6. *Derivation of Distribution of K under H_0 (No-Ties Case).* Let R_i be the rank of X_i in the joint ranking of $X_1, \ldots, X_n$ and let S_i be the rank of Y_i in the joint ranking of $Y_1, \ldots, Y_n$. It is clear that knowledge of the R's and S's is sufficient to calculate K (8.6). (See Problem 2.) We use this fact to illustrate how the null distribution of K can be obtained. Without loss of generality, we take $R_1 = 1, \ldots, R_n = n$; then, since under H_0 (8.1) all possible $n!$ $(S_1, S_2, \ldots, S_n)$ Y-rank configurations are equally likely, we realize that each has probability $(1/n!)$.

Let us consider the case $n = 4$. In the following table we display the $4! = 24$ possible (S_1, S_2, S_3, S_4) configurations, the associated values of K, and the corresponding null probabilities.

(R_1, R_2, R_3, R_4)	(S_1, S_2, S_3, S_4)	Null Probability	K
(1, 2, 3, 4)	(1, 2, 3, 4)	$\frac{1}{24}$	6
(1, 2, 3, 4)	(1, 2, 4, 3)	$\frac{1}{24}$	4
(1, 2, 3, 4)	(1, 3, 2, 4)	$\frac{1}{24}$	4
(1, 2, 3, 4)	(1, 3, 4, 2)	$\frac{1}{24}$	2
(1, 2, 3, 4)	(1, 4, 2, 3)	$\frac{1}{24}$	2
(1, 2, 3, 4)	(1, 4, 3, 2)	$\frac{1}{24}$	0
(1, 2, 3, 4)	(2, 1, 3, 4)	$\frac{1}{24}$	4
(1, 2, 3, 4)	(2, 1, 4, 3)	$\frac{1}{24}$	2
(1, 2, 3, 4)	(2, 3, 1, 4)	$\frac{1}{24}$	2
(1, 2, 3, 4)	(2, 3, 4, 1)	$\frac{1}{24}$	0
(1, 2, 3, 4)	(2, 4, 1, 3)	$\frac{1}{24}$	0
(1, 2, 3, 4)	(2, 4, 3, 1)	$\frac{1}{24}$	-2
(1, 2, 3, 4)	(3, 1, 2, 4)	$\frac{1}{24}$	2
(1, 2, 3, 4)	(3, 1, 4, 2)	$\frac{1}{24}$	0
(1, 2, 3, 4)	(3, 2, 1, 4)	$\frac{1}{24}$	0
(1, 2, 3, 4)	(3, 2, 4, 1)	$\frac{1}{24}$	-2
(1, 2, 3, 4)	(3, 4, 1, 2)	$\frac{1}{24}$	-2
(1, 2, 3, 4)	(3, 4, 2, 1)	$\frac{1}{24}$	-4
(1, 2, 3, 4)	(4, 1, 2, 3)	$\frac{1}{24}$	0
(1, 2, 3, 4)	(4, 1, 3, 2)	$\frac{1}{24}$	-2
(1, 2, 3, 4)	(4, 2, 1, 3)	$\frac{1}{24}$	-2
(1, 2, 3, 4)	(4, 2, 3, 1)	$\frac{1}{24}$	-4
(1, 2, 3, 4)	(4, 3, 1, 2)	$\frac{1}{24}$	-4
(1, 2, 3, 4)	(4, 3, 2, 1)	$\frac{1}{24}$	-6

Thus, for example, the probability is $\frac{5}{24}$ under H_0 that K is equal to 2, since $K = 2$ when any of the five outcomes $(S_1, S_2, S_3, S_4) = $ (1, 3, 4, 2), (1, 4, 2, 3), (2, 1, 4, 3), (2, 3, 1, 4), or (3, 1, 2, 4) occurs and each of these outcomes has null probability $\frac{1}{24}$. Simplifying, we obtain the null distribution

Possible Value of K	Probability under H_0
-6	$\frac{1}{24}$
-4	$\frac{3}{24}$
-2	$\frac{5}{24}$
0	$\frac{6}{24}$
2	$\frac{5}{24}$
4	$\frac{3}{24}$
6	$\frac{1}{24}$

The probability, under H_0, that K is greater than or equal to 2, for example, is therefore

$$P_0(K \geq 2) = P_0(K = 2) + P_0(K = 4) + P_0(K = 6)$$
$$= \tfrac{5}{24} + \tfrac{3}{24} + \tfrac{1}{24} = \tfrac{3}{8} = .375.$$

This agrees with the upper-tail probability entry for $n = 4$ and the value $K = 2$ in Table A.30.

Note that we have derived the null distribution of K without specifying the form of the underlying independent X and Y populations under H_0 beyond the point of requiring that they be continuous. That is why the test procedures based on K are called distribution-free procedures. From the null distribution of K we can determine the critical value k_α and control the probability α of falsely rejecting H_0 when H_0 is true, and this error probability does not depend on the specific forms of the underlying continuous and independent X and Y distributions.

7. *Calculation of the Mean and Variance of K under the Null Hypothesis.* Displays (8.11) and (8.12) present formulas for the mean and variance of K when the null hypothesis is true. In this comment we illustrate a direct calculation of $E_0(K)$ and $\text{var}_0(K)$ in the particular case of $n = 4$, using the null distribution of K obtained in Comment 6. (Later, in Comment 10, we present general derivations of $E_0(K)$ and $\text{var}_0(K)$.) The null mean, $E_0(K)$, is obtained by multiplying each possible value of K with its probability under H_0 and summing the products. Thus

$$E_0(K) = -6\left(\tfrac{1}{24}\right) - 4\left(\tfrac{3}{24}\right) - 2\left(\tfrac{5}{24}\right) + 0\left(\tfrac{6}{24}\right) + 2\left(\tfrac{5}{24}\right) + 4\left(\tfrac{3}{24}\right) + 6\left(\tfrac{1}{24}\right)$$
$$= 0.$$

This is in agreement with the value stated in equation (8.7). A check on the expression for $\text{var}_0(K)$ is also easily performed, using the well-known fact that

$$\text{var}_0(K) = E_0(K^2) - \{E_0(K)\}^2.$$

The value of $E_0(K^2)$, the second moment of the null distribution of K, is again obtained by multiplying possible values (in this case, of K^2) by the corresponding probabilities under H_0

and summing. We find

$$E_0(K^2) = \left[(36+36)\left(\tfrac{1}{24}\right) + (16+16)\left(\tfrac{3}{24}\right) + (4+4)\left(\tfrac{5}{24}\right) + 0\left(\tfrac{6}{24}\right)\right]$$
$$= \tfrac{26}{3}.$$

Thus,

$$\text{var}_0(K) = \tfrac{26}{3} - (0)^2 = \tfrac{26}{3},$$

which agrees with what we obtain using equation (8.12) directly, namely,

$$\text{var}_0(K) = \frac{4(4-1)(2(4)+5)}{18} = \frac{26}{3}.$$

8. *Symmetry of the Distribution of K under the Null Hypothesis.* When H_0 is true, the distribution of K is symmetric about its mean 0. (See Comment 6 for verification of this when $n = 4$.) This implies that

$$P_0(K \le -x) = P_0(K \ge x), \qquad (8.20)$$

for all x. Equation (8.20) is used directly to convert upper-tail probabilities, as presented in Table A.30, to lower tail probabilities. In particular, it follows from equation (8.20) that the lower α percentile for the null distribution of K is $-k_\alpha$, thus the use of $-k_\alpha$ as the critical value in procedure (8.9).

9. *Possible Values for K.* If $n = 4j$ or $n = 4j + 1$, $j = 0, 1, \ldots$, the statistic K (8.6) is always an even integer. Similarly, if $n = 4j + 2$ or $n = 4j + 3$, $j = 0, 1, \ldots$, K is always an odd integer. The fact that K can assume only every other integer follows from the counting procedure used to define K (see (8.5) and (8.6)). The even or odd property of K for specific sample sizes can be deduced from the relation $K = 2K' - \{n(n-1)/2\}$ and the fact that $n(n-1)/2$ is an even integer (the product of an odd and an even integer) when $n = 4j$ or $n = 4j + 1$, $j = 0, 1, \ldots$, and is an odd integer (the product of two odd integers) when $n = 4j + 2$ or $n = 4j + 3$, $j = 0, 1, \ldots$.

10. *Large-Sample Approximation.* From the counting representation for K in (8.5) and (8.6), we see immediately that

$$E(K) = E\left[\sum_{i=1}^{n-1}\sum_{j=i+1}^{n} Q((X_i, Y_i), (X_j, Y_j))\right]$$

$$= \sum_{i=1}^{n-1}\sum_{j=i+1}^{n} E[Q((X_i, Y_i), (X_j, Y_j))].$$

$$= \sum_{i=1}^{n-1}\sum_{j=i+1}^{n} [P\{(Y_2 - Y_1)(X_2 - X_1) > 0\}$$
$$- P\{(Y_2 - Y_1)(X_2 - X_1) < 0\}],$$

8.1. A DISTRIBUTION-FREE TEST FOR INDEPENDENCE BASED ON SIGNS (KENDALL)

which, since the X and Y variables are continuous, yields

$$E(K) = \sum_{i=1}^{n-1} \sum_{j=i+1}^{n} [2P\{(Y_2 - Y_1)(X_2 - X_1) > 0\} - 1]$$

$$= \sum_{i=1}^{n-1} \sum_{j=i+1}^{n} \tau = \binom{n}{2} \tau, \qquad (8.21)$$

from expression (8.2) for τ. Since $\tau = 0$ if X and Y are independent, it follows that the expected value of K under H_0 is 0, as noted in equation (8.11). For the variance of K, we can use a well-known expression for the variance of a sum of random variables to obtain

$$\operatorname{var}(K) = \left[\sum_{i=1}^{n-1} \sum_{j=i+1}^{n} \operatorname{var}(Q_{ij}) + \sum_{i=1}^{n-1} \sum_{j=i+1}^{n} \sum_{\substack{s=1 \\ (i,j) \neq (s,t)}}^{n-1} \sum_{t=s+1}^{n} \operatorname{cov}(Q_{ij}, Q_{st}) \right], \qquad (8.22)$$

where $Q_{uv} = Q((X_u, Y_u), (X_v, Y_v))$, for $1 \leq u < v \leq n$.

After considerable tedious calculation, we can show that equation (8.22) simplifies to

$$\operatorname{var}(K) = [n(n-1)] \left[\frac{1}{2}(1 - \tau^2) + 4(n-2) \left\{ \delta - \left(\frac{\tau + 1}{2} \right)^2 \right\} \right], \qquad (8.23)$$

where τ is given in equation (8.2) and

$$\delta = P\{(Y_2 - Y_1)(X_2 - X_1) > 0 \quad \text{and} \quad (Y_3 - Y_1)(X_3 - X_1) > 0\}. \qquad (8.24)$$

Using a mutually exclusive breakdown of the event in δ (8.24) similar to that in equation (8.2), we see that

$$\begin{aligned}
\delta &= [P\{Y_2 > Y_1, X_2 > X_1, Y_3 > Y_1, X_3 > X_1\} \\
&\quad + P\{Y_2 > Y_1, X_2 > X_1, Y_3 < Y_1, X_3 < X_1\} \\
&\quad + P\{Y_2 < Y_1, X_2 < X_1, Y_3 > Y_1, X_3 > X_1\} \\
&\quad + P\{Y_2 < Y_1, X_2 < X_1, Y_3 < Y_1, X_3 < X_1\}] \\
&= [P\{Y_1 < \min(Y_2, Y_3), X_1 < \min(X_2, X_3)\} \\
&\quad + P\{Y_3 < Y_1 < Y_2, X_3 < X_1 < X_2\} \\
&\quad + P\{Y_2 < Y_1 < Y_3, X_2 < X_1 < X_3\} \\
&\quad + P\{Y_1 > \max(Y_2, Y_3), X_1 > \max(X_2, X_3)\}]. \qquad (8.25)
\end{aligned}$$

When X and Y are independent variables (under H_0), equation (8.25) simplifies to

$$\begin{aligned}
\delta_0 &= [P_0\{Y_1 < \min(Y_2, Y_3)\} P_0\{X_1 < \min(X_2, X_3)\} \\
&\quad + P_0(Y_3 < Y_1 < Y_2) P_0(X_3 < X_1 < X_2) \\
&\quad + P_0(Y_2 < Y_1 < Y_3) P_0(X_2 < X_1 < X_3) \\
&\quad + P_0\{Y_1 > \max(Y_2, Y_3)\} P_0\{X_1 > \max(X_2, X_3)\}]. \qquad (8.26)
\end{aligned}$$

However, X_1, X_2, X_3 are mutually independent, identically distributed random variables, as are Y_1, Y_2, Y_3. Thus, we know that

$$P_0\{Y_1 < \min(Y_2, Y_3)\} = P_0\{X_1 < \min(X_2, X_3)\} = \tfrac{1}{3}, \tag{8.27}$$

$$P_0\{Y_1 > \max(Y_2, Y_3)\} = P_0\{X_1 > \max(X_2, X_3)\} = \tfrac{1}{3}, \tag{8.28}$$

and

$$P_0(X_3 < X_1 < X_2) = P_0(X_2 < X_1 < X_3) = P_0(Y_3 < Y_1 < Y_2)$$
$$= P_0(Y_2 < Y_1 < Y_3) = \tfrac{1}{6}. \tag{8.29}$$

Combining (8.26), (8.27), (8.28), and (8.29), we obtain

$$\delta_0 = \left[\tfrac{1}{3}\left(\tfrac{1}{3}\right) + \tfrac{1}{6}\left(\tfrac{1}{6}\right) + \tfrac{1}{6}\left(\tfrac{1}{6}\right) + \tfrac{1}{3}\left(\tfrac{1}{3}\right)\right] = \tfrac{10}{36}. \tag{8.30}$$

From equations (8.23) and (8.30) and the fact that $\tau = 0$ when X and Y are independent, it follows that the null variance of K is given by

$$\mathrm{var}_0(K) = [n(n-1)]\left[\tfrac{1}{2}(1-0)^2 + 4(n-2)\left\{\tfrac{10}{36} - \left(\tfrac{0+1}{2}\right)^2\right\}\right]$$

$$= [n(n-1)]\left[\tfrac{1}{2} + \tfrac{1}{9}(n-2)\right] = \frac{n(n-1)(2n+5)}{18},$$

as previously noted in equation (8.12).

The asymptotic normality under both H_0 and general alternatives of the standardized form

$$K^* = \frac{K - E_0(K)}{\{\mathrm{var}_0(K)\}^{1/2}} = \frac{K}{\{\tfrac{n(n-1)(2n+5)}{18}\}^{1/2}}$$

follows from Hoeffding's (1948a) U-statistic theorem applied to the bivariate setting. (For additional details, see Example 3.6.12 in Randles and Wolfe 1979.)

11. *Ties within the X-Values and/or Y-Values.* We have recommended dealing with tied X observations and/or tied Y observations by counting a zero in the Q^* (8.17) counts leading to the computation of K (8.6). This approach is satisfactory as long as the number of (X, Y) pairs containing a tied X and/or tied Y observation does not represent a sizable percentage of the total number (n) of sample pairs.

We should, however, point out that methods other than this zero assignment to the Q^* (8.17) counts have been considered for dealing with tied X and/or tied Y observations. One could use individual randomization (e.g., flipping a fair coin) to decide whether each of the tied pairs (X or Y) is to be counted as a $+1$ (i.e., as a concordant pair—see Comment 3) or as a -1 (i.e., as a discordant pair—again, see Comment 3) in the computation of K (8.6). (Although this approach maintains many of the nice properties of K that hold when there are no tied X and/or tied Y observations, it introduces extraneous randomness that could quite easily have a direct effect on the outcome of any subsequent inferences based on such a modified value of K.) A second alternative approach in the case of the one-sided test procedures in (8.8), (8.9), (8.14), and (8.15) is to be conservative about rejecting the null hypothesis H_0; that is, we could count all the tied X and/or tied Y observations as if they were in favor of not rejecting H_0. Thus, for example, in applying either procedure (8.8) or (8.14) to test H_0 against the alternative $\tau > 0$, we would treat *all* the pairs of pairs involving tied X and/or tied Y observations as if they were discordant pairs (in favor of not rejecting H_0) leading to Q (8.5) counts of -1 in the calculation of K. (In the case of procedures (8.9) and (8.15), all the pairs of pairs involving tied X and/or tied Y observations would be considered as concordant pairs—again in favor of not rejecting H_0—leading to Q (8.5) counts of $+1$ in the calculation of K.) Any rejection of H_0 with this conservative approach to dealing with tied X and/or tied Y observations could then be viewed as providing strong evidence in favor of the appropriate alternative. For more detailed discussion of methods for handling tied X and/or tied Y observations, see Sillitto (1947), Smid (1956), Burr (1960), and Kendall (1962).

12. *Some Power Results for the Kendall Test for Independence.* We consider the upper-tail α-level test of H_0 (8.1) versus $H_1 : \tau > 0$ given by procedure (8.8). The power, or probability of correctly rejecting H_0, for τ (8.2) values "near" the null hypothesis value of 0 can be approximated by

$$\text{Power} \doteq \Phi(A_F), \tag{8.31}$$

where $\Phi(A_F)$ is the area under a standard normal density to the left of the point

$$A_F = \{[9n(n-1)/(4n+10)]^{1/2}\tau - z_\alpha\}. \tag{8.32}$$

When $F_{X,Y}$ is the bivariate normal distribution with correlation coefficient ρ, it follows that $\tau = \frac{2}{\pi}\sin^{-1}(\rho)$ (see, for example, Section 12.1 in Gibbons, 1971). Thus, when $F_{X,Y}$ is bivariate normal the approximate power depends only on the value of ρ. For purposes of illustration, suppose that the common underlying distribution is bivariate normal with $\rho = .4$. For the case of $n = 9$ and $\alpha = .060$, the test rejects H_0 if and only if $K \geq 16$. Substituting $\tau = (2/\pi)\sin^{-1}(.4) = (2/\pi)(.4115) = .2620$ in equation (8.32), we obtain

$$A_{\text{BIV NOR}} = \{[9(9)(8)/2(2(9)+5)]^{1/2}.2620 - 1.555\}$$
$$= \{[14.09]^{1/2}(.2620) - 1.555\} = \{.9835 - 1.555\} = -.57.$$

Thus, from Table A.1, the approximate power of this test for a bivariate normal distribution with $\rho = .4$ (and *any* means and variances) is

$$\text{Power} \doteq \Phi(-.57) = 1 - \Phi(.57) = 1 - (1 - .28) = .28.$$

This compares with the simulation estimated exact power of .35 for $n = 9, \rho = .4$, and $\alpha = .05$, as given in Table 3 of Bhattacharyya, Johnson, and Neave (1970). Additional simulation estimated exact power values for the one-sided Kendall test and sample sizes $n = 5, 7, 9$ and significance levels $\alpha = .01$ and $.05$ can be found in Bhattacharyya, Johnson, and Neave (1970) for bivariate normal and bivariate exponential distributions.

13. *Sample-Size Determination.* Noether (1987) shows how to determine an approximate sample size n so that the α-level one-sided test given by procedure (8.8) will have approximate power $1 - \beta$ against an alternative value of τ (8.2) greater than zero. This approximate value of n is

$$n \doteq \frac{4(z_\alpha + z_\beta)^2}{9\tau^2}. \tag{8.33}$$

As an illustration of the use of equation (8.33), suppose we are testing H_0 and we desire to have an upper-tail level $\alpha = .010$ test with power $1 - \beta$ at least .90 against an alternative bivariate distribution for which $\tau = .4$. Since $z_\alpha = z_{.01} = 2.326$ and $z_\beta = z_{.10} = 1.282$, we find that the approximate required sample size for the alternative $\tau = .4$ is

$$n \doteq \frac{4(2.326 + 1.282)^2}{9(.4)^2} = 36.2.$$

To be conservative, we would take $n = 37$.

14. *Trend Test.* If we take $X_i = i, i = 1, \ldots, n$ and consider

$$K = \sum_{i=1}^{n-1} \sum_{j=i+1}^{n} Q((i, Y_i), (j, Y_j))$$

$$= \sum_{i=1}^{n-1} \sum_{j=i+1}^{n} c(Y_j - Y_i),$$

where

$$c(a) = \begin{cases} 1 & \text{if } a > 0, \\ 0 & \text{if } a = 0, \\ -1 & \text{if } a < 0, \end{cases}$$

then K can be used as a test for a time trend in the univariate random sample $Y_1, \ldots, Y_n$. This use of K to test for a time trend was suggested by Mann (1945).

15. *Other Uses for the K Statistic.* Wilcoxon's rank sum test (Section 4.1) and Jonckheere-Terpstra's test (Section 6.2) can be viewed as tests based on K (8.6) (or, equivalently, $\hat{\tau}$ (8.34)). For this interpretation see Jonckheere (1954a) and Kendall (1962, Sections 3.12 and 13.9). Also, Wolfe (1977) has used the K statistic to compare the correlation between variables X_2 and X_1 with that between the variables X_3 and X_1, when both X_2 and X_3 are potential predictors for X_1.

16. *Consistency of the K Test.* Under the assumption that $(X_1, Y_1), \ldots, (X_n, Y_n)$ is a random sample from a continuous bivariate population with joint distribution function $F_{X,Y}(x, y)$, the consistency of the tests based on K depends on the parameter τ (8.2). The test procedures defined by (8.8), (8.9), and (8.10) are consistent against the class of alternatives corresponding to $\tau >, <,$ and $\neq 0$, respectively.

17. *Multivariate Concordance.* Joe (1990) has generalized Kendall's measure of association τ from the bivariate case where τ measures the strength of association between two variables X, Y to the multivariate case where $\mathbf{X} = (X_1, \ldots, X_m)$ is an m-dimensional random variable and one is interested in a measure of the strength of the association between the components $X_1, \ldots, X_m$ of $\mathbf{X}$. Let F denote the joint distribution function of $\mathbf{X}$,

$$F(x_1, \ldots, x_m) = P(X_1 \leq x_1 \text{ and } X_2 \leq x_2 \text{ and } \ldots \text{ and } X_m \leq x_m)$$

and denote the marginal distribution functions as $F_j(x_j) = P(X_j \leq x_j), j = 1, \ldots, m$. The null hypothesis of mutual independence of $X_1, \ldots, X_m$ is

$$H_0 : F(x_1, \ldots, x_m) = \prod_{j=1}^{m} F_j(x_j), \quad \text{for all } (x_1, \ldots, x_m).$$

That is, the joint distribution is equal to the product of the marginals.

Joe has defined a class of measures of the strength of association between $X_1, X_2, \ldots, X_m$. Let $\mathbf{X}_i = (X_{i1}, \ldots, X_{im})$, $i = 1, 2$ be two independent m-dimensional random variables each with joint distribution function F. One member of Joe's class reduces to $\bar{\tau}$, the average of all pairwise τ's. The measure $\bar{\tau}$ was introduced by Hays (1960) and is given by

$$\bar{\tau} = \sum_{u=1}^{m-1} \sum_{v=1}^{m} \frac{\tau_{uv}}{\left\{\frac{m(m-1)}{2}\right\}},$$

where

$$\begin{aligned}\tau_{uv} &= P\{(X_{1u} - X_{1v})(X_{2u} - X_{2v}) > 0\} - P\{(X_{1u} - X_{1v})(X_{2u} - X_{2v}) < 0\} \\ &= 2P\{(X_{1u} - X_{1v})(X_{2u} - X_{2v}) > 0\} - 1.\end{aligned}$$

Joe has also generalized Spearman's measure (see Section 8.5) and a measure due to Blomqvist (1950).

Properties

1. *Consistency.* The tests defined by (8.8), (8.9), and (8.10) are consistent against the alternatives $\tau >, <,$ and $\neq 0$, respectively.
2. *Asymptotic Normality.* See Hoeffding (1948a) or Randles and Wolfe (1979, 108–109).
3. *Efficiency.* See Section 8.7.

PROBLEMS

1. The data in Table 8.3 are a subset of the data obtained by Featherston (1971). Among other things, he was interested in the relationship between the weight of tapeworms *(Taenia hydatigena)* fed to dogs and the weight of the scoleces recovered from the dogs after 20 days. (A scolex is the attachment end of a tapeworm, consisting of the head and neck.) The cysticerci used in the experiment were collected from sheep carcasses and force-fed to 10 dogs via gelatine capsules. The scoleces were

TABLE 8.3. Relation Between Weight of the Cysticerci of *Taenia hydatigena* Fed to Dogs and Weight of Worms Recovered at 20 Days

Dog	Mean Weight (mg)	
	Cysticerci	Worms Recovered
1	28.9	1.0
2	32.8	7.7
3	12.0	7.3
4	9.9	7.9
5	15.0	1.1
6	38.0	3.5
7	12.5	18.9
8	36.5	33.9
9	8.6	28.6
10	26.8	25.0

Source: D. W. Featherston (1971).

recovered from each dog at autopsy, 20 days after the introduction of the tapeworms. Table 8.3 gives the mean weight of the initial cysticerci and the mean weight of the recovered worms for each of the 10 dogs in the study.

Test the hypothesis of independence versus the alternative that the mean weight of introduced cysticerci is positively correlated with the mean weight of worms recovered.

2. Let R_i be the rank of X_i in the joint ranking of $X_1, \ldots, X_n$, and let S_i be the rank of Y_i in the joint ranking of $Y_1, \ldots, Y_n$. Show that knowledge of $R_1, \ldots, R_n$ and $S_1, \ldots, S_n$ is sufficient to calculate K (8.6).

3. The data in Table 8.4 are a subset of the data obtained by Sylvester (1969) in a study concerned with the anatomical and pathological status of the corticospinal and somatosensory tracts and parietal lobes of patients who had had cerebral palsy. Among other things, he was interested in the relationship between brain weights and large fiber ($> 7.5\mu$ in diameter) counts in the medullary pyramid. Table 8.4 gives the mean brain weights (in g) and medullary pyramid large fiber counts for 11 cerebral palsy subjects. Test the hypothesis of independence versus the general alternative that brain weight and large fiber count in the medullary pyramid are correlated in subjects who have had cerebral palsy.

TABLE 8.4. Mean Brain Weights and Medullary Pyramid Large Fiber Counts for Cerebral Palsy Subjects

Subject Number	Brain Weight (g)	Pyramidal Large Fiber Count
1	515	32,500
2	286	26,800
3	469	11,410
4	410	14,850
5	461	23,640
6	436	23,820
7	479	29,840
8	198	21,830
9	389	24,650
10	262	22,500
11	536	26,000

Source: P. E. Sylvester (1969).

4. Let (X_1, Y_1) and (X_2, Y_2) be independent and identically distributed continuous bivariate random variables with joint probability density function

$$f_{X,Y}(x, y) = \begin{cases} e^{-y}, & 0 < x < y < \infty, \\ 0, & \text{elsewhere.} \end{cases}$$

Calculate the value of τ for this bivariate distribution.

5. Let (X_1, Y_1) and (X_2, Y_2) be independent and identically distributed discrete bivariate random variables with joint probability function

$$f_{X,Y}(x, y) = \begin{cases} \frac{x+y}{21}, & x = 1, 2, 3; \ y = 1, 2, \\ 0, & \text{elsewhere.} \end{cases}$$

Calculate the value of τ for this bivariate distribution.

6. Let (X_1, Y_1) and (X_2, Y_2) be independent and identically distributed continuous bivariate random variables with joint probability density function

$$f_{X,Y}(x, y) = \begin{cases} \frac{1}{2}y^2 e^{-x-y}, & 0 < x < \infty, \quad 0 < y < \infty, \\ 0, & \text{elsewhere.} \end{cases}$$

Calculate the value of τ for this bivariate distribution.

7. The data in Table 8.5 are a subset of the data considered by Clark, Vandenberg, and Proctor (1961) in a study concerned with the relationship of scores on various psychological tests and certain physical characteristics of twins. Table 8.5 gives the test scores (totals of a number of different psychological tests) of 13 dizygous (i.e., nonidentical) male twins. Test the hypothesis of independence versus the alternative that the twins' test scores are positively correlated.

TABLE 8.5. Psychological Test Scores of Dizygous Male Twins

Pair i	Twin X_i	Twin Y_i
1	277	256
2	169	118
3	157	137
4	139	144
5	108	146
6	213	221
7	232	184
8	229	188
9	114	97
10	232	231
11	161	114
12	149	187
13	128	230

Source: P. J. Clark, S. G. Vandenberg, and C. H. Proctor (1961).

8. Previously, it was shown that if X and Y are independent random variables, then τ (8.2) has a value of 0. Show that the converse is not true by constructing a joint probability distribution for the pair of random variables X and Y such that $\tau = 0$ but X and Y are not independent.

9. If we have 25 bivariate observations and $F_{X,Y}$ is bivariate normal with correlation coefficient .3, what is the approximate power of the level $\alpha = .045$ test of H_0 (8.1) versus the alternative $\tau > 0$?

10. For an arbitrary number, n, of bivariate observations, what are the smallest and largest values of K? Give examples of data sets where these extremes are achieved.

11. Give an example of a data set of $n \geq 10$ bivariate observations for which K has value 0. (Consider Comment 9.)

12. Suppose $n = 20$. Compare the critical region for the exact level $\alpha = .064$ test of H_0 (8.1) versus $H_2 : \tau < 0$ based on K with the critical region for the corresponding nominal level $\alpha = .064$ based on the large-sample approximation. What is the exact significance level of this .064 nominal level test based on the large-sample approximation?

13. Consider a level $\alpha = .10$ test of H_0 (8.1) versus the alternative $\tau > 0$ based on K. How many bivariate observations (n) will we need to collect in order to have approximate power at least .95 against an alternative for which $\tau = .6$?

14. A question of significance to state legislators working with tight budgets is the spending for secondary education. The data in Table 8.6 are from the Department of Education, National Center for Education Statistics and were considered by Merline (1991) in assessing the relationship between the amount of money spent on secondary education and various performance criteria for high school seniors. Table 8.6 gives the spending ($) per high-school senior and the percentage of those seniors who graduated for each of the 50 states in the 1987–88 school year.

TABLE 8.6. Spending per High-School Senior and the Percentage of Those Seniors Who Graduated during the 1987–1988 School Year

State	$ Per Pupil	% Graduated	State	$ Per Pupil	% Graduated
Alaska	7,971	65.5	Ohio	3,998	79.6
New York	7,151	62.3	Nebraska	3,943	85.4
New Jersey	6,564	77.4	Hawaii	3,919	69.1
Connecticut	6,230	84.9	West Virginia	3,858	77.3
Massachusetts	5,471	74.4	California	3,840	65.9
Rhode Island	5,329	69.8	Indiana	3,794	76.3
Vermont	5,207	78.7	Missouri	3,786	74.0
Maryland	5,201	74.1	Arizona	3,744	61.1
Wyoming	5,051	88.3	New Mexico	3,691	71.9
Delaware	5,017	71.7	Nevada	3,623	75.8
Pennsylvania	4,989	78.4	Texas	3,608	65.3
Oregon	4,789	73.0	North Dakota	3,519	88.3
Wisconsin	4,747	84.9	Georgia	3,434	61.0
Michigan	4,692	73.6	South Carolina	3,408	64.6
Colorado	4,462	74.7	North Carolina	3,368	66.7
New Hampshire	4,457	74.1	South Dakota	3,249	79.6
Minnesota	4,386	90.9	Louisiana	3,138	61.4
Illinois	4,369	75.6	Oklahoma	3,093	71.7
Maine	4,246	74.4	Tennessee	3,068	69.3
Montana	4,246	87.3	Kentucky	3,011	69.0
Washington	4,164	77.1	Arkansas	2,989	77.2
Virginia	4,149	71.6	Alabama	2,718	74.9
Iowa	4,124	85.8	Idaho	2,667	75.4
Florida	4,092	58.0	Mississippi	2,548	66.9
Kansas	4,076	80.2	Utah	2,454	79.4

Source: J. W. Merline (1991).

Use the large-sample approximation to test the hypothesis of independence versus the alternative that spending per high-school senior and the percentage of seniors graduating are positively correlated (Discuss any other social or economic factors that might impact on these data and, thereby, on the conclusion from this statistical analysis.)

15. For the case of $n = 5$ untied bivariate (X, Y) observations, obtain the form of the exact null (H_0) distribution of K. (See Comment 6.)

16. Johnson (1973) studied several different managerial aspects of university associated schools of nursing. Among the data she collected were the "extent of agreement (between the dean and the faculty) on the responsibilities for decision making" and "faculty satisfaction." The ranks on the two variables for the 12 institutions that were involved in Johnson's study are presented in Table 8.7.
  Test the hypothesis of independence versus the alternative that faculty/dean decision-making agreement and faculty satisfaction are negatively correlated in university schools of nursing. (*Note:* Low ranks are associated with poor faculty satisfaction and little faculty/dean decision-making agreement, respectively.)

17. Consider the test of H_0 (8.1) versus $H_1 : \tau > 0$ based on K for the following $n = 10$ (X, Y) observations: (1.5, 6), (1.9, 4), (2.3, 6), (2.7, 12), (1.5, 13), (1.8, 16), (3.6, 16), (4.2, 9), (4.7, 0), and (4.0, 3). Compute the P-values for the competing K-procedures based on either (a) using Q^* (8.17) counts of zero, as recommended in the Ties portion of this section, or (b) dealing with the tied X and tied Y observations in a conservative manner, as presented in Comment 11. Discuss the results.

18. In Comment 4, we noted that we have $K = 2K' - \{n(n-1)/2\}$, where $K' =$ (number of concordant pairs), when there are neither tied X nor tied Y observations. Obtain the corresponding expression for the relationship between K and K' when there are no tied X pairs and a total of $t(\neq 0)$ tied Y

TABLE 8.7. Rankings for Faculty/Dean Decision-Making Agreement and Faculty Satisfaction for Participating Schools of Nursing

School	Rank for Faculty/Dean Decision-Making Agreement	Rank for Faculty Satisfaction
1	8	8
2	9	2
3	6	10
4	12	5
5	1	12
6	11	4
7	10	6
8	2	9
9	4	7
10	5	3
11	7	11
12	3	1

Source: B. M. Johnson (1973).

pairs (among the $\binom{n}{2}$ total Y pairs) and we use the Q^* (8.17) counts of zero to deal with the tied Y pairs. How does this expression change if there are no tied Y pairs and t tied X pairs? Discuss the necessary changes in the expression relating K and K' when there are $s(\neq 0)$ tied X pairs and $t(\neq 0)$ tied Y pairs.

19. Gerstein (1965) studied the long-term pollution of Lake Michigan and its effect on the water supply for the city of Chicago. One of the measurements considered by Gerstein was the annual number of "odor periods" over the period of years 1950–1964. Table 8.8 contains this information for Lake Michigan for each of these 15 years.

TABLE 8.8. Annual Number of Odor Periods for Lake Michigan for the Period 1950–1964

Year	Number of Odor Periods
1950	10
1951	20
1952	17
1953	16
1954	12
1955	15
1956	13
1957	18
1958	17
1959	19
1960	21
1961	23
1962	23
1963	28
1964	28

Source: H. H. Gerstein (1965).

Test the hypothesis that the degree of pollution (as measured by the number of odor periods) had not changed with time against the alternative that there was a general increasing trend in the pollution of Lake Michigan over the period 1950–1964. (See Comment 14.)

8.2. AN ESTIMATOR ASSOCIATED WITH THE KENDALL STATISTIC (KENDALL)

Procedure

The estimator of the Kendall population correlation coefficient τ (8.2), based on the statistic K (8.6), is

$$\hat{\tau} = \frac{2K}{n(n-1)}. \tag{8.34}$$

The statistic $\hat{\tau}$ is known as Kendall's sample rank correlation coefficient and appropriately assumes values between -1 and 1, inclusive.

Example 8.2: *(Continuation of Example 8.1).* For the canned tuna data of Table 8.1, we see from (8.34) that the sample estimate of τ is

$$\hat{\tau} = \frac{2(16)}{9(8)} = \frac{4}{9}. \tag{8.35}$$

Comments

18. *Ties.* In the presence of ties, use $\hat{\tau} = 2K/n(n-1)$, where

$$K = \sum_{i=1}^{n-1} \sum_{j=i+1}^{n} Q^*((X_i, Y_i), (X_j, Y_j)) \tag{8.36}$$

and $Q^*((X_i, Y_i), (X_j, Y_j))$ is defined by (8.17).

19. *Probability Estimation.* For many problems, distribution-free test statistics are used directly to estimate basic probability parameters other than the usual distributional parameters associated with the corresponding normal theory problems. In particular, note that $\hat{\tau} = 2K/[n(n-1)]$ estimates the probability parameter τ (8.2) rather than the usual correlation coefficient for the underlying bivariate population. Estimators of such readily interpretable parameters are very helpful in data analysis. (See Crouse (1966), Wolfe and Hogg (1971), and Comment 4.18.)

Properties

1. *Standard Deviation of $\hat{\tau}$.* For the asymptotic standard deviation of $\hat{\tau}$ (8.34), see Noether (1967a, 78), Fligner and Rust (1983), Samara and Randles (1988), and Comment 25.
2. *Asymptotic Normality.* See Hoeffding (1948a) and Randles and Wolfe (1979, 108–109).

PROBLEMS

20. Estimate τ for the tapeworm data of Table 8.3.
21. What is the maximum possible value of $\hat{\tau}$ when there are no tied X and/or tied Y observations? What is the minimum possible value of $\hat{\tau}$ when there are no tied X and/or tied Y observations? Construct three examples with $n \geq 10$ with no tied X and/or tied Y observations: one in which $\hat{\tau}$ achieves its maximum value, one in which it achieves its minimum value, and one for which $\hat{\tau} = 0$.
22. Estimate τ for the cerebral palsy data of Table 8.4.

23. Estimate τ for the twin data of Table 8.5.
24. Estimate τ for the secondary education data of Table 8.6.
25. Use Comment 17 to redo Problem 21 for the case when there are $t\,(\neq 0)$ tied X pairs (among the total of $\binom{n}{2}$ X pairs) and no tied Y observations. How is the answer affected if there are no tied X observations and $t\,(\neq 0)$ tied Y pairs? if there are $s\,(\neq 0)$ tied X pairs and $t\,(\neq 0)$ tied Y pairs?
26. Estimate τ for the nursing faculty data of Table 8.7.

8.3. AN ASYMPTOTICALLY DISTRIBUTION-FREE CONFIDENCE INTERVAL BASED ON THE KENDALL STATISTIC (SAMARA-RANDLES, FLIGNER-RUST, NOETHER)

Procedure

For an asymptotically distribution-free symmetric two-sided confidence interval for τ, with approximate confidence coefficient $1 - \alpha$, we first compute

$$C_i = \sum_{\substack{t=1 \\ t \neq i}}^{n} Q((X_i, Y_i), (X_t, Y_t)), \quad \text{for } i = 1, \ldots, n, \tag{8.37}$$

where $Q((a, b), (c, d))$ is given by (8.5). Let $\bar{C} = (1/n) \sum_{i=1}^{n} C_i = 2K/n$ and define

$$\hat{\sigma}^2 = \frac{2}{n(n-1)} \left[\frac{2(n-2)}{n(n-1)^2} \sum_{i=1}^{n} (C_i - \bar{C})^2 + 1 - \hat{\tau}^2 \right], \tag{8.38}$$

where $\hat{\tau}$ is given by (8.34). The approximate $100(1 - \alpha)\%$ confidence interval (τ_L, τ_U) for τ that is associated with the point estimator $\hat{\tau}$ (8.34) is then given by

$$\tau_L = \hat{\tau} - z_{\alpha/2}\hat{\sigma}, \quad \tau_U = \hat{\tau} + z_{\alpha/2}\hat{\sigma}. \tag{8.39}$$

With τ_L and τ_U given by display (8.39), we have

$$P_\tau\{\tau_L < \tau < \tau_U\} \approx 1 - \alpha \text{ for all } \tau. \tag{8.40}$$

(For approximate upper or lower confidence bounds for τ associated with $\hat{\tau}$, see Comment 23.)

Example 8.3: *(Continuation of Examples 8.1 and 8.2).* Consider the canned tuna data of Table 8.1. We illustrate how to obtain an approximate 90% symmetric two-sided confidence interval for τ. From equation (8.37) we see that

$$C_5 = \sum_{j \neq 5} Q((X_5, Y_5), (X_j, Y_j))$$

$$= [Q((X_5, Y_5), (X_1, Y_1)) + Q((X_5, Y_5), (X_2, Y_2))$$

$$+ Q((X_5, Y_5), (X_3, Y_3)) + Q((X_5, Y_5), (X_4, Y_4))$$

$$+ Q((X_5, Y_5), (X_6, Y_6)) + Q((X_5, Y_5), (X_7, Y_7))$$

$$+ Q((X_5, Y_5), (X_8, Y_8)) + Q((X_5, Y_5), (X_9, Y_9))].$$

Using the fact that $Q((X_i, Y_i), (X_j, Y_j)) = Q((X_j, Y_j), (X_i, Y_i))$ for every $i \neq j$ and the Q counts for the canned tuna data in Table 8.2, it follows that

$$C_5 = 1 - 1 + 1 + 1 - 1 + 1 - 1 + 1 = 2.$$

Note that C_5 is simply equal to the sum of the Q values in the $j = 5$ row and the $i = 5$ column in Table 8.2. In the same way, we find

$$C_1 = 7 - 1 = 6, \quad C_2 = 6 - 2 = 4, \quad C_3 = 8 - 0 = 8, \quad C_4 = 6 - 2 = 4$$
$$C_6 = 3 - 5 = -2, \quad C_7 = 6 - 2 = 4, \quad C_8 = 6 - 2 = 4, \quad C_9 = 5 - 3 = 2.$$

Thus, we have

$$\bar{C} = \frac{1}{9}\sum_{i=1}^{9} C_i = \frac{1}{9}[6 + 4 + 8 + 4 + 2 - 2 + 4 + 4 + 2] = \frac{32}{9}.$$

Thus,

$$\sum_{i=1}^{9}(C_i - \bar{C})^2 = \sum_{i=1}^{9}\left(C_i - \frac{32}{9}\right)^2$$

$$= \left[\left(\frac{22}{9}\right)^2 + \left(\frac{4}{9}\right)^2 + \left(\frac{40}{9}\right)^2 + \left(\frac{4}{9}\right)^2 + \left(-\frac{14}{9}\right)^2\right.$$

$$\left. + \left(-\frac{50}{9}\right)^2 + \left(\frac{4}{9}\right)^2 + \left(\frac{4}{9}\right)^2 + \left(-\frac{14}{9}\right)^2\right]$$

$$= \frac{484 + 4(16) + 1600 + 2(196) + 2500}{81} = \frac{560}{9}. \tag{8.41}$$

Using the values for $\hat{\tau}$ and $\sum_{i=1}^{9}(C_i - \bar{C})^2$ given in (8.35) and (8.41), respectively, we see from equation (8.38) that

$$\hat{\sigma}^2 = \frac{2}{9(8)}\left[\frac{2(7)}{9(8)^2}\left(\frac{560}{9}\right) + 1 - \left(\frac{4}{9}\right)^2\right]$$

$$= \frac{1}{36}[1.512 + 1 - .198] = .064.$$

With $1 - \alpha = .90$ (so that $\alpha = .10$), we see from Table A.1 that $z_{.05} = 1.65$. Hence, from (39), we obtain

$$\tau_L = \tfrac{4}{9} - 1.65(.064)^{1/2} = .444 - .417 = .027$$

and

$$\tau_U = \tfrac{4}{9} + 1.65(.064)^{1/2} = .444 + .417 = .861.$$

Our approximate 90% symmetric confidence interval for τ is thus $(\tau_L, \tau_U) = (.027, .861)$.

Comments

20. *Interpretation as a Confidence Interval for a Probability.* The confidence interval given by (8.39) is an approximate $1 - \alpha$ confidence interval for a parameter that is a linear function of the probability $P\{(X_1 - X_2)(Y_1 - Y_2) > 0\}$. This is common practice in the field of nonparametric statistics, where probabilities are often natural and easily interpretable parameters. Recall the relation of the Wilcoxon two-sample test of Section 4.1 to the parameter $P(X < Y)$. See Comments 4.7, 4.10, 4.14, and 4.18.

21. *Ties.* In the presence of ties, use $Q^*((X_i, Y_i), (X_j, Y_j))$ defined by (8.17) instead of $Q((X_i, Y_i), (X_j, Y_j))$ given by (8.5) in the computation of $C_1, \ldots, C_n$ and $\hat{\tau}$.

22. *Alternative Method of Calculation.* The following equivalent formula for the term $\sum_{i=1}^{n}(C_i - \bar{C})^2$ in the definition of $\hat{\sigma}^2$ (8.38), namely,

$$\sum_{i=1}^{n}(C_i - \bar{C})^2 = \sum_{i=1}^{n} C_i^2 - \frac{4K^2}{n}, \tag{8.42}$$

is often computationally more convenient.

23. *Concordant/Discordant Pairs Representation for the C_i's.* Let C_i' and C_i'' be the numbers of pairs (X_j, Y_j), $j \neq i$, that are concordant and discordant, respectively, with (X_i, Y_i), for $i = 1, \ldots, n$. Then the C_i (8.37) counts can be expressed as $C_i = C_i' - C_i''$, for $i = 1, \ldots, n$.

24. *Confidence Bounds.* In many settings we are interested only in making one-sided confidence statements about the parameter τ; that is, we wish to assert with specified confidence that τ is no larger (or, in other settings, no smaller) than some upper (lower) confidence bound based on the sample data. To obtain such one-sided confidence bounds for τ, we proceed as follows. For a specified confidence coefficient $1 - \alpha$, find z_α (not $z_{\alpha/2}$, as for the confidence interval) from Table A.1. An approximate $100(1 - \alpha)\%$ *upper* confidence bound τ_U^* for τ is then given by

$$[-1, \tau_U^*) = [-1, \hat{\tau} + z_\alpha \hat{\sigma}), \tag{8.43}$$

where $\hat{\tau}$ and $\hat{\sigma}^2$ are given by equations (8.34) and (8.38), respectively. With τ_U^* given by display (8.43), we have

$$P_\tau\{-1 \leq \tau < \tau_U^*\} \approx 1 - \alpha \text{ for all } \tau. \tag{8.44}$$

The corresponding approximate $100(1 - \alpha)\%$ *lower* confidence bound τ_L^* for τ is given by

$$(\tau_L^*, 1] = (\hat{\tau} - z_\alpha \hat{\sigma}, 1], \tag{8.45}$$

with

$$P_\tau\{\tau_L^* < \tau \leq 1\} \approx 1 - \alpha \text{ for all } \tau. \tag{8.46}$$

25. *Alternative Approximate Confidence Limits.* Samara and Randles (1988) showed that the statistic $\hat{\tau}/\hat{\sigma}$ is itself distribution-free under the null hypothesis (H_0) of independence and they tabled the upper αth percentile of its null distribution, k_α^*, for $\alpha = .005, .01, .025, .05,$ and .10 and sample sizes $n = 6(1)20$. Slightly improved confidence intervals and confidence bounds can be obtained by replacing the normal percentiles $z_{\alpha/2}$ and z_α by $k_{\alpha/2}^*$ and k_α^*, respectively, in equations (8.39), (8.43), and (8.45).

26. *Estimating the Asymptotic Standard Deviation of $\hat{\tau}$.* The statistic $\hat{\sigma}$ (8.38) is chosen to be a consistent estimator for the asymptotic standard deviation of the point estimator $\hat{\tau}$ (8.34). It is not necessary to use all the sample observations in calculating $\hat{\sigma}$. In fact, any fixed percentage subset of the n sample observations can be employed to find the C_i values used in (8.38). For example, 25% of a random sample of n paired observations (namely, $n/4$ observations) could be used to obtain $\hat{\sigma}$.

27. *Asymptotic Coverage Probability.* Asymptotically, the true coverage probability of the interval defined by (8.39) and the bounds in (8.43) and (8.45) will agree with the nominal confidence coefficient $1 - \alpha$. Subject to Assumption A this asymptotic (n infinitely large) result does not depend on the distribution of the underlying bivariate population. Thus the interval given by (8.39) and the bounds in (8.43) and (8.45) have been constructed to have the asymptotically distribution-free property.

28. *Historical Development.* The initial effort at constructing asymptotically distribution-free confidence intervals and bounds for τ was due to Noether (1967a). The approximate $100(1 - \alpha)\%$ confidence interval proposed by Noether is $\hat{\tau} \pm z_{\alpha/2}\hat{\sigma}_N$, where $\hat{\sigma}_N^2$ is a consistent estimator (based on U-statistics theory) of the variance of $\hat{\tau}$. However, it was later pointed out that $\hat{\sigma}_N^2$ can assume negative values, even though it is estimating the nonnegative quantity var($\hat{\tau}$). Although this distressing possibility is more likely to occur in small samples, it can be negative for sample sizes as large as $n = 30$. To avoid this problem, Fligner and Rust (1983) proposed the use of $\hat{\tau} \pm z_{\alpha/2}\hat{\sigma}_{FR}$ as an asymptotically distribution-free $100(1 - \alpha)\%$ confidence interval for τ, where $\hat{\sigma}_{FR}^2$ is a jackknife estimator (different from $\hat{\sigma}_N^2$) of var($\hat{\tau}$) that is consistent and cannot assume negative values. A few years later, Samara and Randles (1988) noted that although the Fligner-Rust jackknife estimator $\hat{\sigma}_{FR}^2$ can never be negative, it can be zero for a variety of rank configurations, including some nonextreme cases. They suggested a final modification leading to the asymptotically distribution-free $100(1 - \alpha)\%$ confidence interval in display (8.39), where $\hat{\sigma}^2$ (8.38) = $\hat{\sigma}_{SR}^2$ is a third consistent estimator of var($\hat{\tau}$). The estimator $\hat{\sigma}_{SR}^2 = \hat{\sigma}^2$ (8.38) is also based on U-statistics methodology (as is $\hat{\sigma}_N^2$), but it can never be negative and is zero only in the two extreme cases where $\hat{\tau} = \pm 1$. For the approximate confidence interval $\hat{\tau} \pm z_{\alpha/2}\hat{\sigma}$ to be simply the singleton point $\hat{\tau}(= +1$ or $-1)$ in such extreme cases is not ideal, but it is also not unreasonable.

29. *Competitor Tests for Independence.* In Section 8.1 we discussed tests of independence (8.1) based on Kendall's statistic K (8.6). Noether (1967a), Fligner and Rust (1983) and Samara and Randles (1988) also proposed distribution-free tests of H_0 (8.1) based on the statistics $\hat{\tau}/\hat{\sigma}_N, \hat{\tau}/\hat{\sigma}_{FR}$, and $\hat{\tau}/\hat{\sigma}_{SR}$, respectively, where $\hat{\tau}$ is given by (8.34) and $\hat{\sigma}_N^2, \hat{\sigma}_{FR}^2$, and $\hat{\sigma}_{SR}^2$ are the various consistent estimators of var($\hat{\tau}$) discussed in Comment 28. Although not generally as powerful as the procedures based on K for testing H_0 (8.1), the tests based on $\hat{\tau}/\hat{\sigma}_N, \hat{\tau}/\hat{\sigma}_{FR}$, and $\hat{\tau}/\hat{\sigma}_{SR}$ all have the advantage (not possessed by the tests based on K) that they are also asymptotically distribution-free procedures for testing the more general null hypothesis $H_0^* : \tau = 0$.

30. *Partial Correlation Coefficients.* Let $(X_1, Y_1, Z_1), \ldots, (X_n, Y_n, Z_n)$ be a random sample from a continuous trivariate distribution. It is often of interest to assess the association between the X and Y variables, controlled for the third variable Z. Gripenberg (1992) proposed measuring this "partial correlation" by the parameter

$$\tau_{XY/Z} = 2P\{(Y_2 - Y_1)(X_2 - X_1) > 0 | Z_1 = Z_2\} - 1$$
$$= E[Q((X_1, Y_1), (X_2, Y_2)) | Z_1 = Z_2], \qquad (8.47)$$

where $Q((X_1, Y_1), (X_2, Y_2))$ is defined by (8.5). To estimate $\tau_{XY/Z}$, Gripenberg arranged the (X_i, Y_i, Z_i) triples in an increasing order with respect to the values of the Z variable. Letting

$Z_{N(1)} \leq \cdots \leq Z_{N(n)}$ denote the order statistics for $Z_1, \ldots, Z_n$, the ordered triples correspond to $(X_{N(1)}, Y_{N(1)}, Z_{N(1)}), \ldots, (X_{N(n)}, Y_{N(n)}, Z_{N(n)})$ (with respect to increasing Z values). Gripenberg's estimator for $\tau_{XY/Z}$ is then given by

$$T_{XY/Z} = \frac{1}{n-1} \sum_{i=1}^{n-1} Q((X_{N(i)}, Y_{N(i)}), (X_{N(i+1)}, Y_{N(i+1)})), \tag{8.48}$$

where, once again, $Q((X_{N(i)}, Y_{N(i)}), (X_{N(i+1)}, Y_{N(i+1)}))$ is given by (8.5). The approximate $100(1-\alpha)\%$ confidence interval for $\tau_{XY/Z}$ (8.47) proposed by Gripenberg is then

$$\frac{T_{XY/Z}}{\left[1 + \frac{bz_{\alpha/2}^2}{n}\right]} \pm \frac{\left\{\frac{bz_{\alpha/2}^2}{n}\left(1 - T_{XY/Z}^2 + \frac{bz_{\alpha/2}^2}{n}\right)\right\}^{1/2}}{\left[1 + \frac{bz_{\alpha/2}^2}{n}\right]}, \tag{8.49}$$

where b is an arbitrary consistent estimator of $\beta = \frac{\sigma^{*2}}{1-\tau_{XY/Z}^2}$, with σ^{*2} representing the asymptotic variance of $n^{1/2}T_{XY/Z}$. Two competing estimators b are considered by Gripenberg.

Properties

1. *Asymptotic Distribution-Freeness.* For populations satisfying Assumption A, equation (8.40) holds. Hence, we can control the coverage probability to be approximately $1-\alpha$ for large sample size n without having more specific knowledge about the form of the underlying bivariate (X, Y) distribution. Thus (τ_L, τ_U) is an asymptotically distribution-free confidence interval for τ over the class of all continuous bivariate distributions.

PROBLEMS

27. For the tapeworm data of Table 8.3, find a confidence interval for τ with approximate confidence coefficient .98.

28. For the cerebral palsy data of Table 8.4, find a confidence interval for τ with approximate confidence coefficient .90.

29. Use only six (X, Y) pairs (those corresponding to the first six lot numbers) in Table 8.1 and compute a new estimator $\hat{\sigma}^2$ (8.38) for the asymptotic variance of $\hat{\tau}$ (see Comment 26). Compare it with the estimator based on all nine observations obtained in Example 8.3.

30. For the twins data in Table 8.5, find a lower confidence bound for τ with approximate confidence coefficient .95. (See Comment 24.)

31. For the educational expense data in Table 8.6, find a lower confidence bound for τ with approximate confidence coefficient .96. (See Comment 24.)

32. For the nursing data of Table 8.7, find an upper confidence bound for τ with approximate confidence coefficient .94.

33. Suppose that $(X_1, Y_1, Z_1) = (7.0, 2.5, 1.9)$, $(X_2, Y_2, Z_2) = (6.3, 9.6, 4.1)$, $(X_3, Y_3, Z_3) = (6.9, 3.7, 12.4)$, $(X_4, Y_4, Z_4) = (3.6, 12.1, 6.5)$, $(X_5, Y_5, Z_5) = (9.0, 6.4, 11.2)$, $(X_6, Y_6, Z_6) = (3.0, 6.2, 7.7)$, and $(X_7, Y_7, Z_7) = (4.2, 0.4, 8.2)$ represent a random sample of size $n = 7$ from a trivariate probability distribution. Estimate the partial correlation $\tau_{XY/Z}$ (8.47) between X and Y conditional on Z. (See Comment 30.)

8.4. AN ASYMPTOTICALLY DISTRIBUTION-FREE CONFIDENCE INTERVAL BASED ON EFRON'S BOOTSTRAP

The asymptotically distribution-free confidence for the parameter τ described in Section 8.3 is based on obtaining a mathematical expression for σ^2, the variance of $\hat{\tau}$. Such an expression depends on the unknown underlying bivariate distribution and so σ^2 must be estimated from the data. $\hat{\sigma}^2$, given by (8.38), is consistent estimator, and it is used to form the confidence interval of Section 8.3. In many problems, however, it will be difficult or impossible to obtain a tractable mathematical expression for the variance of the statistic of interest. Efron's bootstrap is a general method for obtaining estimated standard deviations of estimators $\hat{\theta}$ and confidence intervals for parameters θ without requiring a tractable mathematical expression for the asymptotic variance of $\hat{\theta}$. Efron's technique eliminates the mathematical intractibility obstacle by relying on computing power and is known as a computer-intensive method. It is applicable in a great variety of problems (see Efron (1979), Efron and Gong (1983), Efron and Tibshirani (1993), Davison and Hinkley (1997), DiCiccio and Efron (1996) and Manly (1997)). In this section we apply Efron's bootstrap method to obtain an asymptotically distribution-free confidence interval for the parameter τ (the population measure of association defined by (8.2)) using the estimator $\hat{\tau}$ given by (8.34), where Q (8.5) is replaced by Q^* (8.17) in the definition of K.

Procedure

Denote the observed bivariate sample values as

$$Z_1 = (X_1, Y_1), \quad Z_2 = (X_2, Y_2), \ldots, Z_n = (X_n, Y_n).$$

1. Make n random draws with replacement from the bivariate sample $Z_1, Z_2, \ldots, Z_n$. This is equivalent to doing independent random sampling from the bivariate empirical distribution function $\hat{F}$, which puts probability $1/n$ on each of the data points $Z_i, i = 1, \ldots, n$.

For the canned tuna data of Table 8.1, $n = 9$ and

$$Z_1 = (44.4, 2.6), \; Z_2 = (45.9, 3.1), \; Z_3 = (41.9, 2.5), \; Z_4 = (53.3, 5.0), \; Z_5 = (44.7, 3.6)$$

$$Z_6 = (44.1, 4.0), \; Z_7 = (50.7, 5.2), \; Z_8 = (45.2, 2.8), \; Z_9 = (60.1, 3.8).$$

A possible bootstrap sample of these data is, for example, 1 copy of Z_1, 2 copies of Z_2, 0 copies of Z_3, 0 copies of Z_4, 1 copy of Z_5, 3 copies of Z_6, 0 copies of Z_7, 1 copy of Z_8, and 1 copy of Z_9.

2. Perform step 1 a large number, say, B, of times. For each draw, compute $\hat{\tau}$. Note that in computing $\hat{\tau}$ it will be necessary to use Q^* (8.17) rather than Q (8.5) in the definition of K. This is because ties will occur in most bootstrap samples because we sample with replacement. Denote the B values of $\hat{\tau}$ as $\hat{\tau}^{*1}, \hat{\tau}^{*2}, \ldots, \hat{\tau}^{*B}$. These are called the bootstrap replications of $\hat{\tau}$. Let $\hat{\tau}^{*(1)} \leq \hat{\tau}^{*(2)} \leq \cdots \leq \hat{\tau}^{*(B)}$ denote the ordered values of the bootstrap replications.

An asymptotically distribution-free confidence interval for τ, with approximate confidence coefficient $100(1 - \alpha)\%$, is (τ'_L, τ'_U) where

$$\tau'_L = \hat{\tau}^{*(k)}, \qquad \tau'_U = \hat{\tau}^{*(B+1-k)} \tag{8.50}$$

and

$$k = B\left(\frac{\alpha}{2}\right). \tag{8.51}$$

8.4. BOOTSTRAP CONFIDENCE INTERVAL

If $k = B(\alpha/2)$ is an integer, then τ'_L is the kth-largest bootstrap replication and τ'_U is the $(B + 1 - k)$th-largest replication. For example, if $\alpha = .10$ and $B = 1,000$, $k = 1,000(.05) = 50$, τ'_L is the bootstrap replication occupying position 50 in the ordered list, and τ'_U is the bootstrap replication occupying position 951 in the ordered list. If $B(\alpha/2)$ is not an integer, we follow the convention of Efron and Tibshirani (see Efron and Tibshirani, 1993, 160) and set $k = \langle (B + 1)(\alpha/2) \rangle$, the largest integer that is less than or equal to $(B + 1)(\alpha/2)$. With this value of k, τ'_L is the bootstrap replication occupying position k in the ordered list and τ'_U is the bootstrap replication occupying position $B + 1 - k$ in the ordered list.

Example 8.4: *(Continuation of Examples 8.1, 8.2, and 8.3).* Using an S-plus program, boots.tau, which can be found in Appendix B, we obtained 1,000 bootstrap replications of $\hat{\tau}$. Figure 8.1 is a histogram of the 1,000 bootstrap replications. The 1,000 bootstrap replications are:

$\hat{\tau}^{*(1)} = -.556, \hat{\tau}^{*(2)} = -.500, \quad \hat{\tau}^{*(3)} = \cdots = \hat{\tau}^{*(5)} = -.444, \quad \hat{\tau}^{*(6)} = -.417,$

$\hat{\tau}^{*(7)} = \cdots = \hat{\tau}^{*(9)} = -.361, \quad \hat{\tau}^{*(10)} = \hat{\tau}^{*(11)} = -.306, \quad \hat{\tau}^{*(12)} = -.250,$

$\hat{\tau}^{*(13)} = \hat{\tau}^{*(14)} = -.222, \quad \hat{\tau}^{*(15)} = \hat{\tau}^{*(16)} = -.194, \quad \hat{\tau}^{*(17)} = \hat{\tau}^{*(18)} = -.167,$

$\hat{\tau}^{*(19)} = -.139, \quad \hat{\tau}^{*(20)} = \hat{\tau}^{*(21)} = -.111, \quad \hat{\tau}^{*(22)} = \cdots = \hat{\tau}^{*(32)} = -.083,$

$\hat{\tau}^{*(33)} = \cdots = \hat{\tau}^{*(40)} = -.056, \quad \hat{\tau}^{*(41)} = \cdots = \hat{\tau}^{*(50)} = -.028,$

$\hat{\tau}^{*(51)} = \cdots = \hat{\tau}^{*(70)} = .000, \quad \hat{\tau}^{*(71)} = \cdots = \hat{\tau}^{*(85)} = .028,$

$\hat{\tau}^{*(86)} = \cdots = \hat{\tau}^{*(91)} = .056, \quad \hat{\tau}^{*(92)} = \cdots = \hat{\tau}^{*(113)} = .083,$

$\hat{\tau}^{*(114)} = \cdots = \hat{\tau}^{*(133)} = .111, \quad \hat{\tau}^{*(134)} = \cdots = \hat{\tau}^{*(148)} = .139,$

$\hat{\tau}^{*(149)} = \cdots = \hat{\tau}^{*(173)} = .167, \quad \hat{\tau}^{*(174)} = \cdots = \hat{\tau}^{*(194)} = .194,$

$\hat{\tau}^{*(195)} = \cdots = \hat{\tau}^{*(227)} = .222, \quad \hat{\tau}^{*(228)} = \cdots = \hat{\tau}^{*(264)} = .250,$

$\hat{\tau}^{*(265)} = \cdots = \hat{\tau}^{*(316)} = .278, \quad \hat{\tau}^{*(317)} = \cdots = \hat{\tau}^{*(351)} = .306,$

$\hat{\tau}^{*(352)} = \cdots = \hat{\tau}^{*(393)} = .333, \quad \hat{\tau}^{*(394)} = \cdots = \hat{\tau}^{*(433)} = .361,$

$\hat{\tau}^{*(434)} = \cdots = \hat{\tau}^{*(478)} = .389, \quad \hat{\tau}^{*(479)} = \cdots = \hat{\tau}^{*(530)} = .417,$

$\hat{\tau}^{*(531)} = \cdots = \hat{\tau}^{*(592)} = .444, \quad \hat{\tau}^{*(593)} = \cdots = \hat{\tau}^{*(640)} = .472,$

$\hat{\tau}^{*(641)} = \cdots = \hat{\tau}^{*(687)} = .500, \quad \hat{\tau}^{*(688)} = \cdots = \hat{\tau}^{*(730)} = .528,$

$\hat{\tau}^{*(731)} = \cdots = \hat{\tau}^{*(769)} = .556, \quad \hat{\tau}^{*(770)} = \cdots = \hat{\tau}^{*(808)} = .583,$

$\hat{\tau}^{*(809)} = \cdots = \hat{\tau}^{*(847)} = .611, \quad \hat{\tau}^{*(848)} = \cdots = \hat{\tau}^{*(880)} = .639,$

$\hat{\tau}^{*(881)} = \cdots = \hat{\tau}^{*(912)} = .667, \quad \hat{\tau}^{*(913)} = \cdots = \hat{\tau}^{*(935)} = .694,$

$\hat{\tau}^{*(936)} = \cdots = \hat{\tau}^{*(955)} = .722, \quad \hat{\tau}^{*(956)} = \cdots = \hat{\tau}^{*(969)} = .750,$

$\hat{\tau}^{*(970)} = \cdots = \hat{\tau}^{*(986)} = .778, \quad \hat{\tau}^{*(987)} = \cdots = \hat{\tau}^{*(992)} = .806,$

$\hat{\tau}^{*(993)} = \hat{\tau}^{*(994)} = .833, \quad \hat{\tau}^{*(995)} = \cdots = \hat{\tau}^{*(999)} = .861, \quad \hat{\tau}^{*(1000)} = .889.$

For an approximate 90% confidence interval $\alpha = .1, (\alpha/2) = .05$, and from (8.51) we find $k = 1000(.05) = 50$. Then from (8.50) and the ordered list of 1000 bootstrap replications,

$$\tau'_L = \hat{\tau}^{*(50)} = -.028, \quad \tau'_U = \hat{\tau}^{*(951)} = .722.$$

Recall that the method of Section 8.3 gave an approximate 90% confidence interval of $(.027, .861)$. See Example 8.3.

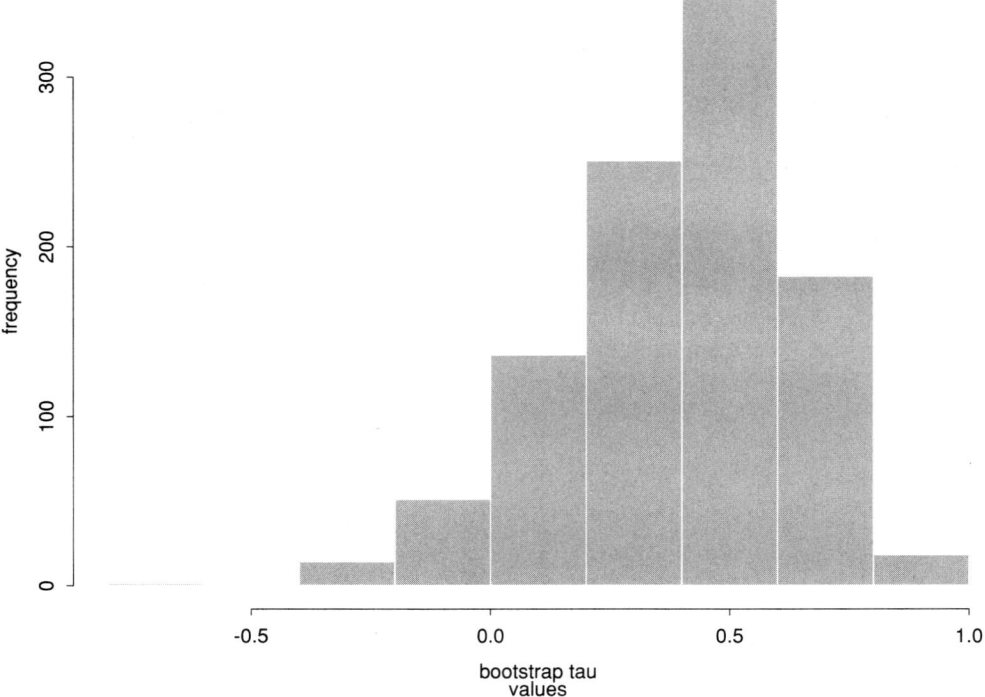

FIGURE 8.1. Histogram of 1,000 bootstrap replications of Kendall's sample correlation coefficient for the canned tuna data of Table 8.1.

Comments

31. *The Bootstrap Estimated Standard Error.* For Kendall's sample correlation coefficient $\hat{\tau}$, the standard deviation of $\hat{\tau}$, which we have denoted thus far as σ, depends on the bivariate distribution function $F_{X,Y}$. We now denote $F_{X,Y}$ as F, dropping the subscripts. We could exhibit the dependence of σ on F by writing σ as $\sigma(F)$. F is unknown, but it can be estimated by the bivariate empirical distribution function $\widehat{F}$, which puts probability $1/n$ on each of the observed data points $Z_i = (X_i, Y_i), i = 1, \ldots, n$. The bootstrap estimate of $\sigma(F)$ is $\sigma(\widehat{F})$, where $\sigma(\widehat{F})$ is the standard deviation of $\hat{\tau}$ when the true underlying distribution is $\widehat{F}$ rather than F. A tractable mathematical expression for $\sigma(\widehat{F})$ is very difficult to obtain. However, $\sigma(\widehat{F})$ can be estimated using the B bootstrap replications, by

$$\hat{\sigma}_B = \left\{ \frac{\sum_{i=1}^{n}(\hat{\tau}^{*i} - \hat{\tau}^{*\cdot})^2}{(B-1)} \right\}^{1/2} \tag{8.52}$$

where

$$\hat{\tau}^{*\cdot} = \frac{\sum_{i=1}^{B} \hat{\tau}^{*i}}{B}. \tag{8.53}$$

As B tends to ∞, $\hat{\sigma}_B$ tends to $\sigma(\widehat{F})$. As n tends to ∞, $\sigma(\widehat{F})$ tends to $\sigma(F)$. Thus $\hat{\sigma}_B$ can be used as an estimate of the standard deviation of $\hat{\tau}$.

8.4. BOOTSTRAP CONFIDENCE INTERVAL

32. *The Bootstrap in the One-Sample Nonparametric Framework*. In this section we applied the bootstrap in a bivariate situation, where the data are bivariate observations $Z_i = (X_i, Y_i)$, $i = 1, \ldots, n$, and the parameter of interest is τ. The bootstrap can be used in a wide variety of situations, including the one-sample problem, the k-sample problem, censored data problems, and complicated multivariate frameworks. See Efron and Gong (1983), Efron and Tibshirani (1993), Davison and Hinkley (1997), and DiCiccio and Efron (1996). In this comment we describe the approach in the context of the one-sample nonparametric framework.

Suppose we are interested in estimating a parameter $\theta = \theta(F)$, when $X_1, \ldots, X_n$ are a random sample from an unknown distribution F. The nonparametric maximum likelihood estimate of F is the statistic $\widehat{\theta} = \theta(F_n)$, where F_n is the sample distribution function. For example, if we are interested in estimating the rth moment of the F distribution, $\theta(F) = E(X^r)$, then $\theta(F_n) = (\sum_{i=1}^{n} X_i^r)/n$.

The bootstrap procedure in the one-sample problem is analogous to the procedure we described for the bivariate situation. The steps are as follows:

1. Make n random draws with replacement from the sample $X_1, \ldots, X_n$.
2. Perform step 1 a large number, say B, of times. For each draw, compute $\widehat{\theta}$. Denote the B values of $\widehat{\theta}$ as $\widehat{\theta}^{*1}, \widehat{\theta}^{*2}, \ldots, \widehat{\theta}^{*B}$. These are the bootstrap replications of $\widehat{\theta}$.

The bootstrap estimate of the standard deviation of $\widehat{\theta}$ is

$$\widehat{\sigma}_B = \left\{ \frac{\sum_{i=1}^{n} (\widehat{\theta}^{*i} - \widehat{\theta}^{*\cdot})^2}{(B-1)} \right\}^{1/2} \tag{8.54}$$

where

$$\widehat{\theta}^{*\cdot} = \frac{\sum_{i=1}^{B} \widehat{\theta}^{*i}}{B}. \tag{8.55}$$

An asymptotically distribution-free confidence interval for θ, with approximate confidence coefficient $100(1-\alpha)\%$, is (θ'_L, θ'_U), where

$$\theta'_L = \widehat{\theta}^{*(k)}, \quad \theta'_U = \widehat{\theta}^{*(B+1-k)}, \tag{8.56}$$

where $\widehat{\theta}^{*(1)} \leq \widehat{\theta}^{*(2)} \leq \cdots \leq \widehat{\theta}^{*(B)}$ are the ordered values for the bootstrap replications and $k = \langle (B+1)(\alpha/2) \rangle$, the largest integer that is less than or equal to $(B+1)(\alpha/2)$.

The confidence interval defined by (8.56) is called the percentile interval. Let $\widehat{G}$ denote the cumulative distribution function of $\widehat{\theta}^*$:

$$\widehat{G}(t) = \frac{\#\{\widehat{\theta}^{*i} < t\}}{B}. \tag{8.57}$$

The endpoints θ'_L, θ'_U are, respectively, the α and $1-\alpha$ percentiles of $\widehat{G}$.

The percentile confience interval is *transformation-respecting*. If $\eta = m(\theta)$ is a monotone transformation, then a confidence interval (η_L, η_U) for the parameter η is obtained directly from the confidence interval (8.56) for θ via $\eta_L = m(\theta'_L)$, $\eta_U = m(\theta'_U)$. For example, a confidence interval for θ^2 is obtained by squaring the endpoints of the confidence interval for θ.

The percentile confidence interval is *range-preserving*. For example, consider the percentile interval based on bootstrapping $\hat{\tau}$. The values of the parameter τ, Kendall's population correlation coefficient, are always between -1 and 1. The possible values of the estimator $\hat{\tau}$ also are in the interval $[-1, 1]$. Thus the bootstrap replications of $\hat{\tau}$ must be in the interval $[-1, 1]$, as must the confidence interval endpoints, because the endpoints are particular bootstrap replications. More generally, the estimator $\hat{\theta}$ of the form $\hat{\theta} = \theta(F_n)$ satisfies the same range restrictions as $\theta = \theta(F)$, and thus the percentile interval based on bootstrapping $\hat{\theta}$ also satisfies the same range restrictions as θ.

33. *The BC_a Confidence Interval.* Efron and Tibshirani (1993, Chapter 14) (see also DiCiccio and Efron, 1996) present a method, called the BC_a method, that gives more accurate confidence limits than does the percentile method of Comment 32. The acronymn BC_a means "bias-corrected and accelerated". The BC_a method depends on a bias-correction z_0 and an acceleration a. In the one-sample nonparametric framework, z_0 can be estimated by

$$\hat{z}_0 = \Phi^{-1} \left\{ \frac{\#\{\hat{\theta}^{*i} < \hat{\theta}\}}{B} \right\} \tag{8.58}$$

where Φ denotes the standard normal cumulative distribution function. Thus $\hat{z}_0$ is Φ^{-1} of the proportion of the bootstrap replications less than $\hat{\theta}$.

The estimate of a is

$$\hat{a} = \frac{\sum_{i=1}^{n} (\hat{\theta}_{\cdot} - \hat{\theta}_{-i})^3}{6\{\sum_{i=1}^{n} (\hat{\theta}_{\cdot} - \hat{\theta}_{-i})^2\}^{3/2}} \tag{8.59}$$

where $\hat{\theta}_{-i}$ is the estimate of θ obtained by deleting X_i and computing $\hat{\theta}$ for the reduced sample $X_1, X_2, \ldots, X_{i-1}, X_{i+1}, \ldots, X_n$ and

$$\hat{\theta}_{\cdot} = \frac{\sum_{i=1}^{n} \hat{\theta}_{-i}}{n}. \tag{8.60}$$

The lower and upper endpoints of the $100(1 - \alpha)\%$ confidence interval are

$$\theta_L'' = \hat{G}^{-1} \Phi \left(\hat{z}_0 + \frac{\hat{z}_0 + z^{(\alpha/2)}}{1 - \hat{a}(\hat{z}_0 + z^{(\alpha)})} \right) \tag{8.61}$$

$$\theta_U'' = \hat{G}^{-1} \Phi \left(\hat{z}_0 + \frac{\hat{z}_0 + z^{(1-\alpha/2)}}{1 - \hat{a}(\hat{z}_0 + z^{(1-\alpha/2)})} \right) \tag{8.62}$$

where in (8.61) and (8.62), Φ is the standard normal distribution function, $z^{(\alpha/2)} = \Phi^{-1}(\alpha/2)$, $z^{(1-\alpha/2)} = \Phi^{-1}(1-\alpha/2)$ (if, for example, $\alpha = .10$, then $z^{(\alpha/2)} = -1.65$, and $z^{(1-\alpha/2)} = 1.65$), $\hat{G}$ is given by (8.57), $\hat{z}_0$ is given by (8.58), and $\hat{a}$ is given by (8.59).

The endpoints θ_L'' and θ_U'' of the BC_a interval are also percentiles of the bootstrap distribution $\hat{G}$ but not necessarily the same ones as given by the percentile interval. If $\hat{a} = \hat{z}_0 = 0$, the BC_a and percentile intervals are the same.

The BC_a interval also enjoys the transformation-respecting and range-preserving properties that hold for the percentile interval. The BC_a interval, however, has an accuracy advantage. The BC_a interval has a second-order accuracy property, whereas the percentile interval is only first-order accurate. See Section 14.3 of Efron and Tibshirani (1993).

The appendix of Efron and Tibshirani (1993) describes some available bootstrap software and contains some programs in the S language, including a program for computing BC_a intervals.

34. *The Choice of B, the Number of Bootstrap Replications.* The choice of B depends to some extent on the particular statistic that is being bootstrapped and the complexity of the situation. Efron and Tibshirani (1993, 52) give some rules of thumb based on their extensive experience with the bootstrap. Roughly speaking, $B = 200$ replications are usually sufficient for estimating a standard error but much larger values of B, such as 1,000 or 2,000, are required for bootstrap confidence intervals.

35. *An Example Where the Bootstrap Fails.* Let $X_1, \ldots, X_n$ be a random sample from the uniform distribution on $(0, \theta)$. The maximum likelihood estimator of θ, the upper endpoint of the interval, is $\widehat{\theta} = \text{maximum}(X_1, \ldots, X_n) = X_{(n)}$. Efron and Tibshirani (1993, 81) point out that the bootstrap does not do well in this situation. Miller (1964) showed that the jackknife estimator of θ also fails in this situation, because it depends not only on $X_{(n)}$, but also on $X_{(n-1)}$, the second largest observation, and the latter contains no additional information about θ when the value of $X_{(n)}$ is available.

36. *Jackknife versus Bootstrap.* For a linear statistic of the form $\widehat{\theta} = \mu + \{\sum_{i=1}^{n} h(X_i)/n\}$, where μ is a constant and h is a function, there is no loss of information in using the jackknife rather than the bootstrap. For nonlinear statistics, there is a loss of information and the bootstrap should be preferred. See Efron and Tibshirani (1993) for a detailed discussion of the relationship between the jackknife and the bootstrap.

One disadvantage of the bootstrap is that two different people bootstrapping the same data will not in general get the same bootstrap estimate of the standard deviation or the same confidence interval. This violates what Gleser (1996) calls "the first law of applied statistics," namely: "Two individuals using the same statistical method on the same data should arrive at the same conclusion." See Gleser (1996) for other disadvantages of the bootstrap.

37. *Development of the Bootstrap.* The bootstrap was formally introduced by Efron (1979). Efron and Tibshirani (1993, 56), however, credit many authors for similar ideas and they designate as "particularly notable" the contributions of Hartigan's typical value theory (1969, 1971, 1975). Hartigan recognized the wide applicability of subsample methods (a subsample of $X_1, \ldots, X_n$ is any subset of the whole sample) as a tool for assessing variability. See Efron and Tibshirani (1993, 56) for references to other papers that contain ideas related to bootstrapping.

Properties

See Bickel and Freedman (1981) for asymptotic consistency. See Efron and Tibshirani (1993) and Davison and Hinkley (1997) for various properties including accuracy, transformation-respecting, range-preserving, and the relationship of the bootstrap to the jackknife. See Hall (1992) for a high-level mathematical treatment of the bootstrap. See Manly (1997) for bootstrap methods in biology.

PROBLEMS

34. For the cerebral palsy data of Table 8.4, use the bootstrap method to find a confidence interval for τ with approximate confidence coefficient .90. Compare your results with those of Problem 28.

35. For the psychological test scores data of Table 8.5, use the bootstrap method to find a confidence interval for τ with approximate confidence coefficient .95.

36. Consider the case $n = 3$ where you have three bivariate observations Z_1, Z_2, Z_3. List the possible bootstrap samples and give the corresponding probability of each being selected on a given bootstrap replication.
37. Consider the case where you have four bivariate observations Z_1, Z_2, Z_3, Z_4. List the possible bootstrap samples and give the corresponding probability of each being selected on a given bootstrap replication.
38. Illustrate by means of an example or show directly that with n observations the number of possible bootstrap samples is $\binom{2n-1}{n}$.
39. Show that if $\hat{a} = \hat{z}_0 = 0$, the BC_a interval given by (8.61) and (8.62), reduces to the percentile interval.

8.5. A DISTRIBUTION-FREE TEST FOR INDEPENDENCE BASED ON RANKS (SPEARMAN)

Hypothesis

Let $(X_1, Y_1), \ldots, (X_n, Y_n)$ be a random sample from a continuous bivariate population (i.e., Assumption A is satisfied) with joint distribution function $F_{X,Y}$ and marginal distribution functions F_X and F_Y. In this section we return to the problem of testing for independence between the X and Y variables corresponding to the null hypothesis H_0 (8.1). Here, however, alternatives to H_0 will no longer be stated in terms of the correlation coefficient τ (8.2). Instead, the alternatives of interest in this section will be less specifically interpretable, corresponding to the quite general (but vague; see Comment 47) concepts of positive or negative association between the X and Y variables.

Procedure

To compute the Spearman rank correlation coefficient r_s, we first order the n X observations from least to greatest and let R_i denote the rank of $X_i, i = 1, \ldots, n$, in this ordering. Similarly, we separately order the n Y observations from least to greatest and let S_i denote the rank of $Y_i, i = 1, \ldots, n$, in this ordering. The Spearman (1904) rank correlation coefficient is defined by

$$r_s = \frac{12 \sum_{i=1}^{n} \left\{ \left[R_i - \frac{n+1}{2} \right] \left[S_i - \frac{n+1}{2} \right] \right\}}{n(n^2 - 1)} \tag{8.63}$$

$$= 1 - \frac{6 \sum_{i=1}^{n} D_i^2}{n(n^2 - 1)}, \tag{8.64}$$

where $D_i = S_i - R_i, i = 1, \ldots, n$.

a. One-Sided Upper-Tail Test. To test the null hypothesis of independence, H_0 (8.1), versus the directional alternative

$$H_1 : [X \text{ and } Y \text{ are positively associated}] \tag{8.65}$$

at the α level of significance,

$$\text{Reject } H_0 \text{ if } r_s \geq r_{s,\alpha}; \quad \text{otherwise do not reject,} \tag{8.66}$$

where the constant $r_{s,\alpha}$ is chosen to make the type I error probability equal to α. Selected values of $r_{s,\alpha}$ are given in Table A.31.

b. One-Sided Lower-Tail Test. To test independence, H_0 (8.1), versus the directional alternative

$$H_2 : [X \text{ and } Y \text{ are negatively associated}] \tag{8.67}$$

at the α level of significance,

$$\text{Reject } H_0 \text{ if } r_s \leq -r_{s,\alpha}; \quad \text{otherwise do not reject.} \tag{8.68}$$

c. Two-Sided Test. To test independence, H_0 (8.1), versus the general dependency alternative

$$H_3 : [X \text{ and } Y \text{ are not independent variables}] \tag{8.69}$$

at the α level of significance,

$$\text{Reject } H_0 \text{ if } |r_s| \geq r_{s,\alpha/2}; \quad \text{otherwise do not reject.} \tag{8.70}$$

This two-sided procedure is the two-sided symmetric test with $\alpha/2$ probability in each tail of the null distribution of r_s.

Large-Sample Approximation

The large-sample approximation is based on the asymptotic normality of r_s, suitably standardized. For this standardization, we need to know the expected value and variance of r_s when the null hypothesis of independence is true. Under H_0, the expected value and variance of r_s are

$$E_0(r_s) = 0 \tag{8.71}$$

and

$$\text{var}_0(r_s) = \frac{1}{n-1}, \tag{8.72}$$

respectively. These expressions for $E_0(r_s)$ and $\text{var}_0(r_s)$ are verified by direct calculations in Comment 42 for the special case of $n = 4$. General derivations of both expressions are discussed in Comment 45.

The standardized version of r_s is

$$r_s^* = \frac{r_s - E_0(r_s)}{\{\text{var}_0(r_s)\}^{1/2}} = (n-1)^{1/2} r_s. \tag{8.73}$$

When H_0 is true, r_s^* has, as n tends to infinity, an asymptotic $N(0, 1)$ distribution. (See Comment 45 for indications of the proof.) The normal theory approximation for procedure (8.66) is

$$\text{Reject } H_0 \text{ if } r_s^* \geq z_\alpha; \quad \text{otherwise do not reject,} \tag{8.74}$$

the normal theory approximation for procedure (8.68) is

$$\text{Reject } H_0 \text{ if } r_s^* \leq -z_\alpha; \quad \text{otherwise do not reject,} \tag{8.75}$$

and the normal theory approximation for procedure (8.70) is

$$\text{Reject } H_0 \text{ if } |r_s^*| \geq z_{\alpha/2}; \quad \text{otherwise do not reject.} \tag{8.76}$$

Ties

If there are ties among the n X observations and/or separately among the n Y observations, assign each of the observations in a tied (either X or Y) group the average of the integer ranks that are associated with the tied group. After computing r_s with these average ranks for tied observations, use procedures (8.66), (8.68), or (8.70) and refer the value of r_s to Table A.31. Note, however, that this test associated with tied X's and/or tied Y's is only approximately, and not exactly, of significance level α. (To get an exact level α test even in this tied setting, see Comment 46.)

When applying the large-sample approximation, however, a change in both the null expectation and null variance for r_s, due to the tied X's and/or tied Y's, must be taken into account. In such tied settings, we replace r_s (8.64) in r_s^* (8.73) by the modified statistic

$$r_s = \frac{n(n^2-1) - 6\sum_{s=1}^{n} D_s^2 - \frac{1}{2}\left\{\sum_{i=1}^{g}[t_i(t_i^2-1)] + \sum_{j=1}^{h}[u_j(u_j^2-1)]\right\}}{\left\{\left[n(n^2-1) - \sum_{i=1}^{g} t_i(t_i^2-1)\right]\left[n(n^2-1) - \sum_{j=1}^{h} u_j(u_j^2-1)\right]\right\}^{1/2}}, \quad (8.77)$$

where in equation (8.77) g denotes the number of tied X groups, t_i is the size of tied X group i, h is the number of tied Y groups, and u_j is the size of tied Y group j. We note that an untied $X(Y)$ observation is considered to be a tied $X(Y)$ group of size 1. In particular, if neither the collection of X nor the collection of Y observations contains tied values, we have $g = h = n, t_i = u_j = 1, i = 1, \ldots, n$ and $j = 1, \ldots, n$. In this case of no tied X's and no tied Y's, each term involving either $(t_i^2 - 1)$ or $(u_j^2 - 1)$ reduces to zero and the "ties" expression for r_s in (8.77) reduces to the "no-ties" form for r_s, as given in equation (8.64).

As a consequence of this effect that ties have on the null distribution of r_s, in order to use the large-sample approximation when there are tied X observations and/or tied Y observations, we first compute r_s^* (8.73) using average ranks and the ties-corrected version of r_s (8.77). Approximations (8.74), (8.75), or (8.76) can then be applied, as appropriate for the problem, with this value of r_s^*.

Example 8.5: *Proline and Collagen in Liver Cirrhosis.* Kershenobich, Fierro, and Rojkind (1970) have studied the relation between the free pool of proline and collagen content in human liver cirrhosis. The data in Table 8.9 are based on an analysis of cirrhotic livers from seven patients, each having a histological diagnosis of portal cirrhosis.

We are interested in assessing whether there is a positive relationship between the total collagen and the free proline in cirrhotic livers. Thus we wish to apply procedure (8.66) to test the hypothesis of independence, H_0 (8.1), versus the alternative, H_1 (8.65), of positive

TABLE 8.9. Free Proline and Total Collagen Contents of Cirrhotic Patients

Patient	Total Collagen, X_i (mg/g Dry Weight of Liver)	Free Proline, Y_i (μ mole/g Dry Weight of Liver)
1	7.1	2.8
2	7.1	2.9
3	7.2	2.8
4	8.3	2.6
5	9.4	3.5
6	10.5	4.6
7	11.4	5.0

Source: D. Kershenobich, F. J. Fierro, and M. Rojkind (1970).

association. For purposes of illustration, we consider the significance level $\alpha = .025$. From Table A.31 we find (with $n = 7$) $r_{s,.025} = .786$ and procedure (8.66) becomes

$$\text{Reject } H_0 \text{ if } r_s \geq .786.$$

Ranking the X (total collagen) values from least to greatest, using average ranks for the tied pair, we obtain $R_1 = (1+2)/2 = 1.5, R_2 = (1+2)/2 = 1.5, R_3 = 3, R_4 = 4, R_5 = 5, R_6 = 6$, and $R_7 = 7$. Similarly, ranking the Y (free proline) values from least to greatest, again using average ranks for the tied pairs, we find $S_1 = (2+3)/2 = 2.5, S_2 = 4, S_3 = (2+3)/2 = 2.5$, $S_4 = 1, S_5 = 5, S_6 = 6$, and $S_7 = 7$. Taking differences, we see that

$$D_1 = 2.5 - 1.5 = 1, \quad D_2 = 4 - 1.5 = 2.5, \quad D_3 = 2.5 - 3 = -.5$$
$$D_4 = 1 - 4 = -3, \quad D_5 = 5 - 5 = 0, \quad D_6 = 6 - 6 = 0,$$
$$D_7 = 7 - 7 = 0.$$

Thus, from (8.64), we obtain

$$r_s = 1 - \frac{6[(1)^2 + (2.5)^2 + (-.5)^2 + (-3)^2 + 0^2 + 0^2 + 0^2]}{7(7^2 - 1)}$$
$$= 1 - \frac{6(16.5)}{7(48)} = .705.$$

Since this value of r_s is not greater than the critical value .786, we do not reject H_0 (independence) at the $\alpha = .025$ level.

Since the one-sided P-value for these data is the smallest significance level at which we can reject H_0 in favor of a positive association between total collagen and free proline in cirrhotic patients with the observed value of the test statistic $r_s = .705$, we see from Table A.31 (with $n = 7$) that the P-value for these data, namely, $P_0(r_s \geq .705)$, is between .05 and .10. Thus there is some marginal evidence that total collagen and free proline are positively associated in subjects with liver cirrhosis.

For the large-sample approximation we need to use the ties-corrected version of r_s as given in (8.77). For this purpose, we note that there are $g = 6$ tied X groups, with $t_1 = 2, t_2 = t_3 = t_4 = t_5 = t_6 = 1$, and $h = 6$ tied Y groups, with $u_2 = 2, u_1 = u_3 = u_4 = u_5 = u_6 = 1$. Thus, for these tied data, the modified value of r_s (8.77) is calculated to be

$$r_s = \frac{7(7^2 - 1) - 6[(1)^2 + (2.5)^2 + (-.5)^2 + (-3)^2 + 3(0)^2] - \frac{1}{2}(2)(2)(2^2 - 1)}{\{[7(7^2 - 1) - 2(2^2 - 1)][7(7^2 - 1) - 2(2^2 - 1)]\}^{1/2}}$$
$$= \frac{7(48) - 6(16.5) - 6}{\{[7(48) - 6][7(48) - 6]\}^{1/2}} = \frac{231}{330} = .700.$$

Using this value of $r_s = .700$ to compute r_s^* (8.73), we obtain

$$r_s^* = (6)^{1/2}(.700) = 1.71.$$

Thus the smallest significance level at which we can reject H_0 in favor of positive association between total collagen and free proline in subjects with liver cirrhosis using the normal approximation is .0436, since $z_{.0436} = 1.71$ in Table A.1.

Comments

38. *Motivation for the Test.* The null hypothesis of this section is that the X and Y variables are independent, which, in the case of no ties, implies that any permutation of the X ranks $(R_1, \ldots, R_n)$ is equally likely to occur with any permutation of the Y ranks $(S_1, \ldots, S_n)$. As a result, under the null hypothesis H_0 (8.1) of independence, the Spearman rank correlation coefficient r_s (8.64) will have a tendency to assume values near zero. However, when the alternative $H_1 : [X$ and Y are positively associated] is true, the rank vectors $(R_1, \ldots, R_n)$ and $(S_1, \ldots, S_n)$ will tend to agree, resulting in small differences $D_i = S_i - R_i, i = 1, \ldots, n$. Thus, when H_1 (8.65) is true, we would expect the value of $\sum_{j=1}^{n} D_j^2$ to be small and the resulting value of r_s (8.64) to be large and positive. This suggests rejecting H_0 in favor of positive association H_1 (8.65) for large positive values of r_s and motivates procedures (8.66) and (8.74). Similar rationales apply to procedures (8.68), (8.70), (8.75), and (8.76).

39. *Minitab Computation for r_s.* The value of r_s (8.63) can also be obtained by applying Minitab's CORRELATION command to the columns $(R_1, \ldots, R_n)$ and $(S_1, \ldots, S_n)$ of X and Y ranks, respectively. Note, however, that prior to using CORRELATION for this purpose we must manually obtain the separate rank vectors $(R_1, \ldots, R_n)$ and $(S_1, \ldots, S_n)$. In addition, CORRELATION does not provide the ties-corrected version of r_s (8.77), even if we use average ranks to break the ties and obtain $(R_1, \ldots, R_n)$ and $(S_1, \ldots, S_n)$.

40. *Pearson's Product Moment Sample Correlation Coefficient.* The classical Pearson product moment sample correlation coefficient for the pair of vectors $(X_1, \ldots, X_n)$ and $(Y_1, \ldots, Y_n)$ is given by

$$r_p = \frac{\sum_{k=1}^{n}(X_k - \bar{X})(Y_k - \bar{Y})}{\left[\sum_{i=1}^{n}(X_i - \bar{X})^2 \sum_{j=1}^{n}(Y_j - \bar{Y})^2\right]^{1/2}}, \tag{8.78}$$

where $\bar{X} = \sum_{s=1}^{n} X_s/n$ and $\bar{Y} = \sum_{t=1}^{n} Y_t/n$. We note that the Spearman rank correlation coefficient r_s is simply the classical correlation coefficient applied to the rank vectors $(R_1, \ldots, R_n)$ and $(S_1, \ldots, S_n)$ instead of the actual X and Y observations, respectively. (See Problem 49.)

41. *Derivation of the Distribution of r_s under H_0 (No-Ties Case).* Without loss of generality, we take $R_1 = 1, \ldots, R_n = n$; then, since under H_0 (8.1) all possible $n!$ $(S_1, S_2, \ldots, S_n)$ Y-rank configurations are equally likely, it follows that each has null probability $1/n!$.

Let us consider the case $n = 4$. In the following table we display the $4! = 24$ possible (S_1, S_2, S_3, S_4) configurations, the associated values of r_s and the corresponding null probabilities.

(R_1, R_2, R_3, R_4)	(S_1, S_2, S_3, S_4)	Null Probability	r_s
(1, 2, 3, 4)	(1, 2, 3, 4)	$\frac{1}{24}$	1
(1, 2, 3, 4)	(1, 2, 4, 3)	$\frac{1}{24}$	.8
(1, 2, 3, 4)	(1, 3, 2, 4)	$\frac{1}{24}$	.8
(1, 2, 3, 4)	(1, 3, 4, 2)	$\frac{1}{24}$	.4
(1, 2, 3, 4)	(1, 4, 2, 3)	$\frac{1}{24}$	.4
(1, 2, 3, 4)	(1, 4, 3, 2)	$\frac{1}{24}$	.2
(1, 2, 3, 4)	(2, 1, 3, 4)	$\frac{1}{24}$	.8
(1, 2, 3, 4)	(2, 1, 4, 3)	$\frac{1}{24}$	.6
(1, 2, 3, 4)	(2, 3, 1, 4)	$\frac{1}{24}$	.4

(R_1,R_2,R_3,R_4)	(S_1,S_2,S_3,S_4)	Null Probability	r_s
(1, 2, 3, 4)	(2, 3, 4, 1)	$\frac{1}{24}$	−.2
(1, 2, 3, 4)	(2, 4, 1, 3)	$\frac{1}{24}$	0
(1, 2, 3, 4)	(2, 4, 3, 1)	$\frac{1}{24}$	−.4
(1, 2, 3, 4)	(3, 1, 2, 4)	$\frac{1}{24}$	.4
(1, 2, 3, 4)	(3, 1, 4, 2)	$\frac{1}{24}$	0
(1, 2, 3, 4)	(3, 2, 1, 4)	$\frac{1}{24}$	.2
(1, 2, 3, 4)	(3, 2, 4, 1)	$\frac{1}{24}$	−.4
(1, 2, 3, 4)	(3, 4, 1, 2)	$\frac{1}{24}$	−.6
(1, 2, 3, 4)	(3, 4, 2, 1)	$\frac{1}{24}$	−.8
(1, 2, 3, 4)	(4, 1, 2, 3)	$\frac{1}{24}$	−.2
(1, 2, 3, 4)	(4, 1, 3, 2)	$\frac{1}{24}$	−.4
(1, 2, 3, 4)	(4, 2, 1, 3)	$\frac{1}{24}$	−.4
(1, 2, 3, 4)	(4, 2, 3, 1)	$\frac{1}{24}$	−.8
(1, 2, 3, 4)	(4, 3, 1, 2)	$\frac{1}{24}$	−.8
(1, 2, 3, 4)	(4, 3, 2, 1)	$\frac{1}{24}$	−1

Thus, for example, the probability is $\frac{3}{24}$ under H_0 that r_s is equal to .8, since $r_s = .8$ when any of the three outcomes $(S_1, S_2, S_3, S_4) = (1, 2, 4, 3), (1, 3, 2, 4),$ or $(2, 1, 3, 4)$ occurs and each of these outcomes has null probability $\frac{1}{24}$. Simplifying, we obtain the null distribution

Possible Value of r_s	Probability under H_0
−1.0	$\frac{1}{24}$
−.8	$\frac{3}{24}$
−.6	$\frac{1}{24}$
−.4	$\frac{4}{24}$
−.2	$\frac{2}{24}$
0	$\frac{2}{24}$
.2	$\frac{2}{24}$
.4	$\frac{4}{24}$
.6	$\frac{1}{24}$
.8	$\frac{3}{24}$
1.0	$\frac{1}{24}$

The probability, under H_0, that r_s is greater than or equal to .6, for example, is therefore

$$P_0(r_s \geq .6) = P_0(r_s = 1.0) + P_0(r_s = .8) + P_0(r_s = .6)$$
$$= \tfrac{1}{24} + \tfrac{3}{24} + \tfrac{1}{24} = \tfrac{5}{24}.$$

Note that we have obtained the null distribution of r_s without specifying the form of the underlying independent X and Y populations under H_0 beyond the point of requiring that they be continuous. That is why the test procedures based on r_s are called distribution-free procedures. From the null distribution of r_s we can determine the critical value $r_{s,\alpha}$ and control the probability α of falsely rejecting H_0 when H_0 is true and this error probability does not depend on the specific forms of the underlying continuous and independent X and Y distributions.

42. *Calculation of the Mean and Variance of r_s under the Null Hypothesis.* In displays (8.71) and (8.72) we presented formulas for the mean and variance of r_s when the null hypothesis is true. In this comment, we illustrate a direct calculation of $E_0(r_s)$ and $\text{var}_0(r_s)$ in the particular case of $n = 4$, using the null distribution of r_s obtained in Comment 41. (Later, in Comment 45, we present general derivations of $E_0(r_s)$ and $\text{var}_0(r_s)$.) The null mean, $E_0(r_s)$, is obtained by multiplying each possible value of r_s with its probability under H_0 and summing the products. Thus,

$$E_0(r_s) = -1\left(\tfrac{1}{24}\right) - .8\left(\tfrac{3}{24}\right) - .6\left(\tfrac{1}{24}\right) - .4\left(\tfrac{4}{24}\right) - .2\left(\tfrac{2}{24}\right) + 0\left(\tfrac{2}{24}\right)$$
$$+ .2\left(\tfrac{2}{24}\right) + .4\left(\tfrac{4}{24}\right) + .6\left(\tfrac{1}{24}\right) + .8\left(\tfrac{3}{24}\right) + 1\left(\tfrac{1}{24}\right)$$
$$= 0.$$

This is in agreement with the value stated in equation (8.71). A check on the expression for $\text{var}_0(r_s)$ is also easily performed, using the well-known fact that

$$\text{var}_0(r_s) = E_0(r_s^2) - \{E_0(r_s)\}^2.$$

The value of $E_0(r_s^2)$, the second moment of the null distribution of r_s, is again obtained by multiplying possible values (in this case, of r_s^2) by the corresponding probabilities under H_0 and summing. We find

$$E_0(r_s^2) = \left[(1 + 1)\left(\tfrac{1}{24}\right) + (.64 + .64)\left(\tfrac{3}{24}\right) + (.36 + .36)\left(\tfrac{1}{24}\right) \right.$$
$$\left. + (.16 + .16)\left(\tfrac{4}{24}\right) + (.04 + .04)\left(\tfrac{2}{24}\right) + 0\left(\tfrac{2}{24}\right)\right]$$
$$= \tfrac{1}{3}.$$

Thus,

$$\text{var}_0(r_s) = \tfrac{1}{3} - 0^2 = \tfrac{1}{3},$$

which agrees with what we obtain using equation (8.72) directly, namely,

$$\text{var}_0(r_s) = \tfrac{1}{4-1} = \tfrac{1}{3}.$$

43. *Symmetry of the Distribution of r_s under the Null Hypothesis.* When H_0 is true, the distribution of r_s is symmetric about its mean 0. (See Comment 41 for verification of this when $n = 4$.) This implies that

$$P_0(r_s \leq -x) = P_0(r_s \geq x), \qquad (8.79)$$

for all x. Equation (8.79) is used directly to convert upper-tail probabilities, as presented in Table A.31, to lower-tail probabilities. In particular, it follows from (8.79) that the lower αth percentile for the null distribution of r_s is $-r_{s,\alpha}$; thus the use of $-r_{s,\alpha}$ as the critical value in procedure (8.68).

44. *Equivalent Form.* Let $(R_1, \ldots, R_n)$ and $(S_1, \ldots, S_n)$ be the vectors of separate ranks for the X and Y observations, respectively. We note that

$$\sum_{i=1}^{n}\left(R_i - \frac{n+1}{2}\right)\left(S_i - \frac{n+1}{2}\right) = \sum_{i=1}^{n} R_i S_i - \frac{n+1}{2}\sum_{i=1}^{n} R_i - \frac{n+1}{2}\sum_{i=1}^{n} S_i + \frac{n(n+1)^2}{4}.$$

However, $\sum_{i=1}^{n} R_i = \sum_{i=1}^{n} S_i = \sum_{i=1}^{n} i = n(n+1)/2$. Thus, we have

$$\sum_{i=1}^{n}\left(R_i - \frac{n+1}{2}\right)\left(S_i - \frac{n+1}{2}\right) = \sum_{i=1}^{n} R_i S_i - \frac{n(n+1)^2}{4}.$$

Combining this fact with the definition of r_s in display (8.63), we obtain an alternative computational expression for r_s, namely,

$$r_s = \frac{12 \sum_{i=1}^{n} R_i S_i}{n(n^2 - 1)} - 3\left(\frac{n+1}{n-1}\right). \tag{8.80}$$

Thus, r_s is a linear function of the statistic $\sum_{i=1}^{n} R_i S_i$. Therefore, the various tests of independence discussed in this section can be as easily based on $\sum_{i=1}^{n} R_i S_i$ as on the more complicated formula for r_s given in (8.63) (or its counterpart in (8.64)).

45. *Large-Sample Approximation.* Under the null hypothesis H_0 (8.1), the rank vectors $(R_1, \ldots, R_n)$ and $(S_1, \ldots, S_n)$ are independent and each is uniformly distributed over the set of $n!$ permutations of $(1, 2, \ldots, n)$. It follows that the random variables $\sum_{i=1}^{n} R_i S_i$ and $\sum_{j=1}^{n} j S_j$ have the same null distribution. Combining this fact with the representation for r_s given in equation (8.80), it follows that

$$E_0(r_s) = E_0\left[\frac{12 \sum_{j=1}^{n} j S_j}{n(n^2 - 1)} - 3\left(\frac{n+1}{n-1}\right)\right]$$

$$= \frac{12 \sum_{j=1}^{n} j E_0(S_j)}{n(n^2 - 1)} - 3\left(\frac{n+1}{n-1}\right).$$

Since each S_j, $j = 1, \ldots, n$, has a probability distribution that is uniform over the set of integers $\{1, 2, \ldots, n\}$, it follows that $E_0(S_j) = \sum_{k=1}^{n} k(1/n) = (n+1)/2$, for $j = 1, \ldots, n$. Thus, we have that

$$E_0(r_s) = \frac{12 \sum_{j=1}^{n} j \left(\frac{n+1}{2}\right)}{n(n^2 - 1)} - 3\left(\frac{n+1}{n-1}\right)$$

$$= \frac{12 \frac{n(n+1)}{2}\left(\frac{n+1}{2}\right)}{n(n^2 - 1)} - 3\left(\frac{n+1}{n-1}\right) = 0,$$

as previously noted in equation (8.71). For the null variance of r_s, we first note that

$$\text{var}_0(r_s) = \text{var}_0\left[\frac{12\sum_{j=1}^n jS_j}{n(n^2-1)} - 3\left(\frac{n+1}{n-1}\right)\right]$$

$$= \frac{144}{n^2(n^2-1)^2}\text{var}_0\left(\sum_{j=1}^n jS_j\right). \tag{8.81}$$

Using a well-known expression for the variance of a sum of random variables, we have that

$$\text{var}_0\left(\sum_{j=1}^n jS_j\right) = \sum_{j=1}^n \text{var}_0(jS_j) + \sum_{j=1}^n\sum_{k=1, k\neq j}^n \text{cov}_0(jS_j, kS_k)$$

$$= \sum_{j=1}^n j^2 \text{var}_0(S_j) + \sum_{j=1}^n\sum_{k=1, k\neq j}^n jk\,\text{cov}_0(S_j, S_k).$$

Since the joint distribution of (S_j, S_k) is the same for every $j \neq k = 1, \ldots, n$ and the marginal distribution of S_j is the same for each $j = 1, \ldots, n$, it follows that

$$\text{var}_0\left(\sum_{j=1}^n jS_j\right) = \text{var}_0(S_1)\sum_{j=1}^n j^2 + \text{cov}_0(S_1, S_2)\sum_{j=1}^n\sum_{k=1, k\neq j}^n jk.$$

Using the facts that

$$\sum_{j=1}^n j^2 = \frac{n(n+1)(2n+1)}{6}$$

and

$$\sum_{j=1}^n\sum_{k=1, k\neq j}^n jk = \left(\sum_{j=1}^n j\right)\left(\sum_{k=1}^n k\right) - \sum_{j=1}^n j^2$$

$$= \left[\frac{n(n+1)}{2}\right]^2 - \frac{n(n+1)(2n+1)}{6}$$

$$= \frac{n(n^2-1)(3n+2)}{12},$$

we obtain

$$\text{var}_0\left(\sum_{j=1}^n jS_j\right) = \left[\frac{n(n+1)(2n+1)}{6}\text{var}_0(S_1) + \frac{n(n^2-1)(3n+2)}{12}\text{cov}_0(S_1, S_2)\right].$$

Moreover, under H_0 (8.1), it can be shown (see Problems 53 and 54) that $\text{var}_0(S_1) = (n^2-1)/12$ and $\text{cov}_0(S_1, S_2) = -(n+1)/12$. Thus, we have

$$\text{var}_0\left(\sum_{j=1}^{n} jS_j\right) = \left[\frac{n(n+1)(2n+1)(n^2-1)}{72}\right.$$

$$\left. - \frac{n(n^2-1)(3n+2)(n+1)}{144}\right]$$

$$= \frac{n(n+1)(n^2-1)}{144}[2(2n+1)-(3n+2)]$$

$$= \frac{n^2(n+1)(n^2-1)}{144}. \tag{8.82}$$

Combining equations (8.81) and (8.82) yields

$$\text{var}_0(r_s) = \frac{144}{n^2(n^2-1)^2}\left[\frac{n^2(n+1)(n^2-1)}{144}\right] = \frac{1}{n-1},$$

as noted in equation (8.72).

The asymptotic normality under H_0 of the standardized form

$$r_s^* = \frac{r_s - E_0(r_s)}{\{\text{var}_0(r_s)\}^{1/2}} = (n-1)^{1/2} r_s$$

follows from the fact that r_s has the same null distribution as

$$\frac{12\sum_{j=1}^{n} jS_j}{n(n^2-1)} - 3\left(\frac{n+1}{n-1}\right)$$

and standard techniques for establishing the asymptotic normality of a linear combination $\left(\sum_{j=1}^{n} jS_j\right)$ of random variables. (For additional details, see Sections 8.4 and 12.3 in Randles and Wolfe (1979).)

46. *Exact Conditional Null Distribution of r_s with Ties among the X- and/or Y-Values.* To have a test with exact significance level even in the presence of tied X and/or Y observations, we must consider all the possible values of r_s corresponding to the fixed observed rank vector $(R_1,\ldots,R_n) = (r_1,\ldots,r_n)$ and every one of the $n!$ permutations of the observed rank vector $(S_1,\ldots,S_n) = (s_1,\ldots,s_n)$, where average ranks have been used to break ties in both of the rank vectors. As in Comment 41, it still follows that under H_0 each of the $n!$ possible outcomes for the ordered configurations $(s_1,\ldots,s_n)$, in conjunction with a fixed value of $(r_1,\ldots,r_n)$, both based on using average ranks to break ties, occurs with probability $1/n!$. For each such $(s_1,\ldots,s_n)$ configuration and fixed $(r_1,\ldots,r_n)$, the value of r_s is computed and the results tabulated. We illustrate this construction for $n = 4$ and the data $(X_1,Y_1) = (2,3.1), (X_2,Y_2) = (3.9,4)$, $(X_3,Y_3) = (2,5.1)$, and $(X_4,Y_4) = (3.6,4)$. Using average ranks to break ties, the associated X and Y rank vectors are $(R_1,R_2,R_3,R_4) = (1.5,4,1.5,3)$ and $(S_1,S_2,S_3,S_4) = (1,2.5,4,2.5)$, respectively. Thus, we have $D_1 = -.5, D_2 = -1.5, D_3 = 2.5, D_4 = -.5$ and an obtained value of $r_s = .1$. To assess the significance of r_s, we obtain its conditional distribution by considering the $4! = 24$ equally likely (under H_0) possible values of r_s for the fixed rank vector $(r_1,r_2,r_3,r_4) = (1.5,4,1.5,3)$ in conjunction with each of the 24 permutations of the rank vector $(s_1,s_2,s_3,s_4) = (1,2.5,4,2.5)$. These 24 permutations of $(1,2.5,4,2.5)$ and associated values of r_s are as follows:

(s_1, s_2, s_3, s_4)	Probability under H_0	Value of r_s
(1, 2.5, 4, 2.5)	$\frac{1}{24}$	.1
(1, 2.5, 2.5, 4)	$\frac{1}{24}$	.55
(1, 2.5, 2.5, 4)	$\frac{1}{24}$	.55
(2.5, 1, 2.5, 4)	$\frac{1}{24}$	−.2
(1, 4, 2.5, 2.5)	$\frac{1}{24}$	.85
(1, 4, 2.5, 2.5)	$\frac{1}{24}$	.85
(1, 2.5, 4, 2.5)	$\frac{1}{24}$	.1
(2.5, 1, 4, 2.5)	$\frac{1}{24}$	−.65
(4, 1, 2.5, 2.5)	$\frac{1}{24}$	−.65
(4, 1, 2.5, 2.5)	$\frac{1}{24}$	−.65
(4, 2.5, 1, 2.5)	$\frac{1}{24}$	.1
(2.5, 4, 1, 2.5)	$\frac{1}{24}$	.85
(4, 2.5, 1, 2.5)	$\frac{1}{24}$	.1
(4, 2.5, 2.5, 1)	$\frac{1}{24}$	−.35
(4, 2.5, 2.5, 1)	$\frac{1}{24}$	−.35
(2.5, 4, 2.5, 1)	$\frac{1}{24}$	.4
(2.5, 1, 4, 2.5)	$\frac{1}{24}$	−.65
(2.5, 1, 2.5, 4)	$\frac{1}{24}$	−.2
(2.5, 2.5, 1, 4)	$\frac{1}{24}$	.55
(2.5, 2.5, 1, 4)	$\frac{1}{24}$	.55
(2.5, 4, 1, 2.5)	$\frac{1}{24}$	.85
(2.5, 4, 2.5, 1)	$\frac{1}{24}$	.4
(2.5, 2.5, 4, 1)	$\frac{1}{24}$	−.35
(2.5, 2.5, 4, 1)	$\frac{1}{24}$	−.35

This yields the null-tail probabilities

$$P_0(r_s \geq .85) = \tfrac{4}{24} \quad P_0(r_s \geq -.2) = \tfrac{16}{24}$$
$$P_0(r_s \geq .55) = \tfrac{8}{24} \quad P_0(r_s \geq -.35) = \tfrac{20}{24}$$
$$P_0(r_s \geq .4) = \tfrac{10}{24} \quad P_0(r_s \geq -.65) = 1.$$
$$P_0(r_s \geq .1) = \tfrac{14}{24}$$

This distribution is called the conditional null distribution or the permutation null distribution of r_s, given the observed sets of tied ranks $(r_1, r_2, r_3, r_4) = (1.5, 4, 1.5, 3)$ and $(s_1, s_2, s_3, s_4) = (1, 2.5, 4, 2.5)$. For the particular observed value $r_s = .1$, we have $P_0(r_s \geq .1) = \tfrac{14}{24}$, so that such a value does not indicate a deviation from H_0 in the direction of positive association between the X and Y variables. (We note that *both* the null expected value and null variance for r_s are different in this case of tied ranks (see Problem 51) than the corresponding expressions given in equations (8.71) and (8.72) for the no ties setting.)

47. *Point Estimation and Confidence Intervals Associated with r_s.* The Kendall statistic K (8.6) is directly associated with the population correlation coefficient τ (8.2). This leads naturally to point estimators and approximate confidence intervals for τ based on K. Such is not the case for the Spearman statistic r_s (8.63). The measure of association linked with the independence tests based on r_s is

$$\eta = \frac{3[\tau + (n-2)\phi]}{n+1},$$

where τ is given by equation (8.2) and

$$\phi = 2P\{(Y_3 - Y_1)(X_2 - X_1) > 0\} - 1.$$

This measure of association η has several undesirable properties, including the facts that it is dependent on the sample size n and it is asymmetric in the X and Y labels. (For more discussion along these lines, see Fligner and Rust (1983).) As a result, point estimators and confidence intervals for η based on r_s are of little practical interest.

48. *Trend Test.* If we take $X_i = i, i = 1, \ldots, n$, and compute r_s, then the procedures based on r_s can be used as tests for a time trend in the univariate random sample $Y_1, \ldots, Y_n$.

49. *Other Uses for the r_s Statistic.* Spearman's rank correlation coefficient r_s also finds use in other settings where association is a primary issue. One such instance is in connection with Page's test for ordered alternatives in a two-way layout (see Section 7.2). Page's L statistic (7.10) is directly related to r_s. For more details, see Comment 7.22.

Properties

1. *Asymptotic Normality.* See Randles and Wolfe (1979, 405–407).
2. *Efficiency.* See Section 8.7.

PROBLEMS

40. In order to study the effects of pharmaceutical and chemical agents on mucociliary clearance, doctors often use the ciliary beat frequency (CBF) as an index of ciliary activity. One accepted way to measure CBF in a subject is through the collection and analysis of an endobronchial forceps biopsy specimen. However, this technique is a rather invasive method for measuring CBF. In a study designed to assess the effectiveness of less invasive procedures for measuring CBF, Low et al. (1984) considered the alternative technique of nasal brushing. The data in Table 8.10 are a subset of the data collected by Low et al. during their investigation.

    The subjects in the study were all men undergoing bronchoscopies for diagnoses of a variety of pulmonary problems. The CBF values reported in Table 8.10 are averages of 10 consecutive measurements on each subject.

    Test the hypothesis of independence versus the alternative that the CBF measurements via nasal brushing and endobronchial forceps biopsy are positively associated (and, therefore, that nasal brushing is an acceptable alternative to the more invasive endobronchial forceps biopsy technique for measuring CBF).

41. Test the hypothesis of independence versus the alternative that the mean weight of introduced cysticerci is positively correlated with the mean weight of worms recovered for the tapeworm data in Table 8.3.

42. Test the hypothesis of independence versus the alternative that spending per high school senior and percentage seniors graduating are positively correlated for the secondary education data in Table 8.6.

TABLE 8.10. Relation between Ciliary Beat Frequency (CBF) Values Obtained through Nasal Brushing and Endobronchial Forceps Biopsy

	CBF (hertz)	
Subject	Nasal Brushing	Endobronchial Forceps Biopsy
1	15.4	16.5
2	13.5	13.2
3	13.3	13.6
4	12.4	13.6
5	12.8	14.0
6	13.5	14.0
7	14.5	16.0
8	13.9	14.1
9	11.0	11.5
10	15.0	14.4
11	17.0	16.0
12	13.8	13.2
13	17.4	16.6
14	16.5	18.5
15	14.4	14.5

Source: P. P. Low, C. K. Luk, M. J. Dulfano, and P. J. P. Finch (1984).

43. Show that the two expressions for r_s in displays (8.63) and (8.64) are equivalent.
44. For arbitrary number of observations, what are the smallest and largest possible values of r_s? Justify your answers.
45. Suppose $n = 5$ and we observe the data $(X_1, Y_1) = (3.7, 9.2)$, $(X_2, Y_2) = (4.3, 9.4)$, $(X_3, Y_3) = (5.0, 9.2)$, $(X_4, Y_4) = (6.2, 10.4)$, and $(X_5, Y_5) = (5.3, 9.2)$. What is the conditional probability distribution of r_s under H_0 (8.1) when average ranks are used to break ties among the Y's? How extreme is the observed value of r_s in this conditional null distribution? Compare this fact with that obtained by taking the observed value of r_s to the (incorrect) unconditional null distribution of r_s given in Table A.31. (See also Problem 48.)
46. Give an example of a data set of $n \geq 10$ bivariate observations for which r_s has value 0.
47. Suppose $n = 25$. Compare the critical region for the level $\alpha = .05$ test of H_0 (8.1) versus H_2 (8.67) based on r_s and Table A.31 with the critical region for the corresponding nominal level $\alpha = .05$ test based on the large-sample approximation.
48. For the case of $n = 5$ untied bivariate (X, Y) observations, obtain the form of the exact null (H_0) distribution of r_s. (See Comment 41.)
49. Let r_p be the Pearson product moment correlation coefficient defined in equation (8.78). Show that r_s (8.63) is simply this Pearson product moment correlation coefficient applied to the rank vectors $(R_1, \ldots, R_n)$ and $(S_1, \ldots, S_n)$ instead of the original $(X_1, \ldots, X_n)$ and $(Y_1, \ldots, Y_n)$ vectors.
50. Use the computer software Minitab to obtain the value of r_s for the secondary education data in Table 8.6, using average ranks to break the ties in the X and Y values. (See Comment 39, noting that this value of r_s obtained from Minitab does not equal the ties-corrected version of r_s given in equation (8.77).)
51. Obtain the values of $E_0(r_s)$ and $\text{var}_0(r_s)$ corresponding to the exact conditional null distribution of r_s for the case of $n = 5$ and the tied data considered in Comment 46. Compare these values with the corresponding values for $E_0(r_s)$ and $\text{var}_0(r_s)$ given in expressions (8.71) and (8.72) for the no ties setting. Discuss a possible reason for the difference in these null variances.
52. Use the Lake Michigan pollution data in Table 8.8 to test the hypothesis that the degree of pollution (as measured by the number of odor periods) had not changed with time against the alternative that

8.5. A DISTRIBUTION-FREE TEST FOR INDEPENDENCE BASED ON RANKS (SPEARMAN)

there was a general increasing trend in the pollution of Lake Michigan over the period of 1950–1964. (See Comment 48.)

53. Let $(S_1, \ldots, S_n)$ be a vector of ranks that is uniformly distributed over the set of all $n!$ permutations of $(1, 2, \ldots, n)$. Show that the marginal probability distribution of each S_i, for $i = 1, \ldots, n$, is uniform over the set $\{1, 2, \ldots, n\}$. Use this fact to show that $E(S_i) = (n + 1)/2$ and $\text{var}(S_i) = (n^2 - 1)/12$, for $i = 1, \ldots, n$.

54. Let $(S_1, \ldots, S_n)$ be a vector of ranks that is uniformly distributed over the set of all $n!$ permutations of $(1, 2, \ldots, n)$. Show that the joint marginal probability distribution of (S_i, S_j), for $i \neq j = 1, \ldots, n$, is given by

$$P(S_i = s, S_j = t) = \begin{cases} \dfrac{1}{n(n-1)}, & s \neq t = 1, \ldots, n \\ 0, & \text{otherwise.} \end{cases}$$

Use this fact to show that $\text{cov}(S_i, S_j) = -(n+1)/12$, for $i \neq j = 1, \ldots, n$.

55. The data in Table 8.11 were considered by Gentry and Pike (1970) in their study of the relationship between the mean rate of return over the period 1956 through 1969 and the 1969 value of common stock portfolios for 32 life insurance companies.

TABLE 8.11. Mean Rate of Return of Common Stock Portfolios over the Period 1956–1969 and the 1969 Value of Each Equity Portfolio for 32 Life Insurance Companies

Company	Mean rate (%) of Return, 1956–1969	Value of Common Stock Portfolio, December 31, 1969 (Millions of Dollars)
1	18.83	96.0
2	16.98	54.6
3	15.36	84.4
4	14.65	251.5
5	14.21	131.8
6	13.68	37.3
7	13.65	109.9
8	13.07	13.5
9	12.99	76.3
10	12.81	72.6
11	11.60	42.1
12	11.51	41.5
13	11.50	56.2
14	11.41	59.3
15	11.26	1,184.0
16	10.67	144.0
17	10.44	111.9
18	10.44	179.8
19	10.33	29.2
20	10.30	279.5
21	10.22	166.6
22	10.05	194.3
23	10.04	40.8
24	9.57	428.4
25	9.50	7.0
26	9.48	485.6
27	9.29	165.3
28	9.21	343.8
29	9.04	35.4
30	8.82	24.7
31	8.78	2.7
32	7.26	8.9

Source: J. Gentry and J. Pike (1970).

Test the hypothesis of independence versus the general alternative that the 1956–1969 mean rate of return for a stock portfolio is correlated in some fashion with its 1969 value.

8.6. A DISTRIBUTION-FREE TEST FOR INDEPENDENCE AGAINST BROAD ALTERNATIVES (HOEFFDING)

Hoeffding (1948b) proposed a test of independence that is able to detect a much broader class of alternatives to independence than the classes of alternatives that can be detected by the tests of Sections 8.1 and 8.5 based on sample correlation coefficients.

Procedure

To test the hypothesis that the X and Y random variables are independent, namely, H_0 given by (8.1), we first rank $X_1, \ldots, X_n$ jointly and let R_i denote the rank of X_i in this joint ranking, $i = 1, \ldots, n$. Then rank $Y_1, \ldots, Y_n$ jointly, and let S_i denote the rank of Y_i in this joint ranking, $i = 1, \ldots, n$. We let c_i denote the number of sample pairs (X_α, Y_α) for which both $X_\alpha < X_i$ and $Y_\alpha < Y_i$; that is,

$$c_i = \sum_{\alpha=1}^{n} \phi(X_\alpha, X_i)\phi(Y_\alpha, Y_i), \quad i = 1, \ldots, n, \tag{8.83}$$

where $\phi(a, b) = 1$ if $a < b$, $= 0$, otherwise.
  We set

$$Q = \sum_{i=1}^{n} (R_i - 1)(R_i - 2)(S_i - 1)(S_i - 2), \tag{8.84}$$

$$R = \sum_{i=1}^{n} (R_i - 2)(S_i - 2)c_i, \tag{8.85}$$

and

$$S = \sum_{i=1}^{n} c_i(c_i - 1), \tag{8.86}$$

and compute

$$D = \frac{Q - 2(n-2)R + (n-2)(n-3)S}{n(n-1)(n-2)(n-3)(n-4)}. \tag{8.87}$$

For a (two-sided) test of H_0 versus the alternative that X and Y are dependent (see Comment 52), at the α level of significance,

$$\text{Reject } H_0 \text{ if } D \geq d_\alpha; \quad \text{otherwise do not reject,} \tag{8.88}$$

where the constant d_α satisfies the equation $P_0(D \geq d_\alpha) = \alpha$. Values of d_α can be obtained from Table A.32.

Large-Sample Approximation

For the large-sample approximation, we use a statistic B, proposed by Blum, Kiefer, and Rosenblatt (1961), that is slightly different than Hoeffding's D statistic. (The tests based on B and D are, however, asymptotically equivalent, since the statistics $nD + (\frac{1}{36})$ and nB have the same asymptotic distribution under H_0. See Comment 53.) Let

$$B = n^{-5} \sum_{i=1}^{n} [N_1(i)N_4(i) - N_2(i)N_3(i)]^2, \tag{8.89}$$

where

$N_1(i) = $ number of sample pairs (X_α, Y_α) lying in the region
$\qquad T_1(i) = \{(x, y) : x \leq X_i \text{ and } y \leq Y_i\}$,

$N_2(i) = $ number of sample pairs (X_α, Y_α) lying in the region
$\qquad T_2(i) = \{(x, y) : x > X_i \text{ and } y \leq Y_i\}$,

$N_3(i) = $ number of sample pairs (X_α, Y_α) lying in the region
$\qquad T_3(i) = \{(x, y) : x \leq X_i \text{ and } y > Y_i\}$,

$N_4(i) = $ number of sample pairs (X_α, Y_α) lying in the region
$\qquad T_4(i) = \{(x, y) : x > X_i \text{ and } y > Y_i\}$. $\tag{8.90}$

That is, for each i, determine the number of sample pairs (X_α, Y_α) lying in each of the regions determined by the horizontal and vertical lines through the point (X_i, Y_i).

Table A.33 gives values of the upper α percentiles b_α of the asymptotic (n tending to infinity) null distribution of $\frac{1}{2}\pi^4 nB$. A large-sample approximation to procedure (8.88) is

$$\text{Reject } H_0 \text{ if } \tfrac{1}{2}\pi^4 nB \geq b_\alpha; \quad \text{otherwise do not reject.} \tag{8.91}$$

Blum, Kiefer, and Rosenblatt suggested that when n is small, the error introduced when utilizing the large-sample approximation may be reduced by substituting $(n-1)B$ for nB in the left-hand side of (8.91). For a different large-sample approximation, see Comment 53.

Ties

Use average ranks and replace (8.83) by

$$c_i = \sum_{\substack{\alpha=1 \\ \alpha \neq i}}^{n} \phi^*(X_\alpha, X_i)\phi^*(Y_\alpha, Y_i), \quad i = 1, \ldots, n, \tag{8.92}$$

where

$$\phi^*(a, b) = \begin{cases} 1, & \text{if } a < b, \\ \tfrac{1}{2}, & \text{if } a = b, \\ 0, & \text{otherwise.} \end{cases} \tag{8.93}$$

Example 8.6: *Continuation of Example 8.5.* We return to the data of Table 8.9 and consider the relation between the free pool of proline and collagen content in human liver cirrosis. We apply Hoeffding's test of independence. From (8.92) we find

$$\begin{aligned}
c_1 &= \phi^*(X_2,X_1)\phi^*(Y_2,Y_1) + \phi^*(X_3,X_1)\phi^*(Y_3,Y_1) + \phi^*(X_4,X_1)\phi^*(Y_4,Y_1) \\
&\quad + \phi^*(X_5,X_1)\phi^*(Y_5,Y_1) + \phi^*(X_6,X_1)\phi^*(Y_6,Y_1) + \phi^*(X_7,X_1)\phi^*(Y_7,Y_1) \\
&= \tfrac{1}{2}(0) + 0\left(\tfrac{1}{2}\right) + 0(1) + 0(0) + 0(0) + 0(0) = 0, \\
c_2 &= \phi^*(X_1,X_2)\phi^*(Y_1,Y_2) + \phi^*(X_3,X_2)\phi^*(Y_3,Y_2) + \phi^*(X_4,X_2)\phi^*(Y_4,Y_2) \\
&\quad + \phi^*(X_5,X_2)\phi^*(Y_5,Y_2) + \phi^*(X_6,X_2)\phi^*(Y_6,Y_2) + \phi^*(X_7,X_2)\phi^*(Y_7,Y_2) \\
&= \tfrac{1}{2}(1) + 0(1) + 0(1) + 0(0) + 0(0) + 0(0) = \tfrac{1}{2}, \\
c_3 &= \phi^*(X_1,X_3)\phi^*(Y_1,Y_3) + \phi^*(X_2,X_3)\phi^*(Y_2,Y_3) + \phi^*(X_4,X_3)\phi^*(Y_4,Y_3) \\
&\quad + \phi^*(X_5,X_3)\phi^*(Y_5,Y_3) + \phi^*(X_6,X_3)\phi^*(Y_6,Y_3) + \phi^*(X_7,X_3)\phi^*(Y_7,Y_3) \\
&= 1\left(\tfrac{1}{2}\right) + 1(0) + 0(1) + 0(0) + 0(0) + 0(0) = \tfrac{1}{2}, \\
c_4 &= \phi^*(X_1,X_4)\phi^*(Y_1,Y_4) + \phi^*(X_2,X_4)\phi^*(Y_2,Y_4) + \phi^*(X_3,X_4)\phi^*(Y_3,Y_4) \\
&\quad + \phi^*(X_5,X_4)\phi^*(Y_5,Y_4) + \phi^*(X_6,X_4)\phi^*(Y_6,Y_4) + \phi^*(X_7,X_4)\phi^*(Y_7,Y_4) \\
&= 1(0) + 1(0) + 1(0) + 0(0) + 0(0) + 0(0) = 0, \\
c_5 &= \phi^*(X_1,X_5)\phi^*(Y_1,Y_5) + \phi^*(X_2,X_5)\phi^*(Y_2,Y_5) + \phi^*(X_3,X_5)\phi^*(Y_3,Y_5) \\
&\quad + \phi^*(X_4,X_5)\phi^*(Y_4,Y_5) + \phi^*(X_6,X_5)\phi^*(Y_6,Y_5) + \phi^*(X_7,X_5)\phi^*(Y_7,Y_5) \\
&= 1(1) + 1(1) + 1(1) + 1(1) + 0(0) + 0(0) = 4, \\
c_6 &= \phi^*(X_1,X_6)\phi^*(Y_1,Y_6) + \phi^*(X_2,X_6)\phi^*(Y_2,Y_6) + \phi^*(X_3,X_6)\phi^*(Y_3,Y_6) \\
&\quad + \phi^*(X_4,X_6)\phi^*(Y_4,Y_6) + \phi^*(X_5,X_6)\phi^*(Y_5,Y_6) + \phi^*(X_7,X_6)\phi^*(Y_7,Y_6) \\
&= 1(1) + 1(1) + 1(1) + 1(1) + 1(1) + 0(0) = 5, \\
c_7 &= \phi^*(X_1,X_7)\phi^*(Y_1,Y_7) + \phi^*(X_2,X_7)\phi^*(Y_2,Y_7) + \phi^*(X_3,X_7)\phi^*(Y_3,Y_7) \\
&\quad + \phi^*(X_4,X_7)\phi^*(Y_4,Y_7) + \phi^*(X_5,X_7)\phi^*(Y_5,Y_7) + \phi^*(X_6,X_7)\phi^*(Y_6,Y_7) \\
&= 1(1) + 1(1) + 1(1) + 1(1) + 1(1) + 1(1) = 6.
\end{aligned}$$

We next compute the values of Q, R, S, and D. Using (8.84) and (8.87) and the R_i's and S_i's found in Example 8.5, we obtain

$$\begin{aligned}
Q &= .5(-.5)(1.5)(.5) + .5(-.5)(3)(2) + 2(1)(1.5)(.5) \\
&\quad + 3(2)(0)(-1) + 4(3)(4)(3) + 5(4)(5)(4) + 6(5)(6)(5) \\
&= 1443.81, \\
R &= -.5(.5)(0) + (-.5)(2)\left(\tfrac{1}{2}\right) + 1(.5)\left(\tfrac{1}{2}\right) \\
&\quad + 2(-1)(0) + 3(3)(4) + 4(4)(5) + 5(5)(6) \\
&= 265.75,
\end{aligned}$$

8.6. A DISTRIBUTION-FREE TEST FOR INDEPENDENCE AGAINST BROAD ALTERNATIVES

$$S = 0(-1) + \tfrac{1}{2}\left(-\tfrac{1}{2}\right) + \tfrac{1}{2}\left(-\tfrac{1}{2}\right) + 0(-1) + 4(3) + 5(4) + 6(5)$$
$$= 61.5$$

and

$$D = \frac{1443.81 - 2(5)(265.75) + 5(4)(61.5)}{7(6)(5)(4)(3)}$$
$$= \frac{16.31}{2520}.$$

Entering Table A.32 at $n = 7$, we find $P_0\{1260 \cdot D \geq 8\} = .0905$. That is, $d_{.0905} = \frac{8}{1260}$. For the data we are analyzing, $1260 \cdot D = 8.16$, and hence we reject the hypothesis of independence at the (two-sided) level of .0905. (The test is only approximate in this case because the ties in the data invalidate the assumption that the underlying bivariate population is continuous.)

We now use these data to illustrate the computations needed to perform the large-sample approximation. For example, for the pair $(X_4, Y_4) = (8.3, 2.6)$, dividing the plane into the four regions defined by (8.90) and counting the number of sample pairs in these regions yields the $N_1(4), N_2(4), N_3(4), N_4(4)$ values defined by (8.90), namely,

$$N_1(4) = 1, \quad N_2(4) = 0, \quad N_3(4) = 3, \quad N_4(4) = 3.$$

Performing similar subdivisions and counts corresponding to the other six sample pairs, we find

$$N_1(1) = 1, \quad N_2(1) = 2, \quad N_3(1) = 1, \quad N_4(1) = 3,$$
$$N_1(2) = 2, \quad N_2(2) = 2, \quad N_3(2) = 0, \quad N_4(2) = 3,$$
$$N_1(3) = 3, \quad N_2(3) = 1, \quad N_3(3) = 1, \quad N_4(3) = 3,$$
$$N_1(5) = 5, \quad N_2(5) = 0, \quad N_3(5) = 0, \quad N_4(5) = 2,$$
$$N_1(6) = 6, \quad N_2(6) = 0, \quad N_3(6) = 0, \quad N_4(6) = 1,$$
$$N_1(7) = 7, \quad N_2(7) = 0, \quad N_3(7) = 0, \quad N_4(7) = 0.$$

From (8.89) we then obtain

$$B = (7)^{-5}\{[3-2]^2 + [6-0]^2 + [6-1]^2 + [3-0]^2 + [10-0]^2 + [6-0]^2 + [0-0]^2\}$$
$$= 7^{-5}(207).$$

Since $n = 7$ is relatively small, we calculate the left-hand side of (8.91) with $(7-1)B$ replacing $7B$. We find

$$\tfrac{1}{2}\pi^4(n-1)B = \tfrac{1}{2}(3.14)^4(6)(207)(7)^{-5} = 3.60.$$

From Table A.33, we find $b_{.0205} = 3.60$. Hence the lowest level at which we reject H_0 using the large-sample approximation is .0205. That it is not in close agreement with the small-sample procedure based on D may be attributed to a combination of factors, including the relatively small sample size, the (small-sample) differences between the tests based on B and D, and the presence of ties.

Recall that in Section 8.5 we applied Spearman's test to the data of Table 8.9 and found the one-sided P-value to be between .05 and .10. Thus the two-sided P-value for the test is between .10 and .20, and this shows some agreement with the exact two-sided P-value of .0905 based on D.

Comments

50. *Motivation for Hoeffding's Test.* Define

$$D^*(x, y) = P(X \leq x \text{ and } Y \leq y) - P(X \leq x)P(Y \leq y). \tag{8.94}$$

We note that $D^*(x, y) = 0$ for all (x, y) if and only if H_0 is true. This fact was used by Hoeffding in devising the test based on D. The statistic D estimates the parameter

$$\Delta_1(F) = E_F\{D^*(X', Y')\}^2, \tag{8.95}$$

where (X', Y') is a random member from the underlying bivariate population with distribution F. In other words, we may think of $D^*(x, y)$ as a measure of the deviation from H_0 at the point (x, y), and $\Delta_1(F)$ as the average value of the square of this deviation.

51. *Null Distribution of D.* In determining the null distribution of D, we can, without loss of generality, take $R_i = i$ and obtain the associated values of D for the $n!$ possible Y rank configurations of the form $(S_1, \ldots, S_n)$. Each of these configurations has probability $[1/(n!)]$ under H_0.

52. *Consistency of D against a Broad Class of Alternatives.* The D test was designed by Hoeffding to detect a broad class of alternatives to the hypothesis of independence, and in this sense its character differs from that of the tests of independence of Sections 8.1 and 8.5 based on sample correlation coefficients. Although Hoeffding (1948b) showed that the D test is not sensitive to all alternatives to H_0, he demonstrated that under mild restrictions on the nature of the underlying bivariate population F, the test is consistent when H_0 is false. Thus the D test detects alternatives where the X's and Y's are positively associated and alternatives where the X's and Y's are negatively associated. Furthermore, there exist populations F where X, Y are dependent and D is consistent, but the tests based on the sample correlation coefficients are not consistent.

53. *Relationship of D and B.* The statistics $nD + (\frac{1}{36})$ and nB have the same asymptotic distribution under H_0. (See Hoeffding (1948b) and Blum, Kiefer, and Rosenblatt (1961).) Thus another large-sample approximation to procedure (8.88) is

$$\text{Reject } H_0 \text{ if } \left(\tfrac{1}{2}\right) \pi^4 \left\{nD + \left(\tfrac{1}{36}\right)\right\} \geq b_\alpha; \quad \text{otherwise do not reject.}$$

where b_α is given by Table A.33.

54. *Development of D Test.* The test based on D was introduced by Hoeffding (1948b). The related test based on B was considered by Blum, Kiefer, and Rosenblatt (1961), who extended the approach to testing for the independence of k ($k \geq 2$) variables. A one-sided test, similar in character to the two-sided B test, was proposed by Crouse (1966). Skaug and Tjøstheim (1993) considered the Blum-Kiefer-Rosenblatt statistic in a time-series setting and established (under mild conditions) consistency against lag one dependent alternatives. Zheng (1997) used smoothing methods to develop a nonparametric test of independence between two variables. His test is consistent against any form of dependence.

Properties

1. *Consistency.* The test defined by (8.88) is consistent against populations for which the parameter $\Delta_1(F)$ defined by (8.95) is positive. For conditions on F that ensure that $\Delta_1(F)$ will be positive, see Hoeffding (1948b) and Yanagimoto (1970).
2. *Asymptotic Distribution.* For the asymptotic distribution of $\{nD + (\frac{1}{36})\}$ see Hoeffding (1948b) and Blum, Kiefer, and Rosenblatt (1961).

PROBLEMS

56. The data in Table 8.12 are a subset of the data obtained by S. Shen et al. (1970) in an experiment concerned with the hypothesis that diabetes mellitus is not simply a function of insulin deficiency and that perhaps insulin insensitivity could play an important role in the hyperglycemia of diabetes. One of the purposes of the study was to investigate the relation between the response to a glucose tolerance test and glucose impedance, a quantity describing the body tissues' resistance to glucose and expected to be constant for a given individual throughout the experimental range of glucose uptake rate in the author's study. The seven subjects represented in Table 8.12 were volunteers recently released from a minimum security prison and characterized by low plasma glucose response to oral glucose. Table 8.12 gives the weighted glucose response to an oral glucose tolerance test (X) and the glucose impedance reading (Y) for each of the seven subjects.

TABLE 8.12. Weighted Glucose Response and Glucose Impedance

Subject	Weighted Glucose Response, X	Glucose Impedance, Y
1	130	26.1
2	116	19.7
3	122	26.8
4	117	23.7
5	108	23.4
6	115	24.4
7	107	16.5

Source: S. Shen, G. M. Reaven, J. W. Farquhar, and R. H. Nakanishi (1970).

Use procedure (8.88) to test for impedance of weighted glucose response and glucose impedance. (Recall that procedure (8.88) is designed to detect all alternatives to the hypothesis of independence. However, if one has prior reasons or evidence to suspect that the weighted glucose response is positively correlated with glucose impedance, it would be more appropriate to focus on alternatives of positive association by using the one-sided procedure, based on Kendall's K, given by (8.8).)

57. Apply Kendall's two-sided test based on (8.10) to the data of Table 8.12. Compare your result with the result of Problem 56.
58. For the case $n = 9$, compare the exact $\alpha = .0407$ test based on D given by (8.88) to the approximate $\alpha = .04$ test based on the large-sample approximation given by (8.91).
59. Apply the large-sample approximation given in Comment 53 to the data in Table 8.9. Compare this approximation with the approximation based on (8.91) that was used in Example 8.6.

8.7. EFFICIENCIES OF INDEPENDENCE PROCEDURES

Investigation of the asymptotic relative efficiencies of tests for independence is made more difficult by our inability to define natural classes of alternatives to the hypothesis of independence. The asymptotic relative efficiencies of the test procedure (one- or two-sided) based on Kendall's statistic K (8.6) with respect to the corresponding normal theory test based on Pearson's product moment correlation coefficient r_p (8.78) have been found by Stuart (1954) and Konijn (1956) for a class of dependence alternatives "close" to the hypothesis of independence. Values of this asymptotic relative efficiency, $e(K, r_p)$, for selected bivariate $F_{X,Y}$ are as follows:

$F_{X,Y}$:	Normal	Uniform	Double Exponential
$e(K, r_p)$:	.912	1.000	1.266

In the normal setting, natural alternatives to independence correspond to bivariate normal distributions with nonzero correlation. In this case, the asymptotic relative efficiency of the test procedure (one- or two-sided) based on Spearman's statistic r_s (8.63) with respect to the corresponding test procedure based on Kendall's K is 1. Moreover, the common asymptotic relative efficiency of either the test procedure based on r_s or the test procedure based on K with respect to the corresponding normal theory test based on r_p is $(3/\pi)^2 = .912$.

The point estimator and confidence interval associated with normality assumptions for the independence problem are concerned with the underlying correlation coefficient, whereas the estimator and confidence intervals based on Kendall's K relate to the parameter τ. In view of this, the estimator $\hat{\tau}$ (8.34) and the approximate confidence intervals given by (8.39) and (8.50) are not easily compared with the normal theory procedures; hence, their asymptotic efficiencies are not presented here.

We do not know of any results for the asymptotic efficiency of Hoeffding's independence test (Section 8.6).

CHAPTER 9

Regression Problems

INTRODUCTION

Among the most common applications of statistical techniques are those involving some sort of regression analysis. Such procedures are designed to detect and interpret stochastic relationships between a dependent (response) variable and one or more independent (predictor) variables. These regression relationships can vary from that of a simple linear relationship between the dependent variable and a single independent variable to complex, nonlinear relationships involving a large number of predictor variables.

In Sections 9.1–9.4, we present nonparametric procedures designed for the simplest of regression relationships, namely, that of a single stochastic linear relationship between a dependent variable and one independent variable. (Such a relationship is commonly referred to as a regression line.) In Section 9.1 we present a distribution-free test of the hypothesis that the slope of the regression line is a specified value. Sections 9.2 and 9.3 provide, respectively, a point estimator and distribution-free confidence intervals and bounds for the slope parameter. In Section 9.4, we complete the analysis for a single regression line by discussing both an estimator of the intercept of the line and the use of the estimated linear relationship to provide predictions of dependent variable responses to additional values of the predictor variable. In Section 9.5, we consider the case of two or more regression lines and describe an asymptotically distribution-free test of the hypothesis that the regression lines have the same slope; that is, that the regression lines are parallel.

In Section 9.6, we present the reader with an introduction to the extensive field of rank-based regression analysis for more complicated regression relationships than that of a straight line. In Section 9.7, we provide short introductions to a number of recent developments in the rapidly expanding area of non-rank-based nonparametric regression, where the goal is to make statistical inferences about the relationship between a dependent variable and one or more independent variables without a priori specification of a formal model describing the regression relationship. These non-rank-based approaches to nonparametric regression are generally more complicated than the level assumed throughout the rest of this text. As a result, our approach in Section 9.7 is simply to give brief descriptions of a variety of statistical techniques that are commonly used to develop such procedures and provide appropriate references for readers interested in more detailed information about them, rather than to concentrate on specific procedures and their application to appropriate data sets.

Finally, in Section 9.8 we consider the asymptotic relative efficiencies of the straight line regression procedures discussed in Sections 9.1–9.3 and 9.5–9.6 with respect to their competitors based on classical least squares estimators.

ONE REGRESSION LINE

Data. At each of n fixed values, $x_1, \ldots, x_n$, of the independent (predictor) variable x, we observe the value of the response random variable Y. Thus we obtain a set of observations $Y_1, \ldots, Y_n$, where Y_i is the value of the response variable when $x = x_i$. The x's are assumed to be distinct and, without loss of generality, we take $x_1 < x_2 < \cdots < x_n$.

Assumptions

A1. Our straight line model is

$$Y_i = \alpha + \beta x_i + e_i, \quad i = 1, \ldots, n, \tag{9.1}$$

where the x's are known constants and α (the intercept) and β (the slope) are unknown parameters.

A2. The random variables $e_1, \ldots, e_n$ are a random sample from a continuous population that has median 0.

9.1. A DISTRIBUTION-FREE TEST FOR THE SLOPE OF THE REGRESSION LINE (THEIL)

Hypothesis

The null hypothesis of interest here is that the slope, β, of the postulated regression line is some specified value β_0, namely,

$$H_0 : \beta = \beta_0. \tag{9.2}$$

Thus, the null hypothesis asserts that for every unit increase in the value of the independent (predictor) variable x, we would expect an increase (or decrease, depending on the sign of β_0) of roughly β_0 in the value of the dependent (response) variable Y.

Procedure

To compute the Theil (1950a) statistic C, we first form the n differences

$$D_i = Y_i - \beta_0 x_i, \quad i = 1, \ldots, n. \tag{9.3}$$

Let

$$C = \sum_{i=1}^{n-1} \sum_{j=i+1}^{n} c(D_j - D_i), \tag{9.4}$$

where

$$c(a) = \begin{cases} -1, & \text{if } a < 0, \\ 0, & \text{if } a = 0, \\ 1, & \text{if } a > 0. \end{cases} \tag{9.5}$$

Thus, for each pair of subscripts (i, j), with $1 \leq i < j \leq n$, score 1 if $D_j - D_i$ is positive, and score -1 if $D_j - D_i$ is negative. The statistic C (9.4) is then just the sum of these 1s and -1s.

a. One-Sided Upper-Tail Test. To test the null hypothesis

$$H_0 : \beta = \beta_0$$

versus the alternative that the slope is larger than the specified β_0 corresponding to

$$H_1 : \beta > \beta_0, \qquad (9.6)$$

at the α level of significance,

$$\text{Reject } H_0 \text{ if } C \geq k_\alpha; \quad \text{otherwise do not reject,} \qquad (9.7)$$

where the constant k_α is chosen to make the type I error probability equal to α. Values of k_α are given in Table A.30. (See Comment 2.)

b. One-Sided Lower-Tail Test. To test the null hypothesis

$$H_0 : \beta = \beta_0$$

versus the alternative that the slope is smaller than the specified β_0 corresponding to

$$H_2 : \beta < \beta_0 \qquad (9.8)$$

at the α level of significance,

$$\text{Reject } H_0 \text{ if } C \leq -k_\alpha; \quad \text{otherwise do not reject.} \qquad (9.9)$$

c. Two-Sided Test. To test the null hypothesis

$$H_0 : \beta = \beta_0$$

versus the alternative that the slope is simply not equal to the specified β_0 corresponding to

$$H_3 : \beta \neq \beta_0 \qquad (9.10)$$

at the α level of significance,

$$\text{Reject } H_0 \text{ if } |C| \geq k_{\alpha/2}; \quad \text{otherwise do not reject.} \qquad (9.11)$$

This two-sided procedure is the two-sided symmetric test with $\alpha/2$ probability in each tail of the null distribution of C.

Large-Sample Approximation

The large-sample approximation is based on the asymptotic normality of C, suitably standardized. For this standardization, we need to know the expected value and variance of C when the null hypothesis H_0 (9.2) is true. Under H_0, the expected value and variance of C are

$$E_0(C) = 0 \qquad (9.12)$$

and

$$\text{var}_0(C) = \frac{n(n-1)(2n+5)}{18}, \tag{9.13}$$

respectively. (See Comment 2.)

The standardized version of C is

$$C^* = \frac{C - E_0(C)}{\{\text{var}_0(C)\}^{1/2}} = \frac{C}{\{n(n-1)(2n+5)/18\}^{1/2}}. \tag{9.14}$$

When H_0 is true, C^* has, as n tends to infinity, an asymptotic $N(0, 1)$ distribution (see Comment 2). The normal theory approximation for procedure (9.7) is

$$\text{Reject } H_0 \text{ if } C^* \geq z_\alpha; \quad \text{otherwise do not reject,} \tag{9.15}$$

the normal theory approximation for procedure (9.9) is

$$\text{Reject } H_0 \text{ if } C^* \leq -z_\alpha; \quad \text{otherwise do not reject,} \tag{9.16}$$

and the normal theory approximation for procedure (9.11) is

$$\text{Reject } H_0 \text{ if } |C^*| \geq z_{\alpha/2}; \quad \text{otherwise do not reject.} \tag{9.17}$$

Ties

If there are ties among the D_i (9.3) differences, C may be computed as described in (9.4), but keep in mind that procedures (9.7), (9.9), and (9.11) are then approximate rather than exact. Sen (1968) suggested a way to deal with ties among the values of the independent variable x.

Example 9.1: *Effect of Cloud Seeding on Rainfall.* Smith (1967) described experiments performed in Australia to investigate the effects of a particular method of cloud seeding on the amount of rainfall. In one experiment that took place in the Snowy Mountains, two areas served as target and control, respectively, and during any one period a random process was used to determine whether clouds over the target area should be seeded. The effect of seeding was measured by the *double ratio* $[T/Q \text{ (seeded)}]/[T/Q \text{ (unseeded)}]$, where T and Q are the total rainfalls in the target and control areas, respectively. Table 9.1 provides the double ratio calculated for each year of a 5-year experiment.

The slope parameter β represents the rate of change in Y per unit change in x. We apply the one-sided lower-tail test (9.9) with β_0 equal zero. This should be viewed as a test of the

TABLE 9.1. Double Ratio for 5 Years in the Snowy Mountains of Australia

Years Seeded, x_i	Double Ratio, Y_i
1	1.26
2	1.27
3	1.12
4	1.16
5	1.03

Source: E. J. Smith (1967).

9.1. A DISTRIBUTION-FREE TEST FOR THE SLOPE OF THE REGRESSION LINE

null hypothesis that the double ratio does not change with time (i.e., the effects of seeding during one year do not overlap into other years) against the alternative that there is a decrease over time, either in the rainfall increases resulting from the seeding or in the ability of the experiments to detect such increases.

From (9.3), with $\beta_0 = 0$, we see that $D_i = Y_i$. We now illustrate the computations required to obtain the value of C (9.4) for these data.

(i, j)	$D_j - D_i$	$c(D_j - D_i)$
(1, 2)	.01	1
(1, 3)	−.14	−1
(1, 4)	−.10	−1
(1, 5)	−.23	−1
(2, 3)	−.15	−1
(2, 4)	−.11	−1
(2, 5)	−.24	−1
(3, 4)	.04	1
(3, 5)	−.09	−1
(4, 5)	−.13	−1

Thus, we find the value of C to be

$$C = \sum_{i=1}^{4} \sum_{j=i+1}^{5} c(D_j - D_i) = -6.$$

Using the fact that the null distribution of C is symmetric about zero (see Comment 2), we find from Table A.30 (entering the table with $n = 5$ and $x = 6$) that the P-value for these data is $P_0(C \leq -6) = P_0(C \geq 6) = .117$. Thus, there is not much evidence of a decrease over time of the rainfall increases resulting from the seeding.

To illustrate the normal approximation (which should not be expected to be highly accurate for a sample size as small as 5), we first find from (9.14) that

$$C^* = \frac{-6}{\{5(4)(15)/18\}^{1/2}} = -1.47.$$

Thus the smallest significance level at which we can reject $H_0 : \beta = 0$ in favor of $\beta < 0$ using the normal approximation is .0708, since $z_{.0708} = -1.47$ in Table A.1. As expected for this small sample size ($n = 5$), this is not in especially good agreement with the exact P-value of .117 found previously.

Comments

1. *Motivation for the Test.* From (9.4) we see that C will be large when $D_j > D_i$ for many (i, j) pairs. Now

$$D_j - D_i = [Y_j - \beta_0 x_j - (Y_i - \beta_0 x_i)] = [Y_j - Y_i + \beta_o(x_i - x_j)].$$

Furthermore, under model (9.1), the median of $Y_j - Y_i = [\beta(x_j - x_i) + (e_j - e_i)]$ is $\beta(x_j - x_i)$. Thus, under model (9.1) the median of $D_j - D_i$ is $[\beta(x_j - x_i) + \beta_0(x_i - x_j)] = (\beta - \beta_0)(x_j - x_i)$.

Hence we tend to obtain positive $D_j - D_i$ differences when $\beta > \beta_0$, and these positive differences lead to large values of C. This serves as partial motivation for procedure (9.7).

2. *Relationship to Kendall's Correlation Statistic K.* The statistic C (9.4) is simply Kendall's correlation statistic K (8.6) computed between the x and $Y - \beta_0 x$ values. In particular, a test of $\beta_0 = 0$ can be interpreted as a test for correlation between the x and Y sequences. Moreover, the null H_0 (9.2) distribution properties of the statistic C (when there are no tied D values) are identical with the corresponding distributional properties of Kendall's statistic K under its null hypothesis of independence (see Section 8.1). This leads immediately to the use of Table A.30 for the critical values k_α in procedures (9.7), (9.9), and (9.11). In addition, the symmetry about zero for the null distribution of C follows from Comment 8.8 and the values of $E_0(C)$ and $\text{var}_0(C)$ are direct consequences of the corresponding values of $E_0(K)$ and $\text{var}_0(K)$, respectively, developed in Comment 8.10. Finally, the asymptotic $(n \to \infty)$ normality for the standardized statistic C^* under the null hypothesis H_0 (9.2) derives from the similar property for the standardized K^*, as discussed in Comment 8.10.

3. *Testing for Trends over Time.* In the special case when the x-values are the time order (as in Example 9.1), the procedures in (9.7), (9.9), and (9.11) (with β_0 set equal to zero) can be viewed as tests against a time trend and have been suggested for this use by Mann (1945). (See also Comment 8.14.)

Properties

1. *Consistency.* The tests defined by (9.7), (9.9), and (9.11) are consistent against the alternatives $\beta >, <,$ and $\neq \beta_0$, respectively.
2. *Asymptotic Normality.* See Comment 2.
3. *Efficiency.* See Sen (1968) and Section 9.8.

PROBLEMS

1. Johnson et al. (1970) considered the behavior of a cenosphere-resin composite under hydrostatic pressure. The authors pointed out that most deep submersible vehicles utilize a buoyancy material, known as syntactic foam, that is a composite of closely packed hollow glass microspheres embedded in a resin matrix. These microspheres are relatively expensive to manufacture, and the cost of the syntactic foam is principally determined by the cost of the microspheres. The authors also noted that the ash from generating stations burning pulverized coal contains a small proportion of hollow glassy microspheres, known as cenospheres, and these have about the right size distribution for use in syntactic foam. The cenospheres can be readily collected from the ash-disposal method used in certain British generating stations. The authors were thus interested in whether the cenospheres would, in some applications, perform as well as the manufactured microspheres.

    In attempting to assess the usefulness of cenospheres as a component of syntactic foam, Johnson et al. investigated the effects of hydrostatic pressure (such as exists in the ocean depths) on the density of a cenosphere-resin composite. The results are given in Table 9.2. What is the P-value for a test of $H_0: \beta = 0$ against the alternative $\beta > 0$ for these data?
2. Explain why the effect of the unknown intercept parameter α (see model (9.1)) is "eliminated" in the application of procedure (9.7) to a set of data.
3. Consider the tapeworm data discussed in Problem 8.1. Using the mean weight of the initial force fed cysticerci as the independent (predictor) variable, test the hypothesis that there was virtually no change in the mean weight of the cysticerci over the 20-day period following introduction into the dogs against the alternative that the typical tapeworm grew in size during the period of the study.
4. Stitt, Hardy, and Nadel (1971) studied the relationship between the surface area and body weight of squirrel monkeys. The data in Table 9.3 represent the total surface areas (in cm^3) and body weights (in

TABLE 9.2. The Effects of Hydrostatic Pressure on the Density of a Cenosphere-Resin Composite

Specimen	Pressure (psi)	Density (g/cm³)
1	0	0.924
2	5,000	0.988
3	10,000	0.992
4	15,000	1.118
5	20,000	1.133
6	25,000	1.145
7	30,000	1.157
8	100,000	1.357

Source: A. A. Johnson, K. Mukherjee, S. Schlosser, and E. Raask (1970).

g) for nine squirrel monkeys. Treating body weight as the independent variable, test for the presence of a linear relationship between these two measurements in squirrel monkeys.

5. Explain the meaning of the intercept parameter α and slope parameter β in model (9.1).
6. Consider the odor periods data of Table 8.8 in Problem 8.19. Test the conjecture that over the period 1950–1964 the number of odor periods for Lake Michigan generally increased at a rate greater than two per year.

TABLE 9.3. Body Weight and Total Surface Area of Squirrel Monkeys

Monkey	Body Weight (g)	Total Surface Area (cm³)
1	660	780.6
2	705	887.6
3	994	1,122.8
4	1,129	1,125.2
5	1,005	1,070.4
6	923	1,039.2
7	953	1,040.0
8	1,018	1,133.4
9	1,181	1,148.0

Source: J. T. Stitt, J. D. Hardy, and E. R. Nadel (1971).

9.2. A SLOPE ESTIMATOR ASSOCIATED WITH THE THEIL STATISTIC (THEIL)

Procedure

To estimate the slope parameter β of model (9.1), compute the $N = n(n-1)/2$ individual sample slope values $S_{ij} = (Y_j - Y_i)/(x_j - x_i)$, $1 \leq i < j \leq n$. The estimator of β (Theil (1950c)) associated with the Theil statistic, C, is

$$\widehat{\beta} = \text{median } \{S_{ij}, 1 \leq i < j \leq n\}. \tag{9.18}$$

Let $S^{(1)} \leq \cdots \leq S^{(N)}$ denote the ordered values of the sample slopes S_{ij}. Then if N is odd, say $N = 2k + 1$, we have $k = (N-1)/2$ and

$$\widehat{\beta} = S^{(k+1)}, \tag{9.19}$$

the value that occupies position $k + 1$ in the list of the ordered S_{ij} values. If N is even, say $N = 2k$, then $k = N/2$ and

$$\widehat{\beta} = [S^{(k)} + S^{(k+1)}]/2. \tag{9.20}$$

That is, when N is even, $\widehat{\beta}$ is the average of the two S_{ij} values that occupy positions k and $k + 1$ in the ordered list of all N sample slopes S_{ij}.

Example 9.2: *Effect of Cloud Seeding on Rainfall—Example 9.1 Continued.* Consider the double ratio data of Table 9.1. The ordered values of the $N = 5(4)/2 = 10$ sample slopes $S_{ij} = (Y_j - Y_i)/(x_j - x_i)$ are $S^{(1)} \leq \cdots \leq S^{(10)}$: $-.150, -.130, -.080, -.070, -.0575, -.055, -.045, -.033, .010,$ and $.040$. Since $N = 10$ is even, we use equation (9.20) with $k = \frac{10}{2} = 5$ to obtain the slope estimate $\widehat{\beta} = [S^{(5)} + S^{(6)}]/2 = [-.0575 - .055]/2 = -.0563$.

Comments

4. *Generalization for Nondistinct x-Values.* Sen (1968) generalized Theil's (1950c) estimator to the case where the x's are not distinct. Let N' denote the number of positive $x_j - x_i$ differences, for $1 \leq i < j \leq n$. (In the case where the x's are distinct, $N' = N$.) Sen's estimator of β is the median of the N' sample slope values that can be computed from the data. In the special case when $x_1 = x_2 = \cdots = x_m = 0$ and $x_{m+1} = x_{m+2} = \cdots = x_{m+q} = 1$ (with $n = m + q$ and $m < n$), Sen's estimator reduces to the median of the mq $(Y_j - Y_i)$ differences, where $i = 1, \ldots, m$ and $j = m + 1, \ldots, m + q$. That is, Sen's estimator reduces to the Hodges-Lehmann two-sample estimator of Section 4.2 applied to the two samples $Y_1, \ldots, Y_m$ and $Y_{m+1}, \ldots, Y_{m+q}$.

Dietz (1989) considers various nonparametric estimators of the slope including Theil's estimator. She finds that Theil's estimator is robust, easy to compute, and competitive in terms of mean squared error with alternative slope estimators. She also considers various nonparametric estimators of the intercept and of the mean response at a given x-value.

5. *Sensitivity to Gross Errors.* The estimator $\widehat{\beta}$ (9.18) is less sensitive to gross errors than is the classical least squares estimator

$$\overline{\beta} = \frac{\sum_{i=1}^{n}(Y_i - \overline{Y})(x_i - \overline{x})}{\sum_{j=1}^{n}(x_j - \overline{x})^2},$$

where $\overline{x} = \sum_{i=1}^{n} x_i/n$ and $\overline{Y} = \sum_{j=1}^{n} Y_j/n$.

6. *Median versus Weighted Average.* The estimator $\widehat{\beta}$ (9.18) is the median of the N individual slope estimators $S_{ij} = (Y_j - Y_i)/(x_j - x_i)$. The least squares estimator $\overline{\beta}$ (see Comment 5) is a weighted average of the S_{ij}'s.

7. *Use of* Minitab *to Compute the Sample Slopes.* The $N = n(n - 1)/2$ individual sample slope values $S_{ij} = (Y_j - Y_i)/(x_j - x_i), 1 \leq i < j \leq n$, can also be obtained using Minitab's WSLOPE command. In addition to their importance in finding the value of the slope estimator $\widehat{\beta}$ (9.18), they are also necessary for determining confidence intervals or bounds for β, as described in Section 9.3. The form of the WSLOPE command is

WSLOPE y in C, x in C, put slopes into C (put row indicies into CC).

After setting the Y-values in Column C1 and the corresponding x-values in Column C2, use

MTB > WSLOPES C1 C2 C3 C4 C5.

For given $i, j \in \{1, 2, \ldots, n\}$, the individual slope $S_{ij} = (Y_j - Y_i)/(x_j - x_i)$ appears in Column C3 corresponding to index j in Column C4 and index i in Column C5. To get the ordered values of the S_{ij} slopes from Minitab, use the SORT command. Thus to put the S_{ij} slopes from Column C3 in order from least to greatest in Column C6, use

$$\text{MTB} > \text{SORT } C3 \ C6.$$

We illustrate the Minitab WSLOPE command with the double-ratio rainfall data of Table 9.1. Here we have $n = 5$ and $N = 10$. The Minitab output for Columns C3 to C6 for these data consist of the following four columns (in agreement with the ordered slopes computed in Example 9.1):

C3 S_{ij}	C4 j	C5 i	C6 Ordered S_{ij}
.01	2	1	−.15
−.07	3	1	−.13
−.033	4	1	−.08
−.0575	5	1	−.07
−.15	3	2	−.0575
−.055	4	2	−.055
−.08	5	2	−.045
.04	4	3	−.033
−.045	5	3	.01
−.13	5	4	.04

Properties

1. *Standard Deviation of $\widehat{\beta}$ (9.18).* For the asymptotic standard deviation of $\widehat{\beta}$ (9.18), see Sen (1968).
2. *Asymptotic Normality.* See Sen (1968).
3. *Efficiency.* See Sen (1968) and Section 9.8.

PROBLEMS

7. Estimate β for the cenosphere-resin data of Table 9.2.
8. Compute the least squares estimator $\overline{\beta}$ (see Comment 5) for the cenosphere-resin data of Table 9.2, and compare $\overline{\beta}$ with the $\widehat{\beta}$ value obtained in Problem 7. In general, which of $\widehat{\beta}$ and $\overline{\beta}$ is easier to compute?
9. Estimate β for the body-weight and surface-area data for squirrel monkeys discussed in Problem 4.
10. Use the Minitab WSLOPE command to obtain the set of 28 ordered individual sample slopes for the cenosphere-resin data of Table 9.2.
11. Estimate β for the tapeworm data discussed in Problems 3 and 8.1.
12. Use the Minitab WSLOPE command to obtain the set of 45 ordered individual sample slopes for the tapeworm data discussed in Problems 3 and 8.1.

9.3. A DISTRIBUTION-FREE CONFIDENCE INTERVAL ASSOCIATED WITH THE THEIL TEST (THEIL)

Procedure

For a symmetric two-sided confidence interval for β, with confidence coefficient $1 - \alpha$, first obtain the upper $(\alpha/2)$th percentile point $k_{\alpha/2}$ of the null distribution of C (9.4) from Table A.30. Let $C_\alpha = k_{\alpha/2} - 2$ and set

$$M = \frac{N - C_\alpha}{2}, \tag{9.21}$$

and

$$Q = \frac{N + C_\alpha}{2} = M + C_\alpha, \tag{9.22}$$

where, once again, $N = n(n-1)/2$. The $100(1-\alpha)\%$ confidence interval (β_L, β_U) for the slope β that is associated with the two-sided Theil test (Section 9.1) is then given by

$$\beta_L = S^{(M)}, \quad \beta_U = S^{(Q+1)}, \tag{9.23}$$

where $S^{(1)} \leq \cdots \leq S^{(N)}$ are the ordered individual sample slopes $S_{ij} = (Y_j - Y_i)/(x_j - x_i)$, $1 \leq i < j \leq n$, used in computing the point estimator $\hat{\beta}$ (9.18). That is, β_L is the sample slope S_{ij} that occupies position M in the list of N ordered sample slopes. The upper endpoint β_U is the sample slope S_{ij} value that occupies position $Q = M + C_\alpha$ in this ordered list. With β_L and β_U given by display (9.23), we have

$$P_\beta(\beta_L < \beta < \beta_U) = 1 - \alpha \text{ for all } \beta. \tag{9.24}$$

(For upper or lower confidence bounds for β associated with appropriate one-sided Theil test procedures, see Comment 9).

Large-Sample Approximation

For large n, the integer C_α may be approximated by

$$C_\alpha \approx z_{\alpha/2} \left\{ \frac{n(n-1)(2n+5)}{18} \right\}^{1/2}. \tag{9.25}$$

In general, the value of the right-hand side of (9.25) is not an integer. To be conservative, take C_α to be the largest integer that is less than or equal to the right-hand side of (9.25) for use in (9.21) and (9.22).

Example 9.3: *Effect of Cloud Seeding on Rainfall—Example 9.1 Continued.* Consider the double-ratio data of Table 9.1. We illustrate how to obtain the 91.6% confidence interval for β. With $1 - \alpha = .916$ (so that $\alpha = .084$), we see from Table A.30 with $n = 5$ that $k_{\alpha/2} = k_{.042} = 8$. Thus $C_{.084} = k_{.042} - 2 = 8 - 2 = 6$. Since $N = 5(4)/2 = 10$, we see from (9.21) and (9.22) that

$$M = \frac{10 - 6}{2} = 2$$

and
$$Q = \frac{10+6}{2} = 8.$$

Using these values of $M = 2$ and $Q = 8$ in display (9.23), we see that
$$\beta_L = S^{(2)}, \quad \beta_U = S^{(9)}$$

provide the endpoints of our 91.6% confidence interval for β. From the ordered list of sample slope values given in Example 9.2, we obtain $S^{(2)} = -.130$ and $S^{(9)} = .010$, so that our 91.6% confidence interval for β is

$$(\beta_L, \beta_U) = (-.130, .010).$$

Comments

8. *Use of* Minitab *to Compute the Endpoints of the Confidence Interval (9.23).* The $N = n(n-1)/2$ individual sample slope values $S_{ij} = (Y_j - Y_i)/(x_j - x_i)$, $1 \le i < j \le n$, can also be obtained using Minitab's WSLOPE command. For details, see Comment 7.

9. *Confidence Bounds.* In many settings we are interested only in making one-sided confidence statements about the parameter β; that is, we wish to assert with specified confidence that β is no larger (or, in other settings, no smaller) than some upper (lower) confidence bound based on the sample data. To obtain such one-sided confidence bounds for β, we proceed as follows. For specified confidence coefficient $1 - \alpha$, find the upper αth [not $(\alpha/2)$th, as for the confidence interval] percentile point k_α of the null distribution of C (9.4) from Table A.30. Let $C_\alpha^* = k_\alpha - 2$ and set

$$M^* = \frac{N - C_\alpha^*}{2} \quad \text{and} \quad Q^* = \frac{N + C_\alpha^*}{2}. \quad (9.26)$$

The $100(1-\alpha)\%$ *lower* confidence bound β_L^* for β is then given by

$$(\beta_L^*, \infty) = (S^{(M^*)}, \infty), \quad (9.27)$$

where, as before, $S^{(1)} \le \cdots \le S^{(N)}$ are the ordered individual sample slopes. With β_L^* given by display (9.27), we have

$$P_\beta(\beta_L^* < \beta < \infty) = 1 - \alpha \quad \text{for all } \beta. \quad (9.28)$$

The corresponding $100(1-\alpha)\%$ *upper* confidence bound β_U^* is given by

$$(-\infty, \beta_U^*) = (-\infty, S^{(Q^*+1)}). \quad (9.29)$$

It follows that

$$P_\beta(-\infty < \beta < \beta_U^*) = 1 - \alpha \quad \text{for all } \beta. \quad (9.30)$$

For large n, the integer C_α^* may be approximated by

$$C_\alpha^* \approx z_\alpha \left\{ \frac{n(n-1)(2n+5)}{18} \right\}^{1/2}. \quad (9.31)$$

In general, the value of the right-hand side of (9.31) is not an integer. To be conservative, take C_α^* to be the largest integer that is less than or equal to the right-hand side of (9.31) for use in display (9.26).

10. *Midpoint of the Confidence Interval as an Estimator.* The midpoint of the confidence interval given by (9.23), namely, $[S^{(M)} + S^{(Q+1)}]/2$, suggests itself as a reasonable estimator of β. (Note that this actually yields a class of estimators depending on the value of α.) In general, this midpoint does not give the same value as $\widehat{\beta}$ (9.18).

Properties

1. *Distribution-Freeness.* For populations satisfying Assumptions A1 and A2, equation (9.24) holds (see Theil (1950b, 1950c)). Hence, we can control the coverage probability to be $1 - \alpha$ without having more specific knowledge about the form of the common underlying distribution of the e_i's. Thus (β_L, β_U) is a distribution-free confidence interval for β over a very large class of populations.
2. *Efficiency.* See Sen (1968) and Section 9.8.

PROBLEMS

13. Obtain a 93.8% confidence interval for β for the cenosphere-resin data in Table 9.2.
14. Obtain a 95.6% confidence interval for β for the body-weight and surface-area data for squirrel monkeys discussed in Problems 4 and 9.
15. Consider a fixed set of data. Show that for $\alpha_2 > \alpha_1$, the symmetric two-sided $(1 - \alpha_1)$ confidence interval for β given by (9.23) is always as long as or longer than the symmetric two-sided $(1 - \alpha_2)$ confidence interval for β.
16. Obtain a 95.8% upper confidence bound (see Comment 9) for β for the double-ratio cloud seeding data in Table 9.1.
17. Obtain a 97.7% lower confidence bound (see Comment 9) for β for the tapeworm data discussed in Problems 8.1, 3, and 11.
18. Obtain a 94.2% confidence interval for β for the Lake Michigan odor periods data discussed in Problems 8.19 and 6.
19. Find the midpoint of the 95.6% confidence interval for β obtained in Problem 14 for the squirrel monkey body-weight and surface-area data. As noted in Comment 10, this midpoint can be used to estimate the value of β. Compare this midpoint estimator with the value of $\widehat{\beta}$ (9.18) obtained in Problem 9.

9.4. AN INTERCEPT ESTIMATOR ASSOCIATED WITH THE THEIL STATISTIC AND USE OF THE ESTIMATED LINEAR RELATIONSHIP FOR PREDICTION (HETTMANSPERGER-MCKEAN-SHEATHER)

Procedure

To estimate the intercept parameter α of model (9.1), we define

$$A_i = Y_i - \widehat{\beta} x_i, \quad i = 1, \ldots, n, \tag{9.32}$$

where $\widehat{\beta}$ is the point estimator of β given in equation (9.18). An estimator associated with the Theil statistic C and suggested by Hettmansperger, McKean, and Sheather (1997) is

$$\widehat{\alpha} = \text{median}\{A_1, \ldots, A_n\}. \tag{9.33}$$

9.4. INTERCEPT ESTIMATOR AND PREDICTION

Let $A^{(1)} \leq \cdots \leq A^{(n)}$ denote the ordered A_i values (9.32). Then if n is odd, say $n = 2k + 1$, we have $k = (n-1)/2$ and

$$\widehat{\alpha} = A^{(k+1)}, \tag{9.34}$$

the value that occupies position $k + 1$ in the list of ordered A_i values. If n is even, say $n = 2k$, then $k = n/2$ and

$$\widehat{\alpha} = \frac{A^{(k)} + A^{(k+1)}}{2}. \tag{9.35}$$

That is, when n is even, $\widehat{\alpha}$ is the average of the two values that occupy positions k and $k + 1$ in the ordered list of all n A_i's.

Employing both the estimator $\widehat{\beta}$ (9.18) for the slope and the estimator $\widehat{\alpha}$ (9.33) for the intercept, our estimated linear relationship between the x and Y variables is then given by

$$\widetilde{\mathrm{med}Y}_{x=x^*} = \widetilde{[\mathrm{median}\ Y\ \mathrm{when}\ x = x^*]} = \widehat{\alpha} + \widehat{\beta}\, x^*. \tag{9.36}$$

That is, we would predict $\widetilde{\mathrm{med}Y}_{x=x^*}$ to be the typical value of the dependent variable Y for a future setting of the independent variable x at x^*. (See also Comment 12.)

Example 9.4: *Effect of Cloud Seeding on Rainfall—Example 9.1 Continued.* Once again, consider the double-ratio data of Table 9.1. From Example 9.2 we see that the slope estimate for these data is $\widehat{\beta} = -.0563$. Combining this value with the (x_i, Y_i) pairs from Table 9.1, the five ordered A (9.32) values are $A^{(1)} \leq \cdots \leq A^{(5)}$: 1.2889, 1.3115, 1.3163, 1.3826, and 1.3852. Since $n = 5$ is odd, we use equation (9.34) with $k = (5-1)/2 = 2$ to obtain the intercept estimate $\widehat{\alpha} = A^{(3)} = 1.3163$.

Combining the slope estimate of $\widehat{\beta} = -.0563$ and this intercept estimate of $\widehat{\alpha} = 1.3163$, our final estimated linear relationship between the x and Y variables is then given by (9.36) to be

$$\widetilde{\mathrm{med}Y}_{x=x^*} = \widetilde{[\mathrm{median}\ Y\ \mathrm{when}\ x = x^*]} = 1.3163 - .0563x^*. \tag{9.37}$$

Thus, for example, we would estimate the median double-ratio value after 4.5 years of the cloud-seeding study to have been

$$\widetilde{\mathrm{med}Y}_{x=4.5} = \widetilde{[\mathrm{median}\ Y\ \mathrm{when}\ x = 4.5\ \mathrm{years}\,]}$$
$$= 1.3163 - .0563(4.5) = 1.06295.$$

Using (9.37) once again, the predicted double-ratio value if the study were to continue for a sixth year would be

$$\widetilde{\mathrm{med}Y}_{x=6} = \widetilde{[\mathrm{median}\ Y\ \mathrm{when}\ x = 6\ \mathrm{years}]}$$
$$= 1.3163 - .0563(6) = 0.9785.$$

One must always exercise caution in using an estimated linear relationship to predict typical values of the dependent variable Y for values of the independent variable x that are too different from the range of x-values used in establishing the estimated linear relationship (see Comment 12). For example, it would make no sense whatsoever to use the relationship (9.37) to estimate the median double-ratio value for negative values of x^*, since they are not possible.

In addition, although $x^* = 20$ years would certainly be a possible value for the independent variable (corresponding to 20 consecutive years of the cloud-seeding study), in order to use equation (9.37) to predict the typical double-ratio value Y after 20 years of the study would require the assumption that the regression relationship (9.1) remains linear for that extended time period. Although this may be a reasonable assumption to make, it is not one that comes automatically. Careful consideration should be given to its validity before using equation (9.37) to predict the double-ratio value that far into the future based solely on the 5 years of available data.

Comments

11. *Competing Estimators*. The intercept estimator given by (9.33) is not the only nonparametric estimator of α that has been studied in the statistical literature. In the case of symmetry of the underlying distribution for the error random variables $e_1, \ldots, e_n$ in Assumption A2, Hettmansperger and McKean (1977) proposed the competing estimator $\tilde{\alpha}$ associated with the median of the $n(n + 1)/2$ Walsh averages of the n individual A_i (9.32) differences. Adichie (1967) proposed and studied the asymptotic properties of an entire class of estimators for α associated with rank tests.

12. *Appropriate Range of Values of the Independent Variable for Purposes of Prediction*. When we choose to use the estimated linear relationship (9.36) to predict typical values of the dependent variable Y for a particular setting of the independent variable x^*, caution must always be the rule. For prediction purposes, we must be relatively confident that the linear relationship holds at least approximately when x assumes the value x^*. This is seldom of concern when x^* is well-situated among the values of the independent variable at which we observed sample dependent variables in obtaining the estimated relationship (9.36) in the first place. However, when we are interested in predicting the typical value of the dependent variable Y for a setting of the independent variable x^* that is outside the range for which sample data had been collected in obtaining the estimated relationship (9.36), we must not automatically assume that the linear relationship (9.1) is still appropriate. Careful consideration must be given to justification of the reasonableness of this relationship for the particular problem of interest prior to using (9.36) for prediction purposes when considering such extended ranges of the independent variable.

PROBLEMS

20. Estimate α for the cenosphere-resin data of Table 9.2.
21. Estimate α for the body-weight and surface-area data for squirrel monkeys discussed in Problem 4.
22. Use the linear relationship (9.36) to estimate the typical density of a cenosphere-resin composite under hydrostatic pressure of 17,500 psi.
23. Use the linear relationship (9.36) to estimate the typical total surface area (cm^3) for a squirrel monkey with a body weight of 1,000 g.
24. Estimate α for the tapeworm data discussed in Problems 8.1, 3, 11, and 17. Use the linear relationship (9.36) to estimate the typical weight of worms recovered from a dog that had been force-fed 20 mg cysticerci of *Taenia hydatigena*.
25. Consider the cenosphere-resin data of Table 9.2. Discuss the reasonableness of using the linear relationship (9.36) established in Problem 20 for these data to estimate the typical density of cenosphere-resin composites under hydrostatic pressures of 35,000 psi, 75,000 psi, and 200,000 psi.
26. Consider the body-weight and surface-area data for squirrel monkeys presented in Problem 4. Discuss the reasonableness of using the linear relationship (9.36) established in Problem 21 for these data to estimate the typical total surface area (cm^3) for squirrel monkeys with body weights of 320 g, 975 g, and 2,500 g.

$k(\geq 2)$ REGRESSION LINES

9.5. AN ASYMPTOTICALLY DISTRIBUTION-FREE TEST FOR THE PARALLELISM OF SEVERAL REGRESSION LINES (SEN, ADICHIE)

In this section we discuss an asymptotically distribution-free procedure to test for parallelism of $k \geq 2$ regression lines. Thus we are concerned with testing equality of the k slope parameters without additional constraints on the corresponding, unspecified intercepts.

Data. For the ith line, $i = 1, \ldots, k$, we observe the value of the ith response random variable Y_i at each of n_i fixed levels, $x_{i1}, \ldots, x_{in_i}$, of the ith independent (predictor) variable x_i. Thus for the ith line, $i = 1, \ldots, k$, we obtain a set of observations $Y_{i1}, \ldots, Y_{in_i}$, where Y_{ij} is the value of the response variable Y_i when $x_i = x_{ij}$.

Assumptions

B1. We take as our straight-line model

$$Y_{ij} = \alpha_i + \beta_i x_{ij} + e_{ij}, \quad i = 1, \ldots, k; \quad j = 1, \ldots, n_i, \qquad (9.38)$$

where the x_{ij}'s are known constants and $\alpha_1, \ldots, \alpha_k$ and $\beta_1, \ldots, \beta_k$ are the unknown intercept and slope parameters, respectively.

B2. The $N = n_1 + \cdots + n_k$ random variables $e_{11}, \ldots, e_{1n_1}, \ldots, e_{k1}, \ldots, e_{kn_k}$ are mutually independent.

B3. The random variables $\{e_{i1}, \ldots, e_{in_i}\}, i = 1, \ldots, k$, are k random samples from a common continuous population with distribution function $F(.)$.

Hypothesis

The null hypothesis of interest here is that the k regression lines in model (9.38) have a common, but unspecified, slope, β, namely,

$$H_0 : [\beta_1 = \cdots = \beta_k = \beta, \text{ with } \beta \text{ unspecified}]. \qquad (9.39)$$

Note that this null hypothesis does not place any conditions whatsoever on the intercept parameters $\alpha_1, \ldots, \alpha_k$. Thus, the assertion in H_0 (9.39) is simply that the k regression lines in model (9.38) are parallel.

Procedure

To construct the Sen-Adichie statistic V, we first align each of the k regression samples. Let $\overline{\beta}$ be the pooled least squares estimator for the common slope β under the null hypothesis H_0 (9.39), as given by

$$\overline{\beta} = \frac{\sum_{i=1}^{k} \sum_{j=1}^{n_i} (x_{ij} - \overline{x}_i) Y_{ij}}{\sum_{i=1}^{k} \sum_{j=1}^{n_i} (x_{ij} - \overline{x}_i)^2}, \qquad (9.40)$$

where

$$\overline{x}_i = \sum_{j=1}^{n_i} \frac{x_{ij}}{n_i}, \quad \text{for } i = 1, \ldots, k. \qquad (9.41)$$

For each of the k regression samples, compute the aligned observations

$$Y_{ij}^* = (Y_{ij} - \overline{\beta} x_{ij}), \quad i = 1, \ldots, k; \quad j = 1, \ldots, n_i. \tag{9.42}$$

Order these aligned observations Y_{ij}^* from least to greatest separately within each of the k regression samples. Let r_{ij}^* denote the rank of Y_{ij}^* in the joint ranking of the aligned observations $Y_{i1}^*, \ldots, Y_{in_i}^*$ in the ith regression sample.

Compute

$$T_i^* = \sum_{j=1}^{n_i} [(x_{ij} - \overline{x}_i) r_{ij}^*]/(n_i + 1), \quad i = 1, \ldots, k, \tag{9.43}$$

where $\overline{x}_i$ is given by (9.41). Setting

$$C_i^2 = \sum_{j=1}^{n_i} (x_{ij} - \overline{x}_i)^2, \quad i = 1, \ldots, k, \tag{9.44}$$

the Sen-Adichie statistic V is then given by

$$V = 12 \sum_{i=1}^{k} \left[\frac{T_i^*}{C_i} \right]^2. \tag{9.45}$$

To test

$$H_0 : [\beta_1 = \cdots = \beta_k = \beta, \text{ with } \beta \text{ unspecified}]$$

versus the general alternative

$$H_1 : [\beta_1, \ldots, \beta_k \text{ not all equal}] \tag{9.46}$$

at the approximate α level of significance,

$$\text{Reject } H_0 \text{ if } V \geq \chi_{k-1,\alpha}^2; \quad \text{otherwise do not reject}, \tag{9.47}$$

where $\chi_{k-1,\alpha}^2$ is the upper α percentile of a chi-square distribution with $k-1$ degrees of freedom. Values of $\chi_{k-1,\alpha}^2$ can be obtained from Chart A.2.

Ties

If there are ties among the n_i aligned observations Y_{ij}^* (9.42) for the ith regression sample, use average ranks to break the ties and compute the weighted sum T_i^* (9.43) contribution to V (9.45) for that sample.

Example 9.5: *Ammonium Flux in Coastal Sediments.* Coastal sediments are an important reservoir for organic nitrogen (ON). The degradation and mineralization of ON in coastal sediments is bacterially mediated and is known to involve several distinct steps. Moreover, it is possible to measure the rates of the processes at each of these steps. During the first stage of ON remineralization, ammonium is generated by heterotrophic bacteria during a process called *ammonification*. Ammonium can then be released to the environment or be microbially transformed to other nitrogenous species.

9.5. AN ASYMPTOTICALLY DISTRIBUTION-FREE TEST FOR PARALLELISM

Mortazavi (1997) collected four sediment cores from Apalachicola Bay, Florida, and analyzed them at Florida State University. The flux of ammonium (μ moles N per square meter of surface area) to the overlying water was measured for each core sample every 90 min during a 6-hour incubation period. These data are presented in Table 9.4 for the four core samples.

We are interested in assessing whether the rate of ammonium flux is similar across these four coastal sediments (at least over the 6-hour period of the study). Thus, if we let β_i correspond to the rate of ammonium flux for the ith coastal sediment core sample, $i = 1, \ldots, 4$, we are interested in testing the null hypothesis H_0 (9.39) against the general alternative (9.46) that the rates are not the same for the four coastal areas in Apalachicola from which the core samples were drawn.

First, we must obtain the pooled least squares estimator $\overline{\beta}$ (9.40). Since the set of x_{ij} values is the same for each of the coastal sediment samples, we see that

$$\overline{x}_1 = \overline{x}_2 = \overline{x}_3 = \overline{x}_4 = \frac{0 + 1.5 + 3 + 4.5 + 6}{5} = 3.$$

Hence, from equation (9.44) we obtain

$$C_1^2 = C_2^2 = C_3^2 = C_4^2 = (0-3)^2 + (1.5-3)^2 + (3-3)^2 + (4.5-3)^2 + (6-3)^2$$
$$= 9 + 2.25 + 0 + 2.25 + 9 = 22.5,$$

TABLE 9.4. Coastal Sediment Ammonium Flux in Apalachicola Bay, Florida

Core Sample, i	Time, x_{ij} (h)	Ammonium Flux, Y_{ij} (μ moles N/m^2)
Core 1	0	0
	1.5	33.019
	3	111.314
	4.5	196.205
	6	230.658
Core 2	0	0
	1.5	131.831
	3	181.603
	4.5	230.070
	6	258.119
Core 3	0	0
	1.5	33.351
	3	97.463
	4.5	196.615
	6	217.308
Core 4	0	0
	1.5	8.959
	3	105.384
	4.5	211.392
	6	255.105

Source: B. Mortazavi (1997).

which, in turn, yields

$$\sum_{i=1}^{4}\sum_{j=1}^{5}(x_{ij}-\bar{x}_i)^2 = \sum_{i=1}^{4}c_i^2 = 4(22.5) = 90.$$

For the numerator of $\bar{\beta}$ (9.40), we see that

$$\sum_{i=1}^{4}\sum_{j=1}^{5}(x_{ij}-\bar{x}_i)(Y_{ij}) = [(0-3)(0+0+0+0)$$

$$+ (1.5-3)(33.019 + 131.831 + 33.351 + 8.959)$$
$$+ (3-3)(111.314 + 181.603 + 97.463 + 105.384)$$
$$+ (4.5-3)(196.205 + 230.070 + 196.615 + 211.392)$$
$$+ (6-3)(230.658 + 258.119 + 217.308 + 255.105)]$$
$$= [0 - 310.74 + 0 + 1251.423 + 2883.57] = 3824.253.$$

Combining these two quantities, we obtain the value of the pooled least squares slope estimator (9.40) to be

$$\bar{\beta} = \frac{3824.253}{90} = 42.49.$$

Next, we create the aligned observations Y_{ij}^* (9.42) for each of the core samples:

Core 1 : $Y_{11}^* = 0 - 42.49(0) = 0$
$Y_{12}^* = 33.019 - 42.49(1.5) = -30.716$
$Y_{13}^* = 111.314 - 42.49(3) = -16.156$
$Y_{14}^* = 196.205 - 42.49(4.5) = 5$
$Y_{15}^* = 230.658 - 42.49(6) = -24.282$

Core 2 : $Y_{21}^* = 0 - 42.49(0) = 0$
$Y_{22}^* = 131.831 - 42.49(1.5) = 68.096$
$Y_{23}^* = 181.603 - 42.49(3) = 54.133$
$Y_{24}^* = 230.070 - 42.49(4.5) = 38.865$
$Y_{25}^* = 258.119 - 42.49(6) = 3.179$

Core 3 : $Y_{31}^* = 0 - 42.49(0) = 0$
$Y_{32}^* = 33.351 - 42.49(1.5) = -30.384$
$Y_{33}^* = 97.463 - 42.49(3) = -30.007$
$Y_{34}^* = 196.615 - 42.49(4.5) = 5.41$
$Y_{35}^* = 217.308 - 42.49(6) = -37.632$

9.5. AN ASYMPTOTICALLY DISTRIBUTION-FREE TEST FOR PARALLELISM

$$\text{Core 4}: \quad Y_{41}^* = 0 - 42.49(0) = 0$$
$$Y_{42}^* = 8.959 - 42.49(1.5) = -54.776$$
$$Y_{43}^* = 105.384 - 42.49(3) = -22.086$$
$$Y_{44}^* = 211.392 - 42.49(4.5) = 20.187$$
$$Y_{45}^* = 255.105 - 42.49(6) = 0.165.$$

Ordering these aligned observations Y_{ij}^* from least to greatest separately within each of the four core samples, we obtain the following within-samples rankings:

$$\text{Core 1}: \quad r_{11}^* = 4, \quad r_{12}^* = 1, \quad r_{13}^* = 3, \quad r_{14}^* = 5, \quad \text{and} \quad r_{15}^* = 2$$
$$\text{Core 2}: \quad r_{21}^* = 1, \quad r_{22}^* = 5, \quad r_{23}^* = 4, \quad r_{24}^* = 3, \quad \text{and} \quad r_{25}^* = 2$$
$$\text{Core 3}: \quad r_{31}^* = 4, \quad r_{32}^* = 2, \quad r_{33}^* = 3, \quad r_{34}^* = 5, \quad \text{and} \quad r_{35}^* = 1$$
$$\text{Core 4}: \quad r_{41}^* = 3, \quad r_{42}^* = 1, \quad r_{43}^* = 2, \quad r_{44}^* = 5, \quad \text{and} \quad r_{45}^* = 4.$$

The values of $T_1^*, \ldots, T_4^*$ are then obtained from (9.43) to be:

$$T_1^* = \frac{[(0-3)(4) + (1.5-3)(1) + (3-3)(3) + (4.5-3)(5) + (6-3)(2)]}{(5+1)} = 0,$$

$$T_2^* = \frac{[(0-3)(1) + (1.5-3)(5) + (3-3)(4) + (4.5-3)(3) + (6-3)(2)]}{(5+1)} = 0,$$

$$T_3^* = \frac{[(0-3)(4) + (1.5-3)(2) + (3-3)(3) + (4.5-3)(5) + (6-3)(1)]}{(5+1)} = -.75,$$

$$T_4^* = \frac{[(0-3)(3) + (1.5-3)(1) + (3-3)(2) + (4.5-3)(5) + (6-3)(4)]}{(5+1)} = 1.5.$$

Combining these T_i^* values with the corresponding values of C_i^2 previously obtained, we see from equation (9.45) that the Sen-Adichie statistic V for these data is given by

$$V = 12 \left\{ \frac{(0)^2}{22.5} + \frac{(0)^2}{22.5} + \frac{(-.75)^2}{22.5} + \frac{(1.5)^2}{22.5} \right\}$$
$$= 12\{0 + 0 + .025 + .1\} = 1.5.$$

For the Sen-Adichie procedure (9.47), we compare the value of V to the chi-square distribution with $k - 1 = 3$ degrees of freedom. From Chart A.2, we see that the observed value of $V = 1.5$ is approximately the .70 upper percentile for the chi-square distribution with 3 degrees of freedom. Thus the approximate P-value for these data and test procedure (9.47) is .70, indicating that there is virtually no sample evidence in support of significant differences in the rates (slopes) of ammonium flux for the four coastal areas sampled.

Comments

13. *Motivation for the Test.* The pooled least squares estimator $\overline{\beta}$ (9.40) estimates some weighted combination, say β^*, of the k individual slopes $\beta_1, \ldots, \beta_k$ (9.38). From Assumptions B1 and B3, it follows that the aligned observations Y_{ij}^* (9.42), $i = 1, \ldots, k$ and $j = 1, \ldots, n_i$,

will tend to have values near

$$\begin{aligned}\text{med}\,(Y_{ij}^*) &= \text{med}\,(Y_{ij} - \overline{\beta}x_{ij})\\ &\approx \text{med}\,(Y_{ij}) - \beta^* x_{ij} = \alpha_i + \beta_i x_{ij} - \beta^* x_{ij} + \text{med}\,(e_{ij})\\ &= \alpha_i + (\beta_i - \beta^*)x_{ij} + \text{med}\,(e_{ij}). \end{aligned} \qquad (9.48)$$

If the null hypothesis H_0 (9.39) is true, then $\beta_1 = \cdots = \beta_k = \beta^* = \beta$ and we would expect each of $Y_{i1}^*, \ldots, Y_{in_i}^*$ to be near $\alpha_i + \text{med}\,(e_{ij})$, for each of the regression samples $i = 1, \ldots, k$. Since the r_{ij}^* ranks are obtained separately within each of the k samples, it follows that under H_0 (9.39) the ranks $r_{i1}^*, \ldots, r_{in_i}^*$ should behave like a random permutation of the integers $1, \ldots, n_i$, and exhibit no additional relationship with the regression constants $x_{i1}, \ldots, x_{in_i}$, for $i = 1, \ldots, k$. Thus the null hypothesis setting should lead to values of T_i^* near zero, for $i = 1, \ldots, k$, and subsequently to small values of the Sen–Adichie test statistic V (9.45). On the other hand, if the null hypothesis H_0 (9.39) is not true, then some of the β_i's will be larger than β^* and some of them will be smaller than β^*. For those regression populations for which β_i is larger than β^*, we see from equation (9.48) that the aligned observations Y_{ij}^* (9.42) will be positively related to the values of the corresponding regression constants x_{ij}. This would tend to produce large positive values for the corresponding T_i^*'s (9.43). For those regression populations for which β_i is smaller than β^*, we see from equation (9.48) that the aligned observations Y_{ij}^* (9.42) will be negatively related to the values of the corresponding regression constants x_{ij}. This would tend to produce large negative values for the corresponding T_i^*'s (9.43). Since each of these T_i^* values is squared in the calculation of the Sen–Adichie test statistic V (9.45), regression populations with either β_i larger or smaller than β^* will tend to produce large contributions to the test statistic V, providing partial motivation for procedure (9.47).

14. *Historical Development.* The general form of the test procedure (9.47), but using a rank estimate for the common value of the slope parameter β under H_0 (9.39) in the construction of the aligned observations Y_{ij}^* (9.42), was first proposed and studied by Sen (1969). The use of the pooled least squares estimator $\overline{\beta}$ (9.40) in the construction of the Y_{ij}^*'s was first suggested by Adichie (1984).

15. *Potthoff's Conservative Test of Parallelism.* For the case $k = 2$, Potthoff (1974) proposed a Wilcoxon-type test of $\beta_1 = \beta_2$. He compares each sample slope that can be computed from line 2 data with each sample slope that can be computed from line 1 data, scoring 1 if the sample 2 slope is larger than the sample 1 slope and 0 otherwise. His statistic is the average of the $n_1(n_1 - 1)(n_2)(n_2 - 1)/4$ such indicators. (To avoid complications, he assumes no two x_{1j}'s are equal and no two x_{2j}'s are equal.) The test associated with his statistic is neither distribution-free nor asymptotically distribution-free. Instead, he uses an upper bound for the null variance of the statistic to produce a conservative test procedure.

16. *Competitor Based on Joint Rankings When the Intercept Is Common.* The Sen–Adichie procedure (9.47) is based on the individual rankings of the aligned observations Y_{ij}^* (9.42) *separately* within each of the k samples. This requires a good deal more computational time than if we could use a single simultaneous ranking of all $N = n_1 + \cdots + n_k$ aligned observations. Although such a joint ranking is not appropriate for the general model (9.38), Adichie (1974) proposed a procedure based on the joint ranking of all N of the aligned observations for settings where it is also reasonable to assume equality of the k intercepts $\alpha_1, \ldots, \alpha_k$ in model (9.38). Thus Adichie's procedure is appropriate for testing H_0 (9.39) under Assumptions B2, B3 and the following more restrictive Assumption B1′ replacing Assumption B1:

B1'. We take as our straight-line model

$$Y_{ij} = \alpha + \beta_i x_{ij} + e_{ij}, \quad i = 1, \ldots, k; j = 1, \ldots, n_i, \tag{9.49}$$

where the x_{ij}'s are known constants, α is the common (unknown) intercept and $\beta_1, \ldots, \beta_k$ are the unknown slope parameters, respectively.

Adichie's (1974) test statistic for this more restrictive setting is quite similar in form to the Sen–Adichie test statistic V (9.45). The major difference is the use of the single simultaneous ranking of all N of the aligned observations, rather than the k separate rankings utilized in constructing V. (We note, in passing, that the assumption of a common intercept α would be quite reasonable for the ammonium flux data considered in Example 9.5.)

17. *Test Procedures for Restricted Alternatives.* The Sen–Adichie procedure (9.47) is designed to test H_0 (9.39) against the class of general alternatives H_1 (9.46). Other authors have proposed nonparametric procedures designed to test H_0 against more restricted classes of alternatives. Adichie (1976) and Rao and Gore (1984) studied asymptotically distribution-free test procedures designed to reach a decision between H_0 and the class of ordered alternatives $H_2 : [\beta_1 \leq \cdots \leq \beta_k$, with at least one strict inequality]. Finally, Kim and Lim (1995) considered asymptotically distribution-free procedures for testing H_0 against umbrella alternatives of the form $H_3 : [\beta_1 \leq \cdots \leq \beta_p \geq \beta_{p+1} \geq \cdots \geq \beta_k$, with at least one strict inequality].

18. *Comparing Several Regression Lines with a Control.* The Sen–Adichie procedure (9.47) is designed to test H_0 (9.39) against the class of general alternatives H_1 (9.46). In this context, the test involves a comparison of each regression line with every other regression line. For settings where one of the regression lines corresponds to a standard line for a control population, we might want to make only the $k - 1$ comparisons between the noncontrol regression lines and this control line. Lim and Wolfe (1997) proposed and studied an asymptotically distribution-free procedure for testing the null hypothesis H_0 (9.39) against the "treatments" versus control alternative $H_4 : [\beta_1 \leq \beta_i, i = 2, \ldots, k$, with at least one strict inequality], where, without loss of generality, the first regression line plays the role of the control line.

Properties

1. *Asymptotic Chi-Squareness.* See Sen (1969).
2. *Efficiency.* See Sen (1969) and Section 9.8.

PROBLEMS

27. Wells and Wells (1967) discussed Project SCUD, an attempt to study the effects of cloud seeding on cyclones. The basic hypothesis of interest was that cloud seeding in areas of cyclogenesis on the east coast of the United States had no measurable effect on the development of storms there. Table 9.5, based on a subset (Experiment 1 of Table I of Wells and Wells (1967)) of the observational data from Project SCUD, gives "*RI*" and "*M*" values for 11 seeded and 10 control units. The quantity *RI* is a measure of precipitation and the quantity *M*, the geostrophic meridional circulation index, was used in predicting cyclogenesis. Cyclones were expected to develop only when *M* was predicted positive. Test that the regression lines of *RI* on *M* for seeded and control units are parallel.

28. Consider the aligned observations Y_{ij}^* (9.42), $i = 1, \ldots, k$ and $j = 1, \ldots, n_i$. Discuss why additional knowledge about the intercept parameters $\alpha_1, \ldots, \alpha_k$ is not necessary in order to use the *separate* within-samples ranks of the Y_{ij}^*'s in the construction of the test statistic V (9.45).

TABLE 9.5. Precipitation Amounts *RI* and Circulation Index *M* for Seeded and Control Units

	Seeded		Control	
Unit, j	$x_{1j}(M)$	$Y_{1j}(RI)$	$x_{2j}(M)$	$Y_{2j}(RI)$
1	24	.180	−7	.138
2	28	.175	10	.081
3	30	.178	17	.072
4	37	.021	25	.188
5	43	.260	44	.075
6	47	.715	51	.435
7	52	.441	53	.423
8	57	.205	63	.339
9	71	.417	75	.519
10	87	.498	90	.738
11	115	.603		

Source: J. M. Wells and M. A. Wells (1967).

29. Consider the aligned observations Y_{ij}^* (9.42), $i = 1, \ldots, k$ and $j = 1, \ldots, n_i$.
    (a) Discuss why it would *not* be appropriate, in general, to use a single simultaneous ranking of all $N = n_1 + \cdots + n_k$ aligned observations in the construction of a statistic for testing H_0 (9.39).
    (b) Under what conditions on the intercept parameters $\alpha_1, \ldots, \alpha_k$ might such a single simultaneous ranking be appropriate for developing a statistic to test H_0 (9.39)? (See Comment 16.)

30. Wardlaw and van Belle (1964) discussed the mouse hemidiaphragm method for assaying insulin. This procedure depends on the ability of the hormone to stimulate glycogen synthesis by the diaphragm tissue, in vitro. Hemidiaphragms are dissected from mice of uniform weight that have been starved for 18 h. The tissues are incubated in tubes, and after incubation the hemidiaphragms are washed with water and analyzed for glycogen content using anthrone reagent. The content is measured in terms of optical density. The procedure makes use of the fact that increasing the concentration of insulin in the incubation medium tends to increase glycogen synthesis by the hemidiaphragms. Specifically, for levels of insulin between .1 to 1.0 μ/ml there is an approximate linear relationship between glycogen content and log concentration of insulin (see Wardlaw and Moloney (1961)).

TABLE 9.6. Glycogen Content of Hemidiaphragms Measured by Optical Density in the Anthrone Test × 1000

	Standard Insulin		Sample 1 Insulin	
j	x_{1j} (Log Dose)	Y_{1j} (Glycogen)	x_{2j} (Log Dose)	Y_{2j} (Glycogen)
1	log (.3)	230	log (.3)	310
2	log (.3)	290	log (.3)	265
3	log (.3)	265	log (.3)	300
4	log (.3)	225	log (.3)	295
5	log (.3)	285	log (.3)	255
6	log (.3)	280	log (.3)	280
7	log (1.5)	365	log (1.5)	415
8	log (1.5)	325	log (1.5)	375
9	log (1.5)	360	log (1.5)	375
10	log (1.5)	300	log (1.5)	275
11	log (1.5)	360	log (1.5)	380
12	log (1.5)	385	log (1.5)	380

Source: A. C. Wardlaw and G. van Belle (1964).

The data in Table 9.6 are the log concentrations of insulin and the glycogen contents for 12 observations each from two varieties of insulin, namely, standard insulin and sample 1 insulin. For both standard and sample 1 lines there are six observations at an insulin volume of .3 ml and six observations at a volume of 1.5 ml. In this insulin assay, and in many bioassays, the question of parallelism is extremely important, since the concept of relative potency (of a test preparation with respect to a standard) depends on the assumption that the dose-response lines are parallel. Using the data in Table 9.6, test the hypothesis that the dose-response lines for standard insulin and sample 1 insulin are parallel.

31. Experimental geneticists use survival under stressful conditions to compare the relative fitness of different species. Dowdy and Wearden (1983) consider data relating to the survival of three species of *Drosophila* under increasing levels of organic phosphorus insecticide. Four batches of medium, identical except for the levels of insecticide they contained, were prepared. One hundred eggs from each of three *Drosophila* species were deposited on each of the four medium preparations and the level of insecticide (x) in parts per million (ppm) and number of *Drosophila* flies that survived to adulthood (y) for each combination are recorded in Table 9.7. Test the hypothesis that the three species of *Drosophila* exhibit the same response to increasing levels of insecticide in the medium studied.

TABLE 9.7. Numbers of *Drosophila* Flies (Three Different Species) That Survive to Adulthood after Exposure to Various Levels (ppm) of an Organic Phosphorus Insecticide

Species	Level of Insecticide (ppm)	Number Survived to Adulthood
Drosophila melanogaster	0.0	91
	0.3	71
	0.6	23
	0.9	5
Drosophila pseudoobscura	0.0	89
	0.3	77
	0.6	12
	0.9	2
Drosophila serrata	0.0	87
	0.3	43
	0.6	22
	0.9	8

Source: S. Dowdy and S. Wearden (1983).

32. Among the pieces of information used to assess the age of primates are measurements of skull, muzzle, and long-bone development. Reed (1973) collected such measurements for *Papio cynocephalus* baboons over a period of 5 years and developed a regression relationship between these attributes and age. A portion of Reed's data (from African colonies existing at the Southwest Foundation for Research and Education in San Antonio, Texas) is presented in Table 9.8 for male and female *Papio cynocephalus* baboons. The recorded data are age, in months, and the sum of skull, muzzle, and long-bone measurements, in millimeters.

Use these data to decide whether there is any difference in the slopes defining the linear relationships between age and the sum of skull, muzzle, and long-bone measurements for male and female *Papio cynocephalus* baboons.

TABLE 9.8. Age, in Months, and Sum of Skull, Muzzle, and Long-Bone Measurements, in Millimeters, for Male and Female *Papio cynocephalus* Baboons

	Male		Female	
j	x_{1j} (Sum)	Y_{1j} (Age)	x_{2j} (Sum)	Y_{2j} (Age)
1	175.0	1.36	175.0	1.58
2	183.0	2.20	183.0	2.48
3	190.0	3.05	190.0	3.40
4	200.0	4.45	200.0	4.92
5	211.0	6.19	211.0	6.87
6	220.0	7.78	220.0	8.66
7	230.0	9.70	230.0	10.86
8	239.5	11.66	239.5	13.14
9	245.5	12.96	245.5	14.67
10	260.0	16.33	260.0	18.68
11	271.5	19.21	271.5	22.14
12	284.0	22.52	284.0	26.18
13	291.0	24.46	291.0	28.57
14	302.5	27.78	302.5	32.68
15	314.0	31.25	314.0	37.03
16	318.5	32.65	318.5	38.80
17	327.0	35.36	327.0	42.22
18	337.0	38.65		
19	345.5	41.52		
20	360.0	46.61		
21	375.0	52.10		
22	384.5	55.69		
23	397.0	60.55		
24	411.0	66.18		
25	419.5	69.68		
26	428.5	73.47		
27	440.0	78.41		
28	454.5	84.81		

Source: O. M. Reed (1973).

GENERAL MULTIPLE LINEAR REGRESSION

9.6. ASYMPTOTICALLY DISTRIBUTION-FREE RANK-BASED TESTS FOR GENERAL MULTIPLE LINEAR REGRESSION (JAECKEL, HETTMANSPERGER-MCKEAN)

The statistical procedures discussed in Sections 9.1 through 9.4 are concerned with the case of a straight-line relationship between a single independent (predictor) variable x and a response random variable Y. In Section 9.5, we presented a test procedure for assessing parallelism of two such straight-line relationships, each with a single independent (predictor) variable. However, in many settings where a regression relationship is of interest there are several independent (predictor) variables that potentially influence the value of a single response random variable. In this section, we present an asymptotically distribution-free rank-based procedure for testing appropriate hypotheses in such a setting, commonly known as *multiple linear regression*.

Data. Let $\mathbf{x}' = (x_1, \ldots, x_p)$ be a row vector of p independent (predictor) variables and let $\mathbf{x}'_1 = (x_{11}, \ldots, x_{p1}), \ldots, \mathbf{x}'_n = (x_{1n}, \ldots, x_{pn})$ denote n fixed values of this vector. At each of these fixed vectors $\mathbf{x}'_1, \cdots \mathbf{x}'_n$, we observe the value of the single response random variable Y.

9.6. RANK-BASED TESTS FOR GENERAL MULTIPLE LINEAR REGRESSION

Thus we obtain a set of observations $Y_1, \ldots, Y_n$, where Y_i is the value of the response variable when $\mathbf{x}' = \mathbf{x}'_i$.

Assumptions

C1. Our model for multiple linear regression is

$$Y_i = \xi + \beta_1 x_{1i} + \beta_2 x_{2i} + \cdots + \beta_p x_{pi} + e_i = \xi + \mathbf{x}'_i \boldsymbol{\beta}, \quad i = 1, \ldots, n, \tag{9.50}$$

where $\mathbf{x}'_1 = (x_{11}, \ldots, x_{p1}), \ldots, \mathbf{x}'_n = (x_{1n}, \ldots, x_{pn})$ are vectors of known constants, ξ is the unknown "intercept" parameter, and $\boldsymbol{\beta}' = (\beta_1, \ldots, \beta_p)$ is a row vector of unknown parameters, commonly referred to as the set of regression coefficients. For convenience later, we also write expression (9.50) in matrix notation. Let $\mathbf{Y}' = (Y_1, \ldots, Y_n)$, $\boldsymbol{\xi}' = (\xi, \ldots, \xi)$, and set

$$\mathbf{X} = \begin{bmatrix} x_{11} & x_{21} & \cdots & x_{p1} \\ x_{12} & x_{22} & \cdots & x_{p2} \\ \vdots & \vdots & & \vdots \\ x_{1,n-1} & x_{2,n-1} & \cdots & x_{p,n-1} \\ x_{1n} & x_{2n} & \cdots & x_{pn} \end{bmatrix}. \tag{9.51}$$

Then, using matrix notation, the multiple linear regression model (9.50) can also be written as

$$\mathbf{Y} = \boldsymbol{\xi} + \mathbf{X}\boldsymbol{\beta}. \tag{9.52}$$

C2. The error random variables $e_1, \ldots, e_n$ are a random sample from a continuous distribution that is symmetric about its median 0, has cumulative distribution function $F(.)$, and has probability density function $f(.)$ satisfying the mild condition that $\int_{-\infty}^{\infty} f^2(t)\,dt < \infty$.

Hypothesis

We are interested in testing the null hypothesis that a specific subset $\boldsymbol{\beta}_q$ of the regression parameters $\boldsymbol{\beta}$ are zero. Without loss of generality (since the ordering of the $(x_1, \beta_1), \ldots, (x_p, \beta_p)$ pairs in model (9.50) is arbitrary), we take this subset $\boldsymbol{\beta}_q$ to be the first q components of $\boldsymbol{\beta}$; that is, we take $\boldsymbol{\beta}'_q = (\beta_1, \ldots, \beta_q)$. Thus we wish to test the null hypothesis

$$H_0 : [\boldsymbol{\beta}'_q = \mathbf{0}; \boldsymbol{\beta}'_{p-q} = (\beta_{q+1}, \ldots, \beta_p) \text{ and } \xi \text{ unspecified}]. \tag{9.53}$$

Thus, the null hypothesis asserts that the independent variables $x_1, \ldots, x_q$ do not play significant roles in determining the value of the dependent variable Y. (In many settings we are interested in assessing the effect of *all* the independent variables simultaneously, which corresponds to taking $q = p$ in H_0 (9.53). Also see Problem 35.)

Procedure

To compute the Jaeckel-Hettmansperger-McKean test statistic HM, we proceed in several distinct steps. First, we obtain an unrestricted estimator for the vector of regression parameters $\boldsymbol{\beta}$. Let $R_i(\boldsymbol{\beta})$ denote the rank of $Y_i - \mathbf{x}'_i \boldsymbol{\beta}$ among $Y_1 - \mathbf{x}'_1 \boldsymbol{\beta}, \ldots, Y_n - \mathbf{x}'_n \boldsymbol{\beta}$, as a function of $\boldsymbol{\beta}$, for $i = 1, \ldots, n$ (see Comment 20). The unrestricted estimator for $\boldsymbol{\beta}$, corresponding to a special case of a class of such estimators proposed by Jaeckel (1972), is then that value of $\boldsymbol{\beta}$, say, $\widehat{\boldsymbol{\beta}}$, that minimizes the measure of dispersion (once again, see Comment 20)

$$D_J(\mathbf{Y} - \mathbf{X}\boldsymbol{\beta}) = (12)^{1/2}(n+1)^{-1}\sum_{i=1}^{n}\left[R_i(\boldsymbol{\beta}) - \frac{n+1}{2}\right](Y_i - \mathbf{x}_i'\boldsymbol{\beta}). \quad (9.54)$$

The estimator $\widehat{\boldsymbol{\beta}}$ does not, in general, have a closed-form expression (see Comment 21 for a special case where such a closed-form expression is available) and iterative computer methods are generally necessary to obtain numerical solutions. Fortunately, as we shall see later, we are able to complete this step as part of the total test procedure by using Minitab's RREG command.

The second step in the computation of the Jaeckel-Hettmansperger-McKean test statistic HM involves repeating the steps leading to $\widehat{\boldsymbol{\beta}}$, except now the minimization of the Jaeckel dispersion measure $D_J(\mathbf{Y} - \mathbf{X}\boldsymbol{\beta})$ is obtained under the condition imposed by the null hypothesis H_0 (9.53), namely, that $\boldsymbol{\beta}_q = \mathbf{0}$, with $\boldsymbol{\beta}_{p-q}$ unspecified. Let $\widehat{\boldsymbol{\beta}}_0$ denote the value of $\boldsymbol{\beta}$ that minimizes $D_J(\mathbf{Y} - \mathbf{X}\boldsymbol{\beta})$ in (9.54) under the null constraint that $\boldsymbol{\beta}_q = \mathbf{0}$. (Once again, $\widehat{\boldsymbol{\beta}}_0$ will generally not be available in a closed-form expression and we will utilize Minitab's RREG to obtain its value.)

Let $D_J(\mathbf{Y} - \mathbf{X}\widehat{\boldsymbol{\beta}})$ and $D_J(\mathbf{Y} - \mathbf{X}\widehat{\boldsymbol{\beta}}_0)$ denote the overall minimum and the minimum under the null constraint that $\boldsymbol{\beta}_q = \mathbf{0}$, respectively, of the Jaeckel dispersion measure $D_J(\mathbf{Y} - \mathbf{X}\boldsymbol{\beta})$ in (9.54) and set

$$D_J^* = D_J(\mathbf{Y} - \mathbf{X}\widehat{\boldsymbol{\beta}}_0) - D_J(\mathbf{Y} - \mathbf{X}\widehat{\boldsymbol{\beta}}). \quad (9.55)$$

We note that D_J^* represents the drop or reduction in Jaeckel dispersion from fitting the full model as opposed to the reduced model corresponding to the null hypothesis H_0 (9.53) constraint that $\boldsymbol{\beta}_q = \mathbf{0}$.

The third and final step in the construction of the Jaeckel-Hettmansperger-McKean test statistic HM is the computation of a consistent estimator (see Comment 23) of the parameter

$$\tau = [12]^{-1/2}\left[\int_{-\infty}^{\infty} f^2(t)\,dt\right]^{-1}. \quad (9.56)$$

Once again, the Minitab procedure RREG will provide this consistent estimator, say, $\widehat{\tau}$, of τ.

Combining the results of these three construction steps, the Jaeckel-Hettmansperger-McKean test statistic HM is given by

$$HM = \frac{2D_J^*}{\widehat{\tau}}. \quad (9.57)$$

When H_0 (9.53) is true, the statistic HM has, as n tends to infinity, an asymptotic chi-square (χ^2) distribution with q degrees of freedom, corresponding to the q constraints placed on $\boldsymbol{\beta}$ under H_0.

To test

$$H_0 : [\boldsymbol{\beta}_q' = \mathbf{0}; \boldsymbol{\beta}_{p-q}' = (\beta_{q+1},\ldots,\beta_p) \text{ and } \xi \text{ unspecified}]$$

against the general alternative

$$H_1 : [\boldsymbol{\beta}_q' \neq \mathbf{0}; \boldsymbol{\beta}_{p-q}' = (\beta_{q+1},\ldots,\beta_p) \text{ and } \xi \text{ unspecified}] \quad (9.58)$$

9.6. RANK-BASED TESTS FOR GENERAL MULTIPLE LINEAR REGRESSION

at the approximate α level of significance,

$$\text{Reject } H_0 \text{ if } HM \geq \chi^2_{q,\alpha}; \quad \text{otherwise do not reject,} \tag{9.59}$$

where $\chi^2_{q,\alpha}$ is the upper α percentile point of a chi-square distribution with q degrees of freedom. Values of $\chi^2_{q,\alpha}$ can be obtained from Chart A.2.

Hettmansperger and McKean (1977) and McKean and Sheather (1991) point out that the chi-square distribution is often too light-tailed for use with small or moderate size samples. They suggest replacing the chi-square percentile $\chi^2_{q,\alpha}$ in (9.59) by $qF_{q,n-p-1,\alpha}$, where $F_{q,n-p-1,\alpha}$ is the upper α percentile of the F distribution with q and $n-p-1$ degrees of freedom.

Ties

If there are ties among $Y_1 - \mathbf{x}_1'\boldsymbol{\beta}, \ldots, Y_n - \mathbf{x}_n'\boldsymbol{\beta}$, use average ranks to break the ties in the computation of the minimum $D_J(\mathbf{Y} - \mathbf{X}\boldsymbol{\beta})$. Similarly, if there are ties among $Y_1 - \mathbf{x}_1'\boldsymbol{\beta}_0$ $\ldots, Y_n - \mathbf{x}_n'\boldsymbol{\beta}_0$, use average ranks to break the ties in the computation of the minimum $D_J(\mathbf{Y} - \mathbf{X}\boldsymbol{\beta}_0)$.

Example 9.6: Wildlife science involves the study of how environmental conditions affect wildlife habits. Freund and Wilson (1997) report data on such a study to assess how a variety of environmental conditions affect the time that lesser snow geese leave their overnight roost sites to fly to their feeding areas. The data in Table 9.9 represent the following observations collected at a refuge near the Texas coast for 36 days of the 1987–88 winter season:

TIME (Y) : minutes before (-) or after (+) sunrise,

TEMP (x_1) : air temperature in degrees Celsius,

HUM (x_2) : relative humidity,

LIGHT (x_3) : light intensity,

CLOUD (x_4) : percent cloud cover.

Here we consider a multiple regression analysis to assess the influence that the environmental conditions temperature (TEMP), relative humidity (HUM), light intensity (LIGHT), and percent cloud cover (CLOUD) have on the departure times (TIME) of lesser snow geese in this region of the country. For illustrative purposes, we use Minitab's RREG command to compute the values of HM (9.57) and provide approximate P-values (based on the appropriate chi-square approximations) for tests of the following three distinct null hypotheses:

$$H_{01} : [\beta_1 = \beta_2 = \beta_3 = \beta_4 = 0; \xi \text{ unspecified}], \tag{9.60}$$

$$H_{02} : [\beta_1 = \beta_2 = 0; \beta_3, \beta_4, \text{ and } \xi \text{ unspecified}], \tag{9.61}$$

and

$$H_{03} : [\beta_2 = 0; \beta_1, \beta_3, \beta_4, \text{ and } \xi \text{ unspecified}]. \tag{9.62}$$

The general Minitab setup for entering the data in preparation for testing these three hypotheses for the snow geese data in Table 9.9 is as follows:

TABLE 9.9. Environmental Conditions Related to Snow Goose Departure Times

DATE	TIME	TEMP	HUM	LIGHT	CLOUD
11/10/87	11	11	78	12.6	100
11/13/87	2	11	88	10.8	80
11/14/87	−2	11	100	9.7	30
11/15/87	−11	20	83	12.2	50
11/17/87	−5	8	100	14.2	0
11/18/87	2	12	90	10.5	90
11/21/87	−6	6	87	12.5	30
11/22/87	22	18	82	12.9	20
11/23/87	22	19	91	12.3	80
11/25/87	21	21	92	9.4	100
11/30/87	8	10	90	11.7	60
12/05/87	25	18	85	11.8	40
12/14/87	9	20	93	11.1	95
12/18/87	7	14	92	8.3	90
12/24/87	8	19	96	12.0	40
12/26/87	18	13	100	11.3	100
12/27/87	−14	3	96	4.8	100
12/28/87	−21	4	86	6.9	100
12/30/87	−26	3	89	7.1	40
12/31/87	−7	15	93	8.1	95
01/02/88	−15	15	43	6.9	100
01/03/88	−6	6	60	7.6	100
01/05/88	−14	2	92	9.0	60
01/07/88	−8	2	96	7.1	100
01/08/88	−19	0	83	3.9	100
01/10/88	−23	−4	88	8.1	20
01/11/88	−11	−2	80	10.3	10
01/12/88	5	5	80	9.0	95
01/14/88	−23	5	61	5.1	95
01/15/88	−7	8	81	7.4	100
01/16/88	9	15	100	7.9	100
01/20/88	−27	5	51	3.8	0
01/21/88	−24	−1	74	6.3	0
01/22/88	−29	−2	69	6.3	0
01/23/88	−19	3	65	7.8	30
01/24/88	−9	6	73	9.5	30

Source: R. J. Freund and W. J. Wilson (1997).

```
MTB>   READ   C10   C1   C2    C3     C4    C5   C6   C7   C8   C9
DATA >         11   11   78   12.6   100    1    0    0    0    1
DATA >          2   11   88   10.8    80    0    1    0    0   -1
DATA >         -2   11  100    9.7    30    0    0    1    0    0
DATA >        -11   20   83   12.2    50    0    0    0    1    0
DATA >         -5    8  100   14.2     0
DATA >          2   12   90   10.5    90
DATA >         -6    6   87   12.5    30
DATA >         22   18   82   12.9    20
DATA >         22   19   91   12.3    80
DATA >         21   21   92    9.4   100
DATA >          8   10   90   11.7    60
DATA >         25   18   85   11.8    40
```

```
DATA >     9    20    93   11.1    95
DATA >     7    14    92    8.3    90
DATA >     8    19    96   12.0    40
DATA >    18    13   100   11.3   100
DATA >   -14     3    96    4.8   100
DATA >   -21     4    86    6.9   100
DATA >   -26     3    89    7.1    40
DATA >    -7    15    93    8.1    95
DATA >   -15    15    43    6.9   100
DATA >    -6     6    60    7.6   100
DATA >   -14     2    92    9.0    60
DATA >    -8     2    96    7.1   100
DATA >   -19     0    83    3.9   100
DATA >   -23    -4    88    8.1    20
DATA >   -11    -2    80   10.3    10
DATA >     5     5    80    9.0    95
DATA >   -23     5    61    5.1    95
DATA >    -7     8    81    7.4   100
DATA >     9    15   100    7.9   100
DATA >   -27     5    51    3.8     0
DATA >   -24    -1    74    6.3     0
DATA >   -29    -2    69    6.3     0
DATA >   -19     3    65    7.8    30
DATA >    -9     6    73    9.5    30
DATA >   END
MTB> NAME C10='TIME' C1='TEMP' C2='HUM' C3='LIGHT' C4='CLOUD'
```

These `Minitab` commands enter the TIME data in Column C10, the TEMP data in Column C1, the HUM data in Column C2, the LIGHT data in Column C3, and the CLOUD data in Column C4. The entries in Columns C5 through C9 will be used to set up the design matrices to enable `Minitab` to test the three specific hypotheses in (9.60), (9.61), and (9.62).

To use `Minitab` to test a general hypothesis of the form H_0 (9.53), we must set up the appropriate design matrix $\mathbf{M}$ so that the conditions in H_0 are equivalent to the constraint $\mathbf{M}\boldsymbol{\beta} = \mathbf{0}$ (see Comment 25). As a first illustration, we consider the null hypothesis H_{01} (9.60). Here the appropriate design matrix $\mathbf{M}_1$ is given by

$$\mathbf{M}_1 = \begin{bmatrix} 1 & 0 & 0 & 0 \\ 0 & 1 & 0 & 0 \\ 0 & 0 & 0 & 0 \\ 0 & 0 & 0 & 1 \end{bmatrix},$$

so that $\mathbf{M}_1\boldsymbol{\beta} = \mathbf{0}$ is equivalent to H_{01} (9.60). The corresponding commands to complete the `Minitab` test of H_{01} are then:

```
MTB > copy C5-C8 into M1
MTB > RREG 'TIME' 4 'TEMP' 'HUM' 'LIGHT' 'CLOUD';
SUBC > HYPOTHESIS M1
```

The relevant output (in addition to some classical least squares information) is, in `Minitab`'s notation:

The regression equation is

TIME = -52.0 + 1.03 TEMP + 0.125 HUM + 2.53 LIGHT + 0.0895 CLOUD

	Coefficient	Std. dev. of coef.
Predictor	Rank	Rank
Constant	−52.000	9.321
TEMP	1.0272	0.2806
HUM	0.1253	0.1207
LIGHT	2.5255	0.7968
CLOUD	0.08946	0.04659

Hodges-Lehmann estimate of $\tau = 8.583$
Anova for hypothesis matrix $M1$

	Dispersion		
	Red. model	Full model	DF
Rank	530.3	248.3	4

Thus, the Minitab output provides the following results:

$$D_J^* = D_J(\mathbf{Y} - \mathbf{X}\widehat{\boldsymbol{\beta}}_0) - D_J(\mathbf{Y} - \mathbf{X}\widehat{\boldsymbol{\beta}}) = 530.3 - 248.3 = 282,$$
$$q = DF = 4, \quad \widehat{\tau} = \text{estimate of } \tau = 8.583.$$

From equation (9.57), the value of the Jaeckel-Hettmansperger-McKean test statistic HM for this setting is

$$HM_1 = \frac{2(282)}{8.583} = 65.711.$$

From Chart A.2, we see that the observed value of $HM_1 = 65.711$ is well above the .001 upper percentile for the chi-square distribution with 4 degrees of freedom, so that the approximate P-value for this test of H_{01} (9.60) is much smaller than .001. Thus, there is strong evidence that temperature, relative humidity, light intensity, and percent cloud cover contribute significantly to determination of the time that lesser snow geese leave their overnight roost sites to fly to their feeding areas. Moreover, the overall estimated regression (predictive) relationship provided by the Minitab output is given by

$$\widehat{Y} = -52.0 + 1.03x_1 + 0.125x_2 + 2.53x_3 + 0.0895x_4.$$

In order to get additional information about potential contributions of some of the individual independent (predictor) variables, we illustrate the use of Minitab to test the two subhypotheses indicated in (9.61) and (9.62). For H_{02} (9.61), the appropriate design matrix $\mathbf{M}_2$ is given by

$$\mathbf{M}_2 = \begin{bmatrix} 1 & 0 & 0 & 0 \\ 0 & 1 & 0 & 0 \end{bmatrix},$$

so that $\mathbf{M}_2\boldsymbol{\beta} = \mathbf{0}$ is equivalent to H_{02} (9.61). The corresponding commands to conduct the Minitab test of H_{02} are:

```
MTB  > copy C5-C6 into M2
MTB  > transpose M2, put in M3
MTB  > RREG 'TIME' 4 'TEMP' 'HUM' 'LIGHT' 'CLOUD';
SUBC > HYPOTHESIS M3
```

9.6. RANK-BASED TESTS FOR GENERAL MULTIPLE LINEAR REGRESSION

The additional `Minitab` output specifically related to this subhypothesis H_{02} (9.61) is then

Hodges-Lehmann estimate of $\tau = 8.583$
Anova for hypothesis matrix $M3$

	Dispersion		
	Red. model	Full model	DF
Rank	301.37	248.27	2

Thus, the `Minitab` output for testing H_{02} (9.61) provides the following results:

$$D_J^* = D_J(\mathbf{Y} - \mathbf{X}\widehat{\boldsymbol{\beta}}_0) - D_J(\mathbf{Y} - \mathbf{X}\widehat{\boldsymbol{\beta}}) = 301.37 - 248.27 = 53.10,$$
$$q = DF = 2, \quad \widehat{\tau} = \text{estimate of } \tau = 8.583.$$

From equation (9.57), the value of the Jaeckel-Hettmansperger-McKean test statistic HM_2 for this setting is

$$HM_2 = \frac{2(53.10)}{8.583} = 12.373.$$

From Chart A.2, we see that the observed value of $HM_2 = 12.373$ is slightly greater than the .002 upper percentile for the chi-square distribution with 2 degrees of freedom, so that the approximate P-value for this test of H_{02} (9.61) is smaller than .002. Thus, there is clear evidence that the first two independent variables, temperature and relative humidity, contribute significantly (over and above the contributions of light intensity and cloud cover) to determination of the time that lesser snow geese leave their overnight roost sites to fly to their feeding areas.

Finally, we turn to the third hypothesis H_{03} (9.62). For this subhypothesis, the appropriate design matrix $\mathbf{M}_3$ is given by

$$\mathbf{M}_3 = [0\ 1\ 0\ 0],$$

so that $\mathbf{M}_3\boldsymbol{\beta} = \mathbf{0}$ is equivalent to H_{03} (9.62). The corresponding commands to conduct the `Minitab` test of H_{03} are

```
MTB  > copy C6 into M4
MTB  > transpose M4, put in M5
MTB  > RREG 'TIME' 4 'TEMP' 'HUM' 'LIGHT' 'CLOUD';
SUBC > HYPOTHESIS M5
```

The additional `Minitab` output specifically related to this subhypothesis H_{03} (9.62) is then:

Hodges-Lehmann estimate of $\tau = 8.583$
Anova for hypothesis matrix $M5$

	Dispersion		
	Red. model	Full model	DF
Rank	255.111	248.270	1

Thus, the `Minitab` output for testing H_{03} (9.62) provides the following results:

$$D_J^* = D_J(\mathbf{Y} - \mathbf{X}\widehat{\boldsymbol{\beta}}_0) - D_J(\mathbf{Y} - \mathbf{X}\widehat{\boldsymbol{\beta}}) = 255.111 - 248.270 = 6.841,$$
$$q = DF = 1, \quad \widehat{\tau} = \text{estimate of } \tau = 8.583.$$

From equation (9.57), the value of the Jaeckel-Hettmansperger-McKean test statistic HM_3 for this setting is

$$HM_3 = \frac{2(6.841)}{8.583} = 1.594.$$

From Chart A.2, we see that the observed value of $HM_3 = 1.594$ is approximately equal to the .20 upper percentile for the chi-square distribution with 1 degree of freedom, so that the P-value for this test of H_{03} (9.62) is approximately .20. Thus, there is no evidence that the relative humidity contributes significantly (over and above the contributions of the other independent variables, temperature, light intensity, and percent cloud cover) to the determination of the time that lesser snow geese leave their overnight roost sites to fly to their feeding areas.

Comments

19. *Motivation for the Test.* Use of a measure of dispersion to assess the effectiveness of a model fit to a set of data is common in regression analysis. The estimators $\widehat{\boldsymbol{\beta}}$ and $\widehat{\boldsymbol{\beta}}_0$ are chosen to minimize the Jaeckel dispersion associated with the differences $Y_i - \mathbf{x}_i'\boldsymbol{\beta}, i = 1,\ldots,n$, under no restrictions on $\boldsymbol{\beta}$ and under the null hypothesis restriction that $\boldsymbol{\beta} = (\boldsymbol{\beta}_q, \boldsymbol{\beta}_{p-q}) = (\mathbf{0}_q, \boldsymbol{\beta}_{p-q})$, respectively. Thus, the numerator of the Jaeckel-Hettmansperger-McKean test statistic $2D_J^* = 2[D_J(\mathbf{Y} - \mathbf{X}\widehat{\boldsymbol{\beta}}_0) - D_J(\mathbf{Y} - \mathbf{X}\widehat{\boldsymbol{\beta}})]$ is twice the drop or reduction in Jaeckel dispersion from fitting the full model as opposed to the reduced null hypothesis model (9.53) with $\boldsymbol{\beta}_q = \mathbf{0}_q$. Large values of this drop in dispersion will lead to large values of HM (9.57) and are indicative of lack of agreement between the collected data and the null hypothesis. This serves as partial motivation for procedure (9.59).

20. *Translation Invariance—"Effect" of the "Intercept" Parameter ξ.* The Jaeckel dispersion measure $D_J(\mathbf{Y} - \mathbf{X}\boldsymbol{\beta})$ is translation-invariant in the sense that it is not affected by the unknown value of the "intercept" parameter, ξ. We note that the rank $R_i(\boldsymbol{\beta})$ of $Y_i - \mathbf{x}_i'\boldsymbol{\beta}$ among $Y_1 - \mathbf{x}_1'\boldsymbol{\beta},\ldots,Y_n - \mathbf{x}_n'\boldsymbol{\beta}$, as a function of $\boldsymbol{\beta}$, is exactly the same as the rank of $Y_i - \xi - \mathbf{x}_i'\boldsymbol{\beta}$ among $Y_1 - \xi - \mathbf{x}_1'\boldsymbol{\beta},\ldots,Y_n - \xi - \mathbf{x}_n'\boldsymbol{\beta}$, for $i = 1,\ldots,n$. It follows that the Jaeckel measure of dispersion is independent of the value of the intercept parameter ξ, since

$$D_J(\mathbf{Y} - \xi - \mathbf{X}\boldsymbol{\beta}) = (12)^{1/2}(n+1)^{-1}\sum_{i=1}^{n}\left[R_i(\boldsymbol{\beta}) - \frac{n+1}{2}\right](Y_i - \xi - \mathbf{x}_i'\boldsymbol{\beta})$$

$$= D_J(\mathbf{Y} - \mathbf{X}\boldsymbol{\beta}) - \xi[(12)^{1/2}(n+1)^{-1}]\sum_{i=1}^{n}\left[R_i(\boldsymbol{\beta}) - \frac{n+1}{2}\right]$$

$$= D_J(\mathbf{Y} - \mathbf{X}\boldsymbol{\beta}), \quad \text{because} \quad \sum_{i=1}^{n}\left[R_i(\boldsymbol{\beta}) - \frac{n+1}{2}\right] = 0.$$

21. *Closed-Form Expression for $\widehat{\boldsymbol{\beta}}$-Special Case of Straight-Line Regression.* One situation where the unrestricted estimator $\widehat{\boldsymbol{\beta}}$ minimizing $D_J(\mathbf{Y} - \mathbf{X}\boldsymbol{\beta})$ in (9.54) has a closed-form expression is when we have only a single independent (predictor) variable so that model (9.50) corresponds to a straight-line regression $Y_i = \xi + \beta_1 x_i + e_i, i = 1,\ldots,n$. For this setting, the Jaeckel (1972) estimator of the slope β_1 is a weighted median of the set of all pairwise slopes $S_{ij} = (Y_j - Y_i)/(x_j - x_i)$, for (i, j) such that $x_i \neq x_j$. This particular estimator was first derived using different criteria and studied by Adichie (1967).

22. *Estimation of the "Intercept" ξ.* Since the Jaeckel dispersion measure $D_J(\mathbf{Y} - \mathbf{X}\boldsymbol{\beta})$ is independent of the unknown value of the "intercept" ξ (see Comment 20), the estimator

$\widehat{\boldsymbol{\beta}}$ obtained by minimizing $D_J(\mathbf{Y} - \mathbf{X}\boldsymbol{\beta})$ does not provide any information relative to ξ. Hettmansperger and McKean (1977) suggested using the full-model residuals $e_i^* = Y_i - \mathbf{x}_i'\widehat{\boldsymbol{\beta}}$, for $i = 1, \ldots, n$, to estimate ξ. In particular, they proposed the estimator

$$\widehat{\xi} = \text{median}\left\{\frac{e_i^* + e_j^*}{2}, 1 \leq i \leq j \leq n\right\}. \qquad (9.63)$$

(We note, in passing, that $\widehat{\xi}$ is simply the Hodges-Lehmann one-sample estimator $\widehat{\theta}$ (3.23) applied to the full-model residuals $e_1^*, \ldots, e_n^*$.)

23. *Estimation of the Parameter τ (9.56)*. Part of the construction of the Jaeckel-Hettmansperger-McKean test statistic, HM (9.57), is the computation of a consistent estimator of the parameter τ (9.56). A variety of approaches leading to a number of competing consistent estimators have been considered in the literature. Hettmansperger and McKean (1977) suggested a consistent estimator for τ based on the length of a Wilcoxon signed rank confidence interval (see Section 3.3) applied to the full-model residuals $e_1^*, \ldots, e_n^*$ discussed in Comment 22. Koul, Sievers, and McKean (1987) recommended an estimator of τ based on the empirical distribution function of the absolute differences of the full-model residuals $e_1^*, \ldots, e_n^*$. An approach to the estimation of τ based on window- or kernel-type estimation of the probability density function $f(.)$ has been considered by Schuster (1974) and Schweder (1975, 1981).

24. *Generalized Score Functions.* The rank-regression procedure discussed in this section is based on the use of the Wilcoxon-type scoring function of the ranks in the construction of the Jaeckel dispersion function $D_J(\mathbf{Y} - \mathbf{X}\boldsymbol{\beta})$ in (9.54). Other scoring functions, such as $\Phi^{-1}(.)$ associated with the van der Waerden test discussed in Comment 4.12, were also considered by Jaeckel (1972) in the construction and study of an entire class of rank-based dispersion measures for the multiple linear regression setting.

25. *Test for a More General Null Hypothesis.* Hettmansperger, McKean, and Sheather (1997) describe a generalization of the null hypothesis presented in H_0 (9.53). They discuss statistical procedures for the more inclusive problem of testing $H_0^* : \mathbf{M}\boldsymbol{\beta} = \mathbf{0}$ versus the general alternative $H_A^* : \mathbf{M}\boldsymbol{\beta} \neq \mathbf{0}$, where $\mathbf{M}$ is an arbitrary full row rank $q \times p$ matrix, for some $q \leq p$. Minitab can be used to test this more general null hypothesis as well as the more specific hypothesis given by H_0 (9.53). We illustrate the use of Minitab in this fashion to test the null hypothesis $H_0^* : [\beta_1 = \beta_2; \beta_3, \beta_4 \text{ and } \xi \text{ unspecified}]$ against the alternative $H_{A1}^* : [\beta_1 \neq \beta_2; \beta_3, \beta_4 \text{ and } \xi \text{ unspecified}]$ for the snow geese data in Example 9.6. For this hypothesis, the appropriate design matrix $\mathbf{M}$ is given by

$$\mathbf{M} = [1\ -1\ 0\ 0],$$

so that $\mathbf{M}\boldsymbol{\beta} = \mathbf{0}$ is equivalent to H_0^*. The corresponding commands to conduct the Minitab test of H_0^* are

```
MTB  > ccpy C9 into M1
MTB  > transpose M1, put in M2
MTB  > RREG 'TIME' 4 'TEMP' 'HUM' 'LIGHT' 'CLOUD';
SUBC > HYPOTHESIS M2
```

The additional Minitab output specifically related to this hypothesis $H_0^* : [\beta_1 = \beta_2; \beta_3, \beta_4 \text{ and } \xi \text{ unspecified}]$ is then

Hodges-Lehmann estimate of $\tau = 8.583$

Anova for hypothesis matrix $M2$

	Dispersion		
	Red. model	Full model	DF
Rank	281.92	248.27	1

Thus, the `Minitab` output for testing H_0^* provides the following results:

$$D_J^* = D_J(\mathbf{Y} - \mathbf{X}\widehat{\boldsymbol{\beta}}_0) - D_J(\mathbf{Y} - \mathbf{X}\widehat{\boldsymbol{\beta}}) = 281.92 - 248.27 = 33.65,$$

$$q = DF = 1, \quad \widehat{\tau} = \text{estimate of } \tau = 8.583.$$

From equation (9.57), the value of the Jaeckel-Hettmansperger-McKean test statistic HM^* for this setting is

$$HM^* = \frac{2(33.65)}{8.583} = 7.841.$$

From Chart A.2, we see that the observed value of $HM^* = 7.841$ is approximately equal to the .005 upper percentile for the chi-square distribution with 1 degree of freedom, so that the P-value for this test of H_0^* is approximately .005. Thus, there is good evidence that the independent variables temperature and relative humidity do not contribute in the same degree to determination of the time that lesser snow geese leave their overnight roost sites to fly to their feeding areas. This finding is in good agreement with what had been established previously in Example 9.6.

26. *Extension to the General Linear Model.* Our discussion of rank-based regression in this section has only touched upon a small portion of a much more extensive rank-based approach to the large class of linear models. Although a discussion of this more general setting is beyond the scope of this text, we recommend that the interested reader take advantage of two excellent survey articles on this topic by Draper (1988) and Hettmansperger, McKean, and Sheather (1997).

Properties

1. *Consistency.* Under certain regularity conditions (see, for example, Hettmansperger, McKean, and Sheather (1997)), the test defined by (9.59) is consistent against the alternatives H_1 (9.58).
2. *Asymptotic Chi-Squareness.* See McKean and Hettmansperger (1976).
3. *Efficiency.* See McKean and Hettmansperger (1976) and Section 9.8.

PROBLEMS

33. In heart catheterization a 3-mm-diameter Teflon catheter (tube) is inserted into a major vein or artery at the femoral region and maneuvered up into the heart itself to assess the heart's physiology and functional ability. Heart catheterizations are sometimes performed on children with congenital heart defects. In such cases, the length of the catheter is often determined by a physician's educated guess. Rice (1988) considered a data set obtained by Weindling (1977) in a preliminary study involving 12 children. For each child, the exact catheter length required was determined by using a fluoroscope to check that the tip of the catheter had reached the pulmonary artery. The 12 catheter lengths (cm) and the heights (in) and weights (lb) for the 12 children in the study are given in Table 9.10.

    Treating length of heart catheter as the independent variable, test for the importance of height and weight in determining the required catheter length.

9.6. RANK-BASED TESTS FOR GENERAL MULTIPLE LINEAR REGRESSION

TABLE 9.10. Required Length of Heart Catheter as a Function of Height and Weight

Child	Height (in)	Weight (lb)	Length of Heart Catheter (cm)
1	42.8	40.0	37.0
2	63.5	93.5	49.5
3	37.5	35.5	34.5
4	39.5	30.0	36.0
5	45.5	52.0	43.0
6	38.5	17.0	28.0
7	43.0	38.5	37.0
8	22.5	8.5	20.0
9	37.0	33.0	33.5
10	23.5	9.5	30.5
11	33.0	21.0	38.5
12	58.0	79.0	47.0

Source: J. A. Rice (1988).

34. Iman (1994) considered data obtained by Leaf et al. (1989) in a study of options for reducing concentrations of total plasma triglycerides. Leaf et al. obtained measurements of the following variables on each of 13 patients:

Y : Total triglyceride level, mmol/ℓ,
x_1 : Sex of the patient (coded as female = 0, male = 1),
x_2 : Whether patient is obese (coded as no = 0, yes = 1),
x_3 : Chylomicrons,
x_4 : Very low-density lipoprotein (VLDL),
x_5 : Low-density lipoprotein (LDL),
x_6 : High-density lipoprotein (HDL),
x_7 : Age of the patient.

These data are presented in Table 9.11.

TABLE 9.11. Blood Plasma Measurements Related to Total Triglyceride Level

Patient	Total Triglyceride Level	Sex/Obese	Chylomicrons	VLDL	LDL	HDL	Age
1	20.19	1/1	3.11	4.51	2.05	0.67	53
2	27.00	0/1	4.90	6.03	0.67	0.65	51
3	51.75	0/0	5.72	7.98	0.96	0.60	54
4	51.36	0/1	7.82	9.58	1.06	0.42	56
5	28.98	1/1	2.62	7.54	1.42	0.36	66
6	21.70	0/1	1.48	3.96	1.09	0.23	37
7	14.40	1/1	0.57	8.60	2.16	0.83	41
8	15.14	1/1	0.60	5.46	1.58	0.85	55
9	50.00	1/1	6.29	13.03	1.48	0.28	43
10	23.73	1/1	1.94	7.12	0.91	0.57	58
11	29.33	0/1	0.52	8.94	1.58	0.88	39
12	19.98	0/1	1.11	5.85	1.19	0.62	41
13	13.28	1/0	1.61	3.73	1.58	0.62	54

Source: D. A. Leaf, W. E. Connor, R. Illingworth, S. P. Bacon, and G. Sexton (1989) and R. L. Iman (1994).

(a) Including all the measured variables, find the approximate P-value for a test of the null hypothesis that obesity does not play a significant role in determination of the total triglyceride level.

(b) Including all the measured variables, find the approximate P-value for a test of the null hypothesis that none of the lipoproteins play significant roles in the determination of the total triglyceride level.

(c) Does age play a significant role in the determination of the total triglyceride level, when all the measured variables are taken into account? Justify your answer.

(d) Find an approximate P-value for a test of the null hypothesis that none of the measured variables contribute significantly to the determination of the total triglyceride level.

35. Consider the multiple linear regression model in (9.50). Often it is the case that we are interested in testing whether *any* of the independent variables $x_1, \ldots, x_p$ have significant effects on the determination of the value of the dependent random variable Y. This corresponds to taking $q = p$ in the statement of the null hypothesis (9.53). For this setting, what would be the form of the null constraint estimator $\widehat{\boldsymbol{\beta}}_0$? Provide a closed form expression for $D_J(\mathbf{Y} - \mathbf{X}\widehat{\boldsymbol{\beta}}_0)$ for this setting in terms of the ordered Y observations, $Y^{(1)} \leq \cdots \leq Y^{(n)}$.

36. Freund and Wilson (1997) present a set of data relating survival times (TIME) of liver-transplant patients to the following information collected from the patients prior to their transplant operations:

CLOT: a measure of the clotting potential of the patient's blood
PROG: a subjective index of the patient's prospect of recovery
ENZ: a measure of a protein present in the body
LIV: a measure relating to white blood cell count.

These data for 54 liver-transplant patients are presented in Table 9.12. Examine the relationship of survival time (TIME) to the four measured preoperation variables. Which of them provide significant input into the determination of survival time for liver-transplant patients?

37. In Section 9.1 we discussed a procedure designed to test the effect of a single independent (predictor) variable x on a dependent random variable Y when the anticipated relationship between x and Y is linear. Sometimes the anticipated relationship between x and Y is better described by a higher order polynomial in x, rather than a simple linear relationship. Discuss how the general procedure presented in this section can be used to test for a relationship between x and Y that is best described by a polynomial of degree $p > 1$.

38. Consider the cenosphere-resin composite data of Problem 1. In that problem, you were asked to assess the significance of a possible linear relationship between hydrostatic pressure, x, and the density of the cenosphere-resin composite, Y. Suppose that someone suggested that the relationship between x and Y might be better represented by a cubic polynomial through the expression

$$E[Y \mid x] = \xi + \beta_1 x + \beta_2 x^2 + \beta_3 x^3,$$

where $\xi, \beta_1, \beta_2,$ and β_3 are unknown parameters. (See Problem 37.)

(a) Find the approximate P-value for an appropriate test of the null hypothesis that there is no (cubic, quadratic, or linear) significant relationship between x and Y.

(b) Find the approximate P-value for an appropriate test of the null hypothesis that the relationship between x and Y is actually quadratic, as opposed to cubic.

(c) Find the approximate P-value for an appropriate test of the null hypothesis that the relationship between x and Y is actually linear, as opposed to either quadratic or cubic.

39. Hettmansperger, McKean, and Sheather (1997) describe the following generalization of the null hypothesis presented in H_0 (9.53). They discuss statistical procedures for the more general problem of testing $H_0^* : \mathbf{M}\boldsymbol{\beta} = \mathbf{0}$ versus the general alternative $H_A^* : \mathbf{M}\boldsymbol{\beta} \neq \mathbf{0}$, where $\mathbf{M}$ is an arbitrary full row rank $q \times p$ matrix, for some $q \leq p$ (see Comment 25). Within this more general setting, what matrix $\mathbf{M}$ corresponds to the special case of the null hypothesis H_0 in (9.53)?

40. In an attempt to gain a better understanding of the complexities of air pollution in general and to predict pollutant levels in particular, the Los Angeles Pollution Control District routinely records the levels of pollutants and several meteorological conditions at various sites around the city. As reported by Rice (1988), the data in Table 9.13 represent the maximum level of an oxidant (a photochemical

TABLE 9.12. Survival Times of Liver Transplant Patients and Related Biological Measurements

Patient	TIME	CLOT	PROG	ENZ	LIV
1	34	3.7	51	41	1.55
2	58	8.7	45	23	2.52
3	65	6.7	51	43	1.86
4	70	6.7	26	68	2.10
5	71	3.2	64	65	0.74
6	72	5.2	54	56	2.71
7	75	3.6	28	99	1.30
8	80	5.8	38	72	1.42
9	80	5.7	46	63	1.91
10	87	6.0	85	28	2.98
11	95	5.2	49	72	1.84
12	101	5.1	59	66	1.70
13	101	6.5	73	41	2.01
14	109	5.2	52	76	2.85
15	115	5.4	58	70	2.64
16	116	5.0	59	73	3.50
17	118	2.6	74	86	2.05
18	120	4.3	8	119	2.85
19	123	6.5	40	84	3.00
20	124	6.6	77	46	1.95
21	125	6.4	85	40	1.21
22	127	3.7	68	81	2.57
23	136	3.4	83	53	1.12
24	144	5.8	61	73	3.50
25	148	5.4	52	88	1.81
26	151	4.8	61	76	2.45
27	153	6.5	56	77	2.85
28	158	5.1	67	77	2.86
29	168	7.7	62	67	3.40
30	172	5.6	57	87	3.02
31	178	5.8	76	59	2.58
32	181	5.2	52	86	2.45
33	184	5.3	51	99	2.60
34	191	3.4	77	93	1.48
35	198	6.4	59	85	2.33
36	200	6.7	62	81	2.59
37	202	6.0	67	93	2.50
38	203	3.7	76	94	2.40
39	204	7.4	57	83	2.16
40	215	7.3	68	74	3.56
41	217	7.4	74	68	2.40
42	220	5.8	67	86	3.40
43	276	6.3	59	100	2.95
44	295	5.8	72	93	3.30
45	310	3.9	82	103	4.55
46	311	4.5	73	106	3.05
47	313	8.8	78	72	3.20
48	329	6.3	84	83	4.13
49	330	5.8	83	88	3.95
50	398	4.8	86	101	4.10
51	483	8.8	86	88	6.40
52	509	7.8	65	115	4.30
53	574	11.2	76	90	5.59
54	830	5.8	96	114	3.95

Source: R. J. Freund and W. J. Wilson (1997).

TABLE 9.13. Maximum Oxidant Level, Wind Speed, Temperature, Humidity, and Insolation for a 30-Day Summer Period in the Los Angeles Pollution Control District

Day	Oxidant Level	Wind Speed	Temperature	Humidity	Insolation
1	15	50	77	67	78
2	20	47	80	66	77
3	13	57	75	77	73
4	21	38	72	73	69
5	12	52	71	75	78
6	12	57	74	75	80
7	12	53	78	64	75
8	11	62	82	59	78
9	12	52	82	60	75
10	20	42	82	62	58
11	11	47	82	59	76
12	17	40	80	66	76
13	20	42	81	68	71
14	23	40	85	62	74
15	17	48	82	70	73
16	16	50	79	66	72
17	10	55	72	63	69
18	11	52	72	61	57
19	11	48	76	60	74
20	9	52	77	59	72
21	5	52	73	58	67
22	5	48	68	63	30
23	4	65	67	65	23
24	7	53	71	53	72
25	18	36	75	54	78
26	17	45	81	44	81
27	23	43	84	46	78
28	23	42	83	43	78
29	24	35	87	44	77
30	25	43	92	35	79

Source: J. A. Rice (1988).

pollutant) and the morning averages of the four meteorological variables: wind speed, temperature, humidity, and insolation (measure of amount of sunlight) over a 30-day period in a single summer.

Ignoring the distinct possibility that there is some degree of correlation between maximum oxidant levels collected on adjacent days (which would violate Assumption C3 regarding the independence of the observations of the dependent variable), examine the relationship of oxidant level to the four meteorological variables. Which of them contribute significantly to the maximum oxidant level on a given day for the Los Angeles Pollution Control District?

41. For the data discussed in Problem 40, consider a multiple linear regression of the maximum oxidant level on the four meteorological measurements. Find the approximate P-value for an appropriate test of the null hypothesis that the regression coefficients for wind speed and humidity are the same, as are the regression coefficients for temperature and insolation. (See Comment 25.)

42. Dowdy and Wearden (1983) considered the relationship between several environmental factors and the number of larvae of the phantom midge, genus *Chaoborus,* which is similar to a mosquito in appearance, but is not blood-sucking. The larva burrows into the sediment at the bottom of a body of water and remains there during the daylight hours. At night it migrates to the surface of the water to feed. The larva is itself eaten by larger animals and therefore plays an important role in the food chain for freshwater fish. A team of biologists studied a recreational lake created by damming a small stream and recorded the following measurements at each of 14 sampling points in the lake:

Y : number of larvae of *Chaoborus* collected in a grab sample of the sediment from an area of approximately 225 cm^2 of lake bottom.
x_1 : depth (meters) of the lake at the sampling point.
x_2 : brackishness (conductivity) of the water at the lake bottom (recorded in mhos per decimeter).
x_3 : dissolved oxygen (milligrams per liter) in the water sampled from the lake bottom.

The data from these 14 sampling points are presented in Table 9.14. Examine the relationship of the number of *Chaoborus* larvae to the three measured water quality variables. Which of them provide significant input into the determination of the number of *Chaoborus* larvae in a lake environment?

TABLE 9.14. Number of *Chaoborus* Larvae and Water Quality of Samples

Sample	Number of Larvae	Depth	Brackishness	Dissolved Oxygen
1	35	8.4	8.0	1.0
2	10	2.0	6.5	8.5
3	9	3.5	6.2	6.5
4	30	10.4	5.0	1.5
5	20	6.5	6.5	7.5
6	23	6.2	7.3	4.5
7	28	12.4	6.4	4.0
8	8	7.0	6.0	10.0
9	29	5.8	6.1	3.0
10	4	3.0	5.4	11.0
11	18	6.0	7.3	4.5
12	14	5.5	6.6	5.5
13	32	9.0	6.5	2.5
14	6	1.1	5.8	7.0

Source: S. Dowdy and S. Wearden (1983).

NONPARAMETRIC REGRESSION ANALYSIS

9.7. AN INTRODUCTION TO NON-RANK-BASED APPROACHES TO NONPARAMETRIC REGRESSION ANALYSIS

In all the previous sections of this chapter, the *modus operandi* has been to consider a specific regression model with associated parameters (e.g., straight line, two straight lines, multiple linear regression) and then to discuss appropriate rank-based procedures for making statistical inferences about the unknown parameters. They have all been nonparametric in nature, in that the inferential procedures were not dependent upon the assumption of a particular underlying distribution for the error terms. Recently, however, there has been considerable research activity in the literature in an arena that has become known generally as *nonparametric regression*. Although it maintains an indifference to the form of the underlying distribution for the error terms, the distinction in this new area of endeavor is that even a specific regression model is no longer stipulated a priori. The data are asked to provide not only the eventual statistical inference but also aid with the development of an appropriate regression relationship between the dependent random variable and the independent predictor variables(s). Thus the intent of these nonparametric regression procedures is to permit the data to aid in both the selection of an appropriate model for the regression relationship and in the inferences eventually drawn from this model.

All the procedures previously discussed in this chapter have also been rank-based, in the sense that some form of ranking was used in arriving at the appropriate inferences. When the model itself is open for data input, however, ranks are no longer sufficient to provide both model selection and inferential procedures. Hence, the procedures associated with this field known as nonparametric regression do not generally utilize rankings in reaching their conclusions. As a result, they are often more complicated and computationally intensive than the level assumed throughout the rest of this text. Consequently, our approach in this section will be to discuss briefly some of the statistical techniques that are commonly used in developing such nonparametric regression procedures and cite appropriate references for readers interested in pursuing more detailed information about them rather than to provide details of specific procedures and their applications to appropriate data sets. More detailed summaries of various aspects of this general area of nonparametric regression can be found in Cleveland (1985), Eubank (1988), Hastie and Tibshirani (1990), and Ryan (1997), for example.

We concentrate here on the setting where we are interested in obtaining information about the relationship between a single dependent random variable Y and a single independent (predictor) variable x. For available procedures in the area of nonparametric regression when there are multiple independent variables, the reader is referred to recent work by Friedman (1991), Stone (1994), and Zhou and Wolfe (1997), for example.

Data. At each of n fixed values, $x_1, \ldots, x_n$, of the independent (predictor) variable x, we observe the value of the response random variable Y. Thus we obtain a set of observations $Y_1, \ldots, Y_n$, where Y_i is the value of the response variable when $x = x_i$.

Assumptions

The most general nonparametric regression relationship between Y_i and x_i is given by

$$Y_i = \mu(x_i) + e_i, \quad \text{for} \quad i = 1, \ldots, n, \tag{9.64}$$

where the random variables $e_1, \ldots, e_n$ are a random sample from a continuous population that has median 0.

The goal, of course, is to make valid statistical inferences about the form of the regression function $\mu(.)$. Depending on the specific approach to nonparametric regression under consideration, a variety of additional *regularity* conditions are often imposed on the form of $\mu(.)$ to enable development of appropriate statistical methodology.

Since there is likely to be a good deal of fuzziness (variability) in the response data $Y_1, \ldots, Y_n$, it is often difficult to describe the relationship between x and Y, as expressed in the median $\mu(x)$, without the aid of a more formal model. Therefore, we search for ways to dampen, or "smooth," the fluctuations present in the Y observations as we move along the various x values. In this section we discuss a variety of ways to approach this smoothing of the data. Ryan (1997) refers to each of these smoothing techniques as a "smoother" and to the associated estimates $\widehat{\mu}(x_i)$ at the n x_i values as a "smooth." All four smoothers discussed in this section are linear smoothers, in the sense that the estimates $\widehat{\mu}(x_1), \ldots, \widehat{\mu}(x_n)$ in a particular smooth are always linear combinations of the observations $Y_1, \ldots, Y_n$. (For more details on any of these approaches, see Chapter 10 in Ryan (1997).)

Running Line Smoother. One of the earliest attempts at nonparametric regression is associated with the running line smoother proposed and studied by Cleveland (1979). For this smoother a moving window of points is utilized and a simple least squares linear regression line is computed each time a point is deleted and another added as the window moves along the x values. The plot of these running lines as a function of the independent variable x is referred to as the running line smoother estimator for $\mu(x)$.

A number of issues are important relative to the construction of running line smoothers. First, one must decide on how many points are to be used in each window (i.e., the window *size* or *size of the neighborhood*) for which the least squares regression line is to be computed. Clearly, a window size that is too small will result in very little smoothing of the data, whereas a window size that is too large will virtually force a single straight-line relationship on the data, regardless of its validity. This choice of window size is discussed in Hastie and Tibshirani (1987), where they indicate that a window size of roughly 10–15% of the data is reasonable. Another matter of concern with running line smoothers is how to deal with the extremes of the data, where symmetric windows are not possible. Statistical inference about $\widehat{\mu}(x)$ associated with running line smoothers is addressed in Hastie and Tibshirani (1990).

Kernel Regression Smoother. As with the running line smoother, the kernel regression smoother utilizes *neighborhood data* to provide its estimate of the regression function $\mu(x)$. In this setting, the neighborhoods are often referred to as *strips* and the size of a strip is called the *bandwidth*. One of the clear distinctions between running line smoothers and kernel regression smoothers is in how they weight the observations in a given window. For a running line smoother, the points in a neighborhood are equally weighted, although, of course, they could have differing influences on the estimation process. On the other hand, for a kernel regression smoother, the distance of the points in a neighborhood from the center of a neighborhood, say, x_0, is used to differentially weight their contributions. Basically, in the process of estimating $\mu(x_0)$, no weight is given to those observations outside of the neighborhood centered at x_0 and the greatest weight in the neighborhood is given to those observations Y_i for which the corresponding x_i are closest to x_0. A *kernel function* is utilized to assign these differential weights to the observations across the various neighborhoods.

Altman (1992) addresses the question of how many strips to use in constructing a kernel regression smoother, as well as some related procedures for statistical inference. The selection of a kernel function and its relationship to the stipulation of both the number of strips and the bandwidth is discussed in Hastie and Tibshirani (1990) and Härdle (1990). One particular shortcoming of kernel regression smoothers is that their performance at the boundaries of the predictor region can be rather poor, as documented by Hastie and Loader (1993) and Fan and Marron (1993).

Local Regression Smoother. Local regression smoothers were first introduced by Cleveland (1979), where he referred to the process as *locally weighted regression*. Local regression smoothers once again use overlapping neighborhoods and, as with the kernel regression smoothers, weight the contributions of points to the estimation of $\mu(x_0)$ in an inverse relationship to their distances from x_0. The estimation in a particular neighborhood is thus like a local weighted least squares fit.

Robust versions of local regression smoothers, which downweight large residuals, have also been proposed (see Cleveland (1985) and Cleveland, Grosse, and Shyu (1992) for example) for the setting where the random errors have a symmetric distribution. Computational methods for local regression smoothers are presented in Cleveland and Grosse (1991). Approaches to statistical inferences for $\mu(x_0)$, as well as diagnostic checks associated with local regression smoothers, are discussed in Cleveland, Grosse, and Shyu (1992).

Spline Regression Smoothers. A *spline* is a curve pieced together from a number of individually constructed curve/line segments; that is, a spline is simply a piecewise polynomial. (Smith (1979) and Eubank (1988) provide nice discussions of this general concept.) When each segment of a spline contains only linear terms, it is called a *linear spline*.

The application of splines to regression problems in a general sense is discussed in Wegman and Wright (1983), where they refer to splines associated with a regression model as *regression splines*. The simplest of these regression spline smoothers are those associated with linear splines. The junctures where these lines are pieced together are known as *knots*. When the positions of these knots are known a priori, the use of linear regression spline smoothers is relatively straightforward. However, when the positions of the knots are also unknown, the problem becomes a good deal more complicated. Applications of higher-order polynomial splines (in particular, quadratic and cubic splines) are discussed in Eubank (1988).

A different approach to the use of splines in regression problems is associated with the development of *smoothing splines*. For these procedures, the regression smooth results from minimization of a sum of squares augmented by a smoothing term related to the order of the desired smoothing spline. For further information on smoothing splines, the reader is referred to Eubank (1988) and Wahba (1990).

General Discussion. As mentioned previously, all the nonparametric regression procedures discussed in this section are considerably more computationally intensive than the material presented elsewhere in the text. As a result, computer software is essential for the implementation of these nonparametric regression smoothers. Such software is available in `Minitab` for running line smoothers and local regression smoothers. Kernel regression smoothers and spline regression smoothers can be obtained via `S-Plus`.

Finally, we note that in determining which of these approaches to nonparametric regression is most appropriate for a given problem, the decision invariably comes down to the relative importance of minimizing bias versus minimizing variance. All these smoothers produce biased estimators for the regression function $\mu(x)$, so that the desired trade-off between the sizes of the variance and bias (along with computational capabilities, of course) often strongly influences the choice of a particular nonparametric regression smoother.

9.8. EFFICIENCIES OF REGRESSION PROCEDURES

The asymptotic relative efficiencies of the Theil procedures of Sections 9.1–9.3 with respect to their normal theory counterparts based on the least squares estimator of β have been found by Sen (1968) to be given by the expression

$$e_F = \epsilon^2 \left[12\sigma_F^2 \left\{ \int_{-\infty}^{\infty} f^2(u)\,du \right\}^2 \right] = \epsilon^2\, e_F^*, \tag{9.65}$$

where σ_F^2 is the variance of the common underlying (continuous) distribution $F(.)$ for the random variables $e_1, \ldots, e_n$ in equation (9.1), $f(.)$ is the probability density function corresponding to F, and ϵ^2 is the limiting value (n tending to infinity) of ϵ_n^2, where ϵ_n is the product moment correlation coefficient between $(x_1, \ldots, x_n)$ and $(1, \ldots, n)$ as given in expression (6.2) of Sen (1968). The parameter $\int_{-\infty}^{\infty} f^2(u)\,du$ is the area under the curve associated with $f^2(.)$, the square of the common probability density function. We note that the expression e_F (9.65) is simply ϵ^2 times the corresponding Pitman efficiencies (e_F^*) in the one-sample, two-sample, and k-sample location settings (see Sections 3.11, 4.5, and 6.10).

We note that ϵ_n clearly depends on the design configuration for the values $(x_1, \ldots, x_n)$ of the independent variable. An important special case where $\epsilon_n = 1$ is the equally spaced, no-replications design, where $x_i = x_1 + (i-1)a$, for some $a > 0$ and $i = 1, \ldots, n$. When $\epsilon^2 = 1$, values of e_F (9.65) correspond to e_F^* and can be obtained from display (3.116).

The asymptotic relative efficiency under a sequence of near alternatives of the Sen-Adichie parallelism test based on V (9.45) with respect to the corresponding normal theory procedure based on least squares estimators was found to be e_F^* (9.65) by Sen (1969). The asymptotic relative efficiency under a sequence of contiguous alternatives of the Jaeckel-Hettmansperger-McKean rank-based multiple linear regression test based on HM (9.57) with respect to the corresponding least squares competitor test procedure was found by McKean and Hettmansperger (1976) to be e_F^* (9.65) as well. Once again, values of e_F^* can be found in display (3.116).

CHAPTER 10

Comparing Two Success Probabilities

INTRODUCTION

In Chapter 2 we described inferential procedures for a single success probability. These procedures are based on the proportion of successes observed in n independent Bernoulli trials. Recall that each observation could be classified a success or failure (depending on whether or not a specified attribute was present). In this chapter the object is to compare two unknown success probabilities, p_1, p_2, on the basis of the corresponding rates of success in independent samples.

Section 10.1 describes approximate tests and confidence intervals for $p_1 - p_2$. The tests and confidence intervals are based on large-sample approximations. Section 10.2 presents Fisher's exact (conditional) test. In Section 10.3, we introduce the odds ratio and present inferential procedures (tests, estimators, confidence intervals) for the odds ratio. Section 10.4 describes tests, estimators, and confidence intervals for analyzing k strata of 2×2 tables. Section 10.5 considers efficiency properties.

Data. We observe the outcomes of $n_{1\cdot}$ independent repeated Bernoulli trials, each with success probability p_1. We also observe the outcomes of $n_{2\cdot}$ independent repeated Bernoulli trials, each with success probability p_2. The data are represented in Table 10.1.

Assumptions

In Section 10.1 we make the following assumptions.

A1. $\mathcal{O}_{11}$ is the number of successes observed in $n_{1\cdot}$ independent Bernoulli trials, each with success probability p_1.

A2. $\mathcal{O}_{21}$ is the number of successes observed in $n_{2\cdot}$ independent Bernoulli trials, each with success probability p_2.

TABLE 10.1. **2 × 2 Table of Outcomes** (10.1)

	Successes	Failures	Totals
Sample 1	$\mathcal{O}_{11}$	$\mathcal{O}_{12}$	$n_{1\cdot}$
Sample 2	$\mathcal{O}_{21}$	$\mathcal{O}_{22}$	$n_{2\cdot}$
Totals:	$n_{\cdot 1}$	$n_{\cdot 2}$	$n_{\cdot\cdot}$

10.1. APPROXIMATE TESTS AND CONFIDENCE INTERVALS

A3. The Bernoulli trials corresponding to sample 1 are independent of the Bernoulli trials corresponding to sample 2.

The hypothesis of interest in Section 10.1 is

$$H_0 : p_1 = p_2 = p \tag{10.2}$$

with the common value p being unspecified.

An exact conditional test of H_0 is achievable but computationally difficult, as follows. The random quantity

$$D = \frac{\mathcal{O}_{11}}{n_{1.}} - \frac{\mathcal{O}_{21}}{n_{2.}} \tag{10.3}$$

is an estimator of $p_1 - p_2$. Suppose, for our data, the observed value of D is D_{obs}. Using the independent binomial distributions of $\mathcal{O}_{11}$ and $\mathcal{O}_{21}$, one can, for a specified value of p, compute $P_p(D \geq D_{\text{obs}})$. Then define the P-value to be

$$P = \max_{0 \leq p \leq 1} P_p(D \geq D_{\text{obs}}).$$

This computationally intensive unconditional test is due to Barnard (1945), but later its originator and others have preferred Fisher's exact conditional test. See Barnard (1945, 1947), Suissa and Shuster (1985), and Haber (1986, 1987) for more on Barnard's unconditional test and Fisher's exact test. See Mehta and Hilton (1993) for a discussion of conditional versus unconditional tests and the computational difficulties incurred by unconditional tests, especially for contingency tables beyond the 2×2 case.

In this chapter, we defer Fisher's exact test to Section 10.2. In Section 10.1, we present approximate tests and confidence intervals for $p_1 - p_2$. These approximate procedures are based on large-sample approximations.

10.1. APPROXIMATE TESTS AND CONFIDENCE INTERVALS FOR THE DIFFERENCE BETWEEN TWO SUCCESS PROBABILITIES (PEARSON)

To test H_0, given by (10.2), we use the following large-sample tests.

Large-Sample Test Procedures

The test statistic is a suitably standardized version of $\widehat{p}_1 - \widehat{p}_2$, where

$$\widehat{p}_1 = \frac{\mathcal{O}_{11}}{n_{1.}}, \quad \widehat{p}_2 = \frac{\mathcal{O}_{21}}{n_{2.}}. \tag{10.4}$$

The standard deviation of $D = \widehat{p}_1 - \widehat{p}_2$ can be estimated by

$$\widehat{SD}(\widehat{p}_1 - \widehat{p}_2) = \sqrt{\frac{\widehat{p}(1-\widehat{p})}{n_{1.}} + \frac{\widehat{p}(1-\widehat{p})}{n_{2.}}}, \tag{10.5}$$

where

$$\widehat{p} = \frac{\mathcal{O}_{11} + \mathcal{O}_{21}}{n_{1.} + n_{2.}} \tag{10.6}$$

is an estimator of the hypothesized common success rate p. Recall that $\mathcal{O}_{11}$ is the number of successes in sample 1, $\mathcal{O}_{21}$ is the number of successes in sample 2, and $n_1 + n_2$ is the total number of trials for both samples. Thus $\widehat{p}$ may be viewed as being obtained by pooling the data in the two samples. The standardized version of $\widehat{p}_1 - \widehat{p}_2$ is

$$A = \frac{\widehat{p}_1 - \widehat{p}_2}{\widehat{SD}(\widehat{p}_1 - \widehat{p}_2)}. \tag{10.7}$$

When H_0 is true, the asymptotic distribution of A is $N(0, 1)$.

a. Approximate One-Sided Upper-Tail Test. We denote the difference in success rates by

$$p_d = p_1 - p_2. \tag{10.8}$$

To test

$$H_0 : p_d = 0$$

versus

$$H_1 : p_d > 0,$$

at the approximate α level of significance,

$$\text{Reject } H_0 \text{ if } A \geq z_\alpha; \quad \text{otherwise do not reject.} \tag{10.9}$$

b. Approximate One-Sided Lower-Tail Test. To test

$$H_0 : p_d = 0$$

versus

$$H_2 : p_d < 0,$$

at the approximate α level of significance,

$$\text{Reject } H_0 \text{ if } A \leq -z_\alpha; \quad \text{otherwise do not reject.} \tag{10.10}$$

c. Approximate Two-Sided Test. To test

$$H_0 : p_d = 0$$

versus

$$H_3 : p_d \neq 0,$$

at the approximate α level of significance,

$$\text{Reject } H_0 \text{ if } |A| \geq z_{\alpha/2}; \quad \text{otherwise do not reject.} \tag{10.11}$$

Approximate confidence intervals for p_d are obtained as follows. For confidence intervals, a different estimator is utilized to estimate the standard deviation of $\widehat{p}_1 - \widehat{p}_2$ than was used in

10.1. APPROXIMATE TESTS AND CONFIDENCE INTERVALS

the testing procedure given earlier. Let

$$\widetilde{SD}(\widehat{p}_1 - \widehat{p}_2) = \sqrt{\frac{\widehat{p}_1(1-\widehat{p}_1)}{n_{1.}} + \frac{\widehat{p}_2(1-\widehat{p}_2)}{n_{2.}}} \tag{10.12}$$

where

$$\widehat{p}_1 = \frac{\mathcal{O}_{11}}{n_{1.}}, \quad \widehat{p}_2 = \frac{\mathcal{O}_{21}}{n_{2.}}.$$

In estimating the standard deviation of $\widehat{p}_1 - \widehat{p}_2$ for use in the confidence interval procedure, we no longer assume $p_1 = p_2$, as in the case when H_0 is true. That is why $\widetilde{SD}(\widehat{p}_1 - \widehat{p}_2)$ given by equation (10.12) differs from $\widehat{SD}(\widehat{p}_1 - \widehat{p}_2)$ given by equation (10.5).

For a symmetric two-sided confidence interval for p_d, with approximate confidence coefficient $1 - \alpha$, set

$$p_{d,L} = \widehat{p}_1 - \widehat{p}_2 - z_{\alpha/2} \cdot \widetilde{SD}(\widehat{p}_1 - \widehat{p}_2) \tag{10.13}$$

and

$$p_{d,U} = \widehat{p}_1 - \widehat{p}_2 + z_{\alpha/2} \cdot \widetilde{SD}(\widehat{p}_1 - \widehat{p}_2). \tag{10.14}$$

With $p_{d,L}$ and $p_{d,U}$ given by displays (10.13) and (10.14), respectively, we have, for all (p_1, p_2) pairs,

$$P_{(p_1,p_2)}(p_{d,L} < p_d < p_{d,U}) \approx 1 - \alpha. \tag{10.15}$$

Example 10.1: *Care Patterns for Black and White Patients with Breast Cancer.* Diehr et al. (1989) point out that it is well known that survival of women with breast cancer tends to be lower in blacks than whites. See the references of their paper for documentation of this fact. Diehr and her colleagues were interested in whether differences seen in survival could be accounted for by differences in diagnostic methods and treatment. Their study sought to determine if there are statistically significant patterns of care and, if so, whether these differences can be attributed to differences between black and white patients in age, stage, type of insurance, type of hospital, or type of physician.

The information used in the Diehr et al. (1989) study concerning management and treatment during the first 4 months of diagnosis was abstracted from a systematic sample of inpatient and outpatient records of female patients in 107 participating hospitals. Diehr and her colleagues reported on a subset of 10 breast cancer patterns. The 10 were chosen because they were applicable to most patients and thus could be assessed for enough black patients to make the study possible. One pattern of interest was *liver scan*. Did patients with local or regional disease have a liver scan or CT scan of the liver? Such scans are not routinely required for a patient with local or regional disease because the likelihood of finding an abnormality with the scan is low in the absense of abnormal liver chemistry or hepatomegaly. In the Diehr et al. study it was found that black patients with local disease were more likely to have a liver scan or a CT scan than were white patients. The percentage of black patients receiving appropriate care was about 10 percentage points lower than that of white patients, even after controlling for other factors. In particular, considering the data for patients in the 19 hospitals that had enough black patients for individual analysis, Diehr and her colleagues found that black patients were more likely than white patients to receive liver scan, and this tendency could not be attributed simply to chance. To see how this conclusion was reached, we first focus on the liver scan data of hospital 8 given in Table 10.2. We will illustrate how, for that hospital, the approximate test

TABLE 10.2. Patients with Local or Regional Disease Receiving Liver Scan in Hospital 8

	Liver Scan		
Patients	Yes	No	Totals
Black	4	8	12
White	1	20	21
Totals:	5	28	33

Source: P. Diehr, J. Yergan, J. Chu, P. Feigl, G. Glaefke, R. Moe, M. Bergner, and J. Rodenbaugh (1989).

based on A can be used to see if there was a significant difference between the chance of a white patient receiving a scan and the chance of a black patient receiving a scan. (We return to this problem in Section 10.2 and apply Fisher's exact test. Later, in Section 10.4, we apply the Mantel-Haenszel test to the data from the 19 hospitals with the most black patients to get an overall conclusion.)

Let p_1 be the unknown probability that a black patient in hospital 8 with local or regional disease will receive a liver scan and let p_2 be the unknown probability that a white patient in hospital 8 with local or regional disease will receive a liver scan. We write the null hypothesis as

$$H_0 : p_1 = p_2,$$

or equivalently, $p_d = 0$, where $p_d = p_1 - p_1$.

Diehr and her colleagues suspected that a deviation from H_0 would be in the direction of the one-sided alternative $p_1 > p_2$. To test against this one-sided alternative we use procedure (10.8). We find, from equations (10.3)–(10.6),

$$\widehat{p}_1 = \frac{4}{12} = .3333, \quad \widehat{p}_2 = \frac{1}{21} = .0476, \quad \widehat{p} = \frac{4+1}{12+21} = \frac{5}{33} = .1515,$$

and

$$\widehat{SD} = \sqrt{\frac{(.1515)(.8485)}{12} + \frac{(.1515)(.8485)}{21}} = .1296.$$

Then, from equation (10.7),

$$A = \frac{.3333 - .0476}{.1296} = 2.20.$$

The approximate $\alpha = .05$ test given by procedure (10.8) is reject H_0 if $A \geq 1.65$, accept H_0 otherwise. Since $A = 2.20$, we reject H_0 at that level. The P-value is found by referring $A = 2.20$ to Table A.1. From Table A.1 we find $P = .014$. This constitutes strong evidence that in hospital 8 the chance that a black patient with local or regional breast cancer receives a liver scan is higher than the corresponding chance that a white patient with local or regional breast cancer receives a liver scan.

To find an approximate 95% confidence interval for $p_1 - p_2$, we first compute, via equation (10.12),

10.1. APPROXIMATE TESTS AND CONFIDENCE INTERVALS

$$\widetilde{SD}(\widehat{p}_1 - \widehat{p}_2) = \sqrt{\frac{.3333(.6667)}{12} + \frac{(.0476)(.9524)}{21}} = .1439.$$

Then, the lower and upper confidence limits, given by equations (10.13) and (10.14), respectively, are:

$$p_{d,L} = .3333 - .0476 - 1.96(.1439) = .004,$$

$$p_{d,U} = .3333 - .0476 + 1.96(.1439) = .568.$$

The 2 × 2 Chi-Squared Test of Homogeneity

The large-sample two-sided test based on A can also be presented via Karl Pearson's famous chi-squared statistic. It is motivated as follows. Suppose the null hypothesis H_0 is true. Then the best estimator of the common success probability is $\widehat{p}$, given by equation (10.6). In the notation of Table 10.1, this can be written as

$$\widehat{p} = \frac{n_{.1}}{n_{..}}. \tag{10.16}$$

Using this estimator, the expected values of the random quantities $\mathcal{O}_{11}, \mathcal{O}_{12}, \mathcal{O}_{21}, \mathcal{O}_{22}$ in Table 10.1 can be estimated, respectively, by $E_{11}, E_{12}, E_{21}, E_{22}$, where

$$E_{11} = n_{1.} \times \widehat{p} = \frac{n_{1.} \times n_{.1}}{n_{..}},$$

$$E_{12} = n_{1.} \times (1 - \widehat{p}) = \frac{n_{1.} \times n_{.2}}{n_{..}},$$

$$E_{21} = n_{2.} \times \widehat{p} = \frac{n_{2.} \times n_{.1}}{n_{..}},$$

$$E_{22} = n_{2.} \times (1 - \widehat{p}) = \frac{n_{2.} \times n_{.2}}{n_{..}}. \tag{10.17}$$

A measure of the discrepancy between the observed frequencies, the $\mathcal{O}$'s, and the estimated expected frequencies under the hypothesis, the E's, is the chi-squared statistic given by

$$\chi^2 = \frac{(\mathcal{O}_{11} - E_{11})^2}{E_{11}} + \frac{(\mathcal{O}_{12} - E_{12})^2}{E_{12}} + \frac{(\mathcal{O}_{21} - E_{21})^2}{E_{21}} + \frac{(\mathcal{O}_{22} - E_{22})^2}{E_{22}}. \tag{10.18}$$

The chi-squared statistic can be written in abbreviated notation as

$$\chi^2 = \sum \frac{(\mathcal{O} - E)^2}{E} \tag{10.19}$$

where we have omitted the subscripts on the $\mathcal{O}$'s and the E's, but it is to be understood that the summation $\sum$ is over the four cells of Table 10.1. Note that the differences "observed minus expected," that is, $\mathcal{O} - E$, are squared, eliminating the balancing out of positive and negative discrepancies. Also, each squared difference is weighted by the inverse of the corresponding E, so that the differences involving small E's assume the greatest importance.

It can be shown that

$$A^2 = \chi^2 \tag{10.20}$$

where A is given by equation (10.7). This implies that the two-sided approximate α-level test of $p_d = 0$ versus $p_d \neq 0$ given by procedure (10.11) is equivalent to the test

$$\text{Reject } H_0 \text{ if } \chi^2 \geq \chi^2_{\alpha,1}; \quad \text{otherwise do not reject,} \tag{10.21}$$

where $\chi^2_{\alpha,1}$ is the upper α percentile point of the chi-squared distribution with 1 degree of freedom. There is a shortcut formula for the calculation of the chi-squared test statistic, namely,

$$\chi^2 = \frac{n_{..}(\mathcal{O}_{11}\mathcal{O}_{22} - \mathcal{O}_{21}\mathcal{O}_{12})^2}{n_{.1} \times n_{.2} \times n_{1.} \times n_{2.}}. \tag{10.22}$$

For the liver scan data of Table 10.2, we can use equation (10.22) to find

$$\chi^2 = \frac{33(4 \times 20 - 1 \times 8)^2}{(5)(28)(12)(21)} = 4.85,$$

agreeing (allowing for round-off error) with the value obtained by squaring $A = 2.20$.

Minitab can be used to calculate χ^2. We illustrate this with the liver scan data. After accessing Minitab, enter the data and give the CHISQUARE command as follows.

```
MTB>READ C1-C2
DATA>4 8
DATA>1 20
DATA>END
MTB>CHISQUARE C1-C2
```

The output will look like the following:

```
Expected counts are printed below observed counts

Expected    C1       C2              Total
   1         4        8               12
            1.82    10.18

   2         1       20               21
            3.18    17.82

 Total       5       28               33
```

ChiSq= 2.618 + 0.468 + 1.496 + 0.267 = 4.849
   df=1
 2 cells with expected counts less than 5.0

We can find the P-value corresponding to $\chi^2 = 4.85$ by using Minitab's CDF command in conjunction with the CHISQUARE subcomand:

```
MTB>SET C1
DATA>4.85
DATA>END
MTB>CDF C1;
SUBC>CHISQUARE v=1.
```

The output is

```
4.8500    0.9724
```

Thus the two-sided P-value is $1 - .9724 = .028$, agreeing with what we found earlier (recall $A = 2.20$ with one-sided P-value .014).

The chi-squared test, as developed here, is called a chi-squared test of *homogeneity*. This is because, for Table 10.1, we have considered the data to be based on a sample of size $n_{1.}$ from one population and a separate independent sample of size $n_{2.}$ from a second population. Thus for the liver scan data of Table 10.2, $n_{1.} = 12$, $n_{2.} = 21$, and the null hypothesis specifies that $p_1 = p_2$, where p_1 denotes the probability that a black patient with local or regional disease will receive a liver scan and p_2 denotes the probability that a white patient with local or regional disease will receive a liver scan. The null hypothesis is called the homogeneity hypothesis because it specifies that the chance of success is the same for both populations.

The 2 × 2 Chi-Squared Test of Independence

In contrast to the homogeneity framework, 2×2 tables also arise when neither of $n_{.1}, n_{.2}, n_{1.}$, or $n_{2.}$ are fixed, but instead when each observation from a general population is cross-classified on the basis of two characteristics (having characteristic C, say, not having characteristic C; having characteristic D, say, not having characteristic D). The question is whether the occurences of the characteristics are *independent*. We now describe how the chi-squared statistic defined by equation (10.18) is also appropriate for a test of independence.

We rewrite Table 10.1 using slightly different notation (see Table 10.3).

TABLE 10.3. 2 × 2 Table of Outcomes

	C	Not C	Totals
D	$\mathcal{O}_{11}$	$\mathcal{O}_{12}$	$n_{1.}$
Not D	$\mathcal{O}_{21}$	$\mathcal{O}_{22}$	$n_{2.}$
Totals:	$n_{.1}$	$n_{.2}$	$n_{..}$

(10.23)

Let $p_{i,j}$, $i = 1, 2$, $j = 1, 2$ denote the true unknown joint probability of falling into cell (i, j) of Table 10.3. Thus

$$p_{11} = P(C \text{ and } D), \quad p_{12} = P(\text{not } C \text{ and } D),$$
$$p_{21} = P(C \text{ and not } D), \quad p_{22} = P(\text{not } C \text{ and not } D). \tag{10.24}$$

The marginal probabilities are

$$p_{1.} = p_{11} + p_{12}, \quad p_{2.} = p_{21} + p_{22},$$
$$p_{.1} = p_{11} + p_{21}, \quad p_{.2} = p_{12} + p_{22}. \tag{10.25}$$

The hypothesis H_I of independence asserts that all joint probabilities are equal to the product of their marginal probabilities, namely,

$$H_I : p_{ij} = p_{i.} \times p_{.j}, \quad i = 1, 2, \quad j = 1, 2. \tag{10.26}$$

If, for example, E_{11} denotes the expected number of observations that fall in cell $(1, 1)$ of Table 10.3, we have

$$E_{11} = n_{..} \times P(C \text{ and } D) = n_{..} \times p_{11},$$

and under H_I,

$$E_{11} = n_{..} \times P(C) \times P(D) = n_{..} \times p_{.1} \times p_{1.}$$

It is natural to estimate $P(C) = p_{1.}$ by $(O_{11} + O_{21})/n_{..}$ and $P(D) = p_{1.}$ by $(O_{11} + O_{12})/n_{..}$. That is, $P(C)$ is estimated by the relative frequency of event C, and $P(D)$ is estimated by the relative frequency of event D. Thus under the hypothesis of independence, the E's are estimated as (we are abusing notation and using the same symbol E_{ij} for the expected number of observations falling into the (i, j) cell and an estimator of that expected number)

$$E_{11} = n_{..} \times \left(\frac{O_{11} + O_{21}}{n_{..}}\right) \times \left(\frac{O_{11} + O_{12}}{n_{..}}\right) = \frac{n_{.1} \times n_{1.}}{n_{..}},$$

and

$$E_{12} = \frac{n_{.2} \times n_{1.}}{n_{..}},$$

$$E_{21} = \frac{n_{.1} \times n_{2.}}{n_{..}},$$

$$E_{22} = \frac{n_{.2} \times n_{2.}}{n_{..}}. \tag{10.27}$$

Note that the E's given by display (10.27) agree with the E's given by display (10.17). It follows that the χ^2 statistic given by equation (10.20) can also be used to test independence. Specifically, an approximate α-level test of H_I, versus alternatives where association holds between the two characteristics, is

$$\text{Reject } H_I \text{ if } \chi^2 \geq \chi^2_{\alpha,1}; \quad \text{otherwise do not reject.} \tag{10.28}$$

We illustrate this test in Example 10.2.

Example 10.2: *Death Penalty and Gun Registration.* The data in Table 10.4 are given by Agresti (1990) and were reported by Clogg and Shockey (1988), whose source was the 1982 General Social Survey.

For these data, we can calculate χ^2 via equation (10.18), the shortcut formula (10.22), or using `Minitab` as we did for the liver scan data. We find $\chi^2 = 5.15$ with a P-value of .023, indicating there is an association between the two characteristics, namely, attitude toward gun registration and attitude toward the death penalty.

TABLE 10.4. Gun Registration and Death Penalty Cross-Classification

Gun Registration	Death Penalty		Totals
	Favor	Oppose	
Favor	784	236	1020
Oppose	311	66	377
Totals:	1095	302	1397

Source: 1982 General Social Survey; see C. C. Clogg and J. W. Shockey (1988).

10.1. APPROXIMATE TESTS AND CONFIDENCE INTERVALS

Although χ^2 measures, via the formal hypothesis test, the significance of association between two characteristics, it does *not* measure the *degree* of association. In Section 10.3 we will discuss a measure of the degree of association based on the odds ratio.

Comments

1. *Sample-Size Determination.* Suppose we want to determine sample sizes so that our estimate $\hat{p}_1 - \hat{p}_2$ of the true difference $p_1 - p_2$ will be within D of the true value, with probability equal to $1 - \alpha$. Equating the desired precision, D, to the actual precision, $(z_{\alpha/2}) \cdot SD(\hat{p}_1 - \hat{p}_2)$, yields the equation

$$D = z_{\alpha/2} \cdot \sqrt{\frac{p_1(1-p_1)}{n_1.} + \frac{p_2(1-p_2)}{n_2.}},$$

Taking $n_1. = n_2. = m$ and solving for m yields

$$m = \frac{(z_{\alpha/2})^2 \cdot [p_1(1-p_1) + p_2(1-p_2)]}{D^2}. \tag{10.29}$$

We cannot use equation (10.29) as it stands, because p_1 and p_2 are not known. (The purpose of the experiment is to obtain information about the unknown values of p_1, p_2, and $p_1 - p_2$.) However, the term in square brackets in the numerator of the right-hand side of (10.29) is largest when $p_1 = p_2 = \frac{1}{2}$. Thus a sufficiently large-sample would be

$$m = \frac{(z_{\alpha/2})^2 \cdot [(\frac{1}{2})(\frac{1}{2}) + (\frac{1}{2})(\frac{1}{2})]}{D^2} = \frac{(z_{\alpha/2})^2}{2D^2}.$$

This sample size assures the desired reliability regardless of the values of p_1 and p_2. In situations in which it is known that p_1 and p_2 are definitely less than some maximum value p^* (say), which is less than $\frac{1}{2}$, p^* can be substituted for p_1 and p_2 in (10.29), yielding

$$m = \frac{(z_{\alpha/2})^2 [2(p^*)(1-p^*)]}{D^2}. \tag{10.30}$$

The same is true if p_1 and p_2 are known to be greater than some value p^*, which is greater than $\frac{1}{2}$.

2. *Testing $p_1 - p_2$ Equals Some Specific Nonzero Value.* In this section, the tests based on A were formulated for the null hypothesis $p_1 = p_2$, or, equivalently, $p_1 - p_2 = 0$. To test $p_1 - p_2 = \delta_0$ (say), where δ_0 is any specified nonzero value between -1 and 1, use the statistic A', defined as

$$A' = \frac{(\hat{p}_1 - \hat{p}_2) - \delta_0}{\widehat{SD}(\hat{p}_1 - \hat{p}_2)}. \tag{10.31}$$

Note that the denominator of A' uses $\widehat{SD}(\hat{p}_1 - \hat{p}_2)$, as given by equation (10.5), rather than the estimator $\widetilde{SD}(\hat{p}_1 - \hat{p}_2)$, given by equation (10.12), based on pooling the two sample proportions together. The statistic A' should be referred to percentiles of the normal distribution. Significantly large values of A' indicate $p_1 - p_2 > \delta_0$; significantly small values of A' indicate $p_1 - p_2 < \delta_0$.

3. *Different Sampling Schemes*. In the framework of the test of homogeneity, we are testing $H_0 : p_1 = p_2$, a comparison of success probabilities using two binomial samples. In the notation of Table 10.1 we observe $\mathcal{O}_{11}$ successes out of $n_{1.}$ observations in sample 1, a sample from an underlying population 1 (say), and we observe $\mathcal{O}_{21}$ successes out of $n_{2.}$ observations in sample 2, a sample from an underlying population 2. Here $n_{1.}$ and $n_{2.}$ are fixed, but $n_{.1}$ and $n_{.2}$ are not fixed but random (although $n_{.1}$ and $n_{.2}$ are constrained to sum to $n_{..}$). In the framework of testing for independence, the sampling scheme is different. It is known as cross-sectional sampling. A total of $n_{..}$ subjects are obtained from an underlying population, and then each subject falls, in the notation of Table 10.3, into the 4 cells of the 2 × 2 table according to whether or not the subject possesses characteristic C and whether or not he has characteristic D. Here none of the row and column totals $n_{.1}$, $n_{.2}$, $n_{1.}$, $n_{2.}$ are fixed; only $n_{..}$ is fixed.

4. *Determining If the Sample Sizes Are Large Enough for the Large-Sample Approximation*. The tests and confidence interval of this section depend on an approximation to exact probabilities. The approximation is close if the samples are large. They should be large enough so that the E's, defined by display (10.17), each are no smaller than five.

5. *Yates' Correction for Continuity*. Yates (1934) proposed a continuity correction for the chi-squared test. The corrected χ^2 statistic is

$$\chi_c^2 = \frac{n_{..}(|\mathcal{O}_{11}\mathcal{O}_{22} - \mathcal{O}_{21}\mathcal{O}_{12}| - n_{..}/2)^2}{n_{.1} \times n_{.2} \times n_{1.} \times n_{2.}}. \tag{10.32}$$

This correction can be used in both the test for independence and the test for homogeneity. To get a P-value, refer χ_c^2 to the chi-squared distribution with 1 degree of freedom. To perform a formal test of the null hypothesis, reject H_0 at the (approximate) type I error probability α if $\chi_c^2 \geq \chi_{\alpha,1}^2$ and accept H_0 if $\chi_c^2 < \chi_{\alpha,1}^2$; otherwise do not reject.

There is disagreement in the statistical literature about the virtue of the continuity correction based on χ_c^2. See Storer and Kim (1990) and the references therein. On the basis of their study, Storer and Kim hold the view that in the context of comparing two binomial samples (i.e., testing homogeneity with two marginals $n_{1.}$ and $n_{2.}$ fixed), Yates' continuity correction should not be used. They also state "... our results suggest that for any reasonable sample size one will not be led far astray by the simple uncorrected χ^2 statistic." Storer and Kim actually compare seven tests of the homogeneity hypothesis $H_0 : p_1 = p_2$, including the approximate test based on χ^2, the approximate test based on χ_c^2, and Fisher's exact test. Fisher's exact test is conditional on all marginal totals $n_{.1}$, $n_{.2}$, $n_{1.}$, $n_{2.}$ being fixed. We present this test in Section 10.2.

6. *McNemar's Test*. Instead of having two independent samples to form the 2 × 2 table, as is the setup for the homogeneity test, there will be experiments in which the categorical data are based on dependent samples. Dependent samples can occur in matched-pair studies. For example, a pair may consist of a sibling and a parent. That is the situation for the Hodgkins tonsillectomy example in this comment. Dependent samples can also occur when the same subject is measured at two different times.

Johnson and Johnson (1972), in a study that was interested in testing the theory that the tonsils protect the body against invasion of the lymph nodes by a Hodgkin's disease virus (also see Problem 3), obtained tonsillectomy data on 85 Hodgkin's cases and a sibling of each case. The data showed 41 tonsillectomies among the Hodgkin's cases and 33 tonsillectomies among the siblings. The pairing of a case with sibling means that the rates for the two groups are not independent. The pairing should be taken into account in the analysis in order to achieve the best chance of detecting a departure from the null hypothesis. The null hypothesis asserts that Hodgkin's cases and their siblings have the same rates of tonsillectomy. A proper way to

10.1. APPROXIMATE TESTS AND CONFIDENCE INTERVALS

TABLE 10.5. Tonsillectomy Rates for Hodgkin's Disease Patients and Siblings

		Sibling		Totals
		Tonsillectomy	No Tonsillectomy	
Hodgkin's	Tonsillectomy	26	15	41
Patients	No Tonsillectomy	7	37	44
	Totals:	33	52	85

Source: S. K. Johnson and R. E. Johnson (1972).

test the null hypothesis is to apply the one-sample binomial test (Section 2.1) to Table 10.5, obtained by Johnson and Johnson.

If there is no association between tonsillectomy and Hodgkin's disease, then the probability is $\frac{1}{2}$ that a patient-sibling pair falls in the upper-right cell and $\frac{1}{2}$ that it falls in the lower-left cell, given that the pair falls off the main diagonal. Since the pairs are independent, the ratio $\frac{15}{22}$ can be compared with $\frac{1}{2}$ by a binomial test, as described in Chapter 2. Specifically, let

$$\widehat{p}_{.1} = \text{proportion of siblings with tonsillectomy} = \frac{26 + 7}{85},$$

$$\widehat{p}_{1.} = \text{proportion of Hodgkin's patients with tonsillectomy} = \frac{26 + 15}{85}.$$

The statistics $\widehat{p}_{.1}$ and $\widehat{p}_{1.}$ are estimates of $p_{.1}$ and $p_{1.}$, respectively. The difference between $\widehat{p}_{1.}$ and $\widehat{p}_{.1}$ is

$$d = \widehat{p}_{1.} - \widehat{p}_{.1} = \frac{26 + 15}{85} - \frac{26 + 7}{85} = \frac{15 - 7}{85}$$

and the estimated standard deviation of this difference is

$$\widehat{SD}(d) = \frac{\sqrt{15 + 7}}{85}.$$

An approximate large-sample test can be applied by referring

$$\frac{d}{\widehat{SD}(d)} = \frac{\frac{15-7}{85}}{\frac{\sqrt{15+7}}{85}} = \frac{15 - 7}{\sqrt{15 + 7}} = 1.71$$

to a $N(0, 1)$ distribution. The approximate one-sided P value is .044, indicating there is evidence that the rate of tonsillectomy is higher for Hodgkin's cases than for their siblings.

The exact one-sided P-value is $\Pr(B \geq 15)$, where B is a binomial random variable with $p = \frac{1}{2}$ and $n = 22$. This is easily found using the Minitab command CDF in conjunction with the subcommand BINOMIAL:

```
MTB>CDF;
SUBC>BINOMIAL N=22,P=0.50.
```

The Minitab output gives $\Pr(B \leq x)$ for each value of x. To find $\Pr(B \geq 15)$ we use

$$\Pr(B \geq 15) = 1 - \Pr(B \leq 14).$$

Minitab gives $\Pr(B \leq 14) = .9931$ and thus

$$\Pr(B \geq 15) = 1 - .9331 = .0669.$$

The one-sided approximate P-value of .044 found by the normal approximation is in reasonable agreement with this exact one-sided P-value of .067. Note, however, that if you were using an $\alpha = .05$ level, you would accept $H_0' : p_{1.} = p_{.1}$ in favor of the alternative $p_{1.} > p_{.1}$ with the one-sided exact McNemar's test, but you would reject H_0 with the normal approximation.

StatXact will give approximate exact P-values for McNemar's test. First create a 2×2 table with the TA command, and enter the four cell values $\mathcal{O}_{11}$, $\mathcal{O}_{12}$, $\mathcal{O}_{21}$ and $\mathcal{O}_{22}$. Then the MC command yields output including the approximate P-value and the exact P-value. For the data of Table 10.5, StatXact gives the one-sided approximate P-value to be .0440 and the one-sided exact P-value to be .0669, agreeing with the values we obtained earlier.

More generally, suppose we are dealing with a retrospective study where each case has been matched with a control. We wish to compare the frequency of an antecedent factor (in the preceding example, tonsillectomy) among the cases with the frequency of the antecedent factor among the controls. The data can be summarized as in Table 10.6.

TABLE 10.6. Data on Matched Pairs

		Controls		
		Factor present	Factor absent	Totals
Cases	Factor present	$\mathcal{O}_{11}$	$\mathcal{O}_{12}$	$n_{1.}$
	Factor absent	$\mathcal{O}_{21}$	$\mathcal{O}_{22}$	$n_{2.}$
	Totals:	$n_{.1}$	$n_{.2}$	$n_{..}$

The null hypothesis is $H_0' : p_{1.} = p_{.1}$, which is equivalent to $p_{12} = p_{21}$.

To test $H_0' : p_{1.} = p_{.1}$, the hypothesis that asserts the cases and controls have the same population proportions of the antecedent factor, refer

$$d = \frac{\hat{p}_{1.} - \hat{p}_{.1}}{\widehat{SD}(d)} = \frac{\mathcal{O}_{12} - \mathcal{O}_{21}}{\sqrt{\mathcal{O}_{12} + \mathcal{O}_{21}}} \qquad (10.33)$$

to a $N(0, 1)$ distribution.

The two-sided test of $H_0' : p_{1.} = p_{.1}$ versus the alternative $p_{1.} \neq p_{.1}$ at the (approximate) α level is reject H_0' if $|d| \geq z_{\alpha/2}$; otherwise do not reject.

The one-sided (approximate) α-level test of H_0' versus the alternative $p_{1.} > p_{.1}$ is reject H_0' if $d \geq z_\alpha$; otherwise do not reject. Similarly, the one-sided (approximate) α level test of H_0' versus the alternative $p_{1.} < p_{.1}$ is reject H_0' if $d \leq -z_\alpha$; otherwise do not reject.

See McNemar (1947), Mosteller (1952), and Agresti (1990) for further details.

7. *Edwards' Correction for Continuity.* Recall McNemar's test (Comment 6) in the matched-pairs situation. The test is based on the statistic

$$d = \frac{\mathcal{O}_{12} - \mathcal{O}_{21}}{\sqrt{\mathcal{O}_{12} + \mathcal{O}_{21}}}.$$

10.1. APPROXIMATE TESTS AND CONFIDENCE INTERVALS

The approximate two-sided α-level test of $H_0' : p_{1.} = p_{.1}$ versus the alternative $p_{1.} \neq p_{.1}$ is reject H_0' if $|d| \geq z_{\alpha/2}$; otherwise do not reject. An equivalent test is to compute

$$d^2 = \frac{(\mathcal{O}_{12} - \mathcal{O}_{21})^2}{\mathcal{O}_{12} + \mathcal{O}_{21}} \tag{10.34}$$

and reject H_0' if $d^2 \geq \chi^2_{\alpha,1}$, do not reject H_0' if $d^2 < \chi^2_{\alpha,1}$, where $\chi^2_{\alpha,1}$ is the upper α percentile point of the chi-squared distribution with 1 degree of freedom. To correct this test for continuity, a correction due to Edwards (1948) is based on computing

$$\chi_e^2 = \frac{(|\mathcal{O}_{12} - \mathcal{O}_{21}| - 1)^2}{\mathcal{O}_{12} + \mathcal{O}_{21}} \tag{10.35}$$

and referring χ_e^2 to tables of the chi-squared distribution with 1 degree of freedom.

Properties

1. *Asymptotic Distribution of Pearson's Chi-Squared Statistic.* See Agresti (1990, Sections 3.3.2 and 12.3.4).
2. *Asymptotic Equivalence of Pearson's Chi-Squared Statistic and the Likelihood Ratio Statistic.* See Agresti (1990, Section 12.3.4).
3. *Power of Chi-Squared Test.* See Agresti (1990, Section 7.6.3).

PROBLEMS

1. Andrews (1995) investigated bodily shame as a possible mediating factor between abusive experiences (sexual and physical) and later depression in a community sample of adult women. One hundred one women, who ranged in age from 32 to 56 years, were selected from an original longitudinal study of 289 women performed between 1980 and 1983 in Islington, an inner-city area of London, England (Brown et al. 1986). The 3-year study was designed to investigate the onset and course of depressive disorder. The investigators concentrated on working-class women with at least one child at home in order to get a group of women who were at high risk of developing clinical depression. Table 10.7, adapted from Andrews (1995), is a 2 × 2 table for the purpose of investigating association between childhood abuse and depression. (Physical and sexual abuse were combined into one category, abuse.)

TABLE 10.7. Abuse in Childhood and Depression in the 8-Year Study Period

Abuse in Childhood	Depression	No Depression	Totals
Yes	17	14	31
No	22	48	70
Totals:	39	62	101

Source: B. Andrews (1995).

From Table 10.7, we see that 17 out of 31 women with childhood abuse had been depressed, whereas 22 out of 70 women who had not expressed childhood abuse had been depressed. Test for independence of childhood abuse and depression against alternatives of association.

2. In the study by Andrews (1995) described in Problem 1, there was also an investigation into the possible association between abuse in adulthood and depression. Table 10.8, adapted from Andrews, gives the results.

TABLE 10.8. Abuse in Adulthood and Depression in the 8-Year Study Period

Abuse in Adulthood	Depression	No Depression	Totals
Yes	23	15	38
No	16	47	63
Totals:	39	62	101

Source: B. Andrews (1995).

Test for independence of adulthood abuse and depression against alternatives of association.

3. Vianna, Greenwald, and Davies (1971) considered a series of 101 Hodgkin's disease patients, with the purpose of testing the theory that the tonsils protect the body against invasion of the lymph nodes by a Hodgkin's disease virus. (The existence of such a virus has not been established.) Among the 101 Hodgkin's cases, they found 67 had had a tonsillectomy, whereas in a control group of 107 patients with other complaints, 43 had had a tonsillectomy. Compute 99% confidence limits for the difference in true rates.

4. R. Goode and D. Coursey, in a study of the theory that the tonsils serve as a reservoir harboring the virus that causes mononucleosis, obtained data on Stanford students seeking treatment for mononucleosis at the Stanford University Student Health Service. The data are given in Miller (1980). Among 46 students 21 years old diagnosed as having mononucleosis, they found that only 8 had had a tonsillectomy, whereas among 139 students of the same age, in the health center for other complaints, 48 had had a tonsillectomy. Compute 95% confidence limits for the difference in rates.

5. Verify directly the equivalence of equations (10.18) and (10.22).

6. Cruess (1989) points out that the error of applying the ordinary chi-squared statistic to paired data occurs frequently in the medical literature. Cruess cites in particular the study of Shen et al. (1988). They compared the results of two tests, ABC-ELISA and standard ELISA, on 101 hydatidosis patients. (Hydatidosis, or hydatid disease, is infestation with echinococcus, a genus of tapeworms.) Shen et al. used the ordinary unpaired chi-squared test and reported a P-value < 0.005. This was inappropriate because each case was tested using both laboratory procedures and thus the data were paired. Instead of the ordinary unpaired chi-square test, McNemar's test should have been performed. Table 10.9 gives the information on the 101 pairs.

TABLE 10.9. ABC-ELISA and Standard ELISA

Standard ELISA		+	−	Totals
ABC-ELISA	+	82	13	95
	−	6	0	6
	Totals:	88	13	101

Source: D. F. Cruess (1989).

Perform McNemar's test. What is the P-value? What do you conclude concerning the hypothesis of equal proportions positive for both ELISA tests?

7. Suppose you are planning an experiment to investigate two success rates p_1, p_2. Determine the value of the sample size m for each sample (in the equal-sample-size-case) so that your estimate, $\widehat{p}_1 - \widehat{p}_2$, of the true difference, $p_1 - p_2$, will be within .2 of the true difference with probability .95.
8. Verify equation (10.20). That is, show χ^2 given by equation (10.18) is equal to the square of A, where A is given by equation (10.7).
9. Astin et al. studied posttraumatic stress disorder (PTSD) and childhood abuse in battered women. PTSD prevalence rates were compared among 50 battered women and 37 maritally distressed women who had not experienced battering. The results are given in Table 10.10.

TABLE 10.10. PTSD Rates

PTSD	Battered Women	Maritally Distressed Women Who Had not Experienced Battering	Totals
Yes	29	7	36
No	21	30	51
Totals:	50	37	87

Source: M. C. Astin, S. M. Ogland-Hand, E. M. Coleman and D. W. Foy (1995).

Is there a significant difference in the PTSD rates for battered women versus maritally distressed women (who had not experienced battering)?

10. Recall Table 10.3 and the definition of independence given by H_I (10.26). We define the conditional probabilities $p_{j|i} = p_{ij}/p_{i.}$, $i = 1, 2$, $j = 1, 2$. Thus, for example, $p_{1|2}$ is the conditional probabililty of the observation landing in column 1 (i.e., has characteristic C) given that the observation has landed in row 2 (i.e., does not have characteristic D). Show that H_I is equivalent to the equalities $p_{1|1} = p_{1|2}$ and $p_{2|1} = p_{2|2}$ being satisfied.
11. Show that the four equalities of H_I (given by (10.26)) are satisfied if and only if $p_{11} = p_{1.} \times p_{.1}$.

10.2. AN EXACT TEST FOR THE DIFFERENCE BETWEEN TWO SUCCESS PROBABILITIES (FISHER)

Recall the basic 2×2 table given in display (10.1). Fisher's (1934) exact test is based on the conditional distribution of $\mathcal{O}_{11}$ given the row and column sums $n_{1.}, n_{2.}, n_{.1}, n_{.2}$. The conditional distribution of $\mathcal{O}_{11}$ is

$$\Pr(\mathcal{O}_{11} = x \mid n_{1.}, n_{2.}, n_{.1}, n_{.2}) = \frac{\binom{n_{1.}}{x}\binom{n_{2.}}{n_{.1} - x}}{\binom{n_{..}}{n_{.1}}}. \qquad (10.36)$$

The range of possible values for x is $n_L \leq x \leq n_U$, where $n_L = \max(0, n_{1.} + n_{.1} - n_{..})$ and $n_U = \min(n_{1.}, n_{.1})$. The conditional probability distribution defined by equation (10.36) is a member of a family of distributions known as *hypergeometric distributions*. Expression (10.36) can be put in a more readily usable form by simplifying the binomial coefficients appearing in the numerator and denominator. Such simplification allows us to rewrite that equation as

$$\Pr(\mathcal{O}_{11} = x \mid n_{1.}, n_{2.}, n_{.1}, n_{.2}) = \frac{n_{.1}! n_{.2}! 1_{1.}! 1_{2.}!}{n_{..}! x! \mathcal{O}_{12}! \mathcal{O}_{21}! \mathcal{O}_{22}!}. \qquad (10.37)$$

Fisher's exact test judges whether $\mathcal{O}_{11}$ is significantly small or significantly large with respect to the conditional distribution defined by equation (10.36). Specifically, to test $H_0 : p_1 = p_2$ versus the alternative $p_1 < p_2$, Fisher's exact test is reject H_0 if $\mathcal{O}_{11} \leq q_\alpha$, otherwise do not reject, where q_α is chosen from the conditional distribution so that $Pr(\mathcal{O}_{11} \leq q_\alpha | n_{1.}, n_{2.}, n_{.1}, n_{.2}) = \alpha$. Similarly, to test $H_0 : p_1 = p_2$ versus the alternative $p_1 > p_2$, Fisher's exact test is reject H_0 if $\mathcal{O}_{11} \geq r_\alpha$, otherwise do not reject, where r_α satisfies $Pr(\mathcal{O}_{11} \geq r_\alpha | n_{1.}, n_{2.}, n_{.1}, n_{.2}) = \alpha$. A two-sided α level test of $H_0 : p_1 = p_2$ versus the alternative $p_1 \neq p_2$ is reject H_0 if $\mathcal{O}_{11} \leq q_{\alpha_1}$ or if $\mathcal{O}_{11} \geq r_{\alpha_2}$, otherwise do not reject, where $q_{\alpha_1}, r_{\alpha_2}$ are chosen to give α_1 probability in the lower tail, and α_2 probability in the upper tail, where $\alpha_1 + \alpha_2 = \alpha$. Critical values for these tests can be obtained from the tables of Finney et al. (1963), and P-values for the test can be obtained from StatXact.

Example 10.3: *Example 10.1 Continued.* Recall the liver scan data of hospital 8. We now illustrate how, for that hospital, Fisher's exact test can be used to see if there was a significant difference between the chance of a white patient receiving a scan and the chance of a black patient receiving a scan. (Later, in Section 10.4, we use a test due to Mantel and Haenszel (1959) to get an overall conclusion based on the data from the 19 hospitals with the most black patients.)

Let p_1 be the unknown probability that a black patient receives a liver scan and let p_2 be the unknown probability that a white patient receives a liver scan. The null hypothesis is $H_0 : p_1 = p_2$. Diehr and her colleagues suspected that a deviation from H_0 would be in the direction of the one-sided alternative $p_1 > p_2$ and thus they reported the one-sided P-value corresponding to large values of $\mathcal{O}_{11}$. To find the P-value corresponding to their observed value $\mathcal{O}_{11} = 4$ (see Table 10.2) for hospital 8, we need to evaluate the probabilities of the tables giving a value as large or larger than the observed value of $\mathcal{O}_{11}$. These tables are

$$\begin{array}{ccc} 4 & 8 & 12 \\ 1 & 20 & 21 \\ \hline 5 & 28 & 33 \end{array} \qquad \begin{array}{ccc} 5 & 7 & 12 \\ 0 & 21 & 21 \\ \hline 5 & 28 & 33 \end{array}$$

with the table on the left corresponding to Table 10.2 and the table on the right being the only more extreme response in the direction $p_1 > p_2$. Next we use equation (10.35) to calculate the probabilities associated with each of these two tables, corresponding to $x = 4$ and $x = 5$.

x	Table	Probability		
4	$\begin{array}{ccc} 4 & 8 & 12 \\ 1 & 20 & 21 \\ \hline 5 & 28 & 33 \end{array}$	$\dfrac{5!28!12!21!}{33!4!8!1!20}$	$= .0438$	(10.38)
5	$\begin{array}{ccc} 5 & 7 & 12 \\ 0 & 21 & 21 \\ \hline 5 & 28 & 33 \end{array}$	$\dfrac{5!28!12!21}{33!5!7!0!21!}$	$= .0033$	(10.39)

Thus the P-value for hospital 8 is

$$P = .0033 + .0438 = .047. \tag{10.40}$$

This constitutes strong evidence that in hospital 8 the chance that a black patient with local or regional breast cancer receives a liver scan is higher than the corresponding chance that a white patient with local or regional breast cancer receives a liver scan.

StatXact can be used to perform Fisher's exact test. After creating the table with the TA command, and entering the four values $\mathcal{O}_{11}, \mathcal{O}_{12}, \mathcal{O}_{21}, \mathcal{O}_{22}$, the FI/EX comand yields output including the exact probability .0438 corresponding to the observed $\begin{smallmatrix} 4 & 8 \\ 1 & 20 \end{smallmatrix}$ (see (10.38)) and the P-value of .0471 corresponding to adding the probability (.0438) of the observed table and the probabilities of the more extreme tables. (In this case, there is only one more extreme table, namely, $\begin{smallmatrix} 5 & 7 \\ 0 & 21 \end{smallmatrix}$ with probability .0033.)

Comments

8. *Justification for Fisher's Exact Test.* Result (10.36) is justified as follows. Suppose that in a group of $n_{..}$ individuals, $n_{1.}$ possess a certain attribute and $n_{..} - n_{1.}$ do not. If a sample of $n_{.1}$ individuals is drawn randomly from the $n_{..}$ individuals, without replacement, the chance that x of these individuals possess the attribute is given by the right-hand side of equation (10.36).

9. *Use of Fisher's Exact Test as a Test of Independence.* In this section we have introduced Fisher's exact test in the context of a test of homogeneity. It can, however, also be used to test independence (just as the approximate chi-squared test of Section 10.1 can be used as a test of homogeneity and also as a test of independence). If, using the notation of Table 10.3, the events C and D are independent, then the conditional distribution of $\mathcal{O}_{11}$ within the restricted set of samples having fixed row and column tables is again given by equation (10.36).

10. *Limited Choice of α Values.* For small sample sizes, the probability distribution of $\mathcal{O}_{11}$, given by expression (10.36), is highly discrete (i.e., has a small number of possible values with corresponding probabilities). Thus the user's choices for α, when performing the formal test, are limited. Equivalently, there will, in small-sample-size cases, be a small number of possible P-values. This is illustrated in Problem 12, where, for the data in Table 10.11, there are only three possible values for $\mathcal{O}_{11}$ with three corresponding P-values.

11. *Use and Misuse of Statistics.* Cruess (1989) reviewed the statistics of the 201 scientific articles published during the calendar year 1988 in *The American Journal of Tropical Medicine and Hygiene*. He determined that 148 of the articles had at least one detectable statistical error; most of the errors involved improper documentation or application of statistical hypothesis testing. Among others, Cruess cites the Mendis, Ihalamulla, and David (1988) article (considered in Problem 12) as one in which the sample sizes were not large enough to justify the large-sample approximation used to compute P-values. The errors cited by Cruess are typical of the uses and misuses of statistics in journals in other areas of medical research. Ironically, the papers containing errors are often the most clearly written. Indeed, it is possible to ascertain errors only if the authors supply sufficient data and detail so that other researchers can check their results. Thus it may be unfair to be overly critical of the papers that contain errors.

Properties

1. *Uniformly Most Powerful Unbiased (UMPU) Property of Fisher's Exact Test.* See Tocher (1950) and Agresti (1990, 70).
2. *Minimal Sample Sizes Required to Achieve a Certain Power for Specified Significance Levels.* See Gail and Gart (1973) and Suissa and Shuster (1985).

PROBLEMS

12. The data in Table 10.11 are from a study by Mendis, Ihalamulla, and David (1988). The researchers compared the reactivity of heterologous human immune sera from patients with multiple malaria attacks to sera from primary-attack patients.

TABLE 10.11. Reactivity of Multiple Malaria Attack Patients and Primary Attack Patients

Patients	Reactivity		Totals
	Low	High	
Multiple Attack	0	5	5
Primary Attack	2	1	3
Totals:	2	6	8

Source: K. N. Mendis, R. I. Ihalamulla, and P. H. David (1988).

We write the null hypothesis as $H_0 : p_1 = p_2$, where p_1 is the unknown probability that a multiple-attack patient will have low reactivity and p_2 is the unknown probability that a primary-attack patient will have low reactivity. What is the P-value achieved by these data if we use Fisher's exact text of H_0 against the alternative $p_1 < p_2$? What is your conclusion?

13. Consider Problem 12 and the reactivity data of Table 10.11. Mendis, Ihalamulla, and David (1988) claimed a "significantly higher incidence of reactivity" in the multiple-attack patients. (In our notation of Problem 12, their conclusion corresponds to the alternative $p_1 < p_2$, or equivalently, $1 - p_1 > 1 - p_2$. Mendis, Ihalamulla, and David (1988) based their conclusion on the χ^2 statistic defined by equation (10.18). Apply the approximate test based on χ^2. Compare their conclusion with your conclusion obtained in Problem 12, recalling that the P-value obtained in Problem 12 is one-sided and the P-value based on χ^2 is two-sided.

14. Return to the reactivity data of Table 10.11. Find the P-value based on Yate's continuity correction to χ^2 (see Comment 5). Compare your result with those of Problems 12 and 13.

15. Show that expressions (10.36) and (10.37), for the conditional distribution of $\mathcal{O}_{11}$ given the row and column sums, are equivalent.

16. For the Diehr et al. (1989) study, the liver scan data for hospital 16 are given in Table 10.12. Let p_1 be the probability that a black patient in hospital 16 with local or regional disease will receive a liver scan and let p_2 be the probability that a white patient in hospital 16 with local or regional disease will receive a liver scan. Test $H_0 : p_1 = p_2$ versus the alternative $p_1 > p_2$ using Fisher's exact test. What is the P-value?

TABLE 10.12. Patients with Local or Regional Disease Receiving Liver Scan in Hospital 16

Patients	Liver Scan		Totals
	Yes	No	
Black	2	3	5
White	3	12	15
Totals:	5	15	20

Source: P. Diehr, J. Yergan, J. Chu, P. Feigl, G. Glaefke, R. Moe, M. Bergner, and J. Rodenbaugh (1989).

17. For the data of Table 10.12, perform two chi-squared approximations, one without a continuity correction and with Yates's continuity correction. Compare the approximate P-values with the exact P-value obtained in Problem 16.

10.3. INFERENCE FOR THE ODDS RATIO (FISHER, CORNFIELD)

Although the χ^2 statistic, given by equation (10.18) of Section 10.1, measures, via the formal hypothesis test, the significance of association between two characteristics, it does *not* measure the *degree* of association. This is because χ^2 has the defect of depending not only on the true probabilities of landing in the four cells of the 2×2 table, but also on the total number of subjects. One commonly used measure for association that does not have this defect and also is readily interpretable is the sample odds ratio. To introduce the sample odds ratio, $\widehat{\theta}$, we first define the corresponding population odds ratio parameter θ.

Recall the notation of Tables 10.1 and 10.3 and the joint probabilities given by display (10.24). The odds, given that the subject is in row 1 of the 2×2 table, that the subject will be in column 1 (instead of column 2) are

$$\theta^{(1)} = \frac{\frac{p_{11}}{p_{11}+p_{12}}}{\frac{p_{12}}{p_{11}+p_{12}}} = \frac{p_{11}}{p_{12}}. \tag{10.41}$$

In terms of the notation of Table 10.3 with characteristics C and D, equation (10.41) can be written as

$$\theta^{(1)} = \frac{P(C \mid D)}{P(\text{not } C \mid D)} = \frac{p_{11}}{p_{12}}. \tag{10.42}$$

Similarly, the odds, given that the subject is in row 2 of the 2×2 table, that the subject will be in column 1 (instead of column 2) are

$$\theta^{(2)} = \frac{P(C \mid \text{not } D)}{P(\text{not } C \mid \text{not } D)} = \frac{p_{21}}{p_{22}}. \tag{10.43}$$

The odds ratio is the parameter

$$\theta = \frac{\theta^{(1)}}{\theta^{(2)}} = \frac{p_{11}p_{22}}{p_{12}p_{21}}. \tag{10.44}$$

The odds ratio can be any number between 0 and ∞. If the cell probabilities p_{11}, p_{12}, p_{21}, p_{22} are all positive, then independence of the characteristics C and D implies $\theta = 1$ and, conversely, $\theta = 1$ implies C and D are independent. If p_{11} or p_{22} is 0 (and p_{12} and p_{21} are not 0), then $\theta = 0$. If p_{12} or p_{21} is 0 (and p_{11} and p_{22} are not 0), then $\theta = \infty$. θ is undefined in each of the four cases (i) $p_{11} = 0$ and $p_{12} = 0$, (ii) $p_{21} = 0$ and $p_{22} = 0$, (iii) $p_{11} = 0$ and $p_{21} = 0$, and (iv) $p_{12} = 0$ and $p_{22} = 0$.

θ measures the strength of the association. A table having $1 < \theta < \infty$ is such that the probability of the subject landing in column 1, given that the subject is known to be in row 1, is higher than the probability that the subject will land in column 1, given that the subject is known to be in row 2. That is, in the notation of the conditional probabilities (see Problem 10), $p_{1|1}$ is greater than $p_{1|2}$. Correspondingly, a table for which $0 < \theta < 1$ has $p_{1|1} < p_{1|2}$.

Unconditional Procedures

Estimator of θ. The unconditional maximum likelihood estimator of θ is the sample odds ratio

$$\widehat{\theta} = \frac{\mathcal{O}_{11}\mathcal{O}_{22}}{\mathcal{O}_{12}\mathcal{O}_{21}}. \tag{10.45}$$

If $\mathcal{O}_{11}$ or $\mathcal{O}_{22}$ is 0 (and $\mathcal{O}_{12}$ and $\mathcal{O}_{21}$ are not zero), then $\widehat{\theta} = 0$. If $\mathcal{O}_{12}$ or $\mathcal{O}_{21}$ is 0 (and $\mathcal{O}_{11}$ and $\mathcal{O}_{22}$ are not zero), then $\widehat{\theta} = \infty$. $\widehat{\theta}$ is undefined in each of the four cases (i) $\mathcal{O}_{11} = 0$ and $\mathcal{O}_{12} = 0$, (ii) $\mathcal{O}_{21}$ and $\mathcal{O}_{22} = 0$, (iii) $\mathcal{O}_{11} = 0$ and $\mathcal{O}_{21} = 0$, and (iv) $\mathcal{O}_{12} = 0$ and $\mathcal{O}_{22} = 0$. To eliminate these difficulties, we may use the adjusted version

$$\widehat{\theta}_a = \frac{(\mathcal{O}_{11} + .5)(\mathcal{O}_{22} + .5)}{(\mathcal{O}_{12} + .5)(\mathcal{O}_{21} + .5)} \tag{10.46}$$

See Comment 20.

Confidence Intervals

We define

$$\nu = \ln(\theta) \tag{10.47}$$

and

$$\widehat{\nu} = \ln(\widehat{\theta}). \tag{10.48}$$

The distribution of $\widehat{\nu}$ converges more rapidly to its asymptotic distribution than does the distribution of $\widehat{\theta}$ to its asymptotic distribution. Thus, it is preferable to base tests and confidence intervals for the odds ratio on the asymptotic distribution of $\widehat{\nu}$.

The standard deviation of $\widehat{\nu}$ can be estimated by

$$\widehat{SD}(\widehat{\nu}) = \sqrt{\frac{1}{\mathcal{O}_{11}} + \frac{1}{\mathcal{O}_{12}} + \frac{1}{\mathcal{O}_{21}} + \frac{1}{\mathcal{O}_{22}}}, \tag{10.49}$$

or the adjusted version

$$\widetilde{SD}(\widehat{\nu}) = \sqrt{\frac{1}{\mathcal{O}_{11} + .5} + \frac{1}{\mathcal{O}_{12} + .5} + \frac{1}{\mathcal{O}_{21} + .5} + \frac{1}{\mathcal{O}_{22} + .5}}. \tag{10.50}$$

The null hypothesis $\theta = 1$ is equivalent to the null hypothesis $\nu = 0$.

a. Approximate One-Sided Upper-Tail Test. To test

$$H_0 : \nu = 0$$

versus

$$H_1 : \nu > 0,$$

10.3. INFERENCE FOR THE ODDS RATIO (FISHER, CORNFIELD)

at the approximate α-level of significance,

$$\text{Reject } H_0 \text{ if } \frac{\widehat{\nu}}{\widehat{SD}(\widehat{\nu})} \geq z_\alpha; \quad \text{otherwise do not reject.} \tag{10.51}$$

b. Approximate One-Sided Lower-Tail Test. To test

$$H_0 : \nu = 0$$

versus

$$H_2 : \nu < 0,$$

at the approximate α-level of significance,

$$\text{Reject } H_0 \text{ if } \frac{\widehat{\nu}}{\widehat{SD}(\widehat{\nu})} \leq -z_\alpha; \quad \text{otherwise do not reject.} \tag{10.52}$$

c. Approximate Two-Sided Test. To test

$$H_0 : \nu = 0$$

versus

$$H_3 : \nu \neq 0,$$

at the approximate α-level of significance,

$$\text{Reject } H_0 \text{ if } \left|\frac{\widehat{\nu}}{\widehat{SD}(\widehat{\nu})}\right| \geq z_{\alpha/2}; \quad \text{otherwise do not reject.} \tag{10.53}$$

For a symmetric two-sided confidence interval for ν, with approximate confidence coefficient $1 - \alpha$, set

$$\nu_L = \widehat{\nu} - z_{\alpha/2}\widehat{SD}(\widehat{\nu}) \tag{10.54}$$

and

$$\nu_U = \widehat{\nu} + z_{\alpha/2}\widehat{SD}(\widehat{\nu}) \tag{10.55}$$

A confidence interval for the odds ratio θ is obtained by exponentiating ν_L and ν_U given by (10.54) and (10.55). That is, a symmetric two-sided confidence interval for θ, with approximate confidence coefficient $1 - \alpha$, is (θ_L, θ_U), where

$$\theta_L = e^{\nu_L}, \tag{10.56}$$

$$\theta_U = e^{\nu_U}. \tag{10.57}$$

Exact conditional tests and confidence intervals for θ, which are more computationally tedious than the unconditional procedures, are presented in Comments 14 and 15, respectively.

Example 10.4: *Example 10.1 Continued.* We return to the liver scan data of Table 10.2. From equation (10.45), we find

$$\widehat{\theta} = \frac{4(20)}{8(1)} = 10.$$

That is, we estimate the odds to be 10 times higher for black patients (with local or regional disease) to receive a liver scan than for white patients (with local or regional disease) to receive a liver scan.

From equations (10.47) and (10.48) we find, respectively,

$$\widehat{\nu} = \ln(\widehat{\theta}) = 2.30,$$

and

$$\widehat{SD}(\widehat{\nu}) = \sqrt{\tfrac{1}{4} + \tfrac{1}{8} + \tfrac{1}{1} + \tfrac{1}{20}} = 1.19.$$

Computing $\widehat{\nu}/\widehat{SD}(\widehat{\nu})$, we obtain

$$\frac{\widehat{\nu}}{\widehat{SD}(\widehat{\nu})} = 1.93.$$

The corresponding approximate one-sided P-value, for testing $H_0 : \nu = 0$ versus $H_1 : \nu > 0$, is .027 (obtained from Table A.1). This is strong evidence that $\nu > 0$ or, equivalently, that the odds ratio θ is greater than 1.

A symmetric two-sided confidence interval for ν, with approximate confidence coefficient .95, is found from (10.54) and (10.55) to be

$$\nu_L = 2.30 - (1.96)(1.19) = -.03,$$
$$\nu_U = 2.30 + (1.96)(1.19) = 4.63.$$

Exponentiating, we find (see (10.56) and (10.57))

$$\theta_L = e^{-.03} = .97,$$
$$\theta_U = e^{4.63} = 102.5.$$

Comments

12. *Odds Ratio and Estimated Odds Ratio under Reversal of Rows and Columns.* If the roles of the rows and columns are reversed, then the odds ratio does not change. Reversing the roles of the rows and columns produces a new 2×2 table with underlying cell probabilities $p'_{11}, p'_{12}, p'_{21}, p'_{22}$, where $p'_{11} = p_{11}, p'_{12} = p_{21}, p'_{21} = p_{12}, p'_{22} = p_{22}$. Thus θ', the odds ratio for the new table, is

$$\theta' = \frac{p'_{11} p'_{22}}{p'_{12} p'_{21}} = \frac{p_{11} p_{22}}{p_{21} p_{12}} = \theta.$$

Similarly, the estimated odds ratio $\widehat{\theta}$ does not change when the roles of the rows and columns are reversed.

13. *Effect of Interchanging the Order of the Rows.* If rows 1 and 2 are interchanged, then the new table has underlying cell probabilities $p'_{11}, p'_{12}, p'_{21}, p'_{22}$, where $p'_{11} = p_{21}, p'_{12} = p_{22}$,

10.3. INFERENCE FOR THE ODDS RATIO (FISHER, CORNFIELD)

$p'_{21} = p_{11}$, $p'_{22} = p_{12}$ and the odds ratio for the new table is

$$\theta' = \frac{p'_{11} p'_{22}}{p'_{12} p'_{21}} = \frac{p_{21} p_{12}}{p_{22} p_{11}} = \frac{1}{\theta},$$

where θ is the odds ratio for the original table. Similarly, if columns 1 and 2 of the original table are interchanged, the new odds ratio is the inverse of the original odds ratio. The same results also hold for the estimated odds ratio. That is, if the rows are interchanged (or if the columns are interchanged), $\widehat{\theta}'$, the estimated odds ratio for the new table, is the inverse of $\widehat{\theta}$, the estimated odds ratio for the original table.

14. *Exact Conditional Test for the Odds Ratio.* The conditional distribution of $\mathcal{O}_{11}$ given $n_{1.}$, $n_{.1}$, and θ (due to Fisher, 1935) is

$$P(\mathcal{O}_{11} = x \mid n_{1.}, n_{.1}, \theta) = \frac{\binom{n_{1.}}{x} \binom{n_{..} - n_{1.}}{n_{.1} - x} \theta^x}{\sum_{y=n_L}^{n_U} \binom{n_{1.}}{y} \binom{n_{..} - n_{1.}}{n_{.1} - y} \theta^y} \qquad (10.58)$$

where (recall that) $n_L = \max(0, n_{1.} + n_{.1} - n_{..})$, $n_U = \min(n_{1.}, n_{.1})$. If $\mathcal{O}_{11}^{\text{obs}}$ denotes the observed value of $\mathcal{O}_{11}$ for your 2×2 table, the P-value for testing $H_0 : \theta = \theta_0$ versus $H_1 : \theta > \theta_0$ is

$$P = \sum_{\{\text{all } x\text{-values} \geq \mathcal{O}_{11}^{\text{obs}}\}} P(\mathcal{O}_{11} = x \mid n_{1.}, n_{.1}, \theta).$$

For testing $H_0 : \theta = \theta_0$ versus $H_2 : \theta < \theta_0$, the P-value is

$$P = \sum_{\{\text{all } x\text{-values} \leq \mathcal{O}_{11}^{\text{obs}}\}} P(\mathcal{O}_{11} = x \mid n_{1.}, n_{.1}, \theta).$$

With the choice $\theta_0 = 1$, these procedures reduce to Fisher's exact test.

The exact conditional test's P-value can be obtained using StatXact's OD/EX command. See Comment 17.

15. *Exact Conditional Confidence Intervals for the Odds Ratio.* Exact confidence intervals for θ can be obtained by inverting the tests of Comment 14. For a confidence interval (θ_L, θ_U), with confidence coefficient $\geq 1 - \alpha$, θ_L is the value of θ_0 for which the P-value is $\alpha/2$ when testing $H_0 : \theta = \theta_0$ versus $H_1 : \theta > \theta_0$.

Similarly, θ_U is the value of θ_0 for which the P-value is $\alpha/2$ when testing $H_0 : \theta = \theta_0$ versus $H_2 : \theta < \theta_0$. StatXact's OD/EX command can be used to obtain the exact conditional confidence interval. See Comment 17.

16. *Conditional Estimator of the Odds Ratio.* The conditional maximum likelihood estimator of θ is the value ($\widetilde{\theta}$, say) of θ that maximizes the right-hand side of (10.38). This yields a different estimator than the unconditional maximum likelihood estimator $\widehat{\theta}$ given by (10.45). The value $\widetilde{\theta}$ can be found by solving the equation $\mathcal{O}_{11} = E_c(\mathcal{O}_{11})$, where E_c denotes expectation with respect to the conditional distribution given by (10.58). The equation has a unique solution (Cornfield, 1956), which can be obtained by using iterative methods. $\widetilde{\theta}$ can be obtained using StatXact's OD/EX command. See Comment 17.

17. *Use of StatXact to Perform the Conditional Procedures.* The conditional methods of Comments 14, 15, and 16 can be performed using StatXact's OD/EX command. For the data of Table 10.2, use of OD/EX gives the one-sided P-value .0471 for testing $\theta = 1$,

agreeing with what we obtained in Example 10.3 using Fisher's exact test. (Recall that the tests of Comment 14 with $\theta_0 = 1$ reduce to Fisher's exact tests.) The OD/EX command also yields exact confidence intervals and the conditional maximum likelihood estimator $\tilde{\theta}$. StatXact gives the 95% confidence interval for $1/\theta$ as $(.0019, 1.306)$. The approximate (unconditional) 95% interval for $1/\theta$ found using the results of Example 10.4 is $(1/102.5, 1/1.03) = (.01, .97)$. The conditional maximum likelihood estimator of $1/\theta$ is found by StatXact to be .1081, and the corresponding conditional maximum likelihood estimator of θ is $(1/.1081) = 9.25$. Recall (see Example 10.4) that the unconditional maximum likelihood estimator of θ is 10. The StatXact procedures are based on the algorithm due to Mehta, Patel, and Gray (1985). For a survey on exact methods for contingency tables, see Agresti (1992).

18. *Yule's Measure of Association.* The odds ratio θ measures association, but it takes values from 0 to ∞. If there is a preference for a measure that lies between -1 and 1, we can use Yule's (1900, 1912) Q, which is defined as

$$Q = \frac{\theta - 1}{\theta + 1}$$

$$= \frac{p_{11}p_{22} - p_{12}p_{21}}{p_{11}p_{22} + p_{12}p_{21}}. \tag{10.59}$$

Its sample analogue for an observed 2×2 table is

$$\widehat{Q} = \frac{\widehat{\theta} - 1}{\widehat{\theta} + 1}$$

$$= \frac{O_{11}O_{22} - O_{12}O_{21}}{O_{11}O_{22} + O_{12}O_{21}}. \tag{10.60}$$

Since Q is an increasing function of θ, confidence limits for Q can be obtained from confidence limits for θ. In particular, an approximate $1 - \alpha$ confidence interval for Q is (Q_L, Q_U), where

$$Q_L = \frac{\theta_L - 1}{\theta_L + 1}, \tag{10.61}$$

$$Q_U = \frac{\theta_U - 1}{\theta_U + 1}, \tag{10.62}$$

where θ_L and θ_U are given by equations (10.56) and (10.57), respectively.

For the liver scan data of Table 10.2, from equation (10.60) we find

$$\widehat{Q} = \frac{(10 - 1)}{(10 + 1)} = .82.$$

From (10.61) and (10.62) and Example 10.4, an approximate 90% confidence interval for Q is (Q_L, Q_U), where

$$Q_L = \frac{(1.40 - 1)}{(1.40 + 1)} = .17,$$

$$Q_U = \frac{(70.8 - 1)}{(70.8 + 1)} = .97.$$

19. *Odds Ratio under Binomial Sampling.* Recall the binomial model of Section 10.1. There $\mathcal{O}_{11}$ is the number of successes in n_1 independent Bernoulli trials, each with success probability p_1, and $\mathcal{O}_{21}$ is the number of successes in n_2 independent Bernoulli trials, each with success probability p_2. (Also, the sample 1 trials are assumed independent of the sample 2 trials.) Then the odds ratio is defined to be

$$\theta = \frac{\dfrac{p_1}{1-p_1}}{\dfrac{p_2}{1-p_2}}. \tag{10.63}$$

20. *Adjusted Unconditional Estimator $\widehat{\theta}_a$.* The denominator of $\widehat{\theta}$ can be 0 with positive probability, and thus in particular, the mean and variance of $\widehat{\theta}$ do not exist. The adjusted estimator $\widehat{\theta}_a$ removes this difficulty. See Haldane (1955) and Gart and Zweiful (1967).

21. *Asymptotic Theory for $\ln(\widehat{\theta})$.* The delta method provides asymptotic normality results for functions g (that satisfy mild regularity conditions) of random variables, which themselves have an asymptotic multivariate normal distribution. The method produces an explicit expression for the variance of g, which in turn can be used to obtain a consistent estimator of that variance. The delta method is frequently used in categorical data analysis, where, under multinomial sampling, the cell entries, suitably standardized, have an asymptotic (singular) multivariate normal distribution (cf. Goodman and Kruskal (1963) and Agresti (1990, 57, 423)). Applying the delta method to the function

$$g(\mathcal{O}_{11}, \mathcal{O}_{12}, \mathcal{O}_{21}, \mathcal{O}_{22}) = \ln\left(\frac{\mathcal{O}_{11}\mathcal{O}_{22}}{\mathcal{O}_{12}\mathcal{O}_{21}}\right) = \ln(\widehat{\theta}) \tag{10.64}$$

provides justification for procedures (10.51)–(10.55). The expression for the variance will depend on the p_{ij}'s that are then replaced by their sample counterparts $\mathcal{O}_{ij}/n_{..}$ to obtain a consistent estimator of the variance. Agresti (1990, 57) provides the elementary details for the case where g is given by (10.64).

Properties

1. *Bias of Odds Ratio Estimators.* See Haldane (1955), Gart and Zweiful (1967), and Agresti (1990, Section 3.4.1).
2. *Invariance Properties of Odds Ratio and Estimated Odds Ratio.* See Edwards (1963) and Comment 12.

PROBLEMS

18. For the data of Table 10.4, estimate θ and compute a 95% confidence interval for θ. Interpret your results.
19. For the data of Table 10.7, estimate θ and compute a 95% confidence interval for θ. Interpret your results.
20. For the data of Table 10.8, estimate θ and compute a 95% confidence interval for θ. Interpret your results.
21. For the data of Table 10.12, compute $\widehat{\theta}$ and obtain an approximate 95% confidence interval for θ.
22. The relative risk is defined to be

$$r = \frac{p_{1|1}}{p_{1|2}},$$

where (recall) $p_{1|1} = p_{11}/(p_{11} + p_{12})$ and $p_{1|2} = p_{21}/(p_{21} + p_{22})$. Show that if $r = 1$, then the hypothesis of independence holds and, conversely, independence implies $r = 1$.

23. Show that the odds ratio θ and the relative risk r satisfy

$$\theta = r \times \frac{1 - p_{1|2}}{1 - p_{1|1}}.$$

24. For the data of Table 10.4, compute $\widehat{Q}$ and obtain an approximate 95% confidence interval for Q.
25. For the data of Table 10.7, compute $\widehat{Q}$ and obtain an approximate 95% confidence interval for Q.
26. For the data of Table 10.8, compute $\widehat{Q}$ and obtain an approximate 95% confidence interval for Q.
27. For the data of Table 10.11, uses StatXact to compute $\tilde{\theta}$ and compare it to $\hat{\theta}$.
28. For the data of Table 10.12, use StatXact to get an exact 95% (conditional) confidence interval for θ.

10.4. INFERENCE FOR k STRATA OF 2×2 TABLES (MANTEL AND HAENSZEL)

Recall that Diehr et al. (1989) study discussed in Example 10.1, Section 10.1. In that example, we computed the respective probabilities, for hospital 8, of white and black patients receiving liver scans. Now we wish to apply an overall test that assesses the respective probabilities across the 19 hospitals (with the most black patients) in the Diehr et al. study.

More formally, suppose our data consist of k strata, and within each statum we have a 2×2 table. The two rows of the 2×2 table in the ith stratum are viewed as data from two independent binomial distributions with respective success probabilities $(p_1^{(i)}, p_2^{(i)})$, $i = 1, 2, \ldots, k$. In the Diehr et al. study $k = 19$ and the strata correspond to hospitals. Let

$p_1^{(i)} =$ probability that a black patient in hospital i (with local or regional disease) will receive a liver scan,

$p_2^{(i)} =$ probability that a white patient in hospital i (with local or regional disease) will receive a liver scan.

The data in the ith 2×2 table can be represented in the notation of Table 10.13.

Mantel and Haenszel (1959) proposed an approximate test of the null hypothesis H_0, which specifies that within each stratum, the success probabilities are equal. That is

$$H_0 : p_1^{(1)} = p_2^{(1)}, p_1^{(2)} = p_2^{(2)}, \ldots, p_1^{(k)} = p_2^{(k)}. \tag{10.65}$$

We let θ_i denote the odds ratio for the ith table. Recall that θ_i is defined as

$$\theta_i = \frac{p_1^{(i)}}{1 - p_1^{(i)}} \bigg/ \frac{p_2^{(i)}}{1 - p_2^{(i)}}. \tag{10.66}$$

TABLE 10.13. 2×2 Table for ith Stratum

	Successes	Failures	Totals
Sample 1	$\mathcal{O}_{11i}$	$\mathcal{O}_{12i}$	$n_{1.i}$
Sample 2	$\mathcal{O}_{21i}$	$\mathcal{O}_{22i}$	$n_{2.i}$
Totals:	$n_{.1i}$	$n_{.2i}$	$n_{..i}$

10.4. INFERENCE FOR k STRATA OF 2×2 TABLES (MANTEL AND HAENSZEL)

H_0 can be rewritten as

$$H_0 : \theta_1 = \theta_2 = \cdots = \theta_k = 1. \tag{10.67}$$

That is, we are testing that there is a common odds ratio and it is equal to 1. Note that H_0 allows for the common success probabilities to differ from hospital to hospital. The alternatives of interest are that $p_1^{(i)} \geq p_2^{(i)}$ for all $i = 1, \ldots, k$ or $p_1^{(i)} \leq p_2^{(i)}$ for all $i = 1, \ldots, k$. In terms of the liver scan problem, the alternatives specify that across the 19 hospitals, the probabilities that black patients receive liver scans are higher than the respective probabilities that white patients receive liver scans or across the 19 hospitals the black patients have lower probabilities than those of the white patients.

Approximate Conditional Procedure

In the ith table, given that the marginal totals $n_{1.i}, n_{2.i}, n_{.1i}, n_{.2i}$ are fixed, the random variable $\mathcal{O}_{11i}$ has a hypergeometric distribution

$$P(\mathcal{O}_{11i} = x) = \frac{\binom{n_{1.i}}{x}\binom{n_{2.i}}{n_{.1i} - x}}{\binom{n_{..i}}{n_{.1i}}}. \tag{10.68}$$

The null mean $E_0(\mathcal{O}_{11i})$ is given by

$$E_0(\mathcal{O}_{11i}) = \frac{(n_{1.i})(n_{.1i})}{n_{..i}}, \tag{10.69}$$

and the null variance $\text{var}_0(\mathcal{O}_{11i})$ is given by

$$\text{var}_0(\mathcal{O}_{11i}) = \frac{(n_{1.i})(n_{2.i})(n_{.1i})(n_{.2i})}{n_{..i}^2(n_{..i} - 1)}. \tag{10.70}$$

The Mantel-Haenszel (1959) statistic is

$$\text{MH} = \frac{\sum_{i=1}^{k}\{\mathcal{O}_{11i} - E_0(\mathcal{O}_{11i})\}}{\sqrt{\sum_{i=1}^{k} \text{var}_0(\mathcal{O}_{11i})}}. \tag{10.71}$$

An approximate α-level one-sided test of H_0 given by (10.65) against the alternatives

$$H_1 : p_1^{(i)} \geq p_2^{(i)}, \quad i = 1, \ldots, k \tag{10.72}$$

(with at least one inequality strict) is

$$\text{Reject } H_0 \text{ if MH} \geq z_\alpha; \quad \text{otherwise do not reject.} \tag{10.73}$$

An approximate α-level one-sided test of H_0 against the alternatives

$$H_2 : p_1^{(i)} \leq p_2^{(i)}, \quad i = 1, \ldots, k \tag{10.74}$$

(with at least one inequality strict) is

$$\text{Reject } H_0 \text{ if MH} \leq -z_\alpha; \quad \text{otherwise do not reject.} \tag{10.75}$$

An approximate α-level two-sided test of H_0 against the alternatives

$$H_3 : p_1^{(i)} \geq p_2^{(i)} \quad \text{for all } k \text{ or } p_1^{(i)} \leq p_2^{(i)} \text{ for all } k \tag{10.76}$$

(with at least one inequality strict) is

$$\text{Reject } H_0 \text{ if (MH)}^2 \geq \chi^2_{\alpha,1}; \quad \text{otherwise do not reject.} \tag{10.77}$$

Example 10.5: *Liver Scan Data for 19 Hospitals.* Table 10.14 gives, for the Diehr et al. (1989) study, the percent of patients with local or regional disease receiving liver scan. The data are for the 19 hospitals with the most black patients.

In Table 10.15 we give the values of $E_0(\mathcal{O}_{11i})$ and $\text{var}_0(\mathcal{O}_{11i})$, obtained using (10.69) and (10.70), respectively.

The 19 2×2 tables formed from Table 10.14 are as follows:

Hospital 1

4	9	13
12	34	46
16	43	59

Hospital 2

4	6	10
34	33	67
38	39	77

Hospital 3

7	2	9
6	7	13
13	9	22

Hospital 4

5	5	10
59	56	115
64	61	125

Hospital 5

7	7	14
22	69	91
29	76	105

Hospital 6

5	6	11
41	80	121
46	86	132

Hospital 7

3	6	9
8	72	80
11	78	89

Hospital 8

4	8	12
1	20	21
5	28	33

Hospital 9

7	2	9
77	38	115
84	40	124

Hospital 10

4	6	10
20	70	90
24	76	100

Hospital 11

1	8	9
16	76	92
17	84	101

Hospital 12

4	10	14
10	91	101
14	101	115

Hospital 13

9	18	27
27	118	145
36	136	172

Hospital 14

3	5	8
35	45	80
38	50	88

Hospital 15

9	5	14
69	20	89
78	25	103

Hospital 16

2	3	5
3	12	15
5	15	20

Hospital 17

6	1	7
45	31	76
51	32	83

Hospital 18

14	10	24
12	70	82
26	80	106

Hospital 19

15	15	30
43	129	172
58	144	202

From (10.71) and Table 10.15 we obtain

$$\text{MH} = \frac{113 - 81.004}{\sqrt{39.383}} = 5.10.$$

TABLE 10.14. Percent of Patients with Local or Regional Disease Receiving Liver Scans in 19 Hospitals with the Most Black Patients

Hospital	% with Scan (Number of Patients Eligible)			
	White		Black	
1	26.1%	(46)	30.8%	(13)
2	50.8%	(67)	40.0%	(10)
3	46.2%	(13)	77.8%	(9)
4	51.3%	(115)	50.0%	(10)
5	24.2%	(91)	50.0%	(14)
6	33.9%	(121)	45.5%	(11)
7	10.0%	(80)	33.3%	(9)
8	4.8%	(21)	33.3%	(12)
9	67.0%	(115)	77.8%	(9)
10	22.2%	(90)	40.0%	(10)
11	17.4%	(92)	11.1%	(9)
12	9.9%	(101)	28.6%	(14)
13	18.6%	(145)	33.3%	(27)
14	43.8%	(80)	37.5%	(8)
15	77.5%	(89)	64.3%	(14)
16	20.0%	(15)	40.0%	(5)
17	59.2%	(76)	85.7%	(7)
18	14.6%	(82)	58.3%	(24)
19	25.0%	(172)	50.0%	(30)

TABLE 10.15. Null Means and Variances of $\mathcal{O}_{11i}$ for Liver Scan Data

i	$\mathcal{O}_{11i}$	$E_0(\mathcal{O}_{11i})$	$\text{var}_0(\mathcal{O}_{11i})$
1	4	3.525	2.038
2	4	4.935	2.204
3	7	5.318	1.347
4	5	5.120	2.317
5	7	3.867	2.449
6	5	3.833	2.307
7	3	1.112	.886
8	4	1.818	1.012
9	7	6.097	1.839
10	4	2.400	1.658
11	1	1.515	1.159
12	4	1.704	1.326
13	9	5.651	3.789
14	3	3.455	1.805
15	9	10.602	2.245
16	2	1.250	.740
17	6	4.301	1.537
18	14	5.887	3.470
19	15	8.614	5.255

$$\sum_{i=1}^{19} \mathcal{O}_{11i} = 113 \qquad \sum_{i=1}^{19} E(\mathcal{O}_{11i}) = 81.004 \qquad \sum_{i=1}^{19} \text{var}_0(\mathcal{O}_{11i}) = 39.383$$

Referring MH = 5.10 to Table A.1 we find $P \ll .0002$. Thus there is very strong evidence that the hospitals do not have a common odds ratio that is 1. Since MH is significantly positive, there is strong evidence that across hospitals the odds that black patients get liver scans are higher than the odds that white patients get liver scans.

Comments

22. *Types of Alternatives the Mantel-Haenszel Test Detects.* Consider the one-sided procedure given by (10.73). This procedure rejects H_0 for significantly large values of MH. This test will be consistent against H_1 (given by (10.72)) if at least one of the inequalities is strict. However, if for some strata $p_1^{(i)} > p_2^{(i)}$ and for others the inequality goes in the other direction, the MH test will have less power because the statistic would then tend to add positive and negative deviations of the form $\{\mathcal{O}_{11i} - E_0(\mathcal{O}_{11i})\}$. Roughly speaking, the one-sided procedure based on (10.73) is consistent against alternatives A for which $E_A(\sum_{i=1}^{k}\{\mathcal{O}_{11i} - E_0(\mathcal{O}_{11i})\}) > 0$, where E_A indicates that the expectation is computed under alternative A. Similar considerations apply to the one-sided procedure given by (10.75) and the two-sided procedure given by (10.77).

23. *The Mantel-Haenszel Test Viewed as a Test of Conditional Independence.* Suppose the k 2×2 tables are tables of cross-classified data and in each stratum we are interested in whether factors C and D are (conditionally) independent. The alternative of positive association would correspond to C and D being positively associated across strata. The one-sided procedure given by (10.73) tests conditional independence against alternatives that C and D are positively associated across strata. As in Comment 22, the MH test will have less power if for some strata C and D are positively associated and, for others, C and D are negatively associated. Similar considerations apply to the one-sided procedure given by (10.75) and the two-sided procedure defined by (10.77).

24. *Zelen's Exact Test for a Common Odds Ratio.* The null hypothesis H_0, given by (10.67) specifies that the strata have a common odds ratio equal to 1. Zelen (1971) developed an exact conditional test of

$$H_0^* : \theta_1 = \theta_2 = \cdots = \theta_n = \theta \text{ (say)}, \tag{10.78}$$

where the common value θ is unspecified. The hypothesis H_0^* is called the hypothesis of homogeneity.

Recall Table 10.13 and let τ denote any collection of k 2×2 tables. Furthermore, define

$$T = \{\tau : \text{table } i \text{ in } \tau \text{ has marginal totals } n_{.1i}, n_{.2i}, n_{1.i}, n_{2.i}\} \tag{10.79}$$

The $\mathcal{O}_{11i}$ values can vary, but the marginal totals are fixed. Zelen's test is based on the restricted reference set

$$T(s) = \left\{\tau : \tau \in T \text{ and } \sum_{i=1}^{k} \mathcal{O}_{11i} = s\right\}. \tag{10.80}$$

Note that $T(s)$ is contained in T. The conditional probability of obtaining a specific set of k 2×2 tables that is a member of $T(s)$ is

$$P(\tau|s) = \frac{\prod_{i=1}^{k} \binom{n_{1.i}}{\mathcal{O}_{11i}} \binom{n_{2.i}}{\mathcal{O}_{21i}} / \binom{n_{..i}}{n_{.1i}}}{\sum_{\tau \in T(s)} \prod_{i=1}^{k} \binom{n_{1.i}}{\mathcal{O}_{11i}} \binom{n_{2.i}}{\mathcal{O}_{21i}} / \binom{n_{..i}}{n_{..i}}}. \tag{10.81}$$

10.4. INFERENCE FOR k STRATA OF 2×2 TABLES (MANTEL AND HAENSZEL)

Zelen's test is conditional on $\sum_{i=1}^{k} \mathcal{O}_{11i}$ as well as the marginal totals $n_{.1i}, n_{.2i}, n_{1.i}, n_{2.i}$ for each table. For all k 2×2 tables having the same marginals as the observed marginals, the probability given by (10.81) is calculated. The tables are then ordered according to those probabilities. The P-value is the sum of those probabilities for those tables whose probabilities are less than or equal to the probability of the observed table. That is, suppose τ_0 is our observed set of k 2×2 tables with marginal totals $n_{.1i}, n_{.2i}, n_{1.i}, n_{2.i}$ for the ith table and suppose that for τ_0, we have $\sum_{i=1}^{k} \mathcal{O}_{11i} = s$. Let

$$T^*(s) = \{\tau : \tau \in T(s) \text{ and } P(\tau \mid s) \leq P(\tau_0 \mid s)\}.$$

Then the P-value for Zelen's test is

$$P = \sum_{\tau \in T^*(s)} P(\tau \mid s).$$

This P-value is two-sided; the test does not indicate the direction of a deviation from H_0^*.

For example, suppose $k = 2$ and our observed set of two tables is

$$\tau_0 = \left\{ \begin{matrix} 2 & 2 \\ 1 & 5, \end{matrix} \begin{matrix} 1 & 4 \\ 5 & 1 \end{matrix} \right\}.$$

Thus, our observed value is $\mathcal{O}_{111} + \mathcal{O}_{112} = 3$. The four tables in the reference set $T(3)$ and their conditional probabilities given by (10.81) are as follows.

Table	Conditional Probability
1. $\left\{ \begin{matrix} 0 & 4 \\ 3 & 3, \end{matrix} \begin{matrix} 3 & 2 \\ 3 & 3 \end{matrix} \right\}$	$\dfrac{\left[\left\{\binom{4}{0}\binom{6}{3}\Big/\binom{10}{3}\right\} \cdot \left\{\binom{5}{3}\binom{6}{3}\Big/\binom{11}{6}\right\}\right]}{\Sigma} = .2841$
2. $\left\{ \begin{matrix} 1 & 3 \\ 2 & 4, \end{matrix} \begin{matrix} 2 & 3 \\ 4 & 2 \end{matrix} \right\}$	$\dfrac{\left[\left\{\binom{4}{1}\binom{6}{2}\Big/\binom{10}{3}\right\} \cdot \left\{\binom{5}{2}\binom{6}{4}\Big/\binom{11}{6}\right\}\right]}{\Sigma} = .6387$
3. $\left\{ \begin{matrix} 2 & 2 \\ 1 & 5, \end{matrix} \begin{matrix} 1 & 4 \\ 5 & 1 \end{matrix} \right\}$	$\dfrac{\left[\left\{\binom{4}{2}\binom{6}{1}\Big/\binom{10}{3}\right\} \cdot \left\{\binom{5}{1}\binom{6}{5}\Big/\binom{11}{6}\right\}\right]}{\Sigma} = .0767$
4. $\left\{ \begin{matrix} 3 & 1 \\ 0 & 6, \end{matrix} \begin{matrix} 0 & 5 \\ 6 & 0 \end{matrix} \right\}$	$\dfrac{\left[\left\{\binom{4}{3}\binom{6}{0}\Big/\binom{10}{3}\right\} \cdot \left\{\binom{5}{0}\binom{6}{6}\Big/\binom{11}{6}\right\}\right]}{\Sigma} = .0004$

where

$$\Sigma = \left\{\binom{4}{0}\binom{6}{3}\bigg/\binom{10}{3}\right\} \cdot \left\{\binom{5}{3}\binom{6}{3}\bigg/\binom{11}{6}\right\}$$

$$+ \left\{\binom{4}{1}\binom{6}{2}\bigg/\binom{10}{3}\right\} \cdot \left\{\binom{5}{2}\binom{6}{4}\bigg/\binom{11}{6}\right\}$$

$$+ \left\{\binom{4}{2}\binom{6}{1}\bigg/\binom{10}{3}\right\} \cdot \left\{\binom{5}{1}\binom{6}{5}\bigg/\binom{11}{6}\right\}$$

$$+ \left\{\binom{4}{3}\binom{6}{0}\bigg/\binom{10}{3}\right\} \cdot \left\{\binom{5}{0}\binom{6}{6}\bigg/\binom{11}{6}\right\}$$

$$= .0722 + .1623 + .0195 + .0001 = .2541.$$

Thus, the four tables have conditional probabilities $.0722/.2541 = .2841$, $.1623/.2541 = .6387$, $.0193/.2541 = .0767$, and $.0001/.2541 = .0004$. Of these four tables, Table 4 has the lowest probability and Table 3 (which is τ_0) has the second lowest. The P-value is

$$P = .0004 + .0767 = .077.$$

Thus, the observed data do not support H_0^*. (The value .077 is also obtained using StatXact's HO/EX command.)

Mehta, Patel, and Gray (1985) developed an algorithm for performing Zelen's test and inverting the test to get a confidence interval for the common odds ratio. StatXact's HO/EX command gives the exact P-value for Zelen's conditional test of H_0^* and an approximate P-value based on a large-sample test of H_0^* proposed by Breslow and Day (1980). For some large data sets, obtaining the exact P-value for Zelen's test is not feasible. StatXact's HO/MO command computes a Monte Carlo sampling estimate of the exact P-value for Zelen's test. The HO/MO command also gives the asymptotic P-value for the Breslow-Day test. The Breslow-Day statistic has, under H_0^*, an asymptotic chi-squared distribution with $k - 1$ degrees of freedom. See Breslow and Day (1980) and Jones et al. (1989). Jones et al. compared seven tests of H_0^*, including the Breslow-Day test. Jones et al. found generally low power for all the tests of H_0^* in the study, especially when the data are sparse. In sparse-data situations a test due to Liang and Self (1985) performed best.

25. *Monte Carlo Sampling to Estimate the Exact P-Value for Zelen's Test.* For some large data sets, computation of the exact P-value for Zelen's test is not feasible. Senchaudhuri, Mehta, and Patel (1995) present a Monte Carlo method of control variates that, with a little extra computational effort, can estimate the exact P-value with greater acuracy than can be obtained by crude Monte Carlo sampling. For precise descriptions of the method of control variates and the method of crude Monte Carlo sampling, see Senchaudhuri, Mehta, and Patel (1995). Both Monte Carlo methods sample a large number N (say) of tables τ_i from the reference set given by (10.80) with respective probabilities $P(\tau_i|s)$ given by (10.81). Senchaudhuri, Mehta, and Patel (1995) present the data set (Dixon v. Margolis 1991) given in Table 10.16. The data may be viewed in the format of 12 2×2 tables. The data relate to promotions, in 1985, 1987, and 1988, of black and white police officers in various ranks.

After entering the data in the 12 2×2 tables, StatXact's HO/MO comand can be used to obatin a confidence interval for Zelen's P-value based on the mothod of control variates. StatXact produces an interval based on a sample of 2000 tables and the user is then given the option of continuing to sample more tables to obtain a narrower interval. We stopped at a sample of 10,000 tables, and the corresponding 99% confidence interval for the P-value

TABLE 10.16. Promotions of Black and White Police Officers

Year	Rank	Black		White	
		Promoted	Total	Promoted	Total
1985	Sgt.	10	84	66	414
	SA Sgt.	3	28	40	110
	Master Sgt.	3	12	37	162
	SA Master Sgt.	0	2	16	62
1987	Sgt.	1	98	32	487
	SA Sgt.	0	29	28	120
	Master Sgt.	3	17	28	176
	SA Master Sgt.	2	5	16	65
1988	Sgt.	4	107	43	591
	SA Sgt.	1	36	4	113
	Master Sgt.	2	20	43	198
	SA Master Sgt.	1	5	18	112

Source: P. Senchaudhuri, C. R. Mehta, and N. R. Patel (1995).

of Zelen's test was (.042, .055). Thus there is strong evidence against H_0^*. That is, the data indicate there is not a common odds ratio. (Senchaurdhuri et al. sampled 100,000 tables and their 99% confidence interval for the P-value was (.0464, .0467).) The Breslow-Day (1980) statistic is also obtained with the HO/MO command. The Breslow-Day statistic, for the data of Table 10.16, is 16.02 with an approximate P-value of .14.

26. *Point and Interval Estimation of a Common Odds Ratio.* If we assume (perhaps on the basis of Zelen's exact test of H_0^*) that there is a common odds ratio θ, it is natural to obtain a point estimate and confidence interval for θ. Mantel and Haenszel (1959) suggested

$$\widehat{\theta}_{MH} = \frac{\sum_{i=1}^{k}(\mathcal{O}_{11i}\mathcal{O}_{22i}/n_{..i})}{\sum_{i=1}^{k}(\mathcal{O}_{12i}\mathcal{O}_{21i}/n_{..i})}. \tag{10.82}$$

Robins, Breslow and Greenland (1986) estimated the variance of $\log(\widehat{\theta}_{MH})$ by

$$\widehat{\sigma}_{RGB}^2 = \frac{\sum_{i=1}^{k}(\mathcal{O}_{11i}+\mathcal{O}_{22i})(\mathcal{O}_{11i}\mathcal{O}_{22i})/n_{..i}^2}{2(\sum_{i=1}^{k}\mathcal{O}_{11i}\mathcal{O}_{22i}/n_{..i})^2}$$
$$+ \frac{\sum_{i=1}^{k}\{(\mathcal{O}_{11i}+\mathcal{O}_{22i})(\mathcal{O}_{12i}\mathcal{O}_{21i}) + (\mathcal{O}_{12i}+\mathcal{O}_{21i})(\mathcal{O}_{11i}\mathcal{O}_{22i})\}/n_{..i}^2}{2(\sum_{i=1}^{k}\mathcal{O}_{11i}\mathcal{O}_{22i}/n_{..i})(\sum_{i=1}^{k}\mathcal{O}_{12i}\mathcal{O}_{21i}/n_{..i})}$$
$$+ \frac{\sum_{i=1}^{k}(\mathcal{O}_{12i}+\mathcal{O}_{21i})(\mathcal{O}_{12i}\mathcal{O}_{21i})/n_{..i}^2}{2(\sum_{i=1}^{k}\mathcal{O}_{12i}\mathcal{O}_{21i}/n_{..i})^2}. \tag{10.83}$$

An approximate $1-\alpha$ confidence interval for $\ln(\theta)$ is (ψ_L, ψ_U), where

$$\psi_L = \ln(\widehat{\theta}_{MH}) - z_{\alpha/2}\widehat{\sigma}_{RGB} \tag{10.84}$$

$$\psi_U = \ln(\widehat{\theta}_{MH}) + z_{\alpha/2}\widehat{\sigma}_{RGB} \tag{10.85}$$

and where $\widehat{\theta}_{MH}$ is given by (10.82) and $\widehat{\sigma}^2_{RGB}$ is given by (10.83). An approximate $1 - \alpha$ confidence interval for the common odds ratio θ is (θ_L, θ_U), where

$$\theta_L = e^{\psi_L}, \tag{10.86}$$

$$\theta_U = e^{\psi_U}, \tag{10.87}$$

where ψ_L, ψ_U are given by (10.84) and (10.85), respectively.

StatXact's OD/EX command will compute $\widehat{\theta}_{MH}$ and the confidence interval. Recall (see Comment 17) that StatXact's format gives estimates and confidence intervals for $1/\theta$ in our notation. For the data of Table 10.16, we have strong evidence from Zelen's test (see Comment 25) that there is not a common odds ratio. Nevertheless, for illustrative purposes we allpy StatXact's OD/EX command to the data. StatXact gives $1/\widehat{\theta}_{MH} = 2.09$ and the approximate 95% confidence interval for $1/\theta$ is $(1.41, 3.09)$.

Properties

1. *Relationship of Mantel-Haenzel Statistic to Cochran's (1954) Statistic.* See Fleiss (1981, 174) and Agresti (1990, 231).
2. *Efficiency of Mantel-Haenzel Estimator of a Common Odds Ratio.* See Hauck (1979), Breslow (1981), Donner and Hauck (1986), Hauck and Donner (1988) and Section 10.5.

PROBLEMS

29. Mittal (1991) considers pooling tables and the paradoxes that could arise from such pooling. For example, with the two 2×2 tables

Table 1		Table 2	
$\mathcal{O}_{111}$	$\mathcal{O}_{121}$	$\mathcal{O}_{112}$	$\mathcal{O}_{122}$
$\mathcal{O}_{211}$	$\mathcal{O}_{221}$	$\mathcal{O}_{212}$	$\mathcal{O}_{222}$

    the pooled 2×2 table is

    Table 3

$\mathcal{O}_{111} + \mathcal{O}_{112}$	$\mathcal{O}_{121} + \mathcal{O}_{122}$
$\mathcal{O}_{211} + \mathcal{O}_{212}$	$\mathcal{O}_{221} + \mathcal{O}_{222}$

    It may happen that the indication obtained by analyzing Table 3 could be different than the individual indications obtained by analyzing Tables 1 and 2. For example, it is possible that the estimated odds ratios $\widehat{\theta}_1$ and $\widehat{\theta}_2$ for Tables 1 and 2 satisfy

    $$\widehat{\theta}_1 = \frac{\mathcal{O}_{111}\mathcal{O}_{221}}{\mathcal{O}_{121}\mathcal{O}_{211}} \geq 1, \quad \widehat{\theta}_2 = \frac{\mathcal{O}_{112}\mathcal{O}_{222}}{\mathcal{O}_{122}\mathcal{O}_{212}} \geq 1$$

    but the estimated odds ratio $\widehat{\theta}_3$ for Table 3 satisfies

    $$\widehat{\theta}_3 = \frac{(\mathcal{O}_{111} + \mathcal{O}_{112}) \times (\mathcal{O}_{221} + \mathcal{O}_{222})}{(\mathcal{O}_{121} + \mathcal{O}_{122}) \times (\mathcal{O}_{211} + \mathcal{O}_{212})} \leq 1.$$

    Or it may be that $\widehat{\theta}_1 \leq 1$ and $\widehat{\theta}_2 \leq 1$ but $\widehat{\theta}_3 \geq 1$. Such anomalies fall into a category known as Simpson's Paradox. Create an example satisfying Simpson's Paradox.

30. Apply the Mantel-Haenszel test to the data of Table 10.16. What is your conclusion? Is the conclusion surprising in view of the test results obtained in Comment 25?

31. Recall the study of R. Goode and D. Coursey reported in Miller (1980) and considered in Problem 4. Goode and Coursey surveyed students seen at Stanford University's Student Health Center between

January 1968 and May 1973. They checked the charts of students treated for infectious mononucleosis (IM) for confirmation of the disease and to determine any history of tonsillectomy. The control group consisted of students seen at the Health Center between April and September 1973, who came in for any ailment and were willing to check on a survey sheet whether or not they had undergone a tonsillectomy. Within the 18–24 age groups, the data are given in Table 10.17. Do the data indicate that tonsillectomy reduces the risk of contracting infectious mononucleosis?

TABLE 10.17. Fractions of Students with Tonsillectomies

Age	18	19	20	21	22	23	24
IM	$\frac{6}{23}$	$\frac{3}{42}$	$\frac{12}{41}$	$\frac{8}{46}$	$\frac{5}{15}$	$\frac{2}{9}$	$\frac{4}{9}$
C	$\frac{17}{49}$	$\frac{26}{96}$	$\frac{34}{112}$	$\frac{48}{139}$	$\frac{45}{118}$	$\frac{29}{66}$	$\frac{36}{75}$

Source: R. G. Miller (1980).

32. Use StatXact to apply Zelen's test (see Comments 24 and 25) to the data of Table 10.17. What is your conclusion?
33. Suppose we have 3 2 × 2 tables of the form

	Successes	Failures
Treatment A	$\mathcal{O}_{11i}$	$\mathcal{O}_{12i}$
Treatment B	$\mathcal{O}_{21i}$	$\mathcal{O}_{22i}$

and the data are as follows:

```
        Table 1      Table 2      Table 3
        
        2   5        4   11       3   11
        1   6        1    7       2    9
```

Using StatXact, apply Zelen's test of H_0^*. What do you conclude?

34. Refer to the three 2 × 2 tables of Problem 33. Assuming a common odds ratio θ, compute $\widehat{\theta}_{\text{MH}}$ (see Comment 26) and obtain an approximate 95% confidence interval for θ.
35. Even if the evidence in a given data set indicates that H_0^* is not true, explain the potential usefulness of computing $\widehat{\theta}_{\text{MH}}$ for that data set.
36. Suppose $k = 2$ and our observed set of two 2 × 2 tables is $\tau_0 = \left\{ \begin{matrix} 3 & 2 \\ 1 & 5, \end{matrix} \begin{matrix} 1 & 4 \\ 5 & 1 \end{matrix} \right\}$

  (a) List the tables in the reference set $T(4)$ defined by (10.80).
  (b) Compute $P(\tau|4)$ for each $\tau \in T(4)$.
  (c) Find the P-value for Zelen's test corresponding to the observed table τ_0.

10.5. EFFICIENCIES

Storer and Kim (1990) studied the size and power of seven tests for the two-sample binomial problem considered in Sections 10.1 and 10.2. The tests compared included the uncorrected chi-squared statistic given by (10.19) in Section 10.1, the chi-squared statistic with Yates' continuity correction (Comment 5) and Fisher's exact test (Section 10.2). In the equal-sample-sizes case, an exact unconditional test due to Suissa and Shuster (1985) does quite well in terms of power. In the general case, the authors find the sample sizes required by Fisher's exact test are 10% to 20% higher than those required by some of the more powerful procedures considered. One of their conclusions is that for any reasonable sample size, one will not be led far astray by using the uncorrected chi-squared statistic.

For estimating the odds ratio (Sections 10.3 and 10.4), Hauck and Donner (1988) studied the asymptotic relative efficiency (ARE) of the Mantel-Haenszel estimator $\widehat{\theta}_{MH}$ (Comment 26) relative to the conditional maximum likelihood estimator proposed by Birch (1964). They assume a common odds ratio $\theta_i = \theta$, $i = 1, \ldots, k$, and $0 < \theta < \infty$. In this situation $p_1^{(i)}$ can be expressed in terms of θ and $p_2^{(i)}$, $p_1^{(i)} = (\theta p_2^{(i)})/\{1 - p_2^{(i)} + \theta p_2^{(i)}\}$ and the ARE can be studied by varying the $p_1^{(i)}$, $= 1, \ldots, k$, and θ. Hauck and Donner (1988) consider the case where the number of strata increases indefinitely for fixed within-stratum sample sizes. For the situations they considered, the ARE does not drop below .9. They point out the ARE decreases monotonically as θ moves away from 1 in either direction. For the less extreme values of θ, $.2 \leq \theta \leq .5$, the smallest ARE is .931. Hauck and Donner (1988) point out that the results they obtained in their 1988 study are similar to ones they obtained (Donner and Hauck 1986) for the fixed-number-of-strata asymptotic case where the ARE was found to be high over a wide range of designs likely to arise in practice. For other efficiency results concerning $\widehat{\theta}_{MH}$ see Breslow (1981) and Tarone, Gart, and Hauck (1983).

CHAPTER 11

Life Distributions and Survival Analysis

INTRODUCTION

In Sections 11.1, 11.2, 11.3, and 11.4 we consider a sample of lifelengths from an underlying life distribution (that is, a distribution that puts all of its probability on nonnegative values). There are many nonparametric classes of life distributions that are used to describe aging. We focus on six natural classes that have easily understood physical interpretations. The classes are the increasing failure rate (IFR) class, the increasing failure rate average (IFRA) class, the new better than used (NBU) class, the new better than used in expectation (NBUE) class, the decreasing mean residual life (DMRL) class, and the initially increasing, then decreasing mean residual life (IDMRL) class. In Sections 11.1, 11.2, 11.3, and 11.4 we describe tests of the null hypothesis of exponentiality. Section 11.1 considers IFR and IFRA alternatives. Section 11.2 considers NBU and NBUE alternatives. Section 11.3 considers DMRL alternatives and also presents confidence bands for the mean residual life function. Section 11.4 considers IDMRL alternatives, and the tests presented are designed to detect a trend change in the mean residual life. Section 11.5 presents a nonparametric confidence band for the distribution function. Sections 11.6 and 11.7 are devoted to censored data. Section 11.6 contains, for censored data, an estimator of the distribution function, confidence bands for the distribution function, and confidence bands for the quantile function. Section 11.7 presents a two-sample test for censored data. Section 11.8 considers asymptotic relative efficiencies.

Data. We obtain n observations, $X_1, \ldots, X_n$.

Assumptions for Sections 11.1, 11.2, 11.3, and 11.4

A1. The observations are a random sample from the underlying continuous population. That is, the X's are independent and identically distributed according to a continuous distribution F.

A2. F is a life distribution; that is, $F(a) = 0$ for $a < 0$. Equivalently, the X's are nonnegative.

11.1. A TEST OF EXPONENTIALITY VERSUS IFR ALTERNATIVES (EPSTEIN)

Hypothesis

We first define the failure rate function, denoted by $r(x)$. If F has a corresponding density f, $r(x)$ is defined as

$$r(x) = \frac{f(x)}{\overline{F}(x)} \qquad (11.1)$$

for those x such that $\overline{F}(x) > 0$. ($\overline{F}(x) = 1 - F(x)$ and $\overline{F}(x)$ is known as the survival function.)

The failure rate has the following physical interpretation. The product $r(x)\delta_x$ is the probability that an item (unit, person, part) alive at age x will fail in the interval $(x, x + \delta_x)$, where δ_x is small. An increasing failure rate correponds to deleterious aging; that is, the failure rate increases as age increases. A decreasing failure rate corresponds to beneficial aging; that is, the failure rate decreases as age increases. A constant failure rate corresponds to a model where the failure rate neither increases nor decreases with age but is, in fact, independent of age.

The null hypothesis is

$$H_0 : r(x) = \lambda, \text{ for some } \lambda > 0, \text{ and all } x > 0. \tag{11.2}$$

The null hypothesis asserts that the failure rate is a constant λ (λ is unspecified). That is, the failure rate does not depend on x. The null hypothesis specifies that F is an exponential distribution. (One characterization of exponential distributions is that F is an exponential distribution if and only if its failure rate is constant.) Thus H_0 can be rewritten as

$$H_0 : F(x) = \begin{cases} 1 - e^{-\lambda x} & (\lambda \text{ unspecified}), \quad x \geq 0, \\ 0, & x < 0. \end{cases} \tag{11.3}$$

Equivalently, H_0 can be expressed as

$$H_0 : \overline{F}(x) = \begin{cases} e^{-\lambda x} & (\lambda \text{ unspecified}), \quad x \geq 0, \\ 1, & x < 0. \end{cases} \tag{11.4}$$

A life distribution F is said to be in the *increasing failure* rate (IFR) class if its failure rate is nondecreasing. Similarly, a life distribution F is said to be in the *decreasing failure rate* (DFR) class if its failure rate is nonincreasing. For mathematically formal definitions of these classes, see, for example, Barlow and Proschan (1981). Whereas the IFR class is used to model deleterious aging, the DFR class is used to model beneficial aging. When the failure rate r exists, r is IFR if

$$r(x) \leq r(y) \quad \text{for all } x < y. \tag{11.5}$$

Similarly, when the railure rate r exists, r is DFR if

$$r(x) \geq r(y) \quad \text{for all } x < y. \tag{11.6}$$

Procedure

Let $X_{(1)} \leq X_{(2)} \leq \cdots \leq X_{(n)}$ denote the order statistics and define $X_{(0)} = 0$. The normalized spacings are $D_1, D_2, \ldots, D_n$, where

$$D_i = (n - i + 1)(X_{(i)} - X_{(i-1)}), \quad i = 1, \ldots, n. \tag{11.7}$$

Let

$$S_i = \sum_{u=1}^{i} D_u, \quad j = 1, \ldots, n \tag{11.8}$$

and define $S_0 = 0$. S_i is called the total time on test at $X_{(i)}$. The total-time-on-test statistic is

$$\mathcal{E} = \frac{\sum_{i=1}^{n-1} S_i}{S_n}. \tag{11.9}$$

a. One-Sided Test against IFR Alternatives. To test

$$H_0 : F \text{ is exponential}$$

versus

$$H_1 : F \text{ is IFR (and not exponential)},$$

at the α-level of significance,

$$\text{Reject } H_0 \text{ if } \mathcal{E} \geq e_\alpha; \quad \text{otherwise do not reject,} \tag{11.10}$$

where the constant e_α is chosen to make the type I error probability equal to α. That is, $P_0\{\mathcal{E} \geq e_\alpha\} = \alpha$. Values of e_α are given in Table A.34.

b. One-Sided Test against DFR Alternatives. To test

$$H_0 : F \text{ is exponential}$$

versus

$$H_2 : F \text{ is DFR (and not exponential)},$$

at the α-level of significance,

$$\text{Reject } H_0 \text{ if } \mathcal{E} \leq \frac{n-1}{2} - e_\alpha; \quad \text{otherwise do not reject.} \tag{11.11}$$

(See Comment 6.)

c. Two-Sided Test against IFR and DFR Alternatives. To test

$$H_0 : F \text{ is exponential}$$

versus

$$H_3 : F \text{ is IFR or DFR (and not exponential)},$$

at the α-level of significance (with equal probabilities $\alpha/2$ in the tails),

$$\text{Reject } H_0 \text{ if } \mathcal{E} \geq e_{\alpha/2} \text{ or if } \mathcal{E} \leq \frac{n-1}{2} - e_{\alpha/2}; \quad \text{otherwise do not reject.} \tag{11.12}$$

Large-Sample Approximation

Let

$$\mathcal{E}^* = \frac{\mathcal{E} - E_0(\mathcal{E})}{\sqrt{\text{var}_0(\mathcal{E})}} = \frac{\mathcal{E} - \frac{(n-1)}{2}}{\sqrt{\frac{n-1}{12}}}. \tag{11.13}$$

Then as $n \to \infty$, the distribution of $\mathcal{E}^*$ tends to the $N(0, 1)$ distribution.

The large-sample approximation to procedure (11.10) is

$$\text{Reject } H_0 \text{ if } \mathcal{E}^* \geq z_\alpha; \quad \text{otherwise do not reject.} \tag{11.14}$$

The large-sample approximation to procedure (11.11) is

$$\text{Reject } H_0 \text{ if } \mathcal{E}^* \leq -z_\alpha; \quad \text{otherwise do not reject.} \tag{11.15}$$

The large-sample approximation to procedure (11.12) is

$$\text{Reject } H_0 \text{ if } |\mathcal{E}^*| \geq z_{\alpha/2}; \quad \text{otherwise do not reject.} \tag{11.16}$$

Example 11.1: *Methylmercury Poisoning.* Van Belle (1972) consulted on an experiment at Florida State University designed to study the effect of methylmercury poisoning on the lifelengths of fish. In the experiment, goldfish were subjected to various dosages of methylmercury. At one dosage level, the ordered times to death (in days) were 42, 43, 51, 61, 66, 69, 71, 81, 82, 82. We will apply the Epstein test of H_0 versus IFR alternatives. The calculations are summarized in Table 11.1.

From the fourth column of Table 11.1, we obtain

$$\sum_{i=1}^{9} S_i = 420 + 429 + 493 + 563 + 593 + 608 + 616 + 646 + 648 = 5{,}016.$$

Then from equation (11.9)

$$\mathcal{E} = \frac{\sum_{i=1}^{9} S_i}{S_{10}} = \frac{5{,}016}{648} = 7.74.$$

Referring $\mathcal{E} = 7.74$ to Table A.34 with $n = 10$, we find $P < .005$. Thus there is strong evidence against the hypothesis of exponentiality in favor of deleterious aging.

To apply the large-sample approximation, we compute, using (11.13),

$$\mathcal{E}^* = \frac{7.74 - 4.5}{\sqrt{\frac{9}{12}}} = 3.74.$$

TABLE 11.1. Calculation of $\mathcal{E}$ for the Methylmercury Data

i	$X_{(i)}$	D_i	S_i
1	42	420	420
2	43	9	429
3	51	64	493
4	61	70	563
5	66	30	593
6	69	15	608
7	71	8	616
8	81	30	646
9	82	2	648
10	82	0	648

From Table A.1 we see $P < .0002$, and thus the large-sample approximation also confirms the strong evidence against H_0 in favor of age deterioration.

Comments

1. *The Total-Time-on-Test Statistic.* In a life-testing situation, n independent items may be put on test to study their survival. Let $X_1, \ldots, X_n$ denote the observed lifelengths and let $X_{(1)} \leq \cdots \leq X_{(n)}$ denote their ordered values, and let $X_{(0)} = 0$. At time $X_{(i)}$, the total time spent on test thus far by the n items is

$$nX_{(1)} + (n-1)(X_{(2)} - X_{(1)}) + \cdots + (n-i+1)(X_{(i)} - X_{(i-1)})$$

$$= \sum_{u=1}^{i} (n - u + 1)(X_{(u)} - X_{(u-1)})$$

$$= \sum_{u=1}^{i} D_u = S_i,$$

where (recall) S_i is given by (11.8). The total-time-on-test transformation transforms $X_{(1)}, \ldots, X_{(n)}$ into $T_1, \ldots, T_n$, where

$$T_i = \frac{\sum_{u=1}^{i} D_u}{\sum_{u=1}^{n} D_u} = \frac{S_i}{S_n}. \qquad (11.17)$$

The quantities $T_1, \ldots, T_n$ are called the total-time-on-test transforms and

$$\mathcal{E} = \sum_{i=1}^{n-1} T_i \qquad (11.18)$$

is known as the total-time-on-test statistic. When H_0 is true, $T_1, \ldots, T_{n-1}$ have the same distribution as the order statistics in a sample of size $n - 1$ from the uniform $(0, 1)$ distribution (see Epstein, 1960). Since the sum of $n - 1$ ordered values equals the sum of $n - 1$ unordered values, it follows that under H_0, $\mathcal{E}$ has the same distribution as $U_1 + U_2 + \cdots + U_{n-1}$, where $U_1, \ldots, U_{n-1}$ are independent and identically distributed $U(0, 1)$ random variables. Since $E(U_1) = \frac{1}{2}$ and $\text{var}(U_1) = \frac{1}{12}$, it follows that

$$E_0(\mathcal{E}) = \frac{n-1}{2} \qquad (11.19)$$

and

$$\text{var}_0(\mathcal{E}) = \frac{n-1}{12}. \qquad (11.20)$$

From the central limit theorem, we obtain the large-sample approximation based on $\mathcal{E}^*$ (see equation (11.13)). The approach to normality is fast and the approximation is very good when $n \geq 9$.

2. *The Barlow-Doksum Class of Monotonic Tests.* Barlow and Proschan (1966) showed that the T's given by (11.17) tend to be larger for an IFR distribution than for an exponential distribution. This led Barlow and Doksum (1972) to consider tests based on statistics that are monotonic in the following sense. A statistic $\mathcal{T}(T_1, \ldots, T_n)$ is monotonic in the T's if

$\mathcal{T}(T_1, \ldots, T_n) \geq \mathcal{T}(T_1', \ldots, T_n')$ whenever $T_i \geq T_i'$, $i = 1, \ldots, n - 1$. Barlow and Doksum studied the particular class of monotonic statistics given by $\mathcal{T}_J = \sum_{i=1}^{n-1} J(T_i)$, where J is a nondecreasing function on $(0, 1)$. The function J is chosen to make the test based on $\mathcal{T}_J$ asymptotically most powerful for a given parametric alternative. The choice $J(u) = u$ yields the total-time-on-test statistic $\mathcal{E}$ that is shown to be asymptotically most powerful for Makeham alternatives. The Makeham parametric family has failure rate function

$$r_m(x) = \lambda\{1 + \theta(1 - \exp(-\lambda x))\}, \quad \theta \geq 0, \quad x > 0, \quad \lambda > 0. \tag{11.21}$$

The failure rate, r_m, is increasing when $\theta > 0$. When $\theta = 0$, $r_m(x) = \lambda$, the failure rate of an exponential distribution with parameter λ.

Barlow (1968) gives a table of percentile points of $\mathcal{E}$ for $n = 2(1)10$ and α in the lower and upper .01, .05, and .10 regions. For other IFR tests see Bickel and Doksum (1969), Problem 3, Klefsjö (1983), Comment 5, and the survey papers by Doksum and Yandell (1984) and Hollander and Proschan (1984).

3. *The Increasing Failure Rate Average (IFRA) Class.* The failure rate, $r(x)$, may have an increasing trend, but it may not be strictly nondecreasing, as is required to be a member of the IFR class. The failure rate may fluctuate due perhaps to seasonal variations. In a medical setting, an early increasing failure rate may decrease for a period due to medical intervention. A distribution F is in the IFRA class if its average failure rate increases. A distribution F is in the DFRA class if its average failure rate decreases. For more formal definitions, see Barlow and Proschan (1981).

The IFR class is contained in the IFRA class and the DFR class is contained in the DFRA class. If F is IFR, then it is IFRA, but the converse is not true. That is, there are IFRA distributions that are not IFR. Similarly, if F is DFR then it is DFRA, but there are DFRA distributions that are not DFR. We may write these containment relations as

$$\text{IFR} \subset \text{IFRA}$$

$$\text{DFR} \subset \text{DFRA}.$$

The exponential distributions form the boundary of the IFR class and the DFR class; that is, the exponential distributions are in the IFR class and in the DFR class, and they are the only distributions that are both IFR and DFR. Similarly, the exponential distributions form the boundary of the IFRA and DFRA class; that is, the exponential distributions are in the IFRA class and in the DFRA class and are the only distributions that are both IFRA and DFRA.

Barlow and Scheuer (1971) and Wang (1987) estimate F when it is known to be in the IFRA class.

4. *Use of $\mathcal{E}$ for Testing IFRA.* Barlow and Proschan (1966) proved that if $X_1, \ldots, X_n$ is a sample from an IFRA distribution and $Y_1, \ldots, Y_n$ is a sample from an exponential distribution, then

$$\mathcal{E}(X_1, \ldots, X_n) \overset{st}{\geq} \mathcal{E}(Y_1, \ldots, Y_n). \tag{11.22}$$

Here $\mathcal{E}(X_1, \ldots, X_n)$ denotes $\mathcal{E}$ computed for the X-sample and $\mathcal{E}(Y_1, \ldots, Y_n)$ denotes $\mathcal{E}$ computed for the Y-sample. The notation $\overset{st}{\geq}$ means *stochastically greater than*. The random variable Z is said to be stochastically greater than the random variable Z' if $P(Z \leq x) \leq P(Z' \leq x)$ for every x.

Motivated by result (11.22), Barlow (1968) suggested rejecting H_0 in favor of IFRA alternatives if $\mathcal{E}$ is large. He provided tables of the null distribution of $\mathcal{E}$ for $n = 2(1)10$ and α in the lower and upper .01, .05, and .10 regions.

Hollander and Proschan (1975) showed that the total-time-on-test statistic $\mathcal{E}$ arises in a natural way for testing exponentiality against NBUE alternatives. See Section 11.2 and Comments 13 and 14. Hollander and Proschan (1975) showed that the consistency class of the test that rejects for large (small) values of $\mathcal{E}$ contains the NBUE (NWUE) distributions. Thus the statistic $\mathcal{E}$ can be more suitably viewed as a test for determining the larger NBUE (NWUE) class.

Other tests of H_0 versus IFRA alternatives have been proposed by Barlow (1968), Barlow and Campo (1975), Bergman (1977), Deshpande (1983) and Klefsjö (1983). Klefsjö's test is presented in Comment 5. IFR and IFRA tests for incomplete data, where some of the items are not observed up to their failure times, have been proposed by Barlow and Proschan (1969). The maximum likelihood estimator ($\widehat{G}_n$, say) of F, when it is known that F is in the IFR class, was obtained by Grenander (1956) and Marshall and Proschan (1965). Also see Barlow et al. (1972) and Robertson, Wright, and Dykstra (1988). Hollander and Proschan (1984) illustrated the calculation of $\widehat{G}_n$ using the methylmercury poisoning data of Table 11.1.

5. *Klefsjö's IFR and IFRA Tests.* Klefsjö (1983) proposed IFR and IFRA tests based on the normalized spacings $D_1, \ldots, D_7$ defined by (11.7). Klefsjö's tests are motivated by a graphical procedure known as the total-time-on-test transform (see Barlow and Campo (1975) and Klefsjö (1983)). Klefsjö rejects H_0 in favor of H_1 for large values of

$$A = \frac{\sum_{j=1}^{n} \alpha_j D_j}{S_n},$$

where S_n is defined by (11.8) and

$$\alpha_j = 6^{-1} \left\{ (n+1)^3 j - 3(n+1)^2 j^2 + 2(n+1) j^3 \right\}.$$

The null distribution of A is determined by using the result that under H_0, $D_1, D_2, \ldots, D_n$ are independent and identically distributed according to the exponential distribution given by (11.3). Klefsjö provides null distribution tables of

$$A^* = A \left(\frac{7560}{n^7} \right)^{1/2}$$

for $n = 5(5)75$. He gives the upper and lower .01, .05, and .10 percentiles. Significantly large values of A^* indicate IFR alternatives; significantly small values of A^* indicate DFR alternatives. Klefsjö shows that under H_0, A^* can be treated asymptotically as a $N(0, 1)$ random variable. Klefsjö also shows that the test which rejects H_0 for large values of A is consistent against the class of continuous IFR distributions.

Klefsjö's (1983) test of H_0 versus F is IFRA (and not exponential) is based on the statistic B, where

$$B = \frac{\sum_{j=1}^{n} \beta_j D_j}{S_n},$$

where

$$\beta_j = 6^{-1} \left\{ 2j^3 - 3j^2 + j(1 - 3n - 3n^2) + 2n + 3n^2 + n^3 \right\}.$$

Klefsjö provides null distribution tables of

$$B^* = B \left(\frac{210}{n^5} \right)^{1/2}$$

for $n = 5(5)75$, giving the upper and lower .01, .05, and .10 percentiles. Significantly large values of B indicate IFRA alternatives; significantly small values of B indicate DFRA alternatives. Klefsjö shows that under H_0, B^* can be treated asymptotically as a $N(0, 1)$ random variable. Klefsjö also shows that the test that rejects H_0 for large values of B is consistent against the class of continuous IFRA distributions.

6. *Symmetry of the Null Distribution of $\mathcal{E}$.* Under H_0, the distribution of $\mathcal{E}$ is symmetric about its mean $(n-1)/2$; that is,

$$P_0\left(\mathcal{E} - \frac{n-1}{2} \geq x\right) = P_0\left(\mathcal{E} - \frac{n-1}{2} \leq -x\right). \tag{11.23}$$

It follows that the lower α percentile point ($e_\alpha^{(2)}$, say) of the null distribution of $\mathcal{E}$ can be obtained from the upper α percentile point $e_\alpha^{(1)}$ via

$$e_\alpha^{(2)} = (n-1) - e_\alpha^{(1)}. \tag{11.24}$$

7. *Some IFR Distributions, Some DFR Distributions.* The exponential distribution has density function

$$f(x) = \lambda e^{-\lambda x}, \quad x > 0, \quad \lambda > 0.$$

Its failure rate is constant, that is,

$$r(x) = \lambda, \quad x > 0.$$

Thus the exponential distribution is both IFR and DFR.

One commonly used generalization of the exponential distribution is the Weibull distribution with density function

$$f(x) = \lambda\alpha(\lambda x)^{\alpha-1} e^{-(\lambda x)^\alpha}, \quad x > 0, \quad \alpha > 0, \quad \lambda > 0.$$

Its failure rate is

$$r(x) = \lambda\alpha(\lambda x)^{\alpha-1}.$$

The failure rate is increasing for $\alpha > 1$ and decreasing for $\alpha < 1$. Thus the Weibull distributions for which $\alpha > 1$ are IFR distributions, and the Weibull distributions with $\alpha < 1$ are DFR distributions. For $\alpha = 1$, the Weibull distribution reduces to the exponential distribution.

Another frequently used family of distributions, which is a generalization of the exponential, is the gamma family. The gamma density function is

$$f(x) = \frac{\lambda^\alpha x^{\alpha-1} e^{-\lambda x}}{\Gamma(\alpha)}, \quad x > 0, \quad \alpha > 0, \quad \lambda > 0$$

where $\Gamma(\alpha)$ is the gamma function (cf. Kalbfleisch and Prentice (1980, 23)). The gamma density does not have a closed-form expression for its failure rate. Its failure rate is increasing when $\alpha > 1$ and is decreasing when $\alpha < 1$. Thus those members of the gamma family for which the parameter α is greater than 1 are IFR distributions, those members corresponding to $\alpha < 1$ are DFR distributions. When $\alpha = 1$, the gamma distribution reduces to the exponential distribution.

For other IFR and DFR distributions, see Barlow and Proschan (1981) and Kalbfleisch and Prentice (1980). Often investigators pool data from different IFR distributions to increase the effective sample size. Gurland and Sethuraman (1995) show that such a pooling may actually reverse the IFR property of the individual samples to a DFR property for the mixture.

Properties

1. *Consistency.* See Hollander and Proschan (1975) for the consistency class of the test defined by (11.10). In particular, if F is continuous and IFR (and not exponential), the test is consistent.
2. *Asymptotic Normality.* See Barlow (1968), Bickel and Doksum (1969), and Doksum and Yandell (1984).
3. *Efficiency.* See Bickel and Doksum (1969), Klefsjö (1983), and Section 11.8.

PROBLEMS

1. Table 11.2 is based on a subset of data considered by Proschan (1963). Proschan investigated the life distribution of the air-conditioning system of a fleet of Boeing 720 jet airplanes. Table 11.2 presents intervals (in hours) between failures of the air-conditioning system of plane 8044.

TABLE 11.2. Intervals in Hours between Failures of the Air-Conditioning System of Plane 8044

i	1	2	3	4	5	6	7	8	9	10	11	12
X_i	487	18	100	7	98	5	85	91	43	230	3	130

Source: F. Proschan (1963).

Using the data of Table 11.2, test H_0 versus IFR alternatives.

2. Bjerkedal (1960) studied the lifelengths of guinea pigs injected with different amounts of tubercle bacilli. One reason for choosing this species is that guinea pigs are known to have a high susceptibility to human tuberculosis. The data in Table 11.3 are a subset of the survival data considered by Bjerkedal.

TABLE 11.3. Ordered Survival Times in Days of Guinea Pigs under Regimen 4.3

i	1	2	3	4	5	6	7	8	9
$X_{(i)}$	10	33	44	56	59	72	74	77	92
i	10	11	12	13	14	15	16	17	18
$X_{(i)}$	93	96	100	100	102	105	107	107	108
i	19	20	21	22	23	24	25	26	27
$X_{(i)}$	108	108	109	112	113	115	116	120	121
i	28	29	30	31	32	33	34	35	36
$X_{(i)}$	122	122	124	130	134	136	139	144	146
i	37	38	39	40	41	42	43	44	45
$X_{(i)}$	153	159	160	163	163	168	171	172	176
i	46	47	48	49	50	51	52	53	54
$X_{(i)}$	183	195	196	197	202	213	215	216	222
i	55	56	57	58	59	60	61	62	63
$X_{(i)}$	230	231	240	245	251	253	254	254	278
i	64	65	66	67	68	69	70	71	72
$X_{(i)}$	293	327	342	347	361	402	432	458	555

Source: T. Bjerkedal (1960).

The data in Table 11.3 correspond to study M (in Bjerkedal's terminology), in which animals in a single cage are under the same regimen. The regimen number is the common logarithm of the number of bacillary units in .5 ml of the challenge solution. The data in Table 11.3 are for regimen 4.3. Regimen 4.3 corresponds to 2.2×10^4 bacillary units per .5 ml ($\log_{10}(2.2 \times 10^4) = 4.342$). There are 72 observations, and Table 11.3 gives the ordered values $X_{(1)} \leq \cdots \leq X_{(72)}$. Using the data of Table 11.3, test H_0 versus IFR alternatives.

3. Bickel and Doksum (1969) considered a class of test statistics based on the ranks of the D's. Let R_i denote the rank of D_i in the joint ranking of $D_1, \ldots, D_n$. One member of the Bickel-Doksum class is

$$W_1 = \sum_{i=1}^{n} i \log \left(1 - \frac{R_i}{n+1}\right). \tag{11.25}$$

When F is IFR (DFR) the spacings tend to show a downward (upward) trend (cf. Doksum and Yandell 1984). Thus H_0 is rejected in favor of H_1 (H_2) for significantly large (small) values of W_1. Bickel and Doksum showed that tests based on W_1 are asymptotically equivalent to those based on $\mathcal{E}$. They found, however, that for finite sample sizes, $\mathcal{E}$ does better than W_1 in terms of power. For finite n, the null distribution of W_1 can be obtained by using the fact that, under H_0, all $n!$ possible outcomes of $(R_1, \ldots, R_n)$ are equally likely (and thus each has probability $1/n!$). Determine the null distribution of W_1 for the case $n = 4$. That is, give the possible values of W_1 and the corresponding probabilities. What is the critical region of the $\alpha = \frac{1}{24}$ test of H_0 versus H_1 based on W_1?

4. Bickel and Doksum (1969) showed that, under H_0,

$$W_1^* = \frac{\sqrt{n}(W_1' + \frac{1}{2})}{s_1} \tag{11.26}$$

tends to a $N(0, 1)$ distribution as $n \to \infty$. In (11.26),

$$W_1' = \frac{W_1}{n(n+1)} \tag{11.27}$$

and

$$s_1^2 = \frac{n-1}{12(n+1)}. \tag{11.28}$$

Thus an approximate α level test of H_0 versus H_1 rejects H_0 if $W_1^* \geq z_\alpha$ and accepts H_0 if $W_1^* < z_\alpha$. Similarly, an approximate α level test of H_0 versus H_2 rejects H_0 if $W_1^* \leq -z_\alpha$ and accepts H_0 if $W_1^* > -z_\alpha$. Apply the large-sample test of H_0 versus H_1, based on W_1^*, to the methylmercury data of Example 11.1.

5. Apply the large-sample test based on W_1^* to the data of Table 11.2.

6. Verify the symmetry of the null distributions of $\mathcal{E}$ as expressed in equation (11.23). (*Hint*: Recall that, under H_0, $\mathcal{E}$ has the same distribution as $U_1 + \cdots + U_{n-1}$, where $U_1, \ldots, U_{n-1}$ are independent $U(0, 1)$ random variables.)

7. Describe a situation in which DFR alternatives (i.e., beneficial aging) might occur.

8. Describe a situation where it is natural to expect the underlying distribution to satisfy the IFRA property but not satisfy the IFR property (that is, where F might be expected to be a member of the IFRA class but not a member of the IFR class).

11.2. A TEST OF EXPONENTIALITY VERSUS NBU ALTERNATIVES (HOLLANDER-PROSCHAN)

Hypothesis

The hypothesis of interest is

$$H_0 : P(X \geq x + y \mid X \geq x) = P(X \geq y), \quad \text{for all } x, y \geq 0. \tag{11.29}$$

11.2. A TEST OF EXPONENTIALITY VERSUS NBU ALTERNATIVES

The vertical bar (|) in the probability statement on the left-hand side of (11.29) is to be read as "given that." The equation in (11.29) asserts that the probability of surviving an additional time period y, given that the item has survived to time x, is equal to the probability that a new item will survive an initial period y. Insisting that the equality holds for all x, y is equivalent to asserting that used items of all ages are no better and no worse than new items. That property is equivalent to the underlying population being an exponential population (see Comment 8). Thus (11.29) is another way of expressing H_0 given by display (11.3) of Section 11.1. Using the survival function $\overline{F}$, (11.29) can be written as

$$H_0 : \frac{\overline{F}(x+y)}{\overline{F}(x)} = \overline{F}(y), \quad \text{all } x, y \geq 0, \tag{11.30}$$

and again as

$$H_0 : \overline{F}(x+y) = \overline{F}(x)\overline{F}(y), \quad \text{all } x, y \geq 0. \tag{11.31}$$

We now turn to the new better than used (NBU) and new worse than used (NWU) alternatives. The distribution F is said to be in the new better than used (NBU) class if

$$\overline{F}(x+y) \leq \overline{F}(x)\overline{F}(y), \quad \text{all } x, y \geq 0. \tag{11.32}$$

Similarly, the distribution F is said to be in the new worse than used (NWU) class if

$$\overline{F}(x+y) \geq \overline{F}(x)\overline{F}(y), \quad \text{all } x, y \geq 0. \tag{11.33}$$

If F is NBU, then new items are better than used items of any age. Similarly, if F is NWU, then new items are worse than used items of any age. The boundary members of the classes are the exponential distributions. An exponential distribution is both NBU and NWU.

Procedure

Let $X_{(1)} \leq \cdots \leq X_{(n)}$ denote the ordered X's. Compute

$$T = \sum_{i>j>k} \psi(X_{(i)}, X_{(j)} + X_{(k)}), \tag{11.34}$$

where

$$\psi(a, b) = \begin{cases} 1, & \text{if } a > b, \\ 0, & \text{if } a < b. \end{cases} \tag{11.35}$$

Note that the summation in (11.34) is over all $n(n-1)(n-2)/6$ ordered triples (i, j, k) with $i > j > k$.

a. One-Sided Test against NBU Alternatives. To test

$$H_0 : F \text{ is exponential}$$

versus

$$H_4 : F \text{ is NBU (and not exponential)}$$

at the α-level of significance,

$$\text{Reject } H_0 \text{ if } T \leq t_{1,\alpha}; \quad \text{otherwise do not reject,} \tag{11.36}$$

where the constant $t_{1,\alpha}$ satisfies $P_0\{T \leq t_{1,\alpha}\} = \alpha$. Approximate values of $t_{1,\alpha}$ are given in Table A.35.

b. One-Sided Test against NWU Alternatives. To test

$$H_0 : F \text{ is exponential}$$

versus

$$H_5 : F \text{ is NWU} \quad \text{(and not exponential)}$$

at the α-level of significance,

$$\text{Reject } H_0 \text{ if } T \geq t_{2,\alpha}; \quad \text{otherwise do not reject,} \tag{11.37}$$

where the constant $t_{2,\alpha}$ satisfies $P_0\{T \geq t_{2,\alpha}\} = \alpha$. Approximate values of $t_{2,\alpha}$ are given in Table A.35.

c. Two-Sided Test against NBU and NWU Alternatives. To test

$$H_0 : F \text{ is exponential}$$

versus

$$H_6 : F \text{ is NBU or NWU} \quad \text{(and not exponential)}$$

at the α-level of significance,

$$\text{Reject } H_0 \text{ if } T \leq t_{1,\alpha_1} \text{ or if } T \geq t_{2,\alpha_2}; \quad \text{otherwise do not reject,} \tag{11.38}$$

where $\alpha_1 + \alpha_2 = \alpha$.

Large-Sample Approximation

Define

$$T^* = \frac{T - E_0(T)}{[\text{var}_0(T)]^{1/2}}$$

$$= \frac{T - \{\frac{n(n-1)(n-2)}{8}\}}{\{(\frac{3}{2})n(n-1)(n-2)[(\frac{5}{2592})(n-3)(n-4) + (n-3)(\frac{7}{432}) + (\frac{1}{48})]\}^{1/2}}. \tag{11.39}$$

When H_0 is true, the statistic T^* has an asymptotic (n tending to infinity) $N(0, 1)$ distribution. The normal theory approximation to procedure (11.36) is

$$\text{Reject } H_0 \text{ if } T^* \leq -z_\alpha; \quad \text{otherwise do not reject.} \tag{11.40}$$

11.2. A TEST OF EXPONENTIALITY VERSUS NBU ALTERNATIVES

The normal theory approximation to procedure (11.37) is

$$\text{Reject } H_0 \text{ if } T^* \geq z_\alpha; \quad \text{otherwise do not reject.} \tag{11.41}$$

The normal theory approximation to procedure (11.38) is

$$\text{Reject } H_0 \text{ if } T^* \leq -z_{\alpha_1} \text{ or if } T^* \geq z_{\alpha_2}; \quad \text{otherwise do not reject,} \tag{11.42}$$

where $\alpha_1 + \alpha_2 = \alpha$.

Ties

If $X_{(i)} = X_{(j)} + X_{(k)}$, compute T by replacing $\psi(X_{(i)}, X_{(j)} + X_{(k)})$ with $\psi^*(X_{(i)}, X_{(j)} + X_{(k)})$, where

$$\psi^*(a, b) = \begin{cases} 1, & \text{if } a > b, \\ \frac{1}{2}, & \text{if } a = b, \\ 0, & \text{if } a < b. \end{cases} \tag{11.43}$$

Example 11.2: *Example 11.1 Continued.* We return to the methylmercury poisoning data of Example 11.1. Recall that the ordered lifelengths $X_{(1)} \leq \ldots \leq X_{(10)}$ are 42, 43, 51, 61, 66, 69, 71, 81, 82, 82. Although in general the ψ functions of equation (11.34) need to be computed for the $n(n-1)(n-2)/6$ $(X_{(i)}, X_{(j)}, X_{(k)})$ triples with $i > j > k$, we note that $X_{(10)} < X_{(1)} + X_{(2)}$; thus for this data set, all 120 ψ functions must be zero. (Since $X_{(10)} < X_{(1)} + X_{(2)}$, there is no (i, j, k) triple with $i > j > k$ satisfying $X_{(i)} > X_{(j)} + X_{(k)}$.) Thus $T = 0$. Hollander and Proschan (1972) showed that for $n \geq 3$,

$$P_0\{T = 0\} = \frac{1}{\binom{2n-2}{n}}. \tag{11.44}$$

Thus with $n = 10$ we find

$$P_0\{T = 0\} = \frac{1}{\binom{18}{10}} = \frac{1}{43758} = .00002.$$

Thus the P-value is .00002 and this is strong evidence against exponentiality in favor of deleterious aging. (Result (11.44) was derived under the assumption that F is continuous. Because we have a tie in the methylmercury poisoning data set, the P-value of .00002 is approximate.)

Example 11.3: *Intervals between Failures for Air-Conditioning System.* Table 11.4 is based on a subset of data considered by Proschan (1963). Proschan investigated the life distribution of the air-conditioning system of a fleet of Boeing 720 jet airplanes. The table presents intervals between failures of the air-conditioning system of plane 7907.

Before testing exponentiality, Proschan considered the following question: Can we view the X's as a random sample from a common population? We are considering a particular plane, and increased operation of the plane might lead to shorter (or longer) intervals between failures. Proschan states: "A trend toward longer intervals, if established, could be the result of greater experience, debugging, or elimination of faulty parts, whereas a trend toward shorter intervals could be the result of wearout, aging, or poor maintenance."

TABLE 11.4. Intervals in Hours between Failures of the Air-Conditioning System of Plane 7907

i	X_i	$X_{(i)}$
1	194	15
2	15	29
3	41	33
4	29	41
5	33	181
6	181	194

Source: F. Proschan (1963).

To see whether it is appropriate to consider the successive intervals from airplane 7907 as values from a common population, we follow Proschan and apply Mann's test for trend. See Comment 8.14 to find that the value of K for the data in Table 11.4 is $K = 1$. Referring this value to Table A.30 we find that there is no significant evidence of a trend; thus we proceed as if the X's are a random sample from a common population. We now apply procedure (11.36) to test the hypothesis of exponentiality against NBU alternatives. In this particular setting, we may interpret the hypothesis as a statement that the air-conditioning system, upon repair, is as good as new. Table 11.5 illustrates the calculations required to compute T.

Thus we have

$$T = \sum_{i>j>k} \psi(X_{(i)}, X_{(j)} + X_{(k)}) = 12.$$

TABLE 11.5. $\psi(X_{(i)}, X_{(j)} + X_{(k)})$ Values Corresponding to the 20 Ordered (i, j, k) Triples with $i > j > k$

(i, j, k)	$(X_{(i)}, X_{(j)}, X_{(k)})$	$X_{(j)} + X_{(k)}$	$\psi(X_{(i)}, X_{(j)} + X_{(k)})$
(3,2,1)	(33,29,15)	44	0
(4,2,1)	(41,29,15)	44	0
(5,2,1)	(181,29,15)	44	1
(6,2,1)	(194,29,15)	44	1
(4,3,1)	(41,33,15)	48	0
(5,3,1)	(181,33,15)	48	1
(6,3,1)	(194,33,15)	48	1
(5,4,1)	(181,41,15)	56	1
(6,4,1)	(194,41,15)	56	1
(6,5,1)	(194,181,15)	196	0
(4,3,2)	(41,33,29)	62	0
(5,3,2)	(181,33,29)	62	1
(6,3,2)	(194,33,29)	62	1
(5,4,2)	(181,41,29)	70	1
(6,4,2)	(194,41,29)	70	1
(6,5,2)	(194,181,29)	210	0
(5,4,3)	(181,41,33)	74	1
(6,4,3)	(194,41,33)	74	1
(6,5,3)	(194,181,33)	214	0
(6,5,4)	(194,181,41)	222	0

11.2. A TEST OF EXPONENTIALITY VERSUS NBU ALTERNATIVES

From Table A.35 we find that the critical constant at the $\alpha = .086$ level is 9. Since $T = 12$ is greater than the critical value, we accept H_0. That is, using procedure (11.36) at the $\alpha = .086$ level, we accept the hypothesis that the underlying distribution is some exponential distribution.

To utilize the normal approximation (which should not be expected to be very accurate for the relatively small sample size $n = 6$), we compute from (11.39),

$$T^* = \frac{12 - \left\{\frac{6(5)(4)}{8}\right\}}{\{(\frac{3}{2})6(5)(4)[(\frac{5}{2592})(3)(2) + 3(\frac{7}{432}) + (\frac{1}{48})]\}^{1/2}} = -.786.$$

The lowest level at which we can reject H_0 in favor of NBU alternatives, using the large-sample approximation, is .22.

Comments

8. *Characterization of the Exponential Distribution.* If (11.31) holds for all $x, y \geq 0$, then it can be shown (cf. Barlow and Proschan (1981), 57) that $P(X \geq y) = e^{-\lambda y}$ for some $\lambda > 0$. That is, the underlying population is exponential. Even though under H_0 the population distributions are restricted to be exponential, we have retained the term *nonparametric* for tests based on T and $\mathcal{E}$, since those tests are designed to detect large nonparametric classes of distributions.

9. *Relationship of NBU to IFR and IFRA.* The NBU class is larger than the IFR and IFRA classes and properly contains those classes. Symbolically,

$$\text{IFR} \subset \text{IFRA} \subset \text{NBU}.$$

The corresponding containment relations for the NWU class in relation to the DFR and DFRA classes are

$$\text{DFR} \subset \text{DFRA} \subset \text{NWU}.$$

Thus, for example, an IFR distribution is also an NBU distribution, but there are NBU distributions that are not IFR or IFRA. For example, the underlying population may be NBU, but its failure rate $r(x)$ may fluctuate (and in particular not be increasing or increasing on average), due perhaps to seasonal variations. The NBU test is designed to detect this larger class. Hollander and Proschan show that the consistency class of the NBU (NWU) test that rejects for small (large) values of T includes the continuous NBU (NWU) distributions.

10. *The NBU Test Employs New Items.* The reader should note that we can test whether new is better (worse) than used employing only new items. That is, we need a sample only of lifelengths of new items to perform the NBU (NWU) test.

11. *An S-plus Program for Computing the NBU Statistic T.* When n is large, computation of T is tedious. Appendix B contains a program, newbet, written in the S-plus language for computing T and determining the approximate one-sided P-value based on the large-sample approximation.

12. *Motivation for the NBU Test.* Define

$$T^*(x, y) = \overline{F}(x)\overline{F}(y) - \overline{F}(x + y). \tag{11.45}$$

Note that $T^*(x, y) = 0$ for all (x, y) if and only if H_0 is true. This fact was used by Hollander and Proschan in devising the test based on T. The statistic $\frac{1}{4} - \{2T/[n(n-1)(n-2)]\}$ estimates the parameter

$$\Delta_{\text{NBU}}(F) = E_F\{T^*(X', Y')\}, \tag{11.46}$$

where X', Y' are independent and each is from the underlying life population with distribution F. We may view $T^*(x, y)$ as a measure of the deviation from H_0 at the point (x, y) and $\Delta_{\text{NBU}}(F)$ as the average value of this deviation. When F is NBU and continuous, the parameter $\Delta_{\text{NBU}}(F)$ is positive. When sampling from such a population the value of $\frac{1}{4} - \{2T/[n(n-1)(n-2)]\}$ tends to be large, or, equivalently, T tends to be small. This partially motivates procedure (11.36). Asymptotic normality of T is directly obtained from Hoeffding's U-statistic theory because $2T/\{n(n-1)(n-2)\}$ is a U-statistic (see Hollander and Proschan 1972).

Chen, Hollander, and Langberg (1983a) extended the NBU test to censored data by estimating the parameter $\Delta_{\text{NBU}}(F)$ using the Kaplan-Meier (1958) estimator of F. See Comment 35 of Section 11.6.

13. *The New Better Than Used in Expectation (NBUE) Class.* The NBUE class is larger than the NSU class. In order to define the NBUE class, we first introduce the mean residual life function. The mean residual life function, corresponding to a distribution F, gives, for each value of $x \geq 0$, the expected remaining life at time x. More formally, the mean residual life (mrl) function corresponding to a distribution F is defined as

$$m(x) = \begin{cases} E_F(X - x \mid X > x), & \text{for those } x \text{ such that } \overline{F}(x) > 0, \\ 0, & \text{for those } x \text{ such that } \overline{F}(x) = 0, \end{cases} \tag{11.47}$$

where X has the distribution F. In (11.47), note that $X - x$ is the residual life of an item with lifelength X, given that it has survived to time x. Also note that $m(0)$ is the mean μ of the distribution F; that is, $m(0) = E(X)$.

A distribution F with finite mean is said to be a member of the new better than used in expectation (NBUE) class if its corresponding mean residual life function m satisfies

$$m(0) \geq m(x) \quad \text{for all } x. \tag{11.48}$$

The NBUE class has the following interpretation. A used NBUE item of any fixed age has a smaller mean residual lifelength than does a new item.

The new worse than used in expectation (NWUE) is similarly defined. A distribution F with finite mean is said to be a member of the NWUE class if its corresponding mean residual life function satisfies

$$m(0) \leq m(x) \quad \text{for all } x. \tag{11.49}$$

The NWUE class has the following interpretation. A used NWUE item of any fixed age has a larger mean residual lifelength than does a new item.

The boundary members of the NBUE and NWUE classes are the exponential distributions. The exponential distributions, given by (11.3), have mean residual life functions $m(x) = (\lambda)^{-1}$, $x \geq 0$. This is another characterization of the exponential distributions, namely, that F is an exponential distribution if and only if its mean residual life function is constant.

The relation of the NBUE class to the smaller classes IFR, IFRA, and NBU is given by the following containment relations:

$$\text{IFR} \subset \text{IFRA} \subset \text{NBU} \subset \text{NBUE}. \tag{11.50}$$

For the dual classes used to model beneficial aging,

$$\text{DFR} \subset \text{DFRA} \subset \text{NWU} \subset \text{NWUE}. \tag{11.51}$$

14. *Using the $\mathcal{E}$ Statistic to Test against NBUE Alternatives.* Let X be a random value from F. Hollander and Proschan (1975) considered the parameter

$$\Delta_{\text{NBUE}}(F) = E_F\{\overline{F}(X)[m(0) - m(X)]\}$$

as a measure of deviation, for a given F, from exponentiality toward "NBUEness." In the definition of $\Delta_{\text{NBUE}}(F)$, $m(x)$ is the mean residual life function defined by (11.47). Hollander and Proschan estimated $\Delta_{\text{NBUE}}(F)$ with its sample counterpart, $\Delta_{\text{NBUE}}(F_n)$, where F_n is the empirical distribution function defined by (11.93). This yields the statistic

$$K = \frac{\sum_{i=1}^{n} d_i X_{(i)}}{n^2},$$

where $X_{(1)} \leq \cdots \leq X_{(n)}$ are the ordered X's and

$$d_i = \frac{3n}{2} - 2i + \frac{1}{2}.$$

Dividing K by $\overline{X}$ to make it scale-invariant, Hollander and Proschan proposed $K^* = K/\overline{X}$ as a statistic for testing H_0 against NBUE alternatives or NWUE alternatives. Significantly large (small) values of K^* lead to rejection of H_0 in favor of NBUE (NWUE) alternatives. Hollander and Proschan showed that

$$nK^* = \mathcal{E} - \frac{n-1}{2} \tag{11.52}$$

where $\mathcal{E}$ is the total-time-on-test statistic defined by (11.9). Thus tests based on K^* are equivalent to tests based on $\mathcal{E}$. Hence, the total-time-on-test statistic, originally proposed to detect IFR (DFR) alternatives and later proposed to detect IFRA (DFRA) alternatives, can be used to detect the larger class of NBUE (NWUE) alternatives. For testing against NBUE (NWUE) alternatives, use Table A.34 or the large-sample approximations given by (11.14), (11.15), and (11.16).

Klefsjö (1983), by considering a graphical method known as the total-time-on-test transform (cf. Barlow and Campo 1975) was also led to the K^* statistic as a test statistic for exponentiality versus NBUE alternatives. Borges, Proschan, and Rodrigues (1984) developed a test of exponentiality versus NBUE alternatives based on the sample coefficient of variation $s/\overline{X}$, where $s^2 = n^{-1} \sum_{i=1}^{n} (X_i - \overline{X})^2$. The test proposed by Borges, Proschan, and Rodrigues is equivalent to a test studied by Lee, Locke, and Spurrier (1980). Under H_0, the distribution of $S' = \sqrt{n}\{(s/\overline{X}) - 1\}$ is asymptotically $N(0, 1)$. Significantly large values of S' indicate NBUE alternatives, and significantly small values (i.e., large negative values) of S' indicate NWUE alternatives.

Let $U_i = S_i/S_n$, where S_i is given by equation (11.8) of Section 11.1. Barlow and Doksum (1972) proposed the statistic $D^+ = \max_{1 \leq i \leq n}\{U_i - i/n\}$ for testing H_0 against IFR alternatives. Koul (1978b) showed that the test that rejects H_0 for significantly large values of D^+ can be more appropriately viewed as a test of H_0 against the larger NBUE class. The null distribution of D^+, tabled by Birnbaum and Tingey (1951), can be used in this life-testing context. Asymptotically, under H_0,

$$P\left\{n^{1/2} D^+ \leq x\right\} = 1 - e^{-2x^2}.$$

Whitaker and Samaniego (1989) derive estimates of F for the situation when it is known that F is a member of the NBUE class.

15. *Koul's NBU Tests.* Koul (1977, 1978a) suggested other statistics for testing H_0 versus NBU alternatives. Koul (1977) proposed the statistic $S = \min_{1 \leq k \leq j \leq n} T_{kj}$, where for $1 \leq k \leq j \leq n$, $T_{kj} = nS_{kj} - (n-k)(n-j)$ and $S_{kj} = \sum_{i=1}^{n} \psi(X_{(i)}, X_{(k)} + X_{(j)})$. S/n^2 estimates the parameter $\alpha(F) = \inf_{x,y \geq 0}\{\bar{F}(x+y) - \bar{F}(x)\bar{F}(y)\}$. The parameter $\alpha(F)$ can be viewed as a measure of the deviation of F from H_0 toward H_4. When F is exponential, $\alpha(F) = 0$; it is negative when F is NBU. Thus the test rejects H_0 in favor of H_4 when S is significantly small. Koul gives critical values of S for $\alpha = .005, .01, .025, .05, .10, .20$ and $n = 3(1)30(5)50$. Koul (1977) does not provide a dual test of H_0 versus H_5 (NWU alternatives). Koul (1978a) suggested a class of tests of H_0 versus NBU alternatives indexed by a function ψ satisfying mild conditions. The Hollander-Proschan NBU test corresponds to the choice $\psi(u) = u$. Koul advocated the choice $\psi(u) = u^{1/2}$.

16. *Boyles-Samaniego Estimator.* Boyles and Samaniego (1984) considered the problem of estimating F when it is known that F is NBU. Their estimator is

$$\widehat{F}_{\text{NBU}}(x) = \max_i \left\{ \frac{\bar{F}_n(x + X_{(i)})}{\bar{F}_n(X_{(i)})} \right\}$$

where $X_{(1)} < \cdots < X_{(n)}$ are the ordered X's and $\bar{F}_n$ is the sample survival function. $\bar{F}_n(x) = 1 - F_n(x)$, where F_n is the sample distribution function defined by equation (11.105). Boyles and Samaniego show that $\widehat{F}_{\text{NBU}}$ is in the NBU class of distributions and that it is relatively easy to compute. It is not, however, a consistent estimator of F for all F in the NBU class. They do show that it is a consistent estimator of F when F is NBU and $T = \inf\{M : F(M) = 1\}$ is well defined and finite.

17. *The NBU Class and Replacement Policy Comparisons.* The NBU class plays an important role in replacement policy comparisons that arise in the study of renewal processes and repairable systems. A renewal process is a sequence of independent, identically distributed, nonnegative random variables that (with probability one) are not all zero. An example of a renewal process is the following. Consider a system operating over an indefinite period of time. Upon failure, the system is repaired or replaced. Assuming negligible time for repair, the successive intervals between failures are independent, identically distributed random variables of a renewal process.

Under an age replacement policy, a unit is replaced upon failure or at age T, whichever comes first. Let $N(t) = $ number of renewals in $[0, t]$ for an ordinary renewal process and $N_A(t, T) = $ number of failures in $[0, t]$ under an age replacement policy with replacement interval T. It can be shown that $N(t)$ is stochastically larger than $N_A(t, T)$ for all $t \geq 0, T \geq 0$, if and only if F is NBU. For this result and related results, see Barlow and Proschan (1981, 179).

Properties

1. *Consistency.* See Hollander and Proschan (1972) for the consistency class of the test defined by (11.36). In particular, if F is continuous, NBU (and not exponential), the test is consistent.
2. *Asymptotic Normality.* See Hollander and Proschan (1972).
3. *Efficiency.* See Hollander and Proschan (1972), Koul (1978b), Klefsjö (1983), and Section 11.8.

PROBLEMS

9. The data in Table 11.6 are from a study discussed by Siddiqui and Gehan (1966) and also considered by Bryson and Siddiqui (1969). The data are survival times (measured from the date of diagnosis) of

TABLE 11.6. Ordered Survival Times (Days from Diagnosis)

7	429	579	968	1877
47	440	581	1077	1886
58	445	650	1109	2045
74	455	702	1314	2056
177	468	715	1334	2260
232	495	779	1367	2429
273	497	881	1534	2509
285	532	900	1712	
317	571	930	1784	

Source: M. M. Siddiqui and E. A. Gehan (1966).

43 patients suffering from chronic granulocytic leukemia. For these data, $T = 8327$. Apply the large-sample approximation (11.40) to test against NBU alternatives. (Note that one might be reluctant to postulate IFR alternatives here because, after the diagnosis of leukemia, medical treatment may cause the failure rate to decrease for a period of time.)

10. Either show directly or illustrate by means of an example that the maximum possible value of T (based on a sample of size n) is $n(n-1)(n-2)/6$ and the minimum possible value is 0.
11. Apply the NBU test to the data of Table 11.2. Compare your result to the result obtained using the test based on $\mathcal{E}$.
12. Let $\mathcal{F}_{a,b}$ denote the class of distributions with support $[a, b]$, where $b < 2a$. (Roughly speaking, each F in $\mathcal{F}_{a,b}$ puts all its probability in the interval $[a, b]$, where $b < 2a$.) Show that if $X_1, \ldots, X_n$ is a random sample from a distribution F that is in the class $\mathcal{F}_{a,b}$, then $P_F(T = 0) = 1$.
13. (Problem 12 Continued) Show that every F in $\mathcal{F}_{a,b}$ is an NBU distribution. (*Hint*: Consider the four cases (i) $x < a$ and $y < a$, (ii) $x \geq a$ and $y \geq a$, (iii) $x < a$ and $y \geq a$, and (iv) $x \geq a$ and $y < a$. Show that in each of these cases, $F(x, y)$ satisfies the inequality of (11.32).)
14. (Problems 12 and 13 Continued) Consider the NBU test that rejects for small values of T. Using the fact that, for $n \geq 3$, $P_0[T = 0] = 1/\binom{2n-2}{n}$, show that when $n \geq 3$ and $\alpha \geq 1/\binom{2n-2}{n}$, the power of the NBU test equals 1 for every F in the class $\mathcal{F}_{a,b}$.
15. Verify directly (or illustrate using an example) equation (11.52).
16. Apply the NBU test to the guinea pig survival data of Table 11.3.

11.3. A TEST OF EXPONENTIALITY VERSUS DMRL ALTERNATIVES (HOLLANDER-PROSCHAN)

Hypothesis

The null hypothesis is

$$H_0 : F \text{ is exponential.} \quad (11.53)$$

The alternatives are expressed in terms of the mean residual life function $m(x)$ defined by equation (11.47), Section 11.2. The alternatives are the decreasing mean residual life (DMRL) alternatives, and the increasing mean residual life (IMRL) alternatives. DMRL distributions model situations where deterioration takes place with age; IMRL distributions model beneficial aging.

The distribution F is said to be a member of the decreasing mean residual life (DMRL) class if $F(0) = 0$ and

$$m(x) \geq m(y), \quad \text{for all } x < y \text{ such that } \overline{F}(x) \text{ and } \overline{F}(y) > 0. \quad (11.54)$$

Similarly, the distribution F is said to be a member of the increasing mean residual life (IMRL) class if $F(0) = 0$ and

$$m(x) \leq m(y), \quad \text{for all } x < y \text{ such that } \overline{F}(x) \text{ and } \overline{F}(y) > 0. \tag{11.55}$$

The distributions that are both DMRL and IMRL are the exponential distributions—that is, the exponentials are the boundary members of the classes.

Procedure

Let $X_{(1)} \leq \cdots \leq X_{(n)}$ denote the ordered X's. Compute

$$V^* = \frac{V}{\overline{X}}, \tag{11.56}$$

where

$$V = \frac{\sum_{i=1}^{n} c_i X_{(i)}}{n^4} \tag{11.57}$$

and

$$c_i = \left(\tfrac{4}{3}\right) i^3 - 4ni^2 + 3n^2 i - \left(\tfrac{1}{2}\right) n^3 + \left(\tfrac{1}{2}\right) n^2 - \left(\tfrac{1}{2}\right) i^2 + \left(\tfrac{1}{6}\right) i.$$

Let

$$V' = \left\{\sqrt{(210)n}\right\} V^*. \tag{11.58}$$

a. One-Sided Test against DMRL Alternatives. To test

$$H_0 : F \text{ is exponential}$$

versus

$$H_7 : F \text{ is DMRL} \quad \text{(and not exponential)},$$

at the α-level of significance,

$$\text{Reject } H_0 \text{ if } V' \geq v_{1,\alpha}; \quad \text{otherwise do not reject}, \tag{11.59}$$

where the constant $v_{1,\alpha}$ is chosen to make the type I error probability equal to α; that is, $P_0\{V' \geq v_{1,\alpha}\} = \alpha$. Values of $v_{1,\alpha}$ are given in Table A.36.

b. One-Sided Test against IMRL Alternatives. To test

$$H_0 : F \text{ is exponential}$$

versus

$$H_8 : F \text{ is IMRL} \quad \text{(and not exponential)},$$

at the α-level of significance,

$$\text{Reject } H_0 \text{ if } V' \leq v_{2,\alpha}; \quad \text{otherwise do not reject}, \tag{11.60}$$

where the constant $v_{2,\alpha}$ is chosen to make the type I error probability equal to α; that is, $P_0\{V' \leq v_{2,\alpha}\} = \alpha$. Values of $v_{2,\alpha}$ are given in Table A.36.

c. *Two-Sided Test against DMRL and IMRL Alternatives.* To test

$$H_0 : F \text{ is exponential}$$

versus

$$H_9 : F \text{ is DMRL or IMRL (and not exponential)}$$

at the α-level of significance,

$$\text{Reject } H_0 \text{ if } V' \geq v_{1,\alpha_1} \text{ or if } V' \leq v_{2,\alpha_2}; \quad \text{otherwise do not reject,} \qquad (11.61)$$

where $\alpha = \alpha_1 + \alpha_2$.

Large-Sample Approximation

Under H_0, the asymptotic distribution of V' tends to the $N(0, 1)$ distribution. Thus, the large-sample approximation to procedure (11.59) is

$$\text{Reject } H_0 \text{ if } V' \geq z_\alpha; \quad \text{otherwise do not reject.} \qquad (11.62)$$

The large-sample approximation to procedure (11.60) is

$$\text{Reject } H_0 \text{ if } V' \leq -z_\alpha; \quad \text{otherwise do not reject.} \qquad (11.63)$$

The (equal-tailed) large-sample approximation to procedure (11.61), with $\alpha_1 = \alpha_2 = \alpha/2$ is

$$\text{Reject } H_0 \text{ if } |V'| \geq z_{\alpha/2}; \quad \text{otherwise do not reject.} \qquad (11.64)$$

Example 11.4: *Methylmercury Poisoning.* We return to the methylmercury poisoning data of Example 11.1 and illustrate the calculation of the DMRL statistic via Table 11.7.

By summing the fourth column of Table 11.7 we obtain

$$\sum_{i=1}^{10} c_i X_{(i)} = 29{,}730.$$

TABLE 11.7. Calculation of V for the Methylmercury Poisoning Data

i	$X_{(i)}$	c_i	$c_i X_{(i)}$
1	42	−189	−7,938
2	43	−1	−43
3	51	122	6,222
4	61	188	11,468
5	66	205	13,530
6	69	181	12,489
7	71	124	8,804
8	81	42	3,402
9	82	−57	−4,674
10	82	−165	−13,530

Thus, from (11.56) we obtain

$$V = \frac{29{,}730}{10{,}000} = 2.9730.$$

Because $\overline{X} = \left(\sum_{i=1}^{10} X_i\right)/10 = 64.8$, from (11.55) and (11.58) we obtain

$$V^* = \frac{2.9730}{64.8} = .0459$$

and

$$V' = \sqrt{2100}\,(.0459) = 2.10.$$

From Table A.36, we find $P \cong .01$ (the .01 critical value for V' is 2.14). Thus the test indicates that a DMRL model is preferable to an exponential model, and there is strong evidence of age deterioration. The large-sample approximation, based on referring $V' = 2.10$ to Table A.1 rather than to Table A.38, yields $P \cong .018$. This is in close agreement with the exact test.

Comments

18. *Implications among Life Distribution Classes.* The relationships between the six classes we have discussed, namely, IFR, IFRA, NBU, NBUE, DMRL and their dual classes DFR, DFRA, NWU, NWUE, IMRL are given in Figure 11.1. Where no implication is shown, no implication exists, as may be demonstrated by counterexample.

Referring to Figure 11.1, we see, for example, that if F is IFR, then F is DMRL. That is, the IFR class is contained in the DMRL class. Similarly, the DMRL class is contained in the NBUE class. That there is no arrow connecting DMRL to the IFRA and NBU classes is meant to signify that a containment relation does not hold between DMRL and IFRA or between DMRL and NBU. Thus, for example, there are IFRA distributions that are not DMRL distributions and there are DMRL distributions that are not IFRA distributions. Similarly, there are NBU distributions that are not DMRL distributions and there are DMRL distributions that are not NBU distributions. See Bryson and Siddiqui (1969).

The classes in Figure 11.1(a) consist of life distributions that can be used to model situations where the lifelengths of items tend to deteriorate with age. The dual classes in Figure 11.1(b) can be used to model situations where the lifelengths tend to improve with age. The boundary members of each class and its dual are the exponential distributions that are used to model situations where lifelengths neither deteriorate or improve with age. Thus, for example, the only distributions that are both DMRL and IMRL are the exponential distributions.

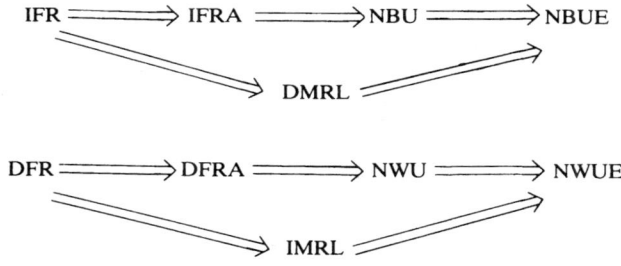

FIGURE 11.1. Implications among life distributions.

11.3. A TEST OF EXPONENTIALITY VERSUS DMRL ALTERNATIVES

19. *Motivation for the DMRL Test.* Let

$$D(x, y) = \overline{F}(x)\overline{F}(y)\{m(x) - m(y)\} \tag{11.65}$$

where $m(x)$ is the mean residual life function (see (11.47)). $D(x, y) = 0$ for all $x \leq y$ if and only if H_0 is true. Let X and Y be independent random variables, each with life distribution F. The parameter

$$\Delta_{\text{DMRL}}(F) = E_F\{I(X < Y)D(X, Y)\} \tag{11.66}$$

can be considered a measure of "DMRLness." In (11.66), $I(X < Y)$ is 1 if $X < Y$, and 0 otherwise. For each $x < y$, $D(x, y)$ is a weighted measure of the deviation from H_0 toward H_7, and $\Delta_{\text{DMRL}}(F)$ is an average value of this deviation. The weights $\overline{F}(x)$ and $\overline{F}(y)$ represent the proportions of the population still alive at times x and y, respectively, thus furnishing comparisons concerning the mean residual lifelengths from x and y respectively. The Hollander-Proschan statistic is obtained by substituting the empirical distribution function F_n for F in (11.66). The asymptotic normality of the statistic follows from Stigler's (1974) results on linear functions of order statistics. See Stigler (1974) and Hollander and Proschan (1975). The exact null distribution of V' is given by Langenberg and Srinivasan (1979) for α in the upper and lower .01, .05 and .10 regions for $n = 2(1)20(5)60$. Their table is the basis for Table A.36. Chen, Hollander, and Langberg (1983b) extended the DMRL test to censored data by estimating $\Delta_{\text{DMRL}}(F)$ using the Kaplan-Meier (1958) estimator.

Aly (1990) derives a test of H_0 versus H_7 using a different parameter to estimate "DMRLness." He finds his test outperforms V' for the three distributions he considered. His test has Pitman efficiencies with respect to V' of 1.219, 1.0714, and 1.4272 for linear failure rate, Makeham, and Weibull alternatives, respectively.

20. *The Empirical Mean Residual Life Function.* The empirical mean residual life function, $\widehat{m}(x)$, for a sample $X_1, \ldots, X_n$ from F is obtained by replacing $\overline{F}$ by the empirical survival function, $\overline{F}_n$, in equation (11.47). The expression for $\widehat{m}(x)$ when this substitution is made reduces to

$$\widehat{m}(x) = \frac{\sum_{j=1}^{S(x)}(X_j^* - x)}{S(x)}, \tag{11.67}$$

where $S(x)$ denotes the number of items at time x, out of the initial sample $X_1, \ldots, X_n$, that exceed x and $X_1^*, \ldots, X_{S(x)}^*$, are those observations that exceed x. Note that $\widehat{m}(x)$ is the average, less x, of the sample values that are greater than x. Yang (1978) and Hall and Wellner (1979) showed strong consistency of $\widehat{m}(x)$ as an estimator of $m(x)$. Hall and Wellner (1979) derived nonparametric simultaneous confidence bands for $m(x)$. See Comment 21. Guess, Hollander, and Proschan (1986) derived tests of exponentiality versus a trend change in the mean residual life function. They considered the situation where the turning point is known. Hawkins, Kochar, and Loader (1992) considered the situation where the turning point is unknown. The turning point procedures are discussed in Section 11.4. Mi (1994) proposed an estimator of $m(x)$ which is continuous and decreasing (increasing) when F is DMRL (IMRL).

21. *Confidence Bands for the Mean Residual Life Function.* Hall and Wellner (1979) developed nonparametric simultaneous confidence bands for the mean residual life function. (Additional bands are presented in Csörgő and Zitikis (1996).) Assume that $E(X^r) < \infty$ for some $r > 2$. The Hall-Wellner bands are

$$\widehat{m}(x) - \frac{D_n}{\overline{F}_n(x)}, \quad \widehat{m}(x) + \frac{D_n}{\overline{F}_n(x)}, \quad 0 \leq x < \infty, \tag{11.68}$$

where

$$\overline{F}_n(x) = \frac{\text{number of } X\text{-values in the sample} > x}{n} \quad (11.69)$$

is the empirical survival function, $\widehat{m}(x)$ is the mean residual life function given by (11.67), and

$$D_n = \frac{a_\alpha S_n}{n^{1/2}}, \quad (11.70)$$

where

$$S_n = \sqrt{\frac{\sum_{i=1}^{n}(X_i - \overline{X})^2}{n-1}} \quad (11.71)$$

is the sample standard deviation. The value a_α is determined so that the coverage probability of the bands is approximately $1 - \alpha$. Values of a_α are given in Table A.37.

The value a_α is chosen from the distribution of $\sup_{0 \le t \le 1} |\mathcal{W}_t|$, where $\mathcal{W}_t$ is the value at t of standard Brownian motion $\mathcal{W}$ (see the Weiner process, Billingsley (1968, 61)). The value a_α satisfies

$$P\left(\sup_{0 \le t \le 1} \mathcal{W}_t \le a_\alpha\right) = 1 - \alpha.$$

Hall and Wellner show that if F is continuous and $E(X^r) < \infty$ for some $r > 2$, then as $n \to \infty$, the limiting value of the probability given by the left-hand side of (11.72) is $1 - \alpha$. The distribution of $Y = \sup_{0 \le t \le 1} \mathcal{W}_t$ is given by (cf. Billingsley 1968, 79)

$$P(Y \le a) \equiv P(a) = \sum_{k=-\infty}^{\infty} (-1)^k \{\Phi((2k+1)a) - \Phi((2k-1)a)\}$$

$$= 1 - 4\{\overline{\Phi}(a) - \overline{\Phi}(3a) + \overline{\Phi}(5a) - \cdots\}$$

where Φ denotes the standard normal cumulative distribution function and $\overline{\Phi} = 1 - \Phi$. Hall and Wellner point out the approximation $P(a) \cong 1 - 4\overline{\Phi}(a)$ gives three-place accuracy for $a > 1.4$.

With the choice of a_α as given in Table A.37, we have

$$P\left(\widehat{m}(x) - \frac{D_n}{\overline{F}_n(x)} \le m(x) \le \widehat{m}(x) + \frac{D_n}{\overline{F}_n(x)}, \text{ for all } x \ge 0\right) \cong 1 - \alpha. \quad (11.72)$$

Thus, for example, to obtain an approximate 95% simultaneous confidence band, use $a_{.05} = 2.2414$ when substituting into equation (11.70).

Figure 11.2 is a plot of $\widehat{m}(x)$ and an approximate 95% simultaneous confidence band for $m(x)$ for the chronic granulocytic leukemia data of Table 11.6. Note that the graph of $\widehat{m}(x)$ changes at each ordered X-value and is a line of slope -1 between adjacent X-values. The same is also true of the band.

An S-plus program, mrl, to compute the bands can be found in Appendix B.

Suppose, rather than simultaneous confidence bands for all $x \ge 0$, we desire tighter bands for $m(x)$ at a specific point. Hall and Wellner (1979) show that if $\overline{F}(x) > 0$, then

$$\frac{n^{1/2}(\widehat{m}(x) - m(x))\{\overline{F}_n(x)\}^{1/2}}{s_n^*(x)}$$

11.3. A TEST OF EXPONENTIALITY VERSUS DMRL ALTERNATIVES

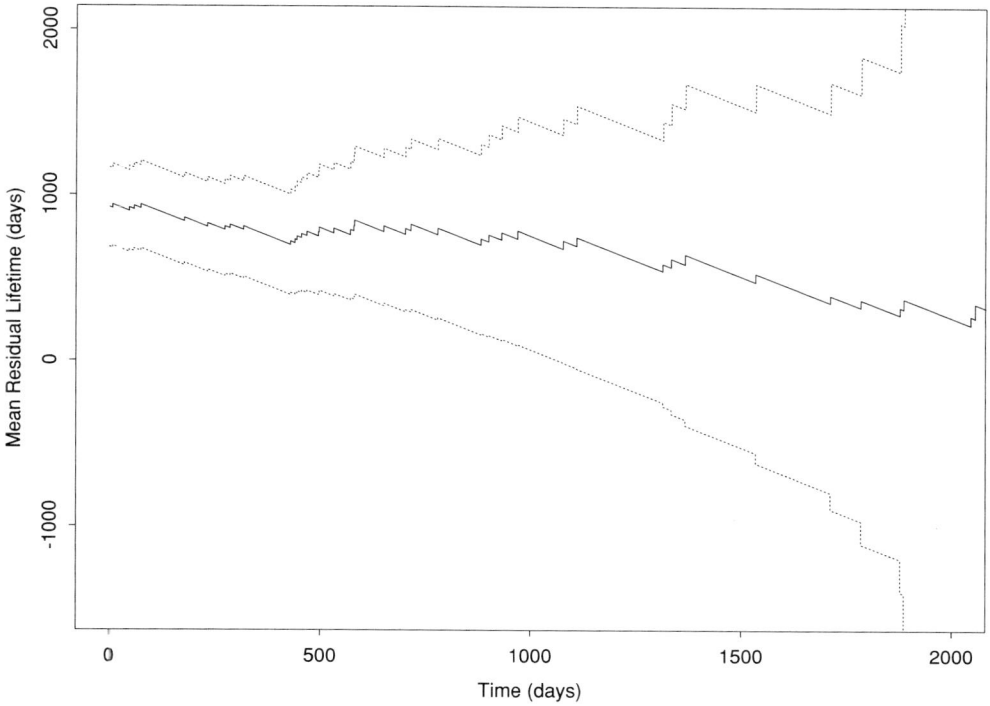

FIGURE 11.2. An approximate 95% confidence band for the mean residual life for the chronic granulocytic leukemia data of Table 11.6.

tends in distribution to a $N(0, 1)$ as $n \to \infty$, where $s_n^*(x)$ is the sample standard deviation of the observations that exceed x. Using the X^* notation of equation (11.67),

$$s_n^*(x) = \left\{ \frac{\sum_{j=1}^{S(x)} (X_j^* - \overline{X}(x))^2}{n - 1} \right\},$$

where $\overline{X}(x) = \sum_{j=1}^{S(x)} X_j^* / S(x)$ is the mean of the observations that exceed x. It follows that

$$\left(\widehat{m}(x) - \frac{z_{\alpha/2} s_n^*(x)}{\{n \overline{F}_n(x)\}^{1/2}},\ \widehat{m}(x) + \frac{z_{\alpha/2} s_n^*(x)}{\{n \overline{F}_n(x)\}^{1/2}} \right)$$

is an approximate $100(1 - \alpha)\%$ confidence interval for $m(x)$ at the point x.

Properties

1. *Consistency.* See Hollander and Proschan (1975) for the consistency class of the test defined by (11.59). In particular, if F is continuous, DMRL (and not exponential), the test is consistent.
2. *Asymptotic Normality.* See Hollander and Proschan (1975).
3. *Efficiency.* See Hollander and Proschan (1975), Klefsjö (1983), and Section 11.8.

PROBLEMS

17. Apply the DMRL test to the chronic granulocytic leukemia data of Table 11.6.
18. Apply the DMRL test to the air-conditioning system failure data of Table 11.2. Compare your result to the results from Problem 11.
19. (a) Calculate the estimated mean residual life function for the air-conditioning system data of Table 11.4.
    (b) Calculate approximate 95% confidence bands for the true mean residual life function.
20. Apply the DMRL test to the guinea pig survival data of Table 11.3. Compare your results to the results in Problems 2 and 16.
21. (a) Calculate the estimated mean residual life function for the guinea pig survival data of Table 11.3.
    (b) Calculate approximate 95% confidence bands for the true mean residual life function.
22. Table 11.8, based on data in Zacks (1992), gives the pneumatic pressure (kg/cm^2) required to break 20 concrete cubes of dimensions $10 \times 10 \times 10$ cm^3.
    (a) Calculate an approximate 92% confidence band for $m(x)$.
    (b) Calculate an approximate 92% confidence interval for $m(229.7)$. Compare the limits of the band at 229.7 with the limits of the interval.

TABLE 11.8. Pneumatic Pressures Required to Break Concrete Cubes

94.9	106.9	229.7	275.7	144.5	112.8	159.3	153.1	270.6	322.0
216.4	544.6	266.2	263.6	138.5	79.0	114.6	66.1	131.2	91.1

Source: S. Zacks (1992).

23. Describe a situation in which it might be expected that the mean residual life function would be initially increasing and then later decreasing.

11.4. A TEST OF EXPONENTIALITY VERSUS A TREND CHANGE IN MEAN RESIDUAL LIFE (GUESS-HOLLANDER-PROSCHAN)

Hypothesis

The null hypothesis is $H_0 : F$ is exponential. The alternatives are specified in terms of two nonparametric classes of distributions defined via the mean residual life function. The initially increasing, then decreasing mean residual life (IDMRL) class models aging that is initially beneficial and then is adverse. The decreasing initially, then increasing mean residual life (DIMRL) class models aging that is initially adverse and then is beneficial. A distribution with finite mean is said to be a member of the increasing initially, then decreasing mean residual life (IDMRL) class if there exists a turning point $\tau \geq 0$ such that

$$m(x) \leq m(y), \quad \text{for } 0 \leq x \leq y < \tau,$$
$$m(x) \geq m(y), \quad \text{for } \tau \leq x \leq y,$$

where $m(x)$ is the mean residual life function. Similarly, a distribution F with finite mean is said to be a member of the decreasing initially, then increasing mean residual life (DIMRL) class if there exists a turning point $\tau \geq 0$ such that

$$m(x) \geq m(y), \quad \text{for } 0 \leq x \leq y < \tau,$$
$$m(x) \leq m(y), \quad \text{for } \tau \leq x \leq y.$$

Procedure

We treat the case where the turning point τ is known, using a procedure due to Guess, Hollander, and Proschan (GHP) (1986). The GHP statistic can be used to detect IDMRL and DIMRL alternatives. In Comment 25 we consider the case where the turning point is not known and describe a procedure due to Hawkins, Kochar, and Loader (HKL) (1992). The HKL statistic can be used to detect IDMRL alternatives, but it is not designed to detect DIMRL alternatives.

To define the GHP test statistic, we set

$$T_1 = \sum_{i=1}^{i^*} B_1\left(\frac{(n-i+1)}{n}\right)(X_{(i)} - X_{(i-1)}) + B_1\left(\frac{(n-i^*)}{n}\right)(\tau - X_{(i^*)})$$
$$+ B_2\left(\frac{(n-i^*)}{n}\right)(X_{(i^*+1)} - \tau) + \sum_{i=i^*+2}^{n} B_2\left(\frac{(n-i+1)}{n}\right)(X_{(i)} - X_{(i-1)}) \quad (11.73)$$

where (letting $X_{(1)} < \cdots < X_{(n)}$ denote the ordered X-values with $X_{(0)} = 0$) the integer i^* is defined by

$$0 < X_{(1)} < \cdots < X_{(i^*)} \le \tau < X_{(i^*+1)} < \cdots < X_{(n)}. \quad (11.74)$$

The functions B_1 and B_2 in (11.73) are defined as

$$B_1(u) = \left[\tfrac{2}{3} - F_n(\tau) + \tfrac{1}{2}F_n^2(\tau)\right]u + \left[-1 + F_n(\tau) - \tfrac{1}{2}F_n^2(\tau)\right]u^2 + \tfrac{1}{3}u^4, \quad (11.75)$$

$$B_2(u) = \left[-\tfrac{1}{6} + \tfrac{1}{2}F_n(\tau) - \tfrac{1}{2}F_n^2(\tau) + \tfrac{1}{3}F_n^3(\tau)\right]u + \left[\tfrac{1}{2} - F_n(\tau) + \tfrac{1}{2}F_n^2(\tau)\right]u^2 - \tfrac{1}{3}u^4, \quad (11.76)$$

where F_n is the empirical distribution function defined by (11.93). For data where there are ties, use

$$T_1 = \sum_{i=1}^{i^*} B_1\left(\frac{S_{i-1}}{n}\right)(\widetilde{X}_{ik} - \widetilde{X}_{(i-1)k}) + B_1\left(\frac{S_{i^*}}{n}\right)(\tau - \widetilde{X}_{i^*k})$$
$$+ B_2\left(\frac{S_{i^*}}{n}\right)(\widetilde{X}_{(i^*+1)k} - \tau) + \sum_{i=i^*+2}^{k} B_2\left(\frac{S_{i-1}}{n}\right)(\widetilde{X}_{ik} - \widetilde{X}_{(i-1)k}), \quad (11.77)$$

where

$$0 = \widetilde{X}_{0k} < \widetilde{X}_{1k} < \cdots < \widetilde{X}_{i^*k} \le \tau < \widetilde{X}_{(i^*+1)k} < \cdots < \widetilde{X}_{kk} \quad (11.78)$$

are the distinct ordered observations,

$$n_i = \text{number of observed deaths at time } \widetilde{X}_{is}, \quad (11.79)$$

$$S_i = n - \sum_{t=0}^{i} n_t, \quad \text{for } i = 0, 1, \ldots, k < n. \quad (11.80)$$

In (11.79), $n_i \ne 0$, $i = 1, \ldots, k$, but n_0 is allowed to be 0.

Under H_0, the distribution of $n^{1/2}T_1$ tends, as $n \to \infty$, to a normal distribution with mean 0 and variance

$$\sigma_{T_1}^2 = \mu^2 \left[-\tfrac{1}{15}F^5(\tau) + \tfrac{1}{6}F^4(\tau) - \tfrac{1}{6}F^3(\tau) + \tfrac{1}{10}F^2(\tau) - \tfrac{1}{30}F(\tau) + \tfrac{1}{210}\right], \quad (11.81)$$

where μ is the mean of F. The test, for the case where τ is known, uses the statistic $n^{1/2}T_1/\widehat{\sigma}_{T_1}$, where

$$\widehat{\sigma}_{T_1}^2 = \bar{X}^2 \left[-\tfrac{1}{15}F_n^5(\tau) + \tfrac{1}{6}F_n^4(\tau) - \tfrac{1}{6}F_n^3(\tau) + \tfrac{1}{10}F_n^2(\tau) - \tfrac{1}{30}F_n(\tau) + \tfrac{1}{210}\right] \quad (11.82)$$

a. One-Sided Test against IDMRL Alternatives. To test

$$H_0 : F \text{ is exponential}$$

versus

$$H_{10} : F \text{ is IDMRL} \quad \text{(and not exponential)},$$

at the approximate α-level of significance,

$$\text{Reject } H_0 \text{ if } \frac{n^{1/2}T_1}{\widehat{\sigma}_{T_1}} \geq z_\alpha; \quad \text{otherwise do not reject.} \quad (11.83)$$

b. One-Sided Test against DIMRL Alternatives. To test

$$H_0 : F \text{ is exponential}$$

versus

$$H_{11} : F \text{ is DIMRL} \quad \text{(and not exponential)}$$

at the approximate α-level of significance,

$$\text{Reject } H_0 \text{ if } \frac{n^{1/2}T_1}{\widehat{\sigma}_{T_1}} \leq -z_\alpha; \quad \text{otherwise do not reject.} \quad (11.84)$$

c. Two-Sided Test against IDMRL and DIMRL Alternatives. To test

$$H_0 : F \text{ is exponential}$$

versus

$$H_{12} : F \text{ is IDMRL or DIMRL} \quad \text{(and not exponential)},$$

at the approximate α-level of significance,

$$\text{Reject } H_0 \text{ if } \frac{n^{1/2}T_1}{\widehat{\sigma}_{T_1}} \geq z_{\alpha_1} \text{ or if } \frac{n^{1/2}T_1}{\widehat{\sigma}_{T_1}} \leq -z_{\alpha_2}; \quad \text{otherwise do not reject,} \quad (11.85)$$

where $\alpha = \alpha_1 + \alpha_2$.

Example 11.5: *Lifelengths of Guinea Pigs Injected with Tubercle Bacilli.* Bjerkedal (1960) studied the lifelengths of guinea pigs injected with different amounts of tubercle bacilli. Guinea pigs are known to have a high susceptibility to human tuberculosis. That is one reason experimenters choose the species. In Bjerkedal's study (M), the animals in a single cage are under the same regimen. The regimen number is the common logarithm of the number

11.4. A TEST OF EXPONENTIALITY VERSUS A TREND CHANGE

TABLE 11.9. Ordered Survival Times in Days of Guinea Pigs under Regimen 5.5

i	1	2	3	4	5	6	7	8	9
$X_{(i)}$	43	45	53	56	56	57	58	66	67
i	10	11	12	13	14	15	16	17	18
$X_{(i)}$	73	74	79	80	80	81	81	81	82
i	19	20	21	22	23	24	25	26	27
$X_{(i)}$	83	83	84	88	89	91	91	92	92
i	28	29	30	31	32	33	34	35	36
$X_{(i)}$	97	99	99	100	100	101	102	102	102
i	37	38	39	40	41	42	43	44	45
$X_{(i)}$	103	104	107	108	109	113	114	118	121
i	46	47	48	49	50	51	52	53	54
$X_{(i)}$	123	126	128	137	138	139	144	145	147
i	55	56	57	58	59	60	61	62	63
$X_{(i)}$	156	162	174	178	179	184	191	198	211
i	64	65	66	67	68	69	70	71	72
$X_{(i)}$	214	243	249	329	380	403	511	522	598

Source: T. Bjerkedal (1960).

of bacillary units in .5ml of the challenge solution. In Table 11.3 (see Problem 2 of Section 11.1) we presented the data for regimen 4.3, which corresponds to 2.2×10^4 bacillary units per .5 ml (because $\log_{10}(2.2 \times 10^4) = 4.342$).

It is natural to postulate DIMRL alternatives in this situation. The motivation is that initially the injection of tubercle bacilli cause an adverse stage of aging, but after the guinea pigs have survived this initial adverse stage, their natural systems recover to yield a beneficial stage.

Hall and Wellner (1981) used regimen 4.3 and fit a parametric distribution that is in the DIMRL class. They estimated the turning point as $\hat{\tau} = 91.9$. We apply the DIMRL test to regimen 5.5 of Table 11.9, using 91.9 as the "known" turning point. This is a reasonable a priori choice because regimens 4.3 and 5.5 are closely related.

For the DIMRL test applied to the data of Table 11.9 with $\tau = 91.9$, we obtain $T_1 = -.6419$, $\hat{\sigma}_{T_1}^2 = 7.1072$, and $n^{1/2}T_1/\hat{\sigma}_{T_1} = -2.04$, yielding a P-value of .02. Thus there is strong evidence to reject H_0 in favor of H_{11}.

Comments

22. *Motivation for the IDMRL Test.* The motivation for the IDMRL test is similar to the motivation (see Comment 19) for the DMRL test of Section 11.3. The test statistic T_1 estimates a parameter $\Delta_{\text{IDMRL}}(F)$ that is a weighted measure of the degree to which F satisfies the IDMRL property. Specifically

$$\Delta_{\text{IDMRL}}(F) = E_F\{I(X < Y < \tau)D(X,Y) - I(\tau < X < Y)D(X,Y)\} \quad (11.86)$$

where D is defined by equation (11.65) and the indicator functions in (11.86) are defined as follows. Let X, Y be independent random variables each with distribution F. Then $I(X < Y < \tau)$ is 1 if $X < Y < \tau$ and 0 otherwise. Similarly $I(\tau < X < Y)$ is 1 if $\tau < X < Y$ and 0 otherwise.

Asymptotic normality of T_1, similarly standardized, is proved directly in Guess, Hollander, and Proschan (1986). Exact null distributions of T_1 can be obtained, but the distributions depend on τ and thus creating tables for different τ values is impractical. There are exact tables, however, for the related problem where one knows the proportion p of the population that "dies" at or before the turning point. See Comment 24.

23. *Some Situations Where Knowledge of τ May Be Available.* Knowledge of τ might be available in a situation where one is studying a biological organism in a physical model of a disease process. In such a situation, it may be the case that the first 3 months (say) constitute an incubation period. As another example, consider a training program for future doctors or a recruiting program for a military service. The value of τ may be known by the length of the intensive stage designed to eliminate the weaker students or recruits.

24. *The IDMRL Test When the Proportion p (of the Population That Dies before or at the Turning Point) Is Known.* In a training program, for example, past experience with earlier classes of recruits may provide knowledge of the proportion p of the population that dies at or before the turning point. Such a p would satisfy $F(\tau) = p$. Guess, Hollander, and Proschan (1986) proposed a statistic (similar to the T_1 statistic) for this situation and provided exact critical values for the case $p = .25$ for the sample sizes $n = 2, \ldots, 30$ in the lower and upper $\alpha = .01, .05,$ and $.10$ regions. Their statistic can be used to detect both IDMRL and DIMRL alternatives. Tables for $p = 0(.1)1, .75, \frac{1}{3},$ and $\frac{2}{3}$ are given in Guess (1984).

25. *Case Where the Turning Point Is Unknown.* Hawkins, Kochar, and Loader (1992) presented two statistics, $T^{(1)}$ and $T^{(2)}$, for the case where the turning point is unknown. Their statistics can be used to test exponentiality versus IDMRL alternatives, but they are not designed to detect DIMRL alternatives. In HKL (1992) Monte Carlo power comparisons showed that $T^{(2)}$ outperformed $T^{(1)}$, and thus we present $T^{(2)}$ here. The HKL statistic is an appropriately standardized estimator of a function of F that is 0 when F is exponential and is positive when F is IDMRL (see HKL (1992) for details). The statistic can be expressed as

$$T^{(2)} = n^{1/2}(\overline{X})^{-1} \max_{0 \le k \le n} \xi_k, \qquad (11.87)$$

where

$$\xi_k = A - 2\left(1 - \frac{k}{n}\right) \sum_{j=k}^{n-1} \left(1 - \frac{j}{n}\right) D_j^* + 4 \sum_{j=k}^{n-1} \left(1 - \frac{j}{n}\right)^2 D_j^*, \qquad (11.88)$$

and

$$A = -X_{(1)} + \sum_{j=1}^{n-1} c_j D_j^*, \qquad (11.89)$$

where

$$c_j = 1 - \frac{j}{n} - 2\left(1 - \frac{j}{n}\right)^2 \qquad (11.90)$$

and

$$D_j^* = X_{(j+1)} - X_{(j)}.$$

The HKL test of H_0 versus H_{10} (F is IDMRL) rejects H_0 for significantly large values of $T^{(2)}$ and accepts H_0 otherwise. (To simplify the notation, in equations (11.87)–(11.90) the dependence on n of $T^{(2)}, \xi_k, A,$ and c_j has been suppressed. In HKL these quantities are called $T_n^{(2)}, \xi_{nk}, A_n,$ and c_{nj}.)

TABLE 11.10. Estimated Large-Sample Percentiles of $T^{(2)}$

$1 - \alpha$	.90	.95	.99
$(1 - \alpha)$th percentile	1.41	1.59	1.93

Source: D. L. Hawkins, S. Kochar, and C. Loader (1992).

HKL do not provide exact tables for the null distribution of $T^{(2)}$ but instead base their test on estimated critical values obtained from an asymptotic approximation given by equation (2.6) of their paper. Table 11.10 contains selected estimated percentiles of the asymptotic distribution of $T^{(2)}$.

From Table 11.10 we see that for large n,

$$P_0\left(T^{(2)} \geq 1.41\right) \cong .10, \qquad P_0\left(T^{(2)} \geq 1.59\right) \cong .05, \qquad P_0\left(T^{(2)} \geq 1.93\right) \cong .01.$$

Thus, for example, the approximate $\alpha = .05$ test of exponentiality versus IDMRL alternatives rejects H_0 if $T^{(2)} \geq 1.59$ and accepts H_0 if $T^{(2)} < 1.59$.

Aly (1990) proposed competitors of the GHP (1986) tests for the case where the turning point τ (or the proportion ρ that dies before or at the turning point) is known and a competitor of the HKL (1992) tests when neither τ or ρ is known.

Properties

1. *Consistency.* The test defined by (11.83) is consistent against those F distributions for which the parameter $T(F)$, given by equation (2.1) of Guess, Hollander, and Proschan (1986), is positive. In particular, if F is continuous, and IDMRL (and not exponential), the test is consistent.
2. *Asymptotic Normality.* See Guess, Hollander, and Proschan (1986).
3. *Efficiency.* See Hawkins, Kochar, and Loader (1992) and Section 11.8. Asymptotic relative efficiencies are unavailable, but HKL give some Monte Carlo power comparisons of their tests with respect to the GHP test.

PROBLEMS

24. The data for Bjerkedal's study M, regimen 6.6, are given in Table 11.11. Test for a trend change in the mean residual life function at $\tau = 89$.

TABLE 11.11. Ordered Survival Times in Days of Guinea Pigs under Regimen 6.6

i	1	2	3	4	5	6	7	8	9
$X_{(i)}$	12	15	22	24	24	32	32	33	34
i	10	11	12	13	14	15	16	17	18
$X_{(i)}$	38	38	43	44	48	52	53	54	54
i	19	20	21	22	23	24	25	26	27
$X_{(i)}$	55	56	57	58	58	59	60	60	60
i	28	29	30	31	32	33	34	35	36
$X_{(i)}$	60	61	62	63	65	65	67	68	70
i	37	38	39	40	41	42	43	44	45
$X_{(i)}$	70	72	73	75	76	76	81	83	84

(continued)

TABLE 11.11. *(continued)*

i	46	47	48	49	50	51	52	53	54
$X_{(i)}$	85	87	91	95	96	98	99	109	110
i	55	56	57	58	59	60	61	62	63
$X_{(i)}$	121	127	129	131	143	146	146	175	175
i	64	65	66	67	68	69	70	71	72
$X_{(i)}$	211	233	258	258	263	297	341	341	376

Source: T. Bjerkedal (1960).

25. Calculate the estimated mean residual life function for the guinea pig survival data of Table 11.9.
26. Calculate the estimated mean residual life function for the guinea pig survival data of Table 11.11.
27. Test for a trend change in the mean residual life using the chronic granulocytic leukemia data of Table 11.6.
28. Describe a situation (different from those mentioned in Comment 23) where one might expect a trend change in the mean residual life.

11.5. A CONFIDENCE BAND FOR THE DISTRIBUTION FUNCTION (KOLMOGOROV)

Assumption

A1. The observations are a random sample from the underlying continuous population. That is, the X's are independent and identically distributed according to a continuous distribution F.

Note that we do not require assumption A2 (used in Sections 11.1–11.4) that F be a life distribution. Here the X's can also assume negative values.

We seek a simultaneous confidence band for the unknown distribution function; that is, we seek random functions (i.e., functions that depend on the observed sample values $X_1, \ldots, X_n$) $\ell(x)$ and $u(x)$ satisfying

$$P\{\ell(x) \leq F(x) \leq u(x), \text{ for all } x\} \geq 1 - \alpha.$$

We then say $\{\ell(x), u(x)\}$ is a simultaneous confidence band (or, more simply, a confidence band) for $F(x)$ with confidence at least $100(1 - \alpha)\%$.

The bands are based on the null distribution of the Kolomogorov statistic (see Comment 27). They are defined as

$$\ell(x) = \begin{cases} F_n(x) - d_\alpha, & \text{if } F_n(x) - d_\alpha \geq 0, \\ 0, & \text{if } F_n(x) - d_\alpha < 0, \end{cases} \quad (11.91)$$

and

$$u(x) = \begin{cases} F_n(x) + d_\alpha, & \text{if } F_n(x) + d_\alpha \leq 1, \\ 1, & \text{if } F_n(x) + d_\alpha > 1. \end{cases} \quad (11.92)$$

In (11.91) and (11.92), $F_n(x)$ is the empirical distribution function of the X's defined by

$$F_n(x) = \frac{\text{number of } X\text{'s in the sample } \leq x}{n}, \quad (11.93)$$

11.5. A CONFIDENCE BAND FOR THE DISTRIBUTION FUNCTION (KOLMOGOROV)

and d_α is the upper α percentile point of the distribution of Kolmogorov's statistic D (see Comments 26 and 5.41), defined as

$$D = \sup_{-\infty < x < \infty} \{|F_n(x) - F(x)|\}; \quad (11.94)$$

that is, d_α satisfies

$$P_F\left(\sup_{-\infty < x < \infty} \{|F_n(x) - F(x)|\} < d_\alpha\right) = 1 - \alpha. \quad (11.95)$$

Values of d_α are given in Table A.38.

Large-Sample Approximation

Kolmogorov (1933) and Smirnov (1939) (see also Feller (1948)) proved that as $n \to \infty$, $P_F(D \leq z/\sqrt{n})$ tends to $L(z)$, where

$$L(z) = 1 - 2\sum_{i=1}^{\infty}(-1)^{i-1}e^{-2i^2z^2}. \quad (11.96)$$

Smirnov (1948) presents a table of values of $L(z)$. For large n, d_α can be approximated by

$$d_\alpha \doteq \frac{z_\alpha^*}{\sqrt{n}}, \quad (11.97)$$

where $L(z_\alpha^*) = 1 - \alpha$. The last row of Table A.38 gives the values $z_\alpha^*/\sqrt{n}$ for $\alpha = .20, .10, .05, .02, .01$. For example, from Table A.38 we find, with $\alpha = .05$, $z_{.05}^*/\sqrt{n} = 1.36/\sqrt{n}$. The large-sample approximation is reasonably good for $n \geq 38$.

Example 11.6: *Example 11.1 Continued.* We use the data of Example 11.1 to determine a confidence band for the distribution of lifelengths. The ordered times to death (in days) are 42, 43, 51, 61, 66, 69, 71, 81, 82, 82. We illustrate the calculation of the 95% band. From Table A.38 we find, with $n = 10$, $d_{.05} = .409$. Table 11.12 illustrates the calculation of $F_n(x)$, $\ell(x)$, and $u(x)$. Figure 11.3 is a plot of $F_n(x)$ and the 95% confidence band.

TABLE 11.12. Calculation of the Confidence Band for $F(x)$ for the Methylmercury Poisoning Data

x	$F_{10}(x)$	$\ell(x)$	$u(x)$
$x < 42$	0	0	.409
$42 \leq x < 43$	.1	0	.509
$43 \leq x < 51$	.2	0	.609
$51 \leq x < 61$	.3	0	.709
$61 \leq x < 66$	.4	0	.809
$66 \leq x < 69$	.5	.091	.909
$69 \leq x < 71$	.6	.191	1
$71 \leq x < 81$	.7	.291	1
$81 \leq x < 82$	.8	.391	1
$x \geq 82$	1	.591	1

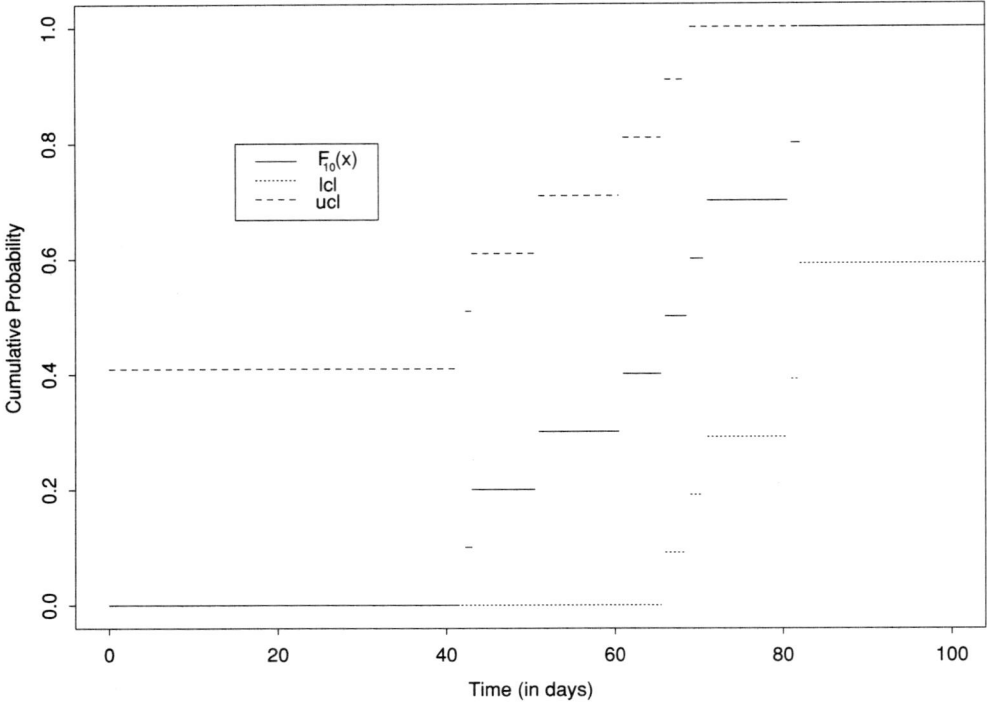

FIGURE 11.3. A 95% confidence band for $F(x)$ for the methylmercury poisoning data.

The null distribution tables for D are derived under the assumption that the underlying distribution is continuous. If there are tied observations (as is the case for the methylmercury poisoning data of this example), the confidence band derived using percentage points of D as given in Table A.38 and the goodness-of-fit test based on D in conjunction with Table A.38 are approximate, not exact.

Comments

26. *Derivation of the Confidence Band.* Using the null distribution of D (see Comment 27 and Table A.38), we can determine a critical value d_α such that

$$P_F\left(|F_n(x) - F(x)| \leq d_\alpha, \text{ for all } x\right) = 1 - \alpha. \tag{11.98}$$

The subscript F in the notation P_F means that the probability is being computed under the assumption that $X_1, \ldots, X_n$ is a random sample from F. We can rewrite (11.98) as

$$P_F\left(-d_\alpha \leq F_n(x) - F(x) \leq d_\alpha, \text{ for all } x\right) = 1 - \alpha. \tag{11.99}$$

Equation (11.99) is equivalent to

$$P_F\left(d_\alpha \geq -F_n(x) + F(x) \geq -d_\alpha, \text{ for all } x\right) = 1 - \alpha. \tag{11.100}$$

Equation (11.100) is equivalent to

$$P_F\left(F_n(x) - d_\alpha \leq F(x) \leq F_n(x) + d_\alpha, \text{ for all } x\right) = 1 - \alpha. \tag{11.101}$$

11.5. A CONFIDENCE BAND FOR THE DISTRIBUTION FUNCTION (KOLMOGOROV)

From (11.101) we can conclude that $\{F_n(x) - d_\alpha, F_n(x) + d_\alpha\}$ is a $1 - \alpha$ confidence band for F. It is, however, possible that the upper boundary $F_n(x) + d_\alpha$ may exceed 1. We know that $F(x)$ itself cannot exceed 1, so we can lower the upper boundary to $u(x) =$ minimum $[F_n(x) + d_\alpha, 1]$. Similarly, it is possible that the lower boundary $F_n(x) - d_\alpha$ may be less than 0. We know that $F(x)$ cannot be less than 0, so we can raise the lower boundary to $\ell(x) =$ maximum $[F_n(x) - d_\alpha, 0]$. These adjustments yield the band given by (11.91) and (11.92).

27. **Goodness-of-Fit Test Based on D.** Suppose we have a random sample $X_1, \ldots, X_n$ from a population with distribution function $F(x)$. Suppose further there is reason to believe (perhaps based on previous experience) that F_0 is some completely specified distribution. For example, F_0 may be specified to be a normal distribution with mean 1 and standard deviation 2 or an exponential distribution (see equation (11.3)) with scale parameter $\lambda = \frac{1}{2}$. Kolmogorov's test of the null hypothesis,

$$H_0^* : F(x) = F_0(x) \quad \text{for all } x, \tag{11.102}$$

against the alternative,

$$H_A^* : F(x) \neq F_0(x) \text{ for at least one } x, \tag{11.103}$$

is based on

$$D = \sup_{-\infty < x < \infty} \{|F_n(x) - F_0(x)|\}, \tag{11.104}$$

where $F_r(x)$, the sample distribution function, is

$$F_n(x) = \frac{\text{number of } X\text{'s in the sample } \leq x}{n}. \tag{11.105}$$

Alternatively, $F_n(x)$ can be expressed as

$$F_n(x) = \begin{cases} 0, & x < X_{(1)}, \\ \dfrac{i}{n}, & X_{(i)} \leq x < X_{(i+1)}, \\ 1, & x > X_{(n)}. \end{cases} \tag{11.106}$$

In (11.104), $\sup_{-\infty<x<\infty}$ denotes the supremum over all x of the absolute value of the difference $F_n(x) - F_0(x)$. If $F_n(x)$ and $F_0(x)$ are plotted as ordinates against x as abscissa, D is the value of the largest vertical distance between F_n and F_0. The supremum occurs at one of the $X_{(i)}$'s (that is, at one of the jump points of F_n) or just to the left of one of the $X_{(i)}$'s.

Formally, Kolmogorov's test, at the α-level of significance, is

$$\text{Reject } H_0^* \text{ if } D \geq d_\alpha; \quad \text{otherwise do not reject}, \tag{11.107}$$

where d_α satisfies $P_{F_0}(D \geq d_\alpha) = \alpha$. Values of d_α are given in Table A.38.

The motivation for the test is as follows. The sample distribution function $F_n(x)$ has many desirable properties as an estimator of the underlying distribution $F(x)$ from which the sample is drawn. In particular $F_n(x)$ converges to $F(x)$. If the hypothesized distribution F_0 is the true distribution F, then $F_n(x)$ should be "close" to $F_0(x)$. The statistic D_n is the largest vertical distance between F_n and F_0 and this largest distance should be small if H_0^* is true.

When all n observations are distinct, D can be computed as

$$D = \max_{i=1,\ldots,n} M_i, \tag{11.108}$$

where

$$M_i = \max\left\{\left|\frac{i}{n} - F_0\left(X_{(i)}\right)\right|, \left|\frac{(i-1)}{n} - F_0\left(X_{(i)}\right)\right|\right\}. \tag{11.109}$$

If there are tied observations, let k denote the number of distinct observations and let $Z_{(1)} < \cdots < Z_{(k)}$ denote the ordered distinct observations. Then D can be computed as

$$D = \max_{i=1,\ldots,n} M'_i, \tag{11.110}$$

where

$$M'_i = \max\left\{\left|F_n\left(Z_{(i)}\right) - F_0\left(Z_{(i)}\right)\right|, \left|F_n\left(Z_{(i-1)}\right) - F_0\left(Z_{(i)}\right)\right|\right\}. \tag{11.111}$$

When F is continuous, the statistic D has a continuous distribution. The statistic D is distribution-free under H_0^* when F_0 is a continuous distribution. To see this let $X_{(0)} < \cdots < X_{(n+1)}$ denote the order statistics, where $X_{(0)} = -\infty$ and $X_{(n+1)} = \infty$. When $X_1, \ldots, X_n$ are independent and identically distributed according to the continuous distribution F_0, it can be shown using the probability integral transformation (cf. Casella and Berger (1990, 52–54)) that $F_0(X_1), \ldots, F_0(X_n)$ are independent and identically distributed according to the uniform distribution on $[0, 1]$. It follows that $F_0(X_{(1)}), \ldots, F_0(X_{(n)})$ have the same joint distribution as that of the order statistics from a sample of size n from the uniform distribution on $[0, 1]$. Therefore, in determining the distribution of D under H_0^*, without loss of generality F_0 can be taken to be the uniform distribution on $[0, 1]$. That is, $F_0(x) = 0$ for $x < 0$, $F_0(x) = x$ for $0 \leq x \leq 1$, $F_0(x) = 1$ for $x > 1$. This simplifies the calculations used to determine the critical values d_α of Table A.38. For further details see Birnbaum (1952) and Miller (1956).

To illustrate the test based on D we return to settling velocity data of Table 3.12. Suppose that we wish to test if the data are from a normal population with mean 14 and standard deviation 2. That is, suppose the hypothesized distribution is $F_0 = N(14, 2)$. The seven ordered values are 12.8, 12.9, 13.3, 13.4, 13.7, 13.8, 14.5. The values $F_0(X_{(1)}), \ldots, F_0(X_{(7)})$ for use in equation (11.109) are calculated as follows. Let Y denote a $N(14, 2)$ random variable. Then

$$F_0(X_{(1)}) = P(Y \leq 12.8) = P\left(\frac{Y-14}{2} \leq \frac{12.8-14}{2}\right) = P(Z \leq -.6) = .2743,$$

where Z has a $N(0, 1)$ distribution. Letting $\Phi(z)$ denote the area under the $N(0, 1)$ curve to the left of z (i.e., $P(Z \leq z)$), we find the value .2743 from Table A.1. Note that, by symmetry, $\Phi(-.6) = 1 - \Phi(.6) = P(Z \geq .6)$. Similarly we find

$$F_0(X_{(2)}) = P(Y \leq 12.9) = P\left(\frac{Y-14}{2} \leq \frac{12.9-14}{2}\right) = P(Z \leq -.55)$$
$$= \Phi(-.55) = .2912,$$

$$F_0(X_{(3)}) = P(Y \leq 13.3) = P\left(\frac{Y-14}{2} \leq \frac{13.3-14}{2}\right) = P(Z \leq -.35)$$
$$= \Phi(-.35) = .3632,$$

11.5. A CONFIDENCE BAND FOR THE DISTRIBUTION FUNCTION (KOLMOGOROV)

$$F_0(X_{(4)}) = P(Y \leq 13.4) = P\left(\frac{Y-14}{2} \leq \frac{13.4-14}{2}\right) = P(Z \leq -.3)$$
$$= \Phi(-.3) = .3821,$$

$$F_0(X_{(5)}) = P(Y \leq 13.7) = P\left(\frac{Y-14}{2} \leq \frac{13.7-14}{2}\right) = P(Z \leq -.15)$$
$$= \Phi(-.15) = .4404,$$

$$F_0(X_{(6)}) = P(Y \leq 13.8) = P\left(\frac{Y-14}{2} \leq \frac{13.8-14}{2}\right) = P(Z \leq -.1)$$
$$= \Phi(-.1) = .4602,$$

$$F_0(X_{(7)}) = P(Y \leq 14.5) = P\left(\frac{Y-14}{2} \leq \frac{14.5-14}{2}\right) = P(Z \leq .25)$$
$$= \Phi(.25) = .5987.$$

Table 11.13 illustrates the calculation of D using equations (11.108) and (11.109).
From Table 11.13 and equations (11.108) and (11.109)

$$D = .4013.$$

From Table A.38 we find, with $n = 7$, $d_{.2} = .381$ and $d_{.1} = .436$. Thus, for our observed value $D = .4013$, the P-value is between .1 and .2. Therefore, there is not sufficient evidence to reject H_0^*. We should not, however, be surprised that H_0^* is not rejected. With the small sample size $n = 7$, Kolmogorov's test will not have good power against reasonable alternatives to H_0^*. Even when the sample sizes are large, there are many types of distributions for which the Kolmogorov statistic D will have low power (cf. Fan 1996).

28. *Goodness-of-Fit Test for a Composite Null Hypothesis.* In Comment 26, D was used to test the simple null hypothesis H_0^*, where the underlying distribution, F, is completely specified under the null hypothesis. If, instead, the underlying distribution, F, is not completely specified under the null hypothesis, but rather the null hypothesis asserts that F is a member of some parametric family with one or more parameters unspecified, this is known as a composite null hypothesis. The statistic D, suitably modified, can be used to test a composite null hypothesis. The modified statistic is D', where

$$D' = \sup_{-\infty < x < \infty} \left|F_n(x) - \widehat{F}_0(x)\right| \tag{11.112}$$

TABLE 11.13. Calculation of D for the Settling Velocity Data in the Case Where $F_0 = N(14, 2)$

i	$X_{(i)}$	$\frac{i}{n} - F_0(X_{(i)})$	$F_0(X_{(i)}) - \frac{(i-1)}{n}$	$\max\{\lvert\frac{i}{n} - F_0(X_{(i)})\rvert, \lvert\frac{(i-1)}{n} - F_0(X_{(i)})\rvert\}$
1	12.8	$\frac{1}{7} - .2743 = -.1314$	$.2743 - 0 = .2743$	.2743
2	12.9	$\frac{2}{7} - .2912 = -.0055$	$.2912 - \frac{1}{7} = .1483$	.1483
3	13.3	$\frac{3}{7} - .3632 = .0654$	$.3632 - \frac{2}{7} = .0775$	.0775
4	13.4	$\frac{4}{7} - .3821 = .1893$	$.3821 - \frac{3}{7} = -.0465$	.1893
5	13.7	$\frac{5}{7} - .4404 = .2739$	$.4404 - \frac{4}{7} = -.1310$	.2739
6	13.8	$\frac{6}{7} - .4602 = .3969$	$.4602 - \frac{5}{7} = -.2541$	.3969
7	14.5	$1 - .5987 = .4013$	$.5987 - \frac{6}{7} = -.2584$	.4013

where $\widehat{F}_0(x)$ is an estimator of F_0 calculated using the method of maximum likelihood estimation to estimate the unspecified parameters of the hypothesized parametric family. When the underlying F is continuous, D' has a continuous distribution. The null distribution and asymptotic null distribution of D for the simple null hypothesis are no longer valid for D' in the composite case, and new tables need to be derived for each parametric family. Lilliefors (1967) used simulation to obtain tables of the null distribution of D' for testing for an underlying normal distribution and more accurate tables based on simulation were provided by Dallal and Wilkinson (1986). For testing for an underlying exponential distribution, see Lilliefors (1969), Stephens (1974) and Durbin (1975), for testing for an underlying logistic distribution see Stephens (1979), and see Chandra, Singpurwalla, and Stephens (1981) for the extreme-value and Weibull families. In Comment 28 we give an illustration of the test for the normal family with the mean and standard deviation unspecified. The test is due to Lilliefors (1967, 1969). For summary articles on Kolmogorov-Smirnov-type tests of fit, see Stephens (1983a, 1983b).

29. *Lilliefors' Test of Normality.* We wish to test if the X's come from a normal distribution (or more realistically, if their distribution can be reasonably approximated by a normal distribution). The maximum likelihood estimators of the mean μ and the standard deviation σ are

$$\widehat{\mu} = \overline{X} = \frac{\sum_{i=1}^{n} X_i}{n} \tag{11.113}$$

and

$$\widehat{\sigma} = s = \sqrt{\frac{\sum_{i=1}^{n}(X_i - \overline{X})^2}{n-1}}. \tag{11.114}$$

The estimator $\widehat{F}_0(x)$ for use in equation (11.112) is the normal distribution with mean $\overline{X}$ and standard deviation s given by (11.113) and (11.114), respectively. That is, for each x,

$$\widehat{F}_0(x) = \Phi\left(\frac{x - \overline{X}}{s}\right). \tag{11.115}$$

Letting $X_{(1)} < \cdots < X_{(n)}$ denote the ordered X's, we need to obtain the n values of $F_n(X_{(i)})$ as well as the n values of

$$\widehat{F}_0\left(Z_{(i)}\right) = \Phi\left(Z_{(i)}\right) = \Phi\left(\frac{X_{(i)} - \overline{X}}{s}\right), \tag{11.116}$$

where

$$Z_{(i)} = \frac{X_{(i)} - \overline{X}}{s}, \tag{11.117}$$

and $Z_{(1)} < \cdots < Z_{(n)}$ are the ordered Z's. Then D' can be written as

$$D' = \underset{i=1,\ldots,n}{\operatorname{maximum}} \left\{ \left|F_n\left(X_{(i)}\right) - \Phi\left(Z_{(i)}\right)\right|, \left|F_n\left(X_{(i)}-\right) - \Phi\left(Z_{(i)}\right)\right| \right\}, \tag{11.118}$$

where, in (11.118), $F_n\left(X_{(i)}-\right)$ denotes the height of the empirical distribution just to the left of $X_{(i)}$. That is, $F_n\left(X_{(i)}-\right)$ is $(i-1)/n$ for $i = 1, \ldots, n$. Table A.39 gives critical values for D' in this case of testing for a normal family.

We illustrate Lilliefors' test using the settling velocities data of Table 3.12. We are testing if the underlying distribution can be reasonably approximated by some unspecified normal

TABLE 11.14. Calculation of D' for the Settling Velocity Data in the Case of Testing for an Underlying Unspecified Normal Distribution

i	$X_{(i)}$	$Z_{(i)} = \frac{(X_{(i)}-\bar{X})}{s}$	$\Phi(Z_{(i)})$	$\left|F_7(X_{(i)}) - \Phi(Z_{(i)})\right|$	$\left|F_7(X_{(i)}-) - \Phi(Z_{(i)})\right|$
1	12.8	−1.1793	.1191	.0237	.1911
2	12.9	−1.0073	.1569	.1288	.0140
3	13.2	−.3194	.3747	.0539	.0890
4	13.4	−.1474	.4414	.1300	.0128
5	13.7	.3685	.6438	.0705	.0723
6	13.8	.5405	.7056	.1516	.0087
7	14.5	1.7444	.9595	.0405	.1023

distribution. Direct calculations yield

$$\bar{X} = 13.486, \quad s = .5815.$$

Table 11.14 illustrates the calculation of D'.

From Table 11.14 and equation (11.118), we find

$$D' = .1516.$$

From Table A.39 we see that at $\alpha = .20$, the critical value for D' is .252. Since $D' = .1516$, the P-value is $> .20$. Thus the hypothesis of normality cannot be rejected. Since $n = 7$ is a small sample size, we should not be surprised that the test based on D' does not lead to rejection of the null hypothesis at a low α value.

30. *Modifications for Discrete Data and for Censored Data.* Pettitt and Stephens (1977) define a Kolmogorov-Smirnov-type statistic for discrete or grouped data. For discrete data the possible outcomes are divided into k cells and the null hypothesis is $H_0 : P$ (an observation falls in cell i) $= p_i$, $i = 1, \ldots, k$, where the p's are specified. For a sample of size N, their statistic is

$$PS = \max_{i=1,\ldots,k} \left|\sum_{i=1}^{j}(O_i - E_i)\right| \qquad (11.119)$$

where O_i is the observed number in the ith cell and $E_i = np_i$ is the expected number for the ith cell. The null hypothesis is rejected for significantly large values of PS. For grouped continuous data, let $x_0 < x_1 < \cdots < x_k$ define the cells in that cell i contains values of the random variable X for which $x_{i-1} < X < x_i$. If O_i and E_i are the observed and expected values in cell i, PS is defined as in (11.119).

Pettitt and Stephens provide null distribution tables of PS (the statistic is S in their notation) for sample sizes 30 or less, giving exact upper-tail probabilities in the equal-cell case (i.e., equal expected values E_i) and good approximations for other situations.

Fleming et al. (1980) provide an uncensored data modification of Kolmogorov's goodness-of-fit test, which they generalize to right-censored data. Chi-squared-type goodness-of-fit tests for censored data are given by Habib and Thomas (1986), Akritas (1988), Hjort (1990a), Hollander and Peña (1992a, 1992b), and Li and Doss (1993). See also Section VI.3.3 of Andersen et al. (1993).

Properties

1. *Asymptotic Distribution.* See Smirnov (1948).
2. *Closeness of Actual Coverage Probability to Nominal Coverage Probability when Using Large-Sample Approximation.* See Nair (1984).
3. *Asymptotic Efficiency.* See Nair (1984) and Hollander and Peña (1989).

PROBLEMS

29. Table 11.15 contains service-time data for a Tallahassee fast-food restaurant. The data were obtained by H. Schonrock (1996). The service time is defined as the time when the car pulled up to the speaker to order to the time when the car left the window with the order. The data were obtained at dinner time on a Thursday evening.

TABLE 11.15. Service Times at a Fast-Food Restaurant

Start	Finish	Time in seconds
5:35:17	5:36:09	54
5:37:33	5:39:21	108
5:38:41	5:40:37	115
5:47:27	5:49:36	129
5:48:34	5:50:06	92
5:49:06	5:51:24	138
5:53:31	5:54:14	43
5:54:36	5:56:57	141
5:55:27	5:57:17	110
5:58:11	6:00:09	118
6:00:42	6:02:00	78
6:01:37	6:03:05	88
6:02:28	6:08:08	340
6:04:59	6:08:49	230
6:08:35	6:10:53	138
6:12:34	6:15:31	177
6:15:18	6:17:48	150
6:19:24	6:21:29	125
6:21:22	6:22:42	80
6:21:39	6:24:07	148
6:22:33	6:25:58	205
6:24:31	6:31:24	413
6:27:05	6:31:41	276
6:31:17	6:33:43	146
6:31:44	6:34:52	188
6:36:03	6:37:42	99
6:38:00	6:40:14	134
6:40:31	6:41:01	30
6:41:02	6:44:04	182
6:41:33	6:45:16	223
6:44:30	6:46:45	135
6:46:14	6:50:43	269
6:47:36	6:51:20	224
6:48:02	6:52:19	257

Source: H. Schonrock (1996).

(a) Do you think the observations can be viewed as a random sample from the service-time distribution? In particular, consider the question of whether the observations are independent.

(b) Compute 90% confidence bands for the distribution of time to service.

30. Refer to Problem 29 and explain why one might, a priori, suspect that the time-to-service distribution is an IFR distribution. Apply a suitable procedure to test exponentiality versus IFR alternatives.

31. With $F_n(x)$ and $F_0(x)$ plotted as ordinates against x as abscissa, the Kolmogorov statistic, D, is the value of the largest vertical distance between them. Justify that the largest vertical distance can be expressed by equations (11.108) and (11.109).

32. Obtain approximate 95% simultaneous confidence bands for the survival function corresponding to the leukemia data of Table 11.6.

33. Obtain approximate 95% simultaneous confidence bands for the survival function corresponding to the guinea pig data of Table 11.9.

11.6. AN ESTIMATOR OF THE DISTRIBUTION FUNCTION WHEN THE DATA ARE CENSORED (KAPLAN-MEIER)

Assumptions

B1. Let $T_1, \ldots, T_n$ be independent each with continuous life distribution F. Let $C_1, \ldots, C_n$ be independent, each with continuous censoring distribution function G. C_i is the censoring time corresponding to T_i.

B2. We observe, for $i = 1, \ldots, n$,

$$X_i = \text{minimum } \{T_i, C_i\}$$

and δ_i where

$$\delta_i = \begin{cases} 1, & \text{if } T_i \leq C_i, \\ 0, & \text{if } T_i > C_i. \end{cases}$$

Thus δ_i is 1 if T_i is uncensored and we observe the true survival time, T_i, rather than the time to censorship. However, δ_i is 0 if T_i is censored and we observe C_i. In this case we know only that the survival time T_i is greater than C_i.

B3. The T's and C's are mutually independent.

Assumption B3 in practical terms means that the censoring times provide no information about the true survival times. This would not be the case, for example, in a methadone study where patients on methadone are given weekly urine analysis tests to see if they have gone back on heroin. A patient who has started using heroin again may censor himself or herself (not show up for the weekly test) in order not to be "caught." Here X, the time to relapse, and C, the time to censorship, would not be independent.

To illustrate the notation of this model, which is known as the randomly right-censored model, consider the radiation of affected node data of Table 11.16. For those data, $X_1 = T_1 = 346$ and $\delta_1 = 1$ because the 346 value is not censored as the true time to relapse is observed. However, for X_4 we have $X_4 = C_4 = 1953$ and $\delta_4 = 0$ because the true number of days to relapse is not observed. The observation is censored and we only know that the true time to relapse is greater than 1953.

In clinical trials, the T's may represent times to the occurrence of an endpoint event. For example, the endpoint event may be relapse or death. The data are typically analyzed before all the subjects have experienced the endpoint event. For example, in a fertility clinic, women may be taking hormones to enhance the chance of becoming pregnant. Suppose, when the data are analyzed, a woman has been undergoing treatment for 418 days and is not yet pregnant. If T denotes the time to pregnancy (measured from the initialization of treatment), we know that $T > 418$, but we do not at this point know the (eventual) true value of T. We call the 418 value a censored observation and denote it by 418^c. Other women in the study may have left town or stopped coming to the clinic and, again, for such women the observations are censored.

Procedure

We let $F(x) = P(T \leq x)$ denote the distribution function and

$$\overline{F}(x) = 1 - F(x) = P(T > x) \tag{11.120}$$

denote the survival function. Furthermore, we let $t_{(1)} < \cdots < t_{(k)}$ denote the ordered distinct failure times. These are the known deaths and the censored values are not in the list of the $t_{(i)}$'s. (If F is continuous, then there will be no tied failure times, but in practice, ties occur.) We let n_i denote the number of patients *at risk* at time $t_{(i)}$. n_i is the number of patients who have not died or been censored before $t_{(i)}$. We let d_i be the number of deaths (failures) at time $t_{(i)}$. (When there are no tied observations, all the d's are 1.) The Kaplan-Meier (1958) estimator of the survival function at time x is

$$\overline{F}_{KM}(x) = \prod_{t_{(i)} \leq x} \left(1 - \frac{d_i}{n_i}\right). \tag{11.121}$$

The estimator of the distribution function at time x is, of course, $1 - \overline{F}_{KM}(x)$. We illustrate the computation of the Kaplan-Meier estimator in Example 11.6.

Example 11.7: *Hodgkin's Disease Data.* The data in Table 11.16 are from a clinical trial in early Hodgkin's disease (i.e., cases in which the disease was detected at an early stage) conducted at the Stanford Medical Center by Drs. H. S. Kaplan and S. A. Rosenberg (1973). The data also appear in Chapter 14 of Brown and Hollander (1977). Hodgkin's disease is a cancer of the lymph system. The two treatments considered by Kaplan and Rosenberg (1973) were (1) radiation treatment of the affected node and (2) radiation treatment of the affected node plus all nodes in the trunk of the body (total nodal radiation). The relapse-free survival times are given in Table 11.16. If a relapse has not occurred by the date of data analysis, that is indicated by an N, and a superscript c is affixed to the observation.

Figure 11.4 is a plot of the Kaplan-Meier estimators for the total nodal treatment and the affected node treatment. Table 11.17 illustrates the calculation of the Kaplan-Meier estimator for the affected node treatment. Because there are 16 known relapses (out of 25 patients), there are $k = 16$ distinct failure times listed in Table 11.17.

Note that the Kaplan-Meier estimator decreases at each distinct failure time and is constant between distinct failure times (it does not decrease at censored observations). If the largest observed value is censored, the survival probability estimate $\overline{F}_{KM}$, as given by (11.121), does not decrease to zero but remains constant from that largest observed value out to ∞. See Comment 34. To illustrate the calculations of Table 11.17 and the use of formula (11.121), consider, for example, the value of $\overline{F}_{KM}$ at $t_{(16)} = 1375$. The right-hand side of formula (11.121) is a product taken over those distinct failure times that are less or equal to 1375—that is, over the times 86, 107, 141, 292, 312, 330, 346, 364, 401, 419, 505, 570, 688, 822, 836, 1375. The product is evaluated as follows:

11.6. AN ESTIMATOR OF THE DISTRIBUTION FUNCTION WHEN THE DATA ARE CENSORED

TABLE 11.16. Relapse-Free Survival Times for Hodgkin's Disease Patients

	Radiation of Affected Node		Total Nodal Radiation
Relapse (Y) or not (N)	Days to Relapse or to Date of Analysis	Relapse (Y) or not (N)	Days to Relapse or to Date of Analysis
Y	346	N	1699^c
Y	141	N	2177^c
Y	296	N	1968^c
N	1953^c	N	1889^c
Y	1375	Y	173
Y	822	N	2070^c
N	2052^c	N	1972^c
Y	836	N	1897^c
N	1910^c	N	2022^c
Y	419	N	1879^c
Y	107	N	1726^c
Y	570	N	1807^c
Y	312	Y	615
N	1818^c	Y	1408
Y	364	N	1763^c
Y	401	N	1684^c
N	1645^c	N	1576^c
Y	330	N	1572^c
N	1540^c	Y	498
Y	688	N	1585^c
N	1309^c	N	1493^c
Y	505	Y	950
N	1378^c	N	1242^c
N	1446^c	N	1190^c
Y	86		

Source: H. S. Kaplan and S. A. Rosenberg (1973)

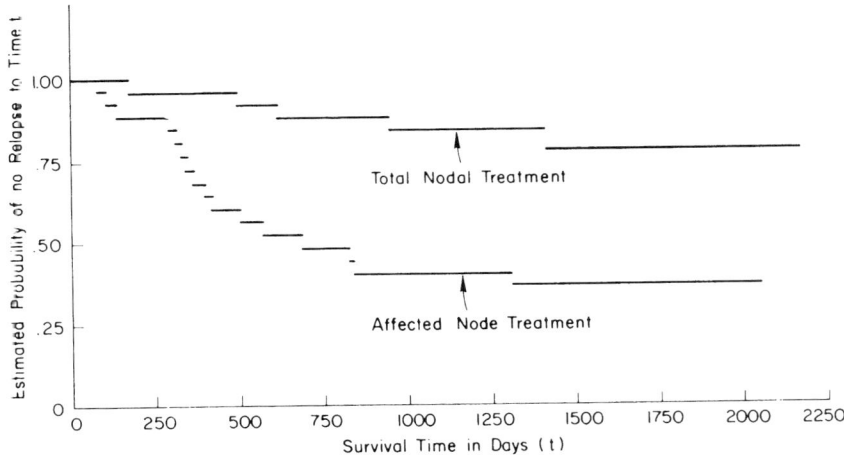

FIGURE 11.4. The Kaplan-Meier estimators for the total nodal and affected node survival distributions.

TABLE 11.17. Calculation of the Kaplan-Meier Estimator for the Affected Node Treatment

i	$t_{(i)}$	n_i	d_i	$\overline{F}_{KM}(t_{(i)})$
1	86	25	1	.960
2	107	24	1	.920
3	141	23	1	.880
4	296	22	1	.840
5	312	21	1	.800
6	330	20	1	.760
7	346	19	1	.720
8	364	18	1	.680
9	401	17	1	.640
10	419	16	1	.600
11	505	15	1	.560
12	570	14	1	.520
13	688	13	1	.480
14	822	12	1	.440
15	836	11	1	.400
16	1375	9	1	.356

$$\overline{F}_{KM}(1375) = \prod_{t_{(i)} \leq 1375} \left(1 - \frac{d_i}{n_i}\right)$$

$$= \left(1 - \tfrac{1}{25}\right)\left(1 - \tfrac{1}{24}\right)\left(1 - \tfrac{1}{23}\right)\left(1 - \tfrac{1}{22}\right)\left(1 - \tfrac{1}{21}\right)$$
$$\times \left(1 - \tfrac{1}{20}\right)\left(1 - \tfrac{1}{19}\right)\left(1 - \tfrac{1}{18}\right)\left(1 - \tfrac{1}{17}\right)\left(1 - \tfrac{1}{16}\right)$$
$$\times \left(1 - \tfrac{1}{15}\right)\left(1 - \tfrac{1}{14}\right)\left(1 - \tfrac{1}{13}\right)\left(1 - \tfrac{1}{12}\right)\left(1 - \tfrac{1}{11}\right)\left(1 - \tfrac{1}{9}\right)$$
$$= .356.$$

The values in Table 11.17 can be obtained by using the S-plus function surv.fit. In addition to the values displayed in Table 11.17, the S-plus printout includes, for each uncensored time, an estimate of the standard deviation of the Kaplan-Meier estimate and asymptotic confidence limits for the true survival probability (you specify the confidence coefficient; .95 is the default value).

In the Kaplan and Rosenberg study, 49 patients were admitted to the study between 1967 and 1970 and randomly assigned to the affected node and the total nodal therapies. The original 49 patients were followed, and the data as of fall 1973 are given in Table 11.16. In the spring of 1970, all subsequent patients were assigned the total nodal therapy because at that point the evidence was mounting that the total nodal therapy was superior to the affected node therapy.

Note that from Table 11.16 we see that by the date of analysis in the fall of 1973, 16 of the 25 affected node patients had relapsed, whereas only 5 of the 24 total nodal patients had relapsed. Furthermore, Figure 11.4 shows that the Kaplan-Meier estimator for the total nodal therapy is always above that of the affected node therapy; that is, the estimated chance of relapse-free survival to any time is higher for the total nodal group than the affected node group. The statistical assessment of this difference can be formally made with a two-sample test for censored data. We describe such a test in Section 11.7.

Comments

31. *Partial Motivation for the Kaplan-Meier Estimator.* The Kaplan-Meier estimator (also known as the product limit estimator) given by equation (11.121) can be motivated as follows: Just before time $t_{(i)}$, there are n_i patients at risk and d_i die at time $t_{(i)}$. Thus it is natural to estimate the probability of death at $t_{(i)}$, given that one has survived to $t_{(i)}$, as the ratio d_i/n_i— i.e., the number of deaths at $t_{(i)}$ divided by the number at risk at $t_{(i)}$. Then $(1 - (d_i/n_i))$ is the estimated conditional probability of surviving past time $t_{(i)}$, given survival up to $t_{(i)}$. The product in (11.121) corresponds to multiplying these conditional probabilities of not dying, for all known death times from zero up to the time of interest. This yields an estimate of the unconditional probability of surviving past the time of interest.

32. *Efron's Redistribute-to-the-Right Algorithm.* Suppose our data consist of the following observations: 4, 5, 5^c, 6^c, 7, 8^c, 9, 11^c, where the superscript c indicates a censored value. If the observations were all uncensored times, the empirical survival function would assign mass $\frac{1}{8}$ to each of the values. At the first censored time, 5^c, a death has not occurred but will occur somewhere to the right of 5. Efron's (1967) redistribute-to-the-right algorithm takes the mass of $\frac{1}{8}$ at 5^c and redistributes it equally among the five times 6^c, 7, 8^c, 9, 11^c to the right of 5^c, adding $\frac{1}{5}(\frac{1}{8})$ to the mass at 6^c, 7, 8^c, 9, 11^c. Now go to the next censored time 6^c and redistribute the new mass $\frac{1}{5}(\frac{1}{8}) + \frac{1}{8}$ equally among the observations to the right of 6^c. Continue this process until you reach the last observation. Efron shows this algorithm yields the Kaplan-Meier estimator. The algorithm is illustrated in the following display. The last value $\overline{F}_{KM}(x) = 0$ at 11^c in the southeast corner of the display is in accordance with Efron's convention (see Comment 34).

Observed Values	Mass at Start	Mass after First Redistribution	Mass after Second Redistribution	Mass after Third redistribution	$\overline{F}_{KM}(x)$
4	$\frac{1}{8}$	.125	.125	.125	.875
5	$\frac{1}{8}$	.125	.125	.125	.750
5^c	$\frac{1}{8}$	0	0	0	.750
6^c	$\frac{1}{8}$	$\frac{1}{8} + (\frac{1}{5})(\frac{1}{8}) = .150$	0	0	.750
7	$\frac{1}{8}$	.150	$.150 + (\frac{1}{4})(.150) = .1875$	.1875	.5625
8^c	$\frac{1}{8}$	.150	.1875	0	.5625
9	$\frac{1}{8}$	.150	.1875	$.150 + (\frac{1}{2})(.1875) = .28125$	.28125
11^c	$\frac{1}{8}$	.150	.1875	.28125	0

33. *Efron's Self-Consistency.* Another way to obtain the Kaplan-Meier estimator is via Efron's self-consistency process. If there is no censoring and we observe $T_1, \ldots, T_n$, the nonparametric estimator of $\overline{F}(x)$ is the empirical survival function

$$\overline{F}_n(x) = \frac{\sum_{i=1}^n \psi(T_i, x)}{n},$$

where $\psi(T_i, x) = 1$ if $T_i > x$, 0 otherwise. Note $n\overline{F}_n(x)$ is a sum of 0's and 1's, where 1 is scored if $T_i > x$ and 0 is scored otherwise. In the censored case, we observe

$$X_i = \min(T_i, C_i), \quad \delta_i = \begin{cases} 1, & \text{if } X_i = T_i, \\ 0, & \text{if } X_i = C_i, \end{cases}$$

and thus for some X's we cannot tell if the corresponding T's will exceed x. If $X_i > x$, we know $T_i > x$, but if $X_i < x$ and $\delta_i = 0$, we do not know if $X_i < T_i \leq x$ or, instead, if $T_i > x$. It is thus reasonable in such a case to score (in place of 1) an estimated conditional probability $\widehat{S}(x)/\widehat{S}(X_i)$, say, an estimated chance that T_i will exceed x, given that $T_i > X_i$. Efron calls an estimator $\widehat{S}$ self-consistent if

$$\widehat{S}(x) = \frac{1}{n}\left[N(x) + \sum_{\delta_i=0, X_i \leq x} \frac{\widehat{S}(x)}{\widehat{S}(X_i)}\right]$$

where $N(x)$ = number of X's $> x$. Start with some estimator (it can, for example, be $\widehat{S}_0(x) = N(x)/n$) on the right-hand side of the preceding defining equation, calculate the left-hand side, plug the calculated value into the right-hand side, and continue this process to form a sequence of estimators

$$\widehat{S}_{j+1}(x) = \frac{1}{n}\left[N(x) + \sum_{\delta_i=0, X_i \leq x} \frac{\widehat{S}_j(x)}{\widehat{S}_j(X_i)}\right].$$

Efron shows $\widehat{S}_j(x)$ converges in a finite number of steps to an estimator that will agree with $\overline{F}_{KM}(x)$ for x less than the largest observation.

34. *Kaplan-Meier Estimated Tail Probabilities.* Let $Z_{(1)} \leq \cdots \leq Z_{(n)}$ denote the ordered values in the combined list of uncensored and censored values. If $Z_{(n)}$ is an uncensored value, then $\overline{F}_{KM}(Z_{(n)}) = 0$ for all $x > Z_{(n)}$. If, however, $Z_{(n)}$ is censored, then $\overline{F}_{KM}(x)$—as defined by (11.121)—remains a nonzero constant from $Z_{(n)}$ to ∞ and thus does not have the property that $\overline{F}_{KM}(x)$ tends to 0 as x tends to ∞, a property that must hold for $\overline{F}(x)$, the true survival function being estimated by $\overline{F}_{KM}$. Some authors leave $\overline{F}_{KM}(x)$ undefined for $x > Z_{(n)}$ when $Z_{(n)}$ is a censored observation (as we have done in Figure 11.4, which gives the Kaplan-Meier estimates for the affected node and total nodal radiation data). Efron (1967) suggested that when $Z_{(n)}$ is censored, $\overline{F}_{KM}$ should be defined to be 0 for all $x > Z_{(n)}$. Gill (1980) suggested that when $Z_{(n)}$ is censored, one should set $\overline{F}_{KM}(x) = \overline{F}_{KM}(Z_{(n)})$ for $x > Z_{(n)}$ so that $\overline{F}_{KM}$ is a nonzero constant from $Z_{(n)}$ to ∞. Efron's convention yields underestimates of the tail survival probabilities, whereas Gill's convention yields overestimates. Brown, Hollander, and Korwar (1974) and Moeschberger and Klein (1985) suggested methods for completing the tail of the Kaplan-Meier estimator, which can be considered intermediate to Efron's and Gill's conventions. Brown, Hollander, and Korwar recommend fitting an exponential survival curve to complete the Kaplan-Meier estimator, namely, the exponential curve that agrees with $\overline{F}_{KM}(Z_{(n)})$ at $Z_{(n)}$. Determining the λ that satisfies $\exp(-\lambda Z_{(n)}) = \overline{F}_{KM}(Z_{(n)})$ yields $\lambda = -\{\ln[\overline{F}_{KM}(Z_{(n)})]\}/Z_{(n)}$, and the tail survival probabilities are defined to be

$$\exp(-\lambda x) = \exp\left(\frac{x\{\ln[\overline{F}_{KM}(Z_{(n)})]\}}{Z_{(n)}}\right) \quad \text{for } x > Z_{(n)}.$$

Moeschberger and Klein (1985) used the Weibull distribution to fit the tail survival probabilities when $Z_{(n)}$ is censored.

In Figure 11.4 we left the two Kaplan-Meier estimates undefined after the respective largest observations (which are censored in both samples). Figure 11.4 as is indicates the superiority of the total nodal treatment and in this case there in not a need to estimate survival probabilities in the tails. In situations where such tail estimates are desired, however, fitting the tail probabilities by the Brown, Hollander, and Korwar method or the Moeschberger and Klein method yields more realistic estimates then the Efron or Gill conventions.

11.6. AN ESTIMATOR OF THE DISTRIBUTION FUNCTION WHEN THE DATA ARE CENSORED 541

35. *Confidence Intervals for the Survival Probability at Time x.* Let

$$V(x) = \text{var}\left(\overline{F}_{KM}(x)\right), \tag{11.122}$$

the variance of the Kaplan-Meier estimator of surviving past time x. $V(x)$ can be estimated by

$$\widehat{V}(x) = \{\overline{F}_{KM}(x)\}^2 \sum_{t_{(i)} \leq x} \frac{d_i}{n_i(n_i - d_i)}. \tag{11.123}$$

The standard error given by (11.123) is known as Greenwood's formula (Greenwood 1926). An asymptotic $100(1-\alpha)\%$ confidence interval for $\overline{F}_{KM}(x)$ is $(\overline{F}_L(x), \overline{F}_U(x))$, where,

$$\overline{F}_L(x) = \overline{F}_{KM}(x) - z_{\alpha/2}\left(\widehat{V}(x)\right)^{1/2}, \quad \overline{F}_U(x) = \overline{F}_{KM}(x) + z_{\alpha/2}\left(\widehat{V}(x)\right)^{1/2}. \tag{11.124}$$

When no censoring and no tied observations occur, the Kaplan-Meier estimator reduces to the empirical survival function

$$\overline{F}_n(x) = \frac{\text{number of observations in the sample} > x}{n}, \tag{11.125}$$

where n is the sample size. Correspondingly, the asymptotic confidence interval given by (11.123) reduces, in the case of no censoring and no tied observations, to the following $100(1-\alpha)\%$ confidence interval for $\overline{F}(x)$:

$$\left(\overline{F}_n(x) - z_{\alpha/2}\left\{\frac{\overline{F}_n(x)(1-\overline{F}_n(x))}{n}\right\}^{1/2}, \quad \overline{F}_n(x) + z_{\alpha/2}\left\{\frac{\overline{F}_n(x)(1-\overline{F}_n(x))}{n}\right\}^{1/2}\right). \tag{11.126}$$

Estimators, confidence intervals, and confidence bands for $\overline{F}$ can also be obtained by exploiting the relationship between the cumulative hazard function of F and the survival function. The cumulative hazard function Λ_F corresponding to the distribution F is defined to be

$$\Lambda_F(x) = -\ln(1 - F(x)).$$

$\overline{F}$ can be expressed in terms of Λ by

$$\overline{F}(x) = e^{-\Lambda(x)}.$$

If one has an estimator $\widehat{\Lambda}$ of Λ, this yields an estimator $\widehat{\overline{F}}$ of $\overline{F}$ via $\widehat{\overline{F}}(x) = e^{-\widehat{\Lambda}(x)}$. The Nelson-Aalen estimator of Λ is

$$\widehat{\Lambda}(x) = \begin{cases} 0, & x < t_{(1)}, \\ \sum_{t_{(i)} \leq x} \frac{d_i}{n_i}, & x \geq t_{(1)}. \end{cases}$$

Nelson defined the estimator in an applied setting (see Nelson 1969, 1972) and Aalen (1978) considered it in a more general theoretical setting.

Kalbfleisch and Prentice (1980) (see also Bie, Borgan, and Liestøl (1987) and Borgan and Liestøl (1990)) used a log transformation of the cumulative hazard function (or, equivalently, the transformation $g(x) = \ln(-\ln(x))$ applied to the survival function) to obtain an asymptotic confidence interval for $\overline{F}(x)$ that is a competitor of the interval given by (11.124). Thomas and Grunkemeier (1975) used an arcsine square root transformation ($g(x) = \arcsin(\sqrt{x})$) to obtain a competitor of the interval given by (11.124). The Kalbfleisch and Prentice asymptotic $100(1 - \alpha)\%$ confidence interval for $\overline{F}(x)$ is

$$(\{\overline{F}_{KM}(x)\}^a, \{\overline{F}_{KM}(x)\}^{1/a})$$

where

$$a = \exp\left[\frac{z_{\alpha/2} \sum_{t_{(i)} \leq x} \frac{d_i}{n_i(n_i - d_i)}}{\ln(\overline{F}_{KM}(x))}\right].$$

Borgan and Liestøl (1990) found that the Thomas and Grunkemeier (1975) and Kalbfleisch and Prentice (1980) confidence intervals do better, in small samples, than the confidence interval given by (11.124).

36. *Exact Moments of the Kaplan-Meier Estimator.* Chen, Hollander, and Langberg (1982) give, for $\alpha > 0$, an exact expression for $E([\overline{F}_{KM}(x)]^\alpha)$, the αth moment of the Kaplan-Meier estimator. They consider the case where the censoring distribution G (the distribution of each C_i) and the survival distribution F (the distribution of each T_i) satisfy a proportional hazards model (see (11.154)). CHL (1982) used the Efron convention (see Comment 34) to define $\overline{F}_{KM}$ in the tail, whereas Wellner (1985) obtained closely related exact results using the Gill convention to define $\overline{F}_{KM}$ in the tail. These exact results enable one to study the exact biases of the Kaplan-Meier estimator and to compare the exact variances of the Kaplan-Meier estimator with its asymptotic variances. Wellner notes that the biases and variances of the Kaplan-Meier estimator based on the Gill convention are almost everywhere smaller than those for the Kaplan-Meier estimator based on the Efron convention. Thus Wellner advocates use of the Gill definition rather than the Efron definition.

37. *Simultaneous Confidence Bands for the Survival Function.* The interval given by (11.124) has approximate coverage probability $1 - \alpha$ at a fixed value x. Suppose, instead, we desire an approximate $1 - \alpha$ simultaneous confidence band for $\overline{F}$. Such a band would have the property

$$P_F\left(\overline{F}_\ell(x) \leq \overline{F}(x) \leq \overline{F}_u(x), \text{ for all } x\right) \doteq 1 - \alpha, \tag{11.127}$$

where $\overline{F}_\ell(x)$ is the lower contour of the band and $\overline{F}_u(x)$ is the upper contour of the band. Display (11.127) is to be interpreted as meaning that simultaneously, for all x, the chance is approximately $1 - \alpha$ that the random contours F_ℓ, F_u contain $\overline{F}$.

Confidence bands based on the Kaplan-Meier estimator have been presented by many authors including Gillespie and Fisher (1979), Gill (1980), Hall and Wellner (1980), Fleming et al. (1980), Nair (1981, 1984), Csörgő and Horváth (1986), Hollander and Peña (1989) and Hollander, McKeague and Yang (1997). (The bands in Fleming et al. (1980) are not explicitly defined in the paper, but they are implicitly given in the appendix of the paper and can be developed as outlined in Exercise 6.5, pp. 397–398, of Fleming and Harrington (1991).) Gulati and Padgett (1996) use the Hollander-Peña (1989) approach but replace the discontinuous Kaplan-Meier estimator by a kernel-based continuous estimator to obtain continuous confidence bands. See also Section IV.3.3 of Andersen et al. (1993) for confidence intervals and simultaneous confidence bands for the survival function. For confidence intervals, also see Thomas and Grunkemeir (1975) and Murphy (1995).

Some of the proposed bands are contained in the family of bands, indexed by a value $c > 0$, presented by Hollander and Peña (1989). The Hollander-Peña family is of the form

$$\left\{ [\overline{F}_\ell(x), \overline{F}_u(x)], \ 0 \le x \le T \right\}, \tag{11.128}$$

where

$$\overline{F}_\ell(x) = \overline{F}_{KM}(x)\{1 - r_n(x; c, \lambda_n)\}, \quad \overline{F}_u(x) = \overline{F}_{KM}(x)\{1 + r_n(x; c, \lambda_n)\}, \tag{11.129}$$

where

$$r_n(x; c, \lambda_n) = \frac{\lambda_n}{[(nc)^{1/2}\{1 - L_n^*(x, c)\}]}, \tag{11.130}$$

$$L_n^*(x, c) = \frac{c d_n^*(x)}{\{1 - c d_n^*(x)\}}, \tag{11.131}$$

$$d_n^*(x) = \sum_{\{t_{(i)} \le x\}} \frac{d_i}{n_i(n_i - d_i)}, \tag{11.132}$$

where, as before, d_i is the number of deaths at $t_{(i)}$ and n_i is the number of risk at $t_{(i)}$. The value of λ_n is obtained by entering Table A.40 (which is adapted from Table 11.1 of Hall and Wellner 1980). Enter Table A.40 with $\beta = 1 - \alpha$ and

$$a = \frac{c d_n^*(T)}{1 + c d_n^*(T)}.$$

Table A.40 gives values of λ such that $G_a(\lambda) = \beta$ for $a = .10, .25, .40, .50, .60, .75, .90, 1.00$ and $\beta = .99, .95, .90, .75, .50, .25$, where

$$G_a(\lambda) = 1 - 2\overline{\Phi}\left[\lambda\{a(1-a)\}^{-1/2}\right] + 2\sum_{k=1}^{\infty}(-1)^k e^{-2k^2\lambda^2}\left[\overline{\Phi}(r(2k-d)) - \overline{\Phi}\{r(2k+d)\}\right], \tag{11.133}$$

where Φ is the standard normal distribution function, $r = \lambda\{(1-a)/a\}^{1/2}$ and $d = (1-a)^{-1}$. $G_a(\lambda) = P\{\sup_{0 \le t \le a} |\mathcal{W}_t^0| \le \lambda\}$, where $\mathcal{W}^0$ is a Brownian bridge process on [0, 1] (cf. Billingsley 1968, 64–65).

The most popular confidence band in use at the time of this writing is the Hall-Wellner (1980) band. It corresponds to $c = 1$ in the Hollander-Peña family. The choice of c is discussed in Section 11.4 of Hollander and Peña (1989). Although the asymptotic coverage probability of all bands in the family is $1 - \alpha$, c controls the width of the bands on various intervals. Bands corresponding to larger values of c are narrower on the left (i.e., for smaller x values). If one desires a band that is narrow at the specific point $x = x_0$, choose $c = 1/d_n^*(x_0)$.

For the affected-node data of Table 11.15, Figures 11.5, 11.6, and 11.7 are, respectively, plots of asymptotic 95% confidence bands for the choices $c = .5, 1$, and 2. The figures also display the Kaplan-Meier estimator.

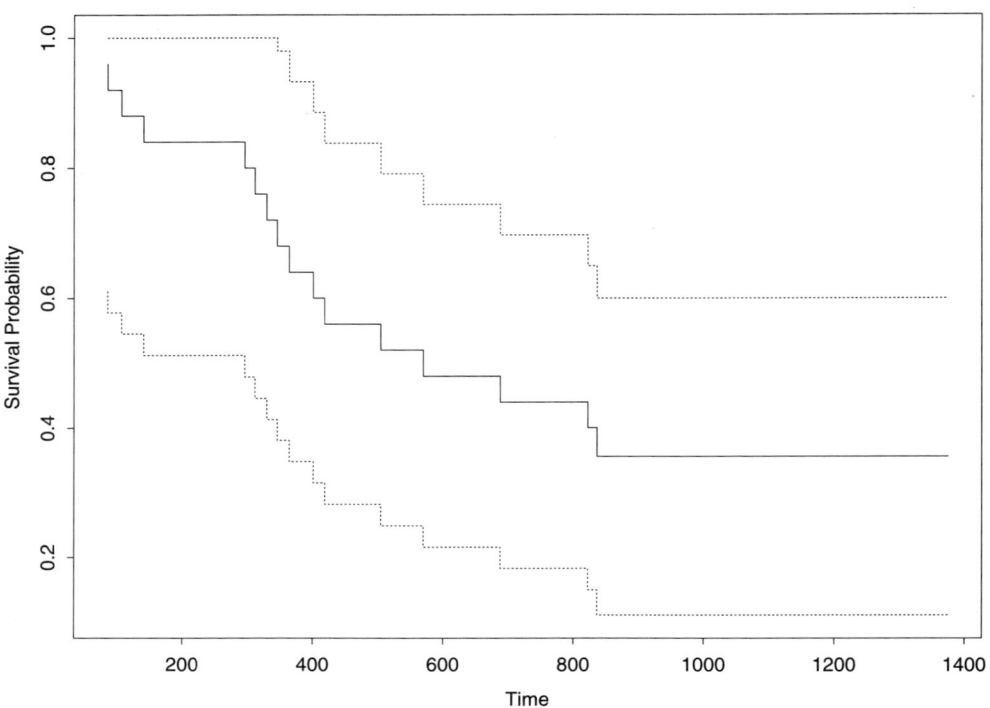

FIGURE 11.5. An approximate 95% confidence band for the affected-node survival distribution ($c = .5$).

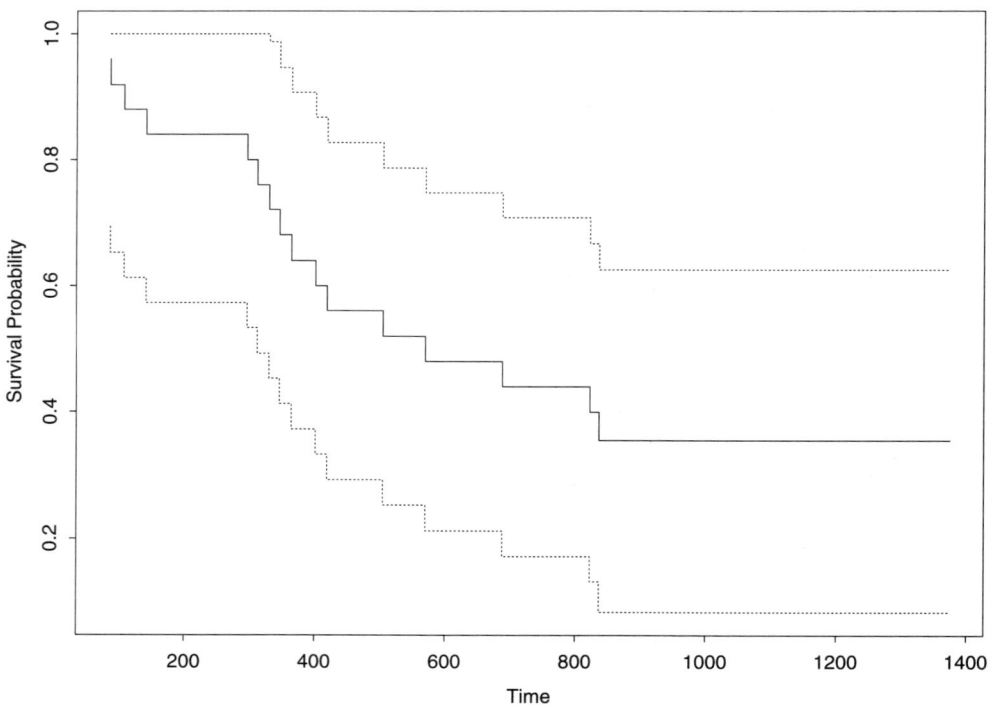

FIGURE 11.6. An approximate 95% confidence band for the affected-node survival distribution ($c = 1$, the Hall-Wellner band).

11.6. AN ESTIMATOR OF THE DISTRIBUTION FUNCTION WHEN THE DATA ARE CENSORED

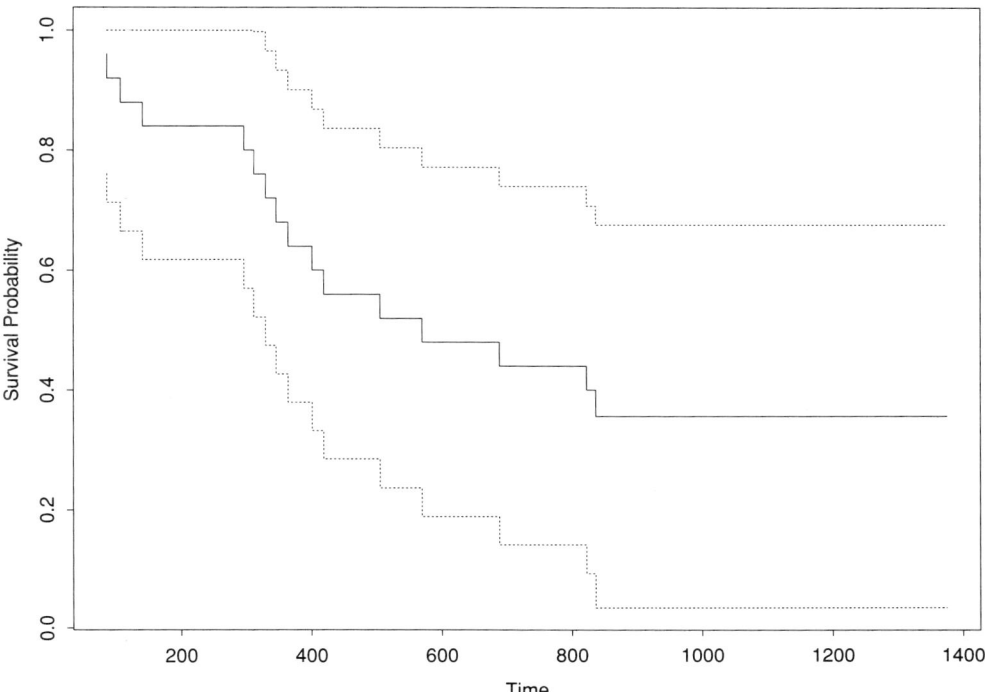

FIGURE 11.7. An approximate 95% confidence band for the affected-node survival distribution ($c = 2$).

38. *Monotonized Bands.* There will be some data sets for which the bands given by (11.129) can be narrowed and yet still retain the same asymptotic coverage probability. The process is called monotonization. Let $0 \leq x < y \leq T$. From (11.129) it can be seen that the ratio of the values of the upper contour of the bands at x and y is

$$\frac{\overline{F}_u(x)}{\overline{F}_u(y)} = \frac{\overline{F}_{KM}(x)\{1 + r_n(x; c, \lambda_n)\}}{\overline{F}_{KM}(y)\{1 + r_n(y; c, \lambda_n)\}}$$

$$= \frac{\overline{F}_{KM}(x)}{\overline{F}_{KM}(y)} \left[1 - \left\{ \frac{r_n(y; c, \lambda_n) - r_n(x; c, \lambda_n)}{1 + r_n(y; c, \lambda_n)} \right\} \right].$$

Because $\overline{F}_{KM}(x) \geq \overline{F}_{KM}(y)$ and $r_n(y; c, \lambda_n) \geq r_n(x; c, \lambda_n)$, it may happen that, for some x and y (with $x \leq y$), the ratio $\overline{F}_u(x)/\overline{F}_u(y)$ will be less than 1. In such a situation the upper contour of the band is not monotone decreasing. This is unacceptable because $\overline{F}$ itself is monotone decreasing. For a confidence band whose upper contour is not monotone decreasing, a narrower band can be formed that will have the monotone property and will also have the same confidence level as the original band. This narrower band is formed by *monotonizing* the upper contour of the original band. In general, suppose that $[m_n(x), M_n(x)]$ is an asymptotic $100(1 - \alpha)\%$ confidence band for $\overline{F}$ on the interval $[0, T]$. Then the monotonized band $[m_n(x), \min_{a \leq x} M_n(a)]$ is also an asymptotic $100(1 - \alpha)\%$ confidence band for $\overline{F}$. This is true because, due to the monotonicity of $\overline{F}$, the events

$$\{m_n(x) \leq \overline{F}(x) \leq M_n(x), \ 0 \leq x \leq T\} \tag{11.134}$$

and

$$\{m_n(x) \leq \overline{F}(x) \leq \min_{a \leq x} M_n(a),\ 0 \leq x \leq T\} \tag{11.135}$$

are identical and thus have the same probability of occurring. Furthermore, because $0 \leq \overline{F}(x) \leq 1$ for all x-values, we may use the confidence band given by

$$\left\{\max[0, m_n(x)],\ \min\left[1, \min_{a \leq x} M_n(a)\right],\ 0 \leq x \leq T\right\}. \tag{11.136}$$

39. *NBU Test for Censored Data.* Chen, Hollander, and Langberg (1983a) extended the NBU test of Section 11.2 to the censored data situation. Their test statistic J_c is obtained by estimating the parameter $\Delta_{\text{NBU}}(F)$ (see equation (11.46)) by $\Delta_{\text{NBU}}(F_{KM})$ where F_{KM} is the Kaplan-Meier estimator. Letting $Z_{(1)} \leq \cdots \leq Z_{(n)}$ denote the ordered values (in the combined list of censored and uncensored values), the statistic can be written as

$$J_c = \sum_{i=1}^n \overline{F}_{KM}\left(2Z_{(i)}\right) d_i^2 + 2\sum_{i<j}^n \overline{F}_{KM}\left(Z_{(i)} + Z_{(j)}\right) d_i d_j, \tag{11.137}$$

where

$$d_i = \overline{F}_{KM}\left(Z_{(i-1)}\right) - \overline{F}_{KM}\left(Z_{(i)}\right) \tag{11.138}$$

and $Z_{(0)}$ is defined to be 0.

The null asymptotic mean of J_c is $\frac{1}{4}$, independent of the scale parameter λ of the exponential distribution corresponding to the null hypothesis and independent of the censoring distribution governing the times to censorship. The null asymptotic variance of $n^{1/2}J_c$ does, however, depend on λ and the censoring distribution and thus must be estimated from the data. The NBU (NWU) test for censored data is based on

$$J_c^* = \frac{\sqrt{n}(J_c - \frac{1}{4})}{\widehat{\sigma}_c}, \tag{11.139}$$

where

$$\widehat{\sigma}_c^2 = (128)^{-1} + \sum_{i=1}^{n-1} n(n-i+1)^{-1}(n-i)^{-1}\{(128)^{-1} - (32)^{-1}Z_{(i)}(\widehat{\lambda}) + (16)^{-1}Z_{(i)}^2(\widehat{\lambda}^2)\}$$

$$\cdot \exp\left\{-4Z_{(i)}\widehat{\lambda}\right\} - n\left\{(128)^{-1} - (32)^{-1}Z_{(n)}\widehat{\lambda} + (16)^{-1}Z_{(n)}^2\widehat{\lambda}^2\right\}\exp\left\{-4Z_{(n)}\widehat{\lambda}\right\}$$

and

$$\widehat{\lambda} = \frac{\text{number of uncensored observations}}{\sum_{i=1}^n Z_i}. \tag{11.140}$$

Under H_0, $\widehat{\lambda}$ consistently estimates λ and $\widehat{\sigma}_c^2$ consistently estimates the asymptotic variance of $\sqrt{n}J_c$. Under H_0, J_c^* is asymptotically $N(0, 1)$ (see Chen, Hollander, and Langberg, 1983a). To test H_0 against NBU alternatives at the approximate α level of significance,

$$\text{Reject } H_0 \text{ if } J_c^* \leq -z_\alpha;\quad \text{otherwise do not reject.} \tag{11.141}$$

To test H_0 against NWU alternatives, at the approximate α level of significance,

$$\text{Reject } H_0 \text{ if } J_c^* \geq z_\alpha; \quad \text{otherwise do not reject.} \tag{11.142}$$

To test H_0 against NBU and NWU alternatives, at the approximate α level of significance,

$$\text{Reject } H_0 \text{ if } |J_c^*| \geq z_{\alpha/2}; \quad \text{otherwise do not reject.} \tag{11.143}$$

40. *Estimation and Confidence Bands for the Quantile Function.* The quantile function is formally defined as $F^{-1}(p) = \sup\{t : F(t) \leq p\}, 0 < p < 1$. Thus, for example, $F^{-1}(.5)$ is the median. Bootstrap confidence bands for the quantile function were given by Doss and Gill (1992). Li et al. (1996) (hereafter, LHMY (1996)) used a likelihood ratio approach to obtain confidence bands for the quantile function. They define the likelihood function

$$L(F) = \prod_{i=1}^n [F(Z_i) - F(Z_i-)]^{\delta_i} [1 - F(Z_i)]^{1-\delta_i},$$

where Z_i is the observed value corresponding to patient i (Z_i may be a censored value or a known death) and δ_i is 1 if Z_i corresponds to a known death and δ_i is 0 if Z_i corresponds to a censored observation. Let Θ be the space of all distribution functions on $[0, \infty)$. Here F is viewed as a parameter taking values in Θ. For any $t \geq 0$ and $0 < p < 1$, LHMY (1996) define

$$R(p,t) = \frac{\sup\{L(F) : F(t) = p \text{ and } F \in \Theta\}}{\sup\{L(F) : F \in \Theta\}}$$

and, for $0 \leq r \leq 1$,

$$C(p,r) = \{t : R(p,t) \geq r\}.$$

A large value of $R(p,t)$ can be considered evidence in favor of the hypothesis $H^* : F(t) = p$. Therefore, $C(p,r)$ can be interpreted, for each fixed p, as the set of times t for which H^* is not rejected by a test based on $R(p,t)$. LHMY (1996) show that $C(p,r)$ is always an interval and their asymptotic confidence band is obtained by pasting together intervals of the form $C(p,r)$, with r chosen appropriately. Furthermore, they obtain an asymptotic $1 - \alpha$ confidence interval for the p-quantile of F. That interval is of the form $C(p, r_\alpha^*)$ where $r_\alpha^* = \exp\{-\chi_{1,\alpha}^2/2\}$ and $\chi_{1,\alpha}^2$ is the upper α-quantile of a chi-square distribution with 1 degree of freedom. Figure 11.8 contains a plot of the Kaplan-Meier quantile function, F_{KM}^{-1}, and an asymptotic 95% confidence band for the true unknown quantile function for the affected node data of Table 11.16.

Properties

1. *Consistency.* See Peterson (1977), Gill (1980), Shorack and Wellner (1986, Section 7.3), Wang (1987), Ying (1989), Fleming and Harrington (1991, Section 3.4), and Andersen et al. (1993, Section IV.3.2).
2. *Asymptotic Distributional Properties.* See Kaplan and Meier (1958), Efron (1967), Breslow and Crowley (1974), Meier (1975), Gill (1980, 1983), Fleming and Harrington (1991, Chapter 6) and Andersen et al. (1993, Section IV.3.2).
3. *Asymptotic Optimality.* See Wellner (1982), van der Vaart (1988, 1991) and Andersen et al. (1993, Chapter VIII).
4. *Efficiency.* See Wellner (1982), Miller (1983), Hollander, Proschan, and Sconing (1985), Gill (1989), and Andersen et al. (1993, Chapter VIII).

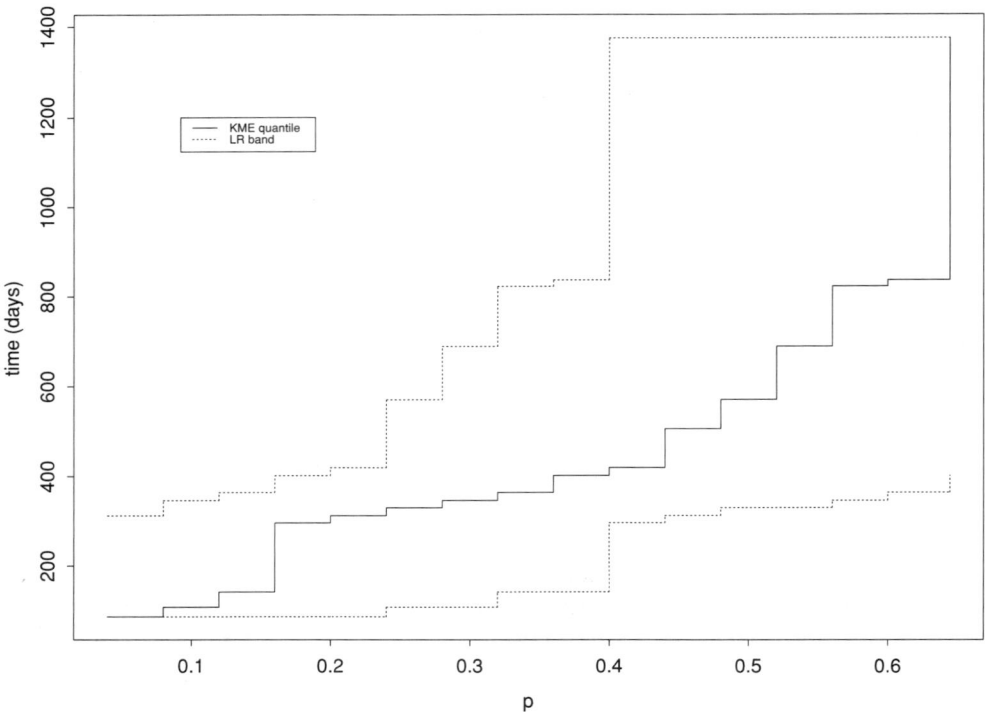

FIGURE 11.8. An approximate 95% confidence band for the affected-node quantile function.

5. *Nonparametric Maximum Likelihood Estimator.* See Johansen (1978) for a proof that the Kaplan-Meier estimator is the nonparametric maximum likelihood estimator of F in the sense of Kiefer and Wolfowitz (1956).

PROBLEMS

34. The data in Table 11.18 are from Hollander, McKeague and Yang (1997) and concern 432 manuscripts submitted for publication to the Theory and Methods Section of the *Journal of the American Statistical Association* in the period January 1, 1994–December 13, 1994. Of interest is the distribution of the time (in days) to first review. When the data were studied on December 13, 1994, 157 papers were still awaiting the first review. Thus there are 157 censored times and 275 uncensored times. In Table 11.18, the variable $X_i = \text{minimum}(T_i, C_i)$, where T_i is the time to first review and C_i is the time to censorship, and the indicator variable δ_i is 1 if the ith observation is uncensored and 0 if it is censored. Compute the Kaplan-Meier estimate of the survival function.
35. For the review times data of Table 11.18, compute asymptotic 95% confidence bands for the survival function.
36. For the review times data of Table 11.18, compute an asymptotic 95% confidence interval for the probability that the time to first review will exceed 150 days.
37. The data in Table 11.19 were provided by Drs. J. A. Koziol and S. B. Green (1978) and can be found in Hollander and Proschan (1979). The data correspond to 211 patients with stage IV prostate cancer who were treated with estrogen in a Veterans Administration Cooperative Urological Research Group (1967) study. The observations over the years 1967 through March 1977. At the March 1977 closing date there were 90 patients who had died of prostate cancer, 105 who had died of other

TABLE 11.18. The Times to First Review of 1994 JASA Theory and Methods Papers

X_i	δ_i	X_i	δ_i	X_i	δ_i	X_i	δ_i	X_i	δ_i	X_i	δ_i	X_i	δ_i	X_i	δ_i	X_i	δ_i	X_i	δ_i	X_i	δ_i
214	1	201	1	28	1	252	0	118	1	187	0	28	1	28	1	76	1	56	0	28	0
184	1	274	1	287	0	96	1	33	1	152	1	21	1	118	0	18	1	21	1	27	0
150	1	265	1	195	1	175	1	69	1	46	1	1	1	40	1	88	0	55	0	27	0
70	1	120	1	86	1	54	1	133	1	103	1	0	1	6	1	85	0	55	0	25	0
16	1	141	1	137	1	167	1	126	1	37	1	144	0	91	1	85	0	54	0	25	0
141	1	48	1	74	1	150	1	84	1	170	1	144	0	34	1	85	0	18	1	22	0
210	1	204	1	71	1	219	1	197	1	64	1	140	0	21	1	20	1	54	0	22	0
132	1	312	0	140	1	86	1	85	1	182	0	14	1	1	1	83	0	53	0	21	0
30	1	220	1	22	1	1	1	15	1	180	0	0	1	111	0	82	0	50	0	21	0
204	1	188	1	120	1	111	1	206	1	176	0	27	1	111	0	81	0	50	0	15	0
84	1	84	1	176	1	128	1	125	1	175	0	23	1	1	1	81	0	1	1	15	0
36	1	84	1	181	1	178	1	57	1	64	1	126	1	48	1	11	1	15	1	15	0
38	1	215	1	155	1	40	1	181	1	42	1	139	0	110	0	77	0	50	0	1	1
69	1	33	1	74	1	131	1	215	0	175	0	55	1	47	1	77	0	47	0	1	1
33	1	55	1	29	1	20	1	3	1	149	1	137	0	68	1	70	1	47	0	14	0
49	1	140	1	100	1	220	1	13	1	158	1	114	1	74	1	74	0	46	0	12	0
203	1	147	1	195	1	84	1	175	1	169	0	56	1	98	1	71	0	16	1	12	0
203	1	41	1	127	1	32	1	37	1	169	0	124	1	105	0	23	1	43	0	12	0
218	1	94	1	34	1	95	1	182	1	22	0	121	1	104	0	28	1	43	0	8	0
267	1	292	1	177	1	188	1	210	0	168	0	1	1	104	0	70	0	18	1	8	0
99	1	131	1	150	1	115	1	92	1	157	1	27	1	103	0	44	1	43	0	8	0
21	1	221	1	265	0	238	0	208	0	89	1	130	0	90	1	69	0	42	0	8	0
78	1	39	1	174	1	1	1	30	1	165	1	130	0	98	0	68	0	42	0	7	0
150	1	3	1	104	1	187	1	28	1	14	1	130	0	98	0	67	0	40	0	7	0
237	1	16	1	203	1	125	1	168	1	161	0	127	0	98	0	64	0	0	1	7	0
91	1	129	1	109	1	110	1	202	0	161	0	100	1	97	0	30	1	12	1	7	0
21	1	210	1	217	1	32	1	114	1	159	0	126	0	96	1	41	1	39	0	6	0
224	1	240	1	238	1	32	1	105	1	91	1	126	0	97	0	62	0	35	0	5	0
126	1	141	1	210	1	228	1	196	0	146	1	28	1	18	1	61	0	35	0	5	0
167	1	231	1	22	1	80	1	195	0	159	0	125	0	96	0	20	1	35	0	4	0
105	1	119	1	148	1	64	1	114	1	134	1	125	0	92	0	57	0	30	1	1	0
146	1	291	0	142	1	231	0	75	1	13	1	125	0	91	0	57	0	35	0	1	0
50	1	199	1	126	1	64	1	194	0	159	0	95	1	91	0	57	0	34	0		
28	1	67	1	220	1	228	0	143	1	18	1	95	1	91	0	57	0	34	0		
288	1	263	1	145	1	18	1	106	1	155	0	123	0	31	1	57	0	34	0		
37	1	155	1	21	1	55	1	128	1	154	0	123	0	27	1	57	0	33	0		
18	1	189	1	256	0	154	1	200	0	124	1	123	0	83	1	42	1	33	0		
113	1	0	1	253	0	139	1	129	1	73	1	109	1	33	1	57	0	29	1		
22	1	209	1	22	1	91	1	138	1	51	1	119	0	11	1	57	0	28	0		
234	1	223	1	80	1	196	1	152	1	21	1	6	1	88	0	57	1	27	0		

Source: M. Hollander, I. W. McKeague, and J. Yang (1997).

diseases, and 16 who were still alive. Those observations corresponding to deaths due to other causes and those corresponding to the 16 survivors are considered censored observations (withdrawals). Compute the Kaplan-Meier estimator of the survival distribution for deaths from cancer of the prostate.

38. For the prostate cancer data of Table 11.19, compute asymptotic 90% confidence bands for the survival function.

39. For the prostate cancer data of Table 11.19, compute an asymptotic 90% confidence interval for the probability of surviving more than 100 months.

40. Verify that, in the case of no censored observations, the confidence interval given by (11.124) reduces to that given by (11.126).

TABLE 11.19. Survival Times and Withdrawal Times in Months for 211 State IV Patients (with Number of Ties Given in Parentheses)

Survival times: 0(3), 2, 3, 4, 6, 7(2), 8, 9(2), 11(3), 12(3), 15(2), 16(3), 17(2), 18, 19(2), 20, 21, 22(2), 23, 24, 25(2), 26(3), 27(2), 28(2), 29(2), 30, 31, 32(3), 33(2), 34, 35, 36, 37(2), 38, 40, 41(2), 42(2), 43, 45(3), 46, 47(2), 48(2), 51, 53(2), 54(2), 57, 60, 61, 62(2), 67, 69, 87, 97(2), 100, 145, 158.

Withdrawal times: 0(6), 1(5), 2(4), 3(3), 4, 6(5), 7(5), 8, 9(2), 10, 11, 12(3), 13(3), 14(2), 15(2), 16, 17(2), 18(2), 19(3), 21, 23, 25, 27, 28, 31, 32, 34, 35, 37, 38(4), 39(2), 44(3), 46, 47, 48, 49, 50, 53(2), 55, 56, 59, 61, 62, 65, 66(2), 72(2), 74, 78, 79, 81, 89, 93, 99, 102, 104(2), 106, 109, 119(2), 125, 127, 129, 131, 133(2), 135, 136(2), 138, 141, 142, 143, 144, 148, 160, 164(3).

Source: J. A. Koziol and S. B. Green (1978).

41. Verify that, due to monotonicity of $\overline{F}$ (i.e., $\overline{F}(x) \geq \overline{F}(y)$ whenever $x < y$), the events given by (11.134) and (11.135) are identical.
42. Apply the NBU test for censored data to the prostate cancer of Table 11.19. What is your conclusion?
43. Apply the NBU test for censored data to the affected node data of Table 11.16. What is your conclusion?

11.7. A TWO-SAMPLE TEST FOR CENSORED DATA (MANTEL)

For the Hodgkin's survival time data of Table 11.15, the two samples, corresponding to the affected node and total nodal radiation treatments, contain censored observations. Figure 11.4 of Section 11.6 indicates that the underlying survival distributions corresponding to the two treatments are not equal and the survival times in the samples are better under total nodal radiation than under radiation of the affected node. That, however, is only a visual assessment, and censoring complicates the picture. A two-sample test for censored data is needed to make an objective assessment. One such test, the most commonly used at the time of this writing, is Mantel's (1966) test, also known as the logrank test.

Assumptions

C1. For sample 1, let $T_1, \ldots, T_m$ be independent each with continuous life distribution function F_1. Let $C_1, \ldots, C_m$ be independent, each with continuous censoring distribution function G_1. C_i is the censoring time corresponding to T_i. For sample 2, let $U_1, \ldots, U_n$ be independent, each with continuous life distribution function F_2. Let $D_1, \ldots, D_n$ be independent, each with continuous censoring distribution function G_2. D_j is the censoring time associated with U_j.

C2. For sample 1, we observe, for $i = 1, \ldots, m$,

$$X_i = \text{minimum }\{T_i, C_i\}$$

and δ_i, where

$$\delta_i = \begin{cases} 1, & \text{if } T_i \leq C_i, \\ 0, & \text{if } T_i > C_i. \end{cases}$$

For sample 2, we observe, for $j = 1, \ldots, n$,

$$Y_j = \text{minimum }\{U_j, D_j\}$$

and

$$\varepsilon_j = \begin{cases} 1, & \text{if } U_j \leq D_j, \\ 0, & \text{if } U_j > D_j. \end{cases}$$

Thus δ_i is 1 if T_i is uncensored and we observe the true survival time T_i rather than the time to censorship C_i. However, δ_i is 0 if T_i is censored and we observe C_i. In this case we only know that the true survival time T_i is greater than C_i. Similarly, ε_j is 1 if U_j is uncensored and we observe the true survival time U_j rather than the time to censorship D_j. However, ε_j is 0 if U_j is censored and we observe D_j. In this case we only know that the true survival time U_j is greater than D_j.

C3. The T's, C's, U's and D's are mutually independent.

To illustrate the notation of this model, which is known as the randomly right-censored model, consider the data of Table 11.16. For those data, where $m = 25$, $n = 24$, we have $X_1 = T_1 = 346$ and $\delta_1 = 1$ because the 346 value is not censored as the true time to relapse is observed. However, for X_4 we have $X_4 = C_4 = 1953$ and $\delta_4 = 0$ because the true number of days to relapse is not observed. The observation is censored and we know only that the true time to relapse is greater than 1953. Similarly, $Y_1 = D_1 = 1699$, $\varepsilon_1 = 0$, and $Y_5 = U_5 = 173$, $\varepsilon_5 = 1$.

Procedure

Combine the two samples and let k denote the number of distinct failure times. Denote these distinct failure times by $w_{(1)} < w_{(2)} < \cdots < w_{(k)}$. Let n_{ij}, $j = 1, 2$, denote the number of patients from sample j at risk at $w_{(i)}$. That is, n_{ij} is the number of patients who have not experienced the endpoint event (death, relapse, etc.) or been censored before time $w_{(i)}$. Let d_{ij}, $j = 1, 2$, be the number of failures from sample j at time $w_{(i)}$. Correspondingly, let n_i be the combined number of patients from both samples who are at risk at $w_{(i)}$ and let d_i be the combined (from both samples) number of failures at $w_{(i)}$.

Mantel's (1966) test statistic M_c for two-sample censored data is

$$M_c = \frac{\sum_{i=1}^{k}(d_{i1} - E_{i1})}{\sqrt{\sum_{i=1}^{k} V_{i1}}}, \tag{11.144}$$

where

$$E_{i1} = \frac{d_i n_{i1}}{n_i} \tag{11.145}$$

and

$$V_{i1} = \frac{d_i(n_i - d_i)n_{i1}n_{i2}}{n_i^2(n_i - 1)} \tag{11.146}$$

are, respectively, the conditional mean and variance of d_{i1} (see Comment 41).

Let F_1 denote the unknown life distribution of group 1 and F_2 the unknown life distribution of group 2. The null hypothesis is

$$H_0 : F_1 = F_2.$$

The test is performed by treating M_c as having an approximate $N(0, 1)$ distribution under H_0.

a. One-Sided Test of H_0 against Alternatives for Which Treatment 2 Is Better. To test

$$H_0 : F_1 = F_2$$

versus

H_1 : Survival times for treatment 2 tend to be longer than those for treatment 1,

at the approximate α-level of significance,

$$\text{Reject } H_0 \text{ if } M_c \geq z_\alpha; \quad \text{otherwise do not reject.} \tag{11.147}$$

b. One-Sided Test of H_0 against Alternatives for Which Treatment 1 is Better. To test

$$H_0 : F_1 = F_2$$

versus

H_2 : Survival times for treatment 1 tend to be longer than those for treatment 2,

at the approximate α-level of significance,

$$\text{Reject } H_0 \text{ if } M_c \leq -z_\alpha; \quad \text{otherwise do not reject.} \tag{11.148}$$

c. Two-Sided Test against Alternatives for Which the Two Treatments Differ. To test

$$H_0 : F_1 = F_2$$

versus

H_3 : Survival times for treatment 2 have a different distribution than that for treatment 1,

at the approximate α-level of significance,

$$\text{Reject } H_0 \text{ if } |M_c| \geq z_{\frac{\alpha}{2}}; \quad \text{otherwise do not reject.} \tag{11.149}$$

Example 11.8: *Example 11.7 Continued.* We return to the Hodgkin's disease data of Table 11.16. Drs. Kaplan and Rosenberg believed the total nodal radiation treatment to be superior to the radiation of affected node treatment. A hypothesis test provides an assessment in terms of P-values. We will apply Mantel's test to the data. In the combined sample there are $k = 21$ distinct failure times $w_{(1)} < \cdots < w_{(21)}$. They are

(1) (1) (1) (2) (1) (1) (1) (1) (1) (1) (1)
86, 107, 111, 173, 296, 312, 330, 346, 364, 401, 419,
(2) (1) (1) (2) (1) (1) (1) (2) (1) (1)
498, 505, 570, 615, 688, 822, 836, 950, 1375, 1408.

In the preceding display, above each w in parentheses we indicate if the failure is from treatment 1 (radiation of affected node) or treatment 2 (total nodal radiation). Table 11.20 illustrates the computation of M_c. Summing columns 2, 7, and 8, respectively, of Table 11.20

11.7. A TWO-SAMPLE TEST FOR CENSORED DATA (MANTEL)

TABLE 11.20. Computation of M_c for the Hodgkin's Disease Data

w_i	d_{i1}	d_i	n_{i1}	n_{i2}	n_i	E_{i1}	V_{i1}
86	1	1	25	24	49	.5102	.2499
107	1	1	24	24	48	.5000	.2500
141	1	1	23	24	47	.4894	.2499
173	0	1	22	24	46	.4783	.2495
296	1	1	22	23	45	.4889	.2499
312	1	1	21	23	44	.4773	.2495
330	1	1	20	23	43	.4651	.2488
346	1	1	19	23	42	.4524	.2477
364	1	1	18	23	41	.4390	.2463
401	1	1	17	23	40	.4250	.2444
419	1	1	16	23	39	.4103	.2420
498	0	1	15	23	38	.3947	.2389
505	1	1	15	22	37	.4054	.2411
570	1	1	14	22	36	.3889	.2377
615	0	1	13	22	35	.3714	.2335
688	1	1	13	21	34	.3824	.2362
822	1	1	12	21	33	.3636	.2314
836	1	1	11	21	32	.3438	.2256
950	0	1	10	21	31	.3226	.2185
1,375	1	1	9	18	27	.3333	.2222
1,408	0	1	8	18	25	.2800	.2016

yields

$$\sum_{i=1}^{21} d_{i1} = 16, \quad \sum_{i=1}^{21} E_{i1} = 8.7220, \quad \sum_{i=1}^{21} V_{i1} = 5.0146.$$

Then, from equation (11.144), we obtain

$$M_c = \frac{16 - 8.7220}{\sqrt{5.0146}} = 3.25.$$

Referring $M_c = 3.25$ to Table A.1 gives an approximate one-sided P-value of .0006. Thus there is strong evidence that total nodal radiation is more effective than radiation of affected nodes in preventing or delaying the recurrence of early stage Hodgkin's disease.

Comments

41. *Motivation for Mantel's Test.* The development is similar to that used in Section 10.4, where success probabilities are compared in k 2×2 tables. Here, k 2×2 tables are formed, one at each known failure time $w_{(i)}$, as in Table 11.21. (Note, however, that whereas k was fixed in Section 10.4, here k is the random number of observed failures in the combined sample.)

Conditioning on the marginal totals in each of the k 2×2 tables, the mean and variance of d_{ij}, are, respectively

$$E_{ij} = \frac{d_i n_{ij}}{n_i} \qquad (11.150)$$

TABLE 11.21. 2 × 2 Table of Failure and Numbers at Risk at Failure Time $w_{(i)}$

	Failures	Not Failures	Totals
Sample 1	d_{i1}	$n_{i1}-d_{i1}$	n_{i1}
Sample 2	d_{i2}	$n_{i2}-d_{i2}$	n_{i2}
Totals	d_i	$n_i - d_i$	n_i

and

$$V_{ij} = \frac{d_i(n_i - d_i)n_{i1}n_{i2}}{n_i^2(n_i - 1)}. \qquad (11.151)$$

The k "observed minus expected" differences $d_{11} - E_{11}, \ldots, d_{1k} - E_{1k}$ are not independent (in contrast to the k "observed minus expected" differences $\mathcal{O}_{111} - E_{111}, \ldots, \mathcal{O}_{11k} - E_{11k}$ of Section 10.4, which are independent because the k 2 × 2 tables of that section are assumed to be independent). Due to the lack of independence, the central limit theorem, which is used to establish asymptotic normality of the MH statistic of Section 10.4, cannot be applied in this case, and other approaches are necessary to establish asymptotic normality of M_c. Proofs of asymptotic normality of M_c, as a consequence of more general results that establish asymptotic normality of classes of two-sample statistics that include M_c, can be found in Gill (1980), Fleming and Harrington (1991), and Andersen et al. (1993).

Mantel's (1966) nonrigorous development of his test for the censored case is via analogy to the Mantel-Haenszel (1959) test for k 2 × 2 tables presented in Section 10.4. Also see Mantel (1963). For further discussion see Miller (1981b), Kalbfleisch and Prentice (1980), Fleming and Harrington (1991), Andersen et al. (1993), and Klein and Moeschberger (1997). Mantel's test is closely related to tests proposed by Peto and Peto (1972) and Cox (1972). (Cox (1972) is a seminal paper on nonparametric regression methods for censored data. For nonparametric regression methods for censored data, also see Kalbfleisch and Prentice (1980), Miller (1981b), Cox and Oakes (1984), Fleming and Harrington (1991), Crowder et al. (1991), Andersen et al. (1993), and Klein and Moeschberger (1997).) Mantel's test, also known as the logrank test (a name first used by Peto and Peto 1972), can be viewed as a generalization of a test due to Savage (1956) for uncensored data (see Kalbfleisch and Prentice 1980). Fleming et al. (1980) provide a two-sample test for censored data that generalizes the Kolmogorov-Smirnov test of Section 5.4. They found that their test tends to do better than Mantel's test and Gehan's test when the failure rates (recall equation (11.1)) of the two underlying distributions cross.

42. *Choice of the Variance for the Logrank Test.* When there are no ties among the k uncensored observations in the combined sample, $d_i = 1, i = 1, \ldots, k$, and M_c can be written as

$$M_c = \frac{M}{\sigma_M},$$

where

$$M = \sum_{i=1}^{k}\left(d_{i1} - \frac{n_{i1}}{n_i}\right)$$

and

$$\sigma_M^2 = \sum_{i=1}^{k} \frac{n_{i1} n_{i2}}{n_i^2}.$$

σ_M^2 is the Mantel variance (also known as the Mantel-Haenszel variance). Under H_0, σ_M^2 is an unbiased estimator of the variance of M, independently of differences between the censoring distributions of the two groups. Brown (1984) shows, however, that with equal sample sizes, when M is large in absolute value, σ_M^2 tends to underestimate the true variance. Large values of $|M|$ tend to be accompanied by small values of σ_M^2. This results in exaggeratedly large values of $|M|/\sigma_M$ and P-values that are too small. This is also discussed by Morton (1978). Brown suggests that when the sample sizes are approximately equal, one should instead base tests on M/σ_P, where σ_P^2 is the permutation variance (see Peto and Peto (1972) and Brown (1984)), which assumes equal censoring distributions. The permutation variance is

$$\sigma_P^2 = \frac{mn}{(m+n)(m+n-1)} \left(k - \sum_{i=1}^{k} n_i^{-1} \right).$$

Brown shows that under equal sample sizes, the permutation variance tends to overstate the true variance when censoring is unequal.

Although Brown's results highlight a weakness of Mantel's procedure in certain situations, the assumption of equal censoring distributions required by σ_P^2 is too restrictive and thus we have used the Mantel variance in our presentation of Mantel's test. See Mantel (1985) for a brief rebuttal to Brown (1984).

43. *Tarone-Ware Tests.* The Mantel test described in this section assigns equal weights to the 2×2 tables formed at each known failure time $w_{(i)}$. Tarone and Ware (1977) define a class of two-sample tests for censored data by allowing different weights for the tables. Their general test statistic, for weights $b_1, \ldots, b_k$, is

$$TW = \frac{\sum_{i=1}^{k} b_i (d_{i1} - E_{i1})}{\sqrt{\sum_{i=1}^{k} b_i^2 V_{i1}}} \tag{11.152}$$

where E_{i1}, V_{i1} are given by (11.145) and (11.146), respectively. Under H_0, TW is asymptotically $N(0, 1)$. Significantly large values of TW indicate survival times for treatment 2 tend to be larger than those for treatment 1. Significantly small values of TW indicate survival times for treatment 1 tend to be larger than those for treatment 2. The choice $b_i = 1$ yields Mantel's statistic. The choice $b_i = n_i$ yields a test due to Gehan (1965) (although Gehan derived his test from a different approach). Tarone and Ware advocate the choice $b_i = \sqrt{n_i}$ based on efficiency considerations.

44. *Other Two-Sample Tests for Censored Data.* There are many two-sample tests for censored data. Gehan (1965) proposed a generalization of Wilcoxon's two-sample test for uncensored data. Efron (1967) proposed a different generalization of Wilcoxon's test. Tarone and Ware (1977) defined a class of tests that include Mantel's test and Gehan's test. Prentice (1978) proposed a family of linear rank tests. Other very general families have been proposed and studied by Gill (1980) and Harrington and Fleming (1982). Leurgans (1983, 1984) gives efficiency and small-sample Monte Carlo power results for many of these tests. Fleming and Harrington (1991, Chapter 7) provide a comprehensive treatment of Gill's 1980 $\mathcal{K}$ class.

45. S-plus *Program for Two-Sample Censored Data Tests.* S-plus provides a function surv.diff that performs two-sample censored data tests and obtains corresponding approximate P-values for statistics in the Harrington and Fleming (1982) $\mathcal{G}^\rho$ class of tests. By setting $\rho = 0$, one gets Mantel's test. To illustrate the use of surv.diff we consider the relapse-free times of Table 11.16. We create three S objects corresponding to (i) the X's of Table 11.16, (ii) the indicators (1 for observed relapse, 0 for censored) of whether the X is a true relapse time or a censored time, and (iii) indicators (1 for sample 1, 2 for sample 2) that designate if X is from sample 1 or sample 2. Letting

rad< −c(346, 141, 296, 1953, 1375, 822, 2052, 836, 1910, 419, 107, 570, 312, 1818, 364, 401, 1645, 330, 1540, 688, 1309, 505, 1378, 1446, 86, 1699, 2177, 1968, 1889, 173, 2070, 1972, 1897, 2022, 1879, 1726, 1807, 615, 1408, 1763, 1684, 1576, 1572, 498, 1585, 1493, 950, 1242, 1190),

cen< −c(1, 1, 1, 0, 1, 1, 0, 1, 0, 1, 1, 1, 1, 0, 1, 1, 0, 1, 0, 1, 0, 1, 0, 0, 1, 0, 0, 0, 0, 1, 0, 0, 0, 0, 0, 0, 1, 1, 0, 0, 0, 0, 1, 0, 0, 1, 0, 0),

smp< −c(1, 1, 1, 1, 1, 1, 1, 1, 1, 1, 1, 1, 1, 1, 1, 1, 1, 1, 1, 1, 1, 1, 1, 1, 1, 0, 0, 0, 0, 0, 0, 0, 0, 0, 0, 0, 0, 0, 0, 0, 0, 0, 0, 0, 0, 0, 0, 0),

the command surv.diff(rad,cen,smp,rho=0) yields the output chisquare=10.6 on 1 degree of freedom, p=0.001153. This agrees (allowing for rounding) with what we obtained in Example 11.7. Recall that in Example 11.7 we found $M_c = 3.25$ with an approximate one-sided P-value of .0006 based on the normal approximation. Note $(3.25)^2 = 10.56$ and $2(.0006) = .0012$.

Properties

1. *Consistency.* See Gill (1980, Section 4.1) and Fleming and Harrington (1991, Section 7.3).
2. *Asymptotic Normality.* See Prentice (1978), Gill (1980), and Fleming and Harrington (1991, Section 7.2).
3. *Efficiency.* See Peto and Peto (1972), Prentice (1978), Gill (1980, Chapter 5), Harrington and Fleming (1982), Leurgans (1983, 1984), and Fleming and Harrington (1991, Section 7.4).
4. *Power and Sample-Size Calculations.* See Latta (1981), Schoenfeld (1981), Leurgans (1983, 1984), Hsieh (1987, 1992), Sposto and Krailo (1987), Lakatos and Lan (1992), and Strawderman (1997).

PROBLEMS

44. The data in Table 11.22, obtained by Dr. P. B. Gregory (1974) of Stanford University, originally appeared in Brown and Hollander (1977). The data are from a clinical trial conducted to study the efficiency of a new drug thought to be helpful for treating patients with a particular type of serious liver disease. Is there evidence that the new drug does significantly better (or significantly worse) than the placebo in terms of survival times?
45. Give an intuitive explanation why the 2 × 2 tables formed at the uncensored times $w_{(i)}, \ldots, w_{(k)}$ are not independent.
46. Apply Gehan's test (see Comment 43) to the hepatitis data of Table 11.22. Compare your results with those of Problem 44.
47. Apply the Tarone-Ware test with weights $b_i = \sqrt{n_i}$ (see Comment 43) to the hepatitis data of Table 11.22. Compare your results with those of Problems 44 and 46.
48. The data in Table 11.23 are from Hollander (1996) and concern 444 manuscripts submitted for publication to the Theory and Methods Section of the *Journal of the American Statistical Association*

11.8. EFFICIENCIES

TABLE 11.22. Severe Viral Hepatitis Study

Patient	Treatment (D=Drug, P=Placebo)	Length of observation (in weeks)	Status (A=Alive, D=Dead)
1	D	6	D
2	P	16	A
3	P	2	D
4	D	1	D
5	D	1	D
6	P	4	A
7	D	16	A
8	D	3	A
9	P	16	A
10	D	8	D
11	P	16	A
12	P	2	D
13	D	4	D
14	P	16	A
15	P	5	A
16	D	1	D
17	D	10	A
18	P	2	A
19	P	2	A
20	D	10	D
21	P	16	A
22	D	1	A
23	P	16	A
24	D	15	A

Source: P. B. Gregory (1974).

in the period January 1, 1995–December 15, 1995. Of interest is the distribution of the time (in days) to first review. When the data were studied on December 15, 1995, 173 papers were still awaiting the first review. Thus there are 173 censored times and 271 uncensored times. In Table 11.23, the variable $X_i = $ minimum (T_i, C_i) where T_i is the time to first review and C_i is the time to censorship, and the indicator variable δ_i is 1 if the ith observation is uncensored and 0 if it is censored. Use the data in Tables 11.18 and 11.23 to test if there is a significant difference between the 1994 times to first review and the 1995 times to first review.

11.8. EFFICIENCIES

The entries in Table 11.24 are given by Klefsjö (1983) and are based on efficiency calculations reported in Bickel and Doksum (1969), Hollander and Proschan (1975) and Klefsjö (1983). The statistics considered in Table 11.24 are the total-time-on-test statistic $\mathcal{E}$ of Section 11.1 given by (11.9) (also see Comment 14 of Section 11.2), the IFR statistic A (see Comment 5), the IFRA statistic B (Comment 5), and the DMRL statistic V' (11.58) of Section 11.3. Table 11.24 gives, for the distributions F_1 (linear failure rate), F_2 (Makeham), F_3 (Pareto), F_4 (Weibull), and F_5 (gamma), the asymptotic relative efficiencies of A, B, V', $\mathcal{E}$ relative to the statistic (among A, B, V', $\mathcal{E}$) having the largest efficacy for that particular F. The c^2_{MAX} column of Table 11.24 gives, for a given F, the largest squared efficacy for the four included statistics.

For the NBU statistic T given by (11.34), Hollander and Proschan (1972) found the asymptotic relative efficiency of T with respect to $\mathcal{E}$ for Weibull and linear failure rate distributions. The values are $e_{F_4}(T, \mathcal{E}) = .937$ and $e_{F_1}(T, \mathcal{E}) = .45$. Other efficiency values for T are given

TABLE 11.23. The Times to First Review of 1995 JASA Theory and Methods Papers

X_i	δ_i	X_i	δ_i	X_i	δ_i	X_i	δ_i	X_i	δ_i	X_i	δ_i	X_i	δ_i	X_i	δ_i	X_i	δ_i	X_i	δ_i
141	1	37	1	16	1	93	1	77	1	134	1	64	1	29	1	88	1	58	0	22	1
140	1	42	1	245	1	240	0	101	1	171	1	56	1	116	0	87	0	56	0	30	0
161	1	126	1	162	1	95	1	175	1	119	1	106	1	115	0	85	0	56	0	30	0
23	1	88	1	42	1	221	1	162	1	155	1	87	1	115	0	15	1	50	0	19	1
14	1	27	1	31	1	130	1	206	0	28	1	103	1	0	1	81	0	50	0	29	0
134	1	64	1	98	1	21	1	132	1	112	1	137	0	114	0	81	0	50	0	29	0
83	1	260	1	127	1	66	1	205	0	169	0	137	0	113	0	81	0	39	1	28	0
145	1	114	1	37	1	90	1	90	1	169	0	38	1	113	0	1	1	33	1	25	0
97	1	114	1	171	1	159	1	19	1	125	1	17	1	113	0	80	1	50	0	25	1
119	1	252	1	118	1	232	0	32	1	162	0	95	1	14	1	79	0	49	0	25	0
255	1	49	1	0	1	70	1	150	1	162	0	133	0	109	0	78	0	22	1	25	0
181	1	154	1	188	1	74	1	190	1	118	1	120	1	109	0	78	0	49	0	25	0
140	1	54	1	257	1	100	1	56	1	96	1	130	0	89	1	71	1	23	1	25	0
209	1	159	1	197	1	188	1	196	0	47	1	53	1	2	1	78	0	45	0	25	0
194	1	38	1	21	1	211	1	82	1	73	1	19	1	45	1	77	0	45	0	25	0
297	1	246	1	263	0	177	1	196	0	150	1	59	1	23	1	77	0	38	1	23	0
46	1	91	1	143	1	221	0	58	1	40	1	93	1	107	0	2	1	44	0	23	0
47	1	197	1	262	0	204	1	191	1	161	0	129	0	107	0	73	0	44	0	16	1
44	1	32	1	197	1	122	1	42	1	14	1	1	1	106	0	73	0	44	0	18	1
85	1	32	1	145	1	14	1	14	1	161	0	119	1	93	1	73	0	43	0	18	0
155	1	104	1	35	1	99	1	21	1	157	0	128	0	105	0	70	0	43	0	18	0
153	1	32	1	45	1	28	1	54	1	157	1	90	1	85	1	70	0	43	1	18	0
48	1	101	1	201	1	218	0	104	1	148	1	127	0	101	0	70	0	31	1	17	0
48	1	291	0	204	1	217	0	151	1	153	1	123	0	101	0	70	0	42	0	17	0
187	1	102	1	126	1	3	1	189	0	157	0	123	1	101	0	70	0	39	0	17	0
99	1	288	0	255	0	28	1	39	1	157	0	123	0	100	0	67	0	39	0	16	0
21	1	225	1	255	0	129	1	78	1	157	0	46	1	100	0	67	0	39	0	16	0
190	1	147	1	99	1	170	1	109	1	41	1	66	1	100	0	64	0	39	0	15	0
319	0	160	1	233	1	93	1	186	0	151	0	43	1	100	0	64	0	14	1	15	0
108	1	284	0	197	1	135	1	158	1	46	1	123	0	0	1	64	0	27	1	11	0
44	1	253	1	42	1	169	1	44	1	151	0	121	0	20	1	47	1	38	0	11	0
127	1	174	1	170	1	28	1	179	1	71	1	34	1	47	1	63	0	38	0	11	0
155	1	180	1	71	1	79	1	64	1	151	0	121	0	99	0	60	0	37	0	9	0
186	1	22	1	76	1	81	1	28	1	150	0	110	1	98	0	59	1	37	0	9	0
215	1	237	1	26	1	211	1	17	1	150	0	120	0	21	1	59	0	33	1	9	0
39	1	218	1	12	1	33	1	177	0	148	0	80	1	95	0	34	1	27	1	8	0
118	1	127	1	187	1	117	1	29	1	12	1	119	0	94	0	59	0	37	0	4	0
22	1	148	1	130	1	41	1	47	1	18	1	31	1	95	0	59	0	18	1	3	0
32	1	100	1	28	1	24	1	23	1	70	1	73	1	93	0	59	0	32	0	3	0
74	1	63	1	1	1	0	1	167	1	47	1	43	1	43	1	59	0	32	0	2	0
														28	1	58	0	31	0	2	0

Source: M. Hollander (1996).

TABLE 11.24. Asymptotic Relative Efficiencies of $A, B, V', \mathcal{E}$

	A	B	V'	$\mathcal{E}$	c^2_{MAX}
F_1 (linear failure rate)	.44	.31	1.00	.91	.820
F_2 (Makeham)	.70	.70	.70	1.00	.083
F_3 (Pareto)	.44	.31	1.00	.91	.820
F_4 (Weibull)	.51	.87	.49	1.00	1.441
F_5 (gamma)	.39	1.00	.28	.90	.498

11.8. EFFICIENCIES

in Koul (1978b) and Deshpande (1983). Other efficiency values for $\mathcal{E}$ are given by Bickel and Doksum (1969) and Borges, Proschan, and Rodrigues (1984).

To our knowledge, asymptotic relative efficiency results have not been obtained for the IDMRL and DIMRL tests of Section 11.4. Hawkins, Kochar, and Loader (1992), however, have performed a limited Monte Carlo power comparison of their IDMRL tests $T^{(1)}$ and $T^{(2)}$ (see Comment 25) for the case where the turning point τ is unknown versus the Guess-Hollander-Proschan IDMRL test for the case where τ is known (given by (11.83) and denoted by HKL as GHP_1) and the Guess-Hollander-Proschan IDMRL test when the proportion $\rho = F(\tau)$ is known (see Comment 24). The latter test is denoted by HKL as GHP_2. The distribution used by HKL in their Monte Carlo study is $F_{\alpha,\beta,\gamma}(x) = 1 - \overline{F}_{\alpha,\beta,\gamma}(x)$, $\alpha > 0$, $\beta > 0$, $\gamma > 0$, where $\overline{F}_{\alpha,\beta,\gamma}(x)$ is given in Section 3 of HKL (1992). The distribution has mean residual life function.

$$m_{\alpha,\beta,\gamma}(x) = \beta + \gamma e^{-\alpha x}(1 - e^{-\alpha x}), \quad x \geq 0. \qquad (11.153)$$

As γ tends to 0, $m_{\alpha,\beta,\gamma}(x)$ tends to the constant mean residual life function of the exponential distribution (11.3) with $\lambda = 1/\beta$. Each member of the $F_{\alpha,\beta,\gamma}$ family is an IDMRL distribution with turning point $\tau = \alpha^{-1}\log(2)$. HKL found their $T^{(1)}$ test and GHP_1 to be slightly conservative. They found GHP_1 generally dominates their $T^{(1)}$ and $T^{(2)}$ tests except when $\rho \leq .5$, where $T^{(2)}$ seems to dominate. When $\rho \leq .75$, $T^{(2)}$ dominates $T^{(1)}$ and compares well with GHP_1 and GHP_2. For ρ in the neighborhood of .90, HKL found $T^{(1)}$ dominates $T^{(2)}$, but both $T^{(1)}$ and $T^{(2)}$ are considerably dominated by GHP_2. As HKL point out, it is not surprising that the GHP tests outperform the HKL in their study, because in their comparisons the GHP tests are allowed to use information about τ, and such information is not required by the HKL tests.

Nair (1984) defined a family of confidence bands for the survival function. Nair's family is indexed by parameters a, b with $0 < a < b < 1$. Each band in his family is an equal-precision band in the sense that its width is proportional to its standard deviation. Nair compared his equal-precision (EP) bands to the Hall-Wellner (HW) band defined in Comment 34, Section 11.6. His efficiency criterion is the ratio of the limiting squared widths of the bands. In the absence of censoring, the HW band reduces to the Kolmogorov band of Section 11.5. Thus Nair's comparisons are relevant to Section 11.5 as well as Section 11.6. Nair concluded that the HW and EP bands are competitive, with the HW band being narrower in the middle and the EP bands narrower in the tails. He also found the relative performance of the EP bands to the HW band gets better as censoring increases. Hollander and Peña (HP) (1989) used the same efficiency criterion to compute the efficiency $e(x; c)$ of the HP band (see Comment 37) for a general c with respect to the HW band (which is the HP band corresponding to $c = 1$). They showed, under certain conditions, $e(x; c)$ tends to c as x tends to 0 and $e(x; c)$ tends to c^{-1} as x tends to infinity.

The Kaplan-Meier estimator (KME) of Section 11.6 is the best estimator of the survival function $\overline{F}$ in the fully nonparametric model when the underlying life distribution F and the censoring distribution G (that governs the censoring patterns) are unspecified. See Andersen et al. (1993, Chapter VIII). Efficiency robustness properties of the KME can be studied by seeing how well the KME does in models for which it is not optimal. Miller (1983) studied the KME's efficiency loss when compared to the maximum likelihood estimator (MLE) in parametric models. Not surprisingly the KME performs poorly relative to the MLE's in a fully parametric setting. For example, when F and G are exponential, Miller (1983) showed that the asymptotic efficiency of the KME $\overline{F}_{KM}(x)$ (see (11.121)) with respect to the MLE tends to zero as x tends to zero and also tends to zero as x tends to infinity.

Hollander, Proschan, and Sconing (1985) considered the efficiency properties of the KME in the proportional hazards model, where the censoring distribution G and the life distribution F satisfy

$$1 - G(x) = \{1 - F(x)\}^\beta, \quad x > 0, \ \beta > 0. \tag{11.154}$$

In this model the cumulative hazard functions $\wedge_G = -\log(1 - G)$ and $\wedge_F = -\log(1 - F)$ are proportional. Also, the expected proportion of uncensored observations is $1/(1 + \beta)$. The model is also known as the Koziol-Green model (see Koziol and Green 1976). Table 11.25 contains asymptotic efficiency values of the KME with respect to the maximum likelihood estimator $\widetilde{F}$ in model (11.154) with β unknown. The estimator $\widetilde{F}$ was independently proposed by Abdushukurov (1987), Cheng and Lin (1987) and Hollander, Proschan and Sconing (1985). (Also see Csörgő (1988) and Csörgő and Faraway (1998). The latter paper bares a paradox that shows the proportional hazards model is "too good to be true." For estimating F, sometimes it is better to have a censored sample than an uncensored sample!) The estimator $\widetilde{F}$ is

$$\widetilde{F}(x) = \{\widehat{H}(x)\}^d \tag{11.155}$$

where, letting $Z_{(1)} \leq Z_{(2)} \leq \cdots \leq Z_{(n)}$ denote the ordered values in the combined list of censored and uncensored values, $\widehat{H}(x) = $ (number of Z's $> x)/n$ and $d = $ (number of uncensored values)$/n$. Table 11.25, part of a larger table in HPS (1985), gives values of the asymptotic efficiency

$$a(x) = e(\overline{F}_{KM}, \widetilde{F})$$

of the KME with respect to $\widetilde{F}$ when F is exponential with parameter 1 and G is exponential with parameter β. For this choice of F, G, it can be shown that $a(x)$ initially increases and then decreases. The value of x for which this change occurs is given by the solution to the equation $(\beta + 1)x = 2[1 - \exp\{-x(\beta + 1)\}]$. Table 11.25 suggests that $a(x)$ decreases as β increases. β increasing is equivalent to censoring increasing stochastically. Thus Table 11.25 suggests that $a(x)$ decreases as censoring increases stochastically.

HPS (1985) also studied, in the fully nonparametric model where the life distribution F and the censoring distribution G are arbitrary, the asymptotic efficiency $b(x)$,

$$b(x) = e(\overline{F}_n, \overline{F}_{KM}),$$

of the empirical survival function (11.125) with respect to the KME; $b(x)$ has the following interpretation. Roughly speaking, the KME requires $nb(x)$ observations in the censored model to do as well as the empirical survival function does with n observations from the noncensored

TABLE 11.25. Values of $a(x)$

x	β:	.1	.2	$\frac{1}{3}$	.5	1.0	2.0
.1		.9186	.8522	.7812	.7130	.5903	.5048
.5		.9466	.9063	.8672	.8345	.7910	.7642
1.0		.9640	.9368	.9091	.8821	.8130	.6477
1.5		.9679	.9403	.9065	.8655	.7358	.4850
2.0		.9639	.9291	.8828	.8239	.6492	.3930

model. HPS (1985) showed (1) as x tends to 0, $b(x)$ tends to 1, (2) as x tends to infinity, $b(x)$ tends to infinity, (3) $b(x)$ is increasing in x, and (4) $b(x)$ increases as censoring increases stochastically. The results show that when x is small, the KME's efficiency loss is small, but for large values of x, the KME should be used with caution, particularly in cases of heavy censoring.

Harrington and Fleming (HF) (1982) obtained, using results of Gill (1980), asymptotic relative efficiencies of their class $\mathcal{G}^\rho$ of tests. The $\mathcal{G}^\rho$ class is a special case of Gill's (1980) $\mathcal{K}$ class. They take the censoring distributions G_1, G_2 to be equal and consider the family of survival functions $\mathcal{H}_\rho(x)$ given by

$$\mathcal{H}_0(x) = e^{-e^x} \qquad (\rho = 0), \qquad (11.156)$$

$$\mathcal{H}_\rho(x) = (1 + \rho e^x)^{-1/\rho} \qquad (\rho > 0). \qquad (11.157)$$

They consider the time-transformed location alternatives of survival functions

$$S^\rho(x) = \mathcal{H}_\rho(g(x) + \theta),$$

where g is an arbitrary monotonically increasing time transformation. The two samples can be viewed as being modeled with the survival functions $\overline{F}_i(x) = \mathcal{H}_\rho(g(x) + \theta_i)$, $i = 1, 2$, with $\Delta = \theta_2 - \theta_1$ and H_0 corresponding to $\Delta = 0$. HF (1982) obtained the asymptotic efficiency of the $\mathcal{G}^\rho$ statistic versus the $\mathcal{G}^{\rho^*}$ statistic for survival alternatives $\overline{F}_i(x) = \mathcal{H}_{\rho^*}(g(x) + \theta_i^N)$. The statistic $\mathcal{G}^{\rho^*}$ is fully efficient against $\mathcal{H}_{\rho^*}$. Expression (11.158) gives the asymptotic efficiencies $e(\mathcal{G}^\rho, \mathcal{G}^{\rho^*})$ in the special case of proportional hazards, where $\overline{G}_1(x) = \overline{G}_2(x) = \{\overline{F}(x)\}^\beta$:

$$e(\mathcal{G}^\rho, \mathcal{G}^{\rho^*}) = \frac{(2\rho + \beta + 1)(2\rho^* + \beta + 1)}{(\rho^* + \rho + \beta + 1)^2}. \qquad (11.158)$$

For $\rho = 0$, $\mathcal{G}^\rho$ is the logrank test, and it is optimal against shifts of the extreme value distribution given by (11.156). See HF (1982), Leurgans (1983, 1984) and Fleming and Harrington (1991, Section 7.4) for further efficiency results concerning Mantel's logrank test and its competitors.

APPENDIX A

Tables and Charts

Table A.1. Upper-tail probabilities for the standard normal distribution

Table A.2. Upper-tail probabilities for binomial distributions: $n = 2(1)10, p = .05(.05).45, \frac{1}{3}$; $n = 2(1)25, p = .50$

Table A.3. Selected confidence limits for the probability of success: $n = 1(1)10$

Table A.4. Selected upper-tail probabilities for the null distribution of the Wilcoxon signed rank T^+ statistic: $n = 3(1)50$

Table A.5. Upper-tail probabilities for the asymptotic conditional null distribution of the Hollander statistic, A_{obs}/n

Table A.6. Upper-tail probabilities for the null distribution of the Wilcoxon rank sum W statistic: $m = 3(1)10, n = 1(1)m; m = 11(1)20, n = 1(1)4$

Table A.7. Selected upper-tail probabilities for the null distribution of the Fligner-Policello $\hat{U}$ statistic: $m = 3(1)12, n = 3(1)m$

Table A.8. Upper-tail probabilities for the null distribution of the Ansari-Bradley C statistic: $2 \le n \le m$, with $(m + n) \le 20$

Table A.9. Selected upper-tail probabilities for the null distribution of the Lepage D statistic:

$$m = 2, n = 3(1)30; m = 3, n = 3(1)26; m = 4, n = 4(1)18;$$
$$m = 5, n = 5(1)13; m = 6, n = 6(1)10; m = 7, n = 7, 8$$

Table A.10. Upper-tail probabilities for the null distribution of the two-sided Kolmogorov-Smirnov J statistic: $n = 1(1)20, m = 1(1)n$

Table A.11. Upper-tail probabilities for the null limiting distribution of the two-sided Kolmogorov-Smirnov J^* statistic

Table A.12. Selected upper-tail probabilities for the null distribution of the Kruskal-Wallis H statistic:

$$k = 3, n_1 = 1(1)6, n_2 = n_1(1)6, 2 \le n_3 = n_2(1)6; n_1 = n_2 = n_3 = 7, 8;$$
$$k = 4, n_1 = 1(1)4, n_2 = n_1(1)4, n_3 = n_2(1)4, 2 \le n_4 = n_3(1)4;$$
$$k = 5, n_1 = 1(1)3, n_2 = n_1(1)3, n_3 = n_2(1)3, n_4 = n_3(1)3; 2 \le n_5 = n_4(1)3$$

Table A.13. Selected critical values for the null distribution of Jonckheere's J statistic: $k = 3, 2 \le n_1 \le n_2 \le n_3 \le 8; k = 4(1)6, n_1 = \cdots = n_k = 2(1)6$

Table A.14. Selected exact critical values for the null distribution of the peak-known Mack-Wolfe A_p statistic:

$$k = 4, p = 3, n_1 = \cdots = n_4 = 2(1)5;$$
$$k = 5, p = 3(1)4, n_1 = \cdots = n_5 = 2(1)5;$$
$$k = 6, p = 4(1)5, n_1 = \cdots = n_6 = 2(1)5$$

Table A.15. Selected critical values for the null distribution of the peak-unknown Mack-Wolfe A_p^* statistic:

$$k = 3, n_1 = n_2 = n_3 = 3(1)10; \ k = 4(1)10, n_1 = \ldots = n_k = 2(1)10$$

Table A.16. Exact critical values for the Steel-Dwass-Critchlow-Fligner two-sided all-treatments multiple comparison procedure:

$$k = 3, 2 \leq n_i \leq 7, i = 1, 2, 3$$

Table A.17. Selected critical values for the range of k independent $N(0, 1)$ variables:

$$k = 2(1)20(2)40(10)100$$

Table A.18. Exact critical values for the Hayter-Stone one-sided all-treatments multiple comparison procedure:

$$k = 3, \ n_i = 3(1)7, i = 1, 2, 3$$

Table A.19. Large-sample approximate critical values for the Hayter-Stone multiple comparison procedure with k treatments and equal sample sizes: $k = 3(1)9, n = n_1 = \cdots = n_k$

Table A.20. Exact critical values for the Nemenyi-Damico-Wolfe one-sided treatments versus control multiple comparison procedure:

$$k = 3, n_1 = 1(1)6, 1 \leq n_2 \leq n_3 \leq 6$$
$$k = 4, n_1 = 1, n_2 = n_3 = n_4 = 5; k = 4, n_1 = 2, n_2 = n_3 = n_4 = 2(1)5;$$
$$k = 4, n_1 = 3, n_2 = n_3 = n_4 = 1(1)4; k = 4, n_1 = 4, n_2 = n_3 = n_4 = 1(1)3;$$
$$k = 4, n_1 = 5, n_2 = n_3 = n_4 = 1(1)4; k = 4, n_1 = 6, n_2 = n_3 = n_4 = 1, 2$$

Table A.21. Cumulative probabilities for the distribution of the maximum of $\ell N(0, 1)$ random variables with common correlation ρ:

$$\ell = 1(1)12; \rho = .1(.1).9, \rho = .125(.125).875, \text{ and } \rho = \tfrac{1}{3}, \tfrac{2}{3}.$$

Table A.22. Selected upper-tail probabilities for the null distribution of the Friedman S statistic:

$$k = 3, n = 2(1)13; k = 4, n = 2(1)8; k = 5, n = 3(1)8; k = 6, n = 2(1)6$$

Table A.23. Selected upper-tail probabilities for the null distribution of the Page L statistic:

$$k = 3, n = 2(1)15; k = 4(1)8, n = 2(1)10$$

Table A.24. Selected critical values for the Wilcoxon-Nemenyi-McDonald-Thompson two-sided all-treatments multiple comparison procedure:

$$k = 3, n = 3(1)15; k = 4(1)15, n = 2(1)15$$

Table A.25. Selected critical values for the Nemenyi-Wilcoxon-Wilcox-Miller one-sided treatments versus control multiple comparison procedure:

$$k = 3, n = 2(1)18; k = 4(1)5, n = 2(1)8; k = 6, n = 2(1)6$$

Table A.26. Selected critical values for the null distribution of the Durbin-Skillings-Mack D statistic: $k = 3(1)6$ and selected BIBD designs

Table A.27. Selected exact critical values for the null distribution of the Skillings-Mack SM statistic: $k = 3(1)6, n = 3(1)11$

Table A.28. Selected exact critical values for the null distribution of the Mack-Skillings MS statistic with the same number (c) of replications in each cell: $k = 2(1)5, n = 2(1)5, c = 2(1)5$

Table A.29. Upper bounds ρ_U^n for the null correlation between two overlapping signed rank statistics: $n = 1(1)50$

Table A.30. Upper-tail probabilities for the null distribution of the Kendall K statistic: $n = 4(1)40$

Table A.31. Selected critical values for the null distribution of the Spearman r_s statistic: $n = 4(1)50$

Table A.32. Upper-tail probabilities for the null distribution of the Hoeffding D statistic: $n = 5(1)9$

Table A.33. Upper-tail probabilities for the null limiting distribution of the Blum-Kiefer-Rosenblatt ($\pi^4 nB/2$) statistic

Table A.34. Selected upper critical values for the null distribution of the Epstein $\mathcal{E}$ statistic: $n = 3(1)12$

Table A.35. Selected critical values for the null distribution of the Hollander-Proschan T statistic: $n = 4(1)20(5)50$

Table A.36. Selected critical values of the Hollander-Proschan decreasing mean residual life statistic V'

Table A.37. Critical values for the Hall-Wellner confidence bands for the mean residual life function

Table A.38. Percentage points of the Kolmogorov statistic

Table A.39. Percentage points of Lilliefors' test for normality

Table A.40. Selected percentiles of G_a for computing confidence bands for the survival function when the data are censored

Chart A.1. The t distribution: degrees of freedom $= n = 1(1)6(2)12(4)20, \infty$

Chart A.2. The χ^2 distribution: degrees of freedom $= n = 1(1)30$

TABLE A.1. Upper-Tail Probabilities for the Standard Normal Distribution

For a given x, the entry is $P\{X \geq x\}$, where X has a $N(0, 1)$ distribution. Thus if x is such that $P\{X \geq x\} = \alpha$, then $z_\alpha = x$.

x	.00	.01	.02	.03	.04	.05	.06	.07	.08	.09
.0	.5000	.4960	.4920	.4880	.4840	.4801	.4761	.4721	.4681	.4641
.1	.4602	.4562	.4522	.4483	.4443	.4404	.4364	.4325	.4286	.4247
.2	.4207	.4168	.4129	.4090	.4052	.4013	.3974	.3936	.3897	.3859
.3	.3821	.3783	.3745	.3707	.3669	.3632	.3594	.3557	.3520	.3483
.4	.3446	.3409	.3372	.3336	.3300	.3264	.3228	.3192	.3156	.3121
.5	.3085	.3050	.3015	.2981	.2946	.2912	.2877	.2843	.2810	.2776
.6	.2743	.2709	.2676	.2643	.2611	.2578	.2546	.2514	.2483	.2451
.7	.2420	.2389	.2358	.2327	.2296	.2266	.2236	.2206	.2177	.2148
.8	.2119	.2090	.2061	.2033	.2005	.1977	.1949	.1922	.1894	.1867
.9	.1841	.1814	.1788	.1762	.1736	.1711	.1685	.1660	.1635	.1611
1.0	.1587	.1562	.1539	.1515	.1492	.1469	.1446	.1423	.1401	.1379
1.1	.1357	.1335	.1314	.1292	.1271	.1251	.1230	.1210	.1190	.1170
1.2	.1151	.1131	.1112	.1093	.1075	.1056	.1038	.1020	.1003	.0985
1.3	.0968	.0951	.0934	.0918	.0901	.0885	.0869	.0853	.0838	.0823
1.4	.0808	.0793	.0778	.0764	.0749	.0735	.0721	.0708	.0694	.0681
1.5	.0668	.0655	.0643	.0630	.0618	.0606	.0594	.0582	.0571	.0559
1.6	.0548	.0537	.0526	.0516	.0505	.0495	.0485	.0475	.0465	.0455
1.7	.0446	.0436	.0427	.0418	.0409	.0401	.0392	.0384	.0375	.0367
1.8	.0359	.0351	.0344	.0336	.0329	.0322	.0314	.0307	.0301	.0294
1.9	.0287	.0281	.0274	.0268	.0262	.0256	.0250	.0244	.0239	.0233
2.0	.0228	.0222	.0217	.0212	.0207	.0202	.0197	.0192	.0188	.0183
2.1	.0179	.0174	.0170	.0166	.0162	.0158	.0154	.0150	.0146	.0143
2.2	.0139	.0136	.0132	.0129	.0125	.0122	.0119	.0116	.0113	.0110
2.3	.0107	.0104	.0102	.0099	.0096	.0094	.0091	.0089	.0087	.0084
2.4	.0082	.0080	.0078	.0075	.0073	.0071	.0069	.0068	.0066	.0064
2.5	.0062	.0060	.0059	.0057	.0055	.0054	.0052	.0051	.0049	.0048
2.6	.0047	.0045	.0044	.0043	.0041	.0040	.0039	.0038	.0037	.0036
2.7	.0035	.0034	.0033	.0032	.0031	.0030	.0029	.0028	.0027	.0026
2.8	.0026	.0025	.0024	.0023	.0023	.0022	.0021	.0021	.0020	.0019
2.9	.0019	.0018	.0018	.0017	.0016	.0016	.0015	.0015	.0014	.0014
3.0	.0013	.0013	.0013	.0012	.0012	.0011	.0011	.0011	.0010	.0010
3.1	.0010	.0009	.0009	.0009	.0008	.0008	.0008	.0008	.0007	.0007
3.2	.0007	.0007	.0006	.0006	.0006	.0006	.0006	.0005	.0005	.0005
3.3	.0005	.0005	.0005	.0004	.0004	.0004	.0004	.0004	.0004	.0003
3.4	.0003	.0003	.0003	.0003	.0003	.0003	.0003	.0003	.0003	.0002

A. TABLES AND CHARTS

TABLE A.2. Upper-Tail Probabilities for Binomial Distributions: $n = 2(1)10$, $p = .05(.05).45, \frac{1}{3}; n = 2(1)25, p = .50$

If B has a binomial distribution with parameters n and p_0, then the tables read as follows:

1. When $p_0 < .5$, the (b, n) entry in the $p = p_0$ table is $P\{B \geq b\}$. Thus if the (b, n) entry in the $p = p_0$ table is α, then $b_\alpha = b$.

2. When $p_0 > .5$, tail probabilities can be obtained by using the equation $P\{B \geq b\} = P\{C \leq (n - b)\}$, where C has a binomial distribution with parameters n and $1 - p_0$. For example, with $n = 10, p_0 = .7$, and $b = 8$, we find $P\{B \geq 8\} = P\{C \leq 2\} = 1 - P\{C \geq 3\} = 1 - .6172 = .3828$.

3. When $p_0 = .5$, the (b, n) entry in the $p = .5$ table is $P\{B \geq b\} = P\{B \leq (n - b)\}$. Thus if the (b, n) entry in the $p = .5$ table is α, then $b_\alpha = b$.

					$p = .05$				
					n				
b	2	3	4	5	6	7	8	9	10
0	1.0000	1.0000	1.0000	1.0000	1.0000	1.0000	1.0000	1.0000	1.0000
1	.0975	.1426	.1855	.2262	.2649	.3017	.3366	.3698	.4013
2	.0025	.0073	.0140	.0226	.0328	.0444	.0572	.0712	.0861
3		.0001	.0005	.0012	.0022	.0038	.0058	.0084	.0115
4			.0000	.0000	.0001	.0002	.0004	.0006	.0010
5				.0000	.0000	.0000	.0000	.0000	.0001
6					.0000	.0000	.0000	.0000	.0000
7						.0000	.0000	.0000	.0000
8							.0000	.0000	.0000
9								.0000	.0000
10									.0000

					$p = .10$				
					n				
b	2	3	4	5	6	7	8	9	10
0	1.0000	1.0000	1.0000	1.0000	1.0000	1.0000	1.0000	1.0000	1.0000
1	.1900	.2710	.3439	.4095	.4686	.5217	.5695	.6126	.6513
2	.0100	.0280	.0523	.0815	.1143	.1497	.1869	.2252	.2639
3		.0010	.0037	.0086	.0159	.0257	.0381	.0530	.0702
4			.0001	.0005	.0013	.0027	.0050	.0083	.0128
5				.0000	.0001	.0002	.0004	.0009	.0016
6					.0000	.0000	.0000	.0001	.0001
7						.0000	.0000	.0000	.0000
8							.0000	.0000	.0000
9								.0000	.0000
10									.0000

(continued)

TABLE A.2. *(continued)*

$p = .15$

b	____	____	____	____	n	____	____	____	____
	2	3	4	5	6	7	8	9	10
0	1.0000	1.0000	1.0000	1.0000	1.0000	1.0000	1.0000	1.0000	1.0000
1	.2775	.3859	.4780	.5563	.6229	.6794	.7275	.7684	.8031
2	.0225	.0608	.1095	.1648	.2235	.2834	.3428	.4005	.4557
3		.0034	.0120	.0266	.0473	.0738	.1052	.1409	.1798
4			.0005	.0022	.0059	.0121	.0214	.0339	.0500
5				.0001	.0004	.0012	.0029	.0056	.0099
6					.0000	.0001	.0002	.0006	.0014
7						.0000	.0000	.0000	.0001
8							.0000	.0000	.0000
9								.0000	.0000
10									.0000

$p = .20$

b	____	____	____	____	n	____	____	____	____
	2	3	4	5	6	7	8	9	10
0	1.0000	1.0000	1.0000	1.0000	1.0000	1.0000	1.0000	1.0000	1.0000
1	.3600	.4880	.5904	.6723	.7379	.7903	.8322	.8658	.8926
2	.0400	.1040	.1808	.2627	.3446	.4233	.4967	.5638	.6242
3		.0080	.0272	.0579	.0989	.1480	.2031	.2618	.3222
4			.0016	.0067	.0170	.0333	.0563	.0856	.1209
5				.0003	.0016	.0047	.0104	.0196	.0328
6					.0001	.0004	.0012	.0031	.0064
7						.0000	.0001	.0003	.0009
8							.0000	.0000	.0001
9								.0000	.0000
10									.0000

$p = .25$

b	____	____	____	____	n	____	____	____	____
	2	3	4	5	6	7	8	9	10
0	1.0000	1.0000	1.0000	1.0000	1.0000	1.0000	1.0000	1.0000	1.0000
1	.4375	.5781	.6836	.7627	.8220	.8665	.8999	.9249	.9437
2	.0625	.1563	.2617	.3672	.4661	.5551	.6329	.6997	.7560
3		.0156	.0508	.1035	.1694	.2436	.3215	.3993	.4744
4			.0039	.0156	.0376	.0706	.1138	.1657	.2241
5				.0010	.0046	.0129	.0273	.0489	.0781
6					.0002	.0013	.0042	.0100	.0197
7						.0001	.0004	.0013	.0035
8							.0000	.0001	.0004
9								.0000	.0000
10									.0000

TABLE A.2. (continued)

$p = .30$

b	n=2	3	4	5	6	7	8	9	10
0	1.0000	1.0000	1.0000	1.0000	1.0000	1.0000	1.0000	1.0000	1.0000
1	.5100	.6570	.7599	.8319	.8824	.9176	.9424	.9596	.9718
2	.0900	.2160	.3483	.4718	.5798	.6706	.7447	.8040	.8507
3		.0270	.0837	.1631	.2557	.3529	.4482	.5372	.6172
4			.0081	.0308	.0705	.1260	.1941	.2703	.3504
5				.0024	.0109	.0288	.0580	.0988	.1503
6					.0007	.0038	.0113	.0253	.0473
7						.0002	.0013	.0043	.0106
8							.0001	.0004	.0016
9								.0000	.0001
10									.0000

$p = 1/3$

b	n=2	3	4	5	6	7	8	9	10
0	1.0000	1.0000	1.0000	1.0000	1.0000	1.0000	1.0000	1.0000	1.0000
1	.5556	.7037	.8025	.8683	.9122	.9415	.9610	.9740	.9827
2	.1111	.2593	.4074	.5391	.6488	.7366	.8049	.8569	.8959
3		.0370	.1111	.2099	.3196	.4294	.5318	.6228	.7009
4			.0123	.0453	.1001	.1733	.2586	.3497	.4407
5				.0041	.0178	.0453	.0879	.1448	.2131
6					.0014	.0069	.0197	.0424	.0766
7						.0005	.0026	.0083	.0197
8							.0002	.0010	.0034
9								.0001	.0004
10									.0000

$p = .35$

b	n=2	3	4	5	6	7	8	9	10
0	1.0000	1.0000	1.0000	1.0000	1.0000	1.0000	1.0000	1.0000	1.0000
1	.5775	.7254	.8215	.8840	.9246	.9510	.9681	.9793	.9865
2	.1225	.2818	.4370	.5716	.6809	.7662	.8309	.8789	.9140
3		.0429	.1265	.2352	.3529	.4677	.5722	.6627	.7384
4			.0150	.0540	.1174	.1998	.2936	.3911	.4862
5				.0053	.0223	.0556	.1061	.1717	.2485
6					.0018	.0090	.0253	.0536	.0949
7						.0006	.0036	.0112	.0260
8							.0002	.0014	.0048
9								.0001	.0005
10									.0000

(continued)

TABLE A.2. *(continued)*

$p = .40$

b	n=2	3	4	5	6	7	8	9	10
0	1.0000	1.0000	1.0000	1.0000	1.0000	1.0000	1.0000	1.0000	1.0000
1	.6400	.7840	.8704	.9222	.9533	.9720	.9832	.9899	.9940
2	.1600	.3520	.5248	.6630	.7667	.8414	.8936	.9295	.9536
3		.0640	.1792	.3174	.4557	.5801	.6846	.7682	.8327
4			.0256	.0870	.1792	.2898	.4059	.5174	.6177
5				.0102	.0410	.0963	.1737	.2666	.3669
6					.0041	.0188	.0498	.0994	.1662
7						.0016	.0085	.0250	.0548
8							.0007	.0038	.0123
9								.0003	.0017
10									.0001

$p = .45$

b	n=2	3	4	5	6	7	8	9	10
0	1.0000	1.0000	1.0000	1.0000	1.0000	1.0000	1.0000	1.0000	1.0000
1	,6975	.8336	.9085	.9497	.9723	.9848	.9916	.9954	.9975
2	.2025	.4253	.6090	.7438	.8364	.8976	.9368	.9615	.9767
3		.0911	-2415	.4069	.5585	.6836	.7799	.8505	.9004
4			.0410	.1312	.2553	.3917	.5230	.6386	.7340
5				.0185	.0692	.1529	.2604	.3786	.4956
6					.0083	.0357	.0885	.1658	.2616
7						.0037	.0181	.0498	.1020
8							.0017	.0091	.0274
9								.0008	.0045
10									.0003

$p = .50$

b	n=2	3	4	5	6	7	8	9
1	.7500							
2	.2500	.5000	.6875					
3		.1250	.3125	.5000	.6563			
4			.0625	.1875	.3438	.5000	.6367	
5				.0313	.1094	.2266	.3633	.5000
6					.0156	.0625	.1445	.2539
7						.0078	.0352	.0898
8							.0039	.0195
9								.0020

TABLE A.2. *(continued)*

				$p = .50$				
				n				
b	10	11	12	13	14	15	16	17
5	.6230							
6	.3770	.5000	.6128					
7	.1719	.2744	.3872	.5000	.6047			
8	.0547	.1133	.1938	.2905	.3953	.5000	.5982	
9	.0107	.0327	.0730	.1334	.2120	.3036	.4018	.5000
10	.0010	.0059	.0193	.0461	.0898	.1509	.2272	.3145
11		.0005	.0032	.0112	.0287	.0592	.1051	.1662
12			.0002	.0017	.0065	.0176	.0384	.0717
13				.0001	.0009	.0037	.0106	.0245
14					.0001	.0005	.0021	.0064
15						.0000	.0003	.0012
16							.0000	.0001
17								.0000

				$p = .50$				
				n				
b	18	19	20	21	22	23	24	25
9	.5927							
10	.4073	.5000	.5881					
11	.2403	.3238	.4119	.5000	.5841			
12	.1189	.1796	.2517	.3318	.4159	.5000	.5806	
13	.0481	.0835	.1316	.1917	.2617	.3388	.4194	.5000
14	.0154	.0318	.0577	.0946	.1431	.2024	.2706	.3450
15	.0038	.0096	.0207	.0392	.0669	.1050	.1537	.2122
16	.0007	.0022	.0059	.0133	.0262	.0466	.0758	.1148
17	.0001	.0004	.0013	.0036	.0085	.0173	.0320	.0539
18	.0000	.0000	.0002	.0007	.0022	.0053	.0113	.0216
19		.0000	.0000	.0001	.0004	.0013	.0033	.0073
20			.0000	.0000	.0001	.0002	.0008	.0020
21				.0000	.0000	.0000	.0001	.0005
22					.0000	.0000	.0000	.0001
23						.0000	.0000	.0000
24							.0000	.0000
25								.0000

TABLE A.3. Selected Confidence Limits for the Probability of Success: $n = 1(1)10$

For given n, α, and the number of successes B, the table entries are $p_L(\alpha)$ and $p_U(\alpha)$, satisfying $P_p\{p_L(\alpha) < p < p_U(\alpha)\} \geq 1 - \alpha$.

		$n = 1$	
B	α	$p_L(\alpha)$	$p_U(\alpha)$
0	.010	.0000	.9950
	.020	.0000	.9900
	.050	.0000	.9750
	.100	.0000	.9500
	.200	.0000	.9000
1	.010	.0050	1.0000
	.020	.0100	1.0000
	.050	.0250	1.0000
	.100	.0500	1.0000
	.200	.1000	1.0000

		$n = 2$	
B	α	$p_L(\alpha)$	$p_U(\alpha)$
0	.010	.0000	.9293
	.020	.0000	.9000
	.050	.0000	.8419
	.100	.0000	.7764
	.200	.0000	.6838
1	.010	.0025	.9975
	.020	.0050	.9950
	.050	.0126	.9874
	.100	.0253	.9747
	.200	.0513	.9487
2	.010	.0707	1.0000
	.020	.1000	1.0000
	.050	.1581	1.0000
	.100	.2236	1.0000
	.200	.3162	1.0000

		$n = 3$	
B	α	$p_L(\alpha)$	$p_U(\alpha)$
0	.010	.0000	.8290
	.020	.0000	.7846
	.050	.0000	.7076
	.100	.0000	.6316
	.200	.0000	.5358
1	.010	.0017	.9586
	.020	.0033	.9411
	.050	.0084	.9057
	.100	.0170	.8647
	.200	.0345	.8042
2	.010	.0414	.9983
	.020	.0589	.9967
	.050	.0943	.9916
	.100	.1353	.9830
	.200	.1958	.9655

		$n = 3$	
B	α	$p_L(\alpha)$	$p_U(\alpha)$
3	.010	.1710	1.0000
	.020	.2154	1.0000
	.050	.2924	1.0000
	.100	.3684	1.0000
	.200	.4642	1.0000

		$n = 4$	
B	α	$p_L(\alpha)$	$p_U(\alpha)$
0	.010	.0000	.7341
	.020	.0000	.6838
	.050	.0000	.6024
	.100	.0000	.5271
	.200	.0000	.4377
1	.010	.0013	.8891
	.020	.0025	.8591
	.050	.0063	.8059
	.100	.0127	.7514
	.200	.0260	.6795
2	.010	.0294	.9706
	.020	.0420	.9580
	.050	.0676	.9324
	.100	.0976	.9024
	.200	.1426	.8574
3	.010	.1109	.9987
	.020	.1409	.9975
	.050	.1941	.9937
	.100	.2486	.9873
	.200	.3205	.9740
4	.010	.2659	1.0000
	.020	.3162	1.0000
	.050	.3976	1.0000
	.100	.4729	1.0000
	.200	.5623	1.0000

		$n = 5$	
B	α	$p_L(\alpha)$	$p_U(\alpha)$
0	.010	.0000	.6534
	.020	.0000	.6019
	.050	.0000	.5218
	.100	.0000	.4507
	.200	.0000	.3690
1	.010	.0010	.8149
	.020	.0020	.7779
	.050	.0051	.7164
	.100	.0102	.6574
	.200	.0209	.5839

TABLE A.3. *(continued)*

	$n = 5$		
B	α	$p_L(\alpha)$	$p_U(\alpha)$
2	.010	.0229	.9172
	.020	.0327	.8944
	.050	.0527	.8534
	.100	.0764	.8107
	.200	.1122	.7534
3	.010	.0828	.9771
	.020	.1056	.9673
	.050	.1466	.9473
	.100	.1893	.9236
	.200	.2466	.8878
4	.010	.1851	.9990
	.020	.2221	.9980
	.050	.2836	.9949
	.100	.3426	.9898
	.200	.4161	.9791
5	.010	.3466	1.0000
	.020	.3981	1.0000
	.050	.4782	1.0000
	.100	.5493	1.0000
	.200	.6310	1.0000

	$n = 6$		
B	α	$p_L(\alpha)$	$p_U(\alpha)$
0	.010	.0000	.5865
	.020	.0000	.5358
	.050	.0000	.4593
	.100	.0000	.3930
	.200	.0000	.3187
1	.010	.0008	.7460
	.020	.0017	.7057
	.050	.0042	.6412
	.100	.0085	.5818
	.200	.0174	.5103
2	.010	.0187	.8564
	.020	.0268	.8269
	.050	.0433	.7772
	.100	.0628	.7287
	.200	.0926	.6668
3	.010	.0663	.9337
	.020	.0847	.9153
	.050	.1181	.8819
	.100	.1532	.8468
	.200	.2009	.7991
4	.010	.1436	.9813
	.020	.1731	.9732
	.050	.2228	.9567
	.100	.2713	.9372
	.200	.3332	.9074

	$n = 6$		
B	α	$p_L(\alpha)$	$p_U(\alpha)$
5	.010	.2540	.9992
	.020	.2943	.9983
	.050	.3588	.9958
	.100	.4182	.9915
	.200	.4897	.9826
6	.010	.4135	1.0000
	.020	.4642	1.0000
	.050	.5407	1.0000
	.100	.6070	1.0000
	.200	.6813	1.0000

	$n = 7$		
B	α	$p_L(\alpha)$	$p_U(\alpha)$
0	.100	.0000	.5309
	.020	.0000	.4821
	.050	.0000	.4096
	.100	.0000	.3482
	.200	.0000	.2803
1	.010	.0007	.6849
	.020	.0014	.6434
	.050	.0036	.5787
	.100	.0073	.5207
	.200	.0149	.4526
2	.010	.0158	.7970
	.020	.0227	.7637
	.050	.0367	.7096
	.100	.0534	.6587
	.200	.0788	.5962
3	.010	.0553	.8823
	.020	.0708	.8577
	.050	.0990	.8159
	.100	.1288	.7747
	.200	.1696	.7214
4	.010	.1177	.9447
	.020	.1423	.9292
	.050	.1841	.9010
	.100	.2253	.8712
	.200	.2786	.8304
5	.010	.2030	.9842
	.020	.2363	.9773
	.050	.2904	.9633
	.100	.3413	.9466
	.200	.4038	.9212
6	.010	.3151	.9993
	.020	.3566	.9986
	.050	.4213	.9964
	.100	.4793	.9927
	.200	.5474	.9851

(continued)

TABLE A.3. *(continued)*

	$n = 7$				$n = 8$		
B	α	$p_L(\alpha)$	$p_U(\alpha)$	B	α	$p_L(\alpha)$	$p_U(\alpha)$
7	.010	.4691	1.0000	8	.010	.5157	1.0000
	.020	.5179	1.0000		.020	.5623	1.0000
	.050	.5904	1.0000		.050	.6306	1.0000
	.100	.6518	1.0000		.100	.6877	1.0000
	.200	.7197	1.0000		.200	.7499	1.0000

	$n = 8$				$n = 9$		
B	α	$p_L(\alpha)$	$p_U(\alpha)$	B	α	$p_L(\alpha)$	$p_U(\alpha)$
0	.010	.0000	.4843	0	.010	.0000	.4450
	.020	.0000	.4377		.020	.0000	.4005
	.050	.0000	.3694		.050	.0000	.3363
	.100	.0000	.3123		.100	.0000	.2831
	.200	.0000	.2501		.200	.0000	.2257
1	.010	.0006	.6315	1	.010	.0006	.5850
	.020	.0013	.5899		.020	.0011	.5440
	.050	.0032	.5265		.050	.0028	.4825
	.100	.0064	.4707		.100	.0057	.4291
	.200	.0131	.4062		.200	.0116	.3684
2	.010	.0137	.7422	2	.010	.0121	.6926
	.020	.0197	.7068		.020	.0174	.6563
	.050	.0319	.6509		.050	.0281	.6001
	.100	.0464	.5997		.100	.0410	.5496
	.200	.0686	.5382		.200	.0608	.4901
3	.010	.0475	.8303	3	.010	.0416	.7809
	.020	.0608	.8018		.020	.0534	.7500
	.050	.0852	.7551		.050	.0749	.7007
	.100	.1111	.7108		.100	.0978	.6551
	.200	.1469	.6554		.200	.1295	.5994
4	.010	.0999	.9001	4	.010	.0868	.8539
	.020	.1210	.8790		.020	.1053	.8290
	.050	.1570	.8430		.050	.1370	.7880
	.100	.1929	.8071		.100	.1687	.7486
	.200	.2397	.7603		.200	.2104	.6990
5	.010	.1697	.9525	5	.010	.1461	.9132
	.020	.1982	.9392		.020	.1710	.8947
	.050	.2449	.9148		.050	.2120	.8630
	.100	.2892	.8889		.100	.2514	.8313
	.200	.3446	.8531		.200	.3010	.7896
6	.010	.2578	.9863	6	.010	.2191	.9584
	.020	.2932	.9803		.020	.2500	.9466
	.050	.3491	.9681		.050	.2993	.9251
	.100	.4003	.9536		.100	.3449	.9022
	.200	.4618	.9314		.200	.4006	.8705
7	.010	.3685	.9994	7	.010	.3074	.9879
	.020	.4101	.9987		.020	.3437	.9826
	.050	.4735	.9968		.050	.3999	.9719
	.100	.5293	.9936		.100	.4504	.9590
	.200	.5938	.9869		.200	.5099	.9392

TABLE A.3. (continued)

	n = 9				n = 10		
B	α	$p_L(\alpha)$	$p_U(\alpha)$	B	α	$p_L(\alpha)$	$p_U(\alpha)$
8	.010	.4150	.9994	5	.010	.1283	.8717
	.020	.4560	.9989		.020	.1504	.8496
	.050	.5175	.9972		.050	.1871	.8129
	.100	.5709	.9943		.100	.2224	.7776
	.200	.6316	.9884		.200	.2673	.7327
9	.010	.5550	1.0000	6	.010	.1909	.9232
	.020	.5995	1.0000		.020	.2183	.9068
	.050	.6637	1.0000		.050	.2624	.8784
	.100	.7169	1.0000		.100	.3035	.8500
	.200	.7743	1.0000		.200	.3542	.8124

	n = 10						
B	α	$p_L(\alpha)$	$p_U(\alpha)$	7	.010	.2649	.9630
					.020	.2971	.9525
0	.010	.0000	.4113		.050	.3475	.9333
	.020	.0000	.3690		.100	.3934	.9127
	.050	.0000	.3085		.200	.4483	.8842
	.100	.0000	.2589				
	.200	.0000	.2057	8	.010	.3518	.9891
					.020	.3883	.9845
1	.010	.0005	.5443		.050	.4439	.9748
	.020	.0010	.5044		.100	.4931	.9632
	.050	.0025	.4450		.200	.5504	.9455
	.100	.0051	.3942				
	.200	.0105	.3369	9	.010	.4557	.9995
					.020	.4956	.9990
2	.010	.0109	.6482		.050	.5550	.9975
	.020	.0155	.6117		.100	.6058	.9949
	.050	.0252	.5561		.200	.6631	.9895
	.100	.0368	.5069				
	.200	.0545	.4496	10	.010	.5887	1.0000
					.020	.6310	1.0000
3	.010	.0370	.7351		.050	.6915	1.0000
	.020	.0475	.7029		.100	.7411	1.0000
	.050	.0667	.6525		.200	.7943	1.0000
	.100	.0873	.6066				
	.200	.1158	.5517				
4	.010	.0768	.8091				
	.020	.0932	.7817				
	.050	.1216	.7376				
	.100	.1500	.6965				
	.200	.1876	.6458				

Adapted from Table 9.6 of D. B. Owen, *Handbook of Statistical Tables*, Addison-Wesley, Reading, MA (1962), with the permission of the author and the publisher. Courtesy of U.S. Atomic Energy Commission.

TABLE A.4. Selected Upper-Tail Probabilities for the Null Distribution of the Wilcoxon Signed Rank T^+ Statistic: $n = 3(1)50$

Additional upper-tail probabilities can be found in Kraft and van Eeden (1968) and Wilcoxon, Katti, and Wilcox (1973).

For a given n, the table entry for the point x is $P_0\{T^+ \geq x\}$. Thus, if x is such that $P_0\{T^+ \geq x\} = \alpha$, then $t_\alpha = x$.

	x	$P_0\{T^+ \geq x\}$		x	$P_0\{T^+ \geq x\}$		x	$P_0\{T^+ \geq x\}$
$n = 3$	3	.625	$n = 8$	22	.320	$n = 10$	40	.116
	4	.375		23	.273		41	.097
	5	.250		24	.230		42	.080
	6	.125		25	.191		43	.065
				26	.156		44	.053
$n = 4$	5	.562		27	.125		45	.042
	6	.438		28	.098		46	.032
	7	.312		29	.074		47	.024
	8	.188		30	.055		48	.019
	9	.125		31	.039		49	.014
	10	.062		32	.027		50	.010
				33	.020		51	.007
$n = 5$	8	.500		34	.012		52	.005
	9	.406		35	.008		53	.003
	10	.312		36	.004		54	.002
	11	.219					55	.001
	12	.156	$n = 9$	23	.500			
	13	.094		24	.455	$n = 11$	33	.517
	14	.062		25	.410		34	.483
	15	.031		26	.367		35	.449
$n = 6$	11	.500		27	.326		36	.416
	12	.422		28	.285		37	.382
	13	.344		29	.248		38	.350
	14	.281		30	.213		39	.319
	15	.219		31	.180		40	.289
	16	.156		32	.150		41	.260
	17	.109		33	.125		42	.232
	18	.078		34	.102		43	.207
	19	.047		35	.082		44	.183
	20	.031		36	.064		45	.160
	21	.016		37	.049		46	.139
				38	.037		47	.120
$n = 7$	14	.531		39	.027		48	.103
	15	.469		40	.020		49	.087
	16	.406		41	.014		50	.074
	17	.344		42	.010		51	.062
	18	.289		43	.006		52	.051
	19	.234		44	.004		53	.042
	20	.188		45	.002		54	.034
	21	.148					55	.027
	22	.109	$n = 10$	28	.500		56	.021
	23	.078		29	.461		57	.016
	24	.055		30	.423		58	.012
	25	.039		31	.385		59	.009
	26	.023		32	.348		60	.007
	27	.016		33	.312		61	.005
	28	.008		34	.278		62	.003
				35	.246		63	.002
$n = 8$	18	.527		36	.216		64	.001
	19	.473		37	.188		65	.001
	20	.422		38	.161		66	<.0005
	21	.371		39	.138			

A. TABLES AND CHARTS

TABLE A.4. *(continued)*

	x	$P_0\{T^+ \geq x\}$		x	$P_0\{T^+ \geq x\}$		x	$P_0\{T^+ \geq x\}$
$n = 12$	39	.515	$n = 13$	74	.024	$n = 15$	77	.180
	40	.485		75	.020		78	.165
	41	.455		76	.016		79	.151
	42	.425		77	.013		80	.138
	43	.396		78	.011		81	.126
	44	.367		79	.009		82	.115
	45	.339		80	.007		83	.104
	46	.311		81	.005		84	.094
	47	.285		82	.004		85	.084
	48	.259		83	.003		86	.076
	49	.235		84	.002		87	.068
	50	.212		85	.002		88	.060
	51	.190		86	.001		89	.053
	52	.170		87	.001		90	.047
	53	.151		88	.001		91	.042
	54	.133		89	<.0005		92	.036
	55	.117					93	.032
	56	.102	$n = 14$	66	.213		94	.028
	57	.088		67	.196		95	.024
	58	.076		68	.179		96	.021
	59	.065		69	.163		97	.018
	60	.055		70	.148		98	.015
	61	.046		71	.134		99	.013
	62	.039		72	.121		100	.011
	63	.032		73	.108		101	.009
	64	.026		74	.097		102	.008
	65	.021		75	.086		103	.006
	66	.017		76	.077		104	.005
	67	.013		77	.068		105	.004
	68	.010		78	.059		106	.003
	69	.008		79	.052		107	.003
	70	.006		80	.045		108	.002
	71	.005		81	.039		109	.002
	72	.003		82	.034		110	.001
	73	.002		83	.029		111	.001
	74	.002		84	.025		112	.001
	75	.001		85	.021		113	.001
	76	.001		86	.018		114	<.0005
	77	<.0005		87	.015			
				88	.012	$n = 16$	93	.106
$n = 13$	58	.207		89	.010		94	.096
	59	.188		90	.008		95	.088
	60	.170		91	.007		96	.080
	61	.153		92	.005		97	.072
	62	.137		93	.004		98	.065
	63	.122		94	.003		99	.058
	64	.108		95	.003		100	.052
	65	.095		96	.002		101	.047
	66	.084		97	.002		102	.042
	67	.073		98	.001		103	.037
	68	.064		99	.001		104	.033
	69	.055		100	.001		105	.029
	70	.047		101	<.0005		106	.025
	71	.040					107	.022
	72	.034	$n = 15$	75	.211		108	.019
	73	.029		76	.195		109	.017

(continued)

TABLE A.4. (continued)

	x	$P_0\{T^+ \geq x\}$		x	$P_0\{T^+ \geq x\}$		x	$P_0\{T^+ \geq x\}$
$n=16$	110	.015	$n=17$	140	.001	$n=19$	152	.0102
	111	.013		141	.001		153	.0090
	112	.011		142	<.0005		157	.0054
	113	.009					158	.0047
	114	.008	$n=18$	115	.106		168	.0010
	115	.007		116	.098			
	116	.006		117	.091	$n=20$	140	.1012
	117	.005		118	.084		141	.0947
	118	.004		119	.077		144	.0768
	119	.003		120	.071		145	.0715
	120	.003		121	.065		149	.0527
	121	.002		122	.059		150	.0487
	122	.002		123	.054		157	.0266
	123	.001		124	.049		158	.0242
	124	.001		125	.045		166	.0107
	125	.001		126	.041		167	.0096
	126	.001		127	.037		172	.0053
	127	.001		128	.033		173	.0047
	128	<.0005		129	.030		184	.0010
				130	.027			
$n=17$	104	.103		131	.024	$n=21$	153	.1015
	105	.095		132	.022		154	.0953
	106	.087		133	.019		157	.0786
	107	.080		134	.017		158	.0735
	108	.073		135	.015		163	.0516
	109	.066		136	.013		164	.0479
	110	.060		137	.012		172	.0251
	111	.054		138	.010		173	.0230
	112	.049		139	.009		181	.0108
	113	.044		140	.008		182	.0097
	114	.040		141	.007		188	.0051
	115	.036		142	.006		189	.0045
	116	.032		143	.005		200	.0011
	117	.028		144	.005		201	.0009
	118	.025		145	.004	$n=22$	166	.1050
	119	.022		146	.003		167	.0991
	120	.020		147	.003		171	.0780
	121	.017		148	.002		172	.0733
	122	.015		149	.002		177	.0527
	123	.013		150	.002		178	.0492
	124	.012		151	.001		187	.0250
	125	.010		152	.001		197	.0104
	126	.009		153	.001		198	.0095
	127	.008		154	.001		204	.0052
	128	.006		155	.001		205	.0046
	129	.006		156	.001		218	.0010
	130	.005		157	<.0005			
	131	.004				$n=23$	181	.1001
	132	.003	$n=19$	127	.1051		182	.0948
	133	.003		128	.0978		186	.0755
	134	.002		131	.0782		187	.0712
	135	.002		132	.0723		192	.0523
	136	.002		136	.0521		193	.0490
	137	.001		137	.0478		202	.0261
	138	.001		143	.0273		203	.0242
	139	.001		144	.0247			

A. TABLES AND CHARTS

TABLE A.4. (continued)

	x	$P_0\{T^+ \geq x\}$		x	$P_0\{T^+ \geq x\}$		x	$P_0\{T^+ \geq x\}$
$n = 23$	213	.0107	$n = 27$	271	.0246	$n = 31$	377	.0052
	214	.0098		285	.0100		378	.0049
	221	.0051		294	.0052		401	.0010
	222	.0046		295	.0048			
	225	.0011		313	.0010	$n = 32$	333	.1016
	226	.0009					334	.0983
			$n = 28$	260	.1008		341	.0772
$n = 24$	195	.1038		261	.0968		342	.0745
	196	.0987		267	.0751		352	.0512
	201	.0758		268	.0719		353	.0492
	202	.0717		275	.0521		368	.0260
	208	.0505		276	.0496		369	.0249
	209	.0475		289	.0252		387	.0103
	218	.0263		290	.0239		388	.0097
	219	.0245		304	.0102		400	.0050
	230	.0106		305	.0096		425	.0010
	231	.0097		314	.0051			
	238	.0053		315	.0048	$n = 33$	353	.1005
	239	.0048		334	.0010		354	.0974
	254	.0010					361	.0774
			$n = 29$	277	.1027		362	.0748
$n = 25$	211	.1001		278	.0988		373	.0503
	212	.0954		284	.0778		374	.0485
	216	.0782		285	.0747		390	.0253
	217	.0742		294	.0504		391	.0242
	224	.0507		295	.0482		409	.0104
	225	.0479		308	.0253		410	.0099
	235	.0258		309	.0240		422	.0052
	236	.0241		324	.0101		423	.0049
	248	.0101		325	.0095		449	.0010
	249	.0094		334	.0053			
	256	.0053		335	.0049	$n = 34$	373	.1013
	257	.0048		356	.0010		374	.0982
	273	.0010					382	.0764
			$n = 30$	295	.1027		383	.0739
$n = 26$	226	.1039		296	.0990		394	.0506
	227	.0994		303	.0759		395	.0488
	233	.0750		304	.0730		412	.0252
	240	.0524		313	.0502		413	.0242
	241	.0497		314	.0481		432	.0103
	252	.0263		327	.0261		433	.0098
	253	.0247		328	.0249		446	.0051
	266	.0102		344	.0104		447	.0048
	267	.0095		345	.0098		474	.0010
	275	.0051		356	.0050			
	276	.0047		378	.0010	$n = 35$	394	.1006
	293	.0010					395	.0977
			$n = 31$	314	.1012		403	.0768
$n = 27$	243	.1010		315	.0977		404	.0744
	244	.0968		322	.0758		416	.0501
	249	.0776		323	.0730		417	.0484
	250	.0741		332	.0512		434	.0257
	258	.0502		333	.0491		435	.0247
	259	.0477		348	.0251		456	.0100
	270	.0260		349	.0239		470	.0051
				365	.0105		471	.0048
				366	.0099		499	.0010

(continued)

TABLE A.4. *(continued)*

	x	$P_0\{T^+ \geq x\}$		x	$P_0\{T^+ \geq x\}$		x	$P_0\{T^+ \geq x\}$
$n = 36$	415	.1015	$n = 40$	555	.0257	$n = 44$	694	.0097
	416	.0987		556	.0249		713	.0051
	425	.0761		582	.0100		714	.0049
	426	.0739		599	.0051		755	.0010
	438	.0505		600	.0049	$n = 45$	632	.1001
	439	.0489		635	.0010		633	.0981
	457	.0258	$n = 41$	530	.1009		645	.0765
	458	.0248		531	.0986		646	.0749
	480	.0100		542	.0759		663	.0510
	480	.0100		543	.0740		664	.0498
	495	.0050		558	.0501		691	.0251
	525	.0010		559	.0488		692	.0244
$n = 37$	437	.1011		581	.0256		722	.0101
	438	.0985		582	.0248		723	.0098
	447	.0767		609	.0100		786	.0010
	448	.0745		627	.0050	$n = 46$	658	.1016
	461	.0503		664	.0010		659	.0997
	462	.0487	$n = 42$	555	.1000		673	.0753
	481	.0254		567	.0759		674	.0737
	482	.0245		568	.0741		691	.0508
	504	.0103		583	.0509		692	.0497
	505	.0099		584	.0496		719	.0256
	520	.0050		608	.0252		720	.0249
	552	.0010		609	.0245		752	.0101
$n = 38$	459	.1022		636	.0102		753	.0098
	460	.0996		637	.0098		774	.0050
	470	.0763		655	.0051		818	.0010
	471	.0742		656	.0049	$n = 47$	686	.1003
	484	.0509		694	.0010		687	.0984
	485	.0493	$n = 43$	580	.1003		700	.0764
	505	.0256		581	.0982		701	.0749
	506	.0247		593	.0752		720	.0501
	529	.0104		594	.0735		721	.0490
	530	.0099		609	.0511		749	.0251
	546	.0050		610	.0498		750	.0245
	579	.0010		635	.0252		782	.0102
$n = 39$	482	.1020		636	.0245		783	.0099
	483	.0996		664	.0102		805	.0050
	494	.0751		665	.0098		851	.0010
	495	.0731		684	.0050	$n = 48$	714	.1000
	508	.0507		724	.0010		729	.0754
	509	.0493	$n = 44$	605	.1017		730	.0739
	530	.0254		606	.0996		749	.0500
	531	.0246		619	.0754		779	.0251
	555	.0103		620	.0737		780	.0244
	556	.0099		636	.0507		813	.0102
	572	.0051		637	.0495		814	.0099
	573	.0049		663	.0250		837	.0050
	607	.0010		693	.0101		884	.0010
$n = 40$	506	.1008						
	507	.0984						
	518	.0750						
	533	.0500						

TABLE A.4. *(continued)*

	x	$P_0\{T^+ \geq x\}$		x	$P_0\{T^+ \geq x\}$		x	$P_0\{T^+ \geq x\}$
$n = 49$	742	.1007	$n = 49$	845	.0100	$n = 50$	809	.0495
	743	.0989		869	.0050		840	.0253
	758	.0752		918	.0010		841	.0247
	759	.0738					877	.0101
	778	.0505	$n = 50$	771	.1005		878	.0098
	779	.0495		772	.0988		902	.0050
	809	.0253		787	.0758		953	.0010
	810	.0247		788	.0744			
				808	.0506			

Adapted, in part, from Table C of C. H. Kraft and C. van Eeden, *A Nonparametric Introduction to Statistics*, Macmillan, New York (1968), with the permission of the authors and the publisher, and, in part, from Table II of F. Wilcoxon, S. K. Katti, and R. A. Wilcox, Critical values and probability levels for the Wilcoxon rank sum test and the Wilcoxon signed rank test, in: *Selected Tables in Mathematical Statistics, Volume 1* (2nd printing with rev.), H. L. Harter and D. B. Owen (eds.), (1973), pp. 171–260, with the permission of the authors and the editors of *Selected Tables in Mathematical Statistics, Volume 1*.

TABLE A.5. Upper-Tail Probabilities for the Asymptotic Conditional Null Distribution of the Hollander Statistic, A_{obs}/n

For a given t, the table entry is $V(t) \approx P_0\{(A_{\text{obs}}/n) \geq t\}$. In particular, if $V(t) = \alpha$, then $t = a_\alpha$, the upper αth percentile for the asymptotic conditional null distribution of A_{obs}/n.

t	$V(t)$
.0581	.95
.0681	.90
.0765	.85
.0842	.80
.0918	.75
.0994	.70
.1073	.65
.1156	.60
.1245	.55
.1341	.50
.1447	.45
.1566	.40
.1701	.35
.1858	.30
.2048	.25
.2284	.20
.2597	.15
.3053	.10
.3870	.05
.4722	.025
.5889	.01
.6794	.005
.8940	.001

Adapted from J. A. Koziol, A test for bivariate symmetry based on the empirical distribution function, *Communications in Statistics—Theory and Methods* **8**, Marcel Dekker, Inc., N.Y. (1979): 207–221, with the permission of the author and the editor of *Communications in Statistics—Theory and Methods*, reprinted by courtesy of Marcel Dekker, Inc.

TABLE A.6. Upper-Tail Probabilities for the Null Distribution of the Wilcoxon Rank Sum W Statistic: $m = 3(1)10, n = 1(1)m; m = 11(1)20, n = 1(1)4$

Additional upper tail probabilities can be found in Wilcoxon, Katti, and Wilcox (1973).
For given m and n, the table entry for the point x is $P_0\{W \geq x\}$. Under these conditions, if x is such that $P_0\{W \geq x\} = \alpha$, then $\omega_\alpha = x$.

					$n = 1$					
x	$m = 3$	$m = 4$	$m = 5$	$m = 6$	$m = 7$	$m = 8$	$m = 9$	$m = 10$	$m = 11$
3	.500	.600							
4	.250	.400	.500	.571					
5		.200	.333	.429	.500	.556			
6			.167	.286	.375	.444	.500	.545	
7				.143	.250	.333	.400	.455	.500
8					.125	.222	.300	.364	.417
9						.111	.200	.273	.333
10							.100	.182	.250
11								.091	.167
12									.083

					$n = 1$					
x	$m = 12$	$m = 13$	$m = 14$	$m = 15$	$m = 16$	$m = 17$	$m = 18$	$m = 19$	$m = 20$
7	.538								
8	.462	.500	.533						
9	.385	.429	.467	.500	.529				
10	.308	.357	.400	.438	.471	.500	.526		
11	.231	.286	.333	.375	.412	.444	.474	.500	.524
12	.154	.214	.267	.312	.353	.389	.421	.450	.476
13	.077	.143	.200	.250	.294	.333	.368	.400	.429
14		.071	.133	.188	.235	.278	.316	.350	.381
15			.067	.125	.176	.222	.263	.300	.333
16				.062	.118	.167	.211	.250	.286
17					.059	.111	.158	.200	.238
18						.056	.105	.150	.190
19							.053	.100	.143
20								.050	.095
21									.048

					$n = 2$					
x	$m = 3$	$m = 4$	$m = 5$	$m = 6$	$m = 7$	$m = 8$	$m = 9$	$m = 10$	$m = 11$
6	.600								
7	.400	.600							
8	.200	.400	.571						
9	.100	.267	.429	.571					
10		.133	.286	.429	.556				
11		.067	.190	.321	.444	.556			
12			.095	.214	.333	.444	.545		
13			.048	.143	.250	.356	.455	.545	
14				.071	.167	.267	.364	.455	.538
15				.036	.111	.200	.291	.379	.462
16					.056	.133	.218	.303	.385
17					.028	.089	.164	.242	.321
18						.044	.109	.182	.256
19						.022	.073	.136	.205
20							.036	.091	.154
21							.018	.061	.115
22								.030	.077
23								.015	.051
24									.026
25									.013

A. TABLES AND CHARTS

TABLE A.6. *(continued)*

					$n = 2$				
x	$m = 12$	$m = 13$	$m = 14$	$m = 15$	$m = 16$	$m = 17$	$m = 18$	$m = 19$	$m = 20$
15	.533								
16	.462	.533							
17	.396	.467	.533						
18	.330	.400	.467	.529					
19	.275	.343	.408	.471	.529				
20	.220	.286	.350	.412	.471	.526			
21	.176	.238	.300	.360	.418	.474	.526		
22	.132	.190	.250	.309	.366	.421	.474	.524	
23	.099	.152	.208	.265	.320	.374	.426	.476	.524
24	.066	.114	.167	.221	.275	.327	.379	.429	.476
25	.044	.086	.133	.184	.235	.287	.337	.386	.433
26	.022	.057	.100	.147	.196	.246	.295	.343	.390
27	.011	.038	.075	.118	.163	.211	.258	.305	.351
28		.019	.050	.088	.131	.175	.221	.267	.312
29		.010	.033	.066	.105	.146	.189	.233	.277
30			.017	.044	.078	.117	.158	.200	.242
31			.008	.029	.059	.094	.132	.171	.212
32				.015	.039	.070	.105	.143	.182
33				.007	.026	.053	.084	.119	.156
34					.013	.035	.063	.095	.130
35					.007	.023	.047	.076	.108
36						.012	.032	.057	.087
37						.006	.021	.043	.069
38							.011	.029	.052
39							.005	.019	.039
40								.010	.026
41								.005	.017
42									.009
43									.004

					$n = 3$				
x	$m = 3$	$m = 4$	$m = 5$	$m = 6$	$m = 7$	$m = 8$	$m = 9$	$m = 10$	$m = 11$
11	500								
12	350	.571							
13	.200	.429							
14	.100	.314	.500						
15	.050	.200	.393	.548					
16		.114	.286	.452					
17		.057	.196	.357	.500				
18		.029	.125	.274	.417	.539			
19			.071	.190	.333	.461			
20			.036	.131	.258	.388	.500		
21			.018	.083	.192	.315	.432	.531	
22				.048	.133	.248	.364	.469	
23				.024	.092	.188	.300	.406	.500
24				.012	.058	.139	.241	.346	.442
25					.033	.097	.186	.287	.385
26					.017	.067	.141	.234	.330
27					.008	.042	.105	.185	.277
28						.024	.073	.143	.228
29						.012	.050	.108	.184
30						.006	.032	.080	.146
31							.018	.056	.113

(continued)

TABLE A.6. (continued)

					$n = 3$				
x	$m = 3$	$m = 4$	$m = 5$	$m = 6$	$m = 7$	$m = 8$	$m = 9$	$m = 10$	$m = 11$
32							.009	.038	.085
33							.005	.024	.063
34								.014	.044
35								.007	.030
36								.003	.019
37									.011
38									.005
39									.003

					$n = 3$				
x	$m = 12$	$m = 13$	$m = 14$	$m = 15$	$m = 16$	$m = 17$	$m = 18$	$m = 19$	$m = 20$
24	.527								
25	.473								
26	.420	.500							
27	.367	.450	.524						
28	.316	.400	.476						
29	.268	.352	.429	.500					
30	.224	.305	.384	.456	.521				
31	.182	.261	.338	.412	.479				
32	.147	.220	.296	.369	.438	.500			
33	.116	.182	.254	.327	.396	.461	.519		
34	.090	.148	.216	.287	.356	.421	.481		
35	.068	.120	.181	.249	.317	.382	.444	.500	
36	.051	.095	.150	.213	.280	.345	.407	.464	.517
37	.035	.073	.122	.180	.244	.308	.370	.429	.483
38	.024	.055	.099	.151	.211	.273	.335	.394	.449
39	.015	.041	.078	.125	.180	.239	.300	.359	.415
40	.009	.029	.060	.102	.152	.208	.267	.325	.382
41	.004	.020	.046	.082	.127	.179	.235	.293	.349
42	.002	.012	.034	.065	.105	.153	.206	.262	.317
43		.007	.024	.050	.086	.129	.178	.232	.286
44		.004	.016	.038	.069	.108	.153	.204	.257
45		.002	.010	.028	.055	.089	.131	.178	.229
46			.006	.020	.042	.073	.111	.154	.202
47			.003	.013	.032	.059	.092	.132	.177
48			.001	.009	.024	.046	.077	.113	.155
49				.005	.017	.036	.062	.095	.134
50				.002	.011	.027	.050	.080	.115
51				.001	.007	.020	.040	.066	.098
52					.004	.014	.031	.054	.083
53					.002	.010	.023	.044	.069
54					.001	.006	.017	.034	.058
55						.004	.012	.027	.047
56						.002	.008	.020	.038
57						.001	.005	.015	.030
58							.003	.010	.023
59							.002	.007	.018
60							.001	.005	.013
61								.003	.009
62								.001	.006
63								.001	.004
64									.002
65									.001
66									.001

TABLE A.6. *(continued)*

				$n = 4$				
x	$m = 4$	$m = 5$	$m = 6$	$m = 7$	$m = 8$	$m = 9$	$m = 10$	$m = 11$
18	.557							
19	.443							
20	.343	.548						
21	.243	.452						
22	.171	.365	.543					
23	.100	.278	.457					
24	.057	.206	.381	.536				
25	.029	.143	.305	.464				
26	.014	.095	.238	.394	.533			
27		.056	.176	.324	.467			
28		.032	.129	.264	.404	.530		
29		.016	.086	.206	.341	.470		
30		.008	.057	.158	.285	.413	.527	
31			.033	.115	.230	.355	.473	
32			.019	.082	.184	.302	.420	.525
33			.010	.055	.141	.252	.367	.475
34			.005	.036	.107	.207	.318	.426
35				.021	.077	.165	.270	.377
36				.012	.055	.130	.227	.330
37				.006	.036	.099	.187	.286
38				.003	.024	.074	.152	.245
39					.014	.053	.120	.206
40					.008	.038	.094	.171
41					.004	.025	.071	.140
42					.002	.017	.053	
43						.010	.038	.089
44						.006	.027	.069
45						.003	.018	.052
46						.001	.012	.039
47							.007	.028
48							.004	.020
49							.002	.013
50							.001	.009
51								.005
52								.003
53								.001
54								.001

				$n = 4$				
x	$m = 12$	$m = 13$	$m = 14$	$m = 15$	$m = 16$	$m = 17$	$m = 18$	$m = 19$	$m = 20$
34	.524								
35	.476								
36	.431	.522							
37	.385	.478							
38	.342	.435	.521						
39	.299	.392	.479						
40	.260	.352	.439	.519					
41	.223	.312	.399	.481					
42	.190	.274	.360	.443	.518				
43	.158	.239	.323	.405	.482				
44	.131	.206	.287	.368	.446	.517			
45	.106	.175	.253	.332	.410	.483			
46	.085	.148	.221	.298	.375	.449	.516		

(continued)

TABLE A.6. (continued)

				$n = 4$					
x	$m = 12$	$m = 13$	$m = 14$	$m = 15$	$m = 16$	$m = 17$	$m = 18$	$m = 19$	$m = 20$
47	.066	.123	.191	.265	.341	.415	.484		
48	.052	.101	.164	.235	.308	.381	.451	.516	
49	.039	.082	.139	.205	.277	.349	.419	.484	
50	.029	.065	.116	.179	.247	.318	.387	.453	.515
51	.021	.051	.096	.154	.219	.287	.356	.422	.485
52	.015	.039	.079	.131	.192	.258	.326	.392	.455
53	.010	.030	.063	.110	.168	.231	.297	.363	.426
54	.007	.022	.051	.092	.145	.205	.269	.334	.397
55	.004	.016	.040	.076	.124	.181	.242	.306	.368
56	.002	.011	.031	.062	.106	.158	.217	.279	.341
57	.001	.008	.023	.050	.089	.138	.193	.253	.314
58	.001	.005	.017	.040	.074	.119	.171	.228	.288
59		.003	.012	.031	.061	.101	.150	.205	.262
60		.002	.009	.024	.050	.086	.131	.183	.239
61		.001	.006	.018	.040	.072	.113	.162	.216
62		.000	.004	.014	.032	.060	.098	.143	.194
63			.002	.010	.025	.049	.083	.125	.174
64			.001	.007	.019	.040	.070	.109	.155
65			.001	.005	.015	.032	.059	.094	.137
66			.000	.003	.011	.026	.049	.081	.120
67				.002	.008	.020	.040	.069	.105
68				.001	.006	.016	.033	.058	.091
69				.001	.004	.012	.027	.049	.079
70				.000	.002	.009	.021	.041	.067
71					.001	.006	.017	.033	.057
72					.001	.005	.013	.027	.048
73					.000	.003	.010	.022	.041
74					.000	.002	.007	.018	.034
75						.001	.005	.014	.028
76						.001	.004	.011	.023
77						.000	.002	.008	.018
78						.000	.002	.006	.015
79							.001	.004	.011
80							.001	.003	.009
81							.000	.002	.007
82							.000	.001	.005
83								.001	.004
84								.000	.003
85								.000	.002
86								.000	.001
87									.001
88									.000
89									.000
90									.000

TABLE A.6. *(continued)*

			$n = 5$			
x	$m = 5$	$m = 6$	$m = 7$	$m = 8$	$m = 9$	$m = 10$
28	.500					
29	.421					
30	.345	.535				
31	.274	.465				
32	.210	.396				
33	.155	.331	.500			
34	.111	.268	.438			
35	.075	.214	.378	.528		
36	.048	.165	.319	.472		
37	.028	.123	.265	.416		
38	.016	.089	.216	.362	.500	
39	.008	.063	.172	.311	.449	
40	.004	.041	.134	.262	.399	.523
41		.026	.101	.218	.350	.477
42		.015	.074	.177	.303	.430
43		.009	.053	.142	.259	.384
44		.004	.037	.111	.219	.339
45		.002	.024	.085	.182	.297
46			.015	.064	.149	.257
47			.009	.047	.120	.220
48			.005	.033	.095	.185
49			.003	.023	.073	.155
50			.001	.015	.056	.127
51				.009	.041	.103
52				.005	.030	.082
53				.003	.021	.065
54				.002	.014	.050
55				.001	.009	.038
56					.006	.028
57					.003	.020
58					.002	.014
59					.001	.010
60					.000	.006
61						.004
62						.002
63						.001
64						.001
65						.000

(continued)

TABLE A.6. (continued)

			n = 6		
x	m = 6	m = 7	m = 8	m = 9	m = 10
39	.531				
40	.469				
41	.409				
42	.350	.527			
43	.294	.473			
44	.242	.418			
45	.197	.365	.525		
46	.155	.314	.475		
47	.120	.267	.426		
48	.090	.223	.377	.523	
49	.066	.183	.331	.477	
50	.047	.147	.286	.432	
51	.032	.117	.245	.388	.521
52	.021	.090	.207	.344	.479
53	.013	.069	.172	.303	.437
54	.008	.051	.141	.264	.396
55	.004	.037	.114	.228	.356
56	.002	.026	.091	.194	.318
57	.001	.017	.071	.164	.281
58		.011	.054	.136	.246
59		.007	.041	.112	.214
60		.004	.030	.091	.184
61		.002	.021	.072	.157
62		.001	.015	.057	.132
63		.001	.010	.044	.110
64			.006	.033	.090
65			.004	.025	.074
66			.002	.018	.059
67			.001	.013	.047
68			.001	.009	.036
69			.000	.006	.028
70				.004	.021
71				.002	.016
72				.001	.011
73				.001	.008
74				.000	.005
75				.000	.004
76					.002
77					.001
78					.001
79					.000
80					.000
81					.000

TABLE A.6. (continued)

	n = 7					n = 8		
x	m = 7	m = 8	m = 9	m = 10	x	m = 8	m = 9	m = 10
53	.500				68	.520		
54	.451				69	.480		
55	.402				70	.439		
56	.355	.522			71	.399		
57	.310	.478			72	.360	.519	
58	.267	.433			73	.323	.481	
59	.228	.389			74	.287	.444	
60	.191	.347	.500		75	.253	.407	
61	.159	.306	.459		76	.221	.371	.517
62	.130	.268	.419		77	.191	.336	.483
63	.104	.232	.379	.519	78	.164	.303	.448
64	.082	.198	.340	.481	79	.139	.271	.414
65	.064	.168	.303	.443	80	.117	.240	.381
66	.049	.140	.268	.406	81	.097	.212	.348
67	.036	.116	.235	.370	82	.080	.185	.317
68	.027	.095	.204	.335	83	.065	.161	.286
69	.019	.076	.176	.300	84	.052	.138	.257
70	.013	.060	.150	.268	85	.041	.118	.230
71	.009	.047	.126	.237	86	.032	.100	.204
72	.006	.036	.105	.209	87	.025	.084	.180
73	.003	.027	.087	.182	88	.019	.069	.158
74	.002	.020	.071	.157	89	.014	.057	.137
75	.001	.014	.057	.135	90	.010	.046	.118
76	.001	.010	.045	.115	91	.007	.037	.102
77	.000	.007	.036	.097	92	.005	.030	.086
78		.005	.027	.081	93	.003	.023	.073
79		.003	.021	.067	94	.002	.018	.061
80		.002	.016	.054	95	.001	.014	.051
81		.001	.011	.044	96	.001	.010	.042
82		.001	.008	.035	97	.001	.008	.034
83		.000	.006	.028	98	.000	.006	.027
84		.000	.004	.022	99	.000	.004	.022
85			.003	.017	100	.000	.003	.017
86			.002	.012	101		.002	.013
87			.001	.009	102		.001	.010
88			.001	.007	103		.001	.008
89			.000	.005	104		.000	.006
90			.000	.003	105		.000	.004
91			.000	.002	106		.000	.003
92				.002	107		.000	.002
93				.001	108		.000	.002
94				.001	109			.001
95				.000	110			.001
96				.000	111			.000
97				.000	112			.000
98				.000	113			.000
					114			.000
					115			.000
					116			.000

(continued)

TABLE A.6. *(continued)*

	n = 9			n = 10
x	m = 9	m = 10	x	m = 10
86	.500		105	.515
87	.466		106	.485
88	.432		107	.456
89	.398		108	.427
80	.365	.516	109	.398
91	.333	.484	110	.370
92	.302	.452	111	.342
93	.273	.421	112	.315
94	.245	.390	113	.289
95	.218	.360	114	.264
96	.193	.330	115	.241
97	.170	.302	116	.218
98	.149	.274	117	.197
99	.129	.248	118	.176
100	.111	.223	119	.157
101	.095	.200	120	.140
102	.081	.178	121	.124
103	.068	.158	122	.109
104	.057	.139	123	.095
105	.047	.121	124	.083
106	.039	.106	125	.072
107	.031	.091	126	.062
108	.025	.078	127	.053
109	.020	.067	128	.045
110	.016	.056	129	.038
111	.012	.047	130	.032
112	.009	.039	131	.026
113	.007	.033	132	.022
114	.005	.027	133	.018
115	.004	.022	134	.014
116	.003	.017	135	.012
117	.002	.014	136	.009
118	.001	.011	137	.007
119	.001	.009	138	.006
120	.001	.007	139	.004
121	.000	.005	140	.003
122	.000	.004	141	.003
123	.000	.003	142	.002
124	.000	.002	143	.001
125	.000	.001	144	.001
126	.000	.001	145	.001
127		.001	146	.001
128		.000	147	.000
129		.000	148	.000
130		.000	149	.000
131		.000	150	.000
132		.000	151	.000
133		.000	152	.000
134		.000	153	.000
135		.000	154	.000
			155	.000

Adapted from Table B of C. H. Kraft and C. van Eeden, *A Nonparametric Introduction to Statistics*, Macmillan, New York (1968), with the permission of the authors and the publisher.

A. TABLES AND CHARTS

TABLE A.7. Selected Upper-Tail Probabilities for the Null Distribution of the Fligner-Policello $\hat{U}$ Statistic: $m = 3(1)12$, $n = 3(1)m$

For given m and n, the table entry for the point x is $P_0\{\hat{U} \geq x\}$. Under these conditions, if x is such that $P_0\{\hat{U} \geq x\} = \alpha$, then $u_\alpha = x$.

n	m	x	$P_0\{\hat{U} \geq x\}$	n	m	x	$P_0\{\hat{U} \geq x\}$
3	3	2.347	.100	4	5	1.500	.095
		∞^*	.050			2.160	.056
3	4	1.732	.114			3.265	.032
		3.273	.057			∞^*	.008
		∞^*	.029	4	6	1.434	.091
3	5	1.632	.089			2.247	.048
		2.324	.071			3.021	.024
		4.195	.036			6.899	.010
		∞^*	.018	4	7	1.428	.100
3	6	1.897	.083			2.104	.052
		2.912	.048			3.295	.021
		5.116	.024			4.786	.012
		∞^*	.012	4	8	1.371	.101
3	7	1.644	.092			2.162	.051
		2.605	.042			2.868	.024
		6.037	.017			4.252	.010
		∞^*	.008	4	9	1.434	.094
3	8	1.500	.097			2.057	.050
		2.777	.042			2.683	.025
		4.082	.024			4.423	.010
		6.957	.012	4	10	1.466	.099
3	9	1.575	.100			2.000	.050
		2.353	.050			2.951	.025
		3.566	.023			4.276	.010
		7.876	.009	4	11	1.448	.100
3	10	1.611	.101			2.067	.049
		2.553	.049			2.776	.026
		3.651	.025			4.017	.011
		8.795	.007	4	12	1.455	.100
3	11	1.638	.099			2.096	.049
		2.369	.055			2.847	.024
		3.503	.028			3.904	.010
		5.831	.011	5	5	1.447	.103
3	12	1.616	.101			2.063	.048
		2.449	.048			2.859	.028
		3.406	.024			7.187	.008
		5.000	.011	5	6	1.362	.102
4	4	1.586	.100			1.936	.056
		2.502	.057			2.622	.026
		4.483	.029			3.913	.011
		∞^*	.014				*(continued)*

TABLE A.7. (continued)

n	m	x	$P_0\{\hat{U} \geq x\}$	n	m	x	$P_0\{\hat{U} \geq x\}$
5	7	1.308	.100	6	11	1.320	.100
		1.954	.051			1.833	.050
		2.465	.025			2.337	.025
		4.246	.009			3.161	.010
5	8	1.378	.099	6	12	1.330	.101
		1.919	.048			1.835	.050
		2.556	.025			2.349	.026
		3.730	.009			3.151	.010
5	9	1.361	.099	7	7	1.333	.099
		1.893	.050			1.804	.050
		2.536	.025			2.331	.025
		3.388	.010			3.195	.010
5	10	1.361	.098	7	8	1.310	.100
		1.900	.049			1.807	.050
		2.496	.025			2.263	.025
		3.443	.011			3.088	.010
5	11	1.340	.100	7	9	1.320	.100
		1.891	.051			1.790	.051
		2.497	.025			2.287	.025
		3.435	.011			2.967	.010
5	12	1.369	.100	7	10	1.313	.100
		1.923	.049			1.776	.050
		2.479	.025			2.248	.025
		3.444	.010			3.002	.010
6	6	1.335	.104	7	11	1.302	.100
		1.860	.051			1.769	.050
		2.502	.028			2.240	.025
		3.712	.011			2.979	.010
6	7	1.326	.100	7	12	1.318	.101
		1.816	.050			1.787	.050
		2.500	.024			2.239	.025
		3.519	.010			2.929	.010
6	8	1.327	.100	8	8	1.295	.101
		1.796	.050			1.766	.050
		2.443	.025			2.251	.026
		3.230	.011			2.954	.010
6	9	1.338	.099	8	9	1.283	.100
		1.845	.050			1.765	.051
		2.349	.024			2.236	.026
		3.224	.010			2.925	.010
6	10	1.339	.100	8	10	1.284	.100
		1.829	.050			1.756	.050
		2.339	.025			2.209	.025
		3.164	.010			2.880	.010

TABLE A.7. (continued)

n	m	x	$P_0\{\hat{U} \geq x\}$	n	m	x	$P_0\{\hat{U} \geq x\}$
8	11	1.290	.100	10	10	1.295	.100
		1.746	.050			1.723	.050
		2.205	.025			2.161	.025
		2.856	.010			2.770	.010
8	12	1.293	.100	10	11	1.284	.100
		1.759	.050			1.726	.050
		2.198	.025			2.152	.025
		2.845	.010			2.733	.010
9	9	1.294	.101	10	12	1.284	.100
		1.744	.050			1.720	.050
		2.206	.025			2.144	.025
		2.857	.010			2.718	.010
9	10	1.304	.099	11	11	1.289	.100
		1.742	.050			1.716	.050
		2.181	.025			2.138	.025
		2.802	.010			2.705	.010
9	11	1.288	.100	11	12	1.290	.100
		1.744	.050			1.708	.050
		2.172	.025			2.127	.025
		2.798	.010			2.683	.010
9	12	1.299	.100	12	12	1.283	.100
		1.737	.050			1.708	.050
		2.172	.025			2.117	.025
		2.770	.010			2.661	.010

*For these cases, the numerator of the $\hat{U}$ statistic achieves its highest value and the denominator of $\hat{U}$ is zero. We have arbitrarily chosen to denote the value of $\hat{U}$ in such cases by $+\infty$.

Adapted from M. A. Fligner and G. E. Policello II, Robust rank procedures for the Behrens-Fisher problem, *Journal of the American Statistical Association* **76** (1981): 162–174. Reprinted with permission from the *Journal of the American Statistical Association*. Copyright ©(1981) by the American Statistical Association. All rights reserved.

TABLE A.8. Upper-Tail Probabilities for the Null Distribution of the Ansari-Bradley C Statistic: $2 \leq n \leq m$ with $(m + n) \leq 20$

For a given m and n, with $n \leq m$, the table entry for the point x is $P_0\{C \geq x\}$. Under these conditions, if x is such that $P_0\{C \geq x\} = \alpha$, then $c_\alpha = x$.

					$n = 2$				
x	m = 2	m = 3	m = 4	m = 5	m = 6	m = 7	m = 8	m = 9	m = 10
2	1.0000	1.0000	1.0000	1.0000	1.0000	1.0000	1.0000	1.0000	1.0000
3	.8333	.9000	.9333	.9524	.9643	.9722	.9778	.9818	.9848
4	.1667	.5000	.6667	.7619	.8214	.8611	.8889	.9091	.9242
5		.2000	.3333	.5238	.6429	.7222	.7778	.8182	.8485
6			.0667	.2381	.3571	.5000	.6000	.6727	.7273
7				.0952	.1786	.3056	.4000	.5091	.5909
8					.0357	.1389	.2222	.3273	.4091
9						.0556	.1111	.2000	.2727
10							.0222	.0909	.1515
11								.0364	.0758
12									.0152

(continued)

TABLE A.8. *(continued)*

				$n = 2$				
x	$m = 11$	$m = 12$	$m = 13$	$m = 14$	$m = 15$	$m = 16$	$m = 17$	$m = 18$
2	1.0000	1.0000	1.0000	1.0000	1.0000	1.0000	1.0000	1.0000
3	.9872	.9890	.9905	.9917	.9926	.9935	.9942	.9947
4	.9359	.9451	.9524	.9583	.9632	.9673	.9708	.9737
5	.8718	.8901	.9048	.9167	.9265	.9346	.9415	.9474
6	.7692	.8022	.8286	.8500	.8676	.8824	.8947	.9053
7	.6538	.7033	.7429	.7750	.8015	.8235	.8421	.8579
8	.5000	.5714	.6286	.6750	.7132	.7451	.7719	.7947
9	.3590	.4286	.5048	.5667	.6176	.6601	.6959	.7263
10	.2308	.2967	.3714	.4333	.5000	.5556	.6023	.6421
11	.1410	.1978	.2667	.3250	.3897	.4444	.5029	.5526
12	.0641	.1099	.1714	.2250	.2868	.3399	.3977	.4474
13	.0256	.0549	.1048	.1500	.2059	.2549	.3099	.3579
14		.0110	.0476	.0833	.1324	.1765	.2281	.2737
15			.0190	.0417	.0809	.1176	.1637	.2053
16				.0083	.0368	.0654	.1053	.1421
17					.0147	.0327	.0643	.0947
18						.0065	.0292	.0526
19							.0117	.0263
20								.0053

				$n = 3$				
x	$m = 3$	$m = 4$	$m = 5$	$m = 6$	$m = 7$	$m = 8$	$m = 9$	$m = 10$	$m = 11$
4	1.0000	1.0000	1.0000	1.0000	1.0000	1.0000	1.0000	1.0000	1.0000
5	.9000	.9429	.9643	.9762	.9833	.9879	.9909	.9930	.9945
6	.7000	.8286	.8929	.9286	.9500	.9636	.9727	.9790	.9835
7	.3000	.5714	.7143	.8095	.8667	.9030	.9273	.9441	.9560
8	.1000	.3429	.5000	.6548	.7500	.8182	.8636	.8951	.9176
9		.1429	.2857	.4643	.5833	.6909	.7636	.8182	.8571
10		.0286	.1071	.2857	.4167	.5455	.6364	.7168	.7747
11			.0357	.1429	.2500	.3939	.5000	.5979	.6703
12				.0595	.1333	.2606	.3636	.4755	.5604
13				.0119	.0500	.1455	.2364	.3497	.4396
14					.0167	.0727	.1364	.2413	.3297
15						.0303	.0727	.1503	.2253
16						.0061	.0273	.0839	.1429
17							.0091	.0420	.0824
18								.0175	.0440
19								.0035	.0165
20									.0055

			$n = 3$			
x	$m = 12$	$m = 13$	$m = 14$	$m = 15$	$m = 16$	$m = 17$
4	1.0000	1.0000	1.0000	1.0000	1.0000	1.0000
5	.9956	.9964	.9971	.9975	.9979	.9982
6	.9868	.9893	.9912	.9926	.9938	.9947
7	.9648	.9714	.9765	.9804	.9835	.9860
8	.9341	.9464	.9559	.9632	.9690	.9737
9	.8857	.9071	.9235	.9363	.9463	.9544
10	.8198	.8536	.8794	.8995	.9154	.9281
11	.7341	.7821	.8206	.8505	.8741	.8930
12	.6374	.6964	.7485	.7892	.8225	.8491
13	.5297	.6000	.6632	.7132	.7575	.7930

TABLE A.8. (continued)

			$n = 3$			
x	$m = 12$	$m = 13$	$m = 14$	$m = 15$	$m = 16$	$m = 17$
14	.4242	.5000	.5735	.6324	.6852	.7281
15	.3209	.4000	.4794	.5441	.6058	.6561
16	.2286	.3036	.3868	.4559	.5232	.5789
17	.1516	.2179	.2985	.3676	.4396	.5000
18	.0945	.1464	.2206	.2868	.3591	.4211
19	.0527	.0929	.1529	.2108	.2817	.3439
20	.0264	.0536	.1015	.1495	.2136	.2719
21	.0110	.0286	.0632	.1005	.1548	.2070
22	.0022	.0107	.0353	.0637	.1073	.1509
23		.0036	.0176	.0368	.0712	.1070
24			.0074	.0196	.0444	.0719
25			.0015	.0074	.0248	.0456
26				.0025	.0124	.0263
27					.0052	.0140
28					.0010	.0053
29						.0018

				$n = 4$				
x	$m = 4$	$m = 5$	$m = 6$	$m = 7$	$m = 8$	$m = 9$	$m = 10$	$m = 11$	$m = 12$
6	1.0000	1.0000	1.0000	1.0000	1.0000	1.0000	1.0000	1.0000	1.0000
7	.9857	.9921	.9952	.9970	.9980	.9986	.9990	.9993	.9995
8	.9286	.9603	.9762	.9848	.9899	.9930	.9950	.9963	.9973
9	.8000	.8889	.9333	.9576	.9717	.9804	.9860	.9897	.9923
10	.6286	.7778	.8571	.9091	.9394	.9580	.9700	.9780	.9835
11	.3714	.6032	.7333	.8242	.8788	.9161	.9401	.9560	.9670
12	.2000	.4286	.5810	.7152	.7980	.8573	.8961	.9238	.9429
13	.0714	.2619	.4190	.5818	.6889	.7762	.8342	.8769	.9066
14	.0143	.1349	.2667	.4424	.5677	.6783	.7542	.8154	.8582
15		.0476	.1429	.3030	.4323	.5650	.6593	.7385	.7951
16		.0159	.0667	.1939	.3111	.4503	.5554	.6520	.7225
17			.0238	.1061	.2020	.3357	.4446	.5546	.6374
18			.0048	.0515	.1212	.2378	.3407	.4564	.5473
19				.0182	.0606	.1538	.2458	.3590	.4527
20				.0061	.0283	.0923	.1658	.2711	.3626
21					.0101	.0490	.1039	.1934	.2775
22					.0020	.0238	.0599	.1319	.2049
23						.0084	.0300	.0821	.1418
24						.0028	.0140	.0484	.0934
25							.0050	.0256	.0571
26							.0010	.0125	.0330
27								.0044	.0165
28								.0015	.0077
29									.0027
30									.0005

		$n = 4$		
x	$m = 13$	$m = 14$	$m = 15$	$m = 16$
6	1.0000	1.0000	1.0000	1.0000
7	.9996	.9997	.9997	.9998
8	.9979	.9984	.9987	.9990
9	.9941	.9954	.9964	.9971
10	.9874	.9902	.9923	.9938

(continued)

TABLE A.8. *(continued)*

	$n = 4$			
x	$m = 13$	$m = 14$	$m = 15$	$m = 16$
11	.9748	.9804	.9845	.9876
12	.9563	.9660	.9732	.9785
13	.9286	.9444	.9561	.9649
14	.8908	.9144	.9324	.9459
15	.8408	.8742	.9002	.9197
16	.7811	.8245	.8599	.8867
17	.7101	.7647	.8101	.8448
18	.6319	.6967	.7528	.7961
19	.5471	.6209	.6873	.7391
20	.4613	.5412	.6166	.6764
21	.3761	.4588	.5413	.6078
22	.2979	.3791	.4654	.5368
23	.2261	.3033	.3896	.4632
24	.1655	.2353	.3189	.3922
25	.1151	.1755	.2531	.3236
26	.0765	.1258	.1953	.2609
27	.0471	.0856	.1450	.2039
28	.0277	.0556	.1042	.1552
29	.0147	.0340	.0712	.1133
30	.0071	.0196	.0470	.0803
31	.0025	.0098	.0289	.0541
32	.0008	.0046	.0170	.0351
33		.0016	.0090	.0215
34		.0003	.0044	.0124
35			.0015	.0062
36			.0005	.0029
37				.0010
38				.0002

	$n = 5$						
x	$m = 5$	$m = 6$	$m = 7$	$m = 8$	$m = 9$	$m = 10$	$m = 11$
9	1.0000	1.0000	1.0000	1.0000	1.0000	1.0000	1.0000
10	.9921	.9957	.9975	.9984	.9990	.9993	.9995
11	.9762	.9870	.9924	.9953	.9970	.9980	.9986
12	.9286	.9610	.9773	.9860	.9910	.9940	.9959
13	.8492	.9156	.9495	.9689	.9800	.9867	.9908
14	.7302	.8420	.9015	.9386	.9600	.9734	.9817
15	.5873	.7446	.8333	.8936	.9291	.9524	.9670
16	.4127	.6147	.7374	.8275	.8821	.9197	.9437
17	.2698	.4805	.6237	.7451	.8212	.8761	.9116
18	.1508	.3463	.5000	.6457	.7423	.8182	.8681
19	.0714	.2294	.3763	.5385	.6523	.7483	.8132
20	.0238	.1342	.2626	.4266	.5514	.6663	.7468
21	.0079	.0693	.1667	.3209	.4486	.5771	.6708
22		.0303	.0985	.2269	.3477	.4832	.5870
23		.0108	.0505	.1507	.2577	.3916	.5000
24		.0022	.0227	.0917	.1788	.3044	.4130
25			.0076	.0513	.1179	.2268	.3292
26			.0025	.0249	.0709	.1608	.2532
27				.0109	.0400	.1086	.1868
28				.0039	.0200	.0686	.1319
29				.0008	.0090	.0406	.0884
30					.0030	.0220	.0563

TABLE A.8. *(continued)*

			$n = 5$			
x	$m = 5$	$m = 6$	$m = 7$	$m = 8$	$m = 9$	$m = 10$	$m = 11$
31					.0010	.0107	.0330
32						.0047	.0183
33						.0017	.0092
34						.0003	.0041
35							.0014
36							.0005

	$n = 5$			
x	$m = 12$	$m = 13$	$m = 14$	$m = 15$
9	1.0000	1.0000	1.0000	1.0000
10	.9997	.9998	.9998	.9999
11	.9990	.9993	.9995	.9996
12	.9971	.9979	.9985	.9988
13	.9935	.9953	.9966	.9974
14	.9871	.9907	.9931	.9948
15	.9767	.9832	.9876	.9907
16	.9601	.9711	.9787	.9840
17	.9368	.9538	.9659	.9743
18	.9047	.9295	.9476	.9604
19	.8633	.8978	.9235	.9417
20	.8116	.8569	.8920	.9171
21	.7508	.8079	.8533	.8861
22	.6810	.7498	.8067	.8483
23	.6054	.6846	.7530	.8038
24	.5254	.6130	.6923	.7523
25	.4449	.5383	.6267	.6950
26	.3662	.4617	.5572	.6329
27	.2928	.3870	.4864	.5673
28	.2262	.3154	.4157	.5000
29	.1690	.2502	.3478	.4327
30	.1214	.1921	.2840	.3671
31	.0835	.1431	.2262	.3050
32	.0546	.1022	.1751	.2477
33	.0339	.0705	.1318	.1962
34	.0197	.0462	.0960	.1517
35	.0107	.0289	.0675	.1139
36	.0052	.0168	.0455	.0829
37	.0023	.0093	.0294	.0583
38	.0008	.0047	.0181	.0396
39	.0002	.0021	.0105	.0257
40		.0007	.0057	.0160
41		.0002	.0028	.0093
42			.0012	.0052
43			.0004	.0026
44			.0001	.0012
45				.0004
46				.0001

(continued)

TABLE A.8. *(continued)*

					$n = 6$				
x	$m=6$	$m=7$	$m=8$	$m=9$	$m=10$	$m=11$	$m=12$	$m=13$	$m=14$
12	1.0000	1.0000	1.0000	1.0000	1.0000	1.0000	1.0000	1.0000	1.0000
13	.9989	.9994	.9997	.9998	.9999	.9999	.9999	1.0000	1.0000
14	.9946	.9971	.9983	.9990	.9994	.9996	.9997	.9998	.9999
15	.9848	.9918	.9953	.9972	.9983	.9989	.9992	.9995	.9996
16	.9632	.9802	.9887	.9932	.9958	.9973	.9982	.9987	.9991
17	.9264	.9592	.9760	.9856	.9910	.9942	.9961	.9973	.9981
18	.8658	.9242	.9547	.9724	.9825	.9887	.9925	.9948	.9964
19	.7846	.8735	.9217	.9518	.9692	.9799	.9865	.9907	.9935
20	.6807	.8048	.8751	.9215	.9487	.9663	.9772	.9843	.9890
21	.5649	.7203	.8139	.8803	.9202	.9469	.9636	.9749	.9823
22	.4351	.6189	.7366	.8260	.8812	.9199	.9445	.9613	.9725
23	.3193	.5122	.6474	.7600	.8322	.8849	.9190	.9431	.9591
24	.2154	.4038	.5501	.6829	.7717	.8407	.8860	.9191	.9413
25	.1342	.3030	.4499	.5984	.7025	.7877	.8451	.8887	.9184
26	.0736	.2133	.3526	.5085	.6246	.7259	.7962	.8514	.8896
27	.0368	.1410	.2634	.4190	.5425	.6574	.7398	.8074	.8549
28	.0152	.0851	.1861	.3323	.4575	.5831	.6765	.7564	.8138
29	.0054	.0484	.1249	.2543	.3754	.5065	.6082	.6996	.7668
30	.0011	.0239	.0783	.1860	.2975	.4292	.5364	.6376	.7139
31		.0105	.0453	.1303	.2283	.3549	.4636	.5723	.6566
32		.0035	.0240	.0859	.1678	.2851	.3918	.5049	.5954
33		.0012	.0113	.0539	.1188	.2226	.3235	.4376	.5322
34			.0047	.0312	.0798	.1678	.2602	.3716	.4678
35			.0017	.0170	.0513	.1226	.2038	.3094	.4046
36			.0003	.0082	.0308	.0859	.1549	.2518	.3434
37				.0036	.0175	.0579	.1140	.2002	.2861
38				.0012	.0090	.0370	.0810	.1550	.2332
39				.0004	.0042	.0226	.0555	.1170	.1862
40					.0017	.0128	.0364	.0855	.1451
41					.0006	.0069	.0228	.0608	.1104
42					.0001	.0033	.0135	.0415	.0816
43						.0015	.0075	.0274	.0587
44						.0005	.0039	.0172	.0409
45						.0002	.0018	.0104	.0275
46							.0008	.0058	.0177
47							.0003	.0031	.0110
48							.0001	.0015	.0065
49								.0007	.0036
50								.0002	.0019
51								.0001	.0009
52									.0004
53									.0001
54									.0000

TABLE A.8. *(continued)*

				$n = 7$			
x	$m = 7$	$m = 8$	$m = 9$	$m = 10$	$m = 11$	$m = 12$	$m = 13$
16	1.0000	1.0000	1.0000	1.0000	1.0000	1.0000	1.0000
17	.9994	.9997	.9998	.9999	1.0000	1.0000	1.0000
18	.9983	.9991	.9995	.9997	.9998	.9999	.9999
19	.9948	.9972	.9984	.9991	.9994	.9996	.9998
20	.9878	.9935	.9963	.9978	.9987	.9992	.9995
21	.9744	.9862	.9921	.9954	.9972	.9982	.9988
22	.9534	.9744	.9851	.9912	.9946	.9966	.9978
23	.9196	.9549	.9734	.9841	.9901	.9937	.9959
24	.8730	.9270	.9559	.9734	.9833	.9893	.9930
25	.8106	.8878	.9306	.9574	.9729	.9826	.9885
26	.7348	.8375	.8965	.9354	.9583	.9730	.9820
27	.6463	.7748	.8523	.9059	.9381	.9595	.9727
28	.5507	.7021	.7981	.8685	.9118	.9415	.9602
29	.4493	.6194	.7336	.8221	.8782	.9181	.9435
30	.3537	.5324	.6608	.7676	.8374	.8889	.9223
31	.2652	.4435	.5820	.7052	.7887	.8532	.8958
32	.1894	.3577	.5000	.6368	.7333	.8111	.8637
33	.1270	.2777	.4180	.5637	.6714	.7626	.8258
34	.0804	.2075	.3392	.4888	.6050	.7085	.7822
35	.0466	.1478	.2664	.4139	.5353	.6494	.7332
36	.0256	.1005	.2019	.3421	.4647	.5869	.6795
37	.0122	.0648	.1477	.2753	.3950	.5220	.6219
38	.0052	.0393	.1035	.2154	.3286	.4568	.5616
39	.0017	.0221	.0694	.1633	.2667	.3925	.5000
40	.0006	.0115	.0441	.1199	.2113	.3311	.4384
41		.0053	.0266	.0847	.1626	.2735	.3781
42		.0022	.0149	.0576	.1218	.2213	.3205
43		.0008	.0079	.0375	.0882	.1749	.2668
44		.0002	.0037	.0233	.0619	.1350	.2178
45			.0016	.0136	.0417	.1014	.1742
46			.0005	.0075	.0271	.0742	.1363
47			.0002	.0038	.0167	.0526	.1042
48				.0017	.0099	.0361	.0777
49				.0007	.0054	.0239	.0565
50				.0003	.0028	.0152	.0398
51				.0001	.0013	.0092	.0273
52					.0006	.0053	.0180
53					.0002	.0029	.0115
54					.0001	.0015	.0070
55						.0007	.0041
56						.0003	.0022
57						.0001	.0012
58						.0000	.0005
59							.0002
60							.0001
61							.0000

(continued)

TABLE A.8. *(continued)*

			$n = 8$		
x	$m = 8$	$m = 9$	$m = 10$	$m = 11$	$m = 12$
20	1.0000	1.0000	1.0000	1.0000	1.0000
21	.9999	1.0000	1.0000	1.0000	1.0000
22	.9996	.9998	.9999	.9999	1.0000
23	.9989	.9994	.9997	.9998	.9999
24	.9974	.9986	.9992	.9996	.9997
25	.9941	.9969	.9983	.9990	.9994
26	.9885	.9938	.9965	.9980	.9988
27	.9789	.9886	.9935	.9962	.9977
28	.9643	.9804	.9887	.9934	.9960
29	.9428	.9680	.9813	.9889	.9932
30	.9133	.9504	.9704	.9823	.9890
31	.8737	.9262	.9551	.9728	.9830
32	.8246	.8947	.9344	.9598	.9745
33	.7650	.8549	.9075	.9423	.9629
34	.6970	.8069	.8738	.9199	.9477
35	.6212	.7508	.8328	.8918	.9281
36	.5413	.6877	.7847	.8578	.9038
37	.4587	.6184	.7296	.8174	.8742
38	.3788	.5457	.6686	.7710	.8392
39	.3030	.4714	.6031	.7189	.7986
40	.2350	.3983	.5347	.6621	.7528
41	.1754	.3281	.4653	.6015	.7022
42	.1263	.2636	.3969	.5386	.6476
43	.0867	.2055	.3314	.4746	.5898
44	.0572	.1557	.2704	.4113	.5302
45	.0357	.1139	.2153	.3500	.4698
46	.0211	.0807	.1672	.2925	.4102
47	.0115	.0548	.1262	.2394	.3524
48	.0059	.0358	.0925	.1919	.2978
49	.0026	.0221	.0656	.1503	.2472
50	.0011	.0131	.0449	.1150	.2014
51	.0004	.0072	.0296	.0856	.1608
52	.0001	.0037	.0187	.0621	.1258
53		.0017	.0113	.0437	.0962
54		.0007	.0065	.0298	.0719
55		.0002	.0035	.0196	.0523
56		.0001	.0017	.0124	.0371
57			.0008	.0075	.0255
58			.0003	.0043	.0170
59			.0001	.0023	.0110
60			.0000	.0012	.0068
61				.0006	.0040
62				.0002	.0023
63				.0001	.0012
64				.0000	.0006
65					.0003
66					.0001
67					.0000
68					.0000

TABLE A.8. (continued)

	n = 9				n = 10
x	m = 9	m = 10	m = 11	x	m = 10
25	1.0000	1.0000	1.0000	30	1.0000
26	1.0000	1.0000	1.0000	31	1.0000
27	.9999	.9999	1.0000	32	1.0000
28	.9996	.9998	.9999	33	.9999
29	.9991	.9995	.9997	34	.9998
30	.9981	.9990	.9995	35	.9996
31	.9963	.9980	.9989	36	.9992
32	.9932	.9964	.9980	37	.9984
33	.9882	.9937	.9964	38	.9971
34	.9805	.9894	.9940	39	.9951
35	.9695	.9831	.9903	40	.9920
36	.9540	.9741	.9849	41	.9874
37	.9332	.9618	.9773	42	.9808
38	.9062	.9453	.9669	43	.9718
39	.8724	.9240	.9532	44	.9597
40	.8313	.8972	.9355	45	.9440
41	.7833	.8646	.9133	46	.9239
42	.7283	.8259	.8862	47	.8993
43	.6677	.7813	.8538	48	.8694
44	.6025	.7310	.8160	49	.8344
45	.5346	.6759	.7731	50	.7940
46	.4654	.6166	.7251	51	.7486
47	.3975	.5548	.6729	52	.6986
48	.3323	.4916	.6173	53	.6449
49	.2717	.4287	.5593	54	.5881
50	.2167	.3673	.5000	55	.5296
51	.1687	.3092	.4407	56	.4704
52	.1276	.2552	.3827	57	.4119
53	.0938	.2064	.3271	58	.3551
54	.0668	.1632	.2749	59	.3014
55	.0460	.1262	.2269	60	.2514
56	.0305	.0952	.1840	61	.2060
57	.0195	.0700	.1462	62	.1656
58	.0118	.0500	.1138	63	.1306
59	.0068	.0347	.0867	64	.1007
60	.0037	.0232	.0645	65	.0761
61	.0019	.0150	.0468	66	.0560
62	.0009	.0093	.0331	67	.0403
63	.0004	.0056	.0227	68	.0282
64	.0001	.0031	.0151	69	.0192
65	.0000	.0017	.0097	70	.0126
66		.0008	.0060	71	.0080
67		.0004	.0036	72	.0049
68		.0002	.0020	73	.0029
69		.0001	.0011	74	.0016
70		.0000	.0005	75	.0008
71			.0003	76	.0004
72			.0001	77	.0002
73			.0000	78	.0001
74			.0000	79	.0000
				80	.0000

Computed by G. A. Mack on the Ohio State University IBM 370/165.

TABLE A.9. Selected Upper-Tail Probabilities for the Null Distribution of the Lepage D Statistic:

$$m = 2, n = 3(1)30; m = 3, n = 3(1)26; m = 4, n = 4(1)18;$$

$$m = 5, n = 5(1)13; m = 6, n = 6(1)10; m = 7, n = 7, 8$$

Additional upper-tail probabilities can be found in Lepage (1973).
For given m and n, the table entry for the point x is $P_0\{D \geq x\}$. Under these conditions, if x is such that $P_0\{D \geq x\} = \alpha$, then $d_\alpha = x$.

m	n	x	$P_0\{D \geq x\}$	m	n	x	$P_0\{D \geq x\}$
2	3	3.429	.2000	2	16	3.046	.1830
						4.611	.0915
2	4	4.366	.1333			7.310	.0261
						8.957	.0131
2	5	3.739	.1429				
		5.146	.0952	2	17	3.052	.1871
						3.983	.0994
2	6	4.200	.1429			6.109	.0468
		5.867	.0714			9.066	.0117
2	7	3.886	.1944				
		4.180	.0833	2	18	3.198	.2000
						4.158	.0842
2	8	2.804	.1778			6.367	.0421
		4.500	.0889			9.237	.0105
		6.895	.0444				
				2	19	4.397	.1000
2	9	3.240	.2000			6.576	.0381
		4.480	.0909			7.879	.0190
		7.226	.0364			9.325	.0095
2	10	3.202	.1818	2	20	3.360	.1732
		4.714	.0909			4.305	.0952
		7.633	.0303			6.799	.0346
						8.069	.0173
2	11	3.193	.1923			9.470	.0087
		4.607	.0897				
		7.868	.0256	2	21	3.266	.1937
						4.302	.0988
2	12	3.419	.1758			5.871	.0474
		4.875	.0659			8.201	.0158
		6.200	.0440			9.543	.0079
2	13	3.403	.1905	2	22	2.982	.1884
		4.860	.0857			4.397	.0942
		6.494	.0381			6.088	.0435
		8.361	.0190			8.364	.0145
						9.667	.0072
2	14	3.699	.1667				
		5.000	.0833	2	23	3.064	.1900
		6.812	.0333			4.430	.0967
		8.615	.0167			6.272	.0400
						8.476	.0133
2	15	3.252	.1912			9.728	.0067
		4.932	.0809				
		7.045	.0294	2	24	3.097	.1969
		8.751	.0147			4.511	.0985
						6.464	.0369
						8.615	.0123
						9.835	.0062

TABLE A.9. (continued)

m	n	x	$P_0\{D \geq x\}$	m	n	x	$P_0\{D \geq x\}$
2	25	3.070	.1937	3	8	3.204	.2000
		4.501	.0912			4.704	.0909
		6.626	.0342			6.172	.0424
		8.719	.0114			8.371	.0121
		9.887	.0057	3	9	4.234	.1000
2	26	3.203	.1799			5.979	.0364
		4.609	.0847			7.053	.0182
		5.896	.0476			9.059	.0091
		8.840	.0106	3	10	3.421	.1888
		9.981	.0053			4.564	.0944
2	27	3.244	.1921			5.496	.0455
		4.521	.0985			7.684	.0140
		5.395	.0468			9.592	.0070
		7.895	.0197	3	11	3.196	.1978
		8.924	.0099			4.214	.0879
2	28	2.947	.1885			5.317	.0440
		4.057	.0966			7.036	.0165
		5.438	.0460			10.146	.0055
		8.030	.0184	3	12	3.182	.1890
		9.034	.0092			4.347	.0989
2	29	3.112	.1892			5.717	.0462
		4.688	.0860			7.228	.0176
		6.376	.0387			8.787	.0088
		8.133	.0172	3	13	3.100	.2000
		9.108	.0086			4.343	.0893
2	30	3.039	.1935			5.298	.0500
		4.397	.0968			7.376	.0179
		5.757	.0484			9.290	.0071
		8.253	.0161	3	14	4.128	.1000
		9.206	.0081			5.746	.0485
3	3	3.857	.1000			7.202	.0162
3	4	3.289	.2000			9.693	.0059
		4.654	.0857	3	15	3.054	.1961
3	5	3.422	.1786			4.594	.0833
		5.129	.0714			5.380	.0466
		5.840	.0357			7.377	.0196
3	6	3.524	.1548			8.675	.0098
		4.724	.0833	3	16	3.198	.1930
		5.600	.0357			4.259	.0970
3	7	3.136	.2000			5.552	.0485
		4.481	.0833			6.785	.0196
		5.546	.0500			9.059	.0083
		7.656	.0167	3	17	3.170	.1860
						4.388	.0982
						5.206	.0491
						7.865	.0140
						9.450	.0070

(continued)

TABLE A.9. *(continued)*

m	n	x	$P_0\{D \geq x\}$	m	n	x	$P_0\{D \geq x\}$
3	18	3.040	.1962	4	5	3.766	.1984
		4.171	.0962			4.400	.0873
		5.606	.0414			5.771	.0397
		7.474	.0188			6.286	.0159
		9.773	.0060	4	6	3.324	.2000
3	19	4.297	.1000			4.597	.0952
		5.417	.0494			5.415	.0476
		6.888	.0195			6.750	.0190
		8.936	.0091			7.296	.0095
3	20	3.114	.1993	4	7	3.465	.2000
		4.180	.0994			4.645	.0939
		5.610	.0429			5.309	.0455
		7.305	.0152			6.922	.0152
		9.247	.0079			7.009	.0091
3	21	2.995	.1986	4	8	3.356	.1818
		4.331	.0988			4.358	.0970
		5.585	.0484			5.656	.0485
		7.129	.0198			7.543	.0162
		9.566	.0069			7.621	.0081
3	22	3.117	.1991	4	9	3.307	.1972
		4.241	.0935			4.450	.0993
		5.609	.0426			5.579	.0490
		7.199	.0170			7.068	.0154
		8.823	.0096			7.710	.0098
3	23	4.246	.1000	4	10	3.300	.1938
		5.480	.0500			4.205	.0959
		7.335	.0154			5.445	.0500
		8.331	.0092			6.820	.0200
3	24	3.073	.1976			8.125	.0080
		4.294	.0926	4	11	3.312	.1941
		5.500	.0472			4.395	.0989
		7.070	.0195			5.531	.0498
		9.385	.0075			6.836	.0198
3	25	3.097	.1972			8.111	.0095
		4.258	.0995	4	12	3.284	.1978
		5.489	.0488			4.370	.0934
		7.191	.0171			5.455	.0495
		8.757	.0098			6.939	.0187
3	26	3.062	.1905			8.438	.0077
		4.354	.0911	4	13	3.309	.1958
		5.454	.0465			4.263	.0987
		7.212	.0200			5.464	.0492
		8.461	.0093			7.001	.0197
4	4	3.483	.2000			8.053	.0097
		5.333	.0571				
		5.600	.0286				

TABLE A.9. (continued)

m	n	x	$P_0\{D \geq x\}$	m	n	x	$P_0\{D \geq x\}$
4	14	3.259	.1993	5	9	3.351	.1998
		4.279	.0993			4.396	.0989
		5.521	.0497			5.444	.0500
		7.134	.0163			7.049	.0200
		8.397	.0078			7.760	.0100
4	15	3.251	.1981	5	10	3.424	.1971
		4.262	.0991			4.498	.0959
		5.462	.0498			5.447	.0486
		6.813	.0193			6.977	.0176
		8.086	.0095			8.044	.0090
4	16	3.156	.1990	5	11	3.335	.1978
		4.390	.0978			4.475	.0989
		5.607	.0495			5.410	.0495
		7.133	.0198			6.965	.0197
		8.367	.0078			8.140	.0096
4	17	3.225	.1978	5	12	3.357	.1973
		4.377	.0979			4.474	.0995
		5.554	.0446			5.503	.0398
		7.077	.0192			6.800	.0194
		8.118	.0099			8.037	.0095
4	18	3.098	.1985	5	13	3.315	.1977
		4.300	.0957			4.388	.0980
		5.516	.0495			5.449	.0497
		7.149	.0200			6.946	.0191
		8.418	.0085			8.236	.0098
5	5	3.333	.1984	6	6	3.317	.1991
		4.773	.0952			4.635	.0996
		5.335	.0476			5.769	.0476
		6.818	.0079			6.669	.0195
						7.628	.0087
5	6	3.604	.1948				
		4.576	.0974	6	7	3.453	.1987
		5.527	.0498			4.672	.0956
		6.551	.0195			5.659	.0490
		7.587	.0065			6.933	.0192
						7.391	.0099
5	7	3.319	.1970				
		4.562	.0960	6	8	3.359	.1951
		5.572	.0480			4.509	.0999
		6.595	.0177			5.671	.0480
		7.950	.0076			6.771	.0200
						7.750	.0100
5	8	3.431	.1943				
		4.503	.0995	6	9	3.349	.1958
		5.504	.0466			4.617	.0997
		6.875	.0194			5.561	.0494
		7.593	.0085			6.819	.0200
						7.841	.0098

(continued)

TABLE A.9. *(continued)*

m	n	x	$P_0\{D \geq x\}$
6	10	3.342	.1966
		4.471	.0989
		5.617	.0500
		6.903	.0200
		8.028	.0100
7	7	3.352	.1987
		4.511	.0985
		5.654	.0484
		6.846	.0198
		7.532	.0099

m	n	x	$P_0\{D \geq x\}$
7	8	3.415	.1967
		4.524	.0999
		5.643	.0497
		6.888	.0196
		7.772	.0099

Adapted from Y. Lepage, A table for a combined Wilcoxon Ansari-Bradley statistic, *Biometrika* **60** (1973): 113–116, by permission of Oxford University Press on behalf of the *Biometrika* Trustees.

TABLE A.10. Upper-Tail Probabilities for the Null Distribution of the Two-Sided Kolmogorov-Smirnov J statistic: $n = 1(1)20$, $m = 1(1)n$

Additional upper-tail probabilities can be found in Kim and Jennrich (1973).
For a given m and n, the table entry for the point x is $P_0\{J \geq x\}$. Under these conditions, if x is such that $P_0\{J \geq x\} = \alpha$, then $j_\alpha = x$.

\multicolumn{2}{c}{$m=1, n=1$}
x	$P_0\{J \geq x\}$
1	1.0000

$m=1, n=2$	
x	$P_0\{J \geq x\}$
2	.6667

$m=2, n=2$	
x	$P_0\{J \geq x\}$
2	.3333

$m=1, n=3$	
x	$P_0\{J \geq x\}$
3	.5000

$m=2, n=3$	
x	$P_0\{J \geq x\}$
6	.2000

$m=3, n=3$	
x	$P_0\{J \geq x\}$
3	.1000

$m=1, n=4$	
x	$P_0\{J \geq x\}$
4	.4000

$m=2, n=4$	
x	$P_0\{J \geq x\}$
4	.1333

$m=3, n=4$	
x	$P_0\{J \geq x\}$
9	.2286
12	.0571

$m=4, n=4$	
x	$P_0\{J \geq x\}$
3	.2286
4	.0286

$m=1, n=5$	
x	$P_0\{J \geq x\}$
5	.3333

$m=2, n=5$	
x	$P_0\{J \geq x\}$
8	.2857
10	.0952

$m=3, n=5$	
x	$P_0\{J \geq x\}$
12	.1429
15	.0357

$m=4, n=5$	
x	$P_0\{J \geq x\}$
15	.1429
16	.0794
20	.0159

TABLE A.10. (continued)

$m=5, n=5$	
x	$P_0\{J \geq x\}$
3	.3571
4	.0794
5	.0079

$m=1, n=6$	
x	$P_0\{J \geq x\}$
6	.2857

$m=2, n=6$	
x	$P_0\{J \geq x\}$
5	.2143
6	.0714

$m=3, n=6$	
x	$P_0\{J \geq x\}$
4	.3333
5	.0952
6	.0238

$m=4, n=6$	
x	$P_0\{J \geq x\}$
8	.1810
9	.0952
10	.0476
12	.0095

$m=5, n=6$	
x	$P_0\{J \geq x\}$
20	.1082
24	.0476
25	.0260
30	.0043

$m=6, n=6$	
x	$P_0\{J \geq x\}$
4	.1429
5	.0260
6	.0022

$m=1, n=7$	
x	$P_0\{J \geq x\}$
7	.2500

$m=2, n=7$	
x	$P_0\{J \geq x\}$
12	.1667
14	.0556

$m=3, n=7$	
x	$P_0\{J \geq x\}$
15	.1667
18	.0667
21	.0167

$m=4, n=7$	
x	$P_0\{J \geq x\}$
20	.1212
21	.0667
24	.0303
28	.0061

$m=5, n=7$	
x	$P_0\{J \geq x\}$
23	.1162
25	.0657
28	.0303
30	.0152
35	.0025

$m=6, n=7$	
x	$P_0\{J \geq x\}$
24	.1469
28	.0909
29	.0676
30	.0385
35	.0152
36	.0082
42	.0012

$m=7, n=7$	
x	$P_0\{J \geq x\}$
4	.2121
5	.0530
6	.0082
7	.0006

$m=1, n=8$	
x	$P_0\{J \geq x\}$
8	.2222

$m=2, n=8$	
x	$P_0\{J \geq x\}$
7	.1333
8	.0444

$m=3, n=8$	
x	$P_0\{J \geq x\}$
18	.1212
21	.0485
24	.0121

$m=4, n=8$	
x	$P_0\{J \geq x\}$
5	.2222
6	.0848
7	.0202
8	.0040

$m=5, n=8$	
x	$P_0\{J \geq x\}$
25	.1259
27	.0793
30	.0420
32	.0202
35	.0093
40	.0016

(continued)

TABLE A.10. *(continued)*

$m = 6, n = 8$	
x	$P_0\{J \geq x\}$
14	.1392
15	.0926
16	.0606
17	.0426
18	.0226
20	.0093
21	.0047
24	.0007

$m = 7, n = 8$	
x	$P_0\{J \geq x\}$
33	.1181
34	.0870
35	.0559
40	.0326
41	.0242
42	.0131
48	.0047
49	.0025
56	.0003

$m = 8, n = 8$	
x	$P_0\{J \geq x\}$
4	.2827
5	.0870
6	.0186
7	.0025
8	.0002

$m = 1, n = 9$	
x	$P_0\{J \geq x\}$
9	.2000

$m = 2, n = 9$	
x	$P_0\{J \geq x\}$
16	.1091
18	.0364

$m = 3, n = 9$	
x	$P_0\{J \geq x\}$
6	.2364
7	.0909
8	.0364
9	.0091

$m = 4, n = 9$	
x	$P_0\{J \geq x\}$
24	.1147
27	.0615
28	.0420
32	.0140
36	.0028

$m = 5, n = 9$	
x	$P_0\{J \geq x\}$
27	.1189
30	.0859
31	.0559
35	.0280
36	.0140
40	.0060
45	.0010

$m = 6, n = 9$	
x	$P_0\{J \geq x\}$
10	.1758
11	.0947
12	.0611
13	.0280
14	.0140
15	.0060
16	.0028
18	.0004

$m = 7, n = 9$	
x	$P_0\{J \geq x\}$
35	.1267
36	.0979
38	.0787

$m = 7, n = 9$	
x	$P_0\{J \geq x\}$
40	.0551
42	.0341
45	.0210
47	.0149
49	.0075
54	.0028
56	.0014
63	.0002

$m = 8, n = 9$	
x	$P_0\{J \geq x\}$
39	.1094
40	.0786
45	.0559
46	.0469
47	.0336
48	.0202
54	.0112
55	.0083
56	.0043
63	.0014
64	.0007

$m = 9, n = 9$	
x	$P_0\{J \geq x\}$
5	.1259
6	.0336
7	.0063
8	.0007

$m = 1, n = 10$	
x	$P_0\{J \geq x\}$
10	.1818

$m = 2, n = 10$	
x	$P_0\{J \geq x\}$
8	.1818
9	.0909
10	.0303

TABLE A.10. (continued)

$m = 3, n = 10$		$m = 7, n = 10$		$m = 1, n = 11$	
x	$P_0\{J \geq x\}$	x	$P_0\{J \geq x\}$	x	$P_0\{J \geq x\}$
21	.1399	50	.0140	11	.1667
24	.0699	53	.0095	$m = 2, n = 11$	
27	.0280	56	.0045	x	$P_0\{J \geq x\}$
30	.0070	60	.0017	18	.1538
$m = 4, n = 10$		63	.0008	20	.0769
x	$P_0\{J \geq x\}$	$m = 8, n = 10$		22	.0256
13	.1259	x	$P_0\{J \geq x\}$	$m = 3, n = 11$	
14	.0839	21	.1203	x	$P_0\{J \geq x\}$
15	.0460	22	.0951	24	.1099
16	.0300	23	.0704	27	.0549
18	.0100	24	.0499	30	.0220
20	.0020	25	.0373	33	.0055
$m = 5, n = 10$		26	.0303	$m = 4, n = 11$	
x	$P_0\{J \geq x\}$	27	.0207	x	$P_0\{J \geq x\}$
6	.1658	28	.0120	28	.1436
7	.0606	30	.0070	29	.0982
8	.0193	31	.0050	32	.0630
9	.0040	32	.0024	33	.0352
10	.0007	35	.0008	36	.0220
$m = 6, n = 10$		$m = 9, n = 10$		40	.0073
x	$P_0\{J \geq x\}$	x	$P_0\{J \geq x\}$	44	.0015
17	.1251	45	.1056	$m = 5, n = 11$	
18	.0922	50	.0839	x	$P_0\{J \geq x\}$
19	.0664	51	.0746	34	.1058
20	.0420	52	.0603	35	.0737
21	.0315	53	.0446	39	.0440
22	.0190	54	.0303	40	.0293
24	.0090	60	.0210	44	.0137
25	.0040	61	.0176	45	.0096
27	.0017	62	.0123	50	.0027
30	.0003	63	.0070	55	.0005
$m = 7, n = 10$		70	.0037	$m = 6, n = 11$	
x	$P_0\{J \geq x\}$	71	.0028	x	$P_0\{J \geq x\}$
39	.1165	72	.0014	37	.1256
40	.0869	80	.0004	38	.0916
42	.0716	$m = 10, n = 10$		42	.0645
43	.0541	x	$P_0\{J \geq x\}$		
46	.0361	5	.1678		
49	.0217	6	.0524		
		7	.0123		
		8	.0021		
		9	.0002		

(continued)

TABLE A.10. (continued)

$m = 6, n = 11$	
x	$P_0\{J \geq x\}$
43	.0478
44	.0297
48	.0213
49	.0133
54	.0060
55	.0027
60	.0011
66	.0002

$m = 7, n = 11$	
x	$P_0\{J \geq x\}$
42	.1053
44	.0828
45	.0709
48	.0489
49	.0357
52	.0244
55	.0143
56	.0111
59	.0062
63	.0028
66	.0011
70	.0005

$m = 8, n = 11$	
x	$P_0\{J \geq x\}$
47	.1012
48	.0815
50	.0653
53	.0468
55	.0329
56	.0275
58	.0201
61	.0132
64	.0074
66	.0045
69	.0031
72	.0014
77	.0005

$m = 9, n = 11$	
x	$P_0\{J \geq x\}$
50	.1134
52	.0893
54	.0696

$m = 9, n = 11$	
x	$P_0\{J \geq x\}$
55	.0577
57	.0501
59	.0391
61	.0279
63	.0187
66	.0136
68	.0111
70	.0074
72	.0041
77	.0023
79	.0016
81	.0008

$m = 10, n = 11$	
x	$P_0\{J \geq x\}$
56	.1061
57	.0915
58	.0747
59	.0578
60	.0432
66	.0339
67	.0303
68	.0242
69	.0173
70	.0112
77	.0075
78	.0064
79	.0044
80	.0024
88	.0012
89	.0009

$m = 11, n = 11$	
x	$P_0\{J \geq x\}$
5	.2115
6	.0747
7	.0207
8	.0044
9	.0007

$m = 1, n = 12$	
x	$P_0\{J \geq x\}$
12	.1538

$m = 2, n = 12$	
x	$P_0\{J \geq x\}$
10	.1319
11	.0659
12	.0220

$m = 3, n = 12$	
x	$P_0\{J \geq x\}$
8	.1890
9	.0879
10	.0440
11	.0176
12	.0044

$m = 4, n = 12$	
x	$P_0\{J \geq x\}$
8	.1121
9	.0484
10	.0165
11	.0055
12	.0011

$m = 5, n = 12$	
x	$P_0\{J \geq x\}$
35	.1312
36	.0963
38	.0795
40	.0537
43	.0326
45	.0210
48	.0100
50	.0068
55	.0019
60	.0003

$m = 6, n = 12$	
x	$P_0\{J \geq x\}$
7	.1146
8	.0463
9	.0149
10	.0041
11	.0008

TABLE A.10. *(continued)*

$m = 7, n = 12$	
x	$P_0\{J \geq x\}$
44	.1265
46	.0978
48	.0745
49	.0648
51	.0503
53	.0344
56	.0241
58	.0169
60	.0098
63	.0073
65	.0042
70	.0018
72	.0008

$m = 8, n = 12$	
x	$P_0\{J \geq x\}$
12	.1496
13	.0907
14	.0557
15	.0320
16	.0182
17	.0086
18	.0048
19	.0020
20	.0009

$m = 9, n = 12$	
x	$P_0\{J \geq x\}$
18	.1138
19	.0785
20	.0606
21	.0414
22	.0260
23	.0180
24	.0120
25	.0072
26	.0046
27	.0025
28	.0014
29	.0010

$m = 10, n = 12$	
x	$P_0\{J \geq x\}$
29	.1113
30	.0927

$m = 10, n = 12$	
x	$P_0\{J \geq x\}$
31	.0736
32	.0617
33	.0489
34	.0370
35	.0277
36	.0226
37	.0198
38	.0153
39	.0105
40	.0067
42	.0048
43	.0039
44	.0026
45	.0014
48	.0007

$m = 11, n = 12$	
x	$P_0\{J \geq x\}$
63	.1085
64	.0906
65	.0735
66	.0589
72	.0498
73	.0460
74	.0394
75	.0314
76	.0235
77	.0169
84	.0131
85	.0118
86	.0093
87	.0064
88	.0040
96	.0026
97	.0022
98	.0015
99	.0008

$m = 12, n = 12$	
x	$P_0\{J \geq x\}$
5	.2558
6	.0995
7	.0314
8	.0079
9	.0015
10	.0002

$m = 1, n = 13$	
x	$P_0\{J \geq x\}$
13	.1429

$m = 2, n = 13$	
x	$P_0\{J \geq x\}$
22	.1143
24	.0571
26	.0190

$m = 3, n = 13$	
x	$P_0\{J \geq x\}$
27	.1250
30	.0714
33	.0357
36	.0143
39	.0036

$m = 4, n = 13$	
x	$P_0\{J \geq x\}$
32	.1328
35	.0891
36	.0664
39	.0378
40	.0294
44	.0126
48	.0042
52	.0008

$m = 5, n = 13$	
x	$P_0\{J \geq x\}$
39	.1008
40	.0868
42	.0609
45	.0399
47	.0247
50	.0154
52	.0075
55	.0049
60	.0014
65	.0002

(continued)

TABLE A.10. (continued)

\multicolumn{2}{c}{$m = 6, n = 13$}		\multicolumn{2}{c}{$m = 9, n = 13$}		\multicolumn{2}{c}{$m = 11, n = 13$}
x	$P_0\{J \geq x\}$		x	$P_0\{J \geq x\}$		x	$P_0\{J \geq x\}$
42	.1151		56	.1153		77	.0390
46	.0862		59	.0980		78	.0341
47	.0709		60	.0808		80	.0310
48	.0509		63	.0645		82	.0259
52	.0341		64	.0556		84	.0201
53	.0294		65	.0423		86	.0146
54	.0195		68	.0374		88	.0106
59	.0106		69	.0284		91	.0085
60	.0069		72	.0220		93	.0075
65	.0029		73	.0178		95	.0058
66	.0021		77	.0119		97	.0039
72	.0005		78	.0079		99	.0024
			81	.0065		104	.0016
\multicolumn{2}{c}{$m = 7, n = 13$}		82	.0048		106	.0013
			86	.0030		108	.0009
x	$P_0\{J \geq x\}$		90	.0015			
49	.1113		91	.0009		\multicolumn{2}{c}{$m = 12, n = 13$}
50	.0941						
51	.0729		\multicolumn{2}{c}{$m = 10, n = 13$}		x	$P_0\{J \geq x\}$
52	.0541					70	.1083
56	.0458		x	$P_0\{J \geq x\}$		71	.0913
57	.0364		61	.1145		72	.0771
58	.0248		64	.0943		78	.0681
63	.0166		65	.0774		79	.0643
64	.0120		67	.0710		80	.0574
65	.0068		68	.0586		81	.0488
70	.0049		70	.0492		82	.0397
71	.0029		71	.0426		83	.0311
77	.0012		74	.0328		84	.0243
78	.0005		77	.0243		91	.0204
			78	.0183		92	.0190
\multicolumn{2}{c}{$m = 8, n = 13$}		80	.0162		93	.0162
			81	.0133		94	.0126
x	$P_0\{J \geq x\}$		84	.0099		95	.0091
52	.1097		87	.0066		96	.0063
54	.0989		90	.0042		104	.0049
56	.0782		91	.0031		105	.0044
57	.0656		94	.0025		106	.0034
59	.0532		97	.0016		107	.0023
62	.0390		100	.0008		108	.0014
64	.0288					117	.0009
65	.0224		\multicolumn{2}{c}{$m = 11, n = 13$}			
67	.0192					\multicolumn{2}{c}{$m = 13, n = 13$}
70	.0124		x	$P_0\{J \geq x\}$			
72	.0087		66	.1106		x	$P_0\{J \geq x\}$
75	.0058		67	.0998		6	.1265
78	.0031		69	.0876		7	.0443
80	.0024		71	.0738		8	.0126
83	.0013		73	.0602		9	.0029
88	.0006		75	.0482		10	.0005

TABLE A.10. (continued)

$m = 1, n = 14$	
x	$P_0\{J \geq x\}$
14	.1333

$m = 2, n = 14$	
x	$P_0\{J \geq x\}$
12	.1000
13	.0500
14	.0167

$m = 3, n = 14$	
x	$P_0\{J \geq x\}$
30	.1029
33	.0588
36	.0294
39	.0118
42	.0029

$m = 4, n = 14$	
x	$P_0\{J \geq x\}$
18	.0159
19	.0719
20	.0523
21	.0301
22	.0229
24	.0098
26	.0033
28	.0007

$m = 5, n = 14$	
x	$P_0\{J \geq x\}$
41	.1084
42	.0789
45	.0666
46	.0475
50	.0303
51	.0191
55	.0115
56	.0057
60	.0036
65	.0010
70	.0002

$m = 6, n = 14$	
x	$P_0\{J \geq x\}$
23	.1131
24	.0863

$m = 6, n = 14$	
x	$P_0\{J \geq x\}$
25	.0661
26	.0531
27	.0373
28	.0255
29	.0217
30	.0139
32	.0077
33	.0049
35	.0021
36	.0014
39	.0004

$m = 7, n = 14$	
x	$P_0\{J \geq x\}$
7	.1762
8	.0827
9	.0331
10	.0118
11	.0034
12	.0008

$m = 8, n = 14$	
x	$P_0\{J \geq x\}$
28	.1151
29	.0910
30	.0727
31	.0564
32	.0464
33	.0381
34	.0280
35	.0201
36	.0176
37	.0134
38	.0086
40	.0058
41	.0040
42	.0021
44	.0016
45	.0009

$m = 9, n = 14$	
x	$P_0\{J \geq x\}$
62	.1006
63	.0821
66	.0706
67	.0583
70	.0455
71	.0418
72	.0320
75	.0262
76	.0214
80	.0149
81	.0108
84	.0081
85	.0071
89	.0043
90	.0030
94	.0019
98	.0010

$m = 10, n = 14$	
x	$P_0\{J \geq x\}$
33	.1071
34	.0913
35	.0758
36	.0623
37	.0498
38	.0424
39	.0339
40	.0265
41	.0225
42	.0164
43	.0148
44	.0108
45	.0083
46	.0066
48	.0043
49	.0027
50	.0023
51	.0016
53	.0010

$m = 11, n = 14$	
x	$P_0\{J \geq x\}$
71	.1060
73	.0901
74	.0802
76	.0684
77	.0575
79	.0512

(continued)

TABLE A.10. *(continued)*

$m = 11, n = 14$	
x	$P_0\{J \geq x\}$
82	.0410
84	.0325
85	.0301
87	.0242
88	.0202
90	.0174
93	.0131
96	.0094
98	.0068
99	.0061
101	.0049
104	.0037
107	.0024
110	.0014
112	.0010
115	.0008

$m = 12, n = 14$	
x	$P_0\{J \geq x\}$
38	.1015
39	.0870
40	.0733
41	.0615
42	.0525
43	.0445
44	.0389
45	.0324
46	.0257
47	.0199
48	.0157
49	.0136
50	.0125
51	.0104
52	.0079
53	.0056
54	.0039
56	.0031
57	.0027
58	.0021
59	.0014
60	.0008

$m = 13, n = 14$	
x	$P_0\{J \geq x\}$
77	.1111
78	.0972
84	.0885

$m = 13, n = 14$	
x	$P_0\{J \geq x\}$
85	.0847
86	.0777
87	.0688
88	.0590
89	.0492
90	.0404
91	.0334
98	.0295
99	.0280
100	.0249
101	.0209
102	.0166
103	.0126
104	.0096
112	.0081
113	.0075
114	.0064
115	.0049
116	.0034
117	.0023
126	.0018
127	.0016
128	.0012
129	.0008

$m = 14, n = 14$	
x	$P_0\{J \geq x\}$
6	.1549
7	.0590
8	.0188
9	.0049
10	.0010
11	.0002

$m = 1, n = 15$	
x	$P_0\{J \geq x\}$
15	.1250

$m = 2, n = 15$	
x	$P_0\{J \geq x\}$
24	.1471
26	.0882
28	.0441
30	.0147

$m = 3, n = 15$	
x	$P_0\{J \geq x\}$
10	.1618
11	.0858
12	.0490
13	.0245
14	.0098
15	.0025

$m = 4, n = 15$	
x	$P_0\{J \geq x\}$
37	.1166
40	.0857
41	.0588
44	.0418
45	.0243
48	.0181
52	.0077
56	.0026
60	.0005

$m = 5, n = 15$	
x	$P_0\{J \geq x\}$
9	.1095
10	.0519
11	.0233
12	.0088
13	.0027
14	.0008

$m = 6, n = 15$	
x	$P_0\{J \geq x\}$
16	.1336
17	.0875
18	.0659
19	.0404
20	.0279
21	.0163
22	.0102
23	.0058
24	.0035
25	.0015
26	.0010
28	.0003

TABLE A.10. (continued)

$m = 7, n = 15$			$m = 9, n = 15$			$m = 12, n = 15$	
x	$P_0\{J \geq x\}$		x	$P_0\{J \geq x\}$		x	$P_0\{J \geq x\}$
55	.1008		30	.0073		27	.1010
56	.0794		31	.0048		28	.0783
60	.0626		32	.0029		29	.0610
61	.0578		33	.0020		30	.0507
62	.0467		34	.0013		31	.0398
63	.0341		35	.0007		32	.0297
68	.0244					33	.0219
69	.0200		$m = 10, n = 15$			34	.0170
70	.0136		x	$P_0\{J \geq x\}$		35	.0131
75	.0085					36	.0096
76	.0075		14	.1181		37	.0068
77	.0048		15	.0774		38	.0051
83	.0024		16	.0498		39	.0035
84	.0015		17	.0296		40	.0025
90	.0006		18	.0181		41	.0018
			19	.0100		42	.0013
$m = 8, n = 15$			20	.0055		43	.0008
			21	.0028			
x	$P_0\{J \geq x\}$		22	.0015		$m = 13, n = 15$	
59	.1039		23	.0006		x	$P_0\{J \geq x\}$
60	.0856						
64	.0777		$m = 11, n = 15$			85	.1016
65	.0678		x	$P_0\{J \geq x\}$		87	.0882
66	.0544					89	.0766
67	.0415		75	.1041		90	.0679
72	.0332		76	.0987		91	.0646
73	.0278		77	.0839		92	.0599
74	.0205		79	.0762		94	.0541
75	.0144		80	.0683		96	.0470
80	.0123		83	.0564		98	.0395
81	.0096		84	.0479		100	.0325
82	.0061		87	.0398		102	.0265
88	.0040		88	.0322		104	.0222
89	.0028		90	.0284		105	.0201
90	.0015		91	.0265		107	.0189
96	.0011		94	.0205		109	.0165
97	.0006		95	.0175		111	.0136
			98	.0136		113	.0105
$m = 9, n = 15$			99	.0105		115	.0079
x	$P_0\{J \geq x\}$		102	.0088		117	.0061
			105	.0062		120	.0053
22	.1006		106	.0056		122	.0048
23	.0733		109	.0040		124	.0040
24	.0590		110	.0031		126	.0030
25	.0418		113	.0024		128	.0021
26	.0297		117	.0015		130	.0014
27	.0223		120	.0009		135	.0011
28	.0148					137	.0010
29	.0104						

(continued)

TABLE A.10. (continued)

$m = 14, n = 15$	
x	$P_0\{J \geq x\}$
91	.1068
92	.0998
93	.0908
94	.0807
95	.0702
96	.0601
97	.0511
98	.0441
105	.0402
106	.0386
107	.0355
108	.0311
109	.0262
110	.0214
111	.0170
112	.0138
120	.0122
121	.0117
122	.0104
123	.0086
124	.0067
125	.0049
126	.0037
135	.0031
136	.0029
137	.0024
138	.0018
139	.0013
140	.0008

$m = 15, n = 15$	
x	$P_0\{J \geq x\}$
6	.1844
7	.0755
8	.0262
9	.0077
10	.0018
11	.0004

$m = 1, n = 16$	
x	$P_0\{J \geq x\}$
16	.1176

$m = 2, n = 16$	
x	$P_0\{J \geq x\}$
13	.1307
14	.0784
15	.0392
16	.0131

$m = 3, n = 16$	
x	$P_0\{J \geq x\}$
33	.1156
36	.0722
39	.0413
42	.0206
45	.0083
48	.0021

$m = 4, n = 16$	
x	$P_0\{J \geq x\}$
10	.1280
11	.0702
12	.0338
13	.0144
14	.0062
15	.0021
16	.0004

$m = 5, n = 16$	
x	$P_0\{J \geq x\}$
45	.1168
48	.0882
49	.0798
50	.0602
54	.0410
55	.0302
59	.0183
60	.0136
64	.0068
65	.0055
70	.0021
75	.0006

$m = 6, n = 16$	
x	$P_0\{J \geq x\}$
26	.1053
27	.0860
28	.0686
29	.0512
30	.0421
31	.0313
32	.0212
33	.0186
34	.0124
36	.0076
37	.0043
39	.0026
40	.0011
42	.0008

$m = 7, n = 16$	
x	$P_0\{J \geq x\}$
57	.1073
59	.0940
61	.0766
63	.0598
64	.0481
66	.0440
68	.0348
70	.0250
73	.0183
75	.0147
77	.0098
80	.0063
82	.0054
84	.0034
89	.0017
91	.0011
96	.0004

$m = 8, n = 16$	
x	$P_0\{J \geq x\}$
8	.1256
9	.0579
10	.0242
11	.0088
12	.0028
13	.0007

TABLE A.10. (continued)

$m = 9, n = 16$	
x	$P_0\{J \geq x\}$
67	.1186
69	.0999
71	.0843
72	.0748
74	.0660
76	.0543
78	.0432
80	.0352
81	.0327
83	.0280
85	.0215
87	.0159
90	.0129
92	.0105
94	.0074
96	.0050
99	.0044
101	.0033
103	.0020
108	.0013
110	.0009

$m = 10, n = 16$	
x	$P_0\{J \geq x\}$
37	.1017
38	.0876
39	.0761
40	.0634
41	.0534
42	.0440
43	.0359
44	.0313
45	.0244
46	.0210
47	.0168
48	.0127
49	.0117
50	.0085
51	.0070
52	.0056
54	.0037
55	.0026
56	.0019
57	.0017
59	.0010

$m = 11, n = 16$	
x	$P_0\{J \geq x\}$
79	.1004
80	.0865
83	.0818
84	.0718
85	.0603
88	.0536
89	.0490
90	.0404
94	.0336
95	.0282
96	.0224
99	.0208
100	.0186
101	.0142
105	.0119
106	.0093
110	.0070
111	.0060
112	.0042
116	.0037
117	.0026
121	.0020
122	.0016
127	.0010

$m = 12, n = 16$	
x	$P_0\{J \geq x\}$
21	.1132
22	.0837
23	.0633
24	.0469
25	.0336
26	.0243
27	.0166
28	.0115
29	.0081
30	.0052
31	.0033
32	.0023
33	.0014
34	.0008

$m = 13, n = 16$	
x	$P_0\{J \geq x\}$
89	.1014
91	.0892
92	.0836

$m = 13, n = 16$	
x	$P_0\{J \geq x\}$
95	.0721
96	.0621
98	.0593
99	.0513
101	.0466
102	.0410
104	.0353
105	.0327
108	.0270
111	.0218
112	.0178
114	.0168
115	.0141
117	.0123
118	.0112
121	.0090
124	.0068
127	.0051
128	.0040
130	.0036
131	.0032
134	.0026
137	.0019
140	.0013
143	.0009

$m = 14, n = 16$	
x	$P_0\{J \geq x\}$
47	.1046
48	.0934
49	.0816
50	.0712
51	.0637
52	.0556
53	.0477
54	.0404
55	.0343
56	.0301
57	.0267
58	.0241
59	.0208
60	.0172
61	.0138
62	.0109
63	.0090
64	.0082
65	.0077
66	.0067
67	.0055

(continued)

TABLE A.10. *(continued)*

$m = 14, n = 16$			$m = 16, n = 16$			$m = 5, n = 17$	
x	$P_0\{J \geq x\}$		x	$P_0\{J \geq x\}$		x	$P_0\{J \geq x\}$
68	.0042		6	.2145		48	.1198
69	.0030		7	.0933		50	.0939
70	.0023		8	.0350		51	.0718
72	.0020		9	.0112		53	.0643
73	.0018		10	.0030		55	.0476
74	.0015		11	.0007		58	.0328
75	.0011					60	.0237
76	.0007		$m = 1, n = 17$			63	.0145
			x	$P_0\{J \geq x\}$		65	.0106
$m = 15, n = 16$			17	.1111		68	.0053
x	$P_0\{J \geq x\}$					70	.0043
100	.1041		$m = 2, n = 17$			75	.0016
101	.0933		x	$P_0\{J \geq x\}$		80	.0005
102	.0825		28	.1170		$m = 6, n = 17$	
103	.0723		30	.0702		x	$P_0\{J \geq x\}$
104	.0633		32	.0351		55	.1062
105	.0563		34	.0117		56	.0842
112	.0525					60	.0674
113	.0509		$m = 3, n = 17$			61	.0545
114	.0476		x	$P_0\{J \geq x\}$		62	.0403
115	.0430		34	.1193		66	.0325
116	.0378		36	.0982		67	.0245
117	.0322		39	.0614		68	.0164
118	.0270		42	.0351		72	.0141
119	.0224		45	.0175		73	.0096
120	.0191		48	.0070		78	.0057
128	.0175		51	.0018		79	.0033
129	.0169					84	.0019
130	.0155		$m = 4, n = 17$			85	.0009
131	.0135		x	$P_0\{J \geq x\}$		$m = 7, n = 17$	
132	.0112		43	.1069		x	$P_0\{J \geq x\}$
133	.0089		44	.0849		60	.1192
134	.0069		47	.0581		61	.0989
135	.0055		48	.0461		63	.0874
144	.0049		51	.0277		64	.0737
145	.0047		52	.0234		67	.0590
146	.0042		56	.0117		68	.0458
147	.0034		60	.0050		70	.0423
148	.0026		64	.0017		71	.0339
149	.0019		68	.0003		74	.0263
150	.0014					77	.0187
160	.0011					78	.0139
161	.0011					81	.0110
162	.0009					84	.0072
						85	.0047

TABLE A.10. (continued)

$m = 7, n = 17$	
x	$P_0\{J \geq x\}$
88	.0040
91	.0024
95	.0013
98	.0008

$m = 8, n = 17$	
x	$P_0\{J \geq x\}$
64	.1136
68	.0974
69	.0927
70	.0813
71	.0674
72	.0537
77	.0439
78	.0391
79	.0315
80	.0235
85	.0180
86	.0168
87	.0134
88	.0093
94	.0064
95	.0052
96	.0034
102	.0020
103	.0018
104	.0011
111	.0005

$m = 9, n = 17$	
x	$P_0\{J \geq x\}$
73	.1051
74	.0911
75	.0764
76	.0633
81	.0551
82	.0494
83	.0409
84	.0323
85	.0257
90	.0236
91	.0205
92	.0159
93	.0115
99	.0091
100	.0076
101	.0053

$m = 9, n = 17$	
x	$P_0\{J \geq x\}$
102	.0036
108	.0031
109	.0024
110	.0014
117	.0009

$m = 10, n = 17$	
x	$P_0\{J \geq x\}$
76	.1137
79	.0974
80	.0837
82	.0760
83	.0666
85	.0563
86	.0535
89	.0436
90	.0362
92	.0320
93	.0283
96	.0225
99	.0173
100	.0140
102	.0119
103	.0111
106	.0082
109	.0059
110	.0049
113	.0039
116	.0026
119	.0017
120	.0016
123	.0011
126	.0007

$m = 11, n = 17$	
x	$P_0\{J \geq x\}$
82	.1049
85	.0921
86	.0884
87	.0775
88	.0660
91	.0600
92	.0548
93	.0460
97	.0385
98	.0334
99	.0270

$m = 11, n = 17$	
x	$P_0\{J \geq x\}$
102	.0239
103	.0228
104	.0184
108	.0146
109	.0128
110	.0097
114	.0082
115	.0066
119	.0048
120	.0044
121	.0032
125	.0025
126	.0020
131	.0013
132	.0009

$m = 12, n = 17$	
x	$P_0\{J \geq x\}$
88	.1057
90	.0927
91	.0856
93	.0763
95	.0655
96	.0580
98	.0536
100	.0455
102	.0385
103	.0368
105	.0318
107	.0260
108	.0233
110	.0211
112	.0169
115	.0140
117	.0115
119	.0089
120	.0084
122	.0074
124	.0055
127	.0046
129	.0035
132	.0026
134	.0022
136	.0015
139	.0013
141	.0009

(continued)

TABLE A.10. *(continued)*

\multicolumn{2}{c}{$m = 13, n = 17$}		\multicolumn{2}{c}{$m = 14, n = 17$}		\multicolumn{2}{c}{$m = 15, n = 17$}
x	$P_0\{J \geq x\}$	x	$P_0\{J \geq x\}$	x	$P_0\{J \geq x\}$
93	.1013	125	.0178	157	.0027
96	.0913	126	.0153	159	.0021
97	.0810	128	.0141	161	.0016
100	.0712	131	.0115	163	.0011
101	.0642	134	.0091	165	.0008
102	.0553	136	.0072		
104	.0530	137	.0068	\multicolumn{2}{c}{$m = 16, n = 17$}
105	.0497	139	.0057	x	$P_0\{J \geq x\}$
106	.0422	140	.0049	108	.1068
109	.0384	142	.0045	109	.0959
110	.0327	145	.0036	110	.0857
113	.0282	148	.0026	111	.0767
114	.0246	151	.0019	112	.0699
117	.0203	153	.0015	119	.0661
118	.0187	154	.0013	120	.0645
119	.0150	156	.0012	121	.0611
122	.0142	159	.0010	122	.0564
123	.0114			123	.0509
126	.0100	\multicolumn{2}{c}{$m = 15, n = 17$}	124	.0450
127	.0083	x	$P_0\{J \geq x\}$	125	.0390
130	.0067	104	.1001	126	.0335
131	.0060	105	.0941	127	.0288
135	.0045	106	.0897	128	.0254
136	.0033	108	.0820	136	.0238
139	.0031	110	.0736	137	.0231
140	.0024	112	.0650	138	.0217
143	.0020	114	.0568	139	.0195
144	.0017	116	.0495	140	.0169
148	.0012	118	.0434	141	.0142
152	.0008	119	.0392	142	.0116
		120	.0378	143	.0095
\multicolumn{2}{c}{$m = 14, n = 17$}	121	.0358	144	.0080
x	$P_0\{J \geq x\}$	123	.0331	153	.0073
98	.1013	126	.0295	154	.0071
100	.0959	127	.0254	155	.0065
102	.0846	129	.0214	156	.0056
103	.0819	131	.0177	157	.0046
105	.0721	133	.0147	158	.0036
106	.0672	135	.0128	159	.0027
108	.0593	136	.0119	160	.0021
109	.0533	138	.0114	170	.0019
111	.0479	140	.0103	171	.0018
112	.0415	142	.0088	172	.0016
114	.0390	144	.0072	173	.0013
117	.0329	146	.0056	174	.0010
119	.0277	148	.0044		
120	.0266	150	.0035		
122	.0225	153	.0032		
123	.0205	155	.0030		

TABLE A.10. *(continued)*

\multicolumn{2}{c}{$m = 17, n = 17$}		\multicolumn{2}{c}{$m = 5, n = 18$}		\multicolumn{2}{c}{$m = 8, n = 18$}
x	$P_0\{J \geq x\}$		x	$P_0\{J \geq x\}$		x	$P_0\{J \geq x\}$

$m = 17, n = 17$	
x	$P_0\{J \geq x\}$
7	.1124
8	.0450
9	.0156
10	.0046
11	.0012
12	.0002

$m = 1, n = 18$	
x	$P_0\{J \geq x\}$
18	.1053

$m = 2, n = 18$	
x	$P_0\{J \geq x\}$
15	.1053
16	.0632
17	.0316
18	.0105

$m = 3, n = 18$	
x	$P_0\{J \geq x\}$
12	.1444
13	.0842
14	.0526
15	.0301
16	.0150
17	.0060
18	.0015

$m = 4, n = 18$	
x	$P_0\{J \geq x\}$
22	.1203
23	.0902
24	.0705
25	.0487
26	.0380
27	.0230
28	.0191
30	.0096
32	.0041
34	.0014
36	.0003

$m = 5, n = 18$	
x	$P_0\{J \geq x\}$
50	.1237
52	.0987
54	.0765
55	.0700
57	.0524
60	.0382
62	.0266
65	.0188
67	.0117
70	.0083
72	.0042
75	.0033
80	.0012
85	.0004

$m = 6, n = 18$	
x	$P_0\{J \geq x\}$
10	.1010
11	.0534
12	.0254
13	.0109
14	.0044
15	.0015
16	.0004

$m = 7, n = 18$	
x	$P_0\{J \geq x\}$
63	.1108
65	.0949
66	.0851
69	.0684
70	.0563
72	.0461
73	.0431
76	.0325
77	.0264
80	.0202
83	.0142
84	.0118
87	.0083
90	.0054
91	.0048
94	.0030
98	.0018
101	.0009

$m = 8, n = 18$	
x	$P_0\{J \geq x\}$
35	.1043
36	.0878
37	.0723
38	.0625
39	.0510
40	.0405
41	.0337
42	.0297
43	.0235
44	.0174
45	.0135
46	.0126
47	.0098
48	.0067
50	.0047
51	.0038
52	.0024
54	.0014
55	.0013
56	.0008

$m = 9, n = 18$	
x	$P_0\{J \geq x\}$
8	.1724
9	.0879
10	.0413
11	.0173
12	.0066
13	.0022
14	.0006

$m = 10, n = 18$	
x	$P_0\{J \geq x\}$
40	.1167
41	.0988
42	.0865
43	.0738
44	.0626
45	.0548
46	.0472
47	.0398
48	.0323
49	.0263
50	.0233
51	.0205

(continued)

TABLE A.10. *(continued)*

$m = 10, n = 18$	
x	$P_0\{J \geq x\}$
52	.0164
53	.0125
54	.0099
55	.0093
56	.0079
57	.0059
58	.0042
60	.0033
61	.0027
62	.0018
63	.0012
65	.0011
66	.0008

$m = 11, n = 18$	
x	$P_0\{J \geq x\}$
86	.1080
88	.0975
89	.0906
90	.0783
92	.0751
93	.0661
96	.0571
97	.0484
99	.0440
100	.0403
103	.0333
104	.0282
107	.0240
108	.0192
110	.0183
111	.0164
114	.0130
115	.0104
118	.0090
121	.0067
122	.0058
125	.0045
126	.0033
129	.0030
132	.0021
133	.0017
136	.0014
140	.0009

$m = 12, n = 18$	
x	$P_0\{J \geq x\}$
15	.1357
16	.0945
17	.0626
18	.0417
19	.0263
20	.0163
21	.0097
22	.0057
23	.0031
24	.0017
25	.0009

$m = 13, n = 18$	
x	$P_0\{J \geq x\}$
97	.1101
99	.0979
100	.0893
102	.0811
104	.0710
105	.0670
107	.0592
108	.0515
110	.0497
112	.0428
113	.0380
115	.0350
117	.0293
118	.0273
120	.0239
123	.0198
125	.0165
126	.0141
128	.0135
130	.0108
131	.0098
133	.0088
136	.0068
138	.0056
141	.0045
143	.0034
144	.0030
146	.0029
149	.0021
151	.0017
154	.0013
156	.0009

$m = 14, n = 18$	
x	$P_0\{J \geq x\}$
51	.1101
52	.0998
53	.0900
54	.0803
55	.0712
56	.0625
57	.0553
58	.0480
59	.0432
60	.0376
61	.0327
62	.0291
63	.0246
64	.0222
65	.0184
66	.0168
67	.0140
68	.0121
69	.0104
70	.0084
71	.0078
72	.0061
73	.0058
74	.0045
75	.0040
76	.0033
77	.0026
78	.0023
80	.0017
81	.0012
82	.0012
83	.0009

$m = 15, n = 18$	
x	$P_0\{J \geq x\}$
36	.1094
37	.0956
38	.0811
39	.0672
40	.0552
41	.0460
42	.0398
43	.0334
44	.0271
45	.0216
46	.0174
47	.0144
48	.0119
49	.0095
50	.0075

TABLE A.10. *(continued)*

$m = 15, n = 18$	
x	$P_0\{J \geq x\}$
51	.0059
52	.0047
53	.0036
54	.0028
55	.0022
56	.0017
57	.0014
58	.0010

$m = 16, n = 18$	
x	$P_0\{J \geq x\}$
57	.1017
58	.0930
59	.0841
60	.0753
61	.0670
62	.0597
63	.0537
64	.0482
65	.0434
66	.0395
67	.0351
68	.0306
69	.0262
70	.0224
71	.0194
72	.0173
73	.0159
74	.0148
75	.0131
76	.0112
77	.0092
78	.0075
79	.0061
80	.0052
81	.0049
82	.0047
83	.0043
84	.0036
85	.0029
86	.0022
87	.0017
88	.0014
90	.0012
91	.0012
92	.0010
93	.0008

$m = 17, n = 18$	
x	$P_0\{J \geq x\}$
117	.1002
118	.0913
119	.0846
126	.0809
127	.0793
128	.0759
129	.0711
130	.0654
131	.0592
132	.0528
133	.0466
134	.0409
135	.0362
136	.0328
144	.0311
145	.0304
146	.0289
147	.0266
148	.0238
149	.0207
150	.0177
151	.0149
152	.0126
153	.0111
162	.0104
163	.0101
164	.0095
165	.0085
166	.0073
167	.0060
168	.0048
169	.0039
170	.0032
180	.0030
181	.0029
182	.0026
183	.0023
184	.0018
185	.0014
186	.0010
187	.0008

$m = 18, n = 18$	
x	$P_0\{J \geq x\}$
7	.1324
8	.0560
9	.0207
10	.0067
11	.0018
12	.0004

$m = 1, n = 19$	
x	$P_0\{J \geq x\}$
19	.1000

$m = 2, n = 19$	
x	$P_0\{J \geq x\}$
30	.1429
32	.0952
34	.0571
36	.0286
38	.0095

$m = 3, n = 19$	
x	$P_0\{J \geq x\}$
39	.1091
42	.0727
45	.0455
48	.0260
51	.0130
54	.0052
57	.0013

$m = 4, n = 19$	
x	$P_0\{J \geq x\}$
48	.1016
49	.0768
52	.0592
53	.0411
56	.0316
57	.0192
60	.0158
64	.0079
68	.0034
72	.0011
76	.0002

$m = 5, n = 19$	
x	$P_0\{J \geq x\}$
55	.1020
56	.0822
57	.0630
60	.0570
61	.0431
65	.0309
66	.0217

(continued)

TABLE A.10. *(continued)*

\$m = 5, n = 19\$		\$m = 8, n = 19\$		\$m = 10, n = 19\$	
x	$P_0\{J \geq x\}$	x	$P_0\{J \geq x\}$	x	$P_0\{J \geq x\}$
70	.0152	72	.1100	84	.1022
71	.0095	74	.0971	85	.0893
75	.0066	76	.0817	90	.0814
76	.0034	77	.0781	91	.0755
80	.0026	79	.0651	92	.0666
85	.0010	80	.0568	93	.0566
		82	.0486	94	.0475
\$m = 6, n = 19\$		85	.0391	95	.0408
x	$P_0\{J \geq x\}$	87	.0310	100	.0387
		88	.0276	101	.0354
60	.1101	90	.0228	102	.0300
64	.0820	93	.0177	103	.0243
65	.0708	95	.0130	104	.0194
66	.0558	96	.0122	110	.0169
70	.0429	98	.0095	111	.0151
71	.0377	101	.0073	112	.0122
72	.0284	104	.0049	113	.0092
76	.0201	106	.0035	114	.0072
77	.0187	109	.0028	120	.0066
78	.0135	112	.0017	121	.0057
83	.0085	114	.0011	122	.0043
84	.0062	117	.0009	123	.0030
89	.0034			130	.0023
90	.0025	\$m = 9, n = 19\$		131	.0019
95	.0011	x	$P_0\{J \geq x\}$	132	.0013
96	.0009			133	.0008
		79	.1067		
\$m = 7, n = 19\$		80	.0917	\$m = 11, n = 19\$	
		81	.0778		
x	$P_0\{J \geq x\}$	86	.0681	x	$P_0\{J \geq x\}$
67	.1048	87	.0633	91	.1044
69	.0880	88	.0552	92	.0935
70	.0795	89	.0458	94	.0847
72	.0676	90	.0371	95	.0735
74	.0541	95	.0313	97	.0708
76	.0439	96	.0301	99	.0610
77	.0413	97	.0263	100	.0565
79	.0338	98	.0211	102	.0491
81	.0254	99	.0161	103	.0426
84	.0202	105	.0129	105	.0390
86	.0157	106	.0115	108	.0323
88	.0109	107	.0091	110	.0269
91	.0089	108	.0065	111	.0244
93	.0064	114	.0048	113	.0211
95	.0040	115	.0045	114	.0175
98	.0036	116	.0036	115	.0167
100	.0023	117	.0024	119	.0131
105	.0013	124	.0016	121	.0106
107	.0007	125	.0013	122	.0093
		126	.0008	124	.0082

TABLE A.10. (continued)

$m = 11, n = 19$	
x	$P_0\{J \geq x\}$
127	.0064
130	.0047
132	.0038
133	.0032
135	.0030
138	.0021
141	.0015
143	.0012
146	.0009

$m = 12, n = 19$	
x	$P_0\{J \geq x\}$
97	.1003
98	.0902
101	.0791
102	.0710
104	.0662
106	.0578
108	.0499
109	.0465
111	.0416
113	.0352
114	.0305
116	.0294
118	.0252
120	.0207
121	.0190
123	.0173
125	.0140
128	.0116
130	.0099
132	.0077
133	.0069
136	.0065
137	.0051
140	.0040
142	.0035
144	.0025
147	.0022
149	.0017
152	.0012
154	.0011
156	.0008

$m = 13, n = 19$	
x	$P_0\{J \geq x\}$
101	.1030
104	.0965
105	.0919
106	.0825
107	.0726
111	.0657
112	.0599
113	.0522
114	.0458
117	.0442
118	.0418
119	.0363
120	.0308
124	.0283
125	.0251
126	.0209
130	.0182
131	.0170
132	.0141
133	.0116
137	.0111
138	.0095
139	.0075
143	.0066
144	.0061
145	.0047
150	.0038
151	.0031
152	.0023
156	.0022
157	.0019
158	.0014
163	.0012
164	.0009

$m = 14, n = 19$	
x	$P_0\{J \geq x\}$
107	.1033
110	.0952
111	.0869
112	.0773
114	.0735
115	.0716
116	.0638
119	.0566
120	.0533
121	.0466
124	.0420
125	.0382
126	.0327
129	.0308
130	.0270
133	.0230
134	.0224

$m = 14, n = 19$	
x	$P_0\{J \geq x\}$
135	.0189
138	.0166
139	.0154
140	.0126
143	.0117
144	.0102
148	.0083
149	.0068
152	.0057
153	.0055
154	.0043
157	.0039
158	.0035
162	.0027
163	.0022
167	.0017
168	.0013
171	.0011
172	.0011
176	.0008

$m = 15, n = 19$	
x	$P_0\{J \geq x\}$
112	.1093
114	.1000
115	.0976
116	.0881
118	.0817
119	.0775
120	.0690
122	.0660
123	.0609
126	.0537
127	.0483
130	.0430
131	.0378
133	.0343
134	.0334
135	.0288
137	.0273
138	.0256
141	.0218
142	.0195
145	.0168
146	.0144
149	.0127
150	.0105
152	.0099
153	.0096

(continued)

TABLE A.10. *(continued)*

$m = 15, n = 19$		$m = 16, n = 19$		$m = 17, n = 19$	
x	$P_0\{J \geq x\}$	x	$P_0\{J \geq x\}$	x	$P_0\{J \geq x\}$
156	.0078	176	.0030	194	.0017
157	.0071	177	.0028	196	.0014
160	.0058	180	.0024	198	.0011
161	.0050	183	.0019	200	.0009
164	.0043	186	.0014		
165	.0034	189	.0011	$m = 18, n = 19$	
168	.0031	190	.0009	x	$P_0\{J \geq x\}$
171	.0024				
172	.0023	$m = 17, n = 19$		126	.1004
175	.0018	x	$P_0\{J \geq x\}$	133	.0968
176	.0016			134	.0951
179	.0012	124	.1045	135	.0917
180	.0010	126	.0953	136	.0869
		128	.0864	137	.0812
$m = 16, n = 19$		130	.0782	138	.0748
x	$P_0\{J \geq x\}$	132	.0710	139	.0681
		133	.0651	140	.0614
119	.1054	134	.0637	141	.0550
120	.0963	135	.0596	142	.0493
122	.0898	136	.0567	143	.0445
123	.0825	137	.0548	144	.0411
125	.0754	139	.0508	152	.0394
126	.0703	141	.0462	153	.0387
128	.0631	143	.0412	154	.0371
129	.0607	145	.0363	155	.0347
132	.0538	147	.0318	156	.0318
133	.0476	149	.0279	157	.0285
135	.0464	151	.0248	158	.0250
136	.0412	152	.0228	159	.0217
138	.0389	153	.0222	160	.0188
139	.0348	154	.0213	161	.0164
141	.0318	156	.0201	162	.0148
142	.0292	158	.0183	171	.0141
144	.0257	160	.0161	172	.0138
145	.0246	162	.0138	173	.0131
148	.0211	164	.0117	174	.0121
151	.0179	166	.0098	175	.0107
152	.0153	168	.0083	176	.0092
154	.0148	170	.0074	177	.0077
155	.0128	171	.0070	178	.0064
157	.0119	173	.0068	179	.0053
158	.0107	175	.0063	180	.0046
160	.0094	177	.0056	190	.0044
161	.0089	179	.0047	191	.0043
164	.0075	181	.0039	192	.0040
167	.0061	183	.0031	193	.0036
170	.0049	185	.0025	194	.0031
171	.0040	187	.0021	195	.0025
173	.0039	190	.0019	196	.0020
174	.0034	192	.0019	197	.0015

TABLE A.10. *(continued)*

$m = 18, n = 19$	
x	$P_0\{J \geq x\}$
198	.0013
209	.0012
210	.0011
211	.0010
212	.0009

$m = 19, n = 19$	
x	$P_0\{J \geq x\}$
7	.1532
8	.0681
9	.0267
10	.0092
11	.0028
12	.0007

$m = 1, n = 20$	
x	$P_0\{J \geq x\}$
19	.1095
20	.0952

$m = 2, n = 20$	
x	$P_0\{J \geq x\}$
16	.1299
17	.0866
18	.0519
19	.0260
20	.0087

$m = 3, n = 20$	
x	$P_0\{J \geq x\}$
40	.1107
42	.0949
45	.0632
48	.0395
51	.0226
54	.0113
57	.0045
60	.0011

$m = 4, n = 20$	
x	$P_0\{J \geq x\}$
12	.1383
13	.0866
14	.0501
15	.0265
16	.0132
17	.0066
18	.0028
19	.0009

$m = 5, n = 20$	
x	$P_0\{J \geq x\}$
11	.1441
12	.0848
13	.0469
14	.0253
15	.0123
16	.0053
17	.0021
18	.0008

$m = 6, n = 20$	
x	$P_0\{J \geq x\}$
32	.1005
33	.0819
34	.0672
35	.0573
36	.0447
37	.0348
38	.0303
39	.0225
40	.0161
41	.0148
42	.0106
44	.0068
45	.0048
47	.0027
48	.0020
50	.0009

$m = 7, n = 20$	
x	$P_0\{J \geq x\}$
71	.1013
72	.0852
73	.0707
77	.0631
78	.0543
79	.0434

$m = 7, n = 20$	
x	$P_0\{J \geq x\}$
80	.0347
84	.0324
85	.0268
86	.0200
91	.0156
92	.0123
93	.0085
98	.0068
99	.0050
100	.0031
105	.0027
106	.0018
112	.0010

$m = 8, n = 20$	
x	$P_0\{J \geq x\}$
19	.1156
20	.0869
21	.0615
22	.0441
23	.0304
24	.0211
25	.0135
26	.0092
27	.0055
28	.0037
29	.0021
30	.0013
31	.0007

$m = 9, n = 20$	
x	$P_0\{J \geq x\}$
82	.1052
84	.0952
86	.0830
88	.0707
90	.0601
91	.0534
93	.0491
95	.0422
97	.0347
99	.0281
100	.0241
102	.0230
104	.0198
106	.0157
108	.0119

(continued)

TABLE A.10. (continued)

\multicolumn{2}{c	}{$m = 9, n = 20$}	\multicolumn{2}{c	}{$m = 11, n = 20$}	\multicolumn{2}{c}{$m = 13, n = 20$}
x	$P_0\{J \geq x\}$	x	$P_0\{J \geq x\}$	x	$P_0\{J \geq x\}$
111	.0097	138	.0034	135	.0184
113	.0086	140	.0026	136	.0152
115	.0067	143	.0025	140	.0129
117	.0047	145	.0021	141	.0126
120	.0036	147	.0015	142	.0107
122	.0033	149	.0010	143	.0086
124	.0026	154	.0008	147	.0078
126	.0017			148	.0072
131	.0011	\multicolumn{2}{c	}{$m = 12, n = 20$}	149	.0057
133	.0009			154	.0046
		x	$P_0\{J \geq x\}$	155	.0039
\multicolumn{2}{c	}{$m = 10, n = 20$}	25	.1148	156	.0030
x	$P_0\{J \geq x\}$	26	.0906	160	.0026
9	.1218	27	.0724	161	.0025
10	.0623	28	.0555	152	.0020
11	.0290	29	.0429	167	.0015
12	.0124	30	.0330	168	.0013
13	.0048	31	.0249	169	.0009
14	.0017	32	.0183		
15	.0005	33	.0137	\multicolumn{2}{c}{$m = 14, n = 20$}
		34	.0100		
\multicolumn{2}{c	}{$m = 11, n = 20$}	35	.0070	x	$P_0\{J \geq x\}$
x	$P_0\{J \geq x\}$	36	.0052	56	.1063
94	.1091	37	.0036	57	.0947
96	.0964	38	.0024	58	.0860
98	.0855	39	.0017	59	.0768
99	.0780	40	.0011	60	.0698
100	.0727	41	.0008	61	.0627
101	.0703			62	.0554
103	.0624	\multicolumn{2}{c	}{$m = 13, n = 20$}	63	.0491
105	.0538	x	$P_0\{J \geq x\}$	64	.0441
107	.0459	107	.1044	65	.0389
109	.0398	108	.0993	66	.0336
110	.0368	109	.0899	67	.0305
112	.0338	110	.0798	68	.0275
114	.0292	114	.0723	69	.0234
116	.0241	115	.0669	70	.0204
118	.0198	116	.0591	71	.0187
120	.0171	117	.0519	72	.0159
121	.0164	120	.0488	73	.0133
123	.0148	121	.0475	74	.0123
126	.0123	122	.0426	75	.0108
127	.0096	123	.0366	76	.0088
129	.0076	127	.0323	77	.0076
132	.0067	128	.0304	78	.0071
134	.0059	129	.0262	79	.0058
136	.0046	130	.0221	80	.0047
		134	.0205	81	.0045
				82	.0038
				83	.0029

TABLE A.10. (continued)

\multicolumn{2}{c}{$m = 14, n = 20$}		\multicolumn{2}{c}{$m = 17, n = 20$}		\multicolumn{2}{c}{$m = 18, n = 20$}
x	$P_0\{J \geq x\}$	x	$P_0\{J \geq x\}$	x	$P_0\{J \geq x\}$
84	.0026	129	.1060	70	.0833
85	.0024	130	.0986	71	.0761
86	.0018	132	.0916	72	.0692
88	.0015	133	.0840	73	.0632
89	.0012	135	.0791	74	.0583
90	.0009	136	.0717	75	.0531
		138	.0694	76	.0479
\multicolumn{2}{c}{$m = 15, n = 20$}	140	.0625	77	.0428
x	$P_0\{J \geq x\}$	141	.0613	78	.0381
		143	.0551	79	.0341
24	.1019	144	.0527	80	.0310
25	.0790	146	.0475	81	.0284
26	.0611	147	.0443	82	.0262
27	.0461	149	.0403	83	.0243
28	.0347	150	.0366	84	.0220
29	.0261	152	.0340	85	.0194
30	.0190	153	.0302	86	.0169
31	.0136	155	.0291	87	.0145
32	.0099	158	.0254	88	.0125
33	.0070	160	.0221	89	.0110
34	.0049	161	.0216	90	.0101
35	.0034	163	.0189	91	.0095
36	.0023	164	.0178	92	.0089
37	.0015	166	.0158	93	.0081
38	.0010	167	.0144	94	.0071
39	.0006	169	.0132	95	.0060
		170	.0115	96	.0050
\multicolumn{2}{c}{$m = 16, n = 20$}	172	.0110	97	.0041
x	$P_0\{J \geq x\}$	175	.0094	98	.0035
		178	.0078	99	.0031
31	.1085	180	.0065	100	.0029
32	.0889	181	.0064	101	.0028
33	.0724	183	.0054	102	.0026
34	.0598	184	.0050	103	.0023
35	.0489	186	.0045	104	.0019
36	.0393	187	.0040	105	.0016
37	.0314	189	.0038	106	.0012
38	.0251	192	.0031	107	.0010
39	.0197	195	.0025		
40	.0154	198	.0020	\multicolumn{2}{c}{$m = 19, n = 20$}
41	.0121	200	.0016	x	$P_0\{J \geq x\}$
42	.0095	201	.0015		
43	.0072	203	.0013	143	.1037
44	.0054	204	.0012	144	.0979
45	.0041	206	.0011	145	.0914
46	.0032	209	.0009	146	.0845
47	.0023			147	.0776
48	.0017	\multicolumn{2}{c}{$m = 18, n = 20$}	148	.0707
49	.0012	x	$P_0\{J \geq x\}$	149	.0642
50	.0009			150	.0584
		67	.1072	151	.0537
		68	.0985	152	.0503
		69	.0904		

(continued)

TABLE A.10. *(continued)*

$m = 19, n = 20$		$m = 19, n = 20$		$m = 19, n = 20$	
x	$P_0\{J \geq x\}$	x	$P_0\{J \geq x\}$	x	$P_0\{J \geq x\}$
160	.0486	185	.0132	221	.0017
161	.0479	186	.0114	222	.0016
162	.0463	187	.0098	223	.0015
163	.0439	188	.0083	224	.0012
164	.0408	189	.0072	225	.0010
165	.0373	190	.0064		
166	.0335	200	.0061	$m = 20, n = 20$	
167	.0298	201	.0060	x	$P_0\{J \geq x\}$
168	.0263	202	.0058		
169	.0232	203	.0053	7	.1745
170	.0208	204	.0047	8	.0811
171	.0192	205	.0040	9	.0335
180	.0184	206	.0033	10	.0123
181	.0182	207	.0027	11	.0040
182	.0174	208	.0022	12	.0011
183	.0163	209	.0019	13	.0003
184	.0148	220	.0018		

Adapted from Table I of P. J. Kim and R. I. Jennrich, Tables of the exact sampling distribution of the two-sample Kolmogorov-Smirnov criterion, $D_{mn}, m \leq n$, in: *Selected Tables in Mathematical Statistics, Volume 1* (2nd printing with rev.), H. L. Harter and D. B. Owen (eds.), (1973), pp. 79–170, with the permission of the authors and the editors of *Selected Tables* in *Mathematical Statistics*, Volume 1.

A. TABLES AND CHARTS

TABLE A.11. Upper-Tail Probabilities for the Null Limiting Distribution of the Two-Sided Kolmogorov-Smirnov J^* Statistic

For a given s, the table entry is

$$Q(s) = 1 - \sum_{k=-\infty}^{\infty} (-1)^k e^{-2k^2 s^2} \approx P_0\{J^* \geq s\}.$$

s	.00	.01	.02	.03	.04	.05	.06	.07	.08	.09
0.3	1.0000	1.0000	1.0000	.9999	.9998	.9997	.9995	.9992	.9987	.9981
0.4	.9972	.9960	.9945	.9926	.9903	.9874	.9840	.9800	.9753	.9700
0.5	.9639	.9572	.9497	.9415	.9325	.9228	.9124	.9013	.8896	.8772
0.6	.8643	.8508	.8368	.8222	.8073	.7920	.7764	.7604	.7442	.7278
0.7	.7112	.6945	.6777	.6609	.6440	.6272	.6104	.5936	.5770	.5605
0.8	.5441	.5280	.5120	.4962	.4806	.4653	.4503	.4355	.4209	.4067
0.9	.3927	.3791	.3657	.3527	.3399	.3275	.3154	.3036	.2921	.2809
1.0	.2700	.2594	.2492	.2392	.2296	.2202	.2110	.2024	.1939	.1857
1.1	.1777	.1701	.1626	.1555	.1486	.1420	.1356	.1294	.1235	.1177
1.2	.1123	.1070	.1019	.0970	.0924	.0879	.0836	.0794	.0755	.0717
1.3	.0681	.0646	.0613	.0582	.0551	.0522	.0495	.0469	.0444	.0420
1.4	.0397	.0375	.0354	.0335	.0316	.0298	.0288	.0266	.0250	.0236
1.5	.0222	.0209	.0197	.0185	.0174	.0164	.0154	.0145	.0136	.0127
1.6	.0120	.0112	.0105	.0098	.0092	.0086	.0081	.0076	.0071	.0066
1.7	.0062	.0058	.0054	.0050	.0047	.0044	.0041	.0038	.0035	.0033
1.8	.0031	.0029	.0027	.0025	.0023	.0021	.0020	.0019	.0017	.0016
1.9	.0015	.0014	.0013	.0012	.0011	.0010	.0009	.0009	.0008	.0007
2.0	.0007	.0006	.0006	.0005	.0005	.0004	.0004	.0004	.0004	.0003
2.1	.0003	.0003	.0003	.0002	.0002	.0002	.0002	.0002	.0001	.0001
2.2	.0001	.0001	.0001	.0001	.0001	.0001	.0001	.0001	.0001	.0001
2.3	.0001	.0000	.0000	.0000	.0000	.0000	.0000	.0000	.0000	.0000

Adapted from Table VIII of M. Fisz, *Probability Theory and Mathematical Statistics* (3rd. ed.), Wiley, New York (1963). Copyright ©John Wiley & Sons Limited. Reprinted with permission.

TABLE A.12. Selected Upper-Tail Probabilities for the Null Distribution of the Kruskal-Wallis H Statistic:

$k = 3, n_1 = 1(1)6, n_2 = n_1(1)6, 2 \leq n_3 = n_2(1)6; n_1 = n_2 = n_3 = 7, 8;$
$k = 4, n_1 = 1(1)4, n_2 = n_1(1)4, n_3 = n_2(1)4, 2 \leq n_4 = n_3(1)4;$
$k = 5, n_1 = 1(1)3, n_2 = n_1(1)3, n_3 = n_2(1)3, n_4 = n_3(1)3; 2 \leq n_5 = n_4(1)3.$

Additional upper-tail probabilities can be found in Kraft and van Eeden (1968) and Iman, Quade, and Alexander (1975).

For given k and sample sizes $n_1, n_2, \ldots, n_k$, the table entry for the point x is $P_0\{H \geq x\}$. Thus if x is such that $P_0\{H \geq x\} = \alpha$, then $h_\alpha = x$.

				$k = 3$						
n_1	n_2	n_3	x	$P_0\{H \geq x\}$	n_1	n_2	n_3	x	$P_0\{H \geq x\}$
1	1	4	3.571	.200	1	2	4	4.018	.114
								4.500	.076
1	1	5	3.857	.143				4.821	.057
1	1	6	4.083	.107	1	2	5	4.050	.119
								4.200	.095
1	2	2	3.600	.200				4.450	.071
								5.000	.048
1	2	3	4.286	.100				5.250	.036

(continued)

TABLE A.12. *(continued)*

				$k = 3$					
n_1	n_2	n_3	x	$P_0\{H \geq x\}$	n_1	n_2	n_3	x	$P_0\{H \geq x\}$
1	2	6	3.822	.127	1	4	4	4.067	.102
			4.200	.095				4.167	.083
			4.289	.087				4.267	.070
			4.356	.079				4.800	.067
			4.622	.063				4.867	.054
			4.822	.048				4.967	.048
			5.400	.032				5.100	.041
			5.600	.024				5.667	.035
								6.000	.029
1	3	3	4.571	.100				6.167	.022
			5.143	.043				6.667	.010
1	3	4	3.889	.129	1	4	5	3.960	.102
			4.056	.093				3.987	.098
			4.097	.086				4.206	.095
			4.208	.079				4.222	.087
			4.764	.071				4.287	.071
			5.000	.057				4.549	.067
			5.208	.050				4.636	.063
			5.389	.036				4.724	.060
			5.833	.021				4.833	.059
								4.860	.056
1	3	5	3.840	.123				4.986	.044
			4.018	.095				5.078	.041
			4.284	.083				5.160	.038
			4.338	.079				5.515	.037
			4.551	.075				5.558	.035
			4.711	.056				5.596	.033
			4.871	.052				5.733	.027
			4.960	.048				5.776	.025
			5.404	.044				5.858	.024
			5.440	.036				5.864	.022
			5.760	.028				5.967	.021
			6.044	.020				6.431	.019
			6.400	.012				6.578	.016
								6.818	.013
1	3	6	3.818	.119				6.840	.011
			3.909	.095				6.954	.008
			3.964	.090				7.364	.005
			4.127	.086					
			4.418	.083	1	4	6	3.864	.110
			4.545	.064				4.038	.094
			4.691	.062				4.106	.092
			4.782	.052				4.197	.088
			4.855	.050				4.273	.086
			5.127	.048				4.341	.081
			5.273	.036				4.356	.079
			5.509	.033				4.402	.067
			5.582	.031				4.538	.063
			5.727	.029				4.583	.061
			5.855	.026				4.818	.059
			5.945	.021				4.841	.057
			6.236	.017				4.924	.055
			6.582	.012				4.947	.047
			6.873	.007				5.023	.044

TABLE A.12. *(continued)*

				$k = 3$					
n_1	n_2	n_3	x	$P_0\{H \geq x\}$	n_1	n_2	n_3	x	$P_0\{H \geq x\}$
1	4	6	5.091	.042	1	5	6	3.921	.104
			5.152	.040				4.128	.093
			5.197	.038				4.167	.091
			5.318	.036				4.221	.089
			5.455	.035				4.269	.086
			5.568	.034				4.344	.080
			5.652	.029				4.374	.070
			5.674	.027				4.385	.068
			5.697	.026				4.497	.066
			5.856	.024				4.590	.063
			5.924	.023				4.782	.062
			6.038	.022				4.823	.060
			6.114	.021				4.836	.051
			6.174	.020				4.990	.047
			6.288	.019				5.028	.046
			6.402	.016				5.090	.043
			6.523	.015				5.151	.042
			6.538	.014				5.336	.040
			6.606	.013				5.359	.039
			6.697	.012				5.400	.034
			7.000	.011				5.459	.032
			7.083	.010				5.562	.031
			7.106	.009				5.574	.030
			7.424	.007				5.728	.029
			7.500	.005				5.767	.028
			7.614	.004				5.769	.027
			7.955	.003				5.862	.026
								5.951	.025
1	5	5	4.036	.105				6.074	.021
			4.109	.086				6.138	.020
			4.182	.082				6.344	.018
			4.400	.076				6.382	.017
			4.546	.074				6.485	.017
			4.800	.056				6.567	.016
			4.909	.053				6.600	.015
			5.127	.046				6.628	.013
			5.236	.039				6.690	.012
			5.636	.033				6.805	.012
			5.709	.030				6.874	.011
			5.782	.027				6.997	.010
			6.000	.022				7.182	.010
			6.146	.019				7.246	.009
			6.509	.018				7.297	.009
			6.546	.015				7.305	.009
			6.582	.014				7.421	.007
			6.727	.012				7.451	.007
			6.836	.011				7.490	.006
			7.309	.009				7.574	.006
			7.527	.008				7.592	.006
			7.746	.005				7.667	.005
			8.182	.002				8.067	.005
								8.077	.004
								8.167	.004
								8.331	.003

(continued)

TABLE A.12. *(continued)*

				$k = 3$					
n_1	n_2	n_3	x	$P_0\{H \geq x\}$	n_1	n_2	n_3	x	$P_0\{H \geq x\}$
1	5	6	8.436	.003	1	6	6	9.308	.001
			8.515	.002				9.692	.001
			8.885	.001					
					2	2	2	3.714	.200
1	6	6	3.978	.107				4.571	.067
			4.000	.098					
			4.077	.094	2	2	3	4.464	.105
			4.209	.088				4.500	.067
			4.308	.082				4.714	.048
			4.352	.077				5.357	.029
			4.593	.069					
			4.648	.067	2	2	4	4.458	.100
			4.692	.064				4.500	.090
			4.769	.057				5.125	.052
			4.857	.051				5.333	.033
			4.945	.048				5.500	.024
			5.220	.047				6.000	.014
			5.231	.043					
			5.264	.037	2	2	5	4.293	.122
			5.352	.035				4.373	.090
			5.451	.034				4.573	.085
			5.626	.032				4.800	.063
			5.736	.031				4.893	.061
			5.791	.027				5.040	.056
			5.912	.025				5.160	.034
			5.923	.024				5.693	.029
			6.055	.022				6.000	.019
			6.088	.021				6.133	.013
			6.286	.020				6.533	.008
			6.352	.017					
			6.407	.017	2	2	6	4.436	.108
			6.626	.015				4.545	.089
			6.637	.014				4.655	.086
			6.769	.014				4.982	.054
			6.802	.013				5.018	.051
			6.879	.012				5.345	.038
			7.066	.010				5.527	.037
			7.121	.009				5.745	.021
			7.374	.009				6.182	.017
			7.407	.008				6.545	.011
			7.495	.008				6.655	.008
			7.516	.007				6.982	.005
			7.593	.006					
			7.769	.006	2	3	3	4.556	.100
			7.934	.005				4.694	.093
			8.165	.005				5.000	.075
			8.198	.005				5.139	.061
			8.220	.004				5.361	.032
			8.264	.004				5.556	.025
			8.429	.003				6.250	.011
			8.516	.003					
			8.923	.003	2	3	4	4.444	.102
			9.000	.002				4.511	.098
			9.077	.002				4.544	.086
								4.611	.083

TABLE A.12. *(continued)*

				$k = 3$					
n_1	n_2	n_3	x	$P_0\{H \geq x\}$	n_1	n_2	n_3	x	$P_0\{H \geq x\}$
2	3	4	4.711	.079	2	3	5	7.182	.004
			4.811	.076				7.636	.002
			4.878	.073					
			4.900	.071	2	3	6	4.545	.101
			4.978	.059				4.682	.086
			5.078	.057				4.742	.079
			5.144	.054				4.803	.077
			5.378	.052				4.848	.075
			5.400	.051				4.909	.072
			5.444	.046				5.015	.061
			5.500	.040				5.045	.059
			5.611	.032				5.076	.056
			5.800	.030				5.136	.055
			6.000	.024				5.167	.054
			6.111	.021				5.227	.052
			6.144	.014				5.348	.046
			6.300	.011				5.379	.045
			6.444	.008				5.394	.044
			7.000	.005				5.500	.041
								5.576	.040
2	3	5	4.494	.101				5.636	.039
			4.651	.091				5.682	.034
			4.694	.089				5.742	.032
			4.724	.087				5.879	.031
			4.727	.085				5.894	.030
			4.814	.071				6.000	.029
			4.869	.067				6.061	.026
			4.913	.063				6.136	.023
			4.942	.062				6.227	.019
			5.076	.060				6.242	.019
			5.087	.053				6.409	.016
			5.106	.052				6.545	.016
			5.251	.049				6.561	.013
			5.349	.046				6.682	.013
			5.513	.044				6.712	.012
			5.524	.043				6.727	.011
			5.542	.041				6.970	.009
			5.727	.037				7.045	.007
			5.742	.034				7.409	.006
			5.786	.033				7.500	.006
			5.804	.033				7.515	.005
			5.949	.026				7.576	.004
			6.004	.025				7.803	.002
			6.033	.024				8.182	.001
			6.091	.021					
			6.124	.020	2	4	4	4.554	.098
			6.294	.017				4.582	.094
			6.386	.016				4.691	.080
			6.414	.015				4.773	.075
			6.818	.012				4.854	.071
			6.822	.010				4.991	.065
			6.909	.009				5.127	.057
			6.949	.006				5.236	.052

(continued)

TABLE A.12. (continued)

				$k = 3$					
n_1	n_2	n_3	x	$P_0\{H \geq x\}$	n_1	n_2	n_3	x	$P_0\{H \geq x\}$
2	4	4	5.454	.046	2	4	5	6.368	.021
			5.509	.044				6.391	.021
			5.536	.042				6.473	.020
			5.646	.039				6.504	.020
			5.727	.034				6.541	.017
			5.946	.028				6.550	.017
			6.082	.025				6.564	.016
			6.327	.024				6.654	.016
			6.409	.022				6.723	.015
			6.546	.020				6.904	.014
			6.600	.017				6.914	.013
			6.627	.016				7.000	.013
			6.873	.011				7.018	.012
			7.036	.006				7.064	.012
			7.282	.004				7.118	.010
			7.854	.002				7.204	.009
								7.254	.009
2	4	5	4.518	.101				7.291	.008
			4.541	.098				7.450	.007
			4.614	.090				7.500	.007
			4.664	.088				7.568	.006
			4.768	.079				7.573	.005
			4.791	.078				7.773	.004
			4.800	.076				7.814	.003
			4.818	.074				8.018	.002
			4.841	.072				8.114	.001
			4.868	.071				8.591	.001
			4.950	.063					
			5.073	.061	2	4	6	4.494	.100
			5.154	.059				4.615	.098
			5.164	.053				4.647	.090
			5.254	.052				4.673	.084
			5.268	.051				4.744	.083
			5.273	.049				4.878	.075
			5.300	.048				4.904	.074
			5.314	.046				4.955	.068
			5.414	.045				4.974	.066
			5.518	.043				5.032	.064
			5.523	.042				5.051	.062
			5.564	.038				5.109	.060
			5.641	.037				5.128	.058
			5.664	.036				5.135	.053
			5.754	.035				5.186	.051
			5.823	.034				5.263	.050
			5.891	.032				5.340	.049
			5.954	.030				5.417	.048
			5.973	.029				5.436	.044
			6.004	.026				5.494	.042
			6.041	.025				5.590	.041
			6.068	.025				5.596	.040
			6.118	.024				5.667	.039
			6.141	.023				5.769	.036
			6.223	.022				5.801	.031

TABLE A.12. *(continued)*

				$k = 3$					
n_1	n_2	n_3	x	$P_0\{H \geq x\}$	n_1	n_2	n_3	x	$P_0\{H \geq x\}$
2	4	6	5.827	.031	2	5	5	5.054	.060
			5.974	.029				5.177	.057
			6.000	.028				5.238	.054
			6.032	.026				5.246	.051
			6.109	.025				5.338	.047
			6.186	.025				5.546	.045
			6.282	.023				5.585	.041
			6.288	.022				5.608	.040
			6.494	.020				5.615	.039
			6.519	.020				5.708	.037
			6.571	.020				5.731	.036
			6.590	.018				5.792	.032
			6.647	.018				5.915	.030
			6.667	.017				5.985	.028
			6.692	.015				6.077	.027
			6.724	.014				6.231	.026
			6.750	.013				6.346	.025
			6.878	.013				6.354	.021
			6.974	.012				6.446	.020
			7.032	.012				6.469	.019
			7.205	.011				6.654	.017
			7.212	.011				6.692	.016
			7.340	.010				6.815	.015
			7.385	.009				6.838	.014
			7.417	.009				6.969	.013
			7.436	.008				7.023	.013
			7.513	.007				7.185	.012
			7.571	.007				7.208	.011
			7.590	.006				7.269	.010
			7.647	.006				7.338	.010
			7.724	.005				7.392	.009
			7.821	.005				7.462	.008
			7.846	.004				7.577	.007
			7.904	.004				7.762	.007
			8.051	.004				7.923	.006
			8.186	.004				8.008	.006
			8.205	.003				8.077	.006
			8.308	.003				8.131	.005
			8.365	.002				8.169	.003
			8.494	.002				8.292	.003
			8.538	.002				8.377	.002
			8.667	.001				8.562	.002
			8.827	.001				8.685	.001
			9.231	<.0005				8.938	.001
								9.423	<.0005
2	5	5	4.508	.100					
			4.623	.097	2	5	6	4.475	.100
			4.685	.092				4.596	.098
			4.754	.084				4.613	.097
			4.808	.081				4.615	.096
			4.846	.073				4.640	.094
			4.877	.068				4.668	.092
			4.992	.066				4.714	.090

(continued)

TABLE A.12. *(continued)*

				$k = 3$					
n_1	n_2	n_3	x	$P_0\{H \geq x\}$	n_1	n_2	n_3	x	$P_0\{H \geq x\}$
2	5	6	4.727	.088	2	5	6	6.613	.018
			4.738	.087				6.646	.018
			4.811	.078				6.657	.018
			4.824	.075				6.673	.017
			4.833	.074				6.690	.016
			4.846	.073				6.771	.015
			4.890	.072				6.811	.015
			4.903	.071				6.824	.015
			4.932	.070				6.954	.014
			4.956	.065				6.987	.014
			5.044	.064				6.989	.013
			5.075	.063				7.042	.013
			5.090	.062				7.068	.013
			5.101	.060				7.119	.012
			5.154	.057				7.132	.012
			5.229	.056				7.185	.011
			5.233	.055				7.218	.011
			5.240	.054				7.299	.010
			5.273	.052				7.376	.010
			5.286	.052				7.382	.010
			5.319	.051				7.404	.009
			5.338	.047				7.462	.009
			5.440	.047				7.481	.009
			5.486	.046				7.640	.008
			5.497	.045				7.646	.007
			5.530	.044				7.673	.007
			5.585	.041				7.701	.007
			5.615	.039				7.738	.007
			5.618	.039				7.760	.007
			5.624	.038				7.804	.007
			5.662	.037				7.833	.006
			5.767	.035				7.870	.006
			5.813	.035				7.910	.006
			5.881	.034				7.956	.006
			5.899	.033				7.958	.005
			5.932	.030				8.068	.005
			5.958	.030				8.167	.005
			6.033	.029				8.187	.005
			6.057	.029				8.196	.005
			6.099	.028				8.200	.005
			6.110	.027				8.240	.004
			6.130	.026				8.273	.004
			6.189	.026				8.299	.004
			6.196	.025				8.332	.004
			6.218	.024				8.354	.004
			6.262	.024				8.404	.004
			6.275	.022				8.503	.004
			6.327	.022				8.530	.004
			6.354	.021				8.571	.003
			6.415	.021				8.615	.003
			6.525	.021				8.662	.003
			6.538	.020				8.727	.003
			6.585	.020				8.747	.003

A. TABLES AND CHARTS

TABLE A.12. *(continued)*

				$k = 3$					
n_1	n_2	n_3	x	$P_0\{H \geq x\}$	n_1	n_2	n_3	x	$P_0\{H \geq x\}$
2	5	6	8.800	.002	2	6	6	7.276	.012
			8.947	.002				7.352	.011
			8.967	.002				7.371	.011
			9.000	.002				7.410	.010
			9.011	.002				7.467	.010
			9.046	.002				7.505	.009
			9.185	.001				7.543	.009
			9.189	.001				7.619	.008
			9.275	.001				7.638	.008
			9.415	.001				7.752	.007
			9.453	.001				7.886	.007
			9.670	<.0005				8.019	.007
								8.038	.006
2	6	6	4.419	.107				8.076	.006
			4.438	.098				8.152	.005
			4.552	.096				8.210	.005
			4.610	.093				8.305	.005
			4.800	.088				8.400	.004
			4.819	.085				8.533	.004
			4.838	.081				8.610	.003
			4.876	.077				8.819	.003
			4.933	.068				8.838	.003
			4.971	.067				8.876	.003
			5.010	.064				8.933	.003
			5.105	.063				9.010	.002
			5.219	.058				9.086	.002
			5.238	.055				9.105	.002
			5.276	.053				9.219	.002
			5.352	.051				9.352	.002
			5.410	.050				9.505	.002
			5.486	.046				9.600	.002
			5.505	.042				9.638	.001
			5.638	.042				9.676	.001
			5.676	.038				9.752	.001
			5.733	.037				9.867	.001
			5.752	.035				9.943	.001
			5.867	.034				10.076	<.0005
			6.019	.032					
			6.038	.030	3	3	3	4.622	.1000
			6.076	.029				5.067	.0857
			6.133	.027				5.422	.0714
			6.171	.026				5.600	.0500
			6.210	.024				5.956	.0250
			6.305	.024				6.489	.0107
			6.343	.023				7.200	.0036
			6.419	.022					
			6.552	.021	3	3	4	4.700	.1010
			6.667	.019				4.709	.0924
			6.705	.018				4.845	.0810
			6.819	.016				5.000	.0743
			6.876	.015				5.727	.0505
			7.010	.015				5.791	.0457
			7.067	.013				6.018	.0267
			7.105	.012				6.155	.0248

(continued)

TABLE A.12. *(continued)*

					$k = 3$				
n_1	n_2	n_3	x	$P_0\{H \geq x\}$	n_1	n_2	n_3	x	$P_0\{H \geq x\}$
3	3	4	6.745	.0100	3	4	5	6.410	.0250
			7.000	.0062				7.395	.0109
			7.318	.0043				7.445	.0097
			8.018	.0014				7.927	.0050
								8.626	.0012
3	3	5	4.412	.1091				8.795	.0009
			4.533	.0970					
			5.042	.0775	3	4	6	4.604	.1000
			5.079	.0693				4.962	.0769
			5.515	.0507				5.033	.0744
			5.648	.0489				5.604	.0504
			6.303	.0255				5.610	.0486
			6.315	.0212				6.538	.0250
			6.982	.0113				7.467	.0101
			7.079	.0087				7.500	.0097
			7.515	.0054				8.033	.0050
			7.636	.0041				9.170	.0010
			8.242	.0011					
			8.727	.0007	3	5	5	4.536	.1020
								4.545	.0997
3	3	6	4.538	.1034				4.993	.0755
			4.590	.0977				5.020	.0720
			4.949	.0770				5.626	.0508
			5.038	.0748				5.705	.0461
			5.551	.0512				6.488	.0254
			5.615	.0497				6.549	.0244
			6.385	.0253				7.543	.0102
			6.436	.0223				7.578	.0097
			7.192	.0102				8.264	.0051
			7.410	.0078				8.316	.0049
			7.615	.0061				9.284	.0010
			7.872	.0043					
			8.692	.0010	3	5	6	4.497	.1002
								4.535	.0993
3	4	4	4.477	.1022				5.008	.0755
			4.545	.0991				5.021	.0748
			5.053	.0781				5.600	.0500
			5.144	.0729				6.621	.0256
			5.576	.0507				6.667	.0245
			5.598	.0487				7.590	.0100
			6.386	.0262				8.297	.0052
			6.394	.0248				8.314	.0048
			7.136	.0107				9.669	.0010
			7.144	.0097					
			7.477	.0062	3	6	6	4.525	.1022
			7.598	.0042				4.558	.0995
			8.326	.0012				5.025	.0759
			8.909	.0005				5.058	.0725
								5.625	.0500
3	4	5	4.523	.1033				6.683	.0251
			4.549	.0989				6.725	.0246
			4.938	.0754				7.683	.0101
			4.953	.0742				7.725	.0099
			5.631	.0504				8.458	.0050
			5.656	.0486				10.150	.0010

TABLE A.12. *(continued)*

				$k = 3$					
n_1	n_2	n_3	x	$P_0\{H \geq x\}$	n_1	n_2	n_3	x	$P_0\{H \geq x\}$
4	4	4	4.500	.1042	4	5	6	4.500	.1011
			4.654	.0966				4.522	.0997
			4.962	.0800				5.023	.0751
			5.115	.0741				5.036	.0746
			5.654	.0546				5.656	.0506
			5.692	.0487				5.661	.0499
			6.577	.0263				6.736	.0251
			6.615	.0242				6.750	.0247
			7.538	.0107				7.936	.0100
			7.654	.0076				8.640	.0050
			7.731	.0066				9.960	.0010
			8.000	.0049					
			8.769	.0012	4	6	6	4.518	.1009
			9.269	.0005				4.548	.0998
								5.000	.0775
4	4	5	4.619	.1000				5.048	.0740
			5.014	.0758				5.721	.0501
			5.024	.0743				5.724	.0495
			5.618	.0503				6.783	.0252
			5.657	.0491				6.812	.0246
			6.597	.0256				8.000	.0100
			6.673	.0243				8.754	.0050
			7.744	.0107				10.283	.0010
			7.760	.0095					
			8.189	.0050	5	5	5	4.500	.1015
			9.129	.0010				4.560	.0995
								4.940	.0807
4	4	6	4.524	.1031				5.040	.0746
			4.595	.0985				5.660	.0509
			5.095	.0753				5.780	.0488
			5.124	.0730				6.720	.0259
			5.667	.0505				6.740	.0248
			5.681	.0488				7.980	.0105
			6.667	.0250				8.000	.0095
			7.724	.0101				8.780	.0050
			7.795	.0099				9.920	.0010
			8.324	.0052					
			8.381	.0048	5	5	6	4.529	.1025
			9.629	.0010				4.547	.0984
								4.993	.0765
4	5	5	4.520	.1009				5.063	.0748
			4.523	.0994				5.699	.0509
			5.023	.0751				5.729	.0497
			5.071	.0738				6.781	.0252
			5.643	.0502				6.788	.0248
			5.666	.0493				8.012	.0100
			6.671	.0254				8.835	.0050
			6.760	.0249				10.216	.0010
			7.791	.0102					
			7.823	.0098					
			8.463	.0050					
			9.506	.0010					

(continued)

TABLE A.12. *(continued)*

				$k = 3$					
n_1	n_2	n_3	x	$P_0\{H \geq x\}$	n_1	n_2	n_3	x	$P_0\{H \geq x\}$
5	6	6	4.541	.1008	7	7	7	4.549	.1007
			4.542	.0999				4.594	.0993
			5.018	.0756				5.076	.0750
			5.054	.0742				5.766	.0506
			5.752	.0504				5.819	.0491
			5.765	.0499				6.909	.0256
			6.838	.0251				6.954	.0245
			6.848	.0249				8.334	.0101
			8.119	.0100				8.378	.0099
			8.982	.0050				9.358	.0051
			10.503	.0010				9.373	.0049
								11.288	.0010
6	6	6	4.538	.1010					
			4.643	.0987	8	8	8	4.580	.1002
			5.064	.0759				4.595	.0993
			5.099	.0742				5.105	.0756
			5.719	.0502				5.120	.0749
			5.801	.0491				5.795	.0507
			6.877	.0259				5.805	.0497
			6.889	.0249				6.980	.0251
			8.187	.0102				6.995	.0249
			8.222	.0099				8.435	.0101
			9.088	.0050				8.465	.0099
			10.819	.0010				9.485	.0051
								9.495	.0049
								11.625	.0010

TABLE A.12. (*continued*)

					$k = 4$						
n_1	n_2	n_3	n_4	x	$P_0\{H \geq x\}$	n_1	n_2	n_3	n_4	x	$P_0\{H \geq x\}$
1	1	1	4	4.929	.1143	1	2	2	4	5.500	.1032
										5.533	.0979
1	1	2	2	4.714	.1333					5.700	.0852
										5.733	.0683
1	1	2	3	4.893	.1429					6.000	.0566
				5.143	.0857					6.133	.0418
				5.464	.0571					6.500	.0270
										6.533	.0206
1	1	2	4	5.208	.1143					6.800	.0127
				5.250	.0905					7.000	.0095
				5.417	.0762					7.200	.0064
				5.458	.0714						
				5.833	.0429	1	2	3	3	5.622	.1008
				6.083	.0286					5.689	.0857
										5.800	.0806
1	1	3	3	5.222	.1107					5.956	.0687
				5.333	.0964					6.156	.0560
				5.444	.0893					6.244	.0425
				5.889	.0643					6.511	.0290
				6.333	.0214					6.689	.0179
										7.044	.0107
1	1	3	4	4.978	.1064					7.200	.0060
				5.067	.0952					7.400	.0048
				5.511	.0825						
				5.644	.0698	1	2	3	4	5.573	.1003
				6.044	.0524					5.591	.0986
				6.178	.0492					5.891	.0795
				6.567	.0286					5.955	.0743
				6.711	.0191					6.300	.0511
				7.067	.0095					6.309	.0494
										6.909	.0273
1	1	4	4	5.127	.1035					6.955	.0232
				5.182	.0997					7.364	.0111
				5.427	.0762					7.455	.0098
				5.564	.0749					7.727	.0052
				5.864	.0571					7.773	.0043
				5.945	.0495					8.182	.0019
				6.927	.0260						
				6.955	.0235	1	2	4	4	5.545	.1003
				7.500	.0114					5.568	.0998
				7.909	.0038					5.955	.0754
										5.977	.0748
1	2	2	2	5.036	.1238					6.364	.0500
				5.357	.0667					7.136	.0261
				5.679	.0381					7.159	.0246
										7.886	.0102
1	2	2	3	5.389	.1095					7.909	.0091
				5.556	.0714					8.227	.0051
				5.583	.0619					8.341	.0032
				5.806	.0571					8.705	.0016
				5.833	.0429					8.909	.0009
				6.056	.0286						
				6.250	.0214	1	3	3	3	5.582	.1129
				6.500	.0143					5.655	.0979
										6.018	.0836

(*continued*)

TABLE A.12. *(continued)*

						$k = 4$					
n_1	n_2	n_3	n_4	x	$P_0\{H \geq x\}$	n_1	n_2	n_3	n_4	x	$P_0\{H \geq x\}$
1	3	3	3	6.164	.0700	2	2	2	2	5.500	.1143
				6.527	.0521					5.667	.0762
				6.600	.0493					6.000	.0667
				6.891	.0343					6.167	.0381
				7.036	.0243					6.667	.0095
				7.327	.0143						
				7.400	.0086	2	2	2	3	5.644	.1000
				7.764	.0057					5.933	.0762
				8.055	.0036					5.978	.0730
				8.345	.0014					6.244	.0540
										6.333	.0476
1	3	3	4	5.667	.1004					6.644	.0270
				5.689	.0960					6.978	.0175
				6.061	.0759					7.000	.0127
				6.114	.0733					7.133	.0079
				6.538	.0515					7.533	.0032
				6.545	.0495						
				7.273	.0251	2	2	2	4	5.673	.1019
				7.326	.0233					5.755	.0930
				7.750	.0119					5.973	.0791
				7.758	.0097					6.082	.0714
				8.121	.0054					6.436	.0524
				8.212	.0042					6.545	.0492
				8.939	.0010					6.982	.0289
										7.064	.0222
1	3	4	4	5.660	.1002					7.309	.0140
				5.692	.0985					7.391	.0089
				6.103	.0752					7.855	.0051
				6.115	.0732					7.964	.0032
				6.615	.0509					8.291	.0013
				6.635	.0498						
				7.481	.0255	2	2	3	3	5.727	.1002
				7.500	.0246					5.745	.0992
				8.218	.0103					6.091	.0757
				8.231	.0096					6.127	.0732
				8.577	.0053					6.473	.0524
				8.583	.0047					6.527	.0492
				9.295	.0010					7.636	.0100
										7.727	.0081
1	4	4	4	5.637	.1009					7.873	.0043
				5.654	.0980					8.455	.0010
				6.165	.0750						
				6.709	.0503	2	2	3	4	5.712	.1012
				6.725	.0498					5.750	.0998
				7.632	.0252					6.121	.0763
				7.648	.0247					6.136	.0748
				8.571	.0101					6.614	.0515
				8.588	.0099					6.621	.0495
				9.000	.0050					7.871	.0100
				9.758	.0010					8.250	.0053
										8.273	.0043
										8.894	.0010

TABLE A.12. (continued)

					$k = 4$						
n_1	n_2	n_3	n_4	x	$P_0\{H \geq x\}$	n_1	n_2	n_3	n_4	x	$P_0\{H \geq x\}$
2	2	4	4	5.769	.1020	2	4	4	4	5.900	.1014
				5.808	.0998					5.914	.0994
				6.135	.0752					6.357	.0751
				6.192	.0741					6.371	.0747
				6.692	.0519					6.943	.0504
				6.731	.0487					6.957	.0496
				7.519	.0250					7.914	.0250
				8.308	.0102					8.857	.0101
				8.346	.0094					8.871	.0099
				8.673	.0051					9.471	.0050
				8.692	.0049					10.386	.0010
				9.462	.0010						
						3	3	3	3	5.974	.1027
2	3	3	3	5.818	.1023					6.026	.0978
				5.879	.0997					6.385	.0788
				6.242	.0755					6.436	.0746
				6.258	.0738					6.897	.0502
				6.682	.0508					7.000	.0435
				6.727	.0495					7.615	.0257
				7.470	.0258					7.667	.0234
				7.515	.0239					8.436	.0108
				7.955	.0112					8.538	.0084
				8.015	.0096					8.744	.0064
				8.318	.0055					8.897	.0044
				8.379	.0038					9.462	.0014
				8.924	.0010					9.513	.0008
2	3	3	4	5.859	.1003	3	3	3	4	6.000	.1001
				5.872	.0993					6.016	.0978
				6.282	.0755					6.456	.0756
				6.288	.0749					6.462	.0741
				6.782	.0501					6.967	.0503
				6.795	.0493					6.984	.0490
				7.558	.0250					7.758	.0252
				8.321	.0106					7.775	.0244
				8.333	.0099					8.654	.0101
				8.718	.0050					8.659	.0099
				9.404	.0011					9.225	.0050
				9.455	.0009					10.000	.0010
2	3	4	4	5.890	.1007	3	3	4	4	6.005	.1003
				5.901	.0995					6.019	.0995
				6.319	.0751					6.481	.0754
				6.330	.0745					6.500	.0747
				6.863	.0507					7.033	.0509
				6.874	.0498					7.038	.0499
				7.747	.0250					7.924	.0250
				8.610	.0100					8.867	.0100
				9.148	.0050					9.490	.0050
				9.896	.0010					10.424	.0010

(continued)

TABLE A.12. *(continued)*

					$k = 4$						
n_1	n_2	n_3	n_4	x	$P_0\{H \geq x\}$	n_1	n_2	n_3	n_4	x	$P_0\{H \geq x\}$
3	4	4	4	6.029	.1008	4	4	4	4	6.066	.1003
				6.042	.0998					6.088	.0990
				6.529	.0750					6.551	.0768
				7.129	.0502					6.574	.0744
				7.142	.0495					7.213	.0507
				8.054	.0251					7.235	.0492
				8.079	.0249					8.206	.0252
				9.067	.0100					8.228	.0248
				9.717	.0050					9.287	.0100
				10.879	.0010					9.949	.0051
										9.971	.0049
										11.338	.0010

						$k = 5$						
n_1	n_2	n_3	n_4	n_5	x	$P_0\{H \geq x\}$	n_1	n_2	n_3	n_4	n_5	x	$P_0\{H \geq x\}$
1	1	1	1	2	4.857	.3333	1	1	2	3	3	6.545	.1012
												6.600	.0993
1	1	1	1	3	5.571	.1429						6.909	.0787
												6.964	.0721
1	1	1	2	2	5.464	.2095						7.200	.0500
					5.786	.0952						7.564	.0255
												7.618	.0245
1	1	1	2	3	6.139	.1000						8.055	.0102
					6.333	.0571						8.073	.0074
					6.583	.0357						8.218	.0071
												8.345	.0033
1	1	1	3	3	6.222	.1191						8.509	.0024
					6.311	.0929							
					6.578	.0810	1	1	3	3	3	6.727	.1016
					6.667	.0667						6.788	.0978
					6.756	.0595						7.152	.0760
					7.111	.0405						7.212	.0669
					7.467	.0119						7.515	.0538
												7.576	.0455
1	1	2	2	2	6.083	.1238						8.000	.0284
					6.250	.0881						8.061	.0233
					6.500	.0524						8.121	.0227
					6.750	.0238						8.364	.0139
												8.424	.0091
1	1	2	2	3	6.511	.1000						8.727	.0051
					6.600	.0825						8.848	.0044
					6.711	.0698						9.212	.0022
					6.800	.0492						9.455	.0007
					7.067	.0294							
					7.200	.0246	1	2	2	2	2	6.533	.1048
					7.400	.0143						6.600	.0889
					7.600	.0079						6.733	.0762
												6.800	.0698
												7.000	.0540

TABLE A.12. (continued)

$k = 5$

n_1	n_2	n_3	n_4	n_5	x	$P_0\{H \geq x\}$	n_1	n_2	n_3	n_4	n_5	x	$P_0\{H \geq x\}$
1	2	2	2	2	7.133	.0413	1	3	3	3	3	8.703	.0240
					7.200	.0349						9.451	.0100
					7.333	.0222						9.802	.0059
					7.533	.0095						9.846	.0049
					7.733	.0053						10.549	.0010
1	2	2	2	3	6.691	.1011	2	2	2	2	2	6.873	.1016
					6.709	.0987						6.982	.0910
					6.982	.0773						7.091	.0794
					7.018	.0735						7.309	.0635
					7.291	.0538						7.418	.0487
					7.309	.0489						7.855	.0254
					7.727	.0275						7.964	.0222
					7.745	.0232						8.073	.0127
					8.018	.0108						8.291	.0095
					8.127	.0094						8.400	.0053
					8.291	.0051						8.727	.0011
					8.327	.0032							
					8.618	.0016	2	2	2	2	3	6.939	.1026
												6.955	.0992
1	2	2	3	3	6.758	.1032						7.303	.0767
					6.788	.0989						7.318	.0729
					7.121	.0779						7.667	.0508
					7.152	.0736						7.682	.0475
					7.576	.0511						8.167	.0256
					7.591	.0492						8.182	.0238
					8.091	.0250						8.667	.0110
					8.561	.0110						8.682	.0096
					8.576	.0098						8.955	.0051
					8.864	.0057						8.985	.0046
					8.924	.0048						9.364	.0010
					9.303	.0010							
							2	2	2	3	3	7.013	.1005
1	2	3	3	3	6.897	.1023						7.026	.0990
					6.910	.0992						7.397	.0761
					7.308	.0755						7.410	.0745
					7.321	.0745						7.897	.0505
					7.731	.0517						7.910	.0493
					7.769	.0489						8.526	.0251
					8.436	.0257						8.538	.0241
					8.449	.0247						9.115	.0100
					9.013	.0101						9.474	.0050
					9.051	.0098						10.000	.0011
					9.372	.0051						10.026	.0009
					9.410	.0047							
					9.974	.0010	2	2	3	3	3	7.099	.1002
												7.121	.0998
1	3	3	3	3	7.033	.1003						7.495	.0759
					7.077	.0984						7.516	.0745
					7.429	.0769						8.022	.0509
					7.473	.0734						8.044	.0492
					7.956	.0505						8.802	.0252
					8.000	.0479						8.813	.0247
					8.659	.0254						9.505	.0100

(continued)

TABLE A.12. *(continued)*

							$k = 5$						
n_1	n_2	n_3	n_4	n_5	x	$P_0\{H \geq x\}$	n_1	n_2	n_3	n_4	n_5	x	$P_0\{H \geq x\}$
2	2	3	3	3	9.890	.0050	3	3	3	3	3	7.300	.1008
					10.637	.0010						7.333	.0992
												7.733	.0767
2	3	3	3	3	7.181	.1020						7.767	.0749
					7.210	.0997						8.300	.0505
					7.629	.0759						8.333	.0496
					7.638	.0745						9.200	.0250
					8.171	.0504						10.167	.0104
					8.200	.0494						10.200	.0099
					9.010	.0252						10.700	.0051
					9.038	.0245						10.733	.0049
					9.848	.0101						11.633	.0010
					9.876	.0097							
					10.305	.0051							
					10.333	.0049							
					11.133	.0010							

Adapted, in part, from Table F of C. H. Kraft and C. van Eeden, *A Nonparametric Introduction to Statistics*, Macmillan, New York (1968), with the permission of the authors and the publisher, and, in part, from R. L. Iman, D. Quade, and D. A. Alexander, Exact probability levels for the Kruskal-Wallis test, in: *Selected Tables in Mathematical Statistics*, Volume 3, H. L. Harter and D. B. Owen (eds.), (1975): pp. 329–384, with the permission of the authors and the editors of *Selected Tables in Mathematical Statistics*, Volume 3.

TABLE A.13. Selected Critical Values for the Null Distribution of the Jonckheere-Terpstra J Statistic: $k = 3, 2 \leq n_1 \leq n_2 \leq n_3 \leq 8; k = 4(1)6, n_1 = \ldots = n_k = 2(1)6$.

For given k, α, and sample sizes $n_1, \ldots, n_k$, the entry in the table is j_α satisfying $P_0\{J \geq j_\alpha\} = \alpha$.

		$k = 3$					$k = 3$		
n_1	n_2	n_3	α	j_α	n_1	n_2	n_3	α	j_α
2	2	2	.5778	6	2	2	7	.5354	16
			.4222	7				.4647	17
			.2889	8				.2121	21
			.1667	9				.1636	22
			.0889	10				.1217	23
			.0333	11				.0879	24
			.0111	12				.0606	25
								.0404	26
2	2	3	.5619	8				.0253	27
			.4381	9				.0152	28
			.2191	11				.0081	29
			.1381	12				.0040	30
			.0762	13					
			.0381	14	2	2	8	.5320	18
			.0143	15				.4680	19
			.0048	16				.2354	23
								.1886	24
2	2	4	.5524	10				.1118	26
			.4476	11				.0822	27
			.2571	13				.0589	28
			.1810	14				.0404	29
			.1167	15				.0269	30
			.0714	16				.0168	31
			.0381	17				.0101	32
			.0191	18				.0054	33
			.0071	19				.0027	34
			.0024	20					
					2	3	3	.5000	11
2	2	5	.5450	12				.4000	12
			.4550	13				.2214	14
			.2156	16				.1518	15
			.1534	17				.0964	16
			.1045	18				.0571	17
			.0661	19				.0304	18
			.0397	20				.0143	19
			.0212	21				.0054	20
			.0106	22				.0018	21
			.0040	23					
					2	3	4	.5429	13
2	2	6	.5397	14				.4571	14
			.4603	15				.2222	17
			.2444	18				.1619	18
			.1849	19				.1119	19
			.1357	20				.0738	20
			.0944	21				.0452	21
			.0635	22				.0262	22
			.0397	23				.0135	23
			.0238	24				.0064	24
			.0127	25				.0024	25
			.0064	26				*(continued)*	
			.0024	27					

TABLE A.13. *(continued)*

	$k = 3$						$k = 3$			
n_1	n_2	n_3	α	j_α		n_1	n_2	n_3	α	j_α
2	3	5	.5000	16		2	4	4	.5375	16
			.4250	17					.4625	17
			.2230	20					.2559	20
			.1694	21					.1981	21
			.1242	22					.1079	23
			.0877	23					.0756	24
			.0591	24					.0502	25
			.0381	25					.0321	26
			.0230	26					.0191	27
			.0131	27					.0108	28
			.0068	28					.0054	29
			.0032	29					.0025	30
2	3	6	.5336	18		2	4	5	.5326	19
			.4665	19					.4674	20
			.2234	23					.2287	24
			.1755	24					.1810	25
			.1340	25					.1049	27
			.0996	26					.0766	28
			.0714	27					.0540	29
			.0496	28					.0368	30
			.0329	29					.0240	31
			.0210	30					.0150	32
			.0126	31					.0088	33
			.0071	32					.0049	34
			.0037	33		2	4	6	.5293	22
2	3	7	.5000	21					.4707	23
			.4400	22					.2086	28
			.2237	26					.1680	29
			.1803	27					.1025	31
			.1096	29					.0774	32
			.0823	30					.0569	33
			.0602	31					.0408	34
			.0427	32					.0282	35
			.0293	33					.0190	36
			.0193	34					.0122	37
			.0123	35					.0076	38
			.0073	36					.0044	39
			.0042	37		2	4	7	.5263	25
2	3	8	.5273	23					.4737	26
			.4727	24					.2321	31
			.2239	29					.1931	32
			.1843	30					.1005	35
			.1183	32					.0780	36
			.0919	33					.0592	37
			.0520	35					.0441	38
			.0377	36					.0319	39
			.0265	37					.0226	40
			.0181	38					.0103	42
			.0119	39					.0066	43
			.0075	40					.0041	44
			.0045	41						

A. TABLES AND CHARTS

TABLE A.13. (continued)

		$k = 3$					$k = 3$		
n_1	n_2	n_3	α	j_α	n_1	n_2	n_3	α	j_α
2	4	8	.5241	28	2	5	7	.0060	50
			.4759	29				.0039	51
			.2150	35	2	5	8	.5215	33
			.1808	36				.4785	34
			.1227	38				.2077	41
			.0988	39				.1776	42
			.0611	41				.1040	45
			.0469	42				.0850	46
			.0259	44				.0546	48
			.0186	45				.0428	49
			.0131	46				.0252	51
			.0089	47				.0189	52
			.0059	48				.0138	53
			.0038	49				.0100	54
2	5	5	.5000	23				.0070	55
			.4423	24				.0048	56
			.2327	28					
			.1900	29	2	6	6	.5234	30
			.1194	31				.4766	31
			.0916	32				.2220	37
			.0501	34				.1882	38
			.0357	35				.1061	41
			.0245	36				.0853	42
			.0105	38				.0526	44
			.0064	39				.0403	45
			.0037	40				.0303	46
2	5	6	.5260	26				.0224	47
			.4740	27				.0114	49
			.2360	32				.0079	50
			.1971	33				.0053	51
			.1046	36				.0034	52
			.0818	37	2	6	7	.5213	34
			.0628	38				.4788	35
			.0472	39				.2109	42
			.0347	40				.1809	43
			.0249	41				.1072	46
			.0118	43				.0880	47
			.0078	44				.0572	49
			.0049	45				.0452	50
2	5	7	.5000	30				.0270	52
			.4530	31				.0204	53
			.2029	37				.0110	55
			.1706	38				.0079	56
			.1159	40				.0055	57
			.0936	41				.0038	58
			.0582	43	2	6	8	.5195	38
			.0448	44				.4805	39
			.0251	46				.2018	47
			.0182	47				.1749	48
			.0129	48				.1080	51
			.0089	49				.0903	52

(continued)

TABLE A.13. *(continued)*

		$k = 3$					$k = 3$		
n_1	n_2	n_3	α	j_α	n_1	n_2	n_3	α	j_α
2	6	8	.0612	54	3	3	3	.5000	14
			.0495	55				.4155	15
			.0314	57				.2595	17
			.0245	58				.1941	18
			.0107	61				.1387	19
			.0079	62				.0946	20
			.0057	63				.0613	21
			.0040	64				.0369	22
								.0208	23
2	7	7	.5000	39				.0107	24
			.4613	40				.0048	25
			.2174	47					
			.1895	48	3	3	4	.5000	17
			.1003	52				.4267	18
			.0836	53				.2283	21
			.0563	55				.1750	22
			.0454	56				.1300	23
			.0286	58				.0931	24
			.0223	59				.0641	25
			.0129	61				.0421	26
			.0096	62				.0264	27
			.0051	64				.0155	28
			.0036	65				.0086	29
								.0043	30
2	7	8	.5178	43					
			.4822	44	3	3	5	.5000	20
			.2229	52				.4353	21
			.1968	53				.2058	25
			.1113	57				.1615	26
			.0947	58				.1235	27
			.0554	61				.0918	28
			.0456	62				.0662	29
			.0299	64				.0462	30
			.0238	65				.0311	31
			.0113	68				.0200	32
			.0086	69				.0123	33
			.0064	70				.0071	34
			.0047	71				.0039	35
2	8	8	.5164	48	3	3	6	.5000	23
			.4836	49				.4421	24
			.2162	58				.2319	28
			.1925	59				.1891	29
			.1139	63				.1185	31
			.0983	64				.0908	32
			.0509	68				.0679	33
			.0423	69				.0495	34
			.0286	71				.0351	35
			.0232	72				.0241	36
			.0117	75				.0102	38
			.0091	76				.0062	39
			.0054	78				.0036	40
			.0040	79					

TABLE A.13. (continued)

		$k = 3$		
n_1	n_2	n_3	α	j_α
3	3	7	.5000	26
			.4476	27
			.2132	32
			.1762	33
			.1145	35
			.0899	36
			.0522	38
			.0385	39
			.0277	40
			.0194	41
			.0132	42
			.0087	43
			.0055	44
			.0034	45
3	3	8	.5000	29
			.4522	30
			.2343	35
			.1984	36
			.1113	39
			.0891	40
			.0545	42
			.0414	43
			.0309	44
			.0226	45
			.0112	47
			.0076	48
			.0050	49
3	4	4	.5322	20
			.4678	21
			.2325	25
			.1853	26
			.1093	28
			.0804	29
			.0576	30
			.0397	31
			.0265	32
			.0169	33
			.0103	34
			.0059	35
			.0032	36
3	4	5	.5000	24
			.4430	25
			.2358	29
			.1933	30
			.1227	32
			.0948	33
			.0528	35
			.0379	36
			.0265	37
			.0179	38
			.0117	39

		$k = 3$		
n_1	n_2	n_3	α	j_α
3	4	5	.0073	40
			.0044	41
3	4	6	.5257	27
			.4743	28
			.2383	33
			.1997	34
			.1072	37
			.0843	38
			.0651	39
			.0492	40
			.0264	42
			.0187	43
			.0128	44
			.0086	45
			.0055	46
			.0034	47
3	4	7	.5000	31
			.4534	32
			.2050	38
			.1728	39
			.1181	41
			.0957	42
			.0600	44
			.0464	45
			.0263	47
			.0193	48
			.0138	49
			.0096	50
			.0066	51
			.0044	52
3	4	8	.5214	34
			.4786	35
			.2095	42
			.1795	43
			.1058	46
			.0867	47
			.0561	49
			.0442	50
			.0262	52
			.0197	53
			.0106	55
			.0075	56
			.0052	57
			.0035	58
3	5	5	.5000	28
			.4491	29
			.2203	34
			.1837	35
			.1220	37
			.0971	38

(continued)

TABLE A.13. *(continued)*

		$k = 3$					$k = 3$		
n_1	n_2	n_3	α	j_α	n_1	n_2	n_3	α	j_α
3	5	5	.0582	40	3	6	6	.5209	36
			.0438	41				.4791	37
			.0323	42				.2151	44
			.0232	43				.1853	45
			.0112	45				.1116	48
			.0074	46				.0923	49
			.0048	47				.0609	51
								.0486	52
3	5	6	.5000	32				.0297	54
			.4541	33				.0227	55
			.2082	39				.0126	57
			.1761	40				.0092	58
			.1214	42				.0066	59
			.0988	43				.0046	60
			.0628	45					
			.0489	46	3	6	7	.5000	41
			.0282	48				.4619	42
			.0209	49				.2209	49
			.0107	51				.1932	50
			.0074	52				.1039	54
			.0050	53				.0870	55
			.0033	54				.0593	57
								.0482	58
3	5	7	.5000	36				.0308	60
			.4581	37				.0242	61
			.2296	43				.0109	64
			.1985	44				.0081	65
			.1002	48				.0059	66
			.0822	49				.0043	67
			.0533	51					
			.0421	52	3	6	8	.5176	45
			.0252	54				.4824	46
			.0190	55				.2258	54
			.0103	57				.1998	55
			.0074	58				.1144	59
			.0052	59				.0977	60
			.0036	60				.0580	63
								.0479	64
3	5	8	.5000	40				.0255	67
			.4615	41				.0203	68
			.2184	48				.0124	70
			.1906	49				.0095	71
			.1014	53				.0054	73
			.0846	54				.0040	74
			.0572	56					
			.0463	57	3	7	7	.5000	46
			.0293	59				.4650	47
			.0228	60				.2137	55
			.0100	63				.1887	56
			.0074	64				.1070	60
			.0053	65				.0911	61
			.0038	66				.0537	64
								.0442	65

TABLE A.13. (continued)

| \multicolumn{5}{c}{$k = 3$} | \multicolumn{5}{c}{$k = 3$} |
n_1	n_2	n_3	α	j_α	n_1	n_2	n_3	α	j_α
3	7	7	.0292	67	4	4	5	.5254	28
			.0234	68				.4747	29
			.0113	71				.2029	35
			.0086	72				.1683	36
			.0065	73				.1105	38
			.0049	74				.0874	39
								.0518	41
3	7	8	.5000	51				.0387	42
			.4677	52				.0283	43
			.2077	61				.0203	44
			.1849	62				.0141	45
			.1095	66				.0096	46
			.0946	67				.0063	47
			.0585	70				.0040	48
			.0492	71					
			.0278	74	4	4	6	.5229	32
			.0227	75				.4771	33
			.0116	78				.2265	39
			.0091	79				.1929	40
			.0054	81				.1109	43
			.0041	82				.0898	44
								.0565	46
3	8	8	.5150	56				.0438	47
			.4850	57				.0334	48
			.2144	67				.0250	49
			.1929	68				.0132	51
			.1054	73				.0093	52
			.0920	74				.0064	53
			.0502	78				.0043	54
			.0425	79					
			.0299	81	4	4	7	.5209	36
			.0248	82				.4791	37
			.0109	86				.2147	44
			.0087	87				.1849	45
			.0054	89				.1112	48
			.0042	90				.0918	49
								.0605	51
4	4	4	.5284	24				.0482	52
			.4716	25				.0294	54
			.2157	30				.0224	55
			.1756	31				.0125	57
			.1099	33				.0091	58
			.0844	34				.0065	59
			.0632	35				.0045	60
			.0463	36					
			.0330	37	4	4	8	.5192	40
			.0229	38				.4808	41
			.0153	39				.2050	49
			.0099	40				.1783	50
			.0062	41				.1114	53
			.0037	42				.0935	54
								.0522	57
								.0420	58

(continued)

TABLE A.13. (continued)

		$k = 3$					$k = 3$		
n_1	n_2	n_3	α	j_α	n_1	n_2	n_3	α	j_α
4	4	8	.0264	60	4	5	8	.5175	46
			.0205	61				.4825	47
			.0119	63				.2013	56
			.0089	64				.1772	57
			.0065	65				.1159	60
			.0047	66				.0992	61
4	5	5	.5000	33				.0592	64
			.4545	34				.0491	65
			.2107	40				.0264	68
			.1787	41				.0210	69
			.1014	44				.0129	71
			.0818	45				.0100	72
			.0509	47				.0057	74
			.0393	48				.0043	75
			.0298	49	4	6	6	.5189	42
			.0222	50				.4811	43
			.0116	52				.2097	51
			.0082	53				.1831	52
			.0056	54				.1161	55
			.0037	55				.0981	56
4	5	6	.5207	37				.0559	59
			.4793	38				.0455	60
			.2172	45				.0291	62
			.1875	46				.0229	63
			.1138	49				.0103	66
			.0944	50				.0077	67
			.0502	53				.0057	68
			.0397	54				.0041	69
			.0310	55	4	6	7	.5173	47
			.0238	56				.4827	48
			.0135	58				.2034	57
			.0099	59				.1794	58
			.0050	61				.1013	62
			.0035	62				.0862	63
4	5	7	.5000	42				.0507	66
			.4622	43				.0417	67
			.2226	50				.0276	69
			.1949	51				.0221	70
			.1057	55				.0107	73
			.0888	56				.0082	74
			.0608	58				.0062	75
			.0496	59				.0046	76
			.0252	62	4	6	8	.5160	52
			.0196	63				.4840	53
			.0115	65				.2217	62
			.0086	66				.1982	63
			.0063	67				.1040	68
			.0046	68				.0897	69
								.0554	72
								.0465	73

A. TABLES AND CHARTS

TABLE A.13. *(continued)*

	k = 3					k = 3			
n_1	n_2	n_3	α	j_α	n_1	n_2	n_3	α	j_α
4	6	8	.0263	76	5	5	5	.5000	38
			.0214	77				.4589	39
			.0110	80				.2032	46
			.0086	81				.1748	47
			.0051	83				.1049	50
			.0039	84				.0867	51
4	7	7	.5000	53				.0572	53
			.4681	54				.0456	54
			.2107	63				.0279	56
			.1880	64				.0214	57
			.1126	68				.0120	59
			.0976	69				.0087	60
			.0515	73				.0063	61
			.0432	74				.0044	62
			.0296	76	5	5	6	.5000	43
			.0243	77				.4625	44
			.0100	81				.2246	51
			.0078	82				.1971	52
			.0061	83				.1078	56
			.0047	84				.0908	57
4	7	8	.5148	58				.0512	60
			.4852	59				.0415	61
			.2170	69				.0264	63
			.1955	70				.0207	64
			.1081	75				.0122	66
			.0945	76				.0092	67
			.0523	80				.0050	69
			.0444	81				.0036	70
			.0262	84	5	5	7	.5000	48
			.0217	85				.4655	49
			.0118	88				.2169	57
			.0095	89				.1920	58
			.0060	91				.1103	62
			.0047	92				.0943	63
4	8	8	.5138	64				.0563	66
			.4862	65				.0467	67
			.2129	76				.0251	70
			.1932	77				.0201	71
			.1116	82				.0124	73
			.0987	83				.0096	74
			.0575	87				.0055	76
			.0497	88				.0041	77
			.0261	92	5	5	8	.5000	53
			.0219	93				.4681	54
			.0102	97				.2104	63
			.0083	98				.1877	64
			.0054	100				.1124	68
			.0043	101				.0973	69
								.0514	73
								.0430	74

(continued)

TABLE A.13. *(continued)*

n_1	n_2	n_3	α	j_α	n_1	n_2	n_3	α	j_α
			$k = 3$					$k = 3$	
5	5	8	.0295	76	5	7	7	.5000	60
			.0241	77				.4707	61
			.0126	80				.2081	71
			.0099	81				.1874	72
			.0060	83				.1032	77
			.0046	84				.0902	78
5	6	6	.5172	48				.0583	81
			.4828	49				.0498	82
			.2052	59				.0300	85
			.1812	59				.0250	86
			.1030	63				.0113	90
			.0879	64				.0090	91
			.0521	67				.0057	93
			.0430	68				.0045	94
			.0286	70	5	7	8	.5000	66
			.0230	71				.4727	67
			.0113	74				.2047	78
			.0087	75				.1856	79
			.0066	76				.1068	84
			.0050	77				.0944	85
5	6	7	.5000	54				.0549	89
			.4683	55				.0474	90
			.2122	64				.0295	93
			.1895	65				.0249	94
			.1141	69				.0119	98
			.0991	70				.0098	99
			.0527	74				.0052	102
			.0443	75				.0041	103
			.0251	78	5	8	8	.5127	72
			.0204	79				.4873	73
			.0105	82				.2117	85
			.0082	83				.1934	86
			.0064	84				.1046	92
			.0049	85				.0932	93
5	6	8	.5147	59				.0564	97
			.4853	60				.0492	98
			.2182	70				.0272	102
			.1968	71				.0232	103
			.1094	76				.0117	107
			.0958	77				.0097	108
			.0533	81				.0054	111
			.0454	82				.0043	112
			.0269	85	6	6	6	.5158	54
			.0224	86				.4842	55
			.0122	89				.2015	65
			.0099	90				.1796	66
			.0062	92				.1072	70
			.0049	93				.0929	71
								.0581	74
								.0490	75

TABLE A.13. (continued)

		$k = 3$					$k = 3$		
n_1	n_2	n_3	α	j_α	n_1	n_2	n_3	α	j_α
6	6	6	.0282	78	6	7	8	.5127	73
			.0231	79				.4873	74
			.0121	82				.2127	86
			.0095	83				.1946	87
			.0058	85				.1057	93
			.0045	86				.0943	94
6	6	7	.5146	60				.0500	99
			.4854	61				.0435	100
			.2196	71				.0279	103
			.1983	72				.0238	104
			.1108	77				.0100	109
			.0972	78				.0083	110
			.0545	82				.0056	112
			.0465	83				.0045	113
			.0278	86	6	8	8	.5118	80
			.0231	87				.4882	81
			.0103	91				.2106	94
			.0083	92				.1936	95
			.0052	94				.1099	101
			.0041	95				.0989	102
6	6	8	.5136	66				.0553	107
			.4864	67				.0487	108
			.2153	78				.0282	112
			.1956	79				.0244	113
			.1010	85				.0110	118
			.0891	86				.0092	119
			.0515	90				.0053	122
			.0444	91				.0044	123
			.0275	94	7	7	7	.5000	74
			.0231	95				.4747	75
			.0110	99				.2041	87
			.0090	100				.1864	88
			.0059	102				.1005	94
			.0047	103				.0894	95
6	7	7	.5000	67				.0540	99
			.4729	68				.0471	100
			.2060	79				.0261	104
			.1869	80				.0223	105
			.1081	85				.0112	109
			.0956	86				.0093	110
			.0560	90				.0052	113
			.0484	91				.0042	114
			.0256	95	7	7	8	.5000	81
			.0215	96				.4764	82
			.0102	100				.2025	95
			.0083	101				.1860	96
			.0054	103				.1048	102
			.0043	104				.0941	103
								.0524	108
								.0461	109

(continued)

TABLE A.13. *(continued)*

	$k = 3$					$k = 4$					
n_1	n_2	n_3	α	j_α	n_1	n_2	n_3	n_4	α	j_α
7	7	8	.0265	113	3	3	3	3	.5276	27
			.0229	114					.4724	28
			.0102	119					.2220	33
			.0086	120					.1823	34
			.0060	122					.1166	36
			.0049	123					.0907	37
									.0515	39
7	8	8	.5111	88					.0374	40
			.4889	89					.0266	41
			.2096	103					.0183	42
			.1938	104					.0123	43
			.1039	111					.0080	44
			.0940	112					.0050	45
			.0544	117						
			.0483	118	4	4	4	4	.5183	48
			.0254	123					.4817	49
			.0221	124					.2172	57
			.0104	129					.1910	58
			.0089	130					.1058	62
			.0053	133					.0895	63
			.0044	134					.0514	66
									.0420	67
8	8	8	.5104	96					.0272	69
			.4896	97					.0215	70
			.2087	112					.0130	72
			.1939	113					.0100	73
			.1085	120					.0056	75
			.0989	121					.0041	76
			.0537	127						
			.0480	128	5	5	5	5	.5132	75
			.0263	133					.4868	76
			.0231	134					.2030	88
			.0115	139					.1846	89
			.0099	140					.1083	94
			.0053	144					.0962	95
			.0045	145					.0574	99
									.0498	100
		$k = 4$							.0271	104
n_1	n_2	n_3	n_4	α	j_α				.0230	105
2	2	2	2	.5492	12				.0113	109
				.4508	13				.0093	110
				.2683	15				.0050	113
				.1929	16				.0040	114
				.1302	17	6	6	6	6	.5101	108
				.0829	18					.4899	109
				.0484	19					.2004	125
				.0262	20					.1863	126
				.0123	21					.1052	133
				.0052	22					.0961	134
				.0016	23					.0529	140
										.0474	141
										.0265	146

TABLE A.13. (continued)

\multicolumn{6}{c}{$k = 4$}
n_1	n_2	n_3	n_4	α	j_α
6	6	6	6	.0234	147
				.0104	153
				.0089	154
				.0057	157
				.0048	158

\multicolumn{7}{c}{$k = 5$}
n_1	n_2	n_3	n_4	n_5	α	j_α
2	2	2	2	2	.5353	20
					.4647	21
					.2110	25
					.1625	26
					.1213	27
					.0878	28
					.0613	29
					.0412	30
					.0265	31
					.0162	32
					.0094	33
					.0051	34
					.0026	35
3	3	3	3	3	.5198	45
					.4802	46
					.2274	53
					.1982	54
					.1049	58
					.0874	59
					.0588	61
					.0475	62
					.0300	64
					.0234	65
					.0102	68
					.0076	69
					.0055	70
					.0039	71
4	4	4	4	4	.5131	80
					.4870	81
					.2059	93
					.1876	94
					.1113	99
					.0991	100
					.0521	105
					.0452	106
					.0288	109
					.0245	110
					.0102	115
					.0084	116
					.0056	118
					.0046	119

\multicolumn{7}{c}{$k = 5$}
n_1	n_2	n_3	n_4	n_5	α	j_α
5	5	5	5	5	.5094	125
					.4906	126
					.2035	143
					.1903	144
					.1039	152
					.0954	153
					.0549	159
					.0497	160
					.0260	166
					.0232	167
					.0110	173
					.0096	174
					.0055	178
					.0047	179
6	6	6	6	6	.5072	180
					.4928	181
					.2075	203
					.1972	204
					.1049	215
					.0984	216
					.0523	225
					.0484	226
					.0251	234
					.0229	235
					.0107	243
					.0097	244
					.0051	250
					.0046	251

\multicolumn{8}{c}{$k = 6$}
n_1	n_2	n_3	n_4	n_5	n_6	α	j_α
2	2	2	2	2	2	.5271	30
						.4729	31
						.2265	36
						.1871	37
						.1215	39
						.0953	40
						.0553	42
						.0408	43
						.0294	44
						.0207	45
						.0142	46
						.0094	47
						.0061	48
						.0038	49
3	3	3	3	3	3	.5000	68
						.4698	69
						.2015	79
						.1806	80

(continued)

TABLE A.13. (continued)

n_1	n_2	n_3	n_4	n_5	n_6	α	j_α
\multicolumn{7}{c}{$k = 6$}	
3	3	3	3	3	3	.1109	84
						.0969	85
						.0533	89
						.0452	90
						.0266	93
						.0220	94
						.0119	97
						.0096	98
						.0060	100
						.0047	101
4	4	4	4	4	4	.5099	120
						.4901	121
						.2049	137
						.1909	138
						.1005	146
						.0918	147
						.0508	153
						.0457	154
						.0257	159
						.0227	160
						.0103	166
						.0089	167
						.0057	170
						.0049	171

n_1	n_2	n_3	n_4	n_5	n_6	α	j_α
\multicolumn{7}{c}{$k = 6$}	
5	5	5	5	5	5	.5000	188
						.4857	189
						.2039	211
						.1938	212
						.1032	223
						.0968	224
						.0515	233
						.0478	234
						.0270	241
						.0248	242
						.0107	251
						.0096	252
						.0051	258
						.0046	259
6	6	6	6	6	6	.5055	270
						.4945	271
						.2007	301
						.1930	302
						.1048	316
						.0998	317
						.0529	329
						.0499	330
						.0252	341
						.0236	342
						.0108	353
						.0100	354
						.0053	362
						.0049	363

Adapted, in part, from R. E. Odeh, On Jonckheere's k-sample test against ordered alternatives, *Technometrics* **13**(1971): 912–918. Reprinted with permission from *Technometrics*. Copyright ©(1971) by the American Statistical Association and the American Society for Quality Control. All rights reserved; and, in part, from A. R. Jonckheere, A distribution-free k-sample test against ordered alternatives, *Biometrika* **41** (1954): 133–145, by permission of Oxford University Press on behalf of the *Biometika* Trustees.

A. TABLES AND CHARTS

TABLE A.14. Selected Exact Critical Values for the Null Distribution of the Peak Known Mack-Wolfe A_p Statistic: k Is the Number of Treatments, p Is the Known Peak of the Umbrella, and $n = n_1 = \cdots = n_k$ Is the Common Sample Size

For given k, n, p, and significance level α, the table entry is $a_{p,\alpha}$ satisfying $P_0\{A_p \geq a_{p,\alpha}\} = \alpha$.

$k = 4$	$k = 5$	$k = 6$
$p = 3, n = 2$	$p = 3, n = 2$	$p = 4, n = 2$
$a_{3,.0893} = 13$	$a_{3,.0654} = 19$	$a_{4,.0885} = 26$
$a_{3,.0476} = 14$	$a_{3,.0377} = 20$	$a_{4,.0400} = 28$
$a_{3,.0060} = 16$	$a_{3,.0080} = 22$	$a_{4,.0075} = 31$
$p = 3, n = 3$	$p = 3, n = 3$	$p = 4, n = 3$
$a_{3,.0768} = 27$	$a_{3,.0882} = 38$	$a_{4,.0962} = 54$
$a_{3,.0364} = 29$	$a_{3,.0385} = 41$	$a_{4,.0419} = 58$
$a_{3,.0080} = 32$	$a_{3,.0086} = 45$	$a_{4,.0078} = 64$
$p = 3, n = 4$	$p = 3, n = 4$	$p = 4, n = 4$
$a_{3,.0836} = 45$	$a_{3,.0946} = 64$	$a_{4,.0993} = 92$
$a_{3,.0415} = 48$	$a_{3,.0475} = 68$	$a_{4,.0448} = 98$
$a_{3,.0091} = 53$	$a_{3,.0099} = 75$	$a_{4,.0094} = 107$
$p = 3, n = 5$	$p = 3, n = 5$	$p = 4, n = 5$
$a_{3,.0943} = 67$	$a_{3,.0944} = 97$	$a_{4,.0997} = 140$
$a_{3,.0497} = 71$	$a_{3,.0449} = 103$	$a_{4,.0469} = 148$
$a_{3,.0094} = 79$	$a_{3,.0090} = 113$	$a_{4,.0097} = 161$
	$p = 4, n = 2$	$p = 5, n = 2$
	$a_{4,.0812} = 21$	$a_{5,.0837} = 31$
	$a_{4,.0301} = 23$	$a_{5,.0419} = 33$
	$a_{4,.0077} = 25$	$a_{5,.0064} = 37$
	$p = 4, n = 3$	$p = 5, n = 3$
	$a_{4,.0943} = 43$	$a_{5,.0880} = 65$
	$a_{4,.0452} = 46$	$a_{5,.0418} = 69$
	$a_{4,.0089} = 51$	$a_{5,.0077} = 76$
	$p = 4, n = 4$	$p = 5, n = 4$
	$a_{4,.0973} = 73$	$a_{5,.0914} = 111$
	$a_{4,.0440} = 78$	$a_{5,.0447} = 117$
	$a_{4,.0083} = 86$	$a_{5,.0099} = 127$
	$p = 4, n = 5$	$p = 5, n = 5$
	$a_{4,.0961} = 111$	$a_{5,.0931} = 169$
	$a_{4,.0490} = 117$	$a_{5,.0473} = 177$
	$a_{4,.0086} = 129$	$a_{5,.0095} = 192$

Adapted from G. A. Mack, A k-sample Wilcoxon rank test for the umbrella alternatives. II. Point of the umbrella unknown, Ph. D. dissertation (1977), Department of Statistics, Ohio State University, with the permission of the author.

TABLE A.15. Selected Critical Values for the Null Distribution of the Peak Unknown Mack-Wolfe $A_{\hat{p}}^*$ Statistic:

$$k = 3, n_1 = n_2 = n_3 = 3(1)10; \ k = 4(1)10, n_1 = \cdots = n_k = 2(1)10.$$

For given k, n, and significance level α, the table entry is $a_{\hat{p},\alpha}^*$ satisfying $P_0\{A_{\hat{p}}^* \geq a_{\hat{p},\alpha}^*\} \approx \alpha$.

k	$n = n_1 = \cdots = n_k$	$a_{\hat{p},.01}^*$	$a_{\hat{p},.05}^*$	$a_{\hat{p},.10}^*$
3	3	2.556	2.324	1.889
	4	2.635	2.196	1.850
	5	2.694	2.166	1.849
	6	2.668	2.102	1.787
	7	2.674	2.158	1.836
	8	2.633	2.082	1.837
	9	2.623	2.111	1.800
	10	2.662	2.112	1.825
4	2	2.554	2.195	1.915
	3	2.700	2.213	1.903
	4	2.708	2.180	1.912
	5	2.738	2.221	1.951
	6	2.646	2.160	1.903
	7	2.744	2.205	1.898
	8	2.756	2.184	1.890
	9	2.794	2.201	1.891
	10	2.771	2.172	1.876
5	2	2.619	2.191	1.894
	3	2.725	2.239	1.969
	4	2.744	2.195	1.963
	5	2.716	2.222	1.960
	6	2.749	2.227	1.972
	7	2.761	2.240	1.951
	8	2.765	2.216	1.937
	9	2.786	2.236	1.925
	10	2.772	2.249	1.943
6	2	2.643	2.226	1.964
	3	2.733	2.242	2.040
	4	2.862	2.265	1.939
	5	2.851	2.251	1.989
	6	2.817	2.242	1.964
	7	2.808	2.257	1.950
	8	2.819	2.256	1.981
	9	2.770	2.266	1.978
	10	2.863	2.278	1.982
7	2	2.756	2.233	1.992
	3	2.782	2.286	1.982
	4	2.802	2.279	1.999
	5	2.831	2.312	2.017
	6	2.785	2.280	1.974
	7	2.823	2.294	1.997
	8	2.889	2.282	1.988
	9	2.826	2.276	1.986
	10	2.919	2.338	2.008

TABLE A.15. (continued)

k	$n = n_1 = \cdots = n_k$	$a^*_{\tilde{p},.01}$	$a^*_{\tilde{p},.05}$	$a^*_{\tilde{p},.10}$
8	2	2.723	2.292	2.016
	3	2.821	2.297	2.021
	4	2.866	2.310	2.039
	5	2.798	2.289	2.022
	6	2.885	2.339	2.027
	7	2.928	2.321	2.034
	8	2.875	2.315	2.034
	9	2.874	2.305	2.021
	10	2.893	2.333	2.028
9	2	2.789	2.287	1.999
	3	2.815	2.283	2.027
	4	2.864	2.305	2.031
	5	2.917	2.310	2.035
	6	2.887	2.325	2.041
	7	2.925	2.341	2.030
	8	2.879	2.325	2.037
	9	2.883	2.293	2.027
	10	2.888	2.340	2.059
10	2	2.818	2.315	2.021
	3	2.802	2.315	2.026
	4	2.910	2.331	2.031
	5	2.874	2.319	2.025
	6	2.912	2.297	2.027
	7	2.922	2.347	2.046
	8	2.895	2.343	2.031
	9	2.948	2.380	2.050
	10	2.905	2.351	2.032

Adapted from G. A. Mack and D. A. Wolfe, *K*-sample rank tests for umbrella alternatives, *Journal of the American Statistical Association* **76** (1981): 175–181. Reprinted with permission from the *Journal of the American Statistical Association*. Copyright © (1981) by the American Statistical Association. All rights reserved.

TABLE A.16. Exact Critical Values for the Steel-Dwass-Critchlow-Fligner Two-Sided All-Treatments Multiple Comparison Procedure:

$$k = 3,\ 2 \leq n_i \leq 7,\ i = 1, 2, 3$$

Additional critical values can be found in Steel (1960, 1961) and Critchlow and Fligner (1991).

For $k = 3$ and a given set of ordered sample sizes $n_1 \leq n_2 \leq n_3$, the entries in the table are the experimentwise error rates α and associated critical values ω_α^* satisfying the relationship

$$P_0\{|W_{uv}^*| < \omega_\alpha^*,\ (u, v) = (1, 2), (1, 3),\ \text{and}\ (2, 3)\} = 1 - \alpha.$$

Since relabeling the samples does not change the value of the statistic maximum $\{|W_{12}^*|, |W_{13}^*|, |W_{23}^*|\}$, the appropriate experimentwise error rates and critical values for an arbitrary sample size configuration (n_1, n_2, n_3) are obtained by simply putting the three sample sizes in increasing order and then entering the table.

n_1	n_2	n_3	ω_α^*	α	n_1	n_2	n_3	ω_α^*	α	n_1	n_2	n_3	ω_α^*	α
2	2	5	2.739	.1720	2	5	5	2.739	.1936	2	7	7	2.898	.1227
								2.807	.0556				2.982	.0379
2	2	6	2.828	.1317				3.102	.0317				3.162	.0262
								3.397	.0159				3.343	.0175
2	2	7	2.898	.1040				3.693	.0079				3.524	.0111
													3.704	.0070
2	3	3	2.778	.1000	2	5	6	2.828	.1152				3.885	.0041
								2.840	.0519				4.066	.0023
2	3	4	2.619	.1730				3.098	.0303				4.247	.0012
			3.000	.0571				3.357	.0173				4.427	.0006
								3.615	.0087					
2	3	5	2.739	.1524				3.873	.0043	3	3	4	3.000	.1043
			2.741	.0714										
			3.162	.0357	2	5	7	2.739	.1608	3	3	5	2.778	.1429
								2.871	.0978				3.162	.0669
2	3	6	2.828	.1113				2.898	.0818					
			2.921	.0476				3.101	.0303	3	3	6	2.778	.1593
			3.286	.0238				3.330	.0177				2.921	.0892
								3.560	.0101				3.286	.0453
2	3	7	2.740	.1149				3.790	.0051					
			2.898	.0843				4.019	.0025	3	3	7	2.778	.1399
			3.062	.0333									3.062	.0634
			3.385	.0167	2	6	6	2.828	.1488				3.385	.0321
								2.944	.0411					
2	4	4	2.858	.0571				3.170	.0260	3	4	4	3.000	.1186
			3.266	.0286				3.397	.0152				3.266	.0286
								3.623	.0087					
2	4	5	2.739	.1457				3.850	.0043	3	4	5	3.000	.1050
			2.771	.0635				4.076	.0022				3.118	.0633
			3.118	.0317									3.162	.0490
			3.464	.0159	2	6	7	2.828	.1454				3.464	.0159
								2.898	.0859					
2	4	6	2.828	.1030				3.031	.0350	3	4	6	3.000	.1015
			3.015	.0381				3.233	.0221				3.015	.0589
			3.317	.0190				3.435	.0140				3.286	.0410
			3.618	.0095				3.637	.0082				3.317	.0190
								3.839	.0047				3.618	.0095
2	4	7	2.673	.1207				4.041	.0023					
			2.898	.0928				4.243	.0012	3	4	7	2.940	.1117
			2.940	.0424									3.000	.0975
			3.207	.0242									3.062	.0550
			3.474	.0121									3.207	.0394
			3.742	.0061									3.385	.0278
													3.474	.0121
													3.742	.0061

TABLE A.16. (continued)

n_1	n_2	n_3	ω_α^*	α	n_1	n_2	n_3	ω_α^*	α	n_1	n_2	n_3	ω_α^*	α
3	5	5	2.807	.1083	3	7	7	3.885	.0041	4	6	6	2.718	.1194
			3.102	.0881				4.066	.0023				2.944	.0995
			3.162	.0753				4.247	.0012				3.015	.0873
			3.397	.0159				4.427	.0006				3.170	.0567
			3.693	.0079									3.317	.0473
					4	4	4	2.858	.1403				3.397	.0313
3	5	6	2.840	.1145				3.266	.0736				3.618	.0253
			2.921	.0967									3.623	.0087
			3.098	.0778	4	4	5	2.858	.1023				3.850	.0043
			3.162	.0669				3.118	.0794				4.076	.0022
			3.286	.0393				3.266	.0532					
			3.357	.0173				3.464	.0302	4	6	7	2.828	.1111
			3.615	.0087									2.940	.0978
			3.873	.0043	4	4	6	2.858	.1119				3.015	.0838
								3.015	.0901				3.031	.0689
3	5	7	2.871	.1001				3.266	.0586				3.207	.0577
			3.062	.0852				3.317	.0363				3.233	.0478
			3.101	.0723				3.618	.0184				3.317	.0405
			3.162	.0616									3.435	.0325
			3.330	.0331	4	4	7	2.858	.1181				3.474	.0272
			3.385	.0259				2.940	.0973				3.618	.0219
			3.560	.0101				3.207	.0672				3.637	.0138
			3.790	.0051				3.266	.0472				3.742	.0105
			4.019	.0025				3.474	.0234				3.839	.0047
								3.742	.0118				4.041	.0023
3	6	6	2.921	.1139									4.243	.0012
			2.944	.0779	4	5	5	2.807	.1019					
			3.170	.0645				3.102	.0819	4	7	7	2.940	.1032
			3.286	.0551				3.118	.0692				2.982	.0754
			3.397	.0152				3.397	.0425				3.162	.0652
			3.464	.0357				3.207	.0578
			3.623	.0087				3.693	.0079				3.343	.0378
			3.850	.0043									3.474	.0320
			4.076	.0022	4	5	6	2.840	.1039				3.524	.0216
								3.015	.0859				3.704	.0178
3	6	7	2.828	.1117				3.098	.0708				3.742	.0151
			2.921	.0981				3.118	.0599				3.885	.0041
			3.031	.0801				3.317	.0466				4.066	.0023
			3.062	.0691				3.357	.0386				4.247	.0012
			3.233	.0554				3.464	.0309				4.427	.0006
			3.286	.0483				3.615	.0176					
			3.385	.0296				3.618	.0135	5	5	5	2.807	.1372
			3.435	.0140				3.873	.0043				3.102	.0819
			3.637	.0082									3.397	.0425
			3.839	.0047	4	5	7	2.871	.1039				3.693	.0218
			4.041	.0023				2.940	.0894					
			4.243	.0012				3.101	.0747	5	5	6	2.840	.1146
								3.118	.0643				3.098	.0796
3	7	7	2.801	.1026				3.207	.0515				3.102	.0586
			2.982	.0895				3.330	.0411				3.357	.0451
			3.062	.0796				3.464	.0344				3.397	.0302
			3.162	.0533				3.474	.0215				3.615	.0232
			3.343	.0453				3.560	.0157				3.693	.0154
			3.385	.0396				3.742	.0108				3.873	.0084
			3.524	.0111				3.790	.0051					
			3.704	.0070				4.019	.0025					

(continued)

TABLE A.16. *(continued)*

n_1	n_2	n_3	ω_α^*	α	n_1	n_2	n_3	ω_α^*	α	n_1	n_2	n_3	ω_α^*	α
5	5	7	2.871	.1082	5	6	7	3.839	.0109	6	6	7	3.623	.0230
			3.101	.0795				3.873	.0087				3.637	.0192
			3.102	.0591				4.019	.0048				3.839	.0128
			3.330	.0457				4.041	.0023				3.850	.0085
			3.397	.0327				4.243	.0012				4.041	.0065
			3.560	.0258									4.076	.0043
			3.693	.0168	5	7	7	2.871	.1118				4.243	.0023
			3.790	.0098				2.982	.0846					
			4.019	.0049				3.101	.0749	6	7	7	2.828	.1171
								3.162	.0546				2.982	.0918
5	6	6	2.840	.1208				3.330	.0470				3.031	.0825
			2.944	.0874				3.343	.0343				3.162	.0618
			3.098	.0749				3.524	.0285				3.233	.0545
			3.170	.0538				3.560	.0249				3.343	.0408
			3.357	.0444				3.704	.0160				3.435	.0352
			3.397	.0297				3.790	.0133				3.524	.0251
			3.615	.0238				3.885	.0087				3.637	.0215
			3.623	.0162				4.019	.0070				3.704	.0152
			3.850	.0122				4.066	.0023				3.839	.0126
			3.873	.0102				4.247	.0012				3.885	.0083
			4.076	.0022				4.427	.0006				4.041	.0067
													4.066	.0045
5	6	7	2.840	.1129	6	6	6	2.944	.1041				4.243	.0034
			2.871	.0966				3.170	.0679				4.247	.0012
			3.031	.0822				3.397	.0408				4.427	.0006
			3.098	.0718				3.623	.0238					
			3.101	.0615				3.850	.0122	7	7	7	2.801	.1311
			3.233	.0508				4.076	.0062				2.982	.0963
			3.330	.0437									3.162	.0684
			3.357	.0374	6	6	7	2.828	.1199				3.343	.0467
			3.435	.0299				2.944	.0943				3.524	.0302
			3.560	.0247				3.031	.0826				3.704	.0194
			3.615	.0202				3.170	.0616				3.885	.0115
			3.637	.0164				3.233	.0528				4.066	.0067
			3.790	.0132				3.397	.0388				4.247	.0034
								3.435	.0333				4.427	.0017

Adapted from D. E. Critchlow and M. A. Fligner, On distribution-free multiple comparisons in the one-way analysis of variance, *Communications in Statistics—Theory and Methods* **20**, Marcel Dekker, Inc., N.Y. (1991): 127–139, with the permission of the authors and the editor of *Communications in Statistics—Theory and Methods*, reprinted by courtesy of Marcel Dekker.

TABLE A.17. Selected Critical Values for the Range of k Independent $N(0, 1)$ Variables: $k = 2(1)20(2)40(10)100$

For a given k and α, the table entry is q_α.

	α								
k	.0001	.0005	.001	.005	.01	.025	.05	.10	.20
2	5.502	4.923	4.654	3.970	3.643	3.170	2.772	2.326	1.812
3	5.864	5.316	5.063	4.424	4.120	3.682	3.314	2.902	2.424
4	6.083	5.553	5.309	4.694	4.403	3.984	3.633	3.240	2.784
5	6.240	5.722	5.484	4.886	4.603	4.197	3.858	3.478	3.037
6	6.362	5.853	5.619	5.033	4.757	4.361	4.030	3.661	3.232
7	6.461	5.960	5.730	5.154	4.882	4.494	4.170	3.808	3.389
8	6.546	6.050	5.823	5.255	4.987	4.605	4.286	3.931	3.520
9	6.618	6.127	5.903	5.341	5.078	4.700	4.387	4.037	3.632
10	6.682	6.196	5.973	5.418	5.157	4.784	4.474	4.129	3.730
11	6.739	6.257	6.036	5.485	5.227	4.858	4.552	4.211	3.817
12	6.791	6.311	6.092	5.546	5.290	4.925	4.622	4.285	3.895
13	6.837	6.361	6.144	5.602	5.348	4.985	4.685	4.351	3.966
14	6.880	6.407	6.191	5.652	5.400	5.041	4.743	4.412	4.030
15	6.920	6.449	6.234	5.699	5.448	5.092	4.796	4.468	4.089
16	6.957	6.488	6.274	5.742	5.493	5.139	4.845	4.519	4.144
17	6.991	6.525	6.312	5.783	5.535	5.183	4.891	4.568	4.195
18	7.023	6.559	6.347	5.820	5.574	5.224	4.934	4.612	4.242
19	7.054	6.591	6.380	5.856	5.611	5.262	4.974	4.654	4.287
20	7.082	6.621	6.411	5.889	5.645	5.299	5.012	4.694	4.329
22	7.135	6.677	6.469	5.951	5.709	5.365	5.081	4.767	4.405
24	7.183	6.727	6.520	6.006	5.766	5.425	5.144	4.832	4.475
26	7.226	6.773	6.568	6.057	5.818	5.480	5.201	4.892	4.537
28	7.266	6.816	6.611	6.103	5.866	5.530	5.253	4.947	4.595
30	7.303	6.855	6.651	6.146	5.911	5.577	5.301	4.997	4.648
32	7.337	6.891	6.689	6.186	5.952	5.620	5.346	5.044	4.697
34	7.370	6.925	6.723	6.223	5.990	5.660	5.388	5.087	4.743
36	7.400	6.957	6.756	6.258	6.026	5.698	5.427	5.128	4.786
38	7.428	6.987	6.787	6.291	6.060	5.733	5.463	5.166	4.826
40	7.455	7.015	6.816	6.322	6.092	5.766	5.498	5.202	4.864
50	7.571	7.137	6.941	6.454	6.228	5.909	5.646	5.357	5.026
60	7.664	7.235	7.041	6.561	6.338	6.023	5.764	5.480	5.155
70	7.741	7.317	7.124	6.649	6.429	6.118	5.863	5.582	5.262
80	7.808	7.387	7.196	6.725	6.507	6.199	5.947	5.669	5.353
90	7.866	7.448	7.259	6.792	6.575	6.270	6.020	5.745	5.433
100	7.918	7.502	7.314	6.850	6.636	6.333	6.085	5.812	5.503

Adapted from H. L. Harter, Tables of range and studentized range, *Annals of Mathematical Statistics* **31** (1960): 1122–1147, with permission of the author, the editor of *The Annals of Mathematical Statistics*, and the Institute of Mathematical Statistics.

TABLE A.18. Exact Critical Values for the Hayter-Stone One-Sided All-Treatments Multiple Comparison Procedure:

$$k = 3, \; n_i = 3(1)7, \; i = 1, 2, 3$$

Additional critical values can be found in Hayter and Stone (1991).
For $k = 3$ and a given set of sample sizes n_1, n_2, n_3, the table entries are the experimentwise error rates α and associated critical values c_α^* satisfying the relationship

$$P_0\{W_{uv}^* < c_\alpha^*, \; (u, v) = (1, 2), (1, 3), \text{ and } (2, 3)\} = 1 - \alpha.$$

By symmetry of the null distribution of maximum $\{W_{12}^*, W_{13}^*, W_{23}^*\}$, the tabled experimentwise error rates and critical values for sample sizes (n_1, n_2, n_3) are also appropriate for sample sizes (n_3, n_2, n_1).

n_1	n_2	n_3	c_α^*	α	n_1	n_2	n_3	c_α^*	α	n_1	n_2	n_3	c_α^*	α
3	3	3	2.778	.1262	3	4	7	2.5	.1106	3	5	7	2.418	.1099
								2.673	.0865				2.641	.0917
3	3	4	2.5	.1388				2.740	.0728				2.740	.0804
			2.778	.0940				2.940	.0602				2.741	.0688
			3	.0524				3	.0516				2.871	.0530
								3.062	.0276				3.062	.0447
3	3	5	2.741	.1060				3.207	.0197				3.101	.0383
			2.778	.0775				3.385	.0139				3.162	.0322
			3.162	.0335				3.474	.0061				3.330	.0166
								3.742	.0030				3.385	.0130
3	3	6	2.556	.1252									3.560	.0051
			2.778	.0872	3	5	3	2.741	.1017				3.790	.0025
			2.921	.0448				2.778	.0736				4.019	.0013
			3.286	.0227				3.162	.0356					
										3	6	3	2.556	.1202
3	3	7	2.740	.1026	3	5	4	2.5	.1043				2.778	.0825
			2.778	.0759				2.741	.0837				2.921	.0474
			3.062	.0318				2.771	.0684				3.286	.0238
			3.385	.0161				3	.0545					
								3.118	.0337	3	6	4	2.5	.1138
3	4	3	2.5	.1381				3.162	.0258				2.556	.0950
			2.778	.0905				3.464	.0079				2.714	.0744
			3	.0569									2.921	.0620
					3	5	5	2.511	.1027				3	.0522
3	4	4	2.5	.1195				2.741	.0852				3.015	.0309
			2.858	.0749				2.807	.0571				3.286	.0214
			3	.0623				3.102	.0461				3.317	.0095
			3.266	.0143				3.162	.0389				3.618	.0048
								3.397	.0079					
3	4	5	2.5	.1076				3.693	.0040	3	6	5	2.556	.1061
			2.741	.0836									2.582	.0866
			2.771	.0706	3	5	6	2.556	.1058				2.741	.0734
			3	.0558				2.582	.0893				2.840	.0604
			3.118	.0317				2.741	.0753				2.921	.0505
			3.162	.0245				2.840	.0602				3.098	.0401
			3.464	.0079				2.921	.0504				3.162	.0343
								3.098	.0409				3.286	.0205
3	4	6	2.5	.1172				3.162	.0348				3.357	.0087
			2.556	.0946				3.286	.0197				3.615	.0043
			2.714	.0765				3.357	.0087				3.873	.0022
			2.921	.0634				3.615	.0043					
			3	.0546				3.873	.0022	3	6	6	2.556	.1066
			3.015	.0295									2.718	.0706
			3.286	.0205									2.921	.0598
			3.317	.0095									2.944	.0406
			3.618	.0048									3.170	.0334

TABLE A.18. (continued)

n_1	n_2	n_3	c_α^*	α	n_1	n_2	n_3	c_α^*	α	n_1	n_2	n_3	c_α^*	α
3	6	6	3.286	.0284	3	7	6	2.424	.1105	4	3	7	2.5	.0118
			3.397	.0076				2.556	.0984				2.673	.0874
			3.623	.0043				2.626	.0822				2.740	.0758
			3.850	.0022				2.740	.0721				2.940	.0603
			4.076	.0011				2.828	.0581				3	.0533
								2.921	.0508				3.062	.0276
3	6	7	2.556	.1009				3.031	.0418				3.207	.0197
			2.626	.0816				3.062	.0358				3.385	.0139
			2.740	.0717				3.233	.0284				3.474	.0061
			2.828	.0593				3.286	.0246				3.742	.0030
			2.921	.0517				3.385	.0153					
			3.031	.0416				3.435	.0070	4	4	4	2.450	.1440
			3.062	.0357				3.637	.0041				2.858	.0753
			3.233	.0288				3.839	.0023				3.266	.0388
			3.286	.0249				4.041	.0012					
			3.385	.0148				4.243	.0006	4	4	5	2.450	.1060
			3.435	.0070									2.771	.0810
			3.637	.0041	3	7	7	2.418	.1344				2.858	.0547
			3.839	.0023				2.440	.0993				3.118	.0419
			4.041	.0012				2.740	.0794				3.266	.0280
			4.243	.0006				2.801	.0537				3.464	.0151
								2.982	.0466				3.118	.0314
3	7	3	2.418	.1364				3.162	.0275					
			2.740	.0966				3.343	.0233	4	4	6	2.450	.1087
			2.778	.0714				3.385	.0203				2.714	.0839
			3.062	.0332				3.524	.0055				2.858	.0602
			3.385	.0166				3.704	.0035				3.015	.0476
								4.066	.0012				3.266	.0309
3	7	4	2.5	.1045				4.247	.0006				3.317	.0182
			2.673	.0846									3.618	.0092
			2.740	.0715	4	3	4	2.5	.1273					
			2.940	.0576				2.858	.0762	4	4	7	2.450	.1134
			3	.0500				3	.0654				2.673	.0891
			3.062	.0287				3.266	.0143				2.858	.0639
			3.207	.0204									2.940	.0515
			3.385	.0144	4	3	5	2.5	.1105				3.207	.0355
			3.474	.0061				2.741	.0854				3.266	.0248
			3.742	.0030				2.771	.0698				3.474	.0117
								3	.0570				3.742	.0059
3	7	5	2.418	.1103				3.118	.0317					
			2.641	.0897				3.162	.0245	4	5	4	2.450	.1019
			2.740	.0788				3.464	.0079				2.771	.0805
			2.741	.0658									2.858	.0532
			2.871	.0521	4	3	6	2.5	.1226				3.118	.0417
			3.062	.0441				2.556	.0978				3.266	.0274
			3.101	.0370				2.714	.0764				3.464	.0159
			3.162	.0314				2.921	.0651					
			3.330	.0172				3	.0543	4	5	5	2.425	.1330
			3.385	.0134				3.015	.0295				2.511	.0971
			3.560	.0051				3.286	.0205				2.771	.0795
			3.790	.0025				3.317	.0095				2.807	.0540
			4.019	.0013				3.618	.0048				3.102	.0430
													3.118	.0359
													3.397	.0220
													3.464	.0183
													3.693	.0040

(continued)

TABLE A.18. (continued)

n_1	n_2	n_3	c_α^*	α	n_1	n_2	n_3	c_α^*	α	n_1	n_2	n_3	c_α^*	α
4	5	6	2.425	.1132	4	6	6	3.317	.0244	4	7	7	2.440	.1039
			2.582	.0928				3.618	.0130				2.620	.0931
			2.714	.0794				3.623	.0043				2.673	.0844
			2.771	.0690				4.076	.0011				2.801	.0610
			2.840	.0550									2.940	.0541
			3.015	.0451	4	6	7	2.424	.1041				2.982	.0394
			3.317	.0242				2.626	.0917				3.207	.0299
			3.464	.0160				2.673	.0820				3.343	.0195
			3.615	.0088				2.714	.0711				3.524	.0110
			3.618	.0067				2.940	.0514				3.704	.0091
			3.873	.0022				3.015	.0444				3.742	.0077
								3.233	.0250				3.885	.0020
4	5	7	2.425	.1112				3.618	.0113				4.066	.0012
			2.641	.0911				3.637	.0069				4.247	.0006
			2.673	.0800				3.742	.0052					
			2.771	.0688				3.839	.0023	5	3	5	2.511	.1043
			2.871	.0550				4.041	.0012				2.741	.0890
			2.940	.0469				4.243	.0006				2.807	.0576
			3.207	.0267									3.102	.0475
			3.330	.0214	4	7	4	2.450	.1072				3.162	.0411
			3.474	.0107				2.673	.0872				3.397	.0079
			3.560	.0079				2.858	.0614				3.693	.0040
			3.742	.0054				2.940	.0510					
			3.790	.0025				3.207	.0346	5	3	6	2.556	.1100
			4.019	.0013				3.266	.0240				2.582	.0893
								3.474	.0121				2.741	.0773
4	6	4	2.450	.1036				3.742	.0061				2.840	.0615
			2.714	.0826									2.921	.0526
			2.858	.0582	4	7	5	2.425	.1067				3.098	.0416
			3.015	.0473				2.641	.0898				3.162	.0362
			3.266	.0301				2.673	.0789				3.286	.0197
			3.317	.0190				2.771	.0664				3.357	.0087
			3.618	.0095				2.871	.0546				3.615	.0043
								2.940	.0466				3.873	.0022
4	6	5	2.425	.1089				3.207	.0266					
			2.582	.0919				3.330	.0211	5	3	7	2.418	.1167
			2.714	.0787				3.474	.0111				2.641	.0942
			2.771	.0668				3.560	.0081				2.740	.0846
			2.840	.0549				3.742	.0056				2.741	.0693
			3.015	.0450				3.790	.0025				2.871	.0535
			3.118	.0308				4.019	.0013				3.062	.0461
			3.317	.0241									3.162	.0330
			3.464	.0158	4	7	6	2.424	.1036				3.330	.0166
			3.615	.0091				2.626	.0912				3.385	.0130
			3.618	.0069				2.673	.0816				3.560	.0051
			3.873	.0022				2.714	.0693				3.790	.0025
								2.940	.0512				4.019	.0013
4	6	6	2.412	.1332				3.015	.0434					
			2.491	.0981				3.207	.0299	5	4	5	2.425	.1366
			2.714	.0857				3.233	.0245				2.511	.0965
			2.718	.0633				3.618	.0111				2.771	.0812
			2.944	.0524				3.637	.0071				2.807	.0535
			3.015	.0455				3.742	.0054				3.102	.0435
			3.170	.0295				3.839	.0023				3.118	.0371
								4.041	.0012				3.397	.0223
								4.243	.0006				3.464	.0188
													3.693	.0040

A. TABLES AND CHARTS

TABLE A.18. *(continued)*

n_1	n_2	n_3	c_α^*	α	n_1	n_2	n_3	c_α^*	α	n_1	n_2	n_3	c_α^*	α
5	4	6	2.425	.1134	5	6	5	2.582	.0964	5	7	6	3.615	.0103
			2.582	.0930				2.807	.0699				3.637	.0084
			2.714	.0816				2.840	.0613				3.839	.0055
			2.771	.0688				3.098	.0416				3.873	.0044
			2.840	.0549				3.102	.0302				4.041	.0012
			3.015	.0460				3.357	.0234				4.243	.0006
			3.118	.0318				3.615	.0119					
			3.317	.0244				3.693	.0078	5	7	7	2.440	.1048
			3.464	.0162				3.873	.0043				2.620	.0941
			3.615	.0088									2.641	.0856
			3.618	.0067	5	6	6	2.491	.1102				2.801	.0656
			3.873	.0022				2.582	.0982				2.871	.0589
								2.718	.0745				2.982	.0445
5	4	7	2.425	.1131				2.840	.0639				3.162	.0284
			2.641	.0925				2.944	.0461				3.330	.0243
			2.673	.0831				3.170	.0281				3.560	.0127
			2.771	.0692				3.357	.0230				3.704	.0082
			2.871	.0553				3.615	.0122				3.790	.0067
			2.940	.0481				3.623	.0083				3.885	.0044
			3.118	.0345				3.873	.0052				4.066	.0012
			3.207	.0271				4.076	.0011				4.247	.0006
			3.474	.0107										
			3.560	.0079	5	6	7	2.424	.1112	6	3	6	2.556	.1146
			3.742	.0054				2.582	.0989				2.718	.0723
			3.790	.0025				2.641	.0767				2.921	.0625
			4.019	.0013				2.828	.0670				2.944	.0411
								2.871	.0506				3.286	.0298
5	5	5	2.511	.1217				3.031	.0434				3.397	.0076
			2.807	.0735				3.233	.0265				3.623	.0043
			3.102	.0432				3.330	.0227				4.076	.0011
			3.397	.0221				3.615	.0105					
			3.693	.0112				3.637	.0084	6	3	7	2.556	.1052
								3.839	.0056				2.626	.0844
5	5	6	2.511	.1121				3.873	.0044				2.740	.0755
			2.582	.0955				4.041	.0012				2.828	.0604
			2.807	.0708				4.243	.0006				2.921	.0536
			2.840	.0606									3.031	.0428
			3.098	.0421	5	7	5	2.511	.1016				3.286	.0259
			3.102	.0308				2.641	.0878				3.385	.0148
			3.357	.0235				2.807	.0659				3.435	.0070
			3.615	.0120				2.871	.0573				3.637	.0041
			3.693	.0079				3.101	.0414				4.041	.0012
			3.873	.0042				3.102	.0303				4.243	.0006
								3.330	.0236					
5	5	7	2.511	.1053				3.560	.0133	6	4	6	2.412	.1398
			2.641	.0885				3.693	.0085				2.491	.0990
			2.807	.0677				3.790	.0050				2.714	.0882
			2.871	.0574				4.019	.0025				2.718	.0633
			3.101	.0421									2.944	.0534
			3.102	.0312	5	7	6	2.424	.1093				3.015	.0473
			3.330	.0239				2.582	.0973				3.317	.0250
			3.560	.0133				2.641	.0772				3.618	.0132
			3.693	.0086				2.828	.0662				3.623	.0043
			3.790	.0049				2.871	.0509				4.076	.0011
			4.019	.0025				3.031	.0430					
								3.233	.0263	6	4	7	2.424	.1052
								3.330	.0226				2.626	.0943

(continued)

TABLE A.18. *(continued)*

n_1	n_2	n_3	c_α^*	α	n_1	n_2	n_3	c_α^*	α	n_1	n_2	n_3	c_α^*	α
6	4	7	2.673	.0861	6	6	7	3.233	.0275	7	4	7	2.940	.0563
			2.714	.0724				3.397	.0202				2.982	.0398
			2.940	.0530				3.623	.0118				3.207	.0311
			3.015	.0447				3.637	.0098				3.343	.0197
			3.207	.0306				3.839	.0065				3.524	.0111
			3.233	.0249				3.850	.0044				3.704	.0092
			3.618	.0114				4.243	.0011				3.742	.0079
			3.637	.0069									3.885	.0020
			3.742	.0052	6	7	6	2.491	.1017				4.066	.0012
			3.839	.0023				2.626	.0917				4.247	.0006
			4.041	.0012				2.718	.0728					
			4.243	.0006				2.828	.0639	7	5	7	2.440	.1058
								2.944	.0495				2.620	.0965
6	5	6	2.582	.1010				3.233	.0276				2.641	.0894
			2.718	.0746				3.397	.0200				2.801	.0669
			2.840	.0654				3.623	.0118				2.871	.0609
			2.944	.0461				3.637	.0099				2.982	.0449
			3.170	.0280				3.839	.0065				3.162	.0286
			3.357	.0233				3.850	.0043				3.330	.0247
			3.615	.0123				4.243	.0012				3.560	.0130
			3.623	.0083									3.704	.0082
			3.873	.0052	6	7	7	2.440	.1054				3.790	.0068
			4.076	.0011				2.620	.0947				3.885	.0044
								2.626	.0863				4.066	.0012
6	5	7	2.582	.1003				2.801	.0686				4.247	.0006
			2.626	.0874				2.828	.0619					
			2.641	.0790				2.982	.0484	7	6	7	2.440	.1049
			2.828	.0677				3.233	.0283				2.620	.0958
			2.871	.0517				3.343	.0212				2.626	.0888
			3.031	.0436				3.637	.0110				2.801	.0692
			3.233	.0265				3.704	.0078				2.828	.0635
			3.330	.0230				3.839	.0064				2.982	.0488
			3.615	.0104				3.885	.0042				3.233	.0288
			3.637	.0084				4.243	.0017				3.343	.0212
			3.839	.0056				4.247	.0006				3.637	.0111
			3.873	.0045									3.704	.0078
			4.041	.0012	7	3	7	2.440	.1026				3.839	.0064
			4.243	.0006				2.620	.0928				3.885	.0042
								2.740	.0850				4.243	.0017
6	6	6	2.491	.1189				2.801	.0551				4.247	.0006
			2.718	.0850				2.982	.0485					
			2.944	.0553				3.162	.0281	7	7	7	2.440	.1238
			3.170	.0356				3.343	.0241				2.620	.0946
			3.397	.0212				3.385	.0213				2.801	.0702
			3.623	.0123				3.524	.0055				2.982	.0510
			3.850	.0062				3.704	.0035				3.162	.0359
			4.076	.0031				4.066	.0012				3.343	.0243
								4.247	.0006				3.704	.0100
6	6	7	2.491	.1041									3.885	.0059
			2.626	.0921	7	4	7	2.440	.1065				4.066	.0034
			2.718	.0743				2.620	.0971				4.247	.0017
			2.944	.0502				2.673	.0897				4.427	.0009
			3.031	.0435				2.801	.0625					

Adapted from A. J. Hayter and G. Stone, Distribution free multiple comparisons for monotonically ordered treatment effects, *Australian Journal of Statistics* **33** (1991): 335–346, with the permission of the authors and the Statistical Society of Australia, Inc.

TABLE A.19. Large-Sample Approximate Critical Values for the Hayter-Stone Multiple Comparison Procedure with k Treatments and Equal Sample Sizes: $k = 3(1)9$ and Approximate Experimentwise Error Rate $\alpha = .01, .05,$ and $.10$

The table entry for k treatments and approximate experimentwise error rate α is the value of d_α for large, equal sample sizes $n_1 = n_2 = \cdots = n_k = n$.

k	$\alpha = .01$	$\alpha = .05$	$\alpha = .10$
3	3.809	2.943	2.491
4	4.112	3.295	2.873
5	4.325	3.539	3.136
6	4.489	3.725	3.335
7	4.621	3.875	3.495
8	4.732	4.000	3.628
9	4.827	4.107	3.742

Adapted from A. J. Hayter, A one-sided studentized range test for testing against a simple ordered alternative, *Journal of the American Statistical Association* **85** (1990): 778–785. Reprinted with permission from the *Journal of the American Statistical Association*. Copyright © (1990) by the American Statistical Association. All rights reserved.

TABLE A.20. Exact Critical Values for the Nemenyi-Damico-Wolfe One-Sided Treatments versus Control Multiple Comparison Procedure:

$$k = 3, n_1 = 1(1)6, 1 \leq n_2 \leq n_3 \leq 6$$
$$k = 4, n_1 = 1, n_2 = n_3 = n_4 = 5; k = 4, n_1 = 2, n_2 = n_3 = n_4 = 2(1)5;$$
$$k = 4, n_1 = 3, n_2 = n_3 = n_4 = 1(1)4; k = 4, n_1 = 4, n_2 = n_3 = n_4 = 1(1)3;$$
$$k = 4, n_1 = 5, n_2 = n_3 = n_4 = 1(1)4; k = 4, n_1 = 6, n_2 = n_3 = n_4 = 1, 2$$

Additional critical values can be found in Damico and Wolfe (1989) and Hollander and Wolfe (1973).
For given k, control sample size n_1, and sample sizes $n_2, \ldots, n_k$ for treatments $2, \ldots, k$, respectively, the table entries are the experimentwise error rates α and associated critical values y_α^* satisfying the relationship

$$P_0\{N^*(R_{.u} - R_{.1}) < y_\alpha^*, u = 2, \ldots, k\} = 1 - \alpha.$$

Since relabeling the noncontrol treatments does not change the value of the statistic maximum $\{N^*(R_{.u} - R_{.1}), u = 2, 3\}$ for the case $k = 3$, the appropriate experimentwise error rates and critical values for an arbitrary treatments sample-size configuration (n_2, n_3) are obtained by simply putting the two treatment sample sizes in increasing order and then entering the table.

							$k = 3$							
n_1	n_2	n_3	y_α^*	α	n_1	n_2	n_3	y_α^*	α	n_1	n_2	n_3	y_α^*	α
1	1	2	6	.0833	1	2	6	30	.1032	1	3	6	14	.1143
								31	.0714				15	.0143
1	1	3	12	.0500				33	.0595				16	.0690
								36	.0317				17	.0512
1	1	4	16	.1000				39	.0198				18	.0369
			20	.0333				42	.0079				19	.0238
								45	.0040				20	.0155
1	1	5	25	.0714									21	.0095
			30	.0238	1	3	3	12	.1143				22	.0048
								13	.0571					
1	1	6	30	.1071				14	.0286	1	4	4	19	.1206
			36	.0536				15	.0905				20	.0889
			42	.0179									21	.0603
					1	3	4	52	.1107				22	.0413
1	2	3	21	.1167				54	.0893				23	.0222
			22	.0667				56	.0714				24	.0127
			24	.0500				57	.0536				25	.0063
			27	.0167				60	.0429				26	.0032
								63	.0214					
1	2	4	16	.1143				66	.0107	1	4	5	105	.1024
			17	.0667				68	.0071				108	.0889
			18	.0571				72	.0036				110	.0778
			20	.0190									112	.0643
			22	.0095	1	3	5	72	.1071				115	.0556
								75	.0933				116	.0460
1	2	5	45	.1071				78	.0615				124	.0254
			46	.0714				80	.0536				125	.0214
			50	.0536				81	.0397				130	.0135
			55	.0298				85	.0298				132	.0087
			60	.0119				87	.0198				136	.0048
			65	.0060				95	.0079				150	.0008
								100	.0040					

A. TABLES AND CHARTS

TABLE A.20. *(continued)*

								$k = 3$						
n_1	n_2	n_3	y_α^*	α	n_1	n_2	n_3	y_α^*	α	n_1	n_2	n_3	y_α^*	α
1	4	6	68	.1091	2	1	3	21	.1000	2	2	5	60	.0106
			69	.0996				24	.0333				65	.0040
			72	.0784				27	.0167					
			74	.0597						2	2	6	26	.1175
			75	.0532	2	1	4	15	.0952				28	.0849
			76	.0442				16	.0857				30	.0786
			80	.0268				18	.0476				31	.0516
			81	.0242				20	.0190				32	.0484
			86	.0104				22	.0095				35	.0254
			87	.0095	2	1	5	41	.1012				36	.0246
			93	.0030				43	.0952				39	.0127
			99	.0009				45	.0893				42	.0063
								50	.0476				45	.0024
1	5	5	28	.1205				55	.0298				48	.0008
			29	.0974				60	.0119					
			30	.0765				65	.0060	2	3	3	22	.1036
			31	.0584									23	.0893
			32	.0433	2	1	6	25	.1389				24	.0786
			33	.0310				28	.0913				25	.0643
			34	.0216				29	.0873				26	.0464
			35	.0144				33	.0556				27	.0393
			36	.0087				36	.0317				28	.0214
			38	.0029				37	.0198				30	.0107
			40	.0007				42	.0079				31	.0071
								45	.0040				33	.0036
1	5	6	185	.1034										
			186	.0945	2	2	2	7	.0667	2	3	4	48	.1040
			192	.0770				8	.0222				50	.0849
			195	.0684									51	.0786
			200	.0534	2	2	3	20	.1190				54	.0667
			204	.0478				22	.0667				56	.0492
			215	.0269				24	.0524				60	.0325
			216	.0236				27	.0190				62	.0222
			228	.0115				30	.0048				66	.0127
			230	.0090									68	.0063
			240	.0049	2	2	4	15	.1048				70	.0048
			255	.0009				16	.0976					
								17	.0571	2	3	5	126	.1024
1	6	6	38	.1192				18	.0500				130	.0964
			40	.0977				19	.0238				138	.0762
			41	.0826				19	.0238				141	.0683
			42	.0683				20	.0214				150	.0567
			43	.0558				22	.0071				153	.0488
			44	.0446				24	.0024				165	.0302
			47	.0205									168	.0214
			49	.0108	2	2	5	39	.1296				180	.0115
			50	.0075				41	.0939				183	.0091
			52	.0032				43	.0886				195	.0048
			55	.0007				46	.0556				215	.0008
								48	.0503					
2	1	2	6	.1000				49	.0489	2	3	6	27	.1141
			7	.0333				51	.0265				28	.0996
								53	.0238				30	.0794
													31	.0682

(continued)

TABLE A.20. *(continued)*

							$k=3$							
n_1	n_2	n_3	y_α^*	α	n_1	n_2	n_3	y_α^*	α	n_1	n_2	n_3	y_α^*	α
2	3	6	33	.0519	2	5	5	63	.0272	3	1	5	62	.1190
			34	.0424				64	.0226				69	.0893
			36	.0277				69	.0105				72	.0635
			37	.0221				70	.0081				75	.0615
			39	.0136				73	.0047				80	.0397
			40	.0093				79	.0010				85	.0258
			43	.0041									90	.0159
			48	.0006	2	5	6	150	.1098				95	.0079
								155	.0984				100	.0040
2	4	4	17	.1048				160	.0888					
			18	.0883				168	.0726	3	1	6	27	.1262
			19	.0686				180	.0553				30	.0952
			20	.0559				183	.0490				31	.0738
			21	.0406				201	.0261				33	.0524
			22	.0286				204	.0239				34	.0512
			23	.0178				220	.0104				36	.0369
			24	.0114				234	.0044				38	.0238
			25	.0057				252	.0010				40	.0155
			26	.0032									42	.0095
			28	.0006	2	6	6	32	.1083				44	.0048
								33	.0978					
2	4	5	90	.1082				34	.0886	3	2	2	19	.1238
			92	.0929				36	.0723				20	.0952
			96	.0789				38	.0562				21	.0857
			102	.0622				39	.0497				22	.0571
			106	.0508				43	.0265				23	.0476
			108	.0486				44	.0216				24	.0286
			118	.0264				47	.0105				25	.0190
			120	.0241				48	.0080				27	.0095
			130	.0104				50	.0043					
			132	.0065				54	.0009	3	2	3	21	.1054
			136	.0039									22	.0929
			146	.0010	3	1	1	9	.1000				23	.0714
													24	.0607
2	4	6	57	.1098	3	1	2	19	.1000				25	.0464
			60	.0963				20	.0833				27	.0250
			63	.0774				21	.0500				29	.0107
			66	.0665				22	.0333				30	.0071
			69	.0512				24	.0167				31	.0036
			70	.0455										
			76	.0257	3	1	3	11	.1071	3	2	4	45	.1032
			78	.0232				12	.0643				47	.0857
			84	.0106				13	.0286				49	.0762
			86	.0069				14	.0143				51	.0611
			88	.0045				15	.0071				54	.0508
			99	.0006									55	.0389
					3	1	4	46	.1143				58	.0317
2	5	5	47	.1093				50	.0786				59	.0238
			49	.0980				54	.0500				64	.0111
			52	.0766				60	.0286				66	.0095
			54	.0659				64	.0143				68	.0048
			57	.0510				68	.0071				78	.0008
			58	.0464				72	.0036					

TABLE A.20. (continued)

$k = 3$

n_1	n_2	n_3	y_α^*	α	n_1	n_2	n_3	y_α^*	α	n_1	n_2	n_3	y_α^*	α
3	2	5	122	.1004	3	3	5	90	.0148	3	4	6	83	.0095
			124	.0992				91	.0094				88	.0050
			132	.0758				96	.0049				97	.0010
			136	.0651				110	.0009					
			145	.0532						3	5	5	66	.1089
			146	.0456	3	3	6	25	.1194				68	.0998
			160	.0286				27	.0950				73	.0788
			162	.0246				29	.0741				76	.0675
			175	.0135				31	.0562				81	.0514
			180	.0099				32	.0528				82	.0480
			195	.0044				33	.0411				92	.0250
			215	.0008				36	.0272				102	.0106
								37	.0196				103	.0093
3	2	6	26	.1100				40	.0112				109	.0046
			27	.0991				41	.0071				119	.0010
			29	.0766				43	.0038					
			30	.0654				48	.0010	3	5	6	140	.1079
			31	.0580									144	.0985
			32	.0483	3	4	4	48	.1051				154	.0787
			35	.0290				49	.0996				160	.0693
			36	.0227				53	.0762				170	.0535
			39	.0121				55	.0651				172	.0490
			40	.0084				58	.0521				194	.0255
			43	.0041				59	.0469				195	.0247
			48	.0006				65	.0268				216	.0101
								66	.0244				232	.0044
3	3	3	11	.1131				72	.0109				254	.0010
			12	.0810				73	.0088					
			13	.0571				77	.0045	3	6	6	30	.1026
			14	.0369				83	.0010				31	.0927
			15	.0214									33	.0749
			16	.0095	3	4	5	117	.1079				36	.0523
			17	.0036				121	.0989				37	.0459
								129	.0797				41	.0257
3	3	4	47	.0993				135	.0680				42	.0216
			50	.0771				145	.0508				46	.0100
			53	.0569				146	.0499				49	.0047
			56	.0510				164	.0263				54	.0009
			57	.0395				165	.0249					
			61	.0248				184	.0102	4	1	1	13	.1333
			68	.0117				185	.0094				14	.0667
			69	.0069				196	.0048					
			74	.0026				214	.0010	4	1	2	15	.0762
			78	.0010									16	.0476
					3	4	6	53	.1096				17	.0190
3	3	5	61	.1011				55	.0979				18	.0095
			62	.0985				59	.0775					
			66	.0765				61	.0685	4	1	3	48	.1000
			70	.0699				66	.0504				50	.0714
			75	.0512				67	.0451				51	.0679
			76	.0403				74	.0266				54	.0464
			84	.0252				75	.0231				57	.0250
			85	.0241				82	.0114				60	.0143
													63	.0071
													66	.0036

(continued)

TABLE A.20. *(continued)*

								$k = 3$					
n_1	n_2	n_3	y_α^*	α	n_1	n_2	n_3	y_α^*	α	n_1	n_2	n_3	y_α^*	α
4	1	4	17	.1127	4	2	4	16	.1035	4	3	5	243	.1087
			18	.0873				17	.0819				252	.0973
			19	.0651				18	.0651				267	.0792
			20	.0460				19	.0489				276	.0693
			21	.0302				20	.0365				300	.0500
			22	.0206				21	.0248				336	.0253
			23	.0111				23	.0098				375	.0110
			24	.0063				25	.0029				378	.0096
			25	.0032				27	.0006				402	.0047
													445	.0010
4	1	5	91	.1040	4	2	5	81	.1079					
			96	.0833				86	.0890	4	3	6	52	.1083
			102	.0635				91	.0714				54	.0964
			105	.0627				100	.0541				57	.0800
			106	.0484				101	.0436				60	.0657
			115	.0333				111	.0238				64	.0496
			120	.0238				116	.0169				72	.0261
			130	.0103				125	.0105				73	.0235
			135	.0056				126	.0072				81	.0104
			140	.0032				135	.0042				82	.0091
			150	.0008				150	.0007				87	.0047
													97	.0009
4	1	6	58	.1160	4	2	6	52	.1100					
			61	.0957				55	.0921	4	4	4	16	.1164
			64	.0788				58	.0766				17	.0956
			68	.0619				61	.0621				18	.0779
			70	.0485				64	.0501				19	.0622
			78	.0273				72	.0297				20	.0495
			81	.0190				73	.0226				22	.0290
			84	.0134				81	.0114				23	.0213
			87	.0087				82	.0082				25	.0102
			93	.0030				90	.0035				27	.0038
			99	.0009				99	.0006				29	.0010
4	2	2	14	.1143	4	3	3	44	.1090	4	4	5	84	.1089
			15	.0857				46	.0952				86	.0963
			16	.0619				49	.0762				91	.0789
			17	.0381				50	.0700				96	.0643
			18	.0238				53	.0533				102	.0505
			19	.0095				54	.0486				115	.0288
			20	.0048				60	.0252				116	.0241
								66	.0105				130	.0115
4	2	3	43	.1071				70	.0038				131	.0089
			46	.0841				75	.0010				141	.0037
			48	.0794									155	.0010
			49	.0627	4	3	4	46	.1076					
			51	.0587				49	.0870	4	4	6	53	.1086
			52	.0460				52	.0701				55	.0940
			57	.0278				53	.0674				58	.0781
			58	.0198				57	.0504				61	.0641
			60	.0175				63	.0281				65	.0508
			62	.0095				64	.0227				66	.0493
			66	.0048				70	.0100				74	.0251
			72	.0008				75	.0048					
								81	.0008					

TABLE A.20. (continued)

					$k = 3$									
n_1	n_2	n_3	y_α^*	α	n_1	n_2	n_3	y_α^*	α	n_1	n_2	n_3	y_α^*	α
4	4	6	84	.0101	5	1	3	69	.0714	5	2	2	53	.0053
			89	.0049				72	.0516				55	.0026
			102	.0007				75	.0357					
								78	.0218	5	2	3	116	.1067
4	5	5	87	.1085				81	.0139				118	.0964
			90	.0983				84	.0079				126	.0782
			96	.0783				87	.0040				129	.0698
			100	.0674									138	.0500
			107	.0504	5	1	4	90	.1095				152	.0246
			121	.0259				93	.0913				165	.0107
			122	.0246				97	.0738				168	.0079
			137	.0099				104	.0563				174	.0044
			146	.0050				105	.0444				189	.0008
			162	.0010				112	.0310					
								116	.0222	5	2	4	81	.1066
4	5	6	280	.1000				124	.0095				83	.0980
			303	.0768				132	.0032				89	.0762
			312	.0683				140	.0008				91	.0683
			336	.0497									98	.0505
			381	.0251	5	1	5	25	.1071				108	.0268
			430	.0103				26	.0909				109	.0245
			432	.0098				27	.0761				120	.0101
			462	.0049				28	.0620				121	.0087
			513	.0010				29	.0498				128	.0040
								31	.0292				140	.0007
4	6	6	57	.1088				32	.0216					
			59	.0978				33	.0155	5	2	5	43	.1089
			63	.0786				34	.0108				45	.0942
			65	.0698				35	.0072				48	.0752
			70	.0512				36	.0043				52	.0528
			71	.0480				39	.0007				53	.0476
			80	.0252									59	.0248
			90	.0105	5	1	6	157	.1073				65	.0107
			91	.0094				163	.0925				66	.0087
			97	.0049				171	.0790				70	.0041
			108	.0010				177	.0664				76	.0010
								186	.0547					
5	1	1	18	.1429				189	.0447	5	2	6	136	.1084
			19	.0952				204	.0278				142	.0946
			20	.0476				216	.0159				151	.0761
								222	.0115				155	.0698
5	1	2	39	.1012				228	.0081				168	.0501
			40	.0952				240	.0036				189	.0260
			41	.0714				258	.0007				190	.0230
			42	.0655									213	.0096
			43	.0476	5	2	2	38	.1005				228	.0041
			45	.0298				41	.0741				252	.0008
			46	.0238				43	.0582					
			48	.0119				44	.0476	5	3	3	58	.1071
			50	.0060				47	.0291				60	.0974
								48	.0212				64	.0749
5	1	3	61	.1171				51	.0106				70	.0506
			64	.0913									78	.0264
			66	.0893										

(continued)

TABLE A.20. *(continued)*

					$k = 3$									
n_1	n_2	n_3	y_α^*	α	n_1	n_2	n_3	y_α^*	α	n_1	n_2	n_3	y_α^*	α
5	3	3	79	.0234	5	4	5	85	.1077	6	1	2	26	.1032
			86	.0104				89	.0933				27	.0833
			87	.0095				93	.0797				28	.0595
			91	.0045				97	.0679				29	.0437
			100	.0006				104	.0531				30	.0278
								105	.0477				31	.0159
5	3	4	240	.1088				118	.0256				32	.0079
			246	.0992				119	.0248				33	.0040
			261	.0795				119	.0248					
			272	.0688				133	.0100	6	1	3	28	.1012
			291	.0504				143	.0049				29	.0845
			292	.0496				159	.0010				30	.0679
			328	.0253									31	.0548
			364	.0105	5	4	6	274	.0998				32	.0405
			366	.0099				294	.0794				33	.0298
			388	.0049				304	.0689				34	.0190
			424	.0010				328	.0498				35	.0131
								372	.0257				36	.0083
5	3	5	64	.1017				374	.0247				37	.0048
			65	.0984				422	.0100					
			70	.0740				452	.0050	6	1	4	59	.1039
			77	.0503				507	.0010				61	.0896
			87	.0262									63	.0758
			88	.0231	5	5	5	23	.1007				68	.0506
			98	.0099				25	.0748				72	.0312
			105	.0047				27	.0540				74	.0234
			115	.0010				28	.0453				78	.0121
								31	.0256				80	.0078
5	3	6	137	.0966				35	.0102				84	.0030
			145	.0797				36	.0078				88	.0009
			151	.0682				38	.0044					
			162	.0515				42	.0009	6	1	5	156	.1080
			163	.0490									161	.0956
			184	.0259	5	5	6	141	.0996				167	.0824
			185	.0245				151	.0770				172	.0711
			208	.0102				157	.0663				184	.0500
			209	.0095				168	.0533				200	.0256
			224	.0047				169	.0483				215	.0110
			250	.0009				192	.0263				220	.0078
								193	.0233				230	.0034
5	4	4	82	.1091				217	.0096				245	.0007
			85	.0970				235	.0043					
			90	.0791				264	.0010	6	1	6	35	.0996
			93	.0696									39	.0583
			100	.0510	5	6	6	146	.0999				40	.0493
			101	.0483				156	.0800				43	.0279
			113	.0257				162	.0695				44	.0223
			114	.0243				175	.0500				47	.0102
			127	.0101				199	.0251				48	.0075
			128	.0094				226	.0099				50	.0037
			136	.0047				243	.0049				53	.0010
			150	.0010				273	.0010					
					6	1	1	25	.1071					
								26	.0714					
								27	.0357					

TABLE A.20. (continued)

$k = 3$

n_1	n_2	n_3	y_α^*	α	n_1	n_2	n_3	y_α^*	α	n_1	n_2	n_3	y_α^*	α
6	2	2	24	.1175	6	2	6	33	.0683	6	4	4	52	.1044
			25	.0984				35	.0527				53	.0976
			26	.0841				36	.0461				57	.0748
			27	.0598				40	.0245				62	.0520
			28	.0571				44	.0111				63	.0480
			29	.0444				45	.0088				70	.0265
			30	.0349				48	.0040				71	.0241
			31	.0238				52	.0010				79	.0103
			33	.0111									80	.0092
			34	.0063	6	3	3	25	.1051				85	.0046
			35	.0032				26	.0910				94	.0009
								27	.0792					
6	2	3	25	.1069				28	.0676	6	4	5	262	.1093
			26	.0926				29	.0578				272	.0970
			27	.0799				30	.0482				290	.0795
			28	.0673				33	.0271				300	.0695
			29	.0567				34	.0212				322	.0499
			30	.0468				36	.0128				365	.0254
			32	.0301				37	.0094				412	.0098
			33	.0232				39	.0047				442	.0048
			35	.0126				42	.0010				492	.0010
			36	.0084										
			38	.0032	6	3	4	52	.0996	6	4	6	55	.1078
			40	.0009				55	.0785				57	.0961
								57	.0679				61	.0758
6	2	4	52	.1076				61	.0497				67	.0513
			53	.0968				69	.0241				68	.0492
			55	.0844				77	.0094				76	.0270
			57	.0732				82	.0049				77	.0241
			62	.0523				90	.0010				87	.0096
			63	.0454									93	.0050
			69	.0247	6	3	5	131	.1044				105	.0009
			76	.0111				136	.0921					
			77	.0083				143	.0798	6	5	5	139	.0994
			81	.0040				148	.0695				149	.0782
			88	.0010				160	.0513				154	.0689
								161	.0461				166	.0496
6	2	5	136	.1034				180	.0268				188	.0251
			141	.0915				181	.0235				213	.0099
			148	.0795				202	.0100				229	.0048
			152	.0700				216	.0046				256	.0010
			165	.0516										
			166	.0455	6	3	6	28	.1051	6	5	6	144	.1000
			185	.0264				29	.0939				154	.0794
			186	.0221				31	.0737				160	.0696
			205	.0109				32	.0650				171	.0498
			208	.0085				34	.0497				195	.0260
			219	.0047				38	.0272				196	.0238
			245	.0006				39	.0231				221	.0098
								43	.0109				238	.0050
6	2	6	29	.1077				44	.0088				269	.0010
			30	.0971				47	.0044					
			32	.0774				52	.0010					

(continued)

TABLE A.20. *(continued)*

	$k=3$					$k=4$						$k=4$				
n_1	n_2	n_3	y_α^*	α	n_1	n_2	n_3	n_4	y_α^*	α	n_1	n_2	n_3	n_4	y_α^*	α
6	6	6	29	.1101	2	4	4	4	33	.0209	4	1	1	1	16	.1143
			30	.0990					35	.0109					17	.0571
			32	.0795					36	.0077					18	.0286
			34	.0627					37	.0049						
			35	.0554					41	.0006	4	2	2	2	19	.1010
			36	.0486											20	.0790
			40	.0277	2	5	5	5	71	.1079					21	.0571
			41	.0238					73	.0981					22	.0406
			46	.0102					77	.0800					23	.0260
			47	.0084					80	.0672					24	.0168
			50	.0046					84	.0524					25	.0086
			56	.0010					85	.0493					26	.0048
									93	.0264					28	.0010
	$k=4$								94	.0237						
n_1	n_2	n_3	n_4	y_α^*	α				102	.0099	4	3	3	3	62	.1088
									108	.0044					64	.0963
1	5	5	5	43	.1061				117	.0010					67	.0792
				44	.0936										69	.0694
				45	.0814	3	1	1	1	11	.1000					73	.0511
				46	.0695					12	.0500					74	.0474
				47	.0587											81	.0252
				48	.0486	3	2	2	2	27	.1032					89	.0099
				51	.0252					28	.0802					94	.0047
				54	.0112					29	.0706					103	.0007
				55	.0083					30	.0532						
				57	.0042					31	.0437	5	1	1	1	22	.1071
				61	.0007					33	.0254					23	.0714
										35	.0119					24	.0357
2	2	2	2	9	.1024					36	.0071					25	.0179
				10	.0548					37	.0048						
				11	.0214							5	2	2	2	49	.1088
				12	.0071	3	3	3	3	16	.0996					50	.0970
										17	.0771					52	.0789
2	3	3	3	32	.1029					18	.0574					54	.0620
				33	.0932					19	.0416					55	.0571
				34	.0790					20	.0282					56	.0479
				35	.0697					21	.0183					60	.0258
				37	.0499					22	.0106					65	.0102
				40	.0251					23	.0058					66	.0069
				43	.0123					24	.0029					68	.0039
				44	.0084					26	.0005					73	.0009
				46	.0039												
				51	.0006	3	4	4	4	70	.1075	5	3	3	3	80	.1083
										72	.0975					82	.0981
2	4	4	4	25	.1070					76	.0791					87	.0760
				26	.0930					79	.0666					89	.0679
				27	.0792					83	.0522					94	.0505
				28	.0673					84	.0491					95	.0475
				29	.0555					93	.0254					104	.0252
				30	.0456					103	.0097					114	.0104
				32	.0281					109	.0047					115	.0092
										120	.0009					121	.0046
																132	.0010

TABLE A.20. (continued)

n_1	n_2	n_3	n_4	y_α^*	α	n_1	n_2	n_3	n_4	y_α^*	α	n_1	n_2	n_3	n_4	y_α^*	α
			$k=4$						$k=4$						$k=4$		
5	4	4	4	117	.1077	6	1	1	1	29	.1071	6	2	2	2	32	.0995
				120	.0981					30	.0714					33	.0843
				127	.0778					31	.0476					34	.0706
				131	.0677					32	.0238					35	.0582
				139	.0503					33	.0119					36	.0474
				155	.0255											38	.0290
				173	.0099											39	.0219
				184	.0049											41	.0111
				204	.0010											42	.0077
																43	.0048
																46	.0009

Adapted from J. A. Damico and D. A. Wolfe, Extended tables of the exact distribution of a rank statistic for treatments versus control multiple comparisons in one-way layout designs, *Communications in Statistics—Theory and Methods* **18**, Marcel Dekker, Inc., N.Y. (1989): 3327–3353, with the permission of the authors and the editor of *Communications in Statistics—Theory and Methods*, reprinted by courtesy of Marcel Dekker, Inc.

TABLE A.21. Cumulative Probabilities for the Distribution of the Maximum of ℓ $N(0, 1)$ Random Variables with Common Correlation ρ:

$$\ell = 1(1)12; \rho = .1(.1).9, \rho = .125(.125).875, \text{ and } \rho = \tfrac{1}{3}, \tfrac{2}{3}$$

For given ℓ and ρ, the entry in the table corresponding to x is the probability that ℓ standard normal random variables with common correlation ρ are simultaneously less than or equal to x. If ℓ, ρ, and x are such that the table entry is $1 - \alpha$, then $m_\alpha^* = x$.

$\rho = .100$

x \ ℓ	1	2	3	4	5	6	7	8	9	10	11	12
0.00	.50000	.26594	.14891	.08709	.05286	.03314	.02137	.01413	.00955	.00659	.00462	.00330
0.10	.53983	.30720	.18271	.11283	.07196	.04720	.03173	.02180	.01528	.01089	.00789	.00580
0.20	.57926	.35089	.22070	.14336	.09574	.06551	.04580	.03263	.02365	.01741	.01299	.00982
0.30	.61791	.39645	.26259	.17875	.12461	.08870	.06431	.04741	.03547	.02689	.02063	.01601
0.40	.65542	.44327	.30791	.21889	.15877	.11724	.08795	.06692	.05157	.04020	.03167	.02519
0.50	.69146	.49068	.35605	.26340	.19820	.15141	.11722	.09186	.07277	.05823	.04701	.03827
0.60	.72575	.53802	.40625	.31173	.24261	.19122	.15244	.12276	.09979	.08180	.06758	.05622
0.70	.75804	.58461	.45769	.36310	.29145	.23642	.19360	.15990	.13310	.11159	.09417	.07995
0.80	.78814	.62984	.50948	.41659	.34393	.28641	.24039	.20320	.17289	.14799	.12737	.11019
0.90	.81594	.67313	.56074	.47120	.39906	.34035	.29216	.25226	.21899	.19105	.16744	.14738
1.00	.84134	.71401	.61064	.52586	.45570	.39715	.34794	.30628	.27080	.24041	.21424	.19159
1.10	.86433	.75211	.65841	.57955	.51267	.45557	.40653	.36416	.32738	.29528	.26716	.24241
1.20	.88493	.78715	.70344	.63131	.56879	.51430	.46657	.42456	.38744	.35451	.32518	.29898
1.30	.90320	.81896	.74522	.68034	.62298	.57205	.52664	.48600	.44951	.41664	.38693	.36002
1.40	.91924	.84747	.78340	.72597	.67430	.62764	.58538	.54698	.51200	.48004	.45078	.42393
1.50	.93319	.87272	.81779	.76774	.72199	.68007	.64155	.60608	.57335	.54307	.51501	.48895

(continued)

TABLE A.21. (continued)

$\rho = .100$

x \ ℓ	1	2	3	4	5	6	7	8	9	10	11	12
1.60	.94520	.89480	.84831	.80534	.76552	.72855	.69415	.66209	.63215	.60415	.57792	.55331
1.70	.95543	.91387	.87503	.83868	.80458	.77255	.74241	.71402	.68723	.66193	.63800	.61534
1.80	.96407	.93016	.89811	.86777	.83902	.81174	.78583	.76118	.73772	.71535	.69402	.67365
1.90	.97128	.94390	.91777	.89281	.86893	.84607	.82417	.80317	.78302	.76367	.74507	.72718
2.00	.97725	.95537	.93431	.91403	.89449	.87563	.85743	.83986	.82288	.80646	.79058	.77520
2.10	.98214	.96483	.94806	.93179	.91601	.90069	.88580	.87135	.85729	.84362	.83032	.81738
2.20	.98610	.97255	.95933	.94645	.93387	.92160	.90961	.89791	.88647	.87529	.86436	.85367
2.30	.98928	.97877	.96848	.95839	.94850	.93880	.92929	.91995	.91079	.90180	.89298	.88431
2.40	.99180	.98374	.97580	.96800	.96031	.95275	.94530	.93797	.93074	.92363	.91662	.90971
2.50	.99379	.98766	.98161	.97564	.96974	.96391	.95816	.95247	.94686	.94131	.93583	.93042
2.60	.99534	.99072	.98616	.98164	.97716	.97273	.96834	.96399	.95969	.95542	.95120	.94702
2.70	.99653	.99309	.98968	.98630	.98294	.97960	.97630	.97301	.96976	.96652	.96332	.96013
2.80	.99744	.99491	.99238	.98987	.98738	.98490	.98244	.97999	.97756	.97514	.97273	.97034
2.90	.99813	.99628	.99443	.99259	.99076	.98894	.98712	.98532	.98352	.98173	.97995	.97818
3.00	.99865	.99730	.99596	.99463	.99330	.99197	.99065	.98934	.98802	.98672	.98541	.98412
3.10	.99903	.99807	.99710	.99614	.99519	.99423	.99328	.99233	.99139	.99044	.98950	.98856
3.20	.99931	.99863	.99794	.99726	.99658	.99590	.99522	.99454	.99387	.99319	.99252	.99185
3.30	.99952	.99903	.99855	.99807	.99759	.99711	.99663	.99615	.99568	.99520	.99472	.99425
3.40	.99966	.99933	.99899	.99865	.99832	.99798	.99765	.99731	.99698	.99665	.99631	.99598
3.50	.99977	.99953	.99930	.99907	.99884	.99861	.99838	.99814	.99791	.99768	.99745	.99722

$\rho = .125$

x \ ℓ	1	2	3	4	5	6	7	8	9	10	11	12
0.00	.50000	.26995	.15492	.09344	.05875	.03825	.02566	.01766	.01244	.00894	.00654	.00486
0.10	.53983	.31117	.18905	.11992	.07885	.05346	.03721	.02650	.01925	.01424	.01070	.00815
0.20	.57926	.35475	.22724	.15106	.10360	.07296	.05258	.03865	.02892	.02198	.01694	.01322
0.30	.61791	.40014	.26919	.18692	.13332	.09729	.07242	.05486	.04220	.03291	.02598	.02074
0.40	.65542	.44673	.31443	.22733	.16815	.12685	.09734	.07583	.05987	.04783	.03863	.03150
0.50	.69146	.49388	.36235	.27192	.20804	.16185	.12777	.10217	.08265	.06755	.05573	.04636
0.60	.72575	.54092	.41221	.32011	.25264	.20223	.16390	.13430	.11114	.09279	.07809	.06619
0.70	.75804	.58719	.46321	.37115	.30142	.24769	.20568	.17239	.14570	.12407	.10637	.09177
0.80	.78814	.63209	.51449	.42414	.35357	.29764	.25274	.21629	.18641	.16167	.14103	.12368
0.90	.81594	.67506	.56519	.47811	.40814	.35121	.30441	.26555	.23300	.20554	.18220	.16223
1.00	.84134	.71564	.61450	.53204	.46404	.40737	.35973	.31935	.28487	.25524	.22961	.20733
1.10	.86433	.75346	.66170	.58495	.52013	.46493	.41756	.37663	.34105	.30995	.28263	.25850
1.20	.88493	.78824	.70618	.63593	.57531	.52264	.47659	.43610	.40032	.36856	.34023	.31487
1.30	.90320	.81983	.74746	.68419	.62853	.57929	.53549	.49637	.46127	.42966	.40108	.37517
1.40	.91924	.84816	.78520	.72912	.67891	.63376	.59299	.55603	.52241	.49174	.46366	.43790
1.50	.93319	.87325	.81920	.77025	.72574	.68512	.64792	.61377	.58230	.55325	.52636	.50141
1.60	.94520	.89520	.84940	.80731	.76850	.73261	.69935	.66843	.63963	.61276	.58762	.56407
1.70	.95543	.91417	.87585	.84018	.80688	.77573	.74653	.71911	.69331	.66900	.64605	.62435
1.80	.96407	.93038	.89871	.86890	.84077	.81418	.78902	.76516	.74252	.72099	.70050	.68097
1.90	.97128	.94406	.91821	.89363	.87022	.84790	.82658	.80621	.78671	.76804	.75013	.73295
2.00	.97725	.95548	.93463	.91463	.89542	.87697	.85921	.84212	.82565	.80976	.79443	.77962
2.10	.98214	.96491	.94827	.93221	.91667	.90164	.88708	.87298	.85931	.84605	.83317	.82067
2.20	.98610	.97260	.95948	.94673	.93433	.92226	.91051	.89907	.88791	.87703	.86642	.85606
2.30	.98928	.97880	.96858	.95858	.94881	.93925	.92990	.92075	.91179	.90302	.89442	.88600
2.40	.99180	.98376	.97587	.96813	.96052	.95305	.94572	.93851	.93142	.92446	.91761	.91087
2.50	.99379	.98768	.98165	.97572	.96987	.96411	.95843	.95283	.94731	.94187	.93649	.93119
2.60	.99534	.99073	.98618	.98169	.97725	.97286	.96852	.96422	.95998	.95578	.95163	.94753
2.70	.99653	.99310	.98970	.98633	.98299	.97968	.97641	.97316	.96994	.96675	.96359	.96046

A. TABLES AND CHARTS

TABLE A.21. *(continued)*

$\rho = .125$

x \ ℓ	1	2	3	4	5	6	7	8	9	10	11	12
2.80	.99744	.99491	.99239	.98989	.98741	.98495	.98251	.98008	.97767	.97528	.97290	.97054
2.90	.99813	.99628	.99443	.99260	.99078	.98897	.98716	.98537	.98359	.98182	.98005	.97830
3.00	.99865	.99731	.99597	.99464	.99331	.99199	.99068	.98937	.98807	.98677	.98548	.98419
3.10	.99903	.99807	.99711	.99615	.99519	.99424	.99330	.99235	.99141	.99047	.98954	.98860
3.20	.99931	.99863	.99794	.99726	.99658	.99590	.99523	.99455	.99388	.99321	.99254	.99187
3.30	.99952	.99903	.99855	.99807	.99759	.99711	.99664	.99616	.99568	.99521	.99473	.99426
3.40	.99966	.99933	.99899	.99865	.99832	.99799	.99765	.99732	.99699	.99665	.99632	.99599
3.50	.99977	.99953	.99930	.99907	.99884	.99861	.99838	.99815	.99792	.99768	.99745	.99722

$\rho = .200$

x \ ℓ	1	2	3	4	5	6	7	8	9	10	11	12
0.00	.50000	.28205	.17307	.11301	.07741	.05508	.04043	.03044	.02343	.01836	.01463	.01182
0.10	.53983	.32317	.20810	.14149	.10033	.07358	.05546	.04277	.03363	.02689	.02181	.01791
0.20	.57926	.36644	.24684	.17430	.12769	.09634	.07448	.05876	.04716	.03842	.03170	.02646
0.30	.61791	.41134	.28893	.21138	.15966	.12375	.09797	.07897	.06463	.05360	.04496	.03810
0.40	.65542	.45728	.33392	.25248	.19624	.15599	.12631	.10390	.08662	.07305	.06224	.05351
0.50	.69146	.50364	.38120	.29720	.23725	.19309	.15970	.13389	.11358	.09734	.08417	.07336
0.60	.72575	.54980	.43008	.34496	.28229	.23486	.19813	.16912	.14584	.12689	.11126	.09824
0.70	.75804	.59513	.47982	.39502	.33077	.28089	.24137	.20952	.18348	.16191	.14385	.12859
0.80	.78814	.63906	.52961	.44658	.38194	.33054	.28895	.25477	.22633	.20240	.18206	.16462
0.90	.81594	.68109	.57869	.49874	.43491	.38301	.34016	.30431	.27398	.24806	.22572	.20632
1.00	.84134	.72076	.62632	.55059	.48871	.43733	.39411	.35733	.32572	.29832	.27439	.25334
1.10	.86433	.75773	.67185	.60127	.54234	.49246	.44975	.41283	.38063	.35234	.32732	.30505
1.20	.88493	.79175	.71472	.64999	.59484	.54732	.50596	.46968	.43760	.40905	.38351	.36052
1.30	.90320	.82266	.75451	.69605	.64532	.60088	.56162	.52669	.49542	.46726	.44177	.41861
1.40	.91924	.85040	.79091	.73891	.69302	.65221	.61565	.58270	.55285	.52568	.50083	.47802
1.50	.93319	.87500	.82373	.77816	.73733	.70051	.66710	.63663	.60872	.58305	.55935	.53740
1.60	.94520	.89654	.85293	.81357	.77782	.74515	.71517	.68753	.66196	.63821	.61609	.59543
1.70	.95543	.91518	.87856	.84504	.81421	.78572	.75929	.73468	.71170	.69017	.66995	.65092
1.80	.96407	.93112	.90074	.87260	.84641	.82197	.79906	.77755	.75728	.73814	.72004	.70287
1.90	.97128	.94460	.91971	.89639	.87448	.85382	.83431	.81582	.79827	.78159	.76569	.75051
2.00	.97725	.95587	.93571	.91664	.89856	.88138	.86502	.84940	.83448	.82019	.80649	.79334
2.10	.98214	.96518	.94904	.93365	.91894	.90486	.89135	.87838	.86590	.85388	.84230	.83111
2.20	.98610	.97279	.96002	.94775	.93594	.92456	.91358	.90297	.89271	.88278	.87315	.86381
2.30	.98928	.97893	.96894	.95928	.94993	.94086	.93206	.92352	.91521	.90714	.89927	.89161
2.40	.99180	.98385	.97612	.96860	.96128	.95416	.94721	.94043	.93381	.92734	.92102	.91484
2.50	.99379	.98773	.98182	.97603	.97038	.96485	.95944	.95413	.94893	.94384	.93884	.93393
2.60	.99534	.99077	.98629	.98189	.97758	.97334	.96918	.96509	.96106	.95710	.95321	.94937
2.70	.99653	.99312	.98977	.98646	.98321	.98000	.97684	.97372	.97065	.96762	.96463	.96168
2.80	.99744	.99492	.99243	.98998	.98755	.98515	.98278	.98044	.97812	.97583	.97357	.97133
2.90	.99813	.99629	.99446	.99265	.99086	.98909	.98733	.98560	.98387	.98217	.98047	.97880
3.00	.99865	.99731	.99598	.99467	.99336	.99207	.99078	.98950	.98824	.98698	.98574	.98450
3.10	.99903	.99807	.99712	.99617	.99523	.99429	.99336	.99243	.99151	.99060	.98969	.98879
3.20	.99931	.99863	.99795	.99727	.99660	.99593	.99526	.99460	.99394	.99329	.99263	.99198
3.30	.99952	.99903	.99856	.99808	.99760	.99713	.99666	.99619	.99572	.99525	.99479	.99433
3.40	.99966	.99933	.99899	.99866	.99833	.99799	.99766	.99733	.99701	.99668	.99635	.99603
3.50	.99977	.99953	.99930	.99907	.99884	.99861	.99838	.99815	.99793	.99770	.99747	.99724

(continued)

TABLE A.21. (continued)

$\rho = .250$

x \ ℓ	1	2	3	4	5	6	7	8	9	10	11	12
0.00	.50000	.29022	.18532	.12648	.09066	.06748	.05176	.04067	.03262	.02660	.02202	.01845
0.10	.53983	.33127	.22090	.15615	.11528	.08800	.06900	.05530	.04513	.03739	.03140	.02666
0.20	.57926	.37435	.25994	.18994	.14418	.11274	.09027	.07370	.06116	.05145	.04380	.03768
0.30	.61791	.41893	.30210	.22771	.17745	.14193	.11593	.09634	.08123	.06933	.05981	.05208
0.40	.65542	.46445	.34691	.26918	.21500	.17569	.14623	.12358	.10578	.09154	.07997	.07045
0.50	.69146	.51030	.39377	.31393	.25660	.21393	.18125	.15563	.13514	.11849	.10477	.09331
0.60	.72575	.55588	.44203	.36137	.30181	.25641	.22089	.19252	.16946	.15044	.13453	.12109
0.70	.75804	.60060	.49095	.41080	.35003	.30265	.26484	.23410	.20870	.18745	.16945	.15405
0.80	.78814	.64391	.53980	.46144	.40054	.35202	.31256	.27994	.25259	.22937	.20946	.19223
0.90	.81594	.68530	.58784	.51245	.45248	.40372	.36337	.32946	.30062	.27582	.25429	.23545
1.00	.84134	.72437	.63439	.56299	.50495	.45686	.41638	.38185	.35208	.32616	.30339	.28326
1.10	.86433	.76077	.67884	.61227	.55704	.51046	.47063	.43617	.40607	.37954	.35600	.33497
1.20	.88493	.79427	.72067	.65954	.60787	.56355	.52509	.49137	.46155	.43498	.41115	.38965
1.30	.90320	.82472	.75947	.70419	.65663	.61520	.57874	.54637	.51742	.49135	.46774	.44625
1.40	.91924	.85205	.79498	.74571	.70263	.66457	.63063	.60014	.57258	.54751	.52460	.50357
1.50	.93319	.87630	.82701	.78374	.74533	.71094	.67991	.65173	.62599	.60236	.58058	.56042
1.60	.94520	.89755	.85552	.81806	.78434	.75378	.72590	.70031	.67672	.65489	.63459	.61566
1.70	.95543	.91595	.88057	.84858	.81943	.79270	.76807	.74525	.72403	.70422	.68568	.66826
1.80	.96407	.93170	.90228	.87533	.85050	.82750	.80609	.78609	.76734	.74971	.73308	.71736
1.90	.97128	.94503	.92086	.89847	.87761	.85812	.83982	.82258	.80630	.79089	.77626	.76234
2.00	.97725	.95618	.93656	.91819	.90093	.88465	.86925	.85464	.84074	.82750	.81486	.80276
2.10	.98214	.96540	.94966	.93478	.92069	.90729	.89453	.88234	.87068	.85950	.84877	.83844
2.20	.98610	.97294	.96046	.94856	.93721	.92634	.91592	.90591	.89628	.88699	.87804	.86938
2.30	.98928	.97904	.96925	.95986	.95083	.94214	.93375	.92565	.91782	.91024	.90288	.89575
2.40	.99180	.98392	.97633	.96900	.96191	.95505	.94840	.94194	.93567	.92957	.92363	.91784
2.50	.99379	.98778	.98196	.97630	.97081	.96547	.96026	.95519	.95024	.94541	.94068	.93607
2.60	.99534	.99080	.98638	.98208	.97787	.97376	.96974	.96591	.96196	.95819	.95449	.95086
2.70	.99653	.99314	.98983	.98658	.98340	.98027	.97721	.97420	.97125	.96835	.96549	.96269
2.80	.99744	.99494	.99247	.99005	.98767	.98533	.98302	.98076	.97852	.97632	.97414	.97200
2.90	.99813	.99630	.99449	.99270	.99094	.98920	.98749	.98580	.98413	.98248	.98085	.97924
3.00	.99865	.99732	.99600	.99470	.99341	.99214	.99088	.98963	.98840	.98718	.98597	.98478
3.10	.99903	.99807	.99713	.99619	.99526	.99433	.99342	.99251	.99162	.99072	.98984	.98897
3.20	.99931	.99863	.99795	.99728	.99662	.99596	.99530	.99465	.99400	.99336	.99272	.99209
3.30	.99952	.99904	.99856	.99808	.99761	.99714	.99668	.99622	.99576	.99530	.99484	.99439
3.40	.99966	.99933	.99899	.99866	.99833	.99800	.99768	.99735	.99703	.99670	.99638	.99606
3.50	.99977	.99954	.99930	.99907	.99885	.99862	.99839	.99816	.99794	.99771	.99749	.99727

$\rho = .300$

x \ ℓ	1	2	3	4	5	6	7	8	9	10	11	12
0.00	.50000	.29849	.19774	.14031	.10453	.08077	.06421	.05221	.04325	.03638	.03101	.02673
0.10	.53983	.33949	.23381	.17109	.13073	.10321	.08358	.06907	.05805	.04949	.04267	.03718
0.20	.57926	.38237	.27314	.20575	.16104	.12975	.10694	.08977	.07650	.06602	.05759	.05070
0.30	.61791	.42664	.31534	.24413	.19547	.16056	.13459	.11466	.09901	.08647	.07625	.06781
0.40	.65542	.47175	.35996	.28591	.23397	.19565	.16662	.14397	.12590	.11122	.09911	.09898
0.50	.69146	.51710	.40641	.33064	.27594	.23487	.20306	.17782	.15738	.14055	.12649	.11461
0.60	.72575	.56213	.45405	.37774	.32124	.27789	.24370	.21613	.19348	.17458	.15861	.14496
0.70	.75804	.60624	.50219	.42653	.36916	.32423	.28817	.25863	.23403	.21325	.19549	.18015
0.80	.78814	.64892	.55011	.47627	.41898	.37325	.33592	.30488	.27869	.25630	.23696	.22008
0.90	.81594	.68969	.59715	.52617	.46991	.42417	.38623	.35424	.32690	.30327	.28263	.26446
1.00	.84134	.72815	.64264	.57545	.52111	.47615	.43829	.40593	.37795	.35349	.33191	.31275

TABLE A.21. *(continued)*

						$\rho = .300$						
x \ ℓ	1	2	3	4	5	6	7	8	9	10	11	12
1.10	.86433	.76398	.68602	.62337	.57172	.52828	.49117	.45905	.43095	.40613	.38404	.36423
1.20	.88493	.79695	.72682	.66925	.62093	.57967	.54394	.51264	.48495	.46026	.43807	.41801
1.30	.90320	.82692	.76465	.71252	.66303	.62949	.59568	.56573	.53895	.51484	.49299	.47308
1.40	.91924	.85383	.79926	.75273	.71240	.67698	.64554	.61737	.59195	.56886	.54776	.52838
1.50	.93319	.87771	.83049	.78954	.75354	.72151	.69275	.66674	.64304	.62132	.60132	.58283
1.60	.94520	.89866	.85830	.82277	.79110	.76260	.73674	.71311	.69140	.67135	.65276	.63544
1.70	.95543	.91681	.88275	.85234	.82489	.79991	.77703	.75594	.73640	.71822	.70124	.68532
1.80	.96407	.93236	.90397	.87828	.85483	.83328	.81335	.79483	.77754	.76135	.74612	.73175
1.90	.97128	.94552	.92214	.90073	.88099	.86267	.84558	.82958	.81454	.80035	.78093	.77420
2.00	.97725	.95654	.93751	.91990	.90350	.88816	.87373	.86013	.84726	.83504	.82342	.81233
2.10	.98214	.96566	.95036	.93606	.92262	.90995	.89795	.88657	.87572	.86538	.85548	.84600
2.20	.98610	.97313	.96096	.94949	.93863	.92831	.91848	.90909	.90010	.89148	.88319	.87521
2.30	.98928	.97917	.96961	.96052	.95186	.94357	.93563	.92800	.92067	.91359	.90676	.90015
2.40	.99180	.98401	.97658	.96947	.96264	.95608	.94975	.94365	.93774	.93203	.92648	.92111
2.50	.99379	.98784	.98213	.97663	.97132	.96619	.96122	.95640	.95172	.94717	.94274	.93843
2.60	.99534	.99084	.98650	.98230	.97822	.97426	.97040	.96665	.96299	.95943	.95594	.95254
2.70	.99653	.99317	.98990	.98673	.98363	.98061	.97766	.97478	.97196	.96920	.96650	.96385
2.80	.99744	.99495	.99252	.99015	.98783	.98555	.98333	.98114	.97900	.97690	.97483	.97280
2.90	.99813	.99631	.99452	.99276	.99104	.98935	.98769	.98605	.98445	.98286	.98130	.97977
3.00	.99865	.99732	.99602	.99474	.99347	.99223	.99101	.98980	.98861	.98743	.98627	.98513
3.10	.99903	.99808	.99714	.99621	.99530	.99439	.99350	.99262	.99175	.99088	.99003	.98919
3.20	.99931	.99863	.99796	.99730	.99664	.99599	.99535	.99471	.99408	.99346	.99284	.99223
3.30	.99952	.99904	.99856	.99809	.99763	.99717	.99671	.99625	.99581	.99536	.99492	.99448
3.40	.99966	.99933	.99900	.99867	.99834	.99802	.99769	.99737	.99706	.99674	.99643	.99612
3.50	.99977	.99953	.99930	.99908	.99885	.99862	.99840	.99818	.99796	.99773	.99752	.99730

						$\rho = 1/3$						
x \ ℓ	1	2	3	4	5	6	7	8	9	10	11	12
0.00	.50000	.30409	.20613	.14974	.11413	.09012	.07311	.06061	.05113	.04375	.03790	.03318
0.10	.53983	.34504	.24251	.18121	.14133	.11377	.09384	.07893	.06744	.05838	.05111	.04516
0.20	.57926	.38780	.28201	.21642	.17250	.14144	.11853	.10109	.08744	.07654	.06768	.06030
0.30	.61791	.43187	.32423	.25516	.20763	.17324	.14739	.12737	.11149	.09863	.08805	.07922
0.40	.65542	.47670	.36871	.29711	.24653	.20913	.18049	.15796	.13983	.12497	.11260	.10217
0.50	.69146	.52173	.41489	.34180	.28887	.24891	.21777	.19288	.17258	.15574	.14157	.12950
0.60	.72575	.56638	.46212	.38866	.33418	.29223	.25897	.23200	.20971	.19100	.17508	.16140
0.70	.75804	.61010	.50974	.43703	.38187	.33858	.30371	.27501	.25100	.23061	.21310	.19789
0.80	.78814	.65236	.55707	.48618	.43124	.38733	.35140	.32143	.29605	.27426	.25536	.23880
0.90	.81594	.69271	.60344	.53536	.48150	.43771	.40135	.37062	.34429	.32145	.30144	.28377
1.00	.84134	.73076	.64824	.58381	.53186	.48892	.45275	.42180	.39499	.37150	.35073	.33223
1.10	.86433	.76621	.69093	.63084	.58150	.54009	.50473	.47412	.44731	.42360	.40245	.38346
1.20	.88493	.79882	.73104	.67581	.62968	.59039	.55641	.52665	.50033	.47683	.45570	.43657
1.30	.90320	.82847	.76822	.71818	.67570	.63902	.60691	.57849	.55310	.53024	.50951	.49062
1.40	.91924	.85509	.80223	.75753	.71900	.68529	.65545	.62876	.60471	.58286	.56291	.54458
1.50	.93319	.87873	.83292	.79354	.75912	.72863	.70134	.67670	.65430	.63380	.61493	.59748
1.60	.94520	.89946	.86026	.82604	.79573	.76858	.74403	.72166	.70115	.68224	.66471	.64841
1.70	.95543	.91743	.88431	.85497	.82866	.80484	.78310	.76312	.74466	.72752	.71153	.69656
1.80	.96407	.93283	.90518	.88036	.85785	.83727	.81831	.80076	.78441	.76913	.75479	.74129
1.90	.97128	.94588	.92307	.90235	.88336	.86583	.84956	.83437	.82013	.80674	.79409	.78213
2.00	.97725	.95681	.93821	.92114	.90533	.89062	.87686	.86392	.85172	.84017	.82921	.81878
2.10	.98214	.96586	.95088	.93698	.92401	.91184	.90037	.88952	.87922	.86942	.86007	.85113
2.20	.98610	.97327	.96134	.95017	.93966	.92973	.92031	.91134	.90279	.89460	.88676	.87922
2.30	.98928	.97927	.96988	.96102	.95261	.94462	.93699	.92969	.92269	.91595	.90947	.90322

(continued)

TABLE A.21. (continued)

$\rho = 1/3$

$x \backslash \ell$	1	2	3	4	5	6	7	8	9	10	11	12
2.40	.99180	.98408	.97677	.96982	.96319	.95684	.95074	.94488	.93923	.93378	.92851	.92341
2.50	.99379	.98789	.98226	.97688	.97171	.96673	.96193	.95729	.95280	.94845	.94423	.94012
2.60	.99534	.99088	.98659	.98247	.97849	.97463	.97090	.96728	.96376	.96034	.95701	.95376
2.70	.99653	.99319	.98997	.98684	.98381	.98087	.97801	.97522	.97250	.96984	.96725	.96471
2.80	.99744	.99497	.99256	.99023	.98795	.98573	.98356	.98144	.97937	.97734	.97535	.97339
2.90	.99813	.99632	.99454	.99281	.99112	.98947	.98784	.98625	.98469	.98316	.98165	.98017
3.00	.99865	.99733	.99604	.99477	.99353	.99231	.99111	.98993	.98877	.98763	.98650	.98540
3.10	.99903	.99808	.99715	.99623	.99533	.99444	.99357	.99270	.99185	.99101	.99018	.98937
3.20	.99931	.99863	.99797	.99731	.99666	.99602	.99539	.99477	.99415	.99354	.99294	.99234
3.30	.99951	.99904	.99857	.99810	.99764	.99718	.99673	.99629	.99585	.99541	.99498	.99455
3.40	.99966	.99933	.99900	.99867	.99835	.99803	.99771	.99739	.99708	.99677	.99647	.99616
3.50	.99977	.99953	.99931	.99908	.99885	.99863	.99841	.99819	.99797	.99775	.99754	.99732

$\rho = .375$

$x \backslash \ell$	1	2	3	4	5	6	7	8	9	10	11	12
0.00	.50000	.31118	.21677	.16179	.12653	.10235	.08493	.07189	.06185	.05392	.04753	.04229
0.10	.53983	.35208	.25352	.19408	.15491	.12745	.10730	.09199	.08002	.07046	.06267	.05622
0.20	.57926	.39468	.29320	.22992	.18711	.15645	.13356	.11590	.10191	.09059	.08128	.07349
0.30	.61791	.43850	.33544	.26907	.22304	.18940	.16385	.14383	.12777	.11464	.10371	.09450
0.40	.65542	.48300	.37975	.31120	.26249	.22620	.19816	.17589	.15779	.14282	.13025	.11955
0.50	.69146	.52763	.42558	.35582	.30511	.26660	.23637	.21202	.19200	.17526	.16105	.14886
0.60	.72575	.57182	.47230	.40237	.35040	.31021	.27817	.25203	.23027	.21189	.19615	.18252
0.70	.75804	.61503	.51929	.45020	.39778	.35652	.32315	.29555	.27234	.25252	.23540	.22045
0.80	.78814	.65678	.56587	.49862	.44655	.40489	.37070	.34200	.31775	.29677	.27840	.26239
0.90	.81594	.69661	.61143	.54690	.49598	.45458	.42015	.39099	.36592	.34411	.32493	.30792
1.00	.84134	.73415	.65537	.59434	.54530	.50483	.47072	.44150	.41613	.39385	.37411	.35646
1.10	.86433	.76911	.69719	.64028	.59376	.55481	.52158	.49279	.46756	.44521	.42524	.40726
1.20	.88493	.80127	.73645	.68413	.64066	.60376	.57190	.54402	.51934	.49730	.47746	.45948
1.30	.90320	.83050	.77282	.72539	.68537	.65095	.62089	.59433	.57061	.54925	.52988	.51221
1.40	.91924	.85676	.80608	.76367	.72736	.69574	.66783	.64293	.62051	.60017	.58159	.56453
1.50	.93319	.88007	.83610	.79868	.76622	.73761	.71211	.68914	.66829	.64924	.63172	.61554
1.60	.94520	.90053	.86284	.83028	.80166	.77616	.75321	.73237	.71330	.69575	.67951	.66442
1.70	.95543	.91827	.88636	.85840	.83352	.81113	.79079	.77217	.75501	.73912	.72431	.71048
1.80	.96407	.93348	.90679	.88309	.86177	.84239	.82464	.80826	.79306	.77889	.76562	.75315
1.90	.97128	.94637	.92431	.90449	.88646	.86993	.85466	.84047	.82722	.81479	.80309	.79204
2.00	.97725	.95718	.93916	.92278	.90775	.89385	.88091	.86880	.85743	.84670	.83654	.82690
2.10	.98214	.96613	.95159	.93823	.92586	.91433	.90352	.89334	.88372	.87460	.86593	.85765
2.20	.98610	.97347	.96186	.95110	.94106	.93162	.92272	.91429	.90628	.89864	.89135	.88436
2.30	.98928	.97942	.97026	.96170	.95364	.94603	.93880	.93192	.92534	.91904	.91300	.90719
2.40	.99180	.98418	.97704	.97031	.96393	.95787	.95208	.94654	.94122	.93610	.93117	.92642
2.50	.99379	.98796	.98246	.97723	.97224	.96747	.96290	.95850	.95425	.95016	.94620	.94236
2.60	.99534	.99093	.98673	.98271	.97886	.97516	.97159	.96815	.96481	.96158	.95844	.95539
2.70	.99653	.99322	.99006	.98701	.98407	.98124	.97849	.97583	.97324	.97072	.96827	.96588
2.80	.99744	.99499	.99263	.99034	.98813	.98598	.98389	.98186	.97988	.97795	.97606	.97422
2.90	.99813	.99633	.99458	.99289	.99124	.98964	.98807	.98654	.98504	.98358	.98215	.98074
3.00	.99865	.99734	.99606	.99482	.99361	.99242	.99126	.99012	.98901	.98791	.98684	.98578
3.10	.99903	.99809	.99717	.99626	.99538	.99451	.99366	.99283	.99201	.99120	.99041	.98962
3.20	.99931	.99864	.99798	.99733	.99670	.99607	.99545	.99485	.99425	.99366	.99308	.99251
3.30	.99952	.99904	.99857	.99811	.99766	.99721	.99677	.99634	.99591	.99549	.99507	.99466
3.40	.99966	.99933	.99900	.99868	.99836	.99805	.99773	.99743	.99712	.99682	.99652	.99623
3.50	.99977	.99954	.99931	.99908	.99886	.99864	.99843	.99821	.99800	.99778	.99758	.99737

TABLE A.21. *(continued)*

$\rho = .400$

$x \backslash \ell$	1	2	3	4	5	6	7	8	9	10	11	12
0.00	.50000	.31549	.22324	.16917	.13419	.10998	.09238	.07909	.06876	.06053	.05385	.04833
0.10	.53983	.35636	.26020	.20193	.16325	.13592	.11570	.10022	.08803	.07821	.07015	.06344
0.20	.57926	.39887	.29999	.23812	.19603	.16569	.14287	.12514	.11100	.09949	.08996	.08194
0.30	.61791	.44255	.34223	.27750	.23241	.19928	.17396	.15401	.13791	.12466	.11358	.10419
0.40	.65542	.48685	.38643	.31971	.27216	.23657	.20894	.18689	.16887	.15390	.14126	.13044
0.50	.69146	.53123	.43205	.36428	.31491	.27730	.24765	.22368	.20388	.18725	.17309	.16087
0.60	.72575	.57515	.47847	.41063	.36017	.32105	.28977	.26416	.24277	.22463	.20904	.19550
0.70	.75804	.61807	.52508	.45814	.40734	.36730	.33484	.30794	.28523	.26579	.24895	.23419
0.80	.78814	.65950	.57122	.50612	.45575	.41542	.38229	.35450	.33080	.31033	.29244	.27666
0.90	.81594	.69902	.61629	.55387	.50468	.46470	.43141	.40319	.37888	.35770	.33904	.32245
1.00	.84134	.73625	.65972	.60071	.55339	.51436	.48147	.45327	.42876	.40721	.38808	.37095
1.10	.86433	.77091	.70102	.64600	.60115	.56364	.53165	.50394	.47963	.45808	.43881	.42144
1.20	.88493	.80280	.73978	.68919	.64729	.61179	.58117	.55438	.53067	.50949	.49040	.47309
1.30	.90320	.83178	.77567	.72979	.69122	.65813	.62927	.60378	.58104	.56056	.54198	.52502
1.40	.91924	.85781	.80847	.76743	.73243	.70205	.67527	.65141	.62994	.61047	.59269	.57636
1.50	.93319	.88093	.83807	.80185	.77055	.74305	.71859	.69660	.67666	.65845	.64171	.62625
1.60	.94520	.90121	.86445	.83290	.80529	.78078	.75876	.73881	.72058	.79383	.68833	.67394
1.70	.95543	.91881	.88765	.86053	.83652	.81498	.79547	.77764	.76124	.74607	.73195	.71877
1.80	.96407	.93390	.90781	.88480	.86420	.84554	.82850	.81281	.79829	.78476	.77212	.76024
1.90	.97128	.94669	.92511	.90583	.88840	.87247	.85780	.84420	.83153	.81967	.80951	.79798
2.00	.97725	.95742	.93977	.92383	.90927	.89585	.88341	.87180	.86091	.95066	.84098	.83180
2.10	.98214	.96631	.95205	.93903	.92703	.91589	.90548	.89571	.88649	.87777	.86949	.86161
2.20	.98610	.97360	.96221	.95170	.94194	.93282	.92423	.91613	.90844	.90113	.89416	.88750
2.30	.98928	.97951	.97051	.96214	.95431	.94693	.93995	.93332	.92700	.92097	.91519	.90964
2.40	.99180	.98425	.97722	.97063	.96442	.95853	.95293	.94759	.94247	.93756	.93284	.92829
2.50	.99379	.98801	.98258	.97746	.97259	.96796	.96352	.95927	.95518	.95124	.94744	.94376
2.60	.99534	.99096	.98681	.98287	.97911	.97551	.97204	.96871	.96548	.96237	.95935	.95642
2.70	.99653	.99325	.99012	.98712	.98425	.98148	.97881	.97623	.97372	.97129	.96893	.96663
2.80	.99744	.99501	.99267	.99042	.98825	.98615	.98411	.98214	.98022	.97835	.97653	.97475
2.90	.99813	.99634	.99461	.99294	.99132	.98975	.98822	.98673	.98528	.98386	.98247	.98111
3.00	.99865	.99734	.99608	.99485	.99366	.99250	.99136	.99025	.98916	.98810	.98706	.98604
3.10	.99903	.99809	.99718	.99629	.99542	.99456	.99373	.99291	.99211	.99133	.99055	.98979
3.20	.99931	.99864	.99799	.99735	.99672	.99610	.99550	.99491	.99432	.99375	.99318	.99263
3.30	.99952	.99904	.99858	.99812	.99768	.99724	.99680	.99638	.99596	.99554	.99513	.99473
3.40	.99966	.99933	.99901	.99869	.99837	.99806	.99775	.99745	.99715	.99686	.99656	.99628
3.50	.99977	.99954	.99931	.99909	.99887	.99865	.99844	.99822	.99801	.99781	.99760	.99740

$\rho = .500$

$x \backslash \ell$	1	2	3	4	5	6	7	8	9	10	11	12
0.00	.50000	.33333	.25000	.20000	.16667	.14286	.12500	.11111	.10000	.09091	.08333	.07692
0.10	.53983	.37408	.28772	.23446	.19823	.17192	.15193	.13620	.12350	.11302	.10422	.09673
0.20	.57926	.41623	.32788	.27192	.23308	.20445	.18240	.16487	.15058	.13869	.12863	.12000
0.30	.61791	.45931	.37006	.31206	.27104	.24032	.21637	.19713	.18129	.16801	.15669	.14692
0.40	.65542	.50282	.41379	.35450	.31177	.27930	.25368	.23287	.21559	.20099	.18845	.17757
0.50	.69146	.54624	.45855	.39874	.35486	.32104	.29403	.27187	.25332	.23750	.22384	.21190
0.60	.72575	.58906	.50376	.44425	.39981	.36509	.33704	.31380	.29417	.27732	.26266	.24977
0.70	.75804	.63079	.54885	.49041	.44605	.41091	.38220	.35820	.33775	.32006	.30458	.29089
0.80	.78814	.67098	.59323	.53661	.49293	.45788	.42894	.40451	.38353	.36525	.34915	.33483
0.90	.81594	.70922	.63638	.58224	.53983	.50536	.47660	.45211	.43090	.41231	.39582	.38107
1.00	.84134	.74520	.67778	.62670	.58608	.55267	.52450	.50030	.47920	.46056	.44394	.42898
1.10	.86433	.77866	.71701	.66944	.63107	.59913	.57195	.54839	.52770	.50930	.49279	.47786

(continued)

TABLE A.21. (continued)

						$\rho = .500$						
x \ ℓ	1	2	3	4	5	6	7	8	9	10	11	12
1.20	.88493	.80941	.75373	.71000	.67424	.64414	.61828	.59568	.57569	.55780	.54166	.52698
1.30	.90320	.83734	.78766	.74799	.71510	.68713	.66287	.64151	.62248	.60535	.58980	.57559
1.40	.91924	.86243	.81864	.78309	.75326	.72763	.70519	.68529	.66744	.65127	.63652	.62297
1.50	.93319	.88471	.84656	.81513	.78843	.76525	.74480	.72652	.71001	.69498	.68119	.66847
1.60	.94520	.90427	.87143	.84398	.82040	.79973	.78134	.76479	.74975	.73598	.72328	.71151
1.70	.95543	.92124	.89331	.86964	.84908	.83090	.81459	.79982	.78631	.77388	.76236	.75163
1.80	.96407	.93581	.91233	.89218	.87448	.85870	.84444	.83144	.81948	.80841	.79811	.78848
1.90	.97128	.94817	.92867	.91172	.89669	.88317	.87087	.85958	.84914	.83943	.83036	.82183
2.00	.97725	.95855	.94253	.92845	.91585	.90442	.89395	.88429	.87531	.86691	.85902	.85159
2.10	.98214	.96717	.95416	.94260	.93217	.92264	.91385	.90569	.89806	.89090	.88415	.87776
2.20	.98610	.97424	.96380	.95443	.94590	.93805	.93077	.92397	.91758	.91156	.90586	.90045
2.30	.98928	.97998	.97169	.96419	.95730	.95092	.94497	.93939	.93411	.92911	.92437	.91984
2.40	.99180	.98459	.97809	.97215	.96665	.96153	.95673	.95220	.94790	.94382	.93992	.93619
2.50	.99379	.98825	.98321	.97856	.97423	.97017	.96635	.96272	.95927	.95597	.95282	.94979
2.60	.99534	.99113	.98726	.98366	.98030	.97712	.97411	.07125	.06851	.96588	.06117	.06094
2.70	.99653	.99336	.99043	.98768	.98509	.98264	.98030	.97807	.97592	.07386	.07188	.06996
2.80	.99744	.99509	.99298	.99081	.98884	.98697	.98517	.98345	.98180	.98020	.97865	.97716
2.90	.99813	.99640	.99476	.09321	.99173	.99032	.98896	.98765	.98639	.98517	.98398	.98283
3.00	.99865	.99738	.99618	.99504	.99394	.99288	.99187	.99089	.98994	.98901	.98812	.98724
3.10	.99903	.99812	.99724	.90641	.99560	.99483	.99408	.99335	.00264	.09195	.90128	.09063
3.20	.99931	.99866	.99803	.99743	.99684	.99628	.99573	.00520	.99468	.99417	.99367	.99319
3.30	.99952	.99905	.99861	.99817	.99776	.99735	.99695	.09657	.09619	.99582	.99546	.09511
3.40	.99966	.99934	.99902	.99872	.99842	.99813	.09785	.99757	.09730	.09704	.00678	.00652
3.50	.99977	.99954	.99932	.99911	.99890	.99870	.99850	.99830	.99811	.99792	.99774	.99756

						$\rho = .600$						
x \ ℓ	1	2	3	4	5	6	7	8	9	10	11	12
0.00	.50000	.35242	.27862	.23345	.20259	.17999	.16263	.14882	.13753	.12810	.12009	.11320
0.10	.53983	.39304	.31701	.26941	.23636	.21185	.19282	.17756	.16499	.15442	.14540	.13759
0.20	.57926	.43482	.35742	.30790	.27297	.24674	.22618	.20954	.19573	.18406	.17404	.16532
0.30	.61791	.47732	.39946	.34859	.31214	.28445	.26252	.24463	.22969	.21698	.20600	.19639
0.40	.65542	.52003	.44266	.39105	.35352	.32466	.30160	.28263	.26668	.25302	.24116	.23073
0.50	.69146	.56248	.48650	.43480	.39665	.36699	.34305	.32321	.30641	.29194	.27930	.26815
0.60	.72575	.60419	.53046	.47931	.44104	.41094	.38644	.36596	.34851	.33339	.32012	.30835
0.70	.75804	.64470	.57400	.52403	.48612	.45599	.43123	.41039	.39251	.37693	.36319	.35095
0.80	.78814	.68360	.61660	.56840	.53133	.50155	.47688	.45595	.43788	.42204	.40801	.39545
0.90	.81594	.72054	.65780	.61187	.57608	.54704	.52277	.50204	.48402	.46815	.45402	.44131
1.00	.84134	.75521	.69715	.65392	.61982	.59186	.56830	.54804	.53032	.51463	.50059	.48791
1.10	.86433	.78740	.73429	.69410	.66202	.63545	.61289	.59334	.57616	.56086	.54710	.53462
1.20	.88493	.81694	.76893	.73203	.70222	.67730	.65598	.63738	.62093	.60621	.59291	.58080
1.30	.90320	.84376	.80085	.76739	.74004	.71697	.69707	.67961	.66407	.65010	.63743	.62584
1.40	.91924	.86782	.82993	.79994	.77516	.75408	.73576	.71958	.70510	.69202	.68010	.66916
1.50	.93319	.88918	.85610	.82955	.80739	.78836	.77171	.75691	.74361	.73152	.72046	.71027
1.60	.94520	.90793	.87938	.85616	.83658	.81963	.80469	.79133	.77926	.76825	.75813	.74876
1.70	.95543	.92420	.89984	.87979	.86269	.84778	.83455	.82266	.81185	.80194	.79281	.78432
1.80	.96407	.93817	.91763	.90050	.88577	.87282	.86125	.85079	.84124	.83245	.82431	.81673
1.90	.97128	.95004	.93291	.91846	.90591	.89481	.88482	.87574	.86742	.85972	.85256	.84587
2.00	.97725	.96000	.94588	.93383	.92328	.91387	.90536	.89759	.89042	.88377	.87756	.87173
2.10	.98213	.96828	.95677	.94684	.93808	.93020	.92304	.91647	.91038	.90470	.89938	.89438
2.20	.98610	.97508	.96580	.95772	.95052	.94402	.93807	.93258	.92747	.92269	.91820	.91396
2.30	.98928	.98061	.97321	.96671	.96087	.95556	.95068	.94615	.94192	.93795	.93421	.93066

TABLE A.21. (continued)

$\rho = .600$

x \ ℓ	1	2	3	4	5	6	7	8	9	10	11	12
2.40	.99180	.98505	.97922	.97405	.96937	.96509	.96114	.95745	.95399	.95073	.94765	.94472
2.50	.99379	.98859	.98405	.97998	.97628	.97287	.96970	.96673	.96394	.96130	.95880	.95641
2.60	.99534	.99137	.98787	.98471	.98181	.97913	.97662	.97427	.97204	.96993	.96792	.96600
2.70	.99653	.99354	.99087	.98844	.98620	.98411	.98216	.98031	.97856	.97689	.97530	.97378
2.80	.99744	.99521	.99319	.99135	.98964	.98803	.98652	.98509	.98373	.98243	.98119	.97999
2.90	.99813	.99648	.99498	.99359	.99230	.99108	.98993	.98884	.98779	.98679	.98583	.98490
3.00	.99865	.99744	.99633	.99530	.99433	.99342	.99255	.99173	.99094	.99017	.98944	.98873
3.10	.99903	.99816	.99735	.99659	.99588	.99520	.99455	.99394	.99334	.99277	.99222	.99169
3.20	.99931	.99868	.99810	.99755	.99703	.99653	.99606	.99560	.99516	.99474	.99433	.99393
3.30	.99952	.99907	.99865	.99826	.99788	.99752	.99718	.99685	.99652	.99621	.99591	.99562
3.40	.99966	.99935	.99905	.99877	.99851	.99825	.99800	.99776	.99753	.99730	.99708	.99687
3.50	.99977	.99955	.99934	.99915	.99896	.99877	.99860	.99843	.99826	.99810	.99794	.99779

$\rho = .625$

x \ ℓ	1	2	3	4	5	6	7	8	9	10	11	12
0.00	.50000	.35745	.28618	.24234	.21224	.19007	.17295	.15926	.14802	.13860	.13056	.12362
0.10	.53983	.39804	.32471	.27864	.24651	.22257	.20391	.18886	.17643	.16594	.15694	.14913
0.20	.57926	.43974	.36518	.31736	.28351	.25800	.23792	.22160	.20803	.19651	.18658	.17791
0.30	.61791	.48208	.40717	.35815	.32294	.29610	.27479	.25733	.24271	.23023	.21942	.20994
0.40	.65542	.54259	.45021	.40059	.36443	.33656	.31423	.29580	.28027	.26693	.25532	.24510
0.50	.69146	.56680	.49381	.44419	.40753	.37897	.35587	.33668	.32040	.30634	.29404	.28316
0.60	.72575	.60822	.53744	.48843	.45173	.42283	.39927	.37954	.36270	.34908	.33523	.32381
0.70	.75804	.64842	.58058	.53276	.49649	.46763	.44390	.42389	.40670	.39170	.37845	.36663
0.80	.78814	.68699	.62273	.57665	.54124	.51280	.48921	.46918	.45187	.43669	.42321	.41114
0.90	.81594	.72359	.66343	.61956	.58543	.55775	.53460	.51482	.49762	.48246	.46894	.45677
1.00	.84134	.75792	.70225	.66100	.62852	.60192	.57950	.56022	.54336	.52842	.51503	.50294
1.10	.86433	.78979	.73886	.70053	.67000	.64477	.62334	.60479	.58847	.57394	.56087	.54902
1.20	.88493	.81901	.77297	.73779	.70945	.68581	.66560	.64798	.63240	.61846	.60586	.59439
1.30	.90320	.84552	.80438	.77248	.74650	.72463	.70580	.68929	.67461	.66141	.64943	.63848
1.40	.91924	.86931	.83296	.80438	.78086	.76089	.74357	.72929	.71464	.70231	.69107	.68076
1.50	.93319	.89043	.85868	.83337	.81233	.79433	.77861	.76465	.75212	.74074	.73034	.72076
1.60	.94520	.90896	.88154	.85941	.84082	.82479	.81069	.79811	.78675	.77640	.76690	.75811
1.70	.95543	.92504	.90164	.88250	.86628	.85218	.83970	.82851	.81835	.80905	.80048	.79253
1.80	.96407	.93985	.91909	.90275	.88877	.87652	.86561	.85577	.84680	.83856	.83093	.82383
1.90	.97128	.95057	.93409	.92029	.90938	.89788	.88846	.87992	.87211	.86489	.95919	.85194
2.00	.97725	.96042	.94682	.93531	.92529	.91639	.90836	.90105	.89432	.88809	.88228	.97694
2.10	.98214	.96860	.95751	.94802	.93968	.93223	.92548	.91929	.91358	.90826	.90329	.89862
2.20	.98610	.97533	.96637	.95864	.95180	.94563	.94002	.93485	.93006	.92558	.92138	.91742
2.30	.98928	.98080	.97365	.96742	.96186	.95683	.95222	.94796	.94399	.94027	.93676	.93345
2.40	.99180	.98519	.97956	.97459	.97014	.96608	.96234	.95886	.95562	.95256	.94968	.94694
2.50	.99379	.98869	.98429	.98039	.97686	.97362	.97062	.96783	.96520	.96273	.96038	.95815
2.60	.99534	.99145	.98805	.98501	.98224	.97969	.97732	.97510	.97301	.97103	.96914	.96735
2.70	.99653	.99359	.99100	.98866	.98652	.98453	.98268	.98093	.97929	.97772	.97623	.97480
2.80	.99744	.99525	.99329	.99151	.98987	.98834	.98691	.98556	.98427	.98305	.98188	.98076
2.90	.99813	.99651	.99505	.99371	.99247	.99130	.99021	.98917	.98819	.98724	.98634	.98547
3.00	.99865	.99746	.99638	.99538	.99445	.99358	.99276	.99197	.99122	.99050	.98981	.98915
3.10	.99903	.99817	.99738	.99665	.99596	.99531	.99470	.99411	.99355	.99301	.99249	.99199
3.20	.99931	.99869	.99812	.99759	.99709	.99661	.99616	.99572	.99531	.99491	.99452	.99414
3.30	.99952	.99908	.99867	.99828	.99792	.99758	.99725	.99693	.99662	.99633	.99604	.99577
3.40	.99966	.99935	.99906	.99879	.99853	.99828	.99805	.99782	.99760	.99738	.99718	.99697
3.50	.99977	.99955	.99935	.99916	.99897	.99880	.99863	.99847	.99831	.99815	.99801	.99786

(continued)

TABLE A.21. *(continued)*

$\rho = 2/3$

$x \backslash \ell$	1	2	3	4	5	6	7	8	9	10	11	12
0.00	.50000	.36614	.29921	.25775	.22902	.20769	.19109	.17771	.16665	.15732	.14931	.14234
0.10	.53983	.40668	.33798	.29459	.26411	.24124	.22328	.20871	.19659	.18630	.17744	.16969
0.20	.57920	.44822	.37852	.33366	.30173	.27751	.25833	.24266	.22955	.21837	.20868	.20018
0.30	.61791	.49031	.42041	.37459	.34154	.31622	.29601	.27937	.26537	.25336	.24292	.23372
0.40	.05542	.53248	.46319	.41696	.38317	.35703	.33599	.31856	.30380	.29108	.27996	.27014
0.50	.09146	.57426	.50637	.46027	.42616	.39950	.37788	.35985	.34449	.33119	.31952	.30916
0.60	.72575	.61520	.54944	.50403	.47001	.44317	.42123	.40281	.38704	.37331	.36121	.35043
0.70	.75804	.65487	.59190	.54770	.51418	.48750	.46551	.44694	.43095	.41697	.40459	.39351
0.80	.78814	.69288	.63328	.59076	.55815	.53194	.51020	.49171	.47570	.46164	.44913	.43790
0.90	.81594	.72890	.67313	.63272	.60137	.57596	.55471	.53654	.52072	.50676	.49429	.48305
1.00	.84134	.76266	.71106	.67311	.64335	.61900	.59851	.58087	.56544	.55175	.53948	.52839
1.10	.86433	.79394	.74676	.71154	.68362	.66059	.64106	.62416	.60929	.59605	.58414	.57332
1.20	.88493	.82263	.77996	.74766	.72179	.70026	.68189	.66590	.65176	.63911	.62769	.61728
1.30	.90320	.84863	.81049	.78122	.75753	.73766	.72059	.70565	.69237	.68044	.66963	.65974
1.40	.91924	.87196	.83824	.81202	.79059	.77248	.75681	.74302	.73072	.71961	.70950	.70023
1.50	.93319	.99264	.86317	.83997	.82081	.80450	.79030	.77774	.76647	.75626	.74693	.73835
1.60	.94520	.91079	.88533	.86502	.84810	.83359	.82088	.80957	.79939	.79012	.78163	.77378
1.70	.95543	.92654	.90478	.88722	.87246	.85971	.84847	.83842	.82933	.82102	.81338	.80630
1.80	.90407	.94006	.92168	.90667	.89394	.88287	.87305	.86423	.85621	.84886	.84207	.83577
1.90	.97128	.95154	.93618	.92350	.91266	.90316	.89469	.88704	.88006	.87363	.86768	.86213
2.00	.97725	.96119	.94849	.93790	.92877	.92072	.91350	.90695	.90095	.89541	.89025	.88543
2.10	.98213	.96920	.95883	.95009	.94249	.93574	.92967	.92413	.91903	.91430	.90990	.90576
2.20	.98610	.97579	.96741	.96027	.95402	.94844	.94339	.93876	.93448	.93050	.92678	.92328
2.30	.98928	.98115	.97445	.96869	.96361	.95905	.95489	.95107	.94753	.94422	.94112	.93819
2.40	.99180	.98546	.98016	.97557	.97149	.96780	.96443	.96131	.95841	.95570	.95314	.95072
2.50	.99379	.98889	.98475	.98113	.97789	.97494	.97224	.96973	.96738	.96518	.96310	.96113
2.60	.99534	.99159	.98839	.98557	.98302	.98070	.97855	.97656	.97468	.97292	.97125	.96966
2.70	.99653	.99370	.99125	.98907	.98710	.98529	.98361	.98203	.98056	.97916	.97783	.97657
2.80	.99744	.99532	.99347	.99181	.99029	.98890	.98760	.98638	.98522	.98413	.98309	.98210
2.90	.99813	.99656	.99517	.99392	.99277	.99171	.99071	.98978	.98889	.98805	.98724	.98647
3.00	.99805	.99750	.99647	.99554	.99467	.99387	.99312	.99241	.99173	.99109	.99047	.98989
3.10	.99903	.99819	.99744	.99675	.99612	.99552	.99496	.99442	.99391	.99343	.99296	.99252
3.20	.99931	.99871	.99816	.99766	.99720	.99676	.99634	.99595	.99557	.99521	.99486	.99452
3.30	.99952	.99909	.99870	.99833	.99800	.99768	.99737	.99708	.99681	.99654	.99628	.99603
3.40	.99966	.99936	.99908	.99883	.99858	.99835	.99813	.99792	.99772	.99753	.99734	.99716
3.50	.99977	.99956	.99936	.99918	.99901	.99884	.99869	.99854	.99839	.99825	.99812	.99799

$\rho = .700$

$x \backslash \ell$	1	2	3	4	5	6	7	8	9	10	11	12
0.00	.50000	.37341	.31011	.27069	.24319	.22265	.20656	.19353	.18269	.17351	.16559	.15868
0.10	.53983	.41390	.34906	.30794	.27891	.25700	.23971	.22561	.21382	.20378	.19509	.18747
0.20	.57926	.45531	.38964	.34727	.31697	.29389	.27554	.26048	.24782	.23698	.22756	.21927
0.30	.61791	.49720	.43144	.38828	.35705	.33304	.31380	.29790	.28448	.27293	.26285	.25395
0.40	.65542	.53909	.47400	.43056	.39875	.37407	.35414	.33758	.32352	.31137	.30072	.29128
0.50	.69146	.58053	.51682	.47362	.44161	.41655	.39617	.37913	.36458	.35196	.34085	.33096
0.60	.72575	.62107	.55943	.51695	.48513	.45999	.43941	.42209	.40724	.39428	.38283	.37261
0.70	.75804	.66030	.60133	.56006	.52880	.50389	.48335	.46598	.45099	.43786	.42622	.41579
0.80	.78814	.69785	.64206	.60244	.57210	.54772	.52747	.51025	.49532	.48218	.47049	.45997
0.90	.81594	.73339	.68121	.64361	.61451	.59093	.57122	.55436	.53967	.52669	.51509	.50463
1.00	.84134	.76667	.71842	.68315	.65557	.63304	.61409	.59778	.58350	.57084	.55948	.54920
1.10	.86433	.79749	.75336	.72067	.69484	.67358	.65557	.63999	.62629	.61409	.60310	.59313

TABLE A.21. (*continued*)

$\rho = .700$

x \ ℓ	1	2	3	4	5	6	7	8	9	10	11	12
1.20	.88493	.82572	.78582	.75586	.73196	.71213	.69523	.68053	.66755	.65593	.64544	.63588
1.30	.90320	.85130	.81563	.78849	.76663	.74836	.73269	.71899	.70684	.69592	.68602	.67698
1.40	.91924	.87422	.84269	.81839	.79863	.78200	.76765	.75504	.74380	.73366	.72444	.71599
1.50	.93319	.89455	.86698	.84547	.82783	.81287	.79988	.78841	.77815	.76885	.76037	.75257
1.60	.94520	.91238	.88854	.86972	.85415	.84085	.82924	.81894	.80968	.80126	.79355	.78644
1.70	.95543	.92785	.90746	.89119	.87760	.86592	.85567	.84652	.83827	.83074	.82382	.81742
1.80	.96407	.94112	.92388	.90997	.89826	.88812	.87917	.87115	.86388	.85723	.85110	.84541
1.90	.97128	.95239	.93798	.92622	.91624	.90754	.89982	.89288	.88656	.88075	.87538	.87039
2.00	.97725	.96186	.94994	.94011	.93170	.92433	.91776	.91181	.90638	.90138	.89673	.89240
2.10	.98214	.96973	.95998	.95186	.94486	.93868	.93315	.92812	.92351	.91925	.91528	.91157
2.20	.98610	.97620	.96831	.96167	.95591	.95080	.94620	.94200	.93813	.93454	.93120	.92806
2.30	.98928	.98146	.97515	.96979	.96510	.96092	.95713	.95367	.95046	.94748	.94469	.94207
2.40	.99180	.98569	.98070	.97642	.97265	.96927	.96619	.96336	.96074	.95829	.95599	.95383
2.50	.99379	.98907	.98515	.98177	.97878	.97608	.97361	.97132	.96920	.96722	.96535	.96358
2.60	.99534	.99172	.98869	.98605	.98370	.98156	.97960	.97779	.97609	.97450	.97300	.97157
2.70	.99653	.99379	.99147	.98943	.98760	.98594	.98440	.98297	.98163	.98037	.97918	.97804
2.80	.99744	.99539	.99363	.99208	.99067	.98938	.98819	.98708	.98603	.98505	.98411	.98322
2.90	.99813	.99661	.99529	.99412	.99305	.99207	.99115	.99030	.98949	.98873	.98800	.98731
3.00	.99865	.99753	.99655	.99567	.99487	.99413	.99344	.99279	.99217	.99159	.99103	.99050
3.10	.99903	.99822	.99750	.99685	.99626	.99570	.99519	.99470	.99423	.99379	.99337	.99297
3.20	.99931	.99873	.99821	.99773	.99729	.99689	.99650	.99614	.99580	.99547	.99515	.99485
3.30	.99952	.99910	.99872	.99838	.99806	.99777	.99749	.99722	.99697	.99672	.99649	.99627
3.40	.99966	.99937	.99910	.99886	.99863	.99842	.99821	.99802	.99783	.99766	.99749	.99732
3.50	.99977	.99956	.99937	.99920	.99904	.99889	.99874	.99860	.99847	.99834	.99822	.99810

$\rho = .750$

x \ ℓ	1	2	3	4	5	6	7	8	9	10	11	12
0.00	.50000	.38497	.32746	.29135	.26594	.24679	.23167	.21932	.20899	.20017	.19252	.18580
0.10	.53983	.42540	.36666	.32920	.30255	.28229	.26618	.25296	.24184	.23231	.22402	.21671
0.20	.57926	.46661	.40728	.36886	.34123	.32005	.30310	.28912	.27730	.26712	.25824	.25038
0.30	.61791	.50817	.44891	.40995	.38164	.35976	.34214	.32752	.31510	.30437	.29497	.28663
0.40	.65542	.54963	.49109	.45204	.42336	.40103	.38292	.36781	.35493	.34375	.33392	.32517
0.50	.69146	.59053	.53335	.49465	.46594	.44341	.42502	.40960	.39639	.38489	.37473	.36567
0.60	.72575	.63046	.57522	.53731	.50890	.48642	.46798	.45242	.43904	.42734	.41698	.40770
0.70	.75804	.66901	.61623	.57951	.55172	.52958	.51129	.49579	.48239	.47063	.46018	.45080
0.80	.78814	.70583	.65597	.62080	.59394	.57236	.55444	.53918	.52593	.51426	.50385	.49447
0.90	.81594	.74062	.69403	.66074	.63506	.61430	.59694	.58208	.56914	.55769	.54744	.53819
1.00	.84134	.77315	.73009	.69892	.67467	.65491	.63830	.62401	.61151	.60041	.59045	.58143
1.10	.86433	.80322	.76387	.73503	.71237	.69379	.67808	.66451	.65258	.64195	.63238	.62369
1.20	.88493	.83074	.79516	.76877	.74786	.73059	.71590	.70315	.69190	.68185	.67276	.66449
1.30	.90320	.85564	.82383	.79995	.78087	.76500	.75144	.73961	.72912	.71972	.71120	.70342
1.40	.91924	.87794	.84981	.82845	.81123	.79682	.78444	.77359	.76393	.75525	.74735	.74012
1.50	.93319	.89769	.87309	.85419	.83884	.82591	.81473	.80489	.79611	.78818	.78095	.77431
1.60	.94520	.91500	.89372	.87719	.86366	.85218	.84221	.83340	.82550	.81835	.81180	.80578
1.70	.95543	.93001	.91180	.89751	.88571	.87564	.86686	.85905	.85204	.84566	.83981	.83440
1.80	.96407	.94289	.92748	.91525	.90508	.89636	.88870	.88187	.87571	.87009	.86492	.86014
1.90	.97128	.95382	.94092	.93058	.92191	.91443	.90784	.90194	.89659	.89169	.88718	.88299
2.00	.97725	.96300	.95231	.94367	.93637	.93003	.92442	.91937	.91478	.91057	.90668	.90306
2.10	.98214	.97063	.96187	.95472	.94864	.94334	.93861	.93435	.93046	.92688	.92356	.92046
2.20	.98610	.97690	.96980	.96396	.95895	.95456	.95063	.94707	.94381	.94080	.93800	.93538
2.30	.98928	.98200	.97631	.97158	.96751	.96391	.96068	.95774	.95504	.95254	.95021	.94803

(*continued*)

TABLE A.21. (continued)

$\rho = .750$

x \ ℓ	1	2	3	4	5	6	7	8	9	10	11	12
2.40	.99180	.98610	.98160	.97781	.97453	.97162	.96899	.96660	.96439	.96233	.96042	.95862
2.50	.99379	.98938	.98584	.98285	.98023	.97790	.97579	.97386	.97207	.97041	.96885	.96738
2.60	.99534	.99195	.98921	.98687	.98481	.98297	.98129	.97975	.97832	.97698	.97573	.97455
2.70	.99653	.99396	.99185	.99004	.98844	.98700	.98568	.98447	.98334	.98228	.98128	.98034
2.80	.99744	.99551	.99391	.99253	.99129	.99018	.98916	.98821	.98733	.98650	.98572	.98497
2.90	.99813	.99670	.99549	.99445	.99351	.99265	.99187	.99114	.99046	.98982	.98921	.98863
3.00	.99865	.99759	.99670	.99591	.99521	.99456	.99397	.99341	.99289	.99240	.99193	.99149
3.10	.99903	.99826	.99760	.99702	.99650	.99601	.99557	.99515	.99476	.99438	.99403	.99369
3.20	.99931	.99876	.99828	.99785	.99746	.99711	.99678	.99647	.99617	.99589	.99563	.99538
3.30	.99952	.99912	.99877	.99847	.99818	.99792	.99768	.99745	.99723	.99703	.99683	.99664
3.40	.99966	.99938	.99914	.99892	.99871	.99852	.99835	.99818	.99802	.99787	.99773	.99759
3.50	.99977	.99957	.99940	.99924	.99910	.99896	.99883	.99871	.99860	.99849	.99839	.99829

$\rho = .800$

x \ ℓ	1	2	3	4	5	6	7	8	9	10	11	12
0.00	.50000	.39758	.34638	.31399	.29100	.27354	.25965	.24823	.23861	.23034	.22314	.21678
0.10	.53983	.43794	.38580	.35238	.32844	.31011	.29545	.28335	.27311	.26428	.25657	.24973
0.20	.57926	.47894	.42643	.39232	.36766	.34866	.33337	.32068	.30990	.30059	.29242	.28516
0.30	.61791	.52016	.46785	.43342	.40831	.38882	.37305	.35990	.34870	.33898	.33042	.32281
0.40	.65542	.56115	.50960	.47524	.44995	.43019	.41411	.40065	.38913	.37910	.37025	.36235
0.50	.69146	.60149	.55124	.51732	.49214	.47233	.45613	.44250	.43079	.42057	.41151	.40342
0.60	.72575	.64075	.59230	.55920	.53440	.51477	.49862	.48498	.47322	.46292	.45377	.44556
0.70	.75804	.67858	.63236	.60040	.57626	.55702	.54112	.52763	.51594	.50568	.49653	.48831
0.80	.78814	.71462	.67102	.64051	.61727	.59863	.58314	.56994	.55847	.54836	.53933	.53119
0.90	.81594	.74861	.70792	.67911	.65698	.63912	.62421	.61145	.60032	.59047	.58166	.57369
1.00	.84134	.78033	.74275	.71585	.69502	.67811	.66391	.65171	.64103	.63156	.62305	.61534
1.10	.86433	.80960	.77528	.75044	.73105	.71521	.70185	.69032	.68019	.67117	.66306	.65568
1.20	.88493	.83634	.80534	.78264	.76480	.75012	.73769	.72692	.71742	.70894	.70129	.69432
1.30	.90320	.86051	.83279	.81229	.79605	.78261	.77117	.76122	.75242	.74454	.73740	.73089
1.40	.91924	.88212	.85761	.83930	.82467	.81251	.80210	.79302	.78495	.77770	.77113	.76511
1.50	.93319	.90124	.87981	.86362	.85060	.83971	.83035	.82215	.81484	.80826	.80227	.79677
1.60	.94520	.91798	.89943	.88529	.87383	.86419	.85587	.84854	.84199	.83608	.83069	.82573
1.70	.95543	.93249	.91661	.90438	.89440	.88597	.87865	.87218	.86638	.86113	.85633	.85191
1.80	.96407	.94492	.93147	.92102	.91243	.90512	.89876	.89312	.88804	.88343	.87921	.87530
1.90	.97128	.95547	.94420	.93536	.92805	.92179	.91632	.91145	.90706	.90306	.89938	.89598
2.00	.97725	.96432	.95498	.94759	.94143	.93614	.93149	.92733	.92357	.92014	.91698	.91404
2.10	.98214	.97168	.96402	.95790	.95277	.94834	.94443	.94093	.93775	.93483	.93214	.92964
2.20	.98610	.97772	.97151	.96650	.96227	.95861	.95536	.95244	.94978	.94733	.94507	.94296
2.30	.98928	.98264	.97765	.97359	.97015	.96715	.96448	.96207	.95987	.95785	.95597	.95421
2.40	.99180	.98660	.98263	.97939	.97661	.97418	.97201	.97005	.96825	.96659	.96504	.96360
2.50	.99379	.98975	.98663	.98406	.98185	.97991	.97816	.97658	.97512	.97378	.97252	.97135
2.60	.99534	.99223	.98981	.98780	.98605	.98451	.98313	.98186	.98070	.97962	.97861	.97767
2.70	.99653	.99417	.99231	.99074	.98939	.98818	.98709	.98610	.98518	.98432	.98352	.98277
2.80	.99744	.99567	.99425	.99305	.99200	.99107	.99022	.98945	.98873	.98806	.98743	.98684
2.90	.99813	.99681	.99574	.99483	.99403	.99332	.99267	.99207	.99151	.99099	.99051	.99004
3.00	.99865	.99767	.99688	.99619	.99559	.99505	.99456	.99410	.99367	.99328	.99290	.99255
3.10	.99903	.99832	.99773	.99723	.99678	.99637	.99600	.99565	.99533	.99503	.99475	.99448
3.20	.99931	.99880	.99837	.99800	.99767	.99736	.99709	.99683	.99659	.99637	.99615	.99595
3.30	.99952	.99915	.99884	.99857	.99833	.99811	.99790	.99771	.99754	.99737	.99721	.99706
3.40	.99966	.99940	.99918	.99899	.99881	.99865	.99850	.99837	.99824	.99811	.99800	.99789
3.50	.99977	.99958	.99943	.99929	.99916	.99905	.99894	.99884	.99875	.99866	.99858	.99850

TABLE A.21. (continued)

$\rho = .875$

x \ ℓ	1	2	3	4	5	6	7	8	9	10	11	12
0.00	.50000	.41957	.37935	.35365	.33521	.32104	.30965	.30020	.29218	.28523	.27913	.27370
0.10	.53983	.45981	.41908	.39278	.37378	.35910	.34726	.33740	.32900	.32171	.31529	.30958
0.20	.57926	.50046	.45963	.43301	.41365	.39861	.38643	.37625	.36755	.35999	.35332	.34736
0.30	.61791	.54109	.50061	.47395	.45442	.43918	.42679	.41639	.40749	.39972	.39286	.38671
0.40	.65542	.58130	.54157	.51516	.49568	.48040	.46792	.45742	.44840	.44052	.43353	.42727
0.50	.69146	.62068	.58210	.55620	.53697	.52182	.50939	.49891	.48987	.48195	.47491	.46860
0.60	.72575	.65883	.62176	.59664	.57787	.56300	.55076	.54039	.53144	.52357	.51656	.51027
0.70	.75804	.69542	.66017	.63606	.61793	.60350	.59157	.58144	.57266	.56493	.55803	.55182
0.80	.78814	.73015	.69698	.67409	.65676	.64291	.63141	.62161	.61309	.60558	.59886	.59280
0.90	.81594	.76276	.73189	.71038	.69400	.68084	.66987	.66050	.65233	.64511	.63864	.63278
1.00	.84134	.79309	.76464	.74465	.72933	.71696	.70662	.69775	.69000	.68313	.67696	.67137
1.10	.86433	.82099	.79505	.77667	.76249	.75099	.74134	.73304	.72576	.71930	.71349	.70821
1.20	.98493	.94639	.82299	.80626	.79328	.78271	.77380	.76611	.75936	.75334	.74792	.74299
1.30	.90320	.86928	.84839	.83333	.82157	.81195	.80391	.79677	.70057	.78503	.79003	.77548
1.40	.91924	.88969	.87124	.85782	.84729	.83863	.83128	.92490	.81926	.81422	.80966	.80550
1.50	.93319	.90771	.89158	.87975	.87041	.86270	.85613	.85041	.84535	.84081	.83669	.93293
1.60	.94520	.92345	.90949	.89918	.89099	.89419	.87839	.87332	.86982	.86477	.86110	.85773
1.70	.95543	.93705	.92511	.91621	.90910	.90318	.89810	.89365	.98970	.88613	.88289	.87991
1.80	.96407	.94870	.93858	.93098	.92488	.91977	.91538	.91152	.90808	.90497	.90213	.89953
1.90	.97128	.95856	.95007	.94366	.93847	.93412	.93036	.92704	.92408	.92140	.91895	.91670
2.00	.97725	.96692	.95978	.95442	.95006	.94639	.94320	.94039	.93787	.93558	.93340	.93156
2.10	.99214	.97368	.96700	.96346	.95984	.05677	.05410	.05174	.04962	.91760	.01502	.04428
2.20	.98610	.97930	.97461	.97099	.96800	.96546	.96325	.96129	.95952	.95791	.05643	.95506
2.30	.98928	.98388	.98010	.97716	.97474	.97267	.97085	.96924	.96778	.96645	.96523	.96409
2.40	.99180	.98756	.98455	.98220	.98024	.97857	.97710	.97578	.97460	.97351	.97251	.97158
2.50	.99379	.99048	.98812	.98625	.98469	.98335	.98217	.98112	.98016	.97928	.97847	.97772
2.60	.99534	.99279	.99095	.98948	.98825	.98719	.98626	.98542	.98465	.98395	.98330	.98270
2.70	.99653	.99459	.99317	.99203	.99107	.99024	.98951	.98884	.98824	.98769	.98717	.98669
2.80	.99744	.99598	.99489	.99402	.99328	.99264	.99206	.99155	.99109	.99064	.99024	.98986
2.90	.99813	.99704	.99622	.99555	.99499	.99450	.99406	.99366	.99329	.99296	.99265	.99235
3.00	.99865	.99784	.99722	.99673	.99630	.99593	.99559	.99529	.99501	.99475	.99451	.99429
3.10	.99903	.99844	.99798	.99761	.99729	.99701	.99676	.99653	.99632	.99613	.99595	.99578
3.20	.99931	.99888	.99855	.99828	.99804	.99783	.99764	.99747	.99732	.99717	.99703	.99691
3.30	.99952	.99921	.99897	.99877	.99859	.99844	.99830	.99818	.99806	.99795	.99785	.99776
3.40	.99966	.99944	.99927	.99913	.99900	.99889	.99879	.99870	.99861	.99853	.99846	.99839
3.50	.99977	.99961	.99949	.99939	.99930	.99922	.99915	.99908	.99902	.99896	.99891	.99886

$\rho = .900$

x \ ℓ	1	2	3	4	5	6	7	8	9	10	11	12
0.00	.50000	.42822	.39233	.36931	.35274	.33996	.32967	.32110	.31380	.30747	.30189	.29693
0.10	.53983	.46841	.43213	.40866	.39165	.37848	.36783	.35894	.35135	.34475	.33892	.33373
0.20	.57926	.50892	.47263	.44894	.43168	.41825	.40734	.39822	.39040	.38359	.37758	.37220
0.30	.61791	.54934	.51341	.48976	.47241	.45886	.44782	.43855	.43059	.42364	.41749	.41198
0.40	.65542	.58925	.55405	.53068	.51344	.49991	.48884	.47953	.47151	.46450	.45827	.45269
0.50	.69146	.62825	.59412	.57127	.55431	.54095	.52998	.52072	.51273	.50572	.49950	.49390
0.60	.72575	.66597	.63323	.61112	.59462	.58155	.57079	.56168	.55381	.54689	.54072	.53517
0.70	.75804	.70209	.67099	.64982	.63393	.62130	.61086	.60200	.59431	.58754	.58151	.57607
0.80	.78814	.73631	.70708	.68703	.67189	.65980	.64979	.64124	.63383	.62728	.62143	.61615
0.90	.91594	.76840	.74122	.72241	.70814	.69670	.69718	.67905	.67197	.66570	.66000	.65502
1.00	.84134	.79918	.77317	.75572	.74241	.73169	.72274	.71509	.70939	.70246	.69714	.09232
1.10	.86433	.82554	.80276	.78674	.77446	.76452	.75620	.74905	.74290	.73725	.73226	.72773

(continued)

TABLE A.21. (continued)

						$\rho = .900$						
ℓ x	1	2	3	4	5	6	7	8	9	10	11	12
1.20	.88493	.85042	.82988	.81533	.80411	.79500	.78734	.78075	.77497	.76982	.76519	.76098
1.30	.90320	.87281	.85449	.84141	.83127	.82300	.81603	.81001	.80472	.80000	.79575	.79188
1.40	.91924	.89275	.87658	.86494	.85587	.84845	.84217	.83674	.83195	.82767	.82380	.82028
1.50	.93319	.91033	.89620	.88596	.87793	.87134	.86574	.86089	.85660	.85276	.84929	.84611
1.60	.94520	.92568	.91345	.90453	.89751	.89171	.88678	.88248	.87868	.87527	.87218	.86936
1.70	.95543	.93893	.92847	.92078	.91469	.90965	.90534	.90159	.89826	.89526	.99254	.89005
1.80	.96407	.95025	.94139	.93483	.92962	.92528	.92156	.91931	.91542	.91282	.91045	.90829
1.90	.97128	.95984	.95241	.94687	.94245	.93875	.93558	.93280	.93032	.92808	.92604	.92417
2.00	.97725	.96786	.96170	.95708	.95336	.95025	.94757	.94521	.94310	.94120	.93946	.93786
2.10	.98214	.97451	.96946	.96563	.96255	.95995	.95771	.95573	.95396	.95236	.95090	.94955
2.20	.98610	.97997	.97586	.97273	.97020	.96906	.96620	.96456	.96309	.96176	.96054	.95941
2.30	.98928	.99440	.98110	.97857	.97650	.97476	.97324	.97199	.97068	.96958	.96859	.96764
2.40	.99180	.98796	.98533	.98331	.98165	.98024	.97901	.97791	.97693	.97603	.97521	.97445
2.50	.99379	.99080	.98873	.98712	.98580	.98467	.98368	.98281	.98201	.98129	.98063	.98001
2.60	.99534	.99303	.99142	.99015	.98911	.98822	.98744	.98674	.98611	.98553	.98500	.98451
2.70	.99653	.99477	.99352	.99254	.99173	.99103	.99042	.98987	.98937	.98892	.98850	.98811
2.80	.99744	.99611	.99516	.99441	.99378	.99324	.99276	.99234	.99195	.99159	.99127	.99096
2.90	.99813	.99714	.99642	.99585	.99537	.99495	.99459	.99426	.99396	.99368	.99343	.99319
3.00	.99865	.99791	.99737	.99694	.99658	.99627	.99599	.99574	.99551	.99530	.99510	.99492
3.10	.99903	.99849	.99809	.99777	.99750	.99727	.99706	.99697	.99669	.99654	.99639	.99625
3.20	.99931	.99892	.99863	.99839	.99819	.99802	.99786	.99772	.99759	.99747	.99736	.99726
3.30	.99952	.99923	.99902	.99885	.99870	.99857	.99846	.99836	.99826	.99817	.99809	.99801
3.40	.99966	.99946	.99931	.99919	.99908	.99899	.99890	.99883	.99876	.99869	.99863	.99858
3.50	.99977	.99963	.99952	.99943	.99935	.99929	.99923	.99917	.99912	.99907	.99903	.99899

Adapted from S. S. Gupta, Probability integrals of multivariate normal and multivariate t, *Annals of Mathematical Statistics* **34** (1963): 792–828, with permission of the author, the editor of *The Annals of Mathematical Statistics*, and the Institute of Mathematical Statistics.

TABLE A.22. Selected Upper-Tail Probabilities for the Null Distribution of the Friedman S Statistic:

$$k = 3, n = 2(1)13; k = 4, n = 2(1)8; k = 5, n = 3(1)8; k = 6, n = 2(1)6$$

Additional upper-tail probabilities can be found in Odeh (1977a) and Hollander and Wolfe (1973).
For given k and n, the table entry for the point x is $P_0\{S \geq x\}$. Under these conditions, if x is such that $P_0\{S \geq x\} = \alpha$, then $s_\alpha = x$.

					$k = 3$				
n	x	$P_0\{S \geq x\}$	n	x	$P_0\{S \geq x\}$	n	x	$P_0\{S \geq x\}$
2	4	.167	5	4.8	.124	6	4.333	.142
				5.2	.093		5.333	.072
3	4.667	.194		6.4	.039		6.333	.052
	6	.028		7.6	.024		7	.029
				8.4	.008		8.333	.012
4	4.5	.125		10	.001		9	.008
	6	.069					9.333	.006
	6.5	.042					10.333	.002
	8	.005					12	<.0005

TABLE A.22. (continued)

$k = 3$

n	x	$P_0\{S \geq x\}$	n	x	$P_0\{S \geq x\}$	n	x	$P_0\{S \geq x\}$
7	4.571	.112	10	4.2	.135	12	6.167	.051
	5.429	.085		5	.092		6.5	.038
	6	.051		5.4	.078		7.167	.027
	7.143	.027		5.6	.066		8	.020
	7.714	.021		6.2	.046		8.167	.017
	8	.016		7.2	.030		8.667	.011
	8.857	.008		7.4	.026		9.5	.007
	10.286	.004		7.8	.018		10.167	.005
	10.571	.003		8.6	.012		10.5	.004
	11.143	.001		9.6	.007		10.667	.003
	12.286	<.0005		9.8	.006		11.167	.002
				10.4	.003		12.167	.002
8	4.75	.120		11.4	.002		12.5	.001
	5.25	.079		12.2	.001		12.667	.001
	6.25	.047		12.6	.001		13.167	.001
	6.75	.038		12.8	.001		13.5	<.0005
	7	.030		13.4	<.0005			
	7.75	.018				13	4.308	.129
	9	.010	11	4.909	.100		4.769	.098
	9.25	.008		5.091	.087		5.538	.073
	9.75	.005		5.636	.062		5.692	.065
	10.75	.002		6.545	.043		6	.050
	12	.001		6.727	.038		6.615	.037
	12.25	.001		7.091	.027		7.385	.028
	13	<.0005		7.818	.019		7.538	.025
				8.727	.013		8	.016
9	4.667	.107		8.909	.011		8.769	.012
	5.556	.069		9.455	.006		9.385	.009
	6	.057		10.364	.004		9.692	.007
	6.222	.048		11.091	.003		9.846	.005
	6.889	.031		11.455	.002		10.308	.004
	8	.019		11.636	.001		11.231	.003
	8.222	.016		12.182	.001		11.538	.002
	8.667	.010		13.273	.001		11.692	.002
	9.556	.006		13.636	<.0005		12.154	.001
	10.667	.004					12.462	.001
	10.889	.003	12	4.667	.108		12.923	.001
	11.556	.001		5.167	.080		14	.001
	12.667	.001		6	.058		14.308	<.0005
	13.556	<.0005						

$k = 4$

n	x	$P_0\{S \geq x\}$	n	x	$P_0\{S \geq x\}$	n	x	$P_0\{S \geq x\}$
2	5.4	.167	4	6.0	.105	4	8.4	.019
	6	.042		6.3	.094		8.7	.014
				6.6	.077		9.3	.012
3	5.8	.148		6.9	.068		9.6	.007
	6.6	.075		7.2	.054		9.9	.006
	7	.054		7.5	.052		10.2	.003
	7.4	.033		7.8	.036		10.8	.002
	8.2	.017		8.1	.033		11.1	.001
	9	.002					12.0	<.0005

(continued)

TABLE A.22. *(continued)*

					$k = 4$			
n	x	$P_0\{S \geq x\}$	n	x	$P_0\{S \geq x\}$	n	x	$P_0\{S \geq x\}$
5	6.12	.107	6	12	.002	8	7.35	.058
	6.36	.093		12.2	.002		7.50	.051
	6.84	.075		12.6	.001		7.65	.049
	7.08	.067		12.8	.001		7.80	.046
	7.32	.055		13	.001		7.95	.042
	7.80	.044		13.2	.001		8.10	.038
	8.04	.034		13.4	.001		8.25	.037
	8.28	.031		13.6	<.0005		8.55	.031
	8.76	.023					8.70	.028
	9	.020	7	6.257	.100		8.85	.025
	9.24	.017		6.429	.093		9	.023
	9.72	.012		6.6	.085		9.15	.022
	9.96	.009		6.943	.073		9.45	.019
	10.20	.007		7.114	.063		9.60	.016
	10.68	.005		7.286	.056		9.75	.015
	10.92	.003		7.629	.052		9.90	.014
	11.16	.002		7.8	.041		10.05	.014
	11.64	.002		7.971	.038		10.20	.011
	11.88	.002		8.314	.035		10.35	.011
	12.12	.001		8.486	.033		10.50	.009
	12.60	.001		8.657	.030		10.65	.009
	12.84	<.0005		9	.023		10.80	.008
				9.171	.020		10.95	.008
6	6.2	.108		9.686	.015		11.10	.006
	6.4	.089		9.857	.013		11.25	.006
	6.6	.088		10.029	.012		11.40	.005
	6.8	.073		10.371	.010		11.55	.005
	7	.066		10.543	.009		11.85	.004
	7.2	.060		10.714	.008		12	.004
	7.4	.056		11.057	.007		12.15	.004
	7.6	.043		11.229	.005		12.30	.003
	7.8	.041		11.4	.004		12.45	.003
	8.0	.037		11.743	.004		12.60	.002
	8.2	.035		11.914	.003		12.75	.002
	8.4	.032		12.086	.003		12.90	.002
	8.6	.029		12.429	.002		13.05	.002
	8.8	.023		12.6	.002		13.20	.002
	9.0	.022		12.771	.002		13.35	.001
	9.4	.017		13.114	.001		13.50	.001
	9.6	.014		13.286	.001		13.65	.001
	9.8	.013		13.457	.001		13.80	.001
	10	.010		13.8	.001		13.95	.001
	10.2	.010		13.971	.001		14.25	.001
	10.4	.009		14.143	.001		14.40	.001
	10.6	.007		14.486	<.0005		14.55	.001
	10.8	.006					14.70	.001
	11	.006	8	6.30	.100		14.85	<.0005
	11.4	.004		6.45	.094			
	11.6	.003		6.60	.081			
	11.8	.003		6.75	.079			
				7.05	.068			
				7.20	.060			

TABLE A.22. *(continued)*

					$k = 5$				
n	x	$P_0\{S \geq x\}$		n	x	$P_0\{S \geq x\}$		n	x	$P_0\{S \geq x\}$
3	7.2	.117		5	7.52	.107		6	7.6	.1025
	7.467	.096			7.68	.094			7.733	.0952
	7.733	.080			7.84	.089			8.133	.0800
	8	.063			8	.082			8.267	.0742
	8.267	.056			8.16	.077			8.933	.0550
	8.533	.045			8.32	.073			9.067	.0491
	8.8	.038			8.48	.066			10.267	.0265
	9.067	.028			8.64	.058			10.4	.0235
	9.333	.026			8.80	.056			11.733	.0109
	9.6	.017			8.96	.049			11.867	.0099
	9.867	.015			9.12	.046			12.933	.0051
	10.133	.008			9.28	.042			13.067	.0044
	10.4	.005			9.44	.038			15.067	.0010
	10.667	.004			9.60	.035				
	10.933	.003			9.76	.032		7	7.657	.1025
	11.467	.001			9.92	.029			7.771	.0941
	12	<.0005			10.08	.026			8.229	.0775
					10.24	.024			8.343	.0742
4	7.4	.113			10.40	.022			9.029	.0527
	7.6	.095			10.56	.019			9.143	.0494
	7.8	.086			10.72	.018			10.400	.0261
	8	.080			10.88	.015			10.514	.0247
	8.2	.072			11.04	.013			12.114	.0100
	8.4	.063			11.20	.012			13.143	.0054
	8.6	.060			11.36	.012			13.257	.0048
	8.8	.049			11.52	.010			15.543	.0010
	9	.043			11.68	.009				
	9.2	.038			11.84	.008		8	7.6	.1039
	9.4	.035			12	.007			7.7	.0999
	9.6	.028			12.16	.006			8.3	.0765
	9.8	.025			12.32	.006			8.4	.0708
	10	.021			12.48	.005			9.1	.0521
	10.2	.019			12.64	.004			9.2	.0499
	10.4	.017			12.80	.004			10.5	.0256
	10.6	.014			12.96	.003			10.6	.0247
	10.8	.011			13.12	.003			12.2	.0104
	11	.010			13.28	.003			12.3	.0099
	11.2	.008			13.44	.002			13.4	.0050
	11.4	.007			13.60	.002			15.9	.0010
	11.6	.006			13.76	.002				
	11.8	.005			13.92	.002				
	12	.004			14.08	.001				
	12.2	.004			14.24	.001				
	12.4	.003			14.40	.001				
	12.6	.002			14.56	.001				
	12.8	.002			14.72	.001				
	13	.001			14.88	.001				
	13.2	.001			15.04	<.0005				
	13.4	.001								
	13.6	.001								
	13.8	<.0005								

(continued)

TABLE A.22. (continued)

					$k = 6$				
n	x	$P_0\{S \geq x\}$	n	x	$P_0\{S \geq x\}$	n	x	$P_0\{S \geq x\}$
2	8	.1208	4	8.857	.1018	6	8.952	.1033
	8.286	.0875		9	.0943		9.048	.0990
	8.571	.0681		9.429	.0760		9.619	.0766
	8.857	.0514		9.571	.0723		9.714	.0733
	9.143	.0292		10.143	.0518		10.476	.0515
	9.429	.0167		10.286	.0474		10.571	.0491
	9.714	.0083		11.286	.0263		11.905	.0251
	10	.0014		11.429	.0237		12	.0238
				12.571	.0109		13.524	.0103
3	8.524	.1122		12.714	.0096		13.619	.0097
	8.714	.0957		12.857	.0085		14.667	.0051
	9.095	.0769					14.762	.0048
	9.286	.0699	5	8.886	.1029		17.048	.0010
	9.667	.0560		9	.0986			
	9.857	.0461		9.571	.0754			
	10.619	.0281		9.686	.0720			
	10.810	.0247		10.371	.0513			
	11.571	.0117		10.486	.0483			
	11.762	.0095		11.629	.0258			
	12.333	.0052		11.743	.0244			
	12.524	.0037		13.114	.0106			
	13.286	.0010		13.229	.0099			
				14.257	.0050			
				16.314	.0010			

Adapted, in part, from R. E. Odeh, Extended tables of the distribution of Friedman's S-statistic in the two-way layout, *Communications in Statistics—Simulation and Computation* **6**, Marcel Dekker, Inc., N.Y. (1977a): 29–48, with the permission of the author and the editor of *Communications in Statistics—Simulation and Computation*, reprinted by courtesy of Marcel Dekker, Inc.

TABLE A.23. Selected Upper-Tail Probabilities for the Null Distribution of the Page L Statistic:

$$k = 3, n = 2(1)15; k = 4(1)8, n = 2(1)10$$

Additional upper-tail probabilities can be found in Page (1963) and Odeh (1977b).
For given k and n, the table entry for the point x is $P_0\{L \geq x\}$. Under these conditions, if x is such that $P_0\{L \geq x\} = \alpha$, then $\ell_\alpha = x$.

					$k = 3$				
n	x	$P_0\{L \geq x\}$	n	x	$P_0\{L \geq x\}$	n	x	$P_0\{L \geq x\}$
2	27	.1389	7	89	.1173	10	126	.1112
	28	.0278		90	.0716		127	.0738
				91	.0409		128	.0466
3	39	.1528		92	.0211		129	.0279
	40	.0880		93	.0097		130	.0157
	41	.0324		94	.0040		131	.0083
	42	.0046		95	.0014		132	.0041
				96	.0004		133	.0018
4	52	.1088					134	.0007
	53	.0563	8	101	.1331			
	54	.0255		102	.0862	11	141	.05
	55	.0069		103	.0521		144	.01
	56	.0008		104	.0295		147	.001
				105	.0154			
5	64	.1412		106	.0073	12	153	.05
	65	.0795		107	.0031		156	.01
	66	.0394		108	.0012		160	.001
	67	.0181		109	.0004			
	68	.0066				13	165	.05
	69	.0014	9	113	.1472		169	.01
	70	.0001		114	.0990		172	.001
				115	.0633			
6	76	.1596		116	.0381	14	178	.05
	77	.0996		117	.0215		181	.01
	78	.0572		118	.0113		185	.001
	79	.0288		119	.0055			
	80	.0131		120	.0024	15	190	.05
	81	.0053		121	.0009		194	.01
	82	.0016					197	.001
	83	.0003						

					$k = 4$				
n	x	$P_0\{L \geq x\}$	n	x	$P_0\{L \geq x\}$	n	x	$P_0\{L \geq x\}$
2	55	.1476	4	108	.1008	5	138	.0253
	56	.1059		109	.0724		139	.0167
	57	.0556		110	.0504		140	.0106
	58	.0313		111	.0337		141	.0065
	59	.0122		112	.0217		142	.0037
	60	.0017		113	.0130		144	.0010
				114	.0075		145	.0005
3	82	.1017		115	.0039			
	83	.0702		116	.0017	6	159	.1176
	84	.0446		117	.0007		160	.0916
	85	.0201					162	.0524
	86	.0148	5	133	.1266		163	.0383
	87	.0070		134	.0966		164	.0273
	88	.0029		136	.0524		165	.0190
	89	.0007		137	.0370		166	.0128

(continued)

TABLE A.23. *(continued)*

					$k = 4$			
n	x	$P_0\{L \geq x\}$	n	x	$P_0\{L \geq x\}$	n	x	$P_0\{L \geq x\}$
6	167	.0084	8	211	.1010	9	245	.0113
	168	.0053		212	.0807		246	.0081
	169	.0032		213	.0636		247	.0057
	171	.0010		214	.0493		248	.0039
	172	.0005		216	.0283		251	.0011
				217	.0208		252	.0007
7	185	.1091		219	.0107			
	186	.0862		220	.0075	10	262	.1054
	188	.0512		221	.0051		263	.0866
	189	.0383		222	.0034		265	.0565
	190	.0282		224	.0014		266	.0448
	191	.0203		225	.0008		268	.0271
	192	.0143					269	.0207
	193	.0098	9	236	.1145		271	.0116
	194	.0066		237	.0935		272	.0085
	195	.0043		239	.0600		273	.0061
	197	.0016		240	.0471		274	.0043
	198	.0010		242	.0279		277	.0014
				243	.0209		278	.0009

					$k = 5$			
n	x	$P_0\{L \geq x\}$	n	x	$P_0\{L \geq x\}$	n	x	$P_0\{L \geq x\}$
2	99	.1231	4	205	.0055	7	332	.1080
	100	.0962		206	.0038		333	.0945
	102	.0545		209	.0011		337	.0524
	103	.0381		210	.0007		338	.0446
	104	.0261					341	.0264
	105	.0168	5	239	.1165		342	.0219
	106	.0096		240	.0997		345	.0120
	107	.0047		243	.0596		346	.0097
	108	.0022		244	.0493		348	.0061
	109	.0006		247	.0265		349	.0048
				248	.0211		354	.0013
3	146	.1177		250	.0129		355	.0009
	147	.0960		251	.0100			
	149	.0611		253	.0057	8	378	.1096
	150	.0476		254	.0042		379	.0968
	152	.0272		258	.0011		383	.0562
	153	.0198		259	.0009		384	.0485
	154	.0141					388	.0254
	155	.0097	6	286	.1050		389	.0213
	156	.0065		287	.0907		392	.0122
	157	.0041		290	.0562		393	.0100
	159	.0014		291	.0473		396	.0053
	160	.0007		294	.0269		397	.0042
				295	.0219		402	.0012
4	193	.1090		298	.0113		403	.0009
	194	.0912		299	.0089			
	196	.0617		301	.0054	9	424	.1102
	197	.0499		302	.0041		425	.0981
	199	.0315		306	.0012		430	.0514
	200	.0245		307	.0009		431	.0446
	203	.0105					434	.0284
	204	.0077					435	.0241

TABLE A.23. (continued)

$k = 5$

n	x	$P_0\{L \geq x\}$	n	x	$P_0\{L \geq x\}$	n	x	$P_0\{L \geq x\}$
9	440	.0100	10	470	.1101	10	486	.0118
	441	.0083		471	.0985		487	.0099
	443	.0056		476	.0537		490	.0057
	444	.0045		477	.0470		491	.0047
	450	.0011		481	.0265		498	.0010
	451	.0009		482	.0227		499	.0005

$k = 6$

n	x	$P_0\{L \geq x\}$	n	x	$P_0\{L \geq x\}$	n	x	$P_0\{L \geq x\}$
2	162	.1008	5	390	.1065	8	617	.1002
	163	.0848		391	.0963		618	.0924
	165	.0589		396	.0553		624	.0547
	166	.0480		397	.0492		625	.0498
	168	.0310		402	.0254		631	.0271
	169	.0241		403	.0220		632	.0243
	172	.0101		408	.0101		639	.0106
	173	.0071		409	.0085		640	.0094
	174	.0049		411	.0060		644	.0055
	177	.0011		412	.0050		645	.0048
	178	.0006		419	.0012		655	.0010
				420	.0009		656	.0008
3	238	.1087						
	239	.0953	6	466	.1023	9	692	.1018
	243	.0531		467	.0932		693	.0944
	244	.0451		473	.0505		700	.0530
	247	.0265		474	.0451		701	.0485
	248	.0218		478	.0280		707	.0273
	251	.0115		479	.0247		708	.0247
	252	.0092		485	.0108		716	.0102
	254	.0055		486	.0093		717	.0090
	255	.0042		489	.0058		721	.0055
	259	.0012		490	.0049		722	.0048
	260	.0008		498	.0011		732	.0011
				499	.0009		733	.0010
4	314	.1093						
	315	.0976	7	541	.1062	10	767	.1026
	320	.0522		542	.0975		768	.0955
	321	.0454		549	.0505		776	.0511
	324	.0291		550	.0456		777	.0469
	325	.0248		555	.0263		783	.0272
	330	.0103		556	.0233		784	.0247
	331	.0085		562	.0109		792	.0107
	333	.0056		563	.0095		793	.0096
	334	.0045		567	.0053		798	.0053
	340	.0010		568	.0046		799	.0047
	341	.0008		576	.0012		810	.0011
				577	.0010		811	.0009

(continued)

TABLE A.23. *(continued)*

					$k = 7$				
n	x	$P_0\{L \geq x\}$	n	x	$P_0\{L \geq x\}$	n	x	$P_0\{L \geq x\}$
2	245	.1074	5	593	.1036	8	938	.1007
	246	.0961		594	.0966		939	.0953
	251	.0515		602	.0526		949	.0524
	252	.0448		603	.0484		950	.0491
	255	.0283		610	.0259		959	.0263
	256	.0241		611	.0235		960	.0244
	260	.0115		619	.0101		971	.0101
	261	.0093		620	.0091		972	.0092
	263	.0059		624	.0056		978	.0053
	264	.0046		625	.0050		979	.0049
	268	.0014		636	.0011		994	.0010
	269	.0010		637	.0009		995	.0009
3	362	.1019	6	708	.1039	9	1052	.1033
	363	.0930		709	.0976		1053	.0981
	369	.0509		718	.0524		1064	.0530
	370	.0455		719	.0486		1065	.0499
	374	.0283		727	.0253		1075	.0260
	375	.0249		728	.0232		1076	.0242
	381	.0107		736	.0109		1087	.0106
	382	.0092		737	.0099		1088	.0097
	385	.0056		743	.0052		1095	.0054
	386	.0047		744	.0047		1096	.0049
	393	.0011		756	.0010		1112	.0010
	394	.0009		757	.0009		1113	.0009
4	478	.1007	7	823	.1028	10	1167	.1000
	479	.0931		824	.0969		1168	.0952
	486	.0509		834	.0511		1180	.0500
	487	.0463		835	.0476		1181	.0472
	493	.0251		843	.0262		1191	.0253
	494	.0225		844	.0242		1192	.0237
	500	.0110		854	.0102		1204	.0100
	501	.0097		855	.0092		1205	.0093
	505	.0056		861	.0051		1212	.0053
	506	.0049		862	.0046		1213	.0049
	515	.0011		875	.0011		1230	.0010
	516	.0009		876	.0009		1231	.0009

					$k = 8$				
n	x	$P_0\{L \geq x\}$	n	x	$P_0\{L \geq x\}$	n	x	$P_0\{L \geq x\}$
2	353	.1064	3	522	.1009	4	689	.1030
	354	.0983		523	.0946		690	.0974
	361	.0528		531	.0534		700	.0528
	362	.0479		532	.0494		701	.0494
	367	.0281		539	.0274		710	.0258
	368	.0250		540	.0250		711	.0238
	374	.0115		549	.0101		721	.0102
	375	.0100		550	.0091		722	.0093
	379	.0054		555	.0050		728	.0052
	380	.0045		556	.0045		729	.0047
	387	.0011		566	.0011		742	.0011
	388	.0009		567	.0009		743	.0009

TABLE A.23. (continued)

					k = 8			
n	x	$P_0\{L \geq x\}$	n	x	$P_0\{L \geq x\}$	n	x	$P_0\{L \geq x\}$
5	856	.1015	7	1188	.1025	9	1519	.1028
	857	.0965		1189	.0983		1520	.0991
	868	.0530		1203	.0517		1536	.0520
	869	.0499		1204	.0491		1537	.0498
	879	.0264		1216	.0258		1551	.0258
	880	.0246		1217	.0244		1552	.0245
	892	.0101		1231	.0102		1568	.0103
	893	.0093		1232	.0096		1569	.0097
	900	.0051		1241	.0051		1579	.0053
	901	.0047		1242	.0047		1580	.0050
	916	.0011		1261	.0010		1602	.0010
	917	.0010		1262	.0009		1603	.0010
6	1022	.1028	8	1354	.1011	10	1685	.1002
	1023	.0982		1355	.0972		1686	.0967
	1036	.0515		1370	.0510		1703	.0503
	1037	.0488		1371	.0487		1704	.0482
	1048	.0257		1384	.0254		1718	.0258
	1049	.0241		1385	.0240		1719	.0246
	1062	.0100		1400	.0101		1736	.0103
	1063	.0093		1401	.0095		1737	.0098
	1071	.0051		1410	.0053		1748	.0052
	1072	.0047		1411	.0050		1749	.0049
	1089	.0010		1432	.0010		1773	.0010
	1090	.0009		1433	.0010		1774	.0009

Adapted, in part, from R. E. Odeh, The exact distribution of Page's L-statistic in the two-way layout, *Communications in Statistics—Simulation and Computation* **6**, Marcel Dekker, Inc., N.Y. (1977b): 49–61, with the permission of the author and the editor of *Communications in Statistics—Simulation and Computation*, reprinted by courtesy of Marcel Dekker, Inc., and, in part, from E. B. Page, Ordered hypotheses for multiple treatments: a significance test for linear ranks, *Journal of the American Statistical Association* **58** (1963): 216–230. Reprinted with permission of the *Journal of the American Statistical Association*. Copyright © (1963) by the American Statistical Association. All rights reserved.

TABLE A.24. Selected Critical Values for the Wilcoxon-Nemenyi-McDonald-Thompson Two-Sided All-Treatments Multiple Comparison Procedure:

$$k = 3, n = 3(1)15; k = 4(1)15, n = 2(1)15$$

For a given k and n, the r_α entry in the table corresponds to $P_0\{|R_u - R_v| < r_\alpha, u = 1, \ldots, k-1; v = u+1, \ldots, k\} \approx 1 - \alpha$.

	$k = 3$			$k = 4$	
n	r_α	α	n	r_α	α
3	6*	.028	2	6*	.083
4	7	.042	3	8*	.049
	8*	.005		9*	.007
5	8*	.039	4	10*	.026
	9*	.008		11*	.005
6	9*	.029	5	11	.037
	10*	.009		12*	.013
7	9*	.051	6	12*	.037
	10	.023		13	.018
	11*	.008		14*	.006
8	10*	.039	7	13*	.037
	11	.018		14	.020
	12*	.007		15*	.008
9	10*	.048	8	14*	.034
	11	.026		15	.019
	12*	.013		16*	.009
10	11	.037	9	15	.032
	12	.019		17*	.010
	13*	.010	10	15	.046
11	11*	.049		16	.029
	12	.028		18*	.010
	14*	.008	11	16	.041
12	12*	.038		17	.026
	13	.022		19	.009
	14*	.012	12	17	.038
13	12*	.049		18	.023
	13	.030		20	.008
	15*	.009	13	18	.032
14	13*	.038		19	.021
	14	.023		21	.008
	16*	.007	14	18	.042
15	13*	.047		19	.028
	14	.028		21	.011
	16*	.010	15	19	.037
				20	.024
				22	.010

TABLE A.24. (continued)

	$k = 5$			$k = 6$	
n	r_α	α	n	r_α	α
2	8	.050	2	10	.033
3	10	.067	3	13	.030
	11	.018		14	.008
	12	.002			
			4	15	.047
4	12	.054		16	.018
	13	.020		17	.006
	14	.006			
			5	17	.047
5	14	.040		18	.022
	16	.006		19	.010
6	15	.049	6	19	.040
	16	.028		20	.021
	17	.013		21	.010
7	16	.052	7	20	.049
	17	.033		21	.032
	19	.009		23	.010
8	18	.036	8	22	.039
	19	.022		23	.026
	20	.012		25	.008
9	19	.037	9	23	.043
	20	.024		24	.030
	22	.008		26	.012
10	20	.038	10	24	.047
	21	.025		26	.023
	23	.009		28	.009
11	21	.038	11	26	.036
	22	.025		27	.026
	24	.010		29	.012
12	22	.038	12	27	.039
	23	.025		28	.028
	25	.011		31	.009
13	23	.035	13	28	.039
	24	.024		29	.028
	26	.011		32	.010
14	24	.034	14	29	.040
	25	.024		30	.030
	27	.011		33	.011
15	24	.045	15	30	.040
	26	.022		32	.023
	28	.010		34	.012

(continued)

TABLE A.24. *(continued)*

	$k = 7$			$k = 8$	
n	r_α	α	n	r_α	α
2	12	.024	2	14	.018
3	15	.048	3	17	.067
	16	.016		18	.027
				19	.009
4	18	.040			
	20	.007	4	21	.036
				23	.007
5	20	.052			
	21	.028	5	23	.057
	22	.014		24	.034
				26	.009
6	22	.050			
	23	.032	6	26	.045
	25	.009		27	.027
				29	.009
7	24	.047			
	25	.032	7	28	.048
	27	.011		29	.032
				31	.012
8	26	.041			
	27	.030	8	30	.046
	29	.011		31	.033
				34	.009
9	27	.050			
	29	.026	9	32	.043
	31	.011		33	.032
				36	.010
10	29	.042			
	30	.031	10	34	.040
	33	.010		35	.031
				38	.010
11	30	.049			
	32	.027	11	35	.048
	35	.009		37	.028
				40	.010
12	32	.040			
	33	.030	12	37	.042
	36	.011		39	.026
				42	.010
13	33	.043			
	35	.025	13	39	.039
	38	.009		40	.030
				44	.009
14	34	.047			
	36	.028	14	40	.042
	39	.011		42	.027
				45	.012
15	36	.038			
	37	.030	15	42	.037
	41	.009		43	.030
				47	.011

TABLE A.24. (continued)

	k = 9			k = 10	
n	r_α	α	n	r_α	α
2	15	.069	2	17	.056
	16	.014		18	.011
3	20	.041	3	22	.057
	22	.005		23	.026
				24	.010
4	23	.064			
	24	.034	4	26	.060
	26	.008		27	.033
				29	.009
5	27	.040			
	28	.023	5	30	.047
	29	.013		31	.029
				33	.010
6	29	.058			
	30	.038	6	33	.051
	33	.008		34	.033
				37	.008
7	32	.046			
	33	.032	7	36	.047
	36	.008		37	.033
				40	.010
8	34	.049			
	36	.026	8	38	.052
	38	.012		40	.031
				43	.010
9	36	.050			
	38	.030	9	41	.046
	41	.010		43	.027
				46	.009
10	38	.050			
	40	.031	10	43	.047
	43	.011		45	.030
				49	.009
11	40	.048			
	42	.030	11	45	.049
	46	.009		47	.032
				51	.010
12	42	.046			
	44	.029	12	48	.040
	48	.009		50	.027
				54	.009
13	44	.042			
	46	.027	13	50	.039
	50	.009		52	.026
				56	.009
14	46	.041			
	48	.026	14	52	.039
	52	.009		54	.026
				58	.010
15	47	.046			
	50	.025	15	53	.045
	54	.009		56	.026
				60	.010

(continued)

TABLE A.24. *(continued)*

	$k = 11$			$k = 12$	
n	r_α	α	n	r_α	α
2	19	.045	2	21	.038
	20	.009		22	.008
3	25	.038	3	27	.053
	27	.007		28	.027
				29	.012
4	29	.057			
	30	.033	4	32	.055
	32	.010		33	.033
				35	.011
5	33	.055			
	34	.035	5	37	.042
	37	.008		38	.027
				40	.011
6	37	.045			
	38	.030	6	40	.059
	41	.008		42	.028
				45	.008
7	40	.049			
	41	.035	7	44	.050
	44	.011		46	.026
				49	.009
8	43	.046			
	44	.035	8	47	.050
	48	.009		49	.030
				52	.011
9	46	.043			
	47	.034	9	50	.048
	51	.009		52	.032
				56	.010
10	48	.047			
	50	.031	10	53	.047
	54	.009		55	.032
				59	.010
11	51	.040			
	53	.027	11	56	.043
	57	.009		58	.029
				62	.011
12	53	.043			
	55	.029	12	58	.048
	59	.011		61	.027
				65	.011
13	55	.046			
	57	.031	13	61	.043
	62	.010		63	.030
				68	.010
14	57	.045			
	60	.026	14	63	.046
	64	.011		66	.027
				71	.009
15	59	.046			
	62	.027	15	66	.040
	67	.009		68	.028
				73	.011

TABLE A.24. (continued)

	k = 13			k = 14	
n	r_α	α	n	r_α	α
2	23	.032	2	25	.027
	24	.006		26	.005
3	30	.038	3	32	.052
	32	.009		33	.028
				35	.006
4	35	.054	4	38	.053
	36	.033		39	.034
	38	.012		41	.013
5	40	.049	5	43	.057
	41	.033		45	.027
	44	.009		47	.012
6	44	.054	6	48	.050
	46	.027		50	.026
	49	.009		53	.009
7	48	.051	7	52	.053
	50	.028		54	.030
	53	.010		57	.012
8	52	.046	8	56	.051
	53	.035		58	.031
	57	.010		62	.010
9	55	.048	9	60	.047
	57	.030		62	.029
	61	.010		66	.010
10	58	.047	10	63	.048
	60	.032		65	.033
	65	.009		70	.010
11	61	.046	11	66	.049
	63	.032		69	.029
	68	.010		74	.009
12	64	.045	12	69	.048
	66	.032		72	.030
	71	.010		77	.010
13	67	.041	13	72	.047
	69	.030		75	.030
	74	.011		80	.011
14	69	.046	14	75	.045
	72	.028		78	.028
	77	.010		84	.009
15	72	.040	15	78	.043
	74	.030		81	.028
	80	.010		87	.010

(continued)

TABLE A.24. *(continued)*

n	r_α	α	n	r_α	α
\multicolumn{3}{c}{$k = 15$}	\multicolumn{3}{c}{$k = 15$}

n	r_α	α	n	r_α	α
2	26	.071	9	64	.052
	27	.024		67	.028
	28	.005		71	.011
3	35	.039	10	68	.049
	37	.010		71	.028
				75	.011
4	41	.053			
	42	.035	11	72	.043
	45	.008		74	.032
				79	.011
5	47	.046			
	48	.033	12	75	.045
	51	.010		78	.028
				83	.010
6	52	.047			
	53	.035	13	78	.046
	57	.009		81	.030
				87	.009
7	56	.055			
	58	.032	14	81	.046
	62	.010		84	.030
				90	.010
8	60	.056			
	63	.027	15	84	.043
	67	.009		87	.029
				94	.009

Adapted, in part, from B. J. McDonald and W. A. Thompson, Jr., Rank sum multiple comparisons in one- and two-way classifications, *Biometrika* **54** (1967): 487–497, by permission of Oxford University Press on behalf of the *Biometrika* trustees, and, in part (the starred values), from P. Nemenyi, Distribution-free multiple comparisons, Ph. D. thesis, Princeton University (1963), with the permission of the author.

A. TABLES AND CHARTS

TABLE A.25. Selected Critical Values for the Nemenyi-Wilcoxon-Wilcox-Miller One-Sided Treatments versus Control Multiple Comparison Procedure:

$$k = 3, n = 2(1)18; k = 4(1)5, n = 2(1)8; k = 6, n = 2(1)6$$

Additional critical values can be found in Odeh (1977c) and Hollander and Wolfe (1973).
For a given k and n with treatment 1 corresponding to the control, the table entries are the experimentwise error rates α and associated critical values r_α^* satisfying the relationship $P_0\{(R_u - R_1) < r_\alpha^*, u = 2, \ldots, k\} = 1 - \alpha$.

					$k = 3$				
n	r_α^*	α	n	r_α^*	α	n	r_α^*	α
2	3	.2222	9	7	.1096	13	8	.1239
	4	.0556		8	.0692		9	.0839
				9	.0392		10	.0554
3	4	.1482		10	.0212		11	.0353
	5	.0648		11	.0106		12	.0209
	6	.0093		12	.0046		13	.0120
				13	.0018		14	.0065
4	5	.1019		14	.0006		15	.0033
	6	.0463		15	.0002		16	.0016
	7	.0139					17	.0007
	8	.0015	10	7	.1265		18	.0003
				8	.0837		19	.0001
5	5	.1407		9	.0504			
	6	.0697		10	.0292	14	8	.1361
	7	.0337		11	.0159		9	.0946
	8	.0131		12	.0078		10	.0643
	9	.0028		13	.0036		11	.0424
	10	.0003		14	.0015		12	.0261
				15	.0005		13	.0156
6	5	.1712		16	.0002		14	.0090
	6	.0992					15	.0048
	7	.0524	11	7	.1424		16	.0025
	8	.0251		8	.0977		17	.0012
	9	.0101		9	.0617		18	.0005
	10	.0031		10	.0377		19	.0002
	11	.0006		11	.0219		20	.0001
				12	.0116			
7	6	.1223		13	.0058	15	9	.1050
	7	.0728		14	.0027		10	.0732
	8	.0398		15	.0011		11	.0496
	9	.0183		16	.0004		12	.0317
	10	.0078		17	.0001		13	.0197
	11	.0028					14	.0118
	12	.0007	12	8	.1111		15	.0066
	13	.0001		9	.0729		16	.0035
				10	.0464		17	.0019
8	6	.1454		11	.0284		18	.0009
	7	.0914		12	.0160		19	.0004
	8	.0544		13	.0087		20	.0002
	9	.0286		14	.0044		21	.0001
	10	.0139		15	.0021			
	11	.0061		16	.0009			
	12	.0023		17	.0003			
	13	.0007		18	.0001			
	14	.0002						

(continued)

TABLE A.25. *(continued)*

	k = 3			k = 4			k = 5	
n	r_α^*	α	n	r_α^*	α	n	r_α^*	α
16	9	.1151	2	4	.2292	2	6	.1217
	10	.0820		5	.0938		7	.0467
	11	.0569		6	.0208		8	.0100
	12	.0374						
	13	.0240	3	6	.1046	3	7	.1421
	14	.0149		7	.0460		8	.0808
	15	.0087		8	.0122		9	.0385
	16	.0050		9	.0017		10	.0137
	17	.0027					11	.0035
	18	.0014	4	6	.1566		12	.0005
	19	.0007		7	.0918			
	20	.0003		8	.0481	4	8	.1386
	21	.0001		9	.0205		9	.0872
	22	.0001		10	.0065		10	.0504
				11	.0013		11	.0258
17	9	.1248		12	.0001		12	.0114
	10	.0906					13	.0041
	11	.0642	5	7	.1334		14	.0011
	12	.0432		8	.0804		15	.0002
	13	.0285		9	.0446			
	14	.0183		10	.0222	5	9	.1321
	15	.0111		11	.0095		10	.0872
	16	.0065		12	.0032		11	.0542
	17	.0037		13	.0008		12	.0315
	18	.0020		14	.0001		13	.0168
	19	.0010					14	.0081
	20	.0005	6	8	.1129		15	.0034
	21	.0002		9	.0701		16	.0012
	22	.0001		10	.0404		17	.0004
				11	.0215		18	.0001
18	9	.1342		12	.0103			
	10	.0990		13	.0044	6	10	.1233
	11	.0715		14	.0015		11	.0842
	12	.0492		15	.0004		12	.0548
	13	.0332		16	.0001		13	.0339
	14	.0218					14	.0198
	15	.0137	7	8	.1424		15	.0108
	16	.0083		9	.0952		16	.0054
	17	.0049		10	.0604		17	.0025
	18	.0028		11	.0360		18	.0010
	19	.0015		12	.0200		19	.0004
	20	.0008		13	.0103		20	.0001
	21	.0004		14	.0048			
	22	.0002		15	.0020	7	11	.1136
	23	.0001		16	.0007		12	.0795
				17	.0002		13	.0535
				18	.0001		14	.0345
							15	.0212
			8	9	.1194		16	.0124
				10	.0805		17	.0069
				11	.0517		18	.0036
				12	.0315		19	.0017
				13	.0181		20	.0008
				14	.0098		21	.0003
				15	.0049		22	.0001
				16	.0023			
				17	.0010			
				18	.0004			
				19	.0001			

TABLE A.25. *(continued)*

	$k = 5$			$k = 6$			$k = 6$	
n	r_α^*	α	n	r_α^*	α	n	r_α^*	α
8	12	.1040	2	7	.1438	5	11	.1324
	13	.0741		8	.0713		12	.0931
	14	.0510		9	.0264		13	.0627
	15	.0339		10	.0056		14	.0402
	16	.0217					15	.0245
	17	.0133	3	9	.1139		16	.0140
	18	.0078		10	.0673		17	.0074
	19	.0044		11	.0348		18	.0036
	20	.0023		12	.0148		19	.0016
	21	.0012		13	.0051		20	.0006
	22	.0005		14	.0013		21	.0002
	23	.0002		15	.0002		22	.0001
	24	.0001						
			4	10	.1274	6	12	.1324
				11	.0847		13	.0965
				12	.0529		14	.0680
				13	.0306		15	.0462
				14	.0160		16	.0302
				15	.0075		17	.0189
				16	.0030		18	.0113
				17	.0010		19	.0064
				18	.0003		20	.0034
				19	.0001		21	.0017
							22	.0008
							23	.0003
							24	.0001

Adapted, in part, from R. E. Odeh, Extended tables of the distributions of rank statistics for treatment versus control in randomized block designs, *Communications in Statistics—Simulation and Computation* **6**, Marcel Dekker, Inc., N.Y. (1977): 101–113, with the permission of the author and the editor of *Communications in Statistics—Simulation and Computation*, reprinted by courtesy of Marcel Dekker, Inc.

TABLE A.26. Selected Critical Values for the Null Distribution of the Durbin-Skillings-Mack D Statistic: k Is the Number of Treatments, n Is the Number of Blocks, s Is the Common Number of Observations in Each Block, p Is the Total Number of Observations Present for Each Treatment, λ Is the Common Number of Blocks in Which any Pair of Treatments Occur Together, and These Parameters Satisfy the Restriction $p(s-1) = \lambda(k-1)$.

For given k, n, s, p, and λ combination and significance level α, the table entry is $d_{\alpha,s}$ satisfying $P_0\{D \geq d_{\alpha,s}\} = \alpha$.

$k = 3, s = 2$
$n = 6, p = 4, \lambda = 2$

$d_{.0938,2} = 5.333$

$n = 9, p = 6, \lambda = 3$

$d_{.0824,2} = 6.222$
$d_{.0114,2} = 8.000$

$n = 12, p = 8, \lambda = 4$

$d_{.0527,2} = 6.000$
$d_{.0208,2} = 8.000$
$d_{.0112,2} = 8.667$

$n = 15, p = 10, \lambda = 5$

$d_{.1005,2} = 4.800$
$d_{.0422,2} = 6.933$
$d_{.0113,2} = 8.533$

$n = 18, p = 12, \lambda = 6$

$d_{.0982,2} = 5.333$
$d_{.0321,2} = 7.111$
$d_{.0086,2} = 9.333$

$n = 21, p = 14, \lambda = 7$

$d_{.1023,2} = 4.952$
$d_{.0461,2} = 6.095$
$d_{.0140,2} = 8.000$

$k = 4, s = 2$
$n = 12, p = 6, \lambda = 2$

$d_{.0996,2} = 7.000$
$d_{.0293,2} = 8.000$
$d_{.0059,2} = 10.000$

$n = 18, p = 9, \lambda = 3$

$d_{.0746,2} = 7.000$
$d_{.0389,2} = 8.333$
$d_{.0077,2} = 11.000$

$k = 4, s = 3$
$n = 4, p = 3, \lambda = 2$

$d_{.0926,3} = 6.000$
$d_{.0741,3} = 6.750$
$d_{.0185,3} = 7.500$

$k = 4, s = 3$
$n = 8, p = 6, \lambda = 4$

$d_{.1023,3} = 6.375$
$d_{.0467,3} = 7.500$
$d_{.0108,3} = 10.125$

$n = 12, p = 9, \lambda = 6$

$d_{.0853,3} = 6.500$
$d_{.0516,3} = 7.750$
$d_{.0106,3} = 10.750$

$n = 16, p = 12, \lambda = 8$

$d_{.0878,3} = 6.562$
$d_{.0486,3} = 7.875$
$d_{.0110,3} = 10.687$

$n = 20, p = 15, \lambda = 10$

$d_{.0997,3} = 6.300$
$d_{.0480,3} = 7.800$
$d_{.0110,3} = 10.950$

$k = 5, s = 2$
$n = 10, p = 4, \lambda = 1$

$d_{.1172,2} = 8.000$

$n = 20, p = 8, \lambda = 2$

$d_{.0685,2} = 8.800$
$d_{.0411,2} = 9.600$
$d_{.0089,2} = 12.000$

$n = 30, p = 12, \lambda = 3$

$d_{.1005,2} = 8.000$
$d_{.0400,2} = 9.600$
$d_{.0100,2} = 12.267$

$k = 5, s = 3$
$n = 10, p = 6, \lambda = 3$

$d_{.1078,3} = 7.600$
$d_{.0499,3} = 9.200$
$d_{.0105,3} = 11.600$

$n = 20, p = 12, \lambda = 6$

$d_{.0975,3} = 7.800$
$d_{.0513,3} = 9.200$
$d_{.0101,3} = 12.600$

$k = 5, s = 3$
$n = 30, p = 18, \lambda = 9$

$d_{.1006,3} = 7.733$
$d_{.0500,3} = 9.467$
$d_{.0102,3} = 12.933$

$k = 5, s = 4$
$n = 5, p = 4, \lambda = 3$

$d_{.0907,4} = 7.680$
$d_{.0450,4} = 8.960$
$d_{.0093,4} = 10.880$

$n = 10, p = 8, \lambda = 6$

$d_{.0972,4} = 7.680$
$d_{.0514,4} = 9.120$
$d_{.0104,4} = 12.320$

$n = 15, p = 12, \lambda = 9$

$d_{.1010,4} = 7.787$
$d_{.0502,4} = 9.280$
$d_{.0101,4} = 12.587$

$n = 20, p = 16, \lambda = 12$

$d_{.0972,4} = 7.840$
$d_{.0492,4} = 9.360$
$d_{.0099,4} = 12.400$

$k = 6, s = 3$
$n = 10, p = 5, \lambda = 2$

$d_{.0815,3} = 9.500$
$d_{.0428,3} = 10.500$
$d_{.0112,3} = 12.500$

$n = 20, p = 10, \lambda = 4$

$d_{.0973,3} = 9.250$
$d_{.0514,3} = 10.750$
$d_{.0105,3} = 14.000$

$n = 30, p = 15, \lambda = 6$

$d_{.1021,3} = 9.167$
$d_{.0498,3} = 11.000$
$d_{.0105,3} = 14.500$

$n = 40, p = 20, \lambda = 8$

$d_{.1002,3} = 9.250$
$d_{.0498,3} = 10.875$
$d_{.0102,3} = 14.750$

TABLE A.26. *(continued)*

$k = 6, s = 4$ $n = 15, p = 10, \lambda = 6$	$k = 6, s = 5$ $n = 6, p = 5, \lambda = 4$	$k = 6, s = 5$ $n = 12, p = 10, \lambda = 8$
$d_{.1020,4} = 9.067$	$d_{.0987,5} = 9.000$	$d_{.1011,5} = 9.083$
$d_{.0487,4} = 10.800$	$d_{.0497,5} = 10.500$	$d_{.0500,5} = 10.833$
$d_{.0100,4} = 14.133$	$d_{.0099,5} = 13.333$	$d_{.0098,5} = 14.083$
$n = 30, p = 20, \lambda = 12$		$n = 18, p = 15, \lambda = 12$
$d_{.1001,4} = 9.067$		$d_{.0986,5} = 9.111$
$d_{.0498,4} = 10.733$		$d_{.0509,5} = 10.667$
$d_{.0101,4} = 14.467$		$d_{.0101,5} = 14.444$

Adapted from J. H. Skillings and G. A. Mack, On the use of a Friedman-type statistic in balanced and unbalanced block designs, *Technometrics* **23** (1981): 171–177. Reprinted with permission from *Technometrics*. Copyright © (1981) by the American Statistical Association and the American Society for Quality Control. All rights reserved.

TABLE A.27. Selected Exact Critical Values for the Null Distribution of the Skillings-Mack SM Statistic for a Randomized Block Design with k Treatments, n Blocks, and a Single Missing Observation: $k = 3(1)6$, $n = 3(1)11$

For given k, n, and significance level α, the table entry is sm_α satisfying $P_0\{SM \geq sm_\alpha\} = \alpha$.

$k = 3$ $n = 3$	$k = 3$ $n = 9$	$k = 4$ $n = 5$
$sm_{.0833} = 4.866$	$sm_{.0988} = 4.689$	$sm_{.1007} = 6.084$
$sm_{.0278} = 4.982$	$sm_{.0484} = 5.703$	$sm_{.0499} = 7.323$
$n = 4$	$sm_{.0095} = 8.524$	$sm_{.0100} = 9.532$
$sm_{.0972} = 5.133$	$n = 10$	$n = 6$
$sm_{.0417} = 5.665$	$sm_{.0986} = 4.714$	$sm_{.0999} = 6.108$
$sm_{.0139} = 6.854$	$sm_{.0508} = 6.037$	$sm_{.0497} = 7.385$
$n = 5$	$sm_{.0103} = 8.648$	$sm_{.0100} = 9.887$
$sm_{.1065} = 4.434$	$n = 11$	$n = 7$
$sm_{.0471} = 6.077$	$sm_{.1018} = 4.559$	$sm_{.1000} = 6.054$
$sm_{.0085} = 7.500$	$sm_{.0481} = 5.894$	$sm_{.0499} = 7.341$
$n = 6$	$sm_{.0103} = 8.712$	$sm_{.0100} = 10.122$
$sm_{.1030} = 4.517$	$k = 4$ $n = 3$	$n = 8$
$sm_{.0445} = 5.731$	$sm_{.1007} = 5.936$	$sm_{.1002} = 6.053$
$sm_{.0096} = 8.568$	$sm_{.0486} = 6.867$	$sm_{.0499} = 7.399$
$n = 7$	$sm_{.0104} = 7.400$	$sm_{.0099} = 10.242$
$sm_{.0956} = 4.813$	$n = 4$	$n = 9$
$sm_{.0487} = 5.868$	$sm_{.1013} = 5.920$	$sm_{.0998} = 6.126$
$sm_{.0105} = 7.984$	$sm_{.0498} = 7.087$	$sm_{.0500} = 7.536$
$n = 8$	$sm_{.0098} = 8.875$	$sm_{.0100} = 10.645$
$sm_{.0965} = 4.551$		$n = 10$
$sm_{.0496} = 5.943$		$sm_{.1000} = 6.323$
$sm_{.0097} = 8.528$		$sm_{.0501} = 7.709$
		$sm_{.0100} = 10.734$

(continued)

TABLE A.27. *(continued)*

$k = 4$
$n = 11$

$sm_{.0999} = 6.250$
$sm_{.0501} = 7.657$
$sm_{.0100} = 10.416$

$k = 5$
$n = 3$

$sm_{.1001} = 7.167$
$sm_{.0498} = 8.133$
$sm_{.0101} = 9.589$

$n = 4$

$sm_{.0999} = 7.438$
$sm_{.0499} = 8.571$
$sm_{.0100} = 10.717$

$n = 5$

$sm_{.0999} = 7.495$
$sm_{.0498} = 8.893$
$sm_{.0100} = 11.438$

$n = 6$

$sm_{.1001} = 7.485$
$sm_{.0500} = 8.861$
$sm_{.0100} = 11.760$

$n = 7$

$sm_{.1000} = 7.687$
$sm_{.0500} = 9.216$
$sm_{.0100} = 12.034$

$n = 8$

$sm_{.0996} = 7.732$
$sm_{.0499} = 9.211$
$sm_{.0100} = 12.088$

$k = 5$
$n = 9$

$sm_{.1000} = 7.720$
$sm_{.0500} = 9.182$
$sm_{.0100} = 12.575$

$n = 10$

$sm_{.1000} = 7.625$
$sm_{.0500} = 9.333$
$sm_{.0100} = 12.407$

$n = 11$

$sm_{.1000} = 7.670$
$sm_{.0500} = 9.193$
$sm_{.0100} = 12.642$

$k = 6$
$n = 3$

$sm_{.1001} = 8.496$
$sm_{.0503} = 9.625$
$sm_{.0099} = 11.398$

$n = 4$

$sm_{.0998} = 8.799$
$sm_{.0500} = 10.087$
$sm_{.0100} = 12.468$

$n = 5$

$sm_{.0999} = 9.004$
$sm_{.0499} = 10.455$
$sm_{.0099} = 13.394$

$k = 6$
$n = 6$

$sm_{.1000} = 8.927$
$sm_{.0499} = 10.447$
$sm_{.0100} = 13.449$

$n = 7$

$sm_{.1000} = 9.024$
$sm_{.0500} = 10.609$
$sm_{.0100} = 13.590$

$n = 8$

$sm_{.0999} = 9.115$
$sm_{.0499} = 10.594$
$sm_{.0100} = 13.795$

$n = 9$

$sm_{.1000} = 9.072$
$sm_{.0500} = 10.595$
$sm_{.0100} = 13.994$

$n = 10$

$sm_{.1000} = 9.041$
$sm_{.0500} = 10.704$
$sm_{.0100} = 13.958$

$n = 11$

$sm_{.1000} = 9.142$
$sm_{.0501} = 10.825$
$sm_{.0100} = 14.352$

Adapted from J. H. Skillings and G. A. Mack, On the use of a Friedman-type statistic in balanced and unbalanced block designs, *Technometrics* **23** (1981): 171–177. Reprinted with permission from *Technometrics*. Copyright © (1981) by the American Statistical Association and the American Society for Quality Control. All rights reserved.

A. TABLES AND CHARTS

TABLE A.28. Selected Critical Values for the Null Distribution of the Mack-Skillings MS Statistic with the Same Number (c) of Replications in Each Cell:

$$k = 2(1)5, n = 2(1)5, c = 2(1)5$$

For given k, n, c and significance level α, the table entry is ms_α satisfying $P_0\{MS \geq ms_\alpha\} = \alpha$.

$k = 2, n = 2$	$k = 2, n = 4$	$k = 3, n = 2$	$k = 3, n = 4$
$c = 2$	$c = 3$	$c = 4$	$c = 5$
$ms_{.0556} = 4.800$	$ms_{.1018} = 3.048$	$ms_{.1004} = 4.635$	$ms_{.0996} = 4.580$
	$ms_{.0623} = 3.857$	$ms_{.0494} = 5.846$	$ms_{.0496} = 5.955$
$c = 3$	$ms_{.0099} = 6.857$	$ms_{.0099} = 8.481$	$ms_{.0101} = 9.155$
$ms_{.0900} = 3.429$	$c = 4$	$c = 5$	$k = 3, n = 5$
$ms_{.0400} = 4.667$			$c = 2$
$ms_{.0150} = 6.095$	$ms_{.0971} = 3.000$	$ms_{.0996} = 4.530$	
	$ms_{.0502} = 4.083$	$ms_{.0496} = 5.880$	$ms_{.0960} = 4.629$
$c = 4$	$ms_{.0100} = 6.750$	$ms_{.0099} = 8.310$	$ms_{.0515} = 5.886$
			$ms_{.0099} = 8.629$
$ms_{.0820} = 3.375$	$c = 5$	$k = 3, n = 3$	
$ms_{.0502} = 4.167$		$c = 2$	$c = 3$
$ms_{.0074} = 7.042$	$ms_{.0847} = 3.153$		
	$ms_{.0525} = 3.938$	$ms_{.1068} = 4.667$	$ms_{.1010} = 4.604$
$c = 5$	$ms_{.0094} = 6.818$	$ms_{.0467} = 6.000$	$ms_{.0502} = 5.920$
		$ms_{.0092} = 8.667$	$ms_{.0099} = 8.871$
$ms_{.0894} = 3.142$	$k = 2, n = 5$		
$ms_{.0441} = 4.276$	$c = 2$	$c = 3$	$c = 4$
$ms_{.0122} = 6.305$			
	$ms_{.1188} = 3.000$	$ms_{.1001} = 4.622$	$ms_{.1007} = 4.669$
$k = 2, n = 3$	$ms_{.0543} = 4.320$	$ms_{.0483} = 5.896$	$ms_{.0506} = 5.977$
$c = 2$	$ms_{.0067} = 7.680$	$ms_{.0100} = 8.622$	$ms_{.0099} = 8.746$
$ms_{.1204} = 3.200$	$c = 3$	$c = 4$	$c = 5$
$ms_{.0370} = 5.000$			
$ms_{.0093} = 7.200$	$ms_{.1190} = 2.752$	$ms_{.0986} = 4.654$	$ms_{.1002} = 4.572$
	$ms_{.0495} = 4.200$	$ms_{.0497} = 5.936$	$ms_{.0500} = 5.844$
$c = 3$	$ms_{.0095} = 6.943$	$ms_{.0102} = 8.705$	$ms_{.0100} = 8.572$
$ms_{.0768} = 3.571$	$c = 4$	$c = 5$	$k = 4, n = 2$
$ms_{.0410} = 4.587$			$c = 2$
$ms_{.0088} = 7.000$	$ms_{.1069} = 2.817$	$ms_{.1000} = 4.587$	
	$ms_{.0444} = 4.267$	$ms_{.0503} = 5.927$	$ms_{.1057} = 6.083$
$c = 4$	$ms_{.0106} = 6.667$	$ms_{.0101} = 8.880$	$ms_{.0509} = 7.250$
			$ms_{.0101} = 9.250$
$ms_{.1142} = 2.778$	$c = 5$	$k = 3, n = 4$	
$ms_{.0539} = 4.000$		$c = 2$	$c = 3$
$ms_{.0078} = 7.111$	$ms_{.0927} = 2.987$		
	$ms_{.0491} = 4.034$	$ms_{.0969} = 4.571$	$ms_{.0995} = 6.128$
$c = 5$	$ms_{.0107} = 6.600$	$ms_{.0498} = 5.786$	$ms_{.0502} = 7.615$
		$ms_{.0093} = 8.643$	$ms_{.0100} = 10.128$
$ms_{.0914} = 3.058$	$k = 3, n = 2$		
$ms_{.0526} = 3.960$	$c = 2$	$c = 3$	$c = 4$
$ms_{.0097} = 6.724$			
	$ms_{.0859} = 5.143$	$ms_{.1006} = 4.622$	$ms_{.0994} = 6.243$
$k = 2, n = 4$	$ms_{.0556} = 5.571$	$ms_{.0504} = 5.956$	$ms_{.0500} = 7.577$
$c = 2$	$ms_{.0111} = 7.429$	$ms_{.0102} = 8.822$	$ms_{.0099} = 10.511$
$ms_{.0787} = 3.750$	$c = 3$	$c = 4$	$c = 5$
$ms_{.0293} = 5.400$			
$ms_{.0077} = 7.350$	$ms_{.1028} = 4.578$	$ms_{.1011} = 4.625$	$ms_{.0997} = 6.160$
	$ms_{.0519} = 5.733$	$ms_{.0503} = 5.894$	$ms_{.0501} = 7.686$
	$ms_{.0105} = 8.133$	$ms_{.0100} = 8.577$	$ms_{.0100} = 10.360$

(continued)

TABLE A.28. *(continued)*

$k = 4, n = 3$	$k = 4, n = 5$	$k = 5, n = 2$	$k = 5, n = 4$
$c = 2$	$c = 2$	$c = 5$	$c = 3$
$ms_{.1005} = 6.167$	$ms_{.0995} = 6.267$	$ms_{.1000} = 7.617$	$ms_{.0998} = 7.700$
$ms_{.0496} = 7.444$	$ms_{.0500} = 7.733$	$ms_{.0500} = 9.338$	$ms_{.0501} = 9.200$
$ms_{.0103} = 10.056$	$ms_{.0099} = 10.767$	$ms_{.0100} = 12.373$	$ms_{.0099} = 12.492$
$c = 3$	$c = 3$	$k = 5, n = 3$	$c = 4$
		$c = 2$	
$ms_{.0998} = 6.231$	$ms_{.1002} = 6.179$	$ms_{.1013} = 7.527$	$ms_{.1001} = 7.761$
$ms_{.0501} = 7.479$	$ms_{.0499} = 7.779$	$ms_{.0497} = 9.018$	$ms_{.0500} = 9.354$
$ms_{.0099} = 10.573$	$ms_{.0100} = 10.672$	$ms_{.0099} = 11.745$	$ms_{.0100} = 12.896$
$c = 4$	$c = 4$	$c = 3$	$c = 5$
$ms_{.0999} = 6.265$	$ms_{.0997} = 6.287$	$ms_{.1000} = 7.633$	$ms_{.0999} = 7.782$
$ms_{.0500} = 7.757$	$ms_{.0500} = 7.747$	$ms_{.0501} = 9.089$	$ms_{.0500} = 9.404$
$ms_{.0099} = 10.949$	$ms_{.0099} = 10.566$	$ms_{.0100} = 12.256$	$ms_{.0100} = 12.958$
$c = 5$	$c = 5$	$c = 4$	$k = 5, n = 5$
			$c = 2$
$ms_{.1000} = 6.211$	$ms_{.0999} = 6.147$	$ms_{.1001} = 7.776$	$ms_{.1003} = 7.702$
$ms_{.0500} = 7.750$	$ms_{.0499} = 7.583$	$ms_{.0499} = 9.300$	$ms_{.0499} = 9.251$
$ms_{.0100} = 10.947$	$ms_{.0100} = 10.822$	$ms_{.0100} = 12.871$	$ms_{.0099} = 12.327$
$k = 4, n = 4$	$k = 5, n = 2$	$c = 5$	$c = 3$
$c = 2$	$c = 2$		
$ms_{.0998} = 6.250$	$ms_{.0998} = 7.418$	$ms_{.1000} = 7.714$	$ms_{.1002} = 7.673$
$ms_{.0500} = 7.625$	$ms_{.0495} = 8.727$	$ms_{.0500} = 9.263$	$ms_{.0500} = 9.347$
$ms_{.0099} = 10.792$	$ms_{.0100} = 10.964$	$ms_{.0100} = 12.736$	$ms_{.0100} = 12.847$
$c = 3$	$c = 3$	$k = 5, n = 4$	$c = 4$
		$c = 2$	
$ms_{.0999} = 6.231$	$ms_{.0993} = 7.500$	$ms_{.0996} = 7.500$	$ms_{.1000} = 7.746$
$ms_{.0502} = 7.667$	$ms_{.0502} = 8.967$	$ms_{.0500} = 8.918$	$ms_{.0500} = 9.317$
$ms_{.0100} = 10.628$	$ms_{.0100} = 11.783$	$ms_{.0100} = 12.300$	$ms_{.0100} = 12.789$
$c = 4$	$c = 4$		$c = 5$
$ms_{.1002} = 6.325$	$ms_{.1000} = 7.664$		$ms_{.1000} = 7.704$
$ms_{.0499} = 7.737$	$ms_{.0500} = 9.036$		$ms_{.0500} = 9.250$
$ms_{.0100} = 11.068$	$ms_{.0100} = 12.336$		$ms_{.0100} = 12.719$
$c = 5$			
$ms_{.0998} = 6.243$			
$ms_{.0500} = 7.717$			
$ms_{.0100} = 11.169$			

Adapted from G. A. Mack and J. H. Skillings, A Friedman-type rank test for main effects in a two-factor ANOVA, *Journal of the American Statistical Association* **75** (1980): 947–951. Reprinted with permission from the *Journal of the American Statistical Association*. Copyright © (1980) by the American Statistical Association. All rights reserved.

TABLE A.29. Upper Bounds ρ_U^n for the Null Correlation between Two Overlapping Signed Rank Statistics Based on n Observations: $n = 1(1)50$

For $1 \leq n \leq 50$, the table entry is ρ_U^n. For $n > 50$, the approximation $\rho_U^n \approx \frac{1}{2}$ is sufficiently accurate for practical purposes.

n	ρ_U^n	n	ρ_U^n
1	.333	26	.488
2	.389	27	.488
3	.416	28	.488
4	.433	29	.489
5	.444	30	.489
6	.452	31	.490
7	.458	32	.490
8	.463	33	.490
9	.467	34	.490
10	.470	35	.491
11	.472	36	.491
12	.474	37	.491
13	.476	38	.491
14	.478	39	.492
15	.479	40	.492
16	.480	41	.492
17	.481	42	.492
18	.482	43	.492
19	.483	44	.493
20	.484	45	.493
21	.485	46	.493
22	.485	47	.493
23	.486	48	.493
24	.487	49	.493
25	.487	50	.493

Adapted from M. Hollander, Rank tests for randomized blocks when the alternatives have an *a priori* ordering, *Annals of Mathematical Statistics* **38** (1967): 867–877, with permission of the author, the editor of *The Annals of Mathematical Statistics*, and the Institute of Mathematical Statistics.

TABLE A.30. Upper-Tail Probabilities for the Null Distribution of the Kendall K Statistic: $n = 4(1)40$

For a given n, the entry in the table for the point x is $P_0\{K \geq x\}$. Under these conditions, if x is such that $P_0\{K \geq x\} = \alpha$, then $k_\alpha = x$. For certain n, the entries are terminated at x_n, where x_n is the smallest possible value of x such that $P_0\{K \geq x\}$ is zero to three decimal places. (For $n = 4(4)40$ or $n = 5(4)37$, all even integers between $-n(n-1)/2$ and $n(n-1)/2$ have positive probability and for $n = 6(4)38$ or $n = 7(4)39$ all odd integers between $-n(n-1)/2$ and $n(n-1)/2$ have positive probability.)

					n				
x	4	5	8	9	12	13	16	17	20
0	.625	.592	.548	.540	.527	.524	.518	.516	.513
2	.375	.408	.452	.460	.473	.476	.482	.484	.487
4	.167	.242	.360	.381	.420	.429	.447	.452	.462
6	.042	.117	.274	.306	.369	.383	.412	.420	.436
8		.042	.199	.238	.319	.338	.378	.388	.411
10		.008	.138	.179	.273	.295	.345	.358	.387
12			.089	.130	.230	.255	.313	.328	.362
14			.054	.090	.190	.218	.282	.299	.339
16			.031	.060	.155	.184	.253	.271	.315
18			.016	.038	.125	.153	.225	.245	.293
20			.007	.022	.098	.126	.199	.220	.271
22			.002	.012	.076	.102	.175	.196	.250
24			.001	.006	.058	.082	.153	.174	.230
26			.000	.003	.043	.064	.133	.154	.211
28				.001	.031	.050	.114	.135	.193
30				.000	.022	.038	.097	.118	.176
32					.016	.029	.083	.102	.159
34					.010	.021	.070	.088	.144
36					.007	.015	.058	.076	.130
38					.004	.011	.048	.064	.117
40					.003	.007	.039	.054	.104
42					.002	.005	.032	.046	.093
44					.001	.003	.026	.038	.082
46					.000	.002	.021	.032	.073
48						.001	.016	.026	.064
50						.001	.013	.021	.056
52						.000	.010	.017	.049
54							.008	.014	.043
56							.006	.011	.037
58							.004	.009	.032
60							.003	.007	.027
62							.002	.005	.023
64							.002	.004	.020
66							.001	.003	.017
68							.001	.002	.014
70							.001	.002	.012
72							.000	.001	.010
74								.001	.008
76								.001	.007
78								.000	.006
80									.005
82									.004
84									.003
86									.002
88									.002

TABLE A.30. (continued)

x	n									
	4	5	8	9	12	13	16	17	20
90									.002
92									.001
94									.001
96									.001
98									.001
100									.000

x	n									
	21	24	25	28	29	32	33	36	37	40
0	.512	.510	.509	.508	.507	.506	.506	.505	.505	.505
2	.488	.490	.491	.492	.493	.494	.494	.495	.495	.495
4	.464	.471	.472	.477	.478	.481	.482	.484	.484	.486
6	.441	.451	.454	.461	.463	.468	.469	.473	.474	.477
8	.417	.432	.436	.446	.448	.455	.457	.462	.464	.468
10	.394	.413	.418	.430	.434	.442	.445	.452	.453	.459
12	.371	.394	.400	.415	.419	.430	.433	.441	.443	.449
14	.349	.375	.382	.400	.405	.417	.421	.430	.433	.440
16	.327	.356	.364	.385	.390	.405	.409	.420	.423	.431
18	.306	.338	.347	.370	.376	.392	.397	.409	.413	.422
20	.285	.320	.330	.355	.362	.380	.385	.399	.403	.413
22	.265	.303	.314	.341	.348	.368	.373	.388	.393	.404
24	.246	.286	.297	.326	.334	.356	.362	.378	.383	.395
26	.228	.270	.282	.312	.321	.344	.350	.368	.373	.386
28	.210	.254	.266	.298	.308	.332	.339	.358	.363	.377
30	.193	.238	.251	.285	.295	.320	.328	.347	.353	.369
32	.177	.223	.237	.272	.282	.309	.317	.338	.344	.360
34	.162	.209	.222	.259	.270	.298	.306	.328	.334	.351
36	.147	.195	.209	.246	.257	.287	.295	.318	.325	.343
38	.134	.181	.196	.234	.246	.276	.285	.308	.315	.334
40	.121	.169	.183	.222	.234	.265	.274	.299	.306	.326
42	.109	.156	.171	.211	.223	.255	.264	.290	.297	.318
44	.098	.145	.159	.200	.212	.244	.254	.280	.288	.309
46	.088	.134	.148	.189	.201	.234	.244	.271	.279	.301
48	.079	.123	.138	.178	.191	.224	.235	.262	.271	.293
50	.070	.113	.128	.168	.181	.215	.225	.254	.262	.285
52	.062	.104	.118	.158	.171	.206	.216	.245	.254	.277
54	.055	.095	.109	.149	.162	.197	.207	.237	.245	.270
56	.049	.087	.101	.140	.153	.188	.199	.228	.237	.262
58	.043	.079	.093	.131	.144	.179	.190	.220	.229	.255
60	.037	.072	.085	.123	.136	.171	.182	.212	.222	.247
62	.032	.066	.078	.115	.128	.163	.174	.204	.214	.240
64	.028	.059	.071	.108	.120	.155	.166	.197	.206	.233
66	.024	.054	.065	.101	.112	.147	.158	.189	.199	.226
68	.021	.048	.059	.094	.105	.140	.151	.182	.192	.219
70	.018	.044	.054	.087	.099	.133	.144	.175	.185	.212
72	.015	.039	.049	.081	.092	.126	.137	.168	.178	.205
74	.013	.035	.044	.075	.086	.119	.130	.161	.171	.199
76	.011	.031	.040	.070	.080	.113	.124	.155	.165	.192
78	.009	.028	.036	.065	.075	.107	.117	.148	.158	.186
80	.008	.025	.032	.060	.070	.101	.111	.142	.152	.180
82	.007	.022	.029	.055	.065	.095	.106	.136	.146	.174
84	.005	.019	.026	.051	.060	.090	.100	.130	.140	.168

(continued)

TABLE A.30. *(continued)*

x	\multicolumn{10}{c}{n}
	21	24	25	28	29	32	33	36	37	40
86	.005	.017	.023	.047	.056	.085	.095	.124	.134	.162
88	.004	.015	.021	.043	.052	.080	.090	.119	.129	.156
90	.003	.013	.018	.039	.048	.075	.085	.114	.123	.151
92	.002	.011	.016	.036	.044	.070	.080	.108	.118	.146
94	.002	.010	.014	.033	.041	.066	.075	.103	.113	.140
96	.002	.009	.013	.030	.037	.062	.071	.099	.108	.135
98	.001	.007	.011	.027	.034	.058	.067	.094	.103	.130
100	.001	.006	.010	.025	.031	.054	.063	.089	.098	.125
102	.001	.006	.009	.023	.029	.051	.059	.085	.094	.121
104	.001	.005	.008	.021	.026	.048	.055	.081	.090	.116
106	.001	.004	.007	.019	.024	.044	.052	.077	.085	.111
108	.000	.003	.006	.017	.022	.041	.049	.073	.081	.107
110		.003	.005	.015	.020	.039	.046	.069	.077	.103
112		.003	.004	.014	.018	.036	.043	.066	.074	.099
114		.002	.004	.012	.017	.033	.040	.062	.070	.095
116		.002	.003	.011	.015	.031	.037	.059	.067	.091
118		.001	.003	.010	.014	.029	.035	.056	.063	.087
120		.001	.002	.009	.012	.027	.032	.053	.060	.083
122		.001	.002	.008	.011	.025	.030	.050	.057	.080
124		.001	.002	.007	.010	.023	.028	.047	.054	.076
126		.001	.001	.006	.009	.021	.026	.044	.051	.073
128		.001	.001	.006	.008	.019	.024	.042	.048	.070
130		.000	.001	.005	.007	.018	.023	.039	.046	.067
132			.001	.004	.007	.016	.021	.037	.043	.064
134			.001	.004	.006	.015	.019	.035	.041	.061
136			.001	.003	.005	.014	.018	.033	.039	.058
138			.000	.003	.005	.013	.017	.031	.037	.055
140				.003	.004	.012	.015	.029	.034	.053
142				.002	.004	.011	.014	.027	.032	.050
144				.002	.003	.010	.013	.025	.031	.048
146				.002	.003	.009	.012	.024	.029	.046
148				.002	.003	.008	.011	.022	.027	.043
150				.001	.002	.007	.010	.021	.025	.041
152				.001	.002	.007	.009	.020	.024	.039
154				.001	.002	.006	.008	.018	.022	.037
156				.001	.002	.006	.008	.017	.021	.035
158				.001	.001	.005	.007	.016	.020	.034
160				.001	.001	.005	.006	.015	.018	.032
162				.001	.001	.004	.006	.014	.017	.030
164				.000	.001	.004	.005	.013	.016	.029
166					.001	.003	.005	.012	.015	.027
168					.001	.003	.004	.011	.014	.026
170					.001	.003	.004	.010	.013	.024
172					.001	.002	.004	.010	.012	.023
174					.000	.002	.003	.009	.011	.022
176						.002	.003	.008	.011	.020
178						.002	.003	.008	.010	.019
180						.002	.002	.007	.009	.018
182						.001	.002	.006	.009	.017
184						.001	.002	.006	.008	.016
186						.001	.002	.006	.007	.015
188						.001	.002	.005	.007	.014
190						.001	.001	.005	.006	.014
192						.001	.001	.004	.006	.013

TABLE A.30. (continued)

					n					
x	21	24	25	28	29	32	33	36	37	40
194						.001	.001	.004	.005	.012
196						.001	.001	.004	.005	.011
198						.001	.001	.003	.005	.011
200						.000	.001	.003	.004	.010
202							.001	.003	.004	.009
204							.001	.003	.004	.009
206							.001	.002	.003	.008
208							.001	.002	.003	.008
210							.000	.002	.003	.007
212								.002	.003	.007
214								.002	.002	.006
216								.001	.002	.006
218								.001	.002	.005
220								.001	.002	.005
222								.001	.002	.005
224								.001	.002	.004
226								.001	.001	.004
228								.001	.001	.004
230								.001	.001	.004
232								.001	.001	.003
234								.001	.001	.003
236								.001	.001	.003
238								.000	.001	.003
240									.001	.002
242									.001	.002
244									.001	.002
246									.001	.002
248									.000	.002
250										.002
252										.002
254										.001
256										.001
258										.001
260										.001
262										.001
264										.001
266										.001
268										.001
270										.001
272										.001
274										.001
276										.001
278										.001
280										.000

(continued)

TABLE A.30. *(continued)*

					n				
x	6	7	10	11	14	15	18	19	22
1	.500	.500	.500	.500	.500	.500	.500	.500	.500
3	.360	.386	.431	.440	.457	.461	.470	.473	.478
5	.235	.281	.364	.381	.415	.423	.441	.445	.456
7	.136	.191	.300	.324	.374	.385	.411	.418	.434
9	.068	.119	.242	.271	.334	.349	.383	.391	.412
11	.028	.068	.190	.223	.295	.313	.354	.365	.390
13	.008	.035	.146	.179	.259	.279	.327	.339	.369
15	.001	.015	.108	.141	.225	.248	.300	.314	.348
17		.005	.078	.109	.194	.218	.275	.290	.328
19		.001	.054	.082	.165	.190	.250	.267	.308
21		.000	.036	.060	.140	.164	.227	.245	.289
23			.023	.043	.117	.141	.205	.223	.270
25			.014	.030	.096	.120	.184	.203	.252
27			.008	.020	.079	.101	.165	.184	.234
29			.005	.013	.063	.084	.147	.166	.217
31			.002	.008	.050	.070	.130	.149	.201
33			.001	.005	.040	.057	.115	.133	.186
35			.000	.003	.031	.046	.100	.119	.171
37				.002	.024	.037	.088	.105	.157
39				.001	.018	.029	.076	.093	.144
41				.000	.013	.023	.066	.082	.131
43					.010	.018	.056	.072	.120
45					.007	.014	.048	.062	.109
47					.005	.010	.041	.054	.099
49					.003	.008	.034	.047	.089
51					.002	.006	.029	.040	.080
53					.002	.004	.024	.034	.072
55					.001	.003	.020	.029	.064
57					.001	.002	.016	.025	.058
59					.000	.001	.013	.021	.051
61						.001	.011	.017	.045
63						.001	.009	.014	.040
65						.000	.007	.012	.035
67							.005	.010	.031
69							.004	.008	.027
71							.003	.006	.024
73							.003	.005	.021
75							.002	.004	.018
77							.001	.003	.015
79							.001	.003	.013
81							.001	.002	.011
83							.001	.002	.010
85							.000	.001	.008
87								.001	.007
89								.001	.006
91								.001	.005
93								.000	.004
95									.003
97									.003
99									.002
101									.002
103									.002
105									.001
107									.001

TABLE A.30. *(continued)*

x	\multicolumn{9}{c}{n}
	6	7	10	11	14	15	18	19	22
109									.001
111									.001
113									.001
115									.000

x	\multicolumn{9}{c}{n}
	23	26	27	30	31	34	35	38	39
1	.500	.500	.500	.500	.500	.500	.500	.500	.500
3	.479	.483	.484	.486	.487	.488	.489	.490	.490
5	.458	.465	.467	.472	.473	.477	.478	.480	.481
7	.438	.448	.451	.458	.460	.465	.466	.470	.472
9	.417	.431	.434	.444	.446	.453	.455	.460	.462
11	.397	.414	.418	.430	.433	.442	.444	.450	.452
13	.377	.397	.402	.416	.420	.430	.433	.440	.443
15	.357	.380	.386	.402	.407	.418	.422	.431	.433
17	.338	.363	.371	.389	.394	.407	.411	.421	.424
19	.319	.347	.355	.375	.381	.396	.400	.411	.414
21	.301	.331	.340	.362	.368	.384	.389	.401	.405
23	.283	.316	.325	.349	.355	.373	.378	.392	.396
25	.265	.300	.310	.336	.343	.362	.368	.382	.387
27	.248	.285	.296	.323	.331	.351	.357	.373	.377
29	.232	.270	.281	.310	.318	.340	.347	.363	.368
31	.216	.256	.268	.298	.306	.329	.336	.354	.359
33	.201	.242	.254	.286	.295	.319	.326	.345	.350
35	.187	.229	.241	.274	.283	.308	.316	.336	.341
37	.173	.216	.228	.262	.272	.298	.306	.327	.333
39	.160	.203	.216	.251	.261	.288	.296	.318	.324
41	.147	.191	.204	.239	.250	.278	.286	.309	.315
43	.135	.179	.192	.228	.239	.268	.277	.300	.307
45	.124	.168	.181	.218	.229	.259	.267	.291	.298
47	.114	.157	.170	.208	.219	.249	.258	.283	.290
49	.104	.147	.160	.198	.209	.240	.249	.274	.282
51	.094	.137	.150	.188	.199	.231	.240	.266	.274
53	.086	.127	.141	.178	.190	.222	.232	.258	.266
55	.078	.118	.132	.169	.181	.213	.223	.250	.258
57	.070	.110	.123	.160	.172	.205	.215	.242	.250
59	.063	.102	.115	.152	.164	.196	.206	.234	.243
61	.057	.094	.107	.144	.155	.188	.198	.227	.235
63	.051	.087	.099	.136	.147	.180	.191	.219	.228
65	.046	.080	.092	.128	.140	.173	.183	.212	.221
67	.041	.073	.085	.121	.132	.165	.176	.205	.214
69	.036	.067	.079	.114	.125	.158	.168	.198	.207
71	.032	.062	.073	.107	.118	.151	.161	.191	.200
73	.028	.057	.067	.100	.112	.144	.154	.184	.193
75	.025	.052	.062	.094	.105	.137	.148	.177	.187
77	.022	.047	.057	.088	.099	.131	.141	.171	.180
79	.019	.043	.052	.083	.093	.125	.135	.165	.174
81	.017	.039	.048	.077	.088	.119	.129	.158	.168
83	.015	.035	.044	.072	.082	.113	.123	.152	.162
85	.013	.032	.040	.067	.077	.107	.117	.147	.156
87	.011	.029	.036	.063	.072	.102	.112	.141	.150
89	.009	.026	.033	.059	.068	.097	.107	.135	.145
91	.008	.023	.030	.054	.063	.092	.101	.130	.139

(continued)

TABLE A.30. (continued)

x	n								
	23	26	27	30	31	34	35	38	39
93	.007	.021	.027	.051	.059	.087	.096	.125	.134
95	.006	.019	.025	.047	.055	.082	.092	.120	.129
97	.005	.017	.022	.043	.052	.078	.087	.115	.124
99	.004	.015	.020	.040	.048	.074	.083	.110	.119
101	.004	.013	.018	.037	.045	.070	.078	.105	.114
103	.003	.012	.016	.034	.041	.066	.074	.101	.109
105	.003	.010	.015	.032	.038	.062	.070	.096	.105
107	.002	.009	.013	.029	.036	.058	.066	.092	.101
109	.002	.008	.012	.027	.033	.055	.063	.088	.096
111	.001	.007	.010	.025	.031	.052	.059	.084	.092
113	.001	.006	.009	.023	.028	.049	.056	.080	.088
115	.001	.005	.008	.021	.026	.046	.053	.076	.084
117	.001	.005	.007	.019	.024	.043	.050	.073	.081
119	.001	.004	.006	.017	.022	.040	.047	.069	.077
121	.000	.004	.006	.016	.020	.038	.044	.066	.074
123		.003	.005	.014	.019	.035	.042	.063	.070
125		.003	.004	.013	.017	.033	.039	.060	.067
127		.002	.004	.012	.016	.031	.037	.057	.064
129		.002	.003	.011	.014	.029	.034	.054	.061
131		.002	.003	.010	.013	.027	.032	.051	.058
133		.001	.003	.009	.012	.025	.030	.049	.055
135		.001	.002	.008	.011	.023	.028	.046	.053
137		.001	.002	.007	.010	.022	.026	.044	.050
139		.001	.002	.006	.009	.020	.025	.041	.048
141		.001	.001	.006	.008	.019	.023	.039	.045
143		.001	.001	.005	.007	.017	.022	.037	.043
145		.001	.001	.005	.007	.016	.020	.035	.041
147		.000	.001	.004	.006	.015	.019	.033	.039
149			.001	.004	.006	.014	.017	.031	.037
151			.001	.003	.005	.013	.016	.029	.035
153			.001	.003	.004	.012	.015	.028	.033
155			.000	.003	.004	.011	.014	.026	.031
157				.002	.004	.010	.013	.025	.029
159				.002	.003	.009	.012	.023	.028
161				.002	.003	.008	.011	.022	.026
163				.002	.003	.008	.010	.021	.025
165				.001	.002	.007	.010	.019	.023
167				.001	.002	.007	.009	.018	.022
169				.001	.002	.006	.008	.017	.021
171				.001	.002	.005	.007	.016	.020
173				.001	.001	.005	.007	.015	.018
175				.001	.001	.005	.006	.014	.017
177				.001	.001	.004	.006	.013	.016
179				.001	.001	.004	.005	.012	.015
181				.000	.001	.003	.005	.011	.014
183					.001	.003	.005	.011	.014
185					.001	.003	.004	.010	.013
187					.001	.003	.004	.009	.012
189					.001	.002	.003	.009	.011
191					.000	.002	.003	.008	.010
193						.002	.003	.008	.010
195						.002	.003	.007	.009
197						.002	.002	.007	.009
199						.001	.002	.006	.008

TABLE A.30. (continued)

					n				
x	23	26	27	30	31	34	35	38	39
201						.001	.002	.006	.007
203						.001	.002	.005	.007
205						.001	.002	.005	.006
207						.001	.001	.004	.006
209						.001	.001	.004	.006
211						.001	.001	.004	.005
213						.001	.001	.004	.005
215						.001	.001	.003	.004
217						.001	.001	.003	.004
219						.000	.001	.003	.004
221							.001	.003	.004
223							.001	.002	.003
225							.001	.002	.003
227							.001	.002	.003
229							.000	.002	.003
231								.002	.002
233								.002	.002
235								.001	.002
237								.001	.002
239								.001	.002
241								.001	.002
243								.001	.001
245								.001	.001
247								.001	.001
249								.001	.001
251								.001	.001
253								.001	.001
255								.001	.001
257								.001	.001
259								.000	.001
261									.001
263									.001
265									.001
267									.001
269									.000

Adapted from L. Kaarsemaker and A. van Wijngaarden, Tables for use in rank correlation, *Statistica Neerlandica* **7** (1953): 41–54, with permission of the authors, the editor of *Statistica Neerlandica*, and Blackwell Publishers.

TABLE A.31. Selected Critical Values for the Null Distribution of the Spearman r_s Statistic: $n = 4(1)50$

Additional critical values can be found in Zar (1972), Owen (1962), Otten (1973a), and De Jonge and Van Montfort (1972). For given n and α, the entry in the table is $r_{s,\alpha}$ satisfying $P_0\{r_s \geq r_{s,\alpha}\} \approx \alpha$.

					α			
n	.10	.05	.025	.01	.005	.0025	.001	.0005
4	1.000	1.000						
5	.800	.900	1.000	1.000				
6	.657	.829	.886	.943	1.000	1.000		
7	.571	.714	.786	.893	.929	.964	1.000	1.000
8	.524	.643	.738	.833	.881	.905	.952	.976
9	.483	.600	.700	.783	.833	.867	.917	.933
10	.455	.564	.648	.745	.794	.830	.879	.903
11	.427	.536	.618	.709	.755	.800	.845	.873
12	.406	.503	.587	.678	.727	.769	.818	.846
13	.385	.484	.560	.648	.703	.747	.791	.824
14	.367	.464	.538	.626	.679	.723	.771	.802
15	.354	.446	.521	.604	.654	.700	.750	.779
16	.341	.429	.503	.582	.635	.679	.729	.762
17	.328	.414	.485	.566	.615	.662	.713	.748
18	.317	.401	.472	.550	.600	.643	.695	.728
19	.309	.391	.460	.535	.584	.628	.677	.712
20	.299	.380	.447	.520	.570	.612	.662	.696
21	.292	.370	.435	.508	.556	.599	.648	.681
22	.284	.361	.425	.496	.544	.586	.634	.667
23	.278	.353	.415	.486	.532	.573	.622	.654
24	.271	.344	.406	.476	.521	.562	.610	.642
25	.265	.337	.398	.466	.511	.551	.598	.630
26	.259	.331	.390	.457	.501	.541	.587	.619
27	.255	.324	.382	.448	.491	.531	.577	.608
28	.250	.317	.375	.440	.483	.522	.567	.598
29	.245	.312	.368	.433	.475	.513	.558	.589
30	.240	.306	.362	.425	.467	.504	.549	.580
31	.236	.301	.356	.418	.459	.496	.541	.571
32	.232	.296	.350	.412	.452	.489	.533	.563
33	.229	.291	.345	.405	.446	.482	.525	.554
34	.225	.287	.340	.399	.439	.475	.517	.547
35	.222	.283	.335	.394	.433	.468	.510	.539
36	.219	.279	.330	.388	.427	.462	.504	.533
37	.216	.275	.325	.383	.421	.456	.497	.526
38	.212	.271	.321	.378	.415	.450	.491	.519
39	.210	.267	.317	.373	.410	.444	.485	.513
40	.207	.264	.313	.368	.405	.439	.479	.507
41	.204	.261	.309	.364	.400	.433	.473	.501
42	.202	.257	.305	.359	.395	.428	.468	.495
43	.199	.254	.301	.355	.391	.423	.463	.490
44	.197	.251	.298	.351	.386	.419	.458	.484
45	.194	.248	.294	.347	.382	.414	.453	.479
46	.192	.246	.291	.343	.378	.410	.448	.474
47	.190	.243	.288	.340	.374	.405	.443	.469
48	.188	.240	.285	.336	.370	.401	.439	.465
49	.186	.238	.282	.333	.366	.397	.434	.460
50	.184	.235	.279	.329	.363	.393	.430	.456

Adapted, in part, from J. H. Zar, Significance testing of the Spearman rank correlation coefficient, *Journal of the American Statistical Association* **67** (1972): 578–580. Reprinted with permission from the *Journal of the American Statistical Association.* Copyright © (1972) by the American Statistical Association. All rights reserved; and, in part, from A. Otten, Note on the Spearman rank correlation coefficient, *Journal of the American Statistical Association* **68** (1973): 585. Reprinted with permission from the *Journal of the American Statistical Association.* Copyright © (1973) by the American Statistical Association. All rights reserved.

A. TABLES AND CHARTS

TABLE A.32. Upper-Tail Probabilities for the Null Distribution of the Hoeffding D Statistic: $n = 5(1)9$

In the notation of (8.88), the d_α values are given by $(x/60)$ in the $n = 5$ table, the d_α values are given by $(x/180)$ in the $n = 6$ table, the d_α values are given by $(x/1260)$ in the $n = 7$ table, the d_α values are given by $(x/1680)$ in the $n = 8$ table, and the d_α values are given by $(x/7560)$ in the $n = 9$ table.

$n = 5$		$n = 8$		$n = 9$	
x	$P_0\{D \geq (x/60)\}$	x	$P_0\{D \geq (x/1680)\}$	x	$P_0\{D \geq (x/7560)\}$
-1	1.0000	-8	.9635	-37	.9942
0	.8667	-7	.9294	-36	.9936
2	.0667	-6	.9278	-35	.9900
		-5	.8119	-34	.9851
$n = 6$		-4	.7389	-33	.9837
x	$P_0\{D \geq (x/180)\}$	-3	.7268	-32	.9773
		-2	.6093	-31	.9750
-2	1.0000	-1	.4728	-30	.9708
-1	.9556	0	.4625	-29	.9611
0	.6444	1	.3704	-28	.9569
1	.2444	2	.2776	-27	.9442
2	.0667	3	.2726	-26	.9263
3	.0556	4	.2028	-25	.9206
6	.0111	5	.1675	-24	.9028
		6	.1659	-23	.8893
$n = 7$		7	.1222	-22	.8744
x	$P_0\{D \geq (x/1260)\}$	8	.1000	-21	.8547
		9	.0992	-20	.8475
-11	1.0000	10	.0706	-19	.8243
-8	.9873	11	.0635	-18	.8014
-7	.9365	12	.0627	-17	.7837
-6	.8857	13	.0522	-16	.7580
-5	.8730	15	.0442	-15	.7349
-4	.8286	16	.0315	-14	.7177
-3	.6889	17	.0272	-13	.6791
-2	.5873	18	.0256	-12	.6631
-1	.4984	19	.0196	-11	.6383
0	.4857	21	.0181	-10	.6214
2	.3460	22	.0149	-9	.6007
3	.2238	24	.0129	-8	.5834
4	.1857	25	.0095	-7	.5489
6	.1794	26	.0079	-6	.5308
8	.0905	28	.0071	-5	.5069
9	.0778	30	.0048	-4	.4934
12	.0714	34	.0040	-3	.4574
14	.0333	36	.0024	-2	.4413
18	.0302	38	.0016	-1	.4140
24	.0111	40	.0014	0	.4088
30	.0079	46	.0010	1	.3856
42	.0016	56	.0002	2	.3700
				3	.3501
$n = 8$		$n = 9$		4	.3390
x	$P_0\{D \geq (x/1680)\}$	x	$P_0\{D \geq (x/7560)\}$	5	.3186
				6	.3106
-12	1.0000	-44	1.0000	7	.2905
-11	.9984	-43	.9992	8	.2816
-10	.9937	-42	.9985	9	.2591
-9	.9929	-39	.9984	10	.2539
		-38	.9977	11	.2400

(continued)

TABLE A.32. *(continued)*

	$n = 9$		$n = 9$		$n = 9$
x	$P_0\{D \geq (x/7560)\}$	x	$P_0\{D \geq (x/7560)\}$	x	$P_0\{D \geq (x/7560)\}$
12	.2332	50	.0459	91	.0107
13	.2090	51	.0427	92	.0104
14	.2017	52	.0407	94	.0084
15	.1926	53	.0386	96	.0078
16	.1864	54	.0384	97	.0071
17	.1762	55	.0367	98	.0069
18	.1721	56	.0359	102	.0067
19	.1589	57	.0333	104	.0062
20	.1552	58	.0329	105	.0061
21	.1455	59	.0315	106	.0060
22	.1417	60	.0313	107	.0059
23	.1329	61	.0296	108	.0058
24	.1311	62	.0294	112	.0052
25	.1207	63	.0279	114	.0043
26	.1175	64	.0274	116	.0041
27	.1140	65	.0266	118	.0037
28	.1118	66	.0264	120	.0037
29	.1066	67	.0258	121	.0035
30	.1050	68	.0241	122	.0033
31	.0962	69	.0215	126	.0030
32	.0928	70	.0214	127	.0028
33	.0859	71	.0207	128	.0026
34	.0836	72	.0205	132	.0023
35	.0819	74	.0181	138	.0018
36	.0782	75	.0171	142	.0016
37	.0737	76	.0170	146	.0016
38	.0725	77	.0152	150	.0014
39	.0688	78	.0151	152	.0014
40	.0676	79	.0144	157	.0010
41	.0637	82	.0142	158	.0009
42	.0628	83	.0136	162	.0008
43	.0580	84	.0135	172	.0006
44	.0565	86	.0130	182	.0005
45	.0543	87	.0128	192	.0003
46	.0539	88	.0119	202	.0002
47	.0533	89	.0110	222	.0001
48	.0515	90	.0110	252	.0000
49	.0468				

Adapted, in part, from W. Hoeffding, A non-parametric test of independence, *Annals of Mathematical Statistics* **19** (1948): 546–557, with permission of the author, the editor of *The Annals of Mathematical Statistics*, and the Institute of Mathematical Statistics. The $n = 8$ and $n = 9$ values were computed by S. P. Leach on the Florida State University CDC 6400.

TABLE A.33. Upper-Tail Probabilities for the Null Limiting Distribution of the Blum-Kiefer-Rosenblatt ($\pi^4 nB/2$) Statistic

For a given y, the table entry is $V(y) \approx P_0\{(\pi^4 nB/2) \geq y\}$. In particular, if $V(y) = \alpha$, then $y = b_\alpha$, the upper α percentile of the asymptotic null distribution of $(\pi^4 nB/2)$.

y	$V(y)$	y	$V(y)$	y	$V(y)$	y	$V(y)$	y	$V(y)$
.30	1.0000	1.35	.3648	2.40	.0864	3.45	.0244	4.50	.0074
.35	.9999	1.40	.3387	2.45	.0812	3.50	.0230	4.55	.0070
.40	.9991	1.45	.3146	2.50	.0762	3.55	.0217	4.60	.0066
.45	.9961	1.50	.2924	2.55	.0716	3.60	.0205	4.65	.0063
.50	.9884	1.55	.2719	2.60	.0673	3.65	.0194	4.70	.0059
.55	.9739	1.60	.2530	2.65	.0633	3.70	.0183	4.75	.0056
.60	.9513	1.65	.2355	2.70	.0595	3.75	.0173	4.80	.0053
.65	.9210	1.70	.2194	2.75	.0560	3.80	.0163	4.85	.0050
.70	.8841	1.75	.2045	2.80	.0527	3.85	.0154	4.90	.0047
.75	.8422	1.80	.1908	2.85	.0496	3.90	.0145	4.95	.0045
.80	.7971	1.85	.1781	2.90	.0467	3.95	.0137	5.00	.0042
.85	.7504	1.90	.1663	2.95	.0440	4.00	.0130	5.50	.0025
.90	.7035	1.95	.1554	3.00	.0414	4.05	.0123	6.00	.0014
.95	.6573	2.00	.1453	3.05	.0390	4.10	.0116	6.50	.0008
1.00	.6127	2.05	.1359	3.10	.0368	4-15	.0110	7.00	.0005
1.05	.5701	2.10	.1273	3.15	.0347	4.20	.0104	7.50	.0003
1.10	.5297	2.15	.1192	3.20	.0327	4.25	.0098	8.00	.0002
1.15	.4918	2.20	.1117	3.25	.0308	4.30	.0093	8.50	.0001
1.20	.4565	2.25	.1047	3.30	.0291	4.35	.0087	9.00	.0001
1.25	.4236	2.30	.0982	3.35	.0274	4.40	.0083	9.50	.0000
1.30	.3930	2.35	.0921	3.40	.0259	4.45	.0078	10.00	.0000

Adapted from J. R. Blum, J. Kiefer, and M. Rosenblatt, Distribution free tests of independence based on the sample distribution function, *Annals of Mathematical Statistics* 32 (1961): 485–498, with permission of the authors, the editor of *The Annals of Mathematical Statistics*, and the Institute of Mathematical Statistics.

TABLE A.34. Selected Upper-Tail Critical Values for the Null Distribution of the Epstein $\mathcal{E}$ Statistic: $n = 3(1)12$

For a given n and α, the table entry is e_α, the value satisfying $P_0\{\mathcal{E} \geq e_\alpha\} = \alpha$. (The lower-tail α critical value is $[(n-1)/2] - e_\alpha$.)

	α				
n	.10	.05	.025	.01	.005
3	1.553	1.684	1.176	1.852	1.900
4	2.157	2.331	2.469	2.609	2.689
5	2.753	2.953	3.120	3.300	3.411
6	3.339	3.565	3.754	3.963	4.097
7	3.917	4.166	4.367	4.610	4.762
8	4.489	4.759	4.988	5.244	5.413
9	5.056	5.346	5.592	5.869	6.053
10	5.619	5.927	6.189	6.487	6.683
11	6.178	6.504	6.781	7.097	7.307
12	6.735	7.077	7.369	7.702	7.924
13	7.289	7.647	7.953	8.302	8.530

Adapted from Table 7.6.1 of T. Robertson, F.T. Wright, and R.L. Dykstra, *Order Restricted Statistical Inference*, John Wiley, New York (1988). Copyright © John Wiley & Sons, Limited. Reproduced with permission.

TABLE A.35. Selected Critical Values for the Null Distribution of the Hollander-Proschan T Statistic: $n = 4(1)20(5)50$

For given n and α, the lower-tail entry is $t_1(\alpha, n)$ satisfying $P_0\{T \leq t_1(\alpha, n)\} \cong \alpha$ and the upper-tail entry is $t_2(\alpha, n)$ satisfying $P_0\{T \geq t_2(\alpha, n)\} \cong \alpha$. (Tail probabilities were computed via Monte Carlo estimation. A parenthetical value adjacent to an upper (lower) critical point x gives the estimated probability $P_0\{T \geq x\}(P_0\{T \leq x\})$. Parenthetical values are included for those estimated tail probabilities that were not within .002 of the nominal α.)

	Lower-Tail				
			α		
n	.01	.025	.05	.075	.10
4				0 (.067)	1 (.103)
5	0 (.018)	1 (.028)	2 (.040)	3 (.072)	4
6	2	5 (.028)	7	8 (.061)	9 (.086)
7	7	11	14 (.045)	16 (.072)	18 (.105)
8	15	21	25	28	30
9	27	34	40	44	47
10	42	52	60	65	69
11	63	76	86	92	97
12	89	105	117	125	131
13	122	141	157	167	174
14	162	185	204	215	223
15	209	236	259	272	282
16	266	298	323	338	350
17	330	368	397	415	429
18	405	446	480	502	518
19	490	538	577	601	619
20	594	642	685	713	732
25	1250	1351	1427	1472	1507
30	2320	2463	2574	2651	2704
35	3850	4064	4215	4316	4394
40	5947	6214	6434	6593	6700
45	8665	9040	9341	9543	9686
50	12170	12661	13020	13255	13439

(continued)

TABLE A.35. *(continued)*

			Upper-Tail		
			α		
n	.01	.025	.05	.075	.10
4					
5					10 (.177)
6			20 (.053)		19 (.167)
7	35		34 (.047)		33 (.103)
8	55	54	53	52 (.083)	52 (.083)
9	81	80 (.019)	78	77	76
10	114	112	110	109 (.067)	107 (.107)
11	155	152 (.028)	150 (.046)	148 (.070)	146
12	205	201	198 (.047)	195	193
13	264	260	255	252 (.072)	249
14	334	328	322	318	315
15	415	408	401	396	392
16	508	499	491	484	480
17	613	603	593	586	580
18	732	720	709	700	693
19	866	852	838	828	820
20	1015	998	982	970	961
25	2018	1986	1956	1934	1918
30	3521	3461	3411	3375	3349
35	5625	5546	5464	5403	5359
40	8418	8301	8189	8113	8049
45	12012	11864	11709	11600	11502
50	16519	16310	16085	15937	15823

Adapted from M. Hollander and F. Proschan, Testing whether new is better than used, *Annals of Mathematical Statistics* **43** (1972): 1136–1146, with permission of the authors, the editor of *The Annals of Mathematical Statistics*, and the Institute of Mathematical Statistics.

TABLE A.36. Selected Critical Values for the Hollander-Proschan Decreasing Mean Residual Life Statistic, V'

For given n and α, the lower-tail entry is $v_{2,\alpha}$ such that $P_0\{V' \leq v_{2,\alpha}\} = \alpha$ and the upper-tail entry is $v_{1,\alpha}$ such that $P_0\{V' \geq v_{1,\alpha}\} = \alpha$.

	Lower-Tail			Upper-Tail		
n	$\alpha = .01$	$\alpha = .05$	$\alpha = .10$	$\alpha = .01$	$\alpha = .05$	$\alpha = .10$
2	−2.511	−2.306	−2.049	2.551	2.306	2.049
3	−3.169	−2.489	−1.979	2.441	2.011	1.689
4	−3.267	−2.372	−1.811	2.247	1.877	1.595
5	−3.227	−2.257	−1.703	2.208	1.795	1.511
6	−3.159	−2.173	−1.639	2.175	1.750	1.465
7	−3.093	−2.115	−1.593	2.162	1.718	1.432
8	−3.037	−2.070	−1.558	2.153	1.697	1.408
9	−2.990	−2.033	−1.531	2.147	1.681	1.389
10	−2.951	−2.003	−1.508	2.144	1.669	1.375
11	−2.916	−1.978	−1.490	2.142	1.660	1.364
12	−2.886	−1.957	−1.475	2.142	1.652	1.354
13	−2.860	−1.939	−1.462	2.142	1.646	1.346
14	−2.836	−1.923	−1.451	2.142	1.641	1.340
15	−2.815	−1.909	−1.441	2.143	1.637	1.334
16	−2.796	−1.897	−1.432	2.144	1.633	1.329
17	−2.779	−1.886	−1.425	2.146	1.630	1.325
18	−2.764	−1.876	−1.418	2.147	1.628	1.321
19	−2.749	−1.867	−1.412	2.148	1.626	1.318
20	−2.736	−1.859	−1.406	2.149	1.642	1.315
25	−2.684	−1.826	−1.385	2.157	1.617	1.304
30	−2.646	−1.804	−1.371	2.164	1.614	1.297
35	−2.618	−1.788	−1.360	2.171	1.612	1.293
40	−2.595	−1.775	−1.352	2.177	1.611	1.289
45	−2.577	−1.765	−1.346	2.182	1.610	1.287
50	−2.562	−1.757	−1.341	2.187	1.610	1.285
55	−2.549	−1.750	−1.337	2.191	1.610	1.283
60	−2.538	−1.744	−1.333	2.195	1.610	1.282

Adapted from P. Langenberg and R. Srinivasan, Null distribution of the Hollander-Proschan statistic for decreasing mean residual life, *Biometrika* **66** (1979): 679–680, by permission of Oxford University Press on behalf of the *Biometrika* Trustees.

TABLE A.37. Critical Values for the Hall-Wellner Confidence Bands for the Mean Residual Life Function

For a given α, the entry is the value a_α such that with D_n defined by (11.70), the probability given by the left-hand side of (11.72) is approximately $1 - \alpha$.

α	a_α	α	a_α	α	a_α
.01	2.8070	.34	1.3721	.67	.9559
.02	2.5758	.35	1.3562	.68	.9452
.03	2.4324	.36	1.3406	.69	.9345
.04	2.3263	.37	1.3253	.70	.9238
.05	2.2414	.38	1.3103	.71	.9132
.06	2.1701	.39	1.2956	.72	.9025
.07	2.1084	.40	1.2812	.73	.8919
.08	2.0537	.41	1.2670	.74	.8812
.09	2.0047	.42	1.2531	.75	.8706
.10	1.9600	.43	1.2394	.76	.8598
.11	1.9189	.44	1.2259	.77	.8491
.12	1.8808	.45	1.2126	.78	.8383
.13	1.8453	.46	1.1995	.79	.8274
.14	1.8119	.47	1.1866	.80	.8164
.15	1.7805	.48	1.1739	.81	.8053
.16	1.7507	.49	1.1614	.82	.7941
.17	1.7224	.50	1.1490	.83	.7828
.18	1.6954	.51	1.1367	.84	.7712
.19	1.6696	.52	1.1246	.85	.7595
.20	1.6448	.53	1.1127	.86	.7475
.21	1.6211	.54	1.1009	.87	.7353
.22	1.5982	.55	1.0892	.88	.7227
.23	1.5761	.56	1.0776	.89	.7098
.24	1.5548	.57	1.0661	.90	.6964
.25	1.5341	.58	1.0547	.91	.6824
.26	1.5141	.59	1.0434	.92	.6677
.27	1.4946	.60	1.0322	.93	.6521
.28	1.4758	.61	1.0211	.94	.6355
.29	1.4574	.62	1.0101	.95	.6173
.30	1.4395	.63	.9992	.96	.5971
.31	1.4220	.64	.9883	.97	.5737
.32	1.4050	.65	.9774	.98	.5450
.33	1.3883	.66	.9666	.99	.5045

Adapted from Table C.5 of J.P. Klein and M.L. Moeschberger, *Survival Analysis: Techniques for Censored and Truncated Data*, Springer-Verlag, New York (1997), with the permission of the authors and the publisher.

TABLE A.38. Percentage Points of the Kolmogorov Statistic

The entries in the table are the values d_α such that $P_F(D \geq d_\alpha) = \alpha$.

n	$1 - \alpha$: .80	.90	.95	.98	.99
1	.900	.950	.975	.990	.995
2	.684	.776	.842	.900	.929
3	.565	.636	.708	.785	.829
4	.493	.565	.624	.689	.734
5	.447	.509	.563	.627	.669
6	.410	.468	.519	.577	.617
7	.381	.436	.483	.538	.576
8	.358	.410	.454	.507	.542
9	.339	.387	.430	.480	.513
10	.323	.369	.409	.457	.489
11	.308	.352	.391	.437	.468
12	.296	.338	.375	.419	.449
13	.285	.325	.361	.404	.432
14	.275	.314	.349	.390	.418
15	.266	.304	.338	.377	.404
16	.258	.295	.327	.366	.392
17	.250	.286	.318	.355	.381
18	.244	.279	.309	.346	.371
19	.237	.271	.301	.337	.361
20	.232	.265	.294	.329	.352
21	.226	.259	.287	.321	.344
22	.221	.253	.281	.314	.337
23	.216	.247	.275	.307	.330
24	.212	.242	.269	.301	.323
25	.208	.238	.264	.295	.317
26	.204	.233	.259	.290	.311
27	.200	.229	.254	.284	.305
28	.197	.225	.250	.279	.300
29	.193	.221	.246	.275	.295
30	.190	.218	.242	.270	.290
31	.187	.214	.238	.266	.285
32	.184	.211	.234	.262	.281
33	.182	.208	.231	.258	.277
34	.179	.205	.227	.254	.273
35	.177	.202	.224	.251	.269
36	.174	.199	.221	.247	.265
37	.172	.196	.218	.244	.262
38	.170	.194	.215	.241	.258
39	.168	.191	.213	.238	.255
40	.165	.189	.210	.235	.252
Approximation for $n > 40$	$\dfrac{1.07}{\sqrt{n}}$	$\dfrac{1.22}{\sqrt{n}}$	$\dfrac{1.36}{\sqrt{n}}$	$\dfrac{1.52}{\sqrt{n}}$	$\dfrac{1.63}{\sqrt{n}}$

Adapted from L.E. Miller, Table of percentage points of Kolmogorov statistics, *Journal of the American Statistical Association* **51** (1956): 111–121. Reprinted with permission from the *Journal of the American Statistical Association*. Copyright © (1956) by the American Statistical Association. All rights reserved.

TABLE A.39. Percentage Points of Lilliefors' Test for Normality

The entries in the table are the values d'_α such that $P(D' \geq d'_\alpha) \doteq \alpha$ when the underlying distribution is a normal distribution.

n	$1 - \alpha$:	.80	.85	.90	.95	.99	.999
4		.303	.321	.346	.376	.413	.433
5		.289	.303	.319	.343	.397	.439
6		.269	.281	.297	.323	.371	.424
7		.252	.264	.280	.304	.351	.402
8		.239	.250	.265	.288	.333	.384
9		.227	.238	.252	.274	.317	.365
10		.217	.228	.241	.262	.304	.352
11		.208	.218	.231	.251	.291	.338
12		.200	.210	.222	.242	.281	.325
13		.193	.202	.215	.234	.271	.314
14		.187	.196	.208	.226	.262	.305
15		.181	.190	.201	.219	.254	.296
16		.176	.184	.195	.213	.247	.287
17		.171	.179	.190	.207	.240	.279
18		.167	.175	.185	.202	.234	.273
19		.163	.170	.181	.197	.228	.266
20		.159	.166	.176	.192	.223	.260
25		.143	.150	.159	.173	.201	.236
30		.131	.138	.146	.159	.185	.217
Approximation for $n > 30$		$\dfrac{.74}{\sqrt{n}}$	$\dfrac{.77}{\sqrt{n}}$	$\dfrac{.82}{\sqrt{n}}$	$\dfrac{.89}{\sqrt{n}}$	$\dfrac{1.04}{\sqrt{n}}$	$\dfrac{1.22}{\sqrt{n}}$

Adapted from G.E. Dallal and L. Wilkinson, An analytical approximation to the distribution of Lilliefors' test statistic, *The American Statistician* **40** (1986): 294–296. Reprinted from *The American Statistician*. Copyright © (1986) by the American Statistical Association. All rights reserved.

TABLE A.40. Selected Percentiles of G_a for Computing Confidence Bands for the Survival Function when the Data Are Censored

The entries are the values of λ such that $G_a(\lambda) = \beta$, where G_a is given by equation (11.133).

β	$a = .10$	$a = .25$	$a = .40$	$a = .50$	$a = .60$	$a = .75$	$a = .90$	$a = 1.00$
.99	.851	1.256	1.470	1.552	1.600	1.626	1.628	1.628
.95	.682	1.014	1.198	1.273	1.321	1.354	1.358	1.358
.90	.599	.894	1.062	1.133	1.181	1.217	1.224	1.224
.75	.471	.711	.854	.920	.967	1.008	1.019	1.019
.50	.356	.544	.663	.720	.765	.809	.827	.828
.25	.272	.420	.518	.567	.608	.652	.675	.676

Adapted from Table 1 of W.J. Hall and J.A. Wellner, Confidence bands for a survival curve, *Biometrika* **67** (1980): 133–143, by permission of Oxford University Press on behalf of the *Biometrika* Trustees.

CHART A.1. The t Distribution: Degrees of Freedom $= n = 1(1)6(2)12(4)20, \infty$.

A point (x, y) on the curve corresponding to n degrees of freedom satisfies $P\{U \geq x\} = y$, where U has a t distribution with n degrees of freedom. Thus, if (x, α) is a point on the curve for n degrees of freedom, then $x = t_{n,\alpha}$ is the upper αth percentile point of the t distribution with n degrees of freedom.

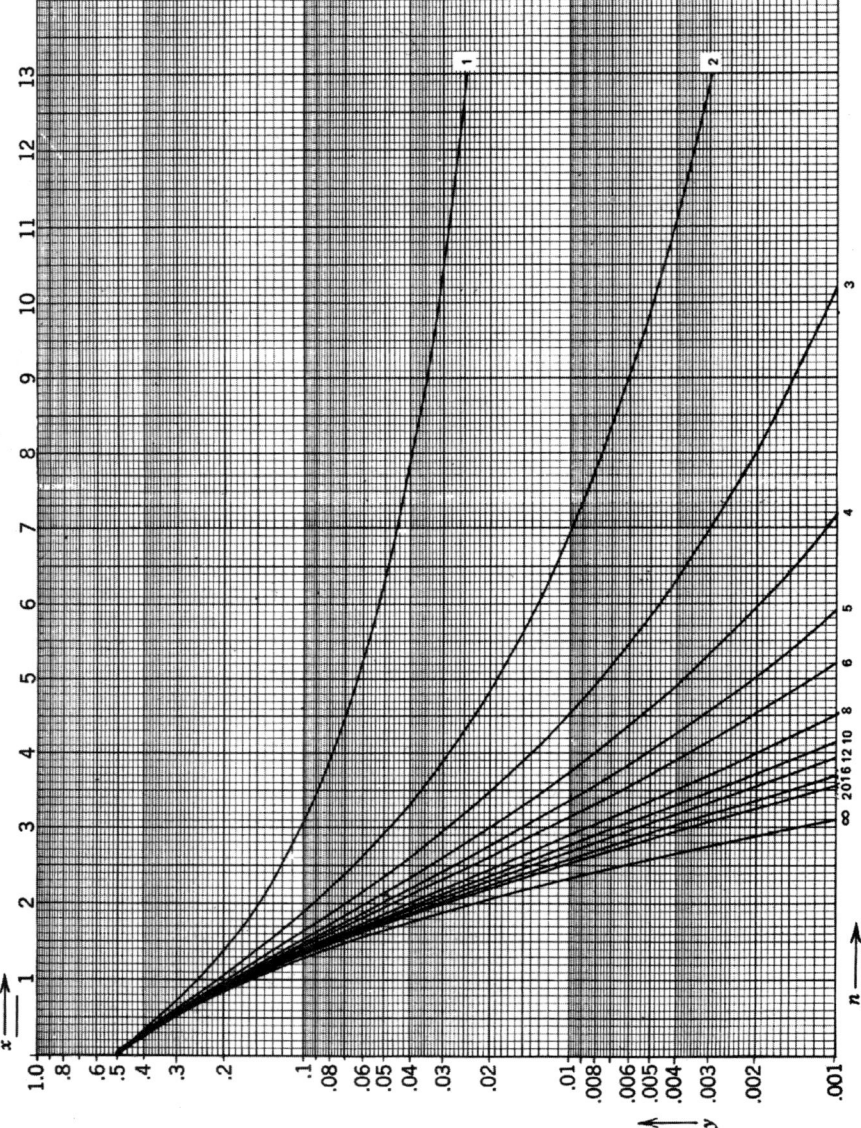

Adapted from Figure 1 of F. Wilcoxon and R. A. Wilcox, *Some Rapid Approximate Statistical Procedures* (2nd ed.), American Cyanamid Company, Lederle Laboratories, Pearl River, NY (1964), with the permission of the authors and Lederle Laboratories (a Division of American Cyanamid Company).

CHART A.2. The χ^2 Distribution: Degrees of Freedom $= n = 1(1)30$.

A point (x, y) on the curve corresponding to n degrees of freedom satisfies $P\{V \geq x\} = y$, where V has a χ^2 distribution with n degrees of freedom. Thus, if (x, α) is a point on the curve for n degrees of freedom, then $x = \chi^2_{n,\alpha}$ is the upper αth percentile point of the χ^2 distribution with n degrees of freedom.

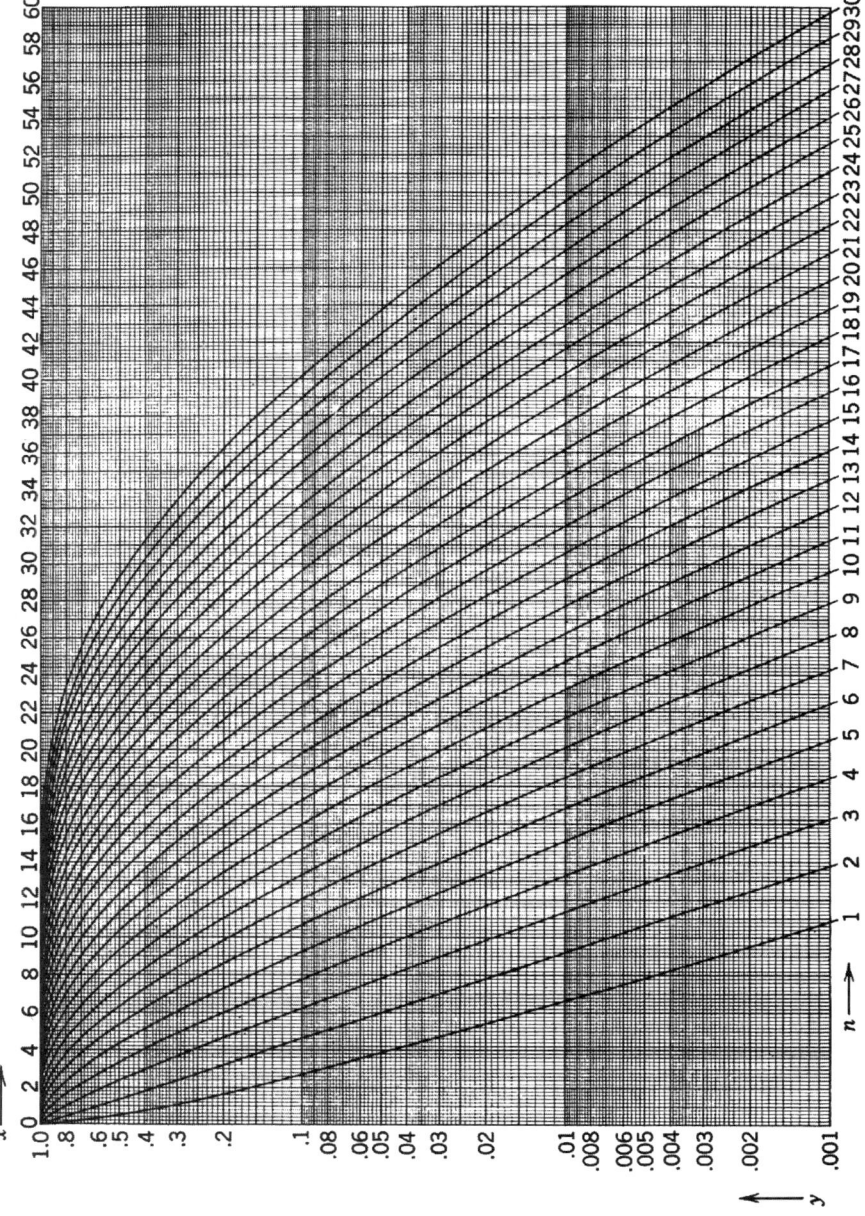

Adapted from C. I. Bliss, A chart of the chi-square distribution, *Journal of the American Statistical Association* **39** (1944): 246–248. Reprinted with permission from the *Journal of the American Statistical Association*. Copyright © (1944) by the American Statistical Association. All rights reserved.

Bibliography

Aalen, O. O. 1978. Nonparametric inference for a family of counting processes. *Ann. Stat.* **6**: 701–726.

Abdushukurov, A. A. 1987. Nonparametric estimation in the proportional hazards model of random censorship. *Akad. Nauk. UzSSR*, Tashkent VINITI No. 3448-V.

Adichie, J. N. 1967. Estimates of regression parameters based on rank tests. *Ann. Math. Stat.* **38**: 894–904.

———. 1974. Rank score comparison of several regression parameters. *Ann. Stat.* **2**: 396–402.

———. 1976. Testing parallelism of regression lines against ordered alternatives. *Comm. Stat. Theory Methods* **5**: 985–997.

———. 1984. Rank tests in linear models. *Handbook in Statistics,* Vol. 4. Edited by P. R. Krishnaiah and P. K. Sen. Amsterdam: Elsevier Science, pp. 229–257.

Agresti, A. 1984. *Analysis of Ordinal Categorical Data.* New York: John Wiley.

———. 1990. *Categorical Data Analysis.* New York: John Wiley.

———. 1992. A survey of exact inference for contingency tables. *Statist. Sci.* **7**: 131–177.

Agresti, A., C. R. Mehta, and N. R. Patel. 1990. Exact inference for contingency tables with ordered categories. *J. Amer. Statist. Assoc.* **85**: 453–458.

Akritas, M. G. 1986. Empirical processes associated with V-statistics and a class of estimators under random censoring. *Ann. Stat.* **14**: 619–637.

———. 1988. Pearson-type goodness-of-fit tests: The univariate case. *J. Amer. Statist. Assoc.* **83**: 222–230.

Altman, N. S. 1992. An introduction to kernel and nearest-neighbor nonparametric regression. *Amer. Statistician* **46**: 175–185.

Aly, E.-E. 1990. Tests for monotonicity properties of the mean residual life function. *Scand. J. Stat.* **17**: 189–200.

Anděl, J. 1967. Local asymptotic power and efficiency of tests of Kolmogorov-Smirnov type. *Ann. Math. Stat.* **38**: 1705–1725.

Andersen, P. K., Ø. Borgan, R. D. Gill, and N. Keiding. 1993. *Statistical Models Based on Counting Processes.* New York: Springer-Verlag.

Anderson, J. D., L. Efron, and S. K. Wong. 1970. Martian mass and earth-moon mass ratio from coherent S-band tracking of Mariners 6 and 7. *Science* **167**: 277–279.

Anderson, T. W., and H. Burstein. 1967. Approximating the upper binomial confidence limit. *J. Amer. Statist. Assoc.* **62**: 857–861.

———. 1968. Approximating the lower binomial confidence limit. *J. Amer. Statist. Assoc.* **63**: 1413–1415. Correction, **64** (1969): 669.

Anderson, V. L., and R. A. McLean. 1974. *Design of Experiments: A Realistic Approach.* New York: Dekker.

Andrews, B. 1995. Bodily shame as a mediator between abusive experiences and depression. *J. Abnorm. Psychol.* **104**: 277–285.

Andrews, F. C. 1954. Asymptotic behavior of some rank tests for analysis of variance. *Ann. Math. Stat.* **25**: 724–736.

Ansari, A. R., and R. A. Bradley. 1960. Rank-sum tests for dispersions. *Ann. Math. Stat.* **31**: 1174–1189.

Anscombe, F. J. 1965. Comments on Kurtz-Link-Tukey-Wallace paper. *Technometrics* **7**: 167–168.

Arbuthnott, J. 1710. An argument for divine providence, taken from the constant regularity observed in the births of both sexes. *Philos. Trans.* **27**: 186–190.

Archambault, W. A. T., Jr., G. A. Mack, and D. A. Wolfe. 1977. K-sample rank tests using pair-specific scoring functions. *Can. J. Stat.* **5**: 195–207.

Arjas, E., and D. Gasbarra. 1994. Nonparametric Bayesian inference for right-censored survival data, using the Gibbs sampler. *Statistica Sinica* **2**: 505–524.

Arnold, H. J. 1965. Small sample power for the one-sample Wilcoxon test for non-normal shift alternatives. *Ann. Math. Stat.* **36**: 1767–1778.

Arnold, B. C., N. Balakrishnan, and H. N. Nagaraja. 1992. *A First Course in Order Statistics*. New York: John Wiley.

Astin, M. C., S. M. Ogland-Hand, E. M. Coleman, and D. W. Foy. 1995. Posttraumatic stress disorder and childhood abuse in battered women: Comparisons with maritally distressed women. *J. Cons. Clin. Psychol.* **63**: 308–312.

August, G. P., W. Hung, and J. C. Houck. 1974. The effects of growth hormone therapy on collagen metabolism in children. *J. Clin. Endocrinol. Metabol.* **39**: 1103–1109.

Barlow, R. E. 1968. Likelihood ratio tests for restricted families of probability distributions. *Ann. Math. Stat.* **39**: 547–560.

Barlow, R. E., D. J. Bartholomew, J. M. Bremner, and H. D. Brunk. 1972. *Statistical Inference Under Order Restrictions*. New York: John Wiley.

Barlow, R. E., and R. Campo. 1975. Total time on test processes and applications to failure data analysis. In *Reliability and Fault Tree Analysis*. Edited by R. E. Barlow, J. Fursell, and N. Singpurwalla. Philadelphia: SIAM, pp. 451–481.

Barlow, R. E., and K. Doksum. 1972. Isotonic tests for convex orderings. *Proc. 6th Berkeley Symp.* **1**: 293–323.

Barlow, R. E., and F. Proschan. 1966. Inequalities for linear combinations of order statistics from restricted families. *Ann. Math. Stat.* **37**: 1574–1592.

———. 1969. A note on tests for monotone failure rate based on incomplete data. *Ann. Math. Stat.* **40**: 595–600.

———. 1981. *Statistical Theory of Reliability and Life Testing*. Second Printing Publisher: To Begin With, 1137 Hornell Drive, Silver Spring, MD 20904.

Barlow, R. E., and E. M. Scheuer. 1971. Estimation from accelerated life tests. *Technometrics* **13**: 145–159.

Barnard, G. A. 1945. A new test for 2 × 2 tables. *Nature* **156**: 177.

———. 1947. Significance tests for 2 × 2 tables. *Biometrics* **34**: 123–138.

Bauer, D. F. 1972. Constructing confidence sets using rank statistics. *J. Amer. Statist. Assoc.* **67**: 687–690.

Becker, R. A. 1994. A brief history of S. In *Computational Statistics*. Edited by P. Dirschedl and R. Osterman, Heidelberg: Physica, pp. 81–110.

Becker, R. A., and J. M. Chambers. 1984. *S: An Interactive Environment for Data Analysis and Graphics*. Pacific Grove, Calif.: Wadsworth and Brooks/Cole.

———. 1985. *Extending the S System*. Pacific Grove, Calif.: Wadsworth and Brooks/Cole.

Becker, R. A., J. M. Chambers, and A. R. Wilds. 1988. *The NEW S-language*. Pacific Grove, Calif.: Wadsworth and Brooks/Cole.

Benard, A., and P. van Elteren. 1953. A generalization of the method of m rankings. *Indag. Math.* **15**: 358–369.

Bergman, B. 1977. Crossings in the total time on test plot. *Scand. J. Stat.* **4**: 171–177.

Bernoulli, J. 1713. *Ars Conjectandi*.

Bhattacharyya, G. K., R. A. Johnson, and H. R. Neave. 1970. Percentage points of some non-parametric tests for independence and empirical power comparisons. *J. Amer. Statist. Assoc.* **65**: 976–983.

Bick, R. L., T. Adams, and W. R. Schmalhorst. 1976. Bleeding times, platelet adhesion, and aspirin. *J. Clin. Path.* **65**: 69–72.

Bickel, P. J. 1965. On some robust estimates of location. *Ann. Math. Stat.* **36**: 847–858.

Bickel, P. J., and K. A. Doksum. 1969. Tests for monotone failure rate based on normalized spacings. *Ann. Math. Stat.* **40**: 1216–1235.

Bickel, P. J., and D. A. Freedman. 1981. Some asymptotic theory for the bootstrap. *Ann. Stat.* **9**: 1196–1217.

Bie, O., Ø. Borgan, and K. Liestøl. 1987. Confidence intervals and confidence bands for the cumulative hazard function and their small sample properties. *Scand. J. Stat.* **14**: 221–233.

Billingsley, P. 1968. *Convergence of Probability Measures*. New York: John Wiley.

Birch, M. W. 1964. The detection of partial correlation I: The 2 × 2 case. *J. Roy. Statist. Soc. B* **26**: 313–324.

Birnbaum, Z. W. 1952. Numerical tabulation of the distribution of Kolmogorov's statistic for finite sample size. *J. Amer. Statist. Assoc.* **47**: 425–441.

———. 1956. On a use of the Mann-Whitney statistic. *Proc. 3rd Berkeley Symp.* **1**: 13–17.

Birnbaum, Z. W., and O. M. Klose. 1957. Bounds for the variance of the Mann-Whitney statistic. *Ann. Math. Stat.* **28**: 933–945.

Birnbaum, Z. W., and R. C. McCarty. 1958. A distribution-free upper confidence bound for Pr$\{Y < X\}$, based on independent samples of X and Y. *Ann. Math. Stat.* **29**: 558–562.

Birnbaum, Z. W., and F. H. Tingey. 1951. One-sided confidence contours for probability distribution functions. *Ann. Math. Stat.* **22**: 592–596.

Bjerkedal, T. 1960. Acquisition of resistance in guinea pigs infected with different doses of virulent tubercle bacilli. *Amer. J. Hygiene* **72**: 130–148.

Bliss, C. I. 1944. A chart of the chi-square distribution. *J. Amer. Statist. Assoc.* **39**: 246–248.

Blomqvist, N. 1950. On a measure of dependence between two random variables. *Ann. Math. Stat.* **21**: 593–600.

Blum, J. R., J. Kiefer, and M. Rosenblatt. 1961. Distribution free tests of independence based on the sample distribution function. *Ann. Math. Stat.* **32**: 485–498.

Blyth, C. 1950. *Notes on the Theory of Estimation*, recorded from lectures by E. L. Lehmann. Berkeley: University of California Press.

Blyth, C. R., and H. A. Still. 1983. Binomial confidence intervals. *J. Amer. Statist. Assoc.* **78**: 108–116.

Bohn, L. L. 1996. A review of nonparametric ranked-set sampling methodology. *Comm. Stat. Theory Methods* **25**: 2675–2685.

Bohn, L. L., and D. A. Wolfe. 1992. Nonparametric two-sample procedure for ranked-set samples data. *J. Amer. Statist. Assoc.* **87**: 552–561.

———. 1994. The effect of imperfect judgment rankings on properties of procedures based on the ranked-set samples analog of the Mann-Whitney-Wilcoxon statistic. *J. Amer. Statist. Assoc.* **89**: 168–176.

Borgan, Ø., and K. Liestøl. 1990. A note on confidence intervals and bands for the survival curve based on transformations. *Scand. J. Stat.* **17**: 35–41.

Borges, W. S., F. Proschan, and J. Rodrigues. 1984. A simple test for new better than used in expectation. *Comm. Stat. Theory Methods* **13**: 3217–3223.

Box, G. E. P. 1953. Non-normality and tests on variances. *Biometrika* **40**: 318–335.

Box, G. E. P., and S. L. Andersen. 1955. Permutation theory in the derivation of robust criteria and the study of departures from assumption. *J. Roy. Statist. Soc. B* **17**: 1–26.

Boyles, R. A., and F. J. Samaniego. 1984. Estimating a survival curve when new is better than used. *Operations Res.* **32**: 732–740.

Bradley, R. A. 1963. Some relationships among sensory difference tests. *Biometrics* **19**: 385–397.

Brady, J. P. 1969. Studies on the metronome effect on stuttering. *Behav. Res. Ther.* **7**: 197–204.

Breslow, N. 1981. Odds ratio estimators when the data are sparse. *Biometrika* **68**: 73–84.

Breslow, N. E., and J. J. Crowley. 1974. A large sample study of the life table and product limit estimates under random censorship. *Ann. Stat.* **2**: 437–453.

Breslow, N., and N. E. Day. 1980. *Statistical Methods in Cancer Research, 1. The Analysis of Case-Control Studies*. Lyon: International Agency for Research on Cancer.

Brown, B. W., Jr., and M. Hollander. 1977. *Statistics: A Biomedical Introduction*. New York: John Wiley.

Brown, B. W., Jr., M. Hollander, and R. M. Korwar. 1974. Nonparametric tests of independence, with applications to heart transplant studies. In *Reliability and Biometry: Statistical Analysis of Lifelength*. Edited by F. Proschan and R. J. Serfling. Philadelphia: SIAM, pp. 327–354.

Brown, G. W., B. Andrews, T. Haris, A. Alder, and L. Bridge. 1986. Social support, self-esteem and depression. *Psychol. Med.* **16**: 182–831.

Brown, M. 1984. On the choice of variance for the log rank test. *Biometrika* **71**: 65–74.

Brunden, M. N., and N. R. Mohberg. 1976. The Benard–van Elteren statistic and nonparametric computation. *Comm. Stat. Simulation Comput.* **5**: 155–162.

Bryson, M. C. 1974. Heavy-tailed distributions: Properties and tests. *Technometrics* **16**: 61–68.

Bryson, M. C., and M. M. Siddiqui. 1969. Some criteria for aging. *J. Amer. Statist. Assoc.* **64**: 1472–1483.

Bugyi, H. L., E. Magnier, W. Joseph, and G. Frank. 1969. A method for measurement of sodium and potassium in erythrocytes and whole blood. *Clin. Chem.* **15**: 712–719.

Burnett, W. C., Jr., and S. B. Jones Jr. 1973. Differential feeding of the southern armyworm on Kentucky and Florida populations of pokeweed. *Amer. Mid. Natur.* **90**: 231–234.

Burr, E. J. 1960. The distribution of Kendall's score S for a pair of tied rankings. *Biometrika* **47**: 151–171.

Byer, A. J., and D. Abrams. 1953. A comparison of the triangular and two-sample taste-test methods. *Food Technol.* **7**: 185–187.

Cain, G. D., G. Mayer, and E. A. Jones. 1970. Augmentation of albumin but not fibrinogen synthesis by corticosteroids in patients with hepatocellular disease. *J. Clin. Invest.* **49**: 2198–2204.

Campbell, J. A., and O. Pelletier. 1962. Determination of niacin (niacinamide) in cereal products. *J. Assoc. Offic. Anal. Chem.* **45**: 449–453.

Capon, J. 1965. On the asymptotic efficiency of the Kolmogorov-Smirnov test. *J. Amer. Statist. Assoc.* **60**: 843–853.

Carroll, R. J., and D. Ruppert. 1988. *Transformation and Weighting in Regression*. London: Chapman & Hall.

Casella, G. 1986. Refining binomial confidence intervals. *Can. J. Stat.* **14**: 113–129.

Casella, G., and R. L. Berger. 1990. *Statistical Inference*. Pacific Grove, Calif.: Wadsworth and Brooks/Cole.

Chambers, J. M., and T. J. Hastie. 1992. *Statistical Models in S*. Pacific Grove, Calif.: Wadsworth and Brooks/Cole.

Chandra, M., N. D. Singpurwalla, and M. A. Stephens. 1981. Kolmogorov statistics for tests of fit for the extreme-value and Weibull distributions. *J. Amer. Statist. Assoc.* **76**: 729–731.

Chen, Y. I., and D. A. Wolfe. 1990a. Modifications of the Mack-Wolfe umbrella tests for a generalized Behrens-Fisher problem. *Can. J. Stat.* **18**: 245–253.

———. 1990b. A study of distribution-free tests for umbrella alternatives. *Biom. J.* **32**: 47–57.

———. 1993. Nonparametric procedures for comparing umbrella pattern treatment effects with a control in a one-way layout. *Biometrics* **49**: 455–465.

Chen, Y. Y., M. Hollander, and N. A. Langberg. 1982. Small-sample results for the Kaplan-Meier estimator. *J. Amer. Statist. Assoc.* **77**: 141–144.

———. 1983a. Testing whether new is better than used with randomly censored data. *Ann. Stat.* **11**: 267–274.

———. 1983b. Tests for monotone mean residual life using randomly censored data. *Biometrics* **39**: 119–127.

Cheng, P. E., and G. D. Lin. 1987. Maximum likelihood estimation of a survival function under the Koziol-Green proportional hazards model. *Stat. Prob. Letters* **5**: 75–80.

Chernoff, H., and I. R. Savage. 1958. Asymptotic normality and efficiency of certain nonparametric test statistics. *Ann. Math. Stat.* **29**: 972–994.

Chew, V. 1971. Point estimation of the parameter of the binomial distribution. *Amer. Statistician* **25**: 47–50.

Clark, P. J., S. G. Vandenberg, and C. H. Proctor. 1961. On the relationship of scores on certain psychological tests with a number of anthropometric characters and birth order in twins. *Human Biol.* **33**: 163–180.

Cleveland, W. S. 1979. Robust locally weighted regression and smoothing scatterplots. *J. Amer. Statist. Assoc.* **74**: 829–836.

———. 1985. *The Elements of Graphing Data*. Pacific Grove, Calif.: Wadsworth.

Cleveland, W. S., and E. Grosse. 1991. Computational methods for local regression. *Stat. and Comput.* **1**: 47–62.

Cleveland, W. S., E. Grosse, and W. M. Shyu. 1992. Local regression models. In *Statistical Models in S*. Edited by J. M. Chambers and T. J. Hastie. Pacific Grove, Calif.: Wadsworth, Chapter 8.

Clogg, C. C., and J. W. Shockey. 1988. Multivariate analysis of discrete data. In *Handbook of Multivariate Experimental Psychology*. Edited by J. R. Nesselroade and R. B. Cattell. New York: Plenum Press, pp. 337–365.

Clopper, C. J., and E. S. Pearson. 1934. The use of confidence or fiducial limits illustrated in the case of the binomial. *Biometrika* **26**: 404–413.

Cochran, W. G. 1937. The efficiencies of the binomial series tests of significance of a mean and of a correlation coefficient. *J. Roy. Statist. Soc.* **100**: 69–73.

Cochran, W. G., and G. M. Cox. 1957. *Experimental Designs,* 2nd ed. New York: John Wiley.

Cole, A. F. W., and M. Katz. 1966. Summer ozone concentrations in southern Ontario in relation to photochemical aspects and vegetation damage. *J. Air Poll. Control Assoc.* **16**: 201–206.

Comroe, J. H., Jr., S. Y. Botelho, and A. B. DuBois. 1959. Design of a body plethysmograph for studying cardiopulmonary physiology. *J. Appl. Physiol.* **14**: 439–444.

Conover, W. J. 1971, 1980. *Practical Nonparametric Statistics,* 1st and 2nd ed. New York: John Wiley.

Cooper, L. M., E. Schubot, S. A. Banford, and C. T. Tart. 1967. A further attempt to modify hypnotic susceptibility through repeated individualized experience. *Int. J. Clin. Exp. Hypn.* **15**: 118–124.

Cornfield, J. 1956. A statistical problem arising from retrospective studies. *Proc. 3rd Berkeley Symp.* **4**: 135–138.

Cox, D. R. 1972. Regression models and life-tables (with discussion). *J. Roy. Statist. Soc. B* **34**: 187–220.

Cox, D. R., and D. Oakes. 1984. *Analysis of Survival Data*. London: Chapman & Hall.

Craig, A. T. 1932. On the distributions of certain statistics. *Amer. J. Math.* **54**: 353–366.

Critchlow, D. E. 1980. *Metric Methods for Analyzing Partially Ranked Data*. Lecture Notes in Statistics, Volume 34. New York: Springer-Verlag.

Critchlow, D. E., and M. A. Fligner. 1991. On distribution-free multiple comparisons in the one-way analysis of variance. *Comm. Stat. Theory Methods* **20**: 127–139.

Crouse, C. F. 1966. Distribution free tests based on the sample distribution function. *Biometrika* **53**: 99–108.

Crow, E. L. 1956. Confidence intervals for a proportion. *Biometrika* **43**: 423–435.

Crowder, M. J., A. C. Kimber, R. L. Smith, and T. J. Sweeting. 1991. *Statistical Analysis of Reliability Data*. London: Chapman & Hall.

Cruess, D. F. 1989. Review of use of statistics in The American Journal of Tropical Medicine and Hygiene for January-December 1988. *Amer. J. Trop. Med. Hyg.* **41**: 619–626.

Csörgő, M., and R. Zitikis. 1996. Mean residual life processes. *Ann. Stat.* **24**: 1717–1739.

Csörgő, S. 1988. Estimation in the proportional hazards model of random censorship. *Statistics* **19**: 437–463.

Csörgő, S., and J. J. Farraway. 1998. The paradoxical nature of the proportional hazards model of random censorship. *Statistics* **31**: 61–78.

Csörgő, S., and L. Horváth. 1986. Confidence bands from censored samples. *Can. J. Stat.* **14**: 131–144.

Dale, M. 1984. Personal communication for report in Statistics 661. Columbus: Ohio State University.

Dallal, G. E., and L. Wilkinson. 1986. An analytical approximation to the distribution of Lilliefors' test statistic. *Amer. Statistician* **40**: 294–296.

Daly, D. A., and E. B. Cooper. 1967. Rate of stuttering adaptation under two electro-shock conditions. *Behav. Res. Ther.* **5**: 49–54.

Damico, J. A., and D. A. Wolfe. 1987. Extended tables of the exact distribution of a rank statistic for all treatments multiple comparisons in one-way layout designs. *Comm. Stat. Theory Methods* **16**: 2343–2360.

———. 1989. Extended tables of the exact distribution of a rank statistic for treatments versus control multiple comparisons in one-way layout designs. *Comm. Stat. Theory Methods* **18**: 3327–3353.

Daniel, W. W. 1990. *Applied Nonparametric Statistics*, 2nd ed. Boston: PWS-Kent.

David, H. A. 1981. *Order Statistics*, 2nd ed. New York: John Wiley.

Davis, C. E., and D. Quade. 1978. U-statistics for skewness or symmetry. *Comm. Stat. Theory Methods* **7**: 413–418.

Davison, A. C., and D. V. Hinkley. 1996. *Bootstrap Methods and Their Application*. New York: Cambridge University Press.

De Jonge, C., and M. A. J. Van Montfort. 1972. The null distribution of Spearman's S when $n = 12$. *Stat. Neerl.* **26**: 15–17.

DeKroon, J., and P. Van der Laan. 1981. Distribution-free test procedures in two-way layouts; a concept of rank interaction. *Stat. Neerl.* **35**: 189–213.

Delse, F. C., and B. W. Feather. 1968. The effect of augmented sensory feedback on the control of salivation. *Psychophysiology* **5**: 15–21.

Deshpande, J. V. 1983. A class of tests for exponentiality against increasing failure rate average alternatives. *Biometrika* **70**: 514–518.

DiCiccio, T. J., and B. Efron. 1996. Bootstrap confidence intervals. *Statist. Sci.* **11**: 189–228.

Dickinson, M. B., F. E. Putz, and C. D. Canham. 1993. Canopy gap closure in thickets of the clonal shrub, *Cornus racemosa*. *Bull. Torrey Bot. Club* **120**: 439–444.

Diehr, P., J. Yergan, J. Chu, P. Feigl, G. Glaefke, M. Moe, M. Bergner, and J. Rodenbaugh. 1989. Treatment modality and quality differences for black and white breast cancer patients treated in community hospitals. *Medical Care* **27**: 942–958.

Dietz, E. J. 1989. Teaching regression in a nonparametric statistics course. *Amer. Statistician* **43**: 35–40.

Dixon v. Margolis. 1991. 56 FEP Cases (N. D. Illinois).

Doksum, K. 1967. Robust procedures for some linear models with one observation per cell. *Ann. Math. Stat.* **38**: 878–883.

———. 1974. Empirical probability plots and statistical inference for nonlinear models in the two-sample case. *Ann. Stat.* **2**: 267–277.

Doksum, K. A., and G. L. Sievers. 1976. Plotting with confidence: Graphical comparisons of two populations. *Biometrika* **63**: 421–434.

Doksum, K. A., and B. S. Yandell. 1984. Tests for exponentiality. In *Handbook of Statistics, Volume 4, Nonparametric Methods*. Edited by P. R. Krishnaiah and P. K. Sen. Amsterdam: North Holland, pp. 579–611.

Donner, A., and W. W. Hauck. 1986. The large-sample relative efficiency of the Mantel-Haenszel estimator in the fixed-strata case. *Biometrics* **42**: 537–545.

Doss, H., and R. D. Gill. 1992. An elementary approach to weak convergence for quantile processes, with applications to survival data. *J. Amer. Statist. Assoc.* **87**: 869–877.

Dowdy, S., and S. Wearden. 1983. *Statistics for Research.* New York: John Wiley.

Draper, D. 1988. Rank-based robust analysis of linear models. I. Exposition and review. *Statist. Sci.* **3**: 239–271.

DuBois, A. B., S. Y. Botelho, G. M. Bedell, R. Marshall, and J. H. Comroe, Jr. 1956. A rapid plethysmographic method for measuring thoracic gas volume: A comparison with a nitrogen washout method for measuring functional residual capacity in normal subjects. *J. Clin. Invest.* **35**: 322–326.

Duffy, D. E., and T. J. Santner. 1987. Confidence intervals for a binomial parameter based on multistage tests. *Biometrics* **43**: 81–93.

Dunn, O. J. 1964. Multiple comparisons using rank sums. *Technometrics* **6**: 241–252.

Dunnett, C. W. 1964. New tables for multiple comparisons with a control. *Biometrics* **20**: 482–491.

Durbin, J. 1951. Incomplete blocks in ranking experiments. *Brit. J. Statist. Psychol.* **4**: 85–90.

———. 1975. Kolmogorov-Smirnov tests when parameters are estimated with applications to tests of exponentiality and tests on spacings. *Biometrika* **62**: 5–22.

Dwass, M. 1960. Some k-sample rank-order tests. In *Contributions to Probability and Statistics.* Edited by I. Olkin, S. G. Ghurye, H. Hoeffding, W. G. Madow, and H. B. Mann. Stanford, Calif.: Stanford University Press, pp. 198–202.

Dykstra, R. L., and P. Laud. 1981. A Bayesian nonparametric approach to reliability. *Ann. Stat.* **9**: 356–367.

Edwards, A. L. 1948. Note on the "correction for continuity" in testing the significance of the difference between correlated proportions. *Psychometrika* **13**: 185–187.

Edwards, A. W. F. 1963. The measure of association in a 2×2 table. *J. Roy. Statist. Soc. A* **126**: 109–114.

Efron, B. 1967. The two-sample problem with censored data. *Proc. 5th Berkeley Symp.* **4**: 831–854.

———. 1971. Forcing a sequential experiment to be balanced. *Biometrika* **58**: 403–417.

———. 1979. Bootstrap methods: another look at the jackknife. *Ann. Stat.* **7**: 1–26.

———. 1982. *The Jackknife, the Bootstrap and other Resampling Plans.* Volume 38 of CBMS-NSF Regional Conference Series in Applied Mathematics. Philadelphia: SIAM.

Efron, B., and G. Gong. 1983. A leisurely look at the bootstrap, the jackknife and cross-validation. *Amer. Statistician* **37**: 36–48.

Efron, B., and R. J. Tibshirani. 1993. *An Introduction to the Bootstrap.* New York: Chapman & Hall.

Ehlers, A. 1995. A 1-year prospective study of panic attacks: Clinical course and factors associated with maintenance. *J. Abnorm. Psychol.* **104**: 164–172.

Epstein, B. 1960a. Tests for the validity of the assumption that the underlying distribution of life is exponential, I. *Technometrics* **2**: 83–101.

———. 1960b. Tests for the validity of the assumption that the underlying distribution of life is exponential, II. *Technometrics* **2**: 167–183.

Eriksen, L., S. Björnstad, and K. G. Götestam. 1986. Social skills training in groups for alcoholics: One-year treatment outcome for groups and individuals. *Addictive Behaviors* **11**: 309–329.

Eubank, R. L. 1988. *Spline Smoothing and Nonparametric Regression.* New York: Dekker.

Everitt, B. S. 1992. *The Analysis of Contingency Tables*, 2nd ed. London: Chapman & Hall.

———. 1994. *A Handbook of Statistical Analyses Using S-PLUS.* London: Chapman & Hall.

Falkner, B., G. Onesti, T. Moshang, Jr., and D. T. Lowenthal. 1981. Growth hormone release in hypertensive adolescents treated with clonidine. *J. Clin. Pharmacol.* **21**: 31–36.

Fan, J. 1996. Tests of significance based on wavelet thresholding and Neyman's truncation. *J. Amer. Statist. Assoc.* **91**: 674–688.

Fan, J., and I. Gijbels. 1996. *Local Polynomial Modelling and Its Applications.* London: Chapman & Hall.

Fan, J., and J. S. Marron. 1993. Comment (discussion of Hastie and Loader). *Statist. Sci.* **8**: 129–134.

Feather, B. W., and D. T. Wells. 1966. Effects of concurrent motor activity on the unconditioned salivary reflex. *Psychophysiology* **2**: 338–343.

Featherston, D. W. 1971. *Taenia hydatigena* II. Evagination of cysticerci and establishment in dogs. *Exp. Parasit.* **29**: 242–249.

Feller, W. 1948. On the Kolmogorov-Smirnov limit theorems for empirical distributions. *Ann. Math. Stat.* **19**: 177–189.

Ferguson, T. S. 1973. A Bayesian analysis of some nonparametric problems. *Ann. Stat.* **1**: 209–230.

Finney, D. J., R. Latscha, B. M. Bennett,, and P. Hsu. 1963. *Tables for Testing Significance in a 2×2 Contingency Table.* New York: Cambridge University Press.

Fisher, R. A. 1925, 1934, 1970. *Statistical Methods for Research Workers* (originally published 1925, 14th edition 1970). Edinburgh: Oliver & Boyd.

———. 1935. The logic of inductive inference (with discussion). *J. Roy. Statist. Soc.* **98**: 39–82.

Fisz, M. 1963. *Probability Theory and Mathematical Statistics*, 3rd ed. New York: John Wiley.

Fleiss, J. L. 1981. *Statistical Methods for Rates and Proportions*, 2nd ed. New York: John Wiley.

Fleming, T. R., and D. P. Harrington. 1991. *Counting Processes and Survival Analysis*. New York: John Wiley.

Fleming, T. R., J. R. O'Fallon, P. C. O'Brien, and D. P. Harrington. 1980. Modified Kolmogorov-Smirnov test procedures with application to arbitrarily right-censored data. *Biometrics* **36**: 607–625.

Fligner, M. A. 1984. A note on two-sided distribution-free treatment versus control multiple comparisons. *J. Amer. Statist. Assoc.* **79**: 208–211.

———. 1985. Pairwise versus joint ranking: Another look at the Kruskal-Wallis statistic. *Biometrika* **72**: 705–709.

Fligner, M. A., and G. E. Policello II. 1981. Robust rank procedures for the Behrens-Fisher problem. *J. Amer. Statist. Assoc.* **76**: 162–174.

Fligner, M. A., and S. W. Rust. 1983. On the independence problem and Kendall's tau. *Comm. Stat. Theory and Methods* **12**: 1597–1607.

Fligner, M. A., and J. S. Verducci. 1993. *Probability Models and Statistical Analyses for Ranking Data*. Lecture Notes in Statistics, Volume 80. New York: Springer-Verlag.

Fligner, M. A., and D. A. Wolfe. 1982. Distribution-free tests for comparing several treatments with a control. *Stat. Neerl.* **36**: 119–127.

Flores, A. M., and L. R. Zohman. 1970. Energy cost of bedmaking to the cardiac patient and the nurse. *Amer. J. Nurs.* **70**: 1264–1267.

Forsman, A., and L. E. Lindell. 1993. The advantage of a big head: swallowing performance in adders, *Vipera Berus*. *Functional Ecology* **7**: 183–189.

Fox, J. R., and J. E. Randall. 1970. Relationship between forearm tremor and the biceps electromyogram. *J. Appl. Physiol.* **29**: 103–108.

Fraser, D. A. S. 1957. *Nonparametric Methods in Statistics*. New York: John Wiley.

Freund, R. J., and W. J. Wilson. 1997. *Statistical Methods*, rev. ed. New York: Academic Press.

Friedman, J. H. 1991. Multivariate adaptive regression splines (with discussion). *Ann. Stat.* **19**: 1–141.

Friedman, Meyer, S. O. Byers, R. H. Rosenman, and R. Neuman. 1971. Coronary-prone individuals (Type A behavior pattern) growth hormone responses. *J. Amer. Med. Assoc.* **217**: 929–932.

Friedman, Meyer, and R. H. Rosenman. 1959. Association of specific overt behavior pattern with blood and cardiovascular findings: Blood cholesterol level, blood clotting time, incidence of arcus senilis, and clinical coronary artery disease. *J. Amer. Med. Assoc.* **169**: 1286–1296.

Friedman, Milton 1937. The use of ranks to avoid the assumption of normality implicit in the analysis of variance. *J. Amer. Statist. Assoc.* **32**: 675–701.

Gabriel, K. R. 1969. Simultaneous test procedures—some theory of multiple comparisons. *Ann. Math. Stat.* **40**: 224–250.

Gail, M., and J. J. Gart. 1973. The determination of sample sizes for use with the exact conditional test in 2×2 comparative trials. *Biometrics* **29**: 441–448.

Gamerman, D. 1991. Dynamic Bayesian models for survival data. *Appl. Stat.* **40**: 63–79.

Gart, J. J., and J. R. Zweifel. 1967. On the bias of various estimators of the logit and its variance with applications to quantal bioassay. *Biometrika* **54**: 181–187.

Gastwirth, J. L., and H. Rubin. 1969. The behavior of robust estimators on dependent data. Mimeo Ser. 197, Department of Statistics, Purdue University.

Gehan, E. A. 1965. A generalized Wilcoxon test for comparing arbitrarily singly-censored samples. *Biometrika* **52**: 203–223.

Gelfand, A. E., and A. F. M. Smith. 1990. Sampling-based approaches to calculating marginal densities. *J. Amer. Statist. Assoc.* **85**: 398–409.

Gentry, J., and J. Pike. 1970. An empirical study of the risk-return hypothesis using common stock portfolios of life insurance companies. *J. Finan. Quant. Anal.* **5**: 179–185.

Gerstein, H. H. 1965. Lake Michigan pollution and Chicago's supply. *J. Amer. Water Works Assoc.* **57**: 841–857.

Gibbons, J. D. 1971. *Nonparametric Statistical Inference*. New York: McGraw-Hill.

———. 1997. *Nonparametric Methods for Quantitative Analysis*, 3rd ed. Syracuse, N. Y.: American Sciences Press.

Gibbons, J. D., and S. Chakraborti. 1992. *Nonparametric Statistical Inference*, 3rd ed. New York: Dekker.

Gilks, W. R., S. Richardson, and D. G. Spiegelhalter. 1996. *Markov Chain Monte Carlo in Practice*. London: Chapman & Hall.

Gill, R. D. 1980. *Censoring and Stochastic Integrals*. Mathematical Centre Tracts 124, Amsterdam: Mathematisch Centrum.

———. 1983. Large sample behaviour of the product-limit estimator on the whole line. *Ann. Stat.* **11**: 49–58.

———. 1989. Non- and semi-parametric maximum likelihood estimators and the von Mises method (Part 1). *Scand J. Stat.* **16**: 97–128.

Gillespie, M. J., and L. Fisher. 1979. Confidence bands for the Kaplan-Meier survival curve estimate. *Ann. Stat.* **7**: 920–924.

Gleser, L. J. 1996. Comment on "Bootstrap confidence intervals" by T. J. DiCiccio and B. Efron. *Statist. Sci.* **11**: 219–221.

Goldsmith, J. R., and J. A. Nadel. 1969. Experimental exposure of human subjects to ozone. *J. Air Poll. Control Assoc.* **19**: 329–330.

Goode, D. J., and H. Y. Meltzer. 1976. Effects of isometric exercise on serum creatine phosphokinase activity. *Arch. Gen. Psychiatry* **33**: 1207–1211.

Goodman, L. A., and W. H. Kruskal. 1963. Measures of association for cross classifications III. Approximate sampling theory. *J. Amer. Statist. Assoc.* **58**: 310–364.

Gottlieb, G. 1965. Prenatal auditory sensitivity in chickens and ducks. *Science* **147**: 1596–1598.

Govindarajulu, Z. 1968. Distribution-free confidence bounds for $P(X < Y)$. *Ann. Inst. Statist. Math.* **20**: 229–238.

Green, P. J. 1995. Reversible jump Markov chain Monte Carlo computation and Bayesian model determination. *Biometrika* **57**: 97–109.

Green, P. J., and B. W. Silverman. 1994. *Nonparametric Regression and Generalized Linear Models: A Roughness Penalty Approach*. London: Chapman & Hall.

Greenberg, V. L. 1966. Robust estimation in incomplete block designs. *Ann. Math. Stat.* **37**: 1331–1337.

Greenwood, M. 1926. The natural duration of cancer. *Reports on Public Health and Medical Subjects* **33**: pp. 1–26. London: Her Majesty's Stationery Office.

Gregory, P. B. 1974. Personal communication.

Grenander, U. 1956. On the theory of mortality measure, Part II. *Skan. Aktuarietidskr* **39**: 125–153.

Gripenberg, G. 1992. Confidence intervals for partial rank correlations. *J. Amer. Statist. Assoc.* **87**: 546–551.

Gross, S. 1966. Nonparametric tests when nuisance parameters are present. Ph.D. diss., University of California, Berkeley.

Guess, F. 1984. Testing whether mean residual life changes trend. Ph.D. diss., Florida State University.

Guess, F., M. Hollander, and F. Proschan. 1986. Testing exponentiality versus a trend change in mean residual life. *Ann. Stat.* **14**: 1388–1398.

Gulati, S., and W. J. Padgett. 1996. Families of smooth confidence bands for the survival function under the general random censorship model. *Lifetime Data Analysis* **2**: 349–362.

Gupta, M. K. 1967. An asymptotically nonparametric test of symmetry. *Ann. Math. Stat.* **38**: 849–866.

Gupta, S. S. 1963. Probability integrals of multivariate normal and multivariate t. *Ann. Math. Stat.* **34**: 792–828.

Gurland, J., and J. Sethuraman. 1995. How pooling failure data may reverse increasing failure rates. *J. Amer. Statist. Assoc.* **90**: 1416–1423.

Haber, M. 1986. An exact unconditional test for the 2×2 comparative trial. *Psychol. Bull.* **99**: 129–132.

———. 1987. A comparison of some conditional and unconditional exact tests for 2 by 2 contingency tables. *Comm. Stat. Simulation Comput.* **16**: 999–1013.

Habib, M. G., and D. R. Thomas. 1986. Chi-squared goodness-of-fit tests for randomly censored data. *Ann. Stat.* **14**: 759–765.

Hájek, J. 1969. *Nonparametric Statistics*. San Francisco: Holden-Day.

Hájek, J., and Z. Šidák. 1967. *Theory of Rank Tests*. New York: Academic Press.

Haldane, J. B. S. 1955. The estimation and significance of the logarithm of a ratio of frequencies. *Ann. Human Genetics* **20**: 309–311.

Hall, P. 1992. *The Bootstrap and Edgeworth Expansion*. New York: Springer-Verlag.

Hall, W. J., and J. A. Wellner. 1979. Estimation of mean residual life. Technical Report, Department of Statistics. Rochester, New York: University of Rochester.

———. 1980. Confidence bands for a survival curve from censored data. *Biometrika* **67**: 133–143.

———. 1981. Mean residual life. In *Statistics and Related Topics*. Edited by M. Csörgő, D. A. Dawson, J. N. K. Rao, and A. K. Md. E. Saleh. Amsterdam: North Holland, pp. 169–184.

Halperin, M., P. R. Gilbert, and J. M. Lachin. 1987. Distribution-free confidence intervals for $\Pr(X_1 < X_2)$. *Biometrics* **43**: 71–80.

Hamilton, M. 1960. A rating scale for depression. *J. Neurol. Neurosurg. Psychiat.* **23**: 56–62.

Härdle, W. 1990. *Applied Nonparametric Regression*. London: Cambridge University Press.

Harrington, D. P., and T. R. Fleming. 1982. A class of rank test procedures for censored survival data. *Biometrika* **69**: 553–566.

Hart, J. D. 1997. *Nonparametric Smoothing and Lack-of-Fit Tests*. New York: Springer-Verlag.

Harter, H. L. 1960. Tables of range and studentized range. *Ann. Math. Stat.* **31**: 1122–1147.

———. 1961. Expected values of normal order statistics. *Biometrika* **48**: 151–165.

Hartigan, J. A. 1969. Using subsample values as typical values. *J. Amer. Statist. Assoc.* **64**: 1303–1317.

———. 1971. Error analysis by replaced samples. *J. Roy. Statist. Soc. B* **33**: 98–110.

———. 1975. Necessary and sufficient conditions for asymptotic joint normality of a statistic and its subsample values. *Ann. Stat.* **3**: 573–580.

Hastie, T., and C. Loader. 1993. Local regression: automatic kernel carpentry. *Statist. Sci.* **8**: 120–129 (discussion: 129–143).

Hastie, T., and R. Tibshirani. 1987. Generalized additive models: some applications. *J. Amer. Statist. Assoc.* **82**: 371–386.

———. 1990. *Generalized Additive Models*. New York: Chapman and Hall.

Hastings, W. K. 1970. Monte Carlo sampling methods using Markov chains and their applications. *Biometrika* **57**: 97–109.

Hauck, W. W. 1979. The large-sample variance of the Mantel-Haenszel estimator of a common odds ratio. *Biometrics* **35**: 817–819.

Hauck, W. W., and A. Donner. 1988. The asymptotic relative efficiency of the Mantel-Haenszel estimator in the increasing-number-of-strata case. *Biometrics* **44**: 379–384.

Hawkins, D. L., S. Kochar, and C. Loader. 1992. Testing exponentiality against IDMRL distributions with unknown change point. *Ann. Stat.* **20**: 280–290.

Hays, W. L. 1960. A note on average tau as a measure of concordance. *J. Amer. Statist. Assoc.* **55**: 331–341.

Hayter, A. J. 1984. A proof of the conjecture that the Tukey-Kramer multiple comparison procedure is conservative. *Ann. Stat.* **12**: 61–75.

———. 1990. A one-sided Studentized range test for testing against a simple ordered alternative. *J. Amer. Statist. Asssoc.* **85**: 778–785.

Hayter, A. J., and G. Stone. 1991. Distribution free multiple comparisons for monotonically ordered treatment effects. *Austral. J. Stat.* **33**: 335–346.

Hebb, D. O., and K. Williams. 1946. A method of rating animal intelligence. *J. Gen. Psychol.* **34**: 59–65.

Hettmansperger, T. P. 1984. *Statistical Inference Based on Ranks*. New York: John Wiley.

Hettmansperger, T. P., and J. W. McKean. 1977. A robust alternative based on ranks to least squares in analyzing linear models. *Technometrics* **19**: 275–284.

———. 1998. *Robust Nonparametric Statistical Methods*. London: Arnold.

Hettmansperger, T. P., J. W. McKean, and S. J. Sheather. 1997. Rank-based analyses of linear models. To appear in *Handbook of Statistics, Volume 15*. Edited by S. Ghosh and C. R. Rao. Amsterdam: Elsevier Science.

Hettmansperger, T. P., and R. M. Norton. 1987. Tests for patterned alternatives in k-sample problems. *J. Amer. Statist. Assoc.* **82**: 292–299.

Hilgard, E. R., L. W. Lauer, and A. H. Morgan. 1963. *Manual for Stanford Profile Scales of Hypnotic Susceptibility, Forms I and II*. Palo Alto, Calif.: Consulting Psychologists Press.

Hjort, N. L. 1990a. Goodness of fit tests in models for life history data based on cumulative hazard rates. *Ann. Stat.* **18**: 1221–1258.

———. 1990b. Nonparametric Bayes estimators based on beta processes in models for life history data. *Ann. Stat.* **18**: 1259–1294.

Hochberg, Y., and A. C. Tamhane. 1987. *Multiple Comparison Procedures*. New York: John Wiley.

Hodges, J. L., Jr., and E. L. Lehmann. 1950. Some problems in minimax point estimation. *Ann. Math. Stat.* **21**: 182–197.

———. 1956. The efficiency of some nonparametric competitors of the t-test. *Ann. Math. Stat.* **27**: 324–335.

———. 1963. Estimates of location based on rank tests. *Ann. Math. Stat.* **34**: 598–611.

———. 1967. On medians and quasimedians. *J. Amer. Statist. Assoc.* **62**: 926–931.

———. 1970. Deficiency. *Ann. Math. Stat.* **41**: 783–801.

———. 1983. Hodges-Lehmann estimators. In *Encyclopedia of Statistical Sciences, Volume 3.* Edited by S. Kotz, N. L. Johnson, and C. B. Read. New York: John Wiley, pp. 463–465.

Hoeffding, W. 1948a. A class of statistics with asymptotically normal distribution. *Ann. Math. Stat.* **19**: 293–325.

———. 1948b. A non-parametric test of independence. *Ann. Math. Stat.* **19**: 546–557.

———. 1951. "Optimum" nonparametric tests. *Proc. 2nd Berkeley Symp.*, pp. 83–92.

———. 1952. The large-sample power of tests based on permutations of observations. *Ann. Math. Stat.* **23**: 169–192.

Hollander, M. 1966. An asymptotically distribution-free multiple comparison procedure—treatments versus control. *Ann. Math. Stat.* **37**: 735–738.

———. 1967a. Rank tests for randomized blocks when the alternatives have an a priori ordering. *Ann. Math. Stat.* **38**: 867–877.

———. 1967b. Asymptotic efficiency of two nonparametric competitors of Wilcoxon's two sample test. *J. Amer. Statist. Assoc.* **62**: 939–949.

———. 1971. A nonparametric test for bivariate symmetry. *Biometrika* **58**: 203–212.

———. 1996. Personal communication.

Hollander, M., and R. M. Korwar. 1982. Nonparametric Bayesian estimation of the horizontal distance between two populations. In *Nonparametric Statistical Inference I.* New York: North Holland, pp. 409–415.

Hollander, M., I. W. McKeague, and J. Yang. 1997. Likelihood ratio-based confidence bands for survival functions. *J. Amer. Statist. Assoc.* **92**: 215–226.

Hollander, M., and E. Peña. 1988. Nonparametric tests under restricted treatment-assignment rules. *J. Amer. Statist. Assoc.* **83**: 1144–1151.

———. 1989. Families of confidence bands for the survival function under the general random censorship model and the Koziol-Green model. *Can. J. Stat.* **17**: 59–74.

———. 1992a. A chi-squared goodness-of-fit test for randomly censored data. *J. Amer. Statist. Assoc.* **87**: 458–463.

———. 1992b. Classes of nonparametric goodness-of-fit tests for censored data: Simple null hypothesis case. In *Nonparametric Statistics and Related Topics.* Edited by A. K. Md. E. Saleh. Amsterdam: North-Holland, 97–118.

Hollander, M., G. Pledger, and P. Lin. 1974. Robustness of the Wilcoxon test to a certain dependency between samples. *Ann. Stat.* **2**: 177–181.

Hollander, M., and F. Proschan. 1972. Testing whether new is better than used. *Ann. Math. Stat.* **43**: 1136–1146.

———. 1975. Tests for the mean residual life. *Biometrika* **62**: 585–593.

———. 1979. Testing to determine the underlying distribution using randomly censored data. *Biometrics* **35**: 393–401.

———. 1984. Nonparametric concepts and methods in reliability. In *Handbook of Statistics, Volume 4, Nonparametric Methods.* Edited by P. R. Krishnaiah and P. K. Sen. Amsterdam: North-Holland, pp. 613–655.

Hollander, M., F. Proschan, and J. Sconing. 1985. Efficiency loss with the Kaplan-Meier estimator. Technical Report M707, Department of Statistics, Tallahassee: Florida State University.

Hollander, M., and D. A. Wolfe. 1973. *Nonparametric Statistical Methods,* 1st ed. New York: John Wiley.

Hotelling, H., and M. R. Pabst. 1936. Rank correlation and tests of significance involving no assumption of normality. *Ann. Math. Stat.* **7**: 29–43.

Høyland, A. 1965. Robustness of the Hodges-Lehmann estimates for shift. *Ann. Math. Stat.* **36**: 174–197.

———. 1968. Robustness of the Wilcoxon estimate of location against a certain dependence. *Ann. Math. Stat.* **39**: 1196–1201.

Hsieh, F. Y. 1987. A simple method of sample size calculation for unequal-sample-size designs that use the logrank or t-test. *Stat. Med.* **6**: 577–581.

———. 1992. Comparing sample size formulae for trials with unbalanced allocation using the logrank test. *Stat. Med.* **11**: 1091–1098.

Hundal, P. S. 1969. Knowledge of performance as an incentive in repetitive industrial work. *J. Appl. Psychol.* **53**: 224–226.

Ijzermans, A. B. 1970. Pitting corrosion and intergranular attack of austenitic Cr-Ni stainless steels in Na SCN. *Corrosion Science* **10**: 607–615.

Iman, R. L. 1994. *A Data-Based Approach to Statistics.* Belmont, Calif.: Duxbury Press.

Iman, R. L., D. Quade, and D. A. Alexander. 1975. Exact probability levels for the Kruskal-Wallis test. In *Selected Tables in Mathematical Statistics, Volume 3*. Edited by H. L. Harter and D. B. Owen. Providence, RI: American Mathematical Society and Institute of Mathematical Statistics, pp. 329–384.

Jaeckel, L. A. 1972. Estimating regression coefficients by minimizing the dispersion of the residuals. *Ann. Math. Stat.* **43**: 1449–1458.

Jamison, H. H. 1971. Development of a gaseous oxygen impact testing method. *Mater. Res. Stand.* **11**: 22–27.

Joe, H. 1990. Multivariate concordance. *J. Mult. Anal.* **35**: 12–30.

Johansen, S. 1978. The product limit estimator as maximum likelihood estimator. *Scand. J. Stat.* **5**: 195–199.

Johnson, A. A., K. Mukherjee, S. Schlosser, and E. Raask. 1970. The behaviour of a cenosphere-resin composite under hydrostatic pressure. *Ocean Engng.* **2**: 45–48.

Johnson, B. 1984. Personal communication for report in Statistics 661, Columbus: Ohio State University.

Johnson, B. M. 1973. Decision making, faculty satisfaction, and the place of the School of Nursing in the university. *Nursing Res.* **22**: 100–107.

Johnson, R. A. 1988. Stress-strength models for reliability. In *Handbook of Statistics, Volume 7*. Edited by P. K. Krishnaiah and C. R. Rao. New York: North Holland, pp. 27–54.

Johnson, S. K., and R. E. Johnson. 1972. Tonsillectomy history in Hodgkin's disease. *N. Engl. J. Med.* **287**: 1122–1125.

Jonckheere, A. R. 1954a. A distribution-free k-sample test against ordered alternatives. *Biometrika* **41**: 133–145.

———. 1954b. A test of significance for the relation between m rankings and k ranked categories. *Brit. J. Statist. Psychol.* **7**: 93–100.

Jones, M. P., T. W. O'Gorman, J. H. Lemke, and R. F. Woolson. 1989. A Monte Carlo investigation of homogeneity tests of the odds ratio under various sample size configurations. *Biometrics* **45**: 171–181.

Jung, D. H., and A. C. Parekh. 1970. A semi-micromethod for the determination of serum iron and iron-binding capacity without deproteinization. *Amer. J. Clin. Path.* **54**: 813–817.

Kaarsemaker, L., and A.van Wijngaarden. 1953. Tables for use in rank correlation. *Stat. Neerl.* **7**: 41–54.

Kalbfleisch, J. D., and R. L. Prentice. 1980. *The Statistical Analysis of Failure Time Data*. New York: John Wiley.

Kaneto, A., K. Kosaka, and K. Nakao. 1967. Effects of stimulation of the vagus nerve on insulin secretion. *Endocrinology* **80**: 530–536.

Kaplan, E. L., and P. Meier. 1958. Nonparametric estimation from incomplete observations. *J. Amer. Statist. Assoc.* **53**: 457–481.

Kaplan, H. S., and S. A. Rosenberg. 1973. Personal communication.

Karpatkin, M., R. F. Porges, and S. Karpatkin. 1981. Platelet counts in infants of women with autoimmune thrombocytopenia. *New Engl. J. Med.* **305**: 936–939.

Kayle, K. A. 1984. Personal communication for report in Statistics 661. Columbus: Ohio State University.

Kendall, M. G. 1938. A new measure of rank correlation. *Biometrika* **30**: 81–93.

———. 1962. *Rank Correlation Methods,* 3rd ed. London: Griffin.

Kendall, M. G., and J. D. Gibbons. 1990. *Rank Correlation Methods,* 5th ed. London: Arnold.

Kershenobich, D., F. J. Fierro, and M. Rojkind. 1970. The relationship between the free pool of proline and collagen content in human liver cirrhosis. *J. Clin. Invest.* **49**: 2246–2249.

Kiefer, J., and J. Wolfowitz. 1956. Consistency of the maximum likelihood estimator in the presence of infinitely many nuisance parameters. *Ann. Math. Stat.* **27**: 887–906.

Kim, D. H., and D. H. Lim. 1995. Rank tests for parallelism of regression lines against umbrella alternatives. *J. Nonpar. Stat.* **5**: 289–302.

Kim, P. J., and R. I. Jennrich. 1973. Tables of the exact sampling distribution of the two-sample Kolmogorov-Smirnov criterion, $D_{mn}, m \leq n$. In *Selected Tables in Mathematical Statistics,* Volume 1 (2nd printing with rev.). Edited by H. L. Harter and D. B. Owen. Providence, RI: American Mathematical Society and Institute of Mathematical Statistics, pp. 79–170.

Klefsjö, B. 1983. Some tests against aging based on the total time on test transform. *Comm. Stat. Theory Methods* **12**: 907–927.

Klein, J. P., and M. L. Moeschberger. 1997. *Survival Analysis: Techniques for Censored and Truncated Data*. New York: Springer-Verlag.

Klotz, J. 1963. Small sample power and efficiency for the one-sample Wilcoxon and normal scores tests. *Ann. Math. Stat.* **34**: 624–632.

———. 1964. On the normal scores two-sample rank test. *J. Amer. Statist. Assoc.* **49**: 652–664.

———. 1967. Asymptotic efficiency of the two sample Kolmogorov-Smirnov test. *J. Amer. Statist. Assoc.* **62**: 932–938.

Kolmogorov, A. N. 1933. Sulla determinazione empirica di una legge di distribuzione. *Giorn. dell' Inst. Ital. degli Att.* **4**: 83–91.

———. 1941. Confidence limits for an unknown distribution function. *Ann. Math. Stat.* **12**: 461–483.

Konijn, H. S. 1956. On the power of certain tests for independence in bivariate populations. *Ann. Math. Stat.* **27**: 300–323. Correction: **29** (1958): 935–936.

Kontula, K., L. C. Andersson, T. Paavonen, G. Myllyla, L. Teerenhovi, and P. Vuopio. 1980. Glucocorticoid receptors and glucocorticoid sensitivity of human leukemic cells. *Int. J. Cancer* **26**: 177–183.

Kontula, K., T. Paavonen, P. Vuopio, and L. C. Andersson. 1982. Glucocorticoid receptors in hairy-cell leukemia. *Int. J. Cancer* **30**: 423–426.

Korwar, R. M. 1971. Personal communication.

Koul, H. L. 1977. A test for new is better than used. *Comm. Stat. Theory Methods* **6**: 563–573.

———. 1978a. A class of tests for testing "new is better than used." *Can. J. Stat.* **6**: 249–271.

———. 1978b. Testing for new is better than used in expectation. *Comm. Stat. Theory Methods* **7**: 685–701.

Koul, H. L., G. L. Sievers, and J. W. McKean. 1987. An estimator of the scale parameter for the rank analysis of linear models under general score functions. *Scand. J. Stat.* **14**: 131–141.

Koziol, J. A. 1979. A test for bivariate symmetry based on the empirical distribution function. *Comm. Stat. Theory Methods* **8**: 207–221.

Koziol, J. A., and S. B. Green. 1976. A Cramër–von Mises statistic for randomly censored data. *Biometrika* **63**: 465–474.

———. 1978. Personal communication.

Kraft, C. H., and C. van Eeden. 1968. *A Nonparametric Introduction to Statistics.* New York: Macmillan.

Kramer, C. Y. 1956. Extension of multiple range tests to group means with unequal numbers of replications. *Biometrics* **12**: 307–310.

———. 1957. Extension of multiple range tests to group correlated adjusted means. *Biometrics* **13**: 13–18.

Krause, A., and M. Olson. 1997. *The Basics of S and S-PLUS.* New York: Springer-Verlag.

Krauth, J. 1988. *Distribution-free Statistics—An Application-Oriented Approach.* Amsterdam: Elsevier.

Kruskal, W. H. 1952. A nonparametric test for the several sample problem. *Ann. Math. Stat.* **23**: 525–540.

———. 1957. Historical notes on the Wilcoxon unpaired two-sample test. *J. Amer. Statist. Assoc.* **52**: 356–360.

Kruskal, W. H., and W. A. Wallis. 1952. Use of ranks in one-criterion variance analysis. *J. Amer. Statist. Assoc.* **47**: 583–621.

Kuehl, R. O. 1994. *Statistical Principles of Research Design and Analysis.* Belmont, Calif.: Duxbury Press.

Kurtz, T. E., R. F. Link, J. W. Tukey, and D. L. Wallace. 1965. Short-cut multiple comparisons for balanced single and double classification, Part I, results. *Technometrics* **7**: 95–161.

Lakatos, E., and G. Lan. 1992. A comparison of sample size methods for the logrank statistic. *Stat. Med.* **11**: 179–191.

Lamp, W. O. 1976. Statistical treatment of a study on the distribution of a stream insect by age. Master's thesis, Ohio State University.

Langenberg, P., and R. Srinivasan. 1979. Null distribution of the Hollander-Proschan statistic for decreasing mean residual life. *Biometrika* **66**: 679–680.

Latta, R. 1981. A Monte Carlo study of some two-sample rank tests with censored data. *J. Amer. Statist. Assoc.* **76**: 713–729.

Leach, C. 1979. *Introduction to Statistics. A Nonparametric Approach for the Social Sciences.* Chichester: John Wiley.

Leach, S. P. 1972. Personal communication.

Leaf, D. A., W. E. Connor, R. Illingworth, S.P. Bacon, and G. Sexton. 1989. The hypolipidemic effects of gemfibrozil in type V hyperlipidemia. *J. Amer. Med. Assoc.* **262**: 3154–3160.

LeCam, L., and J. Neyman, eds. 1965. *Bernoulli Bayes Laplace Anniversary Volume.* New York: Springer-Verlag.

Lee, S. C. S., C. Locke, and J. D. Spurrier. 1980. On a class of tests of exponentiality. *Technometrics* **22**: 547–554.

Lehmann, E. L. 1951. Consistency and unbiasedness of certain nonparametric tests. *Ann. Math. Stat.* **22**: 165–179.

———. 1959. *Testing Statistical Hypotheses.* New York: John Wiley.

———. 1963a. Robust estimation in analysis of variance. *Ann. Math. Stat.* **34**: 957–966.

———. 1963b. Asymptotically nonparametric inference: An alternative approach to linear models. *Ann. Math. Stat.* **34**: 1494–1506.

———. 1963c. Nonparametric confidence intervals for a shift parameter. *Ann. Math. Stat.* **34**: 1507–1512.

———. 1964. Asymptotically nonparametric inference in some linear models with one observation per cell. *Ann. Math. Stat.* **35**: 726–734.

———. 1975. *Nonparametrics: Statistical Methods Based on Ranks.* San Francisco: Holden-Day.

———. 1986. *Testing Statistical Hypotheses,* 2nd ed. New York: John Wiley.

Lepage, Y. 1971. A combination of Wilcoxon's and Ansari-Bradley's statistics. *Biometrika* **58**: 213–217.

———. 1973. A table for a combined Wilcoxon Ansari-Bradley statistic. *Biometrika* **60**: 113–116.

Leurgans, S. 1983. Three classes of censored data rank tests: Strengths and weaknesses under censoring. *Biometrika* **70**: 651–658.

———. 1984. Asymptotic behavior of two-sample rank tests in the presence of random censoring. *Ann. Stat.* **12**: 572–589.

Li, G., and H. Doss. 1993. Generalized Pearson-Fisher chi-square goodness-of-fit tests, with applications to models with life history data. *Ann. Stat.* **21**: 772–797.

Li, G., M. Hollander, I. W. McKeague, and J. Yi. 1996. Nonparametric likelihood ratio confidence bands for quantile functions from incomplete survival data. *Ann. Stat.* **24**: 628–640.

Liang, K. Y., and S. G. Self. 1985. Tests for homogeneity of odds ratio when the data are sparse. *Biometrika* **72**: 353–358.

Lilliefors, H. 1967. On the Kolmogorov-Smirnov test for normality with mean and variance unknown. *J. Amer. Statist. Assoc.* **62**: 399–402.

———. 1969. On the Kolmogorov-Smirnov test for the exponential distribution with mean unknown. *J. Amer. Statist. Assoc.* **64**: 387–389.

Lim, D. H., and D. A. Wolfe. 1997. Nonparametric comparisons of several regression lines with a control. *Far East J. Theor. Stat.* **1**: 51–61.

Livesey, P. J. 1967. The Hebb-Williams elevated pathway test: A comparative study of rat, rabbit and cat performance. *Aust. J. Psychol.* **19**: 55–62.

Lloyd, S. J., K. D. Garlid, R. C. Reba, and A. E. Seeds. 1969. Permeability of different layers of the human placenta to isotopic water. *J. Appl. Physiol.* **26**: 274–276.

Louis, T. A. 1981. Confidence intervals for a binomial parameter after observing no successes. *Amer. Statistician* **35**: 154.

Low, P. P., C. K. Luk, M. J. Dulfano, and P. J. P. Finch. 1984. Ciliary beat frequency of human respiratory tract by different sampling techniques. *Amer. Rev. Respir. Dis.* **130**: 497–498.

Lu, H. H. S., M. T. Wells, and R. C. Tiwari. 1994. Inference for shift functions in the two-sample problem with right-censored data: With applications. *J. Amer. Statist. Assoc.* **89**: 1017–1026.

Mack, G. A. 1971. Personal communication.

———. 1977. A k-sample Wilcoxon rank test for the umbrella alternatives. II. Point of the umbrella unknown. Ph.D. diss., Ohio State University.

———. 1981. A quick and easy distribution-free test for main effects in a two-factor ANOVA. *Comm. Stat. Simulation Comput.* **10**: 571–591.

Mack, G. A., and J. H. Skillings. 1980. A Friedman-type rank test for main effects in a two-factor ANOVA. *J. Amer. Statist. Assoc.* **75**: 947–951.

Mack, G. A., and D. A. Wolfe. 1981. K-sample rank tests for umbrella alternatives. *J. Amer. Statist. Assoc.* **76**: 175–181.

Maesono, Y. 1996. Higher order comparisons of jackknife variance estimators. *J. Nonpar. Stat.* **7**: 35–45.

Manly, B. F. J. 1997. *Randomization, Bootstrap and Monte Carlo Methods in Biology,* 2nd ed. London: Chapman & Hall.

Mann, H. B. 1945. Nonparametric tests against trend. *Econometrica* **13**: 245–259.

Mann, H. B., and D. R. Whitney. 1947. On a test of whether one of two random variables is stochastically larger than the other. *Ann. Math. Stat.* **18**: 50–60.

Mantel, N. 1963. Chi-square tests with one degree of freedom: Extensions of the Mantel-Haenszel procedure. *J. Amer. Statist. Assoc.* **58**: 690–700.

———. 1966. Evaluation of survival data and two new rank order statistics arising in its consideration. *Cancer Chemother. Rep.* **50**: 113–170.

———. 1985. Propriety of the Mantel-Haenszel variance for the log rank test. *Biometrika* **72**: 471–472.

Mantel, N., and W. Haenszel. 1959. Statistical aspects of the analysis of data from retrospective studies of disease. *J. Nat. Cancer Inst.* **22**: 719–748.

Marascuilo, L. A., and M. McSweeney. 1977. *Nonparametric and Distribution-Free Methods for the Social Sciences.* Monterey, Calif.: Brooks/Cole.

Marazzi, A. 1992. *Algorithms, Routines and S Functions for Robust Statistics.* Pacific Grove, Calif.: Wadsworth & Brooks/Cole.

March, G. L., T. M. John, B. A. McKeown, L. Sileo, and J. C. George. 1976. The effects of lead poisoning on various plasma constituents in the Canada goose. *J. Wildl. Dis.* **12**: 14–19.

Maritz, J. S. 1981. *Distribution-Free Statistical Methods.* London: Chapman & Hall.

Marshall, A. W., and F. Proschan. 1965. Maximum likelihood estimation for distributions with monotone failure rate. *Ann. Math. Stat.* **36**: 69–77.

Mather, M. 1984. Personal communication for report in Statistics 661, Columbus: Ohio State University.

Maxson, S. J. 1977. Activity patterns of female ruffed grouse during the breeding season. *Wilson Bull.* **89**: 439–455.

McClave, J. T., and G. Benson. 1978. *Statistics for Business and Economics.* San Francisco: Dellen.

McDonald, B. J., and W. A. Thompson, Jr. 1967. Rank sum multiple comparisons in one- and two-way classifications. *Biometrika* **54**: 487–497.

McKean, J. W., and T. P. Hettmansperger. 1976. Tests of hypotheses of the general linear model based on ranks. *Comm. Stat. Theory Methods* **5**: 693–709.

McKean, J. W., and T. A. Ryan, Jr. 1977. An algorithm for obtaining confidence intervals and point estimates based on ranks in the two-sample location problem. *Trans. Math. Software* **3**: 183–185.

McKean, J. W., and S. J. Sheather. 1991. Small sample properties of robust analyses of linear models based on R-estimates: A survey. In *Directions in Robust Statistics and Diagnostics, Part II.* Edited by W. Stahel and S. Weisberg. New York: Springer-Verlag, pp. 1–19.

McNemar, Q. 1947. Note on the sampling error of the difference between correlated proportions or percentages. *Psychometrika* **12**: 153–157.

Mehta, C. R., and J. F. Hilton. 1993. Exact power of conditional and unconditional tests: Going beyond the 2×2 contingency table. *Amer. Statistician* **47**: 91–98.

Mehta, C. R., and N. R. Patel. 1983. A network algorithm for performing Fisher's exact test in $r \times c$ contingency tables. *J. Amer. Statist. Assoc.* **78**: 427–434.

———. 1986. A hybrid algorithm for Fisher's exact test on unordered $r \times c$ contingency tables. *Comm. Stat. Theory Methods* **15**: 387–403.

———. 1992. *StatXact-Turbo User Manual.* Cambridge, Mass.: CYTEL Software Corporation.

Mehta, C. R., N. R. Patel, and R. Gray. 1985. Computing an exact confidence interval for the common odds ratio in several 2 by 2 contingency tables. *J. Amer. Statist. Assoc.* **80**: 969–973.

Mehta, C. R., N. R. Patel, and P. Senchaudhuri. 1988. Importance sampling for estimating exact probabilities in permutational inference. *J. Amer. Statist. Assoc.* **83**: 999–1005.

Mehta, C. R., N. R. Patel, and A. A. Tsiatis. 1984. Exact significance tests to establish treatment equivalence for ordered categorical data. *Biometrics* **40**: 819–825.

Mehta, C. R., N. R. Patel, and L. J. Wei. 1988. Computing exact significance tests with restricted randomization rules. *Biometrika* **75**: 295–302.

Meier, P. 1975. Estimation of a distribution function from incomplete observations. In *Perspectives in Probability and Statistics.* Edited by J. Gani. Sheffield, England: Applied Probability Trust, pp. 67–87.

Mendenhall, W. 1968. *Introduction to Linear Models and the Design and Analysis of Experiments.* Belmont, Calif.: Wadsworth.

Mendis, K. N., R. I. Ihalamulla, and P. H. David. 1988. Diversity of plasmodium vivax-induced antigens on the surface of infected human erythrocytes. *Amer. J. Trop. Med. Hyg.* **38**: 42–46.

Merline, J. W. 1991. Will more money improve education? *Consum. Res.* **74**: 26–27.

Metropolis, N., A. W. Rosenbluth, M. N. Rosenbluth, A. H. Teller, and E. Teller. 1953. Equation of state calculations by fast computing machines. *J. Chem. Physics* **21**: 1087–1092.

Mi, J. 1994. Estimation related to mean residual life. *J. Nonpar. Stat.* **4**: 179–190.

Miller, L. E. 1956. Table of percentage points of Kolmogorov statistics. *J. Amer. Statist. Assoc.* **51**: 111–121.

Miller, R. G., Jr. 1964. A trustworthy jackknife. *Ann. Math. Stat.* **35**: 1594–1605.

———. 1966. *Simultaneous Statistical Inference.* New York: McGraw-Hill.

———. 1968. Jackknifing variances. *Ann. Math. Stat.* **38**: 567–582.

———. 1974. The jackknife: A review. *Biometrika* **61**: 1–15.

———. 1980. Combining 2 × 2 contingency tables. In *Biostatistics Casebook*. Edited by R. G. Miller, Jr., B. Efron, B. Wm. Brown, Jr., and L. E. Moses. New York: John Wiley, pp. 73–83.
———. 1981a. *Simultaneous Statistical Inference,* 2nd ed. New York: Springer-Verlag.
———. 1981b. *Survival Analysis*. New York: John Wiley.
———. 1983. What price Kaplan-Meier? *Biometrics* **39**: 1077–1082.
Milton, R. C. 1970. *Rank Order Probabilities*. New York: John Wiley.
MINITAB. 1991. *MINITAB Reference Manual, Macintosh Version, Minitab, Inc.* Valley Forge, Penn.: Data Tech Industries, Inc.
Mittal, Y. 1991. Homogeneity of subpopulations and Simpson's paradox. *J. Amer. Statist. Assoc.* **86**: 167–172.
Moeschberger, M. L., and J. P. Klein. 1985. A comparison of several methods of estimating the survival function when there is extreme censoring. *Biometrics* **41**: 253–259.
Molitor, F. T. 1989. Children's acceptance of others' fighting after viewing violent TV and playing cooperatively. M.A. thesis, California State University.
Moore, W., and C. I. Bliss. 1942. A method for determining insecticidal effectiveness using *Aphis rumicis* and certain organic compounds. *J. Econ. Entomol.* **35**: 544–553.
Morton, R. 1978. Regression analysis of life tables and related nonparametric tests. *Biometrika* **65**: 329–333.
Mortazavi, B. 1997. Personal communication.
Moses, L. E. 1963. Rank tests of dispersion. *Ann. Math. Stat.* **34**: 973–983.
———. 1964. One sample limits of some two-sample rank tests. *J. Amer. Statist. Assoc.* **59**: 645–651.
Mosteller, F. 1952. Some statistical procedures in measuring the subjective response to drugs. *Biometrics* **8**: 220–226.
Müller, H. G. 1988. *Nonparametric Regression Analysis of Longitudinal Data*. Lecture Notes in Statistics, Volume 46. New York: Springer-Verlag.
Murphy, S. 1995. Likelihood ratio based confidence intervals in survival analysis. *J. Amer. Statist. Assoc.* **90**: 1399–1405.
Nair, V. 1981. Plots and tests for goodness of fit with randomly censored data. *Biometrika* **68**: 99–103.
———. 1984. Confidence bands for survival functions with censored data: A comparative study. *Technometrics* **26**: 265–275.
Neave, H. R., and P. L. Worthington. 1988. *Distribution-Free Tests*. London: Unwin Hyman.
Nelson, W. 1969. Hazard plotting for incomplete failure data. *J. Qual. Technol.* **1**: 25–72.
———. 1972. Theory and applications of hazard plotting for censored failure data. *Technometrics* **14**: 945–965.
Nelson, W. B. 1982. *Applied Life Data Analysis*. New York: John Wiley.
Nemenyi, P. 1963. Distribution-free multiple comparisons. Ph.D. diss., Princeton University.
Neyman, J. 1937. Outline of a theory of statistical estimation based on the classical theory of probability. *Philos. Trans. A* **236**: 333–380.
Nicholls, G. H., and D. Ling. 1982. Cued speech and the reception of spoken language. *J. Speech Hearing Res.* **25**: 262–269.
Nikitin, Y. 1995. *Asymptotic Efficiency of Nonparametric Tests*. New York: Cambridge University Press.
Noether, G. E. 1955. On a theorem of Pitman. *Ann. Math. Stat.* **26**: 64–68.
———. 1963. Note on the Kolmorogov statistic in the discrete case. *Metrika* **7**: 115–116.
———. 1967a. *Elements of Nonparametric Statistics*. New York: John Wiley.
———. 1967b. Wilcoxon confidence intervals for location parameters in the discrete case. *J. Amer. Statist. Assoc.* **62**: 184–188.
———. 1987. Sample size determination for some common nonparametric tests. *J. Amer. Statist. Assoc.* **82**: 645–647.
Norman, R. D. 1966. A revised formula for the Wechsler adult intelligence scale. *J. Clin. Psychol.* **22**: 287–294.
Obenchain, R. L. 1969. Rank tests invariant only under linear transformations. Mimeo Ser. 617, Institute of Statistics, University of North Carolina.
Odeh, R. E. 1971. On Jonckheere's k-sample test against ordered alternatives. *Technometrics* **13**: 912–918.
———. 1977a. Extended tables of the distribution of Friedman's S-statistic in the two-way layout. *Comm. Stat. Simulation Comput.* **6**: 29–48.
———. 1977b. The exact distribution of Page's L-statistic in the two-way layout. *Comm. Stat. Simulation Comput.* **6**: 49–61.
———. 1977c. Extended tables of the distributions of rank statistics for treatment versus control in randomized block designs. *Comm. Stat. Simulation Comput.* **6**: 101–113.

Oppenheim, R. W. 1968. Light responsivity in chick and duck embryos just prior to hatching. *Anim. Behav.* **16**: 276–280.

Otten, A. 1973a. The null distribution of Spearman's S when $n = 13(1)16$. *Stat. Neerl.* **27**: 19–20.

———. 1973b. Note on the Spearman rank correlation coefficient. *J. Amer. Statist. Assoc.* **68**: 585.

Owen, D. B. 1962. *Handbook of Statistical Tables*. Reading, Mass.: Addison-Wesley.

Owen, D. B., K. J. Craswell, and D. L. Hanson. 1964. Nonparametric upper confidence bounds for $P\{Y < X\}$ and confidence limits for $P\{Y < X\}$ when X and Y are normal. *J. Amer. Statist. Assoc.* **59**: 906–924.

Page, E. B. 1963. Ordered hypotheses for multiple treatments: A significance test for linear ranks. *J. Amer. Statist. Assoc.* **58**: 216–230.

Pan, G. 1996. Distribution-free confidence procedure for umbrella orderings. *Austral. J. Stat.* **38**: 161–172.

Parzen, E. 1962. On estimation of a probability density function and mode. *Ann. Math. Stat.* **33**: 1065–1076.

Pearson, E. S. 1931. The analysis of variance in cases of non-normal variation. *Biometrika* **23**: 114–133.

Pearson, K. 1900. On the criterion that a given system of deviations from the probable in the case of a correlated system of variables is such that it can be reasonably supposed to have arisen from random sampling. *Phil. Mag., Ser. 5* **50**: 157–175.

———. 1911. On the probability that two independent distributions of frequency are really samples from the same population. *Biometrika* **8**: 250–254.

Petersen, A. V. 1977. Expressing the Kaplan-Meier estimator as a function of empirical subsurvival functions. *J. Amer. Statist. Assoc.* **90**: 1399–1405.

Peto, R., and J. Peto. 1972. Asymptotically efficient rank invariant test procedures (with discussion). *J. Roy. Statist. Assoc.* **72**: 854–858.

Pettitt, A. N., and V. Siskind. 1981. Effect of within-sample dependence on the Mann-Whitney-Wilcoxon statistic. *Biometrika* **68**: 437–441.

Pettitt, A. N., and M. A. Stephens. 1977. The Kolmogorov-Smirnov goodness-of-fit statistic with discrete and grouped data. *Technometrics* **19**: 205–210.

Pitman, E. J. G. 1948. Notes on non-parametric statistical inference. Columbia University (duplicated).

Poland, A., D. Smith, R. Kuntzman, M. Jacobson, and A. H. Conney. 1970. Effect of intensive occupational exposure to DDT on phenylbutazone and cortisol metabolism in human subjects. *Clin. Pharmacol. Ther.* **11**: 724–732.

Potthoff, R. F. 1974. A non-parametric test of whether two simple regression lines are parallel. *Ann. Stat.* **2**: 295–310.

Pratt, J. W. 1959. Remarks on zeros and ties in the Wilcoxon signed rank procedures. *J. Amer. Statist. Assoc.* **54**: 655–667.

———. 1964. Robustness of some procedures for the two-sample location problem. *J. Amer. Statist. Assoc.* **59**: 665–680.

Pratt, J. W., and J. D. Gibbons. 1981. *Concepts of Nonparametric Theory*. New York: Springer-Verlag.

Prentice, M. J. 1979. On the problem of m incomplete rankings. *Biometrika* **66**: 167–170.

Prentice, R. L. 1978. Linear rank tests with right censored data. *Biometrika* **65**: 167–179.

Proschan, F. 1963. Theoretical explanation of observed decreasing failure rate. *Technometrics* **5**: 375–383.

Puri, M. L. 1965. Some distribution-free k-sample rank tests of homogeneity against ordered alternatives. *Comm. Pure Appl. Math.* **18**: 51–63.

Puri, M. L., and P. K. Sen. 1968. On Chernoff-Savage tests for ordered alternatives in randomized blocks. *Ann. Math. Stat.* **39**: 967–972.

———. 1971. *Nonparametric Methods in Multivariate Analysis*. New York: John Wiley.

Quenouille, M. H. 1949. Approximate tests of correlation in time-series. *J. Roy. Statist. Soc. B* **11**: 68–84.

Ramachandramurty, P. V. 1966a. On some nonparametric estimates for shift in the Behrens-Fisher situation. *Ann. Math. Stat.* **37**: 593–610.

———. 1966b. On the Pitman efficiency of one-sided Kolmogorov and Smirnov tests for normal alternatives. *Ann. Math. Stat.* **37**: 940–944.

Ramsay, W. N. M. 1957. The determination of iron in blood plasma or serum. *Clin. Chim. Acta* **2**: 214–220.

Randles, R. H., M. A. Fligner, G. E. Policello II, and D. A. Wolfe. 1980. An asymptotically distribution-free test for symmetry versus asymmetry. *J. Amer. Statist. Assoc.* **75**: 168–172.

Randles, R. H., and R. V. Hogg. 1971. Certain uncorrelated and independent rank statistics. *J. Amer. Statist. Assoc.* **66**: 569–574.

Randles, R. H., and D. A. Wolfe. 1979. *Introduction to the Theory of Nonparametric Statistics*. New York: John Wiley.

Rao, K. S. M., and A. P. Gore. 1984. Testing concurrence and parallelism of several sample regressions against ordered alternatives. *Math. Oper.-forsch. und Stat. Series Stat.* **15**: 43–50.

Rasekh, J., A. Kramer, and R. Finch. 1970. Objective evaluation of canned tuna sensory quality. *J. Food Sci.* **35**: 417–423.

Reed, O. M. 1973. *Papio Cynocephalus* age determinations. *Amer. J. Phys. Anthropol.* **38**: 309–314.

Rice, J. A. 1988. *Mathematical Statistics and Data Analysis.* Pacific Grove, Calif.: Wadsworth & Brooks/Cole.

Robertson, T., F. T. Wright, and R. L. Dykstra. 1988. *Order Restricted Statistical Inference.* New York: John Wiley.

Robins, J., N. Breslow, and S. Greenland. 1986. Estimators of the Mantel-Haenszel variance consistent in both spare data and large-strata limiting models. *Biometrics* **42**: 311–323.

Rosenblatt, M. 1956. Remarks on some nonparametric estimates of a density function. *Ann. Math. Stat.* **27**: 832–837.

Rust, S. W., and M. A. Fligner. 1984. A modification of the Kruskal-Wallis statistic for the generalized Behrens-Fisher problem. *Comm. Stat. Theory Methods* **13** 2013–2028.

Ryan, T. P. 1997. *Modern Regression Methods.* New York: John Wiley.

Salsburg, D. S. 1970. Personal communication (with the cooperation of Pfizer and Co., Groton, Conn.).

Samara, B., and R. H. Randles. 1988. A test for correlation based on Kendall's tau. *Comm. Stat. Theory Methods* **17**: 3191–3205.

Santner, T. J. 1998. Teaching large-sample binomial confidence intervals. *Teaching Statistics* **20**: 20–23.

Sauber, S. R. 1971. Approaches to precounseling and therapy training: an investigation of its potential influence on process outcome. Ph.D. diss, Florida State University.

Savage, I. R. 1953. Bibliography of nonparametric statistics and related topics. *J. Amer. Statist. Assoc.* **48**: 844–906. Correction: **53** (1958): 1031.

———. 1956. Contributions to the theory of rank order statistics—the two-sample case. *Ann. Math. Stat.* **27**: 590–615.

———. 1962. *Bibliography of Nonparametric Statistics.* Cambridge: Harvard University Press.

Saxena, K. M. L. 1969. Use of sign statistic in problems concerning $P(Y < X)$. Abstr. in *Ann. Math. Stat.* **40**: 1154.

Scheffé, H. 1943. Statistical inference in the non-parametric case. *Ann. Math. Stat.* **14**: 305–332.

———. 1959. *The Analysis of Variance.* New York: John Wiley.

Scheffé, H., and J. W. Tukey. 1945. Non-parametric estimation. I. Validation of order statistics. *Ann. Math. Stat.* **16**: 187–192.

Schoenfeld, D. 1981. The asymptotic properties of nonparametric tests for comparing survival distributions. *Biometrika* **68**: 316–319.

Schonrock, H. 1996. Personal communication.

Schuster, E. 1974. On the rate of convergence of an estimate of a functional of a probability density. *Scand. Actuar. J.* **1**: 103–107.

Schweder, T. 1975. Window estimation of the asymptotic variance of rank estimators of location. *Scand. J. Stat.* **2**: 113–126.

———. 1981. Correction note. *Scand. J. Stat.* **8**: 55.

Sen, P. K. (1967). A note on asymptotically distribution-free confidence bounds for $Pr(X < Y)$, based on two independent samples. *Sankhya A* **29**: 95–102.

———. 1968. Estimates of the regression coefficient based on Kendall's tau. *J. Amer. Statist. Assoc.* **63**: 1379–1389.

———. 1969. On a class of rank order tests for the parallelism of several regression lines. *Ann. Math. Stat.* **40**: 1668–1683.

———. 1981. *Sequential Nonparametrics: Invariance Principles and Statistical Inference.* New York: John Wiley.

Senchaudhuri, P., C. R. Mehta, and N. R. Patel. 1995. Estimating exact p values by the method of control variates or Monte Carlo rescue. *J. Amer. Statist. Assoc.* **90**: 640–648.

Serfling, R. J. 1968. The Wilcoxon two-sample statistic on strongly mixing processes. *Ann. Math. Stat.* **39**: 1202–1209.

———. 1984. Generalized L-, M-, and R-statistics. *Ann. Stat.* **12**: 76–86.

———. 1992. Nonparametric confidence intervals for generalized quantile parameters in multi-sample contexts. In *Nonparametric Statistics and Related Topics.* Edited by A. K. Md. E. Saleh. New York: North Holland, pp. 121–139.

Shao, J. 1988. Consistency of jackknife estimators of the variances of sample quantiles. *Comm. Stat. Theory Methods* **17**: 3017–3028.

Shao, J., and C. F. J. Wu. 1989. A general theory for jackknife variance estimation. *Ann. Stat.* **17**: 1176–1197.

Shelp, W. D., F. H. Bach, W. A. Kisken, M. Newton, R. E. Rieselbach, and A. B. Weinstein. 1970. Long-term integrity of renal function in cadaver allografts *J. Amer. Med. Assoc.* **213**: 1443–1447.

Shen, S., G. M. Reaven, J. W. Farquhar, and R. H. Nakanishi. 1970. Comparison of impedance to insulin-mediated glucose uptake in normal subjects and in subjects with latent diabetes. *J. Clin. Invest.* **49**: 2151–2160.

Shen, Z. Q., X. H. Feng, Z. X. Qian, R. L. Liu, and C. R. Yang. 1988. Application of biotinadvin system, determination of circulating immune complexes, and evaluation of antibody response in different hydatidosis patients. *Amer. J. Trop. Med. Hyg.* **39**: 93–96.

Sherman, E. 1965. A note on multiple comparisons using rank sums. *Technometrics* **7**: 255–256.

Shlafer, M., and A. M. Karow, Jr. 1971. Ultrastructure-function correlative studies for cardiac cryopreservation II. Hearts frozen to various temperatures without a cryoprotectant. *Cryobiology* **8**: 350–360.

Shorack, G. R. 1965. Nonparametric tests and estimation of scale in the two sample problem. Tech. Rept. 10, Department of Statistics, Stanford University.

———. 1969. Testing and estimating ratios of scale parameters. *J. Amer. Statist. Assoc.* **64**: 999–1013.

Shorack, G. R., and J. A. Wellner. 1986. *Empirical Processes.* New York: John Wiley.

Siddiqui, M. M., and E. A. Gehan. 1966. *Statistical Methodology for Survival Time Studies.* Communication of National Cancer Institute.

Siegel, S., and N. J. Castellan. 1988. *Nonparametric Statistics for the Behavioral Sciences,* 2nd ed. New York: McGraw Hill.

Sillitto, G. P. 1947. The distribution of Kendall's τ coefficient of rank correlation in rankings containing ties. *Biometrika* **34**: 36–40.

Silver, H., N. F. Colovos, J. B. Holter, and H. H. Hayes. 1969. Fasting metabolism of white-tailed deer. *J. Wildl. Mgmt.* **33**: 263–274.

Simpson, D. G., and B. H. Margolin. 1986. Recursive nonparametric testing for dose-response relationships subject to downturns at high doses. *Biometrika* **73**: 589–596.

Skaug, H. J., and D. Tjøstheim. 1993. A nonparametric test of serial independence based on the empirical distribution function. *Biometrika* **80**: 591–602.

Skibinsky, M., and L. Cote. 1963. On the inadmissibility of some standard estimates in the presence of prior information. *Ann. Math. Stat.* **34**: 539–548.

Skillings, J. H. 1980. On the null distribution of Jonckheere's statistic used in two-way models for ordered alternatives. *Technometrics* **22**: 431–436.

Skillings, J. H., and G. A. Mack. 1981. On the use of a Friedman-type statistic in balanced and unbalanced block designs. *Technometrics* **23**: 171–177.

Skillings, J. H., and D. A. Wolfe. 1977. Testing for ordered alternatives by combining independent distribution-free block statistics. *Comm. Stat. Theory Methods* **6**: 1453–1463.

———. 1978. Distribution-free tests for ordered alternatives in a randomized block design. *J. Amer. Statist. Assoc.* **73**: 427–431.

Smid, L. J. 1956. On the distribution of the test statistics of Kendall and Wilcoxon when ties are present. *Stat. Neerl.* **10**: 205–214.

Smirnov, N. V. 1935. Ueber die verteilung des allgemeinen gliedes in der variationsreihe. *Metron* **12**: 59–81.

———. 1939. On the estimation of the discrepancy between empirical curves of distribution for two independent samples. (Russian) *Bull Moscow Univ.* **2**: 3–16.

———. 1948. Table for estimating the goodness of fit of empirical distributions. *Ann. Math. Stat.* **19**: 279–281.

Smith, A. F. M., and G. O. Roberts. 1993. Bayesian computation via the Gibbs sampler and related Markov chain Monte Carlo methods. *J. Roy. Statist. Soc. B* **55**: 3–23.

Smith, E. J. 1967. Cloud seeding experiments in Australia. *Proc. 5th Berkeley Symp.* **V**: 161–176.

Smith, J. D. 1969. Geomorphology of a sand ridge. *J. Geology* **77**: 39–55.

Smith, P. L. 1979. Splines as a useful and convenient statistical tool. *Amer. Statistician* **33**: 57–62.

Spearman, C. 1904. The proof and measurement of association between two things. *Amer. J. Psychol.* **15**: 72–101.

Spector, P. 1994. *An Introduction to S and S-PLUS.* Belmont, Calif.: Duxbury Press.

Spjøtvoll, E. 1968. A note on robust estimation in analysis of variance. *Ann. Math. Stat.* **39**: 1486–1492.

S-PLUS. 1995. S-PLUS Documentation. Seattle: Statistical Sciences, Inc.

Sposto, R., and M. Krailo. 1987. Use of unequal allocation in survival trials. *Stat. Med.* **6**: 119–125.

Sprent, P. 1993. *Applied Nonparametric Statistical Methods,* 2nd ed. London: Chapman & Hall.

Spurrier, J. D. 1991. Improved bounds for the moments of some rank statistics. *Comm. Stat. Theory Methods* **20**: 2603–2608.

Stanton, J. M. 1969. Murderers on parole. *Crime and Delinquency* **15**: 149–155.

Steel, R. G. D. 1959. A multiple comparison rank sum test. Treatments versus control. *Biometrics* **15**: 560–572.

———. 1960. A rank sum test for comparing all pairs of treatments. *Technometrics* **2**: 197–207.

———. 1961. Some rank sum multiple comparisons tests. *Biometrics* **17**: 539–552.

Stephens, M. A. 1974. EDF statistics for goodness of fit and some comparisons. *J. Amer. Statist. Assoc.* **69**: 730–735.

———. 1979. Tests of fit for the logistic distribution based on the empirical distribution function. *Biometrika* **66**: 591–595.

———. 1983a. Kolmogorov-Smirnov statistics. In *Encyclopedia of Statistical Sciences,* Volume 4. Edited by S. Kotz, N. L. Johnson, and C. B. Read. New York: John Wiley, pp. 393–396.

———. 1983b. Kolmogorov-Smirnov-type tests of fit. In *Encyclopedia of Statistical Sciences,* Volume 4. Edited by S. Kotz, N. L. Johnson, and C. B. Read. New York: John Wiley, pp. 398–402.

Sterne, T. E. 1954. Some remarks on confidence or fiducial limits. *Biometrika* **41**: 275–278.

Sternhell, S. 1958. Chemistry of brown coals VI: Further aspects of the chemistry of hydroxyl groups in Victorian brown coals. *Austral. J. Appl. Sci.* **9**: 375–379.

Stigler, S. M. 1974. Linear functions of order statistics with smooth weight functions. *Ann. Stat.* **2**: 676–693.

Stitt, J. T., J. D. Hardy, and E. R. Nadel. 1971. Surface area of the squirrel monkey in relation to body weight. *J. Appl. Physiol.* **31**: 140–141.

Stone, C. 1994. The use of polynomial splines and their tensor products in multivariate function estimation. *Ann. Stat.* **22**: 118–184.

Storer, B. E., and C. Kim. 1990. Exact properties of some exact tests for comparing two binomial proportions. *J. Amer. Statist. Assoc.* **85**: 146–155.

Strawderman, R. L. 1997. An asymptotic analysis of the logrank test. *Lifetime Data Analysis* **3**: 225–249.

Strawderman, R. L., and C. R. Mehta. 1992. On the validation of exact tests for nonparametric inference. *Comput. Stat. Data Anal.* **14**: 263–266.

Stuart, A. 1954. The asymptotic relative efficiencies of tests and the derivatives of their power functions. *Skand. Aktuarietidskr.* **37**: 163–169.

Suissa, S., and J. J. Shuster. 1985. Exact unconditional sample sizes for the 2 by 2 binomial trial. *J. Roy. Statist. Soc. A* **148**: 317–327.

Susarla, V., and J. van Ryzin. 1976. Nonparametric Bayesian estimation of survival curves from incomplete observations. *J. Amer. Statist. Assoc.* **71**: 897–902.

Switzer, P. 1976. Confidence procedures for two-sample problems. *Biometrika* **63**: 13–25.

Sylvester, P. E. 1969. Pyramidal, lemniscal and parietal lobe status in cerebral palsy. *J. Ment. Defic. Res.* **13**: 20–33.

Tanner, M., and W. Wong. 1987. The calculation of posterior distributions by data augmentation (with discussion). *J. Amer. Statist. Assoc.* **82**: 528–550.

Tarone, R. E., J. J. Gart, and W. W. Hauck. 1983. On the asymptotic relative efficiency of certain noniterative estimators of a common relative risk or odds ratio. *Biometrika* **70**: 519–522.

Tarone, R. E., and J. Ware. 1977. On distribution-free tests for equality of survival distributions. *Biometrika* **64**: 156–160.

Tate, M. W., and R. C. Clelland. 1957. *Nonparametric and Shortcut Statistics*. Danville, Ill.: The Interstate.

Tate, M. W., W. L. Cullinan, and A. Ahlstrand. 1961. Measurement of adaptation in stuttering. *J. Speech Hearing Res.* **4**: 321–339.

Terpstra, T. J. 1952. The asymptotic normality and consistency of Kendall's test against trend, when ties are present in one ranking. *Indag. Math.* **14**: 327–333.

Terry, M. E. 1952. Some rank order tests which are most powerful against specific parametric alternatives. *Ann. Math. Stat.* **23**: 346–366.

Theil, H. 1950a. A rank-invariant method of linear and polynomial regression analysis, I. *Proc. Kon. Ned. Akad. v. Wetensch. A* **53**: 386–392.

———. 1950b. A rank-invariant method of linear and polynomial regression analysis, II. *Proc. Kon. Ned. Akad. v. Wetensch. A* **53**: 521–525.

———. 1950c. A rank-invariant method of linear and polynomial regression analysis, III. *Proc. Kon. Ned. Akad. v. Wetensch. A* **53**: 1397–1412.

Thomas, D. R., and G. L. Grunkemeier. 1975. Confidence interval estimation of survival probabilities for censored data. *J. Amer. Statist. Assoc.* **70**: 865–871.

Thomas, H. V., and E. Simmons. 1969. Histamine content in sputum from allergic and nonallergic individuals. *J. Appl. Physiol.* **26**: 793–797.

Thomson, M. L., and M. D. Short. 1969. Mucociliary function in health, chronic obstructive airway disease, and asbestosis. *J. Appl. Physiol.* **26**: 535–539.

Tocher, K. D. 1950. Extension of the Neyman-Pearson theory of tests to discontinuous variates. *Biometrika* **37**: 130–144.

Tukey, J. W. 1949. The simplest signed-rank tests. Memo Rept. 17, Statistical Research Group, Princeton University.

———. 1953. The problem of multiple comparisons. Unpublished manuscript.

———. 1958. Bias and confidence in not-quite large samples. Abstr. in *Ann. Math. Stat.* **29**: 614.

———. 1962. Data analysis and behavioral science. Unpublished manuscript.

Ujah, J. E., and K. B. Adeoye. 1984. Effects of shelterbelts in the Sudan Savanna zone of Nigeria on microclimate and yield of millet. *Agric. Forest Meteor.* **33**: 99–107.

van Belle, G. 1972. Personal communication.

van Dantzig, D. 1951. On the consistency and the power of Wilcoxon's two sample test. *Indag. Math.* **13**: 1–8.

van der Vaart, A. W. 1988. Statistical estimation for large parameter spaces. CWI Tracts 44, Centre for Mathematics and Computer Science, Amsterdam.

———. 1991. Efficiency and Hadamard differentiability. *Scand. J. Stat.* **18**: 63–75.

van der Waerden, B. L., and E. Nievergelt. 1956. *Tafeln zum Vergleich Zweier Stichproben-mittels X-test und Zeichentest.* Berlin: Springer-Verlag, OHG.

van Elteren, P., and G. P. Noether. 1959. The asymptotic efficiency of the χ_r^2-test for a balanced incomplete block design. *Biometrika* **46**: 475–477.

Venables, W. N., and B. D. Ripley. 1997. *Applied Statistics with S-Plus,* 2nd ed. New York: Springer-Verlag.

Veterans Administration Cooperative Urological Research Group. 1967. Treatment and survival of patients with cancer of the prostate. *Surg., Gynec. Obstet.* **124**: 1011–1017.

Vianna, N. J., P. Greenwald, and J. M. P. Davies. 1971. Tonsillectomy and Hodgkin's disease: The lymphoid tissue barrier. *Lancet* **1**: 431–432.

Wahba, G. 1990. *Spline Functions for Observational Data.* CBMS-NSF Regional Conference Series, Volume 59. Philadelphia: SIAM.

Walker, H. M., and J. Lev. 1953. *Statistical Inference,* 1st. ed. New York: Holt, Rinehart & Winston.

Walsh, J. E. 1949. Some significance tests for the median which are valid under very general conditions. *Ann. Math. Stat.* **20**: 64–81.

———. 1962. *Handbook of Nonparametric Statistics.* Princeton, N.J.: Van Nostrand.

———. 1963. Bounded probability properties of Kolmogorov-Smirnov and similar statistics for discrete data. *Ann. Inst. Statist. Math.* **15**: 153–158.

———. 1965. *Handbook of Nonparametric Statistics, II.* Princeton, N.J.: Van Nostrand.

———. 1968. *Handbook of Nonparametric Statistics, III.* Princeton, N.J.: Van Nostrand.

Wand, M. P., and M. C. Jones. 1995. *Kernel Smoothing.* London: Chapman & Hall.

Wang, J. G. 1987. A note on the uniform consistency of the Kaplan-Meier estimator. *Ann. Stat.* **15**: 1313–1316.

Wang, J. L. 1987. Estimators of a distribution function with increasing failure rate average. *J. Statist. Plan. Infer.* **16**: 415–427.

Wardlaw, A. C., and P. J. Moloney. 1961. The assay of insulin with anti-insulin and mouse diaphragm. *Can. J. Biochem. Physiol.* **39**: 695–712.

Wardlaw, A. C., and G. van Belle. 1964. Statistical aspects of the mouse diaphragm test for insulin. *Diabetes* **13**: 622–633.

Wegman, E. J., and I. W. Wright. 1983. Splines in statistics. *J. Amer. Statist. Assoc.* **78**: 351–365.

Weindling, S. 1977. Personal communication to J. A. Rice—statistics report: Math 80B.

Welch, B. L. 1937. The significance of the difference between two means when the population variances are unequal. *Biometrika* **29**: 350–362.

———. 1947. The generalization of "Student's" problem when several different populations are involved. *Biometrika* **34**: 28–35.

Wellner, J. A. 1982. Asymptotic optimality of the product limit estimator. *Ann. Stat.* **10**: 595–602.

———. 1985. A heavy censoring limit theorem for the product limit estimator. *Ann. Stat.* **13**: 150–162.

Wells, J. M., and M. A. Wells. 1967. Note on Project SCUD. *Proc. 5th Berkeley Symp.* **V**: 357–369.

Wells, M. T., and R. C. Tiwari. 1989. Bayesian quantile plots and statistical inference for nonlinear models in the two-sample case with incomplete data. *Comm. Stat. Theory Methods* **18**: 2955–2964.

West, M. 1992. Modelling time-varying hazards and covariate effect. In: *Survival Analysis: State of the Art*. Edited by J. P. Klein and P. K. Goel. Boston: Kluwer, pp. 47–62.

Whitaker, L. R., and F. J. Samaniego. 1989. Estimating a survival curve when new is better than used in expectation. *Nav. Res. Logistics* **36**: 693–707.

Wilcoxon, F. 1945. Individual comparisons by ranking methods. *Biometrics* **1**: 80–83.

Wilcoxon, F., S. K. Katti, and R. A. Wilcox. 1973. Critical values and probability levels for the Wilcoxon rank sum test and the Wilcoxon signed rank test. In *Selected Tables in Mathematical Statistics,* Volume 1 (2nd printing with rev.). Edited by H. L. Harter and D. B. Owen. Providence, R.I.: American Mathematical Society and Institute of Mathematical Statistics, pp. 171–260.

Wilcoxon, F., and R. A. Wilcox. 1964. *Some Rapid Approximate Statistical Procedures,* 2nd ed. Pearl River, N.Y.: American Cyanamid Co., Lederle Laboratories.

Wilks, S. S. 1962. *Mathematical Statistics*. New York: John Wiley.

Windham, B. M. 1971. Personal communication.

Wolfe, B. E., and J. D. Maser. 1994. Treatment of panic disorder: Consensus statement. In *Treatment of Panic Disorder: A Consensus Development Conference*. Edited by B. E. Wolfe and J. D. Maser. Washington, D.C.: American Psychiatric Press, pp. 237–255.

Wolfe, D. A. 1977. A distribution-free test for related correlation coefficients. *Technometrics* **19**: 507–509.

Wolfe, D. A., and R. V. Hogg. 1971. On constructing statistics and reporting data. *Amer. Statistician* **25**: 27–30.

Woodward, W. F. 1970. A comparison of base running methods in baseball. M.Sc. thesis, Florida State University.

Wu, C. F. J. 1990. On the asymptotic properties of the jackknife histogram. *Ann. Stat.* **18**: 1438–1452.

Yanagimoto, T. 1970. On measures of association and a related problem. *Ann. Inst. Statist. Math.* **22**: 57–63.

Yang, G. L. 1978. Estimation of a biometric function. *Ann. Stat.* **6**: 112–116.

Yates, F. 1934. Contingency tables involving small numbers and the χ^2 test. *J. Roy. Statist. Soc. Suppl.* **1**: 217–235.

Ying, Z. 1989. A note on the asymptotic properties of the product-limit estimator on the whole line. *Stat. Prob. Letters* **7**: 311–314.

Yu, C. S. 1971. Pitman efficiencies of Kolmogorov-Smirnov tests. *Ann. Math. Stat.* **42**: 1595–1605.

Yule, G. U. 1900. On the association of attributes in statistics. *Philos. Trans. Roy. Soc. London A* **194**: 257–319.

———. 1912. On the methods of measuring association between two attributes (with discussion). *J. Roy. Statist. Soc.* **75**: 579–642.

Zacks, S. 1992. *Introduction to Reliability Analysis: Probability Models and Statistical Methods*. New York: Springer-Verlag.

Zar, J. H. 1972. Significance testing of the Spearman rank correlation coefficient. *J. Amer. Statist. Assoc.* **67**: 578–580.

Zelen, M. 1971. The analysis of several 2 × 2 contingency tables. *Biometrika* **58**: 129–137.

Zheng, J. X. 1997. A consistent specification test of independence. *J. Nonpar. Stat.* **7**: 297–306.

Zhou, S., and D. A. Wolfe. 1997. On multivariate spline regression. Tech. Rpt., Department of Statistics, Ohio State University.

Answers to Selected Problems

Chapter 2

2.1 $B^* = 4.304$; approximate P-value $< .0002$

2.5 (a) .2898
(b) .7102
(c) .9667

2.10 $\hat{p} = .8615$ and $\hat{sd}(\hat{p}) = .0428$

2.13 $n = 664$

2.14 approximate 96% CI is $(.774, .950)$

2.22 $E_p(\ell) = .6103$

Chapter 3

3.1 $T^+ = 24$; P-value $= .055$

3.8 $n = 59$

3.19 $\hat{\theta} = 80$

3.27 95.4% CI is $(W^{(3)}, W^{(26)}) = (-25, 605)$

3.39 approximate 99% lower CB is $W^{(1)} = -158$

3.43 $B = 5$; P-value $= .1094$

3.48 .1727

3.58 $\bar{\theta} = 2.5$

3.70 96.88% CI is $(Z^{(1)}, Z^{(6)}) = (-1.5, 8)$

3.77 94.61% lower CB is $Z^{(9)} = 7.5$

3.87 $T^+ = 33.5$; P-value $= .734$; $\hat{\theta} = 17.85$; 93.6% CI is $(W^{(16)}, W^{(63)}) = (17.5, 21.25)$

3.94 $B = 4$; P-value $= .1938$; $\tilde{\theta} = .205$; 96.14% CI is $(Z^{(3)}, Z^{(10)}) = (.15, .31)$

3.103 $V = .173$; approximate P-value $= .864$

3.110 $A_{\text{obs}} = \frac{8}{64}$; nonrandomized P-value $= 1$

Chapter 4

4.1 $W^* = 3.17$; approximate P-value $= .0008$

4.6 approximate power $= .3228$

4.15 $\hat{\Delta} = U^{(59)} = 54.3$

4.18 $\hat{\delta} = .9060$; approximate 90% CI is $(.790, 1)$
4.27 approximate 95% CI is $(U^{(29)}, U^{(89)}) = (22.1, 95.8)$
4.28 $\hat{\Delta} = [U^{(9)} + U^{(42)}]/2 = -.305$ (based on a 96% CI)
4.37 $\hat{U} = -1.348$; P-value $> .098$
4.38 $\hat{U} = -5.258$; P-value $< .01$

Chapter 5

5.1 $C = 28$; P-value $= .0686 + .0803 = .1489$
5.9 modified $C^* = .26$; P-value $\approx .7948$
5.13 $Q = .0702$; P-value $\approx .4721$
5.17 $\tilde{\gamma}^2 = 1.0809$
5.24 approximate 91.92% lower CB is $\gamma_L^2 = .2288$
5.25 $D = 2.442$; P-value $= .23$
5.33 $J = 80$; P-value $= .0024$
5.40 $J = 39$; P-value $> .1145$; $J^* = .713$; P-value $\approx .6945$

Chapter 6

6.1 $H' = 4.2645$; P-value $\approx .24$
6.12 $H' = 18.603$; P-value $< .001$
6.13 $J^* = 1.89$; P-value $\approx .0294$
6.22 $J = 73$; P-value $= .0153$
6.28 (a) Maximum number is $k(k-1)/2$ for any k and $p = 1$ or k
    (b) Minimum number is $(k^2 - 1)/4$ if k is odd and $p = (k+1)/2$
    Minimum number is $k^2/4$ if k is even and $p = k/2$
6.36 $\hat{p} = 2$; $A_2^* = 5.21$; P-value $< .01$
6.42 $FW^* = 1.5$; P-value $\approx .0668$
6.46 decide $\tau_{SP} \neq \tau_{BS}, \tau_{SP} \neq \tau_{WP}$, and $\tau_{BS} = \tau_{WP}$ at approximate EER $= .05$
6.54 smallest approximate EER to detect most significant difference is $.0005$
6.56 at EER $= .05$, decide $\tau_u = t_v$ for all $1 \leq u < v \leq 4$
6.63 at EER $= .05747$, decide $\tau_{TR} = \tau_C, \tau_{VTP} = \tau_C$, and $\tau_{RII} > \tau_C$
6.79 $\widehat{\tau_2 - \tau_1} = .62375$; $\widehat{\tau_4 - \tau_1} = .72125$; $\widehat{\tau_5 - \tau_1} = .65406$
6.85 simultaneous approximate 95% CI's are:

$\tau_1 - \tau_2 \in [-5, 16)$; $\tau_1 - \tau_3 \in [-18, -4)$; $\tau_1 - \tau_4 \in [-18, -5)$;
$\tau_2 - \tau_3 \in [-27, -4)$; $\tau_2 - \tau_4 \in [-28, -5)$; $\tau_3 - \tau_4 \in [-4, 3)$

Chapter 7

7.1 P-value $\approx .166$
7.14 P-value $< .00005$
7.17 Reject H_0 at level $\alpha = .01$
7.30 With EER $= .038$, decide $\tau_1 = \tau_2$, $\tau_2 \neq \tau_3$, and $\tau_1 = \tau_3$
7.32 Smallest EER $= .10$

ANSWERS TO SELECTED PROBLEMS

7.39 At EER = .009, decide $\tau_2 = \tau_1$ and $\tau_3 > \tau_1$
7.48 $\tilde{\theta} = 20$
7.58 $D = 12$; do not reject H_0 at approximate level $\alpha = .05$
7.70 At approximate EER = .01, decide $\tau_u = \tau_v$, for every (u, v) pair, $1 \leq u \neq v \leq 7$
7.76 SM = 38.057; Reject H_0 at level $\alpha = .05$
7.77 SM = 15.439; Reject H_0 at level $\alpha = .05$
7.89 MS = 6.540; Approximate P-value = .01
7.104 At approximate EER = .05, decide $\tau_u = \tau_v$, for every (u, v) pair, $1 \leq u \neq v \leq 5$
7.111 $D = 0.9131$; Approximate P-value = .63
7.116 $Q = 4.086$; Approximate P-value < .00002
7.121 With approximate experimentwise error rate $\alpha = .025$, decide $\tau_1 \neq \tau_4$, $\tau_3 \neq \tau_4$, and $\tau_u = \tau_v$, for all other (u, v) pairs, $1 \leq u < v \leq 4$
7.125 With approximate experimentwise error rate $\alpha = .10$, decide $\tau_A < \tau_N$ and $\tau_R < \tau_N$
7.129 $\hat{\theta} = 21$

Chapter 8

8.1 $K = -7$; P-value = .758
8.9 Approximate power = .3685
8.13 $n = 11$
8.20 $\hat{\tau} = -.1556$
8.21 maximum $\hat{\tau}$ is 1; minimum $\hat{\tau}$ is -1
8.27 approximate 98% CI is $(-.8017, .4905)$
8.30 approximate 95% lower CB is $\tau_L^* = -.0222$
8.34 one possible approximate 90% bootstrap CI is $(-.051, .615)$
8.40 $r_S = .8777$; P-value < .0005
8.41 $r_S = -.2$; P-value > .5
8.56 $D = \frac{2}{315}$; P-value = .0905

Chapter 9

9.1 $C = 28$; P-value < .001
9.6 $C = -71$; do not reject H_0
9.7 $\hat{\beta} = .00000555$
9.13 93.8% CI is $(S^{(7)}, S^{(22)}) = (.0000028, .0000102)$
9.17 97.7% lower CB is $\beta_L^* = S^{(12)} = -1.3333$
9.20 $\hat{\alpha} = .9754$
9.22 estimated density is 1.0725
9.27 $V = 1.7011$; P-value $\approx .1921$
9.34 (a) HM = 2.056; P-value $\approx .1516$
    (b) HM = 13.69; P-value $\approx .0034$
    (c) HM = 1.717; P-value $\approx .1901$
    (d) HM = 40.315; P-value < .0001
9.38 (a) HM = 32.362; P-value < .0001
    (b) HM = 1.839; P-value $\approx .1751$
    (c) HM = 9.79; P-value $\approx .0075$

Chapter 10

10.1 $\chi^2 = 4.967$; P-value $\approx .0258$

10.7 $m = 49$

10.12 P-value $= .107$

10.13 $\chi^2 = 4.444$; P-value $\approx .035$

10.18 $\hat{\theta} = .705$; approximate 95% CI is $(.52, .95)$

10.24 $\hat{Q} = -.173$; approximate 95% CI is $(-.316, -.026)$

10.30 MH $= -3.78$; approximate P-value $< .002$

10.34 approximate 95% CI is $(.495, 6.835)$

Chapter 11

11.1 $\mathcal{E} = 4.222$; do not reject H_0 at level $\alpha = .10$

11.4 $W_1^* = 2.28$; P-value $\approx .0113$

11.11 $T = 189$; P-value $\approx .1646$

11.18 $V' = -1.23$; P-value $> .10$

11.25 Use equation (11.67) to estimate the mean residual life function. For example, the mean residual life function estimates are 87.59 at time $t = 84$ and 90.56 at time $t = 100$.

11.37 Use the S-plus function surv.fit to obtain the Kaplan-Meier estimator of the survival distribution. For example, at times $t_1 = 25$ and $t_2 = 42$, the survival estimates are .791 and .604, respectively.

11.44 Use S-plus function surv.diff to obtain Chisq $= 3.9$ and approximate P-value $= .0484$.

Author Index

Aalen, O. O., 541
Abdushukurov, A. A., 560
Abrams, D., 23
Adams, T., 70, 71
Adeoye, K. B., 3, 293, 294, 326
Adichie, J. N., 428, 434, 435, 446
Agresti, A., 10, 12, 46, 67, 466, 470, 471, 475, 482, 483, 492
Ahlstrand, A., 302
Akritas, M. G., 128, 533
Alder, A., 471
Alexander, D. A., 631, 648
Altman, N. S., 455
Aly, E.-E., 517
Anděl, J., 188
Andersen, P. K., 11, 13, 533, 543, 547, 554, 559
Andersen, S. L., 102, 187
Anderson, J. D., 80
Anderson, T. W., 33
Anderson, V. L., 339
Andersson, L. C., 200, 201
Andrews, B., 471, 472
Andrews, F. C., 199, 268
Ansari, A. R., 155, 156, 187
Anscombe, F. J., 244
Arbuthnott, J., 11, 28
Archambault, W. A. T., Jr., 223
Arjas, E., 12
Arnold, B. C., 74
Arnold, H. J., 49
Astin, M. C., 473
August, G. P., 50

Bach, F. H., 97
Bacon, S. P., 449
Balakrishnan, N., 74
Banford, S. A., 70
Barlow, R. E., 496, 499, 500, 501, 502, 503, 509, 511, 512
Barnard, G. A., 459

Bartholomew, D. J., 501
Bauer, D. F., 156
Bedell, G. M., 282
Benard, A., 315, 325
Bennett, B. M., 474
Benson, G., 41
Berger, R. L., 26, 31, 530
Bergman, B., 501
Bergner, M., 4, 461, 462, 476, 484, 486
Bernoulli, J., 28
Bhattacharyya, G. K., 375
Bick, R. L., 70, 71
Bickel, P. J., 55, 393, 500, 503, 504, 557, 559
Bie, O., 542
Billingsley, P., 518, 543
Birch, M. W., 494
Birnbaum, Z. W., 129, 511, 530
Bjerkedal, T., 503, 522, 523, 526
Björnstad, S., 112
Bliss, C. I., 311, 743
Blomqvist, N., 377
Blum, J. R., 409, 412, 413, 735
Blyth, C., 129
Blyth, C. R., 33
Bohn, L. L., 118
Borgan, Ø., 11, 13, 533, 542, 543, 547, 554, 559
Borges, W. S., 511, 559
Botelho, S. Y., 282
Box, G. E. P., 102, 165, 167, 187
Boyles, R. A., 512
Bradley, R. A., 23, 155, 156, 187
Brady, J. P., 293, 322
Bremner, J. M., 501
Breslow, N., 490, 491, 492, 494
Breslow, N. E., 547
Bridge, L., 471
Brown, B. W., Jr., 536, 540, 556
Brown, G. W., 471
Brown, M., 555

Brunden, M. N., 325
Brunk, H. D., 501
Bryson, M. C., 512, 516
Burnett, W. C., Jr., 161, 162
Burr, E. J., 375
Burstein, H., 33
Bugyi, H. L., 168
Byer, A. J., 23
Byers, S. O., 185, 186

Cain, G. D., 103
Campbell, J. A., 331
Campo, R., 501, 511
Canham, C. D., 22
Capon, J., 185, 188
Carroll, R. J., 12
Casella, G., 23, 31, 32, 33, 530
Castellan, N. J., 12
Chakraborti, S., 12
Chandra, M., 532
Chen, Y. I., 210, 224, 233, 239
Chen, Y. Y., 510, 517, 542, 546
Cheng, P. E., 560
Chu, J., 4, 461, 462, 475, 484, 486
Chernoff, H., 11
Chew, V., 31
Clark, P. J., 379
Clelland, R. C., 12
Cleveland, W. S., 454, 455
Clogg, C. C., 466
Clopper, C. J., 32, 33
Cochran, W. G., 105, 286, 492
Cole, A. F. W., 85, 86
Coleman, E. M., 473
Colovos, N. F., 215
Comroe, J. H., Jr., 282
Conney, A. H., 82, 83, 87
Connor, W. E., 449
Conover, W. J., 12, 85
Cooper, E. B., 301, 302
Cooper, L. M., 70
Cornfield, J., 481
Cote, L., 31
Cox, D. R., 11, 13, 554
Cox, G. M., 286
Craig, A. T., 73
Craswell, K. J., 129
Critchlow, D. E., 12, 246, 248, 267, 666, 668
Crouse, C. F., 382, 412
Crow, E. L., 33
Crowder, M. J., 13, 554
Crowley, J. J., 547
Cruess, D. F., 472, 475
Csörgő, M., 517

Csörgő, S., 542, 560
Cullinan, W. L., 301

Dale, M., 211
Dallal, G. E., 532, 741
Daly, D. A., 301, 302
Damico, J. A., 247, 676, 685
Daniel, W. W., 12
David, H. A., 74
David, P. H., 475, 476
Davies, J. M. P., 472
Davis, C. E., 92
Davison, A. C., 12, 388, 391, 393
Day, N. E., 490, 491
De Jonge, C., 732
De Kroon, J., 336
Delse, F. C., 180
Deshpande, J. V., 501, 559
DiCiccio, T. J., 388, 391, 392
Dickinson, M. B., 22
Diehr, P., 4, 461, 462, 476, 484, 486
Dietz, E. J., 422
Doksum, K. A., 122, 308, 343, 347, 361, 362, 499, 500, 503, 504, 511, 557, 559
Donner, A., 492, 494
Doss, H., 533, 547
Dowdy, S., 437, 452, 453
Draper, D., 448
DuBois, A. B., 282
Duffy, D. E., 33
Dulfano, M. J., 3, 405, 406
Dunn, O. J., 248, 255, 259
Dunnett, C. W., 356
Durbin, J., 315, 532
Dwass, M., 246
Dykstra, R. L., 11, 501, 735

Edwards, A. L., 471
Edwards, A. W. F., 483
Efron, B., 1, 12, 120, 388, 389, 391, 392, 393, 539, 540, 547, 555
Efron, L., 80
Ehlers, A., 34
Engelman, W., 28
Epstein, B., 499
Eriksen, L., 112
Eubank, R. L., 12, 454, 455, 456
Everitt, B. S., 12

Falkner, B., 176, 177
Fan, J., 12, 455, 531
Farquhar, J. W., 413
Farraway, J. J., 560
Feather, B. W., 180

AUTHOR INDEX

Featherston, D. W., 377
Feigl, P., 4, 461, 462, 475, 484, 486
Feller, W., 527
Feng, X. H., 472
Ferguson, T. S., 11
Fierro, F. J., 396
Finch, P. J. P., 3, 405, 406
Finch, R., 367
Finney, D. J., 474
Fisher, L., 542
Fisher, R. A., 11, 122, 473, 481
Fisz, M., 74, 631
Fleiss, J. L., 12, 492
Fleming, T. R., 11, 13, 533, 542, 547, 554, 555, 556, 561
Fligner, M. A., 12, 92, 94, 120, 138, 139, 198, 210, 224, 233, 239, 246, 248, 259, 267, 268, 382, 386, 405, 593, 666, 668
Flores, A. M., 87
Forsman, A., 29
Fox, J. R., 349, 355
Foy, D. W., 473
Frank, G., 168
Fraser, D. A. S., 12
Freedman, D. A., 393
Freund, R. J., 441, 442, 450, 451
Friedman, J. H., 454
Friedman, Meyer, 185, 186
Friedman, Milton, 11

Gabriel, K. R., 198, 248, 258, 299
Gail, M., 475
Gamerman, D., 12
Garlid, K. D., 110
Gart, J. J., 475, 483, 494
Gasbarra, D., 12
Gastwirth, J. L., 55
Gehan, E. A., 512, 513, 555
Gelfand, A. E., 12
Gentry, J., 407
George, J. C., 136, 137
Gerstein, H. H., 381
Gibbons, J. D., 12, 375
Gijbels, I., 12
Gilbert, P. R., 130
Gilks, W. R., 12
Gill, R. D., 11, 13, 533, 540, 542, 543, 547, 554, 555, 556, 559, 561
Gillespie, M. J., 542
Glaefke, G., 4, 461, 462, 475, 484, 486
Gleser, L. J., 393
Goldsmith, J. R., 282
Gong, G., 388, 391
Goode, D. J., 283, 327, 328

Goodman, L. A., 483
Gore, A. P., 435
Götestam, K. G., 112
Gottlieb, G., 62
Govindarajulu, Z., 129
Gray, R., 10, 482, 490
Green, P. J., 12
Green, S. B., 548, 550, 560
Greenberg, V. L., 318
Greenland, S., 491
Greenwald, P., 472
Greenwood, M., 541
Gregory, P. B., 556, 557
Grenander, U., 501
Gripenberg, G., 386
Gross, S., 156
Grosse, E., 455
Grunkemeier, G. L., 542, 543
Guess, F., 517, 521, 523, 524, 525
Gulati, S., 542
Gupta, M. K., 74, 93
Gupta, S. S., 698
Gurland, J., 502

Haber, M., 459
Habib, M. G., 533
Haenszel, W., 474, 484, 485, 491, 554
Hájek, J., 12, 155, 180, 188
Haldane, J. B. S., 483
Hall, P., 12, 393
Hall, W. J., 517, 518, 523, 542, 543, 741
Halperin, M., 130
Hamilton, M., 39
Hanson, D. L., 129
Härdle, W., 12, 455
Hardy, J. D., 420, 421
Haris, T., 471
Harrington, D. P., 11, 13, 533, 542, 547, 554, 555, 556, 561
Hart, J. D., 12
Harter, H. L., 122, 581, 630, 648, 669
Hartigan, J. A., 393
Hastie, T. J., 454, 455
Hastings, W. K., 12
Hauck, W. W., 492, 494
Hawkins, D. L., 517, 521, 524, 525, 559
Hayes, H. H., 215
Hays, W. L., 377
Hayter, A. J., 246, 248, 250, 254, 267, 670, 674, 675
Hebb, D. O., 299
Hettmansperger, T. P., 12, 199, 233, 426, 428, 441, 447, 448, 450, 457
Hilgard, E. R., 70

Hilton, J. F., 10, 459
Hinkley, D. V., 12, 388, 391, 393
Hjort, N. L., 11, 533
Hochberg, Y., 247
Hodges, J. L., Jr., 11, 30, 54, 55, 73, 74, 78, 104, 105, 126, 128, 131, 139, 199
Hoeffding, W., 45, 93, 102, 117, 122, 374, 377, 382, 408, 412, 413, 734
Hogg, R. V., 129, 157, 382
Hollander, M., 4, 5, 12, 102, 103, 105, 120, 122, 292, 304, 347, 350, 351, 353, 355, 356, 361, 362, 500, 501, 503, 507, 510, 512, 517, 519, 521, 523, 524, 525, 533, 534, 536, 540, 542, 543, 546, 547, 548, 549, 556, 557, 558, 559, 560, 561, 676, 698, 715, 723, 737
Holter, J. B., 215
Horváth, L., 542
Hotelling, H., 11
Houck, J. C., 50
Høyland, A., 55, 128, 131
Hsieh, F. Y., 556
Hsu, P., 474
Hundal, P. S., 204, 205, 236, 250, 256
Hung, W., 50

Ihalamulla, R. I., 475, 476
Ijzermans, A. B., 82
Illingworth, R., 449
Iman, R. L., 449, 631, 648

Jacobson, M., 82, 83, 87
Jaeckel, L. A., 440, 446, 447
Jamison, H. H., 34
Jennrich, R. I., 606, 630
Joe, H., 376
Johansen, S., 548
John, T. M., 136, 137
Johnson, A. A., 420, 421
Johnson, B., 200
Johnson, B. M., 380, 381
Johnson, R. A., 129, 375
Johnson, R. E., 468, 469
Johnson, S. K., 468, 469
Jonckheere, A. R., 376, 662
Jones, E. A., 103
Jones, M. C., 12
Jones, M. P., 490
Jones, S. B., Jr., 161, 162
Joseph, W., 168
Jung, D. H., 147

Kaarsemaker, L., 731
Kalbfleisch, J. D., 12, 502, 542, 554

Kaneto, A., 50
Kaplan, E. L., 510, 517, 536, 547
Kaplan, H. S., 536, 537
Karow, A. M., Jr., 33
Karpatkin, M., 171, 177
Karpatkin, S., 171, 177
Katti, S. K., 576, 581, 582
Katz, M., 85, 86
Kayle, K. A., 201, 202
Keiding, N., 11, 13, 533, 543, 547, 554, 559
Kendall, M. G., 11, 12, 375, 376
Kershenobich, D., 396
Kiefer, J., 409, 412, 413, 548, 735
Kim, C., 468, 493
Kim, D. H., 435
Kim, P. J., 606, 630
Kimber, A. C., 13, 554
Kisken, W. A., 97
Klefsjö, B., 500, 501, 503, 511, 512, 519, 557
Klein, J. P., 13, 540, 554, 739
Klose, O. M., 129
Klotz, J., 49, 122, 188
Kochar, S., 517, 521, 524, 525, 559
Kolmogorov, A. N., 157, 184, 527
Konijn, H. S., 414
Kontula, K., 200, 201
Korwar, R. M., 74, 122, 540
Kosaka, K., 50
Koul, H. L., 447, 511, 512, 559
Koziol, J. A., 96, 103, 548, 550, 560, 581
Kraft, C. H., 12, 576, 590, 631, 648
Krailo, M., 556
Kramer, A., 367
Kramer, C. Y., 247
Krauth, J., 11, 12
Kruskal, W. H., 113, 197, 199, 483
Kuehl, R. O., 316
Kuntzman, R., 82, 83, 87
Kurtz, T. E., 244

Lachin, J. M., 130
Lakatos, E., 556
Lamp, W. O., 82, 87
Lan, G., 556
Langberg, N. A., 510, 517, 542, 546
Langenberg, P., 517, 738
Latscha, R., 474
Latta, R., 556
Laud, P., 11
Lauer, L. W., 70
Leach, C., 12
Leach, S. P., 259, 734
Leaf, D. A., 449
LeCam, L., 28

Lee, S. C. S., 511
Lehmann, E. L., 11, 12, 30, 48, 54, 55, 57, 58, 59, 73, 74, 78, 102, 104, 105, 117, 119, 122, 123, 126, 128, 129, 131, 133, 134, 139, 199, 210, 211, 262, 263, 268, 281, 344, 347, 353, 360, 362
Lemke, J. H., 490
Lepage, Y., 157, 175, 176, 602, 606
Leurgans, S., 555, 556, 561
Lev, J., 57, 133
Li, G., 533, 547
Liang, K. Y., 490
Liestøl, K., 542
Lilliefors, H., 532
Lim, D. H., 435
Lin, G. D., 560
Lin, P., 120
Lindell, L. E., 29
Link, R. F., 244
Ling, D., 283, 294
Liu, R. L., 472
Livesey, P. J., 299, 300
Lloyd, S. J., 110
Loader, C., 455, 517, 521, 524, 525, 559
Locke, C., 511
Louis, T. A., 33
Low, P. P., 3, 405, 406
Lowenthal, D. T., 176, 177
Lu, H. H. S., 122
Luk, C. K., 3, 405, 406

Mack, G. A., 210, 212, 223, 224, 227, 232, 268, 311, 314, 315, 317, 318, 319, 320, 321, 324, 325, 326, 330, 333, 334, 335, 336, 337, 340, 342, 361, 601, 663, 665, 719, 720, 722
Maesono, Y., 166
Magnier, E., 168
Manly, B. F. J., 12, 388, 393
Mann, H. B., 11, 113, 117, 376, 420
Mantel, N., 474, 484, 485, 491, 550, 551, 554, 555
Marascuilo, L. A., 12
March, G. L., 136, 137
Margolin, B. H., 3, 226, 233
Maritz, J. S., 12
Marron, J. S., 455
Marshall, A. W., 501
Marshall, R., 282
Mather, M., 224, 225
Maxson, S. J., 86
Maser, J. D., 34
Mayer, G., 103
McCarty, R. C., 129

McClave, J. T., 41
McDonald, B. J., 248, 299, 714
McKeague, I. W., 4, 542, 547, 548, 549
McKean, J. W., 12, 128, 426, 428, 441, 447, 450, 457
McKeown, B. A., 136, 137
McLean, R. A., 338
McNemar, Q., 470
McSweeney, M., 12
Mehta, C. R., 10, 120, 459, 482, 490, 491
Meier, P., 510, 517, 536, 547
Meltzer, H. Y., 282, 283, 327, 328
Mendenhall, W., 315
Mendis, K. N., 475, 476
Merline, J. W., 379, 380
Metropolis, N., 12
Mi, J., 517
Miller, L. E., 530, 740
Miller, R. G., Jr., 12, 165, 166, 167, 168, 187, 246, 248, 255, 258, 259, 299, 304, 305, 393, 472, 492, 493, 547, 554, 559
Milton, R. C., 119, 120, 122
Mittal, Y., 492
Moe, R., 4, 461, 462, 475, 484, 486
Moeschberger, M. L., 13, 540, 554, 739
Mohberg, N. R., 325
Molitor, F. T., 124
Moloney, P. J., 436
Moore, W., 312
Morgan, A. H., 70
Morton, R., 555
Mortazavi, B., 4, 431
Moses, L. E., 57, 133, 152, 165, 185
Moshang, T., Jr., 176, 177
Mosteller, F., 470
Mukherjee, K., 420, 421
Müller, H. G., 12
Murphy, S., 543
Myllyla, G., 200, 201

Nadel, E. R., 420, 421
Nadel, J. A., 282
Nagaraja, H. N., 74
Nair, V., 534, 542, 559
Nakanishi, R. H., 413
Nakao, K., 50
Neave, H. R., 12, 375
Nelson, W., 541
Nelson, W. B., 13
Nemenyi, P., 247, 248, 299, 304, 353
Neuman, R., 185, 186
Newton, M., 97
Neyman, J., 28, 33
Nicholls, G. H., 284, 294

Nievergelt, E., 121
Nikitin, Y., 12
Noether, G. E., 12, 49, 69, 78, 120, 180, 281, 315, 361, 375, 382, 386
Norman, R. D., 228
Norton, R. M., 233

Oakes, D., 13, 554
Obenchain, R. L., 353
O'Brien, P. C., 533, 542, 554
Odeh, R. E., 304, 662, 698, 702, 703, 707, 715, 717
O'Fallon, J. R., 533, 542, 554
Ogland-Hand, S. M., 473
O'Gorman, T. W., 490
Onesti, G., 176, 177
Oppenheim, R. W., 62, 63
Otten, A., 732
Owen, D. B., 129, 575, 581, 630, 648, 732

Paavonen, T., 200, 201
Pabst, M. R., 11
Padgett, W. J., 542
Page, E. B., 703, 707
Pan, G., 231
Parekh, A. C., 147
Parzen, E., 74
Patel, N. R., 10, 120, 482, 490, 491
Pearson, K., 11
Pearson, E. S., 32, 33, 167
Pelletier, O., 331
Peña, E., 120, 533, 534, 542, 543, 559
Petersen, A. V., 547
Peto, J., 554, 555, 556
Peto, R., 554, 555, 556
Pettitt, A. N., 120, 533
Pike, J., 407
Pitman, E. J. G., 11, 104, 105, 129, 139
Pledger, G., 120
Poland, A., 82, 83, 87
Policello, G. E., II, 92, 94, 120, 138, 139, 593
Porges, R. F., 171, 177
Potthoff, R. F., 434
Pratt, J. W., 12, 46, 67, 120
Prentice, M. J., 325
Prentice, R. L., 12, 502, 542, 554, 555, 556
Proctor, C. H., 379
Proschan, F., 496, 499, 500, 501, 502, 503, 507, 509, 510, 511, 512, 517, 519, 521, 523, 524, 525, 547, 548, 557, 559, 560, 561, 737
Puri, M. L., 11, 105, 211, 268
Putz, F. E., 22

Qian, Z. X., 472
Quade, D., 92, 631, 648
Quenouille, M. H., 1, 11, 165

Raask, E., 420, 421
Ramachandramurty, P. V., 55, 128, 131, 185, 188
Ramsay, W. N. M., 147
Randall, J. E., 349, 355
Randles, R. H., 12, 45, 49, 67, 92, 93, 94, 117, 155, 156, 157, 174, 210, 211, 290, 292, 374, 377, 382, 385, 386, 403, 405
Rao, K. S. M., 435
Rasekh, J., 367
Reaven, G. M., 413
Reba, R. C., 110
Reed, O. M., 437, 438
Rice, J. A., 337, 340, 448, 449, 450, 452
Richardson, S., 12
Rieselbach, R. E., 97
Roberts, G. O., 12
Robertson, T., 501, 735
Robins, J., 491
Rodenbaugh, J., 4, 461, 462, 475, 484, 486
Rodrigues, J., 511, 559
Rojkind, M., 396
Rosenblatt, M., 74, 409, 412, 413, 735
Rosenberg, S. A., 536, 537
Rosenbluth, A. W., 12
Rosenbluth, M. N., 12
Rosenman, R. H., 185, 186
Rubin, H., 55
Ruppert, D., 12
Rust, S. W., 198, 210, 224, 233, 382, 386, 405
Ryan, T. A., Jr., 128
Ryan, T. P., 454

Salsburg, D. S., 39
Samaniego, F. J., 511, 512
Samara, B., 382, 385, 386
Santner, T. J., 33
Sauber, S. R., 199
Savage, I. R., 11, 554
Saxena, K. M. L., 129
Scheffé, H., 11, 78, 102
Scheuer, E. M., 500
Schlosser, S., 420, 421
Schmalhorst, W. R., 70, 71
Schoenfeld, D., 556
Schonrock, H., 534
Schubot, E., 70
Schuster, E., 447
Schweder, T., 447
Sconing, J., 547, 560, 561

Seeds, A. E., 110
Self, S. G., 490
Sen, P. K., 11, 12, 105, 129, 418, 420, 422, 423, 426, 434, 435, 456, 457
Senchaudhuri, P., 10, 490, 491
Serfling, R. J., 120, 128
Sethuraman, J., 502
Sexton, G., 449
Shao, J., 166
Sheather, S. J., 426, 441, 447, 448, 450
Shelp, W. D., 97
Shen, S., 413
Shen, Z. Q., 472
Sherman, E., 248, 259, 268, 361, 362
Shlafer, M., 33
Shockey, J. W., 466
Shorack, G. R., 165, 167, 187, 547
Short, M. D., 192
Shuster, J. J., 459, 475, 493
Shyu, W. M., 455
Šidák, Z., 12, 155, 180, 188
Siddiqui, M. M., 512, 513, 516
Siegel, S., 12
Sievers, G. L., 122, 447
Sileo, L., 136, 137
Sillitto, G. P., 375
Silver, H., 215
Silverman, B. W., 12
Simmons, E., 123
Simpson, D. G., 3, 226, 233
Singpurwalla, N. D., 532
Siskind, V., 120
Skaug, H. J., 412
Skibinsky, M., 31
Skillings, J. H., 311, 314, 315, 317, 318, 319, 320, 321, 324, 325, 326, 330, 333, 334, 336, 337, 340, 342, 361, 719, 720, 722
Smid, L. J., 375
Smirnov, N. V., 11, 73, 157, 184, 185, 527, 534
Smith, A. F. M., 12
Smith, D., 82, 83, 87
Smith, E. J., 418
Smith, J. D., 83, 84
Smith, P. L., 455
Smith, R. L., 13, 554
Spearman, C., 394
Spiegelhalter, D. G., 12
Spjøtvoll, E., 263, 268
Sposto, R., 556
Sprent, P., 12
Spurrier, J. D., 347, 511
Srinivasan, R., 517, 738
Stanton, J. M., 28

Steel, R. G. D., 246, 259, 666
Stephens, M. A., 532, 533
Sterne, T. E., 33
Sternhell, S., 338
Stigler, S. M., 517
Still, H. A., 33
Stitt, J. T., 420, 421
Stone, C., 454
Stone, G., 250, 254, 267, 670, 674
Storer, B. E., 468, 493
Strawderman, R. L., 10, 556
Stuart, A., 414
Suissa, S., 459, 475, 493
Susarla, V., 11
Sweeting, T. J., 13, 554
Switzer, P., 122
Sylvester, P. E., 378

Tamhane, A. C., 247
Tanner, M., 12
Tarone, R. E., 494, 555
Tart, C. T., 70
Tate, M. W., 12, 301
Teerenhovi, L., 200, 201
Teller, A. H., 12
Teller, E., 12
Terpstra, T. J., 210, 211
Terry, M. E., 122
Theil, H., 416, 421, 422, 426
Thomas, D. R., 533, 542, 543
Thomas, H. V., 123
Thompson, W. A., Jr., 248, 299, 714
Thomson, M. L., 192
Tibshirani, R. J., 12, 388, 389, 391, 392, 393, 454, 455
Tingey, F. H., 511
Tiwari, R. C., 122
Tjøstheim, D., 412
Tocher, K. D., 475
Tsiatis, A. A., 10
Tukey, J. W., 1, 11, 55, 57, 78, 165, 244, 247

Ujah, J. E., 3, 293, 294, 326

van Belle, G., 436, 498
van Dantzig, D., 129
Vandenberg, S. G., 379
van der Laan, P., 336
van der Vaart, A. W., 547
van der Waerden, B. L., 121
van Eeden, C., 12, 576, 590, 631, 648
van Elteren, P., 281, 315, 325, 361
van Montfort, M. A. J., 732
van Ryzin, J., 11

van Wijngaarden, A., 731
Verducci, J. S., 12
Vianna, N. J., 472
Vuopio, P., 200, 201

Wahba, G., 12, 456
Walker, H. M., 57, 133
Wallace, D. L., 244
Wallis, W. A., 197, 199
Walsh, J. E., 12, 54, 180
Wand, M. P., 12
Wang, J. G., 547
Wang, J. L., 500
Wardlow, A. C., 436
Ware, J., 555
Wearden, S., 437, 452, 453
Wegman, E. J., 455
Wei, L. J., 120
Weindling, S., 448
Weinstein, A. B., 97
Welch, B. L., 138
Wellner, J. A., 517, 518, 523, 542, 543, 547, 741
Wells, D. T., 180
Wells, J. M., 435, 436
Wells, M. A., 435, 436
Wells, M. T., 122
West, M., 12
Whitaker, L. R., 511
Whitney, D. R., 11, 113, 117
Wilcox, R. A., 304, 576, 581, 582, 742
Wilcoxon, F., 2, 11, 113, 304, 576, 581, 582, 742
Wilkinson, L., 532, 741
Wilks, S. S., 117, 154
Williams, K., 299
Wilson, W. J., 441, 442, 450, 451
Windham, B. M., 304

Wolfe, B. E., 34
Wolfe, D. A., 5, 12, 45, 49, 67, 92, 93, 94, 117, 118, 129, 155, 156, 157, 174, 210, 211, 212, 223, 224, 227, 232, 233, 239, 247, 268, 290, 292, 304, 336, 347, 351, 374, 376, 377, 382, 403, 405, 435, 454, 665, 676, 685, 698, 715
Wolfowitz, J., 548
Woodward, W. F., 274, 327
Woolson, R. F., 490
Wong, S. K., 80
Wong, W., 12
Worthington, P. L., 12
Wright, F. T., 501, 735
Wright, I. W., 455
Wu, C. F. J., 166

Yanagimoto, T., 413
Yandell, B. S., 500, 503, 504
Yang, C. R., 472
Yang, G. L., 517
Yang, J., 4, 542, 548, 549
Yates, F., 122, 468
Yergan, J., 4, 461, 462, 475, 484, 486
Yi, J., 547
Ying, Z., 547
Yu, C. S., 185, 188
Yule, G. U., 482

Zacks, S., 13, 520
Zar, J. H., 732
Zelen, M., 488
Zheng, J. X., 412
Zhou, S., 454
Zitikis, R., 517
Zohman, L. R., 87
Zweiful, J. R., 483

Subject Index

Ansari-Bradley dispersion test, 142–158
   asymptotic relative efficiency, 187
   consistency, 157
   example, 147–149
   large-sample approximation, 144–145
   motivation, 149
   properties, 157
   tables, 593–601
   ties, 145, 155
Asymptotically distribution-free, 92
Asymptotic relative efficiency:
   independence procedures, 413–414
   odds ratio procedures, 494
   one- and paired-sample procedures, 104–105
   one-way layout procedures, 268–269
   regression procedures, 456–457
   success probability estimators, 30
   survival analysis procedures, 557–561
   two-sample dispersion procedures, 187–188
   two-sample location procedures, 139–140
   two-way layout procedures, 182–184

Balanced incomplete block designs, 310–320
Barlow-Doksum increasing failure rate tests, 499–500
Behrens-Fisher problem:
   k-sample, 198, 223–224, 233
   two-sample, 135–139
Bernoulli trials, 20
Bias, reduction by jackknife, 165
Bickel-Doksum increasing failure rate tests, 504
Binomial confidence interval, 31–34
   example, 31–32
   large-sample approximation, 31
   motivation, 32
   properties, 33
   tables, 572–575
Binomial distribution, 24–25

Binomial estimator, 29–31
   example, 29
   motivation, 29
   properties, 31
   sample-size determination, 30
   standard deviation, estimated, 29
Binomial test, 20–29
   consistency, 28
   example, 22–23
   large-sample approximation, 21–22
   motivation, 24
   power, 26–27
   properties, 28
   tables, 567–571
Bivariate distribution function:
   in bivariate symmetry problem, 95
   in independence problems, 363–364, 376
Bivariate symmetry, test for, 94–104
Blum-Kiefer-Rosenblatt independence test, 409, 412–43
   relation to Hoeffding's test, 412
   tables for large sample approximation, 735
Bootstrap, 388–394
   bias-corrected and accelerated, 392–393
   confidence interval for tau, 388–390
   estimated standard error, 390
   jacknife versus bootstrap, 393
   number of bootstrap replications, 393
   one-sample framework, 391–393
Boyles-Samaniego new better than used estimator, 512

Censored data, 535–561
   confidence bands for survival function, 542–546
   new better than used test, 546–547
   quantile function confidence bands, 547
   survival function estimators, 535–550, 560
   two-sample tests, 550–557

Chen-Hollander Langberg new better than used
   test for censored data, 546–547
Chi-squared test of homogeneity, 458–473
   Minitab, 464
Chi-squared test of independence, 458–473
   Minitab, 464
Concordance, multivariate, 376–377
Concordant pairs, 369
Conditional test:
   balanced incomplete block design, 315–316
   bivariate symmetry, 94–103
   broud alternatives, 183–184
   center of symmetry, 46–48
   common odds ratio, 488–491
   equal means, 115–116
   equal rates, 473–477
   equal variances, 155–156
   independence, 403–404
   odds ratio, 481
   one-way layout, 196–197, 237
   two-way layout, 280–281, 289–290,
      314–315, 325, 333
Contingency tables, 458–494
   2 × 2, 458–484
   k strata of 2 × 2 tables, 484–494
Continuity corrections:
   Edwards, 470–471
   Yates, 468
Contrasts:
   in one-way layout, 260–268
   in two-way layout, 305–309, 357–360
Correlation coefficient:
   Gripenberg partial, 386–387
   Kendall, 382
   Pearson, 398
   Spearman, 394
Critchlow-Fligner simultaneous confidence
      intervals for simple contrasts in one-way
      layout, 264–268
   example, 265–266
   large-sample approximation, 265
   motivation, 266
   properties, 267
   tables, 666–668
Cumulative distribution function, see
      Distribution function
Cumulative hazard function, 541

Decreasing failure rate, see Failure rate
Discordant pairs, 369
Dispersion:
   confidence intervals, 166–167
   estimators, 166–167
   tests, 141–178

Distribution-free, 2
Distribution function:
   confidence bands, censored case, 542–546
   confidence bands, uncensored case, 526–535
   estimation of, censored case, 535–550
   estimation of, uncensored case, 178
Doksum contrast estimator in two-way layout:
   asymptotic relative efficiency, 361
   example, 307–308
   large-sample approximation, 308
   motivation, 308
   properties, 308
Doksum test based on signed ranks for general
      alternatives in two-way layout, 343–348
   asymptotic relative efficiency, 362
   consistency, 347
   example, 345–346
   large-sample approximation, 345
   motivation, 346
   properties, 347
   ties, 345, 347
Durbin, Skillings-Mack test for balanced
      incomplete block design, 309–317
   asymptotic relative efficiency, 361
   example, 311–312
   large-sample approximation, 310–311, 314
   motivation, 312–313
   properties, 315
   tables, 718–719
   ties, 311, 314–315
Dwass, Steel, Critchlow-Fligner one-way layout
      all treatments multiple comparisons,
      240–249
   asymptotic relative efficiency, 268
   example, 242–244
   large-sample approximation, 241, 246–247
   motivation, 244
   properties, 248
   tables, 666–668
   ties, 241–242

Edwards continuity correction, 470–471
Efficiency, see Asymptotic relative efficiency
Efron bootstrap, see Bootstrap
Efron redistribute-to-the-right algorithm, 539
Efron self-consistency property, 539–540
Empirical distribution function, 178
   in goodness-of-fit test, 526–535
   in Kolmogorov-Smirnov test, 178–186
Epstein increasing failure rate test, 495–504
   asymptotic relative efficiency, 557–559
   example, 498–499
   large-sample approximation, 497–499,
   motivation, 499–500

properties, 503
tables, 735
Equivariance, 33
Experimentwise error rate:
  in one-way layout, 244–245
  in two-way layout, 297
Exponentiality, tests of, 495–526

Failure rate, 496
Fisher exact test, 473–477
  example, 474–475
  motivation, 475
  properties, 475
Fisher sign test, for paired replicates, 60–71
  asymptotic relative efficiency, 105
  consistency, 69
  examples, 62–64, 83–84
  large-sample approximation, 40
  Minitab, 63–64
  motivation, 43
  for one-sample data, 83–87
  power, 67–69
  properties, 69–70
  sample-size determination, 69
  tables, 570–571
  ties, 62, 67
Fisher-Yates-Terry-Hoeffding two-sample location test, 122
  asymptotic relative efficiency, 140
  large-sample approximation, 122
Fligner-Policello two-sample test, 135–139
  asymptotic relative efficiency, 140
  consistency, 138–139
  example, 136–138
  large-sample approximation, 136
  motivation, 138
  properties, 139
  tables, 591–593
  ties, 136
Fligner-Wolfe one-way layout treatments versus control test, 234–240
  asymptotic relative efficiency, 268
  consistency, 239
  example, 236–237
  large-sample approximation, 235
  motivation, 237
  properties, 239
  tables, 582–590
  ties, 235–236, 237
Friedman, Kendall-Babington Smith two-way layout test, 272–284
  asymptotic relative efficiency, 361
  consistency, 281
  example, 274–276

large-sample approximation, 273, 281
Minitab, 275–276
motivation, 277
properties, 281
tables, 698–702
ties, 273–274, 280–281

Gehan two-sample test for censored data, 555
Goodness-of-fit tests:
  exponentiality, 495–526
  normality, 532–533
  specified distribution, 529–531
Greenwood formula, 541
Gripenberg estimator and confidence interval for partial correlation, 386–387
Guess-Hollander-Proschan test for trend change in mean residual life, 520–526
  example, 522–523
  motivation, 523–524
  power, 559
  properties, 525
Guide, to procedures, 14–19
  to tables and charts, 563–565

Halperin-Gilbert-Lachin confidence interval for $P(X < Y)$, 130–131
Hall-Wellner mean residual life confidence bands, 517–519
Hall Wellner survival function confidence bands, 542–546
Hawkins-Kochar-Loader tests for trend change in mean residual life, 524–525
Hayter-Stone ordered alternatives multiple comparisons, one-way layout, 249–254
  example, 250–252
  large-sample approximation, 250
  motivation, 252
  properties, 254
  tables, 670–674
  ties, 250
Hettmansperger-McKean-Sheather intercept estimator, 426–429
Hodges-Lehmann one-sample estimator based on Walsh averages, 51–55, 79–83
  asymptotic relative efficiency, 104
  examples, 52–54, 81
  Minitab, 53–54
  motivation, 54
  properties, 55
  standard deviation, estimated, 58
Hodges-Lehmann two-sample estimator, 125–132
  asymptotic relative efficiency, 139–140
  example, 125–126

Hodges-Lehmann two-sample estimator, (*continued*)
　motivation, 126–128
　properties, 131
　standard deviation, estimated, 133
Hoeffding independence test, 408–413
　consistency, 412–413
　example, 410–412
　large-sample approximation, 409–412
　motivation, 412
　properties, 413
　relation to Blum-Kiefer-Rosenblatt test, 412
　tables, 733–734
　ties, 409
Hollander bivariate symmetry test, 94–104
　consistency, 103
　example, 97–102
　large-sample approximation, 96–97
　motivation, 102
　properties, 103
　table for large-sample approximation, 581
　ties, 97
Hollander test based on signed ranks for ordered alternatives in two-way layout, 348–351
　asymptotic relative efficiency, 362
　consistency, 351
　example, 349–350
　large-sample approximation, 349, 351
　motivation, 350
　properties, 351
　ties, 349
Hollander two-way layout treatments versus control multiple comparisons based on signed ranks, 353–357
　asymptotic relative efficiency, 362
　example, 354–355
　large-sample approximation, 354
　motivation, 355
　properties, 356
　ties, 354
Hollander-Peña confidence bands for survival function, 542–546
Hollander-Proschan decreasing mean residual life test, 513–520
　asymptotic relative efficiency, 557–559
　example, 515–516
　large-sample approximation, 515
　motivation, 517
　properties, 519
　tables, 738
Hollander-Proschan new better than used test, 504–513
　asymptotic relative efficiency, 557–559
　consistency, 512

　example, 507–509
　large-sample approximation, 506–507
　motivation, 509–510
　properties, 512
　tables, 736–737
　ties, 507

Incomplete block designs, 309–328
　balanced, 309–319
　arbitrary, 319–328
Increasing failure rate:
　class, 496
　tests for, 495–504
Increasing failure rate average:
　class, 500
　tests for, 500–502
Independence:
　Blum-Kiefer-Rosenblatt test of, 409, 412–413
　Hoeffding test of, 408–413
　in 2×2 contingency tables, 458–477
　Kendall test of, 363–381
　Spearman test of, 394–408
Initially increasing, then decreasing, mean residual life:
　class, 520
　tests for, 520–526
Intercept estimator, 426–429

Jackknife, 165–166
　dispersion confidence interval, 166–167
　dispersion estimator, 166–167
　dispersion test, 158–168
　estimated variance of general estimator, 165
　general confidence interval, 166
　versus bootstrap, 165
Jaeckel-Hettmansperger-McKean test for general multiple linear regression, 438–453
　asymptotic relative efficiency, 457
　example, 441–446
　large-sample approximation, 441
　Minitab, 441–448
　motivation, 446
　properties, 448
　ties, 441
Jonckheere ordered alternatives test, 202–213
　asymptotic relative efficiency, 268
　consistency, 210–211
　example, 204–206
　large-sample approximation, 203, 208–210
　motivation, 206
　power, 210
　properties, 211
　StatXact, 206

SUBJECT INDEX

tables, 649–662
ties, 203–204

Kaplan-Meier estimator of the survival function for censored data, 535–550
  asymptotic relative efficiency, 560–561
  bias, 542
  confidence bands based on, 542–546
  motivation, 539
  properties, 547–548
  redistribute-to-the-right algorithm, 539
  self-consistency property, 539–540
  tail probability estimation, 540
Kendall's test of independence, 363–381
  asymptotic relative efficiency, 413–414
  consistency, 377
  example, 367–368
  large-sample approximation, 365–366
  motivation, 368
  power, 375
  properties, 377
  sample-size determination, 375–376
  tables, 724–731
  ties, 366–367, 374–375
Klefsjö increasing failure rate test, 501–502
Klefsjö increasing failure rate average test, 501–502
Kolmogorov confidence band for distribution function, 526–535
  asymptotic relative efficiency, 559
  example, 527–528
  large-sample approximation, 527
  motivation, 528–529
  properties, 534
  tables, 740
Kolmogorov goodness-of-fit test, 529–531
Kolmogorov-Smirnov test, 178–186
  asymptotic relative efficiency, 187–188
  consistency, 185
  example, 180–181
  large-sample approximation, 179
  properties, 185
  tables, 631
  ties, 180
Koul new better than used tests, 511–512
Kruskal-Wallis one-way layout test, 191–202
  asymptotic relative efficiency, 268
  consistency, 198–199
  example, 192–194
  large-sample approximation, 191, 197–198
  Minitab, 194
  motivation, 199
  properties, 199
  StatXact, 193

tables, 631–648
ties, 191–192, 196–197
k-Sample tests, one-way layout, 189–240
  two-way layout, 270–294, 309–317, 319–340, 343–351

Lehmann contrast estimator in two-way layout, 357–360
  asymptotic relative efficiency, 362
  example, 358–360
  large-sample approximation, 360
  motivation, 360
  properties, 360
Lepage test for location and dispersion, 169–178
  consistency, 176
  example, 171–173
  large-sample approximation, 170
  motivation, 173
  properties, 176
  tables, 602–606
  ties, 170, 175
Li-Hollander-McKeague-Yang quantile function confidence bands, 547
Lillefor normality test, 532–533
Linear regression, 415–453
Location-shift function, 122
Logrank test, see Mantel two-sample test for censored data

Mack-Skillings all treatments multiple comparisons, equal number of replications in each treatment-block configuration, 341–344
  asymptotic relative efficiency, 362
  example, 341–342
  motivation, 342
  properties, 343
  ties, 341
Mack-Skillings test for randomized block design with equal number of replications per treatment-block configuration, 329–341
  asymptotic relative efficiency, 362
  example, 331–332
  large-sample approximation, 330–331, 333–334
  motivation, 333
  properties, 338
  tables, 721–722
  ties, 331, 334
Mack-Wolfe umbrella alternatives test, peak known, 213–226
  asymptotic relative efficiency, 268
  consistency, 224

Mack-Wolfe umbrella alternatives test, peak
  known, (*continued*)
   example, 215–216
   large-sample approximation, 214, 221–223
   motivation, 217
   properties, 224
   tables, 663
   ties, 214–215
Mack-Wolfe umbrella alternatives test, peak
  unknown, 226–234
   example, 228–229
   motivation, 230
   power, 232
   tables, 664
   ties, 227–228
Mann trend test, 376
Mann-Whitney test, *see* Wilcoxon rank sum test
Mann-Whitney U-statistic, 117–118
Mantel-Haenszel estimator of common odds
  ratio, 491–494
   asymptotic relative efficiency, 494
   example, 492
   StatXact, 492
Mantel-Haenszel odds ratio test for k strata of
  2×2 tables, 484–494
   example, 486–488
   motivation, 488
Mantel two-sample test for censored data,
  550–557
   asymptotic relative efficiency, 561
   example, 552–553
   large-sample approximation, 551–552
   motivation, 553–554
   properties, 556
McNemar dependent proportions test, 468–473
   Minitab, 469–470
   StatXact, 470
Mean residual life function, 510
   confidence bands for, 517–519
   decreasing, 513
   estimator, 517
   increasing, 513
Median:
   estimated standard deviation of sample
     median, 74
   of a population, 60
   of a sample, 72
   test for population median being a specified
     value, 42, 64, 79–87
Miller jackknife test for dispersion, 159–169
   asymptotic relative efficiency, 187
   consistency, 168
   example, 161–164
   motivation, 165

   properties, 168
   ties, 161
Minitab, 10–11
Moses confidence interval for location
  differences, 132–135
   asymptotic relative efficiency, 139
   example, 132–133
   large-sample approximation, 132
   Minitab, 126, 133
   motivation, 133
   properties, 134
Moses goodness-of-fit test, 185
Multiple comparisons, one-way layout, 240–260
   all treatments, 240–254
   treatments versus control, 254–260
   two-way layout, 295–305, 317–319,
     340–342, 351–357
   all treatments, 295–300, 317–319, 340–342,
     351–353
   treatments versus control, 300–305, 353–357

Nelson-Aalen cumulative hazard function
  estimator, 541
Nemenyi, Damico-Wolfe one-way layout
  treatment versus control multiple
  comparisons, 254–260
   asymptotic relative efficiency, 268
   example, 256–257
   large-sample approximations, 255–256, 304
   motivation, 257
   properties, 259
   tables, 676–685
   ties, 256
Nemenyi two-way layout all treatments multiple
  comparisons based on signed ranks,
  351–353
   asymptotic relative efficiency, 362
   example, 352–353
   large-sample approximation, 352
   motivation, 353
   properties, 353
   ties, 352
Nemenyi, Wilcoxon-Wilcox, Miller rank sum
  two-way layout treatment versus control
  procedures, 300–305
   asymptotic relative efficiency, 361
   example, 301–302
   large-sample approximation, 301, 304
   motivation, 303
   properties, 305
   tables, 715–717
   ties, 301
New better than used:
   class, 505

estimator, 512
tests for, 504–513
New better than used in expectation:
class, 510
tests for, 510–511
Nonparametric statistical procedures:
advantages of, 1
Normality, tests of, 532–533

Odds ratio, 477–494
confidence interval for common odds ratio, 491–492
estimator, 478–482
estimator of common odds ratio, 478–482
exact conditional test that odds ratio is a specified value, 481
population, 477
StatXact, 481–482, 490
test for a common odds ratio, 488–492
test for a common odds ratio equal to 1, 484–493
One-sample tests, location, 79–81
population symmetry, 87–94
One-way layout, 189–269
Ordered alternatives, one-way layout, 202–213
two-way layout, 284–294, 348–351
Order statistics, in estimation of population median, 73–74

Page test for ordered alternatives in two-way layout, 284–294
asymptotic relative efficiency, 361
consistency, 292
example, 286–287
large-sample approximation, 285–286, 290–292
motivation, 287
properties, 292
tables, 703–707
ties, 286, 289–290
Paired replicates analyses, 35–79, 94–104
Parallelism test, see Sen-Adichie parallelism test
Partial correlation, 386–387
confidence interval, 386–387
estimator, 386–387
Pearson test for comparing two proportions, 459–473
Pitman asymptotic relative efficiency, see Asymptotic relative efficiency
Placements, 135
Probability that $X < Y$, 120, 129–132
estimator, 129
confidence intervals, 129–132
Proportional hazards, 560

Quenouille-Tukey jackknife, 165–166
Quantile function, 547
Quasimedian, 73

Randles-Fligner-Policello-Wolfe, Davis-Quade symmetry test, 87–94
consistency, 93–94
example, 89–91
large-sample approximation, 89, 94
motivation, 91–92
properties, 94
ties, 89
Randomized blocks, 270–309, 343–362
Rank sum test, see Wilcoxon rank sum test
Redistribute-to-the-right algorithm, 539
Regression, 415–457
arbitrary regression function, 453–456
confidence intervals, 424–426
estimators, 421–423, 446–447
intercept estimators, 426–429, 446–447
kernel regression smoother, 455
local regression smoother, 455
multiple linear, 438–453
non-rank based, 453–456
one line, 415–428
running line smoother, 454–455
parallelism, 429–438
several lines, 429–453
slope estimator, 421–423
spline regression smoother, 455–456
tests, 416–423, 429–438, 438–453
Robins-Breslow-Greenland odds ratio confidence interval, 491–492
example, 492
StatXact, 492

Samora-Randles, Fligner-Rust, Noether confidence interval for Kendall's τ, 383–387
example, 383–384
motivation, 385
properties, 387
ties, 385
Scale parameter(s), 142–143
confidence intervals for ratio of, 166–167
estimators for ratio of, 166–167
tests for ratio of, 142–158, 158–169
Self-consistency property, 539–540
Sen-Adichie parallelism test, 429–438
asymptotic relative efficiency, 456–457
example, 430–433
large-sample approximation, 430
motivation, 433–434

Sen-Adichie parallelism test, (*continued*)
  properties, 435
  ties, 430
Sen confidence interval for $P(X < Y)$, 129–130
Signed rank test, *see* Wilcoxon signed rank test
Simpson paradox, 492
Skewness,
  left, 92–93
  right, 92–93
Skillings-Mack multiple comparison procedure
    for balanced incomplete block designs,
    317–319
  asymptotic relative efficiency, 361–362
  example, 317–318
  large-sample approximation, 317
  motivation, 318
  properties, 319
  ties, 317
Skillings-Mack test for arbitrary incomplete
    block design, 319–328
  example, 322–324
  large-sample approximation, 321
  motivation, 323
  properties, 325
  tables, 719–720
  ties, 321
Smoothers, 455–456
Spearman independence test, 394–408
  asymptotic relative efficiency, 414
  example, 396–397
  large-sample approximation, 395
  Minitab, 398
  motivation, 398
  properties, 405
  tables, 732
  ties, 396, 403
Spearman rank correlation coefficient, 394–408
Spjøtvoll contrast estimator in one-way layout,
    260–264
  asymptotic relative efficiency, 268–269
  example, 261–262
  large-sample approximation, 263
  motivation, 262–263
  properties, 263
StatXact, 10
Symmetry, test of bivariate symmetry, 94–104
  test of population symmetry, 87–94

Tables and charts, 563–743
Tarone-Ware two-sample tests for censored
    data, 555
Tau, confidence interval for, 383–394
  estimator of, 382–383
  measure of association, 369

Theil confidence interval for slope, 424–426
  asymptotic relative efficiency, 456
  example, 424–425
  large-sample approximation, 425
  Minitab, 425
  motivation, 425
  properties, 426
Theil slope estimator, 421–423
  asymptotic relative efficiency, 456
  example, 422
  large-sample approximation, 423
  Minitab, 422–423
  motivation, 422
  properties, 423
Theil test for slope, 416–423
  asymptotic relative efficiency, 456
  consistency, 420
  example, 418–419
  large-sample approximation, 417–418
  motivation, 419–420
  properties, 420
  tables, 724–731
  ties, 418
Total-time-on-test statistic, 499
Trend:
  in mean residual life, 513–526
  in one-way layout effects, 202–213
  in sample, 376, 405
  in two-way layout effects, 284–294, 348–351
Triangle test, 23
Triple:
  left, 88
  right, 88
Tukey confidence interval for location, 56–59,
    80–83
  asymptotic relative efficiency, 104
  examples, 56–57, 81
  large-sample approximation, 56
  Minitab, 53, 57
  motivation, 57
  properties, 59
Turning point in mean residual life:
  tests when turning point known, 520–526
  tests when turning point unknown, 524–525
Two-sample tests, broad alternatives, 178–186
  dispersion and location, 169–178
  dispersion differences, 142–158, 158–169
  location differences, 106–125
Two-way layout, 270–362

Umbrella alternatives:
  one-way layout, 213–234
    test for, peak known, 213–226
    test for, peak unknown, 226–234

Van der Waerden two-sample location test, 121–122
    asymptotic relative efficiency, 140
    example, 121
    large-sample approximation, 121
Variance:
    tests for equality of population variances, 141–176, *see also* Dispersion

Walsh average, 54
Wilcoxon, Nemenyi, McDonald-Thompson rank sum two-way layout all treatments multiple comparisons, 295–300
    asymptotic relative efficiency, 361
    example, 296–297
    large-sample approximation, 296, 299
    motivation, 297
    properties, 299
    tables, 708–714
    ties, 296
Wilcoxon rank sum test, 106–125
    asymptotic relative efficiency, 139–140
    consistency, 123
    example, 110–113
    large-sample approximation, 108–109, 116–117
    Minitab, 126
    motivation, 113
    power, 119–120,
    properties, 123
    sample-size determination, 120
    StatXact, 111
    tables, 582–590
    ties, 109, 115–116
Wilcoxon signed rank test, 36–51, 79–81
    asymptotic relative efficiency, 104
    consistency, 49
    examples, 39–42, 80–81
    large-sample approximation, 37–38
    Minitab, 40
    motivation, 42
    power, 48–49
    properties, 49
    sample-size determination, 49
    StatXact, 40, 47–48
    tables, 576–581
    ties, 38–39, 46–48

Yates continuity correction, 468
Yule association measure, 482

Zelen test for common odds ratio, 488–493
    example, 489–491
    StatXact, 490–491

WILEY SERIES IN PROBABILITY AND STATISTICS
ESTABLISHED BY WALTER A. SHEWHART AND SAMUEL S. WILKS

Editors
*Vic Barnett, Noel A. C. Cressie, Nicholas I. Fisher,
Iain M. Johnstone, J. B. Kadane, David G. Kendall, David W. Scott,
Bernard W. Silverman, Adrian F. M. Smith, Jozef L. Teugels;
Ralph A. Bradley, Emeritus, J. Stuart Hunter, Emeritus*

Probability and Statistics Section

*ANDERSON · The Statistical Analysis of Time Series
ARNOLD, BALAKRISHNAN, and NAGARAJA · A First Course in Order Statistics
ARNOLD, BALAKRISHNAN, and NAGARAJA · Records
BACCELLI, COHEN, OLSDER, and QUADRAT · Synchronization and Linearity:
  An Algebra for Discrete Event Systems
BASILEVSKY · Statistical Factor Analysis and Related Methods: Theory and
  Applications
BERNARDO and SMITH · Bayesian Statistical Concepts and Theory
BILLINGSLEY · Convergence of Probability Measures
BOROVKOV · Asymptotic Methods in Queuing Theory
BOROVKOV · Ergodicity and Stability of Stochastic Processes
BRANDT, FRANKEN, and LISEK · Stationary Stochastic Models
CAINES · Linear Stochastic Systems
CAIROLI and DALANG · Sequential Stochastic Optimization
CONSTANTINE · Combinatorial Theory and Statistical Design
COOK · Regression Graphics
COVER and THOMAS · Elements of Information Theory
CSÖRGŐ and HORVÁTH · Weighted Approximations in Probability Statistics
CSÖRGŐ and HORVÁTH · Limit Theorems in Change Point Analysis
DETTE and STUDDEN · The Theory of Canonical Moments with Applications in
  Statistics, Probability, and Analysis
*DOOB · Stochastic Processes
DRYDEN and MARDIA · Statistical Analysis of Shape
DUPUIS and ELLIS · A Weak Convergence Approach to the Theory of Large Deviations
ETHIER and KURTZ · Markov Processes: Characterization and Convergence
FELLER · An Introduction to Probability Theory and Its Applications, Volume 1,
  Third Edition, Revised; Volume II, *Second Edition*
FULLER · Introduction to Statistical Time Series, *Second Edition*
FULLER · Measurement Error Models
GHOSH, MUKHOPADHYAY, and SEN · Sequential Estimation
GIFI · Nonlinear Multivariate Analysis
GUTTORP · Statistical Inference for Branching Processes
HALL · Introduction to the Theory of Coverage Processes
HAMPEL · Robust Statistics: The Approach Based on Influence Functions
HANNAN and DEISTLER · The Statistical Theory of Linear Systems
HUBER · Robust Statistics
IMAN and CONOVER · A Modern Approach to Statistics
JUREK and MASON · Operator-Limit Distributions in Probability Theory
KASS and VOS · Geometrical Foundations of Asymptotic Inference
KAUFMAN and ROUSSEEUW · Finding Groups in Data: An Introduction to Cluster
  Analysis

*Now available in a lower priced paperback edition in the Wiley Classics Library.

Probability and Statistics (Continued)

KELLY · Probability, Statistics, and Optimization
LINDVALL · Lectures on the Coupling Method
McFADDEN · Management of Data in Clinical Trials
MANTON, WOODBURY, and TOLLEY · Statistical Applications Using Fuzzy Sets
MORGENTHALER and TUKEY · Configural Polysampling: A Route to Practical Robustness
MUIRHEAD · Aspects of Multivariate Statistical Theory
OLIVER and SMITH · Influence Diagrams, Belief Nets and Decision Analysis
*PARZEN · Modern Probability Theory and Its Applications
PRESS · Bayesian Statistics: Principles, Models, and Applications
PUKELSHEIM · Optimal Experimental Design
RAO · Asymptotic Theory of Statistical Inference
RAO · Linear Statistical Inference and Its Applications, *Second Edition*
RAO and SHANBHAG · Choquet-Deny Type Functional Equations with Applications to Stochastic Models
ROBERTSON, WRIGHT, and DYKSTRA · Order Restricted Statistical Inference
ROGERS and WILLIAMS · Diffusions, Markov Processes, and Martingales, Volume I: Foundations, *Second Edition;* Volume II: Îto Calculus
RUBINSTEIN and SHAPIRO · Discrete Event Systems: Sensitivity Analysis and Stochastic Optimization by the Score Function Method
RUZSA and SZEKELY · Algebraic Probability Theory
SCHEFFE · The Analysis of Variance
SEBER · Linear Regression Analysis
SEBER · Multivariate Observations
SEBER and WILD · Nonlinear Regression
SERFLING · Approximation Theorems of Mathematical Statistics
SHORACK and WELLNER · Empirical Processes with Applications to Statistics
SMALL and McLEISH · Hilbert Space Methods in Probability and Statistical Inference
STAPLETON · Linear Statistical Models
STAUDTE and SHEATHER · Robust Estimation and Testing
STOYANOV · Counterexamples in Probability
TANAKA · Time Series Analysis: Nonstationary and Noninvertible Distribution Theory
THOMPSON and SEBER · Adaptive Sampling
WELSH · Aspects of Statistical Inference
WHITTAKER · Graphical Models in Applied Multivariate Statistics
YANG · The Construction Theory of Denumerable Markov Processes

Applied Probability and Statistics Section

ABRAHAM and LEDOLTER · Statistical Methods for Forecasting
AGRESTI · Analysis of Ordinal Categorical Data
AGRESTI · Categorical Data Analysis
ANDERSON, AUQUIER, HAUCK, OAKES, VANDAELE, and WEISBERG · Statistical Methods for Comparative Studies
ARMITAGE and DAVID (editors) · Advances in Biometry
*ARTHANARI and DODGE · Mathematical Programming in Statistics
ASMUSSEN · Applied Probability and Queues
*BAILEY · The Elements of Stochastic Processes with Applications to the Natural Sciences
BARNETT and LEWIS · Outliers in Statistical Data, *Third Edition*
BARTHOLOMEW, FORBES, and McLEAN · Statistical Techniques for Manpower Planning, *Second Edition*

*Now available in a lower priced paperback edition in the Wiley Classics Library.

Applied Probability and Statistics (Continued)

BATES and WATTS · Nonlinear Regression Analysis and Its Applications

BECHHOFER, SANTNER, and GOLDSMAN · Design and Analysis of Experiments for Statistical Selection, Screening, and Multiple Comparisons

BELSLEY · Conditioning Diagnostics: Collinearity and Weak Data in Regression

BELSLEY, KUH, and WELSCH · Regression Diagnostics: Identifying Influential Data and Sources of Collinearity

BHAT · Elements of Applied Stochastic Processes, *Second Edition*

BHATTACHARYA and WAYMIRE · Stochastic Processes with Applications

BIRKES and DODGE · Alternative Methods of Regression

BLOOMFIELD · Fourier Analysis of Time Series: An Introduction

BOLLEN · Structural Equations with Latent Variables

BOULEAU · Numerical Methods for Stochastic Processes

BOX · Bayesian Inference in Statistical Analysis

BOX and DRAPER · Empirical Model-Building and Response Surfaces

BOX and DRAPER · Evolutionary Operation: A Statistical Method for Process Improvement

BUCKLEW · Large Deviation Techniques in Decision, Simulation, and Estimation

BUNKE and BUNKE · Nonlinear Regression, Functional Relations and Robust Methods: Statistical Methods of Model Building

CHATTERJEE and HADI · Sensitivity Analysis in Linear Regression

CHOW and LIU · Design and Analysis of Clinical Trials: Concepts and Methodologies

CLARKE and DISNEY · Probability and Random Processes: A First Course with Applications, *Second Edition*

*COCHRAN and COX · Experimental Designs, *Second Edition*

CONOVER · Practical Nonparametric Statistics, *Second Edition*

CORNELL · Experiments with Mixtures, Designs, Models, and the Analysis of Mixture Data, *Second Edition*

*COX · Planning of Experiments

CRESSIE · Statistics for Spatial Data, *Revised Edition*

DANIEL · Applications of Statistics to Industrial Experimentation

DANIEL · Biostatistics: A Foundation for Analysis in the Health Sciences, *Sixth Edition*

DAVID · Order Statistics, *Second Edition*

*DEGROOT, FIENBERG, and KADANE · Statistics and the Law

DODGE · Alternative Methods of Regression

DOWDY and WEARDEN · Statistics for Research, *Second Edition*

DRYDEN and MARDIA · Statistical Shape Analysis

DUNN and CLARK · Applied Statistics: Analysis of Variance and Regression, *Second Edition*

ELANDT-JOHNSON and JOHNSON · Survival Models and Data Analysis

EVANS, PEACOCK, and HASTINGS · Statistical Distributions, *Second Edition*

FLEISS · The Design and Analysis of Clinical Experiments

FLEISS · Statistical Methods for Rates and Proportions, *Second Edition*

FLEMING and HARRINGTON · Counting Processes and Survival Analysis

GALLANT · Nonlinear Statistical Models

GLASSERMAN and YAO · Monotone Structure in Discrete-Event Systems

GNANADESIKAN · Methods for Statistical Data Analysis of Multivariate Observations, *Second Edition*

GOLDSTEIN and LEWIS · Assessment: Problems, Development, and Statistical Issues

GREENWOOD and NIKULIN · A Guide to Chi-Squared Testing

*HAHN · Statistical Models in Engineering

HAHN and MEEKER · Statistical Intervals: A Guide for Practitioners

HAND · Construction and Assessment of Classification Rules

HAND · Discrimination and Classification

*Now available in a lower priced paperback edition in the Wiley Classics Library.

Applied Probability and Statistics (Continued)

HEIBERGER · Computation for the Analysis of Designed Experiments

HINKELMAN and KEMPTHORNE: · Design and Analysis of Experiments, Volume 1: Introduction to Experimental Design

HOAGLIN, MOSTELLER, and TUKEY · Exploratory Approach to Analysis of Variance

HOAGLIN, MOSTELLER, and TUKEY · Exploring Data Tables, Trends and Shapes

HOAGLIN, MOSTELLER, and TUKEY · Understanding Robust and Exploratory Data Analysis

HOCHBERG and TAMHANE · Multiple Comparison Procedures

HOCKING · Methods and Applications of Linear Models: Regression and the Analysis of Variables

HOGG and KLUGMAN · Loss Distributions

HOSMER and LEMESHOW · Applied Logistic Regression

HØYLAND and RAUSAND · System Reliability Theory: Models and Statistical Methods

HUBERTY · Applied Discriminant Analysis

JACKSON · A User's Guide to Principle Components

JOHN · Statistical Methods in Engineering and Quality Assurance

JOHNSON · Multivariate Statistical Simulation

JOHNSON and KOTZ · Distributions in Statistics

Continuous Multivariate Distributions

JOHNSON, KOTZ, and BALAKRISHNAN · Continuous Univariate Distributions, Volume 1, *Second Edition*

JOHNSON, KOTZ, and BALAKRISHNAN · Continuous Univariate Distributions, Volume 2, *Second Edition*

JOHNSON, KOTZ, and BALAKRISHNAN · Discrete Multivariate Distributions

JOHNSON, KOTZ, and KEMP · Univariate Discrete Distributions, *Second Edition*

JUREČKOVÁ and SEN · Robust Statistical Procedures: Aymptotics and Interrelations

KADANE · Bayesian Methods and Ethics in a Clinical Trial Design

KADANE AND SCHUM · A Probabilistic Analysis of the Sacco and Vanzetti Evidence

KALBFLEISCH and PRENTICE · The Statistical Analysis of Failure Time Data

KELLY · Reversability and Stochastic Networks

KHURI, MATHEW, and SINHA · Statistical Tests for Mixed Linear Models

KLUGMAN, PANJER, and WILLMOT · Loss Models: From Data to Decisions

KLUGMAN, PANJER, and WILLMOT · Solutions Manual to Accompany Loss Models: From Data to Decisions

KOVALENKO, KUZNETZOV, and PEGG · Mathematical Theory of Reliability of Time-Dependent Systems with Practical Applications

LAD · Operational Subjective Statistical Methods: A Mathematical, Philosophical, and Historical Introduction

LANGE, RYAN, BILLARD, BRILLINGER, CONQUEST, and GREENHOUSE · Case Studies in Biometry

LAWLESS · Statistical Models and Methods for Lifetime Data

LEE · Statistical Methods for Survival Data Analysis, *Second Edition*

LePAGE and BILLARD · Exploring the Limits of Bootstrap

LINHART and ZUCCHINI · Model Selection

LITTLE and RUBIN · Statistical Analysis with Missing Data

MAGNUS and NEUDECKER · Matrix Differential Calculus with Applications in Statistics and Econometrics

MALLER and ZHOU · Survival Analysis with Long Term Survivors

MANN, SCHAFER, and SINGPURWALLA · Methods for Statistical Analysis of Reliability and Life Data

McLACHLAN and KRISHNAN · The EM Algorithm and Extensions

McLACHLAN · Discriminant Analysis and Statistical Pattern Recognition

McNEIL · Epidemiological Research Methods

*Now available in a lower priced paperback edition in the Wiley Classics Library.

Applied Probability and Statistics (Continued)
  MEEKER and ESCOBAR · Statistical Methods for Reliability Data
  MILLER · Survival Analysis
  MONTGOMERY and PECK · Introduction to Linear Regression Analysis, *Second Edition*
  MYERS and MONTGOMERY · Response Surface Methodology: Process and Product
    in Optimization Using Designed Experiments
  NELSON · Accelerated Testing, Statistical Models, Test Plans, and Data Analyses
  NELSON · Applied Life Data Analysis
  OCHI · Applied Probability and Stochastic Processes in Engineering and Physical
    Sciences
  OKABE, BOOTS, and SUGIHARA · Spatial Tesselations: Concepts and Applications
    of Voronoi Diagrams
  PANKRATZ · Forecasting with Dynamic Regression Models
  PANKRATZ · Forecasting with Univariate Box-Jenkins Models: Concepts and Cases
  PIANTADOSI · Clinical Trials: A Methodologic Perspective
  PORT · Theoretical Probability for Applications
  PUTERMAN · Markov Decision Processes: Discrete Stochastic Dynamic Programming
  RACHEV · Probability Metrics and the Stability of Stochastic Models
  RÉNYI · A Diary on Information Theory
  RIPLEY · Spatial Statistics
  RIPLEY · Stochastic Simulation
  ROUSSEEUW and LEROY · Robust Regression and Outlier Detection
  RUBIN · Multiple Imputation for Nonresponse in Surveys
  RUBINSTEIN · Simulation and the Monte Carlo Method
  RUBINSTEIN and MELAMED · Modern Simulation and Modeling
  RYAN · Statistical Methods for Quality Improvement
  SCHUSS · Theory and Applications of Stochastic Differential Equations
  SCOTT · Multivariate Density Estimation: Theory, Practice, and Visualization
 *SEARLE · Linear Models
  SEARLE · Linear Models for Unbalanced Data
  SEARLE, CASELLA, and McCULLOCH · Variance Components
  SENNOTT · Stochastic Dynamic Programming and the Control of Queueing Systems
  STOYAN, KENDALL, and MECKE · Stochastic Geometry and Its Applications, *Second
    Edition*
  STOYAN and STOYAN · Fractals, Random Shapes and Point Fields: Methods of
    Geometrical Statistics
  THOMPSON · Empirical Model Building
  THOMPSON · Sampling
  TIJMS · Stochastic Modeling and Analysis: A Computational Approach
  TIJMS · Stochastic Models: An Algorithmic Approach
  TITTERINGTON, SMITH, and MAKOV · Statistical Analysis of Finite Mixture
    Distributions
  UPTON and FINGLETON · Spatial Data Analysis by Example, Volume 1: Point
    Pattern and Quantitative Data
  UPTON and FINGLETON · Spatial Data Analysis by Example, Volume II:
    Categorical and Directional Data
  VAN RIJCKEVORSEL and DE LEEUW · Component and Correspondence Analysis
  WEISBERG · Applied Linear Regression, *Second Edition*
  WESTFALL and YOUNG · Resampling-Based Multiple Testing: Examples and
    Methods for *p*-Value Adjustment
  WHITTLE · Systems in Stochastic Equilibrium
  WOODING · Planning Pharmaceutical Clinical Trials: Basic Statistical Principles
  WOOLSON · Statistical Methods for the Analysis of Biomedical Data
 *ZELLNER · An Introduction to Bayesian Inference in Econometrics

*Now available in a lower priced paperback edition in the Wiley Classics Library.

Texts and References Section

AGRESTI · An Introduction to Categorical Data Analysis
ANDERSON · An Introduction to Multivariate Statistical Analysis, *Second Edition*
ANDERSON and LOYNES · The Teaching of Practical Statistics
ARMITAGE and COLTON · Encyclopedia of Biostatistics: Volumes 1 to 6 with Index
BARTOSZYNSKI and NIEWIADOMSKA-BUGAJ · Probability and Statistical Inference
BERRY, CHALONER, and GEWEKE · Bayesian Analysis in Statistics and Econometrics: Essays in Honor of Arnold Zellner
BHATTACHARYA and JOHNSON · Statistical Concepts and Methods
BILLINGSLEY · Probability and Measure, *Second Edition*
BOX · R. A. Fisher, the Life of a Scientist
BOX, HUNTER, and HUNTER · Statistics for Experimenters: An Introduction to Design, Data Analysis, and Model Building
BOX and LUCEÑO · Statistical Control by Monitoring and Feedback Adjustment
BROWN and HOLLANDER · Statistics: A Biomedical Introduction
CHATTERJEE and PRICE · Regression Analysis by Example, *Second Edition*
COOK and WEISBERG · An Introduction to Regression Graphics
COX · A Handbook of Introductory Statistical Methods
DILLON and GOLDSTEIN · Multivariate Analysis: Methods and Applications
DODGE and ROMIG · Sampling Inspection Tables, *Second Edition*
DRAPER and SMITH · Applied Regression Analysis, *Third Edition*
DUDEWICZ and MISHRA · Modern Mathematical Statistics
DUNN · Basic Statistics: A Primer for the Biomedical Sciences, *Second Edition*
FISHER and VAN BELLE · Biostatistics: A Methodology for the Health Sciences
FREEMAN and SMITH · Aspects of Uncertainty: A Tribute to D. V. Lindley
GROSS and HARRIS · Fundamentals of Queueing Theory, *Third Edition*
HALD · A History of Probability and Statistics and their Applications Before 1750
HALD · A History of Mathematical Statistics from 1750 to 1930
HELLER · MACSYMA for Statisticians
HOEL · Introduction to Mathematical Statistics, *Fifth Edition*
HOLLANDER and WOLFE · Nonparametric Statistical Methods, *Second Edition*
JOHNSON and BALAKRISHNAN · Advances in the Theory and Practice of Statistics: A Volume in Honor of Samuel Kotz
JOHNSON and KOTZ (editors) · Leading Personalities in Statistical Sciences: From the Seventeenth Century to the Present
JUDGE, GRIFFITHS, HILL, LÜTKEPOHL, and LEE · The Theory and Practice of Econometrics, *Second Edition*
KHURI · Advanced Calculus with Applications in Statistics
KOTZ and JOHNSON (editors) · Encyclopedia of Statistical Sciences: Volumes 1 to 9 wtih Index
KOTZ and JOHNSON (editors) · Encyclopedia of Statistical Sciences: Supplement Volume
KOTZ, REED, and BANKS (editors) · Encyclopedia of Statistical Sciences: Update Volume 1
KOTZ, REED, and BANKS (editors) · Encyclopedia of Statistical Sciences: Update Volume 2
LAMPERTI · Probability: A Survey of the Mathematical Theory, *Second Edition*
LARSON · Introduction to Probability Theory and Statistical Inference, *Third Edition*
LE · Applied Categorical Data Analysis
LE · Applied Survival Analysis
MALLOWS · Design, Data, and Analysis by Some Friends of Cuthbert Daniel
MARDIA · The Art of Statistical Science: A Tribute to G. S. Watson
MASON, GUNST, and HESS · Statistical Design and Analysis of Experiments with Applications to Engineering and Science

*Now available in a lower priced paperback edition in the Wiley Classics Library.

Texts and References (Continued)

    MURRAY · X-STAT 2.0 Statistical Experimentation, Design Data Analysis, and Nonlinear Optimization
    PURI, VILAPLANA, and WERTZ · New Perspectives in Theoretical and Applied Statistics
    RENCHER · Methods of Multivariate Analysis
    RENCHER · Multivariate Statistical Inference with Applications
    ROSS · Introduction to Probability and Statistics for Engineers and Scientists
    ROHATGI · An Introduction to Probability Theory and Mathematical Statistics
    RYAN · Modern Regression Methods
    SCHOTT · Matrix Analysis for Statistics
    SEARLE · Matrix Algebra Useful for Statistics
    STYAN · The Collected Papers of T. W. Anderson: 1943–1985
    TIERNEY · LISP-STAT: An Object-Oriented Environment for Statistical Computing and Dynamic Graphics
    WONNACOTT and WONNACOTT · Econometrics, *Second Edition*

WILEY SERIES IN PROBABILITY AND STATISTICS
ESTABLISHED BY WALTER A. SHEWHART AND SAMUEL S. WILKS

Editors
Robert M. Groves, Graham Kalton, J. N. K. Rao, Norbert Schwarz, Christopher Skinner

Survey Methodology Section

    BIEMER, GROVES, LYBERG, MATHIOWETZ, and SUDMAN · Measurement Errors in Surveys
    COCHRAN · Sampling Techniques, *Third Edition*
    COUPER, BAKER, BETHLEHEM, CLARK, MARTIN, NICHOLLS, and O'REILLY (editors) · Computer Assisted Survey Information Collection
    COX, BINDER, CHINNAPPA, CHRISTIANSON, COLLEDGE, and KOTT (editors) · Business Survey Methods
 *DEMING · Sample Design in Business Research
    DILLMAN · Mail and Telephone Surveys: The Total Design Method
    GROVES and COUPER · Nonresponse in Household Interview Surveys
    GROVES · Survey Errors and Survey Costs
    GROVES, BIEMER, LYBERG, MASSEY, NICHOLLS, and WAKSBERG · Telephone Survey Methodology
 *HANSEN, HURWITZ, and MADOW · Sample Survey Methods and Theory, Volume 1: Methods and Applications
 *HANSEN, HURWITZ, and MADOW · Sample Survey Methods and Theory, Volume II: Theory
    KISH · Statistical Design for Research
 *KISH · Survey Sampling
    LESSLER and KALSBEEK · Nonsampling Error in Surveys
    LEVY and LEMESHOW · Sampling of Populations: Methods and Applications, *Third Edition*
    LYBERG, BIEMER, COLLINS, de LEEUW, DIPPO, SCHWARZ, TREWIN (editors) · Survey Measurement and Process Quality
    SKINNER, HOLT, and SMITH · Analysis of Complex Surveys

---

*Now available in a lower priced paperback edition in the Wiley Classics Library.